Second Handbook of Research on Mathematics Teaching and Learning

Second Handbook of Research on Mathematics Teaching and Learning

A Project of the National Council of Teachers of Mathematics

Frank K. Lester, Jr.

EDITOR

Information Age Publishing • Charlotte, NC • infoagepub.com

Library of Congress Cataloging-in-Publication Data

Second handbook of research on mathematics teaching and learning : a project of the national council of teachers of mathematics / Frank K. Lester, Jr., editor.
p. cm.
Includes bibliographical references and index.
ISBN-13: 978-1-59311-176-2 (pbk.)
ISBN-13: 978-1-59311-177-9 (hardcover)
1. Mathematics–Study and teaching–Research. I. Lester, Frank K. II. Handbook of research on mathematics teaching and learning
QA11.S365 2007
510.71–dc22

2006033783

Volume 1:
ISBN-13: 978-1-59311-586-9 (pbk.)
ISBN-13: 978-1-59311-587-6 (hardcover)

Volume 2:
ISBN-13: 978-1-59311-588-3 (pbk.)
ISBN-13: 978-1-59311-589-0 (hardcover)

Printed in the United States of America

This handbook is dedicated to the memory of four eminent researchers who were authors of chapters in the first handbook, but who have since died —

Merlyn J. Behr 1932–1995
Robert B. Davis 1923–1997
James J. Kaput 1943–2005
Alba G. Thompson 1946–1996

With their passing, the mathematics education community lost talented, dedicated researchers; those of us who knew them lost valued friends.

CONTENTS

Part IV: Students and Learning

Part V: Assessment

Part VI: Issues and Perspectives

PREFACE

In 1965, a bit more than 40 years ago, in response to growing interest in mathematics education research, the Board of Directors of NCTM appointed a Committee on Research in Mathematics Education (Johnson, Romberg, & Scandura, 1994). One result of the committee's deliberations was the publication in 1967 of a slim volume arguing for the need for more high-quality research (Scandura, 1967). In the preface to this volume, the editor observed that "leading citizens believe in research as never before. . . . [And] many teachers are asking for evidence to support or deny the current crop of claims demanding changes in curriculum and pedagogy. . . . Recommendations for change should be based on research" (Scandura, 1967, p. iii). In the same year, another prominent American mathematics education researcher, Robert Davis, wondered "in a society which has modernized agriculture, medicine, industrial production, communication, transportation, and even warfare as ours has done, it is compelling to ask why we have experienced such difficulty in making more satisfactory improvements in education" (Davis, 1967, p. 53).

The situation is not so different today, but the criticism has become more strident. Presently, in the middle of the first decade of the 21st century, diverse individuals and groups have been promoting a variety of old and new instructional approaches, programs, and policies for mathematics education. Researchers are being exhorted to gather and analyze data for evaluating the efficacy of various instructional approaches and curricula as never before. Individuals both within and outside of the mathematics education research community have begun to promote specific agendas, some largely for political reasons. Thus, a pressing challenge for researchers has been to reach out to its constituents by providing research-based advice about questions that concern that constituent community, and to do so in an unbiased, rational, and convincing manner. Members of the research community have discovered that the general public understands and appreciates very little about *what* researchers do or *why* they do it. They find themselves in the position of having to defend their work. Many factions—from teachers to politicians—are calling for mathematics educators to find out "what works" in classrooms at all levels (see, for example, U.S. Department of Education, What Works Clearinghouse, www.whatworks.ed.gov, retrieved August 6, 2006). This emphasis on studies to document what works presents mathematics education researchers with a dilemma: How can they provide research results showing "what works" and at the same time advance the theoretical and model-building goals that are so important to scientific inquiry in every field?

The purpose of this Handbook is not to answer "what works" questions, although answers to many of these sorts of questions can be found in this volume. Instead, the primary purposes are to coalesce the research that has been done, to provide (perhaps) new conceptualizations of research problems, and to suggest possible research programs to move the field forward. It is only by occasionally stepping back and looking at where mathematics education research has been and suggesting where it should be going that the field is likely to ever be able to answer the questions that practitioners, policy makers, and politicians are asking.

HISTORY OF THE DEVELOPMENT OF THIS HANDBOOK

In May 2002, the Educational Materials Committee (EMC) of NCTM sent the NCTM Board of Directors a request to authorize and fund the development of a new edition of the *Handbook of Research in Mathematics Teaching and Learning*, which had been published in 1992. This request followed a review by the EMC of a proposal from Douglas Grouws and the NCTM Research Advisory Committee (RAC) regarding the revision of the Handbook. Grouws was the editor of the original Handbook and he and the RAC were convinced that, after 10 years, it was time to begin work on a new edition. In July of that year, the Board of Directors approved the EMC's request, and by September NCTM president Johnny Lott had convinced me to serve as editor. In November 2002, I had a very productive discussion with Harry Tunis, Director of Publications at NCTM and George Johnson, President of Information Age Publishing, Inc. (IAP). The result of that discussion was an agreement between NCTM and IAP to publish the new Handbook. A few days later, I met for the first and only time with the advisory board to conceptualize the new Handbook and begin to identify potential authors. The members of the advisory board included some of the most prominent members of the mathematics education community: Doug Grouws, University of Missouri; James Hiebert, University of Delaware; Carolyn Kieran, Université du Québec à Montréal; Judith Sowder, San Diego State University; and past-President of NCTM Lee Stiff, North Carolina State University. My colleague at Indiana University, Paul Kehle, was to serve as assistant editor for the project. At the end of our two-day meeting we had sketched out the overall structure of the Handbook, including identification of chapter topics to emphasize or de-emphasize, and possible authors. Although the topics remained essentially the same throughout the development of the Handbook, some of the authors changed over time for various reasons. During the ensuing four-plus years, the members of the advisory board were also available to give much-needed advice and, in general, to help out in various other ways. During the development process, authors and author teams submitted outlines of their chapters for review by at least two outside reviewers. Having received the reviews, authors proceeded to prepare an initial draft that was closely scrutinized by the same reviewers, who provided valuable constructive criticism of the drafts. Of course, I, as editor, also reviewed each draft and, when appropriate, provided additional commentary. The authors then used the reviewers' feedback to prepare a second draft. I reviewed the second drafts to determine if the authors had adequately addressed the concerns raised by the reviewers. In a few cases, additional work was called for. Once I decided that a draft was acceptable, the editorial staff at Indiana University commenced to prepare it to be sent to the publisher. On average, the process from submission of an outline to submission to the publisher took two full years.

AUDIENCE

The audience remains much the same as for the 1992 Handbook, namely, mathematics education researchers and other scholars conducting work in mathematics education.

> This group includes college and university faculty, graduate students, investigators in research and development centers, and staff members at federal, state, and local agencies that conduct and use research within the discipline of mathematics. The intent of the authors of this volume was to provide useful perspectives as well as pertinent information for conducting investigations that are informed by previous work. The Handbook should also be a useful textbook for graduate research seminars. (Grouws, 1992, p. ix)

In addition to the audience mentioned above, the present Handbook contains chapters that should be relevant to four other groups: teacher educators, curriculum developers, state and national policy makers, and test developers and others involved with assessment. Taken as a whole, the chapters reflect the mathematics education research community's willingness to accept the challenge of helping the public understand what mathematics education research is all about and what the relevance of their research findings might be for those outside their immediate community.

SCOPE

More than 30 years ago, in an effort to define the scope of a reference volume on research in mathematics education, the Publications Committee of NCTM asked mathematics educators to answer two questions about the kind of

> reference on research in mathematics education which you would insist be on your shelf, the shelf of every doctoral student in mathematics education, and the shelf of every person responsible for research in mathematics education. What would you want such a reference to contain? What do you think such a reference need not contain? (Shumway, 1980, p. v).

The advisory board for the present Handbook considered the same two questions. A quick perusal of the previous research compendia published by NCTM (i.e., those published in 1980 [Shumway], and 1992 [Grouws]) shows that the scope of mathematics education research has expanded with each new publication. Recognizing a need "to introduce the beginner to some of the issues and problems in the research process" (Shumway, 1980, p. v), the editorial board of the 1980 volume decided to devote nearly one-third of the volume to this topic, with the balance dealing with the identification of critical, productive problems for researchers, dealing primarily with learning, teaching and curriculum.

Not only was the 1992 handbook considerably longer than Shumway's volume, its scope expanded dramatically. In addition to chapters dealing with specific aspects of mathematics learning (i.e., chapters on research on addition and subtraction, multiplication and division, rational numbers, problem solving, estimation and number sense, algebra, geometry and spatial reasoning, probability and statistics, and advanced mathematical thinking), new avenues of active inquiry appeared. For example, attention to instructional issues was quite prominent (chapters on learning and teaching with understanding, mathematics teaching practices, and grouping for instruction) and research on teachers and teacher education had become part of the mainstream (chapters on teachers' beliefs and conceptions, teachers' knowledge, teacher professionalism, and becoming a mathematics teacher). The topics of the remaining chapters of that handbook indicated that new emphasis had been placed on technology; on assessment and international comparisons; and on philosophical, social, personal, and cultural dimensions of mathematics education (one chapter each on the nature of mathematics, the culture of the mathematics classroom, ethnomathematics, affect, gender, and race, ethnicity, social class, language, and achievement in mathematics).

The current Handbook demonstrates that the research community has sustained its focus on problems of learning, teaching, teacher education, assessment, technology, and social and cultural aspects of mathematics education. But, in addition, some new areas of interest have emerged or been expanded.

The present volume has six sections. The first section, *Foundations*, contains a chapter devoted to two areas not specifically addressed in the 1992 handbook: philosophy (Cobb) and theory (Silver & Herbst). These two chapters plus a chapter on research methods (Schoenfeld) illustrate that the mathematics education research community has taken seriously the importance of thinking hard about the underpinnings of its work.

The second section, *Teachers and Teaching*, contains four chapters focusing on: (1) what knowledge mathematics teachers should have (Hill, Sleep, Lewis, & Ball), (2) how teachers are prepared and the kinds of professional development they receive (J. Sowder), (3) what teachers do in their classrooms and the culture within which teaching takes place (Franke, Kazemi & Battey), and (4) teachers' beliefs and affects (Philipp). Teachers and teaching received considerable attention in the 1992 handbook, but the chapters in section I of the new Handbook demonstrate that new perspectives and new lines of inquiry have emerged.

Section III, *Influences on Student Outcomes*, is also made up of four chapters, one of which focuses on how teaching influences what students learn (Hiebert & Grouws). Among the other three chapters in this section, one focuses on a topic that was not present in the 1992 handbook: how curricula influence student learning (Stein, Remillard, & Smith). The other two chapters comprising the balance of this section—one dealing with the notions of ethnomathematics and everyday cognition (Presmeg), the other with issues of race, class, gender, language, culture, and power (Diversity in Mathematics Education Research Center)—illustrate the ever-expanding scope of research in the field.

More than one-third of the volume (11 chapters) is contained in section IV, *Students and Learning*. Several of the chapters address topics present in the previous handbook: whole number concepts and operations (Verschaffel, Greer, & DeCorte), rational numbers and proportional reasoning (Lamon), algebra in the middle school through college levels (Kieran), problem solving (Lesh & Zawojewski), geometric and spatial thinking (Battista), and post-secondary mathematics thinking and learning (Artigue, Batanero, & Kent). Research on probability and statistics, a single chapter in 1992, has been split into two chapters—one on probability (Jones, Langrall, & Mooney), the other on statistics (Shaughnessy). There also are three chapters representing new areas of research that have burgeoned since the 1992 volume: early childhood (i.e., pre-Grade 1) mathematics learning (Clements & Sarama), early algebra and algebraic reasoning (i.e., pre-middle school; Carraher & Schliemann), and the learning and teaching of proof (Harel & L. Sowder).

The ever-increasing role of assessment in mathematics education made it appropriate to devote an entire section to this topic. The first chapter of Section V, *Assessment*, considers how classroom assessment can actually support learning goals, rather than just measure them (Wiliam). The second chapter (Wilson) deals with high-stakes, standardized tests and considers the influence these tests have had on educational practices, and the third chapter (de Lange) considers issues associated with large-scale, international assessments.

The sixth and final section contains six chapters dealing with a wide range of topics and issues. One chapter (Bishop & Forgasz) focuses on the enduring issues and perspectives that relate to equity and access, another with issues and research associated efforts to understand the impact of computer technologies on learning, teaching, and curriculum (Zbiek, Heid, Blume, & Dick), and a third describes an approach for specifying the content for any assessment of students' achievement (Webb). Two chapters consider issues associated with an area of study not previously within the purview of mathematics education research, namely educational policy. The authors of one of the policy chapters (Ferrini-Mundy & Floden) discuss studies that illustrate the formulation, status, implementation, and effects of educational policy that relate to mathematics education. The authors of the other policy chapter (Tate & Rousseau) provide a review of mathematics education research and relevant policy literature that informs school leaders who are endeavoring to build effective school learning environments for traditionally underserved students. In the final chapter of the volume, the author, Mogens Niss, reflects on the state of mathematics education research, where it seems to be going, and what sort of research would need to be done to get us to the utopian state of being able to provide final, conclusive answers to all the important questions in mathematics education.

ACKNOWLEDGEMENTS

A large number of dedicated individuals helped in the development and publication of this Handbook. As I have already noted, the members of the advisory board (Doug Grouws, James Hiebert, Carolyn Kieran, Judith Sowder, and Lee Stiff) were especially helpful in conceptualizing the volume, providing timely advice, and, on occasion, helping me deal with the inevitable frustrations involved in keeping the project on course and schedule. On behalf of the entire mathematics education community, I wish to express my sincere gratitude to them. I also extend my thanks to Paul Kehle, who, unfortunately, was not able to continue to assist me after the first year—he left Indiana University for another position. He provided valuable help during the early stages of the project. Harry Tunis, Director of Publication Services at NCTM was always helpful and supportive of the project at every stage—even when I had to tell him that we would not be able to stick to our timeline! A special thanks is due to the many reviewers who were willing to spend countless hours poring over drafts of chapters, making insightful and always helpful comments. Indeed, the Handbook exists to a great extent because of their tireless, thoughtful, largely unrecognized efforts on behalf of mathematics education.

A colleague suggested to me that when the Handbook is complete I will be better informed about contemporary mathematics education research than anyone else in the U.S. Perhaps; it is at least arguable! To the extent that it is true, I owe it to the authors and reviewers who worked so hard to make this Handbook the source authority on research in the field. I learned a tremendous amount from reading chapter drafts, considering reviewers' comments, and discussing the focus and organization of chapters. I also gained even more appreciation for my colleagues than I had before—and I already had tremendous respect for them. They are such busy folks—I really don't know how they were able to put together such thoughtful and wonderfully insightful chapters. So, to all authors and reviewers, I extend my sincerest, Thank You!

Closer to home, Indiana University doctoral students Paula Stickles and Andrea McCloskey were absolutely indispensable to the success of the project. Paula helped with the thankless job of formatting the final drafts to make them conform to APA standards. Andrea had an even more tedious task in addition to assisting with formatting: checking every reference of every chapter for correctness and completeness. I am convinced that she now knows as much or more about APA reference conventions as anyone! Both Paula and Andrea assisted me with the tedious job of checking page proofs, as did Zac Rutledge, another very able doctoral student. One other person who played a very important behind-the-scenes role in putting drafts into final form was Cheryl Burkey. Cheryl, an accomplished and very experienced copy editor, knows the *APA Publications Manual* extremely well. She copy edited almost all of the drafts in preparation for sending them to the IAP editors for final production. Thank you Paula, Andrea, and Cheryl—your professionalism, competence, and cheerful dispositions made my job much easier and rewarding. Thanks are also due to the staff of Information Age Publishing. In particular, IAP president, George Johnson, deserves special acknowledgement for his patience and understanding as the Handbook slowly became a reality. I wish also to express my thanks to Scott Suckling of MetroVoice Publishing Services for his work on page design and composition and Ann Salinger who created the indices.

Finally, I wish to thank my wife, Diana, who sometimes wondered if her often-cranky husband would ever bring the Handbook to fruition. Now we can begin to enjoy some time together when I will not have some Handbook issue in the back of my mind.

—Frank K. Lester, Jr.
Bloomington, IN

REFERENCES

Davis, R. J. (1967). The range of rhetorics, scale and other variables. *Proceedings of the National Conference on Needed Research in Mathematics Education* in the *Journal of Research and Development in Education, 1*(1), 51–74.

Grouws, D. A. (Ed.). (1992). *Handbook of research on mathematics teaching and learning.* New York: Macmillan.

Johnson, D. C., Romberg, T. A., & Scandura, J. M. (1994). The origins of *JRME*: A retrospective account. 25th Anniversary Special Issue of the *Journal for Research in Mathematics Education, 25,* 561–582.

Scandura, J. (Ed.). (1967). *Research in mathematics education.* Washington, DC: National Council of Teachers of Mathematics.

Shumway, R. J. (Ed.). (1980). *Research in mathematics education.* Reston, VA: National Council of Teachers of Mathematics.

15

EARLY ALGEBRA AND ALGEBRAIC REASONING

David W. Carraher

TERC

Analúcia D. Schliemann

TUFTS UNIVERSITY

WHY ALGEBRAIC REASONING?

This chapter provides an overview of research about algebraic reasoning among relatively young students (6–12 years). It focuses on mathematics learning and, to a lesser extent, teaching. Issues related to educational policy, epistemology, and curriculum design provide a backdrop for the discussion.

We will adopt as our working definition of *algebra* a description given by Howe (2005) at a meeting about algebraic reasoning in grades K–12:

> [Algebra involves]
>
> i. Working with variables, and in particular, arithmetic with variables, so the formation of polynomial and rational expressions. This also includes representing, or "modeling" concrete situations with expressions, and setting up equations. It is also often extended to include extracting roots. . . . It also includes manipulating expressions and equations, to simplify, solve and interpret.
>
> ii. Algebraic structure, primarily as captured in the Rules of Arithmetic (aka, the field axioms). The Rules of Arithmetic encapsulate the legal manipulations on polynomial or rational expressions. If taking rational powers is allowed, they have to be supplemented by the Law of Exponents and the multiplicativity of fractional powers. These rules, together with the principles for transforming equations (the original techniques which gave rise to the subject known to us as algebra), summarize the basis for algebraic technique, which works on the algebraic expressions described in i), and on equations between them. (p. 1)

The references in Howe's definition to "arithmetic with variables," "modeling," and "Rules of Arithmetic" are especially noteworthy. They reflect the idea that "algebra is inherent to arithmetic." or, as we have expressed it, arithmetic has an "algebraic character" (Carraher, Schliemann, & Brizuela 2000; Schliemann, Carraher, & Brizuela, 2007). They further suggest that arithmetic and elementary algebra are not fully distinct.

We shall treat *arithmetic* as the science of numbers, quantities, and magnitudes. Students' understanding of number develops throughout many years, covering the natural or counting numbers, integers, rational numbers (common fractions and decimal fractions) and, to a very limited extent, real and complex numbers. Operations with numbers vary according to the domain of number. This, together with the fact that students continually learn new conventions for representing numbers and operations on them—e.g., place value and rules for computation (Howe,

2006)—means that arithmetic is a moving target, from the student's and from the teacher's perspectives. Some topics, such as the divisibility of integers, are integral to school mathematics but also belong to special sub-domains of mathematics: divisibility for example is part of elementary number theory or "the higher arithmetic" (Davenport, 1968). We will argue that arithmetic entails functions, but the kinds of functions related to operations and properties suitable for elementary and middle school students are yet to be determined and remain a topic of importance for mathematicians and learning specialists to work out.

Algebraic reasoning refers to psychological processes involved in solving problems that mathematicians can easily express using algebraic notation. This is admittedly a difficult and somewhat subjective matter, one that runs the risk of over-attributing competence to students. However, in the same sense that societies solved problems of algebra before the existence of algebraic notation (Harper, 1987), students may be able to work with variables and the Rules of Arithmetic (i.e., the field axioms) before they have been taught algebra. As we shall see, there is also some compelling evidence (e.g., Davydov, 1991) that students can learn to use algebraic notation and techniques themselves.

We will use the expression *early algebra* (EA) to encompass algebraic reasoning and algebra-related instruction among young learners—from approximately 6 to 12 years of age. Such a definition is useful for getting started, but in order to appreciate the puzzles, challenges and controversies of the field one needs to consider ideas and research findings from recent decades. What emerges, we hope, is a more integrated and challenging view about the early mathematics curriculum, especially with regard to the nature and purposes of arithmetic instruction.

RATIONALE AND STRUCTURE OF CHAPTER

In this section we address the rationale and structure of the chapter. Next, in "A Traveler's Guide To Early Algebra," we present terminological and conceptual distinctions useful for getting around in the field. In the three sections on "Entry Points to EA" we consider various approaches and research findings about elementary school children learning EA. In the final sections, "Special Issues in EA" and "Concluding Thoughts," we summarize the main findings and consider the still open issues about EA and their implications for teacher education, curriculum development and policy.

The idea of writing a chapter on EA may appear unnecessary. After all, handbooks typically do not include chapters on "early calculus," "early geometry," or "early statistics." Besides, textbooks and syllabi from early grades have tended to include relatively little content clearly related to algebra. Furthermore, it is well known that many adolescents have difficulty learning algebra. Could the situation really be more promising for younger students? Would our time and energy not be better spent strengthening the existing early mathematics curriculum? If students had a more solid grounding in arithmetic and elementary geometry, would they not be better prepared when they encountered algebra at its traditional place in the curriculum? These are reasonable questions. However, as we will try to show, there are compelling pragmatic and scientific reasons for paying special attention to the role of algebra in elementary school.

The Recent Focus on Algebraic Reasoning in the Early Grades

Some may react to the idea of introducing algebra in elementary school with puzzlement and skepticism. After all, much prior research highlights the difficulties that middle and high school students have with algebra, showing, for instance, that these students (a) believe that the equals sign[1] only represents a unidirectional operator that produces an output on the right side from the input on the left (Booth, L., 1984; Kieran, 1981; Vergnaud, 1985; Vergnaud, Cortes, A., & Favre-Artigue, P., 1988b); (b) focus on finding particular answers (Booth, 1984); (c) do not recognize the commutative and distributive properties (Boulton-Lewis, 2001; Demana, 1988; MacGregor, 1996); (d) do not use mathematical symbols to express relationships among quantities (Bednarz, 2001; Bednarz & Janvier, 1996; Vergnaud, 1985; Wagner, 1981); (e) do not comprehend the use of letters as generalized numbers or as variables (Booth, L., 1984; Kuchemann, 1981; Vergnaud, 1985); (f) have great difficulty operating on unknowns; and (g) fail to understand that equivalent transformations on both sides of an equation do not alter its truth value (Bednarz, 2001; Bednarz & Janvier, 1996; Filoy & Rojano, 1989; Kieran, 1989; Steinberg, Sleeman, & Ktorza, 1990). Collis (1975), Herscovics & Linchevski (1994), Kuchemann (198), and MacGregor (2001), among others, have attributed students' difficulties to insufficient cognitive

[1] Both "equals sign" and "equal sign" are encountered in the literature. We will adopt the former throughout.

development. And Filoy & Rojano (1989) and Sfard (1995) further relate students' development and their cognitive difficulties with algebra to historical trends in mathematics.

However from time to time in recent decades mathematics educators (Davis, 1967a, 1967b, 1972, 1985, 1989; Davydov, 1991) have discussed the idea of introducing algebra much earlier than its traditional appearance in high school courses. Kaput (1995, 1998) argued that the weaving of algebra throughout the K–12 curriculum could lend coherence, depth, and power to school mathematics, and replace late, abrupt, isolated, and superficial high school algebra courses. Proponents of EA have stressed that the current content of elementary mathematics is not fully distinct from algebra (Mason, 1996; Bass, 1998; Carraher, Schliemann, & Brizuela, 2000, 2005). A deep understanding of arithmetic, for example, requires mathematical generalizations that are algebraic in nature. Some have argued that algebraic notation makes it easier for adults as well as young learners to give expression to such mathematical generalizations (Brizuela & Schliemann, 2004; Carraher, Schliemann, & Brizuela, 2000; 2005; Schliemann, Carraher, & Brizuela, 2007). Bodanskii (1991) suggests that "the algebraic method is the more effective and more 'natural' way of solving problems with the aid of equations in mathematics" than the arithmetical method (p. 276). An Algebra Initiative Colloquium Working Group (Lacampagne, 1995; Schoenfeld, 1995) proposed that algebra should pervade the curriculum instead of appearing in isolated courses in middle or high school. And the 12th International Commission on Mathematical Instruction (ICMI) Study Conference on the Future of the Teaching and Learning of Algebra (Chick, Stacey, Vincent, & Vincent, 2001; Lins & Kaput, 2004; Stacey, Chick, & Kendal, 2004) created a special working group on issues related to EA.

A Decisive Moment

Two events capture the current political significance of EA for mathematics educators in the United States.

Event 1: NCTM's Endorsement

The National Council of Teachers of Mathematics (NCTM, 1989; 2000) recommended that an algebra strand be woven deeply throughout the preK–12 curriculum:

> Algebra encompasses the relationships among quantities, the use of symbols, the modeling of phenomena, and the mathematical study of change. The word algebra is not commonly heard in elementary school classrooms, but the mathematical investigations and conversations of students in these grades frequently include elements of algebraic reasoning. These experiences present rich contexts for advancing mathematical understanding and are an important precursor to the more formalized study of algebra in the middle and secondary grades. For example, when students in grades 3 through 5 investigate properties of whole numbers, they may find that they can multiply 18 by 14 mentally by computing 18×10 and adding it to 18×4, thus using the distributive property of multiplication over addition in a way that contributes to algebraic understanding.
>
> As with number, these concepts of algebra are linked to all areas of mathematics. Much of algebra builds on students' extensive experiences with number. Algebra also is closely linked to geometry and to data analysis. The ideas of algebra are a major component of the school mathematics curriculum and help to unify it. A strong foundation in algebra should be in place by the end of eighth grade, and ambitious goals in algebra should be pursued by all high school students. (c.f. http://standards.nctm.org/document/chapter3/alg.htm)

In brief, teachers are now expected to introduce concepts from algebra in early mathematics instruction. And, given that NCTM is placed "firmly in the driver's seat of the school mathed reform bus" (Howe, 1998b, p. 243), it is not surprising that textbook publishers have been updating their offerings to comply with NCTM's recommendations.

Event 2: The RAND Mathematics Study Panel Report

A 2003 report commissioned by the Rand Corporation argued that "the initial topical choice for focused and coordinated research and development [in K–12 mathematics] should be algebra" because of its fundamental role "for exploring most areas of mathematics, science, and engineering. . . . [and its] unique and formidable gatekeeper role in K–12 schooling" (RAND Mathematics Study Panel, 2003, p. 47).

Although there is some agreement that algebra has a place in the elementary school curriculum, the research basis needed for integrating algebra into the early mathematics curriculum is still emerging, little known, and far from consolidated.

The NCTM directives are general and ambiguous both in terms of mathematical content (Howe, 1998a) and pedagogical implementation. In a sense, the wheels of change have been set in motion before the destination and outcomes are decided upon. By and large the curriculum materials are yet to be developed. Teacher education programs have only recently begun to respond to the call for EA, often without having benefited from reflections on prior work of direct relevance. The mathematics education community still needs to work on establishing a solid research basis for such an undertaking.

Five Key Issues

There is another reason for a chapter specifically about EA. Discussions about EA have raised issues related to the nature of mathematics learning, the structure of mathematics, the roles of teachers, and the feasibility of "algebra for all". Below we mention five major issues in the recurring debate about EA and the mathematics curriculum.

Issue 1: The Relations Between Arithmetic and Algebra

EA goes against the prevailing view in U.S. mathematics education, noted by several (e.g. Chazan, 1996; Gamoran & Hannigan, 2000), that algebra belongs after arithmetic and possibly for only a select minority of students who have the necessary skills and future career needs. Not surprisingly, territorial disputes sometimes arise about the distinctions between arithmetic and algebra. Filoy and Rojano (1984, 1989), for example, claimed there is a clear-cut boundary separating arithmetic from algebra; namely, arithmetical approaches such as "undoing" can handle cases as complex as equations with a variable on one side, but equations with variables placed on each side of the equals sign belong to the realm of algebra.

Whereas EA proponents generally search for continuity between arithmetic and algebra, there have been several well-argued analyses stressing the rupture or discontinuity between the two domains (Balacheff, 2001; Filoy & Rojano, 1984, 1989; Lins, 2001).

Should algebra be treated as a distinct domain with its own methods, objects and perspectives? Or can algebra be developed out of or be 'grafted' onto arithmetic? It might seem that this issue could be straightforwardly resolved by consulting research mathematicians about the relations between arithmetic and algebra. But the question also requires the perspectives of mathematical learning researchers, teacher educators, and historians of mathematics. Such specialists typically ask how arithmetic and algebra curricula can take into account student learning, how teachers can learn to conceive anew the curricula, activities, and goals of mathematics in the early grades, how algebra arose historically, and so on.

Issue 2: Process versus Object

Various authors have contrasted mathematics as, on one hand, operation/process/computation/procedure/algorithm and, on the other, as object/structure/relation (Dubinsky & Harel, 1992a; Otte, 1993; Sfard, 1995; Sfard & Linchevsky, 1994). There is some disagreement over the enduring worth of process approaches. Some authors treat procedural approaches as inherently more primitive and thus in need of replacement by an object orientation; others treat procedural interpretations as different, yet nonetheless desirable even in advanced mathematical thinking. Sfard (1992) claims that students first grasp algebra from an operational perspective and, only later, can develop a structural conception of algebra; therefore, she concludes, the teaching of algebra should start from an operational, instead of from a structural perspective.

Issue 3: The Referential Role of Algebra

For some, algebraic understanding grows out of trying to represent extra-mathematical situations—that is, through modeling (Gravemeijer, 2002; Kirshner, 2001; Lehrer & Schauble, 2000). Others tend to view algebra's referential role as a source of distraction and interference. This tension comes to a head in discussions about the role of patterns in EA instruction.

Issue 4: Symbolic Representation (narrowly defined)

Substantial differences of opinion can be identified regarding the importance, timing, and meanings associated with conventional algebraic notation. Some argue that algebraic (sometimes referred to as "symbolic") notation deserves a prominent place in early instruction (Carraher, Schliemann, & Brizuela, 2000, 2005; Schliemann, Carraher, & Brizuela, 2007). Others believe it should be postponed for several years, on the grounds of being developmentally inappropriate or likely to engage students in meaningless symbol manipulation (Linchevsky, 2001).

The papers presented at the Research Forum on Early Algebra that took place in the 25th International Conference for the Psychology of Mathematics Education provide a good sample of the multiplicity of views on EA (Carraher, D. W., Schliemann, & Brizuela, 2001; Linchevsky, 2001; Radford, 2001; Tall, 2001; Teppo, 2001; Warren & Cooper, 2001). Some, like Linchevsky (2001), believe that algebra should continue to be introduced only after students have spent at least a half-dozen years learning the basics of arithmetic. Even proponents of EA (Blanton & Kaput, 2000, 2001; Carpenter, Franke, & Levi, 2003; Carraher, Schliemann, & Schwartz, 2007; Dougherty, in press; Moss, Beatty, McNab, & Eisenband, 2006; Schifter, Monk, Russell, & Bastable, in press), who regard algebra as deserving a prominent place in the early mathematics curriculum, are likely to disagree about issues so basic as the following:

1. What learning tasks and forms of thinking are algebraic?
2. What kinds of evidence are needed to evaluate the presence of algebraic thinking among young students?
3. What pedagogical approaches, teacher education and policy guidelines ought be encouraged?

Issue 5: Symbolic Representation (broadly defined)

One of the fundamental contributions of math ematics learning research has been to highlight the roles of other symbolic systems—particularly tabular, graphical, and natural language systems–that come into play in algebraic reasoning. There is some diversity of opinion regarding the importance of such systems of representation. Many believe they provide crucial entry points for learning algebra; some consider them to be important even after a certain mastery of symbolic reasoning (in the narrow sense) has been achieved. Still others regard them as "pre-algebraic."

As we will see, in recent years there has been a gradual accumulation of studies of EA (see also Rivera, 2006). One would hope that, as the mathematics education community carefully considers data from empirical studies, beliefs—even ardently held ones—would be reassessed. A conjecture such as, "the more students are confident in their arithmetical competencies, the more they will be reluctant to accept the use of algebra on problems they recognize as pertaining to the field or arithmetic (or computation)" (Balacheff, 2001, p. 252) bears directly on the first issue mentioned above. And it is the sort of conjecture that needs to be assessed in the light of empirical evidence. With these issues in mind, we begin our discussion of the EA research.

A TRAVELER'S GUIDE TO EARLY ALGEBRA

School Algebra and EA

There is a certain agreement in the United States about the algebra students need to know by the end of high school (Bass, 2005). Students are to be familiar with polynomial expressions and certain trigonometric identities. They should be able to solve first order equations in one variable (e.g., $5x - 37 = 2x + 5$), quadratic equations associated with conic sections, and systems of equations in two or more variables. They should be fluent in the use of graphs for representing functions, including elementary trigonometric functions. The should be familiar with mathematical induction as applied, in particular, to the properties of the binomial coefficients. They should be comfortable handling diverse issues about variation and covariation. They should be able to mathematize word problems, use mathematical models, and apply their conclusions to the original problem contexts. High school algebra represents only a tiny sampling of algebra as understood by research mathematicians, and perhaps not the most essential or fundamental part.

Considerably less agreement can be found in the case of early algebra. Does EA include the same topics as high school algebra? If it does not, what topics are 'appropriate for minors'? Does it introduce new topics? Does it introduce the same methods (factoring and expansion of polynomial expressions; solving systems of equations, etc.)? Does it require significant departures in the ordering and handling of topics?

A thorough review of algebraic reasoning in the early grades would scrutinize data from nations outside of the United States and western Europe. And some important lessons could be learned from studying how other countries have dealt with contentious issues facing our own educational system. Curricular materials in wide use in Russia and Singapore reveal that students are being exposed to ideas and techniques from algebra much earlier than in the United States, often with striking success (Milgram, 2006b; Davydov, 1991). This is noteworthy and important. Although it is likely that textbooks "make a difference," we believe that merely adopting course materials from other countries (with translation, of course) is not enough.

Research is gradually providing a sense of young children's capabilities and of the viability of different approaches to EA and to the teaching needs to successfully implement EA activities. The news is generally more encouraging than it was two decades ago, when studies seemed to suggest that algebra was inherently too difficult for many adolescents. In addition, there is a growing recognition that EA is not simply a subset of the high school syllabus; rather, it is a rich sub-domain of mathematics education where special topics and issues are the focus of research and instruction. Furthermore the field now has a better sense of how representations and situations play roles in students' mathematical understanding.

One might be tempted to assume that mathematics has an inherent structure that requires learners to follow the same general pathway regardless of the age at which they first begin to study the subject. According to this view, young learners begin with "the basics" and simply move ahead (perhaps) more slowly than older students. Yet for many centuries adolescents learned the basics in Latin by repeatedly practicing declensions and verb conjugations. It might have appeared obvious to many that relatively young learners would need to learn the basics in like manner—by drilling their declensions and conjugations. But young students' versatility in learning foreign languages does not derive from skills in expressly parsing written sentences. Even though grammar and structural analysis ultimately play important roles in language mastery, they may not constitute the soundest pillars for early language learning. This said, prudence recommends acknowledging the need for a balance between basic skills and conceptual understanding (Wu, 1999).

Competence in algebra requires an ability to derive valid, non-trivial and useful inferences from written forms and their underlying structure. But this in no way suggests that the study of algebraic expressions is the only or even the best point of departure to algebra. As we have stressed before, algebra for young learners may call for somewhat different approaches than those traditionally given prominence in algebra instruction for adolescents, among which are the following (Carraher, Schliemann, & Schwartz, in press):

1. Conventional algebraic notation is not the only vehicle for the expressing algebraic ideas and relations; tables, number sentences, graphs and specialized linguistic structures can also express algebraic ideas.[2]
2. Context and grounding in physical quantities may be desirable in EA instruction even though they can legitimately be curtailed or downplayed in later instruction.
3. Functions provide opportunities for bringing out the algebraic character of many existing early mathematics topics and activities.

EA Versus Pre-Algebra

One can discern two families of approaches related to mathematical learning and teaching before the traditional onset of algebra instruction.

Pre-Algebra Approaches

Pre-algebra approaches (Filoy & Rojano, 1989; Herscovics & Kieran, 1980; Herscovics & Linchevski, 1994) aim to ease the abrupt transition from arithmetic to algebra. They make efforts to augment or redefine the uses and meanings of mathematical symbols such as +, –, ×, ÷, and = as employed in algebraic expressions and equations. The rationale is that careful intervention before algebra instruction proper may mitigate the difficulties that beginning algebra students typically show—difficulties attributed to the inherent differences between arithmetic and algebra. Because young students were presumed to be ill suited for learning algebra, most of the research done in the 1980s, focusing mainly on equations, did not question the idea that algebra should start only in high school.

Herscovics and Kieran (1980) studied the effect of transitional approaches to algebra by attempting to expand seventh and eighth graders' notion of equality and the gradual transformation of arithmetic expressions into algebraic equations. They found that the six participants in the study showed a "clear understanding of arithmetical identities, equations, and algebraic rules" (Herscovics & Kieran, 1980, p. 579)." Vergnaud (1985) and Vergnaud, Cortes, & Favre-Artigue (1988b) used a series of activities based on the behavior of a two-plate balance scale to help seventh to ninth grade students at risk of failing in algebra courses. By the end of the study the at-risk students performed better than the control group generating equations from word problems and then solving them through conventional symbolic methods and syntactic rules. Filloy and Rojano (1989) were less successful in their work with geometrical and balance scale models aimed at promoting use of syntactic rules for solving equations. Students in their study required constant intervention by the teacher to adopt algebraic methods.

Over time, as researchers moved away from equation solving as the principal activity of algebra instruction, transitional approaches switched to generalization, number patterns, variables, and functions. Transitional approaches to algebra as the study of variables and functions have been closely associated with computers in mathematics education. Healy, Hoyles, & Sutherland, (1990) and Ursini (1994, 1997, 2001), worked with Logo environments (Papert, 1980), in carefully structured activities to encourage students to move toward general algebraic representations. Despite the gulf between algebraic symbolism and spreadsheet formulas, which refer to values via their grid location, Sutherland and Rojano (1993), Rojano (1996), and Sutherland (1993) found that, through participation in spreadsheet activities, Mexican and U.K. students as young as 10 years of age, performed relatively well on algebraic tasks working with unknown quantities and operating on them as if they were known. A number of software environments specifically address the teaching and learning of variables and functions. Kieran, Boileau, and Garançon (1996)

---

[2] Some would treat these auxiliary representations as pre-algebraic (Harper, 1987). Regardless of whether one wishes to confer plenary algebraic status on such representations, they deserve recognition as being vital in students' learning trajectories.

found that software-guided support for students' intermediary representations facilitates the adoption of conventional representations. Heid (1996) found that, as they work with software, students develop algebraic thinking that allows them to model real world situations. Software environments designed by Schwartz (1996a, 1996b) and Schwartz & Yerushalmy (1992b, 1992c) give prominence to multiple representations, such as algebra symbolic notation, numbers, graphs, and natural language, and focus on functions as the basic mathematical object and the core of algebra. In Schwartz's approach, the software is a tool for modeling situations and relationships and a medium for students to represent their understandings and to flexibly move among the different kinds of representations. A functions approach to algebra seems to be extremely valuable. But, as was the case with other transitional or introductory approaches, evaluation of students' learning through the proposed activities is still scarce.

It is interesting to note that, even though many of those who developed transitional approaches to algebra recognize that part of the problem originates from students' previous experiences with arithmetic, they generally do not question the sequence of arithmetic first, algebra later.

EA Approaches

EA approaches recognize that mathematical symbols are employed diversely in arithmetic and algebra (Blanton & Kaput, 2000, 2001; Carpenter, Franke, & Levi, 2003; Carraher, Schliemann, & Schwartz, in press; Dougherty, in press; Moss et al., 2006; Schliemann, Carraher, & Brizuela, 2007; Schifter et al., in press). Likewise, they acknowledge the significant and serious difficulties that many adolescents have with algebra. But EA approaches take a different stance regarding the etiology of adolescents' difficulties with algebra. Without denying that algebra moves towards increasingly abstract mathematical objects and relations and depends on ever more elaborate techniques and representational forms, an EA approach attributes difficulties exhibited by adolescent students of algebra as due in large part to shortcomings in how arithmetic and, more generally, elementary mathematics are introduced. As Booth (1988) suggests, "the difficulties that students experience in algebra are not so much difficulties in algebra itself as problems in arithmetic that remain uncorrected" (p. 29).

We develop this view throughout the present chapter. For the moment an example or two may suffice. Carpenter and colleagues have developed teaching activities for exploring number sentences that support the notion that an equals sign (in cases such as $5 + 4 = 9$) means more than "yields" or "produces" (Carpenter & Franke, 2001; Carpenter et al., 2003), a point easily conveyed through examples such as $9 = 5 + 4$ and $3 + 6 = 5 + 4$. Number sentences have the additional virtue of being able to express not only the reflexive property ($a = a$) of the equals sign (which students have no difficulty in accepting), but also the symmetric ($a = b \Rightarrow b = a$) and transitive ($a = b$ and $b = c \Rightarrow a = c$) properties. Remarkably, children as young as 8 and 9 years of age appear to be able to adopt and understand these uses of the equals sign when it is introduced to them through carefully constructed activities (Carpenter, Franke, & Levi, 2003).

Number sentence activities can highlight the *continuity* between arithmetic and algebra. In doing so they raise the possibility that the rupture between arithmetic and algebra is a result of a flaw in the way the early mathematics curriculum has been designed and implemented.

The distinction between pre-algebra and EA is not meant to spark a debate about terminology, much less to separate "believers from non-believers." Some of the most distinguished contributions to EA research have come from people not identified as EA researchers. Kieran (1992, 1996, 2006) for example, has provided seminal overviews of research about the teaching and learning of algebra that have been of tremendous importance for EA work. As we noted earlier, Filloy and Rojano (1989) have recognized the relative difficulty of first-order equations in which the variable appears on both sides of the equation as opposed to only one side. We happen to believe they have taken an extreme stance in arguing that these types of equations demarcate a "cognitive gap" between arithmetic and algebra. Our sense is that, although the variable-on-both-sides case is indeed more challenging, there are a number of intermediary cases for bridging the gap; we would further note that undoing, when performed on each side of an equation, is related to legitimate and important applications of inverse operations.

On the Possibilities of EA

Two questions still under debate are: Can young students really deal with algebra? And, can elementary school teachers teach algebra? Concerning the first question, which is the main focus of this chapter, until recently empirical studies on the teaching and learning of algebra in elementary schools, or of what came to be known as EA, were rather scarce. As we have noted elsewhere (Schliemann, Carraher, & Brizuela, 2006), the lack of empirical data in support of EA led researchers, educators, and policy makers to wonder about issues of possibility (Can teachers teach algebra to young students and can young students learn algebra?), implementation (How can the recommendations for EA be put into effect?), and desirability (Is it important or useful for students to learn algebra early or at all?).

Examples of young children's algebraic reasoning and/or use of algebra notation are described by Ainley (1999), Bellisio and Maher (1998), Brito-Lima's (1996, 1997) Lins Lessa's (1995), Slavitt (1999), Smith (2000), Warren (2004), and Warren and Cooper (2005), among others. Adopting different approaches, these studies, as well as the studies described later in this review, show that elementary school children can successfully learn about the rules, principles, and representations of algebra in the early grades.

Regarding the preparation of teachers to teach EA, some progress has been made (see, for instance (Blanton & Kaput, 2001) and (Schifter, 2005). This however, is beyond the scope of the chapter.

Parsing (Early) Algebra

Authors have parsed algebra and algebraic thinking diversely over the years (see Bednarz & Janvier, 1996; Bednarz, Kieran, & Lee, 1996; Bloedy-Vinner, 2001; Boyer & Merzbach, 1989; Harper, 1987; Kaput, 1995, 1998, in press; Kieran, 1996; Kirshner, 2001; van Amerom, 2002). Most have worked on specific dimensions of interest; relatively few have attempted to characterize the field exhaustively. When they have so tried, the categorical structure occasionally exhibits inconsistencies and overlaps. For example, a breakdown of algebra into generalizing, problem solving, modeling, and functions mixes non-disjoint reasoning processes (generalizing and problem-solving) with a topic of mathematics (functions) and another (modeling) (Bednarz, 1996) that can be understood either as a mathematical topic or a set of reasoning processes. This may reflect the fact that the analysis of algebraic thinking is still in its infancy; or it may reflect the fact that instructional approaches differ on so many dimensions that there may not be a way to parse them into disjoint and exhaustive types. We suspect that approaches to EA will never fit comfortably into a Linnaean taxonomy: a certain degree of eclecticism may even be healthy.

The fact that approaches resist pigeonholing does not mean that the field is conceptually intractable. For instance, there is a widespread acceptance of the historical breakdown of algebra into three stages ranging from an initial rhetoric stage (expressing both the problem and its solution in natural language, through a syncopated stage (in which numbers and letters are used to stand for unknowns or variables) and finally to a symbolic stage (where letters are also used to represent *given* or assigned values) (Boyer & Merzbach, 1989; Harper, 1987; Kieran, 1996; van Amerom, 2002). Consistent with this, most EA theorists and practitioners assign considerable importance to natural language representations of algebra relations in the early stages. Likewise, they tend to agree (e.g., Bloedy-Vinner, 2001) that the symbolic representation of parameters (e.g. $ax + b$) is inherently more advanced than the representation of expressions for which parameters have been assigned values (e.g. $3x + 9$). Although approaches may be conceptually messy, they provide points of access to significant theoretical and empirical work. In upcoming sections we consider approaches that are overlapping and not exclusive.

Kaput, a pioneer EA researcher, worked for many years trying to organize algebra and algebraic thinking according to general approaches (Kaput, 1995, 1998), eventually settling on three approaches, with two additional overarching strands (Kaput, in press). Kaput's three approaches are the following:

1. Algebra as the study of structures and systems abstracted from computations and relations, including those arising in arithmetic (algebra as generalized arithmetic) and in quantitative reasoning;
2. Algebra as the study of functions, relations, and joint variation;
3. Algebra as a cluster of specific modeling languages to express and support reasoning about the situations being modeled (Kaput, 2006, 2007; Kaput, Blanton, & Moreno, 2007).

Kaput's approaches are understood as *ideal types* in Weber's sense (Weber & Shils, 1949/1917, p. 90). The first approach accentuates the use of algebra to capture mathematical structures. The third approach refers to extra-mathematical modeling, that is, using mathematics to describe real-world data, relations and processes. The two approaches correspond generally to Kirshner's *referential* and *structural* "options" (Kirshner, 2001).

Because functions, relations, and joint variation are prominent in both structural and referential approaches, they do not constitute an alternative to the other two. We will deal with functions shortly, treating them as a special entry point into algebra.

Kaput's (2007) core aspects denote reasoning processes considered to flow through each of his approaches; they include:

- Generalization and the expression of generalizations in increasingly systematic, conventional symbol systems and
- Syntactically guided action on symbols within organized systems of symbols

The syntactic guidance referred to by Kaput conveys a notion that Balacheff (2001) has referred more generally to as a "control structure." When a student solves a problem by relying on her knowledge of the quantities manipulated, "the control is left in the reference world of the modelling process (p. 252)." This calls to mind Reed & Lave's (1991) distinction between "manipulation of quantities" and "manipulation of symbols," made in the context of research with Liberian tailors. Balacheff continues:

> The dilemma of controlling the problem-solving process in most situations involving Algebra in the school context is, taking Dettori's (2001) words, the need for students to be able "to associate meanings to the employed symbol, and to manipulate symbols independently of their meaning." Boero (2001) may suggest the overcoming of this dilemma by the ability of "a partial 'suspension of the original meaning' of the transformed expression" during the transformation process. But all authors insist on the tendency of students to come back to a reference world outside Algebra in order to check either the relevance of the actions they performed or the validity of their solution. (p. 253)

Referring to an analysis by Lins (2001), Balacheff (2001) asserts that, even when literals [algebraic letters] are used "to speak about the actions of objects of a referent world of buckets and tanks" (p. 253), quantities-based thinking falls short of being algebraic because "the control on the meaning of the symbolic expressions as well as the control of their manipulation depends on an interpretation in this referent world (p. 253). He uses this as the basis for distinguishing between "symbolic arithmetic" on one hand and algebra on the other.

Let us now propose a set of dimensions for characterizing varieties of algebraic reasoning in broad strokes (see Figure 15.1). The Control Dimension comes directly from Balacheff, with semantically driven inferences on one extreme and syntactically driven inferences on the other. The second dimension, Focus, refers to the domain in which the phenomena to be explained lie.

Quadrant 4, in it's the most extreme form, comprises cases in which algebraic reasoning makes no links to worldly experience or data. Algebra is treated as a self-contained system of structures and symbol manipulation but the symbols do not stand for anything in the real world. Inferences are derived based upon axioms and accepted rules of inference. Cases of modeling that occur within this quadrant are confined to the domain of mathematics itself, for example, when one uses mathematical ideas or representations from a certain mathematical domain to clarify ideas or representations in another mathematical domain. Examples of Quadrant 4 reasoning rely heavily on inferences based on written forms. For example, given the relation, $d = m/v$, where d, m, v are non-negative numbers, one might ask what happens to the value of d, when v is tripled in value.

The extreme reaches of Quadrant 2 comprise cases in which one makes inferences about worldly

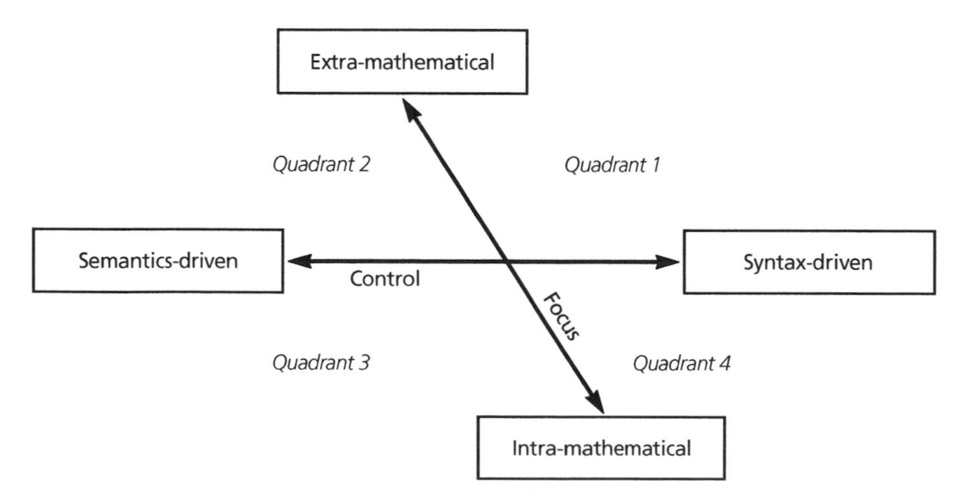

Figure 15.1 Two dimensions underlying approaches to algebra

Note: The *Focus* dimension refers to the domain of phenomena one is attempting to clarify, namely, whether they lie within or outside of mathematics. The *Control* dimension refers to the degree to which inferences are drawn on the basis of the semantics of the problem or the syntax or the representational forms.

[3] We do not wish to suggest that light behaves in contradictory fashion. It may simply be that the metaphors we use to understand light (viz., light-as-particle and light-as-wave) are somewhat limited as models.

phenomena based upon information of a non-mathematical nature. For instance, one might attempt to describe how the crowdedness of gas molecules changes when the gas is moved to a container three times the size of the present container.

Quadrant 1 comprises cases in which mathematical representations can influence the drawing of inferences about extra-mathematical phenomena. For example, upon learning that the density of matter can be described as a ratio of mass to volume, a student may use the formula, $d = m/v$ to conclude that tripling the volume reduces the density to one-third of its original value.

Finally, Quadrant 3 cases correspond to those in which one's use of knowledge or intuitions about worldly phenomena help generate the mathematical representations. For example, one may take measurements of the mass and volume of a set of objects made of the same substance, plot measure of mass by volume, and conclude that there is an invariant relation between the mass and volume of items of the same substance. In other words, the student may conclude that $y = kx$ is a reasonable model for the data.

It should be clear from the present analysis that modeling is not associated with any particular region in Figure 15.1. In fact, modeling is a central, unavoidable process in any aspect of mathematics learning. It is true that the sorts of modeling activities associated with Quadrant 4 are the most challenging to young learners. And tremendous obstacles may be required for students to overcome the intrusion of semantic considerations when issues of mathematical syntax deserve prominence. In any case, the models themselves gradually become objects of study and theorization, an idea brought home by Douady's (1986) *outil-objet* (tool-object) analysis. So, for example, one can use mathematics to keep track of changes in humidity and to model how humidity varies as a function of temperature. In such cases one is preoccupied principally with the data. However, as one becomes increasingly involved in large-scale weather forecasting, the models themselves become more and more prominent. Researchers will often find their attention shifting from the data to the mathematical structures for representing data. This still falls short of pure mathematics, where extra-mathematical modeling may play no role at all, but it is still fairly distant from the extra-mathematical extreme of the continuum.

For purposes of exposition, the two complementarities can be considered independent. In actual practice, there is likely to be a substantial association between them. For this reason we have diagrammed the complementarities as non-orthogonal.

The thing about complementarities, such as the wave-particle tension in Physics, is precisely that the poles do not represent alternatives. Any significant theory of light needs to account for its wave-like and particle-like natures, even though they may seem incompatible. We would claim that any attempt to account for EA needs to take into account each pole of the two complementarities, even though each pole pulls in its own direction, so to speak.

Algebra Is Latent in the Existing Early Mathematics Curriculum

In a symposium on "The Nature of Algebra in the K–14 Mathematics Curriculum," Bass presented his view of school algebra (Bass, 1998):

> School algebra—and *the root of all algebra*—is about the following:
>
> - The basic *number systems*—the *integers* and the *real numbers* and those derived from them, such as the rational and complex numbers.
> - The *arithmetic operations* (+, –, ×, ÷) on these number systems.
> - The *linear ordering* and resulting *geometric structure* defined on the real line. By this I mean the notions of size (whether one number is larger or smaller than another) and of distance between numbers.
> - The study of the *algebraic equations* that arise naturally in these systems. (p. 9, emphasis in original)

The final component, algebraic equations, is fully expected. The other three ingredients—number systems, the arithmetic operations, and the number line—may at first appear to be non-algebraic because they do not require that one use algebraic notation. Algebra nonetheless underlies arithmetic.

In the next three sections of this chapter we examine how conceptions of arithmetic bear on EA. We will consider specifically the cases where arithmetic is understood as the science of (1) numbers (2) numbers, quantities and magnitudes (3) numbers, quantities, magnitudes and functions.

ARITHMETIC AND NUMERICAL REASONING AS AN ENTRY POINT INTO EA

Arithmetic is conventionally understood to be the science of number. It includes the rational operations (addition, subtraction, multiplication, and division) as well as factorization and extraction of roots.

The Field Axioms and Other Properties of Numbers

The question, "What can one do with numbers?" is tantamount to asking, "What are the operations and properties of the number system in which they reside?" The field axioms in Table 15.1 convey some of these properties. They also hint at the algebraic character of arithmetic. The letters *a, b* can be understood as placeholders for particular numbers, but they are intended to stand for *any* and *all* numbers in the domain under consideration. Hence the field axioms assert numerical identities, general properties of numbers. One normally expresses the axioms in the formal language of mathematics—using a limited, precise grammar, and a limited set of symbols (such as the symbols for "for all," "for each," "there exists," and so on). This formal language shares much with algebraic notation itself.

Table 15.1 The Field Axioms (c.f. Weinstein, 1999)

Name	Addition	Multiplication
Commutativity	a + b = b + a	a b = b a
Associativity	(a + b) + c = a + (b + c)	(a b) c = a (b c)
Distributivity	a (b + c) = a b + b c	(a + b) c = a c + b c
Identity	a + 0 = a = 0 + a	a × 1 = 1 × a
Inverses	a + (–a) = 0 = (–a) + a	$a \times a^{-1} = 1 = a^{-1} \times a$ if $a \neq 0$

Of course, what is to be learned about number is not captured by the field axioms. For example, the axioms do not reveal how to compute a + b, that is, how to express a + b without a plus sign once a and b have been assigned values. The same can be said for multiplication, subtraction and division; the latter two are not even explicitly recognized by the axioms. These rational operations depend upon the particular notational system in which numbers are represented. So, for example, column multiplication as we know it depends upon the structure and conventions for our present base 10, place-value system of notation.[4] In ancient Egypt multiplication was carried out by a "halving and doubling" procedure (Menninger, 1992) that also works, but in a very different way. It strikes us as important that there be a formal "grammar" of notation systems[5] akin to the notation-independent grammar of the field axioms.

The field axioms do not even assert that there is a unique value for $a + b$, or that there exist an infinite number of ways to additively compose any number, or that every positive integer is uniquely factorable. Furthermore, division is closed under both the rational and real numbers, but not under the integers because a/b is not defined if b does not divide a or $b = 0$.

Numbers have additional properties that invite expression in algebraic notation. Consider, for instance, the division identity shown below:

> If b is a positive integer and a is a non-negative integer, then there are unique non-negative integers q and r such that,
>
> $$a = qb + r,\ 0 \leq r < b$$

This means that any non-negative integer, a, can be expressed as an integral multiple of any positive integer, b, plus a non-negative remainder, r, less than the second integer. The division identity proves to be fundamental to Euclid's algorithm for finding the greatest factor common to two integers, to continued fractions, to the fundamental theorem of arithmetic and to modular arithmetic (Bass, 1998; Carraher, 1990).

These ideas highlight the connections between arithmetic and algebra for those who are already comfortable with algebraic notation and ideas. They indicate that arithmetic has a *potentially algebraic* character. But surely this does not mean that young students who engage in arithmetic are *de facto* engaging in algebra. Or does it? We will need to give this question careful consideration in the ensuing discussion.

Studies That Introduce Algebra Through Generalizations About Numbers

Let us turn our attention for the moment to problems involving number sentences such as those in Table 15.2.

Table 15.2 Number Sentences from Carpenter et. al. (2003)

8 – 5 = 3
3 × 4 = 15
3 + 5 = 8
8 = 3 + 5
8 = 8
3 + 5 = 3 + 5
3 + 5 = 5 + 3
3 + 5 = 4 + 4
9 + 5 = 14 + 0
9 + 5 = 0 + 14
9 + 5 = 13 + 1
3 + 5 = □ or 3 + □ = 8

[4] Across countries, there occasionally exist variations in how the computations are notated, particularly in the cases of multiplication and division, but these are of little import so long as they do not lead to different answers.

[5] There is a similar need for a grammar of symbolic systems that are not strictly notational (Goodman, 1976), such as graphs.

Teachers working in the project led by Carpenter, Franke, & Levi (2003) initiated discussions about the truth or falsity of number sentences as a means of prodding students to think about the structural relations among the numbers. They found that third to fifth grade students in their project produced and justified generalizations such as:

> "When you add zero to a number you get the number you started with." "When you subtract a number from itself, you get zero." "When multiplying two numbers, you can change the order of the numbers." (Carpenter et al., 2003)

The first generalization is a natural language formulation of the Additive Identity property. The second statement asserts the property of Additive Inverse and the third remark states the Commutative property of addition. Hence, despite the fact that Carpenter's problems did not employ the language of algebra and the students did not use algebraic notation in their answers, the students nonetheless were expressing general, algebraic properties of the number system.

Bastable and Schifter (2007) present an intriguing example given by a teacher. The example concerns a 4th grade homework problem in which students were to represent square numbers on graph paper:

> Knox [the classroom teacher] explained that Adam [a student] had difficulty articulating his idea "even though he knew exactly how to do it." But with help, he was able to share his discovery, which the group then began to explore: If you take two consecutive numbers, add the lower number and its square to the higher number, you get the higher number's square. For example, consider 2 and 3. Adam's rule says add 2 plus the square of 2 plus 3 to get the square of 3, or $2 + 2^2 + 3 = 2 + 4 + 3 = 9$. Knox writes:
>
>> "[The group] had some discussion about whether it would work each time and Fred insisted that it wouldn't work with a higher number than that, so we tried 7 and 8: $7 \times 7 = 49$; $8 \times 8 = 64$; and $7 + 49 + 8 = 64$." Adam then raised an additional question: What would you subtract from a square to find the square below the one you have?

That Adam would find it difficult to explain his theorem is not surprising. Natural language is relatively clumsy for expressing algebraic relations. In addition, Adam stumbled upon his theorem in a geometric context, so he was required to shift between drawings of squares and expressions involving numbers. Algebraic notation offers a more succinct formulation of his theorem in ways that preserve the original number and its successor. Once transcribed into algebraic notation, Adam's reasoning conveys the gist of the expansion of $(n + 1)^2$: for all natural numbers, n, $(n + 1)^2 = n + n^2 + (n + 1)$

Elsewhere Schifter (1999) identified other examples of implicit algebraic reasoning and generalization among elementary school children in classrooms where reasoning about mathematical relations is the focus of instruction. She concluded that:

> Algebraic methods are clearly implicit in the children's work. . . . As they apply different operations to solve a single word problem, they evidence a sense of how the operations are related. For example, as the children come to see that any missing-addend problem can be solved by subtraction, or that any division problem can be solved by finding the missing factor, they acquire experience with the inverse relationships of addition and subtraction, multiplication and division, and thus with equivalent [number sentences]. And as they develop fluency in a variety of computational strategies, they implicitly apply the laws of commutativity, associativity, and distributivity. (p. 75)

Even when children solve "missing addend problems," such as "Maria had 7 marbles and won some more, ending up with 15 marbles; how many did she win?" they may be implicitly dealing with the additive inverse axiom. This corresponds to the number sentence $7 + \square = 15$. This sort of problem is known to be difficult for students who think that "winning" (or the plus sign) indicates that one must add in order to find the result. Further subtleties related to addition are discussed in Vergnaud's (1982) classic treatment of the subject. It is noteworthy that by solving missing addend problems through subtraction one implicitly makes use of the additive inverse axiom.

James Milgram, a reviewer of this chapter, commented that "algebra is a specific TOPIC in mathematics dealing with specific kinds of problems—inverse problems" (Milgram, 2006a). When we asked whether that would include missing addend problems such as the above, he responded:

> YES!!! This is a crucial point. For example, ordinary addition is not algebra, but the missing addend problem IS ALGEBRA. This is a student's first contact with the issues of algebra and it is interesting that in the mathematical instruction in high achieving countries addition and subtraction are introduced at exactly the same time with subtraction being introduced as the inverse problem. In our texts, addition and subtraction are always separate chapters, and the connection between them is not part of typical instruction. This seems to be a part of the reason why algebra is such a hurdle for students in this country.

The above analysis highlights what appears to be an oft-missed digging deeper into arithmetic in order to explore its algebraic nature through the field axioms.

Quasi-Variables

The algebraic character of arithmetic is also witnessed by Fujii and Stephens' (2001) work on *quasi-variables,* a term they use to refer to the implicit variables that students seem to make use of in arithmetical contexts. Quasi-variables, commonly used in Japanese elementary school mathematics, appear "in a number sentence or group of number sentences that indicate an underlying mathematical relationship that remains true *whatever the numbers used are*" (p. 259, emphasis added). For example, a student may note that the expression, 71 – 13, can be solved by taking away 20 from 71 and then adding 7. It is as if the student reasons that 71 – 13 = 71 – (10 + (10 – 7)) = 71 – (20 – 7) = (71 – 20) + 7 = 51 + 7 = 58. Now, this could be mere opportunism on the part of the student. The key issue rests on how the three from 13 becomes re-presented as (10 + (10 – 7)). But there is the opportunity for the student to gain insight into the structure of the number system. To what extent is the student aware of the fact that she can avoid having to subtract a digit, *b*, from another, possibly smaller digit, by first subtracting an additional 10 (not difficult) and *adding* the 10-complement, 10-b? We leave it open for others to decide whether a particular instance is truly algebraic. Suffice it to say here that "quasi-variables" may represent a convenient bridge between arithmetic and algebraic thinking.

Lest the reader think that one cannot have a solid grasp of advanced mathematical ideas until one can articulate them using formal language, consider Thurston's (1994) remarks:

> It is common for excellent mathematicians not even to know the standard formal usage of quantities (for all and there exists), yet all mathematicians certainly perform the reasoning that they encode. (pp. 164–165)

This is a nice example of what Vergnaud (1979) refers to as theorems-in-action. One of the most important questions in EA research is how students learn and can be taught to transform theorems-in-action into explicit expressions.

There are many cases in mathematics where particular instances are reasonably construed as examples of more general mathematical objects and relations. One may use a particular drawing of a triangle to represent a triangle in general. One uses a letter to stand for a particular number or a (possibly infinite) set of numbers. Blanton and Kaput (2000) demonstrated third graders' ability to make robust generalizations and to provide intuitive supporting arguments as they discuss operations on even and odd numbers and consider them as placeholders or as variables. They realized in this analysis that there is a constant shifting back and forth between the specific numbers and the more general classes of odd and even numbers. It is vital that teachers and researchers attempt to discern the scope of validity with which students are effectively operating. It is also helpful to scan current activities for other, possibly broader, contexts. For example, a mathematics teacher can take note of the fact that discussions about even and odd numbers can be construed as examples of modular arithmetic for modulus 2. Students may not appreciate the wider significance of the discussion. And a teacher will need to take into account a host of considerations before deciding whether to explicitly pursue such connections. But if and when she does, some care will be required to distinguish the broadened interpretations from those of her students while encouraging them to partake in the broader view.

The cautious reader will wonder whether we have been "reading algebra into" the cases we have been examining. Is it possible that in our quest for algebra among young students we have been overly zealous in ascribing algebraic interpretations and meanings? In mathematics education, especially with young learners, there is always a risk of over-attributing meaning. But it is a risk one needs to take because of the nature of mathematical reasoning.

Summary: Numerical Reasoning and EA

Several remarks serve to summarize the main points regarding the arithmetic of numbers as an entry point into EA.

1. Arithmetic has an inherently algebraic character and can be usefully regarded as a part of algebra rather than an a domain distinct from algebra;
2. Young students sometimes make algebraic generalizations without using algebraic notation (although natural language is often poorly suited for expressing algebraic relations);
3. Studies of arithmetic as an entry point into algebra are promising, but most of what needs to be known has yet to be investigated.

The first point above may resist appreciation because there are two easily conflated senses of arithmetic: *arithmetic as currently taught* and *arithmetic as it might be taught,* were its algebraic character given greater prominence. For example, narrow approaches to arithmetic conceal the symmetric and transitive properties of equality. Arithmetic need not be taught in such manner, and many of the difficulties students display in learning algebra at adolescence may reflect the misleading ways that symbols have been introduced.

ARITHMETIC AND QUANTITATIVE REASONING AS AN ENTRY POINT INTO EA

The status of *quantities* and *magnitudes* as mathematical objects depends upon one's perspective on modeling. Mathematicians sometimes prefer to think of them as residing outside mathematics, in the realm of science. However, as Whitney (1968a) noted, quantities can be treated as legitimate mathematical objects:

> Let us consider the problem of choosing a model M for masses. An object A has a certain property which we call its "mass"; why not let this property itself be an element of the model? As far as the structure of the model is concerned, we need not theorize on what "mass" really is; we need merely give it certain properties in the model. For instance, if we have distinct objects A and B, with masses m_A and m_B, we may think of the objects as forming a single object C; its mass m_C should then be $m_A + m_B$. Therefore M should contain an operation of addition, and any further properties we choose. (p. 115)

Because modeling plays such a substantial role in elementary mathematics, there are good reasons for treating quantities and magnitudes as belonging within the domain of arithmetic. This does raise the likelihood that lapses in rigor will occur. But it seems to be an unavoidable part of early mathematics teaching.

In this section we will treat quantities and magnitudes (in addition to numbers) as objects of arithmetic and attempt to characterize how this expanded view of arithmetic provides an entry point into EA.

Quantities, Measures, and Magnitudes

Historically, mathematics arose from human enterprises such as commerce (tallying and weighing), surveying (measuring land), manufacturing, astronomy (measuring and predicting position-time values for heavenly bodies), optics, architecture (building, perspective drawing), war (trajectories of projectiles), in which physical quantities are prominent. As Atiyah (2005) observed, "Mathematics originates from the physical world . . . but is organized and developed by the human brain" (00:13:37). He later notes: "In some sense . . . the most fundamental notion of mathematics is that of magnitude, size, that some things are 'bigger than others'" (00:19:13). He refers to the specific case of volume, sound, and light, as instances of "an abstract notion of *magnitude*—mathematics starts off with such abstract notions" (00:19:47).

Significant advances in the history of mathematics are constructed upon the abstract notion of magnitude. The Eudoxian theory of proportion (Euclid & Heath, 1956), for instance, is built upon considerations of the relative lengths of line segments. And Newton's method of fluxions for finding a derivative (Newton, 1967/1671) draws on his reflections about motion.

There are admittedly good reasons for distancing mathematics from quantities. As mathematician, Aleksandrov (1989), put it:

> We operate on abstract numbers without worrying about how to relate them in each case to concrete objects. In school we study the abstract multiplication table, that is, a table for multiplying one abstract number by another, not a number of boys by a number of apples, or a number of apples by the price of an apple. Similarly, in geometry we consider, for example, straight lines and not stretched threads, the concept of a geometric line being obtained by abstraction from all other properties, excepting only extension in one direction. More generally, the concept of a geometric figure is the result of abstraction from all the properties of actual objects except their spatial form and dimensions. (pp. 1–2)

Let us now put forth some basic distinctions related to quantities, drawing on various authors (Fridman, 1991; Lebesgue & May, 1966; Schwartz, 1996a; Thompson, 1988; Whitney, 1968a, 1968b):

- *Quantities* include countable and measurable properties (mass, length, time), as well as derived properties[6] (volume, area, velocity, acceleration, density, coefficient of friction).
- *Abstract numbers* are numbers without any intended reference to concrete situations or referents outside of mathematics. Examples are 3, –0.12, ¾, 4i, π.
- *Concrete numbers* include counts and measures

[6] We are following conventions of the SI base units here. It is of course sometimes the case that quantities such as area and volume (by SI standards) can be measured directly, in which case they are not derived from the observer's point of view.

- *Counts* express the *cardinality of sets of like objects* through a *natural number quantifier* associated with a *unit*; examples are "3 eggs" and "42 donations."
- *Measures* are expressions of the magnitude of a selected property of *individual objects*; they are either the result of measurement that assigns a *number* and *unit of measure* to each respective object, or they are *derived* or composed from other measures, as density is derived from mass and volume.

- *Measurable quantities* (Fridman, 1991) of one dimension[7] are of particular interest; they are by definition
  - Scalar.[8] *Scalar quantities* are those that can be ordered along a single dimension; speed is scalar, but velocity and force are not because they involve direction as well as magnitude
  - Additive. *Additive quantities* are scalar quantities for which the operations of addition/subtraction apply
- *Magnitudes* are general representations of size without explicit reference to extra-mathematical quantities (Euclid & Heath, 1956). Line segments, closed regions in the plane, and closed regions in 3-space commonly represent size via length-distance, area, and volume, respectively. Expressions—for instance, A, B, A–B—can also represent magnitudes. Magnitudes can be thought of as generic physical quantities.

Quantitative Thinking and Number Lines

There is a large body of research about the evolution of young children's understanding of physical quantities, measured as well as unmeasured. The research spans concepts of length, area, volume (Clements & Battista, 1992; Lehrer et al., 1998; Piaget, 1960; Piaget & Inhelder, 1967), mass, time, density, buoyancy and many others (Inhelder & Piaget, 1958; Piaget, 1960, 1970a, 1970b; 1974). However there are two reasons for which it has had a somewhat limited influence on the field. Firstly, given the sparse theoretical groundwork about quantitative thinking, with noteworthy exceptions (e.g., Schwartz, 1996a; Thompson, 1988; Smith & Thompson, in press), it continues to be challenging to relate investigations about quantities, particularly unmeasured and indeterminate quantities, to studies associated with arithmetic, broadly conceived. Secondly, mathematics educators have tended to avoid studies from developmental psychology that seek out universals and downplay the roles of teaching and particular representational forms on students' thinking. Although these concerns are justified, there is much that can be appropriated from developmental studies without endorsing the epistemologies of the researchers. And much work remains to be done analyzing the implications of quantitative reasoning for mathematics education.

Some progress has been made with respect to length. This is encouraging because length and linear ordering take on important symbolic roles (representing magnitudes) in number lines, graphs in Cartesian planes, line segment diagrams and so on. So there exist very special reasons for understanding how students make sense of length and distance.

Bruer (1993) provides an overview of the development of young students' reasoning relating number lines and numerical understanding. This body of work offers important leads for mathematics educators and developers to pursue. It bears directly on Bass' (1998) appeal for grounding the teaching and learning of real numbers—central to EA—on number line representations:

> It is both natural and advantageous to give an early emphasis to the geometric real line model of the real numbers, in which basic arithmetic operations are interpreted geometrically and developed alongside the more algorithmic development rooted in base-10 place-value representation. (p. 15)

The number line is generally compatible with children's learning about natural numbers and integers, despite the "fence post" issue (focusing on numbers as points when intervals are the relevant invariant) and matters of interpretation related to directed numbers (Peled & Carraher, 2007). But as students' concept of number develops, it becomes increasingly difficult for them to reconcile operations on numbers with displacements, intervals, and points on the number line; something that has implications for using number lines to discuss scalar quantities. One manifestation of this is the work on students' belief that "multiplying makes bigger and division makes smaller" (Bell, Fischbein, & Greer, 1984) —a belief that held up well for

---

[7] We shall not consider vector quantities and higher-order tensor quantities—forces, velocities, etc.—here even though they are of considerable importance in secondary mathematics and beyond.

[8] We are uncertain whether there is agreement on the meaning of *scalar*. Some authors seem to treat it as dimensionless, which is to say, not that scalars cannot be ordered, but rather that they are not associated with a particular physical attribute and unit of measure.

the natural numbers but no longer holds for rationals and reals (nor arguably for integers). If $x \times 2/3$ is challenging, it should not be surprising that students will have difficulty in understanding $y = \frac{2}{3}x + b$ as well as its representation as a graph in the Cartesian plane.

Over the past 40 years there have been various attempts to introduce operations on line segments consistent with the Eudoxian theory of proportion (e.g. Braunfeld & Wolfe, 1966; Steiner, 1969). Carraher (1993, 1996) shows how there is room for minimizing the rupture in the transition from natural numbers to rationals by conceiving the numerator of a positive fraction as if it were a natural multiplier and the denominator as if it were a natural divisor.

The work of realistic mathematics education (RME) regarding "empty number lines" (Gravemeijer, 1999; Klein, Beishuizen, & Treffers, 1998) can benefit from and contribute to theory and research about number lines and their relations to EA. This work is predicated on the intriguing notion that there may be advantages to withholding assignment of values to points and line segments so that operations are carried out directly on tokens representing quantities.

Can Students Apply Other Properties of Arithmetic to Magnitudes?

If magnitudes are to be formally treated as objects of arithmetic, they ought to be subject to arithmetical operations and properties, as numbers are. Among other things, this suggests that they should be subject to the field axioms and to some notational grammar. What do expressions such as $A + B$ and $A + (-A) = 0$ mean in the domain of magnitudes? If A and B are line segments, for instance, $A + B$ refers to a particular way of joining them end-to-end. But what does $A + (-A)$ mean for line segments and intervals on the number line? And what about $A + (-B)$? Does it matter if $B < A$ or $B > A$? Multiplication of magnitudes is subject to various representations. $A \times b$, the multiplication of a magnitude times a number is usually depicted as n-folding in a single dimension (Whitney, 1968a, p. 115). When A, B are linear magnitudes, $A \times B$ is generally treated as the area of a rectangle with sides of lengths A and B. But how are students to interpret the product $A \times B$ if A and B stand for quantities of different natures (e.g., an intensive quantity times an extensive quantity)? And how is $A \times b^{-1}$ to be understood on a number line? And what is the multiplicative inverse of a magnitude (or quantity). Furthermore, how is $A \times B^{-1}$ to be interpreted? One cannot simply "ignore the units," that is, fail to consider the relations to physical quantities. As Confrey and Carrejo (2005) discovered with 5th grade students who were asked to express, with a fraction, how 3 pizzas were to be shared equally among 10 students, the answers 3, 3/10, and 1/10 could all be justified, depending upon the unit assignment to the answers. Even if a consensus were to be achieved—say, around the answer 3/10—this might be inconsistent with some students' conceptualization of the magnitudes and underlying quantities.

Referent-Transforming Properties

Quantities raise numerous issues that do not arise in the domain of abstract numbers. Notable among these is what Schwartz (1996a) refers to as the referent-transforming properties of multiplication and division. Consider the simple task of purchasing bottled juices at a vending machine. Let us imagine that each juice costs $.75 and that Jonathan wants to purchase 8 juices. How much will he spend? Presuming that one has decided that the problem calls for multiplication (as opposed to, say, repeated addition), there are several ways of construing the multiplication: (a) multiply .75 by 8, obtaining 6.00 and then attaching the unit of measure, $; (b) multiply $0.75 by 8, obtaining $6.00 straightaway; or (c) multiply two quantities, the intensive quantity, $.75/can, by the extensive quantity, 8 cans, obtaiing $6.00 as the result.

These three approaches may look alike, but entail strikingly different approaches to multiplication. Method (a) is a computation in the realm of abstract numbers, requiring the student to somehow determine the units of the outcome measure. Method (b) involves the multiplication of an extensive quantity by a dimensionless, abstract number; it is somewhat congenial to the repeated addition method. Method (c) involves the multiplication of two quantities, and exemplifies referent-transformation.

The referent transformation character of (c) can be appreciated by considering the juice vending machine as a function machine. Coins are repeatedly inserted but juice comes out. If multiplication were simply a matter of repeated addition, one would expect the inputs and outputs to be of like nature; accordingly, one would expect to put in coins and receive coins instead of cans of juice.

Students will eventually need to deal with problems from science, engineering, or mundane situations in which multiple dimensions are inherent to the situation. The values along the x-axis may be of one nature (e.g., time) whereas the values along the y-axis may be of an entirely different nature (e.g., velocity). At some point the means for converting input values into outputs may require invoking the constant of proportionality of a linear function. For a student to be able to understand how the expression bears upon the problem at hand, she will need to be cognizant of the units of measure associated with constants and variables. The constant of proportionality, a, in $ax + b$, cannot simply sum x's repeatedly, because b and the output will be in units of different natures. The constant of proportionality will not act merely on the number associated with *x*; it also acts as an exchange function, producing a unit different from that of *x* and different from *a*.

Such issues may be too subtle for students who are just beginning to learn about multiplication, but sooner or latter they require addressing. How students eventually come to understand the referent-transforming roles of multiplication is an area that deserves careful empirical research.

Referent-transformation is just the tip of the iceberg when it comes to issues students must negotiate when dealing with quantities. The interested reader may wish to consult Smith & Thompson (2007) for further examples of the challenges elementary and middle-school students face in reconciling computational routines with representations of rich problem situations.

EA Studies that Focus On Magnitudes and Measures

The number line research can be regarded as a specific example of the more general claim that "visualization bootstraps algebraic reasoning and algebraic generalization promotes 'seeing' new spatial structure" (Boester & Lehrer, 2007). This view is consistent with Davydov's (1991) approach to EA, with a significant line of work in Japanese mathematics education (Kobayashi, 1988), and also with the Math Workshop approach to elementary mathematics (Goldenberg & Shteingold, in press; Sawyer, 1964; Wirtz, Botel et al, 1965). By not assigning specific values, magnitudes are left indeterminate and even free to vary. As Freudenthal (1973, pp. 296–298) remarked, in a somewhat more advanced context:

> [In abstract algebra] a polynomial "$a_o x + a_1 x + \ldots + a_n x^n$" is not understood in the sense that $a_o, a_1, \ldots a_n$ are real or complex numbers but rather that they are "indeterminates" which then means that in a fixed ground-field P free transcendants $a_o, a_1, \ldots a_n$ have been adjoined and the given expression is considered as a polynomial over the field extension $P(a_o, a_1, \ldots a_n)$. Of course the substitution of elements of P for $a_o, a_1, \ldots a_n$ is allowed, but this is another thing. (p. 298)

One easily appreciates the relevance of indeterminates for introducing variables: only a slight adjustment in thinking would appear to be needed to shift from treating a letter representing a single, indeterminate value to each and every value in the domain. With regard to numbers, this entails a shift in focus from an arbitrary value to all possible values. With regard to length, there would be a shift from an arbitrary length to all possible lengths. These cases are to be contrasted with unknowns, that is, unique values that, for one reason or another, are not yet known.

Given the importance of variables in algebra, there are good reasons for mathematics educators to treat unknown values as indeterminates or variables. This holds even for equations. For instance, in $8 = 5 + x$, *x* is profitably conceived as a variable (and is thus free to vary) despite the fact that there is only one value, namely 3, for which $8 = 5 + x$ is true. Any other substitution for *x* is legitimate even though it results in a false statement. Such a framing can help students become familiar with variables from early on rather than having to radically overhaul the meaning of symbols that originally had constricted meanings.

One can imagine a line of research to investigate the conditions under which numerical assignment facilitates or interferes with learning, something from which a learning progression[9] in number line understanding and EA might eventually be developed.

Why Quantitative Thinking Is Unavoidable in EA

There is a major reason why we cannot conceive of a pedagogically sound approach to EA that does not give considerable attention to quantitative thinking. This reason is well expressed by Fridman (1991):

> In scientific theory . . . the concept of number is primary and is introduced and constructed independently of the theory of measurement of quantities; the latter is based on the previously constructed system of real numbers. It is impossible to do this in school, however, because the axiomatic method of building the number system may not be used, at least at the elementary level of instruction. (p. 163)

This argument suggests that, despite the promises of arithmetic of abstract numbers as an entry point into EA, one expects that there will be limitations. It does not seem realistic to first introduce youngsters to the algebra of number and then proceed to problems

[9] A learning progression is a conceptual framework for the long-term evolution of understanding in a given domain, taking into account what is known about students' prior understanding as well as the target understanding (c.f. Smith et al, 2006). The approach draws upon learning studies while avoiding the maturationist pitfall common to developmental approaches.

steeped in quantities as "applications" of algebra. One can of course expand the conception of arithmetic to encompass abstract and concrete numbers. But this merely allows arithmetic to save face, so to speak. Issues related to quantities still must be contended with.

The Davydov Approach to EA

In the 1960s, Davydov and colleagues from mathematics and psychology (Davydov, 1991) developed a comprehensive approach to EA where the primary focus was on expressing the basic relationships between explicit and implicit values of quantities. Even before children learned to work with numbers, they were encouraged "[to compare] quantities, to use notation to represent their basic relationships in formulas of equality or inequality (in letter form), and to study the properties of these relationships" (Bodanskii,1991, p. 293). Bodanskii added:

> From the very beginning of instruction, the composition of equations was introduced as a *self-contained* and unique method of solving word problems, and there was no generalization of the arithmetic method. . . . [From the start, children] . . . had to develop an ability to isolate the two equitable quantities and to write down their basic relationship, even before selecting the unknown quantity and searching for other relationships, and then to be guided by their knowledge of this basic relationship. (pp. 289–290)

Table 15.3 provides three examples of the problems children were asked to represent.

Table 15.3 Three EA Problems from Bodanskii (1991).

Problem 1: One brigade built *a* houses; another brigade built *b* houses. It is known that *a* is greater than *b*. What must be done for the number of houses built by the second brigade to be equal to the number of houses built by the first brigade?"(p. 294).

Problem 2: The students in one class made *a* (38 toys) and in another they made *b* (29 fewer toys). How many toys did the students in a third class made if *c* (78) toys altogether were made by the three classes? (p. 295).

Problem 3: In the kindergarten there were 17 more hard chairs than soft ones. When 43 more hard chairs were added there were 5 times more hard chairs than soft. How many hard and soft chairs were there? (p. 302).

Problem 1 requires recognizing that a – b is the missing quantity; problem 2 instantiates the values but is difficult nonetheless, describing the relations among a, b, and c simplifies the problem considerably; problem 3 once again instantiates values but is even more difficult. Note that the first problem entails algebraic relations in a purely additive context, the letters representing unmeasured quantities. The second problem assigns values to the literals but the real challenge lies in the relational calculus (Vergnaud, 1979, 1982) rather than the computation. The third problem involves an even more *relationally complex situation* (Thompson, 1993) however, once the relations among quantities, known and unknown, are understood, finding the answer is fairly straightforward.

The approach described by Bodanskii encouraged the students to represent the relations between givens and unknown values through several representations: (1) diagrams of line segments configured so as to reflect the part-whole relations of quantities; (2) tables of values; and (3) algebraic notation. Students in the experimental classes used tabular notation and geometrical diagrams (line segments) to express relationships among quantities. Equations were written to register that two quantities were equal, something students would seem to have discovered by their diagrams and notations. Once the equation was written, conventional rules for solving were followed until the answer emerged in the form of a statement "x = . . .".

Although the equation solving among Bodanskii's students appears to have been largely syntax-driven, the focus the instruction, particularly in the first years of schooling, fell upon examining the relationships among known and unknown quantities. The initial symbolic notation of a problem thus relied extensively upon semantic-driven inferences (see the Control Dimension in Table 15.1).

Bodanskii (1991) reports results showing that children who received instruction in the multiple representation of verbal problems from grades one to four were able to represent and solve equations to find solutions to algebra problems corresponding to first-order equations. Children in the experimental group used algebraic notation to solve the algebraic word problems, performing substantially better than their control peers. The fourth grade students even performed significantly better than sixth and seventh graders in traditional programs (five years of arithmetic, with algebra instruction beginning in grade six).

The Measure Up Project

The *Measure Up* project (Dougherty, 2007) has implemented in an American setting many ideas originating from the Davydov work. In first grade, children learn to compare objects in terms of length, height, area, volume, or mass.

Comparison situations are created for which students must communicate to others the results of a comparison in the absence of the physical objects. At first children sometimes attempt to use colors to name the physical objects, but this proves unsatisfactory when two objects may have the same color. The teacher guides students to label objects quantities with a letter. Students are also taught to express quantitative relations among quantities through "strips," natural language ("greater than," "less than," "equal to," "not equal to") and eventually literals and comparison operators.

The following dialogue takes place between a teacher and two first grade children who are comparing three volumes, D, K, and P:

> "I think that volume D is greater than volume K," said Caylie.
>
> "How do you know that, Caylie? We didn't directly compare those two volumes," said Mrs. M.
>
> "Well," said Caylie, "we found out that volume D is equal to volume P and volume P is greater than volume K, so volume D must be greater than volume K."
>
> "I agree with Caylie," said Wendy. "Because volume D and volume P are really the same amount so if volume P is greater than volume K, then volume D also has to be greater than volume K." (Dougherty, 2007)

We should not assume, of course, that students have a detailed comprehension of volume; for all we know, they may be treating it as a single, orderable quantity (amount of stuff). But even a one-dimensional notion of quantity may be adequate for the case at hand.

As understanding and representation of the reflexive, symmetric, and transitive properties emerge from the comparisons (we noted transitivity in the above example), numbers and operations enter the picture. For example, using part-whole diagrams or line segments, the children learn to explain that given magnitudes A, B and A > B, A – B is itself a magnitude that can be added to the lesser quantity (B) or subtracted from the greater one (A) to make the two quantities equal.

As operations are carried out on A, B, there appears to be a shift in what the letters refer to. Whereas the students appear to initially understand labels (A, B, C . . .) as representing particular magnitudes, the magnitudes and operations gradually provide a metric and serve as units of measure (Carraher, 1993). There may also be some foreshadowing of the notion of variable, but one must be cautious about such an inference given the well documented and non-trivial confusion between measures or counts (8P as "8 professors") and variables (8P as "8 times the number of professors")—see Clement, Lochhead, & Monk, G. (1981).

Dougherty (2007) described the children's explanations as follows:

> . . . Given $\frac{B}{E} = 5$ (Equation 1) and $\frac{B}{Y} = 8$ (Equation 2), (read as quantity *B* measured by unit *E* is 5 and quantity *B* measured by unit *Y* is 5), they conclude that because the quantities *B* are the same, then unit *Y* must be smaller than unit *E*. They justify this with the explanation that unit *Y* had to be used more times to make *B* than unit *E* was used. This means that unit *Y* is smaller.

Summary: Quantitative Reasoning and EA

Bodanskii's (1991) results demonstrate that young children can learn to represent and correctly find the unknown in algebra problems corresponding to first-degree equations. However, he does not explore the nature of children's understanding of algebraic procedures, especially in what concerns the understanding of the rules for transforming equations, functions, and variables. Even though the potential for exploration of variables and functions exists in his approach, children in his study do not deal with variables and functions, but only with unknowns.

The examples described by Dougherty (2007) suggest that a focus on quantities is a fruitful path towards the development of algebraic reasoning, including work with variables and functions, even among very young children.

ARITHMETIC AND FUNCTIONS AS AN ENTRY POINT INTO EA

The concept of function has played an extremely important role in the history of modern mathematics and has rather recently (Dubinsky & Harel, 1992b) become recognized as having a pivotal role to play in mathematics education at the secondary level and beyond. Schwartz, for example, has proposed that functions, rather than equations, be the fundamental object of algebra instruction (Schwartz, 1999; Schwartz & Yerushalmy, 1992a, b, c; see also, Chazan, 1993; Chazan & Yerushalmy, 2003; Yerushalmy & Chazan, 2002). Placing functions at the center of algebra instruction entails conceiving letters as variables, instead of unknowns, to interpret expressions as rules for functions and the Cartesian coordinate system as a space to display the results of calculations, and to take into account the multiple meanings of the equals sign (Chazan & Yerushalmy, 2003). The idea is that functions can serve to "algebrafy" existing content (Kaput, 1998). We have proposed that arithmetical operations themselves be conceived as functions (Carraher, Schliemann, & Brizuela, 2000, 2005; Schliemann, Carraher, & Brizuela, 2006).

Can Young Students Reason with Functions?

For the purposes of elementary and middle school mathematics,[10] a function is a rule that assigns each element from a domain to a unique element in the co-domain; the domain and co-domains usually are comprised of numbers, but as we suggested in the former section, there are good reasons for treating quantities and magnitudes as fundamental objects; this allows for mappings across sets of like or unlike natures—for example integers to magnitudes or elapsed time values to measures of distance.

By the end of secondary school students are expected to become familiar with a very small part of what is known about functions. What grade school students might be expected to learn is clearly going to be of a much more elementary sort. But we need to ask:

- What evidence exists that young students can reason about functions? What can they understand?
- What evidence, if any, exists suggesting that functions, introduced early, can provide a solid grounding footing for EA?
- And what additional reasons might there be for introducing functions to elementary school students, when functions are known to be challenging even for university level students (Carlson & Oehrtman, 2006)?

Functions As Rules for Generating Collections of Figures

Moss et al. (2006) assessed children's learning about the rules governing visual/geometric and numerical patterns as they participated in an experimental patterning curriculum. In two main studies, one with second graders, the other with fourth graders, in diverse urban settings, Moss and her team of teacher-researchers developed activities aiming at explicitly linking ordinal pattern positions with the number of elements in that position, and to bridge the learning gap between scalar and functional approaches (see Vergnaud, 1983, for the classic analysis of scalar versus functional approaches). The components of Moss' curricula include geometric pattern building with position cards, function machines, t charts (tables), and activities that integrate these activities. Both geometric patterns and numeric patterns were part of the activities. The activities were designed to foster the integration of students' visual and numeric understandings and to move back and forth across these two types of representations of patterns.

Figure 15.2 shows a sequence of geometric forms from Moss et.al.'s (2006) work.

Geometric Patterns and Position Cards

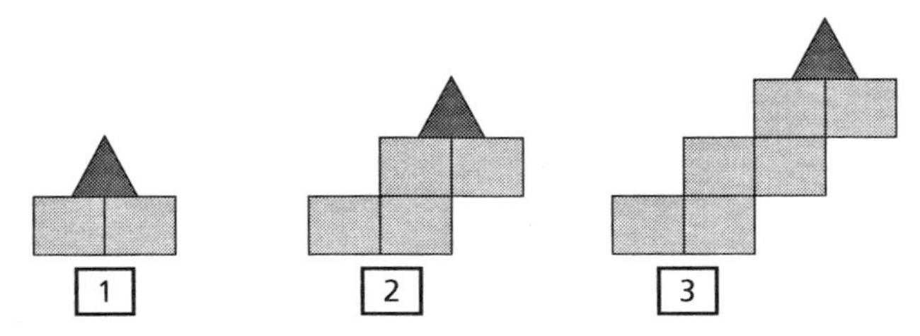

Figure 15.2 A patterns problem from Moss et al (2006). Note the explicit use of the indices corresponding to values of the independent variable.

Note that each step in the pattern sequence is numbered, thus making the order of the produced designs an explicit variable.

The study with second graders was conducted with 78 students in three experimental classrooms in New York and Toronto. On the basis of their results in a number knowledge task, children were assigned to one of three levels (low, medium, high). Twenty-two children in a fourth grade served as a control group. The intervention focused on extending students' prior *numerical* understandings and on helping them integrate their numerical and *visual* knowledge. From this perspective, the merging of the *numerical* and *visual* would provide a new set of powerful insights that would underpin the early learning of a new domain as well as subsequent learning.

A total of 20 lessons were implemented over the period of three months (Moss et al., 2006). Children first worked independently with geometric patterns and numeric patterns. This was followed by lessons designed to foster the integration of visual and numerical understandings. The final set of activities provided opportunities for students to move back and forth across the two types of representations for patterns. The specific activities involved geometric patterns and position cards, multiplicative (number × 2) and composite functions (number × 2 + 1), work with a function machine and tables, among others. The interview administered before and after

[10] We avoid the ordered-pair definition of function that dispenses with a mapping rule. As various authors have pointed out, the ordered-pair approach is likely to be confusing for young students.

the intervention consisted of 10 patterning problems and sub items. Interview results showed that (a) fourth graders in the control group were stronger on number knowledge, (b) the experimental and control groups showed the same rate of improvement on near generalizations, and (c) the experimental group outperformed the control group on far generalizations and explicit reasoning. Children of all ability levels in the experimental group significantly improved from pre to post interview.

Children in the experimental group were able to build geometric patterns based on algebraic representations, to recognize functions from geometric patterns including two-step composite functions, and to use syncopated language to express functions. Even though they had received no formal instruction in multiplication, they displayed understanding of multiplication through use of a variety of invented solution strategies for multiplication problems. They were also more able to apply multiplication than the students in the control group who had received formal instruction in multiplication.

The second study, with fourth graders, involved 50 children in an inner city school and 22 children in a private school, who participated in 18 lessons taught over five weeks and who worked collaboratively over *Knowledge Forum*, an electronic network. Activities included "Guess-my-rule" games, building geometric patterns from rules, deciphering rules from geometric patterns, and word problems involving patterns.

During discussions, all ideas generated by the students were first treated as *improvable*. Discussions took place over a web communication resource (Knowledge Forum), where contributions from participants were incorporated into a process of *idea refinement*.

The following is an example of the kind of rules and explanations produced by fourth graders in the experimental group as they discussed the "Cube Sticker" problem (Figure 15.3).

One fourth grader explained her solution as follows:

> *My theory* is that the rule is "times 4 + 2." I found out the rule right after I made a chart because in the second column there was a pattern going down and the rule was plus 4. Then I thought that the rule was times 4 but when I tried it it didn't work. So I looked at the chart again and I looked across and I saw that 2 was missing. Then I knew that 2 is constant. Then I got the rule times 4 plus 2. (Moss, Beatty, McNab, & Eisenband, 2005, slide 61)

Cube Sticker

A company makes colored rods by joining cubes in a row and using a sticker machine to put "smiley" stickers on the rods. The machine places exactly 1 sticker on each exposed face of each cube. Every exposed face of each cube has to have a sticker. This rod of length 2 (2 cubes) would need 10 stickers.

How many stickers would you need for:

- A rod of 3 cubes
- A rod of 4 cubes
- A rod of 10 cubes
- A rod of 22 cubes
- A rod of 56 cubes
- What is the rule?

Figure 15.3 A problem discussed by fourth graders in Moss et al. (2006)

In this study, children were assigned to low, medium, or high levels, based on their teacher ratings and report cards. Quantitative analyses on pre- and post-test measures revealed that low, medium, and high achiever students' scores were improved significantly as the result of the intervention, with an effect size of 3.2 standard deviations.

Moss et al. (2006) concluded that children found rules for patterns, relationships between patterns, and relationships across different representations, thus moving towards understanding the *relationships between the two variables* in the problem. They also demonstrated that they could use mathematical symbolic language and offer proof and justifications for their conjectures. There are important ideas underlying such a statement and they are suitably challenging, so much so that we will devote the next section of our review to some special topics related to patterns and tables. At the end of that section we will be better prepared to deal with functions as an entry point into EA.

Functions Expressed Through Multiple Representations: The Early Algebra, Early Arithmetic Project

Our approach to EA, developed over a series of interview and longitudinal classroom studies implemented over 10 years, is based on the same premise underlying the work of other EA researchers, namely, that arithmetic and algebra are not fully distinct: a deep understanding of arithmetic requires mathematical generalizations and understanding of basic algebraic principles. We focus on algebra as a generalized arithmetic of numbers and quantities and view the introduction of algebraic activities in elementary school as a move from computations on particular numbers and measures toward thinking about relations among sets of numbers. Specifically, we have proposed (Carraher, Schliemann, & Brizuela, 2000) that the operations of arithmetic should be viewed as functions. Central to our approach is the use of multiple representations, namely, natural language, line segments, function tables, Cartesian graphs, and algebra notation.

In our lessons, problem contexts constitute essential ways to situate and deepen the learning of mathematics and generalizations about quantities and numbers. Contexts are crucial to mathematics educators' concerns since students, particularly young students, learn mathematics through reasoning about various types of situations and activities. A wide range of important research aims to determine how reasoning about situations can foster mathematical understanding (e.g., Moschkovich & Brenner, 2002; T. N. Carraher, Carraher, & Schliemann, 1985, 1987; Nunes, Schliemann, & Carraher, 1993; Schwartz 1996; Smith and Thompson, in press; Verschaffel, Greer, & De Corte 2002). Starting from rich problem contexts and situations one hopes that, at some point, students will be able to derive conclusions directly from a written system of equations or an x-y graph drawn in a plane. The main challenge that contexts pose concerns how abstract knowledge about mathematical objects and structures can result from experience and reasoning in particular situations. While contextualized problems and a focus on quantities help in providing meaning for mathematical relations and structures, algebraic knowledge cannot be fully grounded in thinking about quantities. Important questions for EA research underlie the tension between contextualized situations and mathematical structures and relations (see Carraher & Schliemann, 2002; Schliemann & Carraher, 2002).

We have conducted three longitudinal studies with the aim of documenting how children's algebraic thinking evolves as they are introduced to algebra principles and representations over rather long periods of time (one to three years). The first study was comprised of 16 EA lessons developed over the year in a third grade classroom of 18 children (Carraher, Schliemann & Brizuela 2000, 2005; Schliemann, Carraher, & Brizuela, 2006). The second study was implemented in four classrooms (69 children), with one weekly 90 minutes EA lesson from the second semester in second grade to the end of fourth grade (Brizuela & Earnest, in press; Carraher, Schliemann, Brizuela, & Earnest, 2006; Carraher, Schliemann, and Schwartz, 2007; Schliemann et al., 2003). The third study involved 26 students as they progressed from 3rd through 5th grade. In 3rd and 4th grades, the children participated in two 60 minutes lessons per week (total of 50 lessons in 3rd grade and 36 in 4th grade), each one followed the next day by a 20 minute homework review; in 5th grade they participated in one 90 minutes lesson per week (18 lessons total) followed the next day by 30 minutes of homework review. The students in all three studies came mainly from minority and first generation immigrant families located in the Boston area.

In our classroom work we generally begin not with a mathematical representation, such as a number sentence or a graph, but rather with an open-ended problem or a situation. When working with each problem, we seek to generate a teaching and learning environment that is conducive to children's presentation of their own perspectives, ideas, and ways of representing the problem. Children's first verbal reactions to the problem are brainstormed and they are asked to show, on paper, their ideas about the problem and their suggested solutions. The children in the front of the class then share the notations they produced.

For instance, in a lesson implemented at the beginning of third grade, The Candy Boxes Problem (see Carraher, Schliemann, & Schwartz, 2007), the instructor holds a box of candies in each hand and tells the students that the box in his left hand is John's, and all of John's candies are in that box; the box in his right hand is Mary's, and Mary's candies include those in the box as well as three additional candies resting atop the box; each box has exactly the same number of candies inside. Children are then asked to write or to draw something to compare John and Mary's amounts.

To introduce graphs and the Cartesian space (see Schliemann & Carraher, 2002), we start with a statement like: "Maria has twice as much money as Fred." After holding an initial discussion about the situation, students express their ideas in writing. We discuss their representations and introduce a new representation, often foreshadowed in students' own drawings, such as two parallel number lines drawn on the floor, one for Maria's and one for Fred's amounts, represented as points along the lines. We then ask individual children to represent on the parallel lines the places corresponding to different values Maria and Fred could have, as the relation "twice as much" continues to hold. As the number pairs become larger, it becomes impossible for one single child to touch the two points on the two lines, at the same time. We would then rotate one of the number lines by 90 degrees and ask the students to plot themselves as points at the intersections of ordered pairs for Maria and Fred's amounts.

A "Human Graph" is thus built and becomes the object of discussions and the basis for future work on the graphical representation of functions.

Once children have learned to represent a function via tables and graphs, we introduce problems involving the comparison of functions. For instance, in a problem implemented at the beginning of the second semester in fourth grade, *The Wallet Problem* (see Carraher, Schliemann, & Schwartz, 2007), we describe and ask the children to represent the following situation: "Mike and Robin each have some money. Mike has $8 in his hand and the rest of his money is in his wallet. Robin has altogether exactly three times as much money as Mike has in his wallet. How much money could there be in Mike's wallet? Who has more money?"

Because the *Wallet Problem* involves the comparison of functions, it becomes appropriate to explore when the two functions are equal, even though the problem, as originally stated, does not suggest an equation. But why did we not already present the problem as an equation, at least by telling the students that Mike and Robin have the same amounts?

To assert from the beginning that Mike and Robin have the same amounts overly constrains the functions and encourages the students to think of the variable as a single value. We leave the functions free to vary in order to encourage students to first explore the variation inherent to each function (total amount as a function of the amount in the wallet).

Only when students are informed that the amounts "happen to be equal" does it become appropriate to use an equation such as $w + 8 = 3 \times w$. Until that moment, an inequality such as $3 \times w > w + 8$ or $3 \times w < w + 8$ may hold.

When one treats equations as the setting equal of two functions, there is no need to treat unknowns and variables as fundamentally different. We prefer to think of an *unknown* as a variable that for some reason or other happens to be constrained to a single value, as when $w + 8$ is set equal to $3 \times w$. In this case, the equation holds only for certain values of w—actually, only one value, but it does not transform w from a variable into a single number or instance.

Through experience with problems of this kind, children begin to deal with more than one function at the same time, analyzing the patterns in the changes in the relationships among quantities, variables, and functions, and finding, on the graph, the value that would make the two functions equal. They also deal with the interaction between the multiple systems of representation they are using (see Brizuela & Earnest, in press).

Once children feel comfortable with the above, we begin to address the representation of equations and solution of equations through symbolic manipulation. The following are two examples of the problems 5th graders in our third study were asked to discuss, represent in writing as an algebra equation, and to solve using the manipulation rules of algebra:

> *Problem 1*: Anna went to the arcade with some amount of money. She then spent five dollars playing video games. After that, she won a prize where they doubled her money. The same day, Bobby went to the arcade with ten dollars. When he got there, his mother gave him thirty more dollars. Afterwards, he spent half of all of his money playing video games. At the end of the day, Anna and Bobby discovered that they did have the same amount of money. Write an equation showing that Anna and Bobby had the same amount at the end of the day. Solve the equation.

> *Problem 2*: Elizabeth and Darin each have some money. Elizabeth has $40 in her wallet and the rest of her money is in her piggy bank. Darin has, altogether, exactly five times as much money as Elizabeth has in her piggy bank. Elizabeth's total amount of money is equal to Darin's total amount of money. Write an equation showing that Elizabeth's total amount of money is equal to Darin's total amount of money. Solve the equation. How much money does Elizabeth have in her piggy bank?

The work with equations represented a considerable challenge for us. As we have discussed in this chapter, previous research had highlighted the difficulties that much older, adolescent, students had in dealing with equations. Moreover, the transition from the semantics of the problem to the syntactic rules of algebra, without loss of meaning is a rather challenging endeavor. However, as we will see next, some degree of success was achieved.

Throughout our three longitudinal studies, we have developed qualitative and quantitative analysis of videotaped classroom discussions, written classroom work produced by the children, and children's responses to written assessments and interviews.

Our first study, with 3rd graders, constituted a pilot study for us, in the sense that we could then identify major issues that were to be addressed in the next studies. Three main issues we identified in the first study (Carraher, Schliemann, & Brizuela, 2000; 2005; Schliemann, Carraher, & Brizuela, 2001, 2006) should be mentioned here:

First, reasoning about variable quantities and their interrelations would constitute a natural setting for the discussions about variables and functions. However, using realistic situations to model mathematical ideas and relations presents students with challenging issues. On occasion, children were inclined to assign fixed values to what were meant to be variable quantities, without recognizing their general character. For example, we gave children the following statement: "Tom is 4 inches taller than Maria. Maria is 6 inches shorter than Leslie." To help them focus on the differences between the unspecified heights, we asked three volunteers to enact the problem in front of the class. When the students agreed about the relative order of heights of the three protagonists, we asked them to draw Tom's, Maria's and Leslie's height and to show what the numbers 4 and 6 referred to in their diagrams. As many as 12 of the 18 students in class assigned particular values to each height, while maintaining the differences between the heights in the story consistent with the information given.

Second, when we introduced function tables relating number of items and price, we found that although the children could correctly fill in the tables, they did not attend to the invariant relationship between the values in the first and second columns. For example, to complete the table in Figure 15.4, they would count by ones to find the values in the first column and then count by threes to find the values in the second column one. When a guess-my-rule game was introduced, students were finally able to break away from the isolated column strategies they had been using.

Mary had a table with the prices for boxes of Girl Scout cookies. But it rained and some numbers were wiped out. Let's help Mary fill out her table:

Boxes of cookies	**Price**
	$ 3.00
2	$ 6.00
3	
	$ 12.00
5	
6	
	$ 21.00
8	
9	
10	$ 30.00

Figure 15.4 Table to be completed by third graders.

Third, the introduction of letters to denote any value for the first variable in a function table proved to be a powerful tool for children to focus on the general rule relating the two variables.

The activities in our second study, from 2nd to 4th grade, related to addition, subtraction, multiplication, division, fractions, ratio, proportion, and negative numbers. The project documented, in the classroom and in interviews, how the students worked with variables and functions, using different systems of representation, such as number lines, function tables, graphs, algebraic notation, and equations.

As lessons were implemented during the first year, we witnessed the major role that cognitive development obviously plays in mathematical reasoning: we found, for example, that the idea of "difference," fundamental to the students' understanding of additive structures, takes a long time to evolve. It follows directly from the fact that integers comprise a group that any ordered pair of them, a, b, will have a unique difference, a – b. This idea takes on subtle differences in meaning across the contexts of number lines, measurement, subtraction, tables, graphs and vector diagrams (Carraher, Brizuela, & Earnest, 2001).

We also witnessed the evolution of children's written representation for problems involving variables. In their analysis of the 69 children in our second study, Carraher, Schliemann, & Schwartz (2007) showed children's considerable progress in terms of notation for variables from the beginning of third grade, when they attempted to represent the Candy Boxes Problem, to the middle of fourth grade, when they represented the Wallet Problem (see Figure 15.5).

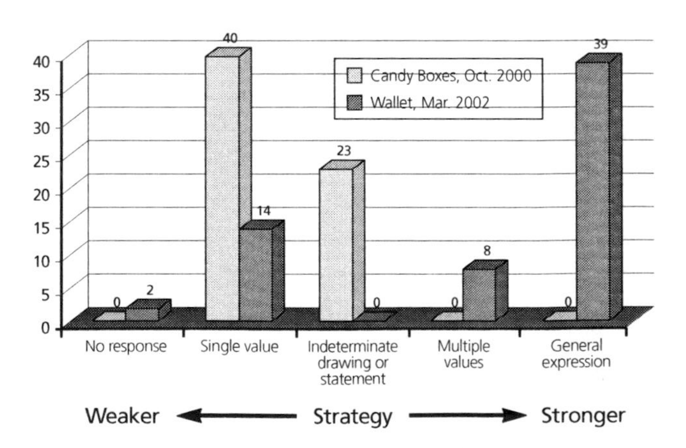

Figure 15.5 Number of children choosing each type of representation for the Candy Boxes problem and the Wallet Problem.

In third grade most children preferred to use a single value for the amount of candies in each box (see Figure 15.6). In contrast, in fourth grade most of them used general notation to represent the amount in the wallet (see Figure 15.7), with only a few still persisting in choosing a specific value.

Concerning the solution of equations, Brizuela and Schliemann (2004) described how, during the

What can you show about John and Mary's candies?
Draw or write something below.

Figure 15.6 A student's representation for the Candy Boxes problem.

Mike	Robin
Mike has \$8 in his hand plus more money in his wallet. N \$8 N+\$8=□	Robin has NX3 money Robin has 3 times as much money as Mike has in his wallet. N N N NX3=3N

Figure 15.7 A student's representation of the Wallet problem.

implementation of the last lesson in fourth grade, children attempted to represent and solve the following problem:

> Two students have the same amount of candies. Briana has one box, two tubes, and seven loose candies. Susan has one box, one tube, and 20 loose candies. If each box has the same amount and each tube has the same amount, can you figure out how much each tube holds? What about each box?

Although some of them still used iconic notations (see Figure 15.8), others generated and solved an equation using letters to represent the unknown amounts (Figure 15.9).

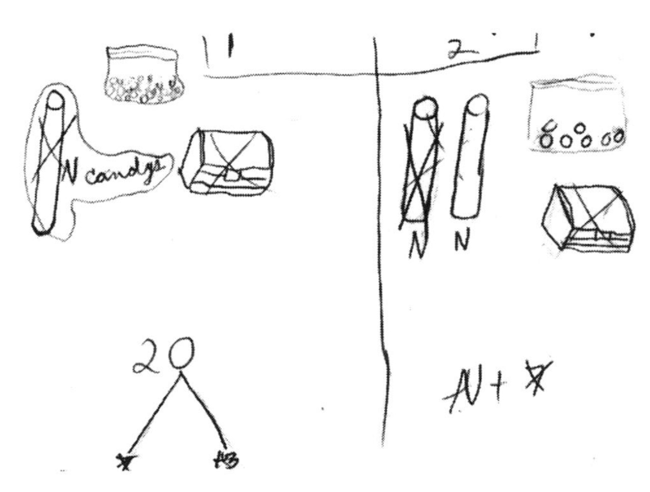

Figure 15.8 An iconic representation of the equation.

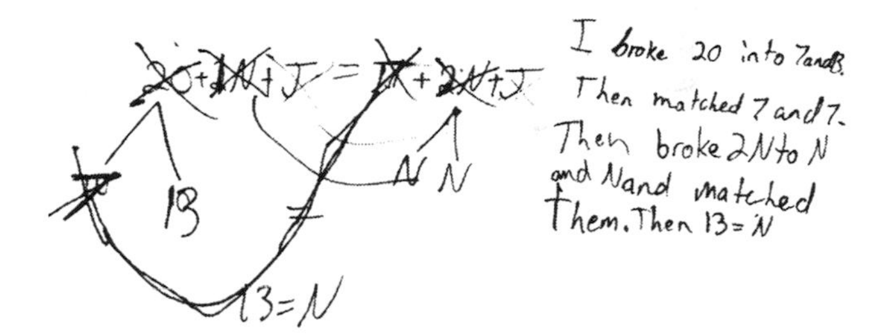

I broke 20 into 7 and 13.
Then matched 7 and 7.
Then broke 2N to N
and N and matched
Them. Then 13 = N
13=N

Figure 15.9 An algebraic representation of the equation and its solution.

At the end of 4th grade, three to four weeks after our last class, the students in the experimental group (EG) were individually interviewed. A control group (CG) of 26 5th graders from the same school were also interviewed. We (Schliemann et al., 2003) found that:

(a) 85% of the 4th graders in the EG correctly stated that the equation $6 + 9 = 7 + 8$ was true while only 65% of the 5th graders in the CG did so.

(b) To represent that Mary had three times as much money as John, 70% of the EG represented John's amount as N and Mary's as $N \times 3$. In the CG, only 29% of the children did so.

(c) When asked which of three linear function graphs showed that "Mary has three times as much money as John," 78% of the EG chose the correct line while only 46% in the CG did so. Of those choosing the correct line, 39% in the EG provided general justifications that took into account any possible pair of numbers. In the CG only 25% did so.

(d) Children were asked to represent in writing and to solve the problem "Harold has some

> money. Sally has four times as much money as Harold. Harold earns $18.00 more dollars. Now he has the same amount as Sally. Can you figure out how much money Harold has altogether? What about Sally?" We found that 56% of children in the EG represented Harold's initial amount as a letter such as N and 49% represented Sally's amount as Nx4. In the EG, 35% of the children wrote N + 18 for Harold's amount after earning 18 more dollars, 17% wrote the full equation $N + 18 = N \times 4$ and 27% correctly solved the problem. However, only 6% (four children) systematically used the algebra method to simplify the equation. In the CG, 23% of the children solved the problem but no one used algebra notation or equations to find the solution.

In the third longitudinal study from 3rd to 5th grade we were conducting at the time of writing this chapter, we have found that participants in the project show significantly better results in the EA assessment we have developed throughout grades 3 to 4, in comparison to control group results. In a written assessment given at the beginning of the second semester in 5th grade, 45% of the children correctly solved the equation $n + 6 = 3n$ and explained what the solution meant in relation to the problem at hand (Goodrow & Schliemann, 2006).

One interesting development in this third study was that the classroom teachers, who were present during the lessons and who had run the review of the homework given after each lesson, as well as the other third to fifth grade teachers, decided to implement all the algebra lessons we had developed with a second and third cohort of students. Analysis of the written assessments of the second cohort of students at the end of third and fourth grade, shows that the students who were taught by the teachers perform as well, and sometimes better, than the students who were taught by the researchers (Goodrow & Schliemann, 2006). These assessments included items from the National Assessment of Educational Progress and from the Massachusetts Comprehensive Assessment System, as well as items developed by our project team.

Summary: What Can Young Students Learn About Functions?

Taken as a whole, Moss et al. (2006) results show that transformation rules are accessible to very young students and that patterns activities, if focused on the transformation rules and numerical and geometrical representations, may constitute a meaningful entry point to EA. Our own results point to the conclusion that young students (8–11 years of age) can benefit from EA activities and learn to (a) think of arithmetical operations as functions rather than as mere computations on particular numbers; (b) learn about negative numbers; (c) grasp the meaning of variables, as opposed to instantiated values; (d) shift from thinking about relations among particular numbers and measures toward thinking about relations among sets of numbers and measures; (e) shift from computing numerical answers to describing and representing relations among variables; (f) build and interpret graphs of linear and non-linear functions; (g) solve algebraic problems using multiple representation systems such as tables, graphs, and written equations; (h) solve equations with variables on both sides of the equality; and (i) are able to inter-relate different systems of representations for functions. We have also found that elementary school teachers can successfully implement the EA activities we have designed. In many of our teacher-researcher meetings, the six teachers we have been working with in our latest study have expressed that (a) they learn mathematics and algebra as they get ready to teach the lessons, (b) the children enjoy and understand the lessons, and (c) the EA lessons help children better deal with other topics in their regular mathematics curriculum. A major research question that we hope to pursue in the next few years concerns the impact of students' EA experiences on their performance in middle and high school mathematics.

CONCLUDING THOUGHTS

Research on EA is still at its infancy. Initial findings are promising, but as we have indicated throughout this review, many basic issues—conceptual and empirical—remain to be clarified and there needs to be a concerted effort by researchers, practitioners, teacher educators and policy makers if the work is to continue productively. The worst possible scenario would be to conclude, "We already know what is needed" and to jump into large-scale implementation. In this final section of the chapter, we share some thoughts about the directions the study of EA and early algebraic reasoning might take in the near future.

What Kinds of Representations Express Algebraic Ideas?

We pointed out earlier that the concept of symbolic representation has undergone a broadening of meaning in recent years. For a long time the concept has included algebraic notation. Most mathematics educators now agree that even arithmetic notation (number sentences) is symbolic insofar as it contains symbols for representing abstract notions and relations as well as a grammar. Indeed four symbolic or representation systems are currently recognized as fundamental in EA (Janvier, 1987): *arithmetical-algebraic notation*, *tables*, *graphs*, and *natural language*.

It is no coincidence that these four fundamental representation systems are the very ones employed by mathematicians for representing functions. We suggest that the principal means for determining whether representations in these systems are conducive to algebraic reasoning is the extent to which they represent functions and capture functional reasoning. We will illustrate this by considering some of the pitfalls in using patterns and function tables. The basic idea is that that patterns, when not taken as instances of functions, tend to encourage the extension of f(n) to f(n+1) without considering the role of the independent variable, n. When tables are treated merely as patterns, the same issues ensue.

Patterns and Functions

Patterns arise in such a wide variety of everyday life contexts—tailoring, choreography, musical composition, floor tiling, DNA matching, fingerprint identification, numerology, voice detection, architectural design, poetic rhyme schemes, code-breaking, and so on—that the power of patterns in developing algebraic reasoning can be overlooked. Even in mathematics, a pattern can refer to many things. To put it bluntly, a pattern is not necessarily a well-defined mathematical object with clear connections to mathematical objects and traditions in mathematics. However, it turns out to be immensely important for establishing a coherent research base in EA. And the fact that NCTM (2000) included "understanding patterns" among the Algebra Standards for grades pre-K through 12, makes it especially urgent to decide whether patterns are a suitable topic for introducing young students to ideas of algebra. It is true that NCTM places patterns in the broad context of relations and functions (see Table 15.4). Even so, there is the implicit suggestion that patterns are bona fide mathematical objects, as relations and functions are, and there is no attempt to clarify how patterns may easily fail to establish connections to those objects.

Consider the collection of triangular dots shown in the top row of Figure 15.10. What item comes next? Most people, children and adults alike, will think of something like the arrangement of 15 dots shown in the figure as the "logical" choice.

Technically, any answer can be defended as long as the universe of discourse has not been clarified, (as Bass and colleagues [2005] noted in their criticism of NAEP algebra problems; see also Schwartz, 1975). Many people find it natural to presume that the

Table 15.4 NCTM's Algebra Standard—Patterns, Relations, & Functions

Grades	Algebra—Understand Patterns, Relations, Functions
Pre-K–2	• Sort, classify, and order objects by size, number, and other properties; • Recognize, describe, and extend patterns such as sequences of sounds and shapes or simple numeric patterns and translate from one representation to another; • Analyze how both repeating and growing patterns are generated.
3–5	• Describe, extend, and make generalizations about geometric and numeric patterns; • Represent and analyze patterns and functions, using words, tables, and graphs.
6–8	• Represent, analyze, and generalize a variety of patterns with tables, graphs, words, and, when possible, symbolic rules; • Relate and compare different forms of representation for a relationship; • Identify functions as linear or nonlinear and contrast their properties from tables, graphs, or equations.
9–12	• Generalize patterns using explicitly defined and recursively defined functions; • Understand relations and functions and select, convert flexibly among, and use various representations for them; • Analyze functions of one variable by investigating rates of change, intercepts, zeros, asymptotes, and local and global behavior; • Understand and perform transformations such as arithmetically combining, composing, and inverting commonly used functions, using technology to perform such operations on more-complicated symbolic expressions; • Understand and compare the properties of classes of functions, including exponential, polynomial, rational, logarithmic, and periodic functions; • Interpret representations of functions of two variables

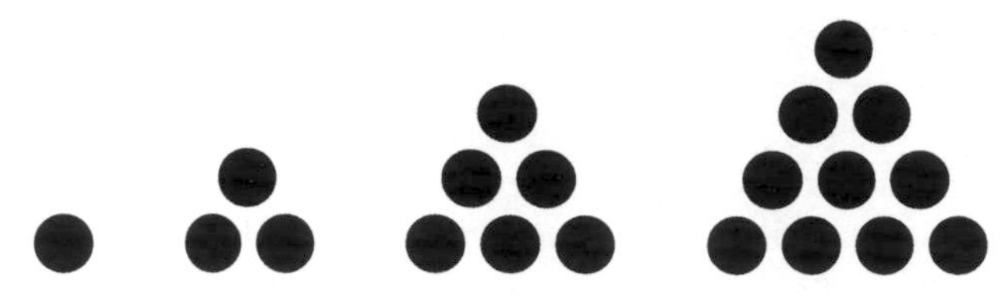

Is the next term uniquely determined?
Does the sequence define a unique function?

Is this the next term? Is it the only possible next term?

Figure 15.10 An example of a growing pattern and a possible next term.

universe is dots—figures corresponding to triangular number—but nothing requires the next item to consist of dots, or that it even be a geometric figure. In addition, even if one takes the universe of discourse for granted, no rule has been advanced to determine how the next item is created. So the fifth term cannot be deduced from the first four unless further information is supplied.

In classroom settings pattern-discovery tasks can be valuable for getting students to make conjectures about the rule that the creator of the series might have used in order to generate each item. This is in fact the rationale behind the "guess my rule" game that Bob Davis (1967) introduced in fifth grade classrooms four decades ago. Now, it is true that different rules can generate the same finite sequence. And when students discover this they face an important question (Carraher & Earnest, 2003): if two rules generate the same finite sequence, are the two rules equivalent, or are they actually different rules that happen to give the same results over a limited number of cases?

Continuing with the example of triangular numbers, the figures themselves change from instance to instance; so does the number of dots in each figure. These are not quite the same: a student may learn to extend the series without paying particular attention to the number of new dots or the number of total dots. For example, if she is working on a computer she might simply copy the last figure, then copy and paste an additional diagonal edge of dots, and then finally add a single dot. This needs to be taken into account if one wishes to use geometric patterns as an entry point to algebra.

Moreover, the dependent variable or output of the function is presumably the total number of dots. However, this is somewhat arbitrary. If one imagines the drawings in Figure 15.10 superimposed as they are produced in succession, it is easy to think of the change, that is, the increase in dots, compared to the pre-existing case, as the output. This shift in focus corresponds to the distinction between a function and its first derivative. Many miscommunications in the classroom arise around a lack of agreement about what the output is—the total or the change. However, there is another essential yet elusive thing that varies, namely, the position or counter associated with each drawing in Figure 15.10: position 0, 1, and 2 and so forth. This corresponds to the independent variable or input of a function. Students often describe serial patterns as if it were necessary only to pay attention to the dependent variable. That is, they treat sequences of geometric figures as if only one variable were involved. This is often inadvertently encouraged by curriculum materials or by teachers themselves, who may view learning how to extend functions as the primary goal of the activity.

Pattern activities can easily be carried out without students giving explicitly attention to the independent variable. This may not be a reason for concern in the grade band pre-K through 2. However, one hopes that students will eventually be able to express the functional dependency, and this entails referring to the independent and dependent variables and their relation.

Issues Common to Patterns and Tables

The first column of a table generally contains values of the independent variable, while the second column is used to hold the respective values of the dependent variable. This would appear to immunize tables against the problems that often plague patterns activities. However, tables are susceptible to the same shortcomings of patterns activities if one fails to keep in mind their role as expressions of functions. Tables of numbers can serve as function tables, that is, as representations of functions. Whether students view them as such is another matter. In the case of linear functions ($f(x) = ax + b$), students may supply additional values by extending each column independently of the other. They may note for example that the first column "grows by 1's" while the next grows by 7's (Schliemann, Carraher, & Brizuela, 2001, 2006). This is not incorrect but it is incomplete. Let us examine why, drawing upon our earlier discussion about serial patterns.

Conceiving a pattern as a "one-variable" sequence restricts attention to how one determines $f(n+1)$ from $f(n)$. In the case of triangular numbers (Fig. 10), one might count the number of dots along the edge of a triangle in any given case, f(n), then increment that number by one and add it to f(n), yielding f(n+1). The answer is correct although it requires that the student already know the value of f(n). The student will typically say something like "it goes up by 3's."

Now, assume that we are not aware of the specific function the student has been working with. Neither the student's technique nor his explanation is explicit enough for us to know what function he is working on. Even if we assume that the three (in "it goes up by 3's") refers to f(n+1) – f(n) and, further, that the independent variable increments by 1 with each row, we would have to examine the given problem in order to determine the precise function underlying the numbers in the table. For all one knows, the function could be any member of the family, $3x + b$; that is, b might be any number.

This failure to fully capture the function is not an inherent shortcoming of "scalar" (Vergnaud, 1983) approaches. Consider the function expressed in closed form as $f(x) = 3x + 7$. The recursive definition of this function is as follows:

For all natural numbers, n,
$f(0) = 7$
$f(n + 1) = f(n) + 3$

This definition of the function contains: (a) a statement that fixes the value of f(x) at x = 0; and (b) a statement describing how to obtain the value of any f(x + 1) from its predecessor, f(x). Students' scalar methods for filling in and extending function tables show features of iteration and recursion but there are also some significant differences.

It is rarely the case that the student needs to begin with the initial condition. If she wishes to extend the table beyond some $f(x_j)$, already known, she can start there paying no attention to f(0). She simply continues to add the numbers 1 and 3, respectively, to the first and second elements of each ordered pair until she reaches her destination, say, $(x_t, f(x_t))$. The student's iterative approach is correct; but it is incomplete insofar as it benefits from the initial condition without explicitly recognizing its role in the production of all values in the table. Furthermore, iteration can inefficient for large values of x: if x = 100, it is difficult to determine f(100).

Chances are, the student will never need to consider the initial condition of the function, a condition upon which the whole table is predicated. He can simply act on f(x) to produce f(x+1).

Is a Scalar Approach Valid?

We noted that students' scalar approaches generally tend not to be fully iterative. This was not to dismiss their approaches. Students notice regularities in patterns and tables of values that are perfectly legitimate. In fact, scalar ways of thinking eventually lead to calculus and differential equations, where great attention is given to delta-y and delta-x.

Two points must be borne in mind (Cuoco, 1990). First, the closed form and recursive approaches to functions do not share the same domain and co-domain: $f(x) = 3x + 7$ can be defined over real numbers, whereas the recursive definition has a domain of only the natural numbers[11] (and an even smaller co-domain). This signifies that a scalar approach will run into difficulties in accommodating the rationals, a stumbling block Kaput & West's (1994) 10–11 year old students repeatedly came across when using (software-generated) tables to determine missing values.

The second point is that scalar thinking does not easily lead to the closed form representation of a function. A closed form expression, such as $f(x) = 3x + 7$, is far easier to reconcile with an input-output approach to functions, that is, one considers x and an input and f(x) as the output. In the context of tables this amounts to thinking how a value in column one can be operated on to produce the value in column two.

Given this analysis, failures to understand the "algebraic meaning of tables" are easier to understand: students tend to be drawn to sequences as they scan tables in a downward, column-wise fashion. The teacher, on the other hand, may be hoping that students will produce a general rule for obtaining f(x) from x. However, it is extremely challenging to produce a closed form algebraic expression of a function based only on information about changes from row n to row $n+1$. (see Table 15.5 for an illustration of this. Note that it takes 7 steps to find f(3) using the piecewise definition given above.)

What Goals Are Achievable in the Short Term, Mid Term, and Long Term (for Students and Teachers)?

The jury is still out on what algebraic material and approaches make sense in the early mathematics curriculum, and whether EA corresponds to a distinct domain of study or whether it is better integrated into the knowledge base about algebra instruction for older students.

---

[11] We follow Hardy and Wright (1979) in including zero among the natural numbers.

Table 15.5 Steps Required to Determine f(3) for the Function Defined by f(0) = 7 and f(x) = f(n – 1) + 3, for all natural numbers, n

Step		Note
1	$f(3) = f(2) + 3$	From $f(n+1) = f(n) + 3$
2	$f(2) = f(1) + 3$	"
3	$f(1) = f(0) + 3$	"
4	$f(0) = 7$	Given
5	$f(1) = 7 + 3 = 10$	From steps 4 & 3.
6	$f(2) = 10 + 3 = 13$	From steps 5 & 2.
7	$f(3) = 13 + 3 = 16$	From steps 6 & 1.

EA is not just about *when*; *what, why,* and *how* are equally as important. It also concerns novel views about arithmetic, algebra, and how they are related. In the United States, arithmetic and algebra have often been treated as distinct subject matters even though certain ideas, techniques, and representations are common to both. Consistent with this view is the notion that arithmetic and algebra need to be *bridged.* The bridging is seen as occurring toward the "end" of arithmetic and the "beginning" of algebra. "Pre-algebra" courses offered one or two years prior to Algebra I in grade 8 or 9 attend to this view of arithmetic and algebra. In these courses, special attention is given to the expanded uses and meanings of symbols from arithmetic. But this *transition* is anything but simple; and part of the reason lies in the unfortunate ways concepts have been treated in the early mathematics curriculum. Consider, for example, the equals sign. In most arithmetic instruction the equals sign has the sense of "yields" or "makes." Students demonstrate that they subscribe to this notion when they read the sentence, "$3 + 5 = 8$" as "three plus five makes eight." They may even cheerfully accept expressions such as "$3 + 5 = 8 + 4 = 12$," but reject "$8 = 3 + 5$" and "$3 + 5 = 7 + 1$." When students begin algebra instruction they are expected to treat the equals sign as a comparison operator expressing an equivalence relation. But, as the preceding examples illustrate, for many students symmetry, and transitivity do not seem to apply to the equals sign.

This makes the difficult task of introducing algebraic equations all the more daunting. Consider, for example, an equation with variables on each side (e.g., $3x = 5x - 14$). Students have to contend with the idea that the expressions on each side of the equals sign can be viewed as functions having variation. The equals sign effectively constrains the values of variable to the solution set: x is free to vary, but the equation is only true for a single value of x. Note that in this case the expressions on the left and right sodes of the equals sign are not interchangeable because they refer to distinct mathematical objects (in the above example, they correspond to the functions $3x$ and $5x - 14$, which are not the same function). This point seems to have gone unnoticed by many mathematics educators, including Freudenthal (1973) who claims that the equals sign means "are the same thing" (p. 302), grounding his point in Leibniz' idea that two things are equal if they are alike in all respects.

There is an alternative to having students relearn mathematics when they take Algebra I. This alternative rests on a strikingly different view about what arithmetic and elementary mathematics are about. The key idea behind this new view is that arithmetic is a *part* of algebra, namely, that part that deals with number systems, the number line, simple functions and so on. Arithmetic deals with the part of algebra in which particular numbers and measures are treated as instances of more general examples.

Isolated examples can always be treated as instances of something more general. The number 327 is an efficient way to represent $(3 \times 100) + (2 \times 10) + (7 \times 1)$, which is just an instance of the more general expression, $(a \times 100) + (b \times 10) + (c \times 1)$, which can be expressed even more generally for an arbitrary radix or base. Opportunities for generalizing, for thinking about functions and variables, for using algebraic notation, abound in elementary mathematics.

To consider arithmetic as a part of algebra encourages us to view isolated examples and topics as instances of more abstract ideas and concepts. Addition, for example, is a computational method. It is also a function with certain general properties. Likewise multiplication by two is a table of number facts ($1 \times 2 = 2$; $2 \times 2 = 4$; $3 \times 2 = 6$; $4 \times 2 = 8$) and also a function that maps a set of input values to unique output values. The latter idea can be expressed algebraically, for instance, through the mapping notation, $x \rightarrow 2x$, the standard notation for functions, $f(x) = 2x$, or the graph on a Cartesian plane of a relation between x and y corresponding to the equation, $y = 2x$.

It will help to bear these points in mind. It may first appear that the problems we gave children to solve are arithmetical. Upon looking more closely, the reader will note their algebraic character. The categories, *arithmetical* and *algebraic*, are not mutually exclusive.

This does not mean that every idea, concept and technique from arithmetic is manifestly algebraic; however, each is *potentially* algebraic. Concepts such as equivalence can and should be treated early on in ways consistent with their usage in more advanced mathematics unless there are compelling reasons not to.

That arithmetic is a part of algebra is neither obvious nor trivial, especially for those of us who followed the "arithmetic first, algebra much later" route through our own mathematics education.

There is another reason that EA deserves special consideration. The mathematics curriculum tends to be a collection of isolated topics. Yet there are good arguments for treating mathematical (and scientific) concepts—particularly the more abstract concepts—as exceedingly relational. That is, much of their meaning derives from connections to other concepts. For example, one of the important meanings of "fraction" stems from the division of one integer by another. Yet division of integers is often treated in the curriculum without any reference to fractions (and vice-versa). It is thus not surprising that many students do not view the expression, "3 ÷ 4" as related to the fraction ¾ (Booth, 1988). In order to correct the fragmented nature of the curriculum it is necessary to take a long-term developmental perspective. In this way, one can more easily see how topics introduced in the early grades foreshadow and set the stage for topics appearing later. Likewise topics appearing relatively late in the curriculum can be more easily understood as building upon, and perhaps departing from, earlier topics.

EA offers a rare opportunity to consider the evolution of mathematical ideas from K–12. That is, it is a special context for examining how early mathematical ideas and methods are related to later ones. This is not to subscribe to the notion that mathematical understanding follows a fixed trajectory, a trajectory laid out entirely by the constraints of psychological development. The course of development depends much on the curricular structure and teaching approaches. That is why we need to ground policy regarding algebra instruction on a careful consideration of the many possible approaches.

REFERENCES

Ainley, J. (1999). Doing Algebra-type stuff: Emergent algebra in the primary school. In O. Zaslavsky (Ed.), *Proceedings of the 23rd Conference of the International Group for the Psychology of Mathematics Education* (Vol. 2, pp. 9–16). Haifa, Israel.

Aleksandrov, A. D. (1989). A general view of mathematics (S. H. Gould, T. Bartha, & K. Kirsh, Trans.). In A. D. Aleksandrov, A. N. Kolmogorov, & M. A. Lavrent'ev (Eds.), Mathematics, its content, methods, and meaning (pp. 1–64). Cambridge, MA: MIT Press.

Atiyah, M. (2005). The nature of space [Quicktime movie, 72 min], KITP Einstein Lecture. Edinburgh, Scotland: Kavli Institute for Theoretical Physics, Oct. 26.

Balacheff, N. (2001). Symbolic arithmetic vs. algebra: The core of a didactical dilemma. Postscript. In R. Sutherland, T. Rojano, A. Bell, & R.C. Lins (Eds.), *Perspectives on school algebra* (pp. 249–260). Dordrecht; The Netherlands: Kluwer.

Bass, H. (1998). Algebra with integrity and reality. *The nature and role of algebra in the K–14 curriculum: Proceedings of a National Symposium* (pp. 9–15). Washington, DC: National Academy Press.

Bass, H., Howe, R., & Milgram, R. J., (2005, September). *Analysis of NAEP items classified under the algebra and functions content strand.* Discussion Panel at the Conference on Algebraic Reasoning: Developmental, Cognitive, and Disciplinary Foundations for Instruction (pp. 14–15). Washington, DC: Brookings Institution.

Bass, H. (2005). Review of the 4th and 8th grade algebra and functions on NAEP [Electronic Version]. Algebraic Reasoning: Developmental, Cognitive, and Disciplinary Foundations for Instruction. Retrieved Feb. 1, 2006 from http://www.brookings.edu/gs/brown/algebraicreasoning.htm.

Bastable, V., & Schifter, D. (in press). Classroom stories: Examples of elementary students engaged in early algebra. In J. Kaput, D. W. Carraher, & M. Blanton (Eds.), *Algebra in the early grades.* Mahwah, NJ: Erlbaum.

Bednarz, N. (1996). Emergence and development of algebra as a problem solving tool: Continuities and discontinuities with arithmetic. In N. Bednarz, C. Kieren, & L. Lee (Eds.), *Approaches to algebra. Perspectives for research and teaching* (pp. 115–136). Dordrecht, The Netherlands: Kluwer.

Bednarz, N. (2001). A problem-solving approach to algebra: Accounting for the reasonings and notations developed by students. In H. Chick, K. Stacey, J. Vincent, & J. Vincent (Eds.), *The future of the teaching and learning of algebra: Proceedings of the 12th ICMI Study Conference* (Vol. 1, pp. 69–78). The University of Melbourne, Australia.

Bednarz, N., & Janvier, B. (1996). Emergence and development of algebra as a problem solving tool: Continuities and discontinuities with arithmetic. In N. Bednarz, C. Kieran, & L. Lee (Eds.), *Approaches to Algebra. Perspectives for Research and Teaching* (pp. 115–136). Dordrecht, The Netherlands: Kluwer.

Bednarz, N., Kieran, C., & Lee, L. (1996). *Approaches to algebra: Perspectives for research and teaching.* Dordrecht, The Netherlands: Kluwer.

Bell, A., Fischbein, E., & Greer, B. (1984). Choice of operation in verbal arithmetic problems: The effects of number size, problem structure and context. *Educational Studies in Mathematics, 15*(2), 129–147.

Bellisio, C., & Maher, C. (1998). What kind of notation do children use to express algebraic thinking? In S. Berenson et al. (Eds.), *Proceedings of the 20th Annual Meeting for the Psychology of Mathematics Education, North American Chapter* (pp. 161–165). Raleigh, NC: Columbus, OH, ERIC Clearinghouse (ED430775-6).

Blanton, M., & Kaput, J. (2000). Generalizing and progressively formalizing in a third grade mathematics classroom: Conversations about even and odd numbers. In M. Fernández (Ed.), *Proceedings of the 20th Annual Meeting of the for the Psychology of Mathematics Education, North American Chapter* (p. 115). Columbus, OH, ERIC Clearinghouse (ED446945).

Blanton, M., & Kaput, J. (2001). Algebrafying the elementary mathematics experience. Part II: Transforming practice on a district wide scale. In H. Chick, K. Stacey, J. Vincent, & J. Vincent (Eds.), *The future of the teaching and learning of algebra Proceedings of the 12th ICMI Study Conference* (Vol. 1, pp. 87–95). The University of Melbourne, Australia.

Bloedy-Vinner, H. (2001). Beyond unknowns and variables-parameters and dummy variables in high school algebra. In R. Sutherland, T. Rojano, A. J. Bishop & R. C. Lins (Eds.), *Perspectives on school algebra* (pp. 177–190). Dordrecht, The Netherlands: Kluwer.

Bodanskii, F. (1991). The formation of an algebraic method of problem solving in primary school. In V. V. Davydov (Ed.), *Psychological abilities of primary school children in learning mathematics* (Vol. 6, pp. 275–338). Reston, VA: National Council of Teachers of Mathematics.

Boero, P. (2001). Transformation and anticipation as key processes in algebraic problem solving. In R. Sutherland, T. Rojano, A. Bell, & R. C. Lins (Eds.), *Perspectives on school algebra* (pp. 99–120). Dordrecht, The Netherlands: Kluwer.

Booth, L. (1984). *Algebra: Children's strategies and errors.* Windsor, UK: NFER-Nelson.

Booth, L. R. (1988). Children's difficulties in beginning algebra. In A. F. Coxford & A. P. Shulte (Eds.), *The ideas of algebra, K–12: 1988 Yearbook* (pp. 20–32). Reston, VA: National Council of Teachers of Mathematics.

Boulton-Lewis, G.M., Cooper, T.J., Atweh, B., Pillay H., & Wilss, L. (2001). Readiness for algebra. In T. Nakahara & M. Koyama (Eds.), *Proceedings of the 24th International Conference for the Psychology of Mathematics Education* (Vol. 2, 89- 96). Hiroshima, Japan.

Boyer, C. B., & Merzbach, U. C. (1989). *History of mathematics* (2nd ed.). New York: Wiley.

Braunfeld, P., & Wolfe, M. (1966). Fractions for low achievers. *The Arithmetic Teacher, 13,* 647–655.

Brito-Lima, A. P. (1996). *Desenvolvimento da representação de igualdades em crianças de primeira a sexta série do primeiro grau.* Unpublished master's thesis, Universidade Federal de Pernambuco, Recife, Brazil.

Brito-Lima, A. P. & da Rocha Falcão, J. T. (1997). Early development of algebraic representation among 6–13 year-old children: The importance of didactic contract. In E. Pehkonen (Ed.), *Proceedings of the 21st Conference of the International Group for the Psychology of Mathematics Education* (Vol. 2, pp. 201- 208). Lahti, Finland.

Brizuela, B. M., & Earnest, D. (2007). Multiple notational systems and algebraic understandings: The case of the "best deal" problem. In J. Kaput, D. Carraher, & M. Blanton (Ed.), *Algebra in the early grades.* Mahwah, NJ: Erlbaum.

Brizuela, B. M., & Schliemann, A. D. (2004). Ten-year-old students solving linear equations *For the Learning of Mathematics, 24*(2), 33–40.

Bruer, J. T. (1993). *Schools for thought: A science of learning in the classroom.* Cambridge, MA: MIT Press.

Carlson, M. P., & Oehrtman, M. (2005). *Key aspects of knowing and learning the concept of function.* Mathematical Association of America [Research Sampler], March 15, 2005, from http://www.maa.org/t_and_l/sampler/rs_9.html

Carpenter, T. & Franke, M. (2001). Developing algebraic reasoning in the elementary school: Generalization and proof. In H. Chick, K. Stacey, J. Vincent, & J. Vincent (Eds.), *The future of the teaching and learning of algebra. Proceedings of the 12th ICMI Study Conference* (Vol. 1, pp. 155–162). The University of Melbourne, Australia.

Carpenter, T. P., Franke, M. L., & Levi, L. (2003). *Thinking mathematically: Integrating arithmetic and algebra in elementary school.* Portsmouth, NH Heinemann.

Carpenter, T., & Levi, L. (2000). Developing conceptions of algebraic reasoning in the primary grades [Electronic Version]. Research Report 00-2. Madison, WI: National Center for Improving Student Learning and Achievement in Mathematics and Science.

Carraher, D. W. (1990). Understanding the division algorithm from different perspectives. In G. Booker, P. Cobb, & T. De Mendicuti (Eds.), *Proceedings of the 14th Conference of the International Group for the Psychology of Mathematics Education* (V. 3, pp. 215–222). Oaxtepec, Mexico.

Carraher, D. W. (1993). Lines of thought: A ratio and operator model of rational number. *Educational Studies in Mathematics, 25*(4), 281–305.

Carraher, D. W. (1996). Learning about fractions. In L. P. Steffe, P. Nesher, G. Goldin, P. Cobb & B. Greer (Eds.), *Theories of mathematical learning* (pp. 241–266). Hillsdale, NJ: Erlbaum.

Carraher, D. W., & Earnest, D. (2003). Guess My Rule Revisited. In N. Pateman, B. Dougherty, & J. Zilliox (Eds.), *Proceedings of the 27th Conference of the International Group for the Psychology of Mathematics Education* (Vol. 2, pp. 173–180). Honolulu: University of Hawai'i.

Carraher, D.W. & Schliemann, A.D. (2002). Is everyday mathematics truly relevant to mathematics education? In J. Moshkovich & M. Brenner (Eds.) Everyday Mathematics. *Monographs of the Journal for Research in Mathematics Education, 11,* 131–153.

Carraher, D. W., Schliemann, A. D., & Brizuela, B. (2000, October). *Early algebra, early arithmetic: Treating operations as functions.* Plenary address at the 22nd Meeting of the Psychology of Mathematics Education, North American Chapter, Tucson, AZ (available on CD-ROM).

Carraher, D. W., Schliemann, A., & Brizuela, B. (2001). Can young students operate on unknowns? In M. v. d. Heuvel-Panhuizen (Ed.), *Proceedings of the 15th Conference of the International Group for the Psychology of Mathematics Education* (Vol. 1, pp. 130–140). Utrecht, The Netherlands: Freudenthal Institute.

Carraher, D. W., Schliemann, A. D., & Brizuela, B. (2005). Treating operations as functions. In D. Carraher & R. Nemirovsky (Eds.), *Monographs of the Journal for Research in Mathematics Education,* XIII, CD-Rom Only Issue.

Carraher, D. W., Schliemann, A. D., Brizuela, B. M., & Earnest, D. (2006). Arithmetic and algebra in early mathematics education. *Journal for Research in Mathematics Education, 37*(2), 87–115.

Carraher, D. W., Schliemann, A. D., & Schwartz, J. L. (2007). Early algebra is not the same as algebra early. In J. Kaput, D. Carraher & M. Blanton (Eds.), *Algebra in the early grades.* Mahwah, NJ: Erlbaum.

Carraher, T. N., Carraher, D. W., & Schliemann, A. D. (1985). Mathematics in the streets and in schools. *British Journal of Developmental Psychology, 3*(21), 21–29.

Carraher, T. N., Carraher, D. W., & Schliemann, A. D. (1987). Written and oral mathematics. *Journal for Research in Mathematics Education, 18*(2), 83–97.

Chazan, D. (1993). F(x) = G(x)?: An approach to modeling with algebra. *For the Learning of Mathematics, 13*(3), 22–26.

Chazan, D. (1996). Algebra for all?: The algebra policy debate. *Journal of Mathematical Behavior, 15*(4), 455–477.

Chazan, D., & Yerushalmy, M. (2003). On appreciating the cognitive complexity of school algebra: Research on algebra learning and directions of curricular change. In J. Kilpatrick, G. W. Martin, & D. Schifter (Eds.), *A research companion to the Principles and Standards for School Mathematics* (pp. 123–135). Reston, VA: National Council of Teachers of Mathematics.

Chick, H., Stacey, K., Vincent, J., & Vincent, J. (Eds.) (2001). *The future of the teaching and learning of algebra. Proceedings of the 12th ICMI Study Conference.* Melbourne, Australia: The University of Melbourne.

Clement, J. J., Lochhead, J., & Monk, G. S. (1981). Translation difficulties in learning mathematics. *American Mathematical Monthly, 88*(4), 286–290.

Clements, D. H., & Batistta, M. (1992). Geometry and spatial reasoning. In D. A. Grouws (Ed.), *Handbook of research on mathematics teaching and learning* (pp. 420–464). New York: Macmillan.

Collis, K. F. (1975). *The development of formal reasoning.* Newcastle, Australia: University of Newcastle.

Confrey, J., & Carrejo, D. (2005). Ratio and fraction: The difference between epistemological complementarity and conflict. In D. Carraher & R. Nemirovsky (Eds.), *Monographs of the Journal for Research in Mathematics Education*, XIII, CD-Rom Only Issue.

Cuoco, A. (1990). *Investigations in algebra.* Cambridge, MA: MIT Press.

Davenport, H. (1968). *The higher arithmetic: An introduction to the theory of numbers* (3rd ed.). London: Hutchinson.

Davis, R. B. (1967a). Mathematics teaching, with special reference to epistemological problems. Athens, GA: College of Education, University of Georgia.

Davis, R. B. (1967b). *Exploration in mathematics: A text for teachers.* Reading, MA: Addison-Wesley.

Davis, R. B. (1972). Observing children's mathematical behavior as a foundation for curriculum planning. *Journal of Children's Mathematical Behavior, 1*(1), 7–59.

Davis, R. B. (1985). ICME-5 Report: Algebraic thinking in the early grades. *Journal of Children's Mathematical Behavior, 4*, 198–208.

Davis, R. B. (1989). Theoretical considerations: Research studies in how humans think about algebra. In S. Wagner & C. Kieran (Eds.), *Research agenda for mathematics education* (Vol. 4, pp. 266–274). Hillsdale, NJ: Erlbaum & National Council of Teachers of Mathematics.

Davydov, V. V. (1991). *Psychological abilities of primary school children in learning mathematics* (Vol. 6). Reston, VA: National Council of Teachers of Mathematics.

Demana, F. & Leitzel, J. (1988). Establishing fundamental concepts through numerical problem solving. In A. F. Coxford & A. P. Shulte (Eds.), *The ideas of algebra, K–12: 1988 Yearbook* (pp. 61–68). Reston, VA: National Council of Teachers of Mathematics.

Douady R. (1986). Jeux de cadres et dialectique outil/objet. *Recherches en Didactique des Mathématiques, 7*(2), 5–31.

Dougherty, B. (2007). Measure up: A quantitative view of early algebra. In J. Kaput, D. Carraher, & M. Blanton (Eds.), *Algebra in the early grades.* Mahwah, NJ: Erlbaum.

Dubinsky, E., & Harel, G. (1992a). The nature of the process conception of function. In E. Dubinsky & G. Harel (Eds.), *The concept of function: Aspects of epistemology and pedagogy* (pp. 85–106). Washington, DC: Mathematical Association of America.

Dubinsky, E., & Harel, G. (Eds.) (1992b). *The concept of function: Aspects of epistemology and pedagogy.* Washington, DC: Mathematical Association of America.

Euclid, & Heath, T. L. (1956). Book V. Propositions I-VII [Related to Eudoxus' theory of proportion]. In The thirteen books of Euclid's Elements (2nd ed., Vol. 2, pp. 138–148). New York: Dover.

Filoy, E., & Rojano, T. (1984). From arithmetical to algebraic thought. In *Proceedings of the 6th Annual Meeting for the Psychology of Mathematics Education, North American Chapter* (pp. 51–56). Madison, WI.

Filoy, E., & Rojano, T. (1989). Solving equations: The transition from arithmetic to algebra. *For the Learning of Mathematics, 2*, 19–25.

Frege, F. L. G. (1953). *The foundations of arithmetic; a logico-mathematical enquiry into the concept of number.* J.L. Austin (Transl.), (German-English, 2nd ed.). Oxford: Basil Blackwell.

Freudenthal, H. (1973). *Mathematics as an educational task.* Dordrecht, The Netherlands: Reidel.

Fridman, L. M. (1991). Features of introducing the concept of concrete numbers in the primary grades. In V. V. Davydov (Ed.), *Psychological abilities of primary school children in learning mathematics. Soviet studies in mathematics education* (Vol. 6, pp. 148–180). Reston, VA: National Council of Teachers of Mathematics.

Fujii, T., & Stephens, M. (2001). Fostering an understanding of algebraic generalization through numerical expressions: The role of quasi-variable. In H. Chick, K. Stacey, J. Vincent, & J. Vincent (Eds.), *The future of the teaching and learning of algebra. Proceedings of the 12th ICMI Study Conference* (Vol. 1, pp. 258–264). The University of Melbourne, Australia.

Gamoran, A., & Hannigan, E. C. (2000). Algebra for everyone? Benefits of college-preparatory mathematics for students with diverse abilities in early secondary school. *Educational Evaluation and Policy Analysis, 22*(3), 241–254.

Goldenberg, P., & Shteingold, N. (in press). Early algebra: The MW Perspective. In J. Kaput, D. W. Carraher & M. Blanton (Eds.), *Algebra in the early grades.* Mahwah, NJ: Erlbaum.

Goodrow, A., & Schliemann, A. D. (2006, April). *The TERC-Tufts Early Algebra Project: Assessment results.* Paper presented at the National Council of Teachers of Mathematics Research Presession. San Francisco.

Gravemeijer, K. (1999). How emergent models may foster the constitution of formal mathematics. *Mathematical Thinking and Learning, 1*(2), 155–177.

Gravemeijer, K. (2002). *Symbolizing, modeling, and tool use in mathematics education.* Boston: Kluwer.

Goodman, N. (1976). The theory of notation. In N. Goodman, *Languages of art; an approach to a theory of symbols* (pp. 127–173). Indianapolis, IN: Hackett.

Hansson, Ö. (2006). *Studying the views of preservice teachers on the concept of function.* Unpublished doctoral dissertation, Luleå University of Technology, Luleå, Sweden.

Hardy, G. H., & Wright, E. M. (1979). *An introduction to the theory of numbers* (5th ed.). Oxford, England: Clarendon.

Harper, E. (1987). Ghosts of Diophantus. *Educational Studies in Mathematics, 18*, 75–90.

Healy, L., Hoyles, C., & Sutherland, R. (1990). Critical decisions in the generalization process: A methodology for researching pupil collaboration in computer and non-computer environments. In G. Booker, P. Cobb, & T. De Mendicuti (Eds.), *Proceedings of the 14th Conference of the*

International Group for the Psychology of Mathematics Education (Vol. 3, pp 83–90). Oxtapec, Mexico.

Heid, M. K. (1996). A technology-intensive functional approach to the emergence of algebraic thinking. In N. Bednarz, C. Kieren, & L. Lee (Eds.), *Approaches to algebra: Perspectives for research and teaching* (pp. 239–255). Dordrecht, The Netherlands: Kluwer.

Herscovics, N., & Kieran, C. (1980). Construction meaning for the concept of equation. *Mathematics Teacher*, 572–580.

Herscovics, N., & Linchevski, L. (1994). A cognitive gap between arithmetic and algebra. *Educational Studies in Mathematics, 27*, 59–78.

Howe, R. (1998a). Reports of AMS association resource group. *Notices of the American Mathematical Society, 45*(2), 270–276.

Howe, R. (1998b). The AMS and mathematics education: The revision of the "NCTM Standards". *Notices of the American Mathematical Society, 45*(2), 243–247.

Howe, R. (2005). Comments on NAEP algebra problems [Electronic version]. *Algebraic Reasoning: Developmental, Cognitive, and Disciplinary Foundations for Instruction.* Retrieved Feb. 1, 2006 from http://www.brookings.edu/gs/brown/algebraicreasoning.htm.

Howe, R. (2006). *Taking place value seriously: Arithmetic, estimation and algebra.* Unpublished manuscript. Department of Mathematics, Yale University.

Hunter, J. and D. Monk (1971). *Algebra and number systems.* London, Blackie.

Inhelder, B., & Piaget, J. (1958). *The growth of logical thinking from childhood to adolescence: An essay on the construction of formal operational structures.* New York: Basic Books.

Jahnke, H. N. (2003). Algebraic analysis in the 18th century. In H. N. Jahnke (Ed.), *A history of analysis* (pp. 105–136). Providence, RI: American Mathematical Society.

Janvier, C. (1987). *Problems of representation in the teaching and learning of mathematics.* Hillsdale, NJ: Erlbaum.

Kaput, J., & West, M. M. (1994). Missing-value proportional reasoning problems: Factors affecting informal reasoning patterns. In G. Harel & J. Confrey (Eds.), *The Development of multiplicative reasoning in the learning of mathematics* (pp. 237–287). Albany, NY: SUNY Press.

Kaput, J. (1995, April). *Transforming algebra from an engine of inequity to an engine of mathematical power by "algebrafying" the K–12 curriculum.* Paper presented at the Annual Meeting of the National Council of Teachers of Mathematics, Boston, MA.

Kaput, J. (1998). Transforming algebra from an engine of inequity to an engine of mathematical power by 'algebrafying' the K–12 Curriculum. Sponsored by the National Council of Teachers of Mathematics and Mathematical Sciences Education Board Center for Science, Mathematics and Engineering Education, National Research Council. *The Nature and Role of Algebra in the K–14 Curriculum* (pp. 25–26). Washington, DC: National Academies Press.

Kaput, J. (2007). What is algebra? What is algebraic reasoning? In J. Kaput, D. W. Carraher, & M. Blanton (Eds.), *Algebra in the early grades.* Mahwah, NJ: Erlbaum.

Kaput, J., Blanton, M., & Moreno, L. A. (2007). Algebra from a symbolization point of view. In J. Kaput, D. W. Carraher & M. Blanton (Eds.), *Algebra in the early grades.* Hillsdale, NJ: Erlbaum & The National Council of Teachers of Mathematics.

Kieran, C. (1981). Concepts associated with the equality symbol. *Educational Studies in Mathematics, 12*, 317–326.

Kieran, C. (1989). The early learning of algebra: A structural perspective. In S. Wagner & C. Kieran (Eds.), *Research issues in the learning and teaching of algebra* (Vol. 4, pp. 33–56). Reston, VA: National Council of Teachers of Mathematics/Erlbaum.

Kieran, C. (1992). The learning and teaching of school algebra. In D. A. Grouws (Ed.), *Handbook of research on mathematics teaching and learning* (pp. 390–419). Reston, VA: National Council of Teachers of Mathematics.

Kieran, C. (1996). The changing face of school algebra. In C. Alsina, B. Alvarez, B. R. Hodgson, C. Laborde, & A. Pérez (Eds.), *8th International Congress on Mathematical Education: Selected lectures* (pp. 271–290). Sevilla, Spain: S.A.E.M. Thales.

Kieran, C. (2006). Research on the learning and teaching of algebra: A broadening of sources of meaning. In A. Gutiérrez & P. Boero (Eds.), *Handbook of research on the psychology of mathematics education: Past, present and future* (pp. 11–49). Rotterdam, The Netherlands: Sense Publishers.

Kieran, C., Boileau, A., & Garançon, A. (1996). Introducing algebra by means of a technology-supported, functional approach. In Bednarz, N., Kieran, C. & L. Lee (Ed.), *Approaches to algebra. Perspectives for research and teaching* (pp. 257–293). Dordrecht, The Netherlands: Kluwer.

Kirshner, D. (2001). The structural algebra option revisited. In R. Sutherland, T. Rojano, A. Bell & R. C. Lins (Eds.), *Perspectives on school algebra* (pp. 83–98). Dordrecht, The Netherlands: Kluwer.

Klein, A. S., Beishuizen, M., & Treffers, A. (1998). The empty number line in Dutch second grades: Realistic versus gradual program design. *Journal for Research in Mathematics Education, 29*(4), 443–464.

Kobayashi, M. (1988). *New ideas of teaching mathematics in Japan.* Tokyo: Chuo University Press.

Kuchemann, D. E. (1981). Algebra. In K. Hart (Ed.), *Children's understanding of mathematics* (pp. 102–119). London: Murray.

Lacampagne, C. B. (1995). *The Algebra Initiative Colloquium.* Washington, DC: U.S. Department of Education, OERI.

Lebesgue, H. L., & May, K. O. (1966). *Measure and the integral.* San Francisco: Holden-Day.

Lehrer, R., & Schauble, L. (2000). Modeling in mathematics and science. In R. Glaser (Ed.), *Advances in instructional psychology: Vol. 5. Educational design and cognitive science.* Mahwah, NJ: Erlbaum.

Lehrer, R., Jacobson, C., Thoyre, G., Kemeny, V., Strom, D., Horvath, J., et al. (1998). Developing understanding of space and geometry in the primary grades. In D. Chazan & R. Lehrer (Eds.) *Designing learning environments for developing understanding of geometry and space* (169–200). Hillsdale, NJ: Erlbaum.

Linchevsky, L. (2001). Operating on the unknowns: What does it really mean? In M. v. d. Heuvel-Panhuizen (Ed.) *Proceedings of the 25th Conference of the International Group for the Psychology of Mathematics Education* (Vol. 4, pp. 141–144). Utrecht, The Netherlands: Freudenthal Institute.

Lins Lessa, M. M. (1995). *A balança de dois pratos versus problemas verbais na iniciação à algebra.* Unpublished master's thesis. Mestrado em Psicologia, Universidade Federal de Pernambuco, Recife, Brazil.

Lins, R. C. (2001). The production of meaning for algebra: A perspective based on a theoretical model of semantic fields. In R. Sutherland (Ed.), *Perspectives on school algebra* (pp. 37–60). Dordrecht, The Netherlands: Kluwer.

Lins, R. C., & Kaput, J. (2004). The early development of algebraic reasoning: The current state of the field. In H. Chick, K. Stacey, J. Vincent, & J. Vincent (Eds.), *The future of the teaching and learning of algebra. Proceedings of the 12th ICMI Study Conference* (Vol. 1, pp. 47–70). The University of Melbourne, Australia.

MacGregor, M. (1996). Curricular aspects of arithmetic and algebra. In R. C. L. J. Gimenez, & B. Gomez (Ed.), *Arithmetics and algebra education: Searching for the future* (pp. 50–54). Tarragona, Spain: Universitat Rovira i Virgili.

MacGregor, M. (2001). Does learning algebra benefit most people? In H. Chick, K. Stacey, J. Vincent, & J. Vincent (Eds.), *The future of the teaching and learning of algebra Proceedings of the 12th ICMI Study Conference* (Vol. 2, pp. 405–411). The University of Melbourne, Australia.

Markie, P. (2004). Rationalism vs. empiricism. In E. N. Zalta (Ed.), *The Stanford Encyclopedia of Philosophy* (Fall 2004 Edition) http://plato.stanford.edu/archives/fall2004/entries/rationalism-empiricism/. Palo Alto, CA: Stanford University.

Mason, J. (1996). Expressing generality and roots of algebra. In N. Bednarz, C. Kieran, & L. Lee (Eds.), *Approaches to algebra: Perspectives for research and teaching* (pp. 65–86). Dordrecht, The Netherlands: Kluwer.

Menninger, K. (1992). *Number words and number symbols: A cultural history of numbers.* New York: Dover Publications.

Milgram, J. (2006a). *Remarks on the manuscript "Early Algebra."* Personal [Email] communication to D. W. Carraher. Boston, MA, July 25.

Milgram, R. J. (2005). Appendix E: The 6th grade treatment of algebra in the Russian program. In R.J. Milgram, *The mathematics pre-service teachers need to know.* (pp. 551–563). Stanford, CA: Stanford University

Moshkovich, J., & Brenner, M. (Eds.) (2002). Everyday mathematics. *Monographs of the Journal for Research in Mathematics Education,* Reston, VA: National Council of Teachers of Mathematics.

Moss, J., Beatty, R., McNab, S. L., & Eisenband, J. (2005). The Potential of Geometric Sequences to Foster Young Students' Ability to Generalize in Mathematics. [Powerpoint file] presented at the Algebraic Reasoning: Developmental, Cognitive and Disciplinary Foundations for Instruction, Sept. 14, 2005. Washington: The Brookings Institution. Retrieved Aug. 1, 2006, from http://www.brookings.edu/gs/brown/algebraicreasoning.htm.

Moss, J., Beatty, R., McNab, S. L., & Eisenband, J. (2006, April). *The potential of geometric sequences to foster young students' ability to generalize in mathematics.* Paper presented at the Annual Meeting of the American Educational Research Association. San Francisco.

National Council of Teachers of Mathematics. (1989). *Curriculum and evaluation standards for school mathematics.* Reston, VA: Author.

National Council of Teachers of Mathematics. (2000). Principles and standards for school mathematics. Reston, VA: Author.

Newton, I. (1967/1671). Methodus Fluxionum et Serierum Infinitarum (1671). In D. T. Whiteside (Ed.), *The mathematical papers of Isaac Newton* (pp. 332–357). Cambridge, England: Cambridge University Press.

Nunes, T., Schliemann, A. D., & Carraher, D. W. (1993). *Street mathematics and school mathematics.* Cambridge, England: Cambridge University Press.

Otte, M. (1993). *O pensamento relacional: Equações [Relational thinking: Equations].* Unpublished manuscript.

Papert, S. (1980). *Mindstorms: Children, computers, and powerful ideas.* New York: Basic Books.

Peled, I., & Carraher, D. W. (2007). Signed numbers and algebraic thinking. In J. Kaput, D. W. Carraher, & M. Blanton (Eds.), *Algebra in the early grades.* Mahwah, NJ: Erlbaum.

Piaget, J. (1960). *The child's conception of geometry.* London: Routledge & Kegan Paul.

Piaget, J. (1970a). *The child's conception of movement and speed.* London: Routledge & Kegan Paul.

Piaget, J. (1970b). *The child's conception of time.* New York: Basic Books.

Piaget, J., & Inhelder, B. (1967). *The child's conception of space.* New York: W. W. Norton.

Piaget, J., & Inhelder, B. (1969). The gaps in empiricism. In A. Koestler & J. Smythies (Eds.), *Beyond Reductionism: New Perspectives In the Life Sciences* (pp. 118–165). Boston: Beacon.

Piaget, J., & Inhelder, B. (1974). *The child's construction of quantities: Conservation and atomism.* London: Routledge & Kegan Paul.

Radford, L. G. (2001). Of course they can! In M. v. d. Heuvel-Panhuizen (Ed.), *Proceedings of the 25th Conference of the International Group for the Psychology of Mathematics Education* (Vol. 4, pp. 145–148). Utrecht, The Netherlands: Freudenthal Institute.

RAND Mathematics Study Panel. (2003). Mathematical proficiency for all students: Toward a strategic research and development program in mathematics education (No. 083303331X). Santa Monica, CA: RAND.

Reed, H. J., & Lave, J. (1981). Arithmetic as a tool for investigating relations between culture and cognition. In R. W. Casson (Ed.), *Language, culture, and cognition: Anthropological perspectives* (pp. 437–455). New York: Macmillan.

Rivera, F. D. (2006, February). Changing the face of arithmetic: Teaching children algebra. *Teaching Children Mathematics,* 306–311.

Rojano, T. (1996). Developing algebraic aspects of problem solving within a spreadsheet environment. In Bednarz, N., Kieran, C. & L. Lee (Ed.), *Approaches to algebra. Perspectives for research and teaching* (pp. 137–145). Dordrecht, The Netherlands: Kluwer.

Sawyer W. W. (1964). *Vision in elementary mathematics.* New York: Penguin.

Schifter, D. (2005). Engaging students' mathematical ideas: Implications for professional development design. *Monographs of the Journal for Research in Mathematics Education,* XIII, CD-Rom Only Issue.

Schifter, D., Monk, S., Russell, S. J., & Bastable, V. (in press). Early algebra: What does understanding the laws of arithmetic mean in the elementary grades? In J. Kaput, D. W. Carraher, & M. Blanton (Eds.), *Algebra in the early grades.* Mahwah, NJ: Erlbaum.

Schliemann, A. D. & Carraher, D. W. (2002). The evolution of mathematical understanding: Everyday versus idealized reasoning. *Developmental Review, 22*(2), 242–266.

Schliemann, A. D., Carraher, D. W. & Brizuela, B. M. (2001). When tables become function tables. In M. v. d. Heuvel-Panhuizen (Ed.), *Proceedings of the 25th Conference of the International Group for the Psychology of Mathematics Education* (Vol. 4, pp. 145–152). Utrecht, The Netherlands: Freudenthal Institute.

Schliemann, A. D., Carraher, D. W., & Brizuela, B. M. (2006). *Bringing out the algebraic character of arithmetic: From children's ideas to classroom practice.* Mahwah, NJ: Erlbaum.

Schliemann, A. D., Carraher, D. W., Brizuela, B. M., Earnest, D., Goodrow, A., Lara-Roth, S., et al. (2003). Algebra in elementary school. In N. Pateman, B. Dougherty, & J. Zilliox (Eds.), *International Conference for the Psychology of Mathematics Education* (Vol. 4, pp. 127–134). Honolulu: University of Hawai'i.

Schoenfeld, A. (1995). Report of Working Group 1. In C. B. Lacampagne (Ed.), *The Algebra Initiative Colloquium: Vol. 2. Working Group Papers* (pp. 11–18). Washington, DC: U.S. Department of Education, OERI.

Schwartz, J. L. (1975). a is to B as c is to anything at all: The illogic of IQ tests. *National Elementary Principal, 54*(4), 38–41.

Schwartz, J. L. (1996a). *Semantic aspects of quantity.* Unpublished manuscript, MIT and Harvard Graduate School of Education, Cambridge, MA.

Schwartz, J. L. (1996b, July). *Can technology help us make the mathematics curriculum intellectually stimulating and socially responsible?* Paper presented at the Eighth International Congress on Mathematics Education. Seville, Spain.

Schwartz, J. L. (1999). Can technology help us make the mathematics curriculum intellectually stimulating and socially responsible? *International Journal of Computers for Mathematical Learning, 4*(2/3), 99–119.

Schwartz, J. & Yerushalmy, M. (1992a). Getting students to function on and with algebra. In E. Dubinsky & G. Harel (Eds.), *The concept of function: Aspects of epistemology and pedagogy* (pp. 261–289). Washington, DC: Mathematical Association of America.

Schwartz, J. L., & Yerushalmy, M. (1992b). *The Function Supposer: Symbols and graphs.* Pleasantville, NY: Sunburst Communications.

Schwartz, J. L., & Yerushalmy, M. (1992c). *The Geometric Supposer Series.* Pleasantville, NY: Sunburst Communications.

Sfard, A. (1995). The development of algebra: Confronting historical and psychological perspectives. *Journal of Mathematical Behavior, 14,* 15–39.

Sfard, A., & Linchevsky, L. (1994). The gains and the pitfalls of reification—The case of algebra. *Educational Studies in Mathematics 26,* 191–228.

Slavitt, D. (1999). The role of operation sense in transitions from arithmetic to algebraic thought. *Educational Studies in Mathematics, 37,* 251–274.

Smith, J. P., & Thompson, P. W. (2007). Quantitative reasoning and the development of algebraic reasoning. In J. Kaput, D. Carraher, & M. Blanton (Ed.), *Algebra in the early grades.* Mahwah, NJ: Erlbaum.

Smith, S. (2000). Second graders discoveries of algebraic generalizations. In M. Fernández (Ed.). *Proceedings of the 22nd Annual Meeting of the North American Chapter of the International Group for the Psychology of Mathematics Education* (pp. 133–139). Tucson, Arizona.

Stacey, K., Chick, H., & Kendal, M. (2004). *The future of the teaching and learning of algebra: The 12th ICMI study.* Boston: Kluwer.

Steinberg, R., Sleeman, D., & Ktorza, D. (1990). Algebra students knowledge of equivalence of equations. *Journal for Research in Mathematics Education, 22*(2), 112–121.

Steiner, H. G. (1969). Magnitudes and rational numbers—A didactical analysis. *Educational Studies in Mathematics, 2*(2), 371–392.

Stern, E., & Mevarech, Z. (1996). Children's understanding of successive divisions in different contexts. *Journal of Experimental Child Psychology, 61*(2), 153–172.

Sutherland, R. (1993). Symbolizing through spreadsheets. *Micromath, 10*(1), 20–22.

Sutherland, R., & Rojano, T. (1993). A spreadsheet approach to solving algebra problems. *Journal of Mathematical Behavior, 12,* 353–383.

Tall, D. (2001). Reflections on early algebra. In M. v. d. Heuvel-Panhuizen (Ed.), *Proceedings of the 25th Conference of the International Group for the Psychology of Mathematics Education* (Vol. 1, pp. 133–139). Utrecht, The Netherlands: Freudenthal Institute.

Teppo, A. (2001). Unknowns or place holders? In M. v. d. Heuvel-Panhuizen (Ed.), *Proceedings of the 25th Conference of the International Group for the Psychology of Mathematics Education* (Vol. 1, pp. 153–155). Utrecht, The Netherlands: Freudenthal Institute.

Thompson, P. W. (1988). *Quantitative concepts as a foundation for algebraic reasoning: sufficiency, necessity, and cognitive obstacles.* Paper presented at the *Annual Conference of the International Group for the Psychology of Mathematics Education,* Dekalb: Northern Illinois University.

Thompson, P. W. (1993). Quantitative reasoning, complexity, and additive structures. *Educational Studies in Mathematics, 25*(3), 165–208.

Ursini, S. (1994). *Ambientes LOGO como apoyo para trabajar las nociones de variación y correspondencia.* Paper presented at the Memoria del primer simposio sobre metodología de la enseñanza de las matemáticas, México.

Ursini, S. (1997). El lenguaje LOGO, los niños y las variables [The LOGO language, children, and variables]. *Educación Matemática, 9*(2), 30–42.

Ursini, S. (2001). General methods: A way of entering the world of algebra In T. R. R. Sutherland, A. Bell, & R. Lins (Ed.), *Perspectives on School Algebra* (pp. 209–230). Dordrecht, The Netherlands: Kluwer.

van Amerom, B. A. (2002). *Reinvention of early algebra: Developmental research on the transition from arithmetic to algebra.* Unpublished doctoral thesis, University of Utrecht, The Netherlands.

Vergnaud, G. (1979). The acquisition of arithmetical concepts. *Educational Studies in Mathematics, 10*(2), 263–274.

Vergnaud, G. (1982). A classification of cognitive tasks and operations of thought involved in addition and subtraction problems. In T. P. Carpenter, J. M. Moser, & T. A. Romberg (Eds.), *Addition and subtraction: A cognitive view* (pp. 39–59). Hillsdale, NJ: Erlbaum.

Vergnaud, G. (1983). Multiplicative structures. In R. A. Lesh & M. Landau (Eds.), *Acquisition of mathematics concepts and processes* (pp. 127–174). New York: Academic Press.

Vergnaud, G. (1985). Understanding mathematics at the secondary-school level. In A. Bell, B. Low, & J. Kilpatrick (Eds.), *Theory, research & practice in mathematical education* (pp. 27–45). Nottingham, UK: University of Nottingham, Shell Center for Mathematical Education.

Vergnaud, G., Cortes, A., & Favre-Artigue, P. (1988). Introduction de l'algèbre auprès des débutants faibles: problèmes épistemologiques et didactiques [Introductory algebra for weak novices: Didactic and epistemological problems]. In

G. Vergnaud, G. Brousseau, & M. Hulin (Eds.), *Didactique et acquisitions des connaissances scientifiques: Actes du Colloque de Sèvres* (pp. 259–280). Sèvres, France: La Pensé Sauvage.

Verschaffel, L., Greer, B., & De Corte, E. (2002). Everyday knowledge and mathematical modeling of school word problems. In K. Gravemeijer, R. Lehrer, B. v. Oers, & L. Verschaffel (Eds.), *Symbolizing, modeling, and tool use in mathematics education* (pp. 249–268). Dodrecht, The Netherlands: Kluwer.

Wagner, S. (1981). Conservation of equation and function under transformations of variable. *Journal for Research in Mathematics Education, 12,* 107–118.

Warren, E. (2004). Generalising arithmetic: Supporting the process in the early years. *Proceedings of the 28th Conference of the International Group for the Psychology of Mathematics Education* (Vol. 4, 417–424). Bergen, Norway.

Warren, E., & Cooper, T. (2001). The unknown that does not have to be known. In M. v. d. Heuvel-Panhuizen (Ed.), *Proceedings of the 25th Conference of the International Group for the Psychology of Mathematics Education* (Vol. 1, pp. 156–162). Utrecht, The Netherlands: Freudenthal Institute.

Warren, E., & Cooper, T. (2005). Introducing functional thinking in year 2: A case study of early algebra teaching. *Contemporary Issues in Early Childhood, 6*(2), 150–162.

Weber, M., & Shils, E. (1949/1917). *The methodology of the social sciences.* Glencoe, IL.: Free Press.

Weinstein, E. W. (1999). Field axioms. Mathworld–A Wolfram Web Resource, July 29, 2006, http://mathworld.wolfram.com/FieldAxioms.html

Whitney, H. (1968a). The mathematics of physical quantities: Part I: mathematical models for measurement. *American Mathematical Monthly, 75*(2), 115–138.

Whitney, H. (1968b). The mathematics of physical quantities, Part 2: Quantity structures and dimensional analysis. *American Mathematical Monthly, 75*(3), 227–256.

Wirtz, R., Botel, M., Beberman, M., & W. W. Sawyer. (1965). *Math workshop.* Chicago: Encyclopaedia Britannica Press.

Wu, H. (1999). Basic skills versus conceptual understanding. A bogus dichotomy in mathematics education. *American Educator,* American Federation of Teachers, retrieved July 29, 2006, http://www.aft.org/pubs-reports/american_educator/fall99/wu.pdf.

Yerushalmy, M. & Chazan, D. (2002). Flux in school algebra: Curricular change, graphing technology, and research on student learning and teacher knowledge. In L. English (Ed.), *Handbook of international research in mathematics education* (pp. 725–755). Hillsdale, NJ: Erlbaum.

AUTHOR NOTE

The National Science Foundation provided us with crucial support for writing this chapter via the grant, "Algebra in Early Mathematics" (NSF-ROLE #0310171).

We wish to thank John Lannin, Illinois State University, and Linda Levi, University of Wisconsin-Madison for their careful and thoughtful reviews of the first draft of this chapter. We are grateful to James Milgram, Stanford University, for his generosity in discussing many points in the manuscript; we assume responsibility for any shortcomings in the exposition. We thank Roger Howe, Yale University, for helping us appreciate the importance of the field axioms in elementary algebra.

We thank Frank Lester for his encouragement throughout the preparation of the chapter as well as his careful feedback on multiple versions of the manuscript.

We thank Judah L. Schwartz, Tufts University, for long and spirited discussions over the years regarding a great many mathematical ideas central to this chapter.

We are indebted to Jim J. Kaput (1942–2005), University of Massachusetts-Dartmouth who—fascinated with mathematics education, early algebra and life itself—was a source of inspiration to us and many others.

16

LEARNING AND TEACHING ALGEBRA AT THE MIDDLE SCHOOL THROUGH COLLEGE LEVELS

Building Meaning for Symbols and Their Manipulation

Carolyn Kieran

UNIVERSITÉ DU QUÉBEC À MONTRÉAL

From the days of al-Khwârizmî, and on through Vieta and Euler, algebra was all about procedures and notations. This view of algebra, as a tool for manipulating symbols and for solving problems, has been reflected in school algebra curricula as they developed and took shape throughout the 1800s and into the 1900s. Like the students of today, those of yesteryear no doubt struggled to adapt to the intricacies of this representational and procedural tool. Thus, research conducted during the first half of the 20th century on the learning of algebra focused, for example, on the relative difficulty of solving various kinds of linear equations (Hotz, 1918; Reeve, 1926), the role of practice (Thorndike et al., 1923), and the errors that first-year algebra students make in applying algorithms (Breslich, 1939). During the 1950s and early 1960s, research related to algebra learning was conducted mostly by psychologists with a behaviorist orientation who used the subject area as a vehicle for studying general questions related to skill development, memory, and the like. In contrast, the research that has been conducted at least since the late 1970s when algebra education researchers began to increase in number and coalesce as a community (Wagner & Kieran, 1989) has tended to focus on the kind of meaning students make of their algebra, as well as on various ways in which to make algebra learning meaningful for students.

It is not surprising that more recent research in algebra learning and teaching has, in general, emphasized pathways to and the development of algebraic meaning. First, Piaget's ideas had begun to circulate among North American researchers during the 1960s and 1970s. The influence of his cognitive development psychology on mathematics education researchers was substantial. Next, studies over several decades had shown that an exclusively skills-based approach to the teaching of algebra did not lead to skilled performance among algebra students (e.g., Monroe, 1915, and the later work of Carry, Lewis, & Bernard, 1980). Nor, according to the ample number of studies of the late 1970s and 1980s, had such approaches led to students' being able to interpret adequately the various ways in which letters are used in algebra (Küchemann, 1981; Matz, 1982), or the structural features of algebraic expressions (Davis, 1975), or equivalence constraints on equations and equation solving (Greeno, 1982). Thirdly, the research on algebra learning since the 1980s has been influenced by socio-democratic forces, such as for example the "Algebra for All" movement in the United States (Chazan, 1996; Edwards, 1990). New ways of thinking about the content of secondary school algebra, as well as the notion that introducing algebraic activity in primary school (Kaput, 1995) could help in removing some of the obstacles associated with secondary school algebra, began to emerge and be discussed. Lastly, the arrival of computing technology in schools in the 1980s led to reshaping conceptions of what could (and should) be taught in school algebra (Fey, 1989)—ideas that were to a large extent both initiated by algebra learning researchers as well as reflected in their subsequent work.

Whereas much of the 1970s and 1980s research focused on the reporting of primarily empirical findings with respect to students' understanding of and activity in algebra, recent theoretical advances in the field of algebra education research, both in North America and abroad (see, e.g., Lerman, Xu, & Tsatsaroni, 2002; Radford, 2000), have widened the community's perspective on the notion of algebraic meaning and have altered its ways of thinking about how students construct meaning for algebraic objects and processes. The theoretical framework of constructivism with its key tenet that knowledge is actively constructed by the cognizing subject, which had been inspired by Piaget's work and had blossomed in the 1980s, had led algebra researchers to move their attention from, for example, the errors made by students to the ways in which students craft their understandings of algebraic concepts and procedures. Although the focus was still cognitive, it tended to be broader than that suggested by the research analyses carried out just a decade or two earlier. Then, during the 1990s, the sociocultural perspective that had developed outside the mathematics education research community began to be felt within. The earlier constructivist/cognitivist orientation shifted for a large number of algebra researchers toward analyses of social factors affecting algebra learning (see, e.g., Cobb, 1994), with an accompanying interest in the mediating role of cultural tools (e.g., Meira, 1998). This shift also led to an increase in classroom-based studies with a focus on teacher-student and student-student discourse (e.g., Bartolini Bussi, 1995; Sfard, 2001). The increasing use of digital tools in learning, in combination with the sociocultural lens being applied to both the design of learning studies and the analysis of the resulting data, led to new questions about the nature of learning with these tools. Dynamic learning environments prompted the development of new theoretical perspectives in which bodily experience was considered to play a role in the creation of meaning for algebraic objects.

More recent research has thus had not only a stronger theoretical component to frame the empirical work, but also one derived from a more diverse set of theories than in the past. Shifting views of the content of school algebra have also been exemplified in this research. In order to give adequate consideration to these shifts, the chapter is divided into two parts, the first part of which is devoted to their discussion. This part also includes the presentation of models of algebraic activity and examines various sources of meaning that shape algebra learning. The second part of the chapter deals with the body of research that has been carried out in the learning and the teaching of algebra, primarily since 1990. Although the research studies conducted prior to 1990 have already been discussed in Kieran (1992), this chapter will at times refer back to this earlier body of research to suggest the ways in which more recent work has been informed by the past.

This review draws upon the considerable number of algebra research resources that exist already in the domain, including research syntheses (Bednarz, Kieran, & Lee, 1996; Filloy & Sutherland, 1996; Kieran, 1990, 1992; Mason & Sutherland, 2002; Rojano, 2002; Sutherland, Rojano, Bell, & Lins, 2001; Wagner & Kieran, 1989), related monographs (Kilpatrick, Martin, & Schifter, 2003; National Research Council, 2001; Wagner, 1993), the Proceedings and final volume of the recent ICMI Algebra Study (Chick, Stacey, Vincent, & Vincent, 2001; Stacey, Chick, & Kendal, 2004), and the algebra research reports found in the Proceedings of both the International Group for the Psychology of Mathematics Education (PME) and the North-American chapter of PME (PME-NA).

In addition to the above sources, the journal literature that was extensively surveyed in preparing this chapter includes articles from the *Journal for Research in Mathematics Education; Educational Studies in Mathematics; Journal of Mathematics Teacher Education; Journal of Mathematical Behavior; Mathematical Thinking and Learning; International Journal of Computers for Mathematics Learning; For the Learning of Mathematics; Canadian Journal of Science, Mathematics, and Technology Education; Recherches en Didactique des Mathématiques; Petit x; Focus on Learning Problems in Mathematics; The Mathematics Educator,* and *Hiroshima Journal of Mathematics Education.* Additional resources, too numerous to mention here but noted in the Reference list, were also consulted.

PART 1. SCHOOL ALGEBRA: BACKGROUND ISSUES AND THEORETICAL CONSIDERATIONS

This part of the chapter deals with some of the issues surrounding current research in algebra learning and teaching, such as the content of school algebra, sources of meaning, and models of algebraic activity. It also includes a description of the model that will serve as a framework for the research studies on learning that are presented in the second part of the chapter.

The Current Content of School Algebra: Competing Views

Since the mid-1980s, the content of school algebra has been experiencing a tug of war between traditional and reformist views. The current situation is such that the content varies not only from country to country, but also within a given country. In the United States, for example, algebra classes might be following either reform-oriented or traditional programs, or something in between (see Lester & Ferrini-Mundy, 2004). Traditional (i.e., nonreform) algebra courses still tend to reflect the content and approaches of the texts written in the 1960s and 1970s by Dolciani and her collaborators (e.g., Dolciani, Berman, & Freilich, 1965). These courses, which are typically 1-year courses and have a strong symbolic orientation, include the simplification of expressions and the solving of equations, inequalities, and systems of equations by formal methods, as well as the factoring of polynomial and rational expressions. Interspersed among the various chapters of typical textbooks are word problems of various types that serve to apply the algebraic techniques just covered. Although functions, along with their graphical, tabular, and letter-symbolic representations, are also treated, they are generally accorded a more minor role—the main thrust of the course being polynomial and rational expressions and their manipulation. According to Saul (1998), this emphasis on polynomial and rational expressions is oriented toward recognizing form, which is considered one of the most important aspects of school algebra within this perspective:

> This level [of computing with rational expressions] is reached when students learn patterns of factoring, for example. Seeing the factorization $A^2 - B^2 = (A + B)(A - B)$, a student who has mastered certain algebraic concepts can factor $x^4 - y^4$, $4x^2 - 1$, or $\cos^2 x - \sin^2 x$. Mathematically, the student is then letting the variables A and B stand for other rational expressions (or rational trigonometric expressions). (p. 138)

Saul is not alone in his view, one that is shared by several other mathematics educators and mathematicians (e.g., Wheeler, 1996). Pimm (1995) similarly argued that, "algebra is about form and about transformation" (p. 88).

In contrast to the position enunciated above, nontraditional, reformist algebra programs tend to give a great deal of weight to functions, to various ways of representing functional situations, and to the solution of "real-world" problems by methods other than manual symbolic manipulation, such as technology-supported methods. These programs reflect the stance formulated by Fey and Good (1985) that practicing manipulative skills should be replaced with the study of families of elementary functions, a shift that "places the function concept at the heart of the curriculum" (p. 49).

The origins of this changed perspective regarding the core elements of school algebra can be traced back to the 1980s when some mathematicians and mathematics educators promoted the idea that computing technology could have significant effects on the content and emphases of school-level and university-level mathematics (Fey, 1984). Early visionaries soon realized that computing technology could be harnessed to more fully integrate the multiple representations of mathematical objects in mathematics teaching. Soon, a *functional perspective* on algebraic activity could be seen in some research projects, a few of them involving pilot programs in schools.[1]

Heid (1988), one of the pioneers of a functional view of algebra (see also Fey et al., 1991; Schwartz & Yerushalmy, 1992), defined it as follows:

> The functional approach to the emergence of algebraic thinking . . . suggests a study of algebra that centers on developing experiences with functions and families of functions through encounters with real world situations whose quantitative relationships can be described by those models. (Heid, 1996, p. 239)

The encouraging results emanating from the research program of Fey, Heid, and their collaborators had an impact not only on the developing projects of other researchers at the time but also on the recommendations to be found in the early *Standards* documents[2] and other related resources published by the National Council of Teachers of Mathematics (NCTM). The movement for function-based approaches to algebra continued throughout the 1990s, becoming more fully elaborated in integrated curricula (see Senk &

---

[1] This was not the first time that functions became a thread of secondary school mathematics programs. During the 1960s, for example, the modern mathematics movement had included similar proposals (see, e.g., the texts authored by Van Engen et al., 1964). However, at that time, the approach being considered was much more formal, emphasizing sets and distinctions between functions and relations.

[2] These documents included *Curriculum and Evaluation Standards for School Mathematics* (NCTM, 1989), *Professional Standards for Teaching Mathematics* (NCTM, 1991), and *Assessment Standards for School Mathematics* (NCTM, 1995).

Thompson, 2003, for examples of National Science Foundation-funded, reform-based curricula that were produced). It culminated in the later publication of the *Principles and Standards for School Mathematics*, where the Algebra Standard for grades pre-K–12 emphasized "relationships among quantities, including functions, ways of representing mathematical relationships, and the analysis of change" (NCTM, 2000, p. 37).

If the number of research studies carried out within this perspective is any gauge, a functional perspective on the teaching of algebra became quite widespread in the United States. However, this perspective is not without its critics. For many researchers, algebra does not consist in the study of functions; algebraic expressions can be used without describing functions, and functions can be expressed without using algebra (Lee, 1997). Pimm (1995) has argued strongly against the position that algebra is about functions:

> Ironically, technology is being used to insist on screen (graphical) interpretations of algebraic forms. There is a strong presumption that symbolic forms are to be interpreted graphically, rather than dealt with directly. . . . There is currently a rapid process of redefinition of algebra, triggered I feel more by the potentialities of these new [technological] systems and the drawbacks of an over-fragmented mathematics curriculum than by any novel epistemological insight. The view of algebraic objects that is being strongly promoted . . . is of algebraic expressions as functions. (p. 104)

One of the by-products of these competing views of algebra content is the development of hybrid versions of programs of study that attempt to include elements of both traditional (rational expressions and equations) and functional orientations to school algebra. However, such mixed approaches can create additional difficulties for algebra learners. As pointed out by Chazan and Yerushalmy (2003), students become confused regarding distinctions between equations and functions, not being able to sort out, for example, how equivalence of equations is different from equivalence of functions. According to Chazan and Yerushalmy, it is not the case that such questions are unanswerable, but rather that combining functional approaches with more standard treatments of school algebra leads to such dilemmas, and that only rarely are opportunities provided for students to inquire into these questions and to attempt to resolve them.

The various curricular approaches to algebra that are currently available in different parts of the United States, ranging from reform to traditional to in-between, are also to be found in the diversity of algebraic content worldwide (see, e.g., Kendal & Stacey, 2004). Sutherland (2002), who conducted a comparative summary of algebra curricula in several countries, noted surprising differences in the ways in which algebra is specified in these various curricula:

> For example, the algebra component of the Australian (Victoria) curriculum is organized around the ideas of "expressing generality," "equations and inequalities," and "function," whereas in the Hungarian curriculum the algebra component refers only to work with systems of equations. . . . The Japanese curriculum is different from the algebra curricula of all the other countries in that there is an emphasis on symbolizing mathematical relationships from elementary school and there is a much more demanding expectation with respect to the use of algebraic symbols for the 15–16-year-old age group than in the other countries studied. . . . The idea of introducing algebra within the context of problem situations is evident within most of the countries studied, although these "problem situations" are sometimes more traditional word problems (e.g., in Italy, Hungary, France, Hong Kong) and are sometimes more "realistic modeling situations." . . . In general where there is more emphasis on solving "realistic problems" there tends to be less emphasis on symbolic manipulation. (p. 29)

This latter observation by Sutherland regarding the greater the emphasis on realistic problems, the less the emphasis on symbolic manipulation, remains a fundamental issue in the arena of competing ideas on algebraic content. The orientation toward the solving of realistic problems, with the aid of technological tools, allows for an algebraic content that is less manipulation oriented. Such orientations also emphasize multi-representational activity with a shift away from the traditional skills of algebra. The logic underpinning this position is that technological tools can step in and assist with problem solving when students do not have the required manipulative skills. This has led to a certain amount of resistance in the United States to the NSF-funded reform-based curricula at the upper secondary level because the technical aspects of algebra would seem to be minimized. The fundamental issue is really one of values—that is, what it is that mathematicians, mathematics educators, and mathematics education researchers consider essential and important in algebra. The next section on sources of meaning takes a second look at some of the same issues, but this time from an epistemological perspective.

Sources of Meaning

This section examines first the general question of where algebraic meaning comes from and then goes on to describe each of four main sources of meaning: (a) meaning from the algebraic structure itself, involving the letter-symbolic form; (b) meaning from other mathematical representations, including multiple representations; (c) meaning from the problem context; and (d) meaning derived from that which is exterior to the mathematics/problem context (e.g., linguistic activity, gestures and body language, metaphors, lived experience, image building, etc.).

Algebraic Meaning: Where Does It Come From?

Several researchers have studied the specific question of meaning in school algebra (e.g., Arzarello, Bazzini, & Chiappini, 2001; Filloy, Rojano, & Rubio, 2001; Kaput, 1989; Kirshner, 2001; Lins, 2001; Sackur, Drouhard, Maurel, & Pécal, 1997). The different perspectives that have been described at various times in the literature have recently been reconceptualized by Radford (2004). For Radford, meaning in school algebra is produced in the "crossroads of diverse semiotic mathematical and non-mathematical systems" (p. 163) and is deemed to come from three primary sources: (a) the algebraic "structure" itself, (b) the problem context, and (c) the exterior of the problem context.

Although the third of Radford's sources of meaning is quite new to algebra research, something akin to it has been discussed by Noss and Hoyles (1996) in their general treatment of "meaning in mathematics education" according to a triple perspective that includes *meanings from mathematical objects, meanings from problem solving,* and *meanings constructed by the individual learner.* The parallelism between the three sources of meaning in algebra from Radford and the corresponding view for mathematics at large from Noss and Hoyles has a certain appeal. However, a strictly threefold approach is not broad enough within the context of current views of algebra content in which graphical and other mathematical representations are considered very important. Thus, a second within-mathematical source of meaning in algebra could be added to Radford's classification, namely *meaning from other mathematical representations, including multiple representations* (see Figure 16.1).

Meaning From the Algebraic Structure Itself, Involving the Letter-Symbolic Form

At the Research Agenda Project conference on algebra in 1987, Booth (1989) argued that

> Our ability to manipulate algebraic symbols successfully requires that we first understand the structural properties of mathematical operations and relations which distinguish allowable transformations from those that are not. These structural properties constitute the *semantic* aspects of algebra. . . . The essential feature of algebraic representation and symbol manipulation, then, is that it should *proceed from* an understanding of the semantics or referential meanings that underlie it. (pp. 57–58)

In other words, "the sense of meaningfulness comes with the ability of 'seeing' abstract ideas hidden behind the symbols" (Sfard & Linchevski, 1994, p. 224), that is, the symbols become transparent.

This structural source of meaning not only links letter-symbolic representations to their numerical foundations but also provides connections among the symbolic forms of algebra, its equivalences, and its property-based manipulation activity (e.g., Cerulli & Mariotti, 2001). Although the algebra research literature often refers to *the structure of expressions,* the latter phrase both shuns definition (see, e.g., Arcavi, 1994; Linchevski & Livneh, 1999; MacGregor & Price, 1999) and proves difficult for students to grasp (e.g., Demby, 1997). Nevertheless, even though the nature of the meaning that students draw from algebraic structure can be elusive, this source of meaning is considered by many mathematics educators and researchers to be fundamental to algebra learning.

Meaning From Other Mathematical Representations, Including Multiple Representations

The diversity of perspectives underlying the content of school algebra suggests that the question of where algebraic meaning comes from is intimately related to the nature of the algebra courses experienced by students. Courses with a traditional focus on literal expressions, equations, equivalent forms, properties

1. Meaning from within mathematics:
    1(a). Meaning from the algebraic structure itself, involving the letter-symbolic form.
    1(b). Meaning from other mathematical representations, including multiple representations.
2. Meaning from the problem context.
3. Meaning derived from that which is exterior to the mathematics/problem context (e.g., linguistic activity, gestures and body language, metaphors, lived experience, image building).

Figure 16.1 Sources of meaning in algebra (adapted from Radford, 2004).

and structure will offer quite different meaning-building experiences from those oriented toward functions and the interplay of tabular, graphical, and symbolic representations. The question then is, what role can be played by the tabular and graphical representations, in conjunction with the letter-symbolic, in algebraic meaning making.

Kaput (1989) has argued that the problem of student learning in algebra is compounded by (a) the inherent difficulties in dealing with the highly concise and implicit syntax of formal algebraic symbols and (b) the lack of linkages to other representations that might provide feedback on the appropriate actions taken. As a consequence, he has promoted the kind of mathematical-meaning building that has its source in *translations between mathematical representations systems.* Although tables, graphs, and equations can all be used to display binary relations, a critical difference, according to Kaput, between tables and graphs is that "a graph engages our gestalt-producing ability, which allows us to consolidate a binary (and especially a functional) quantitative relationship into a single graphical entity—a curve or line" (Kaput, 1989, p. 172). The opportunity to coordinate objects and actions within two different representations, such as the graphical and the letter-symbolic, is considered by many to be crucial in creating meaning in algebra (e.g., Fey, 1989; Romberg, Fennema, & Carpenter, 1993; Yerushalmy & Schwartz, 1993).

Meaning From the Problem Context

In contrast to the internal semantics of algebra as a site for meaning making (Booth, 1989), the external semantics of a problem permit the algebra learner to fuse symbols and notations with events and situations, thereby creating an external meaning for certain objects and processes of algebra. A strongly held belief in algebra education is the notion that problem-solving contexts are foundational to the emergence and evolution of algebraic reasoning (e.g., Bednarz & Janvier, 1996; Bell, 1996). This stance is based to a certain extent on historical grounds whereby problem solving made a major contribution to the development of algebra. As well, algebra grew in status to become the privileged tool for expressing general methods for solving whole classes of problems. Additional arguments for the role of problems in algebraic meaning building are related to notions of relevance and purpose (cf., Lee, 1997). Nevertheless, it has been argued that significant epistemological issues are associated with the whole idea of using word problems for creating algebraic meaning making (e.g., Balacheff, 2001; Pirie & Martin, 1997). Therefore, a broader perspective on problem solving has been advanced by Bell (1996):

> For introducing and developing algebra, I understand problem solving to refer to the solving of problems by the forming and solving of equations; this is the narrow sense of the term. But the essential mathematical activity is that of exploring problems in an open way, extending and developing them in the search for more results and more general ones. Hence [all algebraic learning] . . . is based on problem explorations. This is the broad sense of the term. (p. 167)

Included within this source of algebraic meaning making are "real problems," that is, those that involve *modeling* situations. Much of the current modeling activity that occurs in algebra classes uses physical artifacts or technological modeling tools as an integral part of the activity. In these cases, the contextual aspect of the situation is very much intertwined with the use of gestures and body language in the process of constructing a mathematical model. This activity thus combines two sources of meaning, one derived from the problem context and the other that is exterior to the context and that is presented immediately below.

Meaning Derived From That Which Is Exterior to the Mathematics/Problem Context

A recent approach by algebra researchers to thinking about meaning making concerns sources of meaning that are exterior to both the mathematics and the problem situation. (Note that much of this meaning building that is "exterior" to both the mathematics and the problem situation is related to that which is uniquely human, and thus is clearly "interior" in a very real sense.) This approach focuses on students' processes of meaning production in terms of the way diverse resources such as gestures, bodily movements, words, metaphors, and artifacts become interwoven during mathematical activity (see, e.g., Arzarello & Robutti, 2001; Radford, Demers, Guzmán, & Cerulli, 2003). Past studies of students' ways of thinking and of their errors in algebra had always suggested that students brought more to bear on their learning of algebra than was accounted for by the theories available at the time. Thus, the explicit focus on bodily activity, language, and past lived experience as a source of meaning would seem a natural evolution in research on the learning of algebra.

On the basis of his studies of students' production of oral and written signs and the meanings they ascribe to them as they engage in the construction of expressions of mathematical generality, Radford (2000) has observed that

> Students, at the very beginning, tend to have recourse to other experiential aspects more accessible to them than the structural one. . . . Novice students bring

> meanings from other domains (not necessarily mathematical domains) into the realm of algebra. Hence it seems to us, one of the didactic questions with which to deal is . . . that of the understanding of how those non-algebraic meanings are progressively transformed by the students up to the point to attain the standards of the complex algebraic meanings of contemporary school mathematics. (p. 240)

Radford's closing remark, which implicitly raises the question as to the nature of the algebraic activity whereby students' meanings are gradually transformed and aligned with "the complex algebraic meanings of contemporary school mathematics," leads to the next section dealing with algebraic activity. As will be seen, the section integrates previously elaborated issues of content and sources of meaning in algebra.

Models for Conceptualizing Algebraic Activity

Bell (1996) has described algebra as a means to express generalizations, relations, and formulas; solve problems; denote unknowns; and solve equations. Usiskin (1988) has synthesized four conceptions of algebra: generalized arithmetic, the set of procedures used for solving certain problems, the study of relationships among quantities, and the study of structures. Kaput (1995) has identified five aspects of algebra: generalization and formalization; syntactically guided manipulations; the study of structure; the study of functions, relations, and joint variation; and a modeling language. Other researchers (see, e.g., Gascón, 1994–1995; Lins, 2001; Mason, Graham, & Johnston-Wilder, 2005; Rojano, 2004; Smith, 2003; Star, 2005; Sutherland, 2004) have offered variations on the above, each in accordance with their particular perspectives on algebra, algebraic thinking, or algebraic language.

Participants at an international colloquium on algebra in Montreal in the early 1990s focused on four approaches: generalization of numerical and geometric patterns and of the laws governing numerical relationships; problem solving; functional situations; and modeling of physical and mathematical phenomena (Bednarz et al., 1996). Concurrently, a study by Lee (1997) was in progress, one that adopted an even broader perspective on that which constitutes school algebra. In this study, which explored algebraic understanding, Lee presented the question, "What is algebra?" to a cohort of mathematicians, teachers, students, and mathematics education researchers. Seven themes emerged from her interviews on the question: a school subject, generalized arithmetic, a tool, a language, a culture, a way of thinking, and an activity.

If there was one theme that tended to permeate all others among Lee's interviewees, it was *Algebra is an activity*. Lee (1997) remarked that, "Algebra emerges as an activity, something you do, an area of action, in almost all of the interviews" (p. 187). For example, Pimm, one of Lee's interviewees, is quoted as saying that *action* is the central feature of school algebra: "Algebra . . . is so much more about doing, is actually about action on things, . . . with attention being more on the *transformations* [emphasis added] than on the objects themselves" (p. 187). On the other hand, Kaput, another interviewee, emphasized the alternate side of the coin when he pointed to the importance in algebraic activity of spending a great deal of time in the "building of algebraic objects" (p. 189). And Bell expressed concern about the question of purpose, about students not really "having the experience of what algebra is for" (p. 196).

Building on the idea of algebra as activity (and reflecting the key ideas expressed by Kaput, Pimm, and Bell), Kieran (1996) developed a model that synthesizes the activities of school algebra into three types: *Generational, Transformational,* and *Global/meta-level.* This model, hereafter referred to as the GTG model (see Figure 16.2), will serve in Part 2 as the structuring device for presenting findings from the research involving the algebra learner.

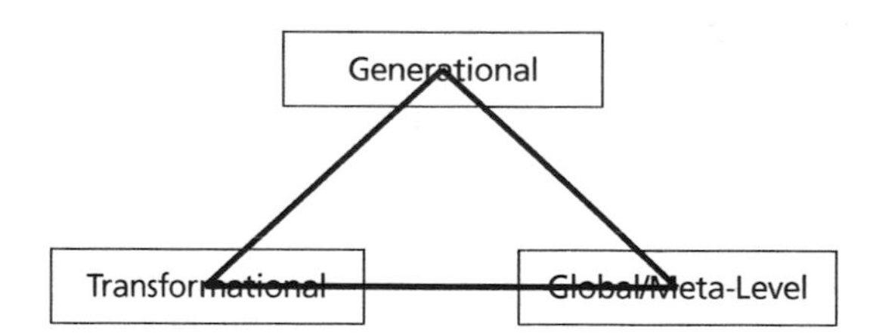

Figure 16.2 GTG model for conceptualizing algebraic activity.

The *Generational* activities of algebra involve the forming of the expressions and equations that are the objects of algebra. Typical examples include (a) equations containing an unknown that represent problem situations (see, e.g., Bell, 1995), (b) expressions of generality arising from geometric patterns or numerical sequences (see, e.g., Mason, 1996), and (c) expressions of the rules governing numerical relationships (see, e.g., Lee & Wheeler, 1987). Much of the meaning building for algebraic objects occurs within the generational activity of algebra. Therefore this activity includes, as well, work with variables, unknowns, and equality; even the notion of that which constitutes the solution to an equation can be initiated within generational activity. Clearly, the meaning that students develop for the notion of, for example, *variable* will depend on whether the algebra content they have experienced has been derived from a polynomial-equation framework or a functions framework, that is, has tended to emphasize either the letter-symbolic or a combination of mathematical representations. These two frameworks each provide a unique transversal thread to the three types of algebraic activity.

Generational activity is the area where, according to Radford (2001), the role of algebra is that of a language to express meaning, and where the *habit of mind* of those who are in "algebra mode" can find expression (Cuoco, Goldenberg, & Mark, 1996). It may be tempting to equate generational activity with the conceptual aspects of algebra and, similarly, transformational activity with the skill-based aspects of algebra, but this parallelism is not intended. Indeed, as is being proposed in this section and argued throughout Part 2, conceptual work and meaning building occur as well within the transformational activity of algebra.

The second type of algebraic activity—the *Transformational* activities (referred to, by some, as the *rule-based* activities)—includes, for instance, collecting like terms, factoring, expanding, substituting one expression for another, adding and multiplying polynomial expressions, exponentiation with polynomials, solving equations and inequalities, simplifying expressions, substituting numerical values into expressions, working with equivalent expressions and equations, and so on. A great deal of this type of activity is concerned with changing the symbolic form of an expression or equation in order to maintain equivalence. In addition to developing meaning for equivalence, this activity also includes meaning building for the use of properties and axioms in the manipulative processes themselves. Even the technique of factoring has its conceptually meaningful aspects: for example, coming to see that, if the exponent in, say, $x^n - 1$ has several divisors, it can generally be factored in more than one way, and thus that this expression can be viewed in terms of multiple possible structures (e.g., seeing $x^6 - 1$ as a difference of squares and also as a difference of cubes). It is in this way that meaning can be said to be present for the transformational processes of algebra. This factoring example also illustrates that a given task can be both transformational and generational. It is generational to the extent that the learning of a new manipulative transformation, such as $x^6 - 1 = (x - 1)(x^5 + x^4 + x^3 + x^2 + x + 1)$, can induce an extended perspective on the algebraic object itself.

In this sense, transformational activity is not just skill-based work. It is much broader than this in that it can include conceptual/theoretical elements, especially during the period in which certain transformations are being learned. According to Lagrange (2002), "Technique has a pragmatic role that permits the production of results; but it also plays an *epistemic* [emphasis added] role in that it constitutes understanding of objects and is the source of new questions" (p. 163, translated from the original).

Lastly, there are the *Global/meta-level*, mathematical activities. These are the activities for which algebra is used as a tool but that are not exclusive to algebra. Note that they are not referred to as *global/meta-level* because they involve taking some distance from the mathematical activity at hand and viewing it comprehensively. Rather, they suggest more general mathematical processes and activity. They also provide the context, sense of purpose, and motivation for engaging in the previously described generational and transformational activity. The global/meta-level activities include problem solving, modeling, working with generalizable patterns, justifying and proving, making predictions and conjectures, studying change in functional situations, looking for relationships or structure, and so on—activities that could indeed be engaged in without using any letter-symbolic algebra at all. These are the kinds of activities that provide the surround for engaging in mathematical activity in general, and algebraic activity in particular—and, thus, for constructing and working with algebraic objects and processes.

However, algebra does not equal problem solving, or modeling, or any of the other global/meta-level activities. Although some may find it helpful to think of these global/meta-level activities in terms of "approaches" to the teaching of algebra—such as a generalization approach to algebra or a problem-solving approach to algebra (see, e.g., Bednarz et al., 1996)—algebra includes as well generational and transformational activity, which may not be a necessary component of the designated approach to algebra. On the other hand, aspects of, for example, generalization and problem solving can be present in much of generational and transformational activity of algebra, especially during the early stages of learning, Thus, the global/meta-level activity of algebra is both broader than and at the same time not quite as broad as algebra.

PART 2. SCHOOL ALGEBRA: FINDINGS FROM RESEARCH STUDIES

This second part, which constitutes the main thrust of the chapter, describes the body of research work on the learning and the teaching of school algebra, emphasizing more recent studies but also integrating some findings from earlier work. The research is presented according to three age ranges: the middle grade and lower secondary level student, the upper secondary and college level student, and the teacher of algebra. Research on the development of algebraic thinking at the primary school level is treated in another chapter (see Carraher & Schliemann, this volume).

Within each of the age-related categories dealing with research on the *learning* of algebra, the GTG model of algebraic activity provides the structural framework. Research where the primary focus is generational activity is treated first, followed by the research dealing with transformational activity, and then that of global/meta-level activity. Because the third type of activity is overarching and cross-cutting, it proved to be more practical, at times, to discuss it within the generational or transformational activity that it served. This was particularly the case for studies involving word problems with the middle grade and lower secondary level student. It is noted that the placement of research studies into the various categories of the GTG model was done according to rather general criteria: For example, if the main emphasis of the study appeared to be transformational activity, then the findings are discussed under that category, even if the study also included some generational activity; if several types of global/meta-level activity were involved, the study is reported in the section related to the major focus. Studies that did not fit this model, such as international assessments, are discussed separately.

Research on the algebra learner occupies the first two sections of this part of the chapter; research that has focused on the algebra teacher and the teaching of algebra is the subject of the third section. The chapter then closes with a fourth section that attempts to draw out that which this body of research on the learning and teaching of school algebra tells us.

The Middle Grade and Lower Secondary Level Student

The research that is reviewed in this section includes those studies carried out with students in the middle grade and lower secondary levels, that is, from Grades 6 to 9 (approximately 11 to 15 years of age). The majority of the research carried out with this age range of pupil has focused on generational activity, with considerably fewer studies dedicated to transformational activity. The studies focusing on global/meta-level activity have been concentrated in the areas of generalizing, proof and proving, and modeling (with problem solving being treated within the context of generational activity).

By way of introduction, this section starts with a brief commentary on results from some recent international assessments. This provides a first glimpse at what students of this age find easy or difficult in algebra. These introductory remarks are followed by the three main sections related to research on the middle grade and lower secondary level student, research categorized according to generational, transformational, and global/meta-level activity. A section on the impact of technology on algebraic activity is then presented, followed by brief concluding remarks.

Background: International Assessments

Two large-scale international studies, the Trends in International Mathematics and Science Study (TIMSS, formerly known as the Third International Mathematics and Science Study) and the Program for International Student Assessment (PISA), announced the results of their 2003 surveys in late 2004 (PISA, 2004; TIMSS, 2004). Forty-eight countries participated in the 2003 TIMSS mathematics achievement assessment for Grade 8 (Gonzales et al., 2004). The number of score points obtained on average in the algebra content area across the 48 countries was 25 out of a possible 53, whereas the U.S. eighth graders in algebra obtained on average 29 score points. Figures 16.3 and 16.4 provide two items from the algebra area of the survey.

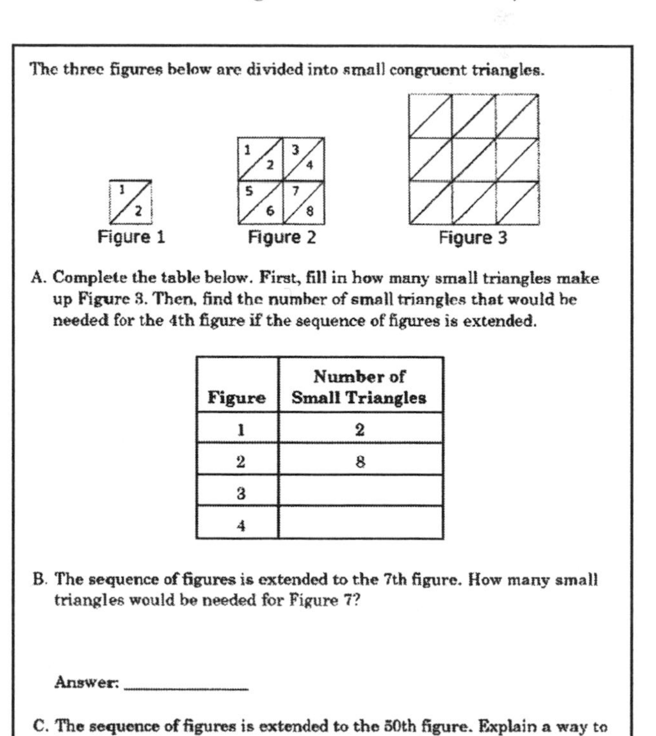

The three figures below are divided into small congruent triangles.

A. Complete the table below. First, fill in how many small triangles make up Figure 3. Then, find the number of small triangles that would be needed for the 4th figure if the sequence of figures is extended.

Figure	Number of Small Triangles
1	2
2	8
3	
4	

B. The sequence of figures is extended to the 7th figure. How many small triangles would be needed for Figure 7?

Answer: ____________

C. The sequence of figures is extended to the 50th figure. Explain a way to find the number of small triangles in the 50th figure that does not involve drawing it and counting the number of triangles.

Figure 16.3 One of the released items from the 2003 TIMSS for eighth grade.

Sam wanted to find three consecutive even numbers that add up to 84.

He wrote the equation $k + (k + 2) + (k + 4) = 84$.

What does the letter k represent?

Ⓐ The least of the three even numbers

Ⓑ The middle even number

Ⓒ The greatest of the three even numbers

Ⓓ The average of the three even numbers

Figure 16.4 Another of the released items from the 2003 TIMSS for eighth grade.

The percentage of correct responses across all TIMSS countries for A, B, and C respectively of the item shown in Figure 16.3 was 36.6%, 22.1%, and 14%. Although the results on this item were universally low, the U.S. students scored higher (50.2%, 29.6%, and 19.4% respectively), perhaps indicating that U.S. students had had more opportunities to examine patterns in this manner than many students internationally. The percentage of correct responses on the item shown in Figure 16.4 (for which the correct answer is A), one with some letter-symbolic content, was 24.7% across all TIMSS countries and 23.1% among U.S. students.

In general, these low results are disturbing—less than 37% correct, worldwide, for each of the four questions shown in the two items of Figures 16.3 and 16.4. Despite the improvement of U.S. eighth graders since the 1995 administration of the test, the fact that only 8 of 48 countries outperformed U.S. eighth graders in algebra in the 2003 assessment suggests that, overall, students are doing quite poorly in algebra. The PISA results, similarly, confirm this situation. The research findings that are presented in this chapter offer a more detailed look at what students know, that is, the nature of the meanings that they attribute to algebraic objects and processes, as well as their approaches to tasks that involve the manipulation of symbols. Some of this research involves the study of alternate learning environments that can lead to more powerful algebraic understandings among students.

Generational Activity: The Middle Grade and Lower Secondary Level Student

The generational activity of algebra involves forming and interpreting the objects of algebra, usually within the context of some global/meta-level activity. This section is one of the larger ones of the chapter dealing as it does with an area where a significant number of research studies on the learning of algebra have been carried out. To render this area more accessible, it has been divided into subthemes according to the source of meaning being invoked in constructing or interpreting the objects of algebra: (a) primarily a letter-symbolic focus, (b) a multiple-representation perspective, or (c) within the context of word problems. The relatively small number of studies that have focused on the fourth source of meaning, that which is "exterior" to both the mathematics and the problem context, usually included as well one of the other three sources of meaning and are thus treated therein.

Generational Activity with a Primary Focus on the Letter-Symbolic Form. The studies that have focused on the kinds of meanings that students of this age range attribute to the letter-symbolic forms of the objects of algebra have been concentrated in the following areas: variables, expressions, and equations; the minus sign and negative numbers; and the beginnings of structure sense.

Variables, Expressions, and Equations. Whereas past research on the ways in which students of this age range interpret algebraic symbols tended to focus on cognitive levels (e.g., Küchemann, 1981), prior arithmetic experience and methods of thinking (Booth, 1984), and difficulty with notation such as brackets (Kieran, 1979) and the equal sign (Behr, Erlwanger, & Nichols, 1976; Kieran, 1981), more recent work suggests additional factors impinging upon students' interpretation of algebraic notation: (a) what one is able to perceive and prepared to notice (Sfard & Linchevski, 1994), (b) difficulties with operation signs (Cooper et al., 1997), (c) the nature of the questions asked and the medium in which they are asked (Warren, 1999), (d) the presence of multiple referents and shifts in the meaning of unknowns (Stacey & MacGregor, 1997), and (e) the nature of the instructional activity (Wilson, Ainley, & Bills, 2003).

More specifically, a large-scale study conducted by MacGregor and Stacey (1997) involving 2000 Australian students (aged 11–15) found that students' interpretations of algebraic notation are based on intuitive assumptions and sensible, pragmatic reasoning about an unfamiliar notation system; analogies with symbol systems used in everyday life, in other parts of mathematics, or in other school subjects; interference from other new learning in mathematics; and poorly designed and misleading teaching materials. The nature of the interpretations brought to bear on algebraic notation by these students suggests the strong role played by the meaning derived from the exterior of the mathematics/problem context. These researchers have argued that recognition of these origins of misunderstanding is necessary for improving the teaching of algebra. In another analysis drawn from the same survey data, MacGregor and Price (1999) assessed metalinguistic awareness[3] and its relation to the use of algebraic notation. The researchers found that very few students with low metalinguistic scores achieved high algebra scores.

The transition from arithmetic to algebra remains an important theme of several studies involving variables, expressions, and equations. Building on earlier research on variables (e.g., Küchemann, 1981; Wagner, 1981), expressions (e.g., Bell, Malone, & Taylor, 1987), and equations (e.g., Kieran, 1988; Whitman, 1976), some of the more recent studies have examined students' abilities to express generalizations and to represent these generalizations in various forms in symbolic expressions (Arzarello, 1992; Lee, 1996; Mason, 1996). Fujii (2003) has introduced younger students to algebraic thinking through generalizable numerical expressions, using numbers as *quasi-variables*—for example, number sentences such as $78 - 49 + 49 = 78$, which are true whatever number is taken away and then added back. Finding that "generalizable numerical expressions can assist students in identifying and discussing algebraic generalizations long before they learn algebraic notation" (p. 59), Fujii claimed that these expressions "allow teachers to build a bridge from existing arithmetic problems to opportunities for thinking algebraically without having to rely on prior knowledge of literal symbolic forms" (p. 62). Other studies (e.g., Graham & Thomas, 2000) have illustrated how calculating technology can be effectively used to develop students' understanding of letters as specific unknowns, generalized number, and variable. In addition, some research studies have focused on the conceptual obstacles involved in moving from simple equations with an unknown on one side to more complex equations with unknowns on both sides and have described phenomena that they have called the *didactic cut* (Filloy & Rojano, 1989) and the *cognitive gap* (Herscovics & Linchevski, 1994).

The Minus Sign and Negative Numbers. In contrast to the findings already accumulated with respect to students' evolving knowledge of the different possible interpretations of the equal sign, we know comparatively little about students' interpretations of the minus sign in algebraic expressions and equations. The subtle distinctions in the ways that the minus sign is used in algebraic expressions and equations were the focus of a study by Vlassis (2004). Of the three possible interpretations of the minus sign, which she described as knowledge of *negativity*, she found that the binary sense predominated among the eighth graders of her study. She noted a high degree of inflexibility in considering either a unary sense (whereby "subtract 3" means "negative 3") or a symmetric sense (where, for example, in $-x = 1$, the $-x$ means "the opposite of x"). Vlassis concluded that the various uses of the minus sign in algebra are counterintuitive to a majority of beginning-algebra pupils and constitute a significant conceptual obstacle in their coming to give meaning to algebraic symbols and processes.

Related research is that involving students' developing notions of negative numbers. Among the few studies in this area is the work of Gallardo (2002; see also Vergnaud, 1989), who applied historical/epistemological analysis to the study of students' extension of the natural-number domain to the integers within word-problem contexts. Gallardo has found that the first three of the four levels of acceptance of negative numbers within historical texts (i.e., subtrahend, relative number, isolated number, and formal negative number) were present among the 12- and 13-year-olds of her study. This suggested to her that students have an intuitive sense of negative numbers long before they are able to formalize this knowledge.

Beginnings of Structure Sense. Linchevski and Livneh (1999) extended some of the earlier research work related to students' structural sense in algebra (e.g., Kieran, 1989; Matz, 1982) in a study that tested whether students' difficulties with aspects of algebraic structure reflect difficulties they already have with numerical structure. They interviewed two classes of end-of-year sixth graders (average age: nearly 12 years) who had not yet learned negative numbers or any algebraic manipulation. They found that, on the whole, students' difficulties with interpreting equations containing several numerical terms and an unknown were a reflection of the same errors that they were making in purely numerical contexts. However, the researchers also noted that some students were inconsistent in that they gave an incorrect answer in one context but a correct one in another. Follow-up interviews with students after they had learned how to operate with negative numbers produced the same results. The researchers suggested that algebra instruction be designed to foster the development of *structure sense* by providing experience with equivalent structures of expressions and with their decomposition and recomposition. They emphasized that more systematic research in this area is needed. One recent thrust in this direction has been taken by Carpenter, Franke, and Levi (2003) whose research with younger students

[3] *Metalinguistic awareness* refers, according to the authors, to the ability to reflect on and analyze spoken or written language, for example, being able to pay attention to sounds and spellings of linguistic signs instead of to their meanings. Both symbol awareness and syntax awareness are considered to be components of the metalinguistic awareness necessary for success in learning to use algebraic notation.

has aimed at having students explain and justify the properties they are using as they carry out their arithmetic calculations. Additional research in this area is discussed in the section on transformational activity.

Generational Activity with Multiple Representations. Most of the approaches to algebra that make use of multiple representations define themselves as *functional approaches*. Tabach and Friedlander (cited in Kieran & Yerushalmy, 2004) suggested that three factors allow for a functional approach in the teaching and learning of algebra: the potential to produce ample numerical tables, the need to use general expressions to express the data in these tables, and the possibility of obtaining a wide variety of corresponding graphs. As an example of such an approach, they referred to the *Compu-Math curriculum* (Hershkowitz et al., 2002), which relies heavily on the use of spreadsheets, in particular. However, not all the research involving multiple representations includes technology environments. In addition, as will be seen, some studies have included multiple representations within word-problem or modeling situations. When the focus of the analysis has been the multi-representational aspect, then the study is included in this section; however, if the word-problem aspect has the higher profile, then the research is treated in the next section. Finally, if the modeling aspect is uppermost, then the study is reserved for the later discussion of global/meta-level activity. This section includes the following themes: tabular representations and the letter-symbolic, graphical representations and the letter-symbolic, and connections among representations.

Tabular Representations and the Letter-Symbolic. Recent studies involving the tabular displays of spreadsheets have investigated whether and how the use of spreadsheet methods lends meaning to algebra. Building on earlier research with spreadsheets as a means to introduce students to algebra-like notation (e.g., Sutherland & Rojano, 1993), Filloy, Rojano, and Rubio (2001) have found that the spreadsheet serves as a bridging tool to algebra because it helps students to create conceptual meaning for algebraic objects and operations, and to move from focusing on a specific example to describing general relationships. For Dettori, Garuti, and Lemut (2001), the functional orientation of spreadsheets allows for their effective use in the investigation of variation but can cause difficulties later when solving equations or inequalities. When these researchers investigated 13- and 14-year-olds working on algebraic problems with spreadsheets, Dettori and colleagues were led to conclude that "spreadsheets can start the journey of learning algebra, but do not have the tools to complete it" (p. 206). As well, Hershkowitz et al. (2002) have described how students tend, in the spreadsheet environment, to generalize recursively rather than explicitly, which makes it more difficult for them to generate closed-form algebraic rules for patterns.

However, other research work that has included tables of values, but not within a spreadsheet environment, has noted that this representation can be used effectively within generational activity. For example, Cedillo and Kieran (2003) carried out a study with 13- and 14-year-olds in which the pattern of input and output values in the given table motivated the search for a computer "program" (algebraic expression) that would reproduce the pattern. Using TI-92 Plus calculators, the students were able to test the algebraic expressions they were generating to see if their expression would yield the same output values as were displayed in the given table. The researchers pointed out the beneficial role played by the calculator in the development of not only students' ability to formulate algebraic expressions but also their emerging notions of variable.

Graphical Representations and the Letter-Symbolic. In a study that focused on the slope-intercept form of a linear function, $y = mx + b$ (Lobato, Ellis, & Muñoz, 2003), the researchers found that students in a first-year reform algebra class developed the notion of slope as a difference (i.e., "it goes up by") rather than as a ratio—as was intended. By means of the theoretical construct of focusing phenomena, Lobato and colleagues analyzed the regularities in the ways in which the teacher, students, artifacts, and curricular materials acted together to direct student attention toward certain mathematical properties over others. Their analysis revealed the presence of four focusing phenomena: the ambiguity of the "goes up by" language, the use of ordered tables of values, the language used when entering into the graphing calculator both the x-values of a function and the scale of the x-axis, and a lack of coordination of differences in x- and y-values when $x \neq 0$. The researchers suggested that a directed focus of attention on the coordination of covarying quantities would be more likely to lead students to generalize slope as a ratio.

Moschkovich (1998) has pointed to students' tendencies to ascribe a role to the x-intercept when learning about linear functions of the form $y = mx + b$. A subgroup of 18 students from a ninth-grade pilot first-year algebra course that included the use of the computer graphing utility, *Superplot*, participated in videotaped discussion sessions with a peer of their choice, using the same graphing utility. Although 13 of these students used the x-intercept at least once at the outset of the group study for either *b* or *m* in an equation, or to describe lines as moving along the *x*-axis as a result of changing *b* in an equation, their thinking nevertheless evolved over the course of the study. Moschkovich has described the resources used by students as they refined their conceptions of the presumed role of the *x*-intercept, such as narrowing the problem contexts in which they applied this conception and making connections between conceptions.

Using the graphing facility of the spreadsheet as a potential learning environment, Ainley (1996) explored some of the ways in which this tool could be used to introduce 12-year-olds to the power of generalizing through algebraic notation. Ainley found that spreadsheet activity allowed students to appreciate the purpose and power of the symbolic notation by providing them with a means to generate more data for the problem at hand and thus work towards a solution. In a study by Kieran and Sfard (1999) that used the software, *Algebra Connections*, students became so skilled at graphing linear functions by focusing on the slope and *y*-intercept that they could mentally approximate the solutions to equations such as $7x + 4 = 5x + 8$ by visualizing the graphs of the two functions and their intersection point. However, in another study, Kieran (2001) found that, for certain nonlinear situations involving rational functions, students experienced considerable difficulty in generating both a graphical and a symbolic representation for the problem situation—the rational function proving to be much more problematic than the linear function.

Researchers have also looked at the ways in which graphical representations are used to give meaning to inequalities. Bazzini, Boero, and Garuti (2001b), who studied the feasibility of a functional approach in the teaching of inequalities to eighth-grade students, have pointed to the positive role that graphical representations can play in helping students to conceptualize the symbolic forms of inequalities, as well as the pitfalls involved in attempting to apply to the solving of inequalities some of the transformational techniques used with equations.

Arzarello and Robutti (2001) have described a study introducing ninth graders to algebra and involving the interpretation of body-motion graphs within an environment that included symbolic-graphing calculators connected to motion sensors (the calculator-based ranger—CBR). Initially students were encouraged to try various running patterns in order to create different graphs. The continuous nature of the CBR graphing allowed students to test conjectures in a direct manner, controllable by their own physical movement. Arzarello and Robutti observed that

> students' cognitive activity passes through a complex evolution, which starts in their bodily experience (namely, running in the corridor), goes on with the evocation of the just lived experience through gestures and words, continues connecting it with the data representation, and culminates with the use of algebraic language to write down the relationships between the quantities involved in the experiment. (p. 39)

Connections Among Representations. Although research continues to show that various technological environments can help in developing students' algebraic thinking by building links among symbolic, tabular, and graphical representations (e.g., Friedlander & Tabach, 2001; Koedinger, Anderson, Hadley, & Mark, 1997; see also Balacheff & Kaput, 1996; Ruthven, 1996), such improvements are not automatic. The quality of the tasks (e.g., Hoyles, 2002), the teaching (e.g., Kubinová & Novotná, 2002), and the general learning environment (e.g., Kramarski, 2000) all play significant roles. Furthermore, as pointed out in the review of graphing calculator research carried out between 1985 and 1995 by Penglase and Arnold (1996), graphing technology clearly helped students improve their understanding of function and graphing concepts; however, several of the studies included in their review suggest that students continued to experience difficulties with seeing the relationship between algebraic and graphical representations of functions.

A partial explanation of this phenomenon has been found to lie in the time needed by students to build an understanding for symbolic notation. For example, Nathan, Stephens, Masarik, Alibali, and Koedinger (2002) investigated seventh and eighth graders' ($N = 90$) abilities to solve problems using tabular, graphical, verbal, and symbolic representations (without the aid of technology) and to translate among these representations before and after instruction that involved bridging activities built on students' invented strategies and representations. Before instruction, students experienced more success solving problems using a given representation than translating among representations, with success strongly influenced by the particular representational formats involved. Results suggested that there are significant gaps in students' abilities to comprehend and to produce representations, and that students may attain more fluency with tables and pointwise graphs before symbolic equations and verbal expressions. (See also Koedinger & Nathan, 2004, whose study is described in a later section.)

Yerushalmy (2000) has studied the long-term impact of a problem-based, functional approach to the teaching of algebra, one that was supported by intensive use of graphing technology, and has reported findings similar to those of Nathan and his colleagues.

By means of classroom observations and clinical interviews with students over their 3 years of following the course, Yerushalmy found that the representation of problem situations evolved as follows: from numbers as the only means of modeling, to intensive work with graphs and tables, to the use of more symbolic representations. Students also moved from analyzing patterns of numbers by watching the behavior of increments to more explicit, closed-form formulations of functional relationships involving pairs of numbers. However, Yerushalmy wondered why it took so long for the students to appreciate algebraic symbols, and why they preferred to use the situation as the source for their answers. From the findings of another study she carried out (Yerushalmy & Shternberg, 2001), and long-term observations by Chazan (2000), Yerushalmy (2000) has argued that "a curricular sequence of algebra based on functions might have to address more subtleties and transitions than initially expected . . . and that the complexity of helping students to value algebraic symbols may require more than bridging between representations" (p. 145). In another related study (Gilead & Yerushalmy, 2001), ninth graders who had been taught by a functional approach to the interpretation of equations (comparison of functions) encountered an obstacle when dealing with a certain type of situation not easily modeled by an equation without introducing another variable or parameter. The researchers were led to suggest that graphical tools may be useful as an exploratory support only for canonical problems (i.e., where y-intercept and slope are both given).

In a study of 1 year's duration, Schwarz and Hershkowitz (1999) investigated ninth graders' learning of the concept of mathematical function in an environment that included problem situations, graphing calculators, and multirepresentational software tools, and where students were encouraged to make their own decisions about which representations to use, when and how to link the representations (especially graphs and tables), and in which medium to work. The researchers compared and contrasted the characteristics of students' concept images of functions that emerged in this environment with the concept images described in previous work involving different environments (e.g., Leinhardt, Zaslavsky, & Stein, 1990; Markovits, Eylon, & Bruckheimer, 1986; Schwarz & Dreyfus, 1995). The questionnaire that was presented to students included questions primarily on the interpretation and generation of graphical representations, and to a limited extent on the relation of certain graphs to algebraic representations. The researchers found that the students of the current study often used the prototypical linear and quadratic functions but did not consider them as exclusive (for persistence of linear models in students' reasoning, see DeBock, Verschaffel, & Janssens, 2002). They suggested further that the numerous manipulative experiences with function examples, which were afforded by the multirepresentational software, influenced positively students' understanding of the functions' attributes. They also claimed that the process of beginning each class activity by small-group work on open-ended situations, followed by full-class discussions that were orchestrated by the teacher, was conducive to students' engaging in a discourse that led to higher levels of mathematical reasoning and conceptual change (see also Yackel & Cobb, 1996).

Generational Activity Within the Context of Word Problems. The solving of word problems has traditionally been considered to involve two phases: the setting-up of an equation to represent the relationships inherent in the word problem, and the actual solving of the equation. Although past studies focused almost exclusively on the equation as the form of problem representation (e.g., Chaiklin, 1989), more recent work has included a much broader range of representations. A further move has been occurring of late that involves an extension of the types of problems being worked on to include the modeling of more open-ended, "real-world" situations. This growing domain of modeling within school mathematics, to the degree that it intersects with the learning of school algebra, is reserved for the later discussion of global/meta-level activity. This particular section is therefore divided into two parts: The first examines the research related to word-problem solving involving multiple representations, and the second, involving the equation representation.

Generating Multiple Representations in the Solving of Word Problems. In a study involving 6 intact classes of seventh and eighth graders at three junior high schools, half of whom were in a treatment group that participated in a representation-based unit on functions and the other half in a comparison group using conventional textbook lessons on algebra that included solving equations, using equations to solve word problems, and a unit on graphing ordered pairs, Brenner et al. (1997) investigated whether instruction on representing word problems by means of multiple representations improved students' abilities to represent and solve problems. The researchers found that instruction on problem-representation skills can work, that is, students learned to create various multiple representations and to use these to find problem solutions. The reform-based instruction was also found to have the same beneficial effects on language-minority students as on the overall school population of the study. Related research by Koedinger et al. (1997), involving the multirepresentational, problem-solving computer environment, *Cognitive Tutor Algebra I* (Carnegie Learning, 1998), and by Nathan et al. (2002), suggests however that students tend to favor tables of values as problem-solving tools before they gain facility with graphs and equations. Multiple representations of word problems have also been used to introduce students to the form of inequalities. In analyzing some of the videotape data from the TIMSS-R eighth-grade algebra lessons (Stigler et al., 2003), Kieran (2004) noted that Japanese students attempted to solve the inequality situation first by using a variety of representations and methods to solve for the corresponding equality, and then by making adjustments to their "solution" in order to provide an answer for the inequality. (Interestingly, none of the students used a graphical representation.) The teacher then used these various "equality" methods, in particular the table-of-values representation, to help students acquire meaning for the form of the algebraic inequality.

Generating Equations in the Solving of Word Problems. Generating equations to represent the relationships found in typical word problems is well known to be an area of difficulty for algebra students. Nevertheless, word problem situations not only continue to be used as a means for infusing algebraic objects with meaning but have also received increased emphasis in reform programs as vehicles for introducing students to algebra. Research in this area continues to provide evidence of students' preferences for arithmetic reasoning and their difficulties with the use of equations to solve word problems (e.g., Bednarz & Janvier, 1996; Cortés, 1998; Swafford & Langrall, 2000). For example, Stacey and MacGregor (1999) found that, at every stage of the process of solving problems by algebra, students were deflected from the algebraic path by reverting to thinking grounded in arithmetic problem-solving methods. Another study dealing with the setting up of equations within the activity of word-problem solving has reported that, for sixth and seventh graders, reasoning and symbolizing appear to develop as independent capabilities (van Amerom, 2003). Although some students could write equations to represent problems, they did not use these equations to find the solution, preferring instead to use more informal methods. Some students did, however, use the equations they had generated as a basis for their reasoning about the problems and for finding their solutions. Similarly, Malara (1999) has observed students using numerical substitution within problem representations in order to arrive at solutions (see also Johanning, 2004). Kutscher and Linchevski (1997) have also noted the beneficial effect of numerical instantiations as mediators in solving word problems. Although contextualized problems are often considered an appropriate avenue for introducing students to algebra (NCTM, 2000), De Bock, Verschaffel, Janssens, and Claes (2000), who explored the influence of authentic and realistic contexts and of self-made drawings on the illusion of linearity in length and area problems, found no beneficial effect of the authenticity factor, nor of the drawing activity, on pupils' performance. Furthermore, the researchers suggested that realistic problems might, in fact, steer pupils away from the underlying mathematical structure of a problem.

However, according to Nathan and Koedinger (2000a), word problems presented in verbal form are easier for students to solve than comparable questions presented in other formats, such as equations or "word-equations." These researchers presented a set of 12 problems to a group of 76 high school students in the following formats: Two were in story-problem format (e.g., "When Ted got home from his waiter job, he multiplied his hourly wage by the 6 hours he worked that day. Then he added the $66 he made in tips and found he earned $81.90. How much per hour did Ted make?"), two were in equation format (e.g., "Solve for x: $x \cdot 6 + 66 = 81.90$"), and two were in word-equation format (e.g., "Starting with some number, if I multiply it by 6 and then add 66, I get 81.90. What did I start with?"). For one problem in each format, the start value was unknown, and for the other, the result was unknown. The equation format was found to be significantly less likely to be correctly solved than either the story-problem or word-equation formats, whereas algebra story problems and algebra word-equation problems were found to be of equal difficulty. In a replication study involving 171 students (Nathan & Koedinger, 2000a), solution success rates for symbolic equations were 25% less than for story problems and nearly 20% less than for word equations. In both of these studies, the start-unknown format was more difficult than the result-unknown problems. Students' difficulties with problems presented in equation format challenge, according to these researchers, the oft-cited view that story problems are inherently harder than symbolic ones. The many possible entry points, as well as the availability of a variety of solving approaches for story problems, are deemed to be factors accounting for the higher success rates with the verbal format.

In a follow-up study, Koedinger and Nathan (2004) further explored their earlier finding that students are more successful solving simple algebra story problems than solving mathematically equivalent equations. They found evidence that this result is not simply a consequence of situated world knowledge facilitating problem-solving performance, but rather a consequence of student difficulties with comprehending the formal symbolic representation of quantitative relations. In fact, students translated story problems into the standard equation format only 5% of the time. According to the researchers, students even after a full year of algebra were particularly challenged by the demands of comprehending the letter symbolic form of the equation: "The language of symbolic algebra presents new demands that are not common in English or in the simpler symbolic arithmetic language of students' past experience" (p. 149). The researchers suggested that more attention be focused on equation-solving instruction; however, they emphasized that this be done after students have had experience with translating back and forth between English and algebra within story problem situations.

Transformational Activity: The Middle Grade and Lower Secondary Level Student

As noted earlier, the transformational activity of algebra includes collecting like terms, factoring, expanding, substituting one expression for another, adding and multiplying polynomial expressions, exponentiation with polynomials, solving equations and inequalities, simplifying expressions, substituting numerical values into expressions, working with equivalent expressions and equations, and so on. Past research in this area tended to focus on the nature of the manipulative processes used in expression simplification (e.g., Davis, Jockusch, & McKnight, 1978; Sleeman, 1984) and in equation solving (e.g., Bell, O'Brien, & Shiu, 1980; Kieran, 1988; Whitman, 1976). The field has recently experienced a change that includes attention to the theoretical foundations of students' manipulative work. In other words, algebraic transformations are not viewed simply as procedural, but also as theoretical. Thus, notions of equivalence figure more prominently in some of this later work. Another movement has been the downward shift of symbol-manipulation technology from the college to the secondary school level, even to the lower secondary level. To reflect the current research emphases, this section is divided into the following three parts: theoretical elements of transformational activity, namely equivalence and theoretical control; technical elements of transformational activity, namely the manipulation of expressions and equations; and the use of concrete manipulatives in transformational activity. The reader is reminded that the age range of the student in this section continues to be approximately 11 to 15 years of age, that is, the sixth to the ninth grader. Comparable research conducted with the older student is reserved for a later section.

Theoretical Elements of Transformational Activity: Equivalence and Theoretical Control. According to Pimm (1995),

> One resource of algebra is a rich plurality of symbolic forms; one core notion, that of equivalence. Equivalence and transformation are linked notions, indicating sameness perceived in difference for some purposes, or indifference with respect to others. The existence of multiple expressions 'for the same thing' (highlighting 'naming') can suggest the very possibility of transforming expressions directly to get from one to another. (p. 89)

Despite the importance of this core idea, only a few of the earlier studies explicitly addressed students' notions of equivalence (e.g., Greeno, 1982; Kieran, 1984a; Linchevski & Vinner, 1990; Steinberg, Sleeman, & Ktorza, 1990).

An example of some of the later research in this area is the work of Cerulli and Mariotti (2001; see also Nicaud, Delozanne, & Grugeon, 2002). Cerulli and Mariotti have described a teaching experiment on equivalence that involved ninth-grade classes and the algebra microworld, *l'Algebrista*, created by Cerulli (2004). Their aim was to develop in students a theoretical perspective on algebraic manipulation, by basing activity on the concept of equivalence relation, and by distinguishing *proof* from *verification*. The equivalence of two expressions is *proved* if one expression is transformed into the other using the axioms. Students were taught that, with literal expressions, the use of axioms is the only way to prove the equivalence of two expressions, whilst numerical *verification* (i.e., substituting the letters with numbers and computing the expressions) is the main way to show that two expressions are not equivalent. Students worked at changing the form of given numerical and literal expressions by selecting axiom buttons of the computer tool or by creating their own "theorem/transformation buttons."

Cerulli and Mariotti's (2001) analysis highlighted two important issues. First, they noted the carryover of the button-icons to students' paper-and-pencil work. Similar iconic traces of the computer environment have been noted in the written algebra work of the participants of Sutherland's (1993) spreadsheet study. According to Cerulli and Mariotti, such signs guide the evolution of meaning in students' algebraic thinking. The second issue concerns the theoretical aspect of students' activity. Cerulli and Mariotti were able to provide evidence to support their claim that they had moved students away from a purely procedural interpretation of algebraic manipulation and had inculcated a theoretical perspective by means of the emphasis on axioms and theorems. The importance of theoretical control in algebraic transformational activity has also been articulated by, among others, Balacheff (2001) and Dettori et al. (2001). However, as noted by Demby (1997), students have a great deal of difficulty in identifying the properties they use when they transform algebraic expressions.

Another element of theoretical control that is considered basic to transformational activity is the knowledge that relates the algebraic domain with the arithmetical. For example, numerical substitution activity within expressions and equations can help students make connections between the arithmetical and the algebraic world (see, e.g., Warren & Pierce, 2004). Graham and Thomas (2000) conducted a teaching experiment with nearly two hundred 13- and 14-year-olds in which students used the letter stores of the graphing calculator as a model of a variable and so evaluated different expressions for a variety of inputs. In the case of equivalent expressions, students came to see that different expressions were being used to represent the same process. This activity also had an impact on students' views of expressions and variables, which suggests that a task that is transformational in nature can simultaneously be related to generational activity if it leads to an evolution of students' conceptions of the objects of algebra.

The kinds of errors that students can make in algebraic transformational activity (e.g., Carry et al., 1980; Lemoyne, Conne, & Brun, 1993; Matz, 1982; Sleeman, 1984) have suggested to some researchers (e.g., Kirshner, 1989) that the issue is not an absence of theoretical control but rather a misperception of form. In a study that extended Kirshner's earlier work on the visual syntax of algebra, Kirshner and Awtry (2004) investigated the role of visual salience[4] in the initial learning of algebra for students in four intact classes of seventh graders (about 12 years old). They found that students did indeed engage with the visual characteristics of the symbol system in their initial learning of algebraic rules: The percentage-correct scores for recognition tasks were significantly higher for visually salient rules than for non-visually-salient rules. Furthermore, visually salient rules turned out to be significantly more difficult to recognize than non-visually-salient rules when presented using tree notation. Similarly, Hewitt (2003), in a study of 40 teachers and a class of 11- to 12-year-olds, found that the inherent mathematical structure, and the visual impact of the notation itself, had an effect on the way in which equations were manipulated.

Technical Elements of Transformational Activity: Expressions, Equations, and Equation Solving. Among the few recent studies dealing with the transformation of algebraic expressions involving this age group of students are a study by Bazzini, Boero, and Garuti (2001a) and another by Ayers (2000) on bracket expansion errors. A greater amount of research has dealt with equations. For instance, Herscovics and Linchevski (1994) conducted a study involving equations of the type $4 + n - 2 + 5 = 11 + 3 + 5$, with various numerical operations on one or both sides of the equal sign. They found that about 50% of the beginning algebra students whom they tested failed to obtain the correct solution, many of them having equated the left side of the equation with only the first member of the right side—a result consistent with Kieran's (1984b) findings with the same kind of equation. In addition, Herscovics and Linchevski observed that several of the unsuccessful solvers incorrectly grouped the left-hand expression to $4 + n - 7$, an error that they have referred to as the *detachment of a term from the indicated operation.* They noted as well that, with the equation $115 - n + 9 = 61$, some students transformed it to $106 - n = 61$, having committed the error of *jumping off with the posterior operation* (see Linchevski & Herscovics, 1996, for additional errors). Similar findings have been reported in research involving eighth-grade Belgian students (Vlassis, 2001).

In a study that focused on the checking of equation solutions, Perrenet and Wolters (1994) found that eighth graders made the same sorts of errors in their checking that they had made while solving the equation, thereby not detecting that their solutions were erroneous. Analysis of students' erroneous checking behavior disclosed their profound structural misunderstandings, uncertainty, and wishful thinking. In another study related to checking solutions, Pawley (1999) noted that students were more successful in both their equation solving and their checking if they were taught how to check sometime later than they were taught how to translate from sentence to equation.

Up until quite recently, researchers have known very little about the ways in which students of this age range approach the solving of systems of equations and the manner in which they think about its underlying concepts. The recent research of Filloy, Rojano,

[4] According to the researchers, "visually salient rules have a visual coherence that makes the left- and right-hand sides of the equations appear naturally related to one another" (p. 229). Although $(x^y)^z = x^{yz}$ is considered a visually salient rule, $x^2 - y^2 = (x - y)(x + y)$ is not.

and Solares (2003, 2004) has focused on the spontaneous approaches of 13- and 14-year-olds, who had already been introduced to the solving of one-unknown linear equations, to problems that could be solved by systems of equations. One of the aims of this research was to document the transition from one-unknown representations and manipulations to the representation and manipulation of one unknown given in terms of the other unknown. It was found that students seemed more inclined to make sense of comparison approaches than substitution methods; however, the researchers noted that manipulation difficulties contributed to making the substitution method less accessible. They also observed that the extension of the notion of transitivity of equality from the numeric to the algebraic domain, as well as the idea of substituting one expression with another, was not at all obvious to students.

Further evidence of students' difficulties with the substitution method for solving systems of equations was found by Drijvers (2003). Within a Computer Algebra System (CAS) environment, 14- and 15-year-olds were asked to solve parametric equations, for example, "Solve $ax + b = 5$ for x." Students experienced difficulty in accepting the expression $(5 - b)/a$ as a solution. According to Drijvers, this required that they conceptualize an expression as an object (Sfard, 1991). Other tasks such as, "Consider the equations $y = a - x$ and $x^2 + y^2 = 10$. Make one equation from these in which y does not appear; you do not need to solve this new equation" (p. 260), which required substituting a variable in one equation by an expression drawn from the other equation, were equally problematic. In some cases, the use of graphical representations helped students to make sense of the equations with which they were working. The obstacles that students encountered in working with these equations led Drijvers to suggest that more attention needs to be paid to the role of paper-and-pencil work throughout CAS activity and to the importance of focused classroom discussions. Friedlander and Stein (2001) found that a significant number of students who could solve equations with the various computer tools at their disposal actually preferred to use paper and pencil.

Within process-object theories (e.g., Sfard, 1991), the transition from process to object tends to be viewed as a leap, a sudden shift in perspective, whereby a process becomes reified as an object. Students who conceive of an expression as a process are considered to be unable to make much sense of, for example, composition of functions (e.g., Ayers, Davis, Dubinsky, & Lewin, 1988), until such time as they make that sudden shift from expression as process to expression as object. However, recent research by Landa (2003) suggests a more gradual learning process, in that composition of function may be preceded conceptually by composition of argument. Landa, who worked with 14- and 15-year-old students in a spreadsheet environment, found that, before students could make sense of, for example, $g(f(x))$, where $f(x) = 2x + 3$ and $g(x) = x^2$, they first needed to be able to create meaning for replacing the argument in a single function by an expression, as in $g(2x + 3)$. The spreadsheet environment, by means of its numerical calculation of functions with composed arguments, was seen to play a vital role in mediating students' passage toward composition of functions.

The Use of Concrete Manipulatives in Transformational Activity. Research has produced mixed results with respect to the use of concrete models and manipulatives in learning about equations and equation solving. Filloy and Rojano (1989) found that the balance scale and geometric models were not effective for helping students construct meaning for the process of operating on both sides of the equation to solve for the unknown. Boulton-Lewis et al. (1997) reported that the eighth graders of their study did not use the concrete manipulatives (cups, counters, and sticks) that were made available for solving linear equations and suggested that the concrete representations increased processing load. In contrast, advocates of such models (e.g., Brown, Eade, & Wilson, 1999; Linchevski & Williams, 1996; Radford & Grenier, 1996) have argued that the balance scale facilitates the understanding of the operation of eliminating the same term from both sides of an equation. In particular, Vlassis (2002) found that, for the two classes of eighth graders whom she observed over a period of 16 lessons, the balance model was an effective tool for conveying the principles of transformation. According to Vlassis, some errors linked to its usage appeared during the first lesson; however, these tended to disappear over time. Nevertheless, Vlassis emphasized that if students have not already extended their numerical range to encompass the negative integers, cancellation errors will be inevitable.

Global/Meta-Level Activity: The Middle Grade and Lower Secondary Level Student

The global/meta-level activities of algebra include problem solving, modeling, working with generalizable patterns, justifying and proving, making predictions and conjectures, studying change in functional situations, looking for relationships or structure, and so on—activities that could indeed be engaged in without using any letter-symbolic algebra at all. Although algebraic activity involving problem solving, more particularly word-problem solving, has been integrated into the previous sections on generational and transformational activity, three types of global/meta-level activity for which a significant body of algebra research exists with this age group of learner have been reserved for this section: generalizing, proof and proving, and modeling.

Global/Meta-Level Activity Involving Generalizing. Generalization activity within algebra has its roots in the use of algebraic notation as a tool for expressing proofs (e.g., Bell, 1976; Fischbein & Kedem, 1982; Mason & Pimm, 1984). The position that generalization is also a route to algebra was developed by Mason, Graham, Pimm, and Gowar (1985; see also Mason, Graham, & Johnston-Wilder, 2005). In some of the pioneering research on the use of algebraic notation as a tool for expressing the general form of geometric and numerical patterns, and for justifying equivalent forms of these patterning relations, Lee (1987; Lee & Wheeler, 1987) found that few students use algebra or appreciate its role in justifying a general statement about numbers. Similar findings have been reported by MacGregor and Stacey (1993), who observed that an additional difficulty was related to students' inability to articulate clearly the structure of a pattern or relationship using ordinary language.

Healy and Hoyles (1999) have pointed out that the visual approaches generated in tasks involving generalization of matchstick patterns can provide strong support for the algebraic representation of sequences and the development of a conceptual framework for functions but emphasized the need to work hard to connect the observed number patterns to symbolic form. Warren (2000), who tested 12- to 15-year-olds ($N = 379$) on tasks related to generalization of patterns, reported that students found this kind of activity to be rather difficult. As a result of follow-up interviews held with some of the students, she noted that the generalization process draws on a variety of skills not traditionally associated with the algebraic domain. In addition, Ainley, Wilson, and Bills (2003), who compared the process of *generalization of the context* with *generalization of the calculation*, found that generalizing the context did not seem to be sufficient to support pupils in moving to a symbolic version of the rule. However, Radford (2000), who has studied the transition from the particular to the general, has argued that such processes take time.

In a study that focused on the role of tabular representations in generalization activity, Sasman, Olivier, and Linchevski (1999) presented eighth graders (14-year-olds) with tasks in which they varied the representation along several dimensions, namely the type of function, the nature of the numbers, the format of tables, and the structure of pictures. Results showed that varying these dimensions had little effect on students' thinking. According to Mason (1996), a scrutiny of established school practice involving generalization in algebra reveals that often, starting from geometric figures or numeric sequences, the emphasis is on the construction of tables of values from which a closed-form formula is extracted and checked with one or two examples. This approach in effect short-circuits all the richness of the process of generalization. In fact, some researchers (e.g., Moss, 2005) have suggested that tabular representations may actually impede students' recognizing the general relationships underlying patterns and representing them algebraically. Mason suggests several possible investigative approaches that can lead to students' construction of algebraic formulas, including visualization and manipulation of the figure on which the generalizing process is based.

From a study of eighth-grade students and the generalization of patterns, Radford (2003) reported that whereas rhythm and movement play a central role in presymbolic generalizations, symbolic algebraic generalizations require a disembodying of the mathematical from the spatial-temporal. This researcher noted that the passage from a nonsymbolic to a symbolic algebraic expression of generality involves two ruptures, one with the sensual geometry of the patterns and the other with their numerical features such as rank. Furthermore, these ruptures are, according to Radford (2000),

> regulated by a socially established mathematical practice where the teacher plays a central role; this role is that of immersing and initiating the students into the peculiarities of the signs and meanings in which the practice of algebra is grounded. (p. 258)

Global/Meta-Level Activity Involving Proof and Proving. A great deal of the research on proof and proving with students of this age includes problems drawn from elementary number theory. For example, Edwards (1998), in her study of 14- and 15-year-olds who had not yet been exposed to instruction in proof in their algebra class, found that none of them used standard algebraic notation to justify or explain their true–false decisions about simple statements involving the combining of odd and even numbers. Miyakawa (2002), who reported on a study involving 14-year-old ninth graders working on a proof problem dealing with the sum of two even numbers, found that the difficulty in constructing this kind of mathematical proof relates not only to algebraic competence and proof conception, but also to general mathematical knowledge. Dreyfus, Hershkowitz, and Schwarz (2001), who observed seventh graders collaboratively constructing an algebraic proof, reported that constructing this proof, which the students were able to do, involved first generalizing the process and then justifying it by means of an algebraic representation.

In contrast to research involving students without prior experience in constructing algebraic proofs, a study by Healy and Hoyles (2000) examined the conceptions of proof held by students who had followed a curriculum that, several years earlier, had mandated activity in formulating and testing conjectures, as well as in explaining and justifying conclusions. The researchers surveyed nearly 2500 high-attaining 14- and 15-year-olds. Consonant with the earlier findings of Lee and Wheeler (1987), Healy and Hoyles found that, when students were asked to pick the argument that they thought would get the best mark, the majority of students chose the ones based on algebraic form. However, students were unlikely to base their own arguments on similar algebraic constructions, feeling that these arguments neither communicated nor illuminated the mathematics involved. Students preferred to give some numerical examples, followed by a narrative explanation—even if they were aware of the limitations of such arguments. According to the researchers, students' responses were influenced mainly by their mathematical competence, but also by curricular factors, their views of proof, and their gender.

Global/Meta-Level Activity Involving Modeling. The majority of the algebra studies involving problem situations have, in general, been carried out with typical word-problem contexts. Nevertheless, a growing number of studies are now including modeling activities that go beyond the traditional problems found in most textbooks. In his plenary address at the closing session of the 12th ICMI Study on The Future of the Teaching and Learning of Algebra, Burkhardt remarked that "modeling as a bridge to algebra" (see below) seemed to turn on its head the idea of algebra as a tool for modeling. However, a number of studies have used real or pseudo-real situations to stimulate algebraic symbolization, for example, the modeling of running with the aid of motion sensors to motivate the development of a symbolic representation of this activity (Arzarello & Robutti, 2001). Stacey and Chick (2004) have commented on the issue of modeling as a bridge to algebra as follows:

> Why, at this point in time, is there an emphasis on "modeling as a bridge to algebra" rather than on instruction designed to help students use their algebra in modeling the world? It may simply reflect the dominance of interest in beginning rather than advanced algebra. It may reflect a feeling of stalemate in improving students' abilities to set up equations for solving word problems, whether presented traditionally or otherwise, and tackling traditional applications. Alternatively, it may reflect an intention to keep algebra and the real world closer together throughout instruction, so that special lessons to close the gap are irrelevant. In any case, it is important to recall the centrality of Kieran's global/meta-level activity. Proving, modeling, and applying are activities that give purpose to learning algebra. (p. 18)

Since the mid-1990s, there has been a noticeable shift toward interest in modeling activity. However, the term *modeling* is interpreted quite broadly in the research literature to include everything from typical word problems to complex physical situations. For Janvier (1996), modeling, in the true sense of the word, is a rare activity in school algebra (see also Doerr & Tripp, 1999). In line with Burkhardt (1981) and de Lange (1987), Janvier has emphasized that modeling involves two phases: formulation and validation. The formulation phase of modeling comprises first the examining of a phenomenon or situation in order to establish some key relationships between the variables involved. After initial assumptions have been investigated, a more or less complex series of mathematical transformations or operations ultimately lead to a model, often expressed as a symbolic expression. The second phase, that of validation, requires testing the validity of the model by going back to the reality that it is supposed to represent. Similarly, according to Zbiek (1998),

> a mathematical model is a combination of one or more mathematical entities and the relationships among them that are chosen to represent aspects of a real-world situation (Niss, 1988); mathematical modeling requires the application of mathematics to unstructured problems in real-life situations (Galbraith & Clatworthy, 1990). (p. 184)

Research in this area includes the work of Lesh, Hoover, Hole, Kelly, and Post (2000) on model-eliciting activities, and Lesh and Doerr (2003) on a theory of models and modeling, as well as, for example, cross-national approaches to research on modeling (Molyneux-Hodgson, Rojano, Sutherland, & Ursini, 1999). (See also the chapter in this volume on problem solving and modeling.) In their research involving both U.S. and Australian middle school students, Doerr and English (2003) have described how students' modeling shifted their focus of thought from finding a solution to a particular problem to creating a system of relationships that is generalizable and reusable.

Dynamic physical models are often an integral part of modeling studies. For example, research with students in the 11- to 15-years-of-age group has included devices with gears of different sizes (Bartolini Bussi, 1995; Meira, 1998), a screen version of a baby duck swimming to catch up with its mother (Kaput & Roschelle, 1997), a simulation of a moving elevator (Noble, Nemirovsky, Wright, & Tierney, 2004), and motion artifacts such as an electronic walking frog (Schorr, 2003). Winch movement was the focus of a case study by Izsák (2003), who analyzed how a pair of eighth graders developed knowledge for representing with algebra the phenomena produced by this device. Izsák reported that, as these students solved problems about the winch, they learned to distinguish equations that are true for any value of the independent variable from equations that constrain the independent variable to a unique value—distinctions that have been shown to be elusive for many algebra students. Izsák's analysis also demonstrated that students have and can use criteria for evaluating algebraic representations, and that they can develop modeling knowledge by coordinating such criteria with knowledge for generating and using algebraic representations of physical situations. Hines (2001), using a similar winch device in her research, has described the learning trajectory of an eighth grader as he created and interpreted tables, equations, and graphs intended to represent functions originating from explorations with this dynamic physical model.

Among the studies on students' modeling that have involved technology environments is the research of Yerushalmy (1997a). She studied seventh-grade Israeli algebra beginners working in a software environment, *The Algebra Sketchbook* (Yerushalmy & Shternberg, 1993), that allowed them to manipulate visual objects and to use two lexical sets (verbal and graphic) to model temporal phenomena related to functions of a single variable. Yerushalmy's descriptions of the linguistic aspects of students' modeling efforts provide substantive examples of student activity in an environment that does not require the use of algebraic symbols (see also the work of Nemirovsky, 1996, on mathematical narratives). In a related study involving situations in two variables, Yerushalmy (1997b) found that students' ability to produce visual representations and to think about and discuss concepts and processes using graphical terminology was an important factor in their modeling success. Visual considerations were a constant feature of students' analyses, even while they were using tables and talking about variables, and before drawing graphs.

The Impact of Technology on Algebraic Activity With the Middle Grade and Lower Secondary Level Student

Since the publication of the last *Handbook* (Grouws, 1992), research on the learning of algebra has experienced a significant increase in the number of technology-related studies. Although this surge must be balanced with the more modest growth in actual classroom usage of technology (e.g., Mullis et al., 2000, report that of the 23 educational systems participating in the 1999 TIMSS, only 21% of eighth graders reported some degree of computer use in their mathematics classes), the results regarding the effects of technology use on the initial learning of algebra have been quite remarkable. The various tools that have been employed include—as has been pointed out in earlier sections—spreadsheets, graphing calculators, computer algebra systems, calculator-based rangers, and specially designed software environments. The aim of this section is to synthesize from the results discussed in the preceding sections the role that technology can play in the learning of algebra with students of the middle grade and lower secondary levels.

The greatest impact of the use of technology in algebra learning has been on students' understanding of graphical representations and the way in which these representations allow a significant number of students to visualize in a graphical way the symbolic form of functions. This has thereby also improved students' understanding of certain functions. However, many students continue to experience difficulties with seeing the relationship between algebraic and graphical representations of function (Penglase & Arnold, 1996). The long-term experience of students in problem-based, functional environments supported by intensive use of graphing technology suggests that their representation of problem situations passes through phases, starting with the use of numbers as the only means of modeling, to working with graphs and tables, and finally to using more symbolic representations (e.g., Yerushalmy, 2000). Several studies, involving various technological environments and an assortment of global/meta-level activities to provide context for algebraic work, have in fact pointed out that supporting students in moving to a symbolic version of a rule, a pattern, or a problem situation takes time. Nevertheless, in environments and with activities specially designed for symbolic work, students can and do develop conceptually with respect to thinking with and acting on symbols (e.g., Cerulli, 2004).

The use of technology, and mainly the use of graphing technology, obviously introduces into school algebra other representations besides symbols and equations. One of the research-tested advantages of this introduction of a variety of representations in technology-supported environments is that such approaches are enabling traditionally unsuccessful students to gain access to the problem-solving aspects of algebra. In fact, "approaches to teaching and learning that emphasize problem solving and exploration, and within which students actively construct and negotiate meaning for the mathematics they encounter, find in this [graphing calculator] technology a natural and mathematically powerful partner" (Penglase & Arnold, 1996, p. 85). More specifically, Huntley, Rasmussen, Villarubi, Sangtong, and Fey (2000), whose study included ninth graders as well as older students, found that students who were not strong in symbol-manipulation skills, when they were able to use context clues and had graphing calculators available, could outperform symbolically capable students in tasks requiring formulation and interpretation of situations. This suggests that weaker students too are deriving a deeper conceptual knowledge of algebraic objects within technological environments. However, research needs to reveal more about how this happens.

There is a perception among some educators that, when powerful technological tools appear to "do all the work," there might be little left to teach in an algebra class. However, as the research in this and a later section shows, the opposite is the case. Nevertheless, the nature of the teaching changes, as does that of the tasks that are presented to students. To take advantage of the potential of technology would seem to require not only tasks that are designed to push students beyond the limits of their current algebraic thinking and encourage further development of that thinking, but also approaches to teaching that facilitate such growth in students. However, a great deal more research is needed in the two areas of task design and teaching within technological environments. Further discussion of the role of the teacher in technology-supported algebra classrooms occurs in a later section.

Concluding Remarks on the Algebra Research Involving the Middle Grade and Lower Secondary Level Student

The middle grade and lower secondary level student has received the bulk of the attention of algebra researchers. This would seem appropriate in that algebra is usually introduced with this age range of student. However, a comparison of the earlier studies of the 1980s involving the middle grade and lower secondary level student with the more current work that has been discussed in this particular section of the chapter discloses the presence of a shift in the nature of the research. Earlier work tended to focus on student difficulties in making the transition from arithmetic to algebra, and on the nature of the algebraic concepts and procedures developed and used by students during their initial attempts at algebra. Later work, while building upon and extending some of this earlier research, has evolved in new directions. Three areas in particular were noted.

One concerns the sources of algebra meaning making. Meaning based on the letter-symbolic form and on the problem situation/context has been extended to include the meaning derived from the use of graphical and tabular representations, as well as the meaning that emerges within dynamic activity. More research, however, needs to be done in this latter area so as to provide further insights into how such meaning develops and becomes disembodied. Another more recent research focus that was signaled in this section, and which will be elaborated more fully in the next section dealing with the upper secondary and college level student, is the role of theorizing within manipulative activity. Although properties and axioms were always an implicit part of symbolic manipulation, they were rarely the theme of algebra research in the 1980s. In contrast, theorizing is now considered to belong not only to the domain of generational activity in algebra, but to transformational activity as well. Finally, the third new direction in algebra learning research is one that was just discussed in the prior section, that of the use of technology. In particular, spreadsheets and graphing technology have played significant roles in this technologically oriented movement in recent algebra research dealing with the middle grade and lower secondary level student.

This section closes with a few additional remarks regarding questions in need of further research. Earlier, it was noted that Radford (2000) had argued that

> One of the didactic questions with which to deal is . . . that of the understanding of how . . . meanings are progressively transformed by the students up to the point to attain the standards of the complex algebraic meanings of contemporary school mathematics. (p. 240)

With the broad scope of generational and global/meta-level activity in current research, including functional approaches with their multirepresentational focus, Radford's question presents a challenge for researchers. School algebra has been taking on so many different forms that one might be led to ask: Are there still commonly accepted "standards of the complex algebraic meanings of contemporary school mathematics"? Although the view of that which constitutes algebraic activity has been considerably enlarged, the answer to this question is, I believe, still *yes*. And, because algebraic symbolism remains foundational to school algebra, it is strongly recommended that further research related to the learning of the letter-symbolic form be carried out. Areas where the field could benefit from additional study include, for example, the question as to how students can be assisted in (a) becoming aware of structure in patterns and in using symbols to express these patterns, (b) seeing relations between graphical representations and the corresponding letter-symbolic forms, and (c) making connections between their verbal problem-solving activity and the generating of equations. With the increased use of technology in algebra learning, another area where further research is needed is in the interaction between paper-and-pencil and machine work with tasks that have been specially designed to take advantage of this interaction. Further discussion of this issue is offered in the next section of the chapter dealing with the upper secondary and college level student.

The Upper Secondary and College Level Student

This section of the chapter examines the algebra research that has been carried out with the upper secondary and college level student, from the 10th grade onwards (about 15 years of age and up). Investigations that have involved students in teacher training programs at the college level are, however, reserved for the later section on teachers. As with the previous section devoted to the middle grade and lower secondary level student, the first subsection below deals with generational activity, whereas the second treats transformational activity and the third, global/meta-level activity. A fourth section discusses research on the effects of technology use, followed by a fifth that offers concluding remarks. Note that the research studies carried out on algebra learning with the upper secondary and college level student are fewer in number than those conducted with the middle grade and lower secondary level student.

Generational Activity: The Upper Secondary and College Level Student

Creating the objects of algebra, and building meaning for them, is at the core of generational activity, not only for the middle grade and lower secondary level student but also for the upper secondary and college level student. This section examines the research bearing on generational activity according to whether the object of study is treated from (a) a primarily letter-symbolic focus or (b) a multiple-representation perspective. The latter of these two sources of meaning is the theme of a larger number of studies than the former, reflecting both the increased attention to functional approaches across reform curricula as well as the traditional emphasis on functions and their various representations with the older student.

Generational Activity With a Primary Focus on the Letter-Symbolic Form. The studies that have focused on generational issues related to the letter-symbolic form among upper secondary and college level students have included research on form and structure, as well as on the kind of meaning building that occurs for the algebraic object, parameter. (A review of the work related to the learning of abstract algebra and of linear algebra is beyond the scope of this chapter.)

Form and structure. Algebra was at one time "all about form," according to Goldenberg (2003, p. 11). Two examples of form-related questions from a test set for 3rd-year Italian secondary students (16-year-olds) are obtained from a study by Menghini (1994):

> Find the value of the following expression:
> $z^4/c^2(b-a) + z^4/c^2(a-b) = \ldots$
>
> For which values of x is the following inequality true?
> $x^2 + x < x^2 + x + 1$.

Menghini reported that, for the first example, most students carried out very detailed calculations, and for the second, the best answers involved a comparison of the two parabolas represented by the two parts of the inequality. What seemed to be missing from the responses to these tasks, according to the researcher, was the ability to recognize more general laws and forms, or to be able to achieve an abstract understanding independent of the specific numerical context. Similarly, in their study of 11th graders, Hoch and Dreyfus (2004) found that very few students used structure sense and those who did were inconsistent. The more advanced students used it more often than the intermediate ones. The presence of brackets seemed to help students see structure, focusing their attention on like terms and breaking up the long string of symbols.

On parameters. While research with the older student has attended to the issue of shifts in the meanings of literal symbols (Bills, 2001)—including those related to the concept of variable (e.g., Ursini & Trigueros, 1997)—particular interest has been focused on parameters and their role in algebraic expressions and equations. According to Bloedy-Vinner (1994, 2001), understanding algebraic language related to parameters means (a) understanding from the context which letters are used as parameters, (b) understanding the role of parameters as opposed to the role of unknowns or variables, and (c) understanding the fact that the meaning of a letter as parameter might change throughout the process of solving a problem. In understanding this difference in roles, implicit quantifiers are involved. From a questionnaire designed to test students' understanding of implicit quantifier structures, administered to 82 Israeli students who had taken matriculation exams in mathematics, she found that most of the questions yielded a very low success rate (ranging from 3% to 69%). Furinghetti and Paola (1994), in their study of 199 Italian students, ages 16 and 17, found that only 20 of these students could adequately describe differences among parameters, unknowns, and variables.

Generational Activity with Multiple Representations. The research interest in multiple representations within functional perspectives is expressed in themes related to the understanding of functions and the interplay between symbolic and graphical representations, with special attention paid to the role of visualization.

Functions and Their Meaning. A function can be regarded as a set of ordered pairs, a correspondence, a graph, a dependent variable, a formula, an action, a process, or an object (Selden & Selden, 1992). When introduced to the formal ordered-pair definition of *function*, students have been found to rely on process-oriented interpretations (Sfard, 1991, 1992) or on their own intuitive concept images (Vinner, 1992; Vinner & Dreyfus, 1989). A significant number of studies have focused on efforts to support students in moving from a process-oriented to an object-oriented view of functions by reifying their process views (see, e.g., Dubinsky & Harel, 1992a). Student misconceptions that have been documented in this area include, for example, identifying functions with just one representation (Vinner, 1992), neglecting domain and range considerations (Sfard, 1992), and relying on the limited information provided by computer screen windows (Goldenberg, 1988). Building on some of the earlier research, more recent studies have explored, for example, the nature of covariational reasoning within the context of exponential functions (e.g., Confrey & Smith, 1994, 1995), the elaboration of a property-oriented view of function that is based on visual aspects of functional growth (Slavit, 1997), the notion of chronicle as a source of error (Janvier, 1998), and the role of concept maps in assessing conceptual understanding of functions (e.g., Williams, 1998).

Technological environments have been widely used in research on students' conceptual understanding of functions, for example, ISETL (Dubinsky & Harel, 1992b), *DynaGraphs* (Goldenberg, Lewis, & O'Keefe, 1992), and *Visualizing Algebra* (Schwartz & Yerushalmy, 1992). Graphing and graphical representations have been a major theme of the research (e.g., Leinhardt et al., 1990; Romberg et al., 1993), with a particular focus on the establishment of connections between the symbolic and graphical representation of function, often with the use of graphing calculators. This latter research is discussed next.

The Interplay Between Symbolic and Graphical Representations and the Role of Visualization. The role of visualization as a representational and interpretational tool has been an ongoing area of interest among algebra researchers (e.g., Arcavi, 2003; Dreyfus, 1991). Earlier work produced evidence that even advanced students demonstrate a reluctance to use visual representations (e.g., Eisenberg & Dreyfus, 1991), preferring letter-symbolic over graphical representations, even when the former are more complicated. More recently, drawing on findings by Schoenfeld, Smith, and Arcavi (1993) and Moschkovich, Schoenfeld, and Arcavi (1993), Knuth (2000) examined 9th- to 12th-grade students' understanding of the concept that the coordinates of any point on a line will satisfy the equation of the line, within the context of problems that require the use of this knowledge. Knuth found an overwhelming reliance on letter-symbolic representations, even on tasks for which a graphical representation seemed more appropriate. The findings indicated to Knuth that for familiar routine problems many students have mastered the connections between the letter-symbolic and graphical representations; however, such mastery appeared to be superficial at best. Similar results were found by Slavit (1998) in a study in which graphing calculators were available. According to Slavit, the students had difficulty in coming to view algebra from a multirepresentational perspective and thus tended to favor symbolic modes of work, using the graphing calculator only on graphing tasks. These findings contrast with those reported for younger students, where the use of symbolic modes of work lagged behind those of graphical representations as problem-solving tools. Visualization in advanced problem solving by both college students (the novices of the study) and mathematics professors (the experts) was the focus of a study reported by Stylianou and Silver (2004). The two groups judged visual representations likely to be useful with different sets of problems; furthermore, when it came to actual usage, the experts constructed visual representations much more often than did the novices.

Dynamic, visual representations were a central feature of a study by Zbiek and Heid (2001), who investigated students' understanding of parameters. They have described an environment involving dynamic control of a slidergraph and provided an example involving a family of functions whose general member is represented by $f(x) = a/(1 + be^{cx}) + d$, where a, b, c, and d are real numbers. At the same time that the slidergraph-control moves along the screen, allowing a selected parameter to take on different values, the graphical representation responds to and reflects the various values that the functional expression takes on. Zbiek and Heid pointed out that the ease with which the slidergraph, unlike static graphing utilities, allows students to vary a single parameter quickly and consistently led to surprising observations and was, in fact, the key to launching student reasoning. This study is an example of one of a growing number related to the use of dynamic environments in algebra in which bodily motion plays a role in meaning building for the functional objects of algebra (see also Kieran & Yerushalmy, 2004).

Zaslavsky, Sela, and Leron (2002) found evidence of much confusion regarding the connection between algebraic and geometric aspects of slope, scale, and angle. Participants, who included 11th-grade students as well as teachers, mathematics educators, and mathematicians, responded to a simple but nonstandard task concerning the behavior of slope under a nonhomogeneous change of scale. Responses revealed two main approaches—analytic and visual—as well as combinations of the two. The researchers recommended that instruction on slope distinguish between the erroneous conception of *visual slope*—the slope of a line (for which the angle is a relevant feature)—and the *analytic slope*—the rate of change of a function. Similar approaches were noted in the Lobato et al. (2003) study involving younger students.

Other related studies have investigated students' interpretations of linear and quadratic graphs as displayed on the graphing screen. Although some studies (e.g., Cavanagh & Mitchelmore, 2000) have suggested that older students continue to experience some of the same difficulties that have been found with younger students, others (e.g., Pierce & Stacey, 2001a) have reported that the use of symbolic, graphical, and numerical representations helps to expand students' previous understandings by allowing them to link visual images to symbolic statements. In particular, Pierce and Stacey observed that college-level students operating in a CAS environment experienced little difficulty in moving between graphical and symbolic representations and in fact seemed to prefer these two modes to the movement between tabular and symbolic representations. Furthermore, studies that have investigated the impact of long-term use of graphing calculators on students' conceptual understanding of function and its representations have reported positive results (e.g., Streun, Harskamp, & Suhre, 2000). The time spent in working with multiple representations has been found to be crucial to their learning.

Transformational Activity: The Upper Secondary and College Level Student

Research that has focused on the transformational activity of algebra with the upper secondary and college level student has investigated the following areas: notions of equivalence and meaning building for equivalence transformations, the solving of equations and inequalities, the factoring of expressions, and the integration of graphical and symbolic representations in transformational work. Before examining this research, a few theoretical remarks are offered to provide a context for some of the more recent work that has been carried out in this area.

Theoretical developments by French researchers have led to a rethinking of the nature of technical activity in mathematics. Combining elements from both anthropological theory (Chevallard, 1999) and cognitive ergonomics (Vérillon & Rabardel, 1995), Artigue (2002b) and Lagrange (2000) have elaborated the notion that, as students develop techniques in response to certain tasks, they engage in a process of theory building. In their theoretical approach, which they have applied to research involving Computer Algebra Systems (CAS), activity with technological tools is considered to promote both conceptual and technical growth in mathematics, as long as the technical aspects are not ignored. An increasing number of research studies on the transformational activity of algebra reflect this newly developed perspective.

Related to this perspective, especially with respect to the emergence of theoretical thinking within the context of CAS use, is the work done by Pierce and Stacey (2001b) on the development of algebraic insight. Algebraic insight, which is considered to consist of both algebraic expectation and the ability to link representations, involves a focus on basic properties, structure, key features, and links.

Transformational Activity Related to Notions of Equivalence. Among the research dealing with students' notions of equivalence and equality has been a study carried out by Sackur et al. (1997). These researchers have described how 10th graders' thinking evolved with respect to these notions within a series of "make false"[5] interviews. The researchers, who used Frege's sense-denotation distinction as their analytical framework, pointed out that,

> When we say that a student must know what the expressions denote, we mean that the student must know the following: (a) that an expression like $y(2x + y)$ has a numerical value, (b) that this value depends on the values of x and y, and (c) that this value is not changed by the algebraic transformation rules to which an expression is subjected, for example, that which transforms $y(2x + y)$ into $2xy + y^2$. (p. 47)[6]

Their analysis led them to conclude that, unless students come to realize that algebra is an arena of sense-making and that they can arrive at rules that will permit them to obtain the same results as their teacher or classmates, they will never be able to control their algebraic work.

Research evidence exists that, without interventions of the type conducted by Sackur et al., students do not spontaneously develop an algebraic notion of equivalence, and that the obstacles associated with this notion cut across the age range. For example, Kieran (1984a) found that 8th graders showed little awareness that the equation-solving transformations they were able to execute proficiently preserved solutions. In a study using a larger sample of 8th- and 9th-grade students, and employing tasks similar to those used by Kieran (1984a), Steinberg et al. (1990) noted that although most students knew how to use transformations to solve simple linear equations, many did not spontaneously relate this knowledge to the production of equivalent expressions. Ball, Pierce, and Stacey (2003) developed an instrument to assess students' abilities to quickly recognize equivalent algebraic forms—the "Algebraic Expectation Quiz." On the basis of the performance of 50 students on this test, administered before and after students' progression from 11th to 12th grade and during which time they participated in algebra instruction involving the use of a computer algebra system, the researchers reported that "recognizing equivalence, even in simple cases, is a significant obstacle for students" (p. 15). Pomerantsev and Korosteleva (2003) presented compelling evidence that difficulties related to understanding algebraic equivalence can extend well beyond the postsecondary level. Their study involved the use of a diagnostic test administered to a large ($N = 416$) sample of students enrolled in different stages of their K–8 teacher preparation program at a major American university. Test items were designed to assess students' abilities to discern and use structural aspects of algebraic expressions; results of this research revealed serious difficulties in doing so, cutting across students in all groups contained in the sample.

The growing increase in the use of CAS in secondary school algebra classes has been accompanied by research interest in the role that this technology might play in the development of students' conceptions of equivalence. Particular attention has been paid to the nature of the tasks that might provoke theoretical growth within the context of transformational activity. For example, Saldanha and Kieran (2005; also Kieran & Saldanha, 2005) reported on a study with 10th graders that involved a combination of CAS- and paper-and-pencil-based tasks on equivalence of expressions. Results suggested that the development of students' theoretical notions of equivalence was guided by the techniques that the tasks invited, and that the most productive learning occurred after the CAS techniques produced an output that conflicted with the students' expectations.

Transformational Activity Related to Equations and Inequalities. While the solving of linear equations has received a great deal of research attention, that of quadratic equations has not. One of the few studies was carried out by Vaiyavutjamai, Ellerton, and Clements (2005). Nearly 500 students from Year 9 classes in Thailand, Year 10 classes in Brunei Darussalam, and 2nd-year university students in the United States attempted to solve the same quadratic equations, all of the form $x^2 = K (K > 0)$ and $(x - a)(x - b) = 0$ (where a and b are any real numbers). All students had already learned to solve such equations before participating in the study. The responses to the second type of equation, in particular, suggested the presence of serious gaps in the theoretical thinking underpinning students' work when solving such equations. For example, in solving $(x - 3)(x - 5) = 0$, several students

[5] A "make false" interview is one that is aimed at having students confront the limits of their own thinking and in which a student is asked to produce an equality that is always false. For example, the researchers presented pupils with tasks such as "Write a false equality for $7x/(7 + x) =$" and then posed questions related to the responses provided by the pupils.

[6] Similarly, according to Arzarello, Bazzini, and Chiappini (1994), "the 'denotation' of a symbolic expression in algebra refers to the number set that is represented by the expression; it is determined by the symbolic expression and by the universe in which the expression is considered (for example the equation $x^2 + 1 = 0$ denotes the empty set when it is considered in **R** and the set {+i, –i} when considered in **C**)" (p. 42).

who correctly solved the equation checked their solutions by substituting $x = 3$ into $(x - 3)$ and $x = 5$ into $(x - 5)$ and concluded that because $0 \times 0 = 0$ their solutions were correct. Related difficulties were encountered by the U.S. students who were asked to respond True or False to the following statement: *This equation* $(x - 3)(x - 5) = 0$ *is equivalent to* $x^2 - 8x + 15 = 0$*, which is a quadratic equation with two solutions. Thus, with* $(x - 3)(x - 5) = 0$*, the x in the first brackets always equals 3, and the x in the second brackets always equals 5*. Fifty-five percent of respondents answered that this was indeed True. To another question, many were unsure whether the x in the "x^2 term" represented the same variable as the x in the "x term" of the equation $x^2 - 8x + 15 = 0$. The finding that most of the students in this study were confused about how the solutions of a quadratic equation related to the equation itself led these researchers to suggest that, if quadratic equations are to remain an important component of middle- and upper-secondary mathematics curricula, then research is needed to guide teachers about how students think about quadratic equations.

Tsamir and Bazzini's (2002) study of 16- and 17-year-old Israeli and Italian students' solutions to algebraic inequalities found that students of both countries intuitively used the balance model to do the same thing to both sides of an inequality and continued to draw on equation-related analogies when dividing both sides by a not necessarily positive value. This reliance on the balance model can be related back to results described in an earlier section of the chapter where the introduction of this model with younger students was found by several researchers to be beneficial, even if some suggested caution. In related research with Israeli students, Tsamir, Almog, and Tirosh (1998) reported that the following methods were used by high school math majors for solving equations and inequalities: algebraic manipulations, drawing a graph, and using the number line. When graphical representations (without technology) were used to solve rational and quadratic inequalities, the solutions were usually correct; however, students often failed to reject the excluded values.

Goodson-Espy (1998) presented 13 college students with a set of eight word problems that involved linear inequalities. Even those participants who were known to be able to solve basic algebraic equations and inequalities when presented in the symbolic format solved the problems using arithmetic forms of reasoning. Similar observations were noted earlier in the research involving younger students who were found to make use of arithmetic forms of reasoning in solving word problems. In all but one case, the participants in the Goodson-Espy study who used algebra to solve the tasks used equations rather than inequalities. This too echoes the findings reported with younger students who, when introduced to problems involving inequalities, chose to think about and solve for the corresponding equality situation and then make adjustments at the end to produce an answer for the inequality.

Transformational Activity Related to Factoring Expressions. With the emergence of CAS in secondary school mathematics classes, the transformational activity of factoring has been receiving research attention (e.g., Artigue, 2002a; Artigue, Defouad, Duperier, Juge, & Lagrange, 1998). For example, Lagrange (2003) has described the activity of classes of 16- to 18-year-olds (reported initially in Mounier & Aldon, 1996) attempting to find and prove general factorization of $x^n - 1$. The students who were in CAS-equipped classes tried to factor $x^n - 1$ for various values of n on their CAS; the machine produced as output complete factorizations of these expressions. This form of output tended to thwart the mathematical aim of the activity as designed by the teachers, which was to have the students arrive at an awareness of general regularities in the factored forms. A related study by Kieran and Drijvers (2006; see also Kieran & Saldanha, 2007) took advantage of the fact that the CAS produced complete factorizations of $x^n - 1$ in order to provoke a confrontation with the existing, but limited, theoretical thinking of the 10th-grade participants with respect to factoring. This confrontation was found to be very productive for most students, but especially so for those who, upon realizing that they could not generate the same factors as had the CAS, insisted on finding out how to do so either from the teacher or from their group-mates during the working part of the activity, or later during the plenary classroom discussion. These CAS-based encounters resulted in the evolution of not only students' paper-and-pencil factoring techniques but also their theoretical views of the structure of these expressions (e.g., seeing that $x^6 - 1$ could be viewed as a difference of squares, or as a difference of cubes, or as an example of the general rule that they had earlier generated, $x^6 - 1 = (x - 1)(x^5 + x^4 + \ldots + x + 1)$).

Another study on factoring is that reported by Nguyen-Xuan, Nicaud, Bastide, and Sander (2002). Two classes of 10th-grade novices (but not beginners) learned factoring techniques in a computer-based tutoring environment for symbol manipulation that provided informative error messages. One class received instruction by means of observing 14 pairs of worked-out examples, each observation being followed by activity on a related example. The second class did not observe any worked-out examples; their activity consisted only of factoring the 14 pairs that the other class had observed being factored. Both classes had from 4 to 6 hour-long sessions with the given activities. The researchers found that the second class learned more. This finding supported their hypothesis that, although observation may be an effective learning approach for rank beginners, it is less effective than learning by doing for students who are trying to become more expert. The researchers also noted that the time taken to move from novice to expert status was significant—a finding already reported in several other studies.

Transformational Activity Involving the Integration of Graphical and Symbolic Work. Ruthven (1990) conducted a study involving advanced-level mathematics students who had access to graphing calculators while working on double-angle trigonometric identity tasks. The most productive work started, according to the researcher, when students began to compare graphs of the different possibilities and used their manipulative skills to establish which of their different algebraic formulations were equivalent. This finding suggests an interesting approach for using the graphical to motivate the manipulation of the symbolic. In another study, Baker, Hemenway, and Trigueros (2001) noted that college students' work with transformations of functions was assisted by graphical representations when these activities were accompanied by written descriptions of what they were seeing on the screen of the graphing calculator.

The integration of graphical and symbolic representations has also been noted in the transformational work reported in studies on (a) finding the roots of quadratic functions (Mourão, 2002), (b) translating quadratic functions (Zazkis, Liljedahl, & Gadowsky, 2003), (c) working with cubic functions (Gómez & Carulla, 2001), (d) solving linear and quadratic equations and inequalities (Asp, Dowsey, & Stacey, 1993; Tsamir et al., 1998), (e) solving problems involving translations of quadratic graphs and parametric curves within a graphing-calculator environment (Drijvers & Doorman, 1996), and (f) the Newton-Raphson method for approximating the zeros of functions in a CAS environment (Hong & Thomas, 2002). Although the research involving CAS technology has focused on either the exclusive use of the letter-symbolic representation or the interplay between graphical and symbolic representations, some CAS research reports have included only graphing tasks, with no reference to the letter-symbolic (Zbiek, 2003; see also Thomas, Monaghan, & Pierce, 2004).

Traditionally, the study of transformations of functions has belonged to the domain of symbolic representation. However, in some technological environments (e.g., Schwartz & Yerushalmy, 1992, 1996), one graph can be transformed into another through direct actions on the graph. Borba (1993), who worked with 16-year-olds in the *Function Probe* environment developed by Confrey (1991), reported that the software environment influenced the mathematics that was studied and produced. He also argued that the nature of the tasks determined whether the students thought about transformations of functions in a static or dynamic way. Kieran and Yerushalmy (2004) have described other environments permitting dynamic transformation of graphical representations, accompanied by simultaneous change in the corresponding symbolic representation; however, they also noted that research on the impact of directly controlling a change in one representation to effect changes in another remains relatively undeveloped.

Global/Meta-Level Activity: The Upper Secondary and College Level Student

As with the middle grade and lower secondary level student, global/meta-level activity with the older student has included work in problem solving and in proving. However, in several of these studies, the two themes have been integrated and are therefore treated in the same section below. Some studies have also been carried out with the older, more mathematically advanced student in the global/meta-level activity of modeling; this research is briefly presented in the second section that follows.

Global/Meta-Level Activity Involving Problem Solving. Much of the research involving problem situations with the older student, as was the case with the younger student, has included technological tools. An example of such research is a study by Hershkowitz and Kieran (2001), who reported on classes of 16-year-olds exploring a situation involving three families of rectangles that increased in size according to a linear, a quadratic, or an exponential relation. In attempting to respond to the question as to which family of rectangles would eventually surpass the others in terms of area, students were found to go from entering lists to graphical representations by means of the regression tool of the graphing calculator without ever seeing or having to examine the algebraic representation of the situation. All of their problem-solving work involved attention to the data in the lists and the resulting graphs, which led the researchers to question whether the use of such tools within problem-solving situations can, in fact, lead students away from the use of letter-symbolic representations.

Much of the research on the global/meta-level activity of problem solving with the older student has tended to integrate number-theoretic work and proving. For example, Arzarello et al. (1994) observed a link between expressing the relations of number-theoretic problems in suitable algebraic code and success in determining the solutions to such problems. According to the researchers, appropriate naming is linked to anticipatory aspects of solving and allows students to orient the solution process toward the aims of the problem. They also pointed out that some students who could express the elements of a problem by using natural language were unable to express them using the algebraic language. In contrast, Douek (1999) found that nonstandard representations of number-theoretic problems did not prevent students from producing valid proofs and problem solutions.

However, this may in fact depend on the methods used by students to generate their proofs. Alcock and Weber (2005) reported that students used both referential (involving some numerical instantiation) and syntactic approaches (involving symbolic manipulation) and that students who used referential approaches may have had a more meaningful understanding of their proofs but may not have been able to complete them. In another study with 10th graders who had been given the task of proving their conjecture that $x + 1$ is always a factor of $x^n - 1$ for even values of n, Kieran and Drijvers (2006) found that most students used in their proofs a combination of syntactic and referential approaches, and that the most successful of the proofs were generic in nature.

Global/Meta-Level Activity Involving Modeling. Dynamic physical models have been an integral part of modeling studies with the upper secondary and college level student. Schnepp and Nemirovsky (2001) reported on a study that involved graphing software and hardware that linked a computer to miniature cars on parallel linear tracks and to a miniature stationary bike. The software allowed the system to work in two ways: either by manipulating the cars (or bike) and then modeling this movement by means of graphical representation, or by constructing a graph on the computer that then moved the mechanical device according to the graphical specifications. The latter, which is the opposite of a modeling perspective, was found effective when combined with the former in the development of conceptions of movement, and its representation, among the 12th-grade calculus students who participated in the study. Related work has been carried out by Rasmussen and Nemirovsky (2003) in an environment involving the physical tool, Water Wheel, with more advanced students.

The Impact of Technology on Algebraic Activity With the Upper Secondary and College Level Student

Just as with the younger algebra student, the presence of technology in the studies involving the upper secondary and college level student has been noteworthy. The main types of technology that have been employed in these studies with older students are the graphing calculator (without symbol-manipulation capabilities) and computer algebra systems, which also include graphing capabilities. Some of the impact of these tools has already been discussed in the individual research studies that have been presented. However, other technology-related studies that cut across the various types of algebraic activity have been reserved for this section. Some of these are large-scale studies, involving meta-analyses or comparison groups; others have focused on the ways in which roles in the algebra classroom change when technology is introduced.

Large-Scale Studies on the Effects of Technology Use. The studies that are presented in this section are categorized as to whether they deal with calculators in general, graphing calculators, or computer algebra systems. The largest set of studies is that having to do with the impact of graphing calculator usage.

Calculators. The first study to be included herein is a meta-analysis that investigated the effects of calculators (basic, scientific, and graphing) on students' achievement and attitude levels by examining 54 studies, 26 of which dealt with the high school level (Ellington, 2003). It was found that, when calculators were included in instruction but not testing, the ability to select the appropriate problem-solving strategies improved for the participating students. Under these conditions, there were no changes in students' computational and conceptual skills. When calculators were part of *both* testing and instruction, the strategic skills, computational and conceptual skills, and problem-solving skills improved for participating students. Under these conditions, there were no changes in students' ability to select the appropriate problem-solving strategies. Students who used calculators while learning mathematics reported more positive attitudes toward mathematics than their non-calculator-using counterparts on surveys taken at the end of the calculator treatment. Ellington also reported that students received the most benefit when calculators had a pedagogical role in the classroom and were not available just for drill and practice or checking work. However, she also noted that in only 11% of the studies analyzed had special curriculum materials been designed for calculator use and that this is thus an area in which more research needs to be conducted.

Graphing Calculators. The next two studies involved primarily the use of the graphing calculator within the computer-intensive algebra (CIA) curriculum (Fey et al., 1991). The first is a study by O'Callaghan (1998) that examined the effects of a one-semester computer-intensive algebra and traditional algebra curricula on college students' understanding of the function concept. He found that the CIA students achieved a better overall understanding of functions and were better at the components of modeling, interpreting, and translating. Further, the CIA students showed significant improvements in their attitudes toward mathematics, were less anxious about mathematics, and rated their classes as more interesting.

Hollar and Norwood (1999) tested four classes of intermediate algebra students (two experimental classes and two control classes), with two instructors each teaching one experimental and one control class in their study of the effects of a graphing-approach curriculum that extended O'Callaghan's (1998) CIA study. The control classes covered the same topics as the experimental classes but did not have access to graphing calculators. In the experimental classes, the calculators were used to explore, estimate, and discover graphically, and to approach problems from a multirepresentational perspective. It was found that students in the graphing-approach classes demonstrated significantly better understanding of functions on all four subcomponents of O'Callaghan's function test than did students in the traditional-approach classes. Additionally, no significant differences were found between the graphing-approach and traditional classes either on a final examination of traditional algebra skills or on an assessment of mathematics attitude.

Further evidence of the benefits of graphing-calculator use was obtained from a study that examined the effects of this technology upon students' understanding in elementary college algebra (Shoaf-Grubbs, 1993, as cited in Penglase & Arnold, 1996). The study involved 37 females from an all-women's liberal arts college, divided into an experimental class and a control class, with the only difference between the two groups being the use of the graphing calculator. Statistical analysis of pretest and posttest results showed that the majority of the students in the experimental group achieved significantly better than the control group in the test results, especially in their understanding of graphing concepts. However, Hall (1993), who compared four treatment classes with four control classes on the use of graphing calculators in a 3-week study of trigonometric functions, found that the graphing calculators had no significant impact on students' achievement.

Another large-scale study involving graphing-calculator use was conducted by Streun et al. (2000) in the Netherlands. These researchers used a three-condition pretest-posttest design to analyze the impact of prolonged use of the graphing calculator throughout the entire school year for all topics of the curriculum (i.e., functions and graphs, change, and exponential and periodic functions). Three experimental classes used the graphing calculator throughout the year; a second set of five experimental classes used the graphing calculator with only one topic for 6 weeks; and four classes, which served as the control group, covered the same subject matter throughout the year but without the graphing calculator. The students who used the calculator throughout the year had enriched solution repertoires and a better understanding of functions. The students who used the graphing calculator for only a short period of time did no better on the posttest than the students in the control group. They merely replaced their symbolic approaches and guess-and-test procedures with graphing methods. Unlike the students who spent more time using the graphing calculator, they had not significantly altered their conceptual understanding of functions. The time factor, which was pivotal in this study, has also been found to be of crucial importance in other technology-related research findings (e.g., Chacon & Soto-Johnson, 1998; Lagrange, Artigue, Laborde, & Trouche, 2001).

Further research involving graphing calculator use is the comparative study of the effects of a 3-year integrated mathematics curriculum for Grades 9–11, the *Core-Plus Mathematics Project*, with more conventional curricula on growth of student understanding, skill, and problem-solving ability in algebra (Huntley et al., 2000). Results indicated that the Core-Plus curriculum was more effective than conventional curricula in developing student ability to solve algebraic problems when those problems were presented in realistic contexts and when students were allowed to use graphing calculators. Conventional curricula were more effective than the Core-Plus curriculum in developing student skills in manipulation of symbolic expressions in algebra when those expressions were presented free of application context and when students were not allowed to use graphing calculators.

Computer Algebra Systems (CAS). The number of large-scale studies involving the use of CAS is much fewer than for the graphing calculator. In one study, Shaw, Jean, and Peck (1997) evaluated the effectiveness of using CAS in an intermediate algebra course by comparing the grade distributions in a follow-up course of three groups: students who went through the CAS technology-based developmental course ($N = 34$), students who went through a traditional developmental course ($N = 109$), and students who were not required to take a developmental course ($N = 1335$). Students in the developmental program that had the technology-based course did at least as well as the students who did not need to take any developmental course. This was considered a real improvement inasmuch as the students who took the traditional intermediate algebra course did not do as well in subsequent courses as students who took the technology-based course.

Technology and the Changing Roles of Students and Teachers. A few studies have investigated the work habits of students using technology. For example, Guin and Trouche (1999) categorized students' use of graphic and symbolic calculators into five profiles of behavior: random, mechanical, rational, resourceful, and theoretical. Doerr and Zangor (2000) classified five modes of graphing calculator use: as a computational, transformational, visualizing, verification, and data collection and analysis tool. These studies suggest the potential of technology for broadening the range of mathematical activity engaged in by students. Other studies have noted the changing roles of both students and teachers when technology is integrated into the mathematics classroom. For instance, Farrell (1996) observed that both teachers and students moved toward the roles of consultant and fellow investigator. Goos, Galbraith, Renshaw, and Geiger (2003) analyzed the impact of an assortment of technological tools on classroom interactions in terms of offering opportunities for students to engage constructively and critically with mathematical ideas, that is, as master, servant, partner, and extension of self. The 3-year Goos et al. study showed how technology can facilitate collaborative inquiry during both small-group interactions and whole-class discussions where students use the computer or calculator and screen projection to share and test their mathematical understanding. This finding is in contrast to that reported by Doerr and Zangor (2000), who observed that the graphing calculator as a private device led to a reduction in small-group interactions. However, Goos et al. (2003) have argued that, when calculators and computers are used as partners or extensions of selves and are permitted to become a part of face-to-face discussions, they facilitate communication and sharing of knowledge.

Similarly, Manouchehri (2004) found that the overall complexity of the mathematical discourse increased from before the technology was introduced to afterward in the senior-level mathematics course that she observed. In particular, she noted that three factors appeared to control the construction of fruitful algebraic classroom discourse: the teacher, the tasks, and the technology. She pointed out that a common tendency in mathematics instruction is to focus on the power of the technology as a presentation device rather than as a discourse participant. In her study, even though the technology served as a powerful information provider, it only became a learning medium when supported by appropriate teacher intervention and tasks. This latter observation has been made time and again by several researchers.

Concluding Remarks on the Algebra Research Involving the Upper Secondary and College Level Student

Although the amount of research conducted with the upper secondary and college level student has been less abundant than that carried out with the middle grade and lower secondary level student, it has produced results that touch upon issues that are both similar to and different from those that emerged from the body of research involving the younger student.

The sources of meaning that have been the object of study in the research with the older student are principally those derived from within mathematics—focusing either on the letter-symbolic or on multiple representations. As was seen, the research involving the letter-symbolic form included studies on structure and on equivalence. With symbol-manipulating technologies being introduced in algebra classes, students' conceptions of equivalence of expressions were a theme of interest. The research involving multirepresentational sources of meaning in algebra, a comparatively larger body of work in relation to that with a letter-symbolic focus, included a broad range of studies. Several of the studies that examined students' notions and usage of symbolic and graphical representations noted students' preference for the symbolic over the graphical. This was in contrast to results from studies involving younger students, which noted their difficulties with using symbolic approaches. Studies dealing with the role of the graphing calculator reported, in general, that sustained use of this technology can provide effective support for multirepresentational thinking among algebra learners. However, the importance of according sufficient time for the conceptual advantages of the technological environment to take hold was noted.

As time has emerged as a key factor in technology-related studies, two other factors have been found to be equally important. One concerns the nature of the tasks in which students engage; the other is the role played by the teacher in orchestrating the development of algebraic thought by means of appropriate classroom discussion. As is borne out by the results of several studies, the technology by itself is clearly insufficient.

The use of CAS technology in the learning of algebra was found to be more widespread in the research with this age group than with the younger student. Although some of the research with this technology makes use of multirepresentational approaches to algebra learning, a significant amount of this work focuses on the letter-symbolic mode. In line with advances by French researchers in the development of theoretical frameworks for studying CAS use, many of the studies reflect these frameworks by their attention to the emergence of theoretical thought within technical activity. Results from this newly developing sphere of research activity provide evidence that meaning building does indeed occur not only within the generational but also within the transformational activity of algebra. However, a great deal more research is needed in this area.

A related issue that arises with the use of technological environments in general, and with CAS in particular, is the question of the role of paper-and-pencil work in the development of understanding of particular concepts. For example, Asp et al. (1993) claimed that physically constructing tables of values for functions was essential to the development of students' understanding of the relationship between graphs and equations. Warren and Pierce (2004), on the basis of their review of Australasian research in algebra learning and teaching during the years 2000 to 2003, have further suggested that for successful CAS use students need to be able to do the following by hand: solve simple equations, use basic algebraic manipulations to rewrite equations in a form they can use, and be aware of standard techniques used to find solutions. However, they admit that further research is needed with respect to the question of the "by hand" skills that students require when working with CAS, as well as the issue of the place of by-hand work in the learning process (see also Forster, Flynn, Frid, & Sparrow, 2004).

The question regarding the by-hand skills needed by students embarking on CAS activity could usefully be researched by means of studies that carefully document the prior experience and incoming knowledge of students (i.e., their current by-hand skills) and then relate these to the theoretical and technical knowledge that develops (as well as to the obstacles encountered) as a result of the technological experience. Such well-documented research might indeed show that the prior by-hand skills needed by students for CAS activity vary according to several yet-to-be-determined dimensions.

The second issue regarding the place of by-hand work in the learning process has been emphasized in several studies. For instance, recent research (e.g., Drijvers, 2003; Kieran & Drijvers, 2006) has suggested that it is in the interaction of paper-and-pencil work and CAS activity, particularly in tasks involving confrontation between students' existing theoretical thinking and CAS output, that students develop both technically and theoretically. However, research has just begun to explore this area in, for example, studies related to equivalence of expressions and factoring. Further research focusing on the interaction of paper-and-pencil work and CAS activity, on a variety of topics, is sorely needed.

Before closing this section, a few remarks on needed research that does not necessarily involve technology are offered. With the current curricular interest in problem situations as a means of motivating and inducting a large majority of students into the study of algebra, along with the use of multiple representations for exploring the domain of problem solutions, several issues arise. For example, the obstacles faced by younger secondary students in using letter-symbolic representations as problem-solving tools versus the oft-found reliance of older secondary students on this same mode of representation leads to several questions: What is the nature of the transition between nonusage and usage of algebraic symbols in problem solving? In particular, what are the factors at play when students shift from using non-letter-symbolic (numerical or graphical) to letter-symbolic representations to solve problems? How do students learn to make connections between graphical images and corresponding symbolic representations? More specifically, what is the nature of the connections that students make between graphical and symbolic representations? The mathematical content involved in the study of these questions also needs to move beyond linear situations to encompass, for example, quadratic and trigonometric situations, as well as systems of equations. Research is needed with the upper secondary and college level student on a much wider variety of algebraic content than has up to now been studied.

Algebra Teaching and the Algebra Teacher

In the algebra chapter of the 1992 *Handbook*, Kieran (1992) remarked

> Even though the research community knows very little about how algebra teachers teach algebra and what

> their conceptions are of their own students' learning, this is not to suggest that there has not been considerable research on new approaches to the teaching of algebra—there has. However, the analysis of results has usually been directed toward detailing learning phenomena. (Kieran, 1992, p. 395)

There are signs that the situation described in the early 1990s has changed in some ways. There has been a noticeable increase in the number of studies related to the teaching or teacher of algebra over the past 15 years or so. Some studies have dealt with the practicing teacher within the algebra classroom; others have been conducted within a context of professional development or inservice training; and yet others have involved the preservice teacher. However, researchers still know relatively little about algebra teaching. As well, there remains a noticeable disconnect between the research on the learning of school algebra and the research on the teaching of school algebra. This latter issue is elaborated throughout the sections that follow.

Research Involving the Practicing Algebra Teacher

One of the first differences one notices in moving from the research focused on the algebra learner to the research focused on the algebra teacher is that the theories used to frame the research tend to be quite different. Whereas studies of algebra learning have included theoretical frameworks based on, for example, constructivism, socioculturalism, semiotic mediation, discourse, embodied cognition, didactical situations, and instrumentation (see also Lerman et al., 2002), the most widely used perspective to frame studies involving the algebra teacher is based on Shulman's (1986) construct of teacher knowledge, which includes content knowledge, general pedagogical knowledge, curriculum knowledge, and pedagogical content knowledge. Shulman identified *pedagogical content knowledge* as the specialized knowledge that teachers need and described it as

> the blending of content and pedagogy into an understanding of how particular topics, problems, or issues are organized, represented, and adapted to the diverse interests and abilities of learners, and presented for inspection . . . the category [of knowledge] most likely to distinguish the understanding of the content specialist from that of the pedagogue. (Shulman, 1987, p. 8)

Ball and Bass (2002) have pointed out that this notion of pedagogical content knowledge suggests that

> even expert personal knowledge of mathematics often may be inadequate for teaching; knowing mathematics for teaching requires a transcendence of the tacit understanding that characterizes much personal knowledge . . . it also requires a unique understanding that intertwines aspects of teaching and learning with content. (p. 4)

This perspective has led Ball and Bass to focus on what it is that teachers of mathematics *do*, rather than on what teachers need to *know*, and to begin to develop a practice-based theory of mathematical knowledge for teaching. This distinction between what it is that algebra teachers know or need to know versus what they do may be helpful in examining the research that has focused on the teacher and the teaching of algebra.

Since the early 1990s, mathematics education researchers, wishing to characterize the components of Shulman's pedagogical content knowledge for various domains of mathematics, have worked at fleshing out this construct. For the domain of school algebra, the specification of pedagogical content knowledge has involved efforts to explore algebra teachers' knowledge and beliefs, as well as analyses of the practice of algebra teaching. Teachers' preexisting views, expectations, and beliefs are generally considered to act as a filter for both their pedagogical content knowledge and teaching practice, but little research exists on the influence of prior conceptions in the learning processes of algebra teachers. Thus, the research that has involved the practicing teacher of algebra has been concentrated in two areas that are treated sequentially below: the first focusing on knowledge and second on teaching, with the research on beliefs interwoven into both sections.

Algebra Teachers' Knowledge and Beliefs. Regarding the domain of school algebra, and functions in particular, Even (1990) has suggested that pedagogical content knowledge needs to include at least the following: different representations, alternative ways of approaching the concept, strength of conceptual knowledge, a basic repertoire of examples, knowledge and understanding of the concept, and knowledge of mathematics. However, as research has recently disclosed, and which will soon be seen, knowledge of students' ways of thinking about particular aspects of school algebra and functions is an equally important component of pedagogical content knowledge. Even (1993) has applied her framework to research involving prospective secondary level teachers and has found their understandings of the concept of function to be weak and fragile.

More recently, a related framework—the Multidimensional Grid for Professional Competence in elementary Algebra (MGPCA)—has been developed by French researchers (Artigue, Assude, Grugeon, & Lenfant, 2001, as cited in Doerr, 2004) with dimensions similar to Shulman's, but they have been elaborated specifically for the teaching of algebra: epistemological, cognitive, and didactic. The *epistemological* dimension includes knowing the following: the content of algebra, the structure of algebra, the role and place of algebra within mathematics, the nature of valuable algebra tasks for learners, and the connections between algebra and other areas of mathematics and to physical phenomena. The *cognitive* dimension includes knowing the following: the development of students' algebraic thinking, students' interpretations of algebraic concepts and notation, students' misconceptions and difficulties in algebra, different approaches taken by learners, ways to motivate learners, and theories of learning. The *didactic* dimension includes knowing the following: the curriculum, the resources, different instructional representations, different practices and approaches taken by other teachers, the connections across the grade levels, and the nature and development of effective classroom discourse. The MGPCA grid has been used as a tool for collecting data and analyzing preservice teachers' initial relationship with algebra, and its evolution, throughout their 1st year of combined teacher training and classroom teaching experience in the field. In general, Artigue et al. have noted that, with respect to the cognitive dimension, teachers begin to become sensitive to students' learning difficulties during their 1st year of fieldwork, but that the length of the training period is too short for the trainees to develop the means for analyzing these difficulties or for generating remedial strategies.

One of the current themes of the existing research inquiring into algebra teachers' pedagogical content knowledge concerns teachers' knowledge of students' algebraic thinking. Recent research suggests discrepancies between teachers' predictions of students' difficulties and students' actual difficulties. Sixty-seven high school mathematics teachers, as well as 35 mathematics education researchers, participated in a study conducted by Nathan and Koedinger (2000a) on teachers' and researchers' beliefs about the development of algebraic reasoning. This group was asked to rank order 12 mathematics problems from easiest to most difficult. Four were in story-problem format, four were in equation format, and four were in word-equation format. A comparison of their predictions with the actual difficulty as experienced by high school students, which was discussed in an earlier section on algebra learning research, shows some important differences. The most salient discrepancy was that teachers and researchers predicted that story-problems and word-equation problems would be more difficult than symbol-equation problems, whereas students found symbolically presented problems most difficult. These differences between students' performance and teachers' predictions could have a significant effect on how teachers perceive students' reasoning and learning in the classroom and in how they teach to these perceptions. In a follow-up study involving 105 K–12 teachers and using similar tasks, Nathan and Koedinger (2000b) found that the middle school teachers came closest to predicting actual student performance. Furthermore, Nathan and Petrosino (2003) have provided evidence to support the argument that it was the well-developed subject-matter knowledge of the high school teachers (i.e., the "expert blind spot" syndrome), which underpinned their view of the equation-before-word-problem sequence, that led them to inaccurately predict student problem-solving difficulty.

Another study (Hadjidemetriou & Williams, 2002) addressing the issue of teachers' knowledge of what students find easy/difficult in algebra was one in which teachers' judgments of the difficulty of items dealing with graphical conceptions, and their awareness of errors and misconceptions, were analyzed and contrasted with learners' difficulty hierarchy. Teachers' judgment of what is difficult was found to be structured by the curriculum and by their mathematical knowledge. Also noted was a gap between pupils' difficulties and teachers' perception of these difficulties: Teachers underestimated technical difficulties of graphing and overestimated the difficulty of interpretive work. Similarly, in a study of 8 upper secondary school teachers in Sweden, Bergqvist (2005) found that the teachers tended to underestimate their students' reasoning levels on conjecture tasks and that the teachers erroneously believed that only a small group of students in each class could use higher level reasoning in mathematics.

Research has also begun to address the question of teachers' knowledge of mathematical links between lessons. Even, Tirosh, and Robinson (1993) compared the connectedness of lessons for two novice secondary teachers and an expert teacher on the topic of equivalence of algebraic expressions. They found that only the expert teacher used connections in mathematical content to guide the unfolding of her lessons. In a later study on teachers' knowledge of students' misconceptions, and the teaching approaches that would address these misconceptions, Tirosh, Even, and Robinson (1998) investigated four 7th-grade teachers' awareness of students' tendency to "finish" algebraic expressions (e.g., to write the expression $2x + 3$ as $5x$ or 5) and their ability to respond on-the-spot to pupils who showed this tendency in class. Although two of the teachers were aware of this tendency, the means they used to address it in class were only partially successful. The authors argued that,

> while one can use the mathematics education literature in order to raise teachers' sensitivity to students' ways of thinking, . . . this literature does not offer enough information and discussion regarding the impact of various approaches and teaching methods related to this conception; . . . it is important for teachers to be acquainted with various teaching methods and to be aware of their pros and cons in different contexts with different teaching aims, and with different students. (pp. 62–63)

Although the interaction between algebra teachers' knowledge and their beliefs is a complex and understudied area, it is widely held that both knowledge and beliefs shape teachers' teaching practice and that teachers' professed beliefs do not always match their instructional practices (Cooney, 1985). It has also been shown that limitations in both content knowledge and pedagogical content knowledge may influence teachers' beliefs and their instructional decisions and actions (Nathan & Koedinger, 2000b). Furthermore, findings by Nathan and Petrosino (2003) have suggested that strong content knowledge in algebra teachers produces beliefs that, in the absence of well-developed pedagogical content knowledge, lead to pedagogical decisions based on the structure of the mathematical domain rather than on the actual ways in which students think. An additional factor that has been found to influence algebra teachers' beliefs is the organization of mathematical material in textbooks (Nathan & Koedinger, 2000b). Cultural factors and related curricular demands also play a role, according to a study by Cai (2004), who found that U.S. and Chinese teachers viewed students' responses involving concrete strategies and visual representations differently. Although both U.S. and Chinese teachers valued responses involving more generalized strategies and symbolic representations equally highly, Chinese teachers expected sixth graders to use the generalized strategies to solve problems whereas U.S. teachers did not.

The existing research on algebra teachers' knowledge and beliefs has, up to now, barely begun to explore the various dimensions of knowledge that teachers need for teaching algebra. If there is one dimension that seems quite pressing, it would be the cognitive dimension of the MGPCA model, that is, the dimension that includes knowing the following: the development of students' algebraic thinking, students' interpretations of algebraic concepts and notation, students' misconceptions and difficulties in algebra, different approaches taken by learners, ways to motivate learners, and theories of learning. These areas of pedagogical content knowledge are considered crucial to the expertise of the practicing teacher of algebra (Nagasaki & Becker, 1993) yet remain largely unresearched. Furthermore, these areas are exactly those where research on the algebra learner has produced abundant findings. Research on the algebra teacher has yet to tap into this resource and make connections between these two bodies of research.

The Practice of Algebra Teaching and Influences on That Practice. Although the above studies on teachers' pedagogical content knowledge have tended to use Shulman's framework, or derivatives of it, the emerging body of research on the practice of algebra teaching reflects a broader set of theoretical perspectives. For example, Boaler (2003) has drawn from Pickering (1995) to analyze the *dance of agency* occurring in reform algebra classes. Within this framework, she described the ways in which a teacher encouraged the interplay between human and disciplinary agency, that is, taught in such a way that students were able to interweave their own thoughts and ideas with standard methods and procedures of the discipline of mathematics. Atweh, Bleicher, and Cooper (1998) have employed a sociosemiotic perspective developed by Halliday and Hasan (1989) to investigate the social context of two 9th-grade mathematics classrooms that differed in the socioeconomic backgrounds and genders of their students. The researchers found that the students in the observed classes were perceived differently by their teachers and that each teacher conducted his classes in a manner that was consistent with his perceptions of the students' needs and abilities. Coulange (2001), a French researcher, has combined the anthropological approach of Chevallard (1999) and the theory of didactical situations of Brousseau (1997) in her study of the ways in which a 9th-grade teacher supported the approaches of students so as to ensure bringing the lesson to closure and achieving her teaching objectives.

Other frameworks that have been adopted in studies of algebra teaching include, for example, Davis's (1996) theory on evaluative, interpretive, and transformative listening, which was used by Coles (2001) to analyze the change that occurred in a given teacher's class as the listening of both students and teachers became transformative. Additional approaches to analyzing teacher change in the classroom have ranged from using (a) perspective taking in middle-school mathematical modeling (English & Doerr, 2003), (b) students' ways of thinking as a motivation for recasting the tasks of teaching about functions (Doerr, 2003), and (c) the occasioning of the growth of teachers' mathematical understandings by focusing on students' mathematics within the classroom collective (Simmt, Davis, Gordon, & Towers, 2003). Another example of a framework for analyzing algebra teaching is presented in a study by Solomon and Nemirovsky (1999). They built upon "Chazan and Ball's (1995) idea that 'any discussion holds the potential for discrepant viewpoints' and that it is the teacher's role to 'manage' these views" (Solomon & Nemirovsky, 1999, p. 217) in their development of a tool for describing the complexity involved in managing the wide range of ideas and choices that emerge from the interactions among students and the teacher.

In some studies on the practice of algebra teaching, the focus has been on the beliefs of the algebra teacher and the role that these beliefs play in teaching. For example, Doerr and Zangor (2000) reported the results of a classroom-based study on the interaction between the role and beliefs of the teacher and the patterns of students' use of graphing calculators in support of their learning of precalculus mathematics. This interaction led to the creation and development of a set of ways the tool was used in the classroom and of related mathematical norms. Doerr and Zangor found that the teacher's confidence, flexibility of use, and awareness of the limitations of the technology led to the establishment of a norm that required results to be justified on mathematical grounds and to the devaluing of regression equations or appeals to the calculator as an authority in a mathematical argument. Simmt (1997), in a study of six 11th- and 12th-grade teachers' use of graphing calculators in the classroom, and the ways in which this use reflected their views of mathematics, found that the graphing calculators were primarily used as a device to provide graphical images from which the students were expected to observe and make generalizations about transformations of the quadratic function. The teachers' philosophies were manifested not so much in their choice of activities but in how they followed up those activities with questions and summary notes.

Another factor influencing teaching practice is the teacher's vision of the role of teacher talk in the algebra classroom. From her analysis of traditional and reform-oriented classroom teaching, Boaler (2003) reported that "the teachers in the traditional classes *gave* students a lot of information, while the teachers of the reform classes chose to *draw* information out of students, by presenting problems and asking students questions" (p. 4). A particularly effective teacher, rather than asking students to present their finished problems, chose to ask students to present their ideas *before* anyone in the class had finished working on the problem, so that students could actively and collaboratively build on each others' ideas. Boaler argued that researchers need to acquire such detailed conceptions of effective teaching if they are to understand and impact practice.

A further example providing detailed descriptions of effective teaching is drawn from the TIMSS-R 1999 video study of eighth-grade algebra lessons around the world (Hiebert et al., 2003). The particular lesson involved a Japanese class (Stigler et al., 2003) where the teacher used a specific problem situation, accompanied by concrete materials to represent it, to help students create meaning for inequalities and for their algebraic form. While the students were working at finding a solution to the problem, the teacher circulated around the class, taking note of the various solution methods being worked at. Afterward, he asked certain students to come to the blackboard to present their solutions in increasing order of mathematical sophistication. The teacher orchestrated the placement of the different student approaches at the board not only so that all the solutions would remain on the board for the class to see and discuss, but also so that he could build upon one method in particular to move ahead with his lesson. The way in which the blackboard was sequentially organized with different representational approaches for the solving of the problem was considered by the teacher to be an important learning tool for the students of his algebra class.

The eclectic nature of the theoretical frameworks of the above studies of algebra teaching reflects the diversity of an emerging field of study where the focus is on what algebra teachers do, rather than on what they know. This diversity can, however, present a challenge in a field where basic knowledge of what constitutes effective algebra teaching is still lacking. The community of researchers of algebra teaching needs to develop a set of appropriate theoretical frameworks for the systematic interpretation of teachers' practice of algebra teaching and their development as learners and teachers of algebra. Although some of the above research has focused on acquiring detailed conceptions of effective teaching practice, the question as to how teachers become effective in their teaching of algebra remains a crucial one. The potential contributing role of professional development and preservice training in this evolutionary process is thus the subject of the next section on research involving algebra teachers in programs of professional or inservice development, as well as in preservice training.

Research Involving the Algebra Teacher Within a Context of Professional Development or Inservice Training

The research that has been conducted within the context of the professional development and inservice training of the algebra teacher has tended to focus on two areas in particular, the integration of new curricula, teaching approaches, and technological environments into the algebra classroom, as well as the nature of the algebra content knowledge and beliefs of teachers involved in these programs.

Integrating New Curricula, Approaches, and Technological Environments Into the Algebra Classroom. The research that has focused on algebra teachers' integration of new curricula into the classroom has highlighted the multiple dimensions involved in studying such phenomena. For instance, Bartolini Bussi and Bazzini (2003) have reported on projects in Italy that include teacher-researchers cooperating with university mathematics educators and with other experts. These projects have typically involved (a) a deep analysis of the nature of the mathematics that is the starting point of the inquiry, (b) the strong involvement of teachers working in their own classrooms, and (c) the observation of both individual and collective learning processes. This study illustrates how, in Italy, teaching and learning have been studied in conjunction one with the other. A related example is the Italian ArAl project (Malara, 2003), an inservice teacher education project that was devoted to the renewal of the teaching of arithmetic and algebra in the sixth to eighth grades. The research explicitly included the classroom teacher as a researcher and emphasized the development of pedagogical content knowledge, the teacher's role in the teaching of mathematics, the impact of his/her personality, as well as social issues within the class group. Chapman (2001) has also pointed to the role played by mathematical belief structures in long-term teacher change involving implementation of new approaches.

Although the practice of situating teachers' professional development within the context of their own classrooms has just been considered above, few studies have also included preservice teachers as additional participants in the learning process. English (2003) has reported on such research, which was based on the four-tiered collaborative model of Lesh and Kelly (2000) and where middle-school teachers and preservice trainees gained powerful insights into students' mathematical development by observing the students as they engaged in modeling problems. Regular meetings between researchers and teachers, which took place both before and after the implementation of each set of modeling activities, were thought to be essential to the success of the collaborative research.

A study (Hallagan, 2004) implementing the reform-based *Connected Mathematics Project* focused on the development of new awarenesses among eighth-grade teachers as a result of their involvement in the project. More specifically, the use of "Ways of Thinking" sheets helped teachers recognize the multiple ways in which their students interpret given tasks, as well as the role of visual images in facilitating students' work on equivalent-expressions tasks. Within another reform-based project, the *Core-Plus Mathematics Project*, involving 40 teachers and their 1466 students in 2 schools, Schoen, Cebulla, Finn, and Fi (2003) analyzed teacher variables related to student achievement. Generally, teaching behaviors that were consistent with the Standards' recommendations and that reflected high mathematical expectations were positively related to growth in student achievement. In another study related to the use of *Core-Plus* curricular materials, Lloyd and Wilson (1998) observed and interviewed an experienced high school teacher during his first implementation of a 6-week unit on functions. The researchers found that the teacher's well-articulated ideas about features of a variety of relationships in different representations supported meaningful discussions with students. The teacher communicated deep and integrated conceptions of functions, dominated by graphical representations and covariation notions.

The increasing implementation of multirepresentational and CAS technology, and curricula involving related new approaches to school algebra, has been reflected in research addressing teacher adaptation and teacher change. Some studies have, for example, examined teachers' attempts at implementing reform-oriented curricula, and their beliefs and attitudes towards different types of curricula (e.g., Heid, Blume, Zbiek, & Edwards, 1999; Stacey, Kendal, & Pierce, 2002). Some teachers have been found to experience tension between familiar methods and the use of technology, or between the curriculum and their personal educational conceptions (e.g., Chazan, 1999; Haimes, 1996; Lloyd, 1999; Slavit, 1996). Others have been found to need support in thinking about how to use symbolic calculators effectively in class and how to reorganize their use of class time when students are involved in CAS work (Guin & Trouche, 2000). In a study of three teachers' implementation of CAS in the classroom, Kendal and Stacey (2001) found that the teachers made very different uses of the CAS even though they had all been involved in planning an agreed common program. The teachers privileged technology use in different ways that were consistent with their beliefs about algebra and about the teaching and learning of algebra. On the other hand, in a 4-year technology-implementation study in Mexico (Rojano, 2003) involving 15 schools of seventh, eighth, and ninth graders, Cedillo and Kieran (2003) observed that, when algebra teachers noticed that the introduction of calculator technology led to improved student learning, as well as a more positive attitude toward mathematics among both stronger and weaker students, they altered their teaching styles to more learner-centered approaches.

The studies in this section, which have focused on integrating new curricula, approaches, and technological environments into the algebra classroom, have presented a diversity of findings that range from descriptions of multilevel support programs that facilitate teacher change to the difficulties experienced by teachers in attempting to implement new curricula and technologies. Several studies have suggested the ways in which teachers' beliefs both shape and are shaped by the implementation process. Although each study has provided its own insights, further research is needed on the long-term impact on algebra teaching practice of the various implementation projects. Few studies have extended beyond the initial implementation to explore the evolution in teachers' practice and its relation to the professional development program that triggered it.

Algebra Content Knowledge and Beliefs. Although most of the professional development studies of implementation described above included some initial indicators regarding teacher change, other research that has involved the professional development or inservice training of algebra teachers has rather focused on algebra teachers' content knowledge. The areas that have been investigated include knowledge of conic sections (Hitt, 1998); the algebraic and geometric aspects of slope, scale, and angle (Zaslavsky et al., 2002); symbolic and graphical representations of rational functions (Arcavi, 2002); and number-theoretic ideas (Barkai, Tsamir, Tirosh, & Dreyfus, 2002; Zazkis, 1998). Zehavi (2004) has reported that the way in which the problems were presented and the prevailing atmosphere of the inservice group influenced the resulting mathematical behaviors and problem-solving products. Another dimension that has been studied concerns teachers' beliefs. Menzel and Clarke (1998) observed that the inservice teachers involved in their study were divided in their views as to whether the generalizable verbal description of a mathematical relationship should be regarded as algebraic—a finding that relates to Nathan and Koedinger's (2000b) results on teachers' beliefs as to what constitutes algebraic activity.

An additional area of inservice teachers' content knowledge that has been researched is that of proof and proving. Knuth (2002) carried out a study with inservice secondary school mathematics teachers (grades 9–12) who were participating in ongoing professional development programs. By means of semistructured interviews with participants in order to examine their conceptions of proof, Knuth found that, although teachers recognized the variety of roles that proof plays in mathematics, the view of proof as a tool for learning mathematics was noticeably absent. With the increasing interest in proof and proving as a potential global/meta-level activity for motivating the work of students with algebraic symbols, the findings of Knuth's study suggest that further teacher development is needed in this area.

The above brief examination of the research involving the algebra teacher within professional development and inservice programs has disclosed that little attention has been paid thus far to the study and development of algebra teachers' pedagogical content knowledge. No studies were found that were similar to the Cognitively Guided Instruction program for elementary school teachers of mathematics (Carpenter, Fennema, Peterson, Chiang, & Loef, 1989). This professional development program with its focus on researched-based evidence regarding children's actual thinking in mathematics has been shown to be effective in changing elementary school teachers' instructional practices. It is suggested that similar programs that focus on the research-based thinking of algebra students be developed. Ironically, for many teachers the need for such knowledge may not be obvious. In a study of 9 high school teachers and their perspectives on what constitutes good mathematics teaching and how it develops, Wilson, Cooney, and Stinson (2005) found that the teachers thought that good teaching requires a sound knowledge of mathematics, promotes mathematical understanding, engages and motivates students, and requires effective management skills. Experience was considered the primary contributor. Nevertheless, many of the studies cited in this chapter and elsewhere (e.g., Krainer, 2005) maintain that knowledge of students' mathematical thinking, in addition to knowledge of mathematics, are necessary preconditions for good teaching.

Research Involving the Preservice Teacher of Algebra

Research involving the future teacher of algebra, although still a relatively undeveloped area of investigation, has been growing steadily. The complexity of the process of educating students to become secondary level mathematics teachers is reflected in programs (e.g., Bednarz, 2001) that do not attempt simply to "deliver" methods of instruction; instead, emphasis is placed on involving future teachers in a process that leads them to construct the knowledge they will need to grapple with, design, and carry out classroom interventions. However, research that informs as to the nature of this complexity is still quite rare, especially with respect to the teaching of algebra. Recent research involving the preservice teacher of algebra has tended to focus on the following themes: the relation between future teachers' subject-matter knowledge and their beliefs, the characteristics of their beliefs and attitudes, and the nature of their subject-matter knowledge.

In a study of the role played by prospective teachers' background knowledge on their beliefs about algebra learning (Nathan & Petrosino, 2003), 48 candidates from an established research and teaching university participated: 35 of them had completed calculus or above and 13 had not. Of the 35 with advanced mathematical knowledge, 16 were in a program for mathematics and science majors and going on to teach mathematics or science at the secondary level; the remaining 19 were teacher education students going on to teach at the elementary level (as were the 13 who had basic mathematical knowledge). The preservice teacher education program in which they were enrolled included the study of subject matter with the use of modern technology, extensive use of the TIMSS video cases, the study of effective classroom interactions, and the development of models of teaching. Participants first ranked six problems in accordance with their expectations of the ease or difficulty that beginning-algebra students would experience when solving them and then responded to a belief survey that broadly addressed issues in mathematical learning, problem solving, and the role of algebra in complex problem solving. The pattern of results, in combination with findings from their previous research, suggested to the researchers that advanced mathematics knowledge, rather than algebra teaching experience, mediates teachers' views of algebra students' development and that highly developed subject-matter knowledge can blind teachers to the actual ways of thinking of algebra students.

Other research related to the subject-matter knowledge of the preservice trainee, as well as the beliefs associated with that knowledge, includes a study by Van Dooren, Verschaffel, and Onghena (2002, 2003). In their investigation of preservice teachers' preferential strategies for solving arithmetic and algebra word problems, they observed that future secondary school mathematics teachers use algebraic methods for solving even very easy problems that could have been handled more appropriately with arithmetic methods. Half of the primary school preservice teachers were found to switch flexibly between arithmetic and algebraic methods, whereas the other half had difficulty with algebraic methods. The authors also noted that the methods the teachers used as individuals were strongly correlated with what they would expect their future students to use and by which they would evaluate student work. Similar results have been reported by Schmidt and Bednarz (1997). In a study conducted by Zazkis and Liljedahl (2002) involving a group of preservice elementary school teachers, participants' attempts to generalize a repeating visual number pattern led to the verbal expression of the generality without necessarily being accompanied by algebraic notation. However, participants perceived that their solutions that did not involve algebraic symbolism, although complete and accurate, were inadequate.

Studies documenting the knowledge of algebra and its teaching that are held by preservice teachers include the research by Even (1998), who presented 152 prospective secondary mathematics teachers with questions related to moving from one functional representation to another. Her results illustrated how knowledge about different representations is interconnected with knowledge about different approaches to functions, knowledge about the context of the presentation, and knowledge of underlying notions. Similarly, Sánchez and Llinares (2003) identified a clear link between preservice teachers' subject-matter knowledge and their pedagogical reasoning for the topic of function and its representations. However, Even and Tirosh (1995) found preservice teachers' subject-matter knowledge to be insufficient for understanding and explaining students' reasoning, as was noted as well in the Nathan and Petrosino (2003) study above.

In a study of the nature of preservice trainees' beliefs and attitudes, Vermeulen (2000) has reported that 1st-year mathematics education students often enter training programs with established concept images in algebra that are not associated with successful mathematics teaching practice. This study, which presented a portrait of the suggested current teaching approaches for eighth-grade algebra in South African schools, contrasted these goals with the orientation of student teachers, many of whom indicated altogether different views on the teaching of algebra and showed their resistance to change. Cooney, Shealy, and Arvold (1998) have noted that preservice programs can seem to have a random effect on the participating teachers being trained, and that such randomness should be expected until researchers begin to better understand the relation between teacher-education activities and the impact of these activities on teachers' belief systems. Similarly, Proulx (2003) found that preservice secondary mathematics teachers were differentially impacted by the teacher training program they were following and that the major factor affecting this impact was their personal frame of reference, which filtered their interpretation of the preservice experience.

Stump (1999), who investigated the notions of slope held by a group of preservice and inservice teachers, found that few thought of slope as a rate of change involving two variables. When questioned about students' conceptual difficulties regarding slope, several teachers answered rather in terms of procedural aspects, such as being able to compute the slope formula. In a related study, Stump (2001) observed preservice teachers teaching a basic algebra course after they had completed a secondary mathematics methods course. She found that the students experienced difficulty in presenting the notion of slope within real-world situations in the classes they were teaching and suggested that methods courses provide experiences that challenge students to construct connections among geometric, algebraic, and real-world representations of slope (see also Zbiek, 1998).

Bowers and Doerr (2001), who used computer-based activities to challenge prospective mathematics teachers' procedural orientations, noted in their findings the intertwined development of subject-matter insights and specific views of teaching. In their research, Chazan, Larriva, and Sandow (1999) asked, "What kind of mathematical knowledge supports teaching for 'conceptual understanding'?" The researchers used tasks from the existing literature on the solving of equations, and questions about teaching students to solve linear equations, to bring to the surface preservice teachers' substantive knowledge in this area. The authors' lack of satisfaction with the nature of the information obtained led them to raise the question of whether finer distinctions might be necessary for describing teachers' substantive knowledge of the mathematics they teach.

The majority of the above studies indicate that research involving the preservice teacher has begun to focus on the complex relation between subject-matter knowledge and pedagogical content knowledge. However, much remains to be done. The finding that teachers with developed subject-matter knowledge may attempt to flesh out underspecified pedagogical content knowledge by drawing on subject-area knowledge suggests first of all that the field needs to develop adequate descriptions of what is considered to be appropriate pedagogical content knowledge for the domain of algebra. Once this is achieved, a great deal more research is needed on the interaction between content knowledge and pedagogical content knowledge for algebra, as well as on how algebra teachers develop both.

Concluding Remarks on the Research Focusing on Algebra Teaching and the Algebra Teacher

Although we now know relatively more about the algebra teacher than we did in the early 1990s, a great deal remains to be researched. There are major areas that have only begun to be studied, for example, the practice of algebra teaching and how teachers learn to be effective teachers of algebra. The above review of research related to studies on algebra teaching and the algebra teacher point to four areas in particular as being in need of focused study.

The first concerns the lack of appropriate models for the observation and analysis of the practice of algebra teaching. The eclectic collection of theoretical frameworks currently being used in research on algebra teaching may work against the systematic building up of a body of coherent knowledge in that area. Second, ways need to be found to incorporate the large body of research findings regarding the learner of algebra into the professional development and preservice and inservice training of algebra teachers. Teachers' insufficient knowledge of students' ways of thinking in algebra is, in part, a reflection of the lack of attention paid to the research base that exists regarding the algebra learner in many programs of teacher training and professional development. Research needs to be conducted on the role played by the gradual integration of this research base into the development of the pedagogical content knowledge of algebra teachers and their practice. Third, the interaction between teachers' content knowledge and pedagogical content knowledge, in conjunction with their understanding of how students' subject-matter-specific knowledge develops, requires major research efforts. Related to this suggestion is the recommendation that research attention focus simultaneously on both the teaching and learning of algebra, and the relation between the two. Fourth, with the steady increase in the use of technology as a tool for algebra learning, considerably more research needs to be done on the importance of the role played by the teacher in maximizing its benefits. Several studies have emphasized that even though technology is a powerful addition to the algebra classroom, it can only become a real learning tool when supported by appropriate teacher intervention. The nature of such intervention remains an understudied area of research.

Two further areas in need of attention have been pointed to by Doerr (2004). One concerns the lack of shared methodologies in the field, which can hinder progress. While the majority of studies use questionnaires, interviews, and observations, and others use concept maps (e.g., Leikin, Chazan, & Yerushalmy, 2001) and varying forms of action research (e.g., Raymond & Leinenbach, 2000), the methods for analyzing the data are almost as varied as the number of researchers. This latter diversity was noted above with respect to theoretical frameworks for the study of algebra teaching. The other point raised by Doerr concerns the fact that

> much of the work of teacher educators, in preparing teachers for practice, is not part of the larger body of research, but rather resides as the uncodified collective wisdom of practitioners and is not subject to the scrutiny of research. (2004, p. 283)

Attention to both of these areas could, according to Doerr, result in the establishment of a professional knowledge base for investigating the teaching of algebra.

To close this section, I return to Tirosh et al. (1998), who remarked:

> The mathematics education literature does not offer enough information and discussion regarding the impact of various approaches and teaching methods . . . it is important for teachers to be acquainted with various teaching methods and to be aware of their pros and cons in different contexts with different teaching aims, and with different students. (pp. 62–63)

We researchers have built a large base with respect to the learner of algebra and have developed an extensive understanding of the nature of algebra learning. We have yet to develop the same understanding of algebra teaching and of the kinds of practice that are effective in bringing about such learning in algebra students.

IN CONCLUSION: WHAT DOES THIS BODY OF RESEARCH TELL US?

This chapter has focused especially on the more recent research in the learning and the teaching of algebra, dating from the early 1990s onward. Nevertheless, explicit links were made with the research that was carried out prior. Students ranging in age from about 11 years and upward, that is to say, from about the sixth grade of middle school up through secondary school and into college, were the subject of the learning research that was examined. Studies involving practicing algebra teachers, as well as those following programs of professional development or inservice or preservice courses, were included in the research that was surveyed regarding the algebra teacher.

The chapter was divided into two parts. The first part looked at the two main competing views of school algebra, at the primary sources of meaning from which the algebra student draws in making sense of the objects and processes of algebra, and at the various models for conceptualizing algebraic activity. The GTG model with its threefold categorization according to the generational, transformational, and global/meta-level activity of algebra served as the framework for presenting the research findings on the learning of algebra within the second part of the chapter. Research on the teaching and teacher of algebra completed the second part.

As the vision of school algebra has widened considerably over the decades—moving from a letter-symbolic and symbol-manipulation view to one that encompasses multiple representations, realistic problem settings, and the use of technological tools—so too has the vision of how algebra is learned. The body of research that was surveyed in this chapter has provided compelling evidence that students derive meaning for algebraic objects from multiple sources. A complex interplay occurs within meaning building among the letter-symbolic and graphical forms, problem situations, and linguistic and gestural activity, as well as past lived experience.

With respect to the manipulative/transformational activity of algebra, the research focus has also expanded. Whereas symbol manipulation has in the past been regarded as a primarily rule-based activity, more recent research has been examining the importance of students' coming to view algebraic transformations as a theory-laden activity. The research involving symbol-manipulation technology has similarly been exploring whether the use of this technology can lead students to embrace a conceptual perspective on symbol manipulation, at the same time that they are developing algebraic techniques. Although this research has begun to show some of the ways in which students think about algebra as theory-based activity, the affordances of such approaches require a great deal more study.

A comparison of the research related to the transformational activity of algebra with that related to the generational activity suggests nevertheless a gap between the two. Within generational activity, students draw upon a wide variety of sources of meaning for algebraic symbols, including multiple representations, word problem situations, and physical and linguistic activity. However, transformational activity narrows the focus primarily to the letter-symbolic. Little is known about how the meaning derived from multiple sources evolves in students and is transformed from embodied and contextual meaning to an understanding of symbols that are subject to formal manipulation. This is an area where further research is especially needed.

Another area where future research might fruitfully be carried out concerns the interaction between functional and nonfunctional approaches to algebra that are included within given programs of instruction. As noted within the chapter, such hybrid programs, which may in fact constitute the majority of algebra programs at the present time, can be a source of confusion to students with respect to the articulation between generational and transformational activity. Teaching experiments designed to assist students in distinguishing function approaches from equation approaches, and in meaningfully integrating them, are strongly recommended.

Research that has investigated the transformational activity of algebra in technological environments such as CAS has signaled the importance of the quality of both the tasks and the classroom discussions during which the teacher attempts to draw out the conceptual aspects underpinning the transformational activity. This research has emphasized that the insertion of such technology into the algebra classroom does not remove the need for paper-and-pencil algebraic techniques; in fact, just the opposite seems true. Further studies that examine the articulation between machine techniques and paper-and-pencil algebraic techniques, as well as the nature of the tasks that enhance the emergence of both conceptual and technical expertise within symbol-manipulation-technology environments, are however needed. CAS research has also noted that the use of this technology as a didactical tool can occasion mathematical discussions that do not normally occur in algebra classrooms. However, for these discussions to be fruitful, the role of the teacher was found to be of crucial importance. A great deal more research is needed concerning the impact of classroom discussion on algebraic growth: what kinds of discussions are most beneficial, how best to encourage students to present their ideas, with what particular kinds of tasks, with which students, at what specific moments, and so on.

The role of word problems in algebra instruction remains somewhat problematic. Although some research reported the advantages of experience with realistic word problems and modeling situations in algebraic activity, others reported the opposite. Furthermore, research on the representation of word-problem situations by algebraic equations continued to show the difficulties that students have with such activity. However, it is not the case that students cannot solve these problems by a variety of arithmetic methods; in fact, they are more likely to solve correctly word problems than either similar-equation or word-equation formats of questions. Research suggests however that word-problem solving, as it is currently engaged in, may not be the most productive route to the generation of equations, particularly with the younger student.

Although generating symbolic representations of problem situations was shown to be fraught with obstacles, the use of graphical representations for both solving problems and interpreting the letter-symbolic was found in many studies to be an effective approach. This is an area where graphing technology has played a significant role. Research has shown that students sometimes require several years to effectively generate equations to model problem situations, but that when that happens, the letter-symbolic becomes a more preferred mode of representation than the graphical. Research that can help to clarify the factors involved in the successful transition to a symbolic mode of representation for problem situations would be most useful.

Research with the older student, which has tended to focus more on the within-mathematical sources of meaning than on the sources of meaning outside of mathematics, indicates that a majority of students continue to experience difficulty in "seeing structure." Just as the younger adolescent needed a great deal of time in coming to see how to model situations with equations, so too has the older student experienced difficulty with what has proved to be another time-demanding phase of symbolic learning, that is, seeing form and structure in those symbols. Although the *Principles and Standards for School Mathematics* (NCTM, 2000) recommends that students in Grades 6 to 8 begin to "represent and analyze mathematical situations and structures using algebraic symbols" (p. 222), the existing evidence suggests that few sixth graders in the United States are actually called upon to engage in symbolic work. Yet, in other countries (e.g., China, Russia, Singapore, and South Korea), students begin a formal study of algebra that involves not only the development of algebraic reasoning and generalization, but also the use of algebraic symbols and the solving of equations as early as the fourth grade (Cai et al., 2005). If, as the findings discussed earlier suggest, students need a great deal of time in becoming comfortable with algebraic symbols and in acquiring the fluency and power that symbols can provide, then serious consideration ought to be given to the notion of having students begin the process at an earlier age.

With respect to the use of technology in algebra learning, some have argued that tools such as computer algebra systems can be made to do all the symbolic work, thus removing the need for students to engage in such learning themselves. However, research has shown that the conceptual does not evolve in the algebra learner if the technical aspects are neglected. The use of technology does not lighten the burden of algebraic work if students have not developed machine techniques to obtain what they want from the machine and conceptual tools for interpreting the machine output. Furthermore, these conceptual tools for interpreting machine output seem to emerge from a complex interplay of paper-and-pencil and machine techniques.

The last section of this chapter involved the algebra teacher and the teaching of algebra. This is a field of study that has been growing during the past 20 years. However, researchers still know relatively little about what makes for effective algebra teaching and how algebra teachers learn to develop their craft. Yet, the current body of research has yielded some notable findings that in turn suggest areas in need of further study. Research has shown that teachers' knowledge of students' actual algebraic thinking may be quite limited. Students can reason beyond what many teachers think they are capable of. In addition, some studies have revealed that the weaknesses that exist in teachers' pedagogical content knowledge may be compensated for by more highly developed subject-matter knowledge that inappropriately maps the structure of the algebraic domain onto students' learning trajectories. This suggests the need for research into the ways in which teachers' knowledge of students' thinking can be further developed, by for example tapping into the research base that exists with respect to the learning of algebra. As emphasized by Stacey and MacGregor (2001),

> Awareness of how students think about . . . relationships and of what they are likely to perceive in mathematical situations will help teachers to decide what aspects of presentation need special emphasis and what follow-up will be necessary, leading to better teaching and better outcomes. (p. 152)

A related component of algebra teachers' pedagogical content knowledge that has been found to be in need of development, especially among beginning algebra teachers, concerns the knowledge of students' misconceptions and the teaching approaches that could address these misconceptions. Researchers in this area have argued that the existing research literature has yet to evolve regarding descriptions not only of teaching approaches that work, but also of those that have been found not to work. It has been proposed that research needs to go beyond documenting what students think and how they interpret various objects and processes to include more information on the advantages and disadvantages of different teaching approaches, the contexts in which they were used, and the nature of the students with whom the approaches were tried. This suggests a closer connection between the study of the learning of algebra and the study of the teaching of algebra.

REFERENCES

Ainley, J. (1996). Purposeful contexts for formal notation in a spreadsheet environment. *Journal of Mathematical Behavior, 15*, 405–422.

Ainley, J., Wilson, K., & Bills, L. (2003). Generalising the context and generalising the calculation. In N. A. Pateman, B. J. Dougherty, & J. Zilliox (Eds.), *Proceedings of the 27th Conference of the International Group for the Psychology of Mathematics Education* (Vol. 2, pp. 9–16). Honolulu, HI.

Alcock, L., & Weber, K. (2005). Referential and syntactic approaches to proof: Case studies from a transition course. In H. L. Chick & J. L. Vincent (Eds.), *Proceedings of the 29th Conference of the International Group for the Psychology of Mathematics Education* (Vol. 2, pp. 33–40). Melbourne, Australia.

Arcavi, A. (1994). Symbol sense: Informal sense-making in formal mathematics. *For the Learning of Mathematics, 14*(3), 24–35.

Arcavi, A. (2002). The everyday and the academic in mathematics. In M. E. Brenner & J. N. Moschkovich (Eds.), *Everyday and academic mathematics in the classroom* (Monograph series by the *Journal for Research in Mathematics Education*, No. 11, pp. 12–29). Reston, VA: National Council of Teachers of Mathematics.

Arcavi, A. (2003). The role of visual representations in the learning of mathematics. *Educational Studies in Mathematics, 52*, 215–241.

Artigue, M. (2002a). L'intégration de calculatrices symboliques à l'enseignement secondaire : les leçons de quelques ingénieries didactiques [The integration of symbolic calculators into secondary education: Some lessons from didactical engineering]. In D. Guin & L. Trouche (Eds.), *Calculatrices symboliques—transformer un outil en un instrument du travail mathématique : un problème didactique* (pp. 277–349). Grenoble: La Pensée sauvage.

Artigue, M. (2002b). Learning mathematics in a CAS environment: The genesis of a reflection about instrumentation and the dialectics between technical and conceptual work. *International Journal of Computers for Mathematical Learning, 7*, 245–274.

Artigue, M., Assude, T., Grugeon, B., & Lenfant, A. (2001). Teaching and learning algebra: Approaching complexity through complementary perspectives. In H. Chick, K. Stacey, J. Vincent, & J. Vincent (Eds.), *Proceedings of the 12th ICMI Study Conference: The Future of the Teaching and Learning of Algebra* (pp. 21–32). Melbourne, Australia: The University of Melbourne.

Artigue. M., Defouad, B., Duperier, M., Juge, G., & Lagrange, J. B. (1998). *Intégration de calculatrices complexes dans l'enseignement des mathématiques au lycée* [The integration of complex calculators into the teaching of high school mathematics]. Paris: Université Denis Diderot, Équipe DIDIREM.

Arzarello, F. (1992). Pre-algebraic problem solving. In J. P. da Ponte, J. F. Matos, J. M. Matos, & D. Fernandes (Eds.), *Mathematical problem solving and new information technologies* (NATO ASI Series F, vol. 89, pp. 155–166). Berlin, Germany: Springer-Verlag.

Arzarello, F., Bazzini, L., & Chiappini, G. (1994). The process of naming in algebraic thinking. In J. P. da Ponte & J. F. Matos (Eds.), *Proceedings of the 18th Conference of the International Group for the Psychology of Mathematics Education* (Vol. 2, pp. 40–47). Lisbon, Portugal.

Arzarello, F., Bazzini, L., & Chiappini, G. (2001). A model for analysing algebraic processes of thinking. In R. Sutherland, T. Rojano, A. Bell, & R. Lins (Eds.), *Perspectives on school algebra* (pp. 61–82). Dordrecht, The Netherlands: Kluwer.

Arzarello, F., & Robutti, O. (2001). From body motion to algebra through graphing. In H. Chick, K. Stacey, J. Vincent, & J. Vincent (Eds.), *Proceedings of the 12th ICMI Study Conference: The Future of the Teaching and Learning of Algebra* (pp. 33–40). Melbourne, Australia: The University of Melbourne.

Asp, G., Dowsey, J., & Stacey, K. (1993). Linear and quadratic graphs with the aid of technology. In B. Atweh, C. Kanes, M. Carss, & G. Booker (Eds.), *Contexts in mathematics education* (proceedings of MERGA 16, pp. 51–56). Brisbane, Australia: MERGA Program Committee.

Atweh, B., Bleicher, R. E., & Cooper, T. J. (1998). The construction of the social context of mathematics classrooms: A sociolinguistic analysis. *Journal for Research in Mathematics Education, 29,* 63–82.

Ayers, P. (2000). An analysis of bracket expansion errors. In T. Nakahara & M. Koyama (Eds.), *Proceedings of the 24th Conference of the International Group for the Psychology of Mathematics Education* (Vol. 2, pp. 25–32). Hiroshima, Japan.

Ayers, T., Davis, G., Dubinsky, E., & Lewin, P. (1988). Computer experiences in learning composition of functions. *Journal for Research in Mathematics Education, 19,* 246–259.

Baker, B., Hemenway, C., & Trigueros, M. (2001). On transformations of functions. In R. Speiser, C. A. Maher, & C. N. Walter (Eds.), *Proceedings of the 23rd Annual Meeting of the North American Chapter of the International Group for the Psychology of Mathematics* (Vol. 1, pp. 91–98). Snowbird, Utah.

Balacheff, N. (2001). Symbolic arithmetic vs. algebra: The core of a didactical dilemma. In R. Sutherland, T. Rojano, A. Bell, & R. Lins (Eds.), *Perspectives on school algebra* (pp. 249–260). Dordrecht, The Netherlands: Kluwer.

Balacheff, N., & Kaput, J. J. (1996). Computer-based learning environments in mathematics. In A. J. Bishop, K. Clements, C. Keitel, J. Kilpatrick, & C. Laborde (Eds.), *International handbook of mathematics education* (pp. 469–501). Dordrecht, The Netherlands: Kluwer.

Ball, D. L., & Bass, H. (2002). Toward a practice-based theory of mathematical knowledge for teaching. In E. Simmt & B. Davis (Eds.), *Proceedings of the Annual Meeting of the Canadian Mathematics Education Study Group* (pp. 3–14). Kingston, Canada: CMESG Program Committee.

Ball, L., Pierce, R., & Stacey, K. (2003). Recognising equivalent algebraic expressions: An important component of algebraic expectation for working with CAS. In N. A. Pateman, B. J. Dougherty, & J. Zilliox (Eds.), *Proceedings of the 27th Conference of the International Group for the Psychology of Mathematics Education* (Vol. 4, pp. 15–22). Honolulu, HI.

Barkai, R., Tsamir, P., Tirosh, D., & Dreyfus, T. (2002). Proving or refuting arithmetic claims: The case of elementary school teachers. In A. D. Cockburn & E. Nardi (Eds.), *Proceedings of the 26th Conference of the International Group for the Psychology of Mathematics Education* (Vol. 2, pp. 57–64). Norwich, UK.

Bartolini Bussi, M. G. (1995). Analysis of classroom interaction discourse from a Vygotskian perspective. In L. Meira & D. Carraher (Eds.), *Proceedings of the 19th Conference of the International Group for the Psychology of Mathematics Education* (Vol. 1, pp. 95–101). Recife, Brazil.

Bartolini Bussi, M. G., & Bazzini, L. (2003). Research, practice and theory in didactics of mathematics: Towards dialogue between different fields. *Educational Studies in Mathematics, 54,* 203–223.

Bazzini, L., Boero, P., & Garuti, R. (2001a). Moving symbols around or developing understanding: The case of algebraic expressions. In M. van den Heuvel-Panhuizen (Ed.), *Proceedings of the 25th Conference of the International Group for the Psychology of Mathematics Education* (Vol. 2, pp. 121–128). Utrecht, The Netherlands.

Bazzini, L., Boero, P., & Garuti, R. (2001b). Revealing and promoting the students' potential in algebra: A case study concerning inequalities. In H. Chick, K. Stacey, J. Vincent, & J. Vincent (Eds.), *Proceedings of the 12th ICMI Study Conference: The Future of the Teaching and Learning of Algebra* (Vol. 1, pp. 53–60). Melbourne, Australia: The University of Melbourne.

Bednarz, N. (2001). Didactique des mathématiques et formation des enseignants: le cas de l'Université du Québec à Montréal [Mathematics education and teacher training: The case of the University of Quebec at Montreal]. *Canadian Journal of Science, Mathematics and Technology Education, 1,* 61–80.

Bednarz, N., & Janvier, B. (1996). Emergence and development of algebra as a problem-solving tool: Continuities and discontinuities with arithmetic. In N. Bednarz, C. Kieran, & L. Lee (Eds.), *Approaches to algebra: Perspectives for research and teaching* (pp. 115–136). Dordrecht, The Netherlands: Kluwer.

Bednarz, N., Kieran, C., & Lee, L. (Eds.). (1996). *Approaches to algebra: Perspectives for research and teaching.* Dordrecht, The Netherlands: Kluwer.

Behr, M., Erlwanger, S., & Nichols, E. (1976). *How children view equality sentences* (PMDC Tech. Rep. No. 3). Tallahassee: Florida State University. (ERIC Document Reproduction Service No. ED144802).

Bell, A. (1976). A study of pupils' proof-explanations in mathematical situations. *Educational Studies in Mathematics, 7,* 23–40.

Bell, A. (1995). Purpose in school algebra. In C. Kieran (Ed.), New perspectives on school algebra: Papers and discussions of the ICME-7 Algebra Working Group [Special issue]. *Journal of Mathematical Behavior, 14,* 41–73.

Bell, A. (1996). Problem-solving approaches to algebra: Two aspects. In N. Bednarz, C. Kieran, & L. Lee (Eds.), *Approaches to algebra: Perspectives for research and teaching* (pp. 167–185). Dordrecht, The Netherlands: Kluwer.

Bell, A., Malone, J. A., & Taylor, P. C. (1987). *Algebra—an exploratory teaching experiment.* Nottingham, UK: Shell Centre for Mathematical Education.

Bell, A., O'Brien, D., & Shiu, C. (1980). Designing teaching in the light of research on understanding. In R. Karplus (Ed.), *Proceedings of the Fourth Conference of the International Group for the Psychology of Mathematics Education* (pp. 119–125). Berkeley, CA.

Bergqvist, T. (2005). How students verify conjectures: Teachers' expectations. *Journal of Mathematics Teacher Education, 8,* 171–191.

Bills, L. (2001). Shifts in the meanings of literal symbols. In M. van den Heuvel-Panhuizen (Ed.), *Proceedings of the 25th Conference of the International Group for the Psychology of Mathematics Education* (Vol. 2, pp. 161–168). Utrecht, The Netherlands.

Bloedy-Vinner, H. (1994). The analgebraic mode of thinking: The case of parameter. In J. P. da Ponte & J. F. Matos (Eds.), *Proceedings of the 18th Conference of the International Group for the Psychology of Mathematics Education* (Vol. 2, pp. 88–95). Lisbon, Portugal.

Bloedy-Vinner, H. (2001). Beyond unknowns and variables—parameters and dummy variables in high school algebra. In R. Sutherland, T. Rojano, A. Bell, & R. Lins (Eds.), *Perspectives on school algebra* (pp. 177–189). Dordrecht, The Netherlands: Kluwer.

Boaler, J. (2003). Studying and capturing the complexity of practice: The case of the dance of agency. In N. A. Pateman, B. J. Dougherty, & J. T. Zilliox (Eds.), *Proceedings of the 27th Conference of the International Group for the Psychology of Mathematics Education* (Vol. 1, pp. 3–16). Honolulu, HI.

Booth, L. R. (1984). *Algebra: Children's strategies and errors.* Windsor, UK: NFER-Nelson.

Booth, L. R. (1989). A question of structure. In S. Wagner & C. Kieran (Eds.), *Research issues in the learning and teaching of algebra* (Vol. 4 of *Research agenda for mathematics education,* pp. 57–59). Reston, VA: National Council of Teachers of Mathematics.

Borba, M. (1993). *Students' understanding of transformations of functions using multi-representational software.* Unpublished doctoral dissertation, Cornell University, Ithaca, NY.

Boulton-Lewis, G., Cooper, T., Atweh, B., Pillay, H., Wilss, L., & Mutch, S. (1997). Processing load and the use of concrete representations and strategies for solving linear equations. *Journal of Mathematical Behavior, 16,* 379–397.

Bowers, J., & Doerr, H. M. (2001). An analysis of prospective teachers' dual roles in understanding the mathematics of change: Eliciting growth with technology. *Journal of Mathematics Teacher Education, 4,* 115–137.

Brenner, M.E., Mayer, R.E., Moseley, B., Brar, T., Durán, R., Smith Reed, B., et al. (1997). Learning by understanding: The role of multiple representations in learning algebra. *American Educational Research Journal, 34,* 663–689.

Breslich, E.R. (1939). Algebra, a system of abstract processes. In C. H. Judd (Ed.), *Education as cultivation of the higher mental processes.* New York: Macmillan.

Brousseau, G. (1997). *Theory of didactical situations in mathematics* (N. Balacheff, M. Cooper, R. Sutherland, & V. Warfield, Eds. & Trans.). Dordrecht, The Netherlands: Kluwer Academic.

Brown, T., Eade, F., & Wilson, D. (1999). Semantic innovation: Arithmetical and algebraic metaphors within narratives of learning. *Educational Studies in Mathematics, 40,* 53–70.

Burkhardt, H. (1981). *The real world and mathematics.* London: Blackie & Son.

Cai, J. (2004). Why do U.S. and Chinese students think differently in mathematical problem solving? Impact of early algebra learning and teachers' beliefs. *Journal of Mathematical Behavior, 23,* 135–167.

Cai, J., Lew, H. C., Morris, A., Moyer, J. C., Ng, S. F., & Schmittau, J. (2005). The development of students' algebraic thinking in earlier grades: A cross-cultural comparative perspective. *Zentralblatt fur Didaktik der Mathematik, 37,* 5–15.

Carnegie Learning, Inc. (1998). Cognitive Tutor® Algebra 1© [Computer software]. Pittsburgh, PA: Carnegie Learning.

Carpenter, T. P., Fennema, E., Peterson, P. L., Chiang, C. P., & Loef, M. (1989). Using knowledge of children's mathematical thinking in classroom teaching: An experimental study. *American Educational Research Journal, 26,* 499–531.

Carpenter, T. P., Franke, M. L., & Levi, L. (2003). *Thinking mathematically: Integrating arithmetic and algebra in elementary school.* Portsmouth, NH: Heinemann.

Carry, L. R., Lewis, C., & Bernard, J. (1980). *Psychology of equation solving: An information processing study* (Final Tech. Rep.). Austin: University of Texas at Austin, Department of Curriculum and Instruction.

Cavanagh, M., & Mitchelmore, M. (2000). Student misconceptions in interpreting basic calculator displays. In T. Nakahara & M. Koyama (Eds.), *Proceedings of the 24th Conference of the International Group for the Psychology of Mathematics Education* (Vol. 2, pp. 161–168). Hiroshima, Japan.

Cedillo, T., & Kieran, C. (2003). Initiating students into algebra with symbol-manipulating calculators. In J. T. Fey (Ed.), *Computer algebra systems in secondary school mathematics education* (pp. 219–239). Reston, VA: National Council of Teachers of Mathematics.

Cerulli, M. (2004). *Introducing pupils to algebra as a theory: L'Algebrista as an instrument of semiotic mediation.* Doctoral dissertation, Università degli Studi di Pisa, Italy. Retrieved on March 30, 2006, from http://www-studenti.dm.unipi.it/~cerulli/tesi/.

Cerulli, M., & Mariotti, M. A. (2001). L'Algebrista: A microworld for symbolic manipulation. In H. Chick, K. Stacey, J. Vincent, & J. Vincent (Eds.), *Proceedings of the 12th ICMI Study Conference: The Future of the Teaching and Learning of Algebra* (pp. 179–186). Melbourne, Australia: The University of Melbourne.

Chacon, P. R., & Soto-Johnson, H. (1998). The effect of CAI in college algebra incorporating both drill and exploration. *International Journal of Computer Algebra in Mathematics Education, 5*(4), 201–216.

Chaiklin, S. (1989). Cognitive studies of algebra problem solving and learning. In S. Wagner & C. Kieran (Eds.), *Research issues in the learning and teaching of algebra* (Vol. 4 of *Research agenda for mathematics education,* pp. 93–114). Reston, VA: National Council of Teachers of Mathematics.

Chapman, O. (2001). Understanding high school mathematics teacher growth. In M. van den Heuvel-Panhuizen (Ed.), *Proceedings of the 25th Conference of the International Group for the Psychology of Mathematics Education* (Vol. 2, pp. 233–240). Utrecht, The Netherlands.

Chazan, D. (1996). "Algebra for all students?" *Journal of Mathematical Behavior, 15,* 455–477.

Chazan, D. (1999). On teachers' mathematical knowledge and student exploration: A personal story about teaching a technologically supported approach to school algebra. *International Journal of Computers for Mathematical Learning, 4*, 121–149.

Chazan, D. (2000). *Beyond formulas in mathematics and teaching: Dynamics of the high school algebra classroom.* New York: Teachers College Press.

Chazan, D., & Ball, D. L. (1995). *Beyond exhortations not to tell: What is the teacher's role in discussion-intensive mathematics classes?* (Craft Paper 95-2). East Lansing: Michigan State University, National Center for Research on Teacher Learning.

Chazan, D., Larriva, C., & Sandow, D. (1999). What kind of mathematical knowledge supports teaching for "conceptual understanding"? In O. Zaslavsky (Ed.), *Proceedings of the 23rd Conference of the International Group for the Psychology of Mathematics Education* (Vol. 2, pp. 193–200). Haifa, Israel.

Chazan, D., & Yerushalmy, M. (2003). On appreciating the cognitive complexity of school algebra: Research on algebra learning and directions of curricular change. In J. Kilpatrick, W. G. Martin, & D. Schifter (Eds.), *A research companion to Principles and Standards for School Mathematics* (pp. 123–135). Reston, VA: National Council of Teachers of Mathematics.

Chevallard, Y. (1999). L'analyse des pratiques enseignantes en théorie anthropologique du didactique [The analysis of teaching practice in the anthropological theory of didactics]. *Recherches en Didactique des Mathématiques, 19*, 221–266.

Chick, H., Stacey, K., Vincent, J., & Vincent, J. (Eds.). (2001). *Proceedings of the 12th ICMI Study Conference: The Future of the Teaching and Learning of Algebra.* Melbourne, Australia: The University of Melbourne.

Cobb, P. (1994). Where is mind? Constructivist and sociocultural perspectives on mathematical development. *Educational Researcher, 23*(7), 13–20.

Coles, A. (2001). Listening: A case study of teacher change. In M. van den Heuvel-Panhuizen (Ed.), *Proceedings of the 25th Conference of the International Group for the Psychology of Mathematics Education* (Vol. 2, pp. 281–288). Utrecht, The Netherlands.

Confrey, J. (1991). Function Probe© [Computer software]. Ithaca, NY: Cornell University.

Confrey, J., & Smith, E. (1994). Exponential functions, rates of change, and the multiplicative unit. *Educational Studies in Mathematics, 26*, 135–164.

Confrey, J., & Smith, E. (1995). Splitting, covariation, and their role in the development of exponential functions. *Journal for Research in Mathematics Education, 26*, 66–86.

Cooney, T. J. (1985). A beginning teacher's view of problem solving. *Journal for Research in Mathematics Education, 16*, 324–336.

Cooney, T. J., Shealy, B. E., & Arvold, B. (1998). Conceptualizing belief structures of preservice secondary mathematics teachers. *Journal for Research in Mathematics Education, 29*, 306–333.

Cooper, T. J., Boulton-Lewis, G. M., Atweh, B., Pillay, H., Wilss, L., & Mutch, S. (1997). The transition from arithmetic to algebra: Initial understanding of equals, operations, and variables. In E. Pehkonen (Ed.), *Proceedings of the 21st Conference of the International Group for the Psychology of Mathematics Education* (Vol. 2, pp. 89–96). Lahti, Finland.

Cortés, A. (1998). Implicit cognitive work in putting word problems into equation form. In A. Olivier & K. Newstead (Eds.), *Proceedings of the 22nd Conference of the International Group for the Psychology of Mathematics Education* (Vol. 2, pp. 208–216). Stellenbosch, South Africa.

Coulange, L. (2001). Enseigner les systèmes d'équations en troisième : une étude économique et écologique [Teaching systems of equations in 9th grade: An economic and ecological study]. *Recherches en Didactique des Mathématiques, 21*, 305–353.

Cuoco, A., Goldenberg, E. P., & Mark, J. (1996). Habits of mind: An organizing principle for mathematics curricula. *Journal of Mathematical Behavior, 15*, 375–402.

Davis, B. (1996). *Teaching mathematics: Toward a sound alternative.* New York: Garland.

Davis, R. B. (1975). Cognitive processes involved in solving simple algebraic equations, *Journal of Children's Mathematical Behavior, 1*(3), 7–35.

Davis, R. B., Jockusch, E., & McKnight, C. (1978). Cognitive processes in learning school algebra. *Journal of Children's Mathematical Behavior, 2*(1), 10–320.

DeBock, D., Verschaffel, L., & Janssens, D. (2002). The effects of different problem presentations and formulations on the illusion of linearity in secondary school students. *Mathematical Thinking and Learning, 4*, 65–89.

DeBock, D., Verschaffel, L., Janssens, D., & Claes, K. (2000). Involving pupils in an authentic context: Does it help them to overcome the "illusion of linearity"? In T. Nakahara & M. Koyama (Eds.), *Proceedings of the 24th Conference of the International Group for the Psychology of Mathematics Education* (Vol. 2, pp. 233–240). Hiroshima, Japan.

de Lange, J. (1987). *Mathematics insight and meaning.* Utrecht, The Netherlands: Rijksuniversiteit, OW&OC.

Demby, A. (1997). Algebraic procedures used by 13- to 15-year-olds. *Educational Studies in Mathematics, 33*, 45–70.

Dettori, G., Garuti, R., & Lemut, E. (2001). From arithmetic to algebraic thinking by using a spreadsheet. In R. Sutherland, T. Rojano, A. Bell, & R. Lins (Eds.), *Perspectives on school algebra* (pp. 191–207). Dordrecht, The Netherlands: Kluwer.

Doerr, H. M. (2003). Using students' ways of thinking to re-cast the tasks of teaching about functions. In N. A. Pateman, B. J. Dougherty, & J. Zilliox (Eds.), *Proceedings of the 27th Conference of the International Group for the Psychology of Mathematics Education* (Vol. 2, pp. 333–340). Honolulu, HI.

Doerr, H. M. (2004). Teachers' knowledge and the teaching of algebra. In K. Stacey, H. Chick, & M. Kendal (Eds.), *The future of the teaching and learning of algebra: The 12th ICMI Study* (pp. 267–290). Dordrecht, The Netherlands: Kluwer.

Doerr, H. M., & English, L. D. (2003). A modeling perspective on students' mathematical reasoning about data. *Journal for Research in Mathematics Education, 34*, 110–136.

Doerr, H. M., & Tripp, J. S. (1999). Understanding how students develop mathematical models. *Mathematical Thinking and Learning, 1*, 231–254.

Doerr, H. M., & Zangor, R. (2000). Creating meaning for and with the graphing calculator. *Educational Studies in Mathematics, 41*, 143–163.

Dolciani, M. P., Berman, S. L., & Freilich, J. (1965). *Modern Algebra.* Boston: Houghton Mifflin.

Douek, N. (1999). Argumentative aspects of proving: Analysis of some undergraduate mathematics students' performances.

In O. Zaslavsky (Ed.), *Proceedings of the 23rd Conference of the International Group for the Psychology of Mathematic Education* (Vol. 2, pp. 273–280). Haifa, Israel.

Dreyfus, T. (1991). On the status of visual reasoning in mathematics and mathematics education. In F. Furinghetti (Ed.), *Proceedings of the 15th Conference of the International Group for the Psychology of Mathematics Education* (Vol. 1, pp. 33–48). Assisi, Italy.

Dreyfus, T., Hershkowitz, R., & Schwarz, B. (2001). The construction of abstract knowledge in interaction. In M. van den Heuvel-Panhuizen (Ed.), *Proceedings of the 25th Conference of the International Group for the Psychology of Mathematics Education* (Vol. 2, pp. 377–384). Utrecht, The Netherlands.

Drijvers, P. (2003). Algebra on screen, on paper, and in the mind. In J. T. Fey (Ed.), *Computer algebra systems in secondary school mathematics education* (pp. 241–267). Reston, VA: National Council of Teachers of Mathematics.

Drijvers, P., & Doorman, M. (1996). The graphics calculator in mathematics education. *Journal of Mathematical Behavior, 15,* 425–440.

Dubinsky, E., & Harel, G. (Eds.). (1992a). *The concept of function: Aspects of epistemology and pedagogy* (MAA Notes, Vol. 25). Washington, DC: Mathematical Association of America.

Dubinsky, E., & Harel, G. (1992b). The nature of the process conception of function. In E. Dubinsky & G. Harel (Eds.), *The concept of function: Aspects of epistemology and pedagogy* (MAA Notes, Vol. 25, pp. 85–106). Washington, DC: Mathematical Association of America.

Edwards, E. L. (Ed.). (1990). *Algebra for everyone.* Reston, VA: National Council of Teachers of Mathematics.

Edwards, L. D. (1998). Odds and evens: Mathematical reasoning and informal proof among high school students. *Journal of Mathematical Behavior, 17,* 489–504.

Eisenberg, T., & Dreyfus, T. (1991). On the reluctance to visualize in mathematics. In W. Zimmermann & S. Cunningham (Eds.), *Visualization in teaching and learning mathematics* (pp. 26–37). Washington, DC: Mathematical Association of America.

Ellington, A. J. (2003). A meta-analysis of the effects of calculators on students' achievement and attitude levels in precollege mathematics classes. *Journal for Research in Mathematics Education, 34,* 433–463.

English, L. D. (2003). Reconciling theory, research, and practice: A models and modelling perspective. *Educational Studies in Mathematics, 54,* 225–248.

English, L. D., & Doerr, H. M. (2003). Perspective-taking in middle-school mathematical modelling: A teacher case study. In N. A. Pateman, B. J. Dougherty, & J. Zilliox (Eds.), *Proceedings of the 27th Conference of the International Group for the Psychology of Mathematics Education* (Vol. 2, pp. 357–364). Honolulu, HI.

Even, R. (1990). Subject matter knowledge for teaching and the case of functions. *Educational Studies in Mathematics, 21,* 521–544.

Even, R. (1993). Subject-matter knowledge and pedagogical content knowledge: Prospective secondary teachers and the function concept. *Journal for Research in Mathematics Education, 24,* 94–116.

Even, R. (1998). Factors involved in linking representations of functions. *Journal of Mathematical Behavior, 17,* 105–121.

Even, R., & Tirosh, D. (1995). Subject-matter knowledge and knowledge about students as sources of teacher presentations of the subject-matter. *Educational Studies in Mathematics, 29,* 1–20.

Even, R., Tirosh, D., & Robinson, N. (1993). Connectedness in teaching equivalent algebraic expressions: Novice versus expert teachers. *Mathematics Education Research Journal, 5*(1), 50–59.

Farrell, A. (1996). Roles and behaviors in technology-integrated precalculus classrooms. *Journal of Mathematical Behavior, 15,* 35–53.

Fey, J. T. (Ed.). (1984). *Computing and mathematics: The impact on secondary school curricula.* Reston, VA: National Council of Teachers of Mathematics.

Fey, J. T. (1989). School algebra for the year 2000. In S. Wagner & C. Kieran (Eds.), *Research issues in the learning and teaching of algebra* (Vol. 4 of *Research agenda for mathematics education,* pp. 199–213). Reston, VA: National Council of Teachers of Mathematics.

Fey, J. T., & Good, R. A. (1985). Rethinking the sequence and priorities of high school mathematics curricula. In C. R. Hirsch & M. J. Zweng (Eds.), *The secondary school mathematics curriculum* (Yearbook of the National Council of Teachers of Mathematics, pp. 43–52). Reston, VA: National Council of Teachers of Mathematics.

Fey, J. T., Heid, M. K., Good, R., Sheets, C., Blume, G., & Zbiek, R. M. (1991). *Computer-intensive algebra.* College Park: University of Maryland; University Park: Pennsylvania State University.

Filloy, E., & Rojano, T. (1989). Solving equations: The transition from arithmetic to algebra. *For the Learning of Mathematics, 9*(2), 19–25.

Filloy, E., Rojano, T., & Rubio, G. (2001). Propositions concerning the resolution of arithmetical-algebraic problems. In R. Sutherland, T. Rojano, A. Bell, & R. Lins (Eds.), *Perspectives on school algebra* (pp. 155–176). Dordrecht, The Netherlands: Kluwer.

Filloy, E., Rojano, T., & Solares, A. (2003). Two meanings of the "equal" sign and senses of comparison and substitution methods. In N. A. Pateman, B. J. Dougherty, & J. T. Zilliox (Eds.), *Proceedings of the 27th Conference of the International Group for the Psychology of Mathematics Education* (Vol. 4, pp. 223–229). Honolulu, HI.

Filloy, E., Rojano, T., & Solares, A. (2004). Arithmetic/algebraic problem solving and the representation of two unknown quantities. In M. J. Hoines & A. B. Fuglestad (Eds.), *Proceedings of the 28th Conference of the International Group for the Psychology of Mathematics Education* (Vol. 2, pp. 391–398). Bergen, Norway.

Filloy, E., & Sutherland, R. (1996). Designing curricula for teaching and learning algebra. In A. J. Bishop, K. Clements, C. Keitel, J. Kilpatrick, & C. Laborde (Eds.), *International handbook of mathematics education* (pp. 139–160). Dordrecht, The Netherlands: Kluwer.

Fischbein, E., & Kedem, I. (1982). Proof and certitude in the development of mathematical thinking. In A. Vermandel (Ed.), *Proceedings of the Sixth Conference of the International Group for the Psychology of Mathematics Education* (pp. 128–131). Antwerp, Belgium.

Forster, P., Flynn, P., Frid, S., & Sparrow, L. (2004). Teaching, learning and assessment with four-function and graphics calculators and computer algebra systems. In B. Perry, G. Anthony, & C. Diezmann (Eds.), *Research in mathematics education in Australasia 2000–2003* (pp. 313–336). Flaxton, Australia: Post Pressed.

Friedlander, A., & Stein, H. (2001). Students' choice of tools in solving equations in a technological learning environment. In M. van den Heuvel-Panhuizen (Ed.), *Proceedings of the 25th Conference of the International Group for the Psychology of Mathematics Education* (Vol. 2, pp. 441–448). Utrecht, The Netherlands.

Friedlander, A., & Tabach, M. (2001). Promoting multiple representations in algebra. In A. A. Cuoco & F. R. Curcio (Eds.), *The roles of representation in school mathematics* (Yearbook of the National Council of Teachers of Mathematics, pp. 173–185). Reston, VA: National Council of Teachers of Mathematics.

Fujii, T. (2003). Probing students' understanding of variables through cognitive conflict problems: Is the concept of variable so difficult for students to understand? In N. A. Pateman, B. J. Dougherty, & J. T. Zilliox (Eds.), *Proceedings of the 27th Conference of the International Group for the Psychology of Mathematics Education* (Vol. 1, pp. 49–65). Honolulu, HI.

Furinghetti, F., & Paola, D. (1994). Parameters, unknowns and variables: A little difference? In J. P. da Ponte & J. F. Matos (Eds.), *Proceedings of the 18th Conference of the International Group for the Psychology of Mathematics Education* (Vol. 2, pp. 368–375). Lisbon, Portugal.

Galbraith, P.L., & Clatworthy, N.J. (1990). Beyond standard models—Meeting the challenge of modeling. *Educational Studies in Mathematics, 21*, 137–163.

Gallardo, A. (2002). The extension of the natural-number domain to the integers in the transition from arithmetic to algebra. *Educational Studies in Mathematics, 49*, 171–192.

Gascón, J. (1994–1995). Un nouveau modèle de l'algèbre élémentaire comme alternative à l'« arithmétique généralisée » [A new model of elementary algebra as an alternative to "generalized arithmetic"]. *Petit x, 37*, 43–63.

Gilead, S., & Yerushalmy, M. (2001). Deep structures of algebra word problems: Is it approach (in)dependent? In M. van den Heuvel-Panhuizen (Ed.), *Proceedings of the 25th Conference of the International Group for the Psychology of Mathematics Education* (Vol. 3, pp. 41–48). Utrecht, The Netherlands.

Goldenberg, E. P. (1988). Mathematics, metaphors, and human factors: Mathematical, technical and pedagogical challenges in the educational use of graphical representations of functions. *Journal of Mathematical Behavior, 7*, 135–173.

Goldenberg, E. P. (2003). Algebra and computer algebra. In J. T. Fey (Ed.), *Computer algebra systems in secondary school mathematics education* (pp. 9–30). Reston, VA: National Council of Teachers of Mathematics.

Goldenberg, P., Lewis, P., & O'Keefe, J. (1992). Dynamic representation and the development of a process understanding of function. In E. Dubinsky & G. Harel (Eds.), *The concept of function: Aspects of epistemology and pedagogy* (MAA Notes, Vol. 25, pp. 235–260). Washington, DC: Mathematical Association of America.

Gómez, P., & Carulla, C. (2001). Students' conceptions of cubic functions. In M. van den Heuvel-Panhuizen (Ed.), *Proceedings of the 25th Conference of the International Group for the Psychology of Mathematics Education* (Vol. 3, pp. 57–64). Utrecht, The Netherlands.

Gonzales, P., Pahlke, E., Guzman, J. C., Partelow, L., Kastberg, D., Jocelyn, L., et al. (2004). *Pursuing excellence: Eighth-grade mathematics and science achievement in the United States and other countries from the Trends in International Mathematics and Science Study (TIMSS) 2003* (NCES 2005-007). Washington, DC: National Center for Education Statistics.

Goodson-Espy, T. (1998). The roles of reification and reflective abstraction in the development of abstract thought: Transitions from arithmetic to algebra. *Educational Studies in Mathematics, 36*, 219–245.

Goos, M., Galbraith, P., Renshaw, P., & Geiger, V. (2003). Perspectives on technology mediated learning in secondary school mathematics classrooms. *Journal of Mathematical Behavior, 22*, 73–89.

Graham, A. T., & Thomas, M. O. J. (2000). Building a versatile understanding of algebraic variables with a graphic calculator. *Educational Studies in Mathematics, 41*, 265–282.

Greeno, J. G. (1982, March). *A cognitive learning analysis of algebra.* Paper presented at the annual meeting of the American Educational Research Association, Boston, MA.

Grouws, D. A. (Ed.). (1992). *Handbook of research on mathematics teaching and learning.* New York: Macmillan.

Guin, D., & Trouche, L. (1999). The complex process of converting tools into mathematical instruments: The case of calculators. *International Journal of Computers for Mathematical Learning, 3*, 195–227.

Guin, D., & Trouche, L. (2000). Thinking of new devices to make viable symbolic calculators in the classroom. In T. Nakahara & M. Koyama (Eds.), *Proceedings of the 24th Conference of the International Group for the Psychology of Mathematics Education* (Vol. 3, pp. 9–16). Hiroshima, Japan.

Hadjidemetriou, C., & Williams, J. (2002). Teachers' pedagogical content knowledge: Graphs from a cognitivist to a situated perspective. In A. D. Cockburn & E. Nardi (Eds.), *Proceedings of the 26th Conference of the International Group for the Psychology of Mathematics Education* (Vol. 3, pp. 57–64). Norwich, UK.

Haimes, D. H. (1996). The implementation of a "function" approach to introductory algebra: A case study of teachers' cognitions, teacher actions, and the intended curriculum. *Journal for Research in Mathematics Education, 27*, 582–602.

Hall, M. K. (1993). Impact of the graphing calculator on the instruction of trigonometric functions in precalculus classes. (Doctoral dissertation, Baylor University, 1992). *Dissertation Abstracts International, 54*(02), 451.

Hallagan, J. E. (2004). A teacher's model of his students' algebraic thinking: "Ways of thinking" sheets. In D. E. McDougall & J. A. Ross (Eds.), *Proceedings of the 26th Annual Meeting of the North American Chapter of the International Group of the Psychology of Mathematics Education* (Vol. 1, pp. 237–244). Toronto, ON.

Halliday, M. A. K., & Hasan, R. (1989). *Language, context, and text: Aspects of language in a social-semiotic perspective.* Geelong, Australia: Deakin University.

Healy, L., & Hoyles, C. (1999). Visual and symbolic reasoning in mathematics: Making connections with computers? *Mathematical Thinking and Learning, 1*, 59–84.

Healy, L., & Hoyles, C. (2000). A study of proof conceptions in algebra. *Journal for Research in Mathematics Education, 31*, 396–428.

Heid, M. K. (1988). Resequencing skills and concepts in applied calculus using the computer as tool. *Journal for Research in Mathematics Education, 19*, 3–25.

Heid, M. K. (1996). A technology-intensive functional approach to the emergence of algebraic thinking. In N. Bednarz, C. Kieran, & L. Lee (Eds.), *Approaches to algebra: Perspectives for research and teaching* (pp. 239–255). Dordrecht, The Netherlands: Kluwer.

Heid, M. K., Blume, G. W., Zbiek, R. M., & Edwards, B. S. (1999). Factors that influence teachers learning to do interviews to understand students' mathematical understandings. *Educational Studies in Mathematics, 37*, 223–249.

Herscovics, N., & Linchevski, L. (1994). A cognitive gap between arithmetic and algebra. *Educational Studies in Mathematics, 27*, 59–78.

Hershkowitz, R., Dreyfus, T., Ben-Zvi, D., Friedlander, A., Hadas, N., Resnick, T., et al. (2002). Mathematics curriculum development for computerized environments: A designer-researcher-teacher-learner activity. In L. D. English (Ed.), *Handbook of international research in mathematics education* (pp. 657–694). Mahwah, NJ: Erlbaum.

Hershkowitz, R., & Kieran, C. (2001). Algorithmic and meaningful ways of joining together representatives within the same mathematical activity: An experience with graphing calculators. In M. van den Heuvel-Panhuizen (Ed.), *Proceedings of the 25th Conference of the International Group for the Psychology of Mathematics Education* (Vol. 1, pp. 96–107). Utrecht, The Netherlands.

Hewitt, D. (2003). Notation issues: Visual effects and ordering operations. In N. A. Pateman, B. J. Dougherty, & J. Zilliox (Eds.), *Proceedings of the 27th Conference of the International Group for the Psychology of Mathematics Education* (Vol. 3, pp. 63–69). Honolulu, HI.

Hiebert, J., Gallimore, R., Garnier, H., Giwin, K. B., Hollingsworth, H., Jacobs, J., et al. (2003). *Teaching mathematics in seven countries: Results from the TIMSS 1999 Video Study* (NCES 2003-013). Washington, DC: National Center for Education Statisitics.

Hines, E. (2001). Developing the concept of linear function: One student's experiences with dynamic physical models. *Journal of Mathematical Behavior, 20*, 337–361.

Hitt, F. (1998). Difficulties in the articulation of different representations linked to the concept of function. *Journal of Mathematical Behavior, 17*, 123–134.

Hoch, M., & Dreyfus, T. (2004). Structure sense in high school algebra: The effects of brackets. In M. J. Hoines & A. B. Fuglestad (Eds.), *Proceedings of the 28th Conference of the International Group for the Psychology of Mathematics Education* (Vol. 3, pp. 49–56). Bergen, Norway.

Hollar, J. C., & Norwood, K. (1999). The effects of a graphing-approach intermediate algebra curriculum on students' understanding of function. *Journal for Research in Mathematics Education, 30*, 220–226.

Hong, Y. Y., & Thomas, M. O. J. (2002). Building Newton-Raphson concepts with CAS. In A. D. Cockburn & E. Nardi (Eds.), *Proceedings of the 26th Conference of the International Group for the Psychology of Mathematics Education* (Vol. 3, pp. 103–112). Norwich, UK.

Hotz, H.G. (1918). First-year algebra scales. *Contributions to education* (No. 90). New York: Columbia University, Teachers College.

Hoyles, C. (2002). From describing to designing mathematical activity: The next step in developing a social approach to research in mathematics education? In C. Kieran, E. Forman, & A. Sfard (Eds.). (2002). *Learning discourse: Discursive approaches to research in mathematics education* (pp. 273–286). Dordrecht, The Netherlands: Kluwer.

Huntley, M. A., Rasmussen, C. L., Villarubi, R. S., Sangtong, J., & Fey, J. T. (2000). Effects of Standards-based mathematics education: A study of the Core-Plus mathematics project algebra and functions strand. *Journal for Research in Mathematics Education, 31*, 328–361.

Izsák, A. (2003). "We want a statement that is always true": Criteria for good algebraic representations and the development of modeling knowledge. *Journal for Research in Mathematics Education, 34*, 191–227.

Janvier, C. (1996). Modeling and the initiation into algebra. In N. Bednarz, C. Kieran, & L. Lee (Eds.), *Approaches to algebra: Perspectives for research and teaching* (pp. 225–236). Dordrecht, The Netherlands: Kluwer.

Janvier, C. (1998). The notion of chronicle as an epistemological obstacle to the concept of function. *Journal of Mathematical Behavior, 17*, 79–103.

Johanning, D. I. (2004). Supporting the development of algebraic thinking in middle school: A closer look at students' informal strategies. *Journal of Mathematical Behavior, 23*, 371–388.

Kaput, J. J. (1989). Linking representations in the symbol systems of algebra. In S. Wagner & C. Kieran (Eds.), *Research issues in the learning and teaching of algebra* (Vol. 4 of *Research agenda for mathematics education*, pp. 167–194). Reston, VA: National Council of Teachers of Mathematics.

Kaput, J. J. (1995). A research base supporting long term algebra reform? In D. T. Owens, M. K. Reed, & G. M. Millsaps (Eds.), *Proceedings of the Seventh Annual Meeting of the North American Chapter of the International Group for the Psychology of Mathematics Education* (Vol. 1, pp. 71–94). Columbus, OH: ERIC Clearinghouse for Science, Mathematics, and Environmental Education. (ERIC Document Reproduction Service No. ED 389 539)

Kaput, J. J., & Roschelle, J. (1997). Developing the impact of technology beyond assistance with traditional formalisms in order to democratize access to ideas underlying calculus. In E. Pehkonen (Ed.), *Proceedings of the 21st Conference of the International Group for the Psychology of Mathematics Education* (Vol. 1, pp. 105–112). Lahti, Finland.

Kendal, M., & Stacey, K. (2001). The impact of teacher privileging on learning differentiation with technology. *International Journal of Computers for Mathematical Learning, 6*, 143–165.

Kendal, M., & Stacey, K. (2004). Algebra: A world of difference. In K. Stacey, H. Chick, & M. Kendal (Eds.), *The future of the teaching and learning of algebra: The 12th ICMI Study* (pp. 329–346). Dordrecht, The Netherlands: Kluwer.

Kieran, C. (1979). Children's operational thinking within the context of bracketing and the order of operations. In D. Tall (Ed.), *Proceedings of the Third Conference of the International Group for the Psychology of Mathematics Education* (pp. 128–133). Coventry, England.

Kieran, C. (1981). Concepts associated with the equality symbol. *Educational Studies in Mathematics, 12*, 317–326.

Kieran, C. (1984a). A comparison between novice and more-expert algebra students on tasks dealing with the equivalence of equations. In J. M. Moser (Ed.), *Proceedings of the Sixth Annual Meeting of the North American Chapter of the International Group for the Psychology of Mathematics Education* (pp. 83–91). Madison, WI.

Kieran, C. (1984b). Cognitive mechanisms underlying the equation-solving errors of algebra novices. In B. Southwell et al. (Eds.), *Proceedings of the Eighth International Conference for the Psychology of Mathematics Education* (pp. 70–77). Sydney, Australia.

Kieran, C. (1988). Two different approaches among algebra learners. In A. F. Coxford (Ed.), *The ideas of algebra, K–12* (Yearbook of the National Council of Teachers of Mathematics, pp. 91–96). Reston, VA: National Council of Teachers of Mathematics.

Kieran, C. (1989). The early learning of algebra: A structural perspective. In S. Wagner & C. Kieran (Eds.), *Research issues in the learning and teaching of algebra* (Vol. 4 of *Research agenda for mathematics education*, pp. 33–56). Reston, VA: National Council of Teachers of Mathematics.

Kieran, C. (1990). Cognitive processes involved in learning school algebra. In P. Nesher & J. Kilpatrick (Eds.), *Mathematics and cognition* (ICMI Study series, pp. 96–112). Cambridge, UK: Cambridge University Press.

Kieran, C. (1992). The learning and teaching of school algebra. In D. A. Grouws (Ed.), *Handbook of research on mathematics teaching and learning* (pp. 390–419). New York: Macmillan.

Kieran, C. (1996). The changing face of school algebra. In C. Alsina, J. Alvarez, B. Hodgson, C. Laborde, & A. Pérez (Eds.), *Eighth International Congress on Mathematical Education: Selected lectures* (pp. 271–290). Seville, Spain: S.A.E.M. Thales.

Kieran, C. (2001). The mathematical discourse of 13-year-old partnered problem solving and its relation to the mathematics that emerges. *Educational Studies in Mathematics, 46*, 187–228.

Kieran, C. (2004). The equation/inequality connection in constructing meaning for inequality situations. In M. J. Hoines & A. B. Fuglestad (Eds.), *Proceedings of the 28th Conference of the International Group for the Psychology of Mathematics Education* (Vol. 1, pp. 143–147). Bergen, Norway.

Kieran, C., & Drijvers, P. (2006). The co-emergence of machine techniques, paper-and-pencil techniques, and theoretical reflection: A study of CAS use in secondary school algebra. *International Journal of Computers for Mathematical Learning, 11*, 205–263.

Kieran, C., & Saldanha, L. (2005). Computer algebra systems (CAS) as a tool for coaxing the emergence of reasoning about equivalence of algebraic expressions. In H. L. Chick & J. L. Vincent (Eds.), *Proceedings of the 29th Conference of the International Group for the Psychology of Mathematics Education* (Vol. 3, pp. 193–200). Melbourne, Australia.

Kieran, C., & Saldanha, L. (2007). Designing tasks for the co-development of conceptual and technical knowledge in CAS activity. In G. Blume & M. K. Heid (Eds.), *Research on technology and the teaching and learning of mathematics: Cases and perspectives* (Vol. 2, pp. 393–414). Greenwich, CT: Information Age.

Kieran, C., & Sfard, A. (1999). Seeing through symbols: The case of equivalent expressions. *Focus on Learning Problems in Mathematics, 21*(1), 1–17.

Kieran, C., & Yerushalmy, M. (2004). Research on the role of technological environments in algebra learning and teaching. In K. Stacey, H. Chick, & M. Kendal (Eds.), *The future of the teaching and learning of algebra: The 12th ICMI Study* (pp. 99–152). Dordrecht, The Netherlands: Kluwer.

Kilpatrick, J., Martin, W. G., & Schifter, D. (Eds.). (2003). *A research companion to Principles and Standards for school mathematics.* Reston, VA: National Council of Teachers of Mathematics.

Kirshner, D. (1989). The visual syntax of algebra. *Journal for Research in Mathematics Education, 20*, 274–287.

Kirshner, D. (2001). The structural algebra option revisited. In R. Sutherland, T. Rojano, A. Bell, & R. Lins (Eds.), *Perspectives on school algebra* (pp. 83–98). Dordrecht, The Netherlands: Kluwer.

Kirshner, D., & Awtry, T. (2004). Visual salience of algebraic transformations. *Journal for Research in Mathematics Education, 35*, 224–257.

Knuth, E. J. (2000). Student understanding of the Cartesian connection: An exploratory study. *Journal for Research in Mathematics Education, 31*, 500–508.

Knuth, E. J. (2002). Secondary school mathematics teachers' conceptions of proof. *Journal for Research in Mathematics Education, 33*, 379–405.

Koedinger, K. R., Anderson, J. R., Hadley, W. H., & Mark, M. A. (1997). Intelligent tutoring goes to school in the big city. *International Journal of Artificial Intelligence in Education, 8*, 30–43.

Koedinger, K. R., & Nathan, M. J. (2004). The real story behind story problems: Effects of representations on quantitative reasoning. *Journal of the Learning Sciences, 13*, 129–164.

Krainer, K. (2005). What is "good" mathematics teaching, and how can research inform practice and policy? *Journal of Mathematics Teacher Education, 8*, 75–81.

Kramarski, B. (2000). The effects of different instructional methods on the ability to communicate mathematical reasoning. In T. Nakahara & M. Koyama (Eds.), *Proceedings of the 24th Conference of the International Group for the Psychology of Mathematics Education* (Vol. 3, pp. 167–174). Hiroshima, Japan.

Kubinová, M., & Novotná, J. (2002). Promoting students' meaningful learning in two different classroom environments. In A. D. Cockburn & E. Nardi (Eds.), *Proceedings of the 26th Conference of the International Group for the Psychology of Mathematics Education* (Vol. 3, pp. 233–240). Norwich, UK.

Küchemann, D. (1981). Algebra. In K. Hart (Ed.), *Children's understanding of mathematics: 11–16* (pp. 102–119). London: John Murray.

Kutscher, B., & Linchevski, L. (1997). Number instantiations as mediators in solving word problems. In E. Pehkonen (Ed.), *Proceedings of the 21st Conference of the International Group for the Psychology of Mathematics Education* (Vol. 3, pp. 168–175). Lahti, Finland.

Lagrange, J.-B. (2000). L'intégration d'instruments informatiques dans l'enseignement : une approche par les techniques [The integration of computer tools into teaching: An approach according to techniques]. *Educational Studies in Mathematics, 43*, 1–30.

Lagrange, J.-B. (2002). Étudier les mathématiques avec les calculatrices symboliques. Quelle place pour les techniques? [Studying mathematics with symbolic calculators: What place for techniques?] In D. Guin & L. Trouche (Eds.), *Calculatrices symboliques. Transformer un outil en un instrument du travail mathématique: un problème didactique* (pp. 151–185). Grenoble, France: La Pensée Sauvage.

Lagrange, J.-B. (2003). Learning techniques and concepts using CAS: A practical and theoretical reflection. In J. T. Fey (Ed.), *Computer algebra systems in secondary school mathematics education* (pp. 269–283). Reston, VA: National Council of Teachers of Mathematics.

Lagrange, J.-B., Artigue, M., Laborde, C., & Trouche, L. (2001). A meta study on IC technologies in education: Towards a multidimensional framework to tackle their integration.

In M. van den Heuvel-Panhuizen (Ed.), *Proceedings of the 25th Conference of the International Group for the Psychology of Mathematics Education* (Vol. 1, pp. 111–122). Utrecht, The Netherlands.

Landa, J. A. (2003). *Mediación de la hoja electrónica de cálculo en la composición de funciones* [Mediation of electronic spreadsheets in composition of functions]. Unpublished doctoral dissertation, Centro de Investigación y de Estudios Avanzados del Instituto Politécnico Nacional, Mexico.

Lee, L. (1987). The status and understanding of generalised algebraic statements by high school students. In J.C. Bergeron, N. Herscovics, & C. Kieran (Eds.), *Proceedings of the 11th Conference of the International Group for the Psychology of Mathematics Education* (Vol. 1, pp. 316–323). Montreal, Quebec.

Lee, L. (1996). An initiation into algebraic culture through generalization activities. In N. Bednarz, C. Kieran, & L. Lee (Eds.), *Approaches to algebra: Perspectives for research and teaching* (pp. 87–106). Dordrecht, The Netherlands: Kluwer.

Lee, L. (1997). *Algebraic understanding: The search for a model in the mathematics education community*. Unpublished doctoral dissertation, Université du Québec à Montréal.

Lee, L., & Wheeler, D. (1987). *Algebraic thinking in high school students: Their conceptions of generalisation and justification* (Research report). Montreal, Quebec: Concordia University, Mathematics Department.

Leikin, R., Chazan, D. I., & Yerushalmy, M. (2001). Understanding teachers' changing approaches to school algebra: Contributions of concept maps as part of clinical interviews. In M. van den Heuvel-Panhuizen (Ed.), *Proceedings of the 25th Conference of the International Group for the Psychology of Mathematics Education* (Vol. 3, pp. 289–296). Utrecht, The Netherlands.

Leinhardt, G., Zaslavsky, O., & Stein, M. (1990). Functions, graphs, and graphing: Tasks, learning, and teaching. *Review of Educational Research, 60*, 1–64.

Lemoyne, G., Conne, F., & Brun, J. (1993). Du traitement des formes à celui des contenus d'écritures littérales : une perspective d'enseignement introductif de l'algèbre [From the treatment of form to that of content in literal symbols: A perspective on the teaching of introductory algebra]. *Recherche en Didactique des Mathématiques, 13*, 333–383.

Lerman, S., Xu, G., & Tsatsaroni, A. (2002). Developing theories of mathematics education research: The ESM story. *Educational Studies in Mathematics, 51*(1–2), 23–40.

Lesh, R., & Doerr, H.M. (2003). *Beyond constructivism: A models and modeling perspective on mathematics problem solving, learning and teaching.* Mahwah, NJ: Erlbaum.

Lesh, R., Hoover, M., Hole, B., Kelly, A., & Post, T. (2000). Principles for developing thought-revealing activities for students and teachers. In A. E. Kelly & R. A. Lesh (Eds.), *Handbook of research design in mathematics and science education* (pp. 591–645). Mahwah, NJ: Erlbaum.

Lesh, R., & Kelly, A. (2000). Multitiered teaching experiments. In A. E. Kelly & R. A. Lesh (Eds.), *Handbook of research design in mathematics and science education* (pp. 197–230). Mahwah, NJ: Erlbaum.

Lester, F. K. Jr., & Ferrini-Mundy, J. (Eds.). (2004). *Proceedings of the NCTM Research Catalyst Conference (Sept. 2003).* Reston, VA: National Council of Teachers of Mathematics.

Linchevski, L., & Herscovics, N. (1996). Crossing the cognitive gap between arithmetic and algebra: Operating on the unknown in the context of equations. *Educational Studies in Mathematics, 30*, 39–65.

Linchevski, L., & Livneh, D. (1999). Structure sense: The relationship between algebraic and numerical contexts. *Educational Studies in Mathematics, 40*, 173–196.

Linchevski, L., & Vinner, S. (1990). Embedded figures and structures of algebraic expressions. In G. Booker, P. Cobb, & T. N. diMendicuti (Eds.), *Proceedings of the 14th Conference of the International Group for the Psychology of Mathematics Education* (Vol. 2, pp. 85–92). Oaxtepec, Mexico.

Linchevski, L., & Williams, J. (1996). Situated intuitions, concrete manipulations and the construction of mathematical concepts: The case of integers. In L. Puig & A. Gutiérrez (Eds.), *Proceedings of the 20th Conference of the International Group for the Psychology of Mathematics Education* (Vol. 3, pp. 265–272). Valencia, Spain.

Lins, R. C. (2001). The production of meaning for algebra: A perspective based on a theoretical model of semantic fields. In R. Sutherland, T. Rojano, A. Bell, & R. Lins (Eds.), *Perspectives on school algebra* (pp. 37–60). Dordrecht, The Netherlands: Kluwer.

Lloyd, G. M. (1999). Two teachers' conceptions of a reform-oriented curriculum: Implications for mathematics teacher development. *Journal of Mathematics Teacher Education, 2*, 227–252.

Lloyd, G. W., & Wilson, M. (1998). Supporting innovation: The impact of a teacher's conception of functions on his implementation of a reform curriculum. *Journal for Research in Mathematics Education, 29*, 248–274.

Lobato, J., Ellis, A. B., & Muñoz, R. (2003). How "focusing phenomena" in the instructional environment support individual student's generalizations. *Mathematical Thinking and Learning, 5*, 1–36.

MacGregor, M., & Price, E. (1999). An exploration of aspects of language proficiency and algebraic learning. *Journal for Research in Mathematics Education, 30*, 449–467.

MacGregor, M., & Stacey, K. (1993). Seeing a pattern and writing a rule. In I. Hirabayashi, N. Nohda, K. Shigematsu, & F.-L. Lin (Eds.), *Proceedings of the 17th Conference of the International Group for the Psychology of Mathematics Education* (Vol. 1, pp. 181–188). Tsukuba, Japan.

MacGregor, M., & Stacey, K. (1997). Students' understanding of algebraic notation: 11–15. *Educational Studies in Mathematics, 33*, 1–19.

Malara, N. (1999). An aspect of a long-term research on algebra: The solution of verbal problems. In O. Zaslavsky (Ed.), *Proceedings of the 23rd Conference of the International Group for the Psychology of Mathematics Education* (Vol. 3, pp. 257–264). Haifa, Israel.

Malara, N. (2003). The dialectics between theory and practice: Theoretical issues and practice aspects from an early algebra project. In N. A. Pateman, B. J. Dougherty, & J. T. Zilliox (Eds.), *Proceedings of the 27th Conference of the International Group for the Psychology of Mathematics Education* (Vol. 1, pp. 31–48). Honolulu, HI.

Manouchehri, A. (2004). Using interactive algebra software to support a discourse community. *Journal of Mathematical Behavior, 23*, 37–62.

Markovits, Z., Eylon, B., & Bruckheimer, M. (1986). Functions today and yesterday. *For the Learning of Mathematics, 6*(2), 18–24, 28.

Mason, J. (1996). Expressing generality and roots of algebra. In N. Bednarz, C. Kieran, & L. Lee (Eds.), *Approaches to algebra: Perspectives for research and teaching* (pp. 65–86). Dordrecht, The Netherlands: Kluwer.

Mason, J., Graham, A., & Johnston-Wilder, S. (2005). *Developing thinking in algebra.* London: Sage.

Mason, J., Graham, A., Pimm, D., & Gowar, N. (1985). *Routes to / roots of algebra.* Milton Keynes, UK: Open University Press.

Mason, J., & Pimm, D. (1984). Generic examples: Seeing the general in the particular. *Educational Studies in Mathematics, 15,* 277–289.

Mason, J., & Sutherland, R. (2002). *Key aspects of teaching algebra in schools.* Report prepared for the Qualifications and Curriculum Authority. London: QCA.

Matz, M. (1982). Towards a process model for high school algebra errors. In D. Sleeman & J. S. Brown (Eds.), *Intelligent tutoring systems* (pp. 25–50). New York: Academic Press.

Meira, L. (1998). Making sense of instructional devices: The emergence of transparency in mathematical activity. *Journal for Research in Mathematics Education, 29,* 121–142.

Menghini, M. (1994). Form in algebra: Reflecting, with Peacock, on upper secondary school teaching. *For the Learning of Mathematics, 14*(3), 9–14.

Menzel, B., & Clarke, D. (1998). Teachers interpreting algebra: Teachers' views about the nature of algebra. In C. Kanes, M. Goos, & E. Warren (Eds.), *Teaching mathematics in new times* (proceedings of MERGA 21, pp. 365–372). Gold Coast, Australia: MERGA.

Miyakawa, T. (2002). Relation between proof and conception: The case of proof for the sum of two even numbers. In A. D. Cockburn & E. Nardi (Eds.), *Proceedings of the 26th Conference of the International Group for the Psychology of Mathematics Education* (Vol. 3, pp. 353–360). Norwich, UK.

Molyneux-Hodgson, S., Rojano, T., Sutherland, R., & Ursini, S. (1999). Mathematical modeling: The interaction of culture and practice. *Educational Studies in Mathematics, 39,* 167–183.

Monroe, W. S. (1915). A test of the attainment of first-year high-school students in algebra. *School Review, 23,* 159–171.

Moschkovich, J. N. (1998). Resources for refining mathematical conceptions: Case studies in learning about linear functions. *Journal of the Learning Sciences, 7,* 209–237.

Moschkovich, J., Schoenfeld, A. H. & Arcavi, A. (1993). Aspects of understanding: On multiple perspectives and representations of linear relations and connections among them. In T. A. Romberg, E. Fennema, & T. C. Carpenter (Eds.), *Integrating research on the graphical representation of function* (pp. 69–100). Hillsdale, NJ: Erlbaum.

Moss, J. (2005, May). *Integrating numeric and geometric patterns: A developmental approach to young students' learning of patterns and functions.* Paper presented at annual meeting of Canadian Mathematics Education Study Group, Ottawa.

Mounier, G., & Aldon, G. (1996). A problem story: Factorizations of x^n–1. *International DERIVE Journal, 3*(3), 51–61.

Mourão, A. P. (2002). Quadratic function and imagery: Alice's case. In A. D. Cockburn & E. Nardi (Eds.), *Proceedings of the 26th Conference of the International Group for the Psychology of Mathematics Education* (Vol. 3, pp. 377–384). Norwich, UK.

Mullis, I., Martin, M., Gonzales, E., Gregory, K., Garden, R., O'Connor, K., et al. (2000). *TIMSS 1999 International Mathematics Report: Findings from IEA's Repeat of the Third International Mathematics and Science Study at the eighth grade.* Chestnut Hill, MA: Boston College.

Nagasaki, E., & Becker, J. P. (1993). Classroom assessment in Japanese mathematics education. In N. L. Webb & A. F. Coxford (Eds.), *Assessment in the mathematics classroom* (Yearbook of the National Council of Teachers of Mathematics, pp. 40–53). Reston, VA: National Council of Teachers of Mathematics.

Nathan, M. J., & Koedinger, K. R. (2000a). Teachers' and researchers' beliefs about the development of algebraic reasoning. *Journal for Research in Mathematics Education, 31,* 168–190.

Nathan, M. J., & Koedinger, K. R. (2000b). An investigation of teachers' beliefs of students' algebra development. *Cognition and Instruction, 18,* 209–237.

Nathan, M. J., & Petrosino, A. (2003). Expert blind spot among preservice teachers. *American Educational Research Journal, 40,* 905–928.

Nathan, M. J., Stephens, A. C., Masarik, K., Alibali, M. W., & Koedinger, K. R. (2002). Representational fluency in middle school: A classroom study. In D. S. Mewborn et al. (Eds.), *Proceedings of the 24th Annual Meeting of the North American Chapter of the International Group for the Psychology of Mathematics Education* (Vol. 1, pp. 464–472). Athens, GA.

National Council of Teachers of Mathematics. (1989). *Curriculum and evaluation standards for school mathematics.* Reston, VA: Author.

National Council of Teachers of Mathematics. (1991). *Professional standards for teaching mathematics.* Reston, VA: Author.

National Council of Teachers of Mathematics. (1995). *Assessment standards for school mathematics.* Reston, VA: Author.

National Council of Teachers of Mathematics. (2000). *Principles and standards for school mathematics.* Reston, VA: Author.

National Research Council. (2001). *Adding it up: Helping children learn mathematics.* J. Kilpatrick, J. Swafford, & B. Findell (Eds.). Mathematics Learning Study Committee, Center for Education, Division of Behavioral and Social Sciences and Education. Washington, DC: National Academy Press.

Nemirovsky, R. (1996). Mathematical narratives, modeling, and algebra. In N. Bednarz, C. Kieran, & L. Lee (Eds.), *Approaches to algebra: Perspectives for research and teaching* (pp. 197–220). Dordrecht, The Netherlands: Kluwer.

Nguyen-Xuan, A., Nicaud, J.-F., Bastide, A., & Sander, E. (2002). Les expérimentations du projet Aplusix [Experiments of the Aplusix project]. *Sciences et Techniques Éducatives, 9,* 63–90.

Nicaud, J.-F., Delozanne, E., & Grugeon, B. (Eds.). (2002). Logiciels pour l'apprentissage de l'algèbre [Software for the learning of algebra]. Special issue of *Sciences et Techniques Éducatives, 9*(1–2).

Niss, M. (Ed.). (1988). Theme group 3: Problem solving, modelling, and applications. In A. Hirst & K. Hirst (Eds.), *Proceedings of the Sixth International Congress on Mathematical Education* (pp. 237–252). Budapest, Hungary: János Bolyai Mathematical Society.

Noble, T., Nemirovsky, R., Wright, T., & Tierney, C. (2001). Experiencing change: The mathematics of change in multiple environments. *Journal for Research in Mathematics Education, 32,* 85–108.

Noss, R., & Hoyles, C. (1996). *Windows on mathematical meanings.* Dordrecht, The Netherlands: Kluwer.

O'Callaghan, B. R. (1998). Computer-intensive algebra and students' conceptual knowledge of function. *Journal for Research in Mathematics Education, 29,* 21–40.

Pawley, D. (1999). To check or not to check? Does teaching a checking method reduce the incidence of the multiplicative reversal error? In O. Zaslavsky (Ed.), *Proceedings of the 23rd Conference of the International Group for the Psychology of Mathematics Education* (Vol. 4, pp. 17–24). Haifa, Israel.

Penglase, M., & Arnold, S. (1996). The graphics calculator in mathematics education: A critical review of recent research. *Mathematics Education Research Journal, 8*, 58–90.

Perrenet, J. C., & Wolters, M. A. (1994). The art of checking: A case study of students' erroneous checking behavior in introductory algebra. *Journal of Mathematical Behavior, 13*, 335–358.

Pickering, A. (1995). *The mangle of practice: Time, agency, and science.* Chicago: University of Chicago Press.

Pierce, R., & Stacey, K. (2001a). Observations on students' responses to learning in a CAS environment. *Mathematics Education Research Journal, 13*, 28–46.

Pierce, R., & Stacey, K. (2001b). A framework for algebraic insight. In J. Bobis, B. Perry, & M. Mitchelmore (Eds.), *Proceedings of 24th MERGA Conference* (Vol. 2, pp. 418–425). Sydney, Australia: MERGA Program Committee.

Pimm, D. (1995). *Symbols and meanings in school mathematics.* London: Routledge.

Pirie, S. E. B., & Martin, L. (1997). The equation, the whole equation and nothing but the equation! One approach to the teaching of linear equations. *Educational Studies in Mathematics, 34*, 159–181.

Pomerantsev, L., & Korosteleva, O. (2003). Do prospective elementary and middle school teachers understand the structure of algebraic expressions? *Issues in the Undergraduate Mathematics Preparation of School Teachers: The Journal, Vol. 1: Content Knowledge*, December 2003 (electronic journal; Retrieved on March 30, 2006, from http://www.k-12prep.math.ttu.edu/journal/journal.shtml).

Program for International Student Assessment (PISA). (2004). (retrieved March 30, 2006, from http://www.pisa.oecd.org).

Proulx, J. (2003). *Pratiques des futurs enseignants de mathématiques au secondaire sous l'angle des explications orales: Intentions sous-jacentes et influences* [The oral-explanation practices of pre-service secondary-level mathematics teachers: Underlying intentions and influences]. Unpublished master's thesis, Université du Québec à Montréal.

Radford, L. (2000). Signs and meanings in students' emergent algebraic thinking: A semiotic analysis. *Educational Studies in Mathematics, 42*, 237–268.

Radford, L. (2001). The historical origins of algebraic thinking. In R. Sutherland, T. Rojano, A. Bell, & R. Lins (Eds.), *Perspectives on school algebra* (pp. 13–36). Dordrecht, The Netherlands: Kluwer.

Radford, L. (2003). Gestures, speech, and the sprouting of signs: A semiotic-cultural approach to students' types of generalization. *Mathematical Thinking and Learning, 5*, 37–70.

Radford, L. (2004). Syntax and meaning. In M. J. Hoines & A. B. Fuglestad (Eds.), *Proceedings of the 28th Conference of the International Group for the Psychology of Mathematics Education* (Vol. 1, pp. 161–166). Bergen, Norway.

Radford, L., Demers, S., Guzmán, J., & Cerulli, M. (2003). Calculators, graphs, gestures and the production of meaning. In N. A. Pateman, B. J. Dougherty, & J. T. Zilliox (Eds.), *Proceedings of the 27th Conference of the International Group for the Psychology of Mathematics Education* (Vol. 4, pp. 55–62). Honolulu, HI.

Radford, L., & Grenier, M. (1996). Les apprentissages mathématiques en situation [Situated mathematical learning]. *Revue des Sciences de l'Éducation, XXII*, 253–275.

Rasmussen, C., & Nemirovsky, R. (2003). Becoming friends with acceleration: The role of tools and bodily activity in mathematical learning. In N. A. Pateman, B. J. Dougherty, & J. T. Zilliox (Eds.), *Proceedings of the 27th Conference of the International Group for the Psychology of Mathematics Education* (Vol. 1, pp. 127–135). Honolulu.

Raymond, A. M., & Leinenbach, M. (2000). Collaborative action research on the learning and teaching of algebra: A story of one mathematics teacher's development. *Educational Studies in Mathematics, 41*, 283–307.

Reeve, W. D. (1926). *A diagnostic study of the teaching problems in high school mathematics.* Boston: Ginn.

Rojano, T. (2002). Mathematics learning in the junior secondary school: Students' access to significant mathematical ideas. In L. English (Ed.), *Handbook of international research in mathematics education* (pp. 143–163). Mahwah, NJ: Erlbaum.

Rojano, T. (2003) Incorporación de entornos tecnológicos de aprendizaje a la cultura escolar : Proyecto de innovación educativa en matemáticas y ciencias en escuelas secundarias públicas de México [Incorporating technological learning environments into the school culture: An educational innovation project in mathematics and science in the public secondary schools of Mexico]. *Revista Iberoamericana de Educación, 33*, 135–165.

Rojano, T. (2004). Local theoretical models in algebra learning: A meeting point in mathematics education. In D. E. McDougall & J. A. Ross (Eds.), *Proceedings of the 26th Annual Meeting of the North American Chapter of the International Group for the Psychology of Mathematics Education* (Vol. 1, pp. 37–53). Toronto, ON.

Romberg, T. A., Fennema, E., & Carpenter, T. P. (Eds.). (1993). *Integrating research on the graphical representation of function.* Hillsdale, NJ: Erlbaum.

Ruthven, K. (1990). The influence of graphic calculator use on translation from graphic to symbolic forms. *Educational Studies in Mathematics, 21*, 431–450.

Ruthven, K. (1996). Calculators in the mathematics curriculum: The scope of personal computing technology. In A. J. Bishop, K. Clements, C. Keitel, J. Kilpatrick, & C. Laborde (Eds.), *International handbook of mathematics education* (pp. 435–468). Dordrecht, The Netherlands: Kluwer.

Sackur, C., Drouhard, J.-Ph., Maurel, M., & Pécal, M. (1997). Comment recueillir des connaissances cachées en algèbre et qu'en faire? [How to get at hidden knowledge in algebra and what to do with it?] *Repères—IREM, 28*, 37–68.

Saldanha, L., & Kieran, C. (2005). A slippery slope between equivalence and equality: Exploring students' reasoning in the context of algebraic instruction involving a computer algebra system. In G. M. Lloyd, M. Wilson, J. L. M. Wilkins & S. L. Behm (Eds.), *Proceedings of the 27th Annual Meeting of the North American Chapter of the International Group for the Psychology of Mathematics Education* [CD-ROM]. Roanoke, VA.

Sánchez, V., & Llinares, S. (2003). Four student teachers' pedagogical reasoning on functions. *Journal of Mathematics Teacher Education, 6*, 5–25.

Sasman, M. C., Olivier, A., & Linchevski, L. (1999). Factors influencing students' generalisation thinking processes. In O. Zaslavsky (Ed.), *Proceedings of the 23rd Conference of the International Group for the Psychology of Mathematics Education* (Vol. 4, pp. 161–168). Haifa, Israel.

Saul, M. (1998). Algebra, technology, and a remark of I. M. Gelfand. In *The nature and role of algebra in the K–14 curriculum: Proceedings of a National Symposium organized by the*

National Council of Teachers of Mathematics, the Mathematical Sciences Education Board, and the National Research Council (pp. 137–144). Washington, DC: National Academy Press.

Schmidt, S., & Bednarz, N. (1997). Raisonnements arithmétiques et algébriques dans un contexte de résolution de problèmes: Difficultés rencontrées par les futurs enseignants [Arithmetic and algebraic reasoning in a problem-solving context: Difficulties encountered by pre-service teachers] *Educational Studies in Mathematics, 32*, 127–155.

Schnepp, M., & Nemirovsky, R. (2001). Constructing a foundation for the fundamental theorem of calculus. In A. A. Cuoco & F. R. Curcio (Eds.), *The roles of representation in school mathematics* (Yearbook of the National Council of Teachers of Mathematics, pp. 90–102). Reston, VA: National Council of Teachers of Mathematics.

Schoen, H. L., Cebulla, K. J., Finn, K. F., & Fi, C. (2003). Teacher variables that relate to student achievement when using a standards-based curriculum. *Journal for Research in Mathematics Education, 34*, 228–259.

Schoenfeld, A. H., Smith, J. P., & Arcavi, A. (1993). The microgenetic analysis of one student's evolving understanding of a complex subject matter domain. In R. Glaser (Ed.), *Advances in instructional psychology* (Vol. 4). Hillsdale, NJ: Erlbaum.

Schorr. R. Y. (2003). Motion, speed, and other ideas that "should be put in books." *Journal of Mathematical Behavior, 22*, 467–479.

Schwartz, J., & Yerushalmy, M. (1992). Getting students to function in and with algebra. In E. Dubinsky & G. Harel (Eds.), *The concept of function: Aspects of epistemology and pedagogy* (MAA Notes, Vol. 25, pp. 261–289). Washington, DC: Mathematical Association of America.

Schwartz, J., & Yerushalmy, M. (1996). *Calculus unlimited* [Computer software]. Tel-Aviv, Israel: Center for Educational Technology.

Schwarz, B. B., & Dreyfus, T. (1995). New actions upon old objects: A new ontological perspective on functions. *Educational Studies in Mathematics, 29*, 259–291.

Schwarz, B. B., & Hershkowitz, R. (1999). Prototypes: Brakes or levers in learning the function concept? The role of computer tools. *Journal for Research in Mathematics Education, 30*, 362–389.

Selden, A., & Selden, J. (1992). Research perspectives on conceptions of functions: Summary and overview. In E. Dubinsky & G. Harel (Eds.), *The concept of function: Aspects of epistemology and pedagogy* (MAA Notes, Vol. 25, pp. 1–16). Washington, DC: Mathematical Association of America.

Senk, S. L., & Thompson, D. R. (2003). *Standards-based school mathematics curricula*. Mahwah, NJ: Erlbaum.

Sfard, A. (1991). On the dual nature of mathematical conceptions: Reflections on processes and objects as different sides of the same coin. *Educational Studies in Mathematics, 22*, 1–36.

Sfard, A. (1992). The case of function. In E. Dubinsky & G. Harel (Eds.), *The concept of function: Aspects of epistemology and pedagogy* (MAA Notes, Vol. 25, pp. 59–84). Washington, DC: Mathematical Association of America.

Sfard, A. (2001). There is more to discourse than meets the ears: Looking at thinking as communication to learn more about mathematical thinking. *Educational Studies in Mathematics, 46*, 13–57.

Sfard, A., & Linchevski, L. (1994). The gains and the pitfalls of reification—the case of algebra. *Educational Studies in Mathematics, 26*, 191–228.

Shaw, N., Jean, B., & Peck, R. (1997). A statistical analysis on the effectiveness of using a computer algebra system in a developmental algebra course. *Journal of Mathematical Behavior, 16*, 175–180.

Shoaf-Grubbs, M.M. (1993). The effects of the graphics calculator on female students' cognitive levels and visual thinking. (Doctoral dissertation, Columbia University, 1992). *Dissertation Abstracts International, 54*(01), 119.

Shulman, L.S. (1986). Those who understand: Knowledge growth in teaching. *Educational Researcher, 15*(2), 4–14.

Shulman, L.S. (1987). Knowledge and teaching: Foundations of the new reform. *Harvard Educational Review, 57*(1), 1–22.

Simmt, E. (1997). Graphing calculators in high school mathematics. *Journal of Computers in Mathematics and Science Teaching, 16*, 269–289.

Simmt, E., Davis, B., Gordon, L., & Towers, J. (2003). Teachers' mathematics: Curious obligations. In N. A. Pateman, B. J. Dougherty, & J. T. Zilliox (Eds.), *Proceedings of the 27th Conference of the International Group for the Psychology of Mathematics Education* (Vol. 4, pp. 175–182). Honolulu, HI.

Slavit, D. (1996). Graphing calculators in "hybrid" Algebra II classroom. *For the Learning of Mathematics, 16*(1), 9–14.

Slavit, D. (1997). An alternate route to the reification of function. *Educational Studies in Mathematics, 33*, 259–281.

Slavit, D. (1998). Three women's understandings of algebra in a precalculus course integrated with the graphing calculator. *Journal of Mathematical Behavior, 17*, 355–372.

Sleeman, D. H. (1984). An attempt to understand students' understanding of basic algebra. *Cognitive Science, 8*, 387–412.

Smith, E. (2003). Stasis and change: Integrating patterns, functions, and algebra throughout the K–12 curriculum. In J. Kilpatrick, W. G. Martin, & D. Schifter (Eds.), *A research companion to principles and standards for school mathematics* (pp. 136–150). Reston, VA: National Council of Teachers of Mathematics.

Solomon, J., & Nemirovsky, R. (1999). "This is crazy, differences of differences!" On the flow of ideas in a mathematical conversation. In O. Zaslavsky (Ed.), *Proceedings of the 23rd Conference of the International Group for the Psychology of Mathematics Education* (Vol. 4, pp. 217–224). Haifa, Israel.

Stacey, K., & Chick, H. (2004). Solving the problem with algebra. In K. Stacey, H. Chick, & M. Kendal (Eds.), *The future of the teaching and learning of algebra: The 12th ICMI Study* (pp. 1–20). Dordrecht, The Netherlands: Kluwer.

Stacey, K., Chick, H., & Kendal, M. (Eds.). (2004). *The future of the teaching and learning of algebra: The 12th ICMI Study*. Dordrecht, The Netherlands: Kluwer Academic.

Stacey, K., Kendal, M., & Pierce, R. (2002). Teaching with CAS in a time of transition. *The International Journal of Computer Algebra in Mathematics Education, 9*, 113–127.

Stacey, K., & MacGregor, M. (1997). Multiple referents and shifting meanings of unknowns in students' use of algebra. In E. Pehkonen (Ed.), *Proceedings of the 21st Conference of the International Group for the Psychology of Mathematics Education* (Vol. 4, pp. 190–197). Lahti, Finland.

Stacey, K., & MacGregor, M. (1999). Learning the algebraic method of solving problems. *Journal of Mathematical Behavior, 18*, 149–167.

Stacey, K., & MacGregor, M. (2001). Curriculum reform and approaches to algebra. In R. Sutherland, T. Rojano, A. Bell, & R. Lins (Eds.), *Perspectives on school algebra* (pp. 141–153). Dordrecht, The Netherlands: Kluwer.

Star, J.R. (2005). Reconceptualizing procedural knowledge. *Journal for Research in Mathematics Education, 36*, 404–411.

Steinberg, R. M., Sleeman, D. H., & Ktorza, D. (1990). Algebra students' knowledge of equivalence of equations. *Journal for Research in Mathematics Education, 22*, 112–121.

Stigler, J., Hiebert, J., Kieran, C., Wearne, D., Seago, N., & Hood, G. (2003). *TIMSS Video Studies: Explorations of algebra teaching.* Intel Corporation.

Streun, A. V., Harskamp, E., & Suhre, C. (2000). The effect of the graphic calculator on students' solution approaches: A secondary analysis. *Hiroshima Journal of Mathematics Education, 8*, 27–40.

Stump, S. (1999). Secondary mathematics teachers' knowledge of slope. *Mathematics Education Research Journal, 11*(2), 124–144.

Stump, S. L. (2001). Developing preservice teachers' pedagogical content knowledge of slope. *Journal of Mathematical Behavior, 20*, 207–227.

Stylianou, D. A., & Silver, E. A. (2004). The role of visual representations in advanced mathematical problem solving: An examination of expert-novice similarities and differences. *Mathematical Thinking and Learning, 6*, 353–387.

Sutherland, R. (1993). Symbolising through spreadsheets. *Micromath, 10*(1), 20–22.

Sutherland, R. (2002). *A comparative study of algebra curricula.* Report prepared for the Qualifications and Curriculum Authority. London: QCA.

Sutherland, R. (2004). A toolkit for analysing approaches to algebra. In K. Stacey, H. Chick, & M. Kendal (Eds.), *The future of the teaching and learning of algebra: The 12th ICMI Study* (pp. 73–96). Dordrecht, The Netherlands: Kluwer.

Sutherland, R., & Rojano, T. (1993). A spreadsheet approach to solving algebra problems. *Journal of Mathematical Behavior, 12*, 353–383.

Sutherland, R., Rojano, T., Bell, A., & Lins, R. (Eds.). (2001). *Perspectives on school algebra.* Dordrecht, The Netherlands: Kluwer.

Swafford, J. O., & Langrall, C. W. (2000). Grade 6 students' preinstructional use of equations to describe and represent problem situations. *Journal for Research in Mathematics Education, 31*, 89–112.

Thomas, M. O. J., Monaghan, J., & Pierce, R. (2004). Computer algebra systems and algebra: Curriculum, assessment, teaching, and learning. In K. Stacey, H. Chick, & M. Kendal (Eds.), *The future of the teaching and learning of algebra: The 12th ICMI Study* (pp. 155–186). Dordrecht, The Netherlands: Kluwer.

Thorndike, E. L., Cobb, M. V., Orleans, J. S., Symonds, P. M., Wald, E., & Woodyard, E. (1923). *The psychology of algebra.* New York: Macmillan.

Tirosh, D., Even, R., & Robinson, N. (1998). Simplifying algebraic expressions: Teacher awareness and teaching approaches. *Educational Studies in Mathematics, 35*, 51–64.

Trends in International Mathematics and Science Study (TIMSS). (2004). (retrieved on April 4, 2006, from http://timss.bc.edu/timss2003.html).

Tsamir, P., Almog, N., & Tirosh, D. (1998). Students' solutions to inequalities. In A. Olivier & K. Newstead (Eds.), *Proceedings of the 22nd Conference of the International Group for the Psychology of Mathematics Education* (Vol. 4, pp. 129–136). Stellenbosch, South Africa.

Tsamir, P., & Bazzini, L. (2002). Algorithmic models: Italian and Israeli students' solutions to algebraic inequalities. In A. D. Cockburn & E. Nardi (Eds.), *Proceedings of the 26th Conference of the International Group for the Psychology of Mathematics Education* (Vol. 4, pp. 289–296). Norwich, UK.

Ursini, S., & Trigueros, M. (1997). Understanding of different uses of variable: A study with starting college students. In E. Pehkonen (Ed.), *Proceedings of the 14th Conference of the International Group for the Psychology of Mathematics Education* (Vol. 4, pp. 254–261). Lahti, Finland.

Usiskin, Z. (1988). Conceptions of school algebra and uses of variable. In A. F. Coxford & A. P. Shulte (Eds.), *The ideas of algebra, K–12* (Yearbook of the National Council of Teachers of Mathematics, pp. 8–19). Reston, VA: National Council of Teachers of Mathematics.

Vaiyavutjamai, P., Ellerton, N. F., & Clements, M. A. (2005). Students' attempts to solve two quadratic equations: A study in three nations. In P. Clarkson et al. (Eds.), *Building connections: Research, theory and practice* (proceedings of MERGA 28, Vol. 2, pp. 735–742). Melbourne, Australia: MERGA.

van Amerom, B. A. (2003). Focusing on informal strategies when linking arithmetic to early algebra. *Educational Studies in Mathematics, 54*, 63–75.

Van Dooren, W., Verschaffel, L., & Onghena, P. (2002). The impact of preservice teachers' content knowledge on their evaluation of students' strategies for solving arithmetic and algebra word problems. *Journal for Research in Mathematics Education, 33*, 319–351.

Van Dooren, W., Verschaffel, L., & Onghena, P. (2003). Preservice teachers' preferred strategies for solving arithmetic and algebra word problems. *Journal of Mathematics Teacher Education, 6*, 27–52.

Van Engen, H., Hartung, M. L. Trimble, H. C., Berger, E. J., Cleveland, R. W., & Evenson, A. B. (1964). *Seeing through mathematics.* Chicago: Scott, Foresman.

Vergnaud, G. (1989). L'obstacle des nombres négatifs et l'introduction à l'algèbre [The obstacle of negative numbers in the introduction to algebra]. In N. Bednarz & C. Garnier (Eds.), *Construction des savoirs : Obstacles et conflits* (pp. 76–83). Ottawa, Canada: Éditions Agence d'Arc.

Vérillon, P., & Rabardel, P. (1995). Cognition and artifacts: A contribution to the study of thought in relation to instrumented activity. *European Journal of Psychology of Education, 10*, 77–101.

Vermeulen, N. (2000). Student teachers' concept images of algebraic expressions. In T. Nakahara & M. Koyama (Eds.), *Proceedings of the 24th Conference of the International Group for the Psychology of Mathematics Education* (Vol. 4, pp. 257–264). Hiroshima, Japan.

Vinner, S. (1992). The function concept as a prototype for problems in mathematics learning. In E. Dubinsky & G. Harel (Eds.), *The concept of function: Aspects of epistemology and pedagogy* (MAA Notes, Vol. 25, pp. 195–213). Washington, DC: Mathematical Association of America.

Vinner, S., & Dreyfus, T. (1989). Images and definitions for the concept of function. *Journal for Research in Mathematics Education, 20*, 356–366.

Vlassis, J. (2001). Solving equations with negatives or crossing the formalizing gap. In M. van den Heuvel-Panhuizen (Ed.), *Proceedings of the 25th Conference of the International Group for the Psychology of Mathematics Education* (Vol. 4, pp. 375–382). Utrecht, The Netherlands.

Vlassis, J. (2002). The balance model: Hindrance or support for the solving of linear equations with one unknown. *Educational Studies in Mathematics, 49,* 341–359.

Vlassis, J. (2004). *Sens et symboles en mathématiques : Étude de l'utilisation du signe « moins » dans les réductions polynomiales et la résolution d'équations du premier degré à une inconnue* [Sense and symbols in mathematics: A study of the use of the "subtraction" sign in polynomial simplifications and the solving of first-degree equations in one unknown]. Liège, Belgium: Université de Liège, Faculté de Psychologie et des Sciences de l'Éducation.

Wagner, S. (1981). Conservation of equation and function under transformation of variable. *Journal for Research in Mathematics Education, 12,* 107–118.

Wagner, S. (Ed.). (1993). *Research ideas for the classroom* (Research Integration Project of the NCTM, Vols. 1–3). New York: Macmillan.

Wagner, S., & Kieran, C. (Eds.). (1989). *Research issues in the learning and teaching of algebra* (Vol. 4 of *Research agenda for mathematics education*). Reston, VA: National Council of Teachers of Mathematics.

Warren, E. (1999). The concept of variable: Gauging students' understanding. In O. Zaslavsky (Ed.), *Proceedings of the 23rd Conference of the International Group for the Psychology of Mathematics Education* (Vol. 4, pp. 313–320). Haifa, Israel.

Warren, E. (2000). Visualisation and the development of early understanding in algebra. In T. Nakahara, & M. Koyama (Ed.), *Proceedings of the 24th Conference of the International Group for the Psychology of Mathematics Education* (Vol. 4, pp. 273–280). Hiroshima, Japan.

Warren, E., & Pierce, R. (2004). A review of research in learning and teaching algebra. In B. Perry, G. Anthony, & C. Diezmann (Eds.), *Research in mathematics education in Australia 2000–2003* (pp. 291–321). Flaxton, Australia: Post Pressed.

Wheeler, D. (1996). Backwards and forwards: Reflections on different approaches to algebra. In N. Bednarz, C. Kieran, & L. Lee (Eds.), *Approaches to algebra: Perspectives for research and teaching* (pp. 317–325). Dordrecht, The Netherlands: Kluwer.

Whitman, B. S. (1976). Intuitive equation solving skills and the effects on them of formal techniques of equation solving (Doctoral dissertation, Florida State University, 1975). *Dissertation Abstracts International, 36,* 5180A. (UMI No. 76-2720)

Williams, C. G. (1998). Using concept maps to assess conceptual knowledge of function. *Journal for Research in Mathematics Education, 29,* 414–421.

Wilson, K., Ainley, J., & Bills, L. (2003). Comparing competence in transformational and generational algebraic activities. In N. A. Pateman, B. J. Dougherty, & J. T. Zilliox (Eds.), *Proceedings of the 27th Conference of the International Group for the Psychology of Mathematics Education* (Vol. 4, pp. 427–434). Honolulu, HI.

Wilson, P. S., Cooney, T. J., & Stinson, D. W. (2005). What constitutes good mathematics teaching and how it develops: Nine high school teachers' perspectives. *Journal of Mathematics Teacher Education, 8,* 83–111.

Yackel, E., & Cobb, P. (1996). Sociomathematical norms, argumentation, and autonomy in mathematics. *Journal for Research in Mathematics Education, 27,* 458–477.

Yerushalmy, M. (1997a). Mathematizing verbal descriptions of situations: A language to support modeling. *Cognition and Instruction, 15,* 207–264.

Yerushalmy, M. (1997b). Designing representations: Reasoning about functions of two variables. *Journal for Research in Mathematics Education, 28,* 431–466.

Yerushalmy, M. (2000). Problem solving strategies and mathematical resources: A longitudinal view on problem solving in a function-based approach to algebra. *Educational Studies in Mathematics, 43,* 125–147.

Yerushalmy, M., & Schwartz, J. L. (1993). Seizing the opportunity to make algebra mathematically and pedagogically interesting. In T. A. Romberg, E. Fennema, & T. C. Carpenter (Eds.), *Integrating research on the graphical representation of function* (pp. 41–68). Hillsdale, NJ: Erlbaum.

Yerushalmy, M., & Shternberg, B. (1993). The Algebra Sketchbook [Computer software]. Tel Aviv, Israel: Center for Educational Technology.

Yerushalmy, M., & Shternberg, B. (2001). Charting a visual course to the concept of function. In A. A. Cuoco & F. R. Curcio (Eds.), *The roles of representation in school mathematics* (Yearbook of the National Council of Teachers of Mathematics, pp. 251–268). Reston, VA: National Council of Teachers of Mathematics.

Zaslavsky, O., Sela, H., & Leron, U. (2002). Being sloppy about slope: The effect of changing the scale. *Educational Studies in Mathematics, 49,* 119–140.

Zazkis, R. (1998). Odds and ends of odds and evens: An inquiry into students' understanding of even and odd numbers. *Educational Studies in Mathematics, 36,* 73–89.

Zazkis, R., & Liljedahl, P. (2002). Generalization of patterns: The tension between algebraic thinking and algebraic notation. *Educational Studies in Mathematics, 49,* 379–402.

Zazkis, R., Liljedahl, P., & Gadowsky, K. (2003). Conceptions of function translation: Obstacles, intuitions, and rerouting. *Journal of Mathematical Behavior, 22,* 437–450.

Zbiek, R. M. (1998). Prospective teachers' use of computing tools to develop and validate functions as mathematical models. *Journal for Research in Mathematics Education, 29,* 184–201.

Zbiek, R. M. (2003). Using research to influence teaching and learning with computer algebra systems. In J. T. Fey (Ed.), *Computer algebra systems in secondary school mathematics education* (pp. 197–216). Reston, VA: National Council of Teachers of Mathematics.

Zbiek, R. M., & Heid, M. K. (2001). Dynamic aspects of function representations. In H. Chick, K. Stacey, J. Vincent, & J. Vincent (Eds.), *Proceedings of the 12th ICMI Study Conference: The Future of the Teaching and Learning of Algebra* (pp. 682–689). Melbourne, Australia: The University of Melbourne.

Zehavi, N. (2004). Symbol sense with a symbolic-graphical system: A story in three rounds. *Journal of Mathematical Behavior, 23,* 183–203.

AUTHOR NOTE

I wish to gratefully acknowledge the helpful comments and feedback on the initial draft of this chapter provided by Helen Doerr, Syracuse University, and Teresa Rojano, Centro de Investigación y Estudios Avanzados del I.P.N., Mexico.

17

PROBLEM SOLVING AND MODELING

Richard Lesh
INDIANA UNIVERSITY

Judith Zawojewski
ILLINOIS INSTITUTE OF TECHNOLOGY

During its relatively short history, research on mathematical problem solving has largely drawn on, and evolved from Polya's (1957) description of problem-solving heuristics. A great deal of research in the 1970s and 1980s focused on processes and strategies thought to be used by mathematicians. Begle (1979) reviewed early stages of this research. He concluded that

> No clear-cut directions for mathematics education are provided by the findings of these studies. In fact, there are enough indications that problem-solving strategies are both problem- and student-specific often enough to suggest that hopes of finding one (or a few) strategies which should be taught to all (or most) students are far too simplistic. (p. 145)

Similarly, Schoenfeld in his 1992 review of the literature concluded that attempts to teach students to use general problem-solving strategies (e.g., draw a picture, identify the givens and goals, consider a similar problem) generally had not been successful. He recommended that better results might be obtained by developing and teaching more *specific problem-solving strategies* (that link more clearly to classes of problems), by studying how to teach *metacognitive strategies* (so that students learn to effectively deploy their problem-solving strategies and content knowledge), and by developing and studying ways to eliminate students' counterproductive beliefs while enhancing productive *beliefs* (to improve students' views of the nature of mathematics and problem solving). However, when Lester (1994) compared his own 1980 list of issues in problem-solving research to those of Schoenfeld's in 1992, he concluded that little had changed. In fact, even a decade later when Lester and Kehle (2003) compared a current list of issues to those described by Lester in 1994, they concluded that, still, little progress had been made in problem-solving research and that the literature on problem solving had little to offer to school practice.

The lack of impact and cumulativeness in the research on mathematical problem solving is unsurprising, given that this area of research has been criticized over the years for its lack of a theoretical base (e.g., Begle, 1979; Kilpatrick, 1969a, 1969b; Lester, 1994; Lester & Kehle, 2003; Silver, 1985). Further, coinciding with the lack of cumulativeness in the research are discouraging cyclic trends in educational policy and practices. In the *Handbook for Research on Mathematics Teaching and Learning*, Schoenfeld (1992) described how, in the United States, the field of mathematics education has been subject to approximately 10-year cycles of pendulum swings between emphases on basic skills and problem solving, and reviews of the literature over these cycles suggest that few knowledge gains have been made from one cycle to the next.

Schoenfeld, whose literature review was written near the end of a cycle that many people considered to be the decade of problem solving, concluded his chapter with some optimism toward a future that would stay focused on problem solving both in school mathematics and in mathematics education research. However, since the 1992 *Handbook* was published, the United States and many other countries have experienced strong shifts back toward curriculum materials' emphasizing basic skills. This trend has been especially strong due to the worldwide emphasis on high-stakes testing of basic competencies, and due to the alignment of state- and local-level standards and curriculum with the goals of these tests.

Assuming that the pendulum of curriculum change again swings back toward an emphasis on problem solving, what directions can be taken that will break the to-and-fro cycles between an emphases on problem solving and basic skills? Different directions for research emerge when connecting problem solving and the learning of mathematics, and when studying problem solving as a developmental process—just as in content areas such as early number, proportional reasoning and geometry are studied developmentally in mathematics education. Therefore, this chapter will focus on the following questions: What new perspectives and directions can lead the field to the next level? What approaches or perspectives can be rejected? What has been learned, and what needs to be learned, in order for the next generation of initiatives to succeed where past ones have not? Will the pendulum swing back?

On the one hand, there is evidence that the amount of research on problem solving per se appears to be on the decline (Lester & Kehle, 2003; Stein, Boaler, & Silver, 2003) and that knowledge accumulation on instruction in problem solving is lagging (Lester & Kehle, 2003). For example, the authors counted the number of articles on research in problem solving in the *Journal for Research in Mathematics Education* (JRME) since 1980[1] and found 31 articles in the 1980s, 22 articles in the 1990s, and 4 articles between 2000 and 2003.[2]

On the other hand, perhaps there are reasons to be optimistic. One indicator that the pendulum of curriculum change may indeed be starting to swing back toward an emphasis on problem solving is at the international level. For example, in a collection of articles in a 2001 issue of the *Journal for Educational Change*, Maclean (2001) described why a number of Asian countries, and especially those that recognize that their prosperity depends on knowledge economies rather than natural resources, have been shifting their curricular emphases toward critical thinking, technological modernity, and mathematical problem solving—and away from a curriculum that emphasizes mainly skill-level instruction. They are doing this despite the fact that students in many of these Asian countries are already reported to be performing at very high levels on international tests of conventional skills.

Nathan (2001) and Tan (2002) provided a specific instance of this trend, describing initiatives in Singapore as emphasizing creativity and innovation, and developing school environments to support the initiatives. The impetus for such change is an increasing demand for future workers who have developed higher-order skills and abilities. Such change is seemingly imminent in North America and coincides with the goal for *all* students to have access to an education that emphasizes creativity, innovation, and problem solving (National Council of Teachers of Mathematics, 2000).

A second indicator of a pendulum swing back toward problem solving is apparent from recent research that emphasizes how mathematics is used in fields such as engineering, medicine and business management—where professionals tend to be heavy users of mathematics, science, and technology (e.g., Hall 1999b; Lesh, Hamilton, & Kaput, in press; Magajna & Monaghan, 2003; Noss & Hoyles, 1996; Noss, Hoyles, & Pozzi, 2002). Experts in such fields report that the nature of problem solving has changed dramatically during the past 20 years (Lesh, Hamilton, et al., in press; Lesh, Zawojewski, & Carmona, 2003; Rosenstein, in press). Furthermore, advisors to future-oriented programs emphasize that powerful new technologies for conceptualization, communication, and computation are leading to fundamental changes in the levels and types of mathematical/scientific understandings and abilities that provide foundations for success beyond school (e.g., Accreditation Board for Engineering and Technology, 2004).

Among mathematics educators, there is a growing recognition that a serious mismatch exists (and is growing) between the low-level skills emphasized in test-driven curriculum materials and the kind of understanding and abilities that are needed for success beyond school. For example, researchers such as Gainsburg (2003a, 2003b), and Hall (1999a) have explicitly noted the differences between the mathematics used by these professionals and the mathematics encountered in schools. These differences include the knowledge and abilities needed to create and modify

---

[1] Titles were identified that included references to problem solving, real world, applications, word problems, problem-solving strategies, problem-solving heuristics, problem-centered programs, verbal problems, and mathematical reasoning

[2] However, the same trend is not evident in *Educational Studies in Mathematics*, since 51 problem-solving articles appeared in the 1980s, 52 articles in the 1990s, and 11 articles between 2000 and 2002.

mathematical models—and to draw upon interdisciplinary knowledge, or knowledge that integrates thinking from a variety of textbook topic areas.

Finally, a third indicator of a pendulum swing back toward problem solving is evident in three current areas of research in mathematics education that are highly relevant to mathematical problem solving. These include research on situated cognition, communities of practice, and representational fluency. The next section of this chapter describes a variety of ways that these three areas provide fresh perspectives that can inform future directions in research in problem solving. Lester and Kehle (2003) observed that "there is general agreement that new perspectives are needed regarding the nature of problem solving and its role in school mathematics" (p. 509).

The purpose of this chapter is to describe trends, perspectives, and key elements of a research agenda that have potential to set a new and different direction for research on mathematical problem solving. A critical characteristic of this agenda for problem-solving research is the clarification of the relationships and connections between the development of mathematical content understandings and the development of problem-solving abilities (Lester & Charles, 2003; Lester & Kehle, 2003; Schoen & Charles, 2003; Silver, 1985; Stein et al., 2003). Clarifying these relationships can inform the development of curriculum and instruction by providing alternatives to treating "problem solving" as an isolated topic, separate from the learning of substantive mathematical concepts. In other words, by clarifying relationships between concept development and the development of problem-solving abilities, our intent is to suggest alternatives to teaching that are based on the assumption that problem-solving abilities develop by first teaching the concepts and procedures, then assigning one-step "story" problems that are designed to provide practice on the content learned, then teaching problem solving as a collection of strategies such as "draw a picture" or "guess and check," and finally, if time, providing students with applied problems that will require the mathematics learned in the first step. When taught in this way, problem solving (and its strategies) is portrayed as concept- and context-independent processes, isolated from important mathematical ideas.

Our alternative perspective treats problem solving as important to developing an understanding of any given mathematical concept or process. This perspective is also based on the recognition that differences between the problem solving of novices and experts go beyond the observed behaviors (i.e., the things they *do*) to the ways in which they interpret and reinterpret a problem situation (i.e., the things they *see*). Therefore, the study of problem solving needs to happen in the context of learning mathematics (and vice versa). Further, research needs to focus on students' interpretations, representations, and reflections—in addition to the computations that they carry out, the deductive reasoning processes that they employ, the skill-level achievements that they develop, or rules and procedures that they learn to execute. In other words, emphasizing synergistic relationships between learning and problem solving portrays a vision of problem solving and learning as being about *mathematical models and modeling*, where a problem situation is interpreted mathematically, and that interpretation is a mathematical model.

AN OVERVIEW OF PAST RESEARCH ON MATHEMATICAL PROBLEM SOLVING

The purpose of this section is to reflect on how several prevailing lines of research related to problem solving have evolved, to uncover underlying assumptions and critical issues, and to raise questions about what it means to learn problem-solving strategies, metacognitive processes, and beliefs and dispositions related to affect and to the development of a productive problem-solving persona. The discussion will introduce opportunities to consider an alternative perspective for future research on mathematical problem solving, in particular, a *models-and-modeling perspective* (Lesh & Doerr, 2003a).

We begin with a discussion of the major areas of research on mathematical problem solving prior to 1990. In the second part of this section, we describe how subsequent research on higher-order thinking (metacognition, habits of mind, beliefs and affect) evolved from this earlier work. In both of these parts, questions are raised, issues identified, and implications for future research described. Later, in the second half of the chapter, we examine current lines of promising research and further describe ways in which a models-and-modeling perspective can provide a productive framework to guide future research directions.

Early Problem-Solving Research

Three major areas of research in problem solving appearing in the *Journal for Research in Mathematics Education* prior to 1990 were determinants of problem difficulty, distinctions between "good" and "poor" problem solvers, and problem-solving instruction (Lester, 1994; Lester & Kehle, 2003). Each of these categories provides important perspectives and information that is used here to identify and raise questions about apparent underlying assumptions of each area.

Task Variables and Problem Difficulty

Goldin and McClintock (1979/1984) wrote the seminal book on task variables in problem solving, complementing a large number of studies that appeared in the *Journal for Research in Mathematics Education* (Lester & Kehle, 2003). Lester and Kehle described this genre of research as focusing "almost exclusively on features of the types of problems students were asked to solve in school" (p. 506). They identified "four classes of variables . . . that contribute to problem difficulty: content and context variables, structure variables, syntax variables and heuristic behavior variables" (p. 506).

Kilpatrick's (1985) description of this early work stated that "problem solving characteristics [were] selected for no special reason other than that they were easily measured and looked like they might have something to do with making a problem tough to solve" (p. 6). Kilpatrick (1985) described how the theme of task variables was revisited in the 1980s by researchers holding an information-processing perspective, emphasizing computer "modeling [of] the processes used in solving problems and predicting the types of errors to be made" (p. 6). He summarized the research on task variables in the following way:

> In the past two decades we have clearly come to a much more sophisticated view of the interactions between task characteristics and the characteristics of the problem solver, and this new view seems likely to yield much useful research (although many of my colleagues in mathematics education appear skeptical on this point). (p. 6)

Skepticism is warranted, given that aptitude-treatment-interaction (ATI) and hypothesis-testing paradigms were long ago described by Cronbach and Snow (1977) as not helpful in establishing "best" instructional practices. In ATI studies, the examination of interaction between educational treatment and student aptitudes was originally thought to have potential, because if a particular treatment could be identified as appropriate for a particular type of student, then clear guidance could be provided for curriculum and instruction of students with the targeted characteristics. However, Cronbach and Snow pointed out major problems with ATI and hypothesis-testing studies:

> It will be evident in what we have said through the book that we have serious reservations about the dominant style of research in instructional psychology. Insofar as educational reality is shaped by multiple parameters of student and treatment, traditional experiments cannot be powerful enough to support adequately complex conclusions. Insofar as variance arises out of the history of a particular class, it will be impossible to catch important causes, in the net of statistical inference. . . . In instructional research, hypothesis-testing turns us toward sterility. (p. 519)

Furthermore, from our perspective, some important variables relevant to problem solving research were not even identified, such as the study of student *response* variables. In particular, task variables alone do not account for the notion that each individual interprets a problem situation differently. That is, the problem solver's interpretation depends not only on external factors (i.e., task variables) but even more so on internal factors (i.e., how one interprets, or "sees," the mathematics problem). Research that moves beyond the simple identification of task variables to more sophisticated considerations of response variables represents a view shifting from problem solvers as simple processors of information toward problem solvers as interpreters and creators of mathematical models, or ideas—where thinking involves the development and use of complex artifacts and complex conceptual systems.

Another issue concerning early studies on task variables is that the studies rarely looked at problem-solving situations outside of school (Lester & Kehle, 2003). Therefore, recent ethnographic research examining arithmetic used by people in everyday situations, and mathematics used in "heavy use" environments such as engineering and architecture, has been very important to continued work in problem-solving research. Studies of the use of mathematics in everyday situations, such as people making decisions in grocery stores (e.g., Lave, 1988), have revealed that mathematics as taught in school is seldom used. Rather, people generally develop and routinize their own algorithms to accomplish everyday tasks—constructing their knowledge locally and situated to the specific context.

Ethnographers, such as Hall (1999a) and Gainsburg (2003b), who have examined problem solving in workplace environments, have found that professionals making "heavy use" of mathematics (such as architects and engineers) do make use of "school" mathematics, but not necessarily as direct applications of procedures and rules. The work of such professionals has been found to often require the interpretation and mathematization of local problem situations where mathematical models are developed to meet the constraints of the problem at hand, and the interpretation and presentation of mathematical solutions are frequently in multimedia formats (i.e., verbal presentations, charts, graphs, written narratives, written directions, etc.). Therefore, the importance of studying the nature of problem solving outside of conventional school mathematics has become evident.

Expert/Novice Problem-Solver Studies

A second emphasis in early research on problem solving was the study of how the behavior and performance of "good" and "poor" problem solvers differed. Lester and Kehle's (2003) summary of findings in this area include (a) "Good problem solvers know more than poor problem solvers and what they know, they know differently—their knowledge is well connected and composed of rich schemas"; and (b) "Good problem solvers tend to focus their attention on structural features of problems, poor problem solvers on surface features" (p. 507).

Krutetskii (1976), Kilpatrick (1985), Schoenfeld (1985), Silver (1985), and others also have suggested that experts may perform better not only because they know more mathematics, but also because they know mathematics *differently* than the novices. An analogy may help clarify the distinction. Larry Bird (arguably one of the most successful basketball players in the history of the game) during a television interview was shown a 3-second video clip of an "amazing move" that he had made during one of his recent games. He was asked to comment on why he did what he did. What was amazing about his comment was that, just by looking at this short video clip, he was able to recount for the interviewer where everybody on the basketball court had come from and where each was going. While most television viewers saw individual players and individual movements as "pieces of information," Bird saw the situation in terms of patterns, organized around his experience in the game. Thus his knowledge was not only more extensive but organized differently. Although his knowledge was situated (in his breadth and depth of personal experience in the game), his knowledge was also flexible and transferable to new situations in new games. Thus, he possessed more than a collection of skills, he possessed a conceptual system of situated, adaptable knowledge that contributed to making him a great player.

Krutetskii's (1976) descriptions of the problem solving of gifted mathematics students have similarities to the description of Bird's expertise. His work provides insights about why descriptions of expert past processes may not serve well as prescriptions to guide novices' next steps during problem solving. Krutetskii reported that gifted mathematics students do not simply have a larger number of ideas and strategies in their repertoire than the novices; they have ideas, strategies, and representations that appear to be organized into a highly sophisticated, elaborate network of knowledge, providing them with powerful ways to approach problem-solving situations. The talented students are described as able to generalize broadly and rapidly from a single instance, curtailing normal chains of reasoning by skipping intermediate steps and moving rapidly from problems to solutions, whereas the average student needs to see and process every logical step. Krutetskii found that gifted students tend to perceive the underlying mathematical structure of problem situations very rapidly, whereas others tend to notice and remember relatively superficial problem characteristics. Notice that the characteristics of expert problem solvers identified by Krutetskii cannot be directly taught to the nonexperts. One cannot directly teach novice problem solvers to "generalize broadly," "identify the underlying structure," or "look ahead to skip steps,"—especially assuming that novices often view mathematical problem solving as keeping track of and carefully processing pieces of information.

A landmark study of expert medical diagnosis by Elstein, Shulman, and Sprafka (1978) demonstrates how problem solving in medical diagnosis taps abilities that are organized more around the individual's experiences (including their own speciality and their own practical experiences) than around a collection of diagnostic heuristics that are used across the field. This was a surprising finding at the time:

> The most startling and controversial aspects of our results have always been the finding of case specificity and the lack of intraindividual consistency over problems, with the accompanying implication that knowledge of content is more critical than mastery of a generic problem-solving process. (p. 292)

Elstein et al.'s findings illuminated the importance of considering the learning of problem-solving processes as embedded in, and linked to, the content and context of the *situation*, rather than existing as a stand-alone processes or skill. Similarly, more recent studies of people in everyday settings (e.g., Carraher, Carraher, & Schliemann, 1985; Carraher & Schliemann, 2002; Lave, 1988; Lave & Wenger, 1991; Saxe, 1988a, 1988b, 1991), as well as studies of experts in workplace environments requiring heavy use of mathematics (e.g., Gainsburg, 2003b; Hall, 1999a, 1999b) have solidified the notion that the knowledge of both experts and novices tends to be organized more around the mathematics of the *situation* than it is around general problem-solving heuristics or conventional mathematical topics.

However, what is known about experts' knowledge is still in very early stages of development. As Lester and Kehle (2003) cautioned,

> Studying how experts think, make decisions, and solve problems does not guarantee that one is studying the experts at what actually makes them experts. For example, one can certainly give expert mathematicians textbook problems to solve, and compare their strategies and mental representations to those of novices. But expert mathematicians do not solve textbook problems for a living. . . . even if cognitive scientists have studied the right people, they may have studied them doing the wrong tasks. (p. 504)

Future study on the *nature* and *development* of expertise needs to go beyond describing typical behaviors of experts to investigating how experts "see" problematic situations: How do they mathematize problematic situations? describe them? systematize them? dimensionalize them? quantify them? Research on the development of expertise also needs to go beyond an assumption that experts first learn the content, then learn the problem-solving strategies, and then learn ways to appropriately select and apply the already-learned mathematics. Rather, the development of expertise seems to involve the holistic co-development of content, problem-solving strategies, higher-order thinking, and affect—all, to varying degrees, and situated in particular contexts (Zawojewski & Lesh, 2003). Future research questions need to address how the development of expertise evolves within problem-solving episodes and over experiences. Furthermore, research on expertise, itself, needs to be viewed developmentally. In particular, the meanings of problem-solving processes and strategies should be assumed to develop similarly to the way concepts develop for topics like rational numbers or proportional reasoning.

The final section of this chapter will describe the potential of research methodologies in which researchers assume that problem-solving expertise evolves in multi-dimensional ways (mathematically, metacognitively, affectively, etc.)—and that researchers' own understandings of expertise also are likely to evolve over the course of a study.

Instruction in Problem-Solving Strategies

In mathematics education, Polya-style problem-solving strategies—such as *draw a picture, work backwards, look for a similar problem,* or *identify the givens and goals*—have long histories of being advocated as important abilities for students to develop. Although experts often use these terms when giving after-the-fact explanations of their own problem-solving behaviors, and researchers find these terms as useful descriptors of the behavior of problem solvers they observe, research has not linked direct instruction in these strategies to improved problem-solving performance. Why do after-the-fact descriptions of past activities not necessarily provide guidelines for next steps during future problem-solving activities? Perhaps one reason is because using a strategy, such as *draw a picture,* assumes that a student would know what pictures to draw when, under what circumstances, and for which type of problems. Therefore, mastering *draw a picture* (in general) depends on interpretation abilities—not just execution abilities. Thus, developing systems for interpreting problem situations is as important, if not more important, than developing processes for doing particular strategies.

Our interpretation of Polya's heuristics is that the strategies are intended to help problem solvers think about, reflect on, and interpret problem situations, more than they are intended to help them decide what to *do* when "stuck" during a solution attempt. To convey our perspective on Polya's heuristics, we first set the stage by considering how the educational use of Polya's problem-solving heuristics has evolved historically. Consider Schoenfeld's (1992) described use of Polya's problem-solving strategies:

> There is no doubt that Pôlya's accounts of problem solving have face validity, in that they ring true to people with mathematical sophistication. Nonetheless, through the 1970s there was little empirical evidence to back up the sense that heuristics could be used as a means to enhance problem solving. . . . In short the critique of the strategies listed in *How to Solve It* and its successors is that the characterizations of them were descriptive rather than prescriptive. That is, the characterizations allowed one to recognize the strategies when they were being used. However, Pôlya's characterizations did not provide the amount of detail that would enable people who were not already familiar with the strategies to be able to implement them. (pp. 352–353)

Schoenfeld went on to describe the amount of detail that would be needed to make problem-solving strategies accessible to those not already familiar with them. Using the *examining special cases* problem-solving heuristic, Schoenfeld illustrated its application to three different mathematical problems (determination of the closed form of a formula for a series represented as a summation; given two polynomials whose coefficients are identical but in "backwards" order, find [and prove] the relationship between the roots of the two functions; given a defined sequence, determine whether the sequence converges and, if so, to what value). In the context of these problems, he described how

> each of these examples typifies a large class of problems and exemplifies a different special-cases strategy. . . . Needless to say, these three strategies hardly

> exhaust 'special cases.' At this level of analysis—the level necessary for implementing the strategies, one could find a dozen more. This is the case for almost all of Pôlya's strategies. (pp. 353–354)

In other words, Schoenfeld claimed that the *examine special cases* problem-solving strategy is, in a sense, "too large" or "too general" a strategy to be useful for prescriptive purposes. He recommended that each conventional heuristic derived from Pôlya's work (e.g., *consider a special case*) should be broken down into longer lists of more specific strategies. An analogy might be made to the process of learning to cook. One can observe and identify when a person is cooking, but learning to cook (as a concept in itself) can be thought of as involving a number of smaller, more specific aspects of cooking (e.g., principles of baking, or making a sauce).

Schoenfeld concluded, describing his own research (1985) and that of others (e.g., Heller & Hungate, 1985; Silver, 1979, 1981), that a knowledge base is available for the careful, prescriptive characterization of problem-solving strategies, which could lead to more detailed specifications of all the problem-solving strategies and heuristics for instruction, which could be linked to various classes of problems. However, with the longer lists of more detailed heuristics to learn and remember comes the need to manage and select from those lists the appropriate problem-solving strategy for a given situation. Therefore "higher order" managerial rules and beliefs would need to be introduced that specify when, why, and how to use the "lower order" prescriptive processes. Thus, Schoenfeld emphasized the need for future research on metacognition and beliefs.

Although Schoenfeld's own classroom-based research (1985, 1987) indicated success using the above-described approach, one needs to keep in mind that the instruction was implemented by a world-class teacher, who was teaching in a complex and lengthy learning environment where many different factors were at work. Thus, even though some indicators of success were achieved, the reasons for success are difficult to sort out. As Silver (1985) pointed out early on, even when a particular problem-solving endeavor has been shown to be successful in improving problem-solving performance, it is not clear *why* performance improved.

The dilemma we see is that, on the one hand, short lists of descriptive processes (i.e., the conventional problem solving strategies) appear to be too general to be meaningful for instructional purposes; on the other hand, long lists of prescriptive processes (i.e., more detailed and specific problem-solving strategies for classes of problems) tend to become so numerous that knowing when to use them becomes the heart of understanding them. Furthermore, adding more metacognitive rules and beliefs to be acquired as another layer for instruction only compounds the difficulty for classroom application.

In contrast, we suggest a perspective on Polya's heuristics as not only prompting ways of selecting and carrying out procedures and rules (i.e., "doing" mathematics[3]), but also as a means of developing systems for interpreting and describing situations (i.e., "seeing" mathematically). Recent research on problem solving in complex problematic situations suggests that the abilities involved in "seeing" are as important as abilities involved in "doing" (Lesh & Doerr, 2003b). Consider, for example, the work of structural engineers, described by Gainsburg (2003b). In particular, for engineers, carrying out the procedures contained in industry-established mathematical models (i.e., "doing the mathematics") is simultaneously both essential and inadequate. The engineers interviewed described how the behavior of the *assumed* buildings (to which the industry-standard models relate) is idealized, whereas actual applications involving more complicated situations and nonideal shapes make the use of conventional models and procedures insufficient for the needed work. Although the industry-standard mathematical models are critical for the engineers' work, the engineers must be aware of the oversimplifications embedded in the models and adapt or combine the models with other models to address the real-world complex problem. As a result the experienced engineer, using what is called *engineering judgment*, draws from many resource areas, including engineering theory, software, experience, and common sense to make judgments about whether to use and trust mathematical methods, or to modify, or to completely override them.

Research on teaching mathematical problem solving needs to move beyond the notion of trying to directly teach what experts apparently "do" to novices, and beyond transforming descriptive information about expert problem-solving behavior into long lists of prescriptive processes for novices to learn. For example, our interpretation of Polya's (1957) heuristics, and the examples that he used to illustrate them, suggests that the heuristics are intended to help stu-

[3] *Doing* is portrayed in this chapter as carrying out procedures, manipulations. This is somewhat in contrast to *doing* as described by Schoenfeld, Goldenberg, and others, who reserve the term for *doing* mathematics as professional mathematicians do—meaning the creation of new mathematics.

dents go *beyond* current ways of thinking about a problem, rather than intended only as strategies to help students function better *within* their current ways of thinking (about givens, goals, and possible solutions steps). Rather than providing a specific list of suggestions (draw a picture, identify a similar/simpler problem) to try out when "stuck," as strategies for plotting specific "next steps," we propose that Polya's heuristics can be thought of as providing a language to help problem solvers think back about their problem-solving experiences. By *describing their own processes*, students can use their reflections to develop flexible prototypes of experiences that can be drawn on in future problem solving.

Using this fresh perspective leads to the emergence of new questions concerning teaching and learning of problem solving: To what extent are students able to particularize a small number of descriptive (heuristic) processes, rather than trying to apply strategies that are recalled from an already-learned list? What collection of descriptive processes seem to be most productive? How do understandings of these descriptive processes develop? What instructional strategies (such as promoting reflection) are useful to facilitate students' development of their descriptive processes?

Research on Higher-Order Thinking

Research and development efforts focused on higher-order thinking in mathematical problem solving have run a course similar to that of research and development efforts focused on problem-solving strategies. Metacognitive behaviors, habits of mind, and dispositions (including affect and beliefs) of expert problem solvers have been described, identified, and labeled and then have been assumed to be productive traits that function across problems, within all stages of solving a given problem, and for all good problem solvers.

Lester and Kehle (2003), in their review of research on problem solving, summarized some of the higher-order thinking characteristics of good problem solvers:

- Good problem solvers are more aware than poor problem solvers of their strengths and weaknesses as problem solvers.
- Good problem solvers are better than poor problem solvers at monitoring and regulating their problem-solving efforts.
- Good problem solvers tend to be more concerned than poor problem solvers about obtaining "elegant" solutions to problems. (p. 507)

Given that characteristics such as these are perceived as accurate and stable, instructional studies in higher-order thinking typically transform such observed metacognitive characteristics of experts directly into instructional goals for novices. However, although students in these studies sometimes seem to gain an awareness of the taught higher-order thinking strategies and dispositions, their knowledge about and attempted use of the strategies has not necessarily led to better problem solving (Lester & Kehle, 2003).

In this section, we break up *higher-order thinking* into three areas: (a) metacognition, (b) habits of mind, and (c) beliefs and dispositions. Each area is used to raise questions and issues about research in higher-order thinking in general, and to pose an alternative perspective. One issue raised concerns the assumption that higher-order-thinking capabilities exist as stable and general traits or capabilities within and across problem situations. Another question raised is whether higher-order thinking exists separately from one's mathematical interpretation of a local problem situation. Finally, other questions are raised about whether descriptions of experts' apparent ways of higher-order thinking should be transformed into prescriptions for novices.

Metacognition

The notion of "thinking about one's own thinking" has been in educational literature at least since Dewey's (1933) discussion of *reflective thinking*. Similarly, Piaget's (1970) reference to *reflective abstraction* preceded Flavell's (1976) introduction of the term *metacognition*, where he stated,

> Metacognition refers, among other things, to the active monitoring and consequent regulation and orchestration of these processes in relation to the cognitive objects or data on which they bear, usually in the service of some concrete goal or objective. (p. 232)

Since the time that Flavell formulated his notions of metacognition, however, a variety of different kinds of operational definitions of metacognition have emerged in the field; and this proliferation of meanings led Lester (1994) to refer to it as an "elusive construct" (p. 666). Similarly, Campione, Brown, and Connell (1989) concluded

> One of the most salient features about metacognition is that the term means different things to different people, with the result that there is considerable confusion in the literature about what is and what is not metacognitive. This confusion leads to apparently contradictory viewpoints, ranging from claims that the concept is too ill-defined or fuzzy to be the object of scientific inquiry

to assertions that things metacognitive are the driving force of learning, and therefore the major aspects of learning we should be studying. (p. 93)

Although definitions for metacognition have not converged to a single meaning, most researchers have described metacognition and cognition as separate entities, and as being hierarchical in nature. For example, Schoenfeld (1992) described metacognition as "self-regulation, or monitoring and control . . . [of] . . . resource allocation during cognitive activity and problem solving" (p. 354). He referred to metacognitive ability (e.g., "executive control," p. 355) as the mechanism that enables problem solvers to break up larger and more complex problems into subtasks, prioritize the subtasks, efficiently sequence each subtask, and finally do each subtask. Another example is in Kuhn's (1999) hierarchical separation, which was apparent in his description: "Metacognitive knowing operates on one's base of declarative knowledge. . . . What do I know, and how do I know it?" (p. 18).

Partly due to the described theoretical formulations, most empirical studies also have operationally defined metacognitive behavior as being separate from cognition. For example, Pintrich (2002) categorized observations of metacognition into strategic knowledge, knowledge about cognitive tasks, and self-knowledge. Wilson and Clark (2004) parsed metacognition into the awareness individuals have of their own thought processes during the development of a solution, the self-evaluation of those thought processes, and the self-regulation of those thought processes. In fact, Wilson and Clark explicitly addressed the hierarchical separation, noting that "the objects on which metacognition acts are cognitive objects" (p. 33). Wilson and Clark, like Pintrich, were successful at *describing* metacognitive actions during problem solving. Their identification and categorization of behaviors were based on students' self-reports, ensuring a high level of validity in the interpretation of students' behaviors. Specifically, students twice watched a videotape of their own problem-solving processes and retroactively described their own awareness, regulation, and evaluation processes (by selecting predetermined statements on cards). However, even though Wilson and Clark obtained conclusive evidence that students did in fact use metacognitive language to describe how they went through cycles of awareness-evaluation-regulation-evaluation during past problem-solving activities, the researchers were unable to relate these metacognitive behaviors directly to improved future problem-solving performance.

A few empirical studies have treated cognition and metacognition as integrated, for example, by embedding the teaching of metacognition into the learning of substantive content. Lester, Garofalo, and Kroll (1989) described the introduction of metacognitive skills in one "regular" and one "advanced" seventh-grade classroom. The teacher served as an external monitor during problem solving within the teaching of a mathematics unit, encouraged discussion of behaviors considered important for internalizing metacognitive skills, and also served as a model of good executive behavior. In their findings, Lester, Garafalo, and Kroll described a dynamic link between the learning of mathematics content and the learning of metacognitive processes. They emphasized that metacognitive processes and skills should be learned in the context of learning substantive mathematics, and that "control processes and awareness of cognitive processes should develop concurrently with the development of an understanding of mathematical concepts" (p. 86). Thus, they hypothesized that "metacognition training is likely to be most effective when it takes place in the context of learning specific mathematical concepts and techniques" (p. 86).

More recently Kramarski, Mevarech, and Arami (2002) also embedded instruction in metacognition within the teaching of specific content—whole and rational number systems. Using a treatment-control design, they taught the same whole and rational number system content to both groups but withheld specific instruction on metacognition from the control group. In the treatment group, the students were taught to generate and answer four types of general questions related to comprehension, connections, strategies, and reflection. The students who were exposed to metacognitive instruction significantly outperformed students who were not on standard tasks (those more customarily used by mathematics teachers—tasks including simplified data and ones that can be solved by applying a straightforward algorithm that was previously taught) and authentic tasks (tasks that portray problem-solving situations, including realistic data, and are not solvable by the straightforward application of a known algorithm). Further, the students in the treatment group were better able to reorganize and process information, and to justify their reasoning.

Why did performance improve in these two cases? Was improvement truly due to learning metacognitive processes? Or did the students learn the mathematics concepts better or differently? Perhaps asking metacognitive questions during instruction on substantive mathematics prompts students to reflect on the content they are learning, and thus superior learning of the content led to better problem-solving performance *in the domain*.

As early as 1985, Silver expressed doubt about the ability to isolate the variables that might lead to improved problem-solving performance and suggested that research would benefit from developing ways of describing problem solving in terms of conceptual systems that influence students' performance (p. 257). Those conceptual systems might be called higher order conceptual systems that are tightly integrated systems of content knowledge, metacognitive capabilities, beliefs, and dispositions. In contrast to a view of metacognition and cognition as hierarchically separate, higher order conceptual systems can be thought of as developing in parallel and interactively with cognition, and as necessarily embedded in the learning of content in problem-based activities.

An assumption that seems to exist in research on metacognition is that the deployment of specific metacognitive behaviors is *always* productive. For example, the research of Schoenfeld (1985, 1987), Kramarski et. al. (2002), Pintrich (2002) and Wilson and Clark (2004) applies the targeted metacognitive strategies across problem-solving tasks, across all stages of the problem-solving process, and across the mathematical content in the problems. Schoenfeld's (1992) description of his early studies (1985, 1987) suggests that students can be trained to ask themselves metacognitive questions throughout a problem-solving episode and across problem-solving activities, and that these particular questions are always productive. He described how when small groups of students were engaged in challenging mathematical problems, he (as their teacher) circulated through the classroom asking specific questions: What (exactly) are you doing? (Can you describe it precisely?) Why are you doing it? (How does it fit into the solution?) How does it help you? (What will you do with the outcome when you obtain it?). Initially students were at a loss to answer the questions. However, recognizing that, despite their discomfort, the professor was going to continue asking those questions, students began to defend themselves by discussing the answers to the questions in advance. By the end of the term, Schoenfeld indicated that this behavior became "habitual" (p. 356). He concluded that his students' use of metacognitive activity after this training was more similar to that of experts than previously, although he said "the students' solution is hardly expert-like in the standard sense, since they found the 'right' approach quite late in the problem session" (1992, p. 357).

A different point of view might be that the use of explicit teacher interventions, intended to help students acquire metacognitive habits, may be disruptive for some students at certain times. Furthermore, drawing explicit attention to a metacognitive activity may be disruptive even in situations where its tacit use might have been productive. For example, consider a metacognitive strategy for problem solving: "Monitor your progress so that you won't go down wrong paths." Problem solvers are indeed likely to explore some paths that are ones they later choose not to take, particularly early in the solution development process. Therefore, verbally cautioning students early in the session to monitor their work in order to avoid going down the wrong path may seem to be a reasonable instructional strategy. However, such a verbal intervention may not only interrupt the problem solvers' thinking processes (whether the problem solver is an individual or a group), but the intervention may be interpreted to mean that the path selected is indeed wrong—simply because of a belief that the teacher would not say anything if one were on the "right" path. An alternative way of thinking is to consider how students can develop an *awareness* that early in complex problem-solving situations, going down "unproductive" paths is common; that something can be learned from the journey down a so-called wrong path; and that part of problem solving is detecting mismatches between the problem situation and trial solutions, leading to iterative cycles of expressing, testing, and revising current ways of thinking. In other words, a productive metacognitive strategy may be more powerful in a tacit and analytical form, rather than in an explicit and deterministic form.

Consider another example of how explicit attention to metacognition can be counterproductive. Human expertise in the performing arts, or in sports such as golf, involves the use of fluidly functioning conceptual or procedural systems. The actions or moves coordinated by the systems are not routine because they are often used to engage in new and unfamiliar situations—much like in problem solving. In such situations, competitors may try to use disruptive techniques to induce "paralysis by analysis" in opponents. For example, John McEnroe, the well-known "bad boy" tennis player, was reported to have won one of his most famous matches by complimenting his opponent on several details about his extraordinary serve (Fein, 2005). Apparently, his opponent started thinking about his own serve, and his explicit metacognitive thinking about his serve started causing him problems.

Lesh, Lester, and Hjalmarson (2003) assumed that the nature of most metacognitive behaviors should be expected to vary in their productivity across problems and over different phases of solving a problem. For example, *brainstorming*, in which group members try to generate a diversity of ideas without being critical of one another's proposals, may be productive in early stages of work on complex and challenging problems. Later in the problem-solving process, however, progress may depend more on detailed monitoring and assessment, whereas continued brainstorming may prevent progress on the solution. Therefore brainstorming, as a behavior or strategy, should be viewed as neither good nor bad across the board. The desirability of brainstorming is dependent on the function that it is intended to serve for the problem at hand and for the stage in problem solving where it is to be used.

Assuming that a particular metacognitive strategy is not necessarily productive all of the time leads to the emergence of new areas of research. For example, Hamilton, Lesh, Lester, and Yoon (in press) reported that students do indeed know they go through multiple cycles of expressing, testing, and revising their own ways of thinking about problems, and the students reported engaging in a revision process when weaknesses were detected in their trial solutions. Thus a number of research questions seem to emerge: How do problem solvers come to recognize inadequacies of their own interpretations, which then lead to further testing and revision of their ideas? What mechanisms facilitate students' recognition that current ways of thinking are unproductive? What mechanisms help students generate ideas for revising their current ways of thinking? How do these mechanisms evolve or develop within and over problem-solving experiences? Along what dimensions do these mechanisms evolve (e.g., from unstable to stable, from "primitive" to mature, from external to internal, and from concrete to abstract)?

Studies on understanding mechanisms for detecting problems in one's way of thinking need to involve developmental investigations of problem-solving and metacognitive processes, similar to how mathematics educators have investigated the nature of children's development of early number concepts, rational number concepts, geometry concepts, and other domains of concepts. With rare exceptions (e.g., Lesh & Zawojewski, 1992; Zawojewski & Lesh, 2003), such developmental perspectives have not been applied to problem-solving processes, metacognition, and other higher-order-thinking areas, such as beliefs, values, and dispositions. Yet, when a developmental perspective is taken on these areas, a vision of problem-solving processes and higher-order thinking can be developed that has the potential to inform teachers' practice. For example, in the above-named sources, we described students' early conceptions of a common problem-solving strategy, *look for a similar problem*, and identified the pitfalls for students who attempted to acquire the strategy as a habit. We suggested that the problem-solving strategy, as an instructional tool, be recast from a developmental perspective: "Look at a particular problem from different perspectives." Over a series of experiences where problem solvers explicitly consider problems from various perspectives, a group (and individuals in a group) may begin to use their experiences to serve as powerful "stories," or metaphors, for thinking about other structurally similar situations. Thus, the problem-solving strategy can be viewed developmentally and as evolving over experiences.

Habits of Mind

Cuoco, Goldenberg, and Mark (1996) described habits of mind as "mental habits that allow students to develop a repertoire of general heuristics and approaches that can be applied in many different situations" (p. 378). For example, they suggest that students should develop general habits of mind such as taking ideas apart and putting them back together and making data-driven conjectures, as well as other mathematics-topic-specific habits (e.g., algebraic and geometric habits of mind). Although students aren't expected to understand the *topics* that mathematicians deal with, Cuoco et. al. "would like students to think about mathematics the way mathematicians do" (p. 377). The notion of habits of mind has become popular in mathematics education rhetoric in recent years, particularly in professional development and more recently in research. Thus, the question addressed in this section is the extent to which habits of mind can enhance directions for research on higher-order thinking in mathematical problem solving.

From our perspective, three issues emerge when considering habits of mind as a potential direction for future research in problem solving. The first issue concerns whether emulation of the apparent behaviors (or habits) of professional mathematicians should serve as a basis of instruction in problem solving for nonexperts. The second issue concerns whether anything new and productive is being added to a research agenda in problem solving by investigating habits of mind compared to investigating problem-solving strategies and metacognitive processes or beliefs. The third issue concerns what the term *habits* may inadvertently communicate to students and teachers, such as the expectation that a habit should be productive across all contexts and all stages of problem solving.

The first issue we raise concerns the apparent assumption that teaching habits of mind (and other behaviors and strategies) of professional mathematicians to students will lead to improved problem-solving performance. For example, the assumption is evident in Cuoco, Goldenberg, and Mark's (1996) suggestion that teaching habits of mind can be a means to help students emulate the problem-solving/research processes of professional mathematicians. Similarly, Schoenfeld (1992) described a number of studies of students engaged in mathematical activity similar to that of mathematicians and as a result suggested that involving students in microcosms of mathematical practice might improve problem-solving performance of students. Little research is available that investigates the direct link between instruction in habits of mind to improved problem-solving performance. We conjecture that although such studies may document that students exhibit desired behaviors (as seen in the section on metacognition), links to improved problem solving will be elusive.

What may explain the lack of empirical support for teaching the behavior of mathematicians to students? Part of the problem may be explained by Lester and Kehle's (2003) observation that little is known about *individual differences* among experts in mathematical problem solving. For example, di Sessa, Hammer, Sherin, and Kolpakowski (1991) found that expert scientists do not all think or work alike: Some work alone, and some in groups, some work from hypothesis to proof, whereas others constantly switch from generalization to specific case perspectives and vice versa. Therefore, a reasonable assumption is that more than one profile of "typical" mathematicians' problem-solving behaviors and habits of mind are likely to exist.

Another questionable aspect of the assumption is the sole use of pure mathematicians as the role models for problem solving. Other professionals, who also engage in creating and modifying mathematics, do so in ways that are different from those of pure mathematicians. For example, Gainsburg (2003b) described how the nature of proof in the work of mathematicians is quite different than the nature of proof in engineering. One difference is in the role of assumptions. For the mathematician, the starting assumptions are assumed to be true, and the work is in the construction of the logic to link the end statement to the assumptions. For the engineer many of the starting assumptions are not identified (e.g., how the design will be simplified, how the design and the environment interact), so identifying the assumptions and justifying their accuracy is a main challenge. Gainsburg (2006) described how in structural engineering there are mathematical methods for analyzing a building with a completely stiff (rigid) floor and another for a completely flexible floor. These extremes describe no real floor, yet the engineer has to choose one (or come up with an entirely different means of analysis) that will satisfy building codes. Selecting the more conservative mathematical method is safe, because of the assumption that the real behavior of the given floor would not be as severe as an ideal floor. However, such assumptions may be modified when resulting in an overly expensive redesign of a building, or resulting in situations where the given design simply cannot carry the loads assumed. Thus, the "proof" that the structure meets building codes is contained in the mathematical argument linking quantities associated with the building to a final set of forces, whereas carrying out the procedures is easy and often accomplished in spreadsheets. The nature of the mathematics of engineers, then, appears to be quite different than the mathematics of mathematicians.

The second issue we raise concerns the potential for the construct, habits of mind, to contribute to research on problem solving beyond what has been learned from investigations on problem-solving strategies and metacognition. The habits of mind in Table 17.1 were drawn from Levasseur and Cuoco (2003) and Goldenberg, Shteingold, and Feurzeig (2003). Most of these habits of mind are easy to map to conventional problem-solving strategies and metacognitive behaviors and therefore seem to simply be new names for old conceptions of problem-solving strategies, heuristics, and metacognitive behaviors. For example, the *guessing is not necessarily bad* habit of mind is virtually the same as the conventional problem-solving strategy, *guess and test*; and the *seeking and using heuristics to solve problems* habit of mind is virtually the same as using executive control to deploy the cognitive resources at hand.

A few of the habits of mind that were cited by the original authors are not listed in Table 17.1. These include *distinguishing between agreement and logical necessity* (Goldenberg et al., 2004, p. 25) and *think algebraically* (Levasseur & Cuoco, 2003, p. 33). We did not include them in Table 17.1 because they are different than the others. They are more closely linked to the learning of specific content (e.g., geometry or algebra). By focusing on such content-specific strategies, Driscoll and colleagues (Driscoll, 1999; Driscoll, Moyer, & Zawojewski, 1998; Driscoll et al., 2001) created professional development materials that more narrowly specified the *think algebraically* habit of mind by addressing *generalize from computation* and *build rules to describe functions.* This specification process is similar to what Schoenfeld suggested in 1992 when he encouraged the development of more detailed and specific problem-solving strategies (previously described). Thus, for the purpose of this chapter, even these kinds of content-specific habits of mind appear to be renamed compilations of problem-solving strategies and metacognitive strategies that have been investigated in the past research. Consequently, the construct of habits of mind seems to have the same strengths and weaknesses as its tra-

Table 17.1 Habits of Mind and Their Counterparts in Problem Solving or Metacognitive Strategy

Habit of Mind	Problem-Solving (PS) or Metacognitive (M) Strategy
Guessing is not necessarily bad (Levasseur & Cuoco, 2003, p. 28)	Guess and test (PS)
Challenge solutions, even correct ones (Levasseur & Cuoco, 2003, p. 29) Analyzing answers, problems, and methods (Goldenberg et al., 2003, p. 26)	Look back (PS) Check the reasonableness of your answer (PS)
Look for patterns (Levasseur & Cuoco, 2003, p. 29) Conserve memory* (Levasseur & Cuoco, 2003, p. 30)	Look for a pattern (PS)
Specialize (Levasseur & Cuoco, 2003, p. 31)	Use simpler numbers (PS)
Use alternative representations (Levasseur & Cuoco, 2003, p. 31) Think about word meanings (Goldenberg et al., 2003, p. 16)	Draw a picture (PS) Act it out (PS) Write an equation (PS)
Carefully classify (Levasseur & Cuoco, 2003, p. 32)	Make a table (PS) Organize the data (PS)
Justifying claims and proving conjectures (Goldenberg et al., 2004, p. 21) Seeking and using heuristics to solve problems (Goldenberg et al., 2004, p. 28)	What are you doing? (M) Why are you doing it? (M) Is it helping? (M) What strategy/tactic/principle can be used in order to solve the problem/task? (M)

* By remembering general patterns—authors' explanation

ditionally named counterparts. Therefore, although such habits of mind are perhaps useful in professional development, where the goal is to stimulate teachers to rethink old practices, they seem to offer no new insights or directions to the existing state of research on problem solving.

Finally, we are concerned that the terminology, habits of mind, conveys a notion of mathematical thinking and problem solving that is not productive. Because in normal everyday language, the word *habit* generally refers to rules or procedures that are executed without much conscious thought, it is usually associated with simple condition-action behaviors. Little thought is given to when an individual habit is "good" or "bad"—generally a particular habit is considered stable and one that should be discontinued (e.g., smoking) or adopted (e.g., saying "please" and "thank you"). On the other hand, there are times when habits of mind, such as *carefully classify* and *challenge solutions, even correct ones* serve productive functions and other times when they can be counterproductive. Therefore, the terminology of *habits of mind* seems to emphasize the automated searching for and doing of explicit heuristics and metacognitive strategies, rather than emphasizing the interpretive, reflective, and generative aspects of problem-solving processes that fit with less behavioristic conceptions of competence.

Beliefs and Dispositions

Beliefs and dispositions related to affect have become an important topic of study related to higher-order thinking. For example, by the late 1980s and early 1990s, the role of metacognition was linked to a wide range of noncognitive factors such as those involved in feelings or other affective variables (e.g., Lester, Garofalo, & Kroll, 1989; Schoenfeld, 1987, 1992). The line of reasoning is that if novice problem solvers can learn how to use various problem-solving heuristics, then the ability to self-regulate, monitor, and call upon control mechanisms often are critically influenced (positively or negatively) by such things as beliefs, attitudes, feelings or dispositions. However, Schoenfeld (1992) pointed out that this picture is complicated by the fact that many students (and often their teachers) need to "unlearn" counterproductive beliefs and dispositions they have developed from past classroom experiences in order to learn and apply the needed metacognitive processes and strategies.

Perceptions about the role of beliefs and dispositions in problem solving have changed little since Schoenfeld's 1992 review of the literature, where he focused on Lampert's (1990) list of common beliefs that have strong (often negative) influences on students' mathematical thinking:

- Mathematics problems have one and only one right answer.

- There is only one correct way to solve any mathematics problem—usually the rule the teacher has most recently demonstrated to the class.
- Ordinary students cannot expect to understand mathematics; they expect simply to memorize it and apply what they have learned mechanically without understanding.
- Mathematics is a solitary activity, done by individuals in isolation.
- Students who have understood the mathematics they have studied will be able to solve any assigned problem in five minutes or less.
- The mathematics learned in school has little or nothing to do with the real world.
- Formal proof is irrelevant to the processes of discovery or invention. (Schoenfeld, 1992, p. 359)

Schoenfeld's (1992) review of research on the beliefs and dispositions of students, in part based on work of McLeod and Adams (1989), led him to conclude that students "abstract their beliefs about formal mathematics—their sense of their discipline—in large measure from their experiences in the classroom" and that "students' beliefs shape their behavior in ways that have extraordinarily powerful (and often negative) consequences" (p. 359). He described how the mathematical practice in the classroom shapes and is shaped by beliefs of students (and teachers, and society). The implications are far-reaching, in as much as what happens in the culture of the classroom determines, in part, future courses taken in college by students, intended majors reported by students, and subsequent approaches to mathematics instruction taken by teachers (Amit, 1988).

The notion of stable, trait-like beliefs and dispositions has been a mainstay in the literature. For example, McLeod (1989) described how emotions that are experienced repeatedly over time transform into relatively stable attitudes and beliefs. Consequently, recommendations for classroom practice naturally focus on eliminating what are believed to be negative beliefs and dispositions—by using direct instruction (providing information and examples counter to given stated beliefs and attitudes), by changing teachers' practice (Schoenfeld, 1992), or by changing the curriculum and the culture of the classroom (McLeod, 1989). For example, consider the first belief listed above—that "mathematics problems have one and only one right answer" (Schoenfeld, 1992, p. 359). Students could receive direct instruction informing them that not all mathematics problems have one and only one right answer, and then they might engage in a collection of activities that result in more than one correct solution (i.e., multiple reasonable solutions). McLeod's suggestion would go even further, encouraging systemic changes in the curriculum and culture by embedding problems with multiple reasonable solutions throughout the curriculum and embedding explicit discussion of the belief whenever possible.

In these described instructional approaches to dealing with debilitating beliefs and dispositions, an implicit message exists that the belief in question (e.g., in this case, the belief that "mathematics problems have only one right answer") is generally "bad." However, students develop most such beliefs precisely because they served some positive function at some point in their school career. For example, the belief that "mathematics problems have only one right answer" tends to be counterproductive in the context of complex problems, where multiple reasonable solution paths and multiple levels and types of solutions exist. But when students take high-stakes tests comprised of timed multiple-choice items (often claiming to include "problem solving" items), then students benefit from the belief that the problems to be encountered do indeed have exactly one correct answer. Even in complex problems, where a real-life situation involves a client who expects only one type of answer as being correct, it would be counterproductive for a student to disregard the notion that "this problem has only one right answer." Therefore, the *problem* of beliefs and dispositions exists not in the trait itself but in students' ability to use the "trait" (or belief) flexibly and appropriately in different circumstances.

When beliefs (or other attributes such as feelings and dispositions) are thought of as all-encompassing rules or simple declarative statements, they may be productive in some situations for some purposes, but they often are counterproductive in others. The goal, therefore, should be to help students recognize the difference and to use such beliefs in ways that are appropriate. The notion of stable trait-like beliefs and dispositions should be abandoned in favor of the notion of developing a productive "problem-solving persona" or "identity" that involves a complex, flexible, and manipulatable profile of attitudes, feelings, dispositions, and beliefs.

The patterns that form a problem-solving identity are complex, involving varied motivational patterns, affective reactions, and cognitive and social engagements in different circumstances both within a given task and across tasks.[4] For example, Hamilton et al.

[4] The notion of flexible problem-solving identity described here goes beyond McLeod's (1989) description of locally dynamic emotion that emerges within a problem-solving episode.

(in press) described how productive problem solvers are not those who adopt some fixed problem-solving personae. Instead, productive problem solvers manipulate their problem-solving personae to fit continually changing circumstances. Instead of trying to get students to learn a list of problem-solving or metacognitive strategies, Hamilton et al. designed opportunities for student reflection following their engagement in problem-solving activities in which students repeatedly expressed, tested, and revised their ways of thinking about the solution. As a result of describing their own problem-solving processes, students gradually developed dynamically changing and manipulatable problem-solving personalities—which integrated metacognitive functions, problem-solving strategies, beliefs, attitudes, feelings, and values.

Middleton and Toluk (1999) also have examined students' problem-solving identity profiles as complex entities involving many conventional variables (e.g., motivation, interest). They found that the conventional "traits" (that are considered to be stable across situations) often change as students progress through various modeling cycles within a specific complex problem-solving episode.[5] They described students' profiles in terms of peaks and valleys of thoughts, feelings, arousal states, and social interactions that played out across the lifespan of the problem-solving activity. For example, in their focus on motivation, they argued that there is no such thing as an "unmotivated child"—where the term *unmotivated* means an absence of will or desire. Middleton and Touluk suggested that what is conventionally perceived as lack of motivation (a stable trait) actually refers to the fact that the child wants to do something other than what the educator desires at a particular moment in time. They argued that research would benefit from a different conception of motivation—motivational models viewed as adaptive systems that interpret and restructure experience in a format that guides future engagement.

More recently, Goldin (in press) also has described affect as being more complex than conventionally viewed. Drawing on the notions of meta-affect previously described in DeBellis and Goldin (1997, 1999) and Goldin (2002), and the previously described constructs of mathematical intimacy and mathematical integrity (Debellis, 1998; Debellis & Goldin, 1997, 1999), Goldin highlighted the need to develop mathematical self-identity, to achieve mathematical integrity, and to experience intimacy with mathematics during problem solving and modeling. He viewed the development of affect as an essential structural part of learning mathematics—rather than as peripheral to cognitive aspects of learning. That is, when students interpret situations mathematically, they do a great deal more than simply engage concepts that are purely logical or mathematical in nature; their interpretations also involve feelings, values, beliefs, and dispositions. Further, Goldin described how mathematically powerful affect involves positive feelings about mathematics *as well as* ambivalent or negative feelings. For example, in mathematical modeling the process of testing and revising often leads to frustration, but one can learn that the frustration is often associated with the anticipation of eventual learning of something new, or the satisfaction of eventually solving the problem.

Middleton, Lesh, and Heger (2003) have developed methods for revealing students' beliefs and dispositions in research settings. They conducted their research with small groups solving thought-revealing, model-eliciting activities. The activities were carefully designed to elicit mathematical models as "answers" to problems—and to reveal students' intermediate and final models as they collaborate during the group problem-solving process (Lesh, Hoover, Hole, Kelly, & Post, 2000). Middleton et al. described how, using these activities, the relationship between students' modeling of mathematics and the self in relation to mathematics can be observed over reasonably brief periods of time. During the problem-solving episodes, students externalize their mathematical thinking to group members by proposing their varying perspectives, and by working to coordinate, negotiate, and sometimes reject ideas. As a result, students overtly assess and revise their mathematical ways of thinking. Using this approach, Middleton et al. have been able to describe how students *simultaneously* construct a framework for understanding classes of mathematical tasks as embodiments of mathematical structure and also construct a framework for understanding their own identity in relationship to mathematics in the problem.

Given a view of beliefs and disposition as flexible and adaptive parts of problem solvers' conceptual system, new research questions need to be posed. Productive questions would seek to understand how successful individuals learns to manipulate their own problem-solving personae to suit the situation, or how they alter a given situation to some degree to fit their own profiles. Research on students' continually adapting problem-solving personae needs to include studies of how students' dispositional components develop along dimensions, such as from external to internal,

[5] Students were given problems in which they worked in small groups to create mathematical models for realistic clients who expressed well-specified needs (i.e., model-eliciting activities as described by Lesh, Hoover, et al. 2000).

and from barren and distorted to increasingly coordinated and adaptable.

Similar to the directions for research on metacognition described earlier, investigations on beliefs and dispositions would benefit from studies that investigate the development of relevant beliefs, feelings, values, and dispositions by involving students in activities where they express, test, and revise their own attributes during post-hoc reflections on problem-solving experiences. For example, in the research described by Hamilton et al. (in press), students developed their own personal models of modeling (i.e., context-specific minitheories about their own learning and problem solving experiences) by using different kinds of reflection tools. A variety of reflection tools were designed to investigate ways to develop more effective learning and problem-solving personae—both as individuals and as members of various problem-solving groups or learning communities. Results of these studies suggest that productive problem-solving identities do not consist of rigid profiles of behaviors or attributes that govern students' behaviors across all situations; they are not a single rigid lists of beliefs, behaviors, dispositions, or values that function in the same ways across all situations—and across all stages of learning or problem solving. Instead, productive profiles are those that students themselves manipulate in ways that are productive in a variety of changing situations.

Summary and Reflections on Research in Higher-Order Thinking

A great deal of research has in common a view of higher-order thinking (e.g., metacognition, motivation, beliefs, attitude, disposition) as separate from the cognitive activity of learning content. However, recent suggestions for research in mathematical problem solving emphasize the parallel and interactive development of both mathematic concepts and higher-order thinking—as well as a variety of attributes that constitute a productive learning and problem-solving persona (e.g., Goldin, in press; Lesh, Lester, & Hjalmarson, 2003; Lester and Kehle, 2003; Middleton, Lesh, & Heger, 2003; Stein et al., 2003). Understanding how students *develop* their own ways of thinking about their own conceptual systems needs to be based on the assumptions that individuals have idiosyncratic conceptual systems influenced by their own personal experiences and that individuals' conceptual systems are dynamically evolving both within and across problem-solving activities. Furthermore, no strategy, process, behavior, or characteristic should be expected to always be productive for every problem, nor for every stage in learning or problem solving.

Research on higher-order thinking in mathematical problem solving needs to be based on the assumption that mathematical ideas and higher-order thinking develop interactively, and the work needs to pursue a greater understanding of the nature of activity that provides opportunities for individuals' problem solving persona to develop. The notion of *flow*, as conceptualized by Csikszentmihalyi (1990), offers a promising direction for such research. *Flow* is a metaphor for a type of experience that is most enjoyable: "like being carried away by a current, everything moving smoothly without effort" (Csikszentmihalyi, 1990, p. xiii). People describe the feelings they experience during flow as including intense "concentration, absorption, deep involvement, joy, a sense of accomplishment . . . as the best moment in their lives" (p. 176), and *flow* experiences are more likely to occur when one is engaged in a difficult enterprise that requires complex skills and leads to a challenging goal than in moments of leisure or during receptive entertainment, such as viewing television (Csikszentmihalyi, 1990). Therefore, the notion of flow seems highly relevant for the study of development of a problem-solving personae, and a close analysis of the characteristic dimensions of flow experiences may help identify the types of experiences that will more readily result in students' co-development of mathematics learning, problem solving, metacognition, and dispositions. For example, Csikszentmihalyi described flow as occurring when the situation provides concrete goals and manageable rules, allows for persons to adjust the opportunities for action to their own capabilities, provides information that makes self-assessment of current progress possible, and makes concentration possible (screening out distractions). Further, the model-eliciting activities briefly described above have these characteristics, making them a potentially useful research site in which to conduct studies related to flow.

Research concerning the development of higher-order thinking should include students reflecting on their own solution processes over a series of model-eliciting activities, identifying their own trends and patterns of feelings related to various phases of problem solving. For example, in post-hoc reflection activities, students can be asked to coordinate landmarks in the solution process with whether time seemed to fly by versus when time seemed to drag, or when they felt "in control" versus felt like they were "floundering." Research questions might include, How are the patterns noticed related to phases of problem solving? How are patterns noticed related to students' problem-solving identity? When students study their own reflections over a series of modeling activities, what personal theories do they develop about what constitutes mathematical problem solving and modeling? The goal for future research should be to better understand how cognition and higher-order thinking *co-develop* in the context of learning mathematics through problem solving.

NEW DIRECTIONS AND PERSPECTIVES FOR RESEARCH IN MATHEMATICAL PROBLEM SOLVING

Lester (1994) and Lester and Kehle (2003) suggested that mathematical problem solving is more complex than previously thought. Concept development and the development of problem-solving ability are highly interdependent and far more socially constructed and contextually situated than traditional theories have supposed. Further, neither concept development nor the development of problem-solving abilities proceeds in the absence of the development of beliefs, feelings, dispositions, values and other components of a complete problem-solving persona. Problem-solving ability cannot be completely separated and studied apart from all of these aspects. Thus, research on problem solving needs to be framed differently than in the past and needs to be based on a definition of problem solving that embraces its complexity in school and beyond. In particular, Lester and Kehle pointed out that that "far too little is known about students' learning in mathematically-rich environments" beyond school (p. 510).

Even though the emerging view of problem solving as a complex endeavor may seem daunting and may create the need for researchers to significantly revise many beliefs that have driven past research, the need for a significant change in perspective should not be a surprise. Historically, as disciplines develop they not only evolve through continuous and incremental steps; they often go through major shifts and revolutions. For example, the 19th century shift from Euclidean geometry to non-Euclidean geometries and the introduction of Gödel's theorem in the 1930s were radical departures from the conventional work of mathematicians in their day. The academic discipline of mathematics, which is more than 300 hundred years old, has gone through a number of such reconceptualizations, and it continues to do so and prospers by doing so.

In contrast, research in mathematics education is very young, and although some of the earliest work was conducted in the mid-1930s (e.g., Brownell, 1970/1935), work directed at mathematical problem solving (in particular, word problems) did not gain much attention until the 1960s. Further, the first research journal devoted entirely to mathematics education in North America, *Journal for Research in Mathematics Education*, did not begin until 1970, and *Educational Studies in Mathematics* (based in Europe) began about 10 years earlier. The field of mathematics education, due to its relative youth, arguably has yet to experience any major and enduring conceptual shifts, and research that clarifies the need for such shifts should be viewed as a major achievement.

The purpose of the second half of this chapter is to articulate a promising shift in perspective concerning future directions for research on mathematical problem solving. A brief overview of the perspective is provided here, and distinguishing characteristics of this perspective are described in greater detail, and with more illustration, in the final section of this chapter. In particular, a models-and-modeling perspective (Lesh & Doerr, 2003a) is proposed for conducting research and interpreting results. In keeping with the American Pragmatists perspective (Thayer, 1982), the proposed shift is not intended to be a "grand theory" of learning or problem solving. Instead, a models-and-modeling perspective can serve as a framework that encourages the integration of ways of thinking drawn from a variety of practical and theoretical perspectives. On the other hand, the proposed perspective is not theory neutral; rather it is generally aligned with extensions of constructivist and sociocultural philosophies about the nature of mathematical concepts and abilities.

One characteristic of research using a models-and-modeling perspective is that the conceptual models researchers use to study and understand mathematical problem solving are expected to be continually under development, or *under design*. Just as students develop mathematical models as they go through iterative cycles of expressing, testing, and revising their solutions to complex problems, researchers' models of students' modeling and problem solving are expected to go through similar cycles. Instead of conducting studies designed to determine whether a particular explanation, predictive model, or hypothesis is "right" or "wrong," research using a models-and-modeling perspective is grounded in the design and production of tangible tools and products for school-based use by applying the principles portrayed by a proposed conceptual model of learning and teaching.

The second characteristic of a models-and-modeling perspective is that the research activity is organized around the production of products and tools that are tested in classrooms. Because the tools educational researchers develop are often intended to help induce changes in the very "subjects" the researcher is trying to study, the expectation is that their tools and their corresponding conceptual models will need continuous modification, extension, and refinement. Therefore, the data gathered informs not only the revision of the tools (through an analysis of how well the tools inform the explanations, predictions, and descriptions in the model) but also the development and design of the researcher's conceptual model of learning and teaching. Thus, a models-and-modeling perspective on research in problem solving has the potential to simultaneously improve learning and teaching while also providing concrete, theory-based, and experience-tested results to inform the continued design of the research model for understanding how the learning of problem solving and mathematics co-develop.

A third characteristic of the proposed perspective is an increased emphasis on theory *development*, and a decreased emphasis on theory borrowing. Mathematics education research needs to shift toward a stronger focus on theory building, including the development of the field's own methodologies, theoretical models, and tools (Kelly & Lesh, 2000; Lesh, Kelly, & Yoon, in press). General theories of learning are useful for guiding the design of worthwhile learning activities and theoretical models, but useful educational models for learning and teaching in specific contexts often need to draw on and build from more than a single practical or theoretical perspective. Similar to the development of realistic solutions to realistically complex problems, model development and theory building in research usually involves trade-offs among competing factors. Therefore, research from a models-and-modeling perspective involves the development of specific methodologies, theoretical models, and tools that are designed in response to the problem being investigated.

A fourth characteristic of a models-and-modeling perspective is that it embraces a goal of producing results that need to be generalizable (i.e., sharable and reusable), but also powerful for local contexts—the context of the school, the context of the students and their teacher, etc.—and, most important, for the context of the mathematics being learned (Lesh, 1985; Silver, 1985; Stein et al., 2003). Thus, the innate tension between generalizability and specificity is constantly at the forefront of a modeling perspective.

The fundamental building blocks for the proposed models-and-modeling perspective are addressed in this section. First, a fresh view of mathematical problem solving, critical to this perspective, is discussed. Then, implications for problem solving are gleaned from current areas of research that are particularly relevant to a models-and-modeling perspective: situated learning, communities of practice, and representational fluency. The chapter ends with an explicit discussion of the critical features and aspects of a models-and-modeling perspective on future directions for problem-solving research.

A Fresh Perspective on Problem Solving

A useful theory of mathematical problem solving probably needs to be sensitive to the fact that the nature of problem solving has not remained unchanged throughout history. For example, in a technology-based *age of information*, powerful tools for computation, conceptualization, and communication are leading to fundamental changes in the levels and types of mathematical understandings and abilities that are needed for success beyond school. As stated by the National Research Council (1999),

> Changing technologies will continue to alter skills and eliminate and create jobs at a rapid rate. Although skill requirements for some jobs may be reduced, the net effects of changing technologies are more likely to raise skill requirements and change them in ways that give greater emphasis to cognitive, communications and interactive skills. (p. 71)

The changes in the nature of mathematics needed for problem solving are remarkable compared to just 50 years ago: Traffic jams are mathematically modeled and used in traffic reports; cell-phone-tower placement is based on mathematical models involving three-dimensional topography of the earth, the availability/cost of property for placing the towers, etc.; mathematical models that form the basis for Internet search engines are based on different assumptions, each one leading to new and more efficient ways to conduct searches. In these new situations, the mathematics used is multitopic, draws on innovative and creative representations and media, and often involves mathematical modeling. Yet, little seems to change for the nature of "problems" encountered by students in mathematics classes. In most school mathematics, problem solving involves thinking about things that are countable or measurable, and the favored problems are those that are readily cast within the conventional topics for teaching—whereas other, multitopic mathematical modeling problems are left for special (expendable) days or chapters at the ends of books.

A fresh perspective of problem solving is needed—one that goes beyond current school curricula and state standards. In particular, a forward-looking view of mathematical problem solving needs to tap research that describes people adapting and creating mathematics for use in everyday environments and in workplaces that require heavy use of mathematics. A fresh view of problem solving needs to view the learning of mathematics and problem solving as integrated, as largely based on modeling activity, and as a construct that is itself continually in need of development. Each of these concerns is addressed in this section, leading to and deriving from a proposed definition of problem solving.

Beyond School Mathematics

In 1989, when Stanic and Kilpatrick reviewed the literature on mathematical problem solving, they identified three main themes—each suggesting a somewhat different perspective on a definition of problem solving. The first theme was "problem solving as context," in which mathematical problems are used to motivate learning mathematics (e.g., justify teaching a specific topic, motivate intrinsic interest in mathematics, provide the development of new skills through carefully sequenced problems, practice learned techniques). This theme presented problem solving as a means to an end, rather than an end in itself. The second theme was a view of problem solving as a set of skills to be learned in their own right. Schoenfeld (1992) suggested that this second theme characterized the mathematical problem-solving curriculum and implementation of the 1980s, in which problem-solving strategies were to be taught as skills to be learned followed by a set of problems that were conducive to practicing the specified strategy. The third theme described by Stanic and Kilpatrick was the "art of solving problems"—meaning what mathematicians *do* as they solve difficult and perplexing problems, formulate proofs, and articulate the new knowledge to the field of mathematics. This last theme is related to how mathematicians select and deploy their mathematical knowledge to solve problems. However, Stanic and Kilpatrick's themes did not refer to human adaptation and creation of mathematics and mathematical models in environments where heavy use is made of mathematics. A fresh perspective on problem solving needs to take into account the situations beyond school mathematics, especially those in which evolving technology plays a large role.

Although traditional topics of school mathematics courses are clearly fundamental in the work of engineers, architects, and business administrators, the method in which the mathematics is deployed and often combined and recreated for the situation is not in the formal "mathematical" fashion learned in school. In fact, one might say that the traditional topics serve as good descriptors of the work, but when used to interpret and understand new and dynamic situations, the mathematics generated and deployed by the users is more complex, situated, and multidisciplinary than the conventional topic descriptions imply. Thus, the users may not even recognize the relationship between the mathematics taught in school and the mathematics they use in problem-solving environments. For example, Gainsburg (2003b), in her study of structural engineers at work, described their use of many of the mathematical topics taught in school, ranging from geometry, algebra, measurement, and graphs to dimensional analysis and probability. However, these topics were not used in the same "formal" way that they appear in mathematics classes, and she described how one of the main ways that the engineers used their knowledge was to describe, interpret, and understand the structure of engineering situations for making decisions or predictions. Of particular note was that the engineers, once submerged in the context of their work, found it difficult to identify school-learned mathematics in their daily work. A similar lack of awareness of the school mathematics used by professionals has been documented for technicians (Magajna & Monaghan, 2003), as well as for architects and engineers (Hall, 1999a, 1999b).

Not surprisingly, a similar gap in perception between school mathematics and the problem solving of the profession may exist for students at the preprofessional stages of education. Capobianco (2003) provided a vivid illustration of the gap in her interviews of first-year engineering students at Purdue University who were making the transition from successful high school science and mathematics accomplishments to college-level engineering. When asked about the purpose of their field (i.e., engineering), most of the students were quick to respond, "problem solving," but when asked to expand, students did not appear to understand very much about the knowledge needed for engineering work and the contexts in which they would need to solve problems.

Identifying and understanding the differences between the mathematics of school and of the workplace is critical to the formulation of a fresh definition for problem solving. A vivid illustration is found in a report by the Mathematics Sciences Education Board ([MSEB], 1998) where they described how employees who make heavy use of mathematics to solve problems effectively must be able to work in dynamic environments with open-ended problems and be able to integrate mathematics with other disciplines and the real world. Such uses of mathematics require that mathematical knowledge be reconstituted or created for the local problem situation, and that useful content knowledge involves the integration of ideas and abilities related to a variety of mathematics topics and other disciplines. For example, Oakes and Rud (2003) described how professionals using technological tools need to adapt rapidly to continually evolving conceptual systems underlying each new generation of technological tools. Another illustration is in the work of Magajna and Monaghan (2003), who have described technicians as using mathematical procedures to not only solve problems but also shape the problems to adapt them to the technology. They also described how technicians sometimes choose to find a technological solution that avoids underlying mathematics problems associated with existing models or methods used in their fields. Parker (1998) also described fields such as engineering as involving problem situations in which the user of mathematics needs to understand and modify mathematical systems rather than acquire piecewise knowledge of components of the system, which is more characteristic of school mathematics.

One notable finding across studies of professionals who make heavy use of mathematics is that mathematical modeling consistently emerges as one of the most important types of activities that needs to be mastered. Gainsburg (2003b) emphasized the role of mathematical modeling in her observations of structural engineers at work. Cross (1994) also described new technologies and newly emerging fields as increasingly using mathematical modeling as the primary form of design. For instance, the new interdisciplinary field of nanotechnology is heavily technology-based and requires the construction of mathematical models, conventions, and procedures to standardize operations across the field. The newest fields are experiencing an increasing need for problem solvers who can create elementary yet powerful constructs and conceptual systems for use in these fields, create mathematical models to use with the technological tools of the trade, and adapt previously developed mathematical models to new problem situations.

A Definition for Problem Solving

Given that the most difficult aspects of newly emerging problem-solving situations involve the development of useful ways to think mathematically about relevant relationships, patterns and regularities, definitions for *problems* and *problem solving* are needed that are consistent with these characteristics—and that do not separate problem solving from concept development and the ways that these concepts are used in "real life" situations beyond school. Therefore, we propose the following definition:

> A task, or goal-directed activity, becomes a problem (or problematic) when the "problem solver" (which may be a collaborating group of specialists) needs to develop a more productive way of thinking about the given situation.

Developing a "productive way of thinking" means that the problem solver needs to engage in a process of interpreting the situation, which in mathematics means modeling. Thus, *problem solving* is defined as the process of interpreting a situation mathematically, which usually involves several iterative cycles of expressing, testing and revising mathematical interpretations—and of sorting out, integrating, modifying, revising, or refining clusters of mathematical concepts from various topics within and beyond mathematics. In contrast to defining problem solving as a search for a procedure to traverse from the "givens" to the "goals" of a problem, the proposed definition views problem solving as iterative cycles of understanding the givens and goals of a problem. Once that understanding is attained, the linking between the two often is almost trivial.

Lester and Kehle (2003) proposed a definition of problem solving similar to the one we pose here. In their discussion, they described mathematics education as moving *away* from a view of "doing mathematics as consisting of performing low-level procedures (i.e., performing skills and routines), recalling facts and formulas, applying conceptual knowledge, and engaging in high-level problem solving (sometimes referred to as critical thinking or creative thinking)" (p. 501) and moving *towards* a view of complex mathematical activity and the mathematics of modeling complex systems. They claimed that such a perspective results in a "more authentic view of students' cognitions as they exist in busy classrooms and in complex realistic settings" (p. 517). We concur, recognizing that in virtually every field where ethnographic studies have been conducted to investigate similarities and differences between experts and novices, results have shown that not only do experts do mathematics differently, but they also see mathematics differently than novices. Thus, mathematical problem solving is about seeing (interpreting, describing, explaining) situations mathematically, and not simply about executing rules, procedures, or skills expertly. The proposed definition also emphasizes the fact that mathematics is, above all, the study of structure. That is, an interpretation is mathematical if the description or explanation focuses on structural characteristics of this situation.

Learning Problem Solving and Learning Mathematics

Our definition for problem solving also embraces the notion that people learn mathematics *through* problem solving and that they learn problem solving *through* creating mathematics (i.e., mathematical models). Embedded in a models-and-modeling perspective is an emphasis on situations in which the problem solver is expected to *create, refine,* or *adapt*

Traditional Perspective on Problem Solving
Applied problem solving is treated as a subset of traditional problem solving.

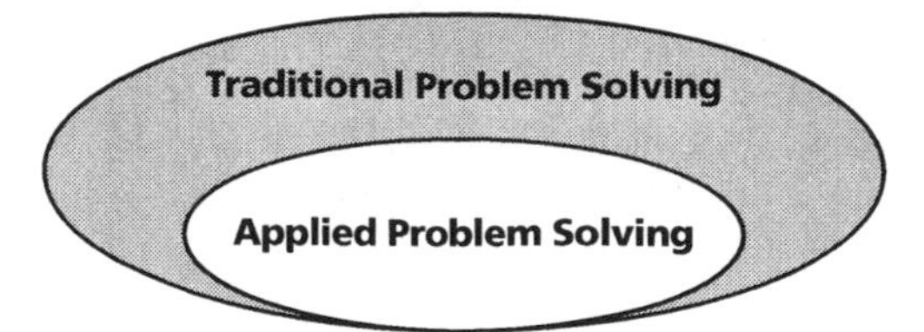

Learning to solve "real life" problems is assumed to involve four steps:

1. First, master the prerequisite ideas and skills in decontextualized situations.
2. Practice the newly mastered ideas and skills on word problems designed to require the use of the learned procedure.
3. Learn general content-independent problem-solving processes and heuristics.
4. Finally (if time permits), learn to use the preceding ideas, skills, and heuristics in messy "real life" situations (i.e., applied problems) where additional information may also be required.

Models-and-Modeling Perspective on Problem Solving
Traditional problem solving is treated as a subset of applied problem solving (i.e., model-eliciting activity).

Applied Problem Solving: As Modeling Activity

Traditional Problem Solving

Solving applied problems involves making mathematical sense of the problem (by paraphrasing, drawing diagrams, and so on) in concert with the development of a sensible solution. Understanding is not thought of as being an all-or-nothing situation, and mathematical ideas and problem-solving capabilities co-develop during the problem-solving process. The constructs, processes, and abilities that are needed to solve "real life" problems (i.e., applied problems) are assumed to be at intermediate stages of development, rather than "mastered" prior to engaging in problem solving.

Figure 17.1 Views of problem-solving: traditional versus modeling (Adapted from Lesh & Doerr, 2003b).

mathematical interpretations, or ways of thinking and procedures.[6] Figure 17.1, adapted from Lesh and Doerr (2003b), compares and contrasts a traditional perspective on problem solving with a models-and-modeling perspective.

As Figure 17.1 suggests, within the boundaries of the traditional school mathematics curriculum, the assumption tends to be made that "real life" applied problems are the most difficult types of problems to solve. Therefore, they are commonly addressed only after computational procedures have been learned, the procedures have been practiced on sets of story problems, and problem-solving strategies have been taught. Thus, only in the final stages of instruction are students engaged in solving realistic and complex applied problems (if time permits). In this traditional perspective, applied problems (requiring mathematical modeling) are a small subset of the problem-solving experiences in which students engage.

From a models-and-modeling perspective, on the other hand, the assumption is that the learning of mathematics takes place *through* modeling. In other words, students *begin* their learning experience by developing conceptual systems (i.e., models) for making sense of real-life situations where it is necessary to create, revise, or adapt a mathematical way of thinking (i.e., a mathematical model). Given model-eliciting activities (described earlier and illustrated below), students are expected to bring their own personal meaning to bear on a problem, and to test and revise their interpretation over a series of modeling cycles. Students are assumed to simultaneously gain an increasing understanding of both the problem situation and their own mathematization of the problem, which can be described as "local concept development" (Lesh & Harel, 2003; Lesh & Zawojewski, 1992), capturing the integration of learning problem solving and learning mathematics. Therefore, students' applied problem-solving experiences (i.e., mathematical modeling) drive the learning in the conventional curriculum, and traditional story problems become a subset of the applied problems through which students learn mathematics.

Model-Eliciting Activities

Our definition for problem solving has embedded in it an assumption that, in response to a realistically complex situation, the problem solvers will engage in "mathematical thinking" as they produce, refine, or

[6] Note the similarity in the process of developing a research model, as described in the introduction to the section, and students' development of a mathematical model in problem solving. In both cases the problem solver/researcher is required to create a model for dealing with new situations for an expressed purpose, and the development of the model takes place *through* the process of expressing, testing, and revising the model *in context*.

adapt complex artifacts or conceptual tools that are needed for some purpose and by some client. Such activities need to be designed to ensure that the solution (artifact, tool) problem solvers create *embodies* the mathematical process they constructed for the situation, and thus these types of problems are called *model-eliciting activities.*[7] For example, one model-eliciting activity, called Camp Placement (Lesh & Zawojewski, 1992), required that a procedure (i.e., mathematical model) be produced for a track-and-field camp director to use in the assignment of campers to groups in order to form competitive teams. Using available data about students' capabilities (i.e., track-and-field best scores, comments from the campers' coaches), the procedure created would not only be applied to a current set of campers, but also to all future sets of campers arriving with similar data. Another model-eliciting activity, called Big Foot (illustrated in Lesh & Harel, 2003), required students to develop a procedure (i.e., a model) for police detectives to use to predict a person's height from a shoe imprint in the soil. The students were told to prepare a procedure that could be used by any police officer in future detective work at a variety of crime scenes.

Both of the preceding model-eliciting activities are quite different from conventional story problems. Instead of requiring students to produce short mathematical answers to questions about premathematized situations, students' solutions (i.e., models) are complex artifacts (or tools) that need to be useful for a given client in a given situation *and* those artifacts need to be sharable and reusable in other situations, for other data sets, or by other people. Because solutions to these types of activities typically evolve over several iterative modeling cycles, the student-generated assumptions and constraints also evolve and change. Therefore, assessing progress on a model-eliciting activity is different than "checking an answer" in traditional story problems. An "end-in-view," as described by English and Lesh (2003), who based their description on Archambault (1964), is used to guide solution development as the problem solver keeps in mind the expressed needs of the client, as the trial solution and assumed conditions change.

Model-eliciting activities require students to focus on the structural and systemic characteristics of the "thing" that is being designed, constructed, or modeled. For example, in the Camp Placement Problem (described above) the procedure developed for the camp director may involve a series of rules, sometimes applied in a stochastic manner, and may be accompanied by graphs, tables, or step-by-step procedures to help communicate how the tool is used. The final solution embodies the system of objects, relationships, and operations that the students thought were important. Thus, the students' work reveals important aspects of their mathematical thinking. For the Big Foot problem, Lesh and Doerr (2003b) reported how one group of students had classmates line up in rank order of height against a wall and then noted something about a "six times" relationship. Although their written response revealed no formal use of proportions, their graph suggested that the group used trends to establish a "six times" relationship between foot length and height of a person. The transcript made from a videotape of the group work confirmed this interpretation of the group's final product; clearly the group used trends, suggested that the foot length be multiplied by six to obtain the height, and did not formally use proportions. Thus, the procedure developed by the students revealed aspects of their thinking not only to the researchers, but also to the teachers and the students themselves.

The iterative modeling cycles that students go through during the solution of model-eliciting problems make it clear how traditional story problems represent only a small part of the modeling process. Figure 17.2 depicts the modeling process as consisting of four steps of interacting processes that do not necessarily occur in any fixed order. *Description* is the creation of a model, which involves mapping from the real world[8] to the modeled world. *Manipulation* of the mathematical model happens within the modeled world and functions to generate predictions (or actions) related to the original problem-solving situation in the real world. *Prediction* carries relevant results from manipulation back into the real world, and *verification* checks the usefulness of the prediction (or action) in the context of the real world. Problems detected in the verification phase often prompt a cycle of expressing, testing, and revising a trial solution.

When students are engaged in model-eliciting activities that involve more than a single cycle of modeling, their *initial* interpretations of these situations tend to be immature, primitive, or unstable compared with the interpretations that underlie their final solutions (Lesh & Harel, 2003). Furthermore, as students develop solutions they generally do more than simply make modifications to a single existing model. For example, small groups of students generally pose a

---

[7] Lesh, Hoover, et al. (2000) and Lesh, Hoover, et al. (1993) described six principles for designing model-eliciting activities.

[8] Note that a "real world" can also be an imagined world, such as the ones in Camp Placement and Big Foot. The critical feature is that the "real world" has enough context and relevance to the student to facilitate meaningful engagement in the mathematical modeling.

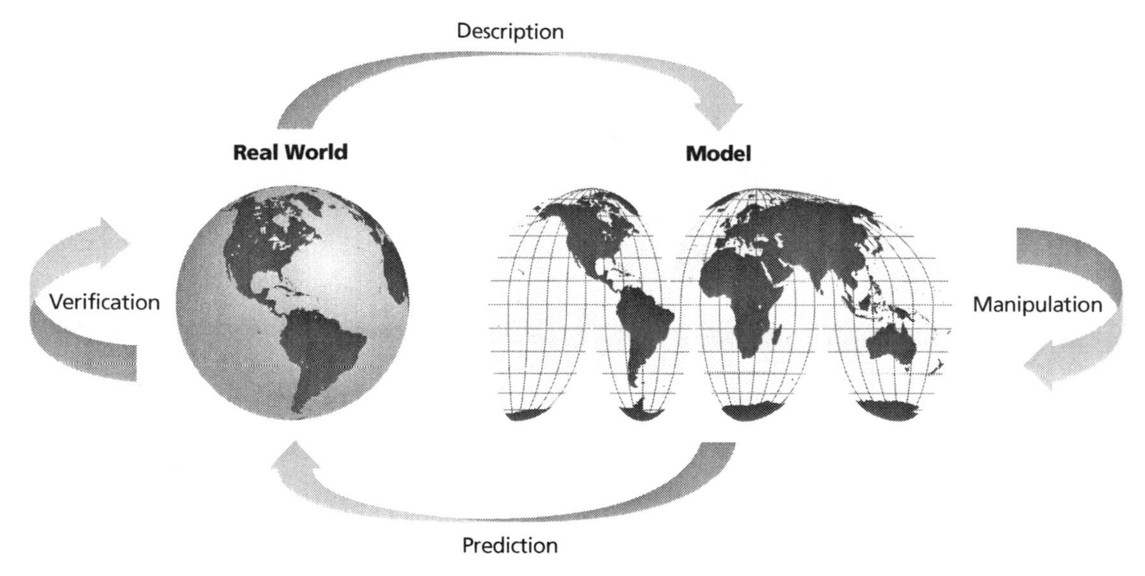

Figure 17.2 Modeling cycles. (Adapted from Lesh & Doerr, 2003b, p. 17).

number of trial solutions that result in the need to sort out, integrate, or discard models (Lesh & Yoon, 2004). Also, trial solutions tend to be expressed using a variety of representational media—including spoken language, written symbols, graphs and graphics, and experience-based metaphors. Each of the representations emphasizes and de-emphasizes somewhat different aspects of the problem-solving situation (Lesh & Doerr, 1998) and therefore also needs to be sorted out, integrated, or discarded. Another source of multiple interpretations results when a variety of technological tools are available, in which multistage solutions tend to require a good deal of planning, monitoring, communicating, and assessing processes (Hamilton et al., in press). Further, when students interpret situations mathematically, their interpretations go beyond logic and mathematics to include feelings, dispositions, values, and beliefs (Goldin, in press), resulting in a need to consider competing models from a dispositional point of view. The breadth and depth of this cyclic process has been described as *making practice mathematical,* which means that the full process of modeling as problem solving is a mathematical practice that needs to be learned through experience. "Making practice mathematical" is in contrast to "making mathematics practical"—which describes the role of traditional story problem in school mathematics (Lesh, Yoon, & Zawojewski, in press).

In contrast to the full modeling situation depicted in Figure 17.2 and described above, traditional story problems generally provide information that is given in prequantified forms. The mathematical question usually requires only short, definitive mathematical answers, rather than the development of a tool (e.g., a procedure, or algorithm) for making decisions based on mathematical analyses and results. Story problems tend to include words that are carefully selected to make apparent the mathematical procedure required to solve the problem, thus the above description—"making mathematics practical." Because the "givens" and "goals" in traditional story problems already occur in premathematized forms, the most problematic aspect involves making meaning out of symbolically described situations. (In contrast, model-eliciting activities require students to create mathematical descriptions of meaningful situations.) Thus, traditional story problems begin in the modeled world with word-based symbolic descriptions in which the key characteristics have already been quantified. Then the word-based symbols are translated to a mathematical equation, computation, or procedure, which is manipulated to yield a mathematical (symbolic) answer, still not leaving the modeled world. If and when students "check their answer for reasonableness," they may map their answer back to the real world. In as much as the problem-solving process begins in the modeled world, upon completing the solution process, students will have only engaged in at most one quarter to one half of a modeling cycle.

An Evolving View of Problem Solving

Just as students' solutions to model-eliciting activities go through cycles of testing and revision, researchers' views of problem solving also can be expected to go through cycles of revision over time. For example, Lesh and Doerr (2003b) described the evolution of

From an Industrial Age	Beyond an Age of Electronic Technologies	Toward an Age of Biotechnologies
Analogies based on *hardware*	Analogies based on computer *software*	Analogies based on *wetware.*
Systems are considered to be no more than the sum of their parts, and the interactions that are emphasized involve no more than simple one-way cause-and-effect relationships.	Silicone-based electronic circuits may involve layers of recursive interactions that often lead to emergent phenomena at higher levels that are not derived from characteristics of phenomena at lower levels.	Neurochemical interactions may involve "logics" that are fuzzy, partly redundant, partly inconsistent, and unstable—as well as living systems that are complex, dynamic, and continually adapting.

Figure 17.3 Recent transitions in models for making (or making sense of) complex systems. (Adapted from Lesh and Doerr, 2003b, p. 30.)

views of learning and problem solving by using an analogy from industry. Figure 17.3 illustrates how the industrial age, or "hardware" age, was characterized by assembly-line and machine metaphors, reflecting an assumption that the whole is simply a sum of the parts, and that causal relationships are simple one-way connections. Behaviorist models of learning mathematics (e.g., Gagne, 1975; Gagne & Briggs, 1974) reflect these assumptions, in which problem solving is viewed as a linking together of pieces of prerequisite knowledge to solve a well-specified problem.

The subsequent age of electronic technologies, or the "software" age, introduced the assumption that more complex systems involve layers of recursive interactions, which in turn involve information management schemes (related to storage and retrieval) as well as metacognitive processes that operate on the lower order skills. The analogy emphasized learning theories that were based on machine metaphors for learning. For example, Newell and Simon (1972) described their theories of learning and problem solving on the basis of their computer simulations of human problem-solving behavior. The goal of their work was to develop a global theory of problem solving that would account for problem-solving performance independent of the task domain (Lester & Kehle, 2003).

More recent descriptions involve metaphors aligned with biotechnologies, or an age of "wetware" (e.g., neurochemical interactions). These descriptions of learning and problem solving introduce additional complexities: logics that are "fuzzy," partly redundant, frequently inconsistent, and often unstable. Systems-as-a-whole are assumed to have properties, referred to as *emergent properties* of the system, that cannot be directly derived from the system's constituent elements. The wetware analogy requires an assumption that the systems involved are highly complex, dynamic, and continually adapting. Therefore scientists (i.e., educational researchers) design situations that lead to the development of living organisms (i.e., cognitive systems) that would not have been likely to evolve naturally. Rather than begin with inert "things" from which life is conjured up, the scientists begin with living systems and cultivate new kinds of living systems. When applied to human learning, problem solving can be modeled as conceptual systems that go on living and functioning, in contrast to a hardware analogy that involves condition-action rules in which systems only "run" when they are triggered by specified conditions.

The assumptions underlying the wetware metaphor are highly reminiscent of Papert's (1991) description of a classroom learning community as a teaming community of conceptual systems (or models for making sense of experience)—each of which is competing for survival in the minds of the resident students, or community of students. In such an environment, survival of the fittest means the survival of models that are useful (for immediate purposes), reusable (for other purposes), sharable (with other people), and durable. Clearly the wetware metaphor suggests the need to consider new definitions, issues, and directions in research on mathematical problem solving and learning.

Promising Lines of Research on Mathematics Learning

Research on mathematical problem solving may be migrating to other areas of research, under different names. An explanation offered by Lester (1994) and Lester and Kehle (2003) was that the introduction of the various *standards documents* by the National Council of Teachers of Mathematics (1989, 1991, 1995, 2000) may have reduced the emphasis on problem-solving research per se by introducing three additional process standards as important for school mathematics: connections, communication, and reasoning. Another explanation offered by Lester (1994) was that constructivism "has replaced problem solving as the dominant 'ideology' driving mathematics education research" (p. 668). He pointed out that some researchers were using the terms *problem solving* and *constructivism* interchangeably and questioned the appropriateness of doing so.

We offer a third possible explanation: a shift from viewing problem solving as a *thing* toward viewing it as related to *a large category of things* that need to be learned. Consider Dark's (2003) description of the skills that are needed for high-performance workplace environments such as engineering and design—and her conclusion that what people need to know about problem solving goes far beyond lists of rules, strategies, or beliefs. Instead, *models* for describing experience are important goals of mathematics teaching and learning, because in an information age, in knowledge economies, in learning organizations, and in global societies, people need to describe and explain the systems within which they operate. Given a view of problem solving as multidimensional and complex, we have identified three areas of current research that are especially relevant to establishing future research directions for mathematical problem solving advanced in this chapter. These are the areas of situated cognition, communities of practice, and representational fluency. Each of these research areas is discussed in the following three sections as they relate to a models-and-modeling perspective on problem solving.

Research on Situated Cognition

Situated cognition refers to learning and problem solving in context (e.g., Greeno, 1998; Greeno & the Middle-School Mathematics Through Applications Project Group, 1997). Studies investigating situated cognition have ranged from those focusing on everyday tasks to those focusing on high-performance work environments that require heavy use of sophisticated mathematics. In everyday contexts, such as street vending (Saxe, 1988a), grocery store shopping (Lave, 1988; Lave & Wenger, 1991), and others (Boaler, 2000; Carraher et al., 1985; Nunes, Schliemann, & Carraher, 1993; Saxe, 1991), significant types of mathematical thinking emerge in response to local problems, and the mathematics that is created to meet local needs tends to be quite different than the knowledge and abilities learned in school. For example, in school-learned mathematics, students' abilities and ideas tend to be organized more around abstractions corresponding to the topics emphasized in textbooks, whereas the conceptual systems that people develop to make sense of real-life experiences tend to integrate ideas and experiences drawn from a variety of textbook topic areas and often making use of a variety of representational media (Greeno, 1991; Noble, Nemirovsky, Wright, & Tierney, 2001).

Similarly, in environments requiring heavy use of substantive mathematics, such as engineering (Gainsburg, 2003a, 2003b, 2006; Parker, 1998), banking (Noss & Hoyles, 1996), nursing (Hoyles, Noss, & Pozzi, 2001; Noss et al., 2002), and other technology-based workplaces (e.g., Magajna & Monaghan, 2003), the users of mathematics tend to organize their mathematical ways of thinking around situations and problem contexts. A main difference from everyday tasks is that these situations often require users to draw on powerful industry-based mathematical models and procedures learned in professional settings (e.g., professional collegiate courses, internships, on the job). However, the problems encountered in these environments frequently do not fit the assumptions underlying the relevant conventional mathematical approaches and therefore are actively adapted or discarded and recreated to meet changing and novel conditions.

Overall, the research from situated cognition consistently demonstrates that nearly all people who are engaged in solving problems in their local contexts are able to develop mathematical concepts and conceptual tools for problem situations that are powerful and reusable. However, these very same people often do not recognize how the mathematics they are using, adapting, or creating is related to the mathematics they learned in school (Gainsburg, 2003b; Lave, 1988). Their perception is that the mathematics used in the real world is significantly different from the corresponding concepts and procedures emphasized in school. Our contention is that the mathematics is indeed different than that learned in school.

While studies in situated cognition have illuminated distinctions between the nature of mathematics needed in "real world" activity and the nature of mathematics typically taught in school, these studies also raise questions about the role of transfer of knowledge to new situations: If people respond to tasks by developing mathematical procedures and models that are situated to the current context, how does one generalize or transfer knowledge? From a models-and-modeling perspective, transfer is based on the idea that in the real world, people seldom take the trouble to develop generalizable conceptual tools if the problem they encounter will only need to be solved once. On the other hand, if the problem requires that a tool be devised to be used more than once or is to be shared with other people, then the problem solver will need to devise a mathematical tool in a way that will move it beyond task-specific knowledge. Consider a problem for which a person needs to develop routines that reduce large classes of tasks to situations that no longer are problematic. When the routines are initially developed to be adapted for large classes of tasks, transfer takes place as the problem solver revises and tests the earlier procedures to meet the new constraints. However, once the routines are established, tasks that were once problematic become little more than exercises in following rules.

The ability to flexibly adapt existing knowledge to new conditions is especially important in environments where technological tools are advancing so rapidly that "the knowledge students gain in college may be considered obsolete within only a few years" (Oakes & Rud, 2003, p. 224). For example, in fields such as industrial engineering, problem solvers are continually encountering and creating *new* kinds of systems for new kinds of problems in their work environments and therefore need to continually adapt or create tools and models to meet the new challenges (Gainsburg, 2006). Further, emerging fields such as nanotechnology, logistics, biotechnology, and computing are places where professionals need to *anticipate* advances and prepare for the future (National Academy of Engineering, 2004).

Lesh, Cramer, Doerr, Post, and Zawojewski (2003) suggested that engaging students in sequences of structurally related, situated modeling activities can facilitate transfer—when discussions and explorations focus on the structural similarities among related mathematical models. The modeling sequences begin with a model-eliciting activity (described earlier), in which groups of students develop their own mathematical model to meet a client's needs for a specified purpose. The initial model-eliciting activity is followed by a model-exploration activity in which students are asked to think *about* the model they have produced and other competing models. The final model-adaptation activity provides opportunities for students to adapt their own, or another model recently explored, to a new situation.

An illustration is drawn from the Small Group Mathematical Modeling Project for Improved Gender Equity in Engineering[9] (SGMM Project; Zawojewski, Diefes-Dux, & Bowman, under development). A modeling sequence was designed using the context of improving the manufacture of the material used to form the surface of artificial hip replacement joints (Diefes-Dux, Hjalmarson, Miller, & Lesh, under development; Diefes-Dux, Moore, Zawojewski, Imbrie, & Follman, 2004; Hjalmarson, Diefes-Dux, & Moore, under development; Moore & Diefes-Dux, 2004;). The roughness of the surface determines how well the hip replacement moves within the hip socket, providing motivation for the development of a mathematical model for determining a measure of roughness.[10] Students were given images of surfaces of material at an atomic level along with a scale that could be used to estimate the heights of the various molecules pictured. The problem was written to require students to design a procedure for measuring roughness that could be applied to other images of materials and thus required generalization. Teams of 4 students were given 1 hour to complete the problem, and their resulting mathematical model was used in the subsequent model-exploration activity.

The model-exploration activity introduced a few conventional mathematical models for measuring roughness in various disciplines. The model-exploration homework activity required students to think *about* the mathematical models by comparing and contrasting the conventional models to each other and to their own models. They used each model to measure roughness of selected surfaces and then identified the relative trade-offs and constraints involved in each model. The final model-adaptation activity was a major project that required the 4-student teams to recast one of the models for measuring roughness as a computer program that would read digitized images of surfaces (where the gray scale of each pixel was quantified to represent a particular height). As an output, the computer would produce the measure of roughness for that image. Because the computer program developed by the students could actually be run on digitized sample images, students and their instructors were readily able to assess progress and final products by running their trial programs on various nano-scale images of materials, which naturally prompted subsequent cycles of revising and again testing their model (i.e., the computer program).

The theoretical framework for model development sequences draws heavily on Dienes's (1960) theory of multiple embodiments for developing mathematical concepts. Dienes emphasized that students need to go beyond thinking with a given embodiment (i.e., model) to also thinking *about* it. To do so, stu-

[9] This was a 4-year NSF-funded project at Purdue University (Award #0120794-HRD) whose goal was to enhance the engineering education experience for both females and males by using mathematical modeling activities. The experiences were designed to address important goals set out by the Accreditation Board for Engineering and Technology ([ABET], 2004) that call for engineering programs to provide more real-world, design-based, team-oriented tasks in order to help students learn what engineering work is like and to help them learn more about the professional work environment of engineering.

[10] This "roughness" task is similar to the "ness" tasks designed by Judah Schwartz and colleagues (Balanced Assessment for the Mathematics Curriculum, 1999) for high school students.

dents need to encounter a concept represented in multiple embodiments (i.e., models) so that they will not solely attend to irrelevant features that are inevitably embedded in specific embodiments. By using multiple embodiments to represent a concept, students begin to recognize the general, abstract concepts that the various embodiments are intended to convey. Similarly, by creating, adapting, and comparing several structurally similar embodiments of a mathematical model, students have opportunities to compare and contrast models, to think about the similarities and differences among them, and to investigate the relationships among alternative models. The generalizing of knowledge from the initial situation to an unfamiliar, yet somewhat close situation is what constitutes *transfer* from a models-and-modeling perspective.

Model development nearly always involves situated cognition: Models are developed for a specific purpose, for specific people, and for specific situations. Nonetheless, model development generally involves much more than task-specific knowledge, because it is seldom worthwhile to develop any significant type of model unless it is intended to be sharable (used by other people) or reusable (in other situations), and therefore generalizable. Situated learning can indeed transfer, when the tasks for which the knowledge is generated require the development of sharable and reusable conceptual tools.

Research on Communities of Practice

Research on *communities of practice* has been closely associated with research on situated cognition. In addition to emphasizing the fact that knowledge is contextually situated, the notion of communities of practice emphasizes the fact that knowledge is also socially situated. Although an emphasis on the social aspects of understanding is not new (Mead, 1962, 1977; Wenger & Snyder, 2000), the resurgence of social perspectives has been fueled, in part, by the increasing expectation in high-performance workplaces, such as engineering, where teams of specialists need to work together in the development of satisfactory solutions to complex problems. Therefore, the ability of individuals to function well within a group is a highly prized attribute of employees (e.g., ABET, 2004), and the characteristics of well-functioning groups also are important to understand and develop. Thus, the "problem solver" is often the *group*, where a diversity of powerful technological and conceptual tools is brought to the solution process, and the trial solutions posed go through cycles of testing and revision.

Another reason for the renewed interest in social perspectives on understanding is that the development of learning theory has increasingly drawn on social aspects to explain how learning occurs. Whether learners act individually or as a group, the assumption is that an essential mechanism for moving the learner (group or individual) beyond current ways of thinking is through the interaction of a *variety* of alternative conceptual systems that are potentially relevant to the interpretation of a given situation.

Consider early research on problem solving, which was largely based on information-processing learning theories (e.g., Newell & Simon, 1972), which tended to rely on computer simulations to model thinking, which generally focused on the learning of individuals in isolation, and which generally treated the mind as if it were disembodied from context. More recent research on situated cognition, however, has provided the stepping stone back to social perspectives, because explanations of how individuals learn (i.e., how their cognitive systems evolve beyond current ways of thinking) have been shown to depend heavily on both cultural capital and interactions among people (e.g., Boaler & Greeno, 2000; Greeno, 2003; Lave & Wenger, 1991; Wenger, 1998; Yackel & Cobb, 1996; Zawojewski, Lesh, & English, 2003).

Although the work of Mead (1952, 1967) and other early American Pragmatists (Thayer, 1982) established many of the most distincitive characteristics of modern research on sociocultural perspectives, Vygotsky's (1978) notion of *zone of proximal development* has more often been used to form the basis for modern theories of learning based on social perspectives. Although originally focusing on how a teacher questions the student, Vygotsky's ideas have been adapted to also consider how student-to-student interactions facilitate learning. Thus, the role of an expert (e.g., a teacher, a tutor, a mentor) has varied in research that draws on a social perspective. One line of research emphasizes learning in an apprenticeship environment (e.g., Boaler & Greeno, 2000; Greeno, 2003; Lave & Wenger, 1991; Wenger, 1998;). Lave and Wenger (1991) coined the phrase "legitimate and peripheral participation" to describe the initial role of apprentices who join a community of practice. In these apprenticeships, the novices do a lot of observing of the experts at work, are given jobs that are "within their zone of proximal development," and work side-by-side with experts, gradually internalizing what was originally only externally observed.

Another line of research emphasizes peers learning from each other under the guidance of a teacher, tutor, or mentor. In these problem-solving situations, groups of students (pairs, small groups, whole classes) may be viewed as communities of practice that are engaged in a task that is considered to be too challenging for any one individual. Consequently, the role of the teacher/tutor/mentor is considered critical to guiding the group experience. For example, one role for the teacher in mathematical problem solving is to ensure that the learning community develops productive sociomathematical norms (Yackel & Cobb, 1996) in order to optimize learning via student-to-student interactions. In another example, one of the tutor's roles in medical problem solving is to help members of the team assign responsibilities (e.g., researching a particular topic, preparing a spreadsheet for the group to use) to individuals, which are to be accomplished prior to their next meeting (Koschmann, Glenn, & Conlee, 2000). In both examples, the assumption is that group-based discourse sets up optimal opportunities for individuals' ideas to be challenged within their zone of proximal development, leading to the further development of those ideas. Instead of solely depending on teacher-student interaction to prompt learning, the main learning is thought to occur through student interaction, as the teacher or tutor orchestrates the experience and ensures that the intended learning takes place.

A third line of research focuses on learning and problem solving in fields like business or engineering, where diverse teams of problem solvers often function as communities of practice in which the role of the teacher/tutor/mentor disappears (e.g. Cook & Yanow, 1993; Wenger, 2000; Wenger & Snyder, 2000; Yanow, 2000;). For example, in new innovative industries (e.g., information technologies, nanotechnology), where members of the workplace community represent a diversity of expertise, the major success of the business depends on the organization-as-a-whole's ability to develop expertise as a community—by adapting to new technologies and innovations, new consumer demands, an evolving workforce, an so on. The lack of a "teacher" or "mentor" in these new industries is in contrast to the traditional physical trades (e.g., carpentry, plumbing) where expertise is passed on from senior members to junior. A challenge for these new companies is that management cannot simply "will" diverse communities of practice to work together productively. Traditional "authority figures" can bring together people who have the desired and varied expertise, and who have interest in a given joint enterprise (i.e., a problem), but the authority figure does not orchestrate the work of the group. Instead, as the diverse members of the group work together, needed leadership roles often must emerge flexibly from within in response to the emerging challenges and opportunities.

If a goal of school mathematics is to prepare students for business environments, how does flexible, as-needed leadership come from within a group, rather than from an external authority figure? How does the role of the classroom teacher change in order to nurture the development of leadership from within the group? How can leadership internal to the group be fostered in classroom settings?

Regardless of the role of a teacher/tutor/mentor, a major contribution made by research in communities of practice is that the learner is not assumed to know *nothing* when encountering a learning or problem-solving situation. Instead, learners generally are thought to bring *some* understanding to the table. Then, interactions among group members provide opportunities for individuals' understandings to be tested, integrated, differentiated, extended, revised, or rejected. Therefore the knowledge (or the solution to the problem) that emerges is viewed as *developing* rather than as being in a state of *learned* versus *not learned*, or *solved* or *not solved*.

Important research questions emerge from this developmental perspective: Along what dimensions does knowledge develop? For example, how does knowledge externally encountered in social problem-solving settings gradually become internalized? How do groups and individuals recognize the need to develop beyond current ways of thinking? What are characteristics of productively functioning groups? How do productive roles of individuals change throughout multistage learning or problem-solving processes?

Some answers to such questions are emerging from transcripts involving small-group problem-solving episodes that are simulations of reasonably complex and realistic decision-making situations. Such studies reveal that successful solutions are typically attained when the group goes through cycles of expressing, testing, and revising their solutions (Lesh & Doerr, 2003a). Furthermore, because significant new ways of thinking often develop within relatively brief periods of time, researchers can often directly observe processes that contribute to local conceptual development (Lesh & Harel, 2003). For example, it certainly is possible to observe students taking on and considering other students' perspectives, comparing and contrasting them to their own, and as a result revising their own ways of thinking about a problem solution. This function of internalizing what was external is one dimension of local concept development. Various other dimensions of knowledge development can also be studied in these contexts, including evolution in thinking from concrete to abstract, from case-specific to general, from simple to complex, or from a collection of uncoordinated, immature ideas to more coordinated, mature, and effective knowledge.

Notice that the focus of the preceding research questions and observations is on the development of *knowledge* more than on the development of *individuals or groups*. Shifting beyond studies of human development toward studies of *idea development* has many practical advantages for mathematics educators. In particular, the researcher can study how *mathematical ideas* evolve in the community by observing the evolution of a concept, idea, or model that is externalized during the group interaction (through talking, note-taking, picture-drawing, etc.). Even though three-person teams may sometimes function in ways that are different than isolated individuals, the opportunity to study problem solving rather than problem solvers emphasizes *idea* development over *child* development. In essence, researchers can use studies of problem-solvers-who-are-a-group to help them understand problem-solvers-who-are-isolated-individuals in much the same way that other researchers use studies of experts to help understand novices—or studies of gifted students to understand average ability students. In each case, no assumptions need to be made that problem-solvers-who-are-groups are no different than problem-solvers-who are-individuals, or that experts or gifted students are no different than novices or average-ability students. The focus is on how mathematical ideas develop, and on mechanisms that prompt rethinking and revision of those ideas.

When Lesh and Yoon (2004) used simulations of complex real-life problem-solving situations to investigate both problem-solvers-who-are-groups and problem-solvers-who-are-isolated-individuals, they found that (a) regardless of whether the problem solver is a group or an individual, the problem solver's thinking often involves a community of alternative ways of thinking, and (b) the evolution of ideas that evolve from this community of minds often resembles the evolution of a community of living, interacting, and evolving biological systems, as described by Dawkins (1990). Regardless of whether the problem solver is a group or an individual, the mathematical model that develops often evolves by sorting out experiences, by adapting to new environments, and by inheriting mathematical traits from multiple lineages (i.e., the various individual ideas of group members). Specifically, for group problem solving, thinking in early phases is based on communities of constructs posed by individuals in the group, which are all most likely at intermediate or immature stages of development. For the model, or *idea*, to develop in productive ways, certain kinds of Darwinian attributes and processes must be present, such as diversity (as described by Minsky, 1987), selection (i.e., survival of stable and productive systems), communication (i.e., spreading ideas throughout the community), and accumulation (i.e., preserving adaptations so they apply to new situations). Applying the wetware analogy (previously described) to a community of minds in problem solving raises new and interesting perspectives for research on problem solving and modeling. For example, in the classroom setting, the class as a whole might be characterized as an ecological system, where communities of constructs and conceptual systems are complex, dynamic, interacting, and continually adapting systems.

The study of knowledge development conducted with the small group as the unit of study benefits from the natural opportunities for internal ideas to be externalized via various representational media (e.g., talking, writing notes, drawing pictures, producing solutions upon which everyone in the group agrees). This is in contrast to studying idea development within an individual, which usually requires researcher probes and prompts in order to get students to reveal their thinking (and students often respond in an effort to please the researcher). In small-group problem solving, idea development often can be observed in "real time." Furthermore, concerns about the validity of data gathered from think-aloud interviews, retroactive self-reports, and surveys is avoided, because all data about the problem-solvers' thinking is generated naturally during the problem-solving process (using field notes, audiotapes, and videotapes).

Research on Representational Fluency

Representations and the tools to produce them are among the most important artifacts that students project into and encounter in the world (Steffe & Thompson, 2000; Lesh & Doerr, 1998). Researchers' interest in representational capabilities is growing as new kinds of representational forms are constantly being introduced with the ever-increasing use of computers and related devices for computation, conceptualization, and communication (Kaput, Hegedus, & Lesh, in press). In particular, newly developed types of expressive media emphasize the importance of graphics-oriented, dynamic, and creative representations (Johnson & Lesh, 2003), as well as multiple-linked representations (e.g., table, graphs, and equations) (Kaput, 1991). Furthermore, problems themselves often occur in multirepresentational forms and require a solution (i.e., a product or a conceptual tool) to be developed that involves a variety of media for the purpose of explanation, simplification, justification, or prediction (Lesh, 1987). Finally, representational fluency is important for professional uses of mathematics, where the problem solver often is a team of diverse specialists communicating from remote sites, and where resources and expertise tend to be distributed in different geographic locations. Thus, representational fluency is critical to enhancing the communication capability and conceptual flexibility that are important to the development of solutions to many real-life problem-solving situations.

In most classrooms, students learn to adopt conventional mathematical language and symbol systems introduced by teachers, textbooks, and technology, with little attention to the dynamic and changing use of representations in the technological world. Further, Thompson and Sfard (1994) cautioned that unless special care is taken, the representational systems that students acquire in school may do little more than represent one another without serving as embodiments for students' internal conceptual systems, and without having the power to describe or explain other systems that occur in students' lives. Therefore, the representational media that students encounter in school need to be more than systems that the students look at; the representational media need to be systems that students develop and use, that is, that the systems become models that the students develop and use to predict, describe, and explain.

Supporting Lesh and Doerr's (2003b) claim that, when average-ability students clearly recognize the need for a specific type of conceptual tool, they often are able to develop (or at least significantly refine or revise) concepts that are both powerful and mathematically important, Lehrer and Schauble (2000) conducted a classroom-based study that engaged elementary students in the development of mathematical models that were intended to serve as classification systems. They found that young students were indeed able to develop and represent mathematical models, and they illustrated their findings using examples of first/second- and fourth/fifth-grade students who created their own languages, notation systems, diagrams, experience-based analogies, and other media to express their thoughts. Lehrer and Schauble's study, building on extensive previous work, not only establishes the possibility that young students can create and represent mathematical models, but also provides an example of research designed to produce a trail of documentation of students' modeling and problem-solving processes. Lehrer and Schuable did so by designing their studies to ensure that students externalize their thinking in representations, which in turn allows important insights to be made concerning the students' thinking, as well as the researchers' own evolving conceptual model of modeling and problem-solving capabilities of children of different age levels.

Another promising line of research looks at the local development of representational ability within the context of a problem-solving session. diSessa and colleagues (diSessa, 2002; diSessa, Hammer, Sherin, & Kolpakowski, 1991; diSessa & Sherin, 2000) have provided a useful description of *metarepresentational competence*: the capabilities students have to construct, critique, and refine external representational forms. Rather than studying the specifically prompted use of standard classroom-taught representations (e.g., the Cartesian coordinate system, or representations of functions), diSessa and colleagues study metarepresentational competence situated in a problem-solving episode where students *need* to develop a representation or *need* to revise a representation in real time—on the spot as they are engaged in the problem. Under these conditions, diSessa et al. (1991) have found that, with minimal intervention from a teacher or other authority, even novice problem solvers reveal local development of metarepresentational competence within the boundaries of the session. Further, they found that this development is most likely to occur if students talk with other students about their representations, which provides opportunities to test representations on peers—usually exposing the weaknesses of early representations. Through continual cycles of expressing, critiquing, and revising representations, students are able to improve their representation until it satisfies a perceived need.

One remarkable finding in this area of research is the lack of relationship between students' mastery of conventional classroom-taught representations and their metarepresentational capability (Izsák, 2003). diSessa et al. (1991) pointed out that much of what students' know about representation appears to exist independent of instruction. Additionally, diSessa (2002) noted that simply presenting one's work or explaining one's thinking has little influence on the development of metarepresentational capability. Instead, the critical feature of classroom activity that contributes to the development of metarepresentational capability is that the activity involves the production, inspection, testing, and revision of representations for a specific purpose. In other words, meta-representational capability involves modeling, and the researchers' descriptions of iterative cycles of expressing, testing, and revising representations are very much in line with the models-and-modeling perspectives posed in this chapter.

Directions for Future Research

Research on situated cognition, communities of practice, and representational fluency all point to new directions for research in not only mathematical problem solving, but other technology-rich domains such as engineering and science education. In particular, Lesh, Hamilton, et al. (in press) have drawn on these areas of research, examined the use of mathematics in technology-rich environments, and as a result proposed a coherent set of research questions that can provide a basis for discussion about future directions for research on mathematics learning and problem solving:

- What is the nature of problem-solving situations in which some type of elementary-but-powerful mathematical (engineering, scientific) constructs and conceptual systems (i.e., models) are needed for success in a technology-based age of information?
- What levels and types of understandings and abilities are needed for success in these situations?
- How do these understandings and abilities develop? Along what dimensions? How are they integrated in students' conceptual models?
- How can development of these understandings and abilities be facilitated, as well as documented and assessed, in teaching and learning?

Important questions, such as these, need to be investigated through research—not simply resolved through political processes such as those that are used in establishing curriculum standards and standardized-test specifications. These research questions focus on the changing nature of mathematics and situations in which mathematics is used rather than focusing on the nature of students, human minds, human information-processing capabilities, or human development. Therefore, research teams need to include those with broad and deep expertise in mathematics and science to play significant roles in the research. Furthermore, these expert participants should include not only the creators of mathematics (i.e., "pure" mathematicians), but also heavy users of mathematics (e.g., "applied" mathematicians, scientists, engineers).

Addressing these research questions requires fresh perspectives that move away from an information-processing perspective introduced to problem-solving research more than 30 years ago by Newell and Simon (1972)—in which problem solving was described as a search in a problem space for applications of known methods to get from givens to goals. In many ways, this view of problem solving still continues to drive a large share of research on mathematical problem solving. Yet, most early proponents of an information-processing perspective now describe this view as inadequate. For example, Greeno (1998) has advocated a "situated perspective" in which he described the problem solving space is as "emerg[ing] in the process of working on the problem" (p. 7), rather than a search for already known procedures. Some of the limitations of an information-processing perspective were recognized by Silver (1985) when he suggested that the field of mathematics education needed to go "beyond process-sequence strings and coded protocols" in research methodologies and beyond "simple procedure-based computer models of performance" to developing ways of describing problem solving in terms of conceptual systems that influence students' performance (p. 257).

The prevalent paradigms adopted in past research on mathematical problem solving have been based on an assumption that the way students learn to solve problems is to first acquire the mathematical knowledge needed, then acquire the problem-solving strategies that will help them decide which already known procedure to deploy, then acquire the metacognitive strategies that will trigger the appropriate use of problem-solving strategies and mathematical knowledge, and finally unlearn beliefs and dispositions that prevent effective use of problem-solving and metacognitive strategies, while also developing productive beliefs and affect. This perspective has led to theories of teaching problem solving in which virtually everything to be learned has been reduced to lists of simple rules or declarative statements (e.g., draw a picture; justify claims and prove conjectures; notice that many mathematical problems have multiple reasonable answers; ask yourself, "Why are you doing that?").

In general, such theories separate concept development from the learning of problem solving by focusing on content-independent problem-solving processes that are assumed to function in relatively invariant ways across diverse sets of tasks and situations. Yet, results from research based on these preceding assumptions have been unimpressive. Consequently, the time seems right to consider alternative assumptions. Many of the more promising recent themes in research—such as those related to situated cognition, communities of practice, and representational fluency—seem to be converging to something that we, and colleagues, are calling a *models-and-modeling perspective* on mathematical learning and problem solving.

Taking a models-and-modeling perspective, and focusing on students' mathematical models and modeling activities, predisposes researchers to take on fresh perspectives on mathematics learning and problem solving. For example,

- Researchers who study students' models and modeling naturally take on integrated approaches to studying the co-development of mathematical concepts, problem-solving processes, metacognitive functions, and beliefs or dispositions.

- Researchers who develop models of students' modeling (thinking and learning) naturally draw on integrated approaches by using a variety of practical and theoretical perspectives—including cognitive and social perspectives, mathematical/epistemological perspectives, developmental psychology perspectives, and so on—to make predictions and explanations.
- Researchers who study how modeling and problem-solving capabilities are learned naturally view problem-solving processes developmentally, in much the same way as those who have investigated the development of mathematical concepts in topic areas such as early number conceptions, rational number concepts, and concepts in geometry, algebra, or calculus.
- Researchers who study students' models and modeling activities naturally gravitate to simulations of "real life" problem-solving situations (e.g., model-eliciting activities).
- Researchers who study the changing nature of mathematics in the real world are drawn to research sites that involve complex multistage projects in which the problem solver is often a diverse team of specialists who have access to powerful conceptual technologies.
- Researchers who study modeling problems assume that mathematical thinking involves creation and interpretation (description, explanation, communication) at least as much as it involves computation, deductive reasoning, and the execution of procedures.

Models-and-modeling perspectives on classroom practice move beyond exclusive reliance on explicitly teaching problem-solving and metacognitive processes that are derived from descriptions of experts' past behaviors. Instead of viewing a novice problem solver as "stuck" and in need of help in the search for a problem space for an appropriate procedure to apply, models-and-modeling perspectives emphasize the fact that students generally are not "stuck" but instead *do* have initial ideas that are relevant, albeit primitive, egocentric, and biased. The goal is to enhance students' ability to use, extend, refine, and develop those mathematical ideas that they *do* bring to bear on the problem they are solving. Thus, other research questions that become important are

- What kind of problem-solving strategies help students use ideas that they *do* have in their initial encounter with a problem-solving situation?
- How can students minimize the debilitating influences associated with the availability of several unstable interpretations of a problem (i.e., conceptual models, trial solutions)?
- How are competing interpretations (i.e., trial solutions, models) differentiated, reconciled, integrated, or discarded altogether?
- How are successively more complex and refined models (i.e., interpretations, trial solutions) gradually constructed?

Zawojewski and Lesh (2003) addressed these types of questions and suggested that the knowledge and use of problem-solving strategies is not black-and-white—learned versus not learned. We argued that, like other forms of mathematical knowledge, students' understanding and use of problem-solving strategies develop and are important components of concept development. Therefore, the role that conventional problem-solving strategies play in instruction needs to be rethought. To illustrate, we described how the problem-solving strategy, "draw a picture," actually plays out in many activities that simulate real-life problem solving. For example, as small groups engage in a model-eliciting activity, group members often use a variety of different forms of representations to explain and defend their early ways of thinking to their teammates. In their early experiences in this type of problem solving, both their interpretations and their representations are often undeveloped and primitive. However, over a series of problem-solving experiences, students often mature in illustrating and explaining their way of thinking to others—and become more proficient at using diagrams and other representations to explore, explain, and converse with others and themselves. They begin to recognize that pictures drawn during early phases of solving a problem may have some inaccuracies, but that with cycles of testing and revising their ideas, the evolving diagrams can lead to enhanced understanding. Therefore, the diagrams themselves are not "right" or "wrong" things to do in order to solve problems; rather, they serve a function of communicating with teammates (or oneself), and of helping to improve the mathematical interpretation as the group engages in cycles of expressing, testing, and revising their thinking. Therefore, the conventional problem-solving strategy, "draw a picture," may be viewed developmentally when considered in the context of communication, and, we propose that the heuristic "draw a picture" can be transformed into "seeking to better understand the situation by explaining your way of thinking to a peer, or a member of a different problem-solving group."

Models-and-modeling perspectives also emphasize the notion that knowledge develops along multiple dimensions. Following the lead of Piaget (e.g., Inhelder & Piaget, 1958; Piaget, 1952; Piaget & Beth, 1966; Piaget & Inhelder, 1975; Piaget, Inhelder & Szeminska, 1960), domain-specific mathematics education research has studied the development in students' understanding of many important mathematical constructs (e.g., early number, rational numbers and proportional reasoning, algebra). As a result, explicit descriptions of stages of concept development have been described for these topics, and have been useful for curriculum and software developers who wish to design textbooks that will guide students towards deeper understandings of important ideas. However, for research and development in problem solving, more complex views are required.

In most of the literature on the development of mathematical knowledge, researchers have studied idea development along only one dimension. For example, Piaget (e.g., Piaget & Beth, 1966) emphasized development from concrete operational systems of thought toward formal operational systems. Vygotsky (1978) emphasized development along the dimensions of external to internal cognitive functions. Bruner (1960) emphasized increasing representational fluency and the development of progressively more powerful representational media in which to express current ways of thinking. Models-and-modeling perspectives adopt more sophisticated conceptions of development based on the observation that when students (or groups) go through a series of modeling cycles in which they integrate, differentiate, revise, and refine their existing relevant ways of thinking, development seldom occurs along a single, one-dimensional, ladder-like sequence. Instead, development occurs along a variety of dimensions, and students' final interpretations often inherit characteristics from a variety of the problem solvers' early interpretations. This multidimensional and integrated view of development requires not only new frameworks for thinking about research on problem solving but also new methodologies and tools for investigating development.

The Need for New Research Tools and Methodologies

Researchers in mature sciences tend to devote significant portions of their energies toward developing tools for their own use. The first hallmark of a high-quality tool is that it does indeed focus on the construct in question; second, the tool is based on assumptions that are consistent with those attributed to the construct being investigated. In research on mathematical problem solving, such tools should range from the development of research sites (or criteria for identifying the sites) to tools that can be used for observations or measurements.

In contrast to research in more mature sciences, few tools exist for measuring constructs claimed to be important in mathematical problem solving. Consequently, researchers have often been forced to measure the problem solving and thinking of students, teachers, and other relevant "subjects" using tools based on assumptions and theories that are inconsistent with the theories and assumptions that the researchers consider to be reasonable. For example, Ridgway, Zawojewski, and Hoover (2000) described how norm-referenced standardized tests that are based on psychometric models for studying problem solving can fall far short of capturing students' problem-solving capabilities.

Understandings and abilities related to problem solving and modeling are especially difficult to measure for a variety of reasons. First, problem solving and modeling are about interpreting (e.g., describing, explaining) situations mathematically. Nontrivial problem situations always require more than a single level or type of interpretation. Furthermore, the very act of interpreting a situation tends to cause the problem solvers' relevant interpretation systems to change. Therefore, to engage students' interpretation systems is to change those systems. In addition, the student models that emerge as potential solutions in the problem-solving process are molded and shaped by the structure of the situation (i.e., problem, task) being interpreted, as well as by the conceptual structures that the problem solvers employ. Thus, what researchers can observe are interactions among the structuring abilities of students and the structure of the tasks, and the data consists of auditable trails of documentation that problem solvers generate as they go through multiple cycles of expressing, testing, and revising their ways of thinking. To study significant types of models and modeling, researchers need to examine interactions between how students structure their experience *and* how teachers and other activity designers structure the problem situations.

Such interactions are captured in multitiered design studies (Kelly & Lesh, 2000; Kelly, Lesh & Baek, in press; Lesh & Kelly, 2000; Lesh, Kelly, & Yoon, in press) and particularized by Zawojewski, Chamberlin, Hjalmarson, and Lewis (in press) in their application of design research to teacher development. Multitiered design studies are based on the assumption that every participant in an evolving complex system (the students, the teachers, the professional development provider, the researchers) will be designing and redesigning their models (i.e., solution, way of thinking, theory, set of principles) in response to information gathered, documented, and analyzed. The notion of *design* draws on the work of research engineers, who create, test, and revise designs rather than conduct classical scientific experiments to determine whether or not "something works."[11] The notion of a *multitiered* study embraces the complexity of school mathematics environments in which students' progress cannot be considered in isolation from teachers' implementation, which in turn cannot be considered without understanding the teachers' conceptual systems, which in turn cannot be considered without understanding the nature of the professional-development experience, and so forth. Given that a change in one part of a complex system can reverberate throughout the whole system, research on problem solving and modeling needs to involve all participants to fully understand the information gathered.

The use of thought-revealing activity at all levels (the students, the teachers, the professional-development providers, and the researcher) is critical to producing a trail of documentation. Thought-revealing activities are designed to reveal the participants' thinking naturally, as a part of their own practice. For students, model-eliciting activities are designed to be thought revealing. Similarly, Doerr, and Lesh (2003) have described a number of teacher-level thought-revealing activities, such as producing a lesson plan, or an evaluation rubric.

Data from the student-level and teacher-level thought-revealing activities provides opportunities not only for researcher analysis, but also self-assessment of progress by the participants who subsequently make decisions about next steps, or subsequent experiences. Designing research studies requires a specific plan for producing a trail of documentation from each participant, which ensures that conceptual models brought to bear are revealed (at least partially), and that changes in the models over time can be detected. The result is that insights into the interacting development of student models, teacher models, and researcher models simultaneously informs the research *and* the subsequent development of the learning experiences at all levels.

Consider, as an illustration, Lesh's multitiered design experiment at Purdue University's Center for Twenty-First Century Conceptual Tools (TCCT). He organized three- to five-person teams of graduate research assistants and faculty colleagues to work together in semester-long sequences of biweekly meetings in which participants collaborated to design thought-revealing (model-eliciting) activities for students. The activities, designed for middle-school students, were intended to be high-fidelity simulations of "real life" situations where mathematical thinking would be needed for success. As the activities were field tested at diverse sites, and as the designers developed tools to document and assess the mathematical understanding and abilities that were most critical for success in the problem-solving activities, the designers repeatedly expressed, tested, and revised their own current ways of thinking about the nature of mathematical thinking needed for success beyond school. Iterative rounds of field testing with students, consensus building among designers and researchers, and redesign of the tasks were built into the design experiment. The kinds of mathematical knowledge and abilities identified by the participants as emphasized in the activities were quite different from what is typically emphasized in traditional tests and textbooks used by the schools, and even what is emphasized in applied mathematics courses (Lesh, 2003). For example, quantifying qualitative information, sorting and selecting data to use when all data is relevant yet not exactly what is needed, and creating and designing mathematical procedures to process information illustrate the types of knowledge that were found to be critical for solving such problems. Thus, the participants learned that mathematical abilities emphasized in model-eliciting activities represent a larger range than those outlined in district-level, state-level, and even national-level documents.

A number of researchers have begun to use multitiered design studies to examine teachers' learning in the context of teachers examining and facilitating their own students' learning. Schorr and Lesh (2003) described a professional-development study in which teachers were asked to interpret their students' work (on thought-revealing, model-eliciting activities) for the purpose of improving implementation and assessment of problem-solving performance. They found enormous changes in the teachers' reflections that corresponded with improved performance on the part of students. Clark and Lesh (2003) described the use of concept maps as a reflective tool in a multitiered design study. Teachers were asked to create a concept map for each student-level model-eliciting activity they had implemented in their classroom. The purpose of

[11] Conducting a series of studies in an attempt to establish "what works" seldom leads to conclusive results. This is because if the investigation indicates that the something "did not work," there exist so many alternative explanations for the lack of findings (e.g., poor implementation, student turnover, lack of resources) that the findings are explained away. Thus, the same research question concerning "what works" gets recycled again and again, resulting in a series of conflicting findings leading to inconclusive results.

the concept map was to provide other teachers with a curricular guide that would identify the concepts and skills elicited in student work on the particular activity, and to identify the state and district standards that were aligned with the expected student work on the problem. The development of the concept maps proved to be a nontrivial problem for teachers and was also thought revealing. Over professional-development cycles, teachers' concept maps evolved from literally describing components found in various standards documents to interpretations that grappled with the notion of *big mathematical ideas*, the differences in skills, concepts, and strategies, and eventually to deep discussions that identified concepts and connections between additive and multiplicative reasoning.

Schorr and Koellner-Clark (2003) conducted a case study of one teacher's development in the context of studying his own students' thinking as the students engaged in model-eliciting activities. Schorr and Koellner-Clark were able to avoid a common problem in which professional development results in teachers who can talk about "reform" but do not actually implement what they are talking about. In this case they documented how a teacher actually began to think *differently* about his student's mathematical thinking—which is essential for making permanent changes in perceptions about teaching and learning mathematics.

A number of recent dissertation studies have also been based on multitiered design experiments. For example, Carmona (2004) focused on developing assessment and evaluation tools for describing and scoring students' work. She organized her dissertation as a multitiered design study to document the teachers' and the assessment team's models for interpreting students' work. She found that as teachers and assessment team members worked on ways to document students' understanding, they noticed more and more about students' thinking.

Another example is Richardson (2004), who was interested in the diffusion of students' knowledge among peers, in contrast to the dissemination of teacher knowledge to students. She facilitated teacher development of mechanisms to enhance student diffusion of knowledge during small-group solutions of model-eliciting activities. For example, one of the teachers provided problem-solving teams with a place in the classroom to go to get needed data, providing an opportunity for cross-group interaction. The teacher expected that interactions across groups would result in solutions that converge along one path. Instead, Richardson and the teacher discovered that although students and groups appropriated the ideas of others, they fit the new ideas into their own line of reasoning. They further developed their own ideas, sometimes eliminating ideas that they found to be useless, and sometimes creating totally new lines of reasoning.

Hjalmarson (2004) also used a multitiered design study in which teachers were motivated to develop "presentation tools" to enhance groups' class presentations of their solutions to model-eliciting activities. Each of the teachers expressed an interest in trying to get students to think more deeply about each other's mathematical models. The teachers each developed their own "tool" (e.g., peer feedback form, planned teacher questions) to enhance the presentation phase of the activity and then revised or changed the tool on the basis of what was learned over a series of model-eliciting activity presentations. The trail of documentation produced by the teachers provided a window into the teachers' ways of thinking about students' learning to present mathematical models to other students.

The main characteristics of multitiered design studies are (a) that activities used by both teachers and their students are designed to be thought-revealing and model-eliciting, (b) that the activities devised require teachers to design a product that is important to them, which results in iterations of expressing, testing, and revising one's own models, and (c) that teachers' models for learning and teaching problem solving are studied as teachers examine the models students design in response to a model-eliciting problem. An assumption underlying this paradigm is that not only will the students and the teachers change as a result of their experience, so too will the professional-development provider and researcher. The role of the latter two participants, then, is to prompt iterative modeling cycles by students and teachers, and *also* to document their own iterative cycles as they revise their own understandings of model development by teachers and students. Clearly, the type of research described is complex and would be enhanced by a cluster of companion studies that use different perspectives, focusing on different participants, and therefore introducing greater diversity of thought into the research process.

An Illustration: The Evolving Models-and-Modeling Perspective

One of the most significant distinguishing characteristics of models-and-modeling perspectives on mathematics learning and problem solving is the assumption that the theory, model, or research perspective being used by the researcher needs to be expressed, tested, and revised systematically as part of the research process—and that these researcher-level models develop along a number of dimensions similarly to the way student-level and teacher-level models develop. In particular, research models should be expected to evolve not only within a particular study, but also over series of studies. For example, the models-and-modeling perspective being proposed in this chapter has gone through a number of major shifts in thinking, and corresponding revisions, based on studies over time. As a close to the chapter, two of the dimensions along which the proposed models-and-modeling perspective has evolved are used to illustrate how the researchers' model is under design while also guiding the work of the research. In particular, we share a glimpse of how the principles for developing model-eliciting activities emerged from early work in understanding the development of rational number ideas and applied problem solving and we share a glimpse of how the researchers' view of students' mathematical knowledge evolved.

Two projects implemented by Lesh and colleagues in the 1980s, the Applied Problem Solving Project (Lesh, Landau, & Hamilton, 1983) and the Rational Numbers Project (Lesh, Post, & Behr, 1985), launched the process of expressing, testing, and revising a perspective on learning and teaching mathematics, which later came to be called a models-and-modeling perspective. The goal of these research projects was to build on the work of Piaget (e.g., Piaget, 1952; Piaget & Inhelder, 1975; Piaget et al., 1960), who studied the evolution of mathematical cognitive structures over long periods of time. Although these projects were designed to do the same, they were designed to look at smaller nuggets—local concept development—over short periods of times, such as in one class period.

In the Applied Problem Solving Project, researchers used complex problem-solving tasks as research sites in which they videotaped small groups of students engaged in collaborative problem solving. One of the early tasks used was the Chicago-to-Tokyo problem, which required students to determine the distance from Chicago (their home town) to Tokyo on different representations of the world (a globe, a Mercador projection map—which is a rectangular map where land masses near the poles look very large, and an Interrupted Mollweide projection map with simplified lobes—which looks like a flattened-out orange peel). The small groups of students who were engaged in solving the problem were usually surprised to find that each map resulted in different distances. Their transcribed conversations were studied and revealed that in successful groups, although students' initial approaches to reconciling the differences could be characterized as a collection of independent, uncoordinated, egocentric ideas, over the course of the problem-solving session, as students tested their ideas on each other, they began to take into account each other's perspectives. They gradually developed a coherent explanation of the discrepancies and decided together on a distance that made sense to them all. Although the process of expressing, testing, and revising their ideas was the goal of the task designed, the students' final mathematical model could only be inferred from the videotaped conversations. The researchers realized that if the problem-solving activities could be designed to capture the group's mathematical way of thinking as a final product, subsequent research studies would be strengthened. Thus, the first principle for task design emerged: that students' mathematical models would be documented in writing as the "answer" to the problem. (This was later called the "model-documentation" principle.)

As researchers grappled with the desire to have students' final written responses be more than the production of a numerical/unitized answer (e.g., the distance between Tokyo and Chicago), they encountered the challenge of designing activities that revealed students' mathematical models as *answers* to complex problems. As a result, a process of expressing, testing, and revising a full set of task design principles began. For example, given that a mathematical model is seldom required as a solution to an isolated problem situation, the model-eliciting activity design needed to require that the students' solution (i.e., model) be shared with other people or used on other sets of data. Thus, the generalizability principle (later called the sharability and reusability principle) was born and was included in an early set of design principles published in Lesh and Lamon (1992). Later iterations of the six principles for task design have been published by Lesh, Hoover, and Kelly (1993) and Lesh, Hoover, et al. (2000).

Another dimension along which models-and-modeling perspectives developed was how students' knowledge of mathematics was viewed. In both the Applied Problem Solving Project and the Rational Numbers Project, activities were designed to reveal information about students' understanding of conventional mathematical topics (e.g., students' rational number conceptual system, students' proportional reasoning system). However, the task designers learned that the more realistic the model-eliciting activity, the more students brought unexpected and contextually based mathematical ideas to bear on the problem. Hindsight is always so clear—because realistic modeling problems almost always have more than one reasonable solution students were bringing their own conceptual systems to bear, and these conceptual systems were largely situated in their own experience and were often multitopic with respect to conventional mathematics. As a result, the researchers' interpretations of students' mathematical models became less oriented around conventional mathematics topics and more associated with the *local conceptual models* students developed for the given problematic situation.

The models-and-modeling perspective on mathematical problem solving and learning has evolved along a number of dimensions and has been developed by a cadre of researchers. Organizing the study of mathematical problem solving around mathematics—as the study of systems, and models of those systems—can result in the simultaneous enhancement of learning and instruction while also increasing the field's understanding of what constitutes mathematical problem solving and learning. Future research needs to embark on investigation of the many questions posed in this chapter, to involve an ever-increasing community of experts who can provide the needed conceptual shift in the field, and to continue to grow and develop the notion of a models-and-modeling perspective on research in mathematical problem solving.

REFERENCES

Accreditation Board for Engineering and Technology (ABET). (2004). Criteria for Accrediting Engineering Programs. Retrieved September 7, 2005, from http://www.abet.org/forms.shtml

Amit, M. (1988). Career choice, gender, and attribution patterns of success and failure in mathematics. In A. Borbas (Ed.), *Proceedings of the 12th annual conference of PME(Psychology of Mathematics Education), Veszprem, Hungary* (Vol. 1, pp. 125–130). (ERIC Document Reproduction Service No. ED411128).

Archambault, R. D. (Ed.). (1964). *John Dewey on education: Selected writings.* Chicago: University of Chicago Press.

Balanced Assessment Project for the Mathematics Curriculum (1999). *Advanced High School Assessment Package 1.* White Plains, NY: Dale Seymour Publications.

Begle, E. G. (1979). *Critical variables in mathematics education.* Washington DC: The Mathematics Association of America and the National Council of Teachers of Mathematics.

Boaler, J. (2000). Exploring situated insights into research and learning. *Journal for Research in Mathematics Education, 31,* 113–119.

Boaler, J. ,& Greeno, J. G. (2000). Identity, agency, and knowing in mathematics worlds. In J. Boaler (Ed), *Multiple perspectives on mathematics teaching and learning.* London: Ablex Publishing.

Brownell, W. A. (1970). Psychological considerations in the learning and teaching of arithmetic. In J. K. Bidwell, & R. G. Clason (Eds.), *Readings in the history of mathematics education* (pp. 504–530). Washington DC: National Council of Teachers of Mathematics. (Reprinted from *Tenth yearbook of the National Council of Teachers of Mathematics: The teaching of arithmetic,* pp. 1–31, 1935. New York: Bureau of Publication, Teachers College, Columbia University.)

Bruner, J. (1960). *The process of education.* New York: Vantage Books.

Campione, J. C., Brown, A. L., & Connell, M. L. (1989). Metacognition: On the importance of understanding what you are doing. In R. I. Charles & E. A. Silver (Eds.), *The teaching and assessing of mathematical problem solving* (pp. 93–114). Reston, VA: National Council of Teachers of Mathematics.

Capobianco, B. (2003). SGMM summary report: ENGR 106 Freshman Engineering interview study fall 2002. West Lafayette, IN: Purdue University, Department of Curriculum and Instruction.

Carmona, G. (2004). Designing an assessment tool to describe students' mathematical knowledge. Unpublished doctoral dissertation, Purdue University, West Lafayette, IN.

Carraher, D.W., & Schliemann, A.D. (2002). Is everyday mathematics truly relevant to mathematics education? In M. Brenner & J. Moschkovich (Eds.), Everyday and academic mathematics in the classroom. *Journal for Research in Mathematics Education Monograph, 11,* 131–153.

Carraher, T. N., Carraher, D. W., & Schliemann, A. D. (1985). Mathematics in the streets and in schools. *British Journal of Developmental Psychology, 3*(1), 21–29.

Clark, K. K., & Lesh, R. (2003) A modeling approach to describe teacher knowledge. In R. Lesh, & H. M. Doerr, (Eds.), *Beyond constructivism: Models and modeling perspectives on mathematics problem solving, learning, and teaching* (pp. 175–190). Mahwah, NJ: Erlbaum.

Cook, S. D. N., & Yanow, D. (1993). Culture and organizational learning. *Journal of Management Inquiry, 2*(4), 373–390.

Cronbach, L. J., & Snow, R. E. (1977). *Aptitudes and instructional methods: A handbook for research on interactions.* New York: Irvington Publishers.

Cross, N. (1994). *Engineering design methods: Strategies for product design.* Chichester, England: John Wiley & Sons.

Csikszentmihalyi, M. (1990). *Flow: The psychology of optimum experience.* New York: Harper Perennial.

Cuoco, A., Goldenberg, E. P., & Mark J. (1996). Habits of mind: An organization principle for mathematics curricula. *Journal of Mathematical Behavior, 15*(4), 375–402.

Dark, M. (2003). A models and modeling perspective on skills for the high performance workplace. In R. Lesh & H. Doerr (Eds.), *Beyond constructivism: Models and modeling perspectives on mathematics problem solving, Learning and teaching* (pp. 297–316). Mahwah, NJ: Erlbaum

Dawkins, R. (1990). *The selfish gene.* Oxford, U.K.: Oxford University Press.

DeBellis, V. A. (1998). Mathematical intimacy: Local affect in powerful problem solvers. In S. Berenson et al. (Eds.), *Proceedings of the 20th annual meeting of PME-NA* (Vol. 2, pp. 435–440). Columbus, OH: ERIC.

DeBellis, V. A., & Goldin, G. A. (1997). The affective domain in mathematical problem solving. In E. Pehkonen (Ed.), *Proceedings of the 21st annual conference of PME* (Vol. 2, pp. 209–216). Helsinki, Finland: University of Helsinki Dept. of Teacher Education.

DeBellis, V. A., & Goldin, G. A. (1999). Aspects of affect: Mathematical intimacy, mathematical integrity. In O. Zaslavsky (Ed.), *Proceedings of the 23rd annual conference of PME* (Vol. 2, pp. 249–256). Haifa, Israel: Technion, Dept. of Education in Technology and Science.

Dewey, J. (1933). *How we think.* Washington D.C.: Heath and Company.

Diefes-Dux, H., Hjalmarson, M., Miller, T, & Lesh, R. (under development). What are model-eliciting activities? In J.

Zawojewski, H. Diefes-Dux & K. Bowman (Eds.), *Models and modeling in engineering education: Designing experiences for all students.*

Diefes-Dux, H., Moore, T., Zawojewski, J. S., Imbrie, P. K., & Follman, D. (2004, October 23). *A framework of posing open-ended engineering problems: Model-eliciting activities.* Proceedings of the 34th ASEE/IEEE Frontiers in Education Conference: Savannah, GA. Retrieved February 7, 2006, from http://fie.engrng.pitt.edu/fie2004/papers/1719.pdf

Dienes, Z. (1960). *Building up mathematics.* London: Hutchinson Educational Ltd.

diSessa, A. A. (2002). Students' criteria for representational adequacy. In K. Gravemeijer, R. Lehrer, B. v. Oers & L. Verschaffel (Eds.), *Symbolizing, modeling and tool use in mathematics education* (pp. 105–129). Dordrect, The Netherlands: Kluwer.

diSessa, A. A., Hammer, D., Sherin, B., & Kolpakowski, T. (1991). Inventing graphing: Meta-representational expertise in children. *Journal of Mathematical Behavior, 10,* 117–160.

diSessa, A. A., & Sherin, B. L. (2000). Meta-representation: An introduction. *Journal of Mathematical Behavior, 19,* 385–398.

Doerr, H. M., & Lesh, R. A. (2003). A modeling perspective on teacher development. In R. A. Lesh & H. M. Doerr (Eds.), *Beyond constructivism: Models and modeling perspectives on mathematics problem solving, learning and teaching* (pp. 125–139). Mahwah, NJ: Erlbaum.

Driscoll, M. (1999). *Fostering algebraic thinking: A guide for teachers in grades 6–10.* Portsmouth, NH: Heinemann.

Driscoll, M., Moyer, J., & Zawojewski, J. S. (1998). Helping teachers implement algebra for all. *National Council of Supervisors of Mathematics Journal of Mathematics Education Leadership,* (1), 3–12.

Driscoll, M., Zawojewski, J. S., Humez, A., Nikula, J., Goldsmith, L., & Hammerman, J. (2001). *Algebraic Thinking Tool Kit.* Portsmouth, NH: Heinemann.

Elstein, A. S., Shulman, L., & Sprafka, S. A. (1978). *Medical problem solving: Analysis of clinical reasoning.* Cambridge, MA: Harvard University Press.

English, L. & Lesh, R. (2003). Ends-in-view problems. In R. Lesh & H. Doerr (Eds.), *Beyond constructivism: Models and modeling perspectives on mathematics problem solving, learning and teaching* (pp. 297–316). Mahwah, NJ: Erlbaum.

Fein, P. (2005). *You can quote me on that: Greatest tennis quips, insights and zingers.* Dulles, VA: Potomac Books.

Flavell, J. H. (1976). Metacognitive aspects of problem solving. In L. Resnick (Ed.), *The nature of learning* (pp. 231–236). Hillsdale, NJ: Erlbaum.

Gagne, R. M. (1975). *Essentials of learning for instruction.* New York: Holt, Reinhardt & Winston.

Gagne, R. M., & Briggs, L. J. (1974). *Principles of instructional design.* New York: Holt, Reinhardt & Winston.

Gainsburg, J. (2003a, April,). *Abstraction and concreteness in the everyday mathematics of structural engineers.* Paper presented at the annual meeting of the American Education Research Association, Chicago, IL.

Gainsburg, J. (2003b). *The mathematical behavior of structural engineers.* (Doctoral dissertation, Stanford University, 2003). Dissertation Abstracts International, A 64, 5.

Gainsburg, J. (2006). The mathematical modeling of structural engineers. *Mathematical Thinking and Learning, 8* (1), 3–36.

Goldin, G. A. (2002). Affect, meta-affect, and mathematical belief structures. In G. C. Leder, E. Pehkonen, & G. Törner (Eds.), *Beliefs: A hidden variable in mathematics education?* (pp. 59–72). Dordrecht, The Netherlands: Kluwer.

Goldin (in press). Aspects of affect and mathematical modeling processes. In R. Lesh, E. Hamilton, & J. Kaput (Eds.) *Models & modeling as foundations for the future in mathematics education.* Mahwah, NJ: Erlbaum.

Goldin, G. A., & McClintock, C. E. (Eds.). (1984). *Task variables in mathematical problem solving.* Hillsdale, NJ: Erlbaum.

Goldenberg, E. P., Shteingold, N., & Feurzeig, N. (2003). Mathematical habits of mind for young children. In F. K. Lester & R. I. Charles (Eds.). *Teaching mathematics through problem solving: Prekindergarten–Grade 6* (pp. 15–29). Reston, VA: National Council of Teachers of Mathematics.

Greeno, J. (1991). Number sense as situated knowing in a conceptual domain. *Journal for Research in Mathematics Education, 22*(3), 170–218.

Greeno, J. (1998). The situativity of knowing, learning, and research. *American Psychologist, 53*(1), 5–26.

Greeno, J. G. (2003). Situative research relevant to standards for school mathematics. In J. Kilpatrick, W. G. Martin & D. Schifter (Eds.), *A Research companion to Principles and Standards for School Mathematics* (pp. 304–332). Reston, VA: The National Council of Teachers of Mathematics.

Greeno, J., & The Middle-School Mathematics Through Applications Project Group (1997). Theories and practices of thinking and learning to think. *American Journal of Education, 106,* 85–126.

Hall, R. (1999a). Following mathematical practices in design-oriented work. In C. Hoyles, C. Morgan, & G. Woodhouse (Eds.), *Studies in mathematics education series: No. 10. Rethinking the mathematics curriculum.* Philadelphia: Falmer Press.

Hall, R. (1999b). *Case studies of math at work: Exploring design-oriented mathematical practices in school and work settings.* (NSF Rep. No. RED-9553648).

Hamilton, E., Lesh, R., Lester, F., & C. Yoon (in press). The use of reflective tools in building personal models of problem solving. In R. Lesh, E. Hamilton, & J. Kaput (Eds.), *Models & modeling as foundations for the future in mathematics education.* Mahwah, NJ: Erlbaum.

Heller, J., & Hungate, H. (1985). Implications for mathematics instruction of research on scientific problem solving. In E. A. Silver (Ed.), *Teaching and learning mathematical problem solving: Multiple research perspectives* (pp. 83–112). Hillsdale, NJ: Erlbaum.

Hjalmarson, M. A. (2004). *Designing presentation tools: A window into mathematics teacher practice.* Unpublished doctoral dissertation, Purdue University. West Lafayette, IN.

Hjalmarson, M. A., Diefes-Dux, H., & Moore, T. (under development). Designing modeling activities. In J. Zawojewski, H. Diefes-Dux & K. Bowman (Eds.), *Models and modeling in engineering education: Designing experiences for all students.*

Hoyles, C., Noss, R., & Pozzi, S. (2001). Proportional reasoning in nursing practice. *Journal for Research in Mathematics Education, 32,* 4–27.

Inhelder, B., & Piaget, J. (1958). *The growth of logical thinking* (A. Parsons & S. Milgram, Trans.) New York: Basic Books.

Izsák, A. (2003). "We want a statement that is always true": Criteria for good algebraic representations and the development of modeling knowledge. *Journal for Research in Mathematics Education, 34,* 191–227.

Johnson, T. & Lesh, R. (2003). A models and modeling perspective on technology-based representational media. In R. Lesh & H. Doerr (Eds.), *Beyond constructivism: Models and modeling perspectives on mathematics problem solving, learning and teaching.* Mahwah, NJ: Erlbaum.

Kaput, J. (1991). Notations and representations as mediators of constructive processes. In E. von Glasersfeld (Ed.), *Constructivism in mathematics education* (pp. 53–74). Dordrecht, The Netherlands: Kluwer.

Kaput, J., Hegedus, S., & Lesh, R. (in press). Technology becoming infrastructural in mathematics education. In R. Lesh, E. Hamilton & J. Kaput (Eds.), *Models & modeling as foundations for the future in mathematics education.* Mahwah, NJ: Erlbaum.

Kelly, E., & Lesh, R. (Eds.). (2000). *The handbook of research design in mathematics and science education.* Hillsdale, NJ: Erlbaum.

Kelly, E. A., Lesh, R. A., & Baek, J. (Eds.). (in press). *Design research in science, mathematics & technology education.* Hillsdale, NJ: Erlbaum.

Kilpatrick, J. (1969a). Problem solving in mathematics. *Review of Educational Research,* 39(4), 523–533.

Kilpatrick, J. (1969b). Problem solving and creative behavior in mathematics. In J. W. Wilson & L. R. Carry (Eds.), *Studies in mathematics; Vol. 19. Reviews of recent research in mathematics education* (pp. 153–187). Stanford, CA: School Mathematics Study Group.

Kilpatrick, J. (1985). A retrospective account of the past twenty-five years of research on teaching mathematical problem solving. In E. A. Silver (Ed.), *Teaching and learning mathematical problem solving: Multiple research perspectives* (pp. 1–16). Hillsdale, NJ: Erlbaum.

Koschmann, T., Glenn, P., & Conlee, M. (2000). When is a problem-based tutorial not tutorial? Analyzing the tutor's role in the emergence of a learning issue. In D. H. Evensen & C. E. Hmelo (Eds.), *Problem-based learning: A research perspective on learning interactions* (pp. 53–74). Mahwah, NJ: Erlbaum.

Kramarski, B., Mevarech, Z. R., & Arami, M. (2002). The effects of metacognitive instruction on solving mathematical authentic tasks. *Educational Studies in Mathematics, 49,* 225–250.

Krutetskii, V. (1976). *The psychology of mathematical abilities in school children.* Chicago, IL: University of Chicago Press.

Kuhn, D. (1999). A developmental model of critical thinking. *Educational Researcher, 28,* 16–25.

Lampert, M. (1990). When the problem is not the question and the solution is not the answer: Mathematical knowing and teaching. *American Educational Research Journal, 27,* 29–63.

Lave, J. (1988). *Cognition in practice: Mind, mathematics and culture in everyday life.* Cambridge, MA: Cambridge University Press.

Lave, J., & Wenger, E. (1991). *Situated learning: Legitimate peripheral participation.* Cambridge, UK: Cambridge University Press.

Lehrer, R., & Schauble, L. (2000). Inventing data structures for representational purposes: Elementary grade students' classification models. *Mathematical Thinking and Learning, 2*(1&2), 49–72.

Lesh, R. (1985). Conceptual analyses of problem solving performance. In E. A. Silver (Ed.), *Teaching and learning mathematical problem solving: Multiple research perspectives* (pp. 309–330). Hillsdale, NJ: Erlbaum.

Lesh, R. (1987). The evolution of problem representation in the presence of powerful conceptual amplifiers. In C. Janvier (Ed.), *Problems of representation in teaching and learning mathematics* (pp. 309–329). Hillsdale, NJ: Erlbaum.

Lesh, R. (Ed.). (2003). *Models and modeling perspectives* [Special issue]. *Mathematical Thinking and Learning, 5*(2&3).

Lesh, R., Cramer, K., Doerr, H. M., Post T., & Zawojewski, J. S. (2003). Model development sequences. In R. Lesh & H. Doerr (Eds.), *Beyond constructivism: Models and modeling perspectives on mathematics problem solving, learning and teaching* (pp. 35–58). Mahwah, NJ: Erlbaum.

Lesh, R., & Doerr, H. (1998). Symbolizing, communicating, and mathematizing: Key components of models and modeling. In P. Cobb & E. Yackel. *Symbolizing, communicating, and mathematizing.* Hillsdale, NJ: Erlbaum.

Lesh, R., & Doerr, H. (Eds.). (2003a). *Beyond constructivism: Models and modeling perspectives on mathematics problem solving, learning and teaching.* Mahwah, NJ: Erlbaum.

Lesh, R., & Doerr, H. (2003b). Foundations of a models and modeling perspective on mathematics teaching, learning, and problem solving. In R. Lesh & H. Doerr (Eds.), *Beyond constructivism: Models and modeling perspectives on mathematics problem solving, learning and teaching* (pp. 3–34). Mahwah, NJ: Erlbaum.

Lesh, R., Hamilton, E., & Kaput, J. (Eds.). (in press). *Models & modeling as foundations for the future in mathematics education.* Mahwah, NJ: Erlbaum.

Lesh, R., & Harel, G. (2003). Problem solving, modeling, and local conceptual development. *Mathematical Thinking and Learning, 5*(2 & 3), 157–190.

Lesh, R., Hoover, M., Hole. B., Kelly, E., & Post, T. (2000). Principles for developing thought-revealing activities for students and teachers. In A. Kelly & R. Lesh (Eds.), *Handbook of research design in mathematics and science education.* Mahwah, NJ: Erlbaum.

Lesh, R., Hoover, M., & Kelly, E. (1993). Equity, assessment, and thinking mathematically: Principles for the design of model-eliciting activities. In I. Wirszup & R. Streit (Eds.), *Developments in school mathematics education around the world: Vol. 3* (pp. 104–130). Reston, VA: National Council of Teachers of Mathematics.

Lesh, R. & Kelly, E. (2000). Multi-tiered teaching experiments. In E. Kelly & R. Lesh (Eds.), *Handbook of research design in mathematics and science education.* Mahwah, NJ: Erlbaum.

Lesh, R., Kelly, A. E. & Yoon, C. (in press). Multi-tier design experiments in mathematics, science and technology education. In A. E. Kelly, R. A. Lesh, & J. Baek (Eds.), *Handbook of design research in mathematics, science and technology education.* Mahwah, NJ: Erlbaum.

Lesh, R., & Lamon, S. (Eds.). (1992). *Assessment of authentic performance in school mathematics.* Washington, DC: American Association for the Advancement of Science.

Lesh, R., Landau, M., & Hamilton, E. (1983). Conceptual models and applied mathematical problem solving research. In R. Lesh & M. Landau (Eds.), *Acquisition of mathematics concepts and processes* (pp. 263–343). New York: Academic Press.

Lesh, R., Lester, F., & Hjalmarson, M. (2003). A models and modeling perspective on metacognitive functioning in everyday situations where problem solvers develop mathematical constructs. In R. Lesh & H. Doerr (Eds.), *Beyond constructivism: Models and modeling perspectives on mathematics problem solving, learning and teaching* (pp. 383–404). Mahwah, NJ: Erlbaum.

Lesh, R., Post, T., & Behr, M. (1985). Representations and translations among representations, in mathematics learning and problem solving. In C. Janvier (Ed.), *Toward a theory of mathematical representations.* Hillsdale, NJ: Erlbaum.

Lesh, R., & Yoon, C. (2004). Evolving communities of mind—in which development involves several interacting and simultaneously developing strands. *Mathematical Thinking and Learning, 6* (2), 205–226.

Lesh, R., Yoon, C., & Zawojewski, J. S. (in press). John Dewey revisited: Making mathematics practical vs. making practice mathematical. In R. Lesh, E. Hamilton, & J. Kaput (Eds.), *Models & modeling as foundations for the future in mathematics education.* Mahwah, NJ: Erlbaum.

Lesh, R., & Zawojewski, J. S. (1992). Problem solving. In T. Post (Ed.), *Teaching mathematics in grades K–8: Research-based methods* (pp. 49–88). Needham Heights, MS: Allyn and Bacon.

Lesh, R., Zawojewski, J. S. & Carmona, G. (2003). What mathematical abilities are needed for success beyond school in a technology-based age of information? In R. Lesh & H. Doerr (Eds.), *Beyond constructivism: Models and modeling perspectives on mathematics problem solving, learning and teaching* (pp. 205–221). Mahwah, NJ: Erlbaum.

Lester, F. (1994). Musings about mathematical problem-solving research: 1970–1994. *Journal for Research in Mathematics Education, 25,* 660–675.

Lester, F. K., & Charles, R. I. (Eds.). (2003). *Teaching mathematics through problem solving: Grades pre-k–6.* Reston, VA: National Council of Teachers of Mathematics.

Lester, F. K., & Kehle, P. E. (2003). From problem solving to modeling: The evolution of thinking about research on complex mathematical activity. In R. Lesh & H. Doerr, (Eds.), *Beyond constructivism: Models and modeling perspectives on mathematics problem solving, learning and teaching* (pp. 501–518). Mahwah, NJ: Erlbaum.

Lester, F., Garofalo, J., & Kroll, D. (1989). Self-confidence, interest, beliefs, and metacognition: Key influences on problem-solving behavior. In D. B. McLeod & V M. Adams (Eds), *Affect and mathematical problem solving: A new perspective* (pp. 75–88). New York: Springer-Verlag.

Levasseur, K., & Cuoco, A. (2003). Mathematical habits of mind. In H. L. Schoen (Ed.), *Teaching mathematics through problem solving.* Reston, VA: National Council of Teachers of Mathematics.

Maclean, R. (2001). Educational change in Asia: An overview. *Journal of Educational Change 2,* 189–192.

Magajna, Z., & Monaghan, J. (2003). Advanced mathematical thinking in a technological workplace. *Educational Studies in Mathematics, 52,* 101–122.

Mathematics Sciences Education Board. (1998). *High school mathematics at work: Essays and examples for the education of all students.* Washington, D.C.: National Academy Press.

McLeod, D. B. (1989). Beliefs, attitudes and emotions: New views of affect in mathematics education. In D. B. McLeod & V. M. Adams (Eds.), *Affect and mathematical problem solving: A new perspective.* New York, NY: Springer-Verlag.

McLeod, D. B., & Adams, V. M. (Eds.). (1989). *Affect and mathematical problem solving: A new perspective.* New York: Springer-Verlag.

Mead, G. H. (1962). *Mind, self, and society: From the standpoint of a social behaviorist.* (C. W. Morris, Ed.). Chicago, IL: University of Chicago Press. (Original work published 1934).

Mead, G. H. (1977). The social psychology of George Herbert Mead: Selected papers. In A. Strauss (Ed.), *Heritage of Sociology Series.* Chicago, IL: University of Chicago Press. (Original work published 1956).

Middleton, J. A., Lesh, R., & Heger, M. (2003). Interest, identify, and social functioning: Central features of modeling activity. In R. Lesh & H. Doerr (Eds.), *Beyond constructivism: Models and modeling perspectives on mathematics problem solving, learning and teaching* (pp. 405–433). Mahwah, NJ: Erlbaum.

Middleton, J. A., & Toluk, Z. (1999). First steps in the development of an adaptive, decision-making theory of motivation. *Educational Psychologist, 34,* 99–112.

Minsky, M. (1987). *The society of mind.* New York: Simon & Schuster.

Moore, T., & Diefes-Dux, H. (2004). Developing model-eliciting activities for undergraduate students based on advanced engineering content. *Proceedings of the 34th ASEE/IEEE Frontiers in Education Conference* (pp. F1A-9 to F1A-14). Savannah, GA.

Nathan, J. M. (2001). Making "thinking schools" meaningful: Creating thinking cultures. In J. Tan, S. Gopinathan, & W. K. Ho (Eds.), *Challenges facing Singapore education system today.* (pp. 35–49). Singapore: Prentice Hall.

National Academy of Engineering (2004). *The engineer of 2020: Visions of engineering in the new century.* Washington DC: The National Academy Press.

National Council of Teachers of Mathematics. (1989). *Curriculum and evaluation standards for school mathematics.* Reston, VA: Author.

National Council of Teachers of Mathematics. (1991). *Professional standards for teaching mathematics.* Reston, VA: Author.

National Council of Teachers of Mathematics. (1995). *Assessment standards for school mathematics.* Reston, VA: Author.

National Council of Teachers of Mathematics. (2000). *Principles and standards for school mathematics.* Reston, VA: Author.

National Research Council. (1999). *The changing nature of work: Implications for occupational analysis.* Washington DC: National Academy Press.

Newell, A., & Simon, H. A. (1972). *Human problem solving.* Englewood Cliffs, NJ: Prentice Hall.

Noble, T., Nemirovsky, R., Wright, T., & Tierney, C. (2001). Experiencing change: The mathematics of change in multiple environments. *Journal for Research in Mathematics Education, 32,* 85–108.

Noss, R., & Hoyles, C. (1996). The visibility of meanings: Modelling the mathematics of banking. *International Journal of Computers for Mathematical Learning,* 1, 3–31.

Noss, R., Hoyles, C., & Pozzi, S. (2002). Abstraction in expertise: A study of nurses' conceptions of concentration. *Journal for Research in Mathematics Education, 33,* 204–229.

Nunes, T., Schliemann, A. D., & Carraher, D. W. (1993). *Street mathematics and school mathematics.* Cambridge, U.K.: Cambridge University Press.

Oakes, W., & Rud, A. (2003). The EPICS model in engineering education: Perspective on problem-solving abilities needed for success beyond schools. In R. Lesh & H. M. Doerr (Eds.), *Beyond constructivism: Models and modeling perspectives on mathematics teaching, learning and problem solving.* (pp. 223–239). Mahwah, NJ: Erlbaum.

Papert, S. (1991). Situating constructivism. In I. Harel & S. Papert (Eds.), *Constructivism.* Norwood, NJ: Ablex.

Parker, A. (1998). SCANS and mathematics—Supporting the transition from schools to careers. In Mathematical Sciences Education Board (Ed.), *High school mathematics at work: Essays and examples for the education of all students.* Washington DC: National Academy Press.

Piaget, J. (1952). *The child's conception of number.* London: Routledge.

Piaget, J. (1970). *Genetic epistemology.* New York: W. W. Norton & Company.

Piaget, J., & Beth, E. (1966). *Mathematical epistemology and psychology.* Dordrecht, The Netherlands: D. Reidel.

Piaget, J., & Inhelder, B., (1975). *The origin of the idea of chance in children.* London: Routledgate & Kegan Paul

Piaget, J., Inhelder, B., & Szeminska, A. (1960). *The child's conception of geometry.* London: Routledge.

Pintrich, P. R. (2002). The role of metacognitive knowledge in learning, teaching and assessing. *Theory into Practice, 41*(4), 219–225.

Pôlya, G. (1957). *How to solve it* (2nd ed.). Princeton, NJ: Princeton University Press.

Richardson, S. (2004). *A design of useful implementation principles for the development, diffusion, and appropriation of knowledge in mathematics classrooms.* Unpublished doctoral dissertation, Purdue University, West Lafayette, IN.

Ridgway, J., Zawojewski, J. S., & Hoover, M. N. (2000). Problematising evidence based policy and practice. *Evaluation and Research in Education, 14* (3/4), 181–192.

Rosenstein, J. G. (in press). Discrete mathematics in 21st century education: An opportunity to retreat from the rush to calculus. In R. Lesh, E. Hamilton, & J. Kaput (Eds.), *Models & modeling as foundations for the future in mathematics education.* Mahwah, NJ: Erlbaum.

Saxe, G. B. (1988a). The mathematics of child street vendors. *Child Development, 5*(59), 1415–1425.

Saxe, G. B. (1988b). Candy selling and math learning. *Educational Researcher, 6*(14), 14–21.

Saxe, G. B. (1991). *Culture and cognitive development: Studies in mathematical understandings.* Hillsdale, NJ: Erlbaum.

Schoen, H. L., & Charles, R. I. (Eds.). (2003). *Teaching mathematics through problem solving: Grades 6–12.* Reston, VA: National Council of Teachers of Mathematics.

Schoenfeld, A. H. (1985). *Mathematical problem solving.* New York: Academic Press.

Schoenfeld, A. H. (1987). What's all the fuss about metacognition? In A. Schoenfeld (Ed.), *Cognitive science and mathematics education* (pp. 189–215). Hillsdale, NJ: Erlbaum.

Schoenfeld, A. H. (1992). Learning to think mathematically: Problem solving, metacognition, and sense making in mathematics. In D. Grouws (Ed.), *Handbook of research on mathematics teaching and learning* (p. 334–370). New York: McMillan.

Schorr, R. Y., & Koellner-Clark, K. (2003). Using a modeling approach to analyze the ways in which teachers consider new ways to teach mathematics. *Mathematical Thinking and Learning, 5*(2/3), 109–130.

Schorr, R. Y., & Lesh, R. (2003). A modeling approach for providing teacher development. In R. Lesh & H. Doerr (Eds.), *Beyond constructivism: Models and modeling perspectives on mathematics problem solving, learning and teaching* (pp. 141–158). Mahwah, NJ: Erlbaum.

Silver, E. A. (1979). Student perceptions of relatedness among mathematical verbal problems. *Journal for Research in Mathematics Education, 10,* 195–210.

Silver, E. A. (1981). Recall of mathematical information: solving related problems. *Journal for Research in Mathematics Education, 12,* 54–64.

Silver, E. A. (Ed.). (1985). *Teaching and learning mathematical problem solving: Multiple research perspectives.* Hillsdale, NJ: Erlbaum.

Stanic, G., & Kilpatrick, J. (1989). Historical perspectives on problem solving in the mathematics curriculum. In R. Charles & E. Silver (Eds.), *The teaching and assessing of mathematical problem solving* (Vol. 3. pp. 1–22). Reston, VA: National Council of Teachers of Mathematics.

Steffe, L, P., & Thompson, P. (2000). Teaching experiment methodology: Underlying principles and essential elements. In E. Kelly & R. Lesh (Eds.), *Handbook of research design in mathematics and science education* (pp. 267–306). Mahwah, NJ: Erlbaum.

Stein, M. K., Boaler, J., & Silver, E. A. (2003). Teaching mathematics through problem solving: Research perspectives. In H. L. Schoen & R. I. Charles (Eds.), *Teaching mathematics through problem solving: Grades 6–12* (pp. 245–256). Reston, VA: National Council of Teachers of Mathematics.

Tan, J. (2002). Education in the twenty-first century: Challenges and dilemmas. In D. da Cunha (Ed.), *Singapore in the new millennium: Challenges facing the city-state.* (pp. 154–186). Singapore: The Institute of Southeast Asian Studies.

Thayer, H. S. (Ed.). (1982). *Pragmatism: The classic writings.* Cambridge, MA: Hackett Publishing.

Thompson, P., & Sfard, A. (1994). Problems of reification: Representations and mathematical objects. In D. Kirschner (Ed.), *Proceedings of the Sixteenth Annual Meeting of the North American Chapter of the Internal Groups for the Psychology of Mathematics Education* (Vol. 1, pp. 3–34). Baton Rouge: Louisiana State University.

Vygotsky, L. S. (1978). *Mind in society: The development of higher psychological processes.* Cambridge, MA: Harvard University Press.

Wenger, E. (1998). *Communities of practice: Learning, meaning, and identity.* Cambridge, U.K.: Cambridge University Press.

Wenger, E. (2000). Communities of practice and social learning systems. *Organization, 7*(2), 225–246.

Wenger, E., & Snyder, W. (2000). Communities of practice: The organizational frontier. [Electronic Version]. *Harvard Business Review. 78*(1), 139–145.

Wilson, J., & Clark, D. (2004). Towards the modeling of mathematical metacognition. *Mathematics Education Research Journal, 16*(2), 25–48.

Yackel, E., & Cobb, P. (1996). Sociomathematical norms, argumentation, and autonomy in mathematics. *Journal for Research in Mathematics Education, 27,* 458–477.

Yanow, D. (2000). Seeing organizational learning: A "cultural" view. *Organization, 7*(2), 247–268.

Zawojewski, J., Chamberlin, M., Hjalmarson, M., & Lewis, C. (in press). The role of design experiments in teacher professional development. In A. E. Kelly, R. A. Lesh & J. Baek (Eds.). *Handbook of design research in mathematics, science and technology education.* Mahwah, NJ: Erlbaum.

Zawojewski, J., Diefes-Dux, H., & Bowman, K. (Eds.). (under development). *Models and modeling in engineering education: Designing experiences for all.*

Zawojewski, J., & Lesh, R. (2003). A models and modeling perspective on problem solving. In R. Lesh & H. Doerr (Eds.), *Beyond constructivism: Models and modeling perspectives on mathematic s problem solving, learning and teaching.* (pp. 317–336). Mahwah, NJ: Erlbaum.

Zawojewski, J., Lesh, R., & English, L. (2003). A models and modeling perspective on small group learning activity. In R. Lesh & H. Doerr (Eds.), *Beyond constructivism: Models and modeling perspectives on mathematics problem solving, learning and teaching* (pp. 337–358). Mahwah, NJ: Erlbaum.

AUTHOR NOTE

Financial support for preparation of this chapter and for the various projects mentioned herein was provided by Lucent Technologies (for the Case Studies for Kids Project) and the National Science Foundation (for the Applied Problem Solving Project, the Rational Numbers Project, and the Small Group Mathematical Modeling Approach to Improve Gender Equity in Engineering Project).

We appreciate the input from many people who read early drafts of this manuscript. In particular, the feedback and support given by Lyn English, Lynn Goldsmith, Eric Hamilton, Margret Hjalmarson, Marsha Landau, Marta Magiera, Terry Wood, and Caroline Yoon were very helpful. In addition, we thank graduate students enrolled in various research classes conducted at Purdue University, Indiana University and the Illinois Institute of Technology, instructed by the authors and Frank Lester, for their input.

18

TOWARD COMPREHENSIVE PERSPECTIVES ON THE LEARNING AND TEACHING OF PROOF

Guershon Harel

UNIVERSITY OF CALIFORNIA, SAN DIEGO

Larry Sowder

SAN DIEGO STATE UNIVERSITY

One of the most remarkable gifts human civilization has inherited from ancient Greece is the notion of mathematical proof. The basic scheme of Euclid's Elements *has proved astoundingly durable over the millennia and, in spite of numerous revolutionary innovations in mathematics, it still guides the patterns of mathematical communication.*

—Babai, 1991, pp. 1–2

This chapter examines proof in mathematics, both informal justifications and the types of justification usually called mathematical proofs. The introductory section below calls for what we label a "comprehensive perspective" toward the examination of the learning and teaching of proof, and identifies the various elements of such a comprehensive perspective. Our viewpoint next centers on students' outlooks on proof, as described by the "proof schemes" evidenced in students' work; the second section elaborates on this proof-scheme notion and includes a description of various proof schemes as well as a listing of the various roles proof can play in mathematics. The third section then gives a brief overview of how the idea of proof in mathematics has evolved historically, and why historical considerations could be a part of educational research on the learning or teaching of proof. The fourth and fifth sections include a look at selected studies[1] dealing with proof, at both the precollege and the college levels, with an effort to show the value of the proof-schemes idea. The final section offers some questions prompted by the earlier sections.

[1] To assure a degree of quality control and for practical reasons, we have restricted our survey to writings that have undergone an external review process, except in the case of a few doctoral dissertations that we examined. Examination of at least one case—the teaching and learning of mathematical induction—was omitted here since it has been treated thoroughly elsewhere (Harel, 2001). We regret that no doubt we have unintentionally overlooked other pieces of research or commentary that could have improved this chapter.

COMPREHENSIVE PERSPECTIVE ON PROOF[2]

No one questions the importance of proof in mathematics, and in school mathematics (Ball & Bass, 2003; Haimo, 1995; Schoenfeld, 1994). Overall, the performance of students at the secondary and undergraduate levels in proof is weak, as the findings reported in this paper will show. Whether the cause lies in the curriculum, the textbooks, the instruction, the teachers' background, or the students themselves, it is clear that the status quo needs, and has needed, improvement. Earlier reviews of research on the teaching and learning of proof (e.g., Battista & Clements, 1992; Hart, 1994; Tall, 1991; Yackel & Hanna, 2003) have informed and inspired more recent studies of proof learning. This chapter argues for "comprehensive perspectives" on proof learning and teaching and provides an example of such a perspective. A comprehensive perspective on the learning and teaching of proofs is one that incorporates a broad range of factors: mathematical, historical-epistemological, cognitive, sociological, and instructional. A unifying and organizing element of our perspective is the construct of "proof scheme." This construct emerged from a long sequence of studies into the concept of proof (Harel & Sowder, 1998), and, in turn, was utilized to address a cohort of foundational questions concerning these factors:

Mathematical and Historical-Epistemological Factors

1. What is proof and what are its functions?
2. How are proofs constructed, verified, and accepted in the mathematics community?
3. What are some of the critical phases in the development of proof in the history of mathematics?

Cognitive Factors

4. What are students' current conceptions of proof?
5. What are students' difficulties with proof?
6. What accounts for these difficulties?

Instructional-Socio-Cultural Factors

7. Why teach proof?
8. How should proof be taught?
9. How are proofs constructed, verified, and accepted in the classroom?
10. What are the critical phases in the development of proof with the individual student and within the classroom as a community of learners?
11. What classroom environment is conducive to the development of the concept of proof with students?
12. What form of interactions among the students and between the students and the teacher can foster students' conception of proof?
13. What mathematical activities—possibly with the use of technology—can enhance students' conceptions of proof?

This list does not purport to be exhaustive or unique; other researchers may choose a different list of questions or formulate those that appear here differently in their attempt to form their comprehensive perspective to proof. This list, however, represents those questions that we have confronted in a decade of investigations into the learning and teaching proof; hence, these questions constitute an essential part of the *content* of our comprehensive perspective on proof. Nor is the classification of these questions as it is outlined here invariable. Clearly, it is difficult—if not impossible—to classify these questions into pairwise disjoint categories of factors. Our use of the labels indicates a loose association rather than a firm classification. In particular, none of the questions should be viewed as a stand-alone question; rather they are all interrelated, constituting a cohort of questions that have guided the comprehensive perspective on proof offered here as an example. Thus, even though the headings of the third, fourth, and fifth sections of the chapter correspond to the three categories outlined in the above list, the content of each section pertains to the other two.

An important character of our perspective is that of *subjectivity:* While the term "proof" often connotes the relatively precise argumentation given by mathematicians, in our perspective "proof" is interpreted subjectively; a proof is what establishes truth for a person or a community. With this interpretation, "proof" connotes an activity that can permeate the whole mathematics curriculum, from kindergarten on as well as throughout the historical development of mathematics. This subjective notion of "proof" is the most central characteristic of the construct of "proof scheme," which will be defined shortly. When we wish

[2] All aspects of proof addressed in this paper must always be understood in the context of the learning and teaching of proof. Even when we address mathematical, historical, or philosophical aspects of proof, the goal is to utilize knowledge of these aspects for the purpose of better understanding the processes of learning and teaching of proof. Thus, phrases such as "research on proof," "perspective on proof," and the like should always be understood in the context of mathematics education, not of mathematics, history, or philosophy per se.

to make clear that we mean the mathematically institutionalized notion of proof, we will write "mathematical proof," otherwise the terms "proof" or "justification" will be used in this subjective sense.

Subjectivity equally applies to how the above questions should be understood. Some of these questions, particularly the first three, might be interpreted as calling for objective, clear-cut answers. Not so. The meaning of proof, its role, and the way it is created, verified, and accepted may vary from person to person and from community to community. One's answers to these questions are greatly influenced by her or his philosophical orientation to the processes of learning and teaching and would reflect her or his answers to questions such as: What bearing, if any, does the epistemology of proof in the history of mathematics have on the conceptual development of proof with students? What bearing, if any, does the way mathematicians construct proofs have on instructional treatments of proof? What bearing, if any, does everyday justification and argumentation have on students' proving behaviors in mathematical contexts? Our emphasis on subjectivity—a motive we will repeat a few times in this paper—stems from the well-known recognition that students' construction of new knowledge is based on what they already know (e.g., Cobb, 1994; Piaget, 1952, 1973a, b, 1978; Vygotsky, 1962, 1978), and hence it is indispensable for teachers to identify students' current knowledge, regardless of its quality, so as to help them gradually refine it. To avoid unnecessary misunderstanding we note here, and we will repeat later, that subjectivity toward the meaning of proof does not imply ambiguous goals in the teaching of this concept. Ultimately, the goal is to help students gradually develop an understanding of proof that is consistent with that shared and practiced in contemporary mathematics.

Why are Comprehensive Perspectives on Proofs Needed?

Comprehensive perspectives on proof are needed in an effort to understand students' difficulties, the roots of the difficulties, and the type of instructional interventions needed to advance students' conceptions of and attitudes toward proof. A single factor usually is not sufficient to account for students' behaviors with proof. For example, while some of the difficulties students have with proof can be accounted for by the cognitive factor, such as the students' lack of logical maturity and understanding of the need for proof, research studies conducted in the last two decades have given evidence that these sources are insufficient to provide a full picture of students' difficulties with mathematics in general and proof in particular. Balacheff's (1991) research, for example, has shown that students "have some awareness of the necessity to prove and some logic" (p. 176) and yet they experience difficulty with proof.

Official documents and research papers on the learning and teaching of proof almost never present their theoretical framework as a comprehensive perspective on proof, but stated curricular goals, instructional recommendations, research design, and so forth, may reveal some elements in the scholars' (potential) perspective on proof. For example, Balacheff (1991) points out that common to mathematics curricula in different parts of the world is the goal of training students in the construction and the formulation of "deductive reasoning," which is defined as "a careful sequence of steps with each step following logically from an assumed or previously proved statement and from previous steps" (National Council of Teachers of Mathematics [NCTM], 1989, p. 144). Balacheff points out that this characterization of proof, which is close to what a logician would formulate, is almost the same in mathematics curricula all over the world. Remarks in these documents on the conceptual development of proof with students stress that proof "has nothing to do with empirical or experimental verification" and "call attention to the move from concrete to abstract" (Balacheff, 1991, p. 177). This suggests a form-driven perspective, according to which the basis for one's answer to the question, how should proof be taught, is the form in which mathematics is organized and accordingly presented in scientific papers or books.

Approaching the concept of proof from an encompassing perspective such as the one suggested here is, in our view, a critical research need. Balacheff, a world-recognized pioneer in the area of the learning and teaching of proof, has addressed this need by raising on several occasions (e.g., 2002) the question: "Is there a shared meaning of 'mathematical proof' among researchers in mathematics education?" Balacheff was not referring to just the standard, more or less formal definition of mathematical proof. Rather, his question, as he puts it, is "whether beyond the keywords, we had some common understanding" (Balacheff, 2002, p. 23). By "common understanding," we believe he means agreed-upon parameters in terms of which one can formulate differences among perspectives into research questions. We agree with Balacheff that without such an understanding it is hard to envision real progress in our field. The comprehensive perspective on proof presented in this paper delineates a set of such parameters.

THE CONCEPT OF PROOF SCHEME

Consistent with our characterization of a comprehensive perspective on proof, several critical factors must be considered in addressing the question, What is proof? First, the construction of new knowledge does not take place in a vacuum but is shaped by existing knowledge. What a learner knows now constitutes a basis for what he or she will know in the future. This fundamental, well documented fact has far-reaching instructional implications. When applied to the concept of proof—our concern in this paper—this fact requires that an answer to the above question takes into account the student as a learner, that is, the cognitive aspects involved in the development of the concept of proof with the individual student. Second, one must maintain the integrity of the concept of proof as has been understood and practiced throughout the history of mathematics. Third, since the concept of proof is social—in that what is offered as a convincing argument by one person must be accepted by others—one must take into account the social nature of the proving process. We see, therefore, that one's answer to the above question attends to a range of factors: cognitive, mathematical, epistemological-historical, and social. The approach we provide here takes into consideration these factors. The conceptual framework of our answer has been formed over a long period of extensive work on students' conception of proof, and it incorporates findings reported in the literature as well as from our own studies. The latter included a range of teaching experiments and a three-year longitudinal study (see, for example, Harel & Sowder, 1998; Harel, 2001; Sowder & Harel, 2003) as well as historical, philosophical, and cultural analyses (e.g., Hanna, 1983, Hanna & Jahnke, 1996, Harel, 1999, Kleiner, 1991). The foundational element of this framework is the concept of "proof scheme."

Obviously, learners' knowledge, in general, and that of proof, in particular, is not homogenous; most commonly, a high-school student's conception of proof is different from that of a college mathematics major student, and a contemporary mathematician's conception of proof is different from that of an ancient mathematician, which, in turn, was different from that of a Renaissance mathematician, etc. "Proof scheme" is a term we use to describe one's (or a community's) conception of proof; it will be defined precisely below. In Harel and Sowder (1998) we offer an elaborate taxonomy of students' proof schemes based on seven teaching experiments with a total of 169 mathematics and engineering majors. The experiments were conducted in classes of linear algebra (elementary and advanced), discrete mathematics, geometry, and real analysis. Later, in Harel (2006), this taxonomy was refined and extended to capture more observations of students' conceptions of proof as well as some developments of this concept in the history of mathematics. In this paper we use our taxonomy of proof schemes to describe or interpret

(a) historical developments and philosophical matters concerning the concept of proof (third section),
(b) results from other studies concerning students' conception of and performance with proof (fourth section), and
(c) curricular and instructional issues concerning proof (fifth section).

Definition of "Proof Scheme"

The definition of "proof scheme" is based on three definitions.

1. ***Conjecture versus fact.*** An assertion can be conceived by an individual either as a *conjecture* or as a *fact*. (A conjecture is an assertion made by an individual who is uncertain of its truth.) The assertion ceases to be a conjecture and becomes a fact in the person's view once he or she becomes certain of its truth.

A critical question, to which we will return shortly, is the following: How in the context of mathematics do students render a conjecture into a fact? That is, how do they become certain about the truth of an assertion in mathematics?

The definitions of "conjecture" and "fact" are the basis for the notion of *proving*.

2. ***Proving.*** *Proving* is the process employed by an individual (or a community) to remove doubts about the truth of an assertion.

The process of proving includes two subprocesses: *ascertaining* and *persuading*.

3. ***Ascertaining versus persuading.*** *Ascertaining* is the process an individual (or a community) employs to remove her or his (or its) own doubts about the truth of an assertion. *Persuading* is the process an individual or a community employs to remove others' doubts about the truth of an assertion.

Mathematics as sense-making means that one should not only ascertain oneself that the particular topic/procedure makes sense, but also that one should be able to convince others through explanation and justification of her or his conclusions. In particular, the convince-others, public side of proof is a social practice (e.g., Bell, 1976), not only for mathematicians but for all students of mathematics. The definition of "persuading" aims at capturing this essential feature of proving. As defined, the process of proving includes two processes: *ascertaining* and *persuading*. Seldom do these processes occur in separation. Among mathematically experienced people and in a classroom environment conducive to intellectual interactions among the students and between the students and the teacher, when one ascertains for oneself, it is most likely that one would consider how to convince others, and vice versa. Thus, proving emerges as a response to cognitive-social needs, rather than exclusively to cognitive needs or social needs—a view consistent with Cobb and Yackel's *emergent perspective* (1996).

As defined, ascertaining and persuading are entirely subjective, for one's proving can vary from context to context, from person to person, from civilization to civilization, and from generation to generation within the same civilization (cf. Harel & Sowder, 1998, Kleiner, 1991, Raman, 2003, Weber, 2001). Thus, we offer the following definition:

4. ***Proof scheme.*** A person's (or a community's) proof scheme consists of what constitutes ascertaining and persuading for that person (or community).

We repeat here what we highlighted in Harel and Sowder (1998): Our definitions of the process of proving and proof scheme are deliberately student-centered. Terms such as "to prove," "to conjecture," "proof," "conjecture," and "fact" must be interpreted in this subjective sense throughout the paper. We emphasize again that despite this subjective definition the goal of instruction must be unambiguous—namely, to gradually refine current students' proof schemes toward the proof scheme shared and practiced by contemporary mathematicians. This claim is based on the premise that such a shared scheme exists and is part of the ground for scientific advances in mathematics.

Taxonomy of Proof Schemes

The taxonomy of proof schemes consists of three classes: the *external conviction proof schemes class*, the *empirical proof schemes class*, and the *deductive proof schemes class*. Below is a brief description of each of these schemes and some of their subschemes. Further elaboration of the different schemes is provided, as needed, throughout the paper. Relations of our taxonomy to other taxonomies and to the functions of proof within mathematics are addressed at the end of this subsection. (For the complete taxonomy, see Harel & Sowder (1998) and Harel (2006).)

External Conviction Proof Schemes

Proving within the *external conviction* proof schemes class depends (a) on an authority such as a teacher or a book, (b) on strictly the appearance of the argument (for example, proofs in geometry must have a two-column format), or (c) on symbol manipulations, with the symbols or the manipulations having no potential coherent system of referents (e.g., quantitative, spatial, etc.) in the eyes of the student (e.g., $(a+b)/(c+b)=(a+\not{b})/(c+\not{b})=a/c$). Accordingly, we distinguish among three proof schemes within the *external conviction* proof schemes class:

- External conviction proof schemes class
  - ➤ Authoritarian proof scheme
  - ➤ Ritual proof scheme
  - ➤ Non-referential symbolic proof scheme

Empirical Proof Schemes

Schemes in the *empirical proof scheme* class are marked by their reliance on either (a) evidence from examples (sometimes just one example) of direct measurements of quantities, substitutions of specific numbers in algebraic expressions, and so forth, or (b) perceptions. Hence, we distinguish between two proof schemes within the *empirical proof scheme* class:

- Empirical proof schemes class
  - ➤ Inductive proof schemes
  - ➤ Perceptual proof schemes

Deductive Proof Schemes

The *deductive proof schemes* class consists of two subcategories, each consisting of various proof schemes:

- Deductive proof schemes class
  - ➤ Transformational proof schemes
  - ➤ Axiomatic proof schemes

All the *transformational* proof schemes share three essential characteristics: *generality, operational thought*, and *logical inference*.

The *generality* characteristic has to do with an individual understanding that the goal is to justify a "for all" argument, not isolated cases and no exception is accepted.

Evidence that *operational thought* is taking place is shown when an individual forms goals and subgoals and attempts to anticipate their outcomes during the evidencing process. Finally, when an individual understands that justifying in mathematics must ultimately be based on logical inference rules, the *logical inference* characteristic is being employed (see also Harel, 2001).

Unlike the proof schemes in the previous two classes (the external conviction proof schemes class and the empirical proof schemes class), the transformation proof schemes require a more elaborate demonstration. Consider the following two responses (taken from Harel, 2001) to the problem:

Prove that for all positive integers n,

$$\log(a_1 \cdot a_2 \cdots a_n) = \log a_1 + \log a_2 + \cdots + \log a_n.$$

Response 1

$\log(4\cdot 3\cdot 7) = \log 84 = 1.924 \quad \log(4\cdot 3\cdot 6) = \log 72 = 1.857$

$\log 4 + \log 3 + \log 7 = 1.924 \quad \log 4 + \log 3 + \log 6 = 1.857$

Since these work, then

$$\log(a_1 \cdot a_2 \cdots a_n) = \log a_1 + \log a_2 + \cdots + \log a_n.$$

A probe into the reasoning of the students who provide responses of this kind reveals that their conviction stems from the fact that the proposition is shown to be true in a few instances, each with numbers that are *randomly* chosen—a behavior that is a manifestation of the empirical proof scheme.

Response 2

(1) $\log(a_1 a_2) = \log a_1 + \log a_2$ by definition

(2) $\log(a_1 a_2 a_3) = \log a_1 + \log a_2 a_3$. Similar to $\log(ax)$ as in step (1), where this time $x = a_2 a_3$.
Then

$$\log(a_1 a_2 a_3) = \log a_1 + \log a_2 + \log a_3$$

(3) We can see from step (2) any $\log(a_1 a_2 a_3 \cdots a_n)$ can be repeatedly broken down to $\log a_1 + \log a_2 + \cdots + \log a_n, n = 1,2,3,\ldots$

It is important to point out that in Response 2 the student recognizes that the process employed in the first and second cases constitutes a pattern that recursively applies to the entire sequence of propositions, $\log(a_1 a_2 \cdots a_n) = \log a_1 + \log a_2 + \cdots + \log a_n$, $n = 1,2,3,\ldots$.

In both responses the generalizations are made from two cases. This may suggest, therefore, that both are empirical. As is explained in Harel (2001), this is not so: Response 2, unlike Response 1, is an expression of the transformational proof scheme. To see why, one needs to examine the two responses against the definitions of the two schemes. While both responses share the first characteristic—i.e., in both the students respond to the "for all" condition in the log-identity problem statement—they differ in the latter two: whereas the mental operations in Response 1 are incapable of anticipating possible subsequent outcomes in the sequence and are devoid of general principles in the evidencing process, the mental operations in Response 2 correctly predict, on the basis of the general rule, $\log(ax) = \log a + \log x$, that the same outcome will be obtained in each step of the sequence. Further, in Response 1 the inference rule that governs the evidencing process is empirical; namely, $(\exists r \in R)(P(r)) \Rightarrow (\forall r \in R)(P(r))$. In Response 2, on the other hand, it is deductive; namely, it is based on the inference rule $(\forall r \in R)(P(r)) \wedge (w \in R) \Rightarrow P(w)$. (Here r is any pair of real numbers a and x, R is the set of all pairs of real numbers, $P(r)$ is the statement "$\log(ax) = \log a + \log x$," and w in step n is a pair of real numbers $a_1 a_2 \cdots a_{n-1}$ and a_n.)

The axiomatic proof scheme too has the three characteristics that define the transformational proof scheme, but it includes others. For now, it is sufficient to define it as a transformational proof scheme by which one understands that in principle any proving process must start from accepted principles (axioms). The situation is more complex, however, as we will show in the section on historical and epistemological considerations (third section). For the purpose of this chapter we will introduce only the Greek axiomatic proof scheme and the modern axiomatic proof scheme—as manifested, for example, in Euclid's *Elements* and Hilbert's *Grundlagen*, respectively. The distinction between these two schemes is further discussed in the third section.

Relations to Other Taxonomies

In broad terms, the empirical proof schemes and the deductive proof schemes correspond to what Bell (1976) calls "empirical justification" and "deductive justification," and Balacheff (1988) calls "pragmatic" justifications and "conceptual" justifications, respectively. Balacheff further divides the pragmatic justifications into three types of justifications: "naïve empiricism" (justification by a few random examples), "crucial experiment" (justification by carefully selected examples), and "generic example" (justification by an example representing salient characteristics of a whole class of cases). "Generic example" in our taxonomy belongs to the deductive proof scheme category, for an analysis similar to that we applied to Response 2 (above) will show that "generic example" satisfies the three characteristics of the transformational proof scheme. Balacheff further classifies conceptual justifications into two types: "thought experiment," where the justification is disassociated from specific examples, and "symbolic calculation," where the justification is based solely on transformation of symbols. In our taxonomy, the latter corresponds to the *referential symbolic* proof scheme. This scheme is the direct opposite of the *non-referential* proof scheme defined above. Recall this is a scheme where neither the symbols nor the operations one performs on them represent a coherent referential reality for the student. Rather, the student thinks and treats symbols and operations on them as if they possess a life of their own without reference to their functional or quantitative meaning. In the referential symbolic proof scheme, to prove or refute an assertion or to solve a problem, one learns to represent the situation algebraically and performs symbol manipulations on the resulting expressions, with the intention to derive information relevant to the problem at hand. We return to this scheme in the next section.

The above definitions and taxonomy are not explicit enough about many critical functions of proof within mathematics. There is a need to point to these functions due to their importance in mathematics in general and to their instructional implications in particular. For this, we point to the work of other scholars in the field, particularly the work by Hanna (1990), Balacheff (1988), Bell (1976), Hersh (1993), and de Villiers (1999). de Villiers, who built on the work of the others mentioned here, raises two important questions about the role of proof: (a) What different functions does proof have within mathematics itself? and (b) How can these functions be effectively utilized in the classroom to make proof a more meaningful activity? (p. 1). According to de Villiers, mathematical proof has six not mutually exclusive roles: verification, explanation, discovery, systematization, intellectual challenge, and communication. At the end of the next section, after the relevant schemes are defined, we show that these functions are describable in terms of the proof scheme construct.

MATHEMATICAL AND HISTORICAL-EPISTEMOLOGICAL FACTORS: SOME PHASES IN THE DEVELOPMENT OF PROOF SCHEMES IN THE HISTORY OF MATHEMATICS[3]

Deductive reasoning is a mode of thought commonly characterized as a sequence of propositions where one must accept any of the propositions to be true if he or she has accepted the truth of those that preceded it in the sequence. This mode of thought was conceived by the Greeks more than 20 centuries ago and is still dominant in the mathematics of today. So remarkable is the Greeks' achievement that their mathematics became an historical benchmark to which other kinds of mathematics are compared. We are here particularly interested in analyzing the proof scheme construct across three periods of mathematics: Greek mathematics, post-Greek mathematics (approximately from the 16th to the 19th century), and modern mathematics (from the late 19th century to today). For the sake of completeness we say a few words about pre-Greek mathematics.

Proof schemes are applied to establish assertions in specific contexts. In this respect, therefore, in addressing proof schemes one must attend to the nature of the context about which the assertions are made. Also of importance is the motive—the intellectual need—that might have brought about the conceptual change from one period to the next. Thus, our discussion will center on three interrelated aspects of the historical-epistemological development: (a) the context of proving, (b) the means of proving (i.e., proof schemes), and (c) the motive for conceptual change. Understanding these elements can shed light on some critical aspects of the learning and teaching of proof, as we will see.

Some Fundamental Differences

Practical World versus Ideal Existence: The Emergence of Structure

Generally speaking, pre-Greek mathematics was concerned merely with *actual* physical entities, particularly with quantitative measurement of different objects. Their "formulas" were in the form of prescriptions providing mostly approximations for measured quantities (area, volume, etc.). Ancient geometry developed in an empirical way through phases of trials and errors, where conjectures were proved by means of empirical—inductive or perceptual—evidence. Even Babylonian mathematics, which, according to Kleiner (1991, p. 291), "is the most advanced and sophisticated of pre-Greek mathematics," lacked the concept of proof as was understood and applied by the Greeks. Accordingly, it is safe to conclude that proving in pre-Greek mathematics is, by and large, governed by the empirical proof schemes.

The Greeks elevated mathematics from the status of practical science to a study of abstract entities. In their mathematics the particular entities under

[3] We are restricting our discussion to classical Greek mathematics and the mathematical developments that grew out of it. The development of deductive reasoning in China, India, and other non-Western cultures is not considered in this paper.

investigation are idealizations of experiential spatial realities and so also are the propositions on the relationships among these entities. The difference between pre-Greek mathematics and Greek mathematics, however, is not just in the nature of the entities considered—actual spatial entities versus their idealizations—but also in the reasoning applied to establish truth about the entities. In Greek mathematics, logical deduction came to be central in the reasoning process, and it alone necessitated and cemented the geometric edifice they created. To construct their geometric edifice the Greeks had to create primary terms—terms admitted without definition—and primary propositions—propositions admitted without proof—what the Greeks called axioms or postulates. In contrast, the mathematics of the civilizations that preceded them established their observations on the basis of empirical measurements, and so their mathematics lacked any apparent structure; it consisted merely of prescriptions of how to obtain measurements of certain spatial configurations.

Constant versus Varying Referential Reality

In constructing their geometry, as is depicted in Euclid's *Elements*, the Greeks had only one model in mind—that of imageries of idealized physical reality. This is supported by the fact that the Greeks, in fact, strove to describe their primitive terms (e.g., "a point is that which has no parts"), which indicates that their sole imagery was that of physical space. Hartshorne (2000) points out that this way of thinking did not start with Euclid. About one hundred years before Euclid, Plato spoke of the geometers:

> Although they make use of the visible forms and reason about them, they are not thinking of these, but of the ideals which they resemble; not the figures which they draw, but of the absolute square and the absolute diameter, and so on . . . (*The Republic*, Book IV).

From the vantage point of modern mathematics, neither the primitive terms nor the primary propositions in Greek mathematics were variables, but constants referring to a single spatial model (Klein, 1968; Wilder, 1967)—as is expressed in the ideal world of Plato's philosophy. This ultimate bond to a real-world context had an impact on the Greeks' proof scheme. Specifically, in his proofs Euclid often uses arguments that are not logical consequences of his initial assumptions but are rooted in humans' intuitive physical experience (the method of superposition, which allows one to move one triangle so that it lies on top of another triangle, is an example). In this respect, while proving in Greek mathematics was governed by the deductive proof scheme and can be characterized as an axiomatic proof scheme, it is different from that of contemporary mathematics where every assertion must be, in principle, derivable from clearly stated assumptions. Wilder (1967) points to a crucial difference between Greek mathematics and modern mathematics: Namely, in modern mathematics the primitive terms are treated as variables, not just undefined; they are free of any referent—real or imagined. In Greek mathematics, on the other hand, they are undefined terms referring to humans' idealized physical reality. Treating primitive terms as undefined is fundamentally different from treating them as variables. Wilder quotes Boole (from 1847) to stress this difference:

> The validity of the processes of analysis does not depend upon the interpretation of the symbols which are employed, but solely upon the laws of their combination. Every system of interpretations which does not affect the truth of the relations supposed, is equally admissible . . . (in Wilder, 1967, p. 116).

Despite the monumental conceptual difference in the referential imageries between Greek mathematics and modern mathematics, the essential condition in applying deductive reasoning in both is the existence of primary terms and primary propositions (axioms).

Content versus Form

The deductive reasoning—or the axiomatic method, as it often is called—of the Greeks dominated the mathematics of the western world until the late 19th century and in essence is still intact today. The difference in referential imageries between Greek mathematics and modern mathematics entails another critical difference: In Greek mathematics proof is valid by virtue of its *content*, not its *form*. Since the Greeks' concern was about relationships among entities in their physical space, the *form* of the proof cannot be completely detached from the *content* of that space. In this conception, one would not consider, for example, the question of whether there exists a consistent geometric model in which the parallel postulate does not hold. In modern mathematics, on the other hand, proof is valid by virtue of its form alone.

In *Greek Mathematical Thought and the Origin of Algebra*, Klein (1968) adds clarity to this distinction. In Greek science concepts are formed in continual dependence on their "natural" foundations, and their scientific meaning is abstracted from "natural," prescientific experience. In modern science, on the other hand, what is intended by the concept is not an object of immediate insight. Rather, it is an object whose scientific meaning can be determined only by its connection to other concepts, by the total edifice to which it belongs, and by its function within this edifice. This difference accounts, for example, for the fact that the Greeks, while interested in idealized entities, were highly selective in their choice of these entities. For instance, what we now call transcendental numbers, like the number π, disturbed the Greeks. "They were unhappy that their ideas of perfection in geometry and arithmetic seemed to be challenged by the existence of such a number" (Ginsburg & Opper, 1969, p. 214–215). In modern mathematics, on the other hand, entities can be quite arbitrary. Upon encountering transcendental numbers modern mathematics incorporated them into a mathematical structure, that of the real numbers, determined by a set of axioms.

It should be highlighted that the idea that the objects are determined by a set of axioms was a revolutionary way of thinking in the development of mathematics. An important manifestation of this revolution is the distinction between Euclid's *Elements* and Hilbert's *Grundlagen.* While the *Elements* is restricted to a single interpretation—namely that its content is a presumed description of human spatial realization—the *Grundlagen* is open to different possible realizations, such as Euclidean space, the surface of a half-sphere, ordered pairs and triples of real numbers, and so forth, including the interpretation that the axioms are meaningless formulas. In other words, the *Grundlagen* characterizes a structure that fits different models. This obviously is not unique to geometry. In algebra, a group or a vector space is defined to be any system of objects satisfying certain axioms that specify the structure under consideration. To reflect this fundamental conceptual difference, we refer to the Greeks' proving means as the *Greek axiomatic proof scheme* and to the proving means of modern mathematics as the *modern axiomatic proof scheme.*

Operation versus Results of an Operation

Klein (1968) argues that the revival and assimilation of Greek mathematics during the 16th century resulted in fundamental conceptual changes that ultimately defined modern mathematics. This conceptual transformation culminated in Vieta's work on symbolic algebra, where the distinction between modern mathematics and Greek mathematics began to crystallize.

The use of symbols in the modern sense is due mainly to Vieta, who was followed in this effort by Descartes and Leibniz. Until then, mathematics had evolved for at least three millennia with hardly any symbols. Symbolic notation was the key to a method of demonstration, and Liebniz—more than anyone else in his generation—advanced the role of symbolism in the process of mathematical proof. He was the first to conceive of proof as a sequence of sentences beginning with identities and proceeding by a finite number of steps of logic and rules of *definitional substitution*—by virtue of symbolic notation—until the theorem was proved.

The works of Vieta that led to the creation of algebra and that of Descartes that led the creation of analytic geometry illustrate another important difference between Greek mathematics and modern mathematics: the focus on "operation" versus "results of an operation." Ginsburg and Opper (1969) discuss the distinction as described by Boutroux, a mathematician, who analyzed the evolution of mathematical thought. According to Boutroux, the Greeks restricted their attention to attributes of spatial configurations but paid no attention to the operations underlying them. As an example, Ginsburg and Opper mention the problems of bisecting and trisecting an angle by means of straightedge-and-compass only. As is well known, the Greeks offered a simple way to solve the bisecting problem but the solution of the trisecting problem had to wait two millennia, until Galois, a French mathematician, solved it in the 19th century. The Greeks attended to geometrical objects (rectilinear angles in this case) by investigating their attributes—whether an angle can be bisected or trisected, for example. The 19th century mathematics, on the other hand, investigated the operation themselves—their algebraic representations and structures. Specifically, the standard Euclidean constructions using only a compass and straightedge were translated in terms of constructibility of the real numbers (i.e., a real number is constructible if its absolute value is the distance between two constructible points). This translation leads to an important observation about the structure of constructible numbers; namely, the constructible numbers form a subfield of the field of the real numbers. A deeper investigation into the theory of fields lead to a proof of the impossibility of certain geometric constructions, including the impossibility of finding a single method of construction for trisecting any given angle with the classical tools. More importantly, with this theory one can understand why certain constructions are possible whereas others are not. The Greeks had no means to build such an understanding. They did not attend to the nature of the operations underlying the Euclidean construction, and hence were unable to understand the difference between bisecting an angle and trisecting an angle and why they were able perform the former but not the latter.

Thus, not until the 17th century with the invention of analytic geometry and algebra did mathematicians begin to shift their attention from the result of mathematical operations to the operations themselves. By means of analytic geometry, mathematicians realized that all Euclidean geometry problems can be solved by a single approach, that of reducing the problems to equations and applying algebraic techniques to solve them. Euclidean straightedge-and-compass constructions were understood to be equivalent to equations, and hence the solvability of a Euclidean problem became equivalent to the solvability of its corresponding equation(s).

Cause versus Reason

Consider Proposition I.32 and its proof in Euclid's *Elements* (slightly adapted from Heath, 1956, pp. 316–317):

> In any triangle, if one of the sides is produced, then the exterior angle equals the sum of the two interior and opposite angles, and the sum of the three interior angles of the triangle equals two right angles.
>
> *Proof:*
>
> Let *ABC* be a triangle, and let one side of it *BC* be produced to *D*.

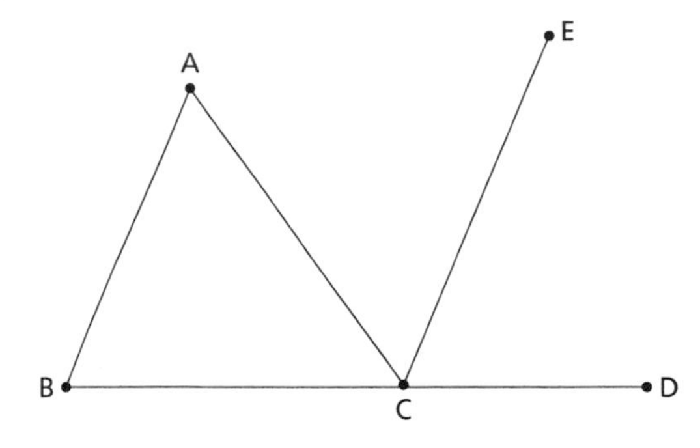

> I say that the exterior angle *ACD* equals the sum of the two interior and opposite angles *CAB* and *ABC,* and the sum of the three interior angles of the triangle *ABC, BCA,* and *CAB* equals two right angles.
>
> Draw *CE* through the point *C* parallel to the straight line *AB* (by Proposition I.31).
>
> Since *AB* is parallel to *CE,* and *AC* falls upon them, therefore the alternate angles *BAC* and *ACE* equal one another (by Proposition I.29).
>
> Again, since *AB* is parallel to *CE,* and the straight line *BD* falls upon them, therefore the exterior angle *ECD* equals the interior and opposite angle *ABC* (by Proposition I.29).
>
> But the angle *ACE* was also proved equal to the angle *BAC.* Therefore the whole angle *ACD* equals the sum of the two interior and opposite angles *BAC* and *ABC.*
>
> Add the angle *ACB* to each. Then the sum of the angles *ACD* and *ACB* equals the sum of the three angles *ABC, BCA,* and *CAB* (by Common Notion 2).
>
> But the sum of the angles *ACD* and *ACB* equals two right angles. Therefore the sum of the angles *ABC, BCA,* and *CAB* also equals two right angles (by Proposition I.13 and Common Notion 1).

This proof appeals to two facts, one about the auxiliary segment CE and the other about the external angle ACD. Note that the property holds whether or not the segment CE is produced and the angle ACD considered. One might then raise the question, what is the true cause of the property proved? This question was a center of debate during the 16th–17th centuries about whether mathematics is a *science.* Philosophers of this period, according to Mancosu (1996), used this proof to demonstrate their argument that mathematics is not a perfect science because "implication" in mathematics is a mere logical consequence rather than a demonstration of the *cause* of the conclusion. Their argument was based on the Aristotelian definition of science, according to which one does not understand something until he or she has grasped the why of it. "We suppose ourselves to possess unqualified scientific knowledge of a thing, . . . , when we think that we know the cause on which the fact depends as the cause of the fact and of no other . . ." (Aristotle, p. 111–112).

Mathematical statements of the form "A if and only if B" provided an additional argument against the scientificness (i.e., causal nature) of mathematics, for—these philosophers claimed—if mathematical proof were scientific (i.e., causal), then such a statement would entail that A is the cause of B and B is the cause of A, which implies A is the cause of itself—an absurdity.

This position entailed rejection of proof by contradiction, for such a proof does not demonstrate the cause of the property that is being argued. When a statement "A implies B" is proved by showing how not B (and A) leads logically to an absurdity, one does not learn anything about the causality relationship between A and B. Nor does one gain any insight into how the result was obtained. Consequently, proof by exhaustion (e.g., Archimedes' known method of proof for calculating volume, area, and parameter of different objects), which is necessarily based on proof by contradiction, also was unsatisfactory to many mathematicians of the 16th and 17th centuries. They argued that the ancients, who broadly used proof by exhaustion to avoid explicit use of infinity, failed to convey their methods of discovery.

Not all philosophers of the time held this position. According to Mancosu (1996), Barozzi, for example, argued that some parts of mathematics are more scientific (causal) than others; but that a proof by contradiction is not a causal proof, and therefore it should be eliminated from mathematics. Others, like Barrow, argued that all mathematics proofs are causal, including proof by contradiction:

> It seems to me . . . that Demonstrations, though some do outdo others in Brevity, Elegance, Proximity to their first Principles, and the like Excellencies, yet are all alike in Evidence, Certitude, Necessity, and the essential Connection and mutual Dependence of the Terms one with another. Lastly, that Mathematical Ratiocinations are the most perfect Demonstrations. (Quoted in Mancosu, 1996, p. 23)

Of particular interest is the position held by Rivaltus on the issue of causal proof in mathematics. Mancosu (1996) illustrates this position by Rivaltus' commentary on Archimedes' proof for the theorem that the area of the surface of the sphere is four times the area of a great circle of the sphere. In this proof, Archimedes inscribes and circumscribes the sphere with auxiliary solids to show that the surface of the sphere can be neither smaller nor greater than four times the great circle—a typical Archimedean proof by exhaustion, which necessarily involves proof by contradiction. So, there are two issues here: the use of proof by contradiction and the use of the auxiliary solids (as in the case of the proof of Proposition I.32). Each of these two features render, in the eyes of some philosophers of the time, Archimedes' proof noncausal. Rivaltus rejects this possibility on the basis of a distinction between "cause" and "reason":

> Ostensive demonstrations in mathematics are not considered more perfect than the ones by contradiction, since in these disciplines it is not made use of the cause of the thing, but of the cause of the knowledge of the thing. . . . The figures drawn are not truly the *cause* [italics added] of that equality but are *reasons* [italics added] from which we know it. From whence it follows that whatever is more fit to knowledge is more appropriate to the mathematician. But we know more easily that absurdities are impossible, false and repugnant by reason that we know the true things. Indeed the truths are concealed and conversely the errors are obvious everywhere. . . . Again it is to be observed that the Geometers do not make use of the cause of a thing, but of the cause from which the thing is known. Indeed it is sufficient to them to show the thing to be so and they do not enquire by which means it is so. (Rivaltus, 1615; quoted in Mancosu, 1996, pp. 26–27)

The Aristotelian theory of science and particularly its appeal to cause and effect manifested itself in another aspect of mathematical practice during the 17th century—that of the use of "genetic definitions." These definitions appeal to the generation of mathematical magnitude by motion; Euclid's definition of a sphere as an object generated by the rotation of a semicircle around a segment taken as axis is an example. An example of a non-genetic definition is that of a circle defined as a set of all points in the plane that are equidistant from a given point in that plane. According to Mancosu (1996), although the insistence on the use of genetic definitions was not universal, important mathematicians of the time emphasized their importance because they were viewed as demonstrating cause, and, hence, conform to Aristotle's epistemological position of what constitutes a science. For example, Barrow stated that of all the possible ways of generating a magnitude, the most important is the method of local movements, and he uses motion as a fundamental concept in his work in geometry" (Mancosu, 1996, p. 96), and Hobbes and Spinoza, "emphasize the role of genetic definitions as the only causal definitions, thereby excluding the nongenetic definitions from the realm of science" (p. 99).

Motives

The Greeks' motive for constructing the remarkable geometric edifice that we now call Euclidean Geometry was their desire to create a consistent system that is free from paradoxes. Avoiding paradoxes constituted, in part, an intellectual need for the transition from Greek to modern mathematics. According to Wilder (1967, based, in part, on Freudenthal, 1962), against the customary view that attributes the change to the introduction of non-Euclidean geometry,

> there was little evidence of excitement or even interest in the mathematical community regarding the work of Bolyai and Lobachevski, for at least 30 years after its publication. . . . The . . . impact of the admissibility of non-Euclidean geometry into mathematics only promoted an evolution already under way (p. 116).

Hacking (1980) argues that the change in perspective occurred much before the late 19th century as is commonly held, but was pioneered by Leibniz. In fact, Hacking insists that Leibniz was aware of the conceptual change and even explained why it did not develop earlier:

> Leibniz himself has a plausible explanation of why the concept of proof emerged at this time. [The idea that proof is independent of its content] is not to be expected when geometry is the standard of rigor. Geometrical demonstrations can appear to rely on their content. Their validity may seem to depend on facts about the very shapes under study, and whose actual

> construction is the aim of the traditional Euclidean theorems. (p. 170)

Here, too, Descartes' contribution was crucial: What brought about Leibniz' new way of understanding the concept of proof was Descartes' algebrization of geometry. By its nature algebra is a way of thinking where one can dispose of or reduce overt attention to content. Together with the recognition that the non-Euclidean geometries are just as consistent as the Euclidean, there seems to have been a general feeling that both the Euclidean and the non-Euclidean systems were still only candidates as sciences of space.

The change in proof scheme from Greek mathematics to modern mathematics was necessitated again by the attempt to establish a consistent foundation for mathematics—just as the Greeks did to construct a consistent system—one that is free from paradoxes. The mathematicians who launched major attempts to form a consistent mathematics, like Zermelo and Russell and Russell and Whitehead, were largely motivated by the desire to meet the crisis in the foundations of mathematics caused by contradictions, primarily those resulting from the Cantorian theory of sets:

> (The) Greek situation, as the latter appears to us through the haze of centuries, is striking. In both situations, crises had developed which threatened the security of mathematics; and in both cases resort was taken to explicit axiomatic statement of the foundations upon which one hoped to build without fear of further charges of inconsistency. (Wilder, 1967, p. 117)

Wilder (1967) also notes that like the Greek mathematicians, the modern mathematicians who launched major attempts to form a consistent mathematics (e.g., Zermelo and Russell) had only one model in mind, albeit a different model. In the case of Zermelo, he focused on "a portion of set theory sufficient for ordinary mathematical purposes, yet carefully limited to avoid the known contradictions" (p. 117). Russell too, like Euclid and Zermelo, "had only one model in mind, and he aimed at making the conditions surrounding his system so stringent that there could be only one model" (p. 118).

Not everyone in the 19th–20th century gave his or her blessing to the new development in mathematics. Both Poincare and Weyl, for example, while acknowledging the legitimacy and accomplishment of the axiomatic method, were disturbed by the widespread preoccupation with it. They called for a return to problems of "mathematical substance" and argued that its role should be limited to giving precision to already created mathematical entities (Wilder, 1967). Nor were all mathematicians of the 17th century enthusiastic about the emergence of a new context of mathematics. The modern notion of number is a case in point. The emergence of negative numbers raised questions as to the utility of symbols without a concrete referent and especially without a geometrical referent. How is it possible, for example, to subtract a greater quantity from a smaller one, where the mental image of "quantity" is nothing else but a physical amount or a spatial capacity? Moreover, how is it possible to understand such a statement as $(-1)/1 = 1/(-1)$, where the quantity 1 is larger than the quantity −1, and therefore, the division of 1 by −1 must be greater than the division of −1 by 1? (See Mancosu, 1996.)

Why Are These Historical-Epistemological Factors Relevant?

What does this brief history account tell us about the learning and teaching of the concept of proof? It is still an open question whether the development of a mathematical concept within an individual student or a community of students parallels the development of that concept in the history of mathematics, though cases of parallel developments have been documented (e.g., Sfard, 1995). If this is the case, one would expect that the path of development would vary from culture to culture. Are there common elements or phases to different paths of development across cultures? Did the development of the concept of proof in, for example, China and India follow a similar path to that of the Western world or was there a leap in time from using perceptual proof schemes to modern axiomatic proof schemes? What are some of the salient social and cultural aspects that might impact the trajectory of these paths? Independent of these particular questions, whose treatment goes beyond the scope and goals of this paper, considerations of historical accounts can evoke other research questions whose answers can potentially direct the development and implementation of instructional treatments of mathematical concepts and ideas. In what follows, we will discuss such questions about the concept of mathematical proof evoked by the historical account outlined above.

Pre-Greek Mathematics versus Greek Mathematics

The Greeks had likely constructed their deductive mathematics on the basis of the mathematics of their predecessors, which was mainly empirical. This mere fact evokes a pedagogical question as to the role of the empirical proof schemes in the development of the deductive proof schemes. The empirical proof schemes are inevitable because natural, everyday thinking utilizes examples so much. Moreover, these schemes have value in the doing and the creating of mathematics. They are even indispensable in enriching one's images by creating examples and non-examples, which, in turn, can help generate ideas and give insights. The question is how to help students utilize their existing proof schemes, largely empirical and external, to help develop deductive proof schemes? As a historian might ask what events—social, cultural, and intellectual—necessitated the transition from pre-Greek mathematics to Greek mathematics, a mathematics educator should ask what instructional interventions can bring students to see a need to refine and alter their existing external and empirical proof schemes into deductive proof schemes.

As was indicated, while pre-Greek mathematics was concerned merely with *actual* physical entities, and its proving was governed by the empirical proof schemes, Greek mathematics dealt with idealizations of spatial and quantitative realities and its proving was deductive. There seems to be a cognitive and epistemological dependency between the nature of the entities considered and the nature of proving applied. To what extent and in what ways is the nature of the entities intertwined with the nature of proving? For example, students' ability to construct an image of a point as a dimensionless geometric entity might impact their ability to develop the Greek axiomatic proof scheme, and vice versa. As far as we know, this interdependency has not been explicitly addressed and its implications for instruction have not been considered.

The motive for the Greeks' construction of their geometric edifice, according to the historical account presented earlier, was their desire to create a consistent system that was free from paradoxes such as those of Zeno. For example, to avoid Zeno's paradoxes the Greeks based their geometric proofs strictly in the context of static concepts. What does this tell us about how to help students see a need for the construction of a geometric structure, particularly that of Euclid? What is the cognitive or social mechanism by which deductive proving can be necessitated for the students?

Greek Mathematics versus Modern Mathematics

As was noted, in constructing their mathematics the Greeks had only one model in mind—that of imageries of idealized physical and quantitative realities. Further, it is these imageries that determined the axioms and postulates on which their geometric and arithmetic edifices stood. In contrast, in modern mathematics entities can be quite arbitrary and one's images of these entities are governed by a set of axioms. This sharp construct is best manifested by Euclid's *Elements*, on the one hand, and Hilbert's *Grundlagen*, on the other: While the former is restricted to a single interpretation, the *Grundlagen* is open to different realizations. Here too the question of intellectual necessity for the transition from Greek mathematics to modern mathematics is of profound pedagogical importance. Historians like Klein and others characterize this transition as revolutionary: It marks a monumental conceptual change in humans' mathematical ways of thinking. More research is needed to systematically document these difficulties to better understand their nature and their implications for instruction.

Another critical difference between Greek mathematics and modern mathematics has to do with *form* versus *content*. In Greek mathematics, the *form* of the proof could not be completely detached from the *content* of the spatial or quantitative context. In contrast, in modern mathematics proof is valid by virtue of its form alone. Here, too, we know of no studies that document systematically students' mathematical behaviors in relation to these fundamental characteristics of Greek mathematics—when students learn Euclidean geometry, for example, as compared to when they learn finite geometry as a case of modern mathematics.

Post-Greek Mathematics

Symbolic algebra, which began with Vieta's work, seems to have played a critical role in the transition from Greek mathematics to modern mathematics, particularly in relation to the reconceptualization of mathematical proof as a sequence of arguments valid by virtue of their form, not content. In the new concept of proof, one would begin with identities and by virtue of rules of symbolic definitional substitutions proceed through a finite number of steps until the theorem is proved. With symbolic algebra, mathematicians shifted their attention from results of operations (e.g., whether and how an angle can be bisected or trisected) to the operations themselves (e.g., the underlying difference between bisecting and trisecting an angle). A critical outcome of this shift was the discovery that all Euclidean geometry problems can be solved by a single approach, that of reducing the problems to equations and applying algebraic techniques to solve them.

The role of symbolic algebra in the reconceptualization of mathematics in general and of proof in particular raises a critical question about the role of symbolic manipulation skills in students' conceptual development of mathematics, in general, and of proof schemes, in particular. Can students develop the modern conception of proof without computational fluency? And in view of the increasing use of electronic technologies in schools, particularly computer algebra systems, one should also ask: Might these tools deprive students of—or, alternatively, provide students with—the opportunity to develop algebraic manipulation skills that might be needed for the development of advanced conception of proof? In addressing this question, it is necessary, we believe, to distinguish between two kinds of symbolic proof schemes: non-referential and referential. As we discussed earlier, in the former scheme, neither the symbols nor the operations one performs on them represent a quantitative reality for the students. Rather, students think of symbols and algebraic operations as if they possess a life of their own without reference to their functional or quantitative meaning. By contrast, in the symbolic referential proof scheme, to prove or refute an assertion or to solve a problem, students learn to represent the statement algebraically and perform symbol manipulations on the resulting expressions. The intention in these symbolic representations and manipulations is to derive relevant information that deepens one's understanding of the statement, and that can potentially lead her or him to a proof or refutation of the assertion or to a solution of the problem. In such an activity, one does not necessarily form referential representations for each of the intermediate expressions and relations that occur in the symbolic manipulation process, but has the ability to attempt to do so in any stage in the process. It is only in critical stages—viewed as such by the person who is carrying out the process—that one forms, or attempts to form, such representations. This ability is potential rather than actual because in many cases the attempt to form quantitative representations may not be successful. Nevertheless, a significant feature of the referential symbolic proof scheme, which is absent from the non-referential symbolic proof scheme, is that one possesses the ability to pause at will to probe into the meaning (quantitative or geometric, for example) of the symbols.

Together with the emergence of symbolic algebra, a new conception of mathematical entity, particularly that of number, began to emerge. A mathematical entity, in this conception, is not necessarily dependent on its "natural" pre-scientific experience but on its connection to other entities within a structure and its function within that structure. For example, while the Greeks were highly selective in their choice of numbers—they rejected irrational numbers, for example—the post-Greek mathematicians began to accept them. This conceptual change was not without difficulty. For example, some mathematicians of the 17th century rejected the utility of negative numbers, which they viewed as symbols without real experiential referents. The conceptual attachment to a context—whether it is the context of intuitive Euclidean space or that of R^n—was dubbed the *contextual proof scheme*. In this scheme, general statements, intended for varying realities, are interpreted and proved in terms of a restricted context. The question of the developmental inevitability of the contextual proof scheme has not been fully addressed in mathematics education research. Some evidence exists to indicate that even students in an advanced stage in their mathematical education have not developed this scheme. For example, it has been shown that many mathematics majors enrolled in advanced geometry courses have major difficulties dealing with any geometric structure except the one corresponding to their spatial imageries, and that mathematics majors enrolled in linear algebra courses interpret and justify general assertions about entities in a general vector space in terms of R^2 and R^3 entities (Harel & Sowder, 1998, Harel, 1999). Such findings have major curricular implications. For example, they raise major doubts as to the wisdom of the practice of starting off college geometry courses with finite geometries or of introducing general vector spaces in the first course of linear algebra.

The debate among philosophers during the Renaissance about the scientificness of mathematics and the mathematics practice that ensued is of particular pedagogical significance. As we have outlined, the question was whether the mathematical practice in which "implication" is a mere logical consequence, rather than a demonstration of the cause of the conclusion, is scientifically acceptable. This, in turn, raises questions about the acceptability of proof by contradiction and proof by exhaustion. Were these issues of marginal concern to the mathematicians of the 16th and 17th centuries, or had they been significantly affected by it? To what extent did the practice of mathematics in the 16th and 17th centuries reflect global epistemological positions that can be traced back to Aristotle's specifications for perfect science? These are important questions, if we are to draw a parallel between the individual's epistemology of mathematics and that of the community. As noted by Mancosu (1996), this debate had a deep and profound impact on the practice of mathematics during the 15th to 18th centuries. For example, the practices of Cavalieri, Guldin, Descartes, and Wallis reflected a deep concern with these issues by, for example, explicitly avoiding proofs by contradiction in order to conform to the Aristotelian position on what constitutes perfect science. This history shows that the modern conception of proof was born out of an intellectual struggle—a struggle in which Aristotelian causality seems to have played a significant role. Is it possible that the development of students' conception of proof includes some of these epistemological obstacles (in the sense of Brousseau, 1997)—obstacles that may be unavoidable, for they are inherent to the meaning of concepts in relation to humans' current schemes?

We conclude this section on the relevance of history and philosophy of mathematics to the learning and teaching of proof with questions pertaining to the idea of "genetic definitions"—mathematical definitions that utilize motion to generate magnitudes—from the 17th century. As we have indicated, the use of such definitions was viewed by some important mathematicians of the time to conform to the Aristotelian epistemological position on the centrality of causality in science. Can this account for the positive impact that dynamic geometry environments might have on advancing students' proof schemes? What is exactly the conceptual basis for the relationship between motion and causal proofs? In a later section we will report on several studies that have examined this effect.

Functions of Proof

Earlier, in the second section, we described a portion of our taxonomy of proof schemes. The discussion that followed brought up several other schemes that emerged in the history of mathematics. In the rest of this section, we depict all the schemes mentioned in this paper (Table 18.1) and discuss their functions within mathematics. This list is not complete; we only depict those schemes that are needed for the discussion in this paper (for the complete taxonomy, see Harel & Sowder, 1998).

Table 18.1 Proof Schemes

External Conviction	Empirical	Deductive
Authoritative	Inductive	Transformational
Ritual	Perceptual	Causality
Non-referential		Greek axiomatic
		Modern axiomatic

As we indicated earlier (see the second section), de Villiers (1999) built on the work of others scholars—particularly Hanna (1990), Balacheff (1988), Bell (1976), Hersh (1993)—to address important questions about the role of proof. Specifically, what different functions does proof have within mathematics itself and how can these functions be effectively utilized in the classroom to make proof a more meaningful activity? de Villiers suggests that mathematical proof has six not mutually exclusive roles:

- verification
- explanation
- discovery
- systematization
- intellectual challenge
- communication

In what follows, we will show that all of these functions but one (intellectual challenge) are describable in terms of the proof scheme construct. Some of the proof schemes used to interpret these functions appeared in the taxonomy presented above. A description of each—with our additions and modifications—follows.

Verification refers to the role of proof as a means to demonstrate the truth of an assertion according to a predetermined set of rules of logic and premises—the *axiomatic proof scheme.*

Explanation is different from verification in that for a mathematician it is usually insufficient to know only that a statement is true. He or she is likely to seek insight into why the assertion is true. We referred to this as the *causality proof scheme.*

Discovery refers to the situations where through the process of proving, new results may be discovered. For example, one might realize that some of the statement conditions can be relaxed, thereby generalizing the statement to a larger class of cases. Or, conversely, through the proving process, one might discover counterexamples to the assertion, which, in turn, would lead to a refinement of the assertion by adding necessary restrictions that would eliminate counterexamples. Lakatos' (1976) thought experiment on the proof of Euler's theorem for polyhedra best illustrates this process. In some cases one may ask whether a certain axiom is needed to establish a certain result, or what form the result would have if a certain axiom is omitted. We considered this as a case of the *axiomatizing proof scheme.*

Systematization refers to the presentation of verifications in organized forms, where each result is derived sequentially from previously established results, definitions, axioms, and primary terms. This too is a case of the axiomatic proof scheme. The difference between systematization and verification is in the extent of formality.

Communication refers to the social interaction about the meaning, validity, and importance of the mathematical knowledge offered by the proof produced. Communication can be viewed in the context of the two subprocesses that define proving: ascertaining and persuading.

Intellectual challenge refers to the mental state of self-realization and fulfillment one can derive from constructing a proof. As we mentioned earlier, this role does not correspond to any of our proof schemes.

With the notion of proof scheme as an organizing concept—appended with these functions—we will now present selected findings reported in the literature that pertain to students' conceptions of proof. Of course, we are unable to describe without speculation most of these findings in terms of the proof scheme construct, because these studies had not been conceptualized or designed with our proof scheme construct in mind. However, as we will see, much can be said about these research findings in relation to proof schemes.

COGNITIVE FACTORS: A LITERATURE REVIEW OF STATUS STUDIES WITH INTERPRETATIONS IN TERMS OF STUDENTS' PROOF SCHEMES

Examining student performance in an area of mathematics is a natural first step in judging whether curricular and teaching efforts in that area seem to be adequate and not to require any special attention. Hence, we first look at several status studies of proof performance. We interpret the results in terms of the (apparent) proof schemes involved in the results, following the proof schemes focus of this chapter. The first two parts of this section will report, respectively, on precollege and college students' conceptions of proof. The findings provide evidence that the pervasive proof schemes among the two populations of students are those belonging to the external conviction proof scheme class and the empirical proof scheme class.

With the deductive proof scheme playing a significant role in mathematics, there is also relevant information from examining performance in logical inferences—a critical characteristic of this scheme. Any description of mathematical proof, the ultimate in justification, will include some mention of logical inferencing ability. Someone not versed in, for example, the mathematical meaning of an "if . . . , then. . . ." statement, or who is not comfortable with reasoning patterns like modus tollens, or who does not understand quantified statements and the role of a counterexample, or who cannot deal correctly at some stage with negating ". . . and . . ." and ". . . or . . ." statements, or who cannot explain the logic of an indirect proof, surely cannot carry out or even understand many mathematical proofs (including disproofs). Hence logical inferencing ability is a basic tool for the process of proving in mathematics and likely enters also into many justifications of a less sophisticated sort. But logic is central to the deductive proof schemes. For example, the transformational proof scheme, which constitutes the essence of the proving process in mathematics and is expected to develop with at least college-bound students and mathematics major students, should be present in students' mathematical behavior. As we have discussed earlier, "logical inference" is one of the three essential characteristics of this scheme. A third part of this section will include findings reported in the literature that seem germane to the characteristics of these deductive proof schemes.

Status Studies: Precollege Students

NAEP Studies

A first place to look for data is the periodic National Assessment of Educational Progress in the United States [NAEP], involving, typically, students at ages 9, 13, and 17 (and now reported by grade: 4, 8, and 12). The wide geographic sampling and the large sample sizes, usually many thousands of students, in a NAEP indicates the U.S. picture, which can be further checked with smaller studies and contrasted with studies outside the U.S. For our purposes, however, one limitation of the NAEP has been that only a few items on proof or logic can be included, because of the scope of the tests. Another limitation is that since all the NAEP questions considered here are multiple-choice items, it is difficult to pinpoint the actual reasoning students employed in answering them.

In the planning stage, the first NAEP mathematics assessment (1972–1973) included mathematical proof and logic as objectives to be evaluated, although only a few items tested these areas (Carpenter, Coburn, Reys, & Wilson, 1978, p. 10). Later NAEPs used different designs for setting objectives to be tested. The fourth NAEP (1985–1986), for example, included items testing "mathematical methods," with a few intended to include "a general understanding of the nature of proof and axiomatic systems, and logic" (Carpenter, 1989, p. 3). Analysts of the results concluded that "most 11th-grade students demonstrated little understanding of the nature and methods of mathematical argumentation and proof" (Silver & Carpenter, 1989, p. 11), citing results on items requiring the recognition of counterexamples (with success rates of 31%–39%—see Figure 18.1 for two items), on items testing understanding of the terms "axiom" and "theorem" (fewer that one-fourth and about half correct, respectively), and on items dealing with an undisclosed but "straightforward" item on indirect proof (about one-third correct) and mathematical induction (similar results) (pp. 17–18). Even when only those students who had taken geometry were considered, results were just slightly better than those for students who had taken mathematics only through first-year algebra. The overall performance led the analysts to conclude that "the generally poor performance on these items dealing with proof and proof-related methods suggests the extent to which students' experiences in school mathematics, even for students in college-preparatory courses, may often fail to acquaint them with the fundamental nature and methods of the discipline" (p. 18).

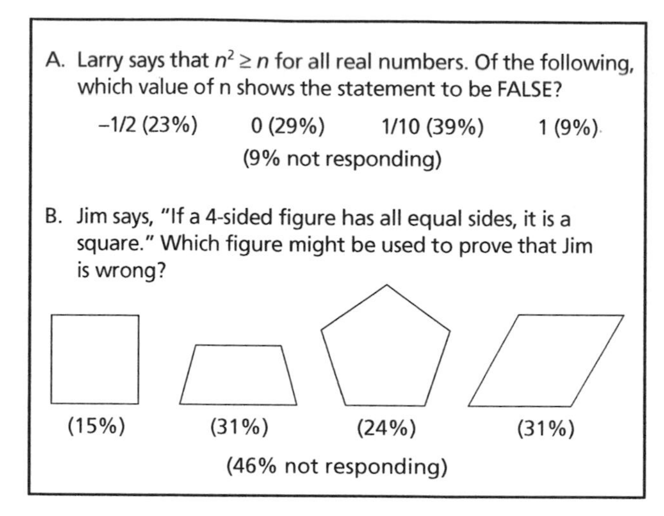

A. Larry says that $n^2 \geq n$ for all real numbers. Of the following, which value of n shows the statement to be FALSE?
−1/2 (23%)
0 (29%)
1/10 (39%)
1 (9%)
(9% not responding)
B. Jim says, "If a 4-sided figure has all equal sides, it is a square." Which figure might be used to prove that Jim is wrong?
(15%)
(31%)
(24%)
(31%)
(46% not responding)

Figure 18.1 Sample NAEP items involving recognition of a counterexample (percents of those responding are in parentheses (after Silver & Carpenter, 1989, p. 17).
Note: Percents may total more than 100 because of rounding.

The test framework for the sixth NAEP (in 1992) did not focus specific attention on proof and logic items. But in commenting on performances on geometry items, the analysts (Strutchens & Blume, 1997, ch. 7) noted that many results could be explained by assuming that the students were basing their responses on the appearances in drawings—the perceptual proof scheme—rather than through reasoning. For example, when asked to choose from a set of triangles one that did not have a particular property, only 21% of the students made the correct choice (p. 180), but when asked whether four given statements applied to an illustrated construction of an angle bisector, 51% were correct on all four, leading the analysts to write, "Often a figure can foster correct reasoning. . . ." (pp. 181–182).

Seventy-seven percent of the Grade 8 students, however, were successful at recognizing a counterexample to a false statement about quadrilaterals. It is interesting to note that the first NAEP (1972–1973) included a multiple-choice item on logic that involved recognizing the logical equivalent of "All good drivers are alert." Only about one-half of the 17-year-olds, 11th graders, chose the correct alternative ("A person who is not alert is not a good driver"). The analysts for those NAEP results noted, however, "The logic exercises were probably answered as much by the semantic context of the problems as by any knowledge of logic" (Carpenter et al., 1978, pp. 126–127). This is in line with the earlier observation that proof schemes can vary from context to context.

Overall, then, the limited picture about proof understanding from the NAEPs is that at best only a small percent of high school students are equipped to deal effectively with the deductive proof schemes, with most apparently relying on the empirical proof schemes.

Other Status Studies in the United States

Studies in which justification and proof have been foci are perhaps more telling than the limited messages from the NAEPs. Some have been large-scale studies and hence are particularly significant. For example, Senk's (1985) study involved 1520 U.S. geometry students in 74 classes in 11 high schools in 5 states within a month of the end of the course. After items in which the student was asked to supply missing reasons or statements, each student was asked to give four proofs, with the first two requiring only one deduction beyond those from the given information, as in the example in Figure 18.2 (72% correct). But only 32% could prove the textbook theorem, The diagonals of a rectangle are congruent.

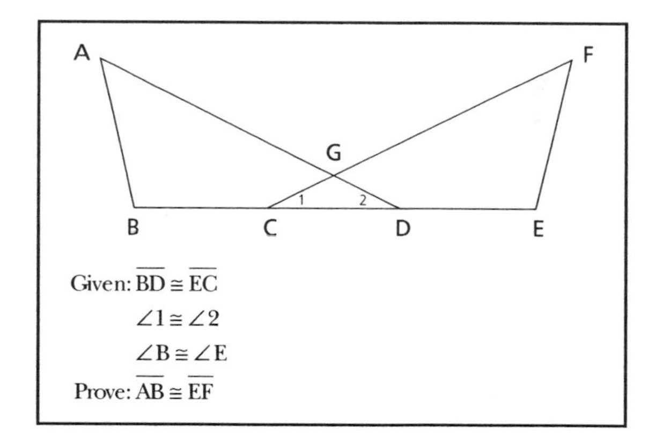

Given: $\overline{BD} \cong \overline{EC}$
$\angle 1 \cong \angle 2$
$\angle B \cong \angle E$
Prove: $\overline{AB} \cong \overline{EF}$

Figure 18.2 A sample geometry proof item from Senk's study (1985, p. 451).

The overall results were dismaying for the course in the U. S. in which deductive proof schemes should be expected to develop: "[The] data suggest that approximately 30 percent of the students in full-year geometry courses that teach proof reach a 75-percent mastery level in proof writing. . . . 29 percent of the sample could not write a single valid proof" (Senk, 1985, p. 453). At the time, about half of high school graduates took a course in geometry, so it is interesting to speculate about what more recent performances might be, when roughly 80% of U. S. high schoolers take geometry (U.S. Department of Education, 2000, p. 122).

Other, smaller studies in the U.S. have also involved students in geometry or later courses, since virtually all formal work with proof has traditionally been introduced in the geometry course (9th or 10th year is most common). But, for example, the interviewees in Tinto's (1988) study felt that proof was used only to verify facts that they already knew—an antithesis to the discovery or explanation functions of proof discussed earlier.

Thompson's study (1991) is of special interest since her subjects were all advanced students taking the last course of a curriculum targeting university-bound students and emphasizing reasoning and proof as a major strand. Yet Thompson expressed concern about the number of students who "proved" a statement by providing a specific example—a manifestation of the inductive proof scheme, and only about one-third of her subjects could find a counterexample to a number theory statement (For all integers a and b, if a^2 is divisible by b then a is divisible by b), in a "prove or disprove" context. Thompson also referred to the "enormous difficulty that students had with indirect proof" (1991, p. 23), with only 3% able to complete one indirect proof (that the sum of a rational number and an irrational number is an irrational number). Difficulties with indirect proof could well be related to the earlier discussion of the causal proof scheme.

Knuth, Slaughter, Choppin, and Sutherland (2002) found that 70% of roughly 350 students in grades 6–8 used examples (the empirical proof scheme) in justifying the truth of two statements (show that the sum of two consecutive numbers is always an odd number; show that when you add any two even numbers, your answer is always even, p. 1696). Only a few students attempted general arguments. On a more positive note, across grade levels, the students did show an increasing sensitivity to adhering to a given definition.

Since the focus of these studies was proof performance, they provide even more striking evidence than the NAEP studies did, that most U. S. students, even those in college-preparatory programs, do not seem to utilize deductive proof schemes.

Status Studies Outside the United States

Another large-scale study of student performance in justification and proof shows that weak performance and reliance on less mature proof schemes is not solely a U.S. phenomenon. Healy and Hoyles (1998, 2000) conducted a study involving 2459 English and Welsh 14–15 year-olds (finishing ninth graders in 94 classes in 90 schools). That the students were in the top 20%–25% on a national test is noteworthy, as is the fact that the English-Welsh national curriculum gives attention to conjecturing and explaining/justifying conclusions and is also, in U.S. terms, integrated with attention to both algebra and geometry, so presumably the items were reasonable for those grade 9 students. The students were asked to describe what proof means and what it is for, and to judge given proofs, as well as to decide on the truth or falsity and to construct their own proofs for two algebra and two geometry items. Over a quarter of these able students had little or no sense of the purpose of proof or its meaning (2000, pp. 417–418). The average score on the constructed-proof items was less than half the maximum, with 14%–62% of the students, depending on the item, not even able to start a proof (1998, p. 2). Of those able to start a proof, 28%–56% could proceed only minimally (1998, p. 2). It is comforting that the students were better at selecting correct proofs, favoring general arguments and clear and explanatory arguments that they found convincing (1998, p. 3). When the stated purpose was to get the best mark, however, the students often felt that more formal—e.g., algebraic—arguments might be preferable to their first choices (1998, p. 3). We interpret this last finding as an indication of the authoritarian proof scheme and the ritual proof scheme, in that in the eyes of these students proof must have a certain appearance (ritual) as determined by the teacher (authority).

It is interesting that, in the large, the findings of this study are consistent with those of Coe and Ruthven (1994), who looked at the proof performances of a much smaller group of "advanced level" British students in university preparation schools, on a project for an end-of-course assessment. These students, like those in the Healy and Hoyles study, relied predominantly on examples and techniques for analyzing numerical data, but showed little feel for the purposes of proof. Sample items from Healy and Hoyles' and Coe and Ruthven's studies are presented in Figure 18.3.

An older study by Reynolds (reported in Lovell, 1971) also examined proof ideas held by British students. Most revealing were the responses of 153 6th formers (final year of secondary school) in mathematics, with about 20% endorsing 16 examples as establishing a given statement—a manifestation of the empirical proof scheme. On the other hand, about 70% of these 6th-formers could complete an indirect argument involving a geometric situation (to show the inequality of a pair of alternate interior angles in a drawing with non-parallel lines) that had been started (p. 75), in contrast to the very weak (3%) performance of Thompson's (1991) U. S. students mentioned above.

Still other studies outside the U.S. also indicate difficulties with certain proof ideas. Galbraith (1981) studied the conceptions of proof held by 12–17-year-old Australian students, most of whom were 13–15

Proof or disproof construction (Healy & Hoyles, 1998)	Prove whether the following statements are true or false. Write down your answer in the way that would get you the best mark you can. 1. When you add any 2 odd numbers, your answer is always even. 2. If p and q are any two odd numbers, (p + q) × (p – q) is always a multiple of 4.
"Starting point" item (Coe & Ruthven, 1994, p. 46)	Any four-digit number is rewritten with its digits arranged in ascending and descending order of size. The smaller is then subtracted from the larger and the process is repeated. Example: 7345 gives 7543 – 3457 = 4086 4086 gives 8640 – 0468 = 8172 Then write 7543–>8640–>8721–>... Complete this chain and then try again with other four-digit numbers. Now try with 3 or 5 or 6 or *n* digits.

Figure 18.3 Sample proof items (adapted from Healy & Hoyles (1998, 2000) and Coe & Ruthven (1994)).

years old. He found that 28% of 130 responses indicated a lack of understanding of what a given counterexample told one about a given statement. On another problem, only 18% of his 73 students felt that one counterexample was sufficient to disprove a conjecture. In commenting on how many counterexamples are sufficient, one student even suggested "'Sometimes one is enough and sometimes it isn't'" (p. 19). The confusion about the role of examples in direct proofs versus in indirect proofs versus in disproofs is consistent with our own findings (Harel & Sowder, 1998), a phenomenon we accounted for in terms of students' empirical proof scheme. Many students in Galbraith's study were guided by their mental pictures of what a geometric term suggested rather than a given definition, and relied on just one diagram without considering other possibilities (p. 23)—a clear manifestation of the perceptual proof scheme.

Porteous (1986, 1990) gave questionnaires to about 400 British students, ages 11–16, and interviewed 50 of them. From the questionnaires, he found that more than 40% of the responses endorsed completely a generalization on the basis of examples only, and only about 10% then offered proofs on their own when asked to explain their decisions (1990, p. 591). Furthermore, 83% of those not offering a proof initially claimed to understand a given proof of an assertion, but only 61% of all the students, including the ones offering a proof on their own, were sure that the author was correct about the assertion after seeing a proof (1996, p. 8). From the students' reactions to given proofs, Porteous concluded that "It is only when a pupil devises a proof for himself that he is convinced of the truth of a statement. A proof provided by a teacher may have some effect, but it is small in comparison with a d(i)scovered proof . . . we need to encourage pupils to investigate relationships for themselves, in order to produce their own reasons, or proof, for general statements. This is not to say, of course, that empirical work has no real part to play in the learning process" (1996, pp. 21–22).

Williams (1980) interviewed 11th grade Canadian students in a college preparatory program and concluded that fewer than 30% showed a grasp of the meaning of mathematical proof, that about half of the students saw no need to prove a statement that they regarded as obvious, roughly 70% did not distinguish between inductive and deductive arguments, and fewer than 20% understood how indirect proof works.

Spanish researchers Recio and Godino (2001) conducted a study involving two groups (n = 429 and n = 193) of beginning university students. They found that about a third of the 429 students and less than a quarter of the 193 students could prove both of two elementary statements (the difference between the squares of consecutive natural numbers is odd and equal to the sum of the numbers, and the bisectors of adjacent supplementary angles are perpendicular). Also, fewer than half of the group of 429 were successful on each individual proof. Roughly 40% used empirical reasoning: "Empirical inductive [proof] schemes were the spontaneous type of argumentation in a high percentage of students when they were confronted with new problems, in which it was necessary to develop new proof strategies, different from the learned procedures" (p. 91).

In Israel, Fischbein and Kedem (1982) found that their high school students did not appear to understand that no further examples need be checked, once a proof was given, a finding confirmed by Vinner (1983), who also noted that many high school students (35%), even high attaining ones (39%), seem to regard a given proof as the method to examine and verify a later particular case.

Similar results were found in a recent Japanese study (cited in Fujita & Jones, 2003). The official curriculum of Japan calls for students to "understand the significance and methodology of proof." However, even though most 14 to 15 year old students are successful at proof writing, "around 70% cannot understand why proofs are needed" (p. 9).

Internationally, overall the most positive conclusion seems to be that the proof glass is not completely empty but that it is by no means even close to full. Even "good" performances may be tainted by little understanding or appreciation of the functions of proof. The prevalence of empirical proof schemes for most students seems to be international.

Status Studies: University Students

One might expect that university students would perform better than secondary students on proof activities. As the Recio and Godino (2001) study indicated, however, performance is disappointingly low in many entering university students. There are only a few formal studies that have examined university students' later conceptions of proof. In one example, up to 80% of prospective elementary school teachers derived the truth of a general statement from examples (Goetting, 1995), confirming a result by Martin and Harel (1989) about the pervasiveness of the empirical proof scheme among students. Martin and Harel had emphasized proof in their course for prospective elementary school teachers (who had had geometry in high school) and then asked them to judge whether particular arguments were mathematical proofs for a statement covered and proved in the course (If the sum of the digits of a whole number is divisible by 3, then the number is divisible by 3) or an unfamiliar one (If a divides b, and b divides c, then a divides c). The arguments were either inductive (based on specific instances) or deductive (assertions via general statements). More than half of the 101 students considered inductive arguments to be mathematical proofs. As many as half of these college students also accepted a false deductive argument as being a mathematical proof, apparently reacting to the appearance of the argument—indicating the dominance of the ritual proof scheme—and over a third found both inductive and deductive arguments to be acceptable mathematical proofs. Morris (2002) echoed this last result, with 40% of her pre-service teachers affirming that both inductive and deductive arguments assured certainty and failing to distinguish between the two types of arguments.

But prospective elementary teachers are often not very sophisticated in mathematics, so one might expect majors in technical fields, especially mathematics majors, to perform much better in proof settings. Such is the case, but only for some students, judging from anecdotes from university mathematics faculty. Furthermore, there are studies that point out areas of weakness, usually with small numbers of students. For example, case studies show that although some students may possess the transformational proof scheme, others seem to enter—and leave—university with little or limited abilities with, and views of, proof, mostly demonstrating external and empirical proof schemes (Sowder & Harel, 2003).

For some university students, even recognizing whether a given argument constitutes a proof cannot be taken as a given. In interviews, Selden and Selden (2003) asked their eight mathematics majors to judge whether four given mathematical arguments were valid, with only one of the arguments valid. The students performed only at a 46% correct level with their first judgments but did improve to 81% correct by the end of the interview with their fourth judgments, with the Seldens crediting the improvement to the added experience and reflection during the interview. Although the study did not examine long-term effects, it suggests that repeated validating activities may be a valuable growth opportunity.

Knowledge about what factors are important in devising proofs is limited. In an expert-novice study, with four graduate students in mathematics as the experts and four undergraduates as the novices, Weber (2001) noticed that "an understanding of mathematical proof and a syntactic knowledge of the facts of a domain are not sufficient for one to be a competent theorem prover" (p. 107); experts have what Weber calls "strategic knowledge," knowledge that undergraduates do not exhibit—knowledge of the domain's proof techniques (p. 111), knowledge of which theorems are important and when they are useful (p. 112), and knowledge of when, and when not, to use strategies based on symbol manipulation rather than deeper knowledge (p. 113). In a similar vein, Raman's (2003) interviews of mathematics students and faculty suggested the importance of a "key idea"—"an heuristic idea which one can map to a formal proof with appropriate sense of rigor" (p. 323). She concluded, "For mathematicians, proof is essentially about key ideas; for many students, it is not" (p. 324). In expert-novice studies, one often does not know how the experts acquired their expertise (or whether there is a selection factor involved), but knowing what differences exist may give ideas for instructional emphases. But in questioning university mathematics faculty about university mathematics majors' proof understanding, a "twice is nice" theme emerged: Exposure to the same material twice allows the student, on the second exposure, to focus on proof methods (Sowder, 2004). Perhaps these second exposures are helpful in attaining other aspects of Weber's strategic knowledge and Raman's key ideas, and in growing in deductive proof schemes. Marty (1991) felt that his explicit attention to proof methods, rather than the common focus on new mathematical content, helped his college students succeed in later mathematics courses.

University students' distinctions among axioms, definitions, and theorems are not sharp. Vinner (1977), for example, found that only about half of a group of Berkeley sophomores and juniors in mathematics could correctly identify all of three statements about exponents as definitions, as opposed to theorems or laws or axioms. Since much of algebra focuses on algorithms and these students had first studied the material in high school, perhaps the lack of such distinctions is expected. Yet, when Brumfiel (1973) questioned a class of University of Michigan juniors and seniors, nearly all of whom had complete a university course in formal geometry, he found that their mastery of similar distinctions were shockingly deficient. For example, collectively the students could recall only one axiom (two points determine a line), about half called a definition of isosceles triangle a theorem, and all were certain that, for two (given) independent postulates about points and lines, one could be deduced from the other. In Israel, Linchevsky, Vinner, and Karsenty (1992) found that only about one-fourth of their university mathematics majors understood that it is possible to have alternate definitions for concepts. These studies speak to about-proof topics. But about-proof topics cannot be mastered without understanding the proof topics themselves. In this case the topics in question involve the meaning and role of axioms and definitions. These studies, then, suggest a weak, or even absent, axiomatic proof scheme among mathematics majors—an observation that is consistent with findings of our study with mathematics majors (Harel & Sowder, 1998).

Findings Germane to the Transformational Proof Scheme

Recall that the deductive proof scheme class consists of two schemes: the transformational proof scheme and the axiomatic proof scheme. The former is a special case of, and a conceptual prerequisite for, the former. The research findings we have just reported paint a gloomy picture as to the quality of students' proof schemes. The failure of mathematics instruction to help many students—even college-bound students and mathematics majors—construct the transformational proof scheme is evidenced in these reports. In this section we provide a closer look at the characteristics of the transformational proof scheme, particularly the deduction process characteristic.

As was defined earlier, the transformational proof scheme has three characteristics: *generality, operational thought,* and *logical inference.* As the language may suggest, our definition of the transformational proof scheme naturally draws one to Piaget's theory of intellectual developmental (for example, Inhelder & Piaget, 1958). Battista and Clements (1992) focused on Piagetian thought in their discussion of proof in an earlier handbook (pp. 439–440), identifying three levels of justification/proof, with the third level being marked by the student being "capable of formal, deductive reasoning based on *any* assumptions" (p. 439)—the generality characteristic—whereas at the second level, the student's reasoning has an empirical reality, with the student not realizing that an argument may cover all cases. The third level of thinking is often identified with Piaget's "formal operational" level—the operational thought characteristic. Although the nominal approximate age for the development of formal operational thought is early adolescence and there are different aspects of formal operational thought (e.g., proportional thinking, ability to control variables, propositional logic), studies suggest that many high school and college students have not reached the formal operational level in some areas important in mathematics. For example, Lawson, Karplus, and Adi (1978) gave tasks in several areas, including propositional logic (one such item is Task 1 in Figure 18.4), to students at grades 6, 8, 10, 12, and 13–14. They found "the lack of clear and substantial improvement with age for the propositional logic items," with percents correct by grade on the Task 1 item being 1.0%, 2.1%, 10.1%, 12.0%, and 16.2% (p. 469). A study of 16-year-old English students found that only 30% tested at a level consistent with formal operational thought (cited in Adey, 1999). It appears that it is risky to assume that logical reasoning at a high level develops automatically in the course of schooling, whether within or outside mathematical contexts.

Psychologists other than Piaget have examined logical thinking for many years, with especially fascinating work done during the last thirty years (for example, Evans, 1982; Rips, 1994; Wason & Johnson-Laird, 1972). More recently, Johnson-Laird (2001) has observed that "deductive reasoning is under intense investigation. The field is fast moving and controversial" (p. 441), with at least some of the controversy coming from different hypothesized mechanisms of reasoning for selected tasks. There does not, however, seem to be controversy about the following findings. Performance on logical inferences involving modus ponens is usually reasonably good, but performance on those tasks involving modus tollens is weak, as is a full understanding of inferences involving if-then statements. Students are too willing to use invalid inference patterns like affirming the consequent (i.e., "if p then q" and "q" yielding "p"). In general, humans do not seem to process negative statements as facilely as affirmative ones, and context is an extremely important variable in performance on deductive reasoning tasks (Wason & Johnson-Laird, 1972). Figure 18.4, for example, gives logically isomorphic tasks on which the success rates of adults are strikingly different. Only 2 out of 24 (about 8%) of the respondents correctly identified the two cases that must be considered for Task 1, but 88% did so on Task 2 (reported in Wason & Johnson-Laird, 1972). Thus, our discussion here focuses primarily on findings and studies dealing with logic in mathematical settings, calling on known work from psychology when it seems particularly pertinent.

Although our focus is mainly on studies dealing directly with mathematics, one sobering study involving everyday contexts deserves special note, because some of the data come from prospective secondary school mathematics teachers. Using items from Eisenberg and McGinty (1974), Easterday and Galloway (1995) compared the performance of last-year university students planning to teach middle school or high school mathematics with those of 7–8th graders and 12th graders on a variety of reasoning tasks. The tasks were based on everyday contexts that should be non-suggestive (e.g., "If John is big, then Jane is big. John is big. Is Jane big? Yes/no/maybe"). In particular, on modus tollens tasks, the college students scored 47%, about the same as the 12th grade calculus students but 20% *less* than the 7–8th graders' 67%. These particular 7–8th graders were studying geometry and may therefore have been exposed to logic, but an earlier comparison in 1986 with 7–8th graders in advanced sections had given a similar result, with the 7–8th graders scoring 62% and the college students only 40% on the modus tollens tasks (Easterday & Galloway, 1995, p. 433). The authors concluded, "College students are barely performing better than children whom they may one day teach" (p. 435), and neither group was by any means topping out in performance.

There seem to be only a few data on students' abilities with specific ideas from logic and within mathematics. Mentioned earlier were the limited results from NAEP, and there the picture was clouded by the use of a non-mathematical context. The large-scale Longitudinal Proof Project in England has, however, looked at students' performance on if-then statements involving number theory ideas (Hoyles & Kuchemann, 2002). Among their findings was that 62% of 14-year-olds thought that a given if-then statement (for example, if the product of two numbers is odd, then the sum of the numbers is even) "said the same thing" as its converse. However, the longitudinal nature of the study also allowed Hoyles and Kuchemann to find a 7% improvement on this item over that of the same students at age 13. (In passing, it may be worth noting that so many theorems, especially in geometry, do have true converses, so it is easy to see how students may be insensitive to the logical difference between an if-then statement and its converse.)

Even advanced students have difficulty with quantifiers. Dubinsky and Yiparaki (2000) found that univer-

Task 1. Given four envelopes with a letter on the front side and a number on the back, select just the envelopes definitely needed to be turned over to find out whether they violate the rule.Separate envelopes show on front: D and C, and on back: 5 and 4.	Task 2. Given four envelopes with a space for a stamp on one side and sealed or not, select just the envelopes definitely need to be turned over to find out whether they violate the rule. Envelopes show (a) back of sealed envelope; (b) unsealed envelope with flap up; (c) front of an envelope with stamp; (d) front of an unstamped envelope.
Rule to test: If a letter has a D on one side, then it has a 5 on the other side. Percent correct (D, 4): 8%	Rule to test: If a letter is sealed, then it has a 5 pence stamp on it. Percent correct (a, d): 88%

Figure 18.4 Two logically isomorphic tasks with "abstract" [Task 1] and "concrete" [Task 2] contexts (after Wason & Johnson-Laird, 1972, pp. 191–192).

sity students at various levels, including some in an abstract algebra course, had much greater trouble giving the mathematical meaning of a doubly quantified statement when the existential quantifier appeared before the universal quantifier ("There is a positive number b such that for every positive number a, $b \leq a$"—19% correct) than when the quantifiers were reversed, universal before existential ("For every positive number a there is a positive number b such that $b < a$"—59%). Selden and Selden (1995) noted that university mathematics students may have difficulties in even restating mathematical statements precisely, with their largely third- and fourth-year students often giving incorrect responses when quantifiers were involved. Thus, although there may be areas of apparent strength in the use of logic (for example, the use of modus ponens), there appear to be many areas of weakness as well, and at a wide gamut of levels of schooling.

An important point is that everyday usage of logical expressions may differ considerably from the precise usage in mathematics. Epp (2003) has summarized the differences in a compelling way, and O'Brien, Shapiro, and Reali (1971) have referred to "child's logic" in describing some of the differences. For example, "or" in everyday usage is most often in the exclusive sense ("I'll wear my sandals or my tennis shoes"), in contrast with the inclusive convention common in mathematics. An everyday if-then statement (for example, "If you finish your work, then you can watch the game") often connotes what would be an if-and-only-if statement in mathematics and to many children seems to be an "and" statement. The disparities between everyday usages and mathematical usages are so marked that explicit instruction in logic as used in mathematics would seem to be necessary, with contrasts to the less precise everyday usages pointed out, yet, as Epp contends (2003), perhaps exploiting nonmathematical usages that do reflect the mathematically precise ones, as exemplars for the latter.

Another interesting disparity between everyday usage and mathematical usage is that of indirect proof. According to Freudenthal (1973) indirect proof is a very common activity. Seven to eight year old children used contradiction in game playing and checking conjectures (Reid & Dobbin, 1998). Antonini (2003) even found that indirect argumentations occurred spontaneously by students in his interviews with them about mathematical assertions. Yet research has shown that students experience difficulties with proof by contradiction in mathematics. Leron (1985), for example, observed that despite the simple and elegant form of certain proofs by contradiction, students experience what seem insurmountable difficulties. Lin, Lee, and Wu Yu (2003) see the ability to negate a statement as a prerequisite ability for succeeding at a proof by contradiction. They found that the difficulty levels of students' negating a statement can be ordered decreasingly as negating statements without quantifiers, negating "some," negating "all," and negating "only one."

In general, then, there are many weak spots in students' likely grasp of the logical reasoning used in advanced proof schemes. Is it a chicken-egg question, or can logical thinking and proof performance grow together? Later in this section we summarize several studies in which explicit instruction in logical principles was incorporated into high school geometry courses.

INSTRUCTIONAL-SOCIAL-CULTURAL FACTORS: EVIDENCE POINTING TO CURRICULUM AND INSTRUCTION

Students' instructional history is without question an important variable in studying their performance in justification/proof settings. Hoyles (1997), in particular, makes a compelling case that studies of advanced learners' performance are relatively meaningless without knowing what curricula (and what teaching emphases) they have received. Two important dimensions of instructional history are the curriculum and the teaching. We begin with studies of teaching and teachers, giving some evidence of the *delivered* curricula. Following this, we review a few studies that highlight the vast differences in curricula, primarily the *intended* curricula as evidenced by national or regional guides. The earlier report of several status studies of students' performance have given a glimpse of the *learned* curricula. We then attend to some studies that give evidence of what might be achieved through *revised* curricula and teaching, as delivered by teachers with a different perspective on teaching, with the section including a review of a few studies that show the possibilities offered by technology. We conclude with several studies that have focused explicitly on logic as a vital component for what we call the deductive proof schemes.

Current Status

Teachers and Teaching

The emphasis that teachers place on justification and proof no doubt plays an important role in shaping students' proof schemes. "You get what you teach" is a common aphorism, and the intended curriculum may differ markedly from the delivered curriculum, especially across classrooms. The essentiality of opportunity to learn must be recognized not only at the intended curriculum level but also in the teachers' enacted curriculum. Yet, a study involving 62 mathematics and science teachers in 18 high schools in six states allowed Porter (1993) to note that for the mathematics teachers, "On average, no instructional time is allocated to students learning to develop proofs, not even in geometry" (p. 4).

Results on items allied to the areas of justification and proof in the 1996 NAEP mathematics assessment indicate not only that students perform much worse on items requiring explanation and justification, but also that at grade 8, the students of teachers who devote more time to "developing reasoning and analytical ability to solve unique problems" have noticeably higher overall scores than other students (Silver & Kenney, 2000). The importance of the role of the teacher (and the curriculum) in fostering justification and proof is further highlighted by the videotape study in the Third International Mathematics and Science Study (TIMSS), in which 30 eighth-grade sessions in each of Germany, Japan, and the U.S. were analyzed. The report reaches the following conclusion.

> The most striking finding in this review of 90 classes was the rarity of explicit mathematical reasoning in the classes. The almost total absence of explicit mathematical reasoning in Algebra and Before Algebra courses raises serious questions about the ways in which those subjects are taught. In order for these courses to help introduce students to mathematical ways of knowing, some of the logical foundations of mathematical knowledge should be explicit. Of course, the total absence of any instances of inductive or deductive reasoning in the analyzed United States classes cries out for curriculum developers to address this aspect of learning mathematics. (Manaster, 1998, p. 803)

It should be clear that only an authoritarian proof scheme is likely to be fostered in these classrooms.

Despite Porter's (1993) finding of little attention to proof in high school, one would certainly expect more explicit attention to proof in the mathematics at those grade levels where proof most often is a conscious part of the curriculum. Senk (1985) noted, however, that there were consistent differences across schools in the geometry students' performances on her proof tasks. Tinto (1988) too noted that one of her four teachers seemed markedly different from the others in his approach to geometry. In their large-scale study of proof in British classrooms, Healy and Hoyles (1998) found that students who had been expected to write proofs and who had classes in which proof was taught as a separate topic performed somewhat better on proof items than other students. Thompson (1991), on the other hand, did not notice differences across her nine teachers and schools; her sample, however, included three private schools and two magnet schools and so was perhaps not representative. Overall, it appears that at least some of the deficiencies in students' acquisition of more sophisticated proof schemes may stem from the lack of opportunity to engage in proof-fostering activities, even in courses where one would expect much attention to proof.

The evidence from the status studies of university students' proof knowledge suggests that some, if not many, precollege teachers are unlikely to teach proof well, perhaps because their own grasp of proof was probably limited in college and may not have grown since then. Knuth (2002a) examined the conceptions of proof of 16 practicing secondary school mathematics teachers, most with backgrounds that would pass a face-validity test for knowledge of mathematics. In interview settings, Knuth asked the teachers to respond to general questions about proof (e.g., What purpose does proof serve in mathematics?), to evaluate given arguments (both proofs and non-proofs), and to identify the arguments that were most convincing. Although all of the teachers endorsed the verification role of proof, none mentioned the explanatory role of proof (see the section on the concept of proof scheme). Six of the 16 thought it might be possible to find contradictory evidence of a (non-specified) statement that had been proved. Four of the 16 tested a statement with a given, endorsed proof with further examples (cf. Fischbein & Kedem, 1982), even though all of the teachers eventually acknowledged that it would not be possible to find a counterexample. Even though the teachers collectively correctly identified 93% of the correct arguments as being proofs, over a third of the non-proof arguments were rated as being proofs! Ten of the 16 accepted the proof of the converse of one statement as a proof of the statement. Thirteen of the 16 teachers found arguments based on examples or visual presentations to be most convincing. Although Knuth felt that their responses may have been directed toward *personally* convincing rather than *mathematically* convincing, that mathematics teachers would be convinced by such arguments more than by a mathematical proof is significant, because it reveals an apparent dominance of the empirical proof schemes among the teachers. Knuth (2002b) further examined these teachers' ideas about proof in the context of school mathematics (versus the earlier just-in-mathematics). In view of the NCTM (2000) recommendation that reasoning and proof be considered fundamental aspects of the study of mathematics at all levels of study, it is disappointing that the teachers in Knuth's study ". . . tended to view proof as an appropriate goal for the mathematics education of a minority of students" (2002b, p. 83), with 14 not considering proof to be a central concern in school mathematics. And, to repeat, none even explicitly mentioned the explanatory role of a proof, although seven did mention its verification role—that a proof shows why a statement is true. Thirteen regarded the development of logical thinking or reasoning as the primary role of proof in school mathematics. Yet, the teachers did not completely reject justifications, but were willing to rely on informal "proofs" (e.g., examples and drawings) to support results, a practice that may mislead students into thinking that such are acceptable mathematical "proofs," and reinforcing the acceptability of their empirical proof schemes.

Curricula

That mathematics curricula differ in their treatments of proof is by no means a recent phenomenon. For example, in his study of students' proof explanations, Bell (1976) found that proof is the topic that shows the greatest variation in approaches internationally. He noted that this variation can be attributed to the tension between the recognition among teachers that deduction is essential to mathematics but that only the most capable students develop a good understanding of it. There is evidence that this condition remains true today as well. For example, Fujita and Jones (2003) compared the textbook treatments of geometry in lower secondary schools in Japan and Scotland and concluded the following.

> Our analysis indicates that . . . Japanese textbooks set out to develop students' deductive reasoning skills through the explicit teaching of proof in geometry, whereas comparative UK [United Kingdom] textbooks tend, at this level, to concentrate on finding angles, measurement, drawing, and so on, coupled with a modicum of opportunities for conjecturing and inductive reasoning. (p. 1)

It is, perhaps, natural to expect great variation in the treatment of curricular topics within countries that do not have national curricular and educational guidelines. In the U. S., for example, how geometry, the primary locus of proof efforts until recently, should be handled has led to vastly different opinions and occasionally to different approaches or emphases in school geometry (cf., e.g., Hoffer, 1981; NCTM, 1973, 1987; Usiskin, 1987).

Examples of Curricula Aimed toward Enhancing Students' Proof Schemes

We do not claim that the studies touched on below use our language of proof schemes, but it will be clear that students who have had such treatments should possess different proof schemes from those that apparently result from traditional teaching. Of particular interest and importance, we think, are the feasibility studies that have involved children in elementary schools.

Earlier Feasibility Studies

Perhaps because of the spirit of the "New Math" times, and the call of the Cambridge Conference Report to move toward more sophisticated mathematics earlier in the curriculum, some studies in the 1970s focused on proof with younger children. To gauge whether and when children seem able to cope with proof or proof-like tasks, Lester (1975) devised a computer-delivered deductive system in a game-like form, with well-defined rules (i.e., postulates) and target configurations (i.e., theorems) to be attained by applying the rules (i.e., with proofs). Hence, the students were dealing implicitly with an axiomatic system.

Lester (1975) sampled 19 students from each of four groups, one group from grades 1–3, a second from grades 4–6, a third from grades 7–9, and the fourth from grades 10–12, gave them practice with the rules, and studied their performances on the target tasks. His grades 7–9 students performed as well as the students from grades 10–12, and the students from grades 4–6 solved about as many tasks as the older students, but took somewhat longer. Lester suggested that "even students in the upper elementary grades can be successful at mathematical activities that are closely related to proof" (p. 23). Indeed, King (1970, 1973) thoroughly developed a 17-day unit dealing with some elementary number theory results (e.g., a number which is a factor of two numbers is also a factor of their sum). He found that a group of 10 above-average sixth graders could reproduce the proofs initially developed with considerable teacher support (in contrast to a non-equivalent control group), but the evidence also suggested that the proofs were given from rote memory.

Fawcett's study of the late 1930s deserves special mention. The title and sub-title give a good summary: *The Nature of Proof, A Description and Evaluation of Certain Procedures Used in a Senior High School to Develop an Understanding of the Nature of Proof.* Whatever the reason—World War II, the usual inertia in curriculum—this study seemed to have had little impact, even though it was reported in a yearbook of the NCTM (Fawcett, 1938/1995). His approach was surprisingly modern in tone. Fawcett's summary includes the following.

> The theorems [of geometry] are not important in themselves. It is the *method* by which they are established that is important, and in this study geometric

theorems are used only for the purpose of illustrating this method. The procedures used are derived from four basic assumptions:

1. That a senior high school student has reasoned and reasoned accurately before he begins the study of demonstrative geometry.
2. That he should have the opportunity to reason about the subject matter of geometry in his own way.
3. That the logical processes which should guide the development of the work should be those of the student and not those of the teacher.
4. That opportunity be provided for the application of the postulational method to non-mathematical material.

Non-mathematical situations of interest to the pupils were used to introduce them to the importance of definition and to the fact that conclusions depend on assumptions, many of which are often unrecognized. To make definitions and assumptions and to investigate their implications is to have firsthand experience with the method of mathematics . . . (p. 117)

Fawcett's teaching experiment, with a non-equivalent control group, continued through two school years, with the report covering just the first year. As the excerpt above suggests, the students eventually composed, collectively, their lists of undefined terms, definitions, and assumptions. The need for such elements arose in discussing everyday situations, such as the importance of definition in discussing how the governor of Ohio handled a particular bill (pp. 51–52). Of course the teacher played a major role in initiating such discussions and in providing fruitful leads for particular results, but in the large the students were responsible for conjecturing results and then proving them. The evaluation of the experiment was based on a state geometry test and a test of the ability to analyze non-mathematical material. Even though the experimental students, after one year of a two-year treatment, had not covered the usual material in the standard course, their performance (although not reported thoroughly) seemed satisfactory on the 80-point state geometry test: Median 52.0, state median 36.5 (p. 102). More telling was the experimental students' performance on the analysis of non-mathematical material, where they out-performed by far the control group (change score of 7.5 vs. a change score of 1.0; maximum possible not given) (p. 103). Fawcett also quoted the laudatory reactions of visitors and of the students themselves, contrasting the students' final remarks with the largely indifferent attitudes expressed at the beginning of the experiment.

One can only conclude from these studies that upper elementary school children can deal with proof ideas or actions, and that high school students can develop meaningful understandings of proof if they are taught appropriately.

More Recent Feasibility Studies

Several researchers have emphasized a sociocultural perspective in investigating and enhancing the development of students' proof schemes. According to this perspective, the development of any higher voluntary form of human knowledge cannot be understood apart from the social context in which it occurs. As such, learning is necessarily a product of social interaction. Key to this perspective is Vygotsky's (1978) notion of "zone of proximal development" (ZPD): the difference between what the students can do under adult guidance or in collaboration with more capable peers and what they can do without guidance. A direct and critical implication of this perspective is *instructional scaffolding* (Mercer, 1995), which refers to the provision of guidance and support that is increased or withdrawn in response to the developing competence of the learner. Blanton, Stylianou, and David (2003) further suggest:

> Since learning is viewed as a product of interaction, it follows that one's development within the ZPD is affected by the intellectual quality and developmental appropriateness of these interactions (Diaz, Neal, & Amaya-Williams, 1999). In other words, the extent of one's development within the ZPD is predicated in part upon how the more knowing other organizes, or scaffolds, the task at hand. Thus, if we intend to understand development within the ZPD, we must be thinking about if and how tasks can be scaffolded to extend one's learning. (p. 114)

Blanton, Stylianou, and David (2003) have investigated the nature and role of this scaffolding in the learning and teaching of proof. Specifically, they addressed two questions: (a) What is the nature and meaning of instructional scaffolding in the classroom in the development of students' proof ability? and (b) How do different types of scaffolding prompts from the teacher affect students' self-regulatory thinking about proof? Their results suggest

> . . . students who engage in whole-class discussions that include metacognitive acts as well as transactive discussions about metacognitive acts make gains in their ability to construct mathematical proofs. Moreover, students' capacity to engage in these types of discussion is a habit of mind that can be scaffolded through the teacher's transactive prompts and facilitative utterances. . . . [This] suggests that students can internalize public argumentation in ways that facilitate private proof construction if instructional scaffolding is appropriately designed to support this. (p. 119)

Some of the most promising work directly or indirectly related to students' growth in elements of proof has been with elementary school children, as early as the primary grades and most often in problem-based settings where questions of proof have come up as intended by-products of the investigations. Evidence exists to indicate that a classroom environment that is conducive to social interactions among students and between the teacher and the students can be productive. Hence, these studies have often incorporated small-group work and whole-class discussions sharing how different children were thinking (that is, justifying their work) rather than just focusing on the numerical result. For example, Zack (2002) had two groups of fifth graders work on the following problem: Find all the squares on an n by n chessboard, first with n = 4, then n = 5, then n = 10. What if it was a 60 by 60 square? Can you prove that you have them all? After working on this problem, her students offered eight different counter-arguments to an answer of 2310 (based on 6 times the established answer of 385 for a 10 by 10). Surely students who experience such instruction will develop different proof schemes than will children whose teachers always judge the correctness of an answer. In some cases (Cobb et al., 1991) researchers worked closely with classroom teachers in designing the tasks and studying the effects of the children's work and their discussions. In other cases (e.g., Carpenter, Franke, Jacobs, Fennema, & Empson, 1998), the teachers were provided with a great deal of background on children's thinking strategies but left to their own in designing and carrying out lessons; however, an earmark of the classroom work was to call for students to give justifications.

Other examples of early appearances of proof ideas come from work aiming toward algebra in the elementary grades (Schifter, Monk, Russell, Bastable, & Earnest, 2003). Some of their instruction deliberately but indirectly deals with such generalizations as commutativity of addition, although that language may not be used in the early exposures. Key are the tasks and teacher questions. Questions such as "Will it always work that way?" or "Why does it work out that way?" are a natural part of the instruction and indeed become an expected part of the lesson. Their work shows that children's conviction about at least some generalizations may follow an interesting path. For example, first graders, when asked "Will it always work that way?" after noting a specific example for commutativity of addition, may be uncertain that the idea will always work. Later, they may be confident that the answer is "Yes" because they have tried lots of examples (an empirical proof scheme), but still later, they may return to uncertainty because they are aware that there are many untested cases and that they have not tried them all. And, not unexpectedly, some students will over-generalize, accepting commutativity of subtraction as well.

Several Italian studies have also examined the nature of students' justifications as encountered in problem-based lessons (e.g., Boero, Chiappini, Garuti, & Sibilla, 1995; Boero, Garuti, & Mariotti, 1996). One example is their grade 8 students' study of shadows. Using data that the students collected over a period of time about the shadows of two vertical sticks, they examined the question of whether the shadows of a vertical stick and an oblique stick can be parallel, and if so, when. Small-group work, with teacher help, led to clearly stated conjectures (e.g., If the sun's rays belong to the vertical plane of the oblique stick, then the shadows are parallel) that were then examined further, with an eye toward establishing them "in general." In the analysis of these subsequent attempts at general arguments (i.e., proofs), the researchers noticed that in the successful proofs, there were connections with key observations made during the conjecture-forming stage. The researchers' collective work has led to their hypothesis of "cognitive unity," emphasizing the close connection between the reasoning during the formation of a conjecture and the reasoning in an eventual proof:

> (D)uring the production of a conjecture, the student progressively works out his/her statement through an intensive argumentative activity functionally intermingling with the justification of the plausibility of his/her choices; during the subsequent statement proving stage, the student links up with this process in a coherent way, organizing some of the justifications ('arguments') produced during the construction of the statements according to a logical chain (Boero, Garuti, Lemut, & Mariotti, 1996, p. 119–120).

They also argue that such a process is followed by many mathematicians—during the conjecturing stage, the mathematician uses arguments that can later be adapted to support his or her mathematical proof—and make the case that much more instruction in mathematics should involve conjecturing.

One series of studies (Maher, 2002; Maher & Martino, 1996; Martino & Maher, 1999) carried out with instruction in a similar problem-based vein, is notable because of its long-term nature (occasional sessions over 14 years, usually separate from the regular mathematics classes) and because of proof behaviors—proof by contradiction, proof by cases, proof by mathematical induction—that arose naturally, if informally, even in elementary school, at least on the part of some students. The nature and flavor of the sessions is communicated by these retrospections of a participant:

> Well, we break up into groups . . . like five groups of three, say, and everyone in their own groups would have their own ideas, and you'd argue within your own group, about what you knew, what I thought the answer was, what you thought the answer was and then from there, we'd all get together and present our ideas, and then this group would argue with this group about who was right with this . . . (Maher, 2002, p. 37). . . . You didn't come in and say, "this is what we were learning today and this is how you're going to figure out the problem." We were figuring out how we were going to figure out the problem. We weren't attaching names to that but we could see the commonness between what we were working on there and maybe what we had done in school at some point in time and been able to put those things together and come up with stuff and to do these problems to come up with, what would be our own formulas because we didn't know that other people had done them before. We were just kind of doing our own thing trying to come up with an answer that was legitimate and that no matter how you tried to attack it, we could still answer it . . . (Maher, 2002, p. 32).

As the excerpts illustrate, all of these student-centered, problem-based studies have involved a way of teaching that is in stark contrast to the stereotype of a mathematics class: Students check homework, teacher illustrates something new, students then do seat-work or homework to practice the new material. The didactical contract (Brousseau, 1997) in the experimental classes was obviously quite different from that in the stereotypical one. In particular, the "social norms" were quite different: Students were expected to work together rather than singly, students were to explain their solution methods, and students were to listen carefully and evaluate the explanations of other students—and hence perhaps learn different proof schemes.

Yackel and Cobb (1996) have sharpened the analysis of social norms to identify "sociomathematical norms," those social norms that refer specifically to *mathematical* activity. For example, coming to accept that an explanation is expected might be a general social norm, whereas what constitutes an acceptable mathematical explanation would be a sociomathematical norm (Yackel & Cobb, 1996). Other sociomathematical norms might include norms dealing with when different explanations are mathematically different, or when a justification is acceptable, or when justifications or explanations convey efficiency or elegance. Indeed, Yackel, Rasmussen, and King (2000) focused on the norms that developed during a problem-based undergraduate differential equations course. They found that students "frequently explained their reasoning without prompting, offered alternative explanations and attempted to make sense of other students' reasoning and explanations, despite the fact that their prior experiences were with traditional approaches to mathematics instruction" (p. 276) and, in particular, there was evidence of the development of such sociomathematical norms as, what is an acceptable mathematical justification and what makes up a mathematically different explanation. That the students, who had likely experienced mathematics classrooms with quite different social norms, developed such norms is particularly encouraging. Yackel and Cobb point out that sociomathematical norms should be examined in teaching modes that are different from the inquiry mode in which they have been studied, because the classroom conduct, whether stereotypic or not, will automatically convey what is an acceptable norm. For example, one can wonder whether some students, young or experienced, might be more comfortable with a more directed approach. Dweck (1999) has identified different goal orientations on the part of students. Some students may be guided primarily by performance goals like grades, parent/teacher approval, high marks, or status with others, whereas others have what Dweck calls "learning goals"—their primary interest is in understanding or mastery. A student's extreme preference for performance goals may be an unfortunate aspect of schooling or society as it exists, of course, but it might also result in resistance to sociomathematical norms that do not obviously support perceived performance. It is an interesting question as to whether different sorts of teaching might shape a student's learning goals.

In most of the studies above, outside support for the teacher was particularly important. The question of what is feasible in classrooms without further teacher preparation or researcher involvement is crucial, as, for example, Yackel and Hanna point out (2003). It is nonetheless exciting to envision learners, starting in the primary grades and continuing through high school and college, developing the social and sociomathematical norms about proof, and the proof schemes, that one might wish, but it is daunting to think of the changes needed in curricula and teachers (and testing programs) to support the development of such norms. As an indication of areas of teacher preparation that are important, Martino and Maher (1999) suggest, based on their videotape analysis of their multi-year study, that there is a "strong relationship between (1) [the teacher's] careful monitoring of students' constructions leading to a problem solution, and (2) the posing of a timely question which can challenge learners to advance their understanding" (p. 53). And there may be useful norm-supporting ideas even for classes in which problem-based or inquiry-based instruction is not the mode. For example, Blanton, Stylianou, and David (2003), based on their work with a college discrete mathematics course and extending the work mentioned earlier (Blanton & Stylianou, 2002), believe that an instructor's attention to likely student understanding (and hence a zone of proximal development) can enable an instructional scaffolding that, along with a careful discussion of (preferably) students' proofs, can support students' understanding of proof.

Again, all of these studies illustrate how appropriate mathematical curricula, together with appropriate teacher intervention, can help students to develop critical elements of deductive reasoning and an openness to the ideas of mathematical proof. As feasibility studies, these efforts have shown that even elementary school children can be remarkably able to make sense of mathematics, if given opportunities. Their explanations not only help them in their communication skills, but the explanations are likely also to help their classmates, to give to their teachers insights into the children's level of understanding, and to engender or foster students' growth in their proof schemes and their grasp of sociomathematical norms.

Use of Technology

The increasing use of dynamic geometry software during the last couple of decades has provoked researchers to look closely into the learning benefits as well as the potential risks of this tool. In this section we focus on studies conducted to investigate the impact of dynamic geometry environments (DGEs) on the learning of proof, primarily because those environments have been involved in several studies.

Generating and measuring many examples, as is now possible and easy with DGEs, would seem only to support the idea that examples prove a result (the empirical proof scheme) and hence interfere with any need for deductive proof. Chazan (1993a,b) noted and corroborated that the ideas that evidence is proof and that proof is merely evidence were widespread. He interviewed 17 students who were taught geometry in a DGE environment (1993b) to see how students fared when they experienced a geometry curriculum that involved both measurement of cases and deductive proof, with the curriculum including attention to the different types of argumentation involved with the two methods. He found both evidence-is-proof (the empirical proof scheme) and proof-is-merely-evidence postures on the part of the students, and hence he noted that the "comparison and contrast of verification and deductive proof certainly deserves explicit attention in mathematics classrooms" (1993b, p. 382), with the teachers involved in the study feeling that more time should have been spent in dealing with students' doubts about deductive proofs (1993a, p. 109). However, he also noticed that "fewer students considered evidence to be proof . . . whereas more students were skeptical about the limits of applicability of deductive proofs. . . . Some students seemed to become more skeptical about deductive proofs as a result of becoming more skeptical about measurement of examples" (1993a, p. 109). He also suggested, based on interviewee comments, "that the explanatory aspect of proofs is a useful starting point for a discussion of the value of deductive proofs" (1993b, p. 383), that some students have no idea of what deductive proofs are intended to do, and that some students resist the idea that a deductive argument can assure that there cannot be counterexamples.

A special issue of the journal, *Educational Studies in Mathematics* (2000, *44*), included four reports on teaching experiments involving DGE. The first four are by Mariotti, by Jones, by Marrades and Gutierrez, and by Hadas, Hershkowitz, and Schwartz. The last paper is by Laborde, and synthesizes the previous four, providing a connecting theme among them by using Brousseau's construct of "milieu." We continue with these studies.

One of the significant results of the four studies is that their findings address a major concern regarding the use of DGEs in the teaching of geometry in school: "the opportunity offered by [DGE] to 'see' mathematical properties so easily might reduce or even kill any need for proof and thus any learning of how to develop a proof" (Laborde, 2000, p. 151).

Jones' study was a teaching experiment with 12-year-old students. The intervention involved students working in pairs or small groups on the classification of quadrilaterals. The instructional activities involved tasks where the students were to reproduce a figure that could not be "messed up" by dragging any of its components (a vertex or segment). More advanced activities included tasks of producing a figure that can be transformed into another specified figure by dragging (e.g, from a rectangle to a square). In each case the students had to explain why their constructed figure was the expected one. Laborde (2000) points out that this type of explanation consists of giving the conditions that imply that the constructed figure is the expected type of quadrilateral, a necessary activity to understand how proof works. From our perspective, these explanations in the context of a DGE involve movement, and, therefore—as we have discussed in earlier in the section—they are causal, and hence deductive. Accordingly, Jones' (2000) study suggests that the DGE does not necessarily eliminate the need for proof in the students' eyes but can enhance students' deductive proof scheme. The developmental path of students' conception in this study started where students were. Initially, according to Jones, students lacked the capability to describe or explain in precise mathematical language. The instructional emphasis in this stage was on description rather than explanation, where students utilized perception rather than mathematical language to describe their observations. As the emphasis extended to explanations, students' language became more precise but was mediated by the DGE terminology (e.g., the term "dragging"). By the end of the teaching experiment, students' explanations related entirely to the mathematical context.

Mariotti's study was a long-term teaching experiment with 15–16-year-old students. Students were engaged in DGE activities through which the students themselves constructed a geometric system of axioms and theorems as a system of Cabri Geometry commands. Similar to Jones, Mariotti emphasized activities where the task is for pairs of students to construct geometric figures, describe the construction procedure, and justify why the procedure produces the expected figure. The basic conceptual change that Mariotti's (2000) study achieved was in students' status of justification, which transitioned from an "intuitive" geometry—a collection of self-evident properties—to a theoretical geometry—a system of statements validated by proof. The theoretical geometry that Mariotti's students had constructed seems to be more than a deductive system, in that students were not only constructing and proving theorems but also establishing the axioms on which these theorems rest, and thereby laying the foundations for their axiomatic proof scheme.

Hadas, Hershkowitz and Schwarz' (2000) study was done in the context of a geometry course that emphasized the concept of proof. They developed instructional activities involving students making assertions about certain geometric relations and later checking them with a DGE. The choice and sequence of activities were such that upon checking their assertions with the DGE the students would find them to be false—a realization that would make the students curious as to the reason for the falsity of the conjecture. For example, in one of the activities the students began with two tasks. The first task was to measure (with the software) the sum of the interior angles in polygons as the number of sides increases, generalizing their observation, and then explaining their conclusion. The second task was to measure (with the software) the sum of the exterior angles of a quadrilateral. Following this, the students were asked to hypothesize the sum of the exterior angles for polygons as the number of sides increases, and to check their hypothesis by measuring (with the software) and explain what they found. Hadas, Hershkowitz and Schwartz succeeded in creating in the students a need to find out the cause for their assertion to be untrue. Laborde (2000) points out that such an achievement would have been impossible without the use of a dynamic geometry system, for "the false conjectures came after students were convinced of other properties thanks to the DG system. . . . [The] interplay of conjectures and checks, of certainty and uncertainty, was made possible by the exploration power and checking facilities offered by the DG environment" (p. 154).

Marrades and Gutierrez investigated how DGEs can help secondary-school students (aged 15–16 years) enhance their proof schemes. As in Hadas, Hershkowitz and Schwarz' study, Marrades and Gutierrez showed that a DGE can help students realize the need for formal proofs in mathematics. By interpreting their results in terms of our taxonomy of proof schemes, an important observation reported in their study is that students' transition to deductive proof schemes is very slow; the total teaching experiment lasted 30 weeks, with two 55 minute class per week. Of particular importance is their finding that for this transition to take place instruction must not ignore students' current empirical proof schemes and must institute a didactical contract that attempts to suppress the authoritative proof scheme. Their method was to repeatedly emphasize "the need to organize justifications by using definitions and results (theorems) previously known and accepted by the class" (p. 120). Finally, another significant finding of this study is that the ability to produce deductive proof evolves hand in hand with students' understanding of subject matter: the concepts and properties related to the topic being studied. This is consistent with other findings. Simon and Blume (1996), for example, illustrated that a learner may not fully understand another's proof because of a limited grasp of the concepts addressed in the proof (p. 29). One can argue that such exposure might lead to a disequilibrium and eventually a greater understanding of both the concepts and the proof.

Hence, DGEs are a promising tool, but they do not automatically or easily lead to improved proof schemes. Accomplishing that sort of growth apparently requires a carefully laid-out curriculum (cf. de Villiers, 1999) and considerable adjustment by a teacher accustomed only to telling as the mode of instruction. Lampert (1993), for example, described some of the difficulties encountered by teachers who allowed conjecturing with a DGE. (Making conjectures, itself, may be difficult for students new to the expectation. Koedinger [1998], with an eye toward possible software activities, noted that with his task of writing a conjecture about kites, given the definition of kite, about a quarter of the roughly 60 geometry students could not come up with a non-trivial conjecture within 20 minutes.) Lampert noted that the change from "sage on the stage" to "guide on the side" required adjustments both for the teacher and the students, since a different sort of didactical contract was involved. Teachers were also concerned about the coverage of a standard body of content (for external testing purposes, or for later courses), as well as the departure from the usual, familiar axiomatic development that often eventuated.

Does Explicit Teaching of Logic Work?

The use of deductive proof schemes, at least implicitly, involves logic. Calls for the explicit teaching of logic are easy to find. More than 40 years ago, consensus at the future-oriented Cambridge Conference on School Mathematics (1963) was that "it is hardly possible to do anything in the direction of mathematical proofs without the vocabulary of logic and explicit recognition of the inference schemes" (p. 39). There have been occasional caveats, as in Suppes' remarks:

> I would not advocate an excessive emphasis on logic as a self-contained discipline. . . . What I do feel is important is that students be taught in an explicit fashion classical rules of logical inference, learn how to use these rules in deriving theorems from given axioms, and come to feel as much at home with simple principles of inference like modus ponendo ponens as they do with elementary algorithms of arithmetic" (1966, p. 72).

Studies of the effects of an explicit attention to logic have not, however, indicated that there is then a pay-off in proof-writing. School geometry has long had a goal of improving students' logical thinking (cf. NCTM, 1970), and several studies have looked at the influence of including an explicit, concentrated treatment of logic in that course (e.g., Deer, 1969; Mueller, 1975; Platt, 1967). Yet, the usual outcome, even when the logic treatment has involved up to four weeks (Platt) and/or a reasonable number of students (Mueller, 146 students; Platt, 12 classes), is a finding of no-significant-differences in proof performance.

It is reasonable to expect that the teacher's emphasis on logical reasoning, even in the absence of explicit treatment in a curriculum, might influence the students' use of logic themselves. In a study of the influence of teachers' language on performance, Gregory and Osborne (1975) explored whether the teachers' use of logic influenced their students' performance on a logic test, by examining audiotapes of five lessons of 20 junior high school mathematics teachers. Although acknowledging some design problems, they found a significant positive correlation between logic test performance for some types of inferences (e.g., when negatives were involved) and the usages of if-then statements by the five teachers with the greatest average use vs. usages of the five teachers with the lowest. Eye-catching, however, was the range in the averages of the number of usages per lesson of if-then sorts of statements: 8.3 to 40.6, with individual lessons giving from 3 to 48 usages (p. 28). Unfortunately, there was no control for content covered (some teachers were teaching number theory, some word problems, and some properties of geometric figures), so the range is only suggestive of what might be an important element in students' growth in deductive reasoning: the frequency of use of logical statements and logical reasoning in the classroom.

Thus there is much evidence that some important elements of deductive reasoning are not natural parts of students' repertoires, at a variety of school levels. How to develop these elements so that they are recognized and utilized in mathematical proofs is an open question, and may not be realized through a single, short unit. As the 2000 NCTM Principles and Standards assert, "Reasoning and proof cannot simply be taught in a single unit on logic, for example, or by 'doing proofs' in geometry" (NCTM, 2000, p. 56).

REFLECTION

In this chapter we have presented a comprehensive perspective on proof learning—a perspective that addresses mathematical, historical-epistemological, cognitive, sociological, and instructional factors. Comprehensive perspectives on proof are needed, we argued, in order to better understand the nature and roots of students' difficulties with proof so that effective instructional treatments can be designed and implemented to advance students' conceptions of and attitudes toward proof. Our perspective grew out of a decade of investigations—empirical as well as theoretical—into students' conceptions of proof. In various periods and stages of these investigations we have repeatedly confronted questions that collectively address a combination of the five factors mentioned above.

The notion of "proof scheme" serves as the main lens of our comprehensive perspective on proof. Through it, for example, we analyze and interpret students' proving behavior—in their individual work as well as in their interaction with others—and understand the development of proof in the history of mathematics. *A proof scheme consists of what constitutes ascertaining and persuading for a person (or community).* This definition was born out of cognitive, epistemological, and instructional considerations. Specifically, a critical observation in our and other scholars' work is that proof schemes vary from person to person and from community to community—in the classroom, with individual students as well as the class as a whole, and throughout history. "Proof," when viewed in this subjective sense, highlights the student as learner. As a result, teachers must take into account what constitutes ascertainment and persuasion for their students and offer, accordingly, instructional activities that can help them gradually refine and modify their proof schemes into desirable ones. This subjective view of proof emerged from our studies and impacted many of the conclusions we drew from them. For example, it influenced our conclusions as to the implications of the epistemology of proof in the history of mathematics to the conceptual development of proof with students, the implications of the way mathematicians construct proofs to instructional treatments of proof, and the implications of the everyday justification and argumentation on students' proving behaviors in mathematical contexts. The subjective notion of proof scheme is not in conflict with our insistence on unambiguous goals in the teaching of proof—namely, to gradually help students develop an understanding of proof that is consistent with that shared and practiced by the mathematicians of today. The question of critical importance is: What instructional interventions can bring students to see an intellectual need to refine and alter their current proof schemes into deductive proof schemes (Harel, 2001)?

The status studies we have reviewed and presented in this paper show the absence of the deductive proof scheme and the pervasiveness of the empirical proof scheme among students at all levels. Students base their responses on the appearances in drawings, and mental pictures alone constitute the meaning of geometric terms. They prove mathematical statements by providing specific examples, not able to distinguish between inductive and deductive arguments. Even more able students may not understand that no further examples are needed, once a proof has been given. Students' preference for proof is ritualistically and authoritatively based. For example, when the stated purpose was to get the best mark, they often felt that more formal—e.g., algebraic—arguments might be preferable to their first choices. These studies also show a lack of understanding of the functions of proof in mathematics, often even among students who had taken geometry and among students for whom the curriculum pays special attention to conjecturing and explaining or justifying conclusions in both algebra and geometry. Students believe proofs are used only to verify facts that they already know, and have no sense of a purpose of proof or of its meaning. Students have difficulty understanding the role of counterexamples; many do not understand that one counterexample is sufficient to disprove a conjecture. Students do not see any need to prove a mathematical proposition, especially those they consider to be intuitively obvious. This is the case even in a country like Japan where the official curriculum emphasizes proof. They view proof as the method to examine and verify a later particular case. Finally, the studies show that students have difficulty writing valid simple proofs and constructing, or even starting, simple proofs. They have difficulty with indirect proofs, and only a few can complete an indirect proof that has been started.

We believe there is a need for more longitudinal studies regarding students' proof schemes. The most difficult studies to carry out, for both financial and design reasons are longitudinal studies. Yet we cannot gain a solid understanding of the effect on students of continued attention to justification and proof throughout their studies in mathematics, except through longitudinal studies. A one- or two-year exposure (let alone a one-semester treatment) to instruction and curricula attentive to reason-giving can be dwarfed by a multiple-year focus on instruction and curricula that, to use an extreme example, emphasize rote skills likely to be useful in some external testing program. In the latter cases, being asked to give reasons and arguments might well be viewed as aberrations and irrelevant to the perceived "really important" side of mathematics.

The findings from studies of teachers' conceptions of proof do not look much better than those with students. Overall, teachers seem to acknowledge the verification role of proof, yet for many the empirical proof schemes seem to be the most dominant, even in dealing with mathematical statements, and they do not seem to understand other important roles of proof, most noticeably its explanatory role. Some teachers tend to view proof as an appropriate goal only for the mathematics education of a minority of students, not considering proof and justification to be a central concern in school mathematics, as has been repeatedly called for by the mathematics education leadership (e.g., NCTM, 1989, 2000). Studies show that little or no instructional time is allocated to the development of the deductive proof schemes, not even in geometry. In the U.S. explicit mathematical reasoning in mathematics classes is rare, and in algebra and pre-algebra courses it is virtually absent. Many teachers are unlikely to teach proof well, since their own grasp of proof is limited. It is important to determine better the extent to which teachers are equipped to deliver a curriculum in which proof is central. Results from studies like those of Knuth (2002a) and Manaster (1998), if indeed typical of a widespread performance of mathematics teachers, demand attention both on the part of university mathematics departments, which have a primary responsibility for the preparation of mathematics teachers, and on school districts, which support the continued development of their existing faculty.

The bright side of the findings is that students who receive more instructional time on developing analytical reasoning by solving unique problems fare noticeably better on overall test scores. Likewise, students who have been expected to write proofs and who have had classes that emphasized proof were somewhat better than other students. It also seems possible to establish desirable sociomathematical norms relevant to proof, through careful instruction, often featuring the student role in proof-giving. There has been a concern that the ease with which technology can generate a large number of examples naturally could undercut any student-felt need for deductive proof schemes. Fortunately, several studies have shown that with careful, non-trivial planning and instruction over a period of time, progress toward deductive proof schemes is possible in technology environments, where such desiderata as making conjectures and definitions occur.

An important element in deductive proof schemes is of course the use of logical reasoning. Yet there is evidence that many students, and possibly even many teachers, do not have a good grasp or appreciation of some important principles of logic. Nor is it clear as to how best to devise instruction to improve performance with logic in mathematical (and non-mathematical) contexts. It is unclear to us how best to prepare students to deal with the logical reasoning essential in mathematical proofs and valuable in even informal justifications—osmosis, or explicit attention? And, in particular, the knowledge of teachers of mathematics about logical reasoning may be a matter of concern (e.g., Easterday & Galloway, 1995). We see a need for the incorporation of items on proof or logic (even multiple-choice ones) into the periodic National Assessments of Education Progress. From the practical viewpoint, the NAEPs exist, and they offer a view of performance across the U.S. Even more pleasing would be to see large-scale efforts devoted explicitly to the study of performance in proof and logic, like those in Great Britain (Healy & Hoyles, 1998, 2000; Hoyles & Kuchemann, 2002). A deep look at the students and teachers' knowledge of proof, on the one hand, and at the development of the deductive proof scheme in the history of mathematics, on the other, has provided us with important insights as to what might account for students' difficulties in constructing this scheme and what instructional approaches can facilitate its construction. In particular, considerations of historical-epistemological developments have led us to new research questions with direct bearing on the learning and teaching of proofs. For example:

1. To what extent and in what ways is the nature of the content intertwined with the nature of proving? In geometry, for example, does students' ability to construct an image of a point as a dimensionless geometric entity impact their ability to develop the Greek axiomatic proof scheme?
2. What are the cognitive and social mechanisms by which deductive proving can be necessitated for the students? The Greek's construction of their geometric edifice seems to have been a result of their desire to create a consistent system that was free from paradoxes. Would paradoxes of the same nature create a similar intellectual need with students?
3. Students encounter difficulties in moving between proof schemes, particularly from the Greek's axiomatic proof scheme (the one they construct in honors high-school geometry, for example) to the modern axiomatic proof scheme (the one they need to succeed in a real analysis course, for example). Exactly what are these difficulties? What role does the emphasis on *form* rather than *content* in modern mathematics (as opposed to Greek mathematics where content is more prominent) play in this transition? Can students develop the modern axiomatic proof without computational fluency? What role does the causality proof scheme play in this transition?

These questions are examples of what we have delineated as important contributions from the history of mathematics to our thinking about students' proof schemes. But there are likely to be other valuable ideas from further study of the growth or development of proof ideas in the history of mathematics. How mathematical proof arose in other cultures—e.g., the basis for the Japanese temple drawings (Rothman & Fukagawa, 1998)—would in itself be fascinating and potentially instructive about how proof ideas might be introduced or developed today. In this respect, our effort to form a comprehensive perspective on proof is an attempt to understand what might be called the "proving conceptual field," a term analogous to Vergnaud's (1983, 1988) "multiplicative conceptual field." Like the multiplicative conceptual field, the proving conceptual field may be thought of as a set of problems and situations for which closely connected concepts, procedures, and representations are necessary.

REFERENCES

Adey, P. (1999). *The science of thinking and science for thinking: A description of logical acceleration through science education.* (INNODATA Monographs-2). Geneva, Switzerland: International Bureau of Education.

Antonini, S. (2003). Non-examples and proof by contradiction. In N. Pateman, B. J. Dougherty, & J. Zilliox (Eds.), *Proceedings of the 27th Annual Meeting of the International Group for Psychology in Mathematics Education (Vol. 2,* pp. 49–55). Honolulu: University of Hawaii.

Aristotle (1941). *The basic works of Aristotle,* R. McKeon (Ed. & Translator). New York: Random House.

Babai, L. (1992). Transparent proofs. *Focus, 12*(3), 1–2.

Balacheff, N. (1988). Aspects of proof in pupils' practice of school mathematics. In D. Pimm (Ed.), *Mathematics, teachers and children* (pp. 216–235). London: Holdder & Stoughton.

Balacheff, N. (1991). The benefits and limits of social interaction: The case of mathematical proof. In A. Bishop, E. Mellin-Olsen, & J. van Dormolen (Eds.), *Mathematical knowledge: Its growth through teaching* (pp. 175–192). Dordrecht, The Netherlands: Kluwer.

Balacheff, N. (2002). A researcher epistemology: A deadlock for educational research on proof. *Proceedings of 2002 International Conference on Mathematics: Understanding Proving and Proving to Understand* (pp. 23–44). Department of Mathematics, National Taiwan Normal University.

Ball, D., & Bass, H. (2003). Making mathematics reasonable in school. In J. Kilpatrick, W. G. Martin, & D. Schifter (Eds.), *A research companion to Principles and Standards for School Mathematics* (pp. 27–44). Reston, VA: National Council of Teachers of Mathematics.

Battista, M., & Clements, D. (1992). Geometry and spatial reasoning. In D. Grouws (Ed.), *Handbook of research on mathematics teaching and learning* (pp. 420–464). New York: Macmillan.

Bell, A. W. (1976). A study of pupils' proof-explanations in mathematical situations. *Educational Studies in Mathematics, 7,* 23–40.

Blanton, M. L., & Stylianou, D. A. (2002). Exploring sociocultural aspects of undergraduate students' transition to mathematical proof. In D. S. Mewborn, P. Sztajn, D. Y. White, H. G. Wiegel, R. L. Bryant, & K. Nooney (Eds.), *Proceedings of the 24th Annual Meeting for Psychology in Mathematics Education—North America, Vol. 4,* (pp. 1673–1680). Athens, GA: University of Georgia.

Blanton, M. L., Stylianou, D. A., & David, M. M. (2003). The nature of scaffolding in undergraduate students' transition to mathematical proof. In N. Pateman, B. J. Dougherty, & J. Zilliox (Eds.), *Proceedings of the 27th Annual Meeting of the International Group for Psychology in Mathematics Education, Vol. 2* (pp. 113–120). Honolulu: University of Hawaii.

Boero, P., & Garuti, R. (1994). Approaching rational geometry: from physical relationships to conditional statements. In J. Ponte & J. Matos (Eds.), *Proceedings of the Eighteenth Annual Meeting of the International Group for Psychology in Mathematics Education, Vol. II* (pp. 96–103). Lisbon: University of Lisbon.

Boero, P., Chiappini, G., Garuti, R., & Sibilla, A. (1995). Towards statements and proofs in elementary arithmetic: An exploratory study about the role of teachers and the behaviour of students. In L. Meira & D. Carraher (Eds.), *Proceedings of the 19th Psychology in Mathematics Education Conference* (Vol. 3, pp. 129–136). Recife, Brazil: Universidade Federal de Pernambuco.

Boero, P., Garuti, R., & Mariotti, M. (1996). Some dynamic mental processes underlying producing and proving conjectures. In L. Puig & A. Gutierrez (Eds.), *Proceedings of the 20th Conference of the International Group for the Psychology of Mathematics Education* (Vol. 2, pp. 121–128). Valencia, Spain: Universitat de Valencia.

Boero, P., Garuti, R., Lemut, E., & Mariotti, M. (1996). Challenging the traditional school approach to the theorems: A hypothesis about the cognitive unity of theorems. In L. Puig & A. Gutierrez (Eds.), *Proceedings of the 20th Conference of the International Group for the Psychology of Mathematics Education* (Vol. 2, pp. 113–120). Valencia, Spain: Universitat de Valencia.

Brousseau, G. (1997). *Theory of didactical situations in mathematics.* N. Balacheff, M. Cooper, R. Sutherland, & V. Warfield (Eds. & Trans.). Dordrecht, The Netherlands: Kluwer.

Brumfiel, C. (1973). Conventional approaches using synthetic Euclidean geometry. In K. Henderson (Ed.), *Geometry in the mathematics curriculum* (pp. 95–115). Thirty-sixth Yearbook of the National Council of Teachers of Mathematics. Reston, VA: National Council of Teachers of Mathematics.

Cambridge Conference on School Mathematics. (1963). *Goals for school mathematics.* Boston: Houghton Mifflin.

Carpenter, T. P. (1989). Introduction. In M. M. Lindquist (Ed.), *Results from the fourth mathematics assessment of the National Assessment of Educational Progress* (pp. 1–9). Reston, VA: National Council of Teachers of Mathematics.

Carpenter, T. P., Coburn, T. G., Reys, R. E., & Wilson, J. W. (1978). *Results from the first mathematics assessment of the National Assessment of Educational Progress.* Reston, VA: National Council of Teachers of Mathematics.

Carpenter, T. P., Franke, M. L., Jacobs, V. R., Fennema, E., & Empson, S. (1998). A longitudinal study of invention and understanding in children's multidigit addition and subtraction. *Journal for Research in Mathematics Education, 29,* 3–20.

Chazan, D. (1993a). Instructional implications of students' understandings of the differences between empirical verification and mathematical proof. In J. Schwartz, M. Yerushalmy, & B. Wilson (Eds.), *The geometric supposer: What is it a case of?* (pp. 107–116). Hillsdale, NJ: Erlbaum.

Chazan, D. (1993b). High school geometry students' justification for their views of empirical evidence and mathematical proof. *Educational Studies in Mathematics, 24,* 359–387.

Cobb, P. (1994). Theories of mathematical learning and constructivism: A personal view. Paper presented at the Symposium on Trends and Perspectives in Mathematics Education, Institute for Mathematics, University of Klagenfurt, Austria.

Cobb, P., Wood, T., Yackel, E., Nicholls, J., Wheatley, G., Trigatti, B., & Perlwitz, M. (1991). Assessment of a problem-centered second-grade mathematics project. *Journal for Research in Mathematics Education, 22,* 3–29.

Cobb, P., & Yackel, E. (1996). Constructivist, emergent, and sociocultural perspectives in the context of developmental research. *Educational Psychologist, 31,* 175–190.

Coe, R., & Ruthven, K.. (1994). Proof practices and constructs of advanced mathematics students. *British Educational Research Journal, 20*(1), 41–53.

Deer, G. W. (1969). The effects of teaching an explicit unit in logic on students' ability to prove theorems in geometry (Doctoral dissertation, Florida State University, 1969). *Dissertation Abstracts International, 30,* 2284–2285B.

de Villiers, M. D. (1999). *Rethinking proof with the Geometer's Sketchpad.* Emeryville, CA: Key Curriculum Press.

Dubinsky, E., & Yiparaki, O. (2000). On student understanding of AE and EA quantification. In E. Dubinsky, A. H. Schoenfeld, & J. Kaput (Eds.), *Research in collegiate mathematics education IV.* (pp. 239–286). Providence, RI: American Mathematical Society.

Dweck, C. S. (1999). *Self-theories: Their role in motivation, personality, and development.* Philadelphia: Psychology Press.

Easterday, K. E., & Galloway, L. L. (1995). A comparison of sentential logic skills: Are teachers sufficiently prepared to teach logic? *School Science and Mathematics, 95,* 431–436.

Eisenberg, T. A., & McGinty, R. L. (1974). On comparing error patterns and the effect of maturation in a unit on sentential logic. *Journal for Research in Mathematics Education, 5,* 225–237.

Epp, S. S. (2003). The role of logic in teaching proof. *American Mathematical Monthly, 110*(10), 886–899.

Evans, J. St. B. T. (1982). *The psychology of deductive reasoning.* London: Routledge & Kegan Paul.

Fawcett, H. P. (1995). *The nature of proof.* Thirteenth Yearbook of the National Council of Teachers of Mathematics. New York: Teachers College, Columbia University. (Original work published 1938).

Fischbein, E., & Kedem, I. (1982). Proof and certitude in the development of mathematical thinking. In A. Vermandel (Ed.), *Proceedings of the Sixth International Conference of the Psychology of Mathematics Education* (128–131). Antwerp, Belgium: Universitaire Instelling Antwerpen.

Freudenthal, H. (1962). The main trends in the foundations of geometry in the 19th century. In E. Nagel, P. Suppes, & A. Tarski, *Logic, methodology, and philosophy of science.* Stanford, CA: Stanford University Press.

Freudenthal, H. (1973). *Mathematics as an educational task.* Dordrecht, The Netherlands: D. Reidel Publishing Company.

Fujita, T., & Jones, K. (2003, July). *Critical review of geometry in current textbooks in lower secondary schools in Japan and the UK.* Paper presented at the 27th annual meeting of the International Group for Psychology of Mathematics Education, Honolulu.

Galbraith, P. (1981). Aspects of proving: a clinical investigation of process. *Educational Studies in Mathematics, 12,* 1–28.

Ginsburg, H. J., & Opper, S. (1969). *Piaget's theory of intellectual development.* Englewood Cliffs, NJ: Prentice-Hall.

Goetting, M. M. (1995). The college students' understanding of mathematical proof (Doctoral dissertation, University of Maryland, 1995). *Dissertation Abstracts International, 56-A,* 3016.

Gregory, J. W., & Osborne, A. R. (1975). Logical reasoning ability and teacher verbal behavior within the mathematics classroom. *Journal for Research in Mathematics Education, 6,* 26–36.

Hacking, I. (1980). Proof and eternal truths: Descartes and Leibniz. In S. Gaukroger (Ed.), *Descartes' philosophy, mathematics and physics* (pp. 169–180). Sussex, U.K.: The Harvester Press.

Hadas, N., Hershkowitz, R., & Schwarz, B. B. (2000). The role of contradiction and uncertainty in promoting the need to prove in dynamic geometry environments. *Educational Studies in Mathematics, 44,* 127–150.

Haimo, D. T. (1995). Experimentation and conjecture are not enough. *American Mathematical Monthly, 102*(2), 102–112.

Hanna, G. (1983). Rigorous proof in mathematics education. Toronto: OISE Press.

Hanna. G. (1990). Some pedagogical aspects of proof. *Interchange, 21,* 6–13.

Hanna, G., & Jahnke, H. N. (1996). Proof and proving. In A. Bishop, K. Clements, C. Keitel, J. Kilpatrick, & C. Laborde (Eds.), *International handbook of mathematics education* (Part 2, pp. 877–908). Dordrecht, The Netherlands: Kluwer.

Harel, G. (1999). Students' understanding of proofs: A historical analysis and implications for the teaching of geometry and linear algebra. *Linear Algebra and Its Applications, 302–303,* 601–613.

Harel, G. (2001). The development of mathematical induction as a proof scheme: A model for DNR-based instruction. In S. Campbell & R. Zazkis (Eds.), *The learning and teaching of number theory* (pp. 185–212). Dordrecht, The Netherlands: Kluwer.

Harel, G. (2006). Students' proof schemes revisited. In P. Boero (Ed.), *Theorems in school: From history, epistemology and cognition to classroom practice* (pp. 61–72). Rotterdam: Sense Publishers.

Harel, G., & Sowder, L. (1998). Students' proof schemes: Results from exploratory studies. In A. Schoenfeld, J. Kaput, & E. Dubinsky (Eds.), *Research in collegiate mathematics education III* (pp. 234–283). Providence, RI: American Mathematical Society.

Hart, E. W. (1994). A conceptual analysis of the proof-writing performance of expert and novice students in elementary group theory. In J. Kaput & E. Dubinsky (Eds.), *Research issues in undergraduate mathematics learning,* MAA Notes No. 33 (pp. 49–157). Washington: Mathematical Association of America.

Hartshorne, R. (2000). *Geometry: Euclid and beyond.* Springer, New York.

Healy, L., & Hoyles, C. (1998). Justifying and proving in school mathematics, executive summary. London: Institute of Education, University of London.

Healy, L. & Hoyles, C. (2000). A study of proof conceptions in algebra. *Journal for Research in Mathematics Education, 31(4),* 396–428.

Heath, T. L. (1956). *The thirteen books of Euclid's Elements* (2nd ed.), Vol. I. New York: Dover.

Hersh, R. (1993). Proving is convincing and explaining. *Educational Studies in Mathematics, 24,* 389–399.

Hoffer, A. (1981). Geometry is more than proof. *Mathematics Teacher, 74,* 11–18.

Hoyles, C. (1997). The curricular shaping of students' approaches to proof. *For the Learning of Mathematics, 17*(1), 7–16.

Hoyles, C., & Kuchemann, D. (2002). Students' understanding of logical implication. *Educational Studies in Mathematics, 51,* 193–223.

Inhelder, B., & Piaget, J. (1958). *The growth of logical thinking from childhood to adolescence* (A. Parsons & S. Milgram, Trans.). New York: Basic Books.

Johnson-Laird, P. N. (2001). Mental models and deduction. *Trends in Cognitive Sciences, 5*, 434–442.

Jones, K. (2000). Providing a foundation for deductive reasoning: Students' interpretations when using dynamic geometry software and their evolving mathematical explanations. *Educational Studies in Mathematics, 44*, 55–85.

King, I. L. (1970). A formative development of a unit on proof for use in the elementary school. (Doctoral dissertation, University of Wisconsin-Madison, 1969). *Dissertation Abstracts, 31*, 680A.

King, I. L. (1973). A formative development of an elementary school unit on proof. *Journal for Research in Mathematics Education, 4*(1), 57–63.

Klein, J. (1968). *Greek mathematical thought and the origin of algebra* (E. Brann, Trans.). Cambridge, MA: MIT Press. (Original work published 1934)

Kleiner, I. (1991). Rigor and proof in mathematics: A historical perspective. *Mathematics Magazine, 64*(5), 291–314.

Knuth, E. J. (2002a). Secondary school mathematics teachers' conceptions of proof. *Journal for Research in Mathematics Education, 33*(5), 379–405.

Knuth, E. J. (2002b). Teachers' conceptions of proof in the context of secondary school mathematics. *Journal of Mathematics Teacher Education, 5*, 61–88.

Knuth, E. J., Slaughter, M., Choppin, J., & Sutherland, J. (2002). Mapping the conceptual terrain of middle school students' competencies in justifying and proving. In S. Mewborn, P. Sztajn, D. Y. White, H. G. Wiegel, R. L. Bryant, & K. Nooney (Eds.), *Proceedings of the 24th Meeting of the North American Chapter of the International Group for the Psychology of Mathematics Education* (Vol. 4, pp. 1693–1700). Athens, GA.

Koedinger, K. R. (1998). Conjecturing and argumentation in high-school geometry students. In R. Lehrer & D. Chazan (Eds.), *Designing learning environments for developing understanding of geometry and space* (pp. 319–347). Mahwah, NJ: Erlbaum.

Laborde, C. (2000). Dynamic geometry environments as a source of rich learning contexts for the complex activity of proving. *Educational Studies in Mathematics, 44*, 151–161.

Lakatos, I. (1976). *Proofs and refutations: The logic of mathematical discovery*. Cambridge: Cambridge University Press.

Lampert, M. (1993). Teachers' thinking about students' thinking about geometry: The effects of new teaching tools. In J. Schwartz, M. Yerushalmy, & B. Wilson (Eds.), *The Geometric Supposer: What is it a case of?* (pp. 107–116). Hillsdale, NJ: Erlbaum.

Lawson, A. E., Karplus, R., & Adi, H. (1978). The acquisition of propositional logic and formal operational schemata during the secondary school years. *Journal of Research in Science Teaching, 15*, 465–478.

Leron, U. (1985). A direct approach to indirect proofs. *Educational Studies in Mathematics, 16*, 321–325.

Lester, F. K. (1975). Developmental aspects of children's ability to understand mathematical proof. *Journal for Research in Mathematics Education, 6*(1), 14–25.

Lin, F-L., Lee, Y-S., & Wu Yu, J-Y. (2003). Students' understanding of proof by contradiction. In N. Pateman, B. J. Dougherty, & J. Zilliox (Eds.), *Proceedings of the 27th Annual Meeting of the International Group for Psychology in Mathematics Education, Vol. 4* (pp. 443–449). Honolulu: University of Hawaii.

Linchevsky, L., Vinner, S., & Karsenty, R. (1992). To be or not to be minimal? Students teachers' views about definitions in geometry. In W. Geeslin & K. Graham (Eds.), *Proceedings of the Sixteenth Psychology in Mathematics Education Conference*, vol. II, (pp. 48–55). Durham, NH: University of New Hampshire.

Lovell, K. (1971). The development of the concept of mathematical proof in abler pupils. In M. Rosskopf, L. Steffe, & S. Taback (Eds.), *Piagetian cognitive-development research and mathematical education* (pp. 66–80). Washington: National Council of Teachers of Mathematics.

Maher, C. A. (2002). How students structure their own investigations and educate us: What we've learned from a fourteen year study. In A. D. Cockburn & E. Nardi (Eds.), *Proceedings of the 26th Annual Meeting of the International Group for the Psychology of Mathematics Education* (Vol. 1, pp. 31–46). Norwich, England: University of East Anglia.

Maher, C. A., & Martino, A. M. (1996). The development of the idea of mathematical proof: A 5-year case study. *Journal for Research in Mathematics Education, 27*, 194–214.

Manaster, A. B. (1998). Some characteristics of eighth grade mathematics classes in the TIMSS videotape study. *American Mathematical Monthly, 108*(9), 793–805.

Mancosu, P. (1996). *Philosophy of mathematical practice in the 17th century*. New York: Oxford University Press.

Marrades, R., & Gutierrez, A. (2000). Proofs produced by secondary school students learning geometry in a dynamic computer environment. *Educational Studies in Mathematics, 44*, 87–125.

Marriotti, M. A. (2000). Introduction to proof: The mediation of a dynamic software environment. *Educational Studies in Mathematics, 44*, 25–53.

Martin, W. G., & Harel, G. (1989). Proof frames of preservice elementary teachers. *Journal for Research in Mathematics Education, 10*(1), 41–51.

Martino, A. M., & Maher, C. A. (1999). Teacher questioning to promote justification and generalization in mathematics: What research practice has taught us. *Journal of Mathematical Behavior, 18*, 53–78.

Marty, R. (1991). Getting to Eureka! Higher order reasoning in math. *College Teaching, 39*(1), 3–6.

Mercer, N. (1995). The guided construction of knowledge: Talk amongst teachers and learners. Philadelphia: Multilingual Matters.

Morris, A. K. (2002). Mathematical reasoning: Adults' ability to make the inductive-deductive distinction. *Cognition and Instruction, 20*(1), 79–118.

Mueller, D. J. (1975). Logic and the ability to prove theorems in geometry (Doctoral dissertation, Florida State University, 1969). *Dissertation Abstracts International, 36*, 851A.

National Council of Teachers of Mathematics. (1970). *A history of mathematics education in the United States and Canada*. Thirty-second Yearbook. Washington, DC: Author.

National Council of Teachers of Mathematics. (1973). *Geometry in the mathematics curriculum*. Thirty-sixth Yearbook. Reston, VA: Author.

National Council of Teachers of Mathematics. (1987). *Learning and teaching geometry, K–12*. 1987 Yearbook. Reston, VA: Author.

National Council of Teachers of Mathematics. (1989). *Curriculum and evaluation standards for school mathematics*. Reston, VA: Author.

National Council of Teachers of Mathematics. (2000). *Principles and standards for school mathematics*. Reston, VA: Author.

O'Brien, T. C., Shapiro, B. J., & Reali, N. C. (1971). Logical thinking—language and context. *Educational Studies in Mathematics, 4,* 201–209.

Piaget, J. (1952). *The origins of intelligence in children.* New York: International Universities Press.

Piaget, J. (1973a). *The child and reality: Problems of genetic psychology.* New York: Grossman.

Piaget, J. (1973b). *The language and thought of the child.* London: Routledge and Kegan Paul.

Piaget, J. (1978). *Success and understanding.* Cambridge, MA: Harvard University Press.

Platt, J. L. (1967). The effect of the use of mathematical logic in high school geometry: An experimental study [Abstract]. *Dissertation Abstracts International, 28,* 4544–4545A.

Porteous, K. (1986, July). *Children's appreciation of the significance of proof.* Paper presented at the (tenth) annual Psychology in Mathematics Education conference, London.

Porteous, K. (1990). What do children really believe? *Educational Studies in Mathematics, 21,* 589–598.

Porter, A. (1993). Opportunity to learn. Brief No. 7. Madison, WI: Center on Organization and Restructuring of Schools.

Raman, M. (2003). Key ideas: What are they and how can they help us understand how people view proof? *Educational Studies in Mathematics, 52,* 319–325.

Recio, A. M., & Godino, J. D. (2001). Institutional and personal meanings of mathematics proof. *Educational Studies in Mathematics, 48,* 83–99.

Reid, D., & Dobbin, J. (1998). Why is proof by contradiction difficult? *Proceedings of the 22nd Psychology in Mathematics Education Conference* (Vol. 4, pp. 41–48), Stellenbosch, South Africa.

Rips, L. J. (1994). *The psychology of proof.* Cambridge, MA: MIT Press.

Rothman, T., & Fukagawa, H. (1998). Japanese temple geometry. *Scientific American, 278*(5), 84–91.

Schifter, D., Monk, S., Russell, S. J., Bastable, V., & Earnest, D. (2003, April). *Early algebra: What does understanding the laws of arithmetic mean in the elementary grades?* Paper presented at the meeting of the National Council of Teachers of Mathematics, San Antonio.

Schoenfeld, A. H. (1994). What do we know about mathematics curricula? *Journal of Mathematical Behavior, 13*(1), 55–80.

Selden, A., & Selden, J. (1995). Unpacking the logic of mathematical statements. *Educational Studies in Mathematics, 29,* 123–151.

Selden, A., & Selden, J. (2003). Validations of proofs considered as texts: Can undergraduates tell whether an argument proves a theorem? *Journal for Research in Mathematics Education, 34,* 4–36.

Senk, S. L. (1985). How well do students write geometry proofs? *Mathematics Teacher, 78*(6), 448–456.

Sfard, A. (1995). The development of algebra—Confronting historical and psychological perspectives. In C. Kieran (Ed.), New perspectives on school algebra: Papers and discussions of the ICME-7 algebra working group (special issue). *Journal of Mathematical Behavior, 14,* 15–39.

Silver, E. A., & Carpenter, T. P. (1989). Mathematical methods. In M. M. Lindquist (Ed.), *Results from the fourth mathematics assessment of the National Assessment of Educational Progress* (pp. 10–18). Reston, VA: National Council of Teachers of Mathematics.

Silver, E. A., & Kenney, P. A. (2000). *Results from the seventh mathematics assessment.* Reston, VA: National Council of Teachers of Mathematics.

Simon, M. A., & Blume, G. W. (1996). Justification in the mathematics classroom: A study of prospective elementary teachers. *Journal of Mathematical Behavior, 15,* 3–31.

Sowder, L. (2004). University faculty views about mathematics majors' understanding of proof. Unpublished manuscript, San Diego State University.

Sowder, L., & Harel, G. (2003). Case studies of mathematics majors' proof understanding, production, and appreciation. *Canadian Journal of Science, Mathematics and Technology Education, 3*(2), 251–267.

Strutchens, M. E., & Blume, G. W. (1997). What do students know about geometry? In P. A. Kenney & E. A. Silver (Eds.), *Results from the sixth mathematics assessment* (pp. 165–193). Reston, VA: National Council of Teachers of Mathematics.

Suppes, P. (1966). The axiomatic method in high school mathematics. In The Conference Board of the Mathematical Sciences, *The role of axiomatics and problem solving in mathematics* (pp. 69–76). Boston: Ginn.

Tall, D. (Ed.) (1991). *Advanced mathematical thinking.* Dordrecht, The Netherlands: Kluwer.

Thompson, D. (1991, April). *Reasoning and proof in precalculus and discrete mathematics.* Paper presented at the meeting of the American Educational Research Association, Chicago.

Tinto, P. (1988, April). *Students' views on learning proof in high school geometry.* Paper presented at the meeting of the American Educational Research Association, New Orleans.

U.S. Department of Education. (2000). *NAEP 1999 trends in academic progress: three decades of student performance,* by J. R. Campbell, C.M. Hombo, & J. Mazzeo. NCES 2000–469. Office of Educational Research and Improvement: Washington, D.C.

Usiskin, Z. (1987). Resolving the continuing dilemmas in school geometry. In M. M. Lindquist, & A. P. Shulte, (Eds.), *Learning and teaching geometry, K–12* (pp. 17–31). 1987 Yearbook, National Council of Teachers of Mathematics. Reston, VA: National Council of Teachers of Mathematics.

Vergnaud, G. (1983). Multiplicative structures. In R. Lesh & M. Landau (Eds.), *Acquisition of mathematics concepts and processes* (pp. 127–174). New York: Academic Press.

Vergnaud, G. (1988). Multiplicative structures. In J. Hiebert & M. Behr (Eds.), *Number concepts and operations in the middle grades* (pp. 141–161). Volume 2 of the Research Agenda for Mathematics Education project. Reston, VA: National Council of Teachers of Mathematics.

Vinner, S. (1977). The concept of exponentiation at the undergraduate level and the definitional approach. *Educational Studies in Mathematics, 8,* 17–26.

Vinner, S. (1983). The notion of proof—some aspects of students' views at the senior high level. In R. Hershkowitz (Ed.), *Proceedings of the Seventh International Conference of the Psychology of Mathematics Education* (pp. 289–294). Rehovot, Israel: Weizmann Institute of Science.

Vygotsky, L. (1962). Thought and language. Cambridge, MA: MIT Press.

Vygotsky, L. (1978). Mind in society: The development of the higher psychological processes. Cambridge, MA: The Harvard University Press.

Wason, P. C., & Johnson-Laird, P. N. (1972). *Psychology of reasoning: Structure and content.* Cambridge, MA: Harvard University Press.

Weber, K. (2001). Student difficulty in constructing proof: The need for strategic knowledge. *Educational Studies in Mathematics, 48*, 101–119.

Wilder, R. (1967). The role of the axiomatic method. *American Mathematical Monthly*, vol. 74, pp. 115–127.

Williams, E. (1980). An investigation of senior high school students' understanding of the nature of mathematical proof. *Journal for Research in Mathematics Education, 11*(3), 165–166.

Yackel, E., & Cobb, P. (1996). Sociomathematical norms, argumentation, and autonomy in mathematics. *Journal for Research in Mathematics Education, 27*, 458–477.

Yackel, E., & Hanna, G. (2003). Reasoning and proof. In J. Kilpatrick, W. G. Martin, & D. Schifter (Eds.), *A research companion to Principles and Standards for School Mathematics* (pp. 227–236). Reston, VA: National Council of Teachers of Mathematics.

Yackel, E., Rasmussen, C., & King, K. (2000). Social and sociomathematical norms in an advanced undergraduate mathematics course. *Journal of Mathematical Behavior, 19*, 275–287.

Zack, V. (2002). Learning from learners: Robust counterarguments in fifth graders' talk about reasoning and proving. In A. Cockburn & E. Nardi (Eds.), *Proceedings of the 26th International Group for the Psychology of Mathematics Education*, (Vol. 4, pp. 434–441). Norwich, UK: University of East Anglia.

AUTHOR NOTE

We wish to acknowledge the helpful comments from Gila Hanna, Carolyn Maher, and Erna Yackel. Preparation of this chapter was supported in part by National Science Foundation Grant No. REC-0310128. The content or opinions expressed herein do not necessarily reflect the views of the National Science Foundation or any other agency of the U.S. government.

19

THE DEVELOPMENT OF GEOMETRIC AND SPATIAL THINKING

Michael T. Battista

MICHIGAN STATE UNIVERSITY

The usefulness of axiomatics . . . goes beyond that of demonstration. . . . It permits us to construct simplified models of reality and thus provides the study of the latter with irreplaceable dissecting instruments. . . . [However] axiomatics cannot claim to be the basis of, and still less to replace, its corresponding experiential science. . . . Axiomatic geometry is incapable of teaching us what the space of the real world is like.

—Piaget, 1950, pp. 30–31

Every spatial representation can depend in part on the geometry of the object and in part on the geometry of the subject—and the dosages of each are hard to determine.

—Piaget, 2001, p. 223

In this chapter I examine geometric and spatial reasoning in a broad context. I describe and analyze recent research on the nature and development of students' thinking about geometry, geometric measurement, and aspects of spatial reasoning that are related to geometry. I concentrate on empirical and theoretical research, not curriculum and technological tool design. I focus on key ideas rather than comprehensively surveying the field. Where possible, I discuss the cognitive roots of geometric reasoning.

DEFINING THE SUBJECT: GEOMETRY AND SPATIAL REASONING

Geometry is a complex interconnected network of concepts, ways of reasoning, and representation systems that is used to conceptualize and analyze physical and imagined spatial environments. Geometric reasoning consists, first and foremost, of the invention and use of formal *conceptual systems* to investigate shape and space (Battista, 2001a, 2001b). For instance, mathematicians employ a property-based conceptual system to analyze and define various types of quadrilaterals and triangles. This system uses concepts such as angle measure, length measure, congruence, and parallelism to conceptualize spatial relationships within and among the shapes. So, defining a square to be a four-sided figure that has four right angles and all sides the same length creates an idealized property-based concept that, with proper spatial grounding, can help people reason more precisely about this special class of shapes.

Underlying most geometric thought is spatial reasoning, which is the ability to "see," inspect, and reflect on spatial objects, images, relationships, and transformations. Spatial reasoning includes generating images, inspecting images to answer questions about them, transforming and operating on images, and maintaining images in the service of other mental operations (Clements & Battista, 1992; Presmeg,

1997; Wheatley, 1997). Thus, spatial reasoning provides not only the "input" for formal geometric reasoning, but critical cognitive tools for formal geometric analyses.

THEORIES OF GEOMETRY LEARNING PRELIMINARIES: DRAWINGS, DIAGRAMS,[1] FIGURES, SHAPES, AND CONCEPTS—TIPS OF AN ENIGMATIC ICEBERG

Before examining theories on geometry learning, it is important to reflect on and clarify the nature of the various "primitive objects" that students operate on in geometric and spatial reasoning.

The Objects of Geometric Analysis

Researchers often distinguish two types of geometric objects: "Drawing refers to the material entity while figure refers to a theoretical object" (Laborde, 1993, p. 49). Laborde argued that teaching often "confuses drawings and the theoretical geometrical objects that the former represent" (1998, p. 115). She further argued that student difficulties in geometry often arise because students reason about material drawings when they are expected to reason about theoretical geometrical objects (Laborde, 1993). (See Laborde, 1993 for further elaboration of these distinctions.) Similarly, Presmeg observed that, "A picture or a diagram is by its nature one concrete case, yet for any but the most trivial mathematical thinking it is necessary to abstract and to generalize" (1997, p. 305). Presmeg claimed that the "one-case concreteness" of drawings and images is the source of many difficulties in visualization-based mathematical reasoning.

However, the complexity of the issue is only partially captured by the drawing-figure dichotomy. Jones, following Holzl, argued that "learners can get 'stuck' somewhere between a drawing and a figure" (2000, p. 58). Furthermore, researchers in psychology and neuroscience have made additional distinctions that illustrate the complexity of identifying the objects of geometric and spatial analysis. For instance, some researchers have characterized the difference between physical and perceptual objects in terms of distal and proximal stimuli: "A distal stimulus is an actual object or event 'out there' in the world; a proximal stimulus is the information our sensory receptors receive about that object" (Coren, Ward, & Enns, 1994, p. 485). Other researchers such as Edelman (1992) have distinguished between perceptual and conceptual objects. He argued that perception is nonconscious and operates on signals from the outside world received by the body's receptors, whereas conception is conscious and operates on the activity of those portions of the brain that record signals from the outside world. Smith (1995) distinguished between a category, which is a group of objects in the world that belong together, and a concept, which is "a mental representation of such a group" (1995, p. 5). Finally, consider the mathematical formalization of an idea: "The concept definition is a form of words used to specify that concept" (Tall & Vinner as quoted in Tall, 1992, p. 496).

Integrating and synthesizing these ideas leads to the identification of five types of basic objects involved in geometric and spatial thought. A *physical object* is an actual physical entity such as a door, box, ball, geoboard figure, picture, drawing, or dynamic draggable computer figure (Coren et al.'s distal stimulus). A *sensory object* is the set of sensory activations evoked when an individual views a physical object (Coren et al.'s proximal stimulus). A *perceptual object* is the mental entity perceived by an individual when viewing a physical object (Edelman's perception). A *conceptual object* or conceptualization (Edelman's conception) is the conscious meaning or way of thinking activated by an individual (a) in response to a perceptual object, memory of a perceptual object, or concept definition; or (b) constructed anew from other conceptual objects. A *concept definition* is an explicit formal mathematical verbal/symbolic specification of a conceptual object.[2]

Another way to distinguish cognitive objects in geometric reasoning is by their use (Battista, in press a). An *object* is a mental entity that is operated on (consciously or unconsciously) during reasoning. A *representation* is something that "stands for" something else (Goldin & Kaput, 1996). In general, in geometric thought, one reasons *about* objects; one reasons *with* representations.

To understand geometric and spatial reasoning, one must understand individuals' cognitions about the various objects described above, including how these cognitions are interrelated. However, in analyzing these cognitions, one must be aware of several complications.

Complication 1: Conception Affects Perception

What one "sees" is affected by what one knows and conceives. For example, in the size-constancy

[1] Although one might distinguish the terms *drawing* (something created freehand by a student or teacher) and *diagram* (a carefully made graphic given in a textbook or other printed material), the two terms seem to be used synonymously in the literature.

[2] Two additional types of objects discussed in the research literature—mental models and concept images—will be discussed later.

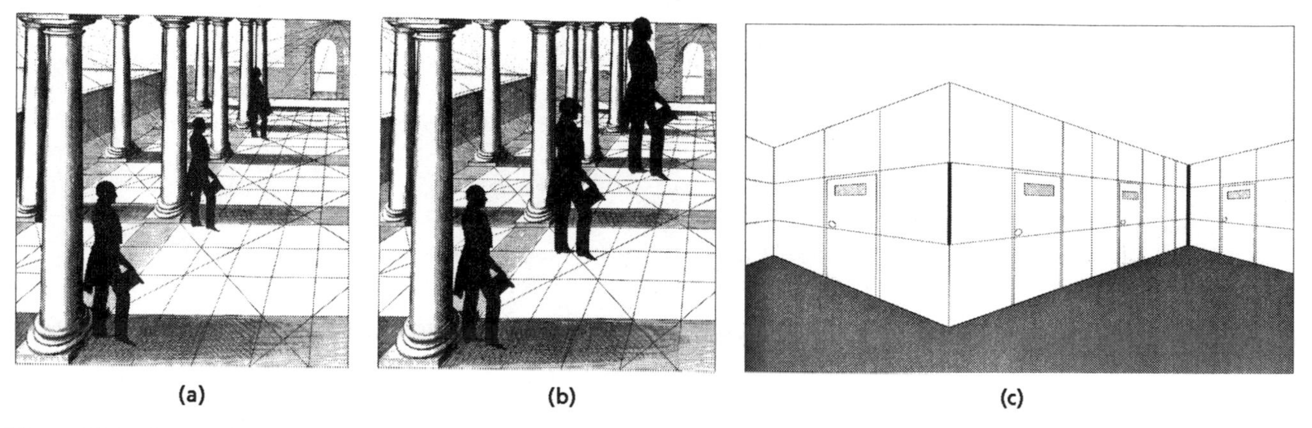

Figure 19.1 Perception of size determined by context (Coren et al., 1994, pp. 488, 497).

phenomenon, the size of an object is conceptualized as constant even though the retinal image it projects varies as the object's distance from the eyes varies; the mind takes into account the apparent distance of the object from the viewer to adjust perceived size (Coren et al., 1994). In Figure 19.1a, the three men appear to be the same size even though the most distant figure is only about a third of the size as the nearest figure. In contrast, in Figure 19.1b, the most distant figure appears much larger than the nearest, even though the actual images are identical. Thus, perception of size is adjusted based on a contextual conceptualization of distance. Figure 19.1c further illustrates how perception is affected by conception and context. Amazingly, the two dark vertical segments are congruent. But our knowledge of the physical world, and the resulting mental models created to represent it, cause us to perceive the two segments as very different in length.

Similarly, verbal labeling can affect conception—an ambiguous figure that is called a broom when presented is drawn to look like a broom, whereas if it is called a rifle when presented, it is drawn as a rifle (Coren et al., 1994). Might a similar phenomenon occur in geometry? Does using a different name for a drawn shape—square, rhombus, diamond—affect a student's perception/conception of the shape? Also, research shows that drawing a figure, in combination with experience and verbal labeling, can make it easier to find when it is embedded in a more complex figure (Coren et al.). Thus again, experience (i.e., current knowledge) affects perception. Finally, research suggests that current knowledge can affect perception of similarities among instances, and thus classification, possibly because such knowledge can influence what stimulus information is attended to (Ross, 1996). For instance, a square with vertical and horizontal sides turned 45° might be perceived as a diamond because previous experience causes a student to attend not to relationships between parts of the shape but to its position, including implicitly, say, how angles line up vertically and horizontally.

Complication 2: Diagrams as Data or Representations?

Diagrams, or more generally instances of physical objects, play two major roles in geometry. On the one hand, instances of physical objects can be thought of as the *data* for geometric conceptualization—it is through analysis of such objects that geometric shape concepts are derived. On the other hand, instances of physical objects, including diagrams, are used to represent formal geometric concepts. However, both curricula and researchers generally neglect the process of forming concepts from physical objects and instead focus on the "representational" perspective. Consider the following statements by researchers: "Diagrams are intended as models . . . meant to be understood as *representing* a class of objects. Every diagram has characteristics that are individual and not *representative* of the class" (Yerushalmy & Chazan, 1993, p. 25). "Students often attribute characteristics of a drawing to the geometric object it *represents*" (Clements & Battista, 1992, p. 448). Chazan and Yerushalmy (1998) argued that this "representational perspective" on diagrams is rooted in the traditional axiomatic approach to geometry instruction. Geometric "objects" are either undefined or are defined formally in terms of undefined terms; "diagrams are aids for intuition and are not the objects of study themselves" (p. 70). And even though Chazan and Yerushalmy stated that geometry "should more explicitly build on students' experience of the world around them" (p. 70), they seemed to conceptualize diagrams and figures in representational terms, citing Aleksandrov: "Geometry has as its object the spatial forms and relations of actual bodies, *removed from their other properties and considered from the purely abstract point of view* [italics added]" (p. 70).

The representational perspective focuses on abstract, general mathematical concepts, which are taken as the objects of study. A diagram is a mere indicator of an abstract mathematical concept; so many characteristics of a particular diagram are deemed *irrelevant distractors*. Consequently, most mathematics education research has treated diagrams as imperfect representations of idealized abstract concepts, neglecting the processes by which formal geometric concepts are abstracted from particular instances. Taking a broader perspective can provide additional insights into the processes by which students construct knowledge about shapes and shape classes. Instead of focusing exclusively on diagrams as representations of formal mathematical concepts, researchers must investigate more carefully and explicitly two complementary areas: (a) the process by which students progress from analysis of particular figures to general abstractions about classes of figures (it is from the analysis of particulars that abstract conceptualizations arise) and (b) the mechanisms that enable students to utilize formal abstract geometric concepts to analyze particular figures. Investigating these areas is critical to understanding how students construct understanding of abstract geometric concepts that are applicable not only within mathematics but to real-world problems.

This new perspective reveals several extremely important avenues for investigation. For instance, how does the ability to abstract what is common among a set of shapes (or what is invariant in a dynamic diagram) develop among children, and how can instruction and curricula promote this ability? How are abstract, general conceptualizations produced? Researchers need to describe how students move from the particular to the general. Investigating such abstractive generalization is an important but neglected area in research on geometric thinking. Although mathematics is about abstract concepts, mathematics education research must focus on exactly how individuals derive meaning for abstract concepts from particular instances.

Difficulties With Drawings and Diagrams

Drawings and diagrams are essential for understanding geometric ideas. In traditional use, drawings and diagrams have been used for two major purposes—(a) to represent classes of shapes (e.g., the *set* of rectangles) and (b) to represent geometric relationships (e.g., the bisectors of the angles of a triangle meet at the incenter of the triangle). Students, however, have difficulty with both uses, often because they do not understand the nature of the objects being considered. In particular, students often attribute irrelevant characteristics of a diagram to the geometric concept it is intended to represent (Clements & Battista, 1992; Yerushalmy & Chazan, 1993). For instance, students might not identify right triangles drawn in nonstandard orientations (having unintentionally abstracted orientation as an attribute). Another difficulty occurs when students use diagrams in proofs (Clements & Battista, 1992; Presmeg, 1997; Yerushalmy & Chazan, 1993). For example, students may assume that sides that look parallel in an accompanying diagram are parallel (adding unintended conditions to hypotheses). Or, students might link a theorem too tightly to the example diagram given with the theorem statement. For instance, if a theorem is originally illustrated by a diagram of an acute triangle, students might believe that the theorem does not apply, or might not think to apply it, to obtuse triangles. Still another type of difficulty can occur when drawings do not capture appropriate geometric relationships—for example, a tangent might be drawn freehand so that it is not perpendicular to the radius it intersects.[3] Finally, students might mistake geometric diagrams for pictures of objects, profoundly changing their interpretation. For instance, a bright fourth grader identified nonrectangular parallelograms as rectangles because he interpreted them as pictures of rectangles viewed "sideways" (Clements & Battista, 1992).

THE VAN HIELE LEVELS: RESEARCH AND REVISIONS

A considerable amount of research has established the van Hiele theory as a generally accurate description of the development of students' geometric thinking (Clements & Battista, 1992). After briefly reviewing the theory, I address recent developments in this important area of research.

The van Hiele Levels of Geometric Thought

According to the van Hiele theory, students progress through discrete, qualitatively different levels of geometric thinking. The levels are sequential and hierarchical, so for students to function adequately at one of the advanced levels, they must have "passed through" the lower levels. Briefly, the levels are as follows (see Clements and Battista, 1992 for a complete

[3] Of course, a sloppy drawing might not interfere with reasoning if the individual creating it is aware of the relationships that *should* exist in it.

description). *Level 0 Pre-recognition:* Because immature perceptual activity causes students to attend to only a subset of a shape's characteristics, students are unable to identify many common shapes. *Level 1 Visual:* Students identify and operate on geometric shapes according to their appearance. They recognize figures as visual gestalts, relying heavily on common prototypes. *Level 2 Descriptive/analytic:* Students recognize and can characterize shapes by their properties. *Level 3 Abstract/relational:* Because one property can signal other properties, students form form definitions, distinguish between necessary and sufficient sets of conditions, and understand and sometimes even provide logical arguments. *Level 4 Formal deduction:* Students establish theorems within an axiomatic system. *Level 5 Rigor/metamathematical:* Students reason formally about mathematical systems.

Recent Developments

New developments will be grouped into three categories: extending the level descriptors beyond 2d shapes; reexamining the nature of the levels; and elaborating the levels phenomenologically and psychologically.

Extending the Level Descriptors Beyond 2d Shapes

Early interpretations of the van Hiele levels focused on reasoning about limited aspects of two-dimensional shapes; that is, the theory was investigated primarily for a well-defined but constrained domain. But a number of researchers have extended the levels to other geometric domains. Although difficulties in interpretation and consistency often arise when researchers attempt to generalize the levels to other domains, such studies have increased our understanding of the levels because generalization typically requires reconceptualization and refinement. As examples of these extensions, Gutiérrez and colleagues extended the van Hiele level descriptions to reasoning about 3d shapes (Gutiérrez, 1992; Gutiérrez, Jaime, & Fortuny, 1991). Johnson-Gentile, Clements, and Battista (1994), Lewellen (1992), and Jaime and Gutiérrez (1989) extended the descriptions to motions/transformations. Clements and Battista (1992) summarized researchers' views on the application of the van Hiele levels to deductive/justification reasoning. Aspects of several of the extensions are discussed below.

Gutiérrez et al.'s (1991) extension of the levels to 3d shapes is straightforward and follows the original conception of the levels as described by Clements and Battista (1992). At Level 1, solids are judged and identified visually and holistically, with no explicit consideration of components or properties. At Level 2, students identify components of solids and informally describe solids using properties, although properties are not logically related. At Level 3, students are able to logically classify solids and understand the logic of definitions. At Level 4, students are able to prove theorems about solids.

In contrast, Gutiérrez and colleagues' extension to 3d visualization is both less straightforward and more ambitious because it attempts to elaborate the relationship between visualization and geometric conceptualization (Gutiérrez, 1992). At Level 1, students compare solids using global perception with no attention given to properties such as angle size, side length, or parallelism. Students are unable to visualize solids, their positions, or motions that they cannot actually see; they manipulate solids by guess and check. At Level 2, students move from global comparisons to visual analysis of components and their properties. Students are able to visualize simple movements from one visible position to another. At Level 3, students compare solids by mathematically analyzing their components. They can visualize movements involving positions that are not visible, and in reasoning about movements, students can match corresponding parts of images and preimages. At Level 4, students can mathematically analyze and formally deduce properties of solids. Visualization is high and is linked to knowledge of properties.

Two observations are relevant to these extensions of the van Hiele levels to 3d figures. First, the descriptors for geometric conceptualization (not visualization) at Levels 3 and 4 in the latter article are a bit less advanced than in the earlier article. One suspects that this may be the case because traditionally there is far less attention given to formal proof in 3d geometry. Second, extending the levels to visualization is complex. On the one hand, visualization is a capability that is not necessarily connected to knowledge of properties. For instance, some students who are not high visualizers may develop analytic (property-based) strategies to help them compensate for a lack of pure visualization skills (Battista, 1990). Other students may possess very high visualization skill well before they develop property-based reasoning about solids. Indeed, some high visualizers can mentally imagine movements of solids so well that, for many problems, they simply have no need to examine components of the solids. On the other hand, Gutiérrez and colleagues have broken new ground by illustrating how visualization and geometric conceptualization can develop hand-in-hand, an idea also proffered by Clements and Battista (2001). However, much more research is needed that clearly describes exactly how geometric conceptualization interacts with visualization during development. For instance, at 3d visualization Level 2, how does progression from global comparison to analysis of components and properties enable improvements in visualizing movement, or vice versa?

Reexamining the Nature of the Levels

Originally, the van Hiele levels were conceptualized as periods of development in geometric reasoning characterized not only by qualitatively different thinking, but by different internal knowledge organization and processing. In this view, students are at Level 1 when their overall cognitive organization and processing disposes them to think about geometric shapes in terms of visual wholes; they are at Level 2 when their overall cognitive organization disposes and enables them to think about shapes in terms of their properties. However, difficulties in classifying students' van Hiele level, oscillation of students between levels, and students being at different levels for different concepts have caused many researchers to question whether the levels are discrete. Indeed, one of the major issues that has arisen is, Should students' thinking be characterized as "at" a single level?

Degrees of Acquisition: A Vector Approach

Arguing that students develop several van Hiele levels simultaneously, Gutiérrez et al. (1991) used a vector with four components to represent the degrees of acquisition of each of van Hiele levels 1 through 4. They identified 5 periods in the acquisition of a level: no, low, intermediate, high, and complete acquisition. For example, in the intermediate acquisition period for a given level, students use reasoning characteristic of the level frequently and often accurately. However, students' lack of mastery of the level causes them to fall back on lower level reasoning when they encounter special difficulties. In scoring student responses, Gutiérrez et al. classified each problem solution according to (a) the nature of the reasoning used via the van Hiele level descriptions, and (b) type. Type was quantified by judging degree of attainment of van Hiele level, correctness of answers, and correctness of reasoning (including justification). The numerical score for degree of acquisition of a van Hiele level was the arithmetic average of type weightings for all items that could be answered at that level. Consequently, for example, a student's degree of acquisition vector might be: 96.67% for Level 1, 82.50% for Level 2, 50.00% for Level 3, and 3.75% for Level 4. Using this vector approach, and focusing on 3d geometry, Gutiérrez et al. described six profiles of level-configurations in students' geometric reasoning. For example, Profile 2 was characterized by complete acquisition of Levels 1 and 2, high acquisition of Level 3, and low acquisition of Level 4.

However, even though level acquisition was described in terms of the vector model, the profiles could easily be interpreted in terms of the original van Hiele levels (as described by Clements and Battista, 1992, with the addition of Level 0). For instance, Profile 2 could be thought of as Level 2 or transition to Level 3. Thus, rather than invalidating the levels, the vector approach merely suggests, as does the research by Lehrer, Jenkins, and Osana (1998), that the original van Hiele levels may need to be elaborated and expanded. Indeed, the profiles suggest a possible decomposition of van Hiele Levels 0–3 into six steps rather than four.[4] Unfortunately, one difficulty with the profiles approach (and perhaps all current elaborations of the van Hiele theory) is that it lacks the conceptual coherence and clarity of the original van Hiele levels. That is, the original levels have a conceptual-logical coherence and eloquence that not only makes them appealing, but has proved difficult to surpass.[5]

Several other components of the Gutiérrez et al. (1991) research should be mentioned. First, by suggesting that the van Hiele levels consist of configurations of different types of reasoning, this research seemed to encourage other researchers to investigate students' geometric reasoning from a "configuration" approach. Second, Gutiérrez et al. and Clements and Battista (2001) both attempted to deal explicitly with the tricky issue of how to classify reasoning that fit the description of a level but was incorrectly used. That is, if a student clearly is attempting to use properties in identifying shapes, but does so incorrectly, is that student thinking at Level 2? Third, and finally, Gutiérrez et al. have provided a careful and detailed attempt to describe the transition between van Hiele levels.

Elaborating the van Hiele Theory by Merging It With the SOLO Taxonomy

Several researchers have attempted to elaborate and reformulate the van Hiele theory using the SOLO taxonomy (Olive, 1991; Pegg, 1992; Pegg & Davey, 1998; Pegg, Gutiérrez, & Huerta, 1998). Although van Hiele stated, "The most distinctive property of the levels of thinking is their discontinuity, the lack of co-

[4] These researchers present the profiles as qualitative descriptions of empirical results, not steps in development (A. Gutiérrez, personal communication, 12/14/04).

[5] Keep in mind that the original levels themselves have had several iterations and interpretations; there have been discrepancies even in the number of these levels.

herence between their networks of relations" (1986, p. 49), Pegg and Davey argued that several groups of researchers have failed to detect these discontinuities and instead have found the levels to be more continuous in nature. Also, similar to Gutiérrez et al. (1991) and Lehrer et al. (1998), Pegg and Davey argued that research suggests that students develop more than one level at a time. They further argued that level discreteness, while appearing valid when the magnification of the analysis is low, seems blurred when the magnification is high. That is, as global descriptions, the levels seem valid, but when one attempts to examine students' thinking more carefully, the level descriptions may be inadequate.

To deal with the van Hiele theory's lack of detail in describing students' progression through the levels, Pegg and Davey (1998) integrated the van Hiele theory with the SOLO theory of Biggs and Collis. In their synthesis of the two theories, Pegg and Davey discussed three major modes of thinking (see also Pegg, 1992). In the ikonic mode, students form and operate on mental images of objects with which they have had contact. In the concrete-symbolic mode, students link concepts and operations to written symbols, as long as the context falls within their personal experience. In the formal mode, students are no longer restricted to concrete referents and can systematically consider principles, theories, and ranges of possibilities and constraints; they become capable of formal proof. Each of these modes is further differentiated by progression through levels of complexity, with progression through each mode requiring at least two cycles (Pegg et al., 1998). At the unistructural (U) level, students focus on one isolated aspect of a situation. At the multistructural (M) level, students focus on two or more unrelated aspects of a situation. At the relational (R) level, students interrelate multiple aspects of a situation.

According to Pegg et al. (1998),[6] van Hiele Level 0 (in the Clements/Battista numbering system) occurs at the ikonic mode, in the first UMR cycle (U1, M1, R1). van Hiele level 1 occurs at the ikonic mode in the second UMR cycle (U2, M2, R2). At U2, the imaging process focuses on one isolated aspect of a situation, such as the sharpness of a vertex. At M2, several unrelated aspects of a situation are considered. At R2, the individual has full control over the imaging process and is able to identify familiar shapes correctly. van Hiele Level 2 occurs at the concrete-symbolic mode in the second UMR cycle, with Levels U2 (focusing on a single property) and M2 (focusing on many properties). (The earlier U1, M1, R1 concrete-symbolic cycle is seen as a transition between van Hiele Levels 1 to 2.) van Hiele Level 3 occurs at the R2 Level in the concrete-symbolic mode. Students seem to achieve van Hiele Level 4 as they progress through two UMR level cycles in the formal mode.

The power of integrating the van Hiele theory with the SOLO theory is that it provides a possible framework for discussing finer delineations of the van Hiele levels. However, many issues and questions remain. For example, (a) The fact that the correspondence between van Hiele level and SOLO taxonomy varies, depending on the article read, indicates that researchers are struggling to make this complex correspondence work. The only agreement seems to be that the original van Hiele levels are too coarse to provide adequate distinctions in students' geometric reasoning. (b) Why should researchers believe that the ability to focus on multiple aspects of a situation is a primary determinant of geometric reasoning? Certainly this ability is important in intellectual development—but is it specific enough to be a primary determinant of geometric reasoning, especially at each van Hiele level? Also, what cognitive mechanisms enable an individual to focus on more than one aspect at a time?

Waves of Acquisition

Several researchers have posited that different types of reasoning characteristic of the van Hiele levels develop simultaneously at different rates, and that at different periods of development, different types or "waves" of reasoning are dominant, depending on the relative competence students exhibit with each type of reasoning (Clements & Battista, 2001; Lehrer et al., 1998; see Figure 19.2). Development depends both on maturation and instruction.

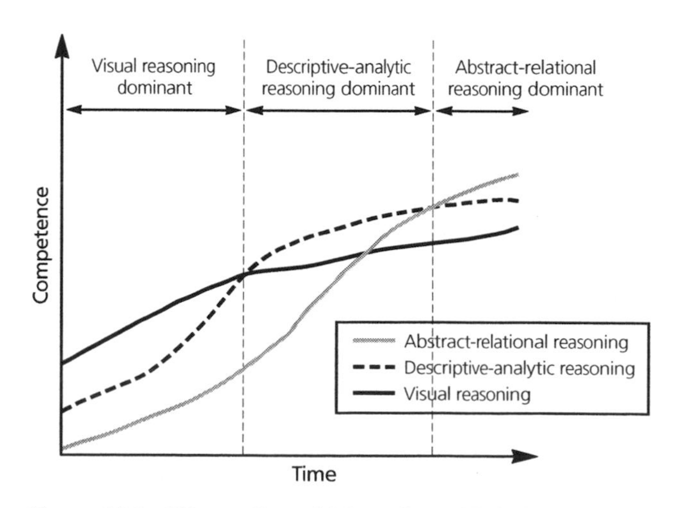

Figure 19.2 Waves of acquisition of van Hiele levels.

---

[6] The correspondence between van Hiele levels and SOLO modes and levels varies by research report. See the cited references in this section.

Based on the administration of a single type of task,[7] Lehrer et al. (1998) argued that "level mixture" was typical for primary-age students, that students' justifications often jumped across nonadjacent van Hiele levels, and that reasoning varied greatly with task. They argued against portraying geometric development in terms of discontinuities among levels, and instead for it being characterized "by which 'waves' or forms of reasoning are most dominant at any single period of time, without necessarily implying the extinction of other forms" (p. 163). They found that children's geometric thinking was accomplished by a range of mental operations including comparing shapes to visual prototypes, mentally "morphing" one shape into another, and finding a variety of attributes of shapes. They argued that these varied types of reasoning should not be described by the single descriptor "visual." However, one can argue that the types of reasoning that Lehrer et al. questioned as being visual are, in fact, examples of visual, or Level 1, reasoning, and that Lehrer et al. have provided us with a richer description of such reasoning. That is, the Lehrer et al. research suggests that Level 1 visual reasoning is not as limited and homogeneous as the original van Hiele levels suggest. Furthermore, the characterization of students' thinking as property-based when they refer to number sides should be questioned (I will return to this issue later in the chapter). And finally, responses the researchers coded as "classification" were not hierarchical classification in the van Hiele sense—students were simply naming figures (it would be a truly exceptional third-grader who used a genuine hierarchical classification). So, as in so many van Hiele studies, the question that arises is whether the "level jumping" was due to students' reasoning or to quirks in the classification system used by the researchers.

Clements and Battista (2001) proposed the view that the van Hiele levels (seen as types of reasoning) develop simultaneously (albeit at different rates), similar to the "overlapping waves" metaphor suggested by Siegler (1996), and by Lehrer et al. (1998). Visual-holistic knowledge, descriptive verbal knowledge, and, to a lesser extent initially, abstract symbolic knowledge grow simultaneously, as do interconnections between levels. However, although these different types of reasoning grow in tandem to a degree, one level tends to become ascendant or privileged in a child's orientation toward geometric problems. Which level is privileged is influenced by age, experience, intentions, tasks, and skill in use of the various types of reasoning. In the early years, while visual reasoning (called syncretic to signify combination without coherent analysis and synthesis) is dominant, descriptive-analytic knowledge begins to emerge and interacts with visual knowledge. The early dominance of visual-holistic reasoning gradually gives way to descriptive-analytic reasoning. As abstract reasoning begins its ascendance, connections among all types of reasoning are strengthened and reformulated. Consistent with this "continuous growth" view, these researchers found little evidence in their own research or that of others of discontinuities between levels as postulated by the van Hiele theory; growth in small increments seemed to be the norm.

Stages or Levels?

Clements and Battista (2001) viewed the van Hiele levels as *levels*, rather than *stages*. They described a stage as a substantive period of time characterized by cognition across a variety of domains that is qualitatively different from that of both the preceding and succeeding stages. In contrast, a level is a period of time of qualitatively distinct cognition, but within a specific domain. In van Hiele research, one of the open questions concerns the size of the "domains" to which the van Hiele levels apply. For instance, Clements and Battista (2001) suggested that many empirical studies indicate that people exhibit behaviors indicative of different levels on different subtopics of geometry (Denis, 1987; Mason, 1989; Mayberry, 1983). This finding, taken together with the observation that students sometimes exhibit different levels on different tasks, suggests that the domains to which levels of thinking apply may be, at most, subtopics within the broad range of geometry.

However, the domain issue may be more complicated than it seems, especially when viewed from a longitudinal rather than a static point of view. For instance, Lewellen (1992) carefully studied 8 students participating in the Logo Geometry Curriculum (Battista & Clements, 1991; Clements & Battista, 2001), which first deals with geometric shapes then with geometric motions. She found that students entering the motions unit all started at Level 1 for motions, but after a short initial learning period, students at Level 2 or in transition to Level 2 for shapes moved up to Level 2 in motions. Lewellen concluded, "These results lend support to the notion that there is a global van Hiele level for each student after they have some initial learning in any new domain" (Lewellen, 1992, p. 279).

In fact, students who entered the motions unit at Level 2 on shapes were able to quickly learn the components and characteristics of objects in this new domain. Rather than limiting themselves to specific

[7] Detailed discussion of this task will be given in a later section.

values and visual feedback, these students looked for underlying properties, components, characteristics, and rules in problem-solving situations. For example, Kelly developed measure-based strategies for verifying whether or not motions were correct, such as measuring the distance of corresponding points from the flip line for flips and the distance of corresponding points from the turn center for turns. She also knew that a slide moves all points of a figure the same distance and direction. Thus, students like Kelly who were at Level 2 for shapes, having seen the value of analytic conceptions in one domain—for instance, using length and angle measure to describe relationships between components of shapes—seemed to seek out similar conceptions in the new domain of motions. Initially as students worked in the new domain, they did not have such conceptions. The key is that they had developed the ability to use such conceptions, and they understood their power, so they looked for properties in the new domain. Of course, an open question is to describe the degree of acquisition of a level of reasoning—along with enabling mental mechanisms—that is required to attempt to and be able to implement that reasoning in another domain.

An Alternative Route: Elaborating the Original van Hiele Levels

To understand the development of reasoning in geometry, Battista has elaborated the van Hiele levels to carefully trace students' progress in moving from informal intuitive conceptualizations of 2d geometric shapes to the formal property-based conceptual system used by mathematicians. The descriptions for Level 2 given below were developed by Battista in several projects; the description for Level 3 was developed by Borrow (2000) and Battista. This new elaboration considerably expands the van Hiele levels in two places—the development of property-based thinking, and the development of inference about properties.

Level 1: Visual-Holistic Reasoning

Students identify, describe, and reason about shapes and other geometric configurations according to their appearance as visual wholes. They may refer to visual prototypes, saying, for example, that a figure is a rectangle because "it looks like a door," or they might judge that two figures are the "same shape" because they "look the same." Students might also say that a square is not a rectangle because rectangles are "long." Students may justify their responses using imagined visual transformations, saying for instance that a shape is a square because if it is turned it looks like a square. Orientation of figures may strongly affect Level 1 students' shape identifications. In fact, Level 1 students tend to identify a square only if its sides are horizontal and vertical.

1.1 Pre-recognition. Students are unable to identify many common shapes.

1.2 Recognition. Students correctly identify many common shapes.

Level 2: Analytic-Componential[8] Reasoning

Students explicitly attend to, conceptualize, and specify shapes by describing their parts and spatial relationships between the parts. However, students' descriptions and conceptualizations vary greatly in sophistication, starting with completely informal and imprecise specifications in Sublevel 2.1 and ending with completely formal geometric specifications in Sublevel 2.3 (which corresponds to original van Hiele Level 2).

Two major interrelated factors contribute to the development of Level 2 reasoning. The first is an increasing ability and inclination to account for the spatial structure of shapes by analyzing their parts and how the parts are related. The second is an increasing ability to understand and apply formal geometric concepts in analyzing relationships between parts of shapes.

2.1 Visual-informal componential reasoning. Students describe parts and properties of shapes informally and imprecisely; they do not possess the formal conceptualizations that enable precise property specifications. Descriptions and conceptualizations are visually based, focusing initially on parts of shapes then on spatial relationships between parts. Examples of references to *parts* are counting the number of sides or angles (usually referred to as "points") in a shape or saying that a figure has straight sides. Examples of descriptions of *relationships between parts* are specifying that "corners" must be "square" or that "all sides must be even." In all cases, students describe parts and their relationships using strictly informal language, that is, language typically learned in everyday experience.

Students' informal language ranges greatly in precision and coherence, from using vague and incompletely formulated conceptualizations to informally describing a conceptualization that corresponds to a formal geometric concept. As an example of the latter case, as students examine a tilted rectangle or square, they might informally refer to the property of having all right angles by saying the shape has "square corners" or "straight sides." Students' descriptions often seem to occur extemporaneously as they inspect shapes (as opposed to describing explicit conceptualizations formulated from previous experiences).

[8] I use the term *analysis* to refer to the process of understanding objects by resolving them into their components.

2.2 Informal and insufficient-formal componential reasoning. As students begin to acquire formal conceptualizations that can be used to "see" and describe spatial relationships between parts of shapes, they use a combination of informal and formal descriptions of shapes. The formal descriptions utilize standard geometric concepts and terms explicitly taught in mathematics curricula. However, the formal portions of students' shape descriptions are insufficient to completely specify shapes. For example, a student might explain that a rectangle has "sides across from each other that are equal [formal] and square corners [informal]." Or, a student might say that "a rectangle has sides across from each other that are equal lengths," making no mention of right angles. Although students often recall properties that they have abstracted for classes of shapes (say, "two long sides and two short sides" for rectangles), their reasoning is still visually based, and most of their descriptions and conceptualizations still seem to occur extemporaneously as they are inspecting shapes.

2.3 Sufficient formal property-based reasoning. Students explicitly and exclusively use formal geometric concepts and language to describe and conceptualize shapes in a way that attends to a sufficient set of properties to specify the shapes. Students have made a decided shift away from visually dominated reasoning because the major criterion for identifying a shape is whether it satisfies a precise set of verbally stated formal properties. So, for example, at Sublevel 2.3, the term *rectangle* refers to a class of shapes that possesses all the properties the student has come to associate with the set of rectangles.

Students can use and formulate formal definitions for classes of shapes. However, their definitions are not minimal because forming minimal definitions requires relating one property to another using some type of inferential reasoning (which occurs at Level 3). Students do not interrelate properties or see that some subset of properties implies other properties. They simply think in terms of unconnected lists of formally described characteristics. The set of properties students give for a class of shapes is a list of all the visual characteristics the student has come to associate with that type of shape, described in terms of formal geometric concepts. (But students seem to recall these property-based specifications rather than discover them on the fly as they inspect shapes.)

Progressing to Sublevel 2.3 requires that formal concepts like side length and angle measure have been raised to a sufficient level of abstraction so that they can be used to form *relational* conceptualizations such as "all sides equal" that describe spatial relationships between shape parts. These relationships, such as opposite sides equal, must have reached the interiorized level of abstraction, detaching them from their original contexts so they are applicable to new situations and available for analyzing shapes.

Level 3: Relational-Inferential Property-Based Reasoning

Students explicitly interrelate and make inferences about geometric properties of shapes. For example, a student might say "If a shape has property X, it also has property Y." However, the sophistication of students' property interrelationships varies greatly, starting with empirical associations (when property X occurs, so does property Y), progressing to componential analyses that explain why one property "causes" another property, then to logically inferring one property from another, and finally to using inference to organize shape conceptualizations into a hierarchical classification system. Because one property can "signal" other properties, students can logically organize sets of properties, correctly specify shapes without naming all their properties, form minimal definitions, and distinguish between necessary and sufficient sets of conditions. At Level 2, students only can use formal properties; at Level 3, students can operate on formal properties.

At Level 3, the spatial relationships described by formal property statements reach the second level of interiorization so that they can be symbolized by the statements, and so that students can reason meaningfully about the statements, in many cases, without having to visually re-present the actual spatial structurings that the statements describe. The verbally-stated properties themselves are interiorized so that they can be meaningfully decomposed, analyzed, and applied to various shapes.

3.1 Empirical relations. Students use empirical evidence to conclude that if a shape has one property, it has another.

3.2 Componential analysis. By analyzing how types of shapes can be built one-component-at-a-time, students conclude that when one property occurs, another property must occur. Students conduct this analysis by making drawings or imagining constructing shapes piece-by-piece. For instance, a student might conclude that if a quadrilateral has four right angles, its opposite sides are equal because if you draw a rectangle by making a sequence of perpendiculars, the sides *must* be equal. (The student explicitly sees visually that making a pair of opposite sides unequal causes some of the angles to *not* be right angles.)

3.3 Logical inference. Students make logical inferences about properties; they mentally operate on property statements, not images. For example, a student might reason that because a square has all sides equal, it has opposite sides equal. Such reasoning enables students to make the inferences needed for hierarchical classification. For instance, a student whose definition for a rectangle is "4 right angles and opposite sides equal" might infer that a square is a rectangle because "a square has 4 right angles, which a rectangle has to have; and because a square has 4 equal sides, it has opposite sides equal, which a rectangle has to have." But students do not use this inferencing ability to logically reorganize their conceptual networks about shapes, so they do not adopt a logical hierarchical shape classification system (i.e., they still resist the notion that a square is a rectangle even though they can follow the logic justifying such a statement).

Students' reasoning is "locally logical" in that they string together logical deductions based on "assumed-true" propositions, that is, propositions that they accept as true based on their experience, intuition, or authority. Thus, students at this level use logic, but they do not question the starting points for their logical analyses.

3.4 Hierarchical shape classification based on logical inference. Students use logical inference *to reorganize* their classification of shapes into a logical hierarchy. They fundamentally restructure their shape classification networks (as opposed to merely making additional connections here and there). It becomes not only clear why a square is a rectangle, but a necessary part of reasoning. Students give logical arguments to justify their hierarchical classifications. Finally, students' use of logic to draw conclusions provides them with a new way to accumulate knowledge. That is, new knowledge can now be generated not merely through empirical or intuitive means, but through logical deduction.

Level 4: Formal Deductive Proof

Students can understand and construct formal geometric proofs. That is, within an axiomatic system, they can produce a sequence of statements that logically justifies a conclusion as a consequence of the "givens." They recognize differences among undefined terms, definitions, axioms, and theorems.

Discussion

One of the major criticisms of the van Hiele levels is that students' progress does not seem to be characterized by jumps from one discrete level to the next, but instead, seems to occur in small incremental steps.[9] To support this criticism, researchers cite the notion of transition between levels and that students often exhibit characteristics of several levels (e.g., Clements & Battista, 1992; Lehrer, 1998). Gutiérrez et al. (1991) dealt with this issue by using vectors to indicate the degree of acquisition of various levels. Clements and Battista (2001) and Lehrer et al. (1998) dealt with the issue by positing that several levels may be developing at the same time, but that different types of reasoning become dominant during learning and development. Although each of these proposed revisions has merit and has moved the field forward, they both face two important issues.

Issue 1: Type versus level of reasoning. Throughout the research on the van Hiele theory, there is a lack of distinction between *type of reasoning* and *qualitatively different levels in the development of reasoning*. That is, sometimes the visual-holistic level is used to refer to a type of reasoning that is strictly visual in nature, and sometimes it is used to refer to a period of development of geometric thinking when an individual's thinking is dominated and characterized by visual-holistic thinking. For instance, Gutiérrez et al. (1991) used vectors to indicate students' "capacity to use each one of the van Hiele levels" (p. 238). This statement makes sense only if van Hiele levels are taken as types of reasoning, not qualitatively different levels of thought. Similarly, Clements and Battista (2001), along with Lehrer et al. (1998), talked about "waves of acquisition" of types of reasoning defined by the van Hiele levels. To complicate matters further, Pegg and Davey used the SOLO taxonomy to describe cognitive processes that enable van Hiele types of reasoning. Thus, broadly speaking, researchers have intermingled and not yet completely sorted out (a) basic enabling cognitive processes, (b) van Hiele levels as types of reasoning, and (c) van Hiele levels as stages of development of geometric reasoning.

Related to this intermingling, several subissues are apparent. First, exactly what are the basic cognitive processes that enable different types of geometric thought? Are they simply the types of thought described by the van Hiele levels? Or are other, more basic, processes implicated? For instance, research suggests several possibilities for basic processes: (a) visualization/imagery; (b) analysis (decomposing shapes into parts and interrelating those parts); (c) conceptualization in terms of formal concepts; (d) ability to attend to, interrelate, and coordinate various aspects of a situation; and (e) inference/logical deduction. Are there other processes? How do each of these interact and produce the types of reasoning characterized by the van Hiele levels?

Second, is the vector or configuration of activations perspective viable? Certainly there is precedent

[9] In fact, longitudinal studies often show students making progress within a van Hiele level, but not moving to the next level (Gutiérrez & Jaime, 1998; Clements & Battista, 2001).

for such a view. For instance, a vector model has been used in describing how the brain represents objects in the perceptual field, not by a set of neurons but by a *configuration* of neuronal firing patterns (Churchland, 2002). Is it then reasonable to think that type of reasoning is enabled by activating a *configuration* of enabling processes? Development of geometric reasoning would then occur (a) as the enabling processes become more sophisticated, (b) as new and more useful configurations of processes form and become stable, and (c) as metacognitive processes regulate the use of configurations. But this approach leads back to defining the enabling processes, distinguishing them from the van Hiele levels as types of reasoning, and explaining how the van Hiele levels *as levels* occur in this activation approach. For instance, how exactly does one type of thinking become ascendant or dominant?

Third, the vector-configuration approach runs the risk of losing the compelling conceptual eloquence attained by the original van Hiele levels. In fact, adopting the vector-configuration approach might regress the field into older "structure of intellect" models, with researchers focused so much on defining and measuring enabling processes that describing the nature of different types of reasoning in conceptually coherent ways is lost.

Issue 2: Assessment. Assessment has long been an issue in the van Hiele theory, both because it operationalizes researchers' conceptions of the levels and because it is necessary for applying the theory to instructional practice. Researchers are still grappling with this issue. For instance, in designing a paper-and-pencil instrument to be used with middle and high school students, Gutiérrez and Jaime (1998) defined four processes (recognition, definition, classification, and proof) and how those processes were implemented by students at different van Hiele levels. *Recognition* refers to recognizing types of geometric figures and identifying components and properties of figures; it includes use of vocabulary to describe properties. *Definition* refers to formulating definitions of concepts during the learning process and using textbook definitions. *Classification* refers to classifying geometric figures into different categories. *Proof* refers to justifying geometric statements. Gutiérrez and Jaime described how each process occurs at several van Hiele levels so that how a student uses a process can be taken as an indicator of the student's level of reasoning. Assessment tasks required students to describe their reasoning.

However, as always with van Hiele level descriptions and assessments, the devil is in the details. Major questions for researchers are (a) the appropriateness of Gutiérrez and Jaime's four characterizing processes and (b) how these processes are defined to occur at each van Hiele level. For instance, their definition of the recognition process focuses on physical attributes at Level 1 but mathematical properties at Level 2, two very different kinds of "recognition" processes that are related but certainly not the same at a basic cognitive level. Thus, it is one thing to devise broad categories of behavioral descriptors; it is another to determine the cognitive processes underlying these categories of behaviors. This has been, and will continue to be, a major challenge facing researchers.

However, despite the difficulty facing researchers attempting to assess students' van Hiele levels, each serious attempt to do so helps us better understand the nature of the levels and students' geometric thinking. For instance, in discussing van Hiele assessments, Gutiérrez and Jaime (1998) claimed that even though researchers most often identify students as belonging to Level 3 if they can understand and use hierarchical classifications of shape families, there are other critical attributes of this level. They argued that students at Level 3 can also establish logical connections between geometric properties (a key component in thinking hierarchically and in using logical definitions), and they can accept and identify nonequivalent definitions of geometric shape classes. Do assessments of Level 3 equally assess all of the characteristics of Level 3, and, if not, how does that affect the validity of the assessments? In fact, how do assessments affect researchers' conceptualizations of the levels?

A totally different approach to assessing van Hiele levels was devised by another group of researchers. In a collaborative effort to find ways to assess elementary students' acquisition of the van Hiele levels in interview situations, Battista, Clements, and Lehrer developed a triad sorting task, that, with variations, both Clements and Battista (2001) and Lehrer et al. (1998) used in separate research efforts. In this task, students are presented with three polygons, such as those shown in Figure 19.3, and are asked, "Which two are most alike? Why?" Choosing B and C and saying that they "look the same, except that B is bent in" was taken as a Level 1 response. Choosing A and B and saying either that they both have two pairs of congruent sides or that they both have four sides was taken as a Level 2 response. The purpose of this task was to determine the type of reasoning used on a task that students had not seen before (so it was unlikely to elicit instructionally programmed responses).

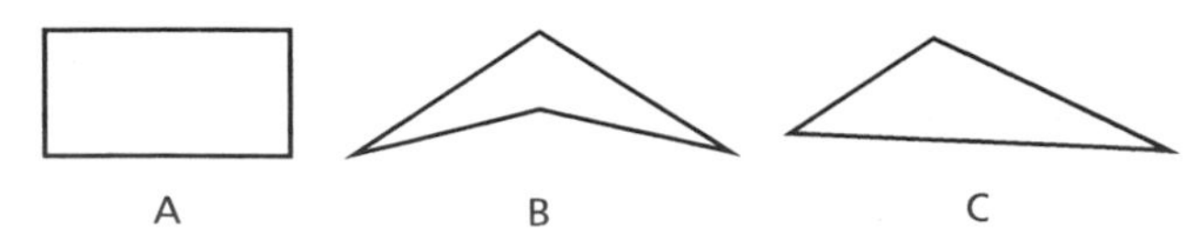

Figure 19.3 Triad polygon sorting task.

One difficulty with this analysis is that giving the number of sides of a polygon is a "low-level" use of properties. That is, there are different types of property use. The simplest property involves describing the number of components in a shape. For instance, a quadrilateral has four sides; a triangle has three angles. A second, somewhat more sophisticated type of property describes global characteristics. For instance, a quadrilateral is a simple closed curve. Describing the straightness of sides also falls in this category. Finally, and most sophisticated, are properties that describe relationships between parts of shapes. For example, the following statements describe *relational* geometric properties: (a) opposite sides of a parallelogram are parallel and congruent; (b) a rectangle has four right angles; (c) adjacent angles of parallelograms are supplementary; (d) this quadrilateral has one line of symmetry. Examples a–c describe relationships between the sides of shapes, (a) explicitly, (b) and (c) indirectly by specifying angle measures. Example (d) describes a relationship between two "halves" of a quadrilateral, which, with further analysis, could be described in terms of line segments and angles.

The distinction in properties described above suggests that students' use of number of sides of a polygon is not a very good indicator of Level 2 thinking, which focuses on relational properties. Thus some jumps in levels observed by Lehrer et al. (1998) may have been caused by coding students' use of number of sides as Level 2. Because a critical factor used in distinguishing van Hiele levels is how students deal with geometric properties, clarifying the meaning of *properties*, as it relates to the van Hiele levels, is important.

Another factor that should be considered with this triad task is that saying Shape B is more like Shape C is not necessarily a less sophisticated response than focusing on number of sides. That is, Shape B is actually more like Shape C if we consider how much movement it takes to transform B into C, compared to B into A. In fact, one could imagine a metric that quantifies the amount of movement required. Thus, the "morphing" response described by Lehrer et al. (1998), and also observed by Clements and Battista (2001), may be an intuitive version of a notion whose mathematization is far beyond the reach of elementary students.

Nevertheless, this being said, the original argument that attending to number of sides is more sophisticated than a visual morphing response may still have some validity. It suggests a shift toward using a more mathematical conceptualization (albeit the primitive one of counting number of sides/vertices).

Another shortcoming of the triad-task approach is pointed out by differences in the ways the researchers used the triads. Lehrer et al. (1998) construed each triad task as an indicator of type of reasoning. So students' use of different types/levels of reasoning on different triads was taken as evidence of jumps in levels. In contrast, Clements and Battista (2001) used a *set* of 9 triad items as an indicator of level of students. To be classified at a given level, a student had to give at least 5 responses at that level. If a student gave 5 responses at one level and at least 3 at a higher level, the student was considered to be in transition to the next higher level. The latter approach, by virtue of its use of multiple items, produces a more reliable indication of level of reasoning. Of course, because it aggregates responses, this approach obscures intertask differences and variability in reasoning. It focuses on determining the ascendant level of students' reasoning on a particular type of task. (So this issue is related to the type-of-reasoning versus level-of-reasoning dichotomy previously discussed.)

Another difference between the researchers' approaches is also important. In analyzing students' reasoning on the triad tasks, Lehrer et al. (1998) classified student responses solely on the basis of the type of reasoning that students employed. In contrast, in classifying students' van Hiele levels, Clements and Battista (2001) attempted to account for the "quality" of students' reasoning—each reason for choosing a pair in a triad was assessed to see if it correctly discriminated the pair that was chosen from the third item in the triad. In this scheme, the van Hiele levels for students were determined based on a complicated algorithm that accounted for both type of reasoning used and discrimination score.

However, although discrimination is a reasonable indication of quality of response in the triad context, it does not address another component that might be important in assessing van Hiele levels, the *validity* of reasoning, which involves the accuracy and precision of students' identifications, descriptions, conceptualizations, explanations, and justifications. That is, a student might validly describe a property of two items in a triad that does not discriminate the items from the third item. Researchers must determine which, if any, of discrimination, validity, or mere use of properties are critical characteristics of van Hiele Level 2.

The last issue in assessment is whether to use individual interviews or paper-and-pencil tests. Because individual interviews provide a more detailed picture of students' geometric reasoning, they provide a more valid assessment of van Hiele level, but are more time-consuming to administer. In contrast, paper-and-pencil instruments are less time-consuming and are thus more practical for many research projects and classroom teachers who must deal with large numbers of students. But paper-and-pencil instruments, even those that ask students to write about their reasoning, rarely give as complete a picture of student reasoning as recorded individual interviews. However, no matter which method is used, creating and interpreting tasks that reliably and validly assess van Hiele levels is extremely challenging.

Final thought on the van Hiele theory. Numerous researchers have concluded that the van Hiele levels accurately, if generally, describe the development of students' geometric thinking, especially about shapes. However, these same researchers, in general, seem to find the theory lacking in depth. They would like a more detailed description of this development. I have already described several proposed alternatives and elaborations. But why does the van Hiele theory, though deemed inadequate in detail, also have such a strong ring of validity?

I suggest that the reason for this feeling of validity is that the van Hiele levels describe a progression of thinking that is part and parcel to scientific/mathematical thinking. That is, we can think of both the levels and the "scientific" method as progressing through four phases: perceptualization (resulting in informal, intuitive, surface-level categories and reasoning), conceptualization (resulting in explicit concepts and analysis), organization (resulting in conceptual organization), axiomatization (resulting in formal logical treatments). For highly educated individuals in scientifically advanced civilizations—like mathematics education researchers—moving through these four phases seems like the natural way for individuals to progress from intuitive, everyday reasoning to formal scientific reasoning.

Psychological Mechanisms Underlying the Levels

To understand the cognitive processes underlying geometric thinking, and movement from one level of thinking to the next, one must understand how individuals construct and mentally represent spatial and geometric knowledge. Because such knowledge originates with perception and imagery, I start with these processes.[10]

Perception of Shape

Kosslyn (1994) has provided a detailed model of the brain processes involved in perception. This model consists of several subsystems. The preprocessing subsystem accepts sensory input from the attention window; it outputs both an image and properties that it extracts from the image. The pattern activation subsystem then matches output from the preprocessing subsystem to stored representations of visual patterns that include extracted properties. The pattern activation subsystem is divided into two additional subsystems. The category pattern activation subsystem classifies a stimulus as an element of a category; the exemplar pattern activation subsystem registers specific examples (e.g., Fido, not dogs in general). The two tasks, "recognizing a stimulus as a specific individual and recognizing it as a member of a category, have incompatible requirements; information that is needed to recognize the individual must be discarded to classify it as a member of the category" (Kosslyn, 1994, p. 182). Thus, even at this elemental level, the two essential geometric tasks of apprehending instances and apprehending categories are differentiated.

The process of abstraction seems to be at the root of construction of entities in the category pattern activation subsystem: "To isolate certain sensory properties of an experience and to maintain them as repeatable combinations, i.e., isolating what is needed to recognize further instantiations of, say, apples, undoubtedly constitutes an empirical abstraction" (von Glasersfeld, 1991, p. 57).[11] An individual comes to recognize triangles as figural patterns of triangular shapes are abstracted and become linked to the word *triangle*.

> Abstraction from the actual sensory material out of which particular configurations or sequences are built up, yields figural patterns that have a certain general applicability and can be semantically associated with specific names. Thus we have, for instance, the notion of 'triangularity' that enables us to *see* spatial configurations as triangles. (von Glasersfeld, 1982, p. 197)

Furthermore, von Glasersfeld argues that perceptual abstraction is active, not passive.

> The experiencing subject attends, not to the specific sensory content of experience, but to the operations that combine perceptual and proprioceptive elements into more or less stable patterns. These patterns are constituted by motion, either physical, or attentional, forming 'scan paths' that link particles of sensory experience. (von Glasersfeld, 1982, p. 196)

Individuals form scan paths that enable them to recognize similar perceptual objects (e.g., triangles that look approximately the same). When several different types of triangular perceptual objects are ab-

[10] There seems to be far less relevant literature about cognitive processes underlying thinking at van Hiele levels above Level 2.

[11] More will be said about the process of abstraction in a later section.

stracted and associated with the word triangle, an early conceptualization of triangle is created. However, at this early stage of recognition, an individual's "conceptualization" of triangle is usually limited to a small subset of typical prototypes and is holistic in nature. A major problem in placing the van Hiele theory on a sound cognitive foundation is answering the question, How do individuals move to finer, parts-based conceptions of shapes?

From Wholes to Parts

Kosslyn argued that to form an image of a shape requires an *implicit* mental construction in a sequential, part-by-part manner not only of the parts of the shape but how they are related; "to represent such shapes is to extract specific properties that will not change when the object assumes a new configuration or a shape-variant appears. One such invariant is the type of spatial relation between parts" (Kosslyn, 1988, p. 1623). Kosslyn (1994) further argued that in object recognition, global shape is processed first, then parts and characteristics. That is, in perceptual processing, the mind first attempts to match the overall pattern that corresponds to an object; if that does not work, it attempts to match input to representations of individual parts. Thus, even when a shape is processed globally, the perceptual system identifies the shape's parts and computes relatively abstract spatial relations among them. Initially, however, the parts are not explicitly accessible.

Kosslyn's research is consistent with that of Smith (1989). She suggested "that represented objects are built in a bottom-up fashion from a feature level of processing" but "that the represented object is a cohesive unit and is experienced as a unit. Represented objects as wholes are given to conscious experience as unitary entities, so that one cannot get to the represented parts without some work" (p. 127). But in addition to Kosslyn's claim that task demands may cause a holistic-to-parts shift in processing, Smith argued that as children get older, they acquire an increased ability to selectively attend to single dimensions, and thus the characteristics of wholes. Clements and Battista (1992, 2001) described the processes by which parts of shapes gradually become accessible to reflection, but their theory focused on experience, reflection, and learning.

Research on brain processes goes even further, separating the identification of shape parts, the creation of spatial relations between parts, and the creation of the whole. Posner and Raichle (1994) found that patients suffering brain damage in either the right or left hemispheres possessed very different types of cognitive deficits. When asked to draw target stimuli, right-hemisphere-damaged patients saw the components of the stimuli but were unable to correctly put the components together; left-hemisphere-damaged patients saw the global picture, but not the components of the target stimuli. The patient drawings in Figure 19.4, especially the top row, clearly illustrate the global-versus-parts nature of patients' cognitive deficits. Patients with right hemisphere damage saw the components of the original triangle—the segments made from rectangles—but were unable to correctly put these components together spatially. They did not perceive or record the overall spatial structure of the triangle. In contrast, left-hemisphere damage seemed to cause little deficit in global spatial processing; subjects correctly captured the shape of the triangle.[12] Left-hemisphere damage did, however, cause important details to be omitted from figures. Such omissions can be important geometrically. For instance, Kosslyn (1994) found that the left hemisphere "is critically involved in detecting target figures when they are embedded in other patterns" (p. 202).

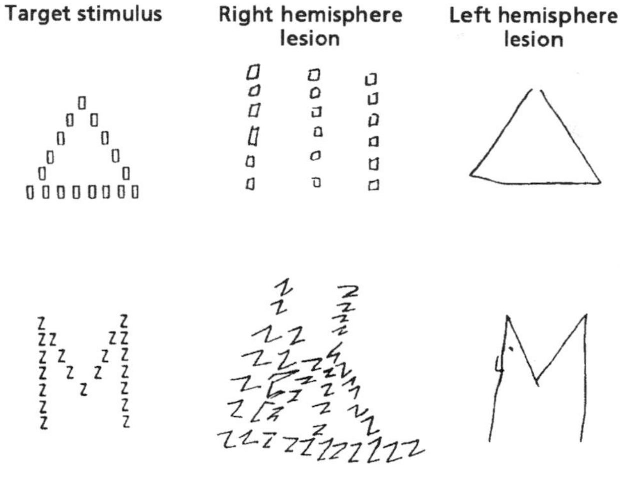

Target stimulus | Right hemisphere lesion | Left hemisphere lesion

Figure 19.4 Global-versus-parts deficits from hemispheric damage (Posner & Raichle, 1994).

Posner and Raichle (1994) explained these hemispheric differences in terms of attentional scale—zooming in or out: "The right hemisphere is more likely to control large or global scales [zoomed out] while the left does the reverse [zoomed in]" (p. 162). Kosslyn (1994) too argued that a visual object "is represented at multiple scales; at one scale the parts are represented, whereas at another they are embedded

[12] Posner and Raichle (1994) reported, however, some research showing that left-hemisphere damage can cause deficits in assembling parts into overall representations.

in an overall pattern" (p. 161). Furthermore, research has found that observers can selectively focus on *either* global or local features in a visual form. "When we are set to attend to one level of detail, our processing of features at the other level is poorer" (Coren et al., 1994, p. 518). Thus the global-local processing separation is built into the architecture of the brain. However, as individuals learn and become more sophisticated in their reasoning, the separate processes get coordinated and integrated into higher level recognition schemes that are sensitive to the whole, its parts, and relationships between parts. Furthermore, this integration is so automatic in educated adults that it is difficult for them to appreciate the world of the child in which the processes remain separate and unintegrated.

In summary, the psychological/neurological research described above sheds light on several findings in research on geometric thinking. It indicates that perceptual wholes are built up from parts-based representations, but that accessing these parts-based representations is initially impossible and only occurs "with some work." Furthermore, there is a separation in perceiving parts of shapes and relationships between parts, and there is a separation in perceiving global or local features of shapes. These findings help explain three important issues in research on the development of geometry reasoning: (a) why students first attend to whole shapes; (b) why students often notice parts of shapes, but not relationships between parts (Clements & Battista, 1992); and (c) how students shift from focusing on wholes to focusing on relationships between parts, as described in the van Hiele theory.

From Images to Explicit Relationships

How does a child progress from viewing shapes merely as visual images to thinking about them primarily in terms of verbal statements about their parts and relationships between parts? First, students must gain mental access to the parts of shapes, explicitly considered as parts of shapes, so that they can mentally construct spatial relationships between the parts. It is not that before this time students cannot notice the parts of shapes; it is that they do not see these parts as parts of wholes—"parts" are merely seen as self-standing entities. In fact, initial processing likely considers parts and wholes on equal footing—the processing system may not detect whether perceived entities are wholes or parts of wholes (Kosslyn, 1994, p. 168). Thus, a student's description that a shape is "pointy" does not imply that the student is attending to a part of the whole; the student may simply be focusing on a salient entity in the visual field. This lack of an inclusion relation between parts and the wholes to which they belong is characteristic of Level 1 thinking and is similar to a type of conceptualization described by Piaget (Gruber & Voneche, 1977), who claimed that "an arm drawn alongside of a manikin is conceived by the child as 'going with' the manikin not as 'forming part of' his body" (p. 102).

Second, students must mentally construct explicit spatial relationships between the parts of a shape that give it its spatial structure. This requires the mind to "break into" the components of perceptual processing for identifying a shape, abstract structural spatial relationships that characterize the shape, and consciously attempt to conceptualize these relationships. According to the theory of abstraction to be discussed later, breaking into the perceptual processes requires intentional reflection on the nature of the shape, and that the perceptual processes be abstracted at the interiorized level. Once spatial relationships become accessible to reflection, they can be verbally described, refined through empirical testing, and later, linked to interiorized formal conceptualizations.

Several factors can engender the move from images to explicit relationships. First, as has been seen, the ability to break into perceptual processes can be affected by development and experience. Second, Kosslyn discussed an imagery-to-verbalization developmental shift. He claimed that even though children rely predominantly upon imagery in accessing and using information, because processing capacity for verbal knowledge increases with development, children increasingly translate observations derived from images into verbal form so facts can be accessed either imagistically or verbally (Kosslyn, 1980, 1983; Kosslyn, Margolis, & Barrett, 1990). In fact, Kosslyn (1980) suggests that the use of imagistic representation may be reduced as children gain experience and repeatedly access the same ideas. That is, a child "may recode information into propositional form as he or she accesses that information repeatedly" (p. 412). Furthermore, because "certain facts are communicated linguistically and presumably memorized in a propositional format. . . . Retrieval . . . should not require consultation of an image" (p. 412).

Comment

Although a number of theories and studies have been reviewed in an attempt to describe the cognitive processes by which students progress through the early van Hiele levels, this area of research is still in its infancy. This is due in great part because researchers are investigating cognitive processes that cannot be observed. To achieve progress in this domain, it is important for mathematics education researchers to heed the work of researchers in other fields such as cognitive science and neuroscience. Such research can provide valuable insights into these difficult-to-observe processes. But we must also continue to conduct studies that expose students to instructional treatments and tasks and carefully investigate phenomenologically how students construct meaning for geometric concepts, always attempting to build cognitive models that explain what we observe. It is through the integration of cognitive and phenomenological investigations that we can make true progress in this vital area of research.

OTHER THEORIES RELEVANT TO GEOMETRIC LEARNING

Several other theories seem particularly relevant for understanding geometric reasoning.

The Theory of Abstraction[13]

Learning occurs as individuals recursively cycle through phases of action (physical and mental), reflection, and abstraction in a way that enables them to develop ever more sophisticated mental models. These mental models consist of abstracted objects and abstracted actions that can be performed on those objects. Once objects and actions have been sufficiently abstracted, they themselves become mental objects that can be mentally operated on (e.g., compared, decomposed, connected). The levels of abstraction described below indicate major landmarks in the process of abstracting mathematical objects, affording increasingly powerful reasoning.

Levels of Abstraction

Perceptual abstraction. At the perceptual/recognition level, abstraction isolates an item in the experiential flow and grasps it as an object: "Focused attention picks a chunk of experience, isolates it from what came before and from what follows, and treats it as a closed entity" (von Glasersfeld, 1991, p. 47). When items are perceptually abstracted, we become conscious of them, entering them into working memory. Perceptual abstraction isolates those sensory properties of an experience that are needed to recognize further instantiations of the experience. However, items that have been abstracted only at the perceptual level cannot be re-presented (visualized) and cannot be operated on unless they are physically present. Furthermore, "the abstracted operational pattern necessary to recognize things of a kind does not automatically turn into an image that can be re-presented" (von Glasersfeld, 1995, p. 93).

Internalization. When material has been sufficiently abstracted so that it can be re-presented in the absence of perceptual input, it has been internalized. Internalization is

> the process that results either in the ability to re-present a sensory item without relevant sensory signals being available in actual perception or in the ability to reenact a motor activity without the presence of the kinesthetic signals from actual physical movement. Internalization leads to 'visualization' in all sensory modalities. (Steffe & Cobb, 1988, p. 337)

According to von Glasersfeld, it is at the internalized level that a concept[14] has been formed, a concept referring "to any structure that has been abstracted from the process of experiential construction as recurrently usable. . . . To be called 'concept' these constructs must be stable enough to be re-presented in the absence of perceptual 'input'" (von Glasersfeld, 1982, p. 194). However, for material that has only been internalized, one cannot reflect upon its re-presentation to consider how it is composed (Steffe & Cobb, 1988). That is, one can only re-present internalized material, not analyze its structure.

Interiorization. "Reflecting upon the results of a re-presentation requires detachment and placing the re-presented activity at a distance in order to analyze its structure and composition" (Steffe & Cobb, 1988, p. 17). To do so requires interiorization, which is "the most general form of abstraction; it leads to the isolation of structure (form), pattern (coordination), and operations (actions) from experiential things and activities" (p. 337). Material has been interiorized when abstraction has disembedded it from its original perceptual context and it can be freely operated on in imagination, including "projecting" it into other perceptual material and utilizing it in novel situations. To use an abstracted item for thinking about a new situation requires that the item be reprocessed, not simply recalled as it has been encountered in the past—reprocessing is the operationalization of interiorization (Steffe, 1998).

Note that researchers should be careful with the notion that abstraction "disembeds" material from its original context. As Hoyles and Healy (1997) have

[13] Other views of abstraction in geometry can be found in (Hollebrands, 2003; Hoyles & Healy, 1997) and (Mitchelmore & White, 2000).
[14] In this case, von Glasersfeld's "concept" is the same as my "conceptual object" or "conceptualization."

noted, abstraction can be *situated* in that, at least at lower levels, it is a process that occurs within, and remains linked to, the context in which it first occurs. That is, at the perceptual and internalized levels of abstraction, context is an inseparable part of an abstraction. At the interiorized level, abstracted material becomes separable from its original context, although links to that context may remain. In fact, these links may be an important part of applying interiorized material in novel situations. Generalized forms of links may form part of the structure of an abstraction that helps individuals recognize new situations in which the abstraction might be applied.

Second level of interiorization. When material reaches the second level of interiorization, operations can be performed on the material without re-presenting it and symbols, acting as "pointers" to the originally abstracted material, can be used to substitute for it.

The Theory of Abstraction Applied to the van Hiele Levels

To apply the theory of abstraction to the van Hiele theory, recall that action-based mental processes enable students to recognize and construct images of shapes—Kosslyn's sequential part-by-part mental constructions, or von Glasersfeld's scan paths. During the first van Hiele level, these processes are abstracted at the perceptual and internalized level to recognize and visualize common shapes, respectively. To move up to the second van Hiele level, shape recognition processes must be raised to the interiorized level, allowing these whole-shape recognition processes to be decomposed, and thus the components of shapes to be recognized, operated on, and interrelated. Because interiorization of whole-shape recognition processes permits relational operations to be performed on shape components, relationships between components of shapes can be noticed, starting another cycle of abstraction, focused not on recognition but on relationships. As these relational operations between components of shapes are abstracted at the perceptual and internalized levels, students can describe relationships between components of individual shapes informally. Subsequently, when relational operations become interiorized, they can be applied to sets of shapes so that the student can see that a class of shapes has a set of characterizing properties. At this point a student is capable Level 2, descriptive/analytic, reasoning. General properties, applied to classes of shapes, can be described. To reach the third van Hiele level, relational operations must achieve "symbol" status (von Glasersfeld, 1995)—that is, they must reach the second level of interiorization. At this level, verbal/symbolic statements of relationships can act as substitutes for, or pointers to, these operations, with individuals able to reason meaningfully about these symbolic statements without having to re-present the actual operations for which they stand.

A Transformational Perspective on Relational Operations

What is the nature of the operations that establish interrelationships between components of shapes? One hypothesis is that the relationships are established with unconscious visual transformations. For example, but differing in detail from Leyton (1992), Battista (in press) posits that in a parallelogram, seeing the relationship that opposite sides are parallel might be based on mentally (but unconsciously) translating one side onto the opposite side. Similarly, seeing that opposite angles in a parallelogram are congruent may be based on an unconscious 180° rotation.

This analysis is consistent with the observation that parallelism is more difficult for students to "see" in typical trapezoids than in rectangles and parallelograms. Seeing parallelism in scalene trapezoids cannot be accomplished using the same mental transformation that establishes this property for rectangles and parallelograms because in a scalene trapezoid, one side is not the translation image of its opposite side. Also consistent with this hypothesis is the common student belief that the property "opposite sides parallel" goes with "opposite sides equal" because the first property is established with a translation that also establishes the second. To disassociate these two properties, parallelism might need to be established, for example, with a translation and a stretch.

Special Types of Abstraction

Three special forms of abstraction are fundamental to geometry learning and reasoning (Battista, 1999a; Battista & Clements, 1996). *Spatial structuring* is the mental act of constructing and abstracting an organization or form for an object or set of objects. It determines an object's nature or shape by identifying its spatial components, combining components into spatial composites, and establishing interrelationships between and among components and composites. *Mental models* are sets of abstractions that are integrated to form nonverbal mental representations that are activated to interpret and reason about situations (more will be said about mental models in a later section). A *scheme* is an organized sequence of actions or operations that has been abstracted from experience and can be applied in response to similar circumstances. It consists of a mechanism for recognizing a situation, a mental model that is activated to interpret actions within the situation, and a set of expectations (usually embedded in the behavior of the model) about the possible results of those actions.

Three Types of Structuring

Geometry learning can be viewed as involving three types of structuring (Battista, in press b). (a) Spatial structuring constructs a spatial organization or form for an object or set of objects. It determines a perception/conception of an object's nature or shape by identifying the object's spatial components, combining components into spatial composites, and establishing interrelationships between and among components and composites. (b) Geometric structuring describes spatial structurings in terms of formal geometric concepts. That is, in geometrically structuring a spatial situation, a person uses geometric concepts such as angles, slope, parallelism, length, rectangle, coordinate systems, and geometric transformations to conceptualize and operate on the situation. In order for a geometric structuring to make sense to a person, it must evoke an appropriate spatial structuring. (c) Logical structuring formally organizes geometric concepts (i.e., geometric structurings) into a system and specifies that interrelationships must be described and established through logical deduction. To accomplish a logical structuring, individuals must logically organize sets of properties. The construct of structuring, construed more generally, provides another perspective on the van Hiele theory: Spatial structuring is the major focus in van Hiele Level 1 and the transition to Level 2, geometric structuring is the major focus at Level 2 (when it has been broadly accomplished, Level 2 is fully attained), and logical structuring is the focus at Levels 3 and 4 ("axiomatic" structuring is the focus at Level 5).

Construction and Use of Mental Models[15,16]

An abundant amount of research indicates that many forms of reasoning are accomplished with mental models (Battista, 1994; Calvin, 1996; English & Halford, 1995; Greeno, 1991; Johnson-Laird, 1983; Markovits, 1993). According to the mental model view of the mind, individuals understand or make sense of a situation, including a set of connected verbal propositions describing a situation, when they activate or construct a mental model to represent the situation (Johnson-Laird, 1983). Mental models are nonverbal recall-of-experience-like mental versions of situations that have structures isomorphic to the perceived structures of the situations they represent (Battista, 1994; Greeno, 1991; Johnson-Laird, 1983, 1998). That is, "parts of the model correspond to the relevant parts of what it represents, and the structural relations between the parts of the model are analogous to the structural relations in the world" (Johnson-Laird, 1998, p. 447). Individuals reason about a situation by activating mental models that enable them to simulate interactions within the situation so that they can explore possible scenarios and solutions to problems. When using a mental model to reason about a situation, a person can mentally move around, move on or into, combine, and transform objects, as well as perform other operations like those that can be performed on objects in the physical world. Individuals reason by manipulating objects in mental models and observing the results. "The behavior of objects in the model is similar to the behavior of objects that they represent, and inferences are based on observing the effects of the operations" (Greeno, 1991, p. 178). Importantly, individuals' use of mental models is constrained by their knowledge and beliefs. That is, much of what happens when we form and manipulate a mental model reflects our underlying knowledge and beliefs about what would happen if we were dealing with the objects they represent. So the properties and behavior of objects in a mental model simulate the properties and behavior we believe the objects they represent possess.

I will now discuss several examples of mental-model-based reasoning in geometry. The first two examples are reformulated from Clements and Battista (2001).

A second-grade student was investigating geometry in a Logo environment. After she had discovered that she needed 90° turns to make a tilted square, she was asked to draw a tilted rectangle. She immediately used 90° turns at the vertices. She was asked, Why?

> **CA:** Because a rectangle is just like a square but just longer, and all the sides are straight. Well, not straight, but not tilted like that (makes an acute angle with her hands). They're all like that (shows a right angle with her hands) and so are the square's.

In later discussions, she also stated that a square is a rectangle.

> **Obs:** Does that make sense to you?
> **CA:** It wouldn't to my [4 year old] sister but it sort of does to me.

---

[15] Researchers employing the theory of mental models have actually created computer models of mental models for specific situations.

[16] Mathematics education researchers using the notion of concept image have reached conclusions similar to researchers embracing the idea of mental models, specifically that students most often do not use definitions of concepts, but rather concept images—a combination of all the mental pictures and properties that have been associated with the concept—to make decisions. (See Clements & Battista, 1992 for more details.) Concept images are similar to mental models.

Obs: How would you explain it to her?
CA: We have these stretchy square bathroom things. And I'd tell her to stretch it out and it would be a rectangle. (Battista & Clements, 1991, p. 18)

It "sort of made sense" that a square could be thought of as a rectangle because CA's mental model of a square enabled her to envision a square being stretched into a rectangle. Because CA had just demonstrated her knowledge that squares and rectangles are similar in having angles made by 90° turns, this response suggests that she may have understood intuitively that all rectangles can be generated from one another by certain transformations, ones that preserve 90° angles. CA's mental models of shapes such as squares and rectangles incorporated allowances and constraints on how they could be acted upon and transformed. In this instance, CA reasoned by performing a simulation of changing a square into a rectangle using a special type of visual transformation, one that incorporated formal geometric constraints—preserving 90° angles—on her mental objects and actions.

Sometimes, however, students' mental models are inadequate to draw valid conclusions. For instance, a second grader was examining her attempt to draw a tilted square in Logo. (She too had successfully taught the turtle to draw squares of various sizes, all of which had sides vertical and horizontal.) Because she did not use 90° turns for the tilted square, her figure was only a crude approximation to a square. In fact, though M had earlier stated that all sides of a square are the same length, in drawing her square, she abandoned this constraint to close the figure (because her angles were not 90°). But M intended to draw a square and concluded that it was a square, reasoning as follows:

Int: How do you know it's a square for sure?
M: It's in a tilt. But it's a square because if you turned it this way it would be a square.

The difference in the validity of CA's and M's reasoning seems to have been due to the degree to which they incorporated relevant knowledge into their mental models. On the basis of a variety of tasks, the researchers found that CA's thinking about shapes was based more on properties than M's, so CA's mental models were more property-based and her imagery more influenced by property-based considerations. M based her thinking on simple visualizations—she, like all students in her class, had seen that squares that are rotated from the vertical-horizontal-side position often do not look like squares. M's mental-model-based transformation was unconstrained by her knowledge that all sides of a square have the same length. Thus, these episodes raise the question of how students' conceptual knowledge gets incorporated into their mental models and visualization.

A different kind of example is provided by a secondary mathematics teacher attempting to decide on the validity of the following statement: If all the angles in a polygon are equal, then all the sides in the polygon are equal. The teacher said that in thinking about the problem he first imagined a regular pentagon, which he attempted to transform so that it had unequal angles. But he could not imagine how to change some but not all of the sides of the pentagon without changing the angle measures. The mental model this teacher activated to think about this problem made the problem impossible for him to solve. When I asked him to think about a rectangle, he immediately solved the problem. Interestingly, had the teacher originally activated a mental model of a square or regular hexagon, it would have been much easier to implement his transformational strategy. This example illustrates that an important question for researchers is, What determines which mental models are activated during problem solving, and what kinds of instruction guide students to activate the most useful models?

Mental Models and Imagery[17]

Images are mental representations "that 'stand in' for (re-present) the corresponding objects" (Kosslyn, 1994, p. 1). Images can be evoked in the absence of perceptual material (re-presentation or visualization) or by perception (e.g., looking at a diagram). Their formation relies on regions of the brain that support *depictive* representations.

> In a depictive representation, each part of an object is represented by a pattern of points, and the spatial relations among these patterns . . . correspond to the spatial relations among the parts themselves. Depictive representations convey meaning via their resemblance to an object, with part of the representation corresponding to parts of the object. (Kosslyn, 1994, p. 5)

A number of researchers have suggested that there are two major types of imagery. Following Lakoff, Wheatley (1998) distinguished between rich images and image schemata. *Rich images* are static, fixed, and contain much visual detail. *Image schemata* repre-

[17] Although I focus on visual imagery, similar discussion can be given for imagery in other modalities such as auditory or kinesthetic.

sent spatial relationships and can be transformed in various ways.[18]

The distinction between rich images and image schemata can be explained by the more general theory of mental models. Johnson-Laird and colleagues argued that it is important to distinguish between visual images (which correspond to rich images) and spatial mental models (which correspond to image schemata). Visual images are derived from perception or generated from mental models, with images being views of mental models from particular viewpoints. Images are mentally manipulable when appropriate mental models have been constructed from them (Johnson-Laird, 1998). Thus, in visual reasoning, sometimes people merely construct and scan visual images, and other times they operate on spatial mental models (Knauff, Fangmeier, Ruff, & Johnson-Laird, 2003). Furthermore, sometimes, when aspects of visual images are irrelevant to a task, the evocation of visual images can actually interfere with construction and use of an appropriate mental model and thus impede reasoning (Knauff et al., 2003).

The distinction between images and mental models can also be interpreted in terms of the theory of abstraction. A re-presentation of a visual image results from an internalization of a perception of a spatial object. A mental model of a spatial entity results from interiorization of the mental processes used to conceive of the entity: "The operations that are carried out in reasoning with models . . . are conceptual and semantic" (Johnson-Laird, 1998, p. 457). Thus, to reason geometrically about an image, one must construct a proper mental model of the image, one that captures the image's relevant spatial structure (and the depth of that structure depends on the sophistication of the concepts used in this structuring). That is, geometric reasoning about figures requires a reconstitution of images in terms of appropriately structured mental models. Indeed, if figures are thought about as pictures (using images), not as sets of spatial/geometric relationships (using mental models), students are not thinking about formal geometry (cf. Laborde, 1993).

Concept Learning

Because concept learning, use, and analysis play a significant role in the development of geometric thinking, research on the formation and use of concepts provides insights into that development, especially for understanding thinking at van Hiele Levels 1 through 3. I will summarize some of the relevant findings of that research.

Not only do human minds naturally categorize objects (Pinker, 1997), categorization is essential for intelligent thought and action (Ross, 1996). What the mind gains from forming categories "is inference. Obviously we can't know everything about every object. But we can observe some of its properties, assign it to a category, and from the category predict properties that we have not observed" (Pinker, 1997, p. 307). For example, if, on seeing round, fist-sized reddish-green objects attached to the branches of a tree, you categorize the objects as apples, you can then infer that they are edible and have seeds.

Research suggests that people form two kinds of categories (Pinker, 1997). Fuzzy or natural categories, such as games, are formed in everyday activity. Such categories generally have no clear definitions, have fuzzy boundaries, and are conceptualized mainly in terms of stereotypes and family-like resemblances—some examples are deemed "better" than others. Identification of instances seems to be the major goal. The second type of category consists of formal, explicitly defined categories, such as odd numbers, in which all instances are logically equivalent as representative of the class. Formulating and studying precise definitions is the goal.

Learning geometry involves the formation of both fuzzy and formal categories. Before geometry instruction, and even in primary-grade teaching, as children encounter shapes in the physical world and those shapes are named by adults, children treat shapes as fuzzy categories—shapes are identified but not defined. Subsequent school instruction demands that students conceptualize shapes in terms of verbally stated, property-based definitions. Difficulties often occur as students attempt to reconcile fuzzy categories with formal concepts. The language used in fuzzy and formal concepts may be the same, but the underlying cognitive entities are vastly different. For instance, logically, a square is as much a rectangle as a nonsquare rectangle. But students who possess fuzzy concepts often resist the idea that a square is a rectangle—they think it is only a square. Progressing from the visual to the descriptive/analytic van Hiele level seems very much like moving from fuzzy to formal categories for a set of concepts in a domain.

There seem to be several competing theories for how people develop fuzzy categories. The most prominent seems to be that people develop prototypes—

[18] The distinction between rich images and image schemata seems related to Piaget's notion of figurative knowledge, which results from empirical abstraction, and operative knowledge, which results from reflective abstraction: "Perception yields a series of static, centration-afflicted snapshots. The knowledge we get from perception is figurative, not operative" (Piaget, 2001, p. 21).

mental images or examples that include, usually implicitly, features that are relatively common among members of a category. People decide whether an object belongs to a category by determining if it is sufficiently similar to prototypes (Smith, 1995).

Researchers describe two ways that prototypes might be formed (Krascum & Andrews, 1993). According to the *abstractionist* model, after experiencing a number of category examples, a perceiver abstracts the category features that occur most frequently into a prototype representation. According to the *exemplar* model, category examples are not decomposed into attributes but instead are abstracted to form a representation that consists of a set of discrete, more-or-less intact instances. Krascum and Andrews (1993) stated that many researchers favor the examplar view because of findings that show that young children are insensitive to component features of objects, show developmental improvement in the ability to selectively attend to single dimensions of stimuli, and perceive objects in global, unanalyzed fashion. However, Krascum and Andrews also cited evidence that is inconsistent with the exemplar model. For instance, they reported that young children do not necessarily fail to decompose objects into their component parts and in fact often fixate on single features. The ability to distribute attention over many stimulus dimensions develops over time. Interestingly, as researchers observe students' learning of geometric shapes, we see many of the behaviors described above—students form prototypes, often limited; they fail to notice important spatial attributes; they sometimes fixate on particular features. Also, Kosslyn's description of perception seems more consistent with the exemplar model and addresses many of the issues raised about it.

In considering how fuzzy categories of (nongeometric) shapes are formed, Gentner and Medina (1998) argued that there is a developmental/experiential shift from perceptual similarity (e.g., recognizing a mobile that is a very close perceptual match to one seen before) to relational similarity (e.g., recognizing the similarity of a mitten covering a hand and a shoe covering a foot). These researchers argued that the perception of similarity is accomplished by a process of "alignment or structure-mapping," and that performing such mappings increases the likelihood of establishing relational rather than purely visual similarities.[19] They hypothesized that giving a common name to shapes in a class can encourage individuals to make comparisons that activate structure mappings that "promote the discovery of deeper commonalities" (p. 276). Indeed, structure mapping is a central learning mechanism that enables individuals to notice and store abstract relational properties, derive abstract knowledge from instances, and extend that knowledge to new cases. This happens as the relations examined in structure mappings are abstracted to become objects in and of themselves.

Because Gentner and Medina's (1998) theory focuses on *structure* mappings, it has the potential to shed light on the mechanisms by which students formulate properties of geometric shapes when formal definitions are not given. The developmental/experiential shift from perceptual to relational similarity seems particularly relevant in progressing from van Hiele Level 1 to 2 because Level 2 reasoning focuses on relationships between parts of shapes. Gentner and Medina's theory suggests that giving common names to like geometric shapes should cause students to start performing structure mappings that can enable them to see, at least informally, spatial relations common to the shapes. But we do not know how spontaneous these are. Perhaps for most students, the mappings occur only after explicit prompts such as, "How are these shapes alike?" Additional research is needed to examine the potential of the structure-mapping theory in explaining the transition to the early phases of van Hiele Level 2.

Vygotsky's (1986) work sheds further light on the relationship between fuzzy and formal concepts. He argued that children become conscious of their spontaneous (i.e., fuzzy) concepts relatively later, with the ability to define them in words appearing long after they have "acquired" the concepts. In contrast, the development of scientific (formal) concepts usually begins with their verbal definition. However,

> though scientific and spontaneous concepts develop in reverse directions, the two processes are closely connected. The development of a spontaneous concept must have reached a certain level for the child to be able to absorb a related scientific concept. . . . In working its slow way upward, an everyday concept clears a path for the scientific concept and its downward development. Scientific concepts, in turn, supply structures for the upward development of the child's spontaneous concepts toward consciousness and deliberate use. Scientific concepts grow downward through spontaneous concepts; spontaneous concepts grow upward through scientific concepts. (p. 194)

Thus, applying this theory to geometric concepts, we see that fuzzy, recognition-based conceptualiza-

[19] A structure mapping is isomorphic-like; it is one-to-one and if two elements are linked in one representation, their images are linked in the corresponding representation.

tions of geometric shapes develop first as the shapes are perceived and named in school and out. For common shapes, and most students, such early experiences "naturally" lead to van Hiele Level 1 reasoning. But getting to Level 2, in which shapes are reasoned about in terms of formal geometric concepts, requires a great deal of work—work that, to be meaningful to students, must connect to the fuzzy concepts developed in Level 1.

Finally, at the formal end of the spectrum on categorization, is the issue of hierarchical classification. According to Jones (2000), following De Villiers, classifications can be hierarchical or partitional. Hierarchical classifications use *inclusive* definitions such as specifying that trapezoid is a quadrilateral with *at least* one pair of sides parallel—which means that a parallelogram is a special type of trapezoid. Partitional classifications use *exclusive* definitions such as specifying that a trapezoid is a quadrilateral with *only* one pair of sides parallel—which *excludes* parallelograms as trapezoids. In general, in mathematics, inclusive definitions (and thus hierarchical classifications) are preferred, although exclusive definitions and partitional classifications are certainly not mathematically incorrect. However, a number of studies show that many students have great difficulty with the hierarchical classification of quadrilaterals (de Villiers, 1994; Jones, 2000). Although research on the van Hiele levels has shed light on this issue, more research is needed that investigates why this difficulty occurs and how to overcome it.

STUDIES OF LEARNING AND TEACHING GEOMETRY

Computer Environments for Learning Geometry[20]

Early research on the use of computer environments for learning geometry focused primarily on investigating predictions about student learning made by environment designers. More recent research, informed by initial studies and extensive experience with the environments, has investigated more closely the learning processes that occur within the environments and how the environments affect and shape student learning.

I will discuss three major types of computer environments for investigating plane shapes (see Clements and Battista, 1992 for a discussion of other types): Logo-based, Geometric Supposers, and Dynamic Geometry Environments (DGEs).[21] Two essential features have been noted for such computer environments. First, they require students to provide *explicit* specifications for geometric shapes—using menu selections in Supposers and DGEs, and command lists in Logo. Thus, unlike using paper and pencil, in these computer environments, students cannot make drawings without some level of *conceptual and representational explicitness*. Researchers have argued that this explicitness promotes and supports reflection on and abstraction of geometric concepts, and movement toward van Hiele Level 2. For instance, Clements and Battista (1992), following Papert, claimed that,

> Writing a sequence of Logo commands, or a procedure, to draw a rectangle . . . obliges the student to externalize intuitive expectations. When the intuition is translated into a program it becomes more obtrusive and more accessible to reflection. . . . Students must analyze the spatial aspects of the rectangle and reflect on how they can build it from parts. (p. 450)

Similarly, Laborde (2001) claimed that when constructing a square with a given side in a DGE,

> With paper and pencil the . . . task is controlled by perception. . . . The same task in Cabri . . . cannot be obtained by eye . . . but uses a circle as a tool for transferring a given distance. The task in Cabri requires more mathematical knowledge about the properties of a square and the characteristic property of a circle. (p. 294)

In the case of DGE research, the amount of explicitness used by students has been, indirectly, a frequent focus of discussion. For instance, in *The Geometer's Sketchpad* (GSP) or Cabri, one can "draw" a rectangle by creating four line segments in a rectangular configuration. Little conceptual explicitness is required, and the drawing does not remain rectangular when its vertices are dragged. Alternately, one can use DGE commands to "construct" a rectangle that preserves its rectangularity with dragging. In this case, it can be argued that each step in the construction requires some conceptual explicitness. The question is, how much? For instance, after creating a segment,

[20] Even though research shows that computer environments have great promise for improving geometry instruction, computer use among mathematics teachers is low. Becker (2000) reported that only 1 of 9 secondary mathematics teachers said that a typical student in their classes used computers on more than 20 occasions during a 30-week period.

[21] Most of the relevant research has been conducted on these three environments. Specialized restricted applets and general drawing programs are not discussed.

students might create a segment perpendicular to it using the perpendicular construction. So they must know, at a minimum, that a specific spatial characteristic of a rectangle is created by the perpendicular construction. Then they must use either perpendicularity or parallelism to construct the remainder of the figure. Because only three right angles are explicitly constructed, how do students know that the last angle must be a right angle? Do they deduce it logically, conduct an informal componential analysis (see the previous description of the Borrow/Battista elaboration of van Hiele Level 3), or merely make a visual assumption? Other properties, such as parallelism or congruence of opposite sides, are not explicitly constructed, so may not be recognized in the construction (but might be visually induced).

Understanding the nature of the conceptual explicitness that occurs in students' constructions of draggable DGE drawings requires much additional research. For instance, are students' conceptualizations of their constructions manifestations of their understanding of abstract mathematical concepts, or are they proceduralized methods for creating visual configurations? That is, when students construct a perpendicular in a DGE, is the perpendicular simply a particular kind of visual configuration, or is it an instantiation of a geometric concept (i.e., two lines intersecting at 90° angles)? Another difficulty with DGEs is that gaining access to commands used in a completed construction is not always straightforward. For instance, to determine how a draggable rectangle was constructed in GSP, one can check all properties of various objects on the rectangle (but the sequence of constructions is not readily available), or one can create a tool for the construction and examine its script. However, if students do not sufficiently understand the syntax of GSP—and learning syntax can be a major barrier for students in DGEs (Laborde, 1992)—students may not be able to use these tools to keep track of what they do, which considerably lessens conceptual explicitness. Similarly, with Logo, although each movement of the turtle requires a command, what the commands mean to students should not be taken for granted. As an example, a student might correctly construct a rectangle, for instance, using trial and error and several turn commands to make each right angle—so there are no explicit commands to make right angles. Without appropriate instructional intervention, the student might not construe the set of RIGHT commands at a vertex as a *set* for which the sum of the inputs determines the size of vertex angle; instead, the commands are taken as separate acts used to visually adjust the turtle's heading. Or, the shape created may not be a rectangle at all, merely a visual approximation of a rectangle; so even if the set of commands is considered, the properties of the rectangle are not accurately embedded in the commands. Thus, the conceptual explicitness that occurs in computer environments for geometry learning depends on a complex interaction between the commands given in constructing figures, the availability of those commands to student inspection, the reasoning of the student, and instruction.

The second feature attributed to computer environments for facilitating geometry learning is the "repeatability" of drawings (Laborde, 1992). "In this kind of software a necessary condition for a construction to be correct is that it produce several (or an infinity of) drawings which preserve the intended properties when variable elements of the figure are modified" (Laborde, 1992, p. 129). Repeatability is achieved very differently in the different programs. In Logo, repeatability requires the use of procedures, and often variables, adding more syntax requirements. In Supposers, repeatability is discrete and constrained. In a DGE, repeatability is continuous and dynamic. Again, however, a major issue for research is not only what kinds of repeatability engender what kinds of learning, but how students use and, more importantly, conceptualize repeatability. Do students really see draggable constructions in DGEs as *representing* multiple elements in a given class of shapes? Or, are such constructions simply another kind of geometric object that can be studied?

This latter point is critical and returns the focus to diagrams. In paper-and-pencil approaches to studying geometry, there are two major types of entities. First, there are diagrams, mere perceptual objects. Second, there are shape conceptualizations—conscious meanings activated for a class of shapes. DGEs introduce a whole new object—draggable drawings, which researchers often view as representations of shape conceptualizations: "A [draggable] figure captures the relationships between the objects in such a way that the figure [i.e., considered as a formal concept] is invariant when . . . the construction is dragged (in other words, that it passes the drag test)" (Jones, 2000, p. 58).

Furthermore, because draggable drawings are constructed to possess specific geometric properties (and consequently possess additional properties necessitated by the given properties), researchers often see all the properties as "included" in the drawings. However, students may not see any of the properties embedded in draggable drawings and may not even think that the drawings have specific properties. Novice students usually reason about draggable drawings as geometric entities in their own right, not explicitly as representations. In analyzing draggable drawings, students first notice movement *constraints*, then later, they might conceptualize these constraints in terms of regularities or invariants, and finally, and often only with great effort, these constraints are conceptualized as formal geometric properties. The question of how students interpret DGE drawings, both draggable and not, is not only critical, but in need of much research attention.

Geometric Supposers

Geometric Supposer construction programs allow students to choose a primitive shape, such as a type of triangle or quadrilateral, and perform measurement operations and geometric constructions on it. The programs record the sequence of constructions and can automatically perform it again on other shapes. The focus of *Supposer* programs, therefore, is to facilitate making, testing, and justifying conjectures, based on examples. Most of the research on *Supposer* programs has taken place in the context of guided-inquiry teaching (Wilson, 1993).

Research generally, but not always, shows that use of *Supposer* programs can enhance geometry learning (Clements & Battista, 1992; Yerushalmy, 1993; Yerushalmy & Chazan, 1993). *Supposer* interacting students were better able to identify non-stereotypical examples of shapes, and their learning often extended beyond standard geometry content, for example, to formulating definitions, making conjectures and arguments, posing and solving significant problems, and devising original proofs. After *Supposer* experience, students more frequently used diagrams in their thinking, treated diagrams flexibly, were less often confused by diagrams in nonstandard orientation, used many diagrams to describe classes of figures, often reasoned about a single diagram as a model for a class of shapes, and understood that *Supposer* constructions can include characteristics that are not shared by all members of the class of figures under consideration. *Supposer* activities can also engender movement away from considering measurement evidence as proof (Clements & Battista, 1992). Indeed, *Supposer* work *can* (but does not have to—see below) help students appreciate the need for proof; unlike for textbook theorems, many students consider *Supposer*-generated conjectures as statements that need to be proved before they can be accepted as true.

Yerushalmy (1993) claimed that three major processes are involved in geometric generalization: "formation of samples of examples to serve as a data base for conjectures, manipulations of the samples, and analysis of ideas in order to form more general ideas" (p. 82). She reported that use of the *Supposer* focused many students' attention on the nature of the set of examples—what information was needed, the variability of the information, and the need to organize information gathering. Students learned to appreciate the value of generating many examples and of formulating a data-collection strategy that would prevent them from getting lost. Students learned "to use extreme cases, negative examples, and non-stereotypic evidence to back up their conjectures" (p. 82). However, Yerushalmy found that some *Supposer* students did very little genuinely independent work, generating many examples, but with little sense of direction—often asking the teacher what to look for in the examples. Students also struggled with knowing the types of examples they should examine (Yerushalmy, 1993), which suggests that they had naïve ideas about general mathematical statements and their validity.

One can pose many questions about the student learning reported by Yerushalmy. For instance, how deliberate is students' use of examples? What is the nature of the reasoning that guides students' choices of examples? How much of the generation of examples is student initiated, and how much teacher generated? If a lot of it is teacher generated, does the generation become interiorized by students, and, if so, how? What do students think as they generate and manipulate examples? What are the elements of students' analyses of examples—how do conceptualizations of relevant ideas or beliefs affect these analyses?

Chazan (1993b) investigated teachers' observations that some students using the *Supposer* did not appreciate that measurement evidence could not prove a proposition true for all cases within an infinite domain. Despite the fact that the instructional unit used in the study argued that measurement of examples is not a valid method for verifying the truth of geometrical statements, instructional "activities in favor of deductive proof were not successful with a sizable portion of the students in the study" (p. 109). Also, by the end of the *Supposer* instructional treatment, although "fewer students considered [empirical] evidence to be proof . . . more students were skeptical about the . . . applicability of deductive proofs" (p. 109). However, Chazan argued that the latter finding may not be as negative as it seems. For instance, such skepticism may be warranted (a) because students realized that they were novices at constructing valid deductive proofs, or (b) because, following Lakatos, being skeptical about proof is actually a mathematically healthy behavior. But there is an even deeper issue here. Almost no geometry curricula actually support students' invention of the deductive method (see Borrow, 2000 for an exception).

That is, even students who are given the opportunity to see the shortcomings of the empirical approach are generally *shown* a "better way" rather than allowed to invent for themselves essential elements of the deductive method. Because students are merely handed the deductive method, not given the opportunity to construct a personal need and meaning for it, they are naturally skeptical of it. Nevertheless, the issue of whether *Supposer* use helps or hinders students' movement away from justifying conjectures with empirical evidence toward the use of deductive proof remains unresolved.

Logo

Instructional Effectiveness

In general, research supports the use of Logo in geometry instruction (Clements & Battista, 1992, 2001; McCoy, 1996). Students can understand and become competent with representing and reflecting on geometry problems using appropriate Logo environments and instructional activities. For instance, in the *Logo Geometry (LG) Project*, 28 teachers (980 K–6 students) used LG, which provided instructional activities and Logo tools to help students progress through the early van Hiele levels (Battista & Clements, 1991; Clements & Battista, 2001). Students in LG scored significantly higher than control students on a paper-and-pencil test that was not directly connected to LG, making about double the gains of control students. Overall, LG helped students progress from van Hiele Level 1, visual thinking, toward Level 2, descriptive/analytic thinking. Observations of LG students also showed that they moved from (a) non-analytical and/or authoritarian-based notions of knowledge, to (b) more autonomous, empirical bases for establishing truth, and finally to (c) knowledge as reasoning about mathematics as a logical system.

Several general comments should be made about this study. First, students spent 4–6 weeks doing LG. This extensive period of time counterbalanced the "investment-time difficulty" required for learning geometry with Logo. That is, to learn geometry with Logo, students must learn the Logo *language*, which can subtract from the time used to learn geometry per se. However, when such a large amount of time is spent in learning geometry, as with LG, the amount of time learning the Logo language is proportionally smaller compared to the amount of time spent learning geometry. Second, LG not only provided a carefully developed sequence of activities, it incorporated into the Logo language numerous supports that lessened the "learning Logo" load on students. Third, participating teachers were given extensive training on the use of the LG curriculum.

Not all research on Logo has been positive, however. Some studies show no significant differences between Logo and control groups or limited transfer of learning (Clements & Battista, 1992, 2001; McCoy, 1996). For example, ninth-grade Logo students did not differ significantly from control students in high school geometry (Olive, 1991).

Logo as Representational Medium

If one of the major reasons for the instructional efficacy of Logo lies in its representational power (Clements & Battista, 1992), it is important to analyze how this representation facility works for students. Is the representational efficacy of Logo due to its capability to support reflection, to its encouragement of measurement-based analysis and conceptualization, or to the fact that multiple representations enrich conceptualizations by enlarging the web of meanings into which an idea fits? Critical in answering this question is how students command the turtle to make shapes, and, more importantly, how they conceptualize these commands. The following episodes illustrate students using Logo commands to support (a) analytic, measure-based reasoning (Episode 1), (b) visual/spatial and trial-and-error reasoning (Episode 2), and (c) the transition from visual-spatial to analytic reasoning (Episode 3). Some researchers claim that one reason that students sometimes do not learn optimally in Logo environments is that students often reason strictly visually despite the fact that Logo allows more analytic and measure-based reasoning (e.g., McCoy, 1996).

Episode 1. Two fifth graders were drawing a tilted square using equal forward commands and 90° turns. When asked if their figure was a square, the students replied, "Yes, a sideways square. . . . It has equal edges and equal turns."

Episode 2. Second grader M was examining her attempt to draw a tilted square in a version of Logo that allowed students to "undo" commands (Clements & Battista, 2001). She used a trial-and-error approach, undoing (signified by E for erase) or inserting additional commands when part of the figure looked incorrect. Her commands were: L 35 F 30 R 45 F 30 E E

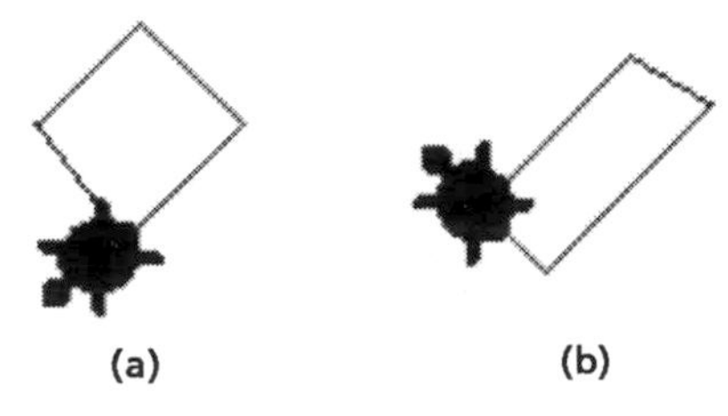

Figure 19.5 M's attempts at making (a) a tilted square and (b) a tilted rectangle in Logo.

R 45 R 35 R 35 E F 30 L 35 R 90 R 35 F 30 L 35 R 90 R 35 F 30 F 5 F 5. Because M did not use all 90° turns, her figure was only a crude approximation of a square (see Figure 19.5a). But she concluded that it was a square, reasoning, "It's in a tilt. But it's a square because if you turned it this way it would be a square."

Episode 3. M was then asked to make a tilted rectangle (previously unreported raw data, Battista & Clements, 1987). She entered the following commands:

M: R 45 F 50 R 45 R 10 R 10 R 10 F 20 L 90 Whoa! R 180 F 50 Whoa! It's supposed to go like this and it goes like this. Because I turned the turtle too far. E E L 25 L 160 F 50 E L 20 R 40 F 50 Phew! R 90 F 20 END [getting the path shown in Figure 19.5b].

Int: Is there anything special about the turns in a rectangle?

M: They're all 90.

Int: But you didn't always use 90s.

M: Because I went like this and like this (motions) and then I didn't turn and I wanted to go like this. And I hadn't turned 90 and I had to make a couple of guesses and I finally got it right.

Int: Could you teach it to make another tilted rectangle?

M: R 45 F 60 R 90 F 10 L 90 R 180 F 60 R 90 F 10 END

In Episode 2 and the first part of Episode 3, M seemed to take a strictly visual, trial-and-error approach. Her thinking was far different than the fifth graders in Episode 1, who used specific Logo commands to ensure their figure had specific properties. But at the end of Episode 3, M seemed to come to a realization that the turtle needed to make 90° turns to make a rectangle. M's work brings up a critical issue in the use of Logo, How exactly does Logo move students to higher levels of reasoning?

One possibility is that Logo, by virtue of requiring a more explicit, analytic, and symbolic representation of geometric ideas, supports and encourages reflection and reformulation of those ideas. Clements and Battista (2001) have argued that programming the Logo turtle to make shapes encourages and facilitates students' reflection on the composition of the shapes. Students analyze the spatial components of shapes and explicitly represent how they can be built from these components, including explicitly conceptualizing how the components are spatially related. This leads to recognition of the shape's properties.

In Episode 2, however, there was no evidence that giving Logo commands enabled or encouraged M to understand the nature of the components of a square or how those components were spatially related. There was no evidence that M consolidated the commands into groups that represented the components of the shape—she merely moved the turtle bit-by-bit, seemingly guided only by visual processes. In fact, even though M had previously stated that all the sides in a square are equal, she explicitly violated this notion with her last set of forward commands. M used the same type of visual reasoning in the first part of Episode 3. However, during the second part of Episode 3, a change occurred. The interviewer's questioning seemed to promote reflection that changed M's conceptualization of how to make a rectangle—she explicitly used 90° turns. What caused this change? Did M explicitly reflect on the movement commands she used in making her first rectangle and draw some specific conclusions from them—that is, was her realization based on unverbalized, even unconceptualized, abstractions from those experiences? This is possible, but, given the number of commands, it seems unlikely that she conducted a command-by-command analysis. Alternately, did M change her conceptualization by interrelating her current experiences in giving Logo commands for a tilted rectangle to the knowledge she exhibited when she stated that all the turns in a rectangle are 90? And, what was the basis for her statement that rectangles have 90° turns? Was it her experience in making rectangles that had vertical and horizontal sides—most students in the class had used 90 for these turns? Or, was it the recollection of statements in class, by students and the teacher, that rectangles have 90° turns? What caused and enabled M to connect the concept of 90° turn to a specific type of spatial configuration—a right angle? Whatever it was, it resulted in an interiorized abstraction of this relationship, an abstraction that allowed the relationship to be applied in this new situation of giving commands to make the final rectangle.

When investigating processes that enable discoveries like those of M, it is important to understand how students use and understand sequences of Logo commands. For instance, M twice used the sequence L 90 R 180 to accomplish a R 90 turn (because her initial left turn was the wrong way). Did M explicitly understand the equivalence between the L 90 R 180 sequence and the R 90 command? That is, did she see this sequence as a substitute for R 90, or was the second command in the sequence merely a visual adjustment, with the sequence not being conceptualized for its net result? On the one hand, because this substitution was a common occurrence among students, it is tempting to think that they saw the equivalence in some way. On the other hand, some students give a sequence of commands such as F 50 F 10 F 10 merely as a way to move the turtle, seemingly unable to integrate the sequence into a length composite equivalent to F 70 (Clements, Battista, Sarama, Swaminathan, & McMillen, 1997). Other students, although able to combine commands when directed to do so, often have difficulty combining commands in more complex situations. For instance, late in an instructional sequence that explicitly attended to encouraging students to combine commands (Clements et al., 1997), a teacher asked if the following commands produced a rectangle:[22] F 40 R 90 F 65 R 90 F 20 F 20 R 90 F 50 F 15 R 90. Monica and Nina agreed that it would be a rectangle until they attempted to draw the resulting path on paper—Nina drew a rectangle; Monica did not. Even when Nina explained to Monica that 20 and 20 sum to 40, which matches the other 40, Monica needed to try the path on the computer before she accepted it. She was not convinced by strictly abstract reasoning.

The next episode shows how the Logo medium can help students develop analytic, measurement-based conceptualizations of shapes.

RE: The rectangle is like a square, except that squares aren't long. But on rectangles, they are long.

Int: What does a shape need to be a rectangle?

RE: All of the sides aren't equal. These two [opposite sides] and these two [other opposite] sides have to be equal.

Int: How about 10 on two sides and 9 on the other two? Would that make a square?

RE: Kind of like a rectangle.

Int: Would it be a square too?

RE: [Shaking head negatively] It's not a square. 'Cause if you make a square, you wouldn't go 10 up, then you turn and it would be 9 this way, and turn and 10 this way. That's not a square.

RE's first comment indicated visual reasoning—rectangles are long. On further questioning, RE talked about opposite, but not all, sides being equal—a property-based response. But we cannot attribute this increase in level of reasoning directly to the Logo context—even students not using Logo make such changes in reasoning in response to appropriate questioning. Finally, however, RE reasoned analytically about side lengths using the action-based command context of Logo. The Logo environment seemed to promote, make accessible, and appropriately support measure-based reasoning for this student.

A final episode further suggests how students' work with Logo can move them toward property-based reasoning about shapes. Prior to the episode, JN had successfully used a variable-input rectangle procedure to draw a tilted rectangle. He was now reflecting on his unsuccessful attempt to use this procedure to make a nonrectangular parallelogram.

JN: Maybe if you used different inputs. [He types in a new initial turn, then stares at the parallelogram on the activity sheet.] No, you can't. Because the lines are slanted. . . .

Tchr: Yes, but this one's slanted [indicating the tilted rectangle that JN had successfully drawn with the Logo procedure].

JN: This one [the parallelogram]—the thing's slanted. This thing [the rectangle] ain't slanted. It looks slanted, but if you put it back [shows how to turn it so that its sides are vertical and horizontal], it wouldn't be slanted. Anyway you move this [the parallelogram], it wouldn't be a rectangle. So, there's no way.

JN was beginning to develop a conceptualization of what is formally captured by the property-based concept that the adjacent sides of a rectangle are perpendicular. In so doing, his reasoning was progressing from the visual reasoning of van Hiele Level 1 to the descriptive-analytic reasoning of Level 2. However, at this time, JN did not possess an interiorized formal concept of perpendicularity (or right angles) that he could apply to the situation.

Zooming in on Conceptual Change

In my final example of Logo research, I summarize a study of Hoyles and Healy (1997), who investigated how one 12-year-old student, Emily, changed her conceptualizations of geometric reflections while working in a Logo-based microworld. On a pretest asking students to draw reflections of figures, two of Emily's answers were correct, three were incorrect. However, in the three incorrect drawings, the reflected images were congruent to their preimages, which is consistent with Emily's description of reflections, "the line . . . has got something on one side and it would be exactly the same on the other." For instance, when drawing the reflection of a vertical line segment through an oblique mirror line, after Emily correctly drew one endpoint of the image, she drew the rest of

[22] For consistency, forward commands are always denoted by F and right commands by R. The actual commands given by students varied by microworld.

the image parallel to the preimage. Hoyles and Healy reported that when Emily did reflection problems, she used a ruler placed orthogonally to the mirror line to measure equal distances between some, but not all, corresponding points. Thus, Emily's initial conceptualization of reflection seemed to have two components: First, a reflection consists of a mirror line along with congruent images and preimages; second, some points on the preimages correspond to points on the image via perpendicularity and equidistance (but this idea was vague and inconsistent).

In the first computer task, a blue turtle was supposed to draw the preimage and a red turtle the reflection image about a vertical mirror line. The two turtles were shown with the same position and heading along the mirror. The commands (but no figure) for the blue turtle were shown; the problem for the students was to give the appropriate commands for the red turtle. The initial idea of Emily and her partner Cheryl was to reverse both movement (F, B) and turn commands (R, L), but they quickly decided, without trying their idea, that they only needed to interchange right and left turn commands, which they did, completing the task correctly.

Because the girls' notions of reflections seemed to be tightly connected to Logo commands, Hoyles and Healy next investigated what happened when Logo commands for figures were absent. Students were given two reflection-congruent turtle paths (with the turtles showing), but no Logo commands, and were asked to find the mirror line. Cheryl simultaneously traced the path of both turtles, stopping where she thought they would meet on the mirror line (so she was using a mental model of two linked and simultaneously moving turtles). After using an experimenter-provided tool for getting the turtles to meet on the mirror line, the girls turned the blue turtle through a series of L 10s until it looked as if it were heading along the mirror line. They then turned the red turtle the same angle to the right, and when they saw that it was facing exactly the same direction as the blue, they were convinced that they had found the direction of the mirror line.

Several questions about these first two tasks arise. Precisely how was the girls' left/right reversal strategy connected to or imbedded in their conceptions of reflections? For instance, had the girls previously abstracted that reflection reverses orientation? Or, was turn reversal an element of Logo procedural knowledge that enabled the girls to produce congruent shapes on both sides of a mirror line? How did they discover this knowledge? How exactly did the girls conceptualize the mirror line? From the second task, we might conjecture that they thought that it was "in the middle" of the two congruent figures and that somehow the turtles drawing the figures made equal angles with the line when they met it.

The next task required students to construct the reflection of a flag when both the flag and mirror line were oblique and no Logo commands were provided. (Colored turtles were used to mark flag vertices so students could find distances between vertices by asking Logo to give distances between specified turtles.) In this task, the girls changed their strategy, trying to set up a correspondence between selected preimage and image points. As shown in Figure 19.6a, Emily projected the endpoints of the flag vertically downward, as if the mirror line were horizontal, making each projected point and its preimage equidistant from the mirror line, in the vertical direction. However, we do not know how the girls arrived at this conceptualization. Is it another manifestation of the vague correspondence idea that Emily used on the pretest?

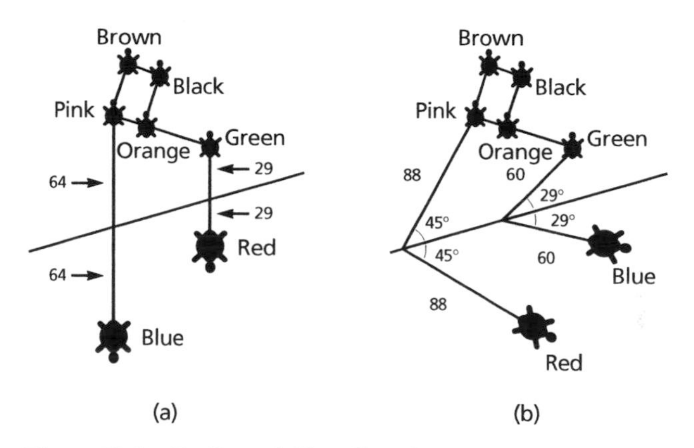

Figure 19.6 Emily and Cheryl's reflection attempts in Logo (Hoyles & Healy, 1997).

When the girls discovered that the distance between the red and blue turtles was unequal to that between the green and pink turtles, they ignored the blue turtle and gave commands to the red turtle. The resulting image flag was congruent to the preimage flag, but when the girls recognized that the image flag's orientation was incorrect, they devised a new strategy for finding the endpoints of the flagpole. They used Logo commands to determine the angle and distance of a turtle on the mirror line to the pink turtle (the preimage turtle); they then turned the turtle the opposite direction from the mirror and went forward the given distance to the image red turtle (see Figure 19.6b). The authors observed,

> Across all the [student] pairs and the many different methods observed, our analysis showed how the microworld offered students a *way of talking about* and *operationalising* the angle properties of reflection, by

> thinking about turtles turning from a heading along the mirror through the same amount in opposite directions. (Hoyles & Healy, 1997, p. 44)

Finally, to investigate how situated the girls' knowledge was in the Logo context, Emily was asked to reflect a triangle in an oblique mirror using only a ruler and a protractor. She was completely puzzled until she drew a turtle on the reflection line to use as a reference point (see Figure 19.7a). She measured the angle and distance of vertex A to the mirror line, then copied this angle and distance to the other side of the line (locating point A'). As she considered the placement of the image of triangle vertex C, as shown in Figure 19.7a, her interiorization of what she did for A enabled her to progress even further, to a generalizing abstraction.

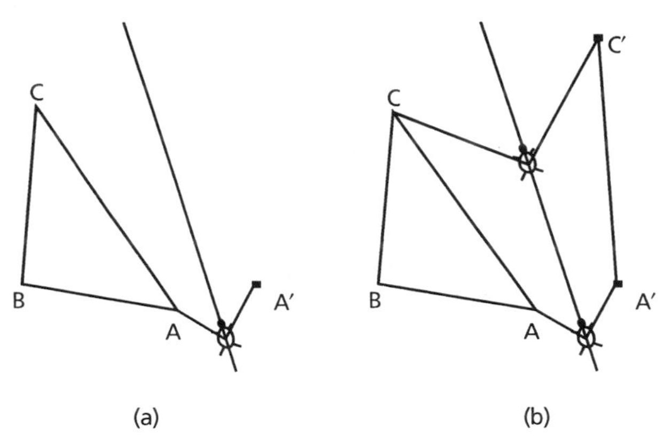

Figure 19.7 Emily's reflection using ruler and protractor and the turtle (Hoyles & Healy, 1997, p. 52).

> Emily: That's one [A], now I suppose I have to do the same (indicates C). Can I (she places protractor on turtle) . . . does the turtle have to be there? It's hard to measure to there so, could I . . . hold on . . . I think I've got it. It doesn't matter does it, you can be anywhere as long as the angle and the distance is the same, it doesn't matter where you actually measure from . . . so say there (draws another turtle), the angle is . . . (measures angle and distance to C) so that like (constructs angle and distance to position C′) [see Figure 19.7b]. (Hoyles & Healy, 1997, p. 52)

Hoyles and Healy noted that "This last vignette illustrates the extent to which, for Emily, the tools used in her previous explorations had become part-and-parcel of her sense making activities" (1997, p. 53). At first, this non-Logo task presented a major obstacle for Emily because it removed the Logo context in which much of her relevant conceptual knowledge resided. Emily could not think of the problem without turtles for reference points. But when she drew a turtle, two things happened. First, placing the turtle enabled her to activate her previous turtle-based scheme for dealing with the situation. She was, through the process of interiorization, able to disembed her abstraction from its original Logo context and apply it in this new, paper-and-pencil situation. This interiorized notion, together with her reflection on the arbitrary placement of the turtle, seemed to enable her to develop an explicit notion of the placement arbitrariness—creating a more general and abstract scheme (and her *aha* moment). However, her new, more general conceptualization still seemed tied to the Logo context—she drew a turtle for C. And the effect of the discovery on Emily's orthogonal-correspondence conception of symmetry is unknown. Was it abandoned, or had it been transformed?

Although the data presented by the authors provide an extremely important view of how Emily's conceptualizations evolved, several questions remain unanswered. For instance, in the initial episodes, both Emily and Cheryl seemed to possess a complex interconnected web of knowledge consisting of holistic visual knowledge of reflections as congruent halves on either side of the mirror line, a vague notion of orthogonal corresponding parts in preimage and image, and developing knowledge of producing reflection paths via Logo commands. By the final episode, and as a result of carefully chosen instructional activities, Emily's conceptualization of reflections seemed to change from a turtle-centric path perspective focusing on Logo movement commands to an exocentric view in which the mirror line served as a reference point for measuring angles and distances. This change may have been mediated by her initial vague notion of finding corresponding parts in reflections. Or, it may have been encouraged by the constant demand to translate between the turtle perspective and an exocentric view of reflections (looking "down" on and comparing figures). Somehow Emily's initial vague notion of corresponding parts became connected to and clarified by the concept of "same angle/same distance."

Emily significantly increased the level of sophistication in her thinking about reflections. Her reasoning progressed from visual to descriptive-analytic; from thinking vaguely and visually, to a synthesis of visual, analytic, and measurement-based reasoning, much of it couched in the language of Logo. The Logo environment, including the instructional activities, seemed to provide Emily with tools for making her conceptions of reflections more analytic, precise, and general. Consistent with what Clements and Battista (2001) found in their Logo studies, this microworld helped Emily and her partner to become more cognizant of and to reshape their thinking in terms of precise, measurement-based properties.

However, as is often asked of geometry learning in Logo, we might ask how Emily's final conceptualization of reflections compares to traditional mathematical descriptions? That is, Emily constructed a nontraditional conceptualization of reflections—that preimage and image points have equal angles and distances to the mirror line. Although Emily's conceptualization is mathematically correct, it is not as conceptually illuminating as the traditional concept because the traditional concept can more easily be related to the intuitively appealing conception of reflection produced by motion. Indeed, thinking of image and preimage points as endpoints of segments bisected by, and perpendicular to, the mirror line, can be directly related to a (nonrigid) motion, with these segments being the paths of preimage to image points. It would have been interesting to see if and how Emily's final conceptualization could have been evolved into the traditional concept. (For a discussion of students' development of comparable reflection concepts in dynamic geometry, see Hollebrands, 2003.)

Finally, let us return to the question, What characteristics of the Logo environment are essential for encouraging and supporting students' learning? As illustrated in the other studies discussed in this chapter, students in the Hoyles and Healy study seemed motivated to work on problems in the Logo context. But beyond that, was it the Logo-command language per se that supported students' learning, or was it one of several other characteristics of Logo that provided such support? For instance, Logo provides quick, accurate, and interesting visual depictions of students' visual-geometric conceptualizations—a type of uninterpreted feedback. It provides and encourages, through its use of measurement specifications in movement and turn commands, an accessible way to analyze shapes in terms of their components' measures—an important step in moving from pure visual to analytic thought. Thus, it might be conjectured that it is not Logo code per se that supports the development of students' geometric understanding but its provision of (a) appropriate motivation; (b) accurate, quick, and uninterpreted visual feedback; and (c) easy-to-use and precise measurement specifications. These same components also appear in dynamic geometry environments.

Dynamic Geometry Environments

Currently, DGEs seem to be one of the most popular types of software used by mathematics teachers (Becker, 2000) and investigated by researchers. On the one hand, it has been claimed that DGEs "provide a revolutionary means for developing geometrical understanding. . . . This software seems to make the exploration of geometrical configurations and the identification of meaningful conjectures more accessible to pupils" (Mariotti, 2001, p. 257). On the other hand, skeptics worry that DGEs weaken the role of proof in high school geometry (Mariotti, 2001). The following discussion deals with all aspects of students' learning in DGEs. As in the section on Logo, the discussion will begin with research on students' work at the elementary school level, then move to the middle school and secondary level.

DGE at the Elementary Level

Because DGE software such as Cabri and GSP were originally designed for the secondary level, use of this software in elementary classrooms requires accommodations. For instance, the *Shape Makers* computer microworld is a special add-on to GSP that provides students with geometric shape-making objects (Battista, 1998a). An example is the Parallelogram Maker that can be used to make any desired parallelogram that fits on the computer screen, no matter what its shape, size, or orientation—but only parallelograms. The appearance of a Parallelogram Maker is changed by dragging its vertices with the mouse. In the *Shape Makers* environment, when measurement is introduced, it is done so by displaying relevant measures on the screen automatically; so students do not have to use GSP measurement tools directly. The major focus of *Shape Makers* units is for students to move from van Hiele Levels 0 and 1 to Levels 2 and 3. To illustrate the type of geometric reasoning and learning exhibited by elementary students in DGEs, I describe several episodes with fifth graders' working with the *Shape Makers* microworld (Battista, 1998a, 2001b, in press b) in their regular classrooms.

Episode 1. Three students were investigating the Square Maker at the beginning of their Shape Maker work: MT, "I think maybe you could have made a rectangle." JD, "No; because when you change one side, they all change." ER, "All the sides are equal."

MT, JD, and ER abstracted different things from their Square Maker manipulations. MT noticed the visual similarity between squares and rectangles, causing him to conjecture that the Square Maker could make a rectangle. JD abstracted a holistic movement regularity—when one side changes length, all sides change (thus, he could not get the sides to be different lengths, which he thought was necessary for a rectangle). Only ER conceptualized the movement regularity with complete precision by expressing it in terms of a formal geometric property.

From the van Hiele perspective, one might say that MT's reasoning on this task was at Level 0, JD's at Level 1, and ER's at Level 2. From the structuring perspective, MT and JD had constructed spatial structurings for the Square Maker, with JD's more sophisticated than MT's. ER, in contrast, had constructed a property-based geometric structuring for the Square Maker by applying a previously interiorized concept of equal sides to a spatial structuring similar to that described by JD. Finally, the abstractions made by MT and JD seemed completely situated in the DGE microworld, whereas that made by ER was not.

Although the above episode illustrates that different levels of reasoning that students can achieve on a task in the Shape Maker environment, it does not show how a student progresses from one level to the next. The following episodes provide insight into this process.

Episode 2. NL was using the seven quadrilateral Shape Makers to make the design shown in Figure 19.8 while a researcher was observing and asking questions.

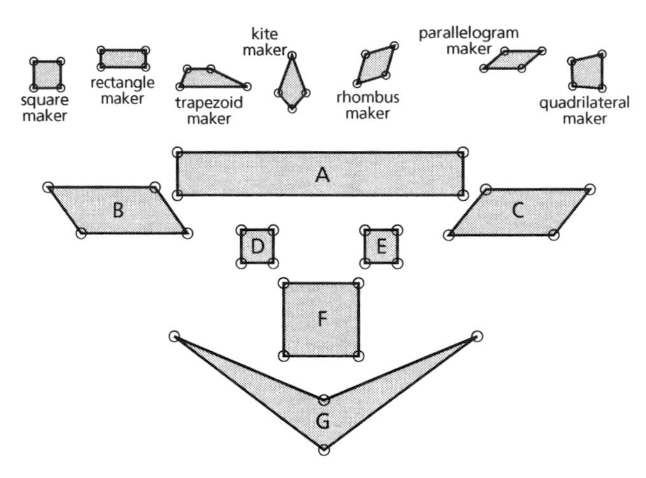

Figure 19.8 Make the design consisting of Shapes A–G with the 7 Shape Makers (Battista, 1998a).

NL: The Rhombus Maker on [Shape] B. It doesn't work. I think I might have to change the Rhombus Maker to [Shape] C.

Res: Why C?

NL: The Rhombus Maker is like leaning to the right. On B, the shape is leaning to the left. I couldn't get the Rhombus Maker to lean to the left, and C leans to the right so I'm going to try it. [After her initial attempts to get the Rhombus Maker to fit exactly on Shape C] I don't think that is going to work.

Res: Why are you thinking that?

NL: When I try to fit it on the shape and I try to make it bigger or smaller, the whole thing moves. It will never get exactly the right size. [Manipulating the Rhombus Maker] Let's see if I can make the square with this. Here's a square. I guess it could maybe be a square. But I'm not sure if this is exactly a square. It's sort of leaning. The lines are a little diagonal. [Continuing to manipulate the Rhombus Maker] Yeah, I think this is a square maybe. . . .

Res: You said the Rhombus Maker could make the same shape as Shape C, what do you mean by that?

NL: It could make the same shape. It could make this shape, the one with 2 diagonal sides and 2 straight sides that are parallel. It could have been almost that shape and it got so close I thought it was that shape.

NL: [Continuing to manipulate the Rhombus Maker] Oh, I see why it didn't work, because the 4 sides are even and this [Shape C] is more of a rectangle.

Res: How did you just come to that?

NL: All you can do is just move it from side to side and up. But you can't get it to make a rectangle. When you move it this way it is a square and you can't move it up to make a rectangle. And when you move this, it just gets a bigger square.

Res: So what made you just notice that?

NL: Well I was just thinking about it. If it [the Rhombus Maker] was the same shape, then there is no reason it couldn't fit into C. But I saw when I was playing with it to see how you could move it and things like that, that whenever I made it bigger or smaller, it was always like a square, but sometimes it would be leaning up, but the sides are always equal.

This episode clearly shows how a student's manipulation of a Shape Maker and reflection on that manipulation can enable the student to move from thinking holistically to thinking about interrelationships between a shape's parts, that is, about its geometric properties. Indeed, NL began the episode thinking about the Rhombus Maker and shapes holistically and vaguely, saying that she was trying to make the Rhombus Maker "lean to the right" and get "bigger or smaller," and that "the whole thing moves." The fact that NL could not make the nonequilateral parallelogram with the Rhombus Maker evoked a perturbation that caused her to reevaluate the spatial structuring contained in her mental model of the Rhombus Maker. Originally, because her model did not include the constraint "all sides equal," her mental simulations of changing the shape of the Rhombus Maker included transforming it into nonequilateral parallelograms. Her subsequent attempts to make a nonequilateral parallelogram with the actual Rhombus Maker tested her model, showing her that it was not viable. As she continued to analyze why the Rhombus Maker would not make the parallelogram—why it would not elongate—her attention shifted to its side lengths. This new focus of attention enabled her to abstract the regularity that all the Rhombus Makers' sides were the same length. As she incorporated this abstraction into her mental model for the Rhombus Maker, she was able to infer that the Rhombus Maker could not make Shape C.

This activity encouraged and enabled NL to progress from a spatial structuring that was incomplete, imprecise, and did not explain the Rhombus Maker's movement (van Hiele Levels 0 and 1) to a property-based geometric structuring that did explain that movement (van Hiele Level 2). A combination of two factors may have supported NL's progress. First, during the episode, NL abstracted the "all sides equal" property to a sufficiently high (interiorized) level that enabled her to apply it in various situations. Previous episodes indicated that NL had already concluded that squares have all sides the same length; in Episode 2 she raised this notion to a higher level of abstraction. Second, it is highly likely that NL's conclusions about the Rhombus Maker came about partly because she had previously made a square with it; she viewed the Rhombus Maker as a transformed square. I conjecture that NL's transformation of the Rhombus Maker into a square, a shape that she conceptualized as having all sides equal in length, made a critical connection that eventually enabled her to transfer the concept of equal side lengths to the new shape of Rhombus Maker (via reaching a higher level of abstraction). In a subsequent section, I will discuss how the transformation facility of the *Shape Makers* is a major component of their instructional efficacy.

Episode 2 also illustrates how students using the *Shape Makers* move toward more sophisticated, property-based conceptions of shapes because of the inherent power these conceptions give to their analyses of spatial phenomena. In the current situation, NL developed a property-based conception of the Rhombus Maker because it enabled her to understand why the Rhombus Maker could not make Shape C—something that truly puzzled her. Only when NL conceptualized the movement of the Rhombus Maker in terms of a geometric structuring did she feel that she really understood its movement. NL acquired this property-based conception not because someone told her to learn it, but because it helped her achieve a goal that she was trying to achieve—understanding why the Rhombus Maker would not make Shapes B or C. (This was a personal goal for NL not only because the problem interested her but because she was a participant in an inquiry-based classroom culture and curriculum.)

Finally, this episode illustrates how geometry learning with the *Shape Makers* is much richer and more powerful than that which occurs in many traditional curricula. First, because NL's learning of properties was purposeful—connected at the outset with attaining a personal goal—it is highly likely that henceforth she would see that knowledge as being applicable and useful. (In fact, NL used this and other properties with increasing regularity in her subsequent work.) Second, because NL constructed her property-based reasoning from her already-existing cognitive structures, the newly constructed knowledge was "well-connected" in the sense that it was firmly anchored in her knowledge web, making it more likely not only to be applied in problem solving but used in further acts of knowledge construction. Third, and finally, it is highly likely that this episode increased NL's overall appreciation for the power of formal geometric reasoning in understanding the environment, again making it more likely that, in the future, NL would seek to use geometry in understanding the world.

Episode 3. The task was to determine which of Shapes 1–7 could be made by the Rectangle Maker (explaining and justifying each conclusion; see Figure 19.9.) Fifth graders M and T predicted that the Rectangle Maker could make shapes 1–3, but not 4–6. They are now checking and discussing their results. (The entire episode took about 25 minutes.)

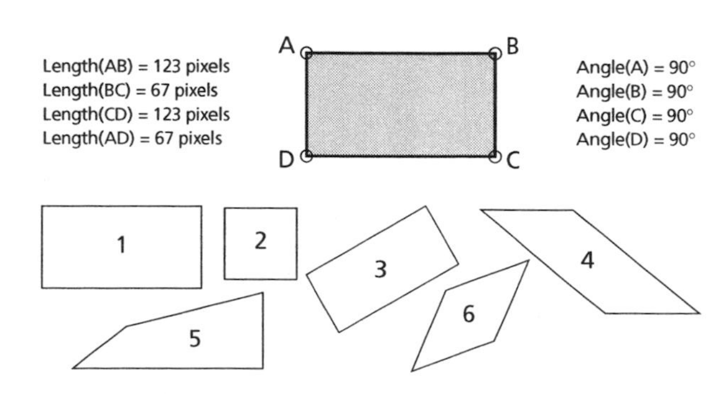

Figure 19.9 Which of Shapes 1–7 can be made by the Rectangle Maker (Battista, 1998a)?

After using the Rectangle Maker to check Shapes 1–3, M and T move on to Shape 4.

T: I'm positive it can't do this one.
M: It [the Rectangle Maker] has no slants. We had enough experience with Number 3, that it can't make a slant.

T: Yes it can, it has a slant in the other one [Shape 3]. It [the Rectangle Maker] has a slant right now. . . .

Tchr: What do you mean by slant?

M: Like this. See how this is shaped like a parallelogram [motioning along the perimeter of Shape 4].

T: This is in a slant right now [points at the Rectangle Maker, which is rotated from the horizontal]. [Note that *slant* means nonperpendicular sides for M, but rotated from the horizontal for T.] . . .

M: It can't make that kind of a shape [pointing to Shape 4]. . . . It can't make something that has a slant at the top and stuff.

T: Do you mean it has to have a *straight* line right here [pointing to Shape 6], like coming across? [T now uses *straight* to mean horizontal.]

M: I know they are *straight*, but they are at a slant and it [the Rectangle Maker] always has lines that aren't at a slant. . . .

Tchr: [After several failed attempts to unobtrusively get students to reconceptualize the *slant* conception in terms of angles] Keep on really looking at what makes these different. And maybe some of the information on the screen will help you. Watch the numbers up there and see if that will help you. . . .

T: [Manipulating the Rectangle Maker after the teacher leaves] Oh, this [the Rectangle Maker] always has to be a 90° angle. And that one [Shape 4] does not have 90° angles. And so this one [Shape 3] has to have a 90° angle, because we made this with that [Rectangle Maker]. So there is one thing different. A 90° angle is a right angle, and this [Shape 4] does not have any right angles.

M and T were trying to find a way to conceptualize and describe the spatial relationship that sophisticated users of geometry would describe as perpendicular. In so doing, and typical of students at van Hiele Level 1, they developed several vague and incomplete spatial structurings, all of which were inadequate. They used familiar terminology and concepts like *slanted* and *straight* that inadequately described the idea with which they were grappling. Progress was finally made, after the teacher left the boys, when T manipulated the Rectangle Maker and focused on its measurements. Through this manipulation, he discovered and abstracted that the Rectangle Maker always has four 90° or right angles. Furthermore, he abstracted this property sufficiently so that he was able to use it to analyze the differences in Shapes 3 and 4. He subsequently conceptualized that the Rectangle Maker could not make Shapes 4, 5, and 6 because they do not have four right angles. In fact, by the end of the class period, the boys saw that the spatial relationship they were attending to could be described in terms of the formal mathematical concept of right angle. They had constructed a geometric structuring, a van Hiele Level 2 analysis, that enabled them to solve the problem they had embraced.

This episode illustrates how difficult it can be for students to reformulate unrefined spatial structurings of Shape Maker movements in terms of traditional geometric concepts.[23] Only after much guidance, reflection, and experimentation can students construct formal, property-based geometric conceptualizations of Shape Maker movement constraints. However, once students meaningfully move from their initial vague spatial structurings to appropriate geometric structurings, they embrace the latter because of their inherent power.

Similar Results

It is instructive to compare *Shape Makers* results to results from other environments. First, note that M and T's reasoning was similar to, but more advanced, than that of JN in Logo. All three students were struggling with conceptualizing the notion of perpendicularity. In fact, Berthelot and Salin (1998) found that only 50% of 10–11-year-olds used right angles to locate the legs of a rectangular bench if moved. The other students explicitly used only information about length. Also, consistent with the Battista results, Sutherland, Godwin, Olivero, and Peel (2002) found that use of DGE constructions similar to the *Shape Makers* helped elementary students learn about properties of shapes.

DGEs at the Middle and High Levels

In examining instructional DGE use by older students, besides having students investigate geometric shapes by manipulating them on the computer screen, two additional major emphases occur. First, students are asked to use DGE tools to construct draggable figures. Second, students are asked to prove conjectures that arise out of exploration.

[23] The difficulty students have with this process points out just how deficient traditional cursory memorization-based approaches to this topic are. It takes a great deal of thought and appropriate experience for students to use the formal geometric conception-based system with genuine understanding.

To promote understanding of quadrilaterals in junior high students in the UK, Jones (2000) gave a series of DGE tasks in which students were asked to construct *draggable* figures. For instance, students were asked to construct a draggable square and explain why the shape is a square. As students progressed through the instructional sequence, Jones found that, initially, students emphasized description rather than explanation, relied on perception rather than mathematical reasoning, and lacked precise mathematical language. At an interim period in the instruction, students' explanations became more mathematically precise but were intertwined with the operation of DGE software. At the end of the instructional unit, students' explanations were entirely mathematical.

To illustrate, when students were asked to explain why all squares are rectangles, early in instruction one pair wrote: "You can make a rectangle into a square by dragging one side shorter . . . until the sides become equal" (Jones, 2000, p. 76). Later in instruction, students were asked to construct a trapezium that could be modified to make a parallelogram and explain why all parallelograms are trapeziums. The same pair of students wrote, "Trapeziums have one set of parallel lines and parallelograms have two sets of parallel lines" (Jones, 2000, p. 76). The apparent argument in the first example is couched in terms of the DGE—"via dragging, you can see that a square is a special type of rectangle." The second argument is apparently more traditional property-based—"parallelograms have all the properties needed for trapeziums." What we do not know, however, is how students made the transition from DGE-based to formal-geometric conceptualizations. More research is needed that investigates this transition.

In another study, conducted by Noss and Hoyles (1996), two female students, C and M (age 14), were given two flags in Cabri—one the reflection of the other—and asked to find the mirror line. After a short period of dragging points on the flags and noticing the effects on the images, C and M visually located the position of the mirror line. In an effort to be more precise about this location, C had the idea of dragging corresponding points on the flags—preimage and reflection image—so that they met. The girls claimed that such points lie on the mirror line. Subsequently, C and M used two pairs of these overlapping corresponding points to draw the mirror line. With the mirror line drawn, the girls then focused on pairs of corresponding points, which led them to see that the mirror line could be constructed by joining the midpoints of segments between corresponding points.

Similar to Emily in Logo, work with Cabri made C and M's conceptualization of reflections more analytic. They reformulated their idea of reflections when they saw that the mirror line was formed by joining midpoints of corresponding points. The "analytic" part of this new conceptualization is the geometric description of the reflection concept in terms of constituent parts (corresponding points, segments, midpoints). Also, C and M, like Emily, did not arrive at the traditional view of reflections—they made no mention of orthogonality, so their midpoints notion was incomplete. It would have been useful to see how further instruction could have extended the girls' notion to include orthogonality. It would also have been enlightening to better understand how C and M made transitions through their successive conceptualizations. For instance, we might conjecture that as C and M focused on a pair of corresponding points while manipulating the flags with the mirror line drawn, the two points formed a perceptual unit, bisected by the mirror line, which may have activated an interiorized version of the concept of segment midpoint (either formal, or situated within Cabri).

Artifacts of DGEs versus Geometric Properties

A fundamental open question is if and how students "distinguish fundamental characteristics of geometry from features that are the result of the particular design of the DGE" (Jones, 2000, p. 59). For instance, similar to Holzl, Healy, Hoyles, and Noss (1994), Jones (2000) argued that students need to understand functional DGE dependencies, such as in Cabri, although basic points, points on objects, and points of intersection look identical on the screen, basic points and points on objects are draggable, whereas intersection points are not. Apparently, students have difficulty developing such understanding (Holzl et al., 1994; Jones, 2000). In fact, students often develop nonmathematical conceptualizations of the draggability of figures. For example, asked why her constructed figure could not be "messed up," one student replied, "They stay together because. . . . It just glued them together" (Jones, 2000, p. 71). Thus, as Holzl et al., Jones, and others (Goldenberg & Cuoco, 1998) have argued, especially in curricula that make significant use of student exploration, research must investigate how students interpret the construction and behavior of DGE drawings. In essence, students working in DGEs need to see beyond the operational syntax of the environment; the DGE representations must become "transparent." But researchers do not yet know how this happens.

What Happens When Constructing a Draggable Dynamic Figure?

Many instructional treatments involving DGEs introduce "a specific criterion of validation for the solution of a construction problem: a solution is *valid* if and only if it is *not* possible to '*mess it up*' by dragging" (Jones, 2000, p. 58). So let us more carefully examine the process of constructing a draggable figure. For instance, two (Year 8) students were attempting to make a draggable rectangle in Cabri (Holzl et al., 1994). After drawing segment AB (see Figure 19.10), they constructed a perpendicular to this segment, passing through B, put point C on the perpendicular and constructed another perpendicular through C, put a point D on that perpendicular and constructed another perpendicular. The students joined points A, B, C, and D and hid the other lines, forming a trapezoid, then dragged A so that the trapezoid looked like a rectangle. The students realized that they had not created a draggable rectangle, but got bogged down in Cabri operations when trying to fix their error. For instance, they attempted to change the nature of point A to be a "point on object [DX]." Finally, when the students were told that this could not be done, they completed their rectangle by creating an intersection point on DX and AB.

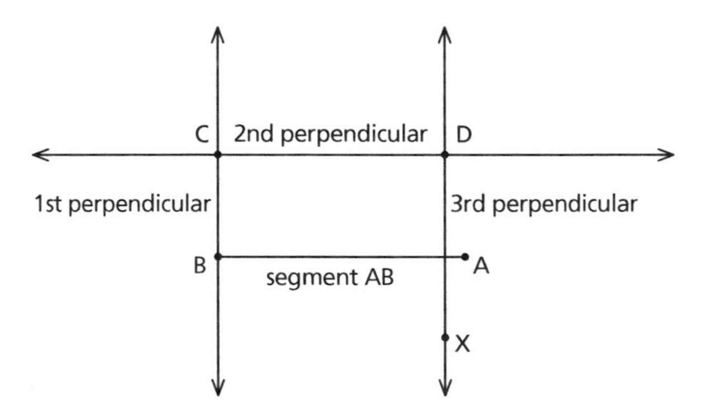

Figure 19.10 Students attempting to make a draggable rectangle in Cabri.

A number of questions must be asked about episodes such as this, which are common in students' DGE constructions. First, Holzl et al. claimed that the students "were clearly aware of the properties of rectangles and the need to construct perpendicular lines to obtain right angles" (1994, p. 9). But exactly which properties of rectangles did the students know? For instance, was their construction a direct result of knowing that the sides of a rectangle are perpendicular? Or, did they know that rectangles have right angles, and connect the concepts of right angle and perpendicularity to make their construction (perhaps making this connection for the first time)? Also, there was no explicit evidence in their construction that they knew that opposite sides were equal and parallel. Of course, they may have known these properties, but because perpendicularity is sufficient to complete the Cabri construction, they did not mention them. Second, if the students already knew the properties of perpendicularity, parallelism, and congruence of opposite sides, what geometry did they learn by working on this problem in Cabri? Perhaps no *new* knowledge was acquired, but instead, the students' knowledge and reasoning were deepened and enriched. The process of applying known concepts in a new situation requires interiorization of the concepts, making them more powerful. Or perhaps connections between properties were newly constructed or extended. Also, if the students noticed parallelism and equality of opposite sides in the constructed figure, they might have seen that perpendicularity implies parallelism and equality of sides (especially if a teacher helped them notice it), moving them toward or reinforcing van Hiele Level 3 thinking. On the flip side of the argument, however, one might ask what the students really understood about the perpendicularity of sides. For instance, the students could have constructed the rectangle visually, using perpendicularity as a procedural tool for making lines meet in a certain way. That is, they knew perpendicularity visually as making things "square," not a geometrically as involving right angles.

The claim by Holzl et al. (1994) that students working in DGEs "ran up against some fundamental aspects of *its [Cabri] geometry*" (p. 11) raises additional issues. First, does Cabri really have its own geometry? If so, what is it and how is it related to traditional mathematical geometry? For instance, as has already been mentioned, unlike mathematical geometry, Cabri "geometry" has different kinds of points that behave differently. As another example, the idea of "point on object," as implemented in a DGE, may promote student conceptualizations that are very different from the corresponding mathematical concept. In Birkoff's formulation of geometry (Moise, 1963), the statement, "let X be a point on segment AB" means that X is any of the points that are elements of the point-set that is segment AB. However, the way the point concept is implemented in a DGE might lead some students to conceptualize a point on a segment as an object "on top of" or separate from the segment, not as part of the segment. In this case, the DGE concept and mathematical concept would be inconsistent. In contrast, in Hilbert's formulation of geometry (Wallace & West, 1992), *point, line,* and *on* are undefined terms, and there is no requirement that individual points be elements of sets of points called lines. In this case, there seems to be no inconsistency between the DGE and mathematical concepts. This recalls the question, What geometric conceptualizations do students develop in DGEs, and how are these conceptualizations related to various formal mathematical concepts?

Moving Toward Justification

At the secondary level, many researchers see the primary purpose for DGE construction activities as a bridge to justification and proof:

> The [instructional] approach to the construction problem within the Cabri environment is expected to introduce first, the need to justify a solution, then the need to negotiate the adequacy of this justification. The dragging test has a fundamental role in this process of justification. The fact that students can see for themselves when and which geometrical properties remain constant under dragging can help them to understand when they have produced something that merits justification. . . . In the subsequent mathematical discussions, it will become necessary to negotiate and introduce explicit criteria for the acceptability of the justification itself. (Mariotti, 2001, p. 262) [Presumably "acceptable" means formal proof.]

Consistent with this claim, but counter to suggestions that students who have explored a conjecture extensively in computer environments often feel no need for further justification, de Villiers (1998) reported that he has been able to encourage deductive thought by asking why DGE results are true. A somewhat different perspective was taken by Hershkowitz et al. (2002) who argued that

> the dragging operation on a geometrical object enables students to apprehend a whole class of shapes in which the conjectured attribute is invariant, *and hence to convince themselves of its truth*. . . The role of proof is then to provide the means to state the conjecture as a theorem . . . *to explain why it is true*, and to enable further generalizations. (p. 676)

So formal proof (a) helps students to symbolically and precisely encapsulate their empirical discoveries and (b) offers corroboration for and explanation of the validity of the discovery.

However, several questions can be raised about these claims. First, does use of DGEs motivate and promote the learning of proof better than traditional instructional contexts? Second, if students already believe that a proposition is true because of DGE empirical evidence, what precisely is added to *students'* reasoning and learning by formally proving it? Although researchers can see much that might be added, what actually happens with students? This is an empirical question. Third, exactly what are the processes by which DGEs promote the development of proof?

A number of researchers have addressed the latter question. For instance, Mariotti (2001) studied students working on DGE instructional activities as part of the regular curriculum. In the first problem, ninth-grade students were asked to construct the bisector of an angle then describe and justify their solutions. In Alex and Gio's first solution attempt (Figure 19.11a), they wrote,

> We transferred a segment AB . . . [to] segment r1 (AB = AC) [despite what they wrote, the students made the two segments AB and AC equal "by eye," so the transference was not draggable]; we drew two circles . . . center in C and point A and center in B and point A. We joined A and [circle intersection] D. . . . [after testing their construction by dragging] Failed! (p. 265)

For their second attempt (Figure 19.11b), Alex and Gio wrote,

> We drew a circle . . . [that] gave us the segments AB and AC belonging to r1 and r2, which are equal because they are radii of the same circle. We drew two circles (centre B and C point A); using the intersection . . . of the two circles, we found the point D that we joined with A, determining the angle bisector. (p. 265, 266)

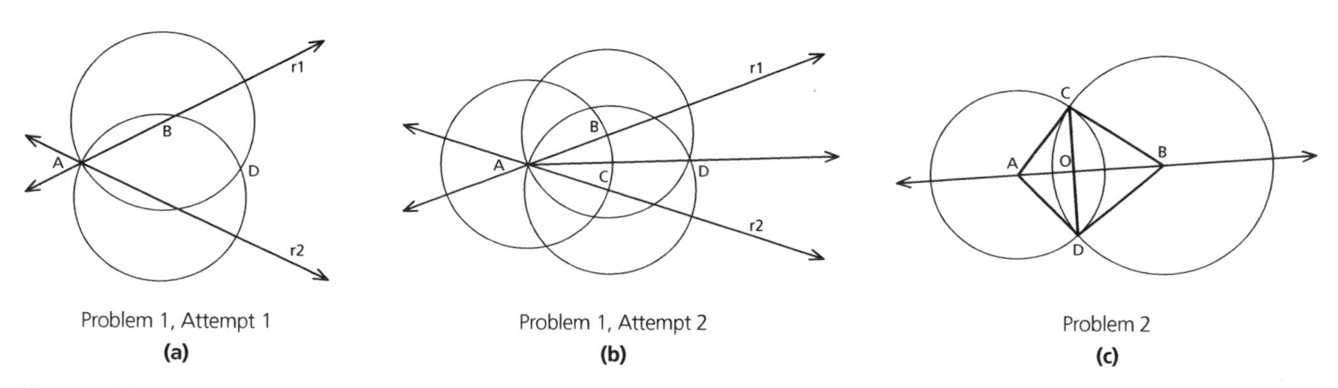

Figure 19.11 Alex and Gio's attempts to construct the bisector of an angle (Mariotti, 2001).

Note that, other than the one statement, "which are equal because they are radii of the same circle," the justification provided by the students is an *explanation* of the DGE constructions that they used. Most of what the students wrote is a description of process, not a justification, or even an explanation for why they did what they did.

On a later problem—constructing a perpendicular to a given line through a given point—Alex and Gio drew two circles having centers on the line and passing through the given point C (Figure 19.11c). Their perpendicular was the line through the intersections of the two circles. To prove the validity of their construction, they used the facts that AD and AC are radii of the same circle as well as that BD and BC are radii of the same circle (and the triangle congruence theorems) to show that angles AOC and AOD are congruent and thus right angles. In this case, the students seemed to link the structure of their proof to their DGE construction; in their proof, they took as given that which was produced by their constructions. So here is an example of using a DGE construction to formulate a proof.

It seems that in Alex and Gio's case, DGE was being used as a problem context—it provided exercises for enriching the proof skills of students who had already been taught formal proof. It helped students learn to use and appreciate geometric proof. It was not being used to evolve students' informal justifications into formal ones (cf., Borrow, 2000).

However, several issues arise in considering Alex and Gio's work. First, how did students move from justifying only a small part of their construction in the first problem to the much more sophisticated solution they provided in the second problem? (In order to prove these ideas, students must have already learned about formal proof.) Second, precisely how were students given, and how did they make sense of, a formal theoretical framework for proof? For instance, was the need to prove the validity of constructions generated by students—did they personally feel a need to remove uncertainty—or was it socially motivated by either the classroom culture or demands of the teacher?

A third issue involves the nature of students' geometric reasoning and how it was supported and furthered by the DGE. For example, what conceptualizations and reasoning (about geometry and Cabri) led students to use equal radii circles in constructing the angle bisector? Were circles seen procedurally as a way of laying off equal segments, or as loci of points in the plane equidistant from a given point? There seems to be evidence of both. Was the students' approach visual or analytic? That is, did the students use the circles analytically as a way of making point D equidistant from the sides of the given angle; or was their approach strictly visual—they simply tried to find a way to put a point D "in the middle" of the angle? Additional research needs to delve more deeply into students' thinking processes.

In contrast to the Alex and Gio episodes in which constructing draggable figures seemed to support students' reasoning, Mariotti (2001) also presented data suggesting a divergence between students' construction of *draggable* figures and their proofs. For instance, Cathy was unable to make a draggable construction for a perpendicular to line though a point P not on the line. However, she was still able to state and correctly prove how to make such a construction. She proved that if one chooses points S and Q on the line so that SP = QP, then the bisector of angle SPQ is perpendicular to line SQ. She derived this proposition by visually manipulating P and observing angle measure changes—her premises, however, were not "constructed" into a draggable drawing. Thus, students can construct valid theorems and proofs even when they cannot construct draggable diagrams. Although this finding is not consistent with claims of the central role that draggability plays in students' work with proof, it is consistent with the view that DGEs provide an exploratory medium that makes students' geometric ideas more analytically manipulable and thus more accessible to reflection and refinement.

Research by Healy and Hoyles (2001) provides additional insights into some of these issues. They conducted a study with above average 14–15-year-olds in England who were studying the national curriculum, which emphasizes proof in a range of contexts, not just geometry. However, despite this emphasis, the researchers reported that "only a few students [out of close to 2500 in a nationwide survey] could actually construct a valid deductive argument" (p. 239). In the experimental treatment, which lasted 6 to 8 weeks, after using Cabri to explore congruent triangles and being introduced to formal proof, students were given tasks that the researchers hoped would integrate DGE and proof experiences. (Before participating in this study, none of the students had used Cabri nor had they created sequences of logically justified geometrical statements.) For example, in one task, students were asked to (a) make and describe a Cabri construction of a rectangle, (b) identify the properties of the rectangle that they considered given in their construction, (c) manipulate their constructed figure to discover new properties, and (d) prove one of their discovered properties.

In response to this task, one student wrote the following (p. 243).

(a) Constructed line between 2 points and put perpendicular line through it. Created a parallel line to the perpendicular line and then a parallel line to the first line. (b) All angles equal 90°. Opposite sides parallel. (c) Opposite sides equal. (d) Used a triangle congruence proof to prove that opposite sides are equal. [For the student's proof that the properties in (b) imply the property in (c), see Figure 19.12.]

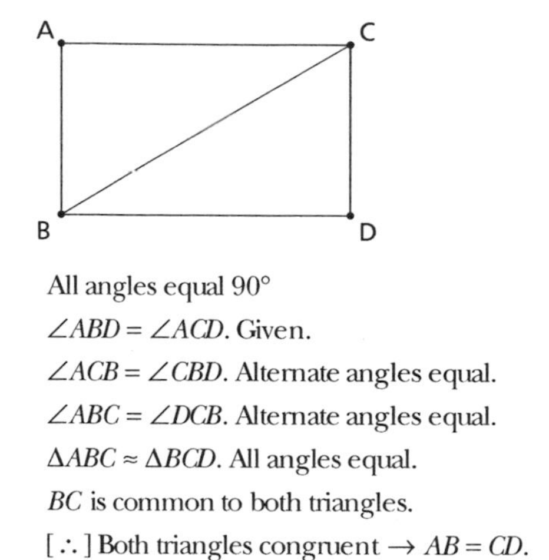

A
C
B
D
All angles equal 90°
$\angle ABD = \angle ACD$. Given.
$\angle ACB = \angle CBD$. Alternate angles equal.
$\angle ABC = \angle DCB$. Alternate angles equal.
$\Delta ABC \approx \Delta BCD$. All angles equal.
BC is common to both triangles.
[$\therefore$] Both triangles congruent $\rightarrow AB = CD$.

Figure 19.12 Student's proof.

This student was able to construct a draggable rectangle and to use a triangle congruence proof to justify his statement that the opposite sides of a rectangle are congruent. However, consider this student's creation of a draggable rectangle in more detail. After creating the first right angle by using a perpendicular, it is unclear how he conceptualized what he was doing. Did he create parallels because he wanted to give the figure the "opposite sides parallel" property about which he wrote? How did he know, or did he discover, that this would produce right angles (which he did not prove, but assumed in his proof)? Was there any explicit connection between the student's DGE construction and his proof (for instance, did his use of parallelism in constructing the rectangle cause him to focus on parallelism in his proof)?

In the final activity of the Healy and Hoyles teaching experiment, students were asked to construct a quadrilateral in which the angle bisectors of two adjacent angles intersect at right angles. They were to discover, then prove, other properties of this quadrilateral. Tim and Richard created quadrilateral ABCD, then the bisectors of angles ABC and BCD (see Figure 19.13). They measured the angle at the intersection of the two bisectors and dragged the vertices of the quadrilateral until this angle measured 90°. They then noticed that BA was parallel to CD and conjectured that whenever the two angle bisectors were at right angles, BA would be parallel to CD. They dragged the vertices for further empirical validation of their conjecture. Then, with the angle bisectors intersecting at right angles, the boys used Cabri's "check-property tool" to test whether BA and CD were parallel. The boys were "disappointed and puzzled that the property that the two lines were parallel was declared *not* to be true in general" (p. 244). However, because in the counterexamples presented by Cabri's check-property tool, the angle bisectors did *not* intersect at right angles, the boys decided that the parallel property had been declared invalid because the condition of 90° between the angle bisectors had not been retained by the construction.

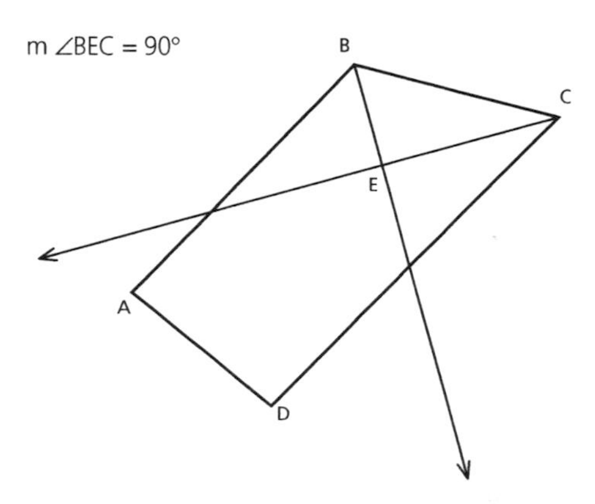

Figure 19.13 Tim and Richard's quadrilateral with angle bisectors of 2 adjacent angles intersecting at right angles.

Note that the boys' realization required sophisticated knowledge about the nature of Cabri constructions and, more importantly, the function of the check-property tool. This latter knowledge is knowledge about Cabri, not geometry. So they had to possess a fairly sophisticated understanding of the nongeometric functioning of Cabri to deal with the instructional task presented to them. Students who do not understand this technology sufficiently would have their geometric investigation thwarted.

Tim and Richard continued investigating their conjecture by constructing a line parallel to CD passing through B. They found that whenever point A was dragged onto this line, the two angle bisectors intersected at right angles, but if A was not on this line, the angle bisectors were no longer perpendicular. They formulated the conjecture, "Whenever the lines are parallel the angle bisectors are perpendicular." In searching for a proof for their conjecture, the boys returned to manipulating the quadrilateral, with A always on the line parallel to segment CD. They measured angles and looked for angles that remained equal when dragging vertices. The boys noticed that the alternate interior angles where the bisector of angle ABC intersected segments BA and CD were always "the same," then realized that proof of this equality would show that the two line segments are parallel.

However, before Tim and Richard moved from their Cabri construction to a strictly formal analysis of the situation, they used their construction to further explore their conjectures. But this was complicated by the fact that the angle bisectors did not always intersect the sides of the quadrilateral, so the relevant angles could not always be measured. The issue was resolved when the teacher suggested that they construct lines through the sides of the quadrilateral, which allowed the boys to measure the relevant angles in all draggable positions. At this point, the boys stopped manipulating the figure. The researchers claimed that the boys had "used the figure they had generated with its specific measurements as a generic example, to help them to formulate a general argument" (Healy & Hoyles, 2001, p. 246).

The DGE construction played two important roles in Tim and Richard's work. First, it provided them with a representation that allowed them to investigate the situation empirically so that they could formulate a valid conjecture about a geometric property. Second, the boys' manipulations of their DGE representation, including the accurate measurements of angles, helped guide their construction of a proof. This measurement process was not only more accurate, it was more general than that which could have been conducted with paper and pencil because it was maintained during dragging. Thus, in this case, for these students, Cabri seemed to form a bridge between empirical exploration and proof. Dynamic geometry helped these students find reasons why their empirical discovery was true. There was no evidence that the DGE discouraged proof activity.

However, on this same task, many students failed to see the parallel relationship between the sides of the resulting quadrilateral. Also, making a draggable figure with the "givens" was not the route successful students took. Like the two students above, all students who were successful on this problem constructed a general quadrilateral with adjacent bisectors and manipulated it to find when the bisectors were parallel. In fact, students who attempted to construct a quadrilateral with the given bisector-perpendicularity property—as the researchers had intended—encountered two problems. The first was reasoning that the given properties were satisfied by a particular quadrilateral, like a parallelogram. Students using this reasoning either became confused about which properties were given and which were to be deduced or did not see that the solution they derived for a special case was incomplete. The second difficulty was that constructing a general quadrilateral that satisfied the given conditions proved too difficult for students. Because they could not use the Cabri tools available to them to construct the quadrilateral, they became frustrated and failed to solve the problem.

The experimenters' solution to the second difficulty was that a more flexible, programmable environment is needed.

> From all our observations of students' interaction with Cabri, we have become increasingly aware that the mediation of students' activities by the software is not necessarily positive for their engagement and for their learning. . . . As we have shown in the work of our less successful students, learners can . . . find themselves in a position where they are unable to use the tools they have in mind, even if they are convinced that their use would make sense mathematically. (Healy & Hoyles, 2001, p. 252)

One can draw several conclusions from the Healy and Hoyles research, though not necessarily those drawn by the authors. First, the DGE apparently helped some students understand and solve posed problems. (Recall, however, that the students who did this seemed to have well-developed proof skills and understanding of associated geometric ideas.) Second, the DGE evidently helped some students develop appropriate proofs.

Just as evident is that other students were stymied in their geometric problem solving. However, it can be argued that the difficulty lies not with the DGE, as claimed by Healy and Hoyles, but with the design of the instructional tasks. For instance, in the last activity, the problem was not chosen so that students could use Cabri to construct, in a reasonably easy way, a draggable figure satisfying the problem hypotheses, even though the researchers, and many students, expected students to do so. If the point of the activity was for students to discover and explain a geometric relationship, there was no reason to expect them to create, a priori, a draggable figure that maintained that relationship. It is much more reasonable to expect students to construct a DGE figure that allowed them to reflect on this relationship (as did Tim and Richard). Alternatively, why not give students a draggable figure with the desired properties and let them use it to explore the problem? Perhaps tasks that do not ask students to make their own constructions are better for some students at some times. In general, then, researchers should be asking what students learn by using specific DGE tools, and what knowledge—both geometric and technological—is required to productively use these tools.

Comparisons of student learning with various task designs could be very revealing. For instance, given the problem of investigating quadrilaterals in which the angle bisectors of two adjacent angles intersect at right angles, researchers might compare the work of students in four different task environments: (a) no DGE, (b) making a draggable figure with the given property, (c) making a draggable figure like Tim and Richard's, and (d) using a preconstructed draggable figure with the given property. The research should compare not only student success rates, but their strategies, required starting knowledge for success, and what they learn from the experience. (Even though the Healy and Hoyles research shows that environment (b) is very difficult for students, it would still be interesting to better understand why it is so difficult.) Related to this suggested research, Arzarello, Micheletti, Olivero, Paola, and Gallirio (1998) attempted to analyze the types of DGE dragging explorations of students during different phases investigation of a geometric problem. Specifically, they investigated types of dragging during empirical activity aimed at trying to understand a problem or generate and explore a conjecture (their "ascending" phase), and activity aimed at building a deductive proof (their "descending" phase). They suggested that different kinds of dragging occur in different phases.

In another study of the use of DGE to support the development of students' proof skills, Marrades and Gutiérrez (2000) described differing levels of sophistication in students' discovery and justification of a theorem about the concurrence of perpendicular bisectors of the sides of a quadrilateral. They stated that "the process of getting conjectures was grounded on the observation of drawings and regularity in the measures of angles" (p. 105). Without prompts for proof, students seemed convinced by empirical validative actions of DGE. However, with teacher prompting, students seemed able to use DGE in concert with proof. When students arrived at a sequence of constructions that enabled them to conceptualize why the theorem was true and their dragging/measuring actions convinced them of the validity of their ideas, they were able to prove the theorem in a way consistent with their construction-based conceptualizations. Thus this research supports the claim that DGE exploration can provide insight and scaffolding for proof, as well as confidence to proceed with a given line of justification (as claimed by de Villiers, 1998, when he cited Polya, "When you have satisfied yourself that the theorem is true, you start proving it").

Unfortunately, many teachers apparently are not taking advantage of DGEs' capability to support the formulation of a proof: "When the students were asked to justify, the teachers did not mention the possibility of using Cabri to find a reason or to elaborate a proof. It is as if there was no interaction between visualisation and proving" (Laborde, 2001, p. 303).

Final Comments on DGE

Several major issues and areas for further research have arisen from the above discussion.

Issue 1. There are two major theoretical perspectives concerning the use of DGEs to promote the development of students' geometric thinking. In the first perspective, dynamic figures, especially draggable ones, are seen as generating numerous examples. For example, Marrades and Gutiérrez claimed that the main advantage of DGEs "is that students can construct complex figures and can easily perform in real time a very wide range of transformations on those figures, so students have access to a variety of examples that can hardly be matched by non-computational or static computational environments" (2000, p. 95).

In the second theoretical perspective, draggable drawings are seen as interesting, manipulable, visual-mechanical objects that have movement constraints that can be conceptualized and analyzed geometrically (Battista, in press b). This approach is taken in Battista's *Shape Makers* environment and Laborde's "black-box" activities in which students are given a DGE figure without being told how it was constructed and are asked to construct a draggable figure that behaves in the same way (Laborde, 1998). In these draggable drawings, geometric properties are visually and mechanically manifested as constraints and allowances of movement and therefore not only can be seen, but can be felt. Thus, rather than examining *a set of figures* to conceptualize how they are the same, with *Shape Makers* or black-box constructions, students investigate how an interesting manipulable "geometric" object works (which can, with proper instructional guidance, support the construction of meaningful conceptualizations of geometric properties). Seemingly consistent with this view, some researchers argue that DGE should not be conceptualized as a device to merely "enable us to do what we already do faster, better or to a larger extent or degree," but as an environment requiring a fundamentally different cognitive view than has been taken in traditional approaches to geometry instruction (Dörfler, 1993, p. 161).

In considering these theoretical perspectives, which guide both research and curriculum development, several questions arise. First, it is important to test the theories and determine empirically how students conceptualize DGE draggable figures. Do students see the various configurations of a dynamic drawing as examples of a class of geometric shapes, or do they see a dynamic drawing as a geometrically interesting manipulable object? How are students' views on this matter affected by instruction?

Second, because much of geometry is about categories of objects, do students who think about DGE figures as manipulable objects naturally move to thinking about classes of shapes? If not, how can such movement be engendered? Does students' reasoning about DGE manipulable objects transfer to thinking about geometric categories? Battista's *Shape Makers* research suggests that instruction can promote such transfer. Are DGE figures viewed differently by students who have already attained strong conceptualizations of the underlying geometric concepts before using a DGE, as opposed to students who initially construct meaning for these concepts in a DGE? That is, for an expert, a dynamic rectangle maker might indeed be a representation of the class of rectangles, whereas for the novice it might be simply an interesting dynamic object whose movement is constrained in particular ways.

Issue 2. How do students conceptualize particular DGE constructions? For instance, what is involved conceptually when a student learns to construct a parallel to a given line through a point not on that line? What does the construction mean to the student? Is it a procedure for constructing a visual configuration, or is it a representation of a geometric concept? How does using the construction change, supplement, or supplant a student's conceptualization of parallelism? That is, learning changes a student's network of cognitive structures—how does learning to use DGE construction tools change students' related conceptual networks? Conversely, what kinds of conceptual networks enable students to productively employ DGE construction tools? Is the nature of conceptualizations created via use of DGE construction tools the same as those learned through verbal definitions or drawing?

Issue 3. In general, much additional qualitative and quantitative research on the use of DGE in geometry learning is needed. Qualitative research is needed to investigate numerous unanswered questions about the nature of students' construction of geometric knowledge in DGE. Quantitative studies are needed to determine whether use of DGE produces greater learning than use of paper and pencil. However, geometry research, and especially its application to classroom practice, is best served by integrated combinations of qualitative and quantitative studies. For instance, as has been shown previously in this chapter, qualitative research results indicate that high quality geometry learning can occur with use of DGE. But quantitative research is needed to determine the generalizability of these results and whether use of DGE is "better" than use of paper-and-pencil techniques. In contrast, when quantitative results indicate that one instructional treatment is more effective than another, qualitative methods should be used to determine why such results were obtained.

Issue 4: The allure of draggable figures. Investigating draggable figures in DGE seems to strongly attract the interest not only of students but of researchers in geometry learning. Why is investigating such figures so appealing? One suggestion for the appeal of draggable figures comes from the theory of psychological essentialism. This theory conjectures that people act as if things have essences or underlying natures that make them what they are, and that when an essence is unknown, it may motivate the search for new meanings (Gelman & Diesendruck, 1999). Thus, one reason that DGE draggable figures may be instructionally useful is that students naturally attempt to determine the essences of these objects—how do they move, why do they move the way they do.

Issue 5: Draggability and invariance. A number of researchers have claimed that one of the reasons that DGE draggable figures are useful for helping students develop understanding of geometric properties is that such properties are invariant under dragging movements. For instance, Laborde claimed that the spatial properties of draggable drawings "may emerge as an invariant in the movement whereas this might not be noticeable in one static drawing" (Laborde, 1998, p. 117). Similarly, Battista (in press b) suggested that properties of shapes are more noticeable as invariants of draggable movements than as commonalities across static examples.

Several clues in the psychological literature point toward a possible explanation of this phenomenon. First, the mind's attentional system is especially alert and sensitive to change (Ornstein, 1991). Thus, when viewing a figure as it is dragged, because the figure maintains its identity as a single entity, changes in its shape are naturally attended to. Second, "[the] search for constancy, the tendency toward certain invariants, constitutes a characteristic feature and immanent function of perception" (Cassirer, 1944, p. 21). Third, comparing a set of static figures and comparing various configurations of a dragged figure should both induce structure-mappings (Gentner & Medina, 1998) that can make properties noticeable. However, transforming draggable figures should facilitate this structural analysis because the continuity of the transformation establishes a structural-mapping correspondence between components of compared figures, whereas when comparing static figures, the perceiver must establish the correspondence (which may not be easy). Of course, if students do not attend to the entire dragging transformation, the original figure and the figure that results from dragging may actually be viewed as two separate entities, losing this facilitative effect. Fourth, a considerable amount of research has shown that objects that move together are seen as part of the same object (e.g. Driver & Baylis, 1989; Regan & Hamstra, 1991). So, dragging motions might create "visual chunking" that facilitates the perception of visual regularities indicative of properties. For instance, because in the Parallelogram Maker the opposite sides move together in a way that maintains their parallelism and congruence, such movement might encourage the formation of two "opposite-sides-parallel" composites or chunks. These composites are a visual manifestation of the geometric properties of opposites sides parallel and congruent (although, in this form, the two properties might not be separable, which is consistent with the observation that children's initial conceptualization of parallel often includes congruence). Thus these composites might form the spatial structuring needed later for students' geometric structuring.

1. In what ways are the figures alike?

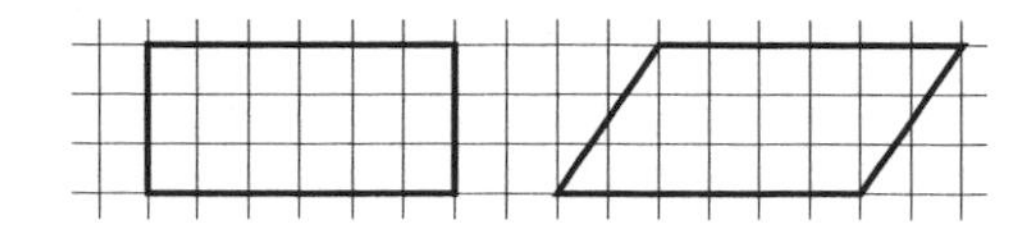

2. Explain how shape N is different from shapes P and Q.

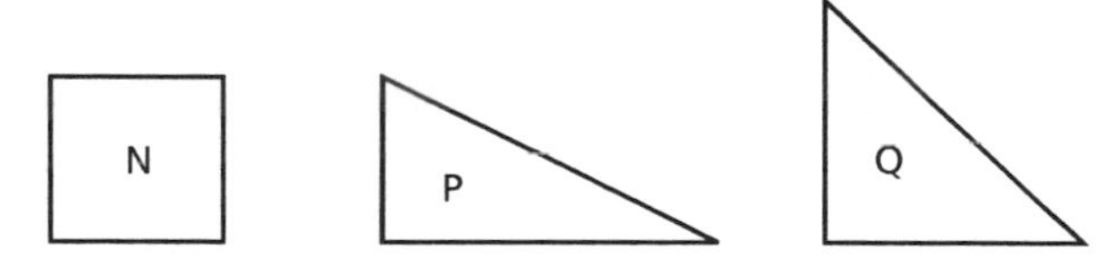

Figure 19.14 NAEP tasks.

RESEARCH ON OTHER GEOMETRIC TOPICS AND NONCOMPUTER ENVIRONMENTS

The National Assessment of Education Progress (NAEP)

Traditionally, the NAEP has been a useful tool for getting an overall picture of U.S. students' mathematics skills. To illustrate, recent administrations of the NAEP in mathematics have shown some improvement in geometry, but students' performance in many areas of geometry and spatial sense is still low (Kloosterman et al., in press; Sowder, Wearne, Martin, & Strutchens, 2004). However, it is difficult to reliably integrate recent NAEP results into the research corpus on geometry learning because of two major problems with the NAEP program. First, in recent years, NAEP assessment tasks have not been released to researchers for analysis, making the results almost impossible to interpret seriously. For instance, it is impossible to make much sense out of a vague NAEP item description such as, "Determine the shape formed under certain conditions."

Second, many NAEP items and scoring rubrics are substantively disconnected from current research on geometry learning. For instance, consider the two tasks shown in Figure 19.14.

Both tasks are intended to assess students' ability to use properties to compare shapes, a critical goal in the geometry strand that is consistent with van Hiele-based research. However, the property focused on in Task 2 is not, as has been previously discussed, a good indicator of van Hiele Level 2; number of sides does not involve relationships between parts of shapes. In contrast, the properties that discriminate the shapes in Task 1 focus on relationships between parts of shapes and thus are good indicators of van Hiele Level 2. Unfortunately, and inconsistent with research on geometry learning, NAEP administered the low-level Task 2 across the grades (4, 8, 12) and the richer Task 1 only at Grade 4 where one would expect to see only minimal development. Furthermore, the scoring guide for Task 1 is not based on a research-derived classification of justifications. For instance, the scoring guide weighs the following properties equally: having 4 sides or angles, having two sets of equal side lengths, having 4 right angles, having equal heights, having equal areas, and "one is more slanted." So informal properties, properties dealing with numbers of components, and relational properties are given equal weights. But research suggests that these properties vary considerably in sophistication. Thus, because the NAEP items and scoring are only loosely connected with research on students' learning of geometry, they provide inadequate information about students' geometry learning for teaching as well as for curriculum evaluation and development.

Proof and Justification in Geometry[24]

In 1992, Clements and Battista (1992) concluded that, in geometry, students are extremely unsuccessful with formal proof and that they are deficient in their ability to establish mathematical truth. More recently, Hershkowitz et al. (2002) drew a similar con-

[24] See (Harel & Sowder, 1998) for a discussion of the proof schemes students use throughout mathematics, not just in geometry.

clusion: "The teaching of mathematical proof appears to be a failure in almost all countries" (p. 675). In a large study in the United Kingdom, Healy and Hoyles (1998) reported that (a) high-achieving Year 10 students showed poor performance in constructing proofs, with empirical verification the most popular form of justification; (b) students' proof performance was considerably better in algebra than in geometry; (c) students recognized that once a statement has been proved, no further work is necessary to check its validity; and (d) students were better at recognizing a valid mathematical argument than constructing one themselves. Hoyles and Jones (1998) reported that (a) students do not distinguish between empirical and deductive arguments and prefer empirical arguments; (b) for many students, deductive proof only *adds evidence* for validity; and (c) most students regard proof as an irrelevant add-on and do not understand its purpose. Similarly, Hershkowitz et al. (2002) stated that students rarely see the point of proving. Battista and Clements (1995) argued that although proof is critically important to mathematicians because it enables them to establish the validity of their mathematical thought, students do not conceive of proof as establishing validity, but see it instead as conforming to a set of formal rules that are unconnected to their personal mathematical activity.

Consistent with earlier research, Koedinger (1998) reported that it was rare for high school students to think that evidence beyond examples was needed to support geometric conjectures. He also stated that many students, when considering conjectures, first said that the one example they drew was enough evidence, then, after prompting, that many examples were needed. Most students had to be explicitly asked to write a proof "like they had seen in class" before they attempted to produce one. Also, most students had difficulty formulating proofs. In fact, Koedinger reported that only 10% of students successfully formulated proof problems and began to work on them. Interestingly, students found it especially difficult to state the givens.

McCrone and Martin (2004) found that high school geometry students' performance was low on almost all items on a proof-construction assessment. Scores were lowest on items in which students had to construct their own proofs, even on items that provided a proof outline (averages ranged from 6–34% correct). Results were a bit better for a fill-in proof (44–50%). Scores were highest, but still low, on an item that asked students to rewrite a conjecture in "if-then" format and specify the givens and what was to be proved (33–53%). These researchers also constructed a questionnaire that probed students' beliefs about essential elements of proofs. About 78% of students agreed with "the principle that one purpose of proof is to provide insight into why a statement is true." About 67% of students agreed with a statement indicating that "knowing that a general statement is true implies that the statement is true for all specific instances." In contrast, only 22% of students found the logical flaw in a fairly simple two-column proof. On a purported proof that the sum of the angles in any triangle is 180 degrees, students were presented with a table of angle measurements for four scalene triangles—3 obtuse, 1 acute. About half the students claimed that this table constituted a proof, with many students saying so because a range of types of triangles had been tested. Some students claimed that the table did not constitute a proof because it omitted right angles.

Chazan's (1993a) investigation of high school students' views of geometry proof elaborates and clarifies some long-standing issues in the research reported above.[25] In interviews, he found that 5 out of 17 high school geometry students believed that measuring examples "proved" assertions, 7 did not believe that such empirical evidence proved assertions, and 5 were unclear or changed their minds about this issue during the interview. Some students who believed that empirical evidence is proof focused on the number of examples examined, whereas others focused on the type of examples examined. As an example of the latter approach, in considering a proposition about triangles, one student said, "Right, obtuse, acute and like isosceles or something like that . . . it's pretty much the same effect as having every one" (p. 369).

Of the students who believed that empirical evidence does not constitute proof, some cited the fact that there may be counterexamples because one cannot examine all examples. Other students cited the inexactness of measurement, a point that raises an important issue. On the one hand, side lengths can be considered as the result of a measuring process (that is, they are obtained by the act of physically measuring and, as such, are only approximations). On the other hand, side lengths can be considered theoretical (that is, they represent the actual length measures of the segments; as such they are exact and without error). In the first case, skepticism about justifications based on lengths is warranted; in the latter

[25] First, use of the *Geometric Supposer* was an integral part of Chazan's students' courses. But because I focus only on what the research indicates about proof, I discuss this research in this section rather than in the section on the *Supposer*. Second, some students were interviewed in January, some in March—so students had not completed their geometry courses.

case, skepticism is unwarranted. It is important for researchers to determine precisely not only how students think about measures given in proof problems (e.g. approximation versus exact), but how instruction and computer geometry programs deal with the measure/approximation issue and how this affects students' conceptualizations.

In contrast to students who believed that empirical evidence was proof, some students thought that proof was merely evidence (Chazan, 1993a). For instance, some students believed that counterexamples were still possible, even if a deductive proof in a textbook was given. However, Chazan's finding could be due to students not understanding the logic of proof, or it could be a reasonable reaction to the possibility of an incorrectly *implemented* proof (especially if the proof is written by a student[26]). This brings up an important issue. Theorems presented in school mathematics textbooks, and their accompanying proofs, are generally assumed by teachers and students to be correct because they appear in the book, which is taken as authoritative, rather than because the proofs are critically examined by students and teachers. This is very different from how mathematicians work—usually, proposed new theorems and proofs are greeted with skepticism, and often with attempts to find counterexamples.[27] So, for some students, skepticism about deductive proofs should not be viewed as a weakness but rather as justified and healthy.

Another reason that students might be skeptical about the generality of a deductive proof occurs when the proof does not really provide insight into why a proposition is true (recall the McCrone and Martin's, 2004 finding that 78% of students agreed that proofs should provide insight into why statements are true). For instance, one can read a proof and believe in its validity because each step in its argument is valid but still not, in the words of Polya, "see it at a glance." In this case, the proof does not provide enough insight to genuinely make sense of the proposition, so one might still be skeptical. In such instances, it is mathematically healthy and reasonable to further explore the proof by examining examples, perhaps reviewing the proof for some of these examples, even developing alternative proofs (think of how many proofs have been given for the Pythagorean theorem). Thus, instead of assuming that student skepticism about proof must be a negative finding, it is important for researchers to investigate whether students' skepticism is due to (a) a lack of understanding of the logic of proof, (b) a healthy critical attitude toward mathematical justification (so students are skeptical not about the logic of proof but about particular implementations), or (c) not being exposed to proofs that help them understand why a theorem is true (that is, they do not believe what does not make sense to them).

According to Chazan (1993a), some students believe that deductive proofs apply only to single diagrams or to single types of shapes. In the latter case, a student might believe that a two-column proof had to be written, say, for an acute, obtuse, and a right triangle (and possibly isosceles and equilateral triangles as well). Chazan noted that the notion of multiple proofs for multiple examples might be a carryover from thinking about proof by examples—students want to see multiple examples to convince themselves.[28] But there may be more to it than this. Progressing from reasoning about a specific example to reasoning about a deductive proof is similar in some sense to moving from reasoning about specific numbers to reasoning about variables. For instance, for a specific diagram, one might say that each of two sides measures 50 so they are equal in length; in a deductive proof of a general proposition, however, we might say that two sides are congruent—that is, algebraically, the measure of both sides both equal some indeterminate number x. Moving from specific to variable thinking requires a significant jump in abstraction. Thus, the multiple proofs for multiple examples conception might indicate a transitional stage in abstraction in moving from empirical to deductive thought. In any event, it is clear that students possessing the multiple proofs for multiple examples view do not genuinely understand the concept of deductive proof. However, this should not be surprising, given that most instructional treatments do not spend significant time specifically helping students understand the nature of mathematical justification and proof. Instead, these treatments focus mostly on having students master formal methods of proof.

In summary, researchers are still grappling with the basic question: Why do students have so much difficulty with geometric proof? Is it because they see little need for proof, because proof is so abstract that it is very difficult to learn, or is it a combination of the two—that because students see proof as difficult and not useful, they do not expend sufficient intellectual effort to master it? What components of proof are

[26] In this research, it is not always clear if the students' statements are about textbook proofs or about student proofs.

[27] In fact, analyzing why counterexamples cannot be found often provides insight into why a proposition is true.

[28] Of course, there are proofs in mathematics texts that are broken down in terms of cases. But students in school would not have been exposed to many such proofs.

difficult for students, and why? How can proof skills best be developed in students? More research is required on students' development of geometric proof skills and their understanding and beliefs about the nature of proof. As was discussed in the section on DGE, use of DGE has shown promise in encouraging and supporting the development of students' geometric proof. To better understand the nature of the facilitative effect of DGE on proof, however, qualitative research comparing students' proof processes in DGE and paper-and-pencil environments is needed.

Apparent contradictions in the research also need to be resolved. For instance, Healy and Hoyles (1998) reported that students recognized that once a statement has been proved, no further work is necessary to check its validity, but Chazan (1993a) reported that some students believed that counterexamples were still possible, even if a deductive proof in a textbook was given. As another instance, Koedinger (1998) reported that students found it especially difficult to state the givens, whereas McCrone and Martin (2004) reported that scores were highest, but still low, on an item that asked students to rewrite a conjecture in if-then format and specify the givens and what was to be proved. It is likely that such contradictions are due to the fact that, in many cases, small numbers of assessment tasks were examined to draw conclusions, and results are task specific. That is, although there are general processes applicable to most geometric proofs, success on most proofs also depends on understanding specific concepts and situations. Research in this area must balance investigating general versus task-specific components of proof.

Understanding Angles

In an investigation of the development of angle conceptualizations in elementary school students, Mitchelmore and White (2000) describe three stages of abstraction. In Stage 1, similarities between angle-related situations that look alike, involve similar actions, and are experienced in similar circumstances lead to abstraction of *situated angle concepts* (*conceptualizations* in my language). For instance, situations involving hills would be a single situated angle concept. In Stage 2, similarities between different situated angle concepts are abstracted to form *contextual angle concepts.* For instance, students might abstract the similarity between hills and roofs into a contextual angle concept of "slope." In Stage 3, similarities between different contextual angle concepts are abstracted to form *abstract angle concepts.* For instance, students might abstract the similarity between intersections of lines, corners, bends in paths, and slopes. Mitchelmore and White suggested that construction of a formal mathematical formulation for the standard abstract angle concept of two rays with common endpoint can be considered the fourth stage of angle concept development. They further argued that, when students are forming their first abstract angle concept, they may be in the process of forming additional contextual angle concepts. Furthermore, they claimed that angle concepts are constantly generalized as new situations and contexts are recognized as similar.

Similar to several other researchers in geometry learning, Mitchelmore and White use a theory of abstraction to analyze students' thinking.

> For example, students often demonstrate the similarity between a tile and a ramp by physically moving a corner of the tile to fit into the space between the ramp and the horizontal base. . . . The recognition of similarities between different angle contexts is therefore a constructive process requiring reflective abstraction. (2000, p. 216)

Could it be that the critical component of this reflective abstraction is the mental operation of matching corresponding parts of angle contexts, and that this matching is the same type of structural mapping discussed by Gentner and Medina (1998)?

Mitchelmore and White (2000) also suggested that an essential part of the process of mentally constructing a general conceptualization of angle is linking together different angle concepts. This returns the discussion to the issue of abstraction as disembedding/decontextualizing versus linking. In fact, the developmental sequence described by Mitchelmore and White provides a nice example of how the sequence of abstractions that lead to abstract angle concepts involves *both* disembedding and linking. For example, to compare the angle concepts embodied by the corner of a tile and the space between a ramp requires the disembedding operation of interiorization because the original abstractions of these embodiments are applied outside of their initial contexts. But the reflective abstraction that compares the tile-and ramp-angle concepts can result in a new abstraction that is linked to both ramps and tiles.

Mitchelmore and White (2000) reported three abstract angle concepts: *point* is sometimes used by young children to relate corners and intersections; *a single sloping line* relates sloping and turning objects; and, most common, *two inclined lines meeting at a point,* which they call the *standard angle concept,* can be used to relate all physical angle contexts. Mitchelmore and White found that more than 80% of students in Grades 2–6 could see similarities between a standard angle (as represented by a folding straw) and physical contexts of scissors, a folding hand fan, a diagram of a road intersection, a four-sided tile, and two intersecting walls. Students had more difficulty relating the standard angle to a wheel or door (turning) and a hill (slope). Less than 30% of students in each of Grades 4, 6, and 8 represented the slope of a hill using a standard angle between the hill and the horizontal, with many, instead, representing the slope as the angle between the vertical and hill. However, this finding must be tempered by the fact that in the representation of the hill used in this study, the angle with the vertical was much more salient (see Figure 19.15).

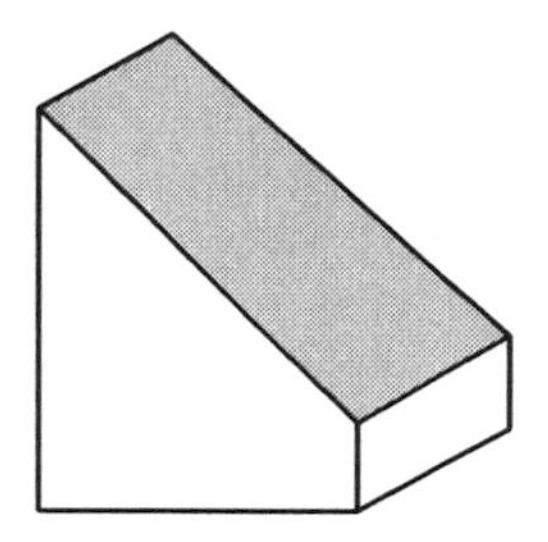

Figure 19.15 Representation of slope of a hill (Mitchelmore & White, 2000, p. 221).

Most importantly, the difficulties that students had relating the standard angle concept to various angle contexts seemed to be directly dependent on the visual availability or salience of structural angle components (the two sides and vertex) in the contexts. For instance, a large portion of students could not identify the two lines that make up a standard angle in the contexts of a wheel or door, and few students saw the standard angle as slope plus horizontal in the hill context illustrated above. Thus, the availability and salience of these structural components seemed to determine students' success in performing the structural mappings required to see similarities between examples from different contexts.

The findings that students have difficulties relating the standard angle concept to inclination and turning are important not only in their own right but because these two contexts are extremely important in postelementary mathematics such as algebra (inclination as slope) and trigonometry (turning). Mitchelmore and White (2000) also suggested that the difficulty that students have in learning to use a protractor might stem from the fact that, on a protractor, several lines can be chosen for the initial side of an angle but the terminal side must be imagined. So again, the cause of students' difficulties seems to be the absence of structural angle components, which leads to failure in establishing appropriate structural mappings.

Because of the difficulty that students have in seeing standard angles in turning contexts, Mitchelmore and White (2000) claim that defining an angle as an amount of turning (about a point) from one line to another is inappropriate for beginning learners. But the notion that understanding angles in terms of turns is *necessarily* bad for children is contraindicated by some (but not all) Logo research. Indeed, Hoffer claimed that "turns or rotations are natural for young children . . . this dynamic way of working with angles should be started early and carried throughout the students' work" (1988, p. 251). In support of this claim, Clements and Battista's research on elementary students' learning in the *Logo Geometry* curriculum shows that learning about angles in the context of turtle turning and paths can be effective for elementary students, *if the relationship between amount of turn and measure of angle produced is explicitly and carefully taught* (Clements & Battista, 2001). However, continued research is needed on how best to instructionally develop general abstract conceptualizations of angle that are adequate for the many manifestations of this concept in mathematics.

Coordinate Systems and Locations

Locating objects in space is a critical process in geometry and everyday life. Geometry formalizes this process through the use of coordinate systems. In this section, I briefly describe some research issues related to the use of coordinate systems.

Types of Coordinate Systems

Students are exposed to two different types of "coordinate-system" conceptualizations, maps and Cartesian coordinate systems. Maps utilize *nonmetric* ordinal coordinates to specify locations: Labels on axes signify ordered locations without metric relationships. To locate a point in 2d space, one must find specified locations on the axes, then move orthogonally to both axes to find the intersection. The key idea here is not distance, but coordinating order on two intersecting lines.[29] On geographic maps, "coordinates" (which can be letters or numbers) often specify intervals, not points. Of course, the difficulty with this conceptualization is dealing with "in-between" locations. In contrast, Cartesian coordinate systems are based on *both* axis-orthogonality and uniform metric axis scales, which ensures that relationships between points are

[29] Although axes are usually perpendicular with equally spaced labels, this concept works even if the lines are not orthogonal and labels are not equally spaced.

uniform throughout the coordinate system and allows the system to be used to represent many important geometric properties.

Research is needed to determine which of these two conceptualizations students are developing in their classroom work with coordinate systems. Such research must account for the fact that mathematical coordinate axes have multiple purposes. One use of coordinate axes is simply to locate points in space. This use does not require metric properties of number lines, only order properties. However, full understanding of mathematical coordinate systems affords students more than the ability to locate points. It also allows them to reason about distances between points using coordinates. This is a critical property of Cartesian coordinate systems and one of their major sources of conceptual/reasoning power. To construct a full understanding of Cartesian coordinate systems, merely having students locate points is not enough; students must also analyze distances between points and how those distances can be determined from coordinates (see, for example, Sarama, Clements, Swaminathan, McMillen, & González Gómez, 2003). However, researchers and teachers must be alert to the fact that students' initial conceptualizations of coordinate systems may be skewed by early map-like uses.

An Example of Research on Student Learning of Coordinate Systems

Sarama et al. (2003) found that, when initially working with numerically labeled axes, some fourth graders ignored the labels. For instance, to locate the point for (8, 9), one student, TA, counted 8 lines over and 9 lines up, even though the grid lines were labeled with numbers (so counting was unnecessary). The authors argued that such students did not take the numeric grid labels as symbols or curtailments of the actual counting process. They conjectured that this is similar to students not being able to count on—they cannot operate on the first number in the count as representing a given number of counting acts—they must produce the counting acts. Such students have not raised their counting process to the symbolic, second level of interiorization. The occurrence of the unnecessary counting behavior on axes is also consistent with the conjecture that such students possess an action-oriented meaning for coordinates—coordinates are considered inputs for a counting- or movement-based procedure for locating an object. Such a proceduralized conception of coordinates would be difficult to apply to nondecade coordinates on grids with only multiples of ten labeled (a difficulty that these researchers observed). Another piece of evidence suggesting that students' initial conceptualizations of coordinates were action based was the finding that some students who did not usually reverse *x* and *y* coordinates would do so when the first coordinate was 0. Such students moved horizontally for the second coordinate (e.g., go over 6) as the first step in locating the point, needing a nonzero movement to start their plotting procedure.

Another difficulty observed by Sarama et al. (2003) was students not seeing that coordinates were given with respect to a particular reference point, the origin (0, 0). For instance, some students, after having plotted one point, plotted the next point relative to that point, not the origin. Of course, this error may have occurred because students were working in a Logo environment in which many commands are relative.

So what does it take to construct a fully developed conceptualization of a mathematical coordinate system? Sarama et al. (2003) suggest that students must be able to "distributively coordinate" two orthogonal "conceptual rulers." This idea is similar to students' conceptualization of 2d arrays of squares (Battista, Clements, Arnoff, Battista, & Van Auken Borrow, 1998). To construct a properly structured mental model of a 2d array, students must iterate a row composite along the elements of a column. To construct a properly structured mental model of a coordinate plane, one must continuously sweep one coordinate axis (conceptualized as a fully metricized number line) along the orthogonal axis. However, the coordinate axis situation is more complicated than the array situation because arrays are discrete, but coordinate axes are continuous. So one cause of students' problems with coordinates axes may be the difficult shift from discrete to continuous objects. Another difficulty that Sarama et al. (2003) observed is that some students fail to interrelate coordinates and locations of multiple points. For instance, asked to plot (20, 35) and (25, 35), less sophisticated students plotted them separately. In contrast, more advanced students saw immediately that the second point was 5 units to the right of the first point. The advanced students had constructed second-level interiorized versions of plotting schemes and properly structured mental models of frames of reference that allowed them to interrelate locations and coordinates without running through the process of locating each point.

Large- and Small-Scale Spaces

Mathematical coordinate systems function the same whether one is locating a point on a sheet of paper or oneself in a city. However, research in psychology suggests that the cognitive processes used in dealing with these situations differ, which might have important implications for research on the learning and teaching of coordinate systems. For instance, according to Siegel (1981), large-scale spaces (e.g., a university campus) cannot be perceived from a single viewpoint, and their structure is mentally constructed from a number of observations over time. In contrast, small-scale spaces (e.g., a sheet of paper) can be mentally constructed from a single vantage point. One acts *in* large-scale spaces, but *on* small-scale spaces. Consequently, it might be difficult for students to transfer their thinking about mapping a classroom or playground to coordinate systems shown on a page or computer screen.

In Siegel's (1981) developmental model for the formation of cognitive maps of large-scale spatial environments, in the first phase, as one moves about in an environment, landmarks are noticed and remembered. In the second phase, routes linking landmarks are formed (and with further experience metricized). In the third phase, as coordinated frames of reference are developed, routes are integrated into overall configurational frameworks or survey maps.

In contrast, for small-scale spaces, Piaget's work on students' construction of coordinate systems is still relevant (Piaget, Inhelder, & Szeminska, 1960). For instance, when students were asked to draw a point on one rectangular sheet that was in exactly the same place as a point on another sheet, several levels were observed. At Level 1, students used global visual estimates to locate the point; measuring devices were either not used or used inappropriately. At Level 2, students used only one measurement (say from a rectangle side or vertex); they were satisfied with one measurement because they could not coordinate two. At Level 3, students recognized that two measurements are needed. For instance, they used a single measurement from a corner and tried to visually preserve the slope of the line when using this measurement. Or, they made two oblique measurements. At Level 4, students successfully used two perpendicular measurement dimensions to locate the point. This perpendicularity served as a conceptual device that enabled students to coordinate the measurement dimensions. However, students' use of two coordinated measurements was not yet interiorized; they tended to discover this approach using trial and error instead of immediately using it. Finally, only at Level 5, did students construct an interiorized mathematical coordinate system.

> [Level 5] subjects do not begin with a single measurement only to discover the need for a second dimension after a period of trial-and-error. From the outset, they realize the logical necessity to take both dimensions into account and the measurements are straightaway coordinated so as to be at right angles to one another. . . . [This] behavior is operational in nature. Before the facts are given in actual experience, they are engendered by structured anticipatory schema. (p. 169)

Presumably it is at this level that relations of order and distance between *objects* are replaced by similar relations between the *positions* themselves. It is as if a space were emptied of objects so as to organize the space itself. In fact, Piaget argued that

> a reference frame is not simply a network composed of relations of order between the various objects themselves. It applies equally to positions within the network as to objects occupying any of these positions and enables the relations between them to be maintained invariant, independent of potential displacement of the objects. (Piaget & Inhelder, 1956, p. 376)

Despite differences in the mental development of large- and small-scale spaces, there are two major commonalities. First, coordination (of actions and objects) is crucial. Second, a frame of reference must be mentally constructed, and actions and objects must be properly integrated into it. Like so many processes in geometry, both the coordination process and the formation of frames of reference depend critically on creating appropriate mental models.

GEOMETRIC MEASUREMENT

The concept of measurement[30] is woven throughout the fabric of geometric conceptualization, reasoning, and application. From the historical beginnings of geometry as "earth measure" to Birkhoff's axioms for Euclidean geometry, measurement plays a central role in reasoning about all aspects of our spatial environment. Measurement is critical for understanding the structure of shapes, using coordinate systems to determine locations in space, specifying transformations, and establishing the size of objects. Geometric

[30] I use the term *geometric measurement* in the broad and abstract mathematical sense to refer to the concept of assigning numbers to geometric entities in accordance with a set of axioms. At the school level, understanding geometric measurement includes (a) conceptualizing that numbers can be used to quantify the amount of an attribute (e.g., length, area, volume) contained in a geometric object by determining the number of attribute-units that fit in the object and (b) being able to implement procedures for assigning measurements to objects (e.g., iterating units, using a ruler, choosing appropriate measurement units). Thus, understanding measurement requires an integration of procedural and conceptual knowledge.

measurement is also embedded in the graphic representation of functions and algebraic equations.

However, despite the importance of geometric measurement, students' performance on measurement tasks is alarmingly low (Kloosterman et al., 2004; Martin & Strutchens, 2000; Sowder et al., 2004). For instance, according to NAEP results, in the year 2000, less than 25% of 4th graders and only about 60% of 8th graders were able to determine the length of an object pictured above a ruler with its end not aligned to that of the ruler (Kloosterman et al., 2004; Sowder et al., 2004). On a similar item in 1996, the percentages of Grade 4, 8, and 12 students answering this item correctly were 22, 63, and 83, respectively. Thus, almost 1 in 5 seniors in high school seemed to lack understanding of measuring length with a ruler. Performance on area and surface area was worse. In 2000, only about 14% of Grade 8 students could determine the number of square tiles it takes to cover a region of given dimensions, and only about 25% could determine the surface area of a rectangular solid (Sowder et al., 2004). In 1996, only 35% of 12th graders and 27% of 8th graders recognized and could explain why a square, and a right triangle with the same height and twice the width of the square, have the same area (Martin & Strutchens, 2000). And on volume, in 1990, only 55% of 12th graders and 41% of 8th graders knew that a measurement of 48 cubic inches for a rectangular box represented volume (NAEP web site).

Strongly implicated in this low performance is the research finding that for many students—perhaps the majority—there is a basic disconnection between spatial and measure-based numerical reasoning (Barrett & Clements, 2003; Battista, 2001a; Clements et al., 1997). That is, many students do not properly maintain the connection between numerical measurements and the process of unit-measure iteration. For instance, students who incorrectly measure the length of an object when one of its ends is not aligned with "zero" on a ruler do not clearly conceptualize how the ruler's numerical markings indicate the iteration of unit lengths. And most students who correctly use the formulas for the area of a rectangle or volume of a right rectangular prism in standard problem contexts, neither understand why the formulas work nor apply the formulas appropriately in nonstandard contexts. For example, a bright eighth grader who was 3 weeks from completing a standard course in high school geometry—so she was 2 years ahead of schedule for college prep students—responded as follows on the problem shown in Figure 19.16 (Battista, 1998b).

S: It's 45 packages. And the way I found it is I multiplied how many packages could fit in the height by the number in the width, which is 3 times 3 equals 9. Then I took that and multiplied it by the length, which is 5, and came up with 9 times 5, which is 45.

Obs: How do you know that is the right answer?

S: Because the equation for the volume of a box is length times width times height.

Obs: Do you know why that equation works?

S: Because you are covering all three dimensions, I think. I'm not really sure. I just know the equation.

Because of an inappropriate connection between her spatial structuring and the numerical procedure, this student did not understand that the mathematical formula she applied was inappropriate for this problem—a common problem for students. Indeed, only 38% of the students in her geometry class answered the item correctly, despite the fact that all of them had scored at or above the 95th percentile in mathematics on a widely used standardized mathematics test in 5th grade. Battista also found that only 19% of Grade 7 and 8 students in a range of classes in the same school, but not including this geometry class, answered this problem correctly (1999b), and that less than 10% of Grade 3–5 students correctly solved a similar problem in which the box was only one layer high (1998b). Simon and Blume (1994) reported preservice elementary teachers making the same error in an analogous area situation. Similarly, in a study by Reynolds and Wheatley (1996), fourth grader Kristin was trying to determine the number of 3-by-5 cards needed to cover a 15-by-30 rectangle. To solve the problem, she divided 15 into 450. However, the researchers found no evidence that Kristin realized that this quotient would be correct only if the two dimensions of the small card divided, respectively, the two dimensions of the large rectangle. In fact, Kristin seemed suspicious of her procedure, saying that she needed to verify it by drawing. Thus, Kristin was struggling to connect the spatial and numeric aspects of this problem.

In general, students working on these types of nonstandard measurement problems must perform two critical processes: (a) they must construct a proper spatial structuring of the situation; (b) they must coordinate their spatial structuring with an appropriate numerical scheme. Too often, students skip the first process and proceed directly to the second. Also, even when students recognize that they must perform the first process, they often have difficulty doing so. That is, because many traditional curricula prematurely teach numerical procedures for geometric measure-

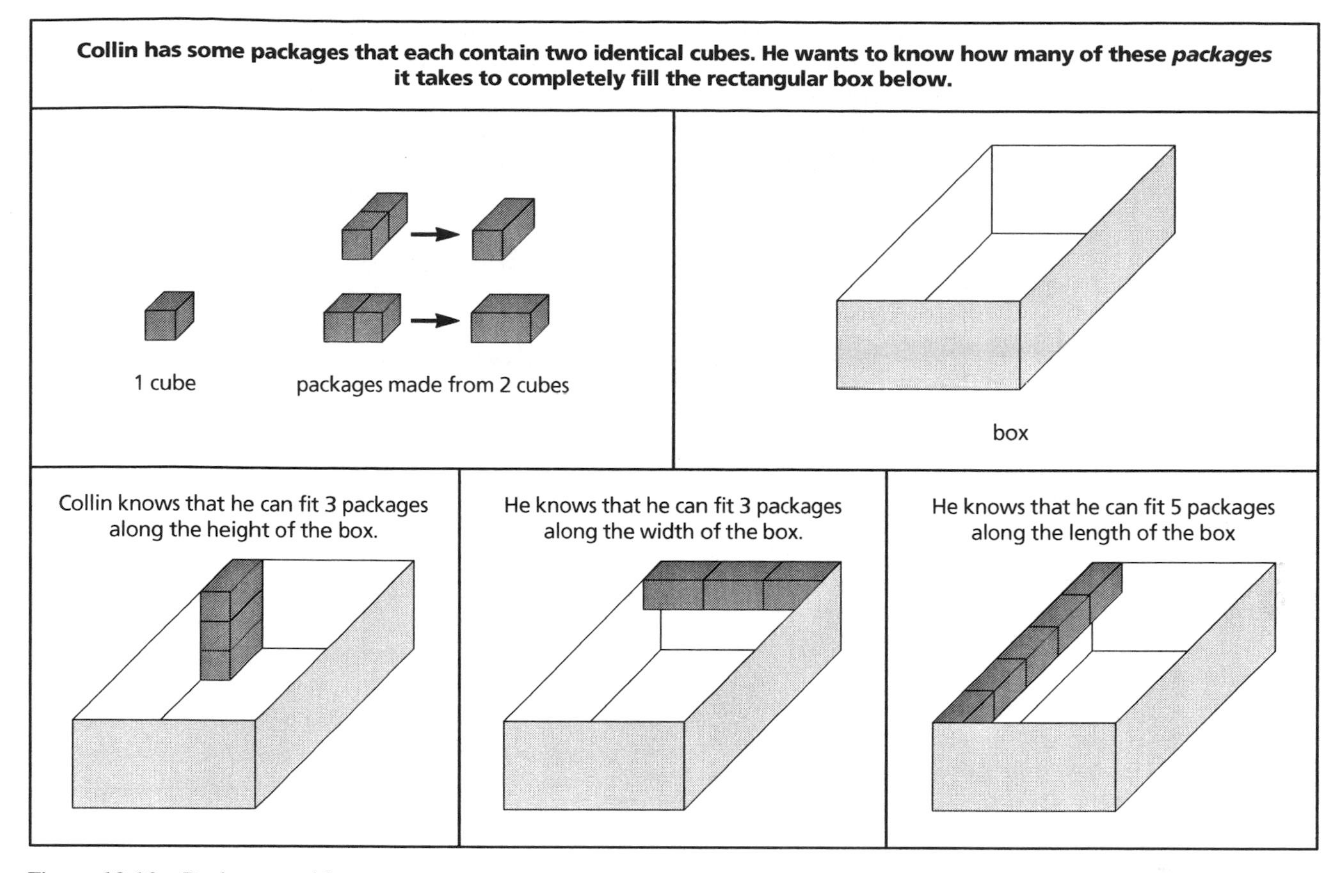

Collin has some packages that each contain two identical cubes. He wants to know how many of these *packages* it takes to completely fill the rectangular box below.
1 cube
packages made from 2 cubes
box
Collin knows that he can fit 3 packages along the height of the box.
He knows that he can fit 3 packages along the width of the box.
He knows that he can fit 5 packages along the length of the box

Figure 19.16 Package problem.

ment, students have little opportunity to think about the appropriateness of the numerical procedures they apply, and they have insufficient opportunities to develop skill in spatially structuring arrays of measurement units. In particular, even when students recognize that the volume formula is inappropriate for the Collin problem, they have significant difficulty constructing an appropriate spatial structuring of the packages because they have had so few opportunities to develop their structuring skills.

Research also shows that students commonly interchange measurement units or computational procedures (Chappell & Thompson, 1999; Nunes, Light, & Mason, 1993; Pesek & Kirshner, 2000; Woodward & Byrd, 1983), and that most students have difficulty understanding how the multiplication of two length units produces units of area (Battista, 2004; Kordaki & Potari, 1998; Nunes et al., 1993). *Overall, then, the research indicates that many students' measurement reasoning is superficial, with poorly understood procedures or formulas substituting for deep understanding.* In fact, the traditional premature instructional focus on computational formulas seems to interfere with students' concept development in geometric measurement.

THE DEVELOPMENT OF STUDENTS' REASONING ABOUT LENGTH

In many ways, research on linear measurement has not answered fundamental questions that were asked 2 decades ago. To illustrate, in a still widely cited study, Hiebert (1981) found that many students who failed Piagetian tasks of conservation and transitivity of length apparently still "learned" measurement strategies. He argued that although

> conservation and transitivity are logical prerequisites for completing many measurement tasks, children do not seem to use this knowledge when they solve the tasks. . . . They use an intermediate measurement to compare two lengths and do not think to ask the transitivity question; they move a unit to measure the length of an object and do not worry about whether the length is being conserved. Simple skills or techniques apparently allow children to bypass the logical structure of many measurement tasks. (p. 208)

Hiebert conjectured, however, that students who possess a particular Piagetian reasoning ability might learn related measurement concepts with a deeper level of understanding than those who do not possess this ability—that, for example, students who do not conserve length learn certain measuring strategies only at a rote level. To partially test this hypothesis, Hiebert made a crooked path with rods and asked students to construct a straight path that had just as far to walk on it (the rods used for the straight path were different in length than those used in the crooked path). All children failed this task before instruction. However, whereas 13 of 16 of the students who passed both Piagetian tasks were at least partially successful on this task after instruction, only 2 of 16 of the students who failed the Piagetian tasks were successful. Thus, it seems that conservation, or more likely, the mental operations that enable conservation, are necessary for a conceptually sound completion of this task. Consistent with Hiebert's hypothesis that simple techniques may substitute for deep understanding, Joram, Subrahmanyam, and Gelman (1998) claimed that even though young children can learn some simple measurement skills, actual physical measurement such as determining lengths with paper clips is challenging for many middle-school students. They also reported poor performance on linear measurement estimation even for high school students and adults. Thus, both old and new research suggests that students' knowledge of linear measurement may be more superficial than it appears in most common assessments.

Levels of Sophistication in Students' Reasoning About Length and Length Measure

Several groups of researchers have attempted to describe levels of sophistication in students' development of the concepts of length and length measurement. Because each group has focused on different aspects of length and used different sets of tasks, comparisons are difficult but also revealing.

Clements et al. (1997) observed three levels of sophistication in students' reasoning about path lengths, which were later extended to four levels by Barrett and Clements (2003). At Barrett and Clements' Level 1, students use gross visual comparison of objects. There is a disconnection between numerical measurement and the extent of line segments; students often estimate or guess measurements. At Level 2a, students connect numbers to iterative movements along a path but often have difficulty keeping track of their unit iterations. "Children at this stage expect that measuring consists in counting, but they are not consistent in their attention to measuring" (p. 55). At Level 2b, students integrate the iteration of a single unit of length into a sequence of units, but restricted to a single direction. So students might correctly iterate a unit along the top of a rectangle but lose track of the unit as they move to an adjacent but perpendicular side. At Level 3, students correctly iterate units around complex multidirectional paths, and they operate on units of units and reason about measurements in the absence of perceptual objects. Number schemes and spatial schemes are properly coordinated.

Battista (2003a, in press c) attempted to integrate previous research and develop a broader characterization of students' construction of meaning for length and length measurement. According to this characterization, there are two fundamentally different types of reasoning about length. *Nonmeasurement* reasoning *does not use numbers.* Instead it involves visual-spatial inferences based on direct or indirect comparisons, imagined transformations, or geometric properties. *Measurement* reasoning involves *unit-length iteration,* that is, determining the *number* of fixed unit lengths that fit end-to-end along the object, with no gaps or overlaps. Measurement reasoning includes not only the process of measuring, but reasoning about numerical measurements (e.g., adding lengths to find the perimeter of a polygon, making inferences about length measurements based on properties of figures). Although students typically develop nonmeasurement strategies before measurement strategies, nonmeasurement reasoning continues to develop in sophistication even after measurement reasoning appears. Furthermore, the most sophisticated reasoning about length involves the integration of nonmeasurement and measurement reasoning.

Nonmeasurement Reasoning

Nonmeasurement Level 0. Students' reasoning about length is appearance-based and holistic. They attend to how things look, focusing on whole shapes rather than systematically on parts within shapes. They often use direct and indirect comparison.

Nonmeasurement Level 1. Students systematically use decomposing/recomposing to compare lengths. At first, students rearrange (physically, by drawing, or in imagination) some or all path pieces, and directly and visually compare the rearranged paths as wholes. Later, students compare two paths by matching, one-by-one, pairs of pieces that they think are the same length—they do not transform one path into another.

Nonmeasurement Level 2. Students compare path lengths by sliding, turning, and flipping shape parts in ways that allow them to infer, *based on shape and motion properties*, that one transformed shape is congruent to another.

Measurement Reasoning

Measurement Level 0. Students use counting to find lengths; however, their counting does not represent the iteration of a fixed unit length. For instance, students might recite numbers as they continuously move their fingers along a path. Or they might count dots along a path but not as true indicators of unit lengths. (See Sarama et al., 2003, for similar findings.)

Measurement Level 1. Students attempt to iterate what they consider to be a unit length along an object or path. However, because students (a) do not fully understand what a length unit is, or (b) do not properly coordinate iterated units with each other, their iterations contain gaps, overlaps, or different length units, *and are incorrect.*

Measurement Level 2. Students not only understand what a length unit is, when iterating unit-lengths, they properly coordinate the position of each unit with the position of the unit that precedes it so that gaps, overlaps, and variations in unit lengths are eliminated.

Measurement Level 3. Students determine some length measurements without explicitly iterating every unit length. That is, after iterating some length units, students operate on the results of their iterations numerically or logically to find other lengths.

Measurement Level 4. Students numerically and inferentially operate on length measurements *without iterating any unit lengths.* They make complex, property-based visual inferences about measurements using properties of geometric shapes and transformations. At this highest level of measurement reasoning, students fully integrate and apply the processes from Nonmeasurement Level 2 with their measurement reasoning.

Cognitive Processes for the Conceptualization of Length Measurement

Research suggests that students construct meaningful understanding of length measurement as they abstract and reflect on the process of iterating unit lengths (Barrett & Clements, 2003; Carpenter & Lewis, 1976; Hiebert, 1981; Kamii, 1995; Cobb, Stephan, McClain, & Gravemeijer, 2001; Piaget, Inhelder, & Szeminska, 1960). For instance, Joram et al. (1998) cited research indicating that, in making linear measurement estimates, iterating a unit is the most commonly reported strategy. In one study, 81% of elementary students reported using unit iteration in estimation, with about one third of these students physically marking off units, whereas the other two thirds were observed marking off units with their eyes. Furthermore, consistent with the unit-counting strategy hypothesis, Joram et al. cited one study in which the time it took adults to estimate the length of line segments increased linearly with the length of the segments and another study in which the times taken to estimate the number of 1-cm units in a 6-cm segment were the same as those to estimate the number of 2-cm segments in a 12-cm segment.

But how exactly do the conceptualizations of length and length measurement develop? What are the basic cognitive operations required? Joram et al. (1998) argued that the mind mentally represents counting numbers as magnitudes using a continuous accumulator mechanism (like putting water into a beaker). This mechanism includes a bidirectional mapping between verbal/written number symbols and accumulated mental magnitudes. In this view, counting a set of 5 objects activates the corresponding magnitude on a mental number line (via the accumulator), and, inversely, when a quantity on the mental number line is activated, a portion of the mental number line is subdivided and intervals assigned appropriate numbers.

Joram et al. (1998) discussed how this counting process can be applied to linear measurement estimation and how the process explains why linear estimation is more difficult than counting. According to the accumulator model, the mental representation activated when counting a set of discrete items is a continuous magnitude. During object counting, this continuous magnitude is made discrete by single bursts "of impulses to the accumulator for each item in the set" (p. 422). The perceptual discreteness of the counted objects provides the signals for these bursts. Joram et al. hypothesized that what makes linear measurement estimation more difficult than counting is that, when estimating, there are no perceptual cues to trigger the bursts. Such burst activation requires an individual to mentally segment the linear object into units, which requires (a) recalling and maintaining an image of the length unit, (b) visualizing the iteration of the unit along the object to be estimated, (c) keeping track of where the last unit ended and where the next one must begin, and (d) maintaining a running count of units while performing tasks a–c. Further complicating matters, the burst interval (interval size) changes, depending on the length unit employed (e.g., inches, feet).

Given that both linear extent and Joram et al.'s (1998) hypothesized mental representation of counting are based on continuous mental models, it is interesting that linear measurement estimation is more difficult to master than counting (Barrett & Clements, 2003; Piaget et al., 1960). In some sense, it would seem that the fact that both length and the counting-accumulator are continuous would make linear quantification easier than discrete quantification. And perhaps, at some level, it is. That is, in some contexts linear extent seems easier to comprehend than counting (thus, for example, the widespread use of bar graphs). However, Joram et al.'s burst-theory explanation suggests that difficulties in applying counting to measuring length reside in the transition from discrete to continuous quantities; individuals must establish an isomorphism between the discrete counting system and linear quantities. The episode below illustrates a student having difficulty with this isomorphism (Battista, in progress; cf., also Clements et al., 1997).

In this episode, when asked how she knew the rectangle she drew was 40 units around, JAK (grade 2) drew irregularly spaced dots along its inside edge, stopping when her count reached 40 (see Figure 19.17). There was no indication that JAK used dots as indicators of fixed unit lengths. She was unable to properly connect the continuous quantity that was the perimeter with the discrete items of the verbal counting sequence.

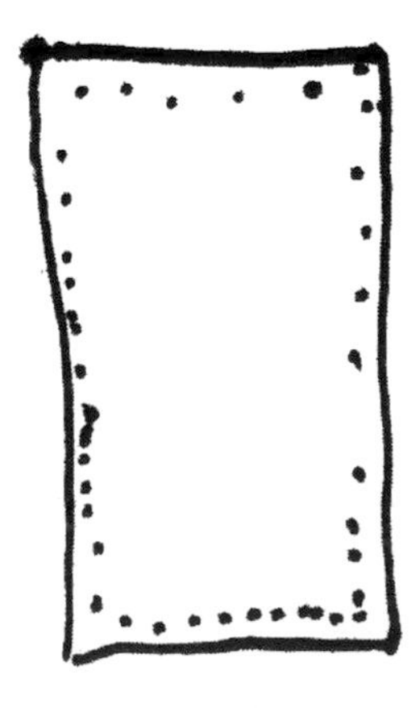

Figure 19.17 JAK's rectangle.

A somewhat different description of the process by which the mind constructs length and length units is given by Steffe (in press), which I will liberally summarize and interpret. According to this theory, individuals abstract continuous experiential items as they move a finger along an item or scan the item with their eyes. Because continuous motion is contained in the abstraction of the object, the perceived object is experienced as continuous. Furthermore, because of the way that an individual becomes aware of a continuous path—the perceptual abstraction contains its endpoints, the *motion* that produces the path, and the duration of the motion—the individual also becomes aware of the path's linear extent. "The property of a segment that we call length is an awareness of the scanning action over an image of the segment along with an awareness of the duration of the scanning action" (p. 3). Extending this analysis, for a segment to be used as a unit in iteration, it must be abstracted at an even higher level. The scanning motion used to perceive the unit length must be interiorized, then applied in steps a–d above.

Cognitive Processes for Conserving Length

What kinds of cognitive processing are required to conserve length? In dealing with conservation of length tasks, Piaget et al. (1960) found that students at the lowest levels (1 and 2a) of sophistication on length conservation failed to simultaneously account for both ends of the sticks, which the researchers saw as an indication that the students were not attending to the intervals of length between the endpoints. So the objects were not compared as continuous entities. Perhaps the reasoning used by students at this level is based on abstraction of experiences with comparisons in which only one set of endpoints are examined, for instance, comparing who is taller when two people are standing next to each other. (Although the *logic* of comparing single endpoints in this context seems obvious to adults, it is complex and may not be at all clear to young children.)

At the next highest level (2b), coordination of parts is clearly evident—student responses indicate "the beginning of a relationship between the two paired extremities (i.e., the four extremities [of the two sticks] taken in pairs" (Piaget et al., 1960, p. 100). This relationship arises as a consequence of students' abstraction and comparison of two scanning motions, one for the extent of each stick. In discussing this level, Piaget et al. made an important observation about operational reasoning. They stated that some students at Stage 2b are close to operational reasoning (which they claimed happens at Stage 3) but not quite there. For instance, one student noted that the sticks were equal when aligned, but when the sticks were moved out of alignment, he had to realign them to convince himself that they were still equal. Piaget et al. identified this reasoning as empirical, not operationally reversible.

Level 3 is achieved only when students see conservation of length as logically necessary. Piaget et al. (1960) claimed that this happens when objects are viewed in terms of a reference system "which provides a common medium for all objects" (p. 103). Within this medium exists a system of sites that exist independent of positional changes because they are fixed in relation to an invariant backdrop: "When an object undergoes a change of position the empty sites which have previously been occupied are equivalent to sites which were previously empty and are now filled, so that the overall length of the object remains constant" (p. 103). That is, students coordinate the scanning motions performed on objects within a system of possible scannings, which is the "space" in which these motions exist (which suggests that the scannings have reached the second level of interiorization).

In contrast to the Piagetian view, a simpler explanation of the mechanism for length conservation is possible. As children repeatedly compare abstracted scanning motions for nonaligned equal-length objects by placing them in appropriate end-to-end alignments, they abstract a general reversible alignment process. This process, coupled with exposure to adults who interpret the process as indicating that the length of an object does not change when it moves, builds the reasoning structure that supports conservation.[31]

THE DEVELOPMENT OF STUDENTS' REASONING ABOUT AREA AND VOLUME

Genuine understanding of area/volume measurement requires comprehending (a) what the attribute of area/volume is and how it behaves (i.e., conserving it as it is moved about and decomposed/recomposed), (b) how area/volume is measured by iterating units of area/volume, (c) how numerical processes can be used to determine area/volume measures for special classes of shapes, and (d) how these numerical processes are represented with words and algebra.

A core idea in developing competence with measuring area and volume in standard measurement systems is understanding how to *meaningfully* enumerate rectangular 2d and 3d arrays of squares and cubes, respectively. Previous research studies have described separate levels of sophistication in students' enumerating squares and cubes in such arrays (Battista, 1999; Battista & Clements, 1996; Battista et al., 1998). More recently, Battista (2003b, 2004) has integrated the separate models into a more general model for area and volume, which is described below.

Underlying Mental Processes

Battista claims that five basic cognitive processes are essential for meaningful enumeration of arrays of squares and cubes: abstraction, forming and using mental models, spatial structuring, units-locating, and organizing-by-composites (the first three processes have already been discussed).[32] The units-locating process locates squares and cubes by coordinating their locations along the dimensions that frame an array. For instance, to understand the location of a square in a rectangular array, an individual must "see" the square in terms of a two-dimensional coordinate-like system—for example, it is in the fourth column and the second row, or it is the fourth unit to the right and the second unit down. The organizing-by-composites process combines an array's basic spatial units (squares or cubes) into more complicated composite units[33] that can be repeated or iterated to generate the whole array. For instance, in a 2d array, a student might mentally unite the squares in a row to form a spatial composite unit that can be iterated in the direction of a column to generate the array. Students can meaningfully enumerate arrays of squares and cubes only if they have developed properly structured mental models that enable them to correctly locate and organize the squares and cubes.

Levels of Sophistication in Students' Structuring and Enumeration of Arrays

Level 1: Absence of Units-Locating and Organizing-by-Composites Processes

Students do not organize units into spatial composites, and, because they do not properly coordinate spatial information, their mental models are insuffi-

[31] Objects actually change length as they are moved from one temperature to another (which was a major problem in establishing stable units of measure in human history). So, when we say that length is preserved, we are implicitly utilizing one particular simplified model of how the world works. Thus, although it might seem that abstracting conservation of length is empirically based, it also depends on social consensus.

[32] See Battista (2003b, 2004) for further details. Also, see (Wheatley, 1997) for additional examples of students' difficulties with these processes in 3d situations.

[33] A *composite unit* is a cognitive entity (an abstraction) that results from mentally uniting and taking as one thing a collection of perceived or re-presented objects.

cient to locate all the units in arrays. Enumeration of squares or cubes seems almost random. The "double-counting" error is ubiquitous.

Level 2: Beginning Use of the Units-Locating and the Organizing-by-Composites Processes

Students not only start to spatially structure arrays in terms of composite units, their emerging development of the units-locating process produces mental models sufficient for them to recognize equivalent composites. For instance, after counting the cubes visible on one side of a building, a student *infers* the number of cubes on the opposite side.

Level 3: Units-Locating Process Sufficiently Coordinated to Eliminate Double-Counting

Students' units-locating process coordinates single-dimension views (e.g., top, side) into a mental model that is sufficient to recognize the same unit from different views, enabling students to eliminate double-counting errors. However, coordination is still insufficient to build a mental model that properly locates interior squares or cubes.

Level 4: Use of Maximal Composites, But Insufficient Coordination for Iteration

Students structure arrays in terms of maximal composites (rows or columns for area; layers for volume). But, due to insufficient coordination, the student cannot precisely locate these composites, instead estimating their locations.

Level 5: Use of Units-Locating Process Sufficient to Correctly Locate All Units, But Less-Than-Maximal Composites Employed

Students' mental models correctly locate all squares/cubes in an array. However, although students sometimes get correct answers, because they inefficiently organize arrays into composites, they frequently lose their place in enumeration. Furthermore, students' structuring and enumeration strategies are not generalizable and are inadequate for large arrays.

Level 6: Complete Development and Coordination of Both the Units-Locating and the Organizing-by-Composites Processes

Students' interiorized mental models fully incorporate a row-by-column or layer structuring so that they can accurately reflect on and enumerate an array without perceptual or concrete material for the individual units within composites.

Level 7: Numerical Procedures Connected to Spatial Structurings, Generalization

Students' spatial structuring *and* enumeration schemes reach a level of abstraction at which they can be reflected on and analyzed, thus enabling students to explicitly understand the connection between an enumeration strategy and the spatial structuring on which it is based. Students' mental models incorporate row-by-column or layer structuring that is abstract and general enough to apply to situations in which the basic units are not cubes (e.g., packages made from two cubes; e.g., see Figure 19.16).

Additional Reasoning

In developing sophisticated reasoning about area and volume measurement, there are additional types of reasoning that students must attain, beyond the levels listed above. For instance, students must generalize the reasoning so that they can reason about fractional units in area and volume measurement, a move that can be difficult for students (Kordaki & Potari, 2002). Students must also generalize their thinking to curved or irregular shapes in which it is not easy to visualize square or cube tilings; in a sense, students must be able to think about deformable units. Students must also generalize their thinking from the small scale spaces in which area/volume are initially studied to large scale spaces such as rooms and buildings. Also, consider the ability to spatially structure the decomposition of, say, a rectangular prism into smaller, non-square units—as might be needed in carpentry (Millroy, 1992). Conceptually, such reasoning may be available at Level 7, but spatially, such reasoning can be extremely demanding and is unlikely to be learned in traditional school settings.

Researchers must carefully investigate students' attainment of these additional types of reasoning. How, for example, are these types of reasoning related to the seven levels already described? Students must also be able to interrelate area and volume. For instance, advanced thinking about volume (e.g., that needed in calculus) depends on understanding a multiplicative relationship between the height of an object and its cross-sectional area. How do students make the transition from a conceptualization based on iterating layers of cubes to this more sophisticated idea?

Additional Research

Transitions to Abstract Tools

The research seems pretty clear that students need appropriate preparation before learning about the use of standard measuring tools such as rulers and the use of formulas. Students need to develop understanding of iteration of appropriate units of length, area, and volume. However, research has not yet clearly described how this transition takes place. Although Battista has described levels of sophistication in the development of students' reasoning about length, area, and volume, other than for volume, he has not followed students during extended periods of instruction. More longitudinal research is needed.

The Connection Between Linear and Area Measurement

Several studies have indicated that students have difficulty properly relating and separating the concepts of length, area, and volume. For instance, preservice and inservice teachers frequently believe that whenever the perimeter of a figure increases, so does its area (Ma, 1999; Tierney, Boyd, & Davis, 1990). Students and teachers also often believe that when the lengths of the sides of a square or cube double, so does the area and volume (e.g., Tierney, Boyd, & Davis). Nunes et al. (1993) found that when 9–10-year-olds were asked to compare the amount of paint needed to cover two rectangles, far more students answered correctly when they used physical squares than when they used rulers. In fact, 26 out of 29 students who used rulers either added the measures of the sides or did not know what do to do with the rulers to make an area comparison.

At an even more fundamental level, Outhred and Mitchelmore (2000) conjectured that some elementary students do not relate the number of squares in the rows and columns of a rectangle to the lengths of the sides. Corroboration for this conjecture comes from Battista (1996) who found that some fifth graders, who correctly enumerated the number of cubes in a box when different types of pictures depicting boxes and cubes were shown, were unable to determine the number of cubic centimeters in an unmarked box when they were given a ruler (students were shown a cubic centimeter and told that it measured 1 cm on each edge). However, this idea has not yet been fully and carefully investigated. For instance, Outhred and Mitchelmore's students' errors may have been caused by a lack of spatial structuring of the squares in a rectangular array, or by deficiency in linear measurement. Battista's students presumably were able to structure 3d arrays for cubes correctly, but he did not separately assess their understanding of linear measurement, so there is no way of knowing if the cause of the problem for the students could be attributed to their deficiencies in conceptualizing length. In fact, given the difficulty that many students have with the concept of length measurement, it is not surprising that some of them struggle with building on this concept and properly relating it to area and volume measurement.

Understanding, Making, and Drawing Tilings

A number of researchers have investigated students' understanding of unit-area tilings of larger shapes and the plane, especially as revealed by students' drawings. For instance, Outhred & Mitchelmore (2000) found several levels of sophistication in elementary students' drawings of unit-square coverings of rectangles. At Level 0, the unit squares drawn by students did not completely cover rectangles without gaps or overlaps. At Level 1, unit squares completely covered rectangles without overlap, but the organization was unsystematic; units varied considerably in size and shape. At Level 2, drawings of unit squares exhibited correct array structure, with equal numbers of units in rows, and in columns, but the size of the units was determined visually, not from the dimensions of the rectangle. At Level 3, rectangles' dimensions were used to iterate rows; one dimension was used to find the number of units in a row, while the other dimension was used to determine the number of rows. At Level 4, students did not need to draw units; the numbers of units in rows and columns were used to calculate the total number of squares.

Investigating students' tilings of different kinds of shapes (not just squares) has shed additional light on the cognitive processes underlying tiling. As a case in point, Owens and Outhred (1997) suggested that, to be successful, students have to understand the geometry of the tile and how it affects gaps and overlaps in tiling. Students also have to understand the structure of the pattern of the tiling. Owens and Outhred found that many students were unable to see or draw appropriate tilings. Wheatley and Reynolds (1996) argued that drawing a tessellation of a shape requires students to develop an image of the shape that can be used in drawing it then developing a plan for covering the plane. Wheatley and Reynolds found that some students drew shapes one at a time, seemingly without understanding the overall structure of the tiling. Other students used an overall structure, including the use of composite units, to guide their activity. Thus, tiling a shape requires two critical components. Component 1 consists of constructing a mental model *of the shape* that (a) incorporates critical features of the shape's geometry and (b) can be mentally manipulated. (Some students are able to visualize exactly where a tiled shape should be placed, while other students must use trial and error with physical materials.) Component 2 consists of developing an appropriately structured mental model *of the array* of shapes in the tiling.

Research by Clements, Wilson, and Sarama (2004) that describes levels of sophistication in young children's ability to make a larger shape with pattern blocks helps elaborate these components. For instance, in Clements et al.'s scheme of seven levels, at the second level, children employ trial and error to make simple given shapes, have limited ability to use turns or flips to help them make these shapes, view shapes only as wholes, and see few geometric relationships between shapes. At the third level, shapes are chosen using a gestalt configuration or one component such as side length, vertices may be matched but without a concept of angle size, and turning and flipping are used, usually by trial and error. At the fourth level, children combine shapes with growing intentionality and anticipation, angles are matched by size, and turning and flipping are used with anticipation to select and place shapes. Thus, in relation to Component 1, students gradually come to use the geometry of shapes, moving from holistic to parts-based reasoning, attending to side length before angle measure, and using motion first physically then in imagery, a progression that harkens back to the van Hiele levels and the previous discussion of mental models. The fact that Clements et al.'s last three levels deal with increasing sophistication in the use of composite units is consistent with the notion that students develop Component 2, structuring the tiling array, after Component 1.

Also related to Component 2, Outhred and Mitchelmore (2000) made an important observation about the difference between drawing and using physical shapes to tile the inside of a larger shape. In the case of squares tiling the inside of a rectangle, Outhred and Mitchelmore argued that because of the way that wooden squares fit together, using wooden squares to cover a rectangle "prestructures" the array—"the array structure is inherent in the materials and does not need to be apprehended by the learner" (p. 146). (They also cited research suggesting that such use of squares may not be effective instructionally.) When drawing, however, students must explicitly create an appropriate structure. Furthermore, "the attempt to represent a covering in a drawing can help children to examine their experience in new ways and lead to new insights" (p. 165).

An episode from Battista et al. (1998) corroborates and extends Outhred and Mitchelmore's claim about material squares pre-structuring arrays. CS was asked to predict how many squares would cover the rectangle shown in Figure 19.18a (making her prediction without drawing). CS pointed and counted as in Figure 19.18b, predicting 30. So clearly she was having difficulty properly structuring the array. When checking her answer with plastic squares, she correctly covered the rectangle with squares. So, consistent with Outhred and Mitchelmore's claim, the placement of the squares was structured by the physical materials. However, CS initially enumerated her correctly placed squares as shown in Figure 19.18c, getting 30, then counted the squares again, first getting 24, then 27. So, although CS was able to correctly place physical squares to cover the rectangle, she was still unable to structure the squares so that they could be correctly enumerated—CS had not mentally constructed the structuring contained in the physical materials.

Let me now address Outhred and Mitchelmore's claim that drawing tilings can help students examine their experience in new ways, leading to new insights into the structure of an array. This claim is consistent with the theory that spatial structuring is a matter of abstracting one's actions on an array, not directly abstracting the array itself. That is, structuring is a reflective not empirical abstraction. However, to gain additional insights from one's drawing, more than abstraction of one's structuring actions is required. For example, typical of young students, CS not only drew an inadequately structured covering as shown in Figure 19.18d, she failed to recognize the inadequacy of her structuring (Battista et al., 1998). Thus, drawing may not promote better structuring than placing

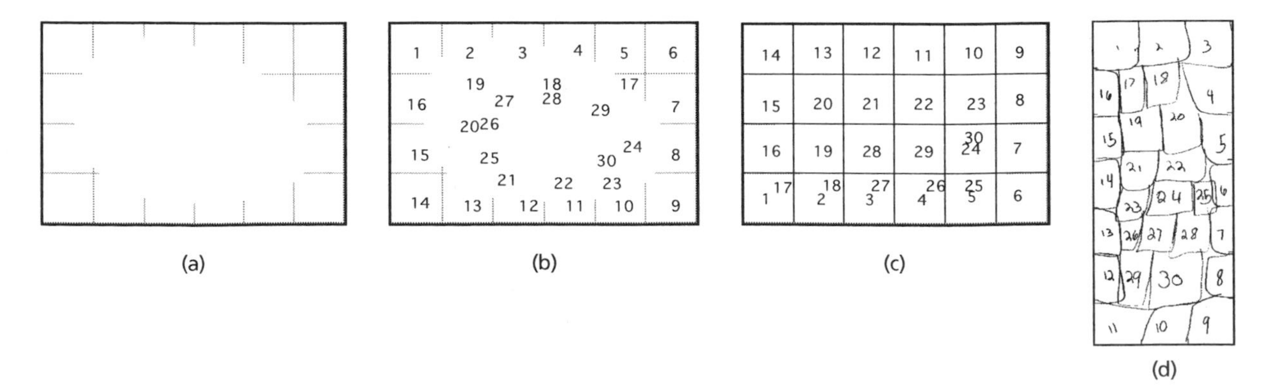

Figure 19.18 CS' structuring of arrays of squares (Battista et al., 1998).

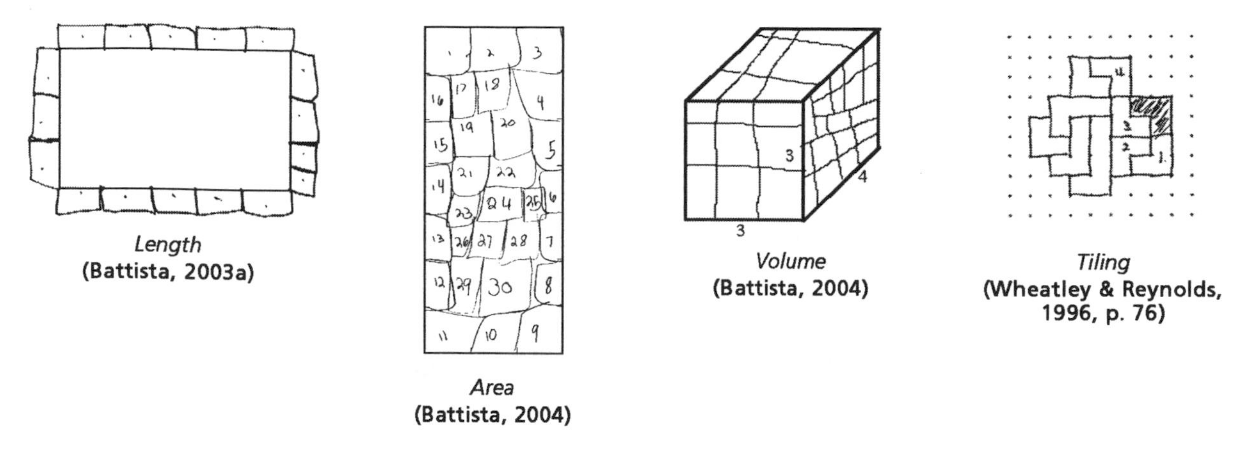

Figure 19.19 Stage 1 of the process of abstracting unit iterations.

physical squares unless students recognize difficulties in their drawings, creating cognitive perturbations. With proper instructional guidance, such perturbations can occur and cause refinements in students' structuring. That is, making drawings and checking drawings using physical squares can cause students to reflect on and refine the structuring contained in their mental models of arrays. In fact, an analogous "predict and check" approach was shown to be very effective in promoting the development of students' structuring and enumeration of 3d arrays of cubes (Battista, 1999). This approach is effective because it focuses students' attention and analysis, not on arrays, but on their *mental models* of arrays.

SYNTHESIS

Cognitive Processes Underlying Geometric Measurement: The Many Faces of Abstraction

Several kinds of abstractions are critical in geometric measurement.

Abstracting the Attribute

First, students must abstract the attribute to be measured (say, length), distinguishing it from all the other spatial attributes possessed by an object. This might happen through reflection initiated by (a) dealing with various kinds of problematic situations (e.g., comparing objects, fitting an object through an opening) or (b) interacting with people who already possess the concept and draw attention to it. Initially, what students abstract during these experiences may not match the formal concept. However, during these or subsequent experiences, students might notice inadequacies in their conceptualizations, causing perturbations to occur. These perturbations may cause them to refocus their attention and abstract conceptualizations of the attribute that are close enough to the formal attribute that social interactions or problem-solving contexts do not cause further perturbations.

Abstracting Unit Iterations

The process of abstracting unit iterations sufficient to measure various geometric attributes is extremely complex. But, based on current research, we might conjecture the following stages in the abstraction process.[34]

Stage 1, incomplete abstraction of unit iterations. Because units and iterations are abstracted at most at the internalized level, iterations are uncoordinated and have gaps, overlaps, and nonequivalence of units. The student work shown in Figure 19.19 illustrates this level for the contexts of length, area, volume, and two-dimensional tiling.

Stage 2, units interiorized and coordinated. To meaningfully and correctly perform unit iteration, students must, at the very least, develop an interiorized mental model *of the unit*. This mental model incorporates relevant elements of the geometry of the unit and permits *coordinated* mental manipulation of the unit so that gaps, overlaps, and nonequivalence are eliminated in the iteration process.

Stage 3, structure of iterations interiorized. Students develop interiorized mental models of the *structure of unit iterations* that make it possible (a) to see a particular unit in relation to the iterated sequence of units, enabling

[34] Each step described below must be properly coordinated with numerical operations.

students to understand a unit's location; (b) to structure iterations in terms of composite units; and (c) to establish a part-whole relationship between the iterated units and the whole by reconstituting the whole as the sequence of iterations of the unit (Barrett & Clements, 2003; Steffe, in press; see also Cobb et al. 2001).[35,36]

Stage 4, numerical measurements become symbols. Mental models of structured unit iterations are raised to the second level of interiorization so that the resulting enumerations become symbols. That is, numbers, construed as measurements, act as symbols for the unactualized process of iteration. This enables students to meaningfully reason about measurements without iterating units.

Integrating Perspectives on Length, Area, and Volume Measurement

Because individual research studies usually have focused on only one type of geometric measurement at a time (length, area, or volume), research has not yet produced a comprehensive theory of geometric measurement. Battista (2003b, 2004) integrated models for the development of area and volume measurement but did not extend this integrated model to length. However, comparing Battista's levels for length to those for area/volume suggests directions and issues for integration. In particular, the question of how nonmeasurement and measurement reasoning are related must be addressed. That is, Battista's levels for area/volume address only one component of understanding these concepts—enumerating units. Nonmeasurement ideas are not addressed, nor is the critical relationship between nonmeasurement and measurement reasoning. A first attempt at developing a general theory of geometric measurement is shown in Table 19.1 (which, at this point, can serve only as a conjecture). In this theory, nonmeasurement and measurement reasoning develop in parallel (though not simultaneously). However, in contrast to this parallel-development hypothesis, Kordaki and Potari (2002) argued that first, students should spatially explore areas; second, they should physically or pictorially transform and compare areas without the use of numbers; third, they should use measurement units; and finally they should use area formulas. Much more research is needed to tease out the interrelationships between nonmeasurement and measurement reasoning.

Final Comments on Measurement

Although students' measurement difficulties are alarming because measure is essential for most real-world applications of geometry, these difficulties may represent only the tip of a huge learning-difficulty iceberg—an iceberg that research has not yet investigated. Indeed, given the pervasiveness of measure in geometric and graphical contexts, poor understanding of measure might be a major cause of learning problems for numerous advanced mathematical concepts.[37] For instance, students' learning of high school geometry and analytic geometry is endangered by poor understanding of geometric measure, especially by misunderstandings of length. Furthermore, given students' disconnection between numerical measurements and the process of unit iteration, researchers must question the depth of students' understanding of coordinate systems, function graphing, locus problems, and theorems about lengths or distances.

Specific examples of these endangered topics abound: (a) Comprehending common proofs of the Pythagorean theorem requires understanding length and area and the distinction between the two. (b) Hollebrands (2003) found high school students unable to generalize their conceptualizations of linear magnitude in a way that enabled them to construct the notions of vectors and translations defined by vectors. (c) Almost half of a set of very advanced eighth graders and more than 80% of above average seventh and eighth graders had great difficulty conceptualizing common geometric shapes (rectangles and squares) in terms of measurement-based properties (Battista, 1998a). (d) A number of researchers have documented students' difficulties in understanding the connection between linear equations and their graphs (Dugdale, 1993; Knuth, 2000; Moschkovich, Schoenfeld, & Arcavi, 1993). Knuth, for instance, found that students failed to recognize that the coordinates of graph points used to calculate the slope of a line are solutions to the equation of that line, a finding that could be directly traceable to an insufficiently developed conceptualization of how coordinates are related to lengths. That is, students might see coordinates not as lengths but as markers on a map (just like margin letters on road maps).

---

[35] It is unlikely, however, that a–c occur at the same time.

[36] Structuring unit-length iterations increases in complexity when paths are situated in 2d space. For instance, Barrett and Clements (2003) found that maintaining unit lengths was more difficult along nonstraight than straight paths.

[37] The notion of poor understanding of geometric measure endangering later mathematics learning was developed in collaboration with Jack Smith.

Table 19.1 Reasoning About Measurable Geometric Quantities

Nonmeasurement Reasoning	Measurement Reasoning
1. Holistic Visual Comparison of Shapes Students' reasoning is appearance-based and holistic. It is appearance-based because students focus strictly on the appearance, or "how things look," of objects. It is holistic because students focus on whole shapes or objects, not parts within shapes. Students' strategies are imprecise and often vague. Conservation and transitivity gradually develop. **2. Visual Comparison of Shapes by Decomposing/Recomposing** Students decompose and recompose spatial regions into smaller regions either physically or mentally, then compare the recomposed regions directly. There is no thought about enumeration of units or measured quantities, so no measurement reasoning occurs. When decomposing shapes, students might compare recomposed whole shapes, or they might compare shapes region-by-region. **3. Comparison of Shapes by Property-Preserving Transformations/Decompositions** Students compare shapes by decomposing them then sliding, turning, or flipping shape parts in a way that allows them to infer, based on shape or transformation properties, that one transformed shape is congruent to another. Students do not necessarily explicitly mention geometric properties (congruence of shapes, segments, angles) or transformations (slide, flip, or turn), but their reasoning is consistent with such properties and transformations.	**1. Use of Numbers Unconnected to Unit Iteration** Students use counting to determine measures; however, their counting does not represent the iteration of an appropriate unit. **2. Unit Iteration and Enumeration** Students attempt to iterate and enumerate appropriate units of measure to fill the shape being measured. At first, because students do not properly coordinate these units with either each other or the whole object being measured, their iterations contain gaps, overlaps, or different units—students often lose track of units while enumerating. Gradually, students overcome these difficulties. *Cognitive Milestones* • Units properly maintained, coordinated, and located • Units properly organized into composites • Iterations properly structured • Operating on iterations • Iteration schemes generalized to fractional units • Iteration schemes generalized to non-segment/-square/-cube units **3. Operating on Numerical Measurements** Students numerically operate on measurements *without iterating units*. At this level, iteration seems to have fallen in the background. It is as if iteration is taken as being done.
4. Integrated Measurement and Nonmeasurement Reasoning Students inferentially operate on measurements *without iterating units*. They can make complex, property-based visual inferences about measurements, which are often accomplished with transformations or by using properties of geometric shapes. At this highest level of measurement reasoning, students fully integrate and apply the processes from Nonmeasurement Level 3 with their measurement reasoning. The difference between Nonmeasurement Level 3 and Measurement Level 4 is that in level 4 students make inferences about numerical measurements, not shapes. For instance, students can develop, meaningfully use, and justify procedures or formulas for finding the areas/volumes of nonrectangular shapes.	

CONCLUSIONS

Despite geometry's importance in mathematical theory and application, students continue to have difficulty learning it with genuine depth. There is much confusion and conceptually shallow, procedural learning. The research described in this chapter discusses in great detail many student difficulties, often suggesting avenues for instructional improvements. But much additional research is needed that delves even more deeply into students' difficulties and carefully examines the effects and effectiveness of alternate instructional approaches.

Moving Beyond the Cognitive Perspective

This chapter has discussed a wide range of research relevant to understanding students' geometric reasoning. Because of the major research foci in this domain, the research has been both heavily phenomenological—how students construct meaning for geometric concepts—as well as cognitive—what cognitive processes are necessary for students' construction of geometric meaning. Far fewer studies have examined geometry learning from a sociocultural perspective. This perspective would focus on how the geometric meanings students construct are affected by their participation in various communities in which certain mathematical practices, interaction patterns, conceptions, beliefs, and reasoning are shared.

The focus of geometry research on cognitive rather than social processes has been both good and bad for the field. On the positive side, the cognitive approach has produced significant understandings of the development of students' geometric reasoning. Although there is still much to understand about the cognitive processes that underlie geometric reasoning, the cognitive approach not only is critical for adequately investigating these processes, it helps connect the field to important related fields such as psychology. It is critical that this approach continue. However, on the negative side, not enough has been done to understand sociocultural and affective factors in geometry learning. Researchers do not know nearly enough about how various components of social practice affect students' construction of geometric concepts and reasoning. For instance, what kinds of classroom practice effectively encourage students to personally embrace (rather than mimic) increasingly sophisticated types of justification in geometry, and what social processes enable such progress? (See Borrow, 2000 and Quinn, 1997 for discussions of this issue that combine social and cognitive perspectives.) We also need additional research on how affective factors are interrelated with geometric learning. For instance, how do students "feel" about being required to give formal proofs when they understand neither the need for such proofs nor their logic, and how do these feelings affect students' cognition and learning? Of course, in all but classes dedicated specifically to geometry, classroom practices generally transcend specific mathematical topics; so investigating some of these issues from a geometric perspective may be difficult. However, such investigations are possible (e.g., Quinn, 1997).

Another sociocultural aspect that has been neglected, but not ignored (e.g., Battista, 1999), includes ways that face-to-face interactions in collaborative small-group work affect geometry learning. For instance, how do students make sense of ideas communicated by other students when those ideas are (a) discrepant from their own views or (b) consistent with their views but expressed in different language or reasoning? In class discussions and small-group work, how do vague and inadequate verbal descriptions of spatial ideas (which are inevitable) affect student learning and participation in the classroom culture? What are the social dynamics of small-group work and how do they affect students' geometric thinking and knowledge construction? How does classroom discourse affect students' geometry learning? Thus, although the current cognitive approach to geometry research should continue, numerous important questions should be investigated from other perspectives.

The Attraction of Computer Environments

A great deal of modern research on geometry learning and teaching has focused on the use of computer environments. Why have geometry teaching and research embraced technology more enthusiastically perhaps than any other area of mathematics education? First, many geometry computer environments are interesting to students. For instance, elementary students often get really excited when first exposed to Logo or the GSP *Shape Makers*. But the pedagogical and research interest in these environments goes well beyond motivational factors. Environments such as Cabri and GSP are tools that enhance the process of doing geometry for everyone, not just students. That is, DGEs allow individuals to explore geometric ideas in ways that are very different, and arguably better, than explorations with paper and pencil. DGEs significantly extend the ability to examine large sets of precisely drawn examples and to infuse dynamic motion into investigations. So it is no wonder that researchers spend so much time investigating the use of this new and powerful tool in instruction. And, as has been discussed earlier in this chapter, this research is getting progressively more sophisticated and providing major insights in geometry learning and teaching.

However, researchers still need to better understand how these technological enhancements affect students' learning. For instance, even though numerous studies show that DGEs can encourage and support significant learning in students, we do not know the effects of scaling up this use, and we do not know how, in very precise terms, learning in DGEs differs from learning in paper-and-pencil environments. Thus we need more comparison studies, both quantitative, to investigate generality, and qualitative, to investigate differences in cognitive processes.

REFERENCES

Arzarello, F., Micheletti, C., Olivero, F., Paola, D., & Gallirio, G. (1998, July). *Dragging in Cabri and modalities of transition from conjectures to proofs in geometry*. Paper presented at the 22nd PME Conference. Stellenbosch, South Africa.

Barrett, J. E., & Clements, D. H. (2003). Quantifying length: Fourth-grade children's developing abstractions for measures of linear quantity. *Cognition and Instruction, 21*(4), 475–520.

Battista, M. T. (1990). Spatial visualization and gender differences in high school geometry. *Journal for Research in Mathematics Education, 21*, 47–60.

Battista, M. T. (1994). On Greeno's environmental/model view of conceptual domains: A spatial/geometric perspective. *Journal for Research in Mathematics Education, 25*(1), 86–94.

Battista, M. T. (1996). Unpublished raw data.

Battista, M. T. (1998a). *Shape Makers: Developing geometric reasoning with the Geometer's Sketchpad.* Berkeley, CA: Key Curriculum Press.

Battista, M. T. (1998b). How many blocks? *Mathematics Teaching in the Middles Grades, 3*(6), 404–411.

Battista, M. T. (1999a). Fifth graders' enumeration of cubes in 3D arrays: Conceptual progress in an inquiry-based classroom. *Journal for Research in Mathematics Education, 30*(4), 417–448.

Battista, M. T. (1999b, April). *Mathematics Education Reform at Lafayette Middle School.* Paper presented at the annual meeting of the American Educational Research Association, Montreal, Canada.

Battista, M. T. (2001a). A research-based perspective on teaching school geometry. In J. Brophy (Ed.), *Advances in research on teaching: Subject-specific instructional methods and activities* (pp. 145–185). New York: JAI Press.

Battista, M. T. (2001b). Shape Makers: A computer environment that engenders students' construction of geometric ideas and reasoning. *Computers in the Schools, 17*(1), 105–120.

Battista, M. T. (2003a, July). *Levels of sophistication in elementary students' reasoning about length.* Paper presented at the 27th annual conference of the International Group for the Psychology of Mathematics Education, Honolulu, HI.

Battista, M. T. (2003b). Understanding students' thinking about area and volume measurement. In D. H. Clements (Ed.), *2003 Yearbook: Learning and teaching measurement* (pp. 122–142). Reston, VA: National Council of Teachers of Mathematics.

Battista, M. T. (2004). Applying cognition-based assessment to elementary school students' development of understanding of area and volume measurement. *Mathematical Thinking and Learning, 6*(2), 185–204.

Battista, M. T. (in press a). Representations and cognitive objects in modern school geometry. In K. Heid & G. Blume (Eds.), *Research on technology in the learning and teaching of mathematics: Syntheses and perspectives.* Greenwich, CT: Information Age Publishing Inc.

Battista, M. T. (in press b). Development of the *Shape Makers* geometry microworld: Design principles and research. In K. Heid & G. Blume (Eds.), *Research on technology in the learning and teaching of mathematics: Syntheses and perspectives.* Greenwich, CT: Information Age Publishing Inc.

Battista, M. T. (in press c). A cognitive perspective on formative assessment. *New England Mathematics Journal.*

Battista, M. T., & Clements, D. H. (1991). *Logo geometry.* Morristown, NJ: Silver Burdett & Ginn.

Battista, M. T., & Clements, D. H. (1995). Geometry and proof. *Mathematics Teacher, 88*(1), 48–54.

Battista, M. T., & Clements, D. H. (1996). Students' understanding of three-dimensional rectangular arrays of cubes. *Journal for Research in Mathematics Education, 27*(3), 258–292.

Battista, M. T., Clements, D. H., Arnoff, J., Battista, K., & Van Auken Borrow, C. (1998). Students' spatial structuring of 2D arrays of squares. *Journal for Research in Mathematics Education, 29*(5), 503–532.

Becker, H. J. (2000, July). *Findings from the Teaching, Learning, and Computing Survery: Is Larry Cuban Right?* Paper presented at the 2000 School Technology Leadership Conference of Council of Chief State School Officers, Washington, D. C.

Berthelot, R., & Salin, M. H. (1998). The role of pupil's spatial knowledge in the elementary teaching of geometry. In C. Mammana & V. Villani (Eds.), *Perspectives on the teaching of geometry for the 21st century* (pp. 71–78). Dordrecht: Kluwer.

Borrow, C. (2000). *An investigation of the development of 6th grade students' geometric reasoning and conceptualizations of geometric polygons in a computer microworld.* Unpublished doctoral dissertation, Kent State University.

Calvin, W. H. (1996). *How brains think.* New York, NY: Basic Books.

Carpenter, T. P., & Lewis, R. (1976). The development of the concept of a standard unit of measure in young children. *Journal for Research in Mathematics Education, 7*(1), 53–58.

Cassirer, E. (1944). The concept of group and the theory of perception. *Philosophy and Phenomenological Research, 5*(1), 1–35.

Chappell, M. F., & Thompson, D. R. (1999). Perimeter or area? Which measure is it? *Mathematics Teaching in the Middle School, 5*(1), 20–23.

Chazan, D. (1993a). High school students' justification for their views of empirical evidence and mathematical proof. *Educational Studies in Mathematics, 24,* 359–387.

Chazan, D. (1993b). Instructional implications of students' understandings of the difference between empirical verification and mathematical proof. In J. L. Schwartz, M. Yersushalmy & B. Wilson (Eds.), *The Geometric Supposer: What is it a case of?* (pp. 107–116). Hillsdale, NJ: Erlbaum.

Chazan, D., & Yerushalmy, M. (1998). Charting a course for secondary geometry. In R. Lehrer & D. Chazan (Eds.), *Designing learning environments for developing understanding of geometry and space* (pp. 67–90). Mahwah, NJ: Erlbaum.

Churchland, P. S. (2002). *Brain-Wise: Studies in neurophilosophy.* Cambridge, MA: MIT Press.

Clements, D. H. (1999). Teaching length measurement: Research challenges. *School Science and Mathematics, 99*(1), 5–11.

Clements, D. H., & Battista, M. T. (1992). Geometry and spatial reasoning. In D. A. Grouws (Ed.), *Handbook of research on mathematics teaching and learning* (pp. 420–464). New York, NY: National Council of Teachers of Mathematics/ Macmillan Publishing Co.

Clements, D. H., & Battista, M. T. (2001). *Logo and Geometry. Journal for Research in Mathematics Education Monograph.* Reston, VA: National Council of Teachers of Mathematics.

Clements, D. H., Battista, M. T., Sarama, J., Swaminathan, S., & McMillen, S. (1997). Students' development of length concepts in a Logo-based unit on geometric paths. *Journal for Research in Mathematics Education, 28*(1), 70–95.

Clements, D. H., Wilson, D. C., & Sarama, J. (2004). Young children's composition of geometric figures: A learning trajectory. *Mathematical Thinking and Learning, 6*(2), 163–184.

Cobb, P., Stephan, M., McClain, K., & Gravemeijer, K. (2001). Participating in classroom mathematical practices. *The Journal of the Learning Sciences, 10*(1 & 2), 113–163.

Coren, S., Ward, L. M., & Enns, J. T. (1994). *Sensation and perception.* Fort Worth, TX: Harcourt Brace College.

de Villiers, M. D. (1994). The role and function of hierarchical classification of quadrilaterals. *For the Learning of Mathematics, 14*(1), 11–18.

de Villiers, M. (1998). An alternative approach to proof in dynamic geometry. In R. Lehrer & D. Chazan (Eds.), *Designing learning environments for developing understanding of geometry and space* (pp. 369–394). Mahwah, NJ: Erlbaum.

Denis, L. P. (1987). Relationships between stages of cognitive development and van Hiele level of geometric thought among Puerto Rican adolescents. *Dissertation Abstracts International,* 48, 859A.

Dorfler, W. (1993). Computer use and views of the mind. In C. Keitel & K. Ruthven (Eds.), *Learning from computers: Mathematics education and technology* (pp. 161–186). Grenoble Cedex, France: NATO ASI Series, Computer and Systems Sciences.

Driver, J., & Baylis, G. C. (1989). Movement and visual attention: The spotlight metaphor breaks down. *Journal of Experimental Psychology: Human Perception and Performance, 15*, 448–456.

Dugdale, S. (1993). Functions and graphs: Perspectives on students' thinking. In T. A. Romberg, E. Fennema & T. P. Carpenter (Eds.), *Integrating research on the graphical representation of functions* (pp. 101–130). Hillsdale, NJ: Erlbaum.

Edelman, G. M. (1992). *Bright air, brilliant fire.* New York, NY: Basic Books.

English, L. D., & Halford, G. S. (1995). *Mathematics education: Models and processes.* Mahwah, NJ: Erlbaum.

Gelman, S. A., & Diesendruck, G. (1999). What's in a concept? Context, variability, and psychological essentialism. In I. E. Siegel (Ed.), *Theoretical perspectives in the concept of representation* (pp. 87–111). Mahwah, NJ: Erlbaum.

Gentner, D., & Medina, J. (1998). Similarity and the development of rules. *Cognition, 65*(2/3), 263–297.

Goldenberg, E. P., & Cuoco, A. A. (1998). What is dynamic geometry? In R. Lehrer & D. Chazan (Eds.), *Designing learning environments for developing understanding of geometry and space* (pp. 351–367). Mahwah, NJ: Erlbaum.

Goldin, G. A., & Kaput, J. J. (1996). A joint perspective on the idea of representation in learning and doing mathematics. In L. P. Steffe, P. Nesher, P. Cobb, G. A. Goldin & B. Greer (Eds.), *Theories of mathematical learning* (pp. 397–430). Mahwah, NJ: Erlbaum.

Greeno, J. G. (1991). Number sense as situated knowing in a conceptual domain. *Journal for Research in Mathematics Education, 22*(3), 170–218.

Gruber, H. E., & Voneche, J. J. (Eds.). (1977). *The essential Piaget.* New York: Basic Books.

Gutiérrez, A. (1992). Exploring the links between van Hiele levels and 3-dimensional geometry. *Structural Topology, 18*, 31–48.

Gutiérrez, A., & Jaime, A. (1998). On the assessment of the van Hiele levels of reasoning. *Focus on Learning Problems in Mathematics, 20*(2/3), 27–47.

Gutiérrez, A., Jaime, A., & Fortuny, J. M. (1991). An alternative paradigm to evaluate the acquisition of the van Hiele levels. *Journal for Research in Mathematics Education, 22*(3), 237–251.

Harel, G., & Sowder, L. (1998). Students' proof schemes: Results from exploratory studies. *CBMS Issues in Mathematics Education, 7*, 234–283.

Healy, L., & Hoyles, C. (1998). *Technical report on the nationwide survey: Justifying and proving in school mathematics.* London: Institute of Education, University of London.

Healy, L., & Hoyles, C. (2001). Software tools for geometrical problem solving: Potentials and pitfalls. *International Journal of Computers for Mathematical Learning, 6*, 235–256.

Hershkowitz, R., Dreyfus, T., Ben-Zvi, D., Friedlander, A., Hadas, N., Resnick, T., et al. (2002). Mathematics curriculum development for computerized environments: A designer–researcher–teacher–learner activity. In L. D. English (Ed.), *Handbook of international research in mathematics education* (pp. 657–694). Mahwah, NJ: Erlbaum.

Hiebert, J. (1981). Cognitive development and learning linear measurement. *Journal for Research in Mathematics Education, 12*, 197–211.

Hoffer, A. R. (1988). Geometry and visual thinking. In T. R. Post (Ed.), *Teaching mathematics in grades K–8: Research based methods* (pp. 232–261). Boston: Allyn and Bacon.

Hollebrands, K. F. (2003). High school students' understandings of geometric transformations in the context of a technological environment. *Journal of Mathematical Behavior, 22*, 55–72.

Holzl, R., Healy, L., Hoyles, C., & Noss, R. (1994). Geometrical relationships and dependencies in Cabri. *Micromath, 10*(3), 8–11.

Hoyles, C., & Healy, L. (1997). Unfolding meanings for reflective symmetry. *International Journal of Computers for Mathematical Learning, 2*, 27–59.

Hoyles, C., & Jones, K. (1998). Proof in dynamic geometry contexts. In C. Mammana & V. Villani (Eds.), *Perspectives on the teaching of geometry for the 21st century* (pp. 121–128). Dordrecht, The Netherlands: Kluwer.

Jaime, A., & Gutiérrez, A. (1989, July). *The learning of plane isometries from the viewpoint of the van hiele model.* Paper presented at the 13th PME Conference. Paris.

Jaime, A., & Gutiérrez, A. (1995). Guidelines for teaching plane isometries in secondary school. *Mathematics Teacher, 88*(7), 591–597.

Johnson-Gentile, K., Clements, D. H., & Battista, M. T. (1994). The effects of computer and noncomputer environment on students' conceptualizations of geometric motions. *Journal of Educational Computing Research, 11*, 121–140.

Johnson-Laird, P. N. (1983). *Mental models: Towards a cognitive science of language, inference, and consciousness.* Cambridge, MA.: Harvard University Press.

Johnson-Laird, P. N. (1998). Imagery, visualization, and thinking. In J. Hochberg (Ed.), *Perception and cognition at century's end* (pp. 441–467). San Diego, CA: Academic Press.

Jones, K. (2000). Providing a foundation for deductive reasoning: Students' interpretations when using dynamic geometry software and their evolving mathematical explanations. *Educational Studies in Mathematics, 44*(1/2), 55–85.

Joram, E., Subrahmanyam, K., & Gelman, R. (1998). Measurement estimation: Learning to map the route from number to quantity and back. *Review of Educational Research, 68*(4), 413–449.

Kamii, C. (1995). *Why is the use of a ruler so hard?* Paper presented at the Proceedings of the seventeenth annual meeting of the North American chapter of the international group Psychology in Mathematics Education, Columbus, OH.

Kloosterman, P., Warfield, J., Wearne, D., Koc, Y., Martin, G., & Strutchens, M. (2004). Knowledge of mathematics and perceptions of learning mathematics of fourth-grade students. In P. Klooosterman & F. K. Lester (Eds.), *Results and interpretations of the 2003 mathematics assessment of the National Assessment of Educational Progress.* Reston, VA: National Council of Teachers of Mathematics.

Knauff, M., Fangmeier, T., Ruff, C. C., & Johnson-Laird, P. N. (2003). Reasoning, models, and images: Behavioral measures and cortical activity. *Journal of Cognitive Neuroscience, 15*(4), 559–573.

Knuth, E. J. (2000). Student understanding of the Cartesian connection: An exploratory study. *Journal for Research in Mathematics Education, 31*(4), 500–508.

Koedinger, K. R. (1998). Conjecturing and argumentation in high-school geometry students. In R. Lehrer & D. Chazan (Eds.), *Designing learning environments for developing understanding of geometry and space* (pp. 319–347). Mahwah, NJ: Erlbaum.

Kordaki, M., & Potari, D. (1998). Children's approaches to area measurement through different contexts. *Journal of Mathematical Behavior, 17*(3), 303–316.

Kordaki, M., & Potari, D. (2002). The effect of area measurement tools on student strategies: The role of a computer microworld. *International Journal of Computers for Mathematical Learning, 7*, 65–100.

Kosslyn, S. M. (1980). *Image and mind.* Cambridge, MA: Harvard University Press.

Kosslyn, S. M. (1983). *Ghosts in the mind's machine.* New York: W. W. Norton.

Kosslyn, S. M., Margolis, J. A., & Barrett, A. M. (1990). Age differences in imagery abilities. *Child Development*, 61(4), 995–1010.

Kosslyn, S. M. (1988). Aspects of a cognitive neuroscience of mental imagery. *Science, 240*(4859), 1621–1626.

Kosslyn, S. M. (1994). *Image and brain.* Cambridge, MA: MIT Press.

Krascum, R. M., & Andrews, S. (1993). Feature-based versus exemplar-based strategies in preschoolers' category learning. *Journal of Experimental Child Psychology, 56*, 1–48.

Laborde, C. (1992). Solving problems in computer-based geometry environments: The influence of the features of the software. *Zentralblatt fur didaktick der mathematik* (*International Reviews on Mathematical Education*), *4*, 128–135.

Laborde, C. (1993). The computer as part of the learning environment: The case of geometry. In C. Keitel & K. Ruthven (Eds.), *Learning from computers: Mathematics education and technology* (Vol. 121, 48–67). Grenoble Cedex, France: NATO ASI Series, Computer and Systems Sciences.

Laborde, C. (1998). Visual phenomena in the teaching/learning of geometry in a computer-based environment. In C. Mammana & V. Villani (Eds.), *Perspectives on the teaching of geometry for the 21st century* (pp. 113–121). Dordrecht, The Netherlands: Kluwer.

Laborde, C. (2001). Integration of technology in the design of geometry tasks with Cabri-geometry. *International Journal of Computers for Mathematical Learning, 6*(3), 283–317.

Lehrer, R., Jenkins, M., & Osana, H. (1998). Longitudinal study of children's reasoning about space and geometry. In R. Lehrer & D. Chazan (Eds.), *Designing learning environments for developing understanding of geometry and space* (pp. 137–167). Mahwah, NJ: Erlbaum.

Lewellen, H. (1992). *Description of van Hiele levels of geometric development with respect to geometric motions.* Unpublished doctoral dissertation, Kent State University, OH.

Leyton, M. (1992). *Symmetry, causality, mind.* Cambridge, MA: MIT Press.

Ma, L. (1999). *Knowing and teaching elementary mathematics.* Mahwah, NJ: Erlbaum.

Mariotti, M. A. (2001). Justifying and proving in the Cabri environment. *International Journal of Computers for Mathematical Learning, 6*, 257–281.

Markovits, H. (1993). The development of conditional reasoning: A Piagetian reformulation of mental models theory. *Merrill-Palmer Quarterly, 39*(1), 131–158.

Marrades, R., & Gutiérrez, A. (2000). Proofs produced by secondary school students learning geometry in a dynamic computer environment. *Educational Studies in Mathematics, 44*(1/2), 87–125.

Martin, G., & Strutchens, M. (2000). Geometry and measurement. In E. A. Silver & P. A. Kenney (Eds.), *Results from the seventh mathematics assessment of the National Assessment of Educational Progress* (pp. 193–234). Reston, VA: National Council of Teachers of Mathematics.

Mason, M. M. (1989). *Geometric understanding and misconceptions among gifted fourth through eighth graders*, Unpublished doctoral dissertation, Northern Illinois University, IL.

Mayberry, J. (1983). The van Hiele levels of geometric thought in undergraduate preservice teachers. *Journal for Research in Mathematics Education, 14*, 58–69.

McCoy, L. P. (1996). Computer-based mathematics learning. *Journal of Research on Computing in Education, 28*, p. 438–460.

McCrone, S. M. S., & Martin, T. S. (2004). Assessing high school students' understanding of geometric proof. *Canadian Journal of Science, Mathematics and Technology Education, 4*(2), 221/240.

Millroy, W. L. (1992). *An ethnographic study of the mathematical ideas of a group of carpenters, Journal for Research in Mathematics Education Monograph. Reston*, VA: National Council of Teachers of Mathematics.

Mitchelmore, M. C., & White, P. (2000). Development of angle concepts by progressive abstraction and generalisation. *Educational Studies in Mathematics, 41*(3), 209–238.

Moise, E. E. (1963). *Elementary geometry from an advanced standpoint.* Reading, MA: Addison-Wesley.

Moschkovich, J., Schoenfeld, A. H., & Arcavi, A. (1993). Aspects of understanding: On multiple perspectives and representations of linear relations and connections among them. In T. A. Romberg, E. Fennema & T. P. Carpenter (Eds.), *Integrating research on the graphical representation of functions* (pp. 41–68). Hillsdale, NJ: Erlbaum.

NAEP web site: http://nces.ed.gov/nationsreportcard/itmrls/.

Noss, R., & Hoyles, C. (1996). *Windows on mathematical meanings: Learning cultures and computers* (Vol. 17). Dordrecht, The Netherlands: Kluwer Academic.

Nunes, T., Light, P., & Mason, J. (1993). Tools for thought: The measurement of length and area. *Learning and Instruction, 3*, 39–54.

Olive, J. (1991). Logo programming and geometric understanding: An in-depth study. *Journal for Research in Mathematics Education, 22*(2), 90–111.

Outhred, L., & Mitchelmore, M. (2000). Yound children's intuitive understanding of rectangular area measurement. *Journal for Research in Mathematics Education, 31*(2), 144–167.

Ornstein, R. (1991). *The evolution of consciousness.* New York: Touchstone.

Owens, K., & Outhred, L. (1997). Early representations of tiling areas. *Proceedings of the 21st Conference of the International Group for the Psychology in Mathematics Education, Helsinki, 3*, 312–319.

Pegg, J. (1992, July). *Students' understanding of geometry: Theoretical perspectives.* Paper presented at the 15th Annual Conference of the Mathematics Education Research Group of Australasia, Sydney.

Pegg, J., & Davey, G. (1998). Interpreting student understanding in geometry: A synthesis of two models. In R. Lehrer & D. Chazan (Eds.), *Designing learning environments for developing understanding of geometry and space* (pp. 109–133). Mahwah, NJ: Erlbaum.

Pegg, J., Gutiérrez, A., & Huerta, P. (1998). Assessing reasoning abilities in geometry. In C. Mammana & V. Villani (Eds.), *Perspectives on the teaching of geometry for the 21st century* (pp. 275–295). Dordrecht, The Netherlands: Kluwer.

Pesek, D. D., & Kirshner, D. (2000). Interference of instrumental instruction in subsequent relational learning. *Journal for Research in Mathematics Education, 31*(5), 524–540.

Piaget, J. (1950). *The psychology of intelligence*. London: Routledge.

Piaget, J. (2001). *Studies in reflecting abstraction*. Philadelphia: Psychology Press.

Piaget, J., & Inhelder, B. (1956). *The child's conception of space*. London: Routledge and Kegan Paul.

Piaget, J., Inhelder, B., & Szeminska, A. (1960). *The child's conception of geometry*. New York: Routledge and Kegan Paul/Basic Books, Inc.

Pinker, S. (1997). *How the mind works*. New York: W. W. Norton.

Posner, M. I., & Raichle, M. E. (1994). *Images of mind*. New York: Scientific American Library.

Presmeg, N. C. (1997). Generalization using imagery in mathematics. In L. D. English (Ed.), *Mathematical reasoning* (pp. 299–312). Mahwah, NJ: Erlbaum.

Quinn, A. (1997). *Justifications, argumentations, and sense making of preservice elementary teachers in a constructivist mathematics classroom*. Unpublished doctoral dissertation, Kent State University, OH.

Regan, D., & Hamstra, S. (1991). Shape discrimination for motion-defined and contrast-defined form: Squareness is special. *Perception, 20*, 315–336.

Reynolds, A. M., & Wheatley, G. H. (1996). Elementary students' construction and coordination of units in an area setting. *Journal for Research in Mathematics Education, 27*(5), 564–581.

Ross, B. H. (1996). Category representations and the effects of interacting with instances. *Journal of Experimental Child Psychology: Learning, Memory and Cognition, 22*(5), 1249–1265.

Sarama, J., Clements, D. H., Swaminathan, S., McMillen, S., & González Gómez, R. M. (2003). Development of mathematical concepts of two-dimensional space in grid environments: An exploratory study. *Cognition and Instruction, 21*(3), 285–324.

Siegel, A. W. (1981). The externalization of cognitive maps by children and adults: In search of ways to ask better questions. In L. S. Liben, A. H. Patterson & N. Newcombe (Eds.), *Spatial representation and behavior across the lifespan* (pp. 167–194). New York: Academic Press.

Siegler, R. S. (1996). *Emerging minds: The process of change in children's thinking*. New York: Oxford University Press.

Simon, M. A., & Blume, G. W. (1994). Building and understanding multiplicative relationships: A study of prospective elementary teachers. *Journal for Research in Mathematics Education, 25*(5), 472–494.

Smith, E. E. (1995). Concepts and categorization. In D. N. Osherson & L. R. Gleitman (Eds.), *An invitation to cognitive science. Vol. 3, Thinking*. Cambridge, MA: MIT Press.

Smith, L. B. (1989). A model of perceptual classification in children and adults. *Psychological Review, 96*(1), 125–144.

Sowder, J. T., Wearne, D., Martin, G., & Strutchens, M. (2004). What grade 8 students know about mathematics: Changes over a decade. In P. Kloosterman & F. Lester (Eds.), *The 1990 through 2000 mathematics assessment of the National Assessment of Educational Progress: Results and interpretations*. (pp. 105–143). Reston, VA: National Council of Teachers of Mathematics.

Steffe, L., & Cobb, P. (1988). *Construction of arithmetical meanings and strategies*. New York: Springer-Verlag.

Steffe, L. P. (1998, April). *Principles of design and use of TIMA software*. Paper presented at the NCTM Research Presession, Washington, D.C.

Steffe, L. P. (in press). Articulation of the reorganization hypothesis. In L. P. Steffe (Ed.) Children's fractional knowledge: A constructive odyssey. *The Journal of Mathematical Behavior* Monograph.

Sutherland, R., Godwin, S., Olivero, F., & Peel, P. (2002). Design Initiatives for Learning: ICT and Geometry in the Primary School. *BERA Conference*, University of Exeter, UK.

Tall, D. (1992). The transition to advanced mathematical thinking: Functions, limits, infinity, and proof. In D. A. Grouws (Ed.), *Handbook of research on mathematics teaching* (pp. 495–511). Reston, VA: National Council of Teachers of Mathematics/Macmillan.

Tierney, C., Boyd, C., & Davis, G. (1990, July). *Prospective primary teachers' conceptions of area*. Paper presented at the 14th PME Conference, Mexico City, Mexico.

van Hiele, P. M. (1986). *Structure and insight*. Orlando, FL: Academic Press.

von Glasersfeld, E. (1982). Subitizing: The role of figural patterns in the development of numerical concepts. *Archives de Psychologie, 50*, 191–218.

von Glasersfeld, E. (1991). Abstraction, re-presentation, and reflection: An interpretation of experience and Piaget's approach. In L. P. Steffe (Ed.), *Epistemological foundations of mathematical experience* (pp. 45–67). New York: Springer-Verlag.

von Glasersfeld, E. (1995). *Radical constructivism: A way of knowing and learning*. London: The Falmer Press.

Vygotsky, L. (1986). *Thought and language*. Cambridge, MA: The MIT Press.

Wallace, E. C., & West, S. F. (1992). *Roads to geometry*. Englewood Cliffs, NJ: Prentice Hall.

Wheatley, G. H. (1997). Reasoning with images in mathematical activity. In L. D. English (Ed.), *Mathematical reasoning* (pp. 281–298). Mahwah, NJ: Erlbaum.

Wheatley, G. H. (1998). Imagery and mathematics learning. *Focus on Learning Problems in Mathematics, 20*(2/3), 65–77.

Wheatley, G. H., & Reynolds, A. M. (1996). The construction of abstract units in geometric and numeric settings. tiling the plane. *Educational Studies in Mathematics, 30*, 67–83.

Wilson, B. (1993). The Geometric Supposer in the classroom. In J. L. Schwartz, M. Yerushalmy & B. Wilson (Eds.), *The Geometric Supposer: What is it a case of?* (pp. 17–22). Hillsdale, NJ: Erlbaum.

Woodward, E., & Byrd, F. (1983). Area: Included topic, neglected concept. *School Science and Mathematics, 83*(4), 343–347.

Yerushalmy, M. (1993). Generalization in geometry. In J. L. Schwartz, M. Yerushalmy & B. Wilson (Eds.), *The Geometric Supposer: What is it a case of?* (pp. 57–83). Hillsdale, NJ: Erlbaum.

Yerushalmy, M., & Chazan, D. (1993). Overcoming visual obstacles with the aid of the Supposer. In J. L. Schwartz, M. Yersushalmy & B. Wilson (Eds.), *The Geometric Supposer: What is it a case of?* (pp. 25–56). Hillsdale, NJ: Erlbaum.

AUTHOR NOTE

Time to prepare portions of this material was partially supported by the National Science Foundation under Grant Nos. ESI 0099047and 0352898. Opinions, findings, conclusions, or recommendations, however, are those of the author and do not necessarily reflect the views of the National Science Foundation.

I would like to thank Angel Gutiérrez for his careful and thoughtful reading of this manuscript.

20

RESEARCH IN PROBABILITY

Responding to Classroom Realities

Graham A. Jones

GRIFFITH UNIVERSITY, GOLD COAST CAMPUS

Cynthia W. Langrall and Edward S. Mooney

ILLINOIS STATE UNIVERSITY

There is a broad consensus that the teaching of probability must begin early in order to build sound intuitions and that such efforts must be grounded in experience. ... However, mere experience is not sufficient, as the research on faulty intuitions and fallacious reasoning amply shows, and a major challenge to the field is harvesting what is known from this research to inform teaching.

—Greer & Mukhopadhyay, 2005, pp. 314–315

Greer and Mukhopadhyay's caveat concerning the importance of research in the teaching of probability is timely as we look back more than 16 years since statistics and probability was introduced as a mainstream strand in the school mathematics curriculum. Consistent with curriculum reforms of the period (e.g., National Council of Teachers of Mathematics [NCTM], 1989, 2000), Greer and Mukhopadhyay acknowledge the need to commence probability experientially in the early grades while highlighting the problematic nature of children's probabilistic reasoning—a picture that is replete in the literature. Lastly, their statement inspires educators to look more deeply at the probability research in order to create effective environments for probability learning. This latter suggestion resonates with one of the key intents of this chapter: to inform the teaching and learning of probability by synthesizing research in the field, especially research that has been published since Shaughnessy's (1992) review of stochastics in the first *Handbook of Research on Mathematics Teaching and Learning* (Grouws, 1992).

The period since Shaughnessy's review (1992) has been a time of curriculum reform in mathematics for the United States (NCTM, 1989, 2000) and for other countries (e.g., Australian Education Council [AEC], 1991; Department of Education and Science and the Welsh Office [DES], 1991). Among other elements of change, national documents like the *Curriculum and Evaluation Standards for School Mathematics* (NCTM, 1989) harbored the emergence of probability and statistics as a mainstream strand that commenced in the primary grades and continued through the high school years. The pervasive and coherent emergence of statistics and probability in curriculum documents and the potential for it in classrooms at all grade levels has had a profound effect on research in statistics and probability education during this period. The curriculum movements created a refreshing urgency for probability research, and much of the research in this period has been designed to support perceived needs in classroom teaching and learning or has been stimulated by classroom experiences that manifest the challenges faced by researchers, teachers, and students in dealing with "knowledge of uncertainty."

Although this *Handbook* devotes separate chapters to research on statistics and probability education, it is critical that we recognize not only the mathematical interrelatedness of the two areas but also the interrelatedness of research in statistics education and research in probability education. Hence we will outline some of the key conceptual bridges between statistics and probability before turning to our main task of reviewing and analyzing research in probability education during the period since Shaughnessy's (1992) review.

The study of statistics and probability focuses on real-world phenomena that involve uncertainty, that is, cannot be predicted with certainty. For example, a statistician involved in health-related research may be considering the question, "Which of the available medications is preferable for treating the common cold in young children?" When a statistician addresses such a problem, he or she collects, organizes, analyzes, and interprets data (Wild & Pfannkuch, 1999). This data might be the number of days each of the medications takes to remedy the common cold. As such, the data are influenced by both the effect of the medication (causal variation) as well as chance variation or random effects associated with the medications or the subjects. As Moore (1990) explained, "'Random' is not a synonym for 'haphazard,' but a description of a kind of order different from the deterministic" (p. 98). It is an order that is not attributable to a specific cause or chain of events (determinism) but one that operates in a phenomenon that has uncertain individual outcomes but a regular pattern of outcomes over many repetitions (e.g., the number of days to cure a common cold for many repeats of the same medication on a particular subject).

These random effects or random processes lie at the very heart of statistics and probability. On the one hand, *statistics* uses the characteristics of random processes and probability models of such processes to make inferences about problems involving data; for example, the most preferred medication for the common cold. On the other hand, *probability* focuses directly on describing, quantifying, modeling, and illuminating random processes. For example, one or more approaches to probability measurement may be used to estimate the probability or likelihood of an event such as "medication A cures the common cold within 5 days or less."

This brief outline of statistics and probability highlights the close connection between the two areas. It also underscores the value of undertaking research in probability, both because of its complementary role in informing statistics and its special place in describing and quantifying the numerous chance phenomena that operate in our world. Accordingly, in writing this chapter we will concentrate on research into the learning and teaching of probability; however, where appropriate we will make connections between ideas in probability and statistics and timely references to research that has addressed commonalities between the two areas. We begin by briefly revisiting research on probability education in the decades prior to the current period.

SETTING THE STAGE: THE EARLIER REVIEW OF PROBABILITY RESEARCH

In Shaughnessy's (1992) review he stated that, up to that point, there had not been much involvement by North American mathematics educators in research on the teaching and learning of probability. He observed that, "Since very little probability or statistics has been systematically taught in our schools in the past, there has been little impetus to carry out research on the problems that students have in learning it" (p. 465). Although it would be difficult to take issue with the overall thrust of Shaugnessy's statement, a number of North American mathematics educators, in the period from the 1950s to the late 1980s, acted as trailblazers for teaching and learning research on probability (e.g., Davies, 1965; Doherty, 1965; Garfield & Ahlgren, 1988; Konold, 1983, 1989; Leake, 1965; Leffin, 1971; Mullenex, 1968; Page, 1959; Shaughnessy, 1977; Shepler, 1970). These North American researchers, like those who charted the way in Europe and other places (e.g., Engel, 1966; Falk, 1981, 1988; Fischbein, 1975; Fischbein & Gazit, 1984; Gillis & Haraud, 1972; Green, 1979, 1983; Jones, 1974; Steinbring, 1984; Varga, 1969a, 1969b) built their research largely on the theoretical models of cognitive-development psychologists like Piaget (e.g., Piaget & Inhelder, 1951/1975) or the work of other psychologists who were interested in people's learning behavior and decision making when faced with situations involving uncertainty (e.g., Kahneman, Slovic, & Tversky, 1982; Offenbach, 1965; Siegel & Andrews, 1962; Stevenson & Zigler, 1958; Tversky & Kahneman, 1974).

During this time frame, psychologists were interested in probability from a cognitive, epistemological, or decision-making perspective, while mathematics educators harbored strong sentiments that their research had implications for the learning and teaching of probability and an earlier and more extensive inclusion of it in the school mathematics curriculum.

Fischbein, Pampu, and Mînzat (1970) echoed this view in the following conclusion from their research:

> In any case, the finding that 9- to 10-year-old Ss [subjects], after a brief instruction, became able to perform chance estimates by comparing numerical ratios and to understand the concept of proportionality might be an argument for starting to teach probabilities while children are still in primary school. (p. 388)

Other early advocates such as Engel (1966), Page (1959), and Varga (1969a, 1969b) reflected similar views to Fischbein's as a result of their own research and their development of exploratory curriculum programs in probability for elementary, intermediate, and secondary students. Steinbring (1984, 1991) even developed an instructional theory that took cognizance of the nature of probability and its different perspectives. However, the efforts of these and other mathematics education researchers in the period prior to Shaughnessy's review (1992) went largely unheralded, as probability received almost no attention in elementary and middle schools and limited exposure in secondary schools. Moreover, when it was taught in secondary schools the focus was largely on "balls-in-urn problems," that is, combinatoric applications.

The protracted gestation finally ended when probability and statistics emerged as a major strand in several national curriculum documents (e.g., AEC, 1991; DES, 1991; NCTM, 1989). Although mathematics educators who had undertaken research on probability thinking and learning in the 3 decades prior to the emergence of these curriculum documents may have felt frustration with the delay, they can take some solace in the fact that knowledge about the teaching and learning of probability was already available before these curriculum programs were introduced across the broad spectrum from kindergarten through secondary grades. There is little doubt that this prior research influenced national curriculum documents in a number of ways: how the aims and "big ideas" in probability were incorporated, how the big ideas were articulated through the grades, and how the design of instruction provided a scaffold for learning.

The research in this prior period also provided an infrastructure and a lighthouse for probability research undertaken during the current period. This infrastructure included a problematique, potential research questions, conceptual frameworks, research designs, assessment instruments, and data-analytic techniques. Notwithstanding the value and significance of this prior research, our thesis is that the research on probability education during the last 15 years has been inescapably driven by the goals and reform directions reflected in early national curriculum documents (e.g., AEC, 1991; DES, 1991; NCTM, 1989). More specifically, we claim that documents like the *Curriculum and Evaluation Standards for School Mathematics* set in train both general directions for mathematics learning and teaching and particular directions for the content and pedagogy associated with probability and statistics. These reform directions and their influence on schools acted as a catalyst for much of the research on probability education that we will review. In fact, gleaning support from the reform documents' espousal of socioconstructivist (Cobb & Bauersfeld, 1995) orientations to learning and instruction that is grounded in students' intuitions (NCTM, 1989), the research agenda has highlighted the profiling of probabilistic thinking that students bring to the classroom and the development of students' individual and collective probabilistic reasoning during various kinds of instruction.

Because of their importance for probability research in the period since Shaughnessy's 1992 review, an examination of some *national curriculum documents* and their impact on probability research will constitute the first of our five goals in writing this chapter. As our second goal, we will discuss and appraise the key research on probability *learning and teaching* during the period and will make connections where appropriate between the research of this period and that of the previous one. In identifying this learning and teaching research we have endeavored to take an international perspective by capturing research from different parts of the globe. Third, under *curriculum research*, we examine the notion of "probability literacy" and its implications for curriculum. Indeed, if the implementation of probability is to be sustainable, research on probability literacy may well be critical. Fourth, we will present *reflections* that evaluate the research activities of this period against the "wish list" for this period that Shaughnessy foreshadowed in the 1992 review. Fifth and finally, mindful of the refinements espoused for probability in more recent curriculum documents like the *Principles and Standards for School Mathematics* (NCTM, 2000), we will create our own agenda that hopefully will serve as a benchmark for future endeavors in probability research and for the next review.

NATIONAL CURRICULA: A BACKDROP FOR PROBABILITY RESEARCH

As we have observed, the advent of national curriculum documents in the late 1980s and early 1990s spawned statistics and probability as a mainstream strand within the mathematics curriculum. In order to expose this new mainstream strand and its potential for impacting research in this period, we will examine briefly key elements of probability in the NCTM *Curriculum and Evaluation Standards for School Mathematics* (1989) and in curriculum documents from Australia and the United Kingdom respectively: *A National Statement on Mathematics for Australian Schools* (AEC, 1991) and *National Curriculum: Mathematics for Ages 5–16* (DES, 1989). We chose these latter two curriculum documents because they appeared at almost the same time as the NCTM *Standards* and, with it, seemed to set the scene for much of the curriculum development in probability and statistics worldwide.

In surveying probability in these three curricula documents we will focus on the content and pedagogical directions at the elementary, middle, and high school levels. Although these curriculum documents were revised in the decade following their initial publication (e.g., AEC, 1994; DfEE, 1999; NCTM, 2000), our survey will concentrate on the original documents as they were the first to impact researchers interested in the teaching and learning of probability.

The Probability Curriculum in the Elementary Grades

Table 20.1 presents the "big ideas" in probability that were introduced at the elementary school level (approximately Grades K through 6) in the three curriculum documents (AEC, 1991; DES, 1989; NCTM, 1989). In all three countries, elementary students were expected to gain an understanding of *chance* (uncertainty) and *random* events through experimentation with simple probability generators (e.g., dice, coins, and spinners) and in some cases through cultural activities (e.g., rainfall, traffic, weather). Students were to use language associated with chance (e.g., *equally likely, fair,* and *unfair*) and to recognize the possibility of several outcomes. However, at this level, understanding of randomness did not extend to long-term patterns.

Each of the three countries also incorporated elements of both the *frequentist* (experimental or *a-posteriori*) and the *classical* (theoretical or *a-priori*) approaches to probability measurement. With respect to the frequentist approach (von Mises, 1928/1952), the probability of an event is defined as the ratio of the number of trials favorable to the event to the total number of trials. In the classical approach (Laplace, 1825/1995), the probability of an event is determined by the ratio of the number of favorable outcomes to the total number of outcomes, where outcomes are assumed to be equally likely. Hence, the frequentist approach deals with experimental trials and statistical or *a-posteriori* evidence whereas the classical approach deals with numerical or geometrical measures and *a-priori* assumptions of symmetry or equal like-

Table 20.1 Probability for the Elementary Grades

Country	Big ideas in probability
Australia	• use, with clarity, everyday language associated with chance events (*very likely, unlikely, more likely,* and *equally likely*) and use appropriately terms like *chance* and *probability*
	• list systematically possible outcomes for familiar random events including one-stage experiments and some two-stage experiments; deduce the order of probability of the outcomes and test predictions experimentally
	• place outcomes for familiar events and one-stage experiments in order from those least likely to happen to those most likely to happen (events arising from random generators like the tossing of a coin and from social situations like the weather)
	• make and interpret empirically based predictions about simple situations (e.g., a drawing pin falling point up or the risk of accidents per 1,000 km traveled for different modes of travel)
United Kingdom	• recognize a degree of uncertainty about the outcomes of some events and that other events are certain or impossible; recognize the possible outcomes of random events (events arising in social situations and from probability generators such as coins and dice)
	• place events in order of "likelihood" and use appropriate words to identify the chance (of such events); distinguish "fair" and "unfair"
	• list all the possible outcomes of an event including the outcomes of two combined events that are independent; use a probability scale from 0 to 1; give and justify subjective estimates of probabilities
	• distinguish between estimates of probabilities based on statistical evidence and those based on assumptions of symmetry; know that different outcomes may result from repeating an experiment and that if each of *n* events is assumed to be equally likely, the probability of one occurring is *1/n*
United States	• begin to recognize the nature of random processes and explore concepts of chance through games and experiments
	• use concepts like *certain, uncertain, likely,* and *luck*
	• compare the likelihood of events theoretically and experimentally using both qualitative and quantitative descriptors

lihood. Although references to these two approaches at the elementary school level are largely nascent, all three curriculum documents make comments like the following: order the outcomes by probability and test predictions experimentally (AEC, 1991), distinguish between estimates of probabilities based on statistical evidence and those based on assumptions of symmetry (DES, 1989) and compare the likelihood of events both theoretically and experimentally (NCTM, 1989).

Not only do the three curriculum documents encourage the use of both frequentist and classical approaches, they also encourage students to make connections between the two. The NCTM *Standards* (1989, p. 56) provided a typical example of the kind of problem that young children could be expected to explore with respect to the two approaches to probability measurement:

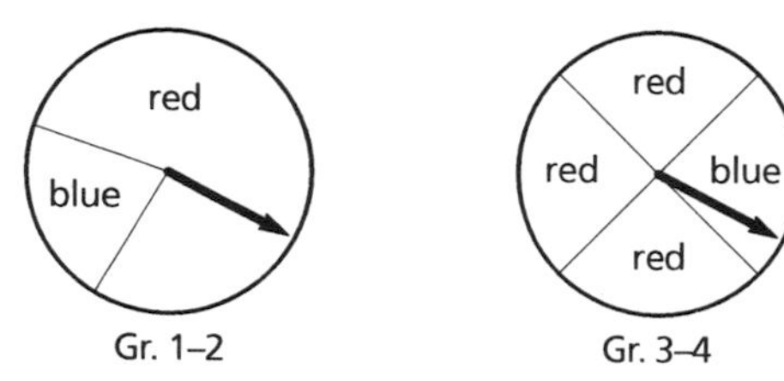

Is red or blue more likely? How likely is yellow? How likely is getting either red or blue? If we spin twelve times, how many blues might we expect to get?

The third approach to chance measurement, *subjective probability*, describes probability as a degree of belief, based on personal judgment and information about an outcome. Thus it depends on two things: the event whose uncertainty is contemplated and the knowledge of the person estimating the probability (Batanero, Henry, & Parzysz, 2005; Lindley, 1980). In the three elementary curricula children have opportunities to use qualitative (subjective) probability descriptors such as certain, more likely, less likely, and impossible; however, the documents are largely silent on the kind of knowledge base students are expected to use in making such judgments. Some evidence (e.g., Huber & Huber, 1987; Metz, 1998a) suggests that children can use knowledge or beliefs about levels of confidence in dealing with subjective probability.

Although it is not included in Table 20.1, the curriculum documents also focused on teaching aspects of probability. In particular the documents advocated a problem-solving focus, tasks that resonated with children's experiences, and approaches that revealed students' intuitions and confronted their misconceptions. Except for the last approach, the pedagogical directions were largely generic in that they espoused the visions in the mathematics curricula as a whole.

The Probability Curriculum in the Middle Grades

The big ideas in probability for the middle grades (roughly Grades 7 through 9) are presented in Table 20.2. All three curriculum documents focused strongly on having students use numerical probabilities; however, there is still continuing consideration of random processes emphasizing both their short-term unpredictability and their long-term stability. The stability aspect was generally considered in association with experimental (empirical) probability.

The strongest feature in all three middle school curricula is the focus on both the classical and the frequentist approaches to probability. Building on the informal start in the elementary grades, the curricula highlight the determination of theoretical and experimental probabilities and also foreshadow the relationship between the two. For example, the United States document uses the phrase, "by comparing experimental results with mathematical expectations" (NCTM, 1989, p. 109). In dealing with the classical (theoretical) approach to probability, all three documents refer to sample space and the systematic listing of outcomes in simple and multistage experiments. There is also a strong thrust on modeling and simulation of real–world phenomena in the Australian and United States documents. This thrust is illustrated in the following problem (Figure 20.1) taken from the NCTM *Standards* (1989, p. 111). The problem is interesting in that it provides data for modeling and hence makes a nice connection between probability and statistics.

Joan Dyer Data and Questions

Home runs	9
Triples	2
Doubles	16
Singles	24
Walks	11
Outs	38
Total	100

The table gives the record for Joan Dyer's last 100 times at bat during the softball season. She is now coming up to bat. Use the data to answer the following questions:

- What is the probability that Joan will get a home run?
- What is the probability that she will get a hit?
- How many times can she expect to get a walk in her next 14 times at bat?

Figure 20.1 Softball data problem.

With respect to subjective probability, the Australian and United Kingdom documents relate it to social contexts. For example, students are expected to explore the odds appearing in gambling situations

Table 20.2 Probability for the Middle School Grades

Country	Big ideas in probability
Australia	• understand and explain social uses of chance processes (e.g., investigate the use of probability in insurance, gambling, and games)
	• construct sample spaces to analyze and explain possible outcomes of simple experiments and calculate probabilities by analysis of equally likely events (including one-, two-, and three-stage experiments)
	• estimate probability using the long-run relative frequency (i.e., empirical probabilities)
	• model situations and devise and carry out simulations
United Kingdom	• understand and use relative frequency as an estimate of probability
	• appreciate, when assigning probabilities, that relative frequency and equally likely considerations may not be appropriate and "subjective" estimates need to be made
	• understand and apply the addition of probabilities for mutually exclusive events
	• understand that when dealing with two independent events, the probability of them both happening is less than the probability of either of them happening (unless the probability is 0 or 1)
	• calculate the probability of a combined event given the probability of each of two independent events and illustrate combined event probabilities of several events using tabulation or tree-diagrams
United States	• model situations by devising and carrying out experiments or simulations to determine probabilities
	• model situations by using a sample space to determine probabilities
	• appreciate the power of using a probability model by comparing experimental results with mathematical expectations
	• make predictions that are based on experimental or theoretical probabilities
	• develop an appreciation for the pervasive use of probability in the real world

(e.g., horse racing) and to recognize that these odds reflect statements of subjective probability (AEC, 1991, p. 175). Moreover, it is also noted that in some situations only subjective estimates of probability are possible; for example, the probability that a cure for cancer will be found in the next 2 years cannot be based on data or *a-priori* assumptions of equal likelihood (DES, 1989, p. 44).

Once again the three documents make limited reference to pedagogical insights that refer specifically to probability. However, reference is made to the fact that "students have many misconceptions and poor intuitions about probabilistic situations" (NCTM, 1989, p. 110), and hence teachers should encourage students to make conjectures (based on these intuitions) and test these conjectures by experimentation. In addition, the emphasis on real-world modeling and experimentation in the probability sections of the curricula is consistent with the prime place given to problem solving throughout the three curriculum documents.

The Probability Curriculum in the High School

Table 20.3 presents an overview of probability in the three curriculum documents for the high school years (approximately Grades 10 through 12). It reveals the big ideas intended for students who take *core* programs in mathematics. All documents note that some students will take specialist mathematics courses that may include more rigorous work in probability.

The study of random processes continues in the high school years but is more implicit than explicit in the United Kingdom (DES, 1989) and the United States documents (NCTM, 1989). The fine detail of the Australian document (AEC, 1991, p. 182) actually suggests having students examine "run lengths" as a means of expanding their understanding of randomization. Developing knowledge and understanding of sample space and probability measurement involving compound events is a growing feature of the three curriculum documents at this level. The emphasis is almost totally on classical (theoretical) and frequentist (experimental) approaches to probability measurement; only the Australian document alludes to subjective probability. Modeling real-world problems using simulation to estimate probabilities is a strong feature of the Australian and United States programs.

The major new conceptual developments at this level are the inclusion of random variable and probability distribution. All documents focus on the notion of probability distribution and also consider some specific probability distributions such as the normal distribution. Although probability distributions play a key role in comparing populations and in making statistical inferences, we will not review research in this area, leaving it to the chapter on statistics (Shaughnessy, this volume).

With respect to pedagogical considerations, the documents recommend that the work should continue to be practical and investigative. Teachers are expected to encourage students to use both theoretical and experimental approaches, especially those involving simulation.

Key Ideas for Research Emanating from the National Curriculum Documents

In undertaking this brief review of three national curriculum documents, we have unashamedly focused on the commonalities of the probability programs rather than their differences. Notwithstanding this perspective, the big ideas in probability in the three curriculum documents are remarkably consistent in scope and evolution from the elementary through the high school grades. We summarize these big ideas below but, in passing, observe that their presence in all three national curriculum documents adds further grist to the argument that researchers across the international arena were influenced in comparable ways. Further, researchers interacted with school classrooms that advocated similar pedagogy.

With respect to probability content, the big ideas that have emerged in the three curriculum programs are the *nature of chance and randomness, sample space, probability measurement* (classical, frequentist, and subjective), and *probability distributions.* Although many of these conceptual ideas were the subject of research in the period prior to Shaughnessy's review (1992), they became more critical and better articulated during the current period as researchers and teachers grappled with the mainstream introduction of probability and statistics.

With respect to pedagogy, the curriculum documents adopted a number of general reform principles (e.g., problem solving) as well as pedagogical suggestions specific to probability. The major thrust in the pedagogy specific to probability dealt with children's misconceptions about chance and the need for teachers to confront these misconceptions through experimental activity and reflection. On the one hand these new directions highlighted the need for research concerning *teachers' knowledge of probability* and their knowledge about *student cognitions of probability*; on the other hand, they stimulated research on *instructional environments.*

RESEARCH INTO THE LEARNING AND TEACHING OF PROBABILITY: A TIME FOR INFORMING CURRICULUM AND INSTRUCTION

In the wake of our discussion of the national curriculum documents of the late 1980s and early 1990s, we

Table 20.3 Probability for the High School Grades

Country	Big ideas in probability
Australia	• use experimental and theoretical approaches to investigate situations involving chance and determine the likelihood of particular outcomes (including random variables)
	• generate, use, interpret, and investigate properties of probability models (including normal, binomial, geometric, and Poisson probability distributions)
	• use simulations to model uncertain events
United Kingdom	• produce a tree-diagram to illustrate the combined probability of several events that are not independent
	• understand the probability of any two events not happening
	• consider different shapes of histograms representing distributions (including the *normal* distribution) with special reference to their mean and dispersion
United States	• use experimental or theoretical probability, as appropriate, to represent and solve problems involving uncertainty
	• use simulations to estimate probabilities
	• understand the concept of random variable; create and interpret discrete probability distributions
	• describe, in general terms, the normal curve and its properties to answer questions about sets of data that are assumed to be normally distributed
	• apply the concept of random variable to generate and interpret probability distributions including binomial, uniform, normal, and chi square

will review research on the learning and teaching of probability during the current period. As noted in the previous section, with the advent of probability in the school curriculum, researchers and teachers required a better understanding of the probabilistic thinking that students of all school ages brought to the classroom. Moreover, there was an even more critical need for pedagogical knowledge that teachers could use to inform probability instruction.

Our review is organized into two parts. In the first, we present key themes of the research on probability thinking and learning, emphasizing the more recent research of the last decade while recognizing the earlier research on which it was built. In the second part, we highlight research on teachers' probabilistic knowledge and beliefs and research on instructional environments that support probability learning.

Research on Probability Learning

Although Shaughnessy's (1992) review characterized one component of earlier probability research as describing *how* people think, we would argue that the research of that period placed greater emphasis on describing *what* people thought. Reflecting the research of the time, the bulk of Shaughnessy's chapter on probability referred to studies about judgment heuristics, biases, and misconceptions in which most of the data were collected from students' forced-choice responses (p. 476). Apart from the developmental studies on probabilistic thinking and intuitions (e.g., Fischbein, 1975; Piaget & Inhelder, 1951/1975), few studies in that era used methodologies that probed students' reasoning or investigated the meanings students attributed to chance situations. For related reviews and analyses of the probability research of this and earlier periods we refer the reader to Borovcnik and Peard (1996), Jones (2005), and Kapadia and Borovcnik (1991).

Since the 1990s, however, research has followed the lead set by Piaget and Fischbein and has become more intent on addressing the ways in which people think about situations involving chance. The research has expanded beyond describing judgmental heuristics, biases, and misconceptions to include the cognitive development of a wider range of conceptual topics in probability. These conceptual topics are reflected in our analysis of the three national curriculum documents and include students' conceptions of chance and randomness, students' understanding of sample space as it relates to both simple and compound events, and students' thinking about probability measurement in its various forms: classical, frequentist, and subjective.

In the sections that follow, we review research on students' learning of these conceptual ideas, often linking it back to the research of Piaget, Fischbein, and others of the earlier period. We also analyze several developmental models of probabilistic reasoning that can be used to inform instruction in probability. These models, developed in this period but standing on the seminal work of Piaget and Inhelder (1951/1975) and Fischbein (1975), presage our review of research on the teaching of probability.

Chance and Randomness

This section commences with a review of the extensive research on *chance*. The *Shorter Oxford English Dictionary* (Trumble, 2002) defined chance as "a possibility or probability: as distinct from a certainty" (p. 312). This usage of chance envisages it as a language for describing events that are characterized by unpredictability or possibility rather than certainty. Chance language is also used in measuring the likelihood of some event. For example, the likelihood of obtaining a "3" in the roll of a die can be referred as a 1-in-6 chance (Gal, 2005). Consequently, chance has become synonymous with probability, albeit in an informal or even colloquial sense.

The Language of Chance

The language that students use to interpret situations involving chance provides insight into the variety of perceptions that this complex construct evokes. Such variety is illustrated by the work of Fischbein, Nello, and Marino (1991) with elementary (Aged 9–11) and junior high school (Aged 11–14) students in Italy. In a written assessment they posed two questions, each of which asked students to indicate whether the event of obtaining a given number was impossible, possible, or certain: (a) when tossing a regular die and (b) when using a uniform random generator with numbers from 1 to 90. The majority of students at both age levels adequately identified impossible, possible, and certain events; however, some students' reasoning revealed nonnormative use of chance language. In the excerpts below, students are identified by grade level (e.g., E4 represents elementary, Grade 4 and M2 represents junior high, Grade 2).

> Will one obtain 31 [in (b)]?
>
> "It is impossible", answers the child, "because the probability is very small" (E4).
>
> "It is impossible, because among 90 numbers there is only one 31" (M1).
>
> "It is certain because in the tombola [game] there is a 31" (E5).

> Is it possible to get 5 when rolling a die?
>
> "It is impossible . . . because it is only one probability among 6" (M3).
>
> "It is impossible because someone believes that he will win and afterwards it does not come out" (M2). (p. 528)

These excerpts show the gallimaufry of meanings that students attribute to the concepts of possible, impossible, and certain. Green (1983), and more recently Moritz, Watson, and Pereira-Mendoza (1996) and Watson and Moritz (2003b), have described similar misinterpretations and nonnormative conceptions of these fundamental descriptors of chance phenomena.

Working in Israel, Amir and Williams (1999) examined students' conceptions of the term *chance*. They reported that the majority of the students held multiple meanings for the construct. The most common interpretation referred to chance as *something that just happens*. As one student explained, "It's like walking down the road and a stone fell out . . . and you fell over. That would be chance. Just happens by accident" (p. 97). Other interpretations included *unexpected* or *unusual* events—"100 tosses of a coin? It would be chance getting 100 out of 100, but possible to get 55 on one, 45 on the other" (p. 97).

When students quantify chance situations, evidence suggests that they misuse language. The most common example is use of the phrase *50–50* to describe the chance of a particular event. In situations in which an event is *possible*, students frequently use the expression *50–50 chance* to indicate the presence of uncertainty rather than a specific measurement of chance (Watson, 2005). In the excerpt below from Amir and Williams's study (1999, p. 101), observe the student's indiscriminate use of the phrase *50–50* to quantify the chance of any outcome that he considered probable.

> **I:** What is the chance of getting a "4" on a 10-sided die?
> **J:** Probable.
> **I:** Could you give me the chances in a number form?
> **J:** 50–50. Even chance of getting a 6 or a 4 or a 3.
> **I:** What is the chance of getting a number bigger than 6?
> **J:** Probable. 50–50.
> **I:** And of getting an even number?
> **J:** Probable. 50–50.

Tarr (2002) highlighted two distinct ways in which the *50–50* phrase is used inappropriately. One involves the use of 50–50 chance to describe a set of more than two equally likely outcomes. This is evident in the excerpt above, when the student used *50–50* with an "even chance of getting a 6 or a 4 or a 3." The other misuse of the phrase involves the use of 50–50 chance to describe two outcomes that are not equally likely. In the excerpt above the student might have applied this reasoning by assigning a *50–50* chance to the two events *getting a number bigger than 6* and *getting a number not bigger than 6*. A generalized version of this reasoning, known as the *equiprobability bias*, refers to situations in which all events are treated as equiprobable, even when they are not. Use of the equiprobability bias has been exhibited by students at all levels of schooling and across various cultures (e.g., Batanero, Serrano, & Garfield, 1996; Konold, Pollatsek, Well, Lohmeier, & Lipson, 1993; Lecoutre, 1992; Li & Pereira-Mendoza, 2002; Moritz & Watson, 2000).

Language and beliefs also play an important role when students talk about "fair" probability generators. Watson and Moritz (2003b) used an interview protocol to investigate 108 elementary and middle school students' beliefs about the fairness of dice and how they would determine whether each of a number of given dice (one of which was "loaded") was fair. Watson and Moritz identified four hierarchical levels of student beliefs about the fairness of dice: (a) the dice are *unfair*, based on idiosyncratic reasons; (b) the dice are *fair*, no theoretical or experiential justification; (c) the dice are fair, qualified by referring to the rolling conditions or the physical manufacturing of the dice; and (d) the outcomes are *fair* in the long term, but *short-term variation* is recognized. The researchers also identified four strategies used by students to determine the fairness of dice: (a) relied on *intuitive and idiosyncratic* beliefs (e.g., luck); (b) asserted dice were *fair* but *no test* was necessary; (c) observed the *physical features*, cited *symmetry*, or carried out a few *unsystematic trials*; and (d) performed *systematic trials* of the dice using *small* and *large* samples to varying degrees. Interestingly, Watson and Moritz noted that the association between students' beliefs and their strategies for judging fairness of dice ($r = 0.28$) was statistically significant but not strong. With respect to the longitudinal aspect of their study, they used the same protocol in interviewing 44 of the participants 3 or 4 years later. The results showed that approximately 50% remained at the same level for both beliefs and strategies, almost 50% responded at a higher level, and a very small number responded at a lower level. The association between beliefs and strategies was almost the same as in the initial study. The authors concluded that student beliefs and strategies for determining fairness may produce dilemmas for teachers in using probability generators like dice and in dealing with language, especially the use of the word *chance* in different contexts.

The diversity of interpretations of chance and the language used to describe it point to the difficulties many students have with the notion of uncertainty. Although there is evidence that children as young as 4 years of age exhibit some understanding of the concept of uncertainty and are able to recognize that some events cannot be predicted (Byrnes & Beilin, 1991; Fay & Klahr, 1996; Fischbein, 1975; Kuzmak & Gelman, 1986), there is considerable empirical data showing that students, as well as adults, overattribute deterministic or causal explanations to situations involving chance (Kahneman & Tversky, 1982; Konold, 1989, 1991b; Metz, 1998a, 1998b; Shaughnessy, 1992). Many misconceptions associated with probabilistic reasoning can be attributed to *causal* or *deterministic* modes of reasoning; that is, to reasoning that assumes that every state of affairs has a cause and that nothing can be attributed to chance or other contingencies (Batanero, Henry, & Parzysz, 2005; Rohmann, 1999).

Causal reasoning is one feature of the *outcome approach* that Konold (1989, 1991b) has introduced to explain the nonstandard reasoning frequently exhibited by school and college students in situations involving chance. For example, when asked to compare the probabilities of a head or tail after a sequence of four heads, students using outcome-oriented reasoning choose "tails" even though they are aware that heads and tails are equally likely. According to Konold, students predict the outcome of the next trial rather than considering probabilities. Moreover, their prediction is often based on a causal or deterministic perspective—"it's tails' turn."

Konold (1989) argued that the use of a causal model for dealing with uncertainty is the most significant difference between novice and expert reasoning in chance situations. He concluded that "as long as students believe that there is some way they can 'know for sure' whether a specific hypothesis is correct, the better part of statistical logic and all of probability theory will evade them" (p. 92). This sentiment resonates with Fischbein's (1975; Greer, 2001) contention that a cultural bias exists toward deterministic thinking. Such a bias may be nurtured by school experiences that emphasize causal explanations and ipso facto strengthen intuitions rooted in deterministic modes of thinking.

Characteristics of Random Phenomena

Random phenomena are characterized by having short-term unpredictability and long-term stability. Gal (2005), in considering short-term unpredictability, identified two perspectives: (a) randomness is a *property* of an outcome, in the sense that a string of outcomes appears unordered; and (b) randomness is a *process* by which a string of outcomes occur and cannot be predicted according to some basic cause. With respect to the long-term stability of randomness, Moore (1990) highlighted the regular pattern associated with the relative frequencies of outcomes over many repetitions. For example, the event of tossing a coin is a random process because we cannot predict with certainty the properties of either a single toss or a sequence of tosses based on the physics of the coin. However, there is a long-term expectation that 50% of tosses will result in a head and 50% in a tail.

Research on students' thinking about randomness has occurred over 4 decades and has focused largely on three facets: random mixtures, random distributions, and random selection. Because Piaget and Inhelder (1951/1975) have laid the foundation for research in this area, we present a brief overview of their research on random processes so as to provide a seamless transition into research from the current period. A fuller review of their work has been presented by Jones and Thornton (2005) and Langrall and Mooney (2005).

Piaget and Inhelder's (1951/1975) research classified students' conceptions of randomness into three stages of development: preoperational (4 to 7 years), concrete operational (8 to 11 years), and formal operational (beyond 11 years). To examine students' understanding of *random mixtures* they used the familiar tilt box task; to study *random distributions* they simulated raindrops falling on squares of pavement; and to investigate students' reasoning on *random selection tasks* they picked objects from a jar, flipped coins, or rotated spinners.

Piaget and Inhelder concluded that students in the *preoperational stage* did not recognize the random mixing of balls in the tilt box. Rather, they predicted regular change patterns and often expected the balls to return to their original position. These students typically placed raindrops evenly on the squares and demonstrated little of the irregularity that is a feature of randomness. On random selection tasks, preoperational students often predicted the most frequently occurring event but exhibited no regard for proportionality. In contrast, students at the *concrete operational stage* predicted a progressive mixing of the balls and also recognized that it was very unlikely that the balls would return to their original positions. Displaying greater understanding of randomness, they constructed irregular distributions that gradually progressed toward regularity as the number of raindrops grew. They also demonstrated more complete quantitative reasoning in selection tasks. Piaget and Inhelder concluded that these students' actions were based on "qualitative intuitions of proportionality" (p. 54) rather than a sense of the law of large numbers. In the *formal operational stage* students used permutations to deal with random mixing and constructed progressively uniform distributions. They also determined probabilities on the basis of their understanding of proportionality, combinatorics, and the law of large numbers.

Much of the research that built upon and extended the work of Piaget and Inhelder (1951/1975) largely corroborated their findings in relation to students' understanding of various facets of randomness. Two notable exceptions were the studies of Paparistodemou, Noss, and Pratt (2002) on random mixtures and Green (1988) on random distributions.

Paparistodemou et al. (2002) took a contemporary approach to the random mixture task by engaging children (Ages 6–8 years) first in the regular tilt box task and then in a microworld task that allowed them to simulate the random movement of balls in a two-dimensional space. On the regular tilt box task they predicted overregulated arrangements for the balls and exhibited little understanding of randomness, similar to the children in Piaget and Inhelder (1951/1975). By way of contrast, children in the microworld environment reflected an emerging understanding of random mixture manifested in their dealings with haphazard movement, complex movement, symmetry of placement, and size of balls. According to the researchers, children moved beyond describing the random behavior of the balls to actually constructing random behavior.

In a replication of the raindrop task, Green (1988) did not find the extent of growth by age that was reported by Piaget and Inhelder (1951/1975). Subsequently, Green changed the raindrop task to one in which random selections of numbered counters were recorded by marking a cross in the appropriate square on a numbered grid. Using a large sample of students, Aged 7 to 16, Green (1983, 1988) found that children's facilities increased relative to age on the more transparent items representing regular and random patterns. However, on less salient items depicting "semi-random" patterns there was no significant improvement by age. Green's items have been used more recently by Dessart (1995) with middle school (Ages 10 to 13) and college students and by Batanero and Serrano (1999) with secondary students in Spain. In accord with Green's findings, these researchers found no significant differences by age in students' ability to identify random distributions. Clearly the notion of random distribution is a complex one for students of all ages.

Researchers (Batanero & Serrano, 1999; Falk, 1981; Green, 1988; Schilling, 1990; Toohey, 1995) have also studied conceptions of randomness in terms of sequences of outcomes. In a study that typifies the approach and findings of these studies, Batanero and Serrano asked students whether sequences of coin tosses, like the following, were generated by actually tossing a coin or whether the results were contrived:

- HTTTHTTHTHTTTHTTTTHHTTTHTTHT-THTTTTHTTTHT
- HTHTTHHTHTHHTTHTTHHTTHTHHTT HTHTHTHTHTHT

The researchers found that "students had greater difficulty in recognizing run properties than frequency properties . . . [and they concluded that] the similarity between the observed and expected frequencies may be more important than run lengths" (p. 562) in students' decisions. For example, most students stated that the second sequence shown above was randomly generated, even though the number of alternations between heads and tails was biased. However, 74% of the students correctly identified the first sequence as nonrandom because the frequency of heads differed from the expected theoretical frequency. To investigate the properties of randomness that students used to make judgments about sequences of outcomes and also random distributions (e.g., modified raindrop task), Batanero and Serrano (1999) asked students to justify their responses. They concluded that students referred to *unpredictability* and *irregular* patterns as indicators of random sequences and distributions and *regular* patterns and *long* runs as indicators of a lack of randomness. Batanero and Serrano interpreted students' difficulties with runs and distributions as an indicator that independence was not intuitive for these students.

Building on Piaget and Inhelder's earlier research, Metz (1998b) investigated randomness from a developmental and nondevelopmental perspective. She examined the reasoning of 5- and 8-year-old children and college students as they confronted classical probability tasks incorporating an experimental component. The experimental component was intended to encourage learning through prediction, experimentation, and reflection. Tasks included the marble tilt box, selection of the spinner most favorable to a target color, and an urn task in which students used sampling to infer the composition of a hidden collection of marbles.

Metz (1998b) found that few kindergarten children (Age 5) were able to interpret random phenomena and those who did failed to do it consistently. She concluded that kindergarten children had a predisposition for attributing order or regularity to chance situations as evidenced in the children's belief that the marbles in the tilt box would return to their original arrangement and in their assumption that the contents of the urn could be inferred from a small sample of marbles. In contrast, the majority of Grade 3 (Age 8) children exhibited some understanding of uncertainty and randomness. However, task characteristics limited their consistency in interpreting random situations as demonstrated by the fact that about one half of the third graders identified the most favorable spinner whereas none of these children consistently recognized the inherent presence of randomness in the urn sampling task. Although all of the undergraduate students in Metz's (1998b) study interpreted the tasks in terms of random phenomena, they sometimes reverted to deterministic interpretations, and this constrained their ability to recognize uncertainty. Metz concluded that deficiencies in children's performance could not be attributed solely to developmental shortcomings given that undergraduate students were also challenged by the probability tasks in her study.

Given the research findings in this section, it is apparent that the concepts of chance and randomness present an enduring challenge to learners of all ages. By way of summary of the critical elements of the research on chance and randomness for teachers, we draw on Metz's (1998b) synthesis of the special challenges these concepts pose for learners: (a) failure to interpret uncertainty in patterns that emerge over many repetitions of an event; (b) a belief that a person or device (e.g., spinner) can exert control over an event; and (c) a belief that some sort of order, purpose, or reason underlies events.

Sample Space

The concept of sample space is a fundamental part of the process of mathematizing random phenomena. It is an important descriptor of the outcomes of random phenomena and also provides the base for measuring the probability of events. As we have noted earlier, random phenomena are characterized by having multiple outcomes, and the sample space is defined as the set of all those possible outcomes. For example, in the simple random experiment of tossing a coin, the sample space comprises two outcomes: a "head" and a "tail," whereas in a compound random experiment, such as tossing a coin and rolling a die, the sample space can be represented as follows: {(H,1) (H,2) (H,3) (H,4) (H,5) (H,6) (T,1) (T, 2) (T,3) (T,4) (T,5) (T,6)}.

Although the concept of sample space appears to be a relatively straightforward aspect of the mathematics of random phenomena, it is more subtle and elusive than it appears. Reflecting on the teaching of probability, Shaughnessy (2003) observed that data from the 1996 National Assessment of Educational Progress (NAEP) indicated that students were weak on the concept of sample space, and he highlighted the value of having students build sample spaces for probability experiments. In this section, we analyze research on the concept of sample space and attempt to interpret and clarify the conflicting results in this research. According to Horvath and Lehrer (1998), understanding and using the notion of sample space requires the coordination of several cognitive skills: (a) recognizing the different ways of obtaining an outcome, (b) being able to systematically and exhaustively generate all possible outcomes, and (c) being able to "map the sample space onto the distribution of outcomes" (p. 123). These cognitive skills will be used as headings to examine and codify the salient features of students' conceptions of sample space.

Recognizing Ways of Obtaining an Outcome

Identifying the set of possible outcomes for even a simple (one-stage) random experiment such as tossing a regular die is nontrivial for many students. Jones, Langrall, Thornton, and Mogill (1999) specifically examined students' (Aged 8 to 9 years) ability and disposition to identify the sample space in random situations. They found that, prior to instruction, 15 of the 37 students in their study did not demonstrate that *all* outcomes could occur in simple event situations. The 15 students exhibited what Jones et al. called the *sample space misconception* in that they either stated a single outcome or eliminated some outcomes because they had occurred on a previous trial. They appeared to think about sample space in a predictive or deterministic way, attributing certainty to situations where it did not exist. A test of the tenacity of this deterministic thinking is reflected in the fact that this misconception remained an intractable problem for 2 of the 15 students in spite of extended instructional experiences with continuous and discrete random generators.

Piaget and Inhelder (1951/1975) did not document sample space as a problematic area for young children. Although subsequent researchers (Borovcnik & Bentz, 1991; Green, 1988; Jones, 1974; Jones, Langrall, Thornton, & Mogill, 1997; Schroeder, 1988) raised issues about it, they made relatively little attempt to explain the cognitive mechanisms that students use in recognizing possible outcomes. Jones, Langrall, et al. (1999) revealed some new insights into young children's conceptions and dispositions towards sample space, but further research is needed.

Generating All Possible Outcomes

As noted above with simple random experiments, determining the sample space necessitates that one be able to systematically and exhaustively generate all possible outcomes. Generating the outcomes for a compound (two-stage) random situation (e.g., flipping two coins) is even more problematic than for a simple one. Part of the difficulty in determining the sample space for compound random situations has been attributed to students' lack of combinatorial reasoning (Batanero, Navarro-Pelayo, & Godino, 1997; Fischbein & Grossman, 1997), a topic typically underrepresented in the school curriculum (English, 2005). Although research on combinatorial reasoning is limited, several studies have produced findings that enrich our knowledge of students' use of combinatorics in dealing with sample space.

In two studies conducted by English (1991, 1993), children (Aged 7 to 12) developed systematic and efficient strategies for generating solutions to meaningful two-stage and even three-stage combinatorial problems. Using the dressing of toy bears to model different combinations, the children in English's studies were able to "develop and modify their solution strategies, detect and correct their errors, and develop generative procedures on their own" (1993, p. 270). The most efficient strategy, coined the *odometer strategy*, was adopted by the majority of children Aged 7 years and older. For a two-stage situation, it involved holding one variable constant (e.g., blue top), matching it with all items of the second variable (e.g., black, white, and blue pants) until all possible combinations had been formed, and then repeating the cycle with a new constant item (e.g., yellow top). In order to develop powerful ideas such as the odometer strategy, English noted that, rather than having direct instruction from the teacher, children needed experiences grappling with meaningful, hands-on combinatorial problems. These kinds of instructional experiences were provided in the Jones, Langrall, et al. (1999) study and likely contributed to the success many students had with regard to compound sample space problems.

Batanero et al. (1997) produced a hierarchical classification of combinational problems after investigating the effects of instruction on the combinatorial reasoning of 14- to 15-year-old Spanish students. This classification, ordered from easiest to most difficult, is presented below.

- *Selection* problems involve determining how many samples of n elements could be obtained from a set of m objects. *Example:* A box contains 10 counters numbered from 1 through 10. A counter is drawn, its numeral is written down and the counter is returned. The process is repeated until a three-digit number is formed. How many three-digit numbers can be obtained using this process?
- *Distribution* problems involve distributing a set of n objects into m cells. *Example:* A toy has three identical red balls and four tubes colored green, yellow, blue, and violet. If only one ball can be placed in each tube, how many different ways can a child place the balls in the tubes?
- *Partition* problems involve splitting a set of n objects in m subsets. *Example:* Sam and Ryan have four cars colored red, orange, blue, and green. How many ways can they share the cars if each gets two?

Selection problems were easiest for students because they could apply a combinatorial formula. Distribution and partition problems were more difficult because students did not recognize how combinatorial operations could be applied to such problems or were unaware of how to translate the problems into the form of a selection problem. Batanero et al. also identified several combinatorial errors in relation to students' construction of sample spaces:

- *Error of order*: Distinguishing the order of elements when it is irrelevant or failing to consider order when it is essential.
- *Error of repetition*: Repeating elements when it is not possible or failing to consider the repeating of elements when it is possible.
- *Nonsystematic listing*: Using trial and error without a recursive procedure to assure identification of all possibilities
- *Faulty interpretation of the tree diagram*: Constructing an inadequate diagram or incorrectly interpreting a diagram.

On the basis of these findings, the researchers suggested that instruction should emphasize "the translation of combinatorial problems into the different models [problem types], recursive reasoning and systematic listing procedures, instead of merely centering on algorithmic aspects and on definitions of combinatorial operations" (p. 196). In essence, the research on combinatorial reasoning underscores the complexities involved in identifying a sample space and the need for instruction that is informed by research-based knowledge of the kinds of errors that students make in listing sample spaces involving compound events.

Mapping the Sample Space on the Distribution of Outcomes

Although nontrivial, the ability to list the sample space outcomes is of little value unless students recognize the importance of sample space in probability measurement. Yet, the research is replete with evidence that students of all ages do not necessarily consider sample space when determining probabilities or examining outcome frequencies (e.g., Ayres & Way, 2000; Fischbein & Schnarch, 1997; Shaughnessy & Ciancetta, 2002). The work of Fischbein et al. (1991) highlights this point.

Fischbein et al. (1991) examined students' (Aged 9 to 14 years) misconceptions about compound events in specific and general contexts. The *specific* context concerned the tossing of two dice where students were asked about the likelihood of two events, "a 5 with one die and a 6 with the other" or a "6 with both dice." The majority of students responded that the likelihood of these compound events was equal and based their justifications on two modes of reasoning:

- the situation involves chance, hence there is no reason to expect one (compound) outcome over the other
- each simple outcome is equally likely and each toss is independent, hence each (compound) outcome is equally likely.

The first mode of reasoning reflects another aspect of Konold's (1991b) outcome approach and the belief that "anything can happen." The second mode of reasoning is more sophisticated and was characteristic of students who had received instruction in probability. The following response illustrates this mode: "Each die is independent from the other. The probability that with one die, one will obtain a certain number is 1/6 and it is the same probability that one will obtain the same number with the other" (p. 535). Even though the second mode of reasoning is more sophisticated than the first it fails to take any cognizance of the sample space associated with these compound events. This lack of focus on the sample space was also true for many students who responded correctly for the wrong reason. Instead of considering the sample space composition, they proffered a belief that "same results" (e.g., 6, 6) usually occur less often than "different results" (e.g., 5, 6).

The other context investigated in the Fischbein et al. (1991) study was *general* in the sense that students were asked whether it is more probable, when tossing two dice, to obtain the same numbers or different numbers. Interestingly, the findings revealed that across these age levels the percentages of correct answers for this problem and a parallel one on the tossing of two coins were greater for the generalized contexts than for the specific contexts. Many of the students, who had provided inappropriate justifications for the specific dice problem, explained their reasoning for the general context in terms of sample space composition. This is illustrated in the following response from a junior high student: "It is more likely to obtain different numbers because, in order to get equal numbers, we have 6 possibilities but for obtaining two different numbers one has 30 possibilities" (pp. 537–538). Although it seems that the more general context motivated students to focus more sharply on sample space, this was not the case for the parallel item involving the tossing of two coins. In the two-coin item, the percentage of correct responses was relatively high, but most students' justifications belied any recognition of the sample space. These baffling differences in students' justifications for the dice and coin problems were attributed by Fischbein et al. (1991) to the richer sample space of the dice context. On the one hand, they claimed that some students exhibit an intuitive capacity to evaluate the sample space in finding the probability of a compound event. On the other hand, they concluded that students do not exhibit a natural understanding that outcomes like (a, b) and (b, a) need to be distinguished and counted separately (p. 547).

As well as providing insights into students' thinking about sample space, the work of Fischbein et al. (1991) and others also highlighted the problematic nature of this research and consequently has seeded many subsequent studies. More recently researchers have begun to observe students' thinking about sample space during teaching experiments and instructional programs.

Horvath and Lehrer (1998) examined the effect of "notational assistance" on Grade 2 students' understanding of the relationship between sample space and distributions. The student tasks involved making predictions with 6-, 8-, and 12-sided dice (e.g., most and least favorable outcomes), experimenting, collecting data, and explaining relationships between their predictions and the data. The study revealed that most children did not relate sample space composition to their experimental dice data, and this was especially true for compound events involving two dice. However, when sample space composition was introduced with a notational system (e.g., bar graphs), the children were able to infer some relationships between sample space composition and experimental data. For example, in the rolling of two 6-sided dice, they came to realize that a sum of 3 was more likely than a sum of 2 because there are two ways of getting a 3 and only one way of getting a 2. Nevertheless, when their experimentation differed from their predictions, students typically changed their predictions to match past experiences rather than use sample space composition. This preference for experimental data over sample space composition was more pronounced when the context was changed and the notational system was not provided. As part of the same study, Horvath and Lehrer (1998) found similar results with a small sample of fourth- and fifth-grade students. However, after having the benefit of notational assistance, these older students readily incorporated the sample space into their thinking even after the assistance was removed. Moreover, they came to accept experimental variation as characteristic of randomness and based their predictions on sample space composition. In summary, whereas the second-grade students in this study predicted likelihood on the basis of subjective judgments (e.g., my favorite number) and established experimental data as the key criterion for determining likelihood, the older students, albeit only 4 of them, accepted variation in experimental data and seemed to forge an ongoing link between sample space composition and outcome likelihood.

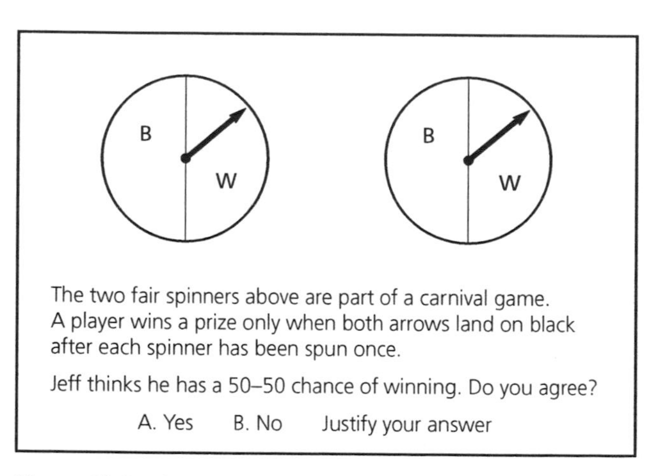

B
W
B
W
The two fair spinners above are part of a carnival game.
A player wins a prize only when both arrows land on black after each spinner has been spun once.
Jeff thinks he has a 50–50 chance of winning. Do you agree?
A. Yes B. No Justify your answer

Figure 20.2 Spinner task from the 1996 NAEP (Shaughnessy & Ciancetta, 2002).

Shaughnessy and Ciancetta (2002) also examined students' understanding of sample space in both classical and experimental contexts. Using the task in Figure 20.2, they found that for a large sample of students in Grades 6–12, (a) only 20% of those in Grades 6–8 responded that the chances of winning the game were less than one half, (b) the performance of Grades 9–11 students enrolled in introductory-level mathematics was similar to the NAEP sample of Grade 12 students (28% gave the correct answer and only 8% gave the correct reason), and (c) about 90% of Grades 10–12 students enrolled in upper level courses answered the problem correctly. The researchers observed that a high percentage of students in Grades 6–11 did not consider the sample space when justifying their response. In a follow-up interview with students in Grades 8–12, the researchers investigated whether actually playing the game would encourage students to attend to the sample space. Students were asked to predict the results of 10 rounds of the game, to collect data using the actual spinners, and to reconsider their response. This cycle was repeated before students made a final response to the question. Prior to playing the game, 13 of 28 students responded correctly, but only 5 provided a correct justification. Of these 5 students, 4 listed the sample space and 1 used the multiplication principle to demonstrate that the probability of winning was one fourth. After the game cycles, 11 of the 15 students who had initially responded incorrectly reported that the game was not 50-50, and 8 of them specifically listed the sample space. The authors claimed that, by playing the game and seeing the variation in repeated trials, some students were prompted to construct the sample space for the task. They drew the following conclusion:

> The conceptual root of the pedagogical power that we gain from having students conduct simulations is the connection that they can make between the observed variation in data in repeated trials of an experiment, and the outcomes that they expect based on knowledge of the underlying sample space or probability distribution. (p. 5)

The instructional benefits of having students link empirical data with sample space composition are also evident in the studies of Horvath and Lehrer (1998), Jones, Langrall, et al. (1999), and Nisbet, Jones, Langrall, and Thornton (2000).

In a series of cross-cultural teaching experiments in Brazil (Kaufman-Fainguelernt & Bolite-Frant, 1998), Israel (Amit, 1998), and the United States (Maher, Speiser, Friel, & Konold, 1998; Speiser & Walter, 1998), researchers investigated the ways that students (ranging in age from elementary to college) represent and deal with sample space in a two-dice game task. These researchers concluded that the game triggered two sample spaces for students: sample space A with 36 outcomes, the Cartesian product of the set {1, 2, 3, 4, 5, 6}; and sample space B with 21 outcomes that can be characterized as {(1, 1) (1,2), (1,3) (2, 3) . . . (6, 6)}. In essence, sample space A recognizes pairs like (1, 2) and (2, 1) as *distinguishable*, whereas sample space B does not and includes just one representative, say (1, 2). Nevertheless, students needed to use sample space A, whose 36 outcomes are equally likely, to compute the unequal probabilities of sample points in B. They did this by essentially using a mapping that is two-to-one for points like (1, 2) and (2, 1) and one-to-one for points like (1, 1). The research suggests that both sample spaces and the map between them are necessary for students to be able to build pervasive solutions for the dice game task. It also highlights that in dealing with compound sample spaces, students need to come to grips with the notion of what constitutes an outcome of a sample space.

From a different cultural perspective, Polaki (2002a) conducted a teaching experiment with fourth- and fifth-grade students (Aged 9 to 10) in Lesotho, Africa, to forge links between probability and sample space composition. The students in Polaki's study were assigned to two instructional groups: one group generated a small set of experimental data in a game situation, then focused on sample space composition to determine the probabilities; the other group examined large-sample computer-generated data prior to considering sample space composition. Both instructional methods positively impacted the students' probabilistic reasoning and their ability to link probability and sample space composition. However, it is interesting to note that there were no significant differences between the two methods, and Polaki conjectured that students might have had difficulty because they were not familiar with computer-generated data.

In contrast to Polaki's study (2002a), students in Pratt's (1998, 2000) microworld studies saw the value of conducting a large number of trials in order to produce a more transparent relationship between sample space composition and relative frequencies. The different findings between the two studies might be attributed to the stronger focus on variation in Pratt's study. Given the importance ascribed to variation by Shaughnessy and Ciancetta (2002), it is not surprising that Pratt's students benefited greatly by observing variation in both small and large samples. In fact, they essentially learned to control variation by preferring large samples over small samples.

Probability modeling presents a different, but related approach to focusing students' attention on sample space. By *modeling* we mean that one "selects a probability generator whose sample space outcomes and their probabilities match with the corresponding outcomes and probabilities of a contextual problem" (Benson & Jones, 1999, p. 2). Benson (2000), using a teaching experiment and tasks like the one in Figure 20.3, found that 6 students in Grades 3 and 4 could construct or select a probability generator that corresponded correctly to the sample space of a contextual problem. All students utilized one-to-one correspondence to make connections between sample space outcomes and their corresponding probabilities; a few also used *n*-to-one or *n*-to-*n* correspondences. In addition, some students recognized when probability generators were *equivalent* and could identify more than one probability generator for a given situation.

Class Draw
Six children, John, Ken, Louis, Sue, Cathy, and Beth, enter a drawing to win a prize. Only one name will be selected. How could you model to find out who wins the drawing?

[Note: Colored bears, dice, two-colored chips, one six-segment and one two-segment spinner were available].

Figure 20.3 Contextual problem on probability modeling (Benson & Jones, 1999).

The modeling tasks in Benson's study seemed to assess students' understanding of sample space more deeply than tasks requiring the listing of all possible outcomes. As she observed,

> It is one thing to identify the outcomes and determine the event probabilities associated with a probability generator; it is another to identify or construct a probability generator that will faithfully represent the probability distribution embodied in a contextual task. (p. 3)

Shaughnessy (2003) highlighted the critical need for instructional programs to involve students in building sample spaces for probability experiments. However, as exemplified by the research in this section, teachers need to be aware that the development of students' thinking about sample space takes many twists and turns: (a) an inability or unwillingness on the part of younger children to list the outcomes of even simple experiments, (b) a lack of systematic processing skills to generate or differentiate between outcomes in compound experiments, and (c) a reluctance to link sample space composition and outcome probabilities (probability distributions) in both classical and experimental situations.

Probability Measurement

Batanero, Henry, et al. (2005) provided an insightful historical exposition on approaches to measuring the probability of random events. In particular they underscored three key approaches to probability measurement: classical (Laplace, 1825/1995), frequentist (von Mises, 1928/1952), and subjective (Lindley, 1980). Notwithstanding the importance of these three approaches, which are described in our earlier section on national curricula, research on probability measurement has focused largely on the classical approach with some recent research beginning to emerge on the frequentist approach. We were not able to locate cognitive research on the subjective approach to probability measurement.

Our review of cognitive research on the classical or theoretical approach to probability measurement has several parts. The first part deals with students' (mostly younger children's) qualitative view of probability in terms of *most likely/least likely events*. Subsequently we examine students' quantitative reasoning on theoretical probability according to the categories: probability of an event, probability comparisons, probability adjustments, and conditional probabilities.

We have already discussed some learning research (e.g., Horvath & Lehrer, 1998; Polaki, 2002a; Pratt, 1998, 2000; Shaughnessy & Ciancetta, 2002) that incorporates to varying degrees a frequentist approach to probability measurement. These studies set the scene for an examination of recent research that focuses more explicitly on experimental probability measurement and the relationship between experimental probability, theoretical probability and sample size.

HT-sequence problem

Part 1. Which of the following is the most likely result of five flips of a fair coin?

a) HHHTT
b) THHTH
c) THTTT
d) HTHTH
e) All four sequences are equally likely.

Part 2. Which of the above sequences would be least likely to occur?

Figure 20.4 Coin-tossing problems (Konold et al., 1993, p. 397).

Most Likely/Least Likely Events

Using the task shown in Figure 20.4, Konold et al. (1993) asserted that high school and college-age students perceived *most likely* and *least likely* outcomes very differently. With respect to the HT-sequence problem, most students responded correctly that the *most likely* result was that all four sequences were equally likely. However, in the *least likely* case, only 38% of the students stated that the four sequences were equally likely. Konold et al. argued that this inconsistency for approximately 60% of students could be attributed to reasoning based on the *outcome approach*. That is, students interpreted the problem as asking them to predict what *would* happen. Hence in predicting equally likely for the *most likely* case, they were not arguing that the sequences were equally likely in the sense of having the same numerical probability; rather, they were using equally likely to mean that "anything could happen" (p. 399). By way of contrast, in the *least likely* case, the majority of students selected item (c) basing their response on the representativeness heuristic (Kahneman & Tversky, 1972) in the sense that this sequence was the least representative.

In a related study, Watson, Collis, and Moritz (1997) asked students in Grades 3, 6, and 9 whether certain events were more likely or equally likely. They found that even when students correctly identified events as more likely or equally likely, their reasoning was not normative. In essence, their findings were similar to those of Konold et al. (1993) in that many students interpreted equally likely as meaning that "anything can happen." The findings of these two studies raise a number of questions about what students of various ages are responding to when they identify the most or least likely event. This caveat is important for studies like those in the next section.

Probability of an Event

A number of studies with younger children (e.g., Acredolo, O'Connor, Banks, & Horobin, 1989; Jones, Langrall, et al., 1997, 1999; Polaki, 2002a) have focused on the construct, probability of an event. In these studies, students were generally asked to identify the most or least favorable outcome usually based on a classical interpretation of probability. In an attempt to avoid the problems raised by Konold et al. (1993) and Watson et al. (1997), they have asked students to justify their responses using both qualitative and quantitative reasoning. The approach taken by Acredolo et al. set the stage for such studies when they examined the reasoning of children in Grades 1, 3, and 5 with respect to the following task: Given a clear plastic bag containing three red jellybeans, two green jellybeans, and one yellow jellybean, indicate the likelihood of randomly drawing a yellow jellybean by sliding a marker along a scale with a sad face at one end and a happy face at the other. They found that students used one of three strategies: (a) a numerator strategy in which they only examined the part of the set that corresponded to the target event, (b) an incomplete denominator strategy in which they examined the complement of the target event, and (c) an integrating strategy in which they related the number of target elements to the total number of elements. More generally, these strategies can be described in terms of part-part and part-whole reasoning as the following studies reveal.

On the basis of their work with students in Grade 3, Jones, Langrall, et al. (1999) concluded that the recognition and use of part-whole relationships, in conjunction with part-part relationships, facilitated students' movement toward quantitative reasoning in probability. Moreover, they found that the part-whole schema, rather than a precise use of fractions, appeared to be a critical element in supporting students' reasoning. A number of other researchers have also identified the importance of part-whole reasoning in younger students' thinking about probability (Jansem, 2005; Metz; 1998a; Polaki, 2002a, 2002b; Ritson, 1998; Tarr, 1997). In fact, Jansem and Polaki have undertaken a detailed analysis of the mechanisms by which elementary school students use different part-part and part-whole schemata in dealing with the probabilities of events.

Probability Comparisons

Some probability measurement situations involve two sets of elements and require students to select the set that is more favorable for a target event (binary-choice task). For example, suppose children are asked which of the following urns is more favorable for red or are they equally favorable: one containing four red marbles and eight green marbles or one containing four red and six green marbles? The consistent finding for this kind of task across several studies (e.g., Falk, 1983; Fischbein et al., 1991; Jones et al., 1997; Piaget & Inhelder, 1951/1975; Way, 1996) is that many students, especially young children, base their choice on idiosyncratic judgments such as "their favorite color" or on restrictive reasoning that focuses on the number of target elements. These students do not consider the number of nontarget elements or the ratio of winning to losing items. Falk, whose studies are seminal in this area, has identified three strategies that students use to identify which set is more favorable: (a) set with more target event, (b) set with less nontarget event, and (c) set with greater difference in favor of target event. Falk observed that as students understood proportionality, they began to use strategies that recognized relationships between the number of winning and losing elements, or between the number of winning/losing elements and the total. In essence, this continuum of strategies resonates with the part-part and part-whole strategies alluded to in the previous section.

The longitudinal study of Watson et al. (1997) illustrates more contemporary research in this area. They surveyed a large sample of students in Grades 3, 6, and 9 on the following problem:

> Box A and Box B are filled with red and blue marbles as follows. Each box is shaken. You want to get a blue marble, but you are only allowed to pick out one marble without looking.
>
> Which box would you choose? Please explain your answer.
>
> (A) Box A (with 6 red and 4 blue)
> (B) Box B (with 60 red and 40 blue)
> (=) It doesn't matter (p. 65)

Their interpretation of students' responses was more complex than that of Falk (1983) and was guided by Biggs and Collis's (1991) development model. Watson et al. described students' responses across four levels for each of two cycles of development that increased in complexity. At the lowest level of the first cycle (*acquisition of the concept*), responses were subjective (pick (=): red and blue are my best colors). At the second level, responses indicated recognition of the colors in the boxes, but no reference to the frequency of colors (pick (=) because there is blue in each). At the third level, responses attended to frequency, but it was inappropriately applied to the problem (pick Box B because there are more marbles to pick from). At the fourth level, responses recognized the number of marbles in each box and compared the two colors (pick (=) because there are more reds in both boxes). In the second cycle (*application of the concept*), more complex explanations were provided. First, the equivalence of the boxes was recognized according to a basic similarity (pick (=) because they are both filled alike). Second, justifications were more mathematical and often used ratios (pick (=) because Box B has ten times the amount in Box A). Finally, responses utilized ratios or percentages for within-box comparisons and then comparisons between boxes (pick (=) because both boxes are 40% blue). Most students in Grades 3 and 6 gave responses at the third level of the first cycle while the majority of Grade 9 students gave responses in the second cycle of development.

The Watson et al. (1997) categorization of responses, based as it is on an extensive age range, has considerable potential for informing probability instruction in both elementary and middle school. It also suggests that students will think differently during the acquisition or building of probability concepts than they do when applying probability concepts.

Probability Adjustments

Citing limitations with tasks typically used in probability of an event and probability comparisons research, Falk and Wilkening (1998) promoted the use of *probability-adjustment tasks* like the one shown in Figure 20.5. In this task students were asked to add white balls to the second urn so that the likelihood of drawing a white was the same for each urn. Falk and Wilkening claimed that these tasks call for "the assessment of the magnitude of the target probability and at the same time [retain] the need to compare two probabilities" (p. 1341).

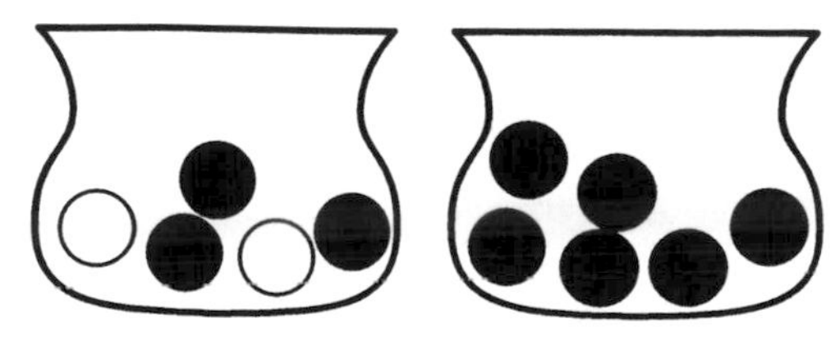

Figure 20.5 Example of a Falk and Wilkening (1998) probability-adjustment task.

The general findings of Falk and Wilkening's (1998) study indicated that students (Ages 6–7) were unable to generate the target probability using notions of proportionality. Although their strategies were often unsystematic, they did reflect a tendency to focus on the number of either target or nontarget balls in the full urn. Students, Aged 9–10, attempted to integrate the two dimensions but tended to do so by making decisions based on the difference between the number of target and nontarget balls in the full urn. By Age 13, students were utilizing proportionality, albeit at less than a perfect level of performance. Falk and Wilkening claimed that the most notable lesson learned from their research is that "children's spontaneous indications . . . show that, in a given setup, binary-choice mastery precedes the ability to adjust probabilities" (p. 1353).

Probability adjustment tasks provide a valuable addition to the network of tasks used to illuminate students' thinking about probability measurement. This pool acts as a fillip for further research, as do the various caveats of Falk and Wilkening (1998), Konold et al. (1993), and Watson et al. (1997) in relation to *most likely/least likely* probability tasks.

Conditional Probability

Drawing heavily on earlier research of Falk (e.g., 1983, 1988), Shaughnessy (1992) described three salient misconceptions in the literature on conditional probability. The first, referred to as the "Falk phenomenon" or *the fallacy of the time axis*, highlights the difficulty students have in dealing with a conditional probability, P(A|B), where the Event A precedes Event B (see also Batanero & Sanchez, 2005). The second misconception reported by Shaughnessy pertains to the difficulty of identifying the conditioning event (see the card problem in Shaughnessy and the coin problem in Batanero & Sanchez). The third relates to the difficulty in distinguishing between a conditional statement and its inverse, that is, between events like "I have measles, given that I have a rash" and "I have a rash, given that I have measles" (see Pollatsek, Well, Konold, & Hardiman, 1987). Given the narrow focus of this earlier research, Shaughnessy (1992) called for "*systematic* investigation of students' conceptions about conditional probabilities" (p. 475). In the current period, one study specifically addressed this need whereas others extended research on students' conditional probability reasoning.

In a teaching experiment designed to increase Grade 5 students' understanding of conditional probability and independence, Tarr (1997) approached conditional probability via a sequence of random tasks that involved *with and without replacement* situations. He detected significant growth in students' understanding following instruction and characterized aspects of students' reasoning about conditional probability. One of the key findings in Tarr's study related to the shift in students' reasoning from part-part to part-whole comparisons and its impact on their ability to deal with conditional probabilities. Prior to instruction, students relied on part-part comparisons in nonreplacement situations and often failed to recognize that the probabilities of *all* events changed. By way of example, consider the following problem:

> A candy jar contains an assortment of flavors: 4 grape, 3 cherry, 2 apple, and 1 lemon candies. A grape candy is drawn and eaten. Has the chance of drawing another grape candy from the jar changed or is it the same chance as it was before? Has the chance of drawing a cherry candy changed? An apple candy? A lemon candy? Explain. (Tarr & Lannin, 2005, p. 223)

Students using part-part reasoning responded that the chance of drawing a grape candy had decreased because there was now one less grape candy in the jar, whereas the other flavors had the same chance as before because their numbers had not changed. By way of contrast, students using part-whole reasoning explained that the lemon candy, for example, now had a better chance because there were fewer candies in the jar. Some even stated that the lemon candy now had a probability of 1 out of 9 compared with 1 out of 10 previously. Tarr also claimed that his practice of having students focus on the sample space and the assigning of numerical probabilities after each draw facilitated students' use of part-whole reasoning.

Many students in Tarr's (1997) study did not use conventional notation when describing conditional probabilities. Tarr characterized four invented representations, the first three of which appeared mostly before instruction. The first representation was a verbal description using chance as a unit of measurement. For example in the candy problem, a student might typically respond that drawing a grape candy has "gone down one chance" because one grape candy has been removed. Tarr reported that students who used this representation focused on part-part relationships between the number of target objects and its complement. A second representation involved the use of a frequency comparison that also incorporated part-part reasoning. For example, the chance of drawing a grape candy has decreased because now there is the same number of grape and cherry candies whereas previously there were more grape candies. A third representation, observed for sample space with two outcomes, involved a verbal comparison of the number of objects of the target event (x) with the total number of objects (y) using the phrase "x of y chance." The fourth representation, appearing after the instructional program, involved the reporting of probabilities as a ratio of percents. For example in the candy problem, the probability of drawing a cherry candy might be described as 30% out of 90% where 30% represented the percentage of cherry candies initially in the jar and 90% represented the percentage of candies remaining after the removal of a grape candy. Although mathematically incorrect, Tarr reported that this invented notation appeared to make sense to the students.

In this section, we have already made reference to the work of Pollatsek et al. (1987) that revealed college students' confusion about a conditional statement and its inverse. Watson and Mortiz (2002) revisited this work with students in Grades 5 to 11. Using similar items, they examined students' estimations of probabilities or frequencies for a conditional event and its inverse (e.g., X|Y and Y|X). Rather than relying on forced-choice responses, as Pollatsek et al. had, they allowed students to choose the form in which they expressed their estimates:

> Please estimate:
>
> (a) Out of 100 men, how many are left-handed?
> (b) Out of 100 left-handed adults, how many are men?
>
> Please estimate:
>
> (a) The probability that a woman is a school teacher.
> (b) The probability that a school teacher is a woman. (p. 66)

Watson and Moritz's results showed that, although the number of correct responses for both items increased by grade level, there were more correct responses for the conditional frequencies item than for the conditional probability item. The researchers stated that "school students rarely had difficulty expressing frequencies but often had difficulties expressing probabilities" (p. 79). They offered two interpretations for the difference in performance on the items: (a) students' experience with the two contexts, and (b) the fact that the phrase, *out of*, in the frequency item cued students to the relevant conditions. Concurring with the findings of Pollatsek et al. (1987), they claimed that the grammatical structure of the statements contributed most strongly to students' performance.

Tarr (1997) stressed the importance of having students build conditional probabilities by sorting and counting sample space outcomes, whereas Watson and Moritz (2002) concluded that students handled conditional frequency items better than conditional probability items. In both studies the research is suggesting that "count" or frequency data is important in developing notions of conditional probability. This finding is also consistent with recent psychological research (Gigerenzer, 1994; Hopfsenberger, Kranendonk, & Scheaffer, 1999) that highlights the value of students using data (frequencies) to build probabilities. In fact, arguing from a more pervasive pedagogical perspective, Shaughnessy (2003) suggested that "the teaching of probability should actually start with data, and that probability questions should be raised from data sets" (p. 224). This data orientation provides an appropriate backdrop for the next part of our review that deals with data generation and experimental probability measurement.

Experimental Probability, Theoretical Probability and Sample Size

Given the importance of experimental (empirical) probability in national curriculum programs (AEC, 1991; DES, 1989, 1991; DfEE, 1999; NCTM, 1989, 2000), it is surprising that so little research has been undertaken on students' conceptions of experimental probability, and "the bidirectional relationship between empirical and theoretical probability and the role of sample size in that relationship" (Stohl & Tarr, 2002, p. 314). Studies reviewed previously (Horvath & Lehrer, 1998; Jones, Langrall, et al., 1999; Polaki, 2002a; Shaughnessy and Ciancetta, 2002) have highlighted the difficulty experienced by students of all ages in making links between the composition of the sample space outcomes of a probability generator and the distribution of outcomes the generator produces; in essence, these studies have underscored the cognitive challenge faced by students in making connections between theoretical and empirical probabilities.

In addition to the demanding task of building and analyzing the sample space (Shaughnessy, 2003), the key element in elucidating the connection between empirical and theoretical probability is the size of the data sample produced by the probability generator. Pratt (2005) claimed that children had a natural proclivity for small samples over large samples in dealing with random phenomena. Nevertheless, the findings of his microworld studies (1998, 2000) suggested that, in an appropriate instructional environment, even 10-year-old students were able to discern the importance of conducting a large rather than a small number of trials in confronting empirical probability and revealing a more transparent relationship between sample space composition and relative frequencies. Studies by Aspinwall and Tarr (2001) and Stohl and Tarr (2002) provide further insights on students' thinking about experimental probability and its relationship to theoretical probability.

During a 5-day instructional program, Aspinwall and Tarr (2001) examined sixth-grade students' conceptions of experimental probability and their understanding of the relationship between experimental probability and sample size. In particular, they investigated whether students understood that, as the number of trials increases, the experimental probability is more likely to reflect the parent population (theoretical probability), and conversely that small samples are more likely to produce nonrepresentative probabilities. Prior to instruction, 5 of their 6 target students showed little or no awareness of the relationship between experimental probability and sample size; typically, they thought that any size sample should reflect the parent distribution. Although students demonstrated a more normative conception of experimental probability and its relationship to sample size following instruction, their understanding was restricted to a realization that events with small probabilities were more likely to occur with a small number of trials and less likely as the number of trials increases. Their responses fell short of a unified understanding of the scope and power of the law of large numbers.

In a related study, Stohl and Tarr (2002) built on earlier research by Stohl (Drier, 2000a, 2000b). They used the *Probability Explorer* (PE) microworld (Stohl, 1999–2002) to explore 2 sixth-grade students' thinking about the relationship between experimental and theoretical probability and the role that sample size plays in that relationship. Using an instructional environment that incorporated the PE tools' facility to perform small- and large-sample simulations and produce multiple data representations (pie graphs, bar graphs, and data tables), the researchers traced students' thinking and reactions to various perturbations associated with the researcher-generated probability tasks. Stohl and Tarr claimed that these two average students were able to make appropriate inferences based on the data, and, in particular, made valid connections between experimental probabilities and their corresponding theoretical probabilities. Moreover, consistent with Pratt (2000) and Drier (2000a), the students recognized the importance of using large samples when making inferences in probability environments.

Although all of these studies involved only a small number of students, the studies do outline the scope of elementary and middle school students' thinking about experimental probability and its relationship with theoretical probability. Although the evidence is nascent, the research reveals that, with appropriately designed learning resources, students can move beyond their predilection for small samples or a *law of small numbers* (Kahneman et al., 1982), and begin to appreciate the power of the *law of large numbers* in dealing with the duality of experimental and theoretical probabilities.

As we have seen in this section, probability measurement remains the key aspect of any quantitative understanding of random processes. Hence, it is critical that teachers become aware of the salient features of research on probability measurement that have been outlined in this review: (a) the need for concurrent and related development of experimental and theoretical probability, (b) the value of notational assistance and invented language in describing probabilities, (c) the role of simulation and modeling as a means of measuring probabilities, (d) the confounding nature of sample size in measuring experimental probability (e) the need for learners to confront misconceptions, and (f) the importance of integrating various kinds of mathematical reasoning (e.g., numerical and proportional reasoning) in probability measurement experiences.

Cognitive Models of Probabilistic Reasoning

Perhaps the most notable shift in research during this period has been towards the use of neo-Piagetian cognitive models (e.g., Biggs & Collis, 1982, 1991; Case, 1985) to describe different levels of complexity in students' probabilistic reasoning. Most researchers working on cognitive models of probabilistic thinking have drawn upon the general developmental model of Biggs and Collis to profile student growth from naïve conceptions to more formal or normative ones.

The Biggs and Collis Model of Cognitive Development

The Biggs and Collis model has been an evolutionary one beginning with the SOLO Taxonomy (Biggs & Collis, 1982) that postulated the existence of five modes of functioning (*sensorimotor* from birth, *ikonic* from around 18 months, *concrete-symbolic* from around 6 years, *formal* from around 14 years, and *postformal* from around 20 years). According to Biggs and Collis (1991), the emergence of each of the five modes incorporates the continuing development of earlier modes. This multimodal functioning also includes, within each mode, a cycle of learning that has five hierarchical levels. At the *prestructural* (P) level, students engage a task but are distracted by an irrelevant aspect; they actually exhibit reasoning indicative of an earlier mode. For the *unistructural* (U) level, the student focuses on the relevant domain and takes up one aspect of the task. At the *multistructural* (M) level the student identifies several relevant aspects of a task but does not integrate them. In the *relational* (R) level, the student integrates the various aspects and produces a more coherent understanding of the task. Finally, at the *extended abstract* (EA) level, the student generalizes the structure to incorporate more abstract features that represent thinking in a higher mode. Within any mode, the middle three levels are of most importance, because as Biggs and Collis noted prestructural responses belong in the previous mode and extended abstract responses belong in the next. A fuller description of the use of the Biggs and Collis model in stochastic processes is found in Jones, Langrall, Mooney, and Thornton (2004) and Langrall and Mooney (2005).

Jones et al. Probability Framework

In order to present a coherent view of students' probabilistic reasoning, Jones et al. (1997) generated and validated a framework that characterizes elementary students' reasoning on four previously reviewed probability constructs: sample space, probability of an event, probability comparisons, and conditional probability. The framework, based on a synthesis of the probability literature, was supported by the researchers' empirical studies with Grades 1–4 children over a period of 2 years. For each probability construct, the framework describes four levels of reasoning that correspond to four levels of thinking within the ikonic and concrete-symbolic modes of the Biggs and Collis (1991) model. An abstracted version of the framework descriptors is shown in Figure 20.6.

At Level 1, *subjective* thinking, students are narrowly and consistently bound to reasoning that is based on idiosyncratic notions such as a student's favorite color. This level relates to the Biggs and Collis (1991) *prestructural* level within the ikonic mode. The Level 1 descriptors presented in Figure 20.6 illustrate students' proclivity for subjective notions in all constructs, especially sample space.

Students exhibiting Level 2, *transitional* thinking, recognize the significance of quantitative measures, but their responses are often naïve and inflexible. Their thinking is indicative of Biggs and Collis's (1991) *unistructural* level in that they begin to function in the concrete-symbolic mode and focus on one quantitative aspect of a task. The descriptors for Level 2 capture the main thrust of their responses in listing outcomes and quantifying probabilities; however, conflicts in their thinking may result in reversions to subjective judgments.

At Level 3, students' thinking involves the use of *informal quantitative* reasoning. This incorporates aspects of both the *multistructural and relational* levels of Biggs and Collis (1991) in that students recognize more than one feature of a task and integrate some features such as attending to the sample space when quantifying the probability of an event. More specifically, as the Level 3 descriptors illustrate, students begin to list outcomes systematically and to use quantitative reasoning and representations to describe probabilities. Finally, at Level 4, students exhibit *numerical* thinking in the sense that they use valid numerical measures to describe probabilities. Moreover, consistent with the characteristics of *relational* thinking (Biggs & Collis), they connect sample space composition and event probabilities.

Although Jones et al. (1997) argued that students' reasoning grows over time, they did not claim that students necessarily follow an ordered progression through the four framework levels. Rather, they saw the framework as presenting a coherent picture of students' reasoning that informs instructional design, implementation, and evaluation (Jones, Thornton, Langrall, & Tarr, 1999). The Jones et al. probability framework has spawned research studies at other grade levels and in other cultures.

Tarr and Jones (1997) expanded the elementary framework into the middle grades, with specific emphasis on conditional probability and independence. Subsequently, a more comprehensive framework (Jones, Thornton, et al., 1999) that merged the elementary and middle school frameworks was published in order to provide teachers with a wider spectrum of students' probabilistic thinking. In a different cultural setting, Polaki, Lefoka, and Jones (2000) used the elementary framework as a springboard to validate a framework that characterized students' probabilistic thinking in Lesotho, Africa. More recently, Volkova (2003) and Skoumpourdi (2004) concluded that the framework validly characterized the probabilistic thinking of elementary students in Russia and Greece, respectively.

	Level 1 (Subjective)	**Level 2** (Transitional)	**Level 3** (Informal Quantitative)	**Level 4** (Numerical)
Sample Space	• does not list all outcomes in dealing with *simple* events	• lists outcomes for *simple* but not for *compound* events	• uses systematic strategies for listing outcomes of compound events	• links sample space and event probabilities
Probability of an Event	• recognizes certain and impossible events	• identifies most or least likely events	• uses quantitative reasoning and measures to describe the likelihood of events	• assigns numerical probabilities
Probability Comparisons	• is unable to distinguish between "fair" and "unfair" situations	• distinguishes between "fair" and "unfair" situations • reverts to subjective judgments	• compares likelihood of events using invented numerical representations	• uses ratios or numerical measures to compare probabilities
Conditional Probability	• uses *subjective* judgments	• recognizes that *some* event probabilities change in nonreplacement situations	• recognizes that *all* event probabilities change in nonreplacement situations	• distinguishes *independent* and *dependent* events

Figure 20.6 Some aspects of the Jones et al. (1997) framework.

Watson et al. Model of Chance Measurement

Watson, Collis, and Moritz (1997; Watson & Moritz, 1998) also used the Biggs and Collis (1991) model to characterize the development of students' understanding of chance measurement. Their work differs from that of Jones et al. (1997) with regard to the levels of thinking exhibited by the students in their study. Watson et al. posited that students' thinking could be characterized according to two hierarchical Unistructural-Multistructural-Relational [U-M-R] cycles "associated with the complexity of responses required to successfully complete tasks" (Watson & Moritz, p. 105).

The Watson et al. (1997) model was based on students' (Grades 3, 6, and 9) responses to three classically oriented probability tasks. The first involved comparing the likelihood of one or six when tossing a six-sided die. The second task required students to explain whether a boy's or a girl's name was more likely or equally likely to be drawn from a hat with 13 boys' and 16 girls' names. The third task was a probability comparison situation: Box A contained 6 red and 4 blue marbles, and Box B contained 60 red and 40 blue marbles. Students were asked to explain which box to choose if they wanted a blue marble. The researchers postulated that the first two tasks could be answered within the first cycle of the concrete-symbolic mode (U_1-M_1-R_1) whereas the third task was more cognitively complex and required a second cycle (U_2-M_2-R_2). Watson et al. developed the following model to characterize students' responses across the two U-M-R cycles.

P: Decisions are made based on non-mathematical beliefs.

U_1: At the unistructural level in the first cycle there is the recognition that the outcome of a physical action such as rolling a die is uncertain. This is not seen as equal likelihood or quantified but seems to be based on concrete experiences rather than beliefs or feelings alone.

M_1: At the multistructural level in the first cycle, the simple uncertainty of the U_1 level becomes qualified in some way; in some contexts a rudimentary numerical comparison may occur.

R_1: At the relational level in the first cycle, a correct quantification of outcomes (chance) is achieved in a straightforward setting, and a relationship established in a more complex setting.

U_2: At the unistructural level in the second cycle, consolidation occurs and in a more complex setting, responses offer single constructs without further justification; any potential conflict is not recognized.

M_2: At the multistructural level in the second cycle, chance measurement involving multiple settings is sought by using informal ratio notions; the need for resolution of the conflict of settings is realized but not addressed.

R_2: At the relational level in the second cycle, a correct mathematical quantitative comparison is made and the result is justified by using ratios and/or percentages. (p. 75)

The researchers reported that their first cycle described the *development* of the concept of chance measurement whereas their second cycle represented *consolidation* of the concept (in terms of using ratio concepts) and *application* of it to more complex problem-solving contexts. Watson and her colleagues have also investigated other probability constructs such as luck (Watson, Collis, & Moritz, 1995) and fairness (Watson & Moritz, 2003b). Similar to Jones et al. (1997), the models they constructed for these latter constructs include only one U-M-R cycle. As researchers continue to build and refine models of development in probabilistic thinking they should remain alert to the possibility of multiple learning cycles within a particular mode of operation. In addition, more research is needed to investigate the formal mode of reasoning about probability.

One further cognitive model on probabilistic thinking (Li & Pereira-Mendoza, 2002) is worthy of mention in that it was built on the written responses of a very large sample of Chinese students ($n = 567$) and the interview responses of 64 of them. The scope of the written questionnaire items was also extensive and incorporated four different categories of probabilistic thinking: identification of impossible, possible, and certain events; interpretation of chance values; chance comparison in one-stage experiments; and chance comparison in two-stage experiments. Li and Pereira-Mendoza's developmental model described students' hierarchical thinking according to one cycle of five SOLO (Biggs & Collis, 1991) levels: prestructural, unistructural, multistructural, relational, and extended abstract. The probability descriptors were similar to those of Jones et al. (1997) and Watson, Collis, and Moritz (1997), even though they related to a wider age range (Grades 6, 8, & 12).

The section on cognitive models of probabilistic reasoning provides a culminating summary of the key research we have discussed on probabilistic thinking and learning: chance and random processes, sample space, and probability measurement. It also provides a timely transition to our review of teaching and instructional environments. Cognitive models of probabilistic reasoning, such as those described above, provide the theoretical background that is needed to inform teachers' planning, implementation, and assessment of instruction in probability. Moreover, the need for cognitive models of stochastic thinking has been highlighted by several researchers (e.g., Cobb et al., 1991) and by the advent of instructional models like *cognitively guided instruction* ([CGI], Fennema et al., 1996) and *realistic mathematics education* (RME) developed at the Freudenthal Institute (Gravemeijer, 1995).

RESEARCH ON PROBABILITY TEACHING

Stohl (2005) observed that "the success of any probability curriculum for developing students' probabilistic reasoning depends greatly on teachers' understanding of probability as well as a much deeper understanding of issues such as students' misconceptions" (p. 351). Her caveat immediately raises questions about research on *teacher knowledge,* and we note that researchers (e.g., Shulman, 1987) have identified a number of important components of teacher knowledge: mathematical content knowledge, pedagogical content knowledge, and knowledge of student cognitions. These components will guide our review on teachers' probability knowledge and set the stage for examining research on instructional environments in probability.

Mathematical content knowledge refers to knowledge of the concepts, procedures, and processes related to the organization and structure of mathematics. It also refers to the relationships of mathematics with other content areas (Shulman, 1987). Linking this to probability, Steinbring (1991) stated that teachers need to have a simultaneous conceptual and theoretical understanding of probability; further they must be cognizant of probabilistic modeling and the implicit assumptions underlying such models. Kvatinsky and Even (2002) identified three critical areas of teachers' understanding of probability content. First, teachers need to understand the essential features that make probability different from other mathematical fields (e.g., its focus on uncertainty and chance). Second, teachers should understand the aspects of mathematics that support probabilistic thinking and those that inhibit it. Third, teachers must understand the power of probability in dealing with everyday situations. This last understanding resonates with Greer and Mukhopadhyay's (2005) proposition that probability is essential for economic competitiveness in commercial and financial endeavors.

Pedagogical content knowledge is knowledge of effective strategies, representations, models, and examples for presenting mathematical content (Shulman,

1987). According to Steinbring (1991), successful learning of probability is dependent on the representations and activities teachers use. For example, he recommended that teachers use "task systems," a well-connected series of tasks, to develop probability concepts. Echoing Steinbring's comments, Kvatinsky and Even (2002) identified various representations and models for probability teaching (e.g., tree diagrams, Venn diagrams, and tables) and advocated that teachers know when each representation or model is appropriate and how they are connected. Kvatinsky and Even also asserted that teachers need a basic repertoire of examples to illustrate probabilistic concepts, properties, and theorems. Such a repertoire should utilize inter alia ideas connected to set theory and combinatorics. Finally, Steinbring and Kvatinsky and Even strongly advocated teachers' use of both the classical and the frequentist approach. In fact, Steinbring recommended the simultaneous use of both approaches in teaching probability concepts.

Knowledge of student cognitions encompasses an understanding of how students think about and learn mathematics. This includes an awareness of the conceptions students have in relation to the content (Shulman, 1987). With respect to probability, knowledge of student cognitions embraces familiarity with different forms of probabilistic understanding including formal, relational, or intuitive thinking (Kvatinsky & Even, 2002). Greer (2001), reflecting on Fischbein's (1975) work, stressed the importance of teachers' dealing with students' probabilistic intuitions, and Kvatinsky and Even warned that "the use of intuitive knowledge in probability [can] lead, in many cases, to wrong answers" (p. 5). In essence, teachers not only need to be familiar with students' intuitions they need to confront them in the classroom (Steinbring, 1991).

The dialogue in this section reveals that teaching probability goes well beyond "balls in urn" problems. On the one hand, teachers need to understand various theoretical and conceptual aspects of probability; on the other hand, in deciding on the tasks, activities, representations, and technology to be used in instruction, teachers need to understand students' intuitions and the means by which probabilistic learning can be facilitated. In subsequent sections, we look at what research tells us about teachers' probabilistic knowledge and instructional environments.

Teachers' Probabilistic Knowledge

Although the work of Fischbein (1975), Kvatinsky and Even (2002), and Steinbring (1991) has produced insightful theory on the knowledge teachers need in teaching probability, research on teachers' mathematical content knowledge, pedagogical content knowledge, and knowledge of student learning is limited. This dearth of research may be due to the fact that the introduction of probability, especially at the elementary level, is still relatively new.

Probabilistic Content Knowledge

In a recent review Stohl (2005) examined four studies that have investigated teachers' beliefs and content knowledge in probability. Working in New Zealand, Begg and Edwards (1999) found that 22 practicing and 12 preservice elementary teachers had a weak knowledge of probability with only about two thirds understanding equally likely events and even fewer understanding independence. Carnell (1997) investigated 13 preservice middle school teachers' understanding of conditional probability. He concluded that each of the teachers showed evidence of holding one or more misconceptions described earlier in this chapter: fallacy of the time axis, finding the conditioning event, and confusing conditionality with causality (Falk, 1988). In a broader study in Australia, Watson (2001) developed a profile instrument to gather information about elementary ($n = 15$) and secondary ($n = 28$) teachers' content and pedagogical knowledge in stochastics. She reported that secondary teachers were significantly more confident than elementary teachers in their ability to teach equally likely outcomes, basic probability measurement, and sampling. In addition, she observed that the majority of teachers had particular difficulty in interpreting 7:2 correctly as "odds" and were unable to transition between part-part odds (7:2) and part-whole probability (7/9 or 2/9). In a related area, Stohl (2005) also noted that mathematics teachers have a *deterministic mindset* in that they experience difficulty in moving mathematically from situations that deal with certainty to situations that deal with uncertainty and nondeterministic reasoning. This finding is corroborated in research on secondary teachers by Nicholson and Darnton (2003) and with elementary teachers by Pereira-Mendoza (2002).

In a study that focused on cultural and language differences, Zaslavsky, Zaslavsky, and Moore (2001) examined 33 Hebrew-speaking preservice teachers' knowledge of independence and mutually exclusive events. With respect to the 9 tasks given by the researchers, nearly 70% could not explain what it means for events to be mutually exclusive and almost half could not give examples of mutually exclusive events. In addition, few teachers could determine whether two given events were mutually exclusive. In a similar way, 70% of the teachers could not explain when events were independent and more than 40% of them could not determine whether two events were independent. Not surprisingly, nearly 90% of them did not understand the relationship between independent/dependent events and mutually exclusive events. Despite their inability to deal with independence, about 80% of the prospective teachers could find the probability of selecting consecutive balls from an urn without replacement. Although Zaslavsky et al.'s findings on teachers' content knowledge were largely consistent with studies reviewed by Stohl (2005), their study produced unexpected results when they gave their tasks to Arabic-speaking, English-speaking, and Russian-speaking preservice teachers. Zaslavsky et al. found that English-speaking and Russian-speaking teachers performed better at identifying mutually exclusive events than their Hebrew- and Arabic-speaking counterparts. Proximity of probabilistic language to everyday language seems to act in a perverse or unhelpful way on teachers' probabilistic thinking in the sense that Hebrew and Arabic words for *mutually exclusive events* are used in everyday language whereas equivalent expressions in English and Russian are not.

Working in Mexico, Sanchez (2002) examined 6 secondary teachers' beliefs about using simulation in the classroom after they had experienced the simulation capabilities of Fathom software (Erickson, 2001) over 8 weekly 3-hour sessions. Responding to a follow-up survey, 3 teachers stated that they understood variability better after solving simulation problems with the software. One even mentioned acquiring a better understanding of conditional probability and the law of large numbers. In addition, most teachers, when asked about the usefulness of simulation, focused on ideas associated with a frequentist view of probability. However, not one teacher referred to the value of simulation activities in building students' understanding of randomness, distribution, or the process of validating models of probability.

Although research on teachers' content knowledge in probability is sobering at best, research on teachers' beliefs about the power and value of probability is even more worrisome. Greer and Ritson (1994) surveyed primary and secondary teachers in Northern Ireland vis-a-vis stochastical issues. Most primary teachers considered probability to be relatively unimportant compared to other mathematics topics, and more than 50% of the primary and secondary teachers rarely used probability experiments in their teaching. Similar results were revealed in an Italian study by Gattuso and Pannone (2002). They reported that 91 secondary teachers considered stochastics worthwhile, but the overwhelming majority stated that it took time away from other aspects of the mathematics curriculum.

Pedagogical Content Knowledge and Knowledge of Student Cognitions

Although Stohl (2005) observed that research on teachers' pedagogical content knowledge and their knowledge of student cognition is limited, she did identify some emergent research that is summarized below. Only two of the studies incorporated teachers' classroom practice.

Studies by Haller (1997) and Dugdale (2001) focused to some extent on pedagogical content knowledge and teachers' actions during classroom instruction. Haller observed 4 middle school teachers' probability lessons in a follow-up to a summer program that included experiences with probability content, misconceptions, and pedagogical issues linked to probability instruction. Haller's classroom observations indicated that teachers at the lower end of the probability knowledge spectrum revealed content errors and misconceptions, relied largely on textbooks, and missed opportunities to forge relationships with fractions, decimals, and percents during probability lessons. By way of contrast, teachers with higher probability knowledge made no content errors, enhanced textbook activities, and exploited opportunities to make connections between probability and decimals, fractions and percents. The researcher observed that teaching experience did not appear to have a great impact on the teachers' instruction.

In a study with preservice teachers, Dugdale (2001) used computer simulation to reveal pedagogical content knowledge. During the process of designing a fair game with a pair of dice, teachers came to see the value of considering relative frequencies for a large number of trials. The enabling aspects of the computer software not only convinced them that they had created a fair game, it also motivated them to find theoretical probabilities as a way of verifying computer-generated results. In a real sense, the computer experience enabled teachers to have a salient pedagogical experience that highlighted Steinbring's (1991) advocacy of making connections between frequentist and classical approaches to probability.

The previously mentioned research of Watson (2001) examined teachers' knowledge of student cognitions. She developed a comprehensive protocol to assess among other things elementary and secondary teachers' knowledge of difficulties students might experience with data and chance. In her documentation of student difficulties, only 2 of the 15 elementary teachers mentioned procedural aspects like "finding probabilities," whereas 13 of the 28 secondary teachers referred to calculating probabilities, permutations, and tree diagrams. Although the data as a whole revealed that teachers could identify students' difficulties with probability, their responses, especially those of the secondary teachers, suggested that procedural approaches to teaching tended to dominate. Watson also claimed that there was little evidence that secondary teachers used activity-based approaches like simulation and sampling to reinforce theory. By way of contrast, elementary teachers used activity-based lessons, but their instructional programs in probability seemed to lack coherence.

Although our review in this section has identified some research on teachers' content knowledge in probability, the research on teachers' knowledge of instructional aspects is embryonic at best. This is a serious situation given Greer and Ritson's (1994) finding that almost all the primary and half the secondary teachers in their survey reported not having studied probability in their initial teacher education course. Clearly, researchers need better ways of obtaining data about teachers' pedagogical knowledge, their knowledge of student cognitions, and their instructional practices in probability. Moreover, such research needs to be classroom-based rather than self-reporting data or information on the number of courses or professional development programs in probability they have received (Watson, 2001).

Instructional Environments for Probability

As discussed above, teachers create instructional environments based on (a) their understanding of how students learn, (b) their knowledge of what their students do and do not know, and (c) their repertoire of instructional strategies and tools. With respect to the first two points, we have already highlighted a substantial corpus of research pertaining to cognitive models (e.g., Jones, Langrall et al., 1997, 1999; Watson et al., 1997) that can be used by teachers in developing, implementing, and assessing instructional activities in probability. The research is less robust, however, in studies that have investigated specific instructional strategies for teaching probability (e.g., Fast, 1999) and in the use of calculator and computer environments (e.g., Kissane, 1997a, 1997b; Pratt, 2000; Stohl & Tarr, 2002; Zimmermann, 2002).

Cognitive Frameworks

If we accept the "pedagogic challenge" (Pratt, 2005, p. 175) of designing instruction to build on students' existing views, be they formal knowledge structures or faulty intuitions, cognitive frameworks of students' probabilistic development can serve as useful guides for teachers. In fact, teaching research (Aspinwall & Tarr, 2001; Jones, Langrall, et al., 1999; Polaki, 2002a; Tarr, 1997) has demonstrated that these frameworks play a key role in instruction because they provide teachers with a coherent overview of the various forms of probabilistic reasoning students exhibit and equip teachers with specific domain knowledge that can be used in the design, implementation, and assessment of instruction in probability.

In terms of the *design of instruction*, cognitive frameworks highlight the complex nature of probability and heighten teachers' awareness of the effects those complexities have on students' reasoning. In turn, this can assist teachers in planning a meaningful sequence of instruction rather than a sequence of potentially disconnected probability activities (Stohl, 2005; Watson, 2001). The use of cognitive frameworks in designing instruction can be amplified through Simon's (1995) notion of *hypothetical learning trajectory*, that is, the formulation of learning goals, learning activities, and a conjectured learning process. Tarr's (1997) teaching experiment on conditional probability and independence serves as an example. His work was informed by domain-specific knowledge drawn from a cognitive framework (Tarr & Jones, 1997), and he used the four-level framework descriptors (ranging from subjective to numerical) to establish learning goals for the students, both individually and as a whole class. With respect to learning activities and the conjectured learning process, Tarr's cognizance of the different levels of students' reasoning guided his selection of instructional tasks and questions. For example, being aware that some students would fail to recognize the changing probabilities in nonreplacement situations, Tarr designed a series of activities that included a task "where one Milky Way bar was drawn without replacement from a bag of 3 Milky Way, 2 Butterfinger, and 1 Snickers candy bars" (Tarr & Lannin, 2005, p. 230). In using this task, Tarr focused discussion on the probability of drawing a Snickers bar, given that the Milky Way bar had been removed from the bag. This drew student attention to the change in the total number of objects in the bag and the consequent changes in probability. The wealth of tasks and instructional activities documented by Tarr have widespread potential for other instructional settings.

With respect to the *implementation of instruction*, frameworks characterize students' conceptual development in a way that helps teachers accommodate the diversity of reasoning reflected by students in their class. The framework serves as a filter for analyzing and classifying students' oral and written responses during instruction and guides teachers in framing questions and written tasks that are accessible to students at different levels of reasoning (see the above illustration for Tarr). Such accommodation and sensitivity on the part of the teacher may enable children to develop more mature levels of reasoning. Cognitive frameworks can also assist teachers in *assessing* students' performance over time and evaluating the effectiveness of instruction. For example, the candy bar task used by Tarr could just as easily have been used to assess students' progress in relation to the framework descriptors. Teachers can also use such assessment to evaluate their own performance in promoting students' probabilistic learning.

Instructional Strategies

Greer and Mukhopadhyay (2005) depicted instruction in probability as impoverished in that curriculum materials treat probability as the "exposition and routine application of a set of formulas to stereotyped problems" (p. 314). They perceived the contexts of many instructional tasks and assessment items as uninteresting, unrealistic, and often contrived. Although Greer and Mukhopadhyay's (2005) perspective resonates with the "balls in urns" curriculum materials that have dominated much of the teaching of probability, recent research has begun to generate instructional strategies that reflect the visions of the NCTM *Principles and Standards* (2000). In a study that revisited a 1970's teaching experiment in France, Brousseau, Brousseau, and Warfield (2002) examined pedagogy in a primary classroom during 18 lessons on probability. The key strategies that emerged were the need to have children involved in empirical exploration and the collection and analysis of data as a base for building meaning in probability. Even though the data were from another era, these findings set the stage for current research.

Castro (1998) explored the impact of two instructional orientations, *traditional* and *conceptual change.* The traditional instructional orientation focused on a clear linear presentation of mathematical ideas but did not relate to students' misconceptions. The conceptual-change orientation emphasized the eliciting of students' thinking and the encouragement of reflection on probabilistic ideas. Castro demonstrated that, compared with the traditional orientation, the conceptual-change orientation significantly improved students' skills in probability calculation and their intuitive reasoning about probability. In discussing this result he highlighted three instructional imperatives: (a) the need to take into account students' preconceptions and beliefs regarding probability, (b) the need to provide an experimental focus to the teaching of probability, as such a focus provoked "the necessary cognitive conflict in the student to perform the transition from the personal model to the scientific model of chance" (p. 252), and (c) the need to incorporate specific training in probability reasoning to challenge misconceptions. This third imperative was consistent with Castro's finding that misconceptions in probability were more resilient among those receiving the traditional instruction, and he attributed this to the fact that the traditional approach did not actually confront students' misconceptions. The notion of confronting misconceptions and engaging students in reflection is not new (e.g., Fischbein, 1975) but is significant in the teaching of probability, as this study and the next two studies reveal.

In a related study, Fast (1999) examined the use of analogies as an instructional strategy for helping students confront probability misconceptions. Forty-one high school seniors were assessed on two parallel sets of tasks. The first set consisted of 10 probabilistic tasks recognized in the research (Kahneman & Tversky, 1972; Konold et al., 1993) as stimulating misconceptions. In the second set, administered immediately after the first set, there were 10 analogous tasks. Figure 20.7 shows an original task and its analogous counterpart. Fast found that students performed better on the analogous tasks (72% success) than on the original tasks (49% success). In interview settings, he observed that many students saw connections between the two versions as indicated by their desire to change their answers to the original tasks. Fast concluded that the use of analogies enabled

Original Task	**Analogous Task**
In families with five children, which birth order, if either, occurs more often: BGGBG or BBBBB? B = Boy and G = Girl a) BGGBG occurs more often. b) BBBBB occurs more often. c) They occur equally.	In a lottery called Pick 4, a 4-digit number like 2798 is generated. To win, the participant must have chosen the same 4-digit number. Albert has chosen the number 2222 and Bill has chosen the number 2332. Compare their chances of winning. a) Albert has a better chance of winning. b) Bill has a better chance of winning. c) They both have the same chance of winning.

Figure 20.7 Sample task and its analogous task.

students to reconstruct their probabilistic thinking, and he claimed that an analogies approach aids teachers to identify tasks that will help students to face misconceptions. Such an approach aligns with suggestions that teachers should confront students' subjective thinking (Steinbring, 1991) and develop their secondary intuitions (Greer, 2001).

In a European study that utilized simulation and focused on confronting misconceptions, Batanero, Biehler, Maxara, Engel, and Vogel (2005) offered a theoretical position that may have implications for the teaching of probability at all levels. These researchers highlighted the importance of simulation in providing teachers with stochastic experience and knowledge to help them confront probabilistic misconceptions. In addition, they claimed that their study provided teachers with an instructional sequence that incorporated an exemplary modeling approach: (a) introduce a real-world problem involving data analysis, (b) build a simulation model in a technological environment, (c) generate data and build simulation-based inferences, (d) critically evaluate the conclusions in terms of any assumptions, and (e) formally analyze the problem on the basis of mathematical probability and statistics.

Zimmermann (2002) used a similar modeling approach to that of Batanero, Biehler, et al. (2005) in examining 23 high school students' understanding of probability simulation. The students participated in a 12-day teaching experiment that involved whole-class and small-group activities. Following instruction students showed significant improvement in their ability to evaluate and construct a probability simulation and in their understanding of the effect of repeated trials on the empirical probability. However, they did not always understand the implicit assumptions underlying a probability model. Zimmermann's study revealed fresh insights on Steinbring's (1991) caveat concerning the implicit assumptions underlying probability models.

Zimmermann's (2002) study is exemplary in its use of graphing calculators as a simulation tool. Although these calculators have become commonplace in many classrooms, research is limited on the role they play in probability teaching and learning. In contrast, research involving computers has received greater attention even though their use in classrooms is less pervasive.

Computer Environments

During this period of research, Konold (1995b) led the way in exploring the use of computers in the teaching and learning of probability. In particular, his study highlighted the need for software designed to promote sense making and the enrichment of students' intuitions. Konold's work pre-empted the development of probability software and subsequent research into its effectiveness. The studies of Pratt (1998, 2000; Pratt & Noss, 2002) and Stohl (Drier, 2000b; Stohl & Rider, 2003; Stohl & Tarr, 2002) are noteworthy in this respect.

Working from the assumption that students hold multiple, even competing, intuitions, Pratt (1998) designed the *Chance-Maker* microworld "in which individuals would meet the consequences of their beliefs" (p. 4). *Chance-Maker* comprises random generator *gadgets* (electronic coins, spinners, dice), some of which are broken in the sense that they produce nonrandom outcomes. Students have access to tools to help them make sense of and edit a gadget's behavior. One tool is the workings box, which lists all possible outcomes for the gadget; other tools include a repeat button to produce large numbers of trials and different options for displaying results numerically and graphically. Pratt found that this computer environment enabled students to generate and examine the long-term behaviors of gadgets vis-à-vis their sample space and data distribution. This in turn supported their understanding of randomness.

Based on similar theoretical assumptions, Stohl's (1999–2002) *Probability Explorer* provides an open-ended learning environment in which students can test their probability intuitions. Probability Explorer allows students to randomly generate a variety of icons (e.g., coins, dice, marbles) that can be represented in graphical or tabular forms. Students determine how much data to generate for the simulation and how to display it. Multiple graphical representations can be presented simultaneously, and numerical results can be presented as odds, probabilities, and decimals. As with Chance-Maker, students can change the likelihood of an event and control the sample space and theoretical probabilities for the simulations they construct. Stohl (Drier, 2000a, 2000b) reported that students using Probability Explorer discovered that greater numbers of trials produced distributions that more closely resembled expected theoretical probabilities. She drew the following conclusion:

> Simulation tools that give students control over designing experiments, running as many trials as they desire, and viewing graphical representations of results may help in the development of deeper understandings of how theoretical probability, empirical probability and sample size can be used to make inferences. (Stohl & Tarr, 2002, p. 322)

The promise of computer microworlds for probability instruction has yet to be realized in school classrooms; both Chance-Maker and Probability Explorer have been used mainly in research and development.

It remains to be seen how classroom teachers would incorporate these technology-rich learning environments into their repertoire of instructional strategies.

Pedagogical Guidelines

Although the research on instructional environments is relatively sparse, the literature does offer some guidelines. On the basis of his work with Chance-Maker, Pratt (2005) generated four pedagogical guidelines for probability instruction. Aspects of these guidelines resonate strongly with the theoretical perspectives of Fischbein (1975), Kvatinsky and Even (2002), and Steinbring (1991) that were presented at the beginning of this section.

The first guideline pertains to task design and involves two constructs: *purpose and utility*. A purposeful task is one whose solution holds some meaning for the learner, and a utilitarian task is one whose context enables the learner to see the usefulness of a mathematics concept or skill. As Pratt (2005) contended, "The difficulty in planning lies in linking purpose to utility in such a way that there is a high probability that the learner will stumble across the utility of the mathematical concept as they engage in the purposeful activity" (pp. 182–183). Both Chance-Maker and Probability Explorer were specifically designed to link purpose and utility, and given their success it would seem desirable for this link to be a feature of probability tasks in all environments.

The second guideline, *testing personal conjectures*, relates to Pratt's (2005) theoretical stance that "old pieces of knowledge coexist with newer pieces of knowledge, either in a connected way or perhaps isolated from each other" (p. 187). From this perspective, instruction is aimed at connecting students' existing conceptions with new knowledge. This occurs through the use of tasks and questions that encourage students to become aware of the limits ("lack of explanatory power") of their current ways of thinking in comparison to alternative views. Feedback is the crucial aspect of this process, and Pratt admitted that "the testing of personal conjectures is an especially difficult aim to achieve without the use of technology" (p. 186).

The third guideline, *large-scale experiments*, suggests that students should determine for themselves how many trials to run when collecting data in probability explorations. With that said, tasks should be designed to encourage students to try increasingly greater numbers of trials, and tools should be able to facilitate the collection of large quantities of data. This is another case where the power of technology could be vital as vindicated in the work of Pratt and Stohl.

The final guideline, *systematic variation of the context*, is closely related to testing personal conjectures. As students experience the lack of explanatory power of their old ideas and come to recognize the power of alternative views, it is important that they experience the reliability of this new knowledge across different contexts. In this way, new knowledge structures are strengthened and eventually supersede earlier intuitions. Thus, students need to be exposed to tasks whose contexts vary while the embedded mathematical structure remains constant and some surface features remain similar to cue the reuse of recently learned knowledge (Pratt, 2005, p. 187).

In this section we have noted that, for the teaching of probability to be effective, teachers need to understand probability from both a theoretical and a conceptual point of view, as well as understand how students' think (Steinbring, 1991). Having research-based knowledge of students' intuitions is critical for teachers in deciding on the tasks, representations, and tools to be used in instruction. To this end, several developmental models have been created to describe students' probabilistic reasoning and to inform instruction (e.g., Tarr & Jones, 1997; Watson et al., 1997). In addition a number of distinctive instructional strategies and environments for the teaching and learning of probability are beginning to emerge (e.g., Fast, 1999; Pratt, 2005).

CURRICULUM RESEARCH

As shown in our discussions of national curriculum documents in Australia, the United Kingdom, and the United States (AEC, 1991; DES, 1989; NCTM, 1989, 2000), data analysis and probability have emerged during this period as a mainstream strand in mathematics curricula. Moreover, these national documents took the bold step of incorporating probability into the mathematics curriculum across all levels of schooling from primary through high school.

Notwithstanding these national visions for data analysis and probability, a very different perspective can be gleaned from the cross-national investigation of curricular intentions in school mathematics conducted as part of the Third International Mathematics and Science Study (Schmidt, McKnight, Valverde, Houang, & Wiley, 1997). The results of this investigation revealed that probability showed the greatest variation among all mathematics topics in terms of when it was introduced in the curriculum: (a) 24 of the 48 countries surveyed introduced probability in Grade 9; (b) 16 countries introduced probability by Grade 6 apparently using an integrated approach from the early grades; and (c) 5 countries "deferred its introduction to a more formal treatment at the end of secondary school" (p. 71). Moreover, although one third claimed that they introduced probability by Grade 6, closer inspection of each country's data revealed that probability was seldom a significant curricular focus prior to Grade 9.

This international survey suggests that the implementation of probability in schools may not be keeping pace with the expectations of national curriculum documents. Similar evidence is also reflected in research cited earlier where it was noted that primary teachers deemed the teaching of probability unimportant (Greer & Ritson, 1994) and that secondary teachers considered stochastics worthwhile but felt it took time away from other aspects of the mathematics curriculum (Gattuso & Pannone, 2002). There may well be a need for a stronger philosophical orientation to curriculum research in probability, one with the power to materialize H. G. Wells' prophetic vision: "Statistical thinking will one day be as necessary for efficient citizenship as the ability to read and write" (as cited in Kranzler, 2003, p. 104). Pursuing this orientation, we examine research that focuses on what constitutes *statistical literacy* and more particularly *probability literacy* (Gal, 1994, 2002, 2005; Watson, 1997a; Watson & Callingham, 2003).

Probability Literacy

Literacy and numeracy, although complex to define, have become fundamental concepts of curriculum research (Steen, 1990, 1997, 2001). For example in the United Kingdom, the Numeracy Strategy appears to play a parallel role with the National Curriculum documents (DfEE, 1999) in articulating what mathematics is taught in school. There is also an expanding meaning for domain-specific literacy (e.g., computer literacy, scientific literacy), and it is within the context of domain-specific literacy that research on statistical literacy and probability literacy has found a place and a theoretical base. Gal (2002) took the position that *probability literacy* cannot be conceived of in isolation, which is in accord with the epistemological position of other researchers (Rychen & Salganic, 2003; Stein, 2000) who argue that probabilistic knowledge and dispositions are part of a broader network of skills, knowledge, and beliefs that individuals possess or need to activate in different life settings. In Gal's more precise terms, probability literacy is closely linked to literacy, numeracy, and his own (Gal, 2002) and Watson's (1997a; Watson & Callingham, 2003) conceptualization of statistical literacy.

It is helpful to start with a brief discussion of research on *statistical literacy* inasmuch as statistical literacy includes components that belong to probability literacy. Watson's (1997a) statistical literacy framework incorporates three increasingly sophisticated tiers: (a) basic understanding of probabilistic and statistical terminology, (b) comprehension of probabilistic and statistical language and concepts when they are embedded in contexts of social communication, and (c) a questioning attitude with respect to claims that are made without proper statistical justification. In an empirical study linked to this framework, Watson and Callingham (2003) used Rasch analysis to show that statistical literacy is a hierarchical construct incorporating six levels of understanding: idiosyncratic, informal, inconsistent, consistent noncritical, critical, and critical mathematical. Watson's work in statistical literacy not only encompasses chance and data but also aspects of variation. Clearly it has implications for the school probability curriculum.

Gal (2002) argued that statistical literacy involves both cognitive and dispositional components and that some of these components are held in common with literacy and numeracy whereas others are unique to statistical literacy. More specifically, Gal referred to statistical literacy as the knowledge and dispositions that people use to critically evaluate and, where relevant, express their opinions regarding statistical information or data-related arguments. For Gal, the knowledge component of statistical literacy involves five bases: literacy skills, statistical knowledge, mathematical knowledge, contextual knowledge, and a list of critical questions. The dispositional component includes a belief in the power of statistical thinking, a belief in your own capability to undertake statistical thinking, and a belief in the legitimacy of adopting a critical perspective on "information one receives from presumably 'official' sources or from experts" (Gal, 2005, p. 45).

Although Gal made only brief reference to probability in his model of statistical literacy, he generated a specific model of probability literacy in his 2005 book chapter. In this chapter, which incorporated a number of allusions to others in literacy and numeracy research (e.g., Moore, 1990; Scheaffer, Watkins, & Landwehr, 1998; Stein, 2000; Utts, 2003), Gal conceptualized *probability literacy* in a similar way to statistical literacy; that is, it contains both cognitive and dispositional components including five classes of knowledge and three dispositional elements. We consider first Gal's knowledge classes: the big ideas of probability, the figuring of probabilities, the terminology and language of probabilities, the critical questions about reasonableness, and the context within which probability issues are considered.

The notion of *big ideas* was suggested by Moore (1990) and refers to foundational elements that students need in order to understand and interpret probabilistic statements. With respect to probability Gal (2005) identified the big ideas as variation (cf. Metz's, 1997, use of the term *chance variation*), randomness, and independence. Moreover, he argued that these ideas have to be understood in tandem with their complementary "alter egos": stability, regularity, and co-occurrence (p. 47). For Gal, *figuring probabilities* is an important aspect of being able to understand probabilistic statements and communicate with others concerning the likelihood of events. It is also a key element in understanding the classical, frequentist, and subjective approaches to probability. In discussing *terminology*, Gal stressed the need for "probability literate" people to be sensitive to the language of chance in dealing with the many verbal and numerical ways in which probability is communicated (p. 50). Terms like *impossible*, *possible*, and *certain* occur in our everyday lives, and hence it is appropriate that they are incorporated in curriculum documents. Gal's (2005) notion of *critical questions or criticalness* occurs in both statistical and probability literacy. It deals with questions of reasonableness and with being able to make critical responses or to ask critical questions about chance phenomena (cf. Watson, 1997a). More specifically, Utts (2003) and Gal have addressed the need for learners to be aware of flaws, problems, and biases that occur in the derivation, reporting, and interpretation of probability information. This discussion about reasonableness has strong links with previous research on intuitions and misconceptions (e.g., Fischbein, 1975; Kahneman & Tversky, 1972) and unsurprisingly is already represented in international curriculum documents as a caveat on the misconceptions children bring to the classroom. Finally, Gal included, in the knowledge component, an element that deals with *context*. This refers to the importance of understanding the role of probabilistic thinking and language in decisions and dialogue associated with personal and public contexts. For example, people need to understand chance variation in random processes as a means of making sense of media statements about group differences, likelihood of results, significance of differences, and other claims related to statistical inference.

Gal's (2005) dispositional elements of probabilistic literacy include personal sentiments regarding uncertainty and risk, critical stance, and beliefs and attitudes. With respect to *personal sentiments*, Gal noted that people react in different ways when confronted with a need to estimate probabilities of certain events or to deal with risk situations. In essence, emotions like risk aversion or anticipated success affect people's decisions in dealing with chance phenomena in social, financial, and political situations. With respect to what Gal called *criticalness*, it is one thing to possess the knowledge to adopt a critical approach in probabilistic situations, it is another to actually hold a questioning attitude to messages that may be misleading, one-sided, biased, or incomplete. Children and adults need to be both able and willing to invoke a set of critical questions when faced with arguments that purport to be based on data or empirical research. According to Gal, several *beliefs and attitudes* underlie people's willingness to take risks vis-à-vis critical evaluations. In other words, critical disposition is dependent on people developing a belief in "the need to apply 'critical-response' skills to real world statistical messages" (Gal, 2005, p. 54). Moreover, this belief in the legitimacy of critical action should apply whether the messages come from official or other sources.

It is not surprising that we have noted instances of probability literacy impinging on our earlier overview of national curriculum documents (AEC, 1991; DES, 1989; NCTM, 1989). In the next section we analyze in more depth what research on probability literacy implies for both extant and future curriculum development in probability.

Curriculum Implications From a Probability Literacy Perspective

Both the knowledge elements and the dispositional elements of Gal's (2005) probability literacy model and Watson's (1997a) statistical literacy framework have implications for the probability curriculum. Using their theoretical perspectives we draw implications for the probability curriculum with respect to both content and pedagogical directions.

Content directions. Within the area of *knowledge*, Gal's (2005) model of probability literacy includes big ideas, figuring probabilities, terminology and language, critical questions, and context. Watson (1997a) also identified language and critical questioning as part of her three-tier framework, and her discussion was consistent with Gal in that it revolved around key ideas like chance, probability measures, and risk taking. The writings of Gal and Watson on probability literacy suggest some new content emphases for the probability curriculum and also provide support for much of the content that is already contained in current curriculum documents.

In addition to having students build a robust concept of randomness, Gal (2005) and Watson (1997a; Watson & Callingham, 2003) have called for a strong emphasis on chance variation and independence. We have already noted that national curriculum documents were limited in the extent to which they dealt with chance variation. Even though one Australian document (AEC, 1994) used the term *chance variation*, the documents, as a whole, fail to build any notion about whether the variation manifest in a set of data has a causal base or whether it is simply due to chance (Metz, 1997; Pfannkuch, 2005). Like Gal and Watson, Moore (1990) contended that it is "crucial for students to grasp the idea that 'chance variation' rather than deterministic variation explains many aspects of the world" (p. 99). The notion of independence is incorporated into national curriculum documents but is not given the salience that Gal (2005) advocated. He argued that people cannot fully understand independence without understanding dependence. In a similar way Gal highlighted stability as the complement of chance variation and regularity as the complement of randomness. The addition of these complementary processes (*alter-egos*) to the curriculum via appropriate experiential activities has the potential to greatly enrich key ideas like chance variation, randomness, and independence. Moreover, it means that these key ideas are accessible, albeit through informal experiences, from an early age.

The emphasis in probability literacy on having students find or estimate probabilities is already strongly represented in national curriculum documents. However, I. Gal (personal communication, October 1, 2003) provided fresh insights on how stronger links might be forged among classical, frequentist, and subjective approaches to probability and on how probabilities in the real world can be estimated:

> Real world probabilities are usually not computed but *estimated* or *judged* [author's emphasis], and in ways that do not fit neatly any one of the three views of probability. Usually, information from multiple sources (classical, frequentist) is used, but heavy reliance is also made of a "subjective" or personal judgment process that is affected by a person's degree of experience in deriving probabilities.

Although the Australian *National Statement* (AEC, 1991) focused on all three approaches to probability measurement and investigated subjective probabilities in social contexts, Gal's work encourages an even stronger link by using all three approaches to estimate real-world probabilities. Investigations such as the one implied by Gal would also seem timely in future revisions of national curriculum documents.

Gal (2005), Moore (1990), and Watson (1997a) placed strong emphasis on the role of language and terminology in communications about chance. Their research supports the strong emphasis on language that is already present in national curriculum documents. Moreover, in harmony with the theoretical positions espoused by these three writers, curriculum documents make strong claims about the need for chance language to evolve and be refined as students build up their knowledge of and experience with probability through the grades.

The notion of *critical questions* or the need for people to become familiar with flaws, problems, and biases in stochastic information is a key aspect of Gal's probability literacy model, Watson's (1997a) statistical literacy framework, and the writings of Rutherford and Ahlgren (1990) and Utts (2003). In addition, Watson (1997a) and Watson and Moritz (2000) have undertaken valuable research that invited students to challenge claims in media extracts. Watson (1997a) concluded that, at the highest level of statistical thinking, students possessed the confidence to challenge what they read in the media so long as teachers made them "aware of the expectation that they must constantly question conclusions" (p. 110). In the following example, students were asked whether the claim made could be justified on the basis of the statistical processes used.

> Some 96 percent of callers to youth radio station Triple J have said marijuana use should be decriminalised in Australia. The phone-in listener poll, which closed yesterday, showed 9,924-out of the 10,000 plus callers-favoured decriminalisation, the station said. Only 389 believed possession of the drug should remain a criminal offence. Many callers stressed they did not smoke marijuana but still believed in decriminalizing its use, a Triple J statement said. (Watson, 1997a, p. 110)

This question involves mostly issues about sampling, and it is true that critical questions about chance are more rare. However, as a result of Watson's influence on the Tasmanian *Mathematics Guidelines K–8* (Department of Education and the Arts, Tasmania [DEA], 1993), there are several challenge-type questions on chance, one of which is illustrated below.

> Peter buys a Tattslotto ticket every week. He never selects 6 consecutive numbers because he believes his chances are better by selecting a number from each line of the entry form. What do you think about this line of reasoning? (DEA, 1993, p. 11)

Watson's research indicates that appropriate critical questions can be extracted from media, financial, and scientific reports. Moreover, these questions should be introduced in the primary school curriculum, as there is certainly a need for children to begin to question statistical and probabilistic data from an early age.

The final knowledge component deals with context and, in particular, probabilistic issues in social activities, public discourse, and personal decisions. In essence, Gal (2005) suggested that his earlier-stated knowledge components (big ideas, figuring probabilities, terminology, and criticalness) need to be applied in various social and human contexts. All three national curriculum documents that we visited earlier do this in regard to games of chance and the language of chance. The Australian documents (AEC, 1991, 1994) also incorporated social contexts like gambling where subjective probability, in addition to classical and frequentist probabilities, is needed to interpret the situation. More can be built into the curriculum documents in order to meet the contextual goal that Gal is envisaging. For example, additional activities involving critical questions and the kinds of cultural and media-related tasks generated by Watson (1997a) would bring students nearer to this aspect of probability literacy.

The *dispositional* elements of Gal's (2005) probability literacy model and Watson's (1997a) statistical literacy framework relate more to pedagogical directions than they do to content directions; hence we will discuss these more in the next section. However, dispositional elements such as personal sentiments, critical stance, and beliefs and attitudes also need to be considered in the development of probability content so that learners will have the opportunity to experience situations in which these dispositions are challenged. Sentiments like risk aversion and anticipated success and failure will not generally be felt in decontextualized problems but can occur in games, personal estimates of likelihood, and in problems children pose for themselves. A critical disposition or a willingness to hold a questioning attitude can only be revealed and enhanced by having children from an early age engage in tasks that require them to challenge data or judgments. Similarly, if children are to develop beliefs such as "it is legitimate to be critical about probabilistic messages," these kinds of messages need to be built into the content of tasks and experiences that the students face.

In summary, the knowledge and dispositional elements of Gal's probability literacy model (2005) and Watson's (1997a; Watson & Callingham, 2003) statistical literacy framework have implications for the kind of probability content that should be incorporated in the Pre-K through Grade 12 curriculum. More specifically, notions like chance variation and stability, independence and co-occurrence, figuring likelihood from multiple sources (classical, frequentist, and subjective), language of chance, critical questions, and contexts need to be examined more deeply when future curriculum documents are developed. Curriculum programs also need to take cognizance of children's dispositions towards situations involving chance.

Pedagogical directions. Gal's (2005) probability literacy model and Watson's statistical literacy framework (1997a) are relatively silent about pedagogical issues. However, a number of pedagogical directions can be inferred from their writings. Both of them have a strong thrust on context or the "context of wider social discussion" (p. 108) in Watson's case, and both support a contextual problem-solving approach to probability especially when dealing with critical questions and scrutinizing media statements. This strong contextual orientation is also supported by Rutherford (1997), who noted that the starting point for deciding what constitutes quantitative literacy is less mathematics itself and more the contexts in which people are likely to encounter the need for mathematical skills (p. 62). Hence, pedagogical approaches that focus on problem-solving within cultural, social, and personal contexts are indicative of what Gal, Rutherford and Watson seem to be advocating for the teaching and learning of probability.

Somewhat in contrast with the almost exclusive emphasis of national curriculum programs on classical and frequentist approaches to probability, Gal and Watson emphasize the importance of students using multiple approaches to estimating probability. Even though inserting multiple approaches to probability into curriculum content is relatively easy, it is critical that teachers feel comfortable with the kinds of activities that support the combined use of classical, frequentist, and subjective approaches. The following "Roulette Investigation" from *Activity Book 3* of the Tasmanian Mathematics Guidelines K–8 (DEA, 1993) is indicative of the kind of task that could be used by a teacher to promote all three approaches to probability.

> Some people believe that if a particular colour comes up on the [roulette] wheel a number of times in succession then the likelihood of the other colour coming up on the next spin is greater than one-half. For example, if red has come up three times in succession, the chance that the next spin will be black is greater than the chance that the next spin will be red. Investigate this using a toy roulette wheel. (DEA, 1993, p. 11)

In using such a task, teachers can encourage students to estimate and justify their own subjective probabilities as well as to use logical reasoning and data collection to generate theoretical and experimental probabilities, respectively. Moreover, such an investigation affords an opportunity for teachers to confront representativeness (Kahneman & Tversky, 1972).

Our final pedagogical implication from the work on probability literacy deals with Gal's (2005) and Moore's (1990) strong emphasis on the big ideas:

chance variation, randomness, independence, and their complementary elements stability, regularity, and co-occurrence. Contrary to the present pedagogical style in which the teaching of probability seems to focus rapidly on qualitative and numerical likelihood judgments, there is a need for teachers to provide opportunities for students, especially young children, to experience the following: the stability associated with causal action and the variation inherent in chance, the regularity in deterministic phenomena and the short-term unpredictability and long-term predictability of randomness, and the co-occurrence of some events and independence of others. For example, teachers might have students compare and contrast activities like the following:

- the day that follows Monday each week (stability)
- the number that follows a "six" when rolling a die (chance variation)
- the order of the days of the week (deterministic phenomena)
- the percentages of heads and tails when you toss a fair coin a small number of times versus a large number of times (randomness)
- getting a six on your first and second throw of a die (independence)
- throwing a six and an even number when rolling a die (co-occurrence).

Activities such as these, especially if they are located in technology environments, can provide informal experiences for children that involve conjecturing, experimenting, collecting data, building arguments, and discussion of results. As a consequence students may be enabled to build increasingly more sophisticated understanding of chance variation, randomness, and independence and hopefully a firm base for subsequent mathematical approaches to probability.

Our review suggests that curriculum research in probability is in its infancy, not surprising given the brief time that probability has been a mainstream part of the school mathematics curriculum. As current curricula on probability are evaluated, more extensive research is undertaken on the teaching and learning of probability, and further work on probability literacy emerges, the probability strand will undergo further evolution as will the instructional environments that support it.

REFLECTIONS AND FUTURE DIRECTIONS

In his chapter on stochastics in the first *Handbook of Research on Mathematics Teaching and Learning*, Shaughnessy (1992) presented a "wish list" of research for the next decade. The topics in his wish list were directed towards mathematics educators; however, they were also pertinent to researchers in other fields like psychology and sociology. As we near the end of this present review, it is appropriate to reflect on the research presented and appraise it in terms of Shaughnessy's wish list. We will also generate our own research agenda in probability for the next 10–15 years by suggesting directions and uncharted waters that need to be addressed.

Reflections

Although Shaughnessy's wish list included research dealing with both statistics and probability, we will concern ourselves with needed research in probability. Shaughnessy's need areas included the following: the development of assessment instruments, studies on secondary school students' conceptions and misconceptions, cross-cultural research studies, studies on teachers' conceptions, studies that examine the effects of instruction on students' conceptual knowledge and intuitions, studies that examine the effects of computer software on learning and teaching, and the role of metacognition in decision making under uncertainty. For each of these areas, we will evaluate the corpus of research that has been presented in this chapter according to the criteria implicit in Shaughnessy's wish list.

Development of Assessment Instruments

The key thrust in Shaughnessy's comments on assessment instruments focused on the need to develop standard and reliable tools, in both written and interview format, that assessed students' probability conceptions. He also underscored the importance of developing consistent instruments across a wide range of grade levels and a variety of contexts.

As a starting point, it is noteworthy that the current period has seen promising growth in the formulation of *assessment frameworks*. Some of these frameworks (e.g., Garfield, 1994; Jolliffe, 1991), although set in the context of statistics and probability, have focused on more general assessment constructs such as what to assess, the purpose of assessment, who will do the assessment, the method of assessment, and the nature of feedback and follow-up action (see Garfield, 1994; Jolliffe, 2005). Other frameworks (e.g., Jones, Thornton, et al., 1999; Tarr & Jones, 1997; Watson et al., 1997; Watson & Moritz, 2003a, 2003b) have produced characterizations of students' probabilistic thinking and beliefs in terms of hierarchical levels. Such characterizations, as well as being useful for monitoring student growth during instruction, have also resulted in the production of written and interview assessments in probability.

These research instruments together with the NAEP assessment material (Shaughnessy & Zawojewski, 1999) have gone some way to addressing Shaughnessy's appeal for standard and reliable probability instruments that cover a wide range of grade levels and incorporate a variety of contexts. In particular, Watson and her colleagues (Watson et al., 1997; Watson & Moritz, 1998, 2003a, 2003b) have produced written survey instruments that measure various aspects of chance measurement, cover a wide range of ages (Grades 3 through 9), and have the advantage that they have been used in longitudinal research and validated on large samples. By way of contrast, Jones and his colleagues (Jones, Langrall, et al., 1997, 1999; Jones, Thornton, et al., 1999; Tarr & Jones, 1997) have generated *interview* assessments. Their instruments cover a range of children's probabilistic reasoning (Grades 1 through 8); however, they have not been subjected to the large-scale population scrutiny that is a powerful feature of Watson's assessments. In addition to these assessment instruments, we underscore Batanero and Serrano's (1999) assessment of randomization and combinatorics with high school students and Amir and Williams's (1999) instruments that examined cultural differences in children's understanding of chance and luck.

Notwithstanding the promising assessment activities adumbrated above, greater coherence is still needed in the way that multiple perspectives of probability (classical, frequentist, and subjective) are assessed. These multiple perspectives of probability have been highlighted for some time as deficit areas in research for both instruction and assessment (Metz, 1997; Steinbring, 1991). We will return to them in discussing our agenda for future research.

Secondary Students' Conceptions and Misconceptions

Shaughnessy (1992) asserted that most of the probability studies have been undertaken with elementary or college students. He called for research on secondary students' thinking about chance, random events, and decision making under uncertainty. In addition, he advocated studies that examine the influence of instruction on secondary students' probabilistic thinking.

This period has seen a much stronger research base on the probabilistic reasoning of secondary students. As described earlier, Batanero and her colleagues have been especially prominent with their research on the following topics: perception of randomness (Batanero & Serrano, 1999), use of heuristics and biases (Batanero, Serrano, et al., 1996), and combinatorial reasoning (Batanero et al., 1997). Other researchers have also investigated heuristics and misconceptions that secondary students bring to the classroom (e.g., Benson & Jones, 1999; Gras & Totohasina, 1995; Sanchez & Hernandez, 2003; Shaughnessy & Ciancetta, 2002; Zimmermann & Jones, 2002), and, in a related area, Fischbein and his colleagues (e.g., Fischbein et al., 1991; Fischbein & Schnarch, 1997) have examined the evolution of probabilistic intuitions across age ranges that included secondary students. Instructional studies such as Zimmermann's (2002) classroom investigation of secondary students' individual and collective thinking about simulation are beginning to emerge. In fact, the extent of recent research involving secondary students is impressive, as can be readily gleaned from Batanero and Sanchez's (2005) review.

Although Shaughnessy's call for additional studies on secondary students' probabilistic conceptions has been heeded, much of the research has focused on their probabilistic reasoning prior to instruction. There is still a void in the kinds of classroom studies that Shaughnessy advocated, that is, studies that investigate the effect of instruction on secondary students' probability learning.

Absence of Cross-Cultural Studies

Shaughnessy raised concerns about the lack of probability research outside of western countries. He advocated large- and small-scale studies that investigate the influence of different cultures on students' thinking in decision making and probability estimation tasks.

During this period probability research has been undertaken outside of western culture. Jansem (2005) and Lamprianou and Lamprianou (2003) studied the nature of primary students' probabilistic thinking in Thailand and Cyprus, respectively; Li (2000) analyzed probabilistic understanding of middle and high school students in the Republic of China; and Polaki (2002a, 2002b) undertook clinical studies and teaching experiments on the probabilistic reasoning of elementary students in Lesotho. However, as Greer and Mukhopadhyay (2005) noted, these studies relate strongly to English or American research and do not focus on "whether the culture of the participants might have any effect on their way of thinking about probability" (p. 317).

By way of contrast, Amir and Williams (1999) and Peard (1995) have investigated the influence of culture on probabilistic thinking. Probing elementary students' religious beliefs and experiences with random phenomena, Amir and Williams concluded that cultural factors influenced the nature of the students' probabilistic knowledge and in turn their probability learning. Working with high school students, Peard investigated the effect of social background. He reported markedly different probabilistic intuitions for students whose social or family background included extensive familiarity with "track" gambling. Although these studies whet our appetite for research into the influence of culture on probabilistic knowledge, they did not incorporate the strong ethnic influences that seemed to be part of Shaughnessy's agenda.

Teachers' Conceptions of Probability

Given the important role that teachers play in probability learning, Shaughnessy advocated research on the knowledge and beliefs that preservice and practicing teachers exhibit toward probability. He also recommended research and development activities that focus on teachers' knowledge of students' conceptions and misconceptions in probability.

Stohl's (2005) review concluded that there had been limited response to Shaughnessy's call for research on teachers' knowledge and beliefs about probability. Nevertheless, we have seen some research in these areas: Begg and Edwards (1999) and Carnell (1997) have documented primary and middle school teachers' misconceptions, and Watson (2001) has profiled primary and secondary teachers' content and pedagogical knowledge in probability and statistics. This research has been sufficient to raise concerns about teachers' lack of confidence in their knowledge of probability, their anxiety about teaching probability, and their disquiet about the time that probability and statistics take away from other strands in the mathematics curriculum (e.g., Gattuso & Pannone, 2002). With regard to Shaughnessy's appeal for research on teachers' knowledge of students' conceptions of probability, we found only one study (Watson, 2001) that addressed this question. The fact that Watson concluded that both secondary and primary teachers were more familiar with students' procedural knowledge than with their conceptual knowledge highlights the need for teacher development programs in probability. However, Stohl (2005) observed that, although many projects have focused on teachers' understanding of data analysis (e.g., *Teach-Stat,* Friel & Joyner, 1997; *Alabama Quantitative Literacy Workshop,* Yarbrough, Daane, & Vessel, 1998), there has been little action on programs in probability. Watson's (1997b) program *Chance and Data for Luddites* may be an exception.

Effects of Instruction

Shaughnessy emphasized the need for research on instruction, especially instruction that relates to students' conceptual knowledge of probability. He called for partnerships between researchers and teachers (e.g., through teaching experiments) that would trace changes in students' probabilistic conceptions and intuitions during instruction.

With the emergence of more sophisticated teaching-experiment designs (Cobb, 2000; Lesh & Kelly, 2000; Simon, 2000), there has been a marked increase in instructional research on probability. At the elementary, middle, and high school levels, several studies (e.g., Aspinwall & Tarr, 2001; Jones, Langrall, et al., 1999; Polaki, 2002a, 2002b; Tarr, 1997; Zimmermann, 2002) have traced students' individual and collective thinking in probability during instruction and have identified pedagogical processes that seem to impact learning. In addition, Polaki and Zimmermann have documented and analyzed classroom mathematical norms and practices that may provide valuable pedagogical content knowledge for teachers. Although these teaching experiments involved teacher-researcher partnerships, none of them constituted the kinds of long-term collaborative interventions that Shaughnessy had in mind. The research seems to have reached a stage requiring teaching experiments that use more robust experimental designs and statistical methods to evaluate the effectiveness of instructional interventions.

Effects of Computer Software

This agenda item is clearly related to the previous one. In identifying issues associated with technology, Shaughnessy called for the development of research into computer software that would change students' probabilistic conceptions by taking advantage of the speed, graphics, and simulation potential of microcomputers. He especially underscored the need for researchers and teachers to be able to provide feedback during the developmental phase of such software.

Although there has been manifest support for the use of technology in the teaching and learning of probability (Ben-Zvi, 2000; Biehler, 1991; Dörfler, 1993; Konold, 1991a) and considerable development in probability-oriented computer software and graphics-calculator capability (Edmark Corporation, 1996; Erickson, 2001; NCTM 2002a, 2002b, 2003), research on the influence of technology on students' probabilistic conceptions is only beginning to emerge. The most promising research involves the use of microworlds and their influence on changing and expanding students' probabilistic thinking (Konold, 1995a; Paparistodemou et al., 2002; Pratt, 1998; Pratt & Noss, 2002; Stohl, 1999–2002). In particular, the work of Pratt and Stohl shows that microworlds can be designed to forge vital links between experimental and theoretical probability and to build cognitive mechanisms that enable children to challenge their own intuitions. There is also evidence of case studies that document the effect of computer software and graphics calculators on probability learning (Kissane, 1997a, 1997b; Konold, 1991a).

The Role of Metacognition in Decision Making Under Uncertainty

Shaughnessy specifically advocated research on metacognitive aspects of probabilistic thinking. He claimed that such research may shed light on students' probabilistic misconceptions and on ways to use metacognitive ideas to challenge such misconceptions.

Regrettably we were unable to locate, in this period, research that has focused on metacognitive aspects of the teaching and learning of probability. The situation in probability seems to reflect Lester's (1994) more general concern about the lack of mathematics education research on metacognition. Perhaps Goos and her colleagues' metacognitive research on mathematical modeling (e.g., Goos, Galbraith, & Renshaw, 2002) might act as a fillip for future research on the role of metacognitive processes in probabilistic modeling.

These reflections on probability research in the last 15 years reveal that gaps still exist in the research outlined in Shaughnessy's (1992) wish list. These extant gaps, together with areas where a need for new research has emerged more recently, will be taken up in the next section when we propose our own agenda for future research.

Future Research Directions in Probability: A Continuing Agenda

Although some aspects of Shaughnessy's (1992) wish list have been addressed during the past 15 years, critical areas remain: developmental research into assessment instruments; long-term teaching experiments, especially in secondary classrooms; studies examining the effects of cultural, social, and ethnic contexts on probabilistic reasoning; research into teachers' knowledge of student conceptions; evaluations of teacher enhancement programs; research into the effect of microworlds on students' probabilistic reasoning; and classroom research using metacognition to examine students' individual and collective thinking in probability. In generating a new agenda for probability research in the next 15 years, we will revisit these critical areas.

Students' Conceptions of Theoretical and Experimental Probability

Our first agenda item relates to cognitive research that is grounded in both classical and frequentist perspectives of probability. Although there has been considerable research at all age levels on students' conceptions of theoretical probability (e.g., Batanero et al., 1996; Fischbein & Schnarch, 1997; Jones, Langrall, et al., 1997, 1999; Watson & Moritz, 1998), there has been relatively little research on students' thinking about experimental probability and even less on students' understanding of the connections between theoretical and experimental probability (Aspinwall & Tarr, 2001; Stohl & Tarr, 2002).

We believe that there is a need to explore students' thinking about classical and frequentist perspectives of probability across the complete range of ages from early childhood to high school. As an illustration of what we intend, consider the following investigative questions in relation to a probability measurement task like the following: "What color is most likely to come up if a cube with 1 green, 2 red, and 3 blue faces is rolled?"

- How do students address such a task unassisted?
- How do students respond to the suggestion that it might be helpful to experiment by rolling the cube a number of times?
- How do they deal with the experimental data including the size of the sample of data?
- How do they respond to the suggestion that it might be helpful to examine the symmetry of the cube?
- How do students deal with symmetry and its implications for theoretical probability?
- Do students perceive any relationship between likelihood estimates based on experimentation and those based on symmetry?

Research questions and tasks like the ones illustrated above can be used to explore students' thinking during clinical interviews and to trace students' individual and collective thinking during classroom instruction. Although curriculum documents (e.g., AEC, 1994; NCTM, 2000) have stressed the importance of focusing on experimental probability, and some researchers (e.g., Aspinwall & Tarr, 2001; Garfield & Ahlgren, 1988; Jones, Langrall, et al., 1999; Shaughnessy, 1977; Steinbring, 1991; Stohl & Tarr, 2002) have identified tasks and learning sequences that have incorporated the simultaneous development of both theoretical and experimental probability, only Aspinwall and Tarr and Stohl and Tarr have documented how students responded to these approaches. We believe that cognitive research under the rubric of classical and frequentist probability is a key endeavor for the next period as such research goes well beyond questions of likelihood. It provides a frame for investigating students' conceptions of key ideas such as the following: chance, random phenomena, conditional probability, independence, the law of large numbers, and probability distributions.

The area of subjective probability is more problematic. Even though it is commonly used by people in everyday reasoning, the fact that it is not widely represented in mathematics curricula means that there is less immediate interest on the part of researchers to investigate this aspect of probability measurement. We think it is timely for researchers in mathematics education to examine subjective probability and the way that students conceptualize it.

Adopting a Critical Stance to Stochastical Information

One of the important goals for teaching probability in school mathematics is to prepare students to deal with the statistical and probabilistic information that increasingly impacts their lives. More specifically, criticalness (Gal, 1994), the ability to take an evaluative stance with respect to stochastical flaws and biases contained in media, marketing, and financial information, is of paramount importance in the quest for probability literacy. In spite of its precedence there has been little research on criticalness and little injection of it into current curriculum programs.

Watson's research (e.g., Watson, 1997a, Watson & Moritz, 2000) has made a valuable beginning in this area, especially her studies in which students are challenged to evaluate media claims. Of particular importance for the future was her use of longitudinal designs and large samples to build models of the evolution of students' cognitions on criticalness. Notwithstanding Watson's efforts, the research on criticalness is in its infancy, and there is a need for empirical studies that investigate students' ability to assess social information that involves probabilistic processes like the following: experimental probabilities, the law of large numbers, conditional probability, simulation, sampling, and applications of probability in statistical inference.

The Development of Instructional Theory

Shaughnessy's (1992) wish list included a recommendation for long-term teaching experiments, especially with high school students. We go further by highlighting the need for teaching experiments that investigate diverse approaches to the teaching of probability. In reality, there has not been substantial work on the development of an instructional theory for probability since Steinbring's (1991) groundbreaking work more than a decade ago. Although Steinbring advocated an approach to probability that incorporated both classical and frequentist probabilities (e.g., through *task systems*), he did not attempt to embellish the instructional strategies incorporated in his theory or provide a great deal of empirical data to support his theory.

We claim that there is a critical need for researchers to document instructional strategies for the teaching and learning of probability at all age levels. For example, as noted earlier, recent research (e.g., Gigerenzer, 1994; Hopfsenberger et al., 1999) suggests that probability should be introduced through frequency data tasks rather than probability tasks per se. Studies are needed to build hypothetical learning trajectories (Simon, 1995) for such tasks so as to support teachers' use of the strategy. This data-oriented strategy might also be the basis for forging connections between probability and statistical research. Accordingly, this *frequency* strategy and emerging strategies like those linked to microworld environments (e.g., Pratt, 2000; Pratt & Noss, 2002) need to be trialed and monitored to examine their effects on students' learning and beliefs.

As part of the emphasis on research involving teaching experiments and innovative instructional strategies, we believe that there is a need for studies that use carefully controlled experimental designs and statistical analyses (Kilpatrick, 2001). We are not suggesting a blanket return to scientific positivism but rather a balanced approach that uses mixed designs similar to those suggested by Tashakkori and Teddlie (1998).

The Development of Assessment Instruments

In maintaining assessment as a continuing research endeavor for the next period we wish to propose different directions from those suggested by Shaughnessy (1992). First, in accord with our advocacy of systematic research on students' cognitions vis-à-vis classical and frequentist approaches to probability, we claim that research is needed to develop and validate written and interview instruments that examine students' reasoning about experimental and theoretical probability and their relationship. Moreover, these instruments should be able to examine students' reasoning in experimental and theoretical probability over an extended age range similar to the written instruments used by Watson and her colleagues (Watson, 1997a; Watson et al., 1997; Watson & Moritz, 1998, 2003a, 2003b). Notwithstanding the importance of written instruments, we also underscore Greer and Mukhopadhyay's (2005) caveat that written assessments do not lend themselves to assessing students' understanding of the relationship between theoretical and experimental probability.

Second, in accord with Metz's (1997) call for chance to be assessed along three dimensions: *cognitive* (conceptual understanding), *epistemological* (beliefs), and *cultural* (especially the culture of the classroom), we believe that there is a need for research that evaluates assessment material according to Metz's dimensions. Such assessments need to be used in tandem with research involving teaching experiments. In particular, it is vital that researchers trace students' conceptual understanding of randomness and chance variation and also monitor changes in their "beliefs about the place of chance and uncertainty in the world" (p. 225).

Third, the development of assessment instruments that embody the *cultural* dimension of Metz's (1997) framework have the potential to stimulate sharper research on the effect of different cultures on students' probabilistic reasoning. As Greer and Mukhopadhyay (2005) have noted, there is a need for "a sensitivity to the cultures of others when teaching mathematics to multicultural classes and when teaching a curriculum that is not indigenously grounded" (p. 318). Moreover, with respect to probability, these two researchers declare that there are few studies that bear on the cultural issues endemic to curriculum, instruction, or assessment.

Large-Scale Curriculum and Instructional Research

Investigations that are integrated into a cluster of school-based longitudinal projects in culturally and economically diverse settings have not, for various reasons (e.g., limited awareness of the importance of stochastics), included probability research. Such projects need to include stochastics, and similar to QUASAR (Silver & Stein, 1996), their manifesto should embrace curriculum development, teacher enhancement, computer resources, and assessment materials.

These large-scale research projects will be appropriately situated to use designs such as multitiered teaching experiments (Lesh & Kelly, 2000) and educational development and developmental research (Gravemeijer, 1998). As such they will have the potential to trace the growth of students' probabilistic conceptions and intuitions from elementary through secondary school. They will also generate increased knowledge about students' collective thinking in probability and document sociomathematical norms and classroom mathematical practices (Cobb, 2000) that will help teachers build appropriate learning environments for probability instruction. Moreover, they will provide an ongoing base for teacher support and opportunities to build and monitor teacher knowledge of students' conceptions and misconceptions in probability.

In a related area, this kind of developmental research will be ideally suited to address Shaughnessy's (1992) concerns about the lack of probability research that focuses on metacognition. Given the impetus of research by Goos and her colleagues (Goos & Galbraith, 1996; Goos et al., 2002) that has documented the metacognitive characteristics of successful and unsuccessful problem solvers, there is infrastructure that can be applied to probability research. More specifically, the theoretical base and methodological design associated with Goos's research seems apposite for future metacognitive research in probability, especially instructional research.

CONCLUSION

This chapter presents an examination of research on cognitive, pedagogical, and curriculum aspects of probability in the wake of its introduction as a mainstream topic in the mathematics curricula of various countries (e.g., AEC, 1991, 1994; DES, 1991; NCTM, 1989, 2000). In our analysis of cognitive and pedagogical research in probability, we have compared and contrasted research in this period with research reviewed by Shaughnessy (1992) in terms of philosophical underpinnings, purpose and direction, context, and methodological features. The research in this period reveals a stronger thrust on documenting and characterizing the probabilistic knowledge that students bring to the classroom (e.g., Batanero et al., 1997; Jones et al., 1997; Metz, 1997; Watson et al., 1997) and on tracing students' individual and collective reasoning during instruction in probability (e.g., Batanero et al., 1997; Jones, Langrall, et al., 1999; Polaki, 2002a, 2002b). Researchers in this period have also generated valuable knowledge about learning environments (e.g., Amir & Williams, 1999; Castro, 1998; Horvath & Lehrer, 1998; Jones, Thornton, et al., 1999) including technology-supported environments (e.g., Pratt, 2000; Stohl, 1999–2002; Zimmermann, 2002) that appear to have the facility to change students' probabilistic intuitions and beliefs.

With respect to curriculum research, we have examined the new literature on probability literacy in the wake of evidence suggesting that the implementation of probability in schools may not be as pervasive or as sustainable as national curriculum documents portray. The literature on probability literacy (e.g., Gal, 2002, 2005; Watson, 1997) has identified a need for fresh directions, greater coherence, and an enlargement of ideas that prepare children to deal with the vagaries of variation. In particular, this research advocates a richer understanding of chance variation and randomization; a more unified development of the classical, frequentist, and subjective approaches to probability; and an emphasis on critical thinking that enables children to confront the potential exploitation of media marketing.

We have also generated a research agenda for the next decade or more. In addition to the curriculum research foreshadowed above, this agenda promotes endeavors like the following: the development of instructional theory in probability especially for secondary students, characterizations of student cognitions in relation to experimental and theoretical probability and their connections, investigations into teachers' pedagogical knowledge and knowledge of students' probabilistic cognitions, and explorations of the role of metacognitive processes in the learning of probability especially classroom learning.

In looking back over this period, one of the striking aspects is the emergence of a robust research base on the probability knowledge and conceptions that students bring to the classroom. There is also evidence of a stronger research literature in pedagogical and instructional aspects of probability teaching. Although Greer and Mukhopadhyay's (2005) caveat, "a major challenge to the field is harvesting what is known from this [probability] research to inform teaching" (p. 315), still has currency, the research in this period has made significant contributions to the teaching and learning of probability in schools.

REFERENCES

Acredolo, C., O'Connor, J., Banks, L., & Horobin, K. (1989). Children's ability to make probability estimates: Skills revealed through application of Anderson's functional measurement methodology. *Child Development, 60,* 933–945.

Amir, G. S., & Williams, J. S. (1999). Cultural influences on children's probabilistic thinking. *Journal of Mathematical Behavior, 18*(1), 85–107.

Amit, M. (1998). Learning probability concepts through games. In L. Pereira-Mendoza, L. S. Kea, T. W. Kee, & W. Wong (Eds.), *Proceedings of the Fifth International Conference on Teaching Statistics* (Vol. 1, pp. 45–48). Voorburg, The Netherlands: International Statistical Institute.

Aspinwall, L., & Tarr, J. E. (2001). Middle school students' understanding of the role sample size plays in experimental probability. *Journal of Mathematical Behavior, 20,* 229–245.

Australian Education Council. (1991). *A national statement on mathematics for Australian Schools.* Carlton, VIC: Curriculum Corporation.

Australian Education Council. (1994). *Mathematics: A curriculum profile for Australian schools.* Carlton, VIC: Curriculum Corporation.

Ayres, P., & Way, J. (2000). Knowing the sample space or not: The effects on decision making. In T. Nakahara & M. Koyama (Eds.), *Proceedings of the 24th Conference of the International Group for the Psychology of Mathematics Education* (Vol.2, pp. 33–40). Hiroshima University, Japan.

Batanero, C., Biehler, R., Maxara, C., Engel, J., & Vogel, M. (2005, May). *Using simulation to bridge teachers' content and pedagogical knowledge in probability.* Paper presented at the meeting of ICMI 15, Aguas de Lindoia, Brazil.

Batanero, C., Henry, M., & Parzysz, B. (2005). The nature of chance and probability. In G. A. Jones (Ed.), *Exploring probability in school: Challenges for teaching and learning* (pp. 15–37). New York: Springer.

Batanero, C., Navarro-Pelayo, V., & Godino, J. D. (1997). Effect of the implicit combinatorial model on combinatorial reasoning in secondary school pupils. *Educational Studies in Mathematics, 32,* 181–199.

Batanero, C., & Sanchez, E. (2005). What is the nature of high school students' conceptions and misconceptions about probability? In G. A. Jones (Ed.), *Exploring probability in school: Challenges for teaching and learning* (pp. 241–266). New York: Springer.

Batanero, C., & Serrano, L. (1999). The meaning of randomness for secondary students. *Journal for Research in Mathematics Education, 30,* 558–567.

Batanero, C., Serrano, L., & Garfield, J. B. (1996). Heuristics and biases in secondary school students' reasoning about probability. In L. Puig & A. Gutiérrez (Eds.), *Proceedings of the 20th Conference of the International Group for the Psychology of Mathematics Education* (Vol.2, pp. 51–58). University of Valencia, Spain.

Begg, A., & Edwards, R. (1999, December). *Teachers' ideas about teaching statistics.* Paper presented at the combined annual meeting of the Australian Association for Research in Education and the New Zealand Association for Research in Education. Melbourne, Australia.

Benson, C. T. (2000). *Assessing students' thinking in modeling probability contexts.* Unpublished doctoral dissertation, Illinois State University, Normal.

Benson, C. T., & Jones, G. A. (1999). Assessing students' thinking in modeling probability contexts. *The Mathematics Educator, 4*(2), 1–21.

Ben-Zvi, D. (2000). Toward understanding the role of technological tools in statistical learning. *Mathematical Thinking and Learning, 2,* 127–155.

Biehler, R. (1991). Computers in probability education. In R. Kapadia & M. Borovcnik (Eds.), *Chance encounters: Probability in education* (pp. 169–211). Amsterdam: Kluwer.

Biggs, J. B., & Collis, K. F. (1982). *Evaluating the quality of learning: The SOLO taxonomy.* New York: Academic.

Biggs, J. B., & Collis, K. F. (1991). Multimodal learning and the quality of intelligent behaviour. In H. A. H. Rowe (Ed.), *Intelligence: Reconceptualization and measurement* (pp. 57–76). Hillsdale, NJ: Erlbaum.

Borovcnik, M., & Bentz, H. J. (1991). Empirical research in understanding probability. In R. Kapadia & M. Borovcnik (Eds.), *Chance encounters: Probability in education* (pp. 73–106). Amsterdam: Kluwer.

Borovcnik, M., & Peard, R. (1996). Probability. In A. Bishop, K. Clements, C. Keitel, J. Kilpatrick, & C. Laborde (Eds.), *International handbook of mathematics education* (Part 1, pp. 239–288). Dordrecht, The Netherlands: Kluwer.

Brousseau, G., Brousseau, N., & Warfield, V. (2002). An experiment on the teaching of statistics and probability. *Journal of Mathematical Behavior, 20*, 363–411.

Byrnes, J. P., & Beilin, H. (1991). The cognitive basis of uncertainty. *Human Development, 34*, 189–203.

Carnell, L. J. (1997). *Characteristics of reasoning about conditional probability (preservice teachers).* Unpublished doctoral dissertation, University of North Carolina-Greensboro.

Case, R. (1985). *Intellectual development: A systematic reinterpretation.* New York: Academic Press.

Castro, C. S. (1998). Teaching probability for conceptual change. *Educational Studies in Mathematics, 35*, 233–254.

Cobb, P. (2000). Conducting teaching experiments in collaboration with teachers. In E. Kelly & R. A. Lesh (Eds.), *Handbook of research design in mathematics and science education*, (pp. 307–334). Mahwah, NJ: Erlbaum.

Cobb, P., & Bauersfeld, H. (1995). *The emergence of mathematical meaning: Interaction in classroom cultures.* Hillsdale, NJ: Erlbaum.

Cobb, P., Wood, T., Yackel, E., Nicholls, J., Wheatley, G., Trigatti, B., et al. (1991). Assessment of a problem-centered second-grade mathematics project. *Journal for Research in Mathematics Education, 22*, 3–29.

Davies, C. M. (1965). Development in probability concepts in children. *Child Development, 36*, 779–788.

Department of Education and the Arts, Tasmania (1993). *Chance and data (Activity Booklet 3): Mathematics Guidelines K-8.* Hobart, Tasmania: Author.

Department of Education and Science and the Welsh Office. (1989). *National curriculum: Mathematics for ages 5 to 16.* York, UK: Central Office of Information.

Department of Education and Science and the Welsh Office. (1991). *National curriculum: Mathematics for ages 5 to 16.* York, UK: Central Office of Information.

Dessart, D. J. (1995). Randomness: A connection to reality. In P. A. House & A. F. Coxford (Eds.), *Connecting mathematics across the curriculum* (pp. 177–181). Reston, VA: National Council of Teachers of Mathematics.

DfEE. (1999). *The national curriculum: Mathematics.* London: DfEE Publication.

Doherty, J. (1965). *Level of four concepts of probability possessed by children of fourth, fifth, and sixth grade before formal instruction.* Unpublished doctoral dissertation, University of Missouri, Columbia.

Dörfler, W. (1993). Computer use and views of the mind. In C. Keitel & K. Ruthven (Eds.), *Learning from computers: Mathematics education and technology* (pp. 159–186). Berlin, Germany: Springer Verlag.

Drier, H. S. (2000a). *Children's probabilistic reasoning with a computer microworld.* Unpublished doctoral dissertation, University of Virginia.

Drier, H. S. (2000b). Children's meaning-making activity with dynamic multiple representations in a probability microworld. In M. Fernandez (Ed.), *Proceedings of the 22nd Annual Meeting of the North American chapter of the International Group for the Psychology of Mathematics Education* (pp. 691–696). Columbus, OH: ERIC Clearinghouse of Science, Mathematics, and Environmental Education.

Dugdale, S. (2001). Pre-service teachers' use of computer simulation to explore probability. *Computers in the Schools, 17*(1/2), 173–182.

Edmark Corporation. (1996). *Mighty math: Number heroes* [Computer software]. Redmond, WA: Author.

Engel, A. (1966). Initiation la theorie des probabilities [Initiation of the theory of probabilities]. In *New trends in mathematics teaching* (Vol. 1, pp. 159–174). Paris: UNESCO.

English, L. D. (1991). Young children's combinatoric strategies. *Educational Studies in Mathematics, 22*, 451–474.

English, L. D. (1993). Children's strategies for solving two- and three-dimensional combinatorial problems. *Journal for Research in Mathematics Education, 24*, 255–273.

English, L. D. (2005). Combinatorics and the development of children's combinatorial reasoning. In G. A. Jones (Ed.), *Exploring probability in school: Challenges for teaching and learning* (pp. 121–141). New York: Springer.

Erickson, T. (2001). *Data in depth: Exploring mathematics with Fathom.* Emeryville, CA: Key Curriculum Press.

Falk, R. (1981). The perception of randomness. In *Proceedings of the Fifth Conference of the International Group for the Psychology of Mathematics Education* (pp. 222–229). Grenoble, France: University of Grenoble.

Falk, R. (1983). Children's choice behaviour in probabilistic situations. In D. R. Grey, P. Holmes, V. Barnett, & G. M. Constable (Eds.), *Proceedings of the First International Conference on Teaching Statistics* (pp. 714–716). Sheffield, England: Teaching Statistics Trust.

Falk, R. (1988). Conditional probabilities: Insights and difficulties. In R. Davidson & J. Swift (Eds.), *The Proceedings of the Second International Conference on Teaching Statistics* (pp. 714–716). Victoria, BC, Canada: University of Victoria.

Falk, R., & Wilkening, F. (1998). Children's construction of fair chances: Adjusting probabilities. *Developmental Psychology, 34*(6), 1340–1357.

Fast, G. R. (1999). Analogies and reconstruction of probability knowledge. *School Science and Mathematics, 99*, 230 – 240.

Fay, A. L., & Klahr, D. (1996). Knowing about guessing and guessing about knowing: Preschoolers understanding of indeterminacy. *Child Development, 67*, 689–716.

Fennema, E., Carpenter, T. P., Franke, M. L., Levi, L., Jacobs, V. R., & Empson, S. B. (1996). A longitudinal study of learning to use children's thinking in mathematics instruction. *Journal for Research in Mathematics Education, 27*, 403–434.

Fischbein, E. (1975). *The intuitive sources of probabilistic thinking in children.* Dordrecht, The Netherlands: Reidel.

Fischbein, E., & Gazit, A. (1984). Does the teaching of probability improve probabilistic intuitions? *Educational Studies in Mathematics, 15*, 1–24.

Fischbein, E., & Grossman, A. (1997). Schemata and intuitions in combinatorial reasoning. *Educational Studies in Mathematics, 34*, 27–47.

Fischbein, E., Nello, M. S., & Marino, M. S. (1991). Factors affecting probabilistic judgments in children in adolescence. *Educational Studies in Mathematics, 22*, 523–549.

Fischbein, E., Pampu, I., & Mînzat, I. (1970). Comparison of ratios and the chance concept in children. *Child Development, 41,* 377–389.

Fischbein, E., & Schnarch, D. (1997). The evolution with age of probabilistic, intuitively based misconceptions. *Journal for Research in Mathematics Education, 28,* 96–105.

Friel, S. N., & Joyner, J. (1997). *Teach-Stat for teachers: Professional development manual.* Palo Alto, CA: Dale Seymour Publication.

Gal, I. (1994, September). *Assessment of interpretive skills.* Summary of a Working Group at the Conference on Assessment Issues in Statistics Education, Philadelphia.

Gal, I. (2002). Adult statistical literacy: Meanings, components, responsibilities. *International Statistical Review, 70*(1), 1–25.

Gal, I. (2005). Towards "probability literacy" for all citizens: Building blocks and instructional dilemmas. In G. A. Jones (Ed.), *Exploring probability in school: Challenges for teaching and learning* (pp. 39–64). New York: Springer.

Garfield, J. (1994). Beyond testing and grading: using assessment to improve student learning. *Journal of Statistics Education, 2*(1). Retrieved March 8, 2004, from http://www.amstat.org/publications/jse/v2n1/garfield.html

Garfield, J. B., & Ahlgren, A. (1988). Difficulties in learning basic concepts in probability and statistics: Implications for research. *Journal for Research in Mathematics Education, 19,* 44–63.

Gattuso, L., & Pannone, M. A. (2002). Teacher's training in a statistics teaching experiment. In B. Phillips (Ed.), *Proceedings of the Sixth International Conference on the Teaching of Statistics* [CD-ROM], Hawthorn, VIC, Australia: International Statistical Institute.

Gigerenzer, G. (1994). Why the distinction between single event probabilities and frequencies is important for psychology (and vice versa). In G. Wright & P. Ayton (Eds.), *Subjective probability* (pp. 129–161). Chichester, England: Wiley.

Gillis, D., & Haraud, B. (1972). Introduction des probabilities a l'elementaire. *Journal of Structural Learning, 3,* 41–59.

Goos, M., & Galbraith, P. (1996). Do it this way! Metacognitive strategies in collaborative mathematical problem solving. *Educational Studies in Mathematics, 30,* 229–260.

Goos, M., Galbraith, P., & Renshaw, P. (2002). Socially mediated cognition: Creating collaborative zones of proximal development in small group problem solving. *Educational Studies in Mathematics, 49,* 193–223.

Gras, R., & Totohasina, A. (1995). Chronologie et causalite, conception source d'obstacles epistemologiques a la notion de probabilite conditionnelle [Chronology and causality, conceptions sources of epistemological obstacles in the notion of conditional probability]. *Recherches en Didactique des Mathematiques, 15*(1), 49–95.

Gravemeijer, K. (1995). *Developing realistic mathematics instruction.* Utrecht, The Netherlands: Freudenthal Institute.

Gravemeijer, K. (1998). Developmental research as a research method. In A. Sierpinska & J. Kilpatrick (Eds.), *Mathematics education as a research domain: A search for identity* (pp. 277–296). Dordrecht, The Netherlands: Kluwer.

Green, D. R. (1979). The chance and probability concepts project. *Teaching Statistics, 1*(3), 66–71.

Green, D. R. (1983). A survey of probability concepts in 3000 pupils aged 11-16 years. In D. R. Grey, P. Holmes, V. Barnett, & G. M. Constable (Eds.), *Proceedings of the First International Conference on Teaching Statistics* (pp. 766–783). Sheffield, England: Teaching Statistics Trust.

Green, D. R. (1988). Children's understanding of randomness: Report of a survey of 1600 children aged 7-11 years. In R. Davidson & J. Swift (Eds.), *Proceeding of the Second International Conference on Teaching Statistics* (pp. 287–291). Victoria, BC, Canada: University of Victoria.

Greer, B. (2001). Understanding probabilistic thinking: The legacy of Efraim Fischbein. *Educational Studies in Mathematics, 45,* 15–33.

Greer, B., & Mukhopadhyay, S. (2005). Teaching and learning the mathematization of uncertainty: Historical, cultural, social and political contexts. In G. A. Jones (Ed.), *Exploring probability in school: Challenges for teaching and learning* (pp. 297–324). New York: Springer.

Greer, B., & Ritson, R. (1994, July). *Readiness of teachers in Northern Ireland to teach data handling.* Paper presented at the Fourth International Conference on Teaching Statistics, Marrakech, Morocco.

Grouws, D. A. (1992). *Handbook of research on mathematics teaching and learning.* New York: Macmillan.

Haller, S. K. (1997). *Adopting probability curricula: The content and pedagogical content knowledge of middle grades teachers.* Unpublished doctoral dissertation, University of Minnesota.

Hopfsenberger, P., Kranendonk, H., & Scheaffer, R. (1999). *Data driven mathematics: probability through data.* Palo Alto, CA: Dale Seymour Publications.

Horvath, J. K., & Lehrer, R. (1998). A model-based perspective on the development of children's understanding of chance and uncertainty. In S. P. Lajoie (Ed.), *Reflections in statistics: Learning, teaching, and assessment in grades K–12* (pp. 121–148). Mahwah, NJ: Erlbaum.

Huber, B. L., & Huber, O. (1987). Development of the concept of comparative subjective probabilities. *Journal of Experimental Child Psychology, 44,* 304–316.

Jansem, S. (2005). *Using instruction to trace thinking and beliefs in probability of primary students.* Unpublished doctoral dissertation, Srinakharinwirot University, Bangkok, Thailand.

Jolliffe, F. (1991). Assessment of the understanding of statistical concepts. In D. Vere-Jones (Ed.), *Proceedings of the Third International Conference on the Teaching of Statistcs* (Vol. 1, pp. 461–466). Voorburg, The Netherlands: International Statistical Institute.

Jolliffe, F. (2005). Assessing probabilistic thinking and reasoning. In G. A. Jones (Ed.), *Exploring probability in school: Challenges for teaching and learning* (pp. 325–344). New York: Springer.

Jones, G. A. (1974). *The performance of first, second, and third grade children on five concepts of probability and the effects of grade, I.Q., and embodiments on their performance.* Unpublished doctoral dissertation, Indiana University, Bloomington.

Jones, G. A. (Ed.). (2005). *Exploring probability in school: Challenges for teaching and learning.* New York: Springer.

Jones, G. A., Langrall, C. W., Mooney, E. S., & Thornton, C. A. (2004). Models of development in statistical reasoning. In D. Ben-Zvi & J. Garfield (Eds.), *The challenge of developing statistical literacy, reasoning and thinking* (pp. 97–117). Dordrecht, The Netherlands: Kluwer.

Jones, G. A., Langrall, C. W., Thornton, C. A., & Mogill, A. T. (1997). A framework for assessing and nurturing young children's thinking in probability. *Educational Studies in Mathematics, 32,* 101–125.

Jones, G. A., Langrall, C. W., Thornton, C. A., & Mogill, A. T. (1999). Students' probabilistic thinking in instruction. *Journal for Research in Mathematics Education, 30,* 487–519.

Jones, G. A., & Thornton, C. A. (2005). An overview of research into the learning and teaching of probability. In G. A. Jones (Ed.), *Exploring probability in school: Challenges for teaching and learning* (pp. 65–92). New York: Springer.

Jones, G. A., Thornton, C. A., Langrall, C. W., & Tarr, J. E. (1999). Understanding students' probabilistic reasoning. In L. V. Stiff & F. R. Curcio (Eds.), *Developing mathematical reasoning in grades K–12* (1999 yearbook, pp. 146–155). Reston, VA: National Council of Teachers of Mathematics.

Kahneman, D., Slovic, P., & Tversky, A. (1982). *Judgment under uncertainty: Heuristics and biases.* Cambridge, England: Cambridge University Press.

Kahneman, D., & Tversky, A. (1972). Subjective probability: A judgment of representativeness. *Cognitive Psychology, 3,* 430–454.

Kahneman, D., & Tversky, A. (1982). Variants of uncertainty. *Cognition, 11,* 143–157.

Kapadia, R., & Borovcnik, M. (1991). *Chance encounters: Probability in education.* Dordrecht, The Netherlands: Kluwer.

Kaufman-Fainguelernt, E., & Bolite-Frant, J. (1998). The emergence of statistical reasoning in Brazilian school children. In L. Pereira-Mendoza, L. S. Kea, T. W. Kee, & W. Wong (Eds.), *Proceedings of the Fifth International Conference on Teaching Statistics* (Vol. 1, pp. 49–52). Voorburg, The Netherlands: International Statistical Institute.

Kilpatrick, J. (2001). Where's the evidence? *Journal for Research in Mathematics Education, 32,* 421–427.

Kissane, B. (1997a). The graphics calculator: Imagine the probabilities. In D. Clarke, P. Clarkson, D. Gronn, M. Horne, L. Lowe, M. Mackinlay, et al. (Eds.), *Mathematics: Imagine the possibilities* (pp. 478–497). Melbourne: Mathematical Association of Australia.

Kissane, B. (1997b). The graphics calculator and the curriculum: The case of probability. In N. Scott & H. Hollingsworth (Eds.), *Mathematics: Creating the future* (pp. 397–404). Melbourne: Australian Association of Mathematics Teachers.

Konold, C. (1983). *Conceptions about probability: Reality between a rock and a hard place.* Unpublished doctoral dissertation, University of Massachusetts, Boston.

Konold, C. (1989). Informal conceptions of probability. *Cognition and Instruction, 6,* 59–98.

Konold, C. (1991a). *ChancePlus: A computer based curriculum for probability and statistics. (Second year report).* Amherst: University of Massachussets Scientific Reasoning Research Institute.

Konold, C. (1991b). Understanding students' beliefs about probability. In E. von Glasersfeld (Ed.), *Radical constructivism in mathematics education* (pp. 139–156). Dordrecht, The Netherlands: Kluwer.

Konold, C. (1995a). Issues in assessing conceptual understanding in probability and statistics. *Journal of Statistics Education* (Online), *3*(1). Retrieved September 25, 2003, from http://www.amstat.org/publications/jse/v3n1/konold.html.

Konold, C. (1995b). Confessions of a coin flipper and would-be instructor. *The American Statistician, 49*(2), 203–209.

Konold, C., Pollatsek, A., Well, A., Lohmeier, J., & Lipson, A. (1993). Inconsistencies in students' reasoning about probability. *Journal for Research in Mathematics Education, 24,* 392–414.

Kranzler, J. H. (2003). *Statistics for the terrified.* Upper Saddle River, NJ: Prentice Hall.

Kuzmak, S., & Gelman, R. (1986). Young children's understanding of random phenomena. *Child Development, 57,* 559–566.

Kvatinsky, T. & Even, R. (2002). Framework for teacher knowledge and understanding of probability. *Proceedings of the Sixth International Conference on the Teaching of Statistics* [CD-ROM], Hawthorn, VIC, Australia: International Statistical Institute.

Lamprianou, I., & Lamprianou, T. A. (2003). *The nature of students' probabilistic thinking in primary school in Cyprus.* Retrieved December 6, 2003, from http://www.education.man.ac.uk/Ita/tal/tall.pdf.

Langrall, C. W., & Mooney, E. S. (2005). Characteristics of elementary school students' probabilistic thinking. In G. A. Jones (Ed.), *Exploring probability in school: Challenges for teaching and learning* (pp. 95–120). New York: Springer.

Laplace, P. S. (1995). Philosophical essay on probabilities (A. I. Dale, Trans.). New York: Springer-Verlag. (Original work published in 1825).

Leake, L. (1965). The status of three concepts of probability in children of seventh, eighth, and ninth grades. *Journal of Experimental Education, 34,* 78–84.

Lecoutre, M. P. (1992). Cognitive models and problem spaces in "purely random" situations. *Educational Studies in Mathematics, 23,* 557–568.

Leffin, W. W. (1971). *A study of three concepts of probability possessed by children in the fourth, fifth, sixth, and seventh grades.* Unpublished doctoral dissertation, University of Wisconsin, Madison.

Lesh, R. A., & Kelly, A. E. (2000). Multitiered teaching experiments. In A. E. Kelly & R. A. Lesh (Eds.), *Handbook of research design in mathematics and science education* (pp. 197–230). Mahwah, NJ: Erlbaum.

Lester, F. K., Jr. (1994). Musings about problem-solving research. *Journal for Research in Mathematics Education, 25,* 660–675.

Li, J. (2000). *Chinese students' understanding of probability.* Unpublished doctoral dissertation, Nanyang Technological University, Singapore.

Li, J., & Pereira-Mendoza, L. (2002). Misconceptions in probability. In B. Phillips (Ed.), *Proceedings of the Sixth International Conference on Teaching Statistics* [CD-ROM]. Hawthorn, VIC, Australia: International Statistical Institute.

Lindley, D. V. (1980). *Introduction to probability and statistics from the Bayesian viewpoint.* Cambridge, England: Cambridge University Press.

Maher, C. A., Speiser, R., Friel, S., & Konold, C. (1998). Learning to reason probabilistically. In S. B. Berenson, K. R. Dawkins, M. Blanton, W. N. Coulombe, J. Kolb, K. Norwood, et al. (Eds.), *Proceedings of the 20th Annual Meeting of the North American Chapter of the International Group for the Psychology of Mathematics Education* (pp. 82–87). Columbus, OH: ERIC Clearinghouse for Science, Mathematics, and Environmental Education.

Metz, K. E. (1997). Dimensions in the assessment of students' understanding and application of chance. In I. Gal & J. B. Garfield (Eds.), *The assessment challenge in statistics education* (pp. 223–238). Amsterdam: IOS Press.

Metz, K. E. (1998a). Emergent ideas of chance and probability in primary-grade children. In S. P. Lajoie (Ed.), *Reflections on statistics: Learning, teaching, and assessment in grades K–12* (pp. 149–174). Mahwah, NJ: Erlbaum.

Metz, K. E. (1998b). Emergent understanding and attribution of randomness: Comparative analysis of reasoning of primary grade children and undergraduates. *Cognition and Instruction, 16*, 285–365.

Mises, R. von (1952). *Probability, statistics and truth* (J. Neyman, O. Scholl, & E. Rabinovitch, Trans.). London: William Hodge. (Original work published 1928).

Moore, D. S. (1990). Uncertainty. In L. A. Steen (Ed.), *On the shoulders of giants: New approaches to numeracy* (pp. 95–137). Washington, DC: The Mathematical Association of America.

Moritz, J. B. & Watson, J. M. (2000). Reasoning and expressing probability in students' judgments of coin tossing. In J. Bana & A. Chapman (Eds.), *Mathematics education beyond 2000: Proceedings of the 23rd annual conference of the Mathematics Education Research Group of Australasia* (Vol. 2, pp. 448–455). Perth, WA: MERGA.

Moritz, J. B., Watson, J. M., & Pereira-Mendoza, L. (1996, November). *The language of statistical understanding: An inverstigation in two countries.* Paper presented at the Joint ERA/AARE Conference, Singapore.

Mullenex, J. L. (1968). *A study of the understanding of probability concepts by selected elementary school children.* Unpublished doctoral dissertation, University of Virginia, Charlottesville.

National Council of Teachers of Mathematics. (1989). *Curriculum and evaluation standards for school mathematics.* Reston, VA: Author.

National Council of Teachers of Mathematics. (2000). *Principles and standards for school mathematics.* Reston, VA: Author.

National Council of Teachers of Mathematics (2002a). *Navigating through data analysis and probability in Prekindergarten–grade 2.* Reston, VA: Author.

National Council of Teachers of Mathematics (2002b). *Navigating through data analysis and probability in grades 3–5.* Reston, VA: Author.

National Council of Teachers of Mathematics (2003). *Navigating through data analysis and probability in grades 6–8.* Reston, VA: Author.

Nicholson, J. R., & Darnton, C. (2003). Mathematics teachers teaching statistics: What are the challenges for the classroom teacher? *In Proceedings of the 54th Session of the International Statistical Institute.* Voorburg, The Netherlands: International Statistical Institute.

Nisbet, S., Jones, G. A., Langrall, C. W., & Thornton, C. A. (2000). A dicey strategy to get your M&Ms. *Australian Primary Mathematics Classroom, 5*(3), 19–22.

Offenbach, S. I. (1965). Studies of children's probability learning behavior II: Effect of method and event frequency. *Child Development, 36*, 952–961.

Page, D. A. (1959). Probability. In P. S. Jones (Ed.), *The growth of mathematical ideas: Grades K-12: Yearbook of the National Council of Teachers of Mathematics* (pp. 229–271). Washington, DC: National Council of Teachers of Mathematics.

Paparistodemou, E., Noss, R., & Pratt, D. (2002). Developing young children's knowledge of randomness. In B. Phillips (Ed.), *Proceedings of the Sixth International Conference on the Teaching of Statistics* [CD-ROM]. Cape Town, South Africa: International Statistical Institute.

Peard, R. (1995). The effect of social background on the development of probabilistic reasoning. In A. J. Bishop (Ed.), *Regional collaboration in mathematics education* (pp. 561–570). Melbourne, VIC, Australia: Monash University.

Pereira-Mendoza, L. (2002). Would you allow your accountant to perform surgery? Implications for the education of primary teachers. In B. Phillips (Ed.), *Proceedings of the Sixth International Conference on the Teaching of Statistics* [CD-ROM), Hawthorn, VIC, Australia: International Statistical Institute.

Pfannkuch, M. (2005). Probability and statistical inference: How can teachers enable students to make the connection. In G. A. Jones (Ed.), *Exploring probability in school: Challenges for teaching and learning* (pp. 267–294). New York: Springer.

Piaget, J., & Inhelder, B. (1975). *The origin of the idea of chance in students* (L. Leake, Jr., P. Burrell, & H. D. Fischbein, Trans). New York: Norton (Original work published 1951).

Polaki, M. V. (2002a). Using instruction to identify key features of Basotho elementary students' growth in probabilistic thinking. *Mathematical Thinking and Learning, 4*, 285–314.

Polaki, M. V. (2002b). Using instruction to identify mathematical practices associated with Basotho elementary students' growth in probabilistic reasoning. *Canadian Journal for Science, Mathematics and Technology Education, 2*, 357–370.

Polaki, M. V., Lefoka, P. J., & Jones, G. A. (2000). Developing a cognitive framework for describing and predicting Basotho students' probabilistic thinking. *Boleswa Educational Research Journal, 17*, 1–21.

Pollatsek, A., Well, A. D., Konold, C., & Hardiman, P. (1987). Understanding conditional probabilities. *Organizational Behavior and Human Decision Processes, 40*, 255–269.

Pratt, D. (1998). The co-ordination of meanings for randomness. *For the Learning of Mathematics, 18*(3), 2–11.

Pratt, D. (2000). Making sense of the total of two dice. *Journal of Research in Mathematics Education, 31*, 602–625.

Pratt, D. (2005). How do teachers foster students' understanding of probability? In G. A. Jones (Ed.), *Exploring probability in school: Challenges for teaching and learning* (pp. 171–189). New York: Springer.

Pratt, D. & Noss, R. (2002). The micro-evolution of mathematical knowledge: The case of randomness. *Journal of the Learning Sciences, 11*(4), 453–488.

Ritson, R. (1998). *The development of primary school children's understanding of probability.* Unpublished thesis, Queen's University, Belfast, Ireland.

Rohmann, C. (1999). *A world of ideas: A dictionary of important theories, concepts, beliefs, and thinkers.* New York: Ballantine Books.

Rutherford, J. F. (1997). Thinking quantitatively about science. In L. A. Steen (Ed.), *Why numbers count: Quantitative literacy for tomorrow's America* (pp. 60–74). New York: The College Board.

Rutherford, J. F., & Ahlgren, A. (1990). *Science for all Americans.* New York: Oxford University Press.

Rychen, D. S., & Salganic, L. H. (Eds.) (2003). *Key competencies for a successful life and a well-functioning society.* Gottingen, Germany: Hogrefe & Huber.

Sanchez, E. S. (2002). Teachers' beliefs about usefulness of simulations with the educational software Fathom for developing probability concepts in statistics classroom. In B. Phillips (Ed.), *Proceedings of the Sixth International Conference on the Teaching of Statistics* [CD-ROM]. Hawthorn, VIC, Australia: International Statistical Institute.

Sanchez, E. S., & Hernandez, R. (2003). Variables de tareaen problemas asociados a la regal de producto en probabilidad [Task variables in product rule problems in probability]. In E. Filloy (Ed.), *Mathematica educativa, aspectos de la investigacion actual* (pp. 295–313). Mexico City: Fondo de Cultura Economica.

Scheaffer, R. L., Watkins, A. E., & Landwehr, J. M. (1998). What every high school graduate should know about statistics. In S. P. Lajoie, (Ed.), *Reflections on statistics: Learning, teaching, and assessment in grades K–12* (pp. 3–31). Mahwah, NJ: Erlbaum.

Schilling, M. F. (1990). The longest run of heads. *College Mathematics Journal, 21*(3), 196–207.

Schmidt, W. H., McKnight, C. C., Valverde, G. A., Houang, R. T., & Wiley, D. E. (1997). *Many visions, many aims: A cross-national investigation of curricular intentions in school mathematics* (Vol. 1). Dordrecht, The Netherlands: Kluwer.

Schroeder, T. L. (1988). Elementary school children's use of strategy in playing microcomputer probability games. In R. Davidson & J. Swift (Eds.), *Proceedings of the Second International Conference on the Teaching of Statistics* (pp. 51–56). Victoria, BC, Canada: University of Victoria.

Shaughnessy, J. M. (1977). Misconceptions of probability: An experiment with a small-group, activity-based, model building approach to introductory probability at the college level. *Educational Studies in Mathematics, 8*, 285–316.

Shaughnessy, J. M. (1992). Research in probability and statistics. In D. A. Grouws (Ed.), *Handbook of research on mathematics teaching and learning* (pp. 465–494). New York: Macmillan.

Shaughnessy, J. M. (2003). Research on students' understandings of probability. In J. Kilpatrick, W. G. Martin, & D. Schifter (Eds.), *A research companion to Principles and Standards for School Mathematics* (pp. 216–226). Reston, VA: National Council of Teachers of Mathematics.

Shaughnessy, J. M., & Ciancetta, M. (2002) Students' understanding of variability in a probability environment. In B. Phillips (Ed.), *Proceedings of the Sixth International Conference on the Teaching of Statistics* [CD-ROM]. Cape Town, South Africa: International Statistical Institute.

Shaughnessy, J. M., & Zawojewski, J. S. (1999). Data and chance. In E. A. Silver & P. A. Kenney (Eds.), *Results from the Seventh Mathematics Assessment of the National Assessment of Educational Progress* (pp. 235–268). Reston, VA: National Council of Teachers of Mathematics.

Shepler, J. L. (1970). Parts of a systems approach to the development of a unit in probability and statistics for the elementary school. *Journal for Research in Mathematics Education, 1*, 197–205.

Shulman, L. S. (1987). Knowledge and teaching: Foundations of the new reform. *Harvard Educational Review, 57*, 1–22.

Siegel, S., & Andrews, J. M. (1962). Magnitude of reinforcement and choice behavior in children. *Journal of Experimental Psychology, 63*, 337–341.

Silver, E., & Stein, M. K. (1996). The QUASAR project: The "revolution of the possible" in mathematics instructional reform in urban middle schools. *Urban Education, 30*, 476–521.

Simon, M. A. (1995). Reconstructing mathematics from a constructivist pedagogy. *Journal for Research in Mathematics Education, 26*, 146–149.

Simon, M. A. (2000). Research on mathematics teacher development. In A. E. Kelly & R. A. Lesh (Eds.), *Handbook of research design in mathematics and science education* (pp. 335–359). Mahwah, NJ: Erlbaum.

Skoumpourdi, C. (2004, July). *The teaching of probability theory as a new trend in Greek primary education.* Paper presented at the 10th International Congress on Mathematical Education (Topic Study Group 1). Copenhagen, Denmark.

Speiser R., & Walter, C. (1998). Two dice, two sample spaces. In L. Pereira-Mendoza, L. S. Kea, T. W. Kee, & W. Wong (Eds.), *Proceedings of the Fifth International Conference on Teaching Statistics* (Vol. 1, pp. 61–66). Voorburg, The Netherlands: International Statistical Institute.

Steen, L. (Ed.). (1990). *On the shoulders of giants: New approaches to numeracy.* Washington, DC: National Academy Press.

Steen, L. (Ed.). (1997). *Why numbers count: Quantitative literacy for tomorrow's America.* New York: The College Board.

Steen, L. A. (2001). *Mathematics and democracy: The case for quantitative literacy.* Washington, DC: Woodrow Wilson National Fellowship Foundation.

Stein, S. (2000). *Equipped for the future content standards: What adults need to know and be able to do in the 21st century.* Washington, DC: National Institute for Literacy. Retrieved October 29, 2004, from http://www.nifl.gov/lincs/collections/eff/eff_publications.html.

Steinbring, H. (1984). Mathematical concepts in didactical situations as complex systems: The case of probability. In H. Steiner & N. Balacheff (Eds.), *Theory of mathematics education (TME: ICME 5): Occasional paper 54* (pp. 56–88). Bielefeld, Germany: IDM.

Steinbring, H. (1991). The theoretical nature of probability in the classroom. In R. Kapadia & M. Borovcnik (Eds.), *Chance encounters: Probability in education* (pp. 135–168). Dordrecht, The Netherlands: Kluwer.

Stevenson, H. W., & Zigler, E. F. (1958). Probability learning in children. *Journal of Experimental Psychology, 56*, 185–192.

Stohl, H. (1999-2002). Probability Explorer [Computer software]. Software application distributed by author. Retrieved February 8, 2005, from http://www.probexplorer.com.

Stohl, H. (2005). Probability in teacher education and development. In G. A. Jones (Ed.), *Exploring probability in school: Challenges for teaching and learning* (pp. 345–366). New York: Springer.

Stohl, H., & Rider, R. (2003). Are these die fair? An analysis of students' technology based exploration. In N. Pateman, B. Dougherty & J. Zilliox (Eds.), *Proceedings of the 27th annual meeting of the International Group for the Psychology of Mathematics Education* (Vol. 1, p. 325). Honolulu: Center for Research and Development Group, University of Hawaii.

Stohl. H., & Tarr. J. E. (2002). Developing notions of inference using probability simulation tools. *Journal of Mathematical Behavior, 21*, 319–337.

Tarr, J. E. (1997). *Using research-based knowledge of students' thinking in conditional probability and independence to inform instruction.* Unpublished doctoral dissertation, Illinois State University, Normal.

Tarr, J. E. (2002). The confounding effects of "50–50 chance" in making conditional probability judgments. *Focus on Learning Problems in Mathematics, 24*, 35–53.

Tarr, J. E., & Jones, G. A. (1997). A framework for assessing middle school students' thinking in conditional probability and independence. *Mathematics Education Research Journal, 9*, 39–59.

Tarr, J. E., & Lannin, J. K. (2005). How can teachers build notions of conditional probability and independence? In G. A. Jones (Ed.), *Exploring probability in school: Challenges for teaching and learning* (pp. 215–238). New York: Springer.

Tashakkori, A., & Teddlie, C. (1998). *Mixed methodology: Combining qualitative and quantitative approaches.* Thousand Oaks, CA: Sage.

Toohey, P. G. (1995). *Adolescent perceptions of the concept of randomness.* Unpublished master's thesis, The University of Waikato, New Zealand.

Trumble, W. R. (Ed.). (2002). *The shorter Oxford English dictionary* (5th ed.).New York: Oxford University Press.

Tversky, A., & Kahneman, D. (1974). Judgment under uncertainty: Heuristics and biases. *Science, 185,* 1124–1131.

Utts, J. (2003). What educated citizens should know about statistics and probability. *The American Statistician, 57*(2), 74–79.

Varga, T. (1969a). Combinatorials and probability for the young (Part 1). *Journal for Structural Learning, 1,* 49–99.

Varga, T. (1969b). Combinatorials and probability for the young (Part 2). *Journal for Structural Learning, 1,* 139–161.

Volkova, T. (2003). *Assessing Russian children's thinking in probability.* Unpublished master's thesis, Illinois State University, Normal.

Watson, J. M. (1997a). Assessing statistical literacy through the use of media surveys. In I. Gal & J. Garfield (Eds.), *The assessment challenge in statistics education* (pp. 107–121). Amsterdam: IOS Press.

Watson, J. M. (1997b). Chance and data for luddites. *Australian Mathematics Teacher, 53*(3), 24–29.

Watson, J. M. (2001). Profiling teachers' competence and confidence to teach particular mathematics topics: The case of chance and data. *Journal of Mathematics Teacher Education, 4*(4), 305–337.

Watson, J. M. (2005). The probabilistic reasoning of middle school students. In G. A. Jones (Ed.), *Exploring probability in school: Challenges for teaching and learning* (pp. 145–168). New York: Springer.

Watson, J., & Callingham, R. (2003). Statistical literacy: A complex hierarchical construct. *Statistics Education Research Journal, 2,* 3–46.

Watson, J. M., Collis, K. F., & Moritz, J. B. (1995). Children's understanding of luck. In B. Atweh & S. Flavel (Eds.), *Proceedings of the 18th Annual Conference of the Mathematics Education Research Group of Australasia* (pp. 550–556). Darwin, NT, Australia: MERGA.

Watson, J. M., Collis, K. F., & Moritz, J. B. (1997). The development of chance measurement. *Mathematics Education Research Journal, 9,* 60–82.

Watson, J. D., & Moritz, J. B. (1998). Longitudinal development of chance measurement. *Mathematics Education Research Journal, 10,* 103–127.

Watson, J. M., & Moritz, J. B. (2000). Developing concepts of sampling. *Journal for Research in Mathematics Education 31,* 44–70.

Watson, J. D., & Moritz, J. B. (2002). School students' reasoning about conjunction and conditional events. *International Journal of Mathematics Education in Science and Techonology, 33*(1), 59–84.

Watson, J. M., & Moritz, J. B. (2003a). The development of comprehension of chance language: Evaluation and interpretation. *School Science and Mathematics, 103,* 65–80.

Watson, J. M., & Moritz, J. B. (2003b). Fairness of dice: A longitudinal study of students' beliefs and strategies for making judgments. *Journal for Research in Mathematics Education, 34,* 270–304.

Way, J. (1996). Children's strategies for comparing two types of random generators. In L. Puig & A. Gutierrez (Eds.), *Proceedings of the 20th conference of the International Group for the Psychology of Mathematics Education* (Vol. 4, pp. 419–526). Valencia, Spain: Universitat de Valencia.

Wild, C., & Pfannkuch, M. (1999). Statistical thinking in empirical enquiry. *International Statistical Review, 67,* 223–265.

Yarbrough, S. J., Daane, C. J., & Vessel, A. M. (1998, November). *An investigation of ten elementary teachers' quantitative literacy instruction as a result of participation in the Alabama Quantitative Literacy Workshop.* Paper presented at the annual meeting of the Mid-South Educational Research Association. New Orleans, LA.

Zaslavsky, T., Zaslavsky, O. & Moore, M. (2001). Language influence on prospective mathematics teachers' understanding of probabilistic concepts. *Focus on Learning Problems in Mathematics, 23,* 23–40.

Zimmermann, G. M. (2002). *Using research-based knowledge of students' thinking in simulation to inform instruction.* Unpublished doctoral dissertation, Illinois State University, Normal.

Zimmermann, G. M., & Jones, G. A. (2002). Probability simulation: What meaning does it have for high school students? *Canadian Journal of Science, Mathematics, and Technology Education, 2,* 221–237.

AUTHOR NOTE

We would like to thank the many colleagues who contributed to this chapter in special ways and made our task so much easier. We would especially like to thank James Schultz, University of Ohio, for his insightful comments on every draft, Cliff Konold, University of Massachusetts-Amherst who made numerous suggestions to an earlier draft, and Michael Shaughnessy for his ongoing collaborative support and comments. In addition, we would like to extend our appreciation to Carmen Batanero, Lyn English, Iddo Gal, Brian Greer, Michel Henry, Flavia Jolliffe, John Lannin, Swapna Mukhopadhyay, Bernard Parzysz, Maxine Pfannkuch, Mokaeane Polaki, Dave Pratt, Ernesto Sanchez, Hollylynne Stohl, James Tarr, Carol Thornton, and Jane Watson: their recent reviews of various parts of the stochastics literature contributed both directly and indirectly to our writing of this chapter.

21

RESEARCH ON STATISTICS LEARNING AND RESEARCH

J. Michael Shaughnessy

PORTLAND STATE UNIVERSITY

At the beginning of the 1990s a rather unique professional opportunity came my way when the National Council of Teachers of Mathematics (NCTM) asked me to author a chapter on research on the teaching and learning of probability and statistics for the *Handbook of Research on the Teaching and Learning of Mathematics* (Shaughnessy, 1992). At that time, research on students' understanding of probability and statistics was just beginning to blossom in the United States though research in stochastics had been conducted for several decades in other countries, principally in Europe. During the 1980s, interest in research in probability and statistics in the United States was being fueled by interactions with researchers from other countries at conferences such as the International Conference on Teaching of Statistics (ICOTS) and the International Group for the Psychology of Mathematics Education (PME). Towards the end of the 1990s, the long efforts of a core group of mathematics and statistics educators in the United States to promote an increased emphasis on statistics in school mathematics programs finally took root. Probability and statistics were included among the content standards in NCTM's groundbreaking document, *Curriculum and Evaluation Standards for School Mathematics* (1989). This was the first time that a national organization in the United States, in fact *the* national organization for mathematics teachers, placed statistics on an equal footing with number sense, algebra, geometry, and measurement as a critical foundation stone for school mathematics. Suddenly statistics became a more attractive area in which to conduct research on student learning. State curriculum leaders followed the lead of the NCTM *Curriculum Standards* and began to build statistics into their state mathematics frameworks and likewise began to assess students' growth in learning statistical concepts. Consequently, curriculum developers began to pay more than lip service to statistics. Prior to the *Standards* statistics had been a lost stepchild in mathematics curriculum frameworks, the mere frosting on any mathematics program if there was time at the end of the school year. Now statistics is here to stay as a major strand in school mathematics programs in the United States.

This book represents the second effort by NCTM to survey, analyze, and compile research in mathematics education, and I have another unique opportunity to author a chapter, this time focusing on statistics (maybe they are hoping I will get it right this time). However, this time around a synthesis and analysis of the research in data and chance is far more challenging than it was 15 years ago. Research in probability and statistics is no longer a fledgling discipline, and reviewing all of the relevant literature is no longer possible as it was for the first *Handbook*. There has been an amazing boom in research, curriculum development, and assessment in statistics education. A recent review of the research literature in data and chance carried out just in Australasia over a 4-year period from 2000 to 2003 turned up over 150 citations in statistics education (Pfannkuch & Watson, 2005). Compare that to around 150 international references that were in the 1992 *Research Handbook* chapter that covered a 30-year period of research in statistics all over the world, and the magnitude of the current task becomes apparent. Contributions to the field have come from a very wide range of investigators, including mathematics and statistics educators, statisticians, cognitive psychologists, educational psychologists, and science educators.

Satellite roundtable conferences on statistics education have been held at the last three meetings of International Congress of Mathematics Education (ICME) in Granada (1996), Hiroshima (2000), and Lund (2004), and these have continued to catalyze research in statistics education. There is also now an international group of statistics educators who have been holding semi-annual conferences on Statistics Reasoning, Thinking, and Literacy (SRTL) in Israel (1999), Australia (2001), the United States (2003), and New Zealand (2005). The field is now expanding so rapidly that keeping abreast of everything is quite impossible. However, I do not think that the growth in research in probability and statistics is well known among the general community of mathematics educators. Recently at a conference a famous and very widely read colleague in mathematics education asked me, "Are you still working in statistics? There hasn't been very much going on there recently, has there?" I was really amazed that he was unaware of what has been going on in the area, and he is probably typical of many mathematics educators.

The sheer volume of research in both probability and statistics has in fact necessitated two separate chapters in this edition of the *Handbook*, one on probability (see Jones, Langrall, & Mooney, this volume) and one on statistics. In one sense, this is rather unfortunate because there are natural connections between statistics concepts and probability concepts, for example in sampling distribution, confidence intervals, and significance tests. Furthermore, researchers must continue to clarify the learning connections between statistics and probability for teachers and students. However, reviews of research are more useful if they focus on particular issues, rather than trying to do everything. In this chapter I will focus primarily on research on students' learning and reasoning in statistics from the past 15 years or so, making connections to probability along the way. Most of the research discussed in this chapter concerns students' understanding of descriptive statistics and data analysis, as there has not yet been much research devoted to students' understanding of inferential statistics—confidence intervals, hypothesis testing, p-values, and so on.

STATISTICS IN THE SCHOOLS TODAY

When the 1989 *Standards* document was updated in NCTM's *Principles and Standards for School Mathematics* (2000) (PSSM), statistics continued to hold a prominent position in NCTM's vision for school mathematics. The two editions of the *Standards* documents (1989, 2000) have had counterparts in other countries that have proclaimed similar messages, touting the important role that statistics plays in the education of our citizens, both as learners of statistics at the school level and as consumers of statistics at the adult level. For example, *A National Statement on Mathematics for Australian Schools* (Australian Educational Council, 1991) and *Mathematics in the New Zealand Curriculum* also called for considerable statistics to be taught as part of school mathematics programs, in part to empower students to critically evaluate data and claims made about data.

National calls for increased attention to statistics have played an important role in catalyzing research in the learning of statistics. However, national calls themselves will not amount to much unless good curriculum materials for students and accompanying guides for classroom teachers are available to implement the suggestions of those calls. Statistics has been embedded in a prominent way in many of the recent curriculum development projects in the United States and Canada throughout the 1990s. These curricula, partly inspired by the development and implementation of the *Standards* documents, have provided another major impetus for research on statistics. The *Standards* documents have influenced curriculum projects such as *Mathematics in Context* (1994), *Core-Plus Mathematics* (1997), *Data Driven Mathematics* (1999), *Connected Mathematics* (1998), and *Investigations Into Number, Data and Space* (1998) just to point out a few examples. The round of curriculum materials in the 1990s systematically wove statistics throughout all their targeted grade levels and incorporated material to help meet the recommendations of *PSSM* for probability and statistics in K–12 that say that students should:

- Formulate questions that can be addressed with data and collect, organize, and display relevant data to answer them.
- Select and use appropriate statistical methods to analyze data.
- Develop and evaluate inferences and predications that are based on data.
- Understand and apply basic concepts of probability.

The *Standards* documents and the curriculum development projects of the '90s have both helped to increase research interest in how students think about data and chance, and how teachers might address statistical concepts. The mere existence of good curriculum materials for statistics does not mean that they are being uniformly and faithfully implemented everywhere throughout the United States and Canada. Our teaching force is undernourished in statistical experience, as statistics has not often been a part of many teachers' own school mathematics programs. In many schools there is a tremendous need for professional development in the area of statistics.

At the same time that many teachers need more work with statistics themselves, the number of students taking Advanced Placement (AP) statistics courses in secondary school has greatly increased in the United States. The AP exams, constructed by Educational Testing Service (ETS), provide a mechanism for secondary school students to obtain college-level credit for coursework taken in high school. Growth in the number of students who register for the AP statistics exam in the United States has been the fastest of any AP course in the history of the Advanced Placement Program. In 1997, the first time it was administered, 7,500 students took the AP statistics test. According to the chair of the AP statistics grading committee, the number of students taking the exam grew to 37,000 students in 2000, then to 50,000 students in 2002, and to as many as 65,000 in 2004 (Roxy Peck, personal communication, March, 2004). The AP statistics option has been another catalytic force in the growth of attention to statistics in K–12 schools. Advanced Placement Statistics is a very popular option in secondary schools because most students need to take some sort of statistics course in college, whatever their major. In fact, introductory statistics has recently outgrown calculus for the largest enrollment in any mathematics or statistics class at many colleges and universities.

Spreading the word on the importance of statistics for students and getting good statistics materials adopted and faithfully implemented in K–12 classrooms requires a long-term effort that must be coupled with research on student learning and assessment of student progress. In that regard, perhaps an overall snapshot of student progress in learning statistics is a good starting point for our research journey in this chapter. In the next section I provide a brief overview of students' knowledge of statistical concepts and skills as documented by the *National Assessment of Educational Progress* (NAEP) in the United States.

NATIONAL ASSESSMENT OF EDUCATIONAL PROGRESS: A SHORT OVERVIEW ON STATISTICS

Statistics items have been included in each of the NAEP frameworks since 1973. The variety and depth of items have continued to grow over the past 30 years in NAEP. The percentage of NAEP items classified as data analysis, statistics and probability has more than tripled at Grade 12 (from 6% in 1986 to 20% in 1996) and almost doubled in Grade 8 (from 8% to 15% over the same time period). In the 1996 administration, the NAEP content strand called Data Analysis, Statistics and Probability was designed to emphasize "the appropriate methods for gathering data, the visual exploration of data, various ways of representing data, and the development and evaluation of arguments based on data analysis." (National Assessment Governing Board, 1994, p. 77).

In the past, NAEP statistics items have included calculating measures of center (mean, median, mode), as well as making inferences from data, reading and interpreting graphical representations of data, predicting beyond the data given in graphs, identifying which statistic (e.g., mean or median) is more appropriate in various situations, and answering questions about samples and sampling. Information from NAEP tasks can help to identify growth in students' understanding of statistical concepts as well as potential trouble spots for students in understanding statistics. Therefore, NAEP data can be a good source for mathematics educators to formulate researchable questions about student growth and understanding. Some summary highlights of the 1996, and the 2000 and 2003 NAEP findings about students' knowledge of statistical concepts provide a backdrop for the rest of the chapter. In an analysis of the 1996 NAEP statistics items, Zawojewski & Shaughnessy (2000) noted that:

- Over half of 8th-grade students read information from tables, charts, and graphs but experienced difficulty in using the information for other substantive purposes such as drawing conclusions based on data.
- Based on performance on two NAEP items, about three-fourths of students in Grade 12 can successfully read line graphs, yet only 37 percent could read a box plot.
- Comparing 1990 results to 1996 results, there was significant growth in 8th- and 12th-grade students' performance on NAEP items that required they find the mean and median for particular data sets.
- When given a choice, 8th-grade and 12th-grade students tend to select the mean over the median, regardless of the distribution of the data.
- Given their performance on items that assessed the appropriateness of survey samples, over half of the students in Grade

8 appropriately considered the potential for bias and the number of data points.

A subsequent analysis by Tarr and Shaughnessy (in press) of the results on the statistics items from the 2000 and 2003 NAEP administrations found that:

- There was significant growth from 1996 to 2003 in 8th-grade students' performance on NAEP items that required them to determine the mean and median of given data sets. However, when asked to select the most appropriate summary statistic, most selected the mean, regardless of the distribution of the data; only about 1 in 5 students were able to justify their choice of the median.
- Although about 4 in 7 8th-grade students could identify a general formula for the mean, only about 1 in 4 students were able to determine the mean from a grouped frequency distribution, and even fewer demonstrated an understanding of the mean as representative of the entire data set.
- Less than half of the students in Grade 8 considered the potential for bias in sampling, but more than half recognized limitations associated with drawing inferences from small samples.
- Student performance on creating data displays was strong and may have reached ceiling levels in Grade 8 on bar graphs pictographs, and circle graphs.
- Performance was poor on complex items that involved interpretation or application of information in tables and graphs. There have been little or no gains from 2000 to 2003, and performance on such items may in fact be slightly eroding.

The picture from these selected NAEP highlights is a good news and bad news situation. On the good news side, students' ability to read and understand tables, charts, and graphs, and their success on procedures such as finding the mean and median, are on the rise in the United States. On the bad news side, conceptual abilities to interpret and draw conclusions from graphs, and to make decisions on which calculation is more appropriate (e.g., mean or median) languish behind their procedural counterparts. Student performance on the statistics items on the NAEP has been steadily improving since 1990, although student performance on data and chance items on the NAEP started out at such a rock-bottom level that this improvement might be somewhat misleading. On the other hand, the steady growth of success on NAEP items does suggest that statistics concepts have been slowly making their way into school mathematics programs. There is still an enormous amount of work to be done to get statistics into every student's school mathematics program throughout the K–12 school years.

PLAN OF THE REST OF THIS CHAPTER

The two principal areas where most of the recent research in statistics education has occurred are students' knowledge and reasoning about statistics and teachers' knowledge of and teaching practices in statistics. To date, far more research has been conducted in students' understanding of statistics concepts than in teachers' knowledge and practices, and so the bulk of the literature reviewed and analyzed in this chapter is on research into student learning and understanding of statistics. A considerable amount of the research on students' understanding has focused on particular concepts or big ideas in statistics, such as centers (averages), variation, comparing data sets, information from samples, students' statistical learning with technology, and students' understanding of graphs. This chapter contains sections on research on students' understanding and reasoning about each of these big ideas. A section on research and development of teachers' understanding of statistics follows the research on student understanding. The chapter will conclude with sections on recommendations for the teaching of statistics, and some recommendations for future research in statistics education.

Before launching into a discussion of the research on student understanding of statistics I wish to begin with an overview and analysis of several models and research frameworks that provide descriptive or interpretive lenses for researchers in statistics education. These models and frameworks include models of *statistical thinking*, models of *statistical literacy*, and cognitive and developmental frameworks that can help to interpret students' *statistical reasoning* as they work on statistical tasks.

Models and Frameworks

For over 3 decades research in probability and statistics has been conducted in a tradition that relies on theoretical frameworks, some built, others borrowed. Earlier frameworks discussed in the first edition of this *Handbook* (Shaughnessy, 1992) included Kahneman and Tversky's theory of heuristics such as representativeness, availability, anchoring, and conjunction in decision making under uncertainty (Kahneman

&Tversky, 1972, 1973a, 1973b; Tversky & Kahneman, 1974, 1983). Other authors (e.g., Shaughnessy, 1977; Konold, 1989; Konold, Pollatsek, Well, Lohmeier, & Lipson, 1993) have applied or extended the work of Kahneman and Tversky. Konold (1989) accounted for types of student reasoning on tasks that did not fall neatly into Kahneman and Tversky's categories of heuristic reasoning. A model for reasoning under uncertainty has turned out to be much more complicated than some of the initial models suggested, as pointed out in several recent interpretive reviews of research in statistics (Konold & Higgens, 2003) and probability (Shaughnessy, 2003a).

Much of the earlier research in probability and statistics was concerned with misconceptions that students had about probability and statistics, a line of research that has persisted even until fairly recently (Fischbein & Schnarck, 1997). However, researchers have begun to pay more careful attention to the details of the development of students' understanding of and thinking about statistics. More mature representations of student thinking have arisen over the past decade. Much of this growth in new models and frameworks has come about from careful documentation and analysis of student thinking by researchers employing a qualitative research methodology while students work on statistical tasks that can elicit an array of responses, from rather naive to somewhat sophisticated responses. As a result, most researchers in statistics (and probability) now hunt for a spectrum of student thinking in their research work, rather than taking an approach that students either do or do not "have it." This alteration of perspectives on student thinking from a misconceptions viewpoint to more of a transitional conceptions position is partly due to an increased attention to and acceptance of a constructivist epistemology that places students at the active center of their learning. It is also partly due to the types of models and frameworks themselves that have been employed in response to a constructivist theory of knowledge.

In reviewing the literature for this second round of the *Handbook*, a new crop of frameworks and models for research in statistics is surfacing. A discussion of some of these models and frameworks is in order early on in this chapter, as they will help provide lenses for discussion and comparison of research results.

STATISTICAL THINKING, STATISTICAL LITERACY, AND STATISTICAL REASONING

The literature on judgment and decision making under uncertainty that was reviewed and analyzed in the probability and statistics chapter in the first edition of this *Handbook* (Shaughnessy, 1992) included a distinction among *normative*, *prescriptive*, and *descriptive* models that may help to distinguish among the current constructs of statistical thinking, statistical literacy, and statistical reasoning. Garfield (2002) summarized a number of general perspectives on what she called statistical reasoning, including correct and incorrect ways that students reason about statistics, and ways of assessing statistical reasoning. In this chapter I take a perspective that attempts to distinguish between statistical thinking and statistical reasoning, as well as to discuss statistical literacy.

Models of *statistical thinking* help both researchers and teachers to attend to the important concepts and processes in the teaching and learning of statistics. These models reflect what we want learners, consumers, and producers of statistics to know. Thus, models of statistical thinking are primarily *normative* models of what statisticians feel are the important concepts and processes of their discipline.

Models of *statistical literacy* help to identify critical statistical survival skills for both school students and adults. Students are primarily learners of statistics, but also they can be consumers of statistics in making decisions on what to buy, or possibly even producers of statistics if they are working on a research project themselves. Adults are often in job situations in which they are producers of statistics. Models of statistical literacy often have a *prescriptive* tone, suggesting what students and life-long learning adults need to *do* in order to be well informed, or to make good decisions, or to take advantage of the data that are available to them. Statistical literacy may also include recommendations for the development of students' and adults' critical thinking skills, so that claims made with data can be questioned and analyzed.

Finally, cognitive and developmental research frameworks can provide interpretive lenses to help us identify and track students' and adults' *statistical reasoning* and their conceptual development. Models of statistical reasoning are primarily *descriptive* models that help clarify how people are thinking about statistics, what they seem to know and understand, and where they have difficulty. The descriptions of student thinking obtained from models of statistical reasoning can also point out opportunities for scaffolding statistical ideas in the teaching of statistics.

Each of these three important realms, statistical thinking, statistical literacy, and statistical reasoning, while overlapping, are potential focal points for research, teaching, and curriculum development in statistics. They have been the motivating force behind the creation of an ongoing international working group in statistics education that has spawned the SRTL conferences and produced a book on statistical thinking, literacy, and reasoning (Garfield & Ben Zvi, 2004).

A Model of Statistical Thinking

How do statisticians generate statistical questions? How do they decide upon a design for a study, which data to collect, and what to take into account when analyzing the data? Are particular ways of thinking germane to statistics? Wild and Pfannkuch (1999) presented a four-dimensional model of statistical thinking consisting of two dimensions they called cycles of thinking activity—an Interrogative cycle and an Investigative cycle—and two more dimensions called Types of statistical thinking and Dispositions, respectively (See Figure 21.1).

According to Wild and Pfannkuch when a statistician works on a statistical problem, parts of these four dimensions are continually, and simultaneously, in use (Pfannkuch & Wild, 2000, 2004). They claim that

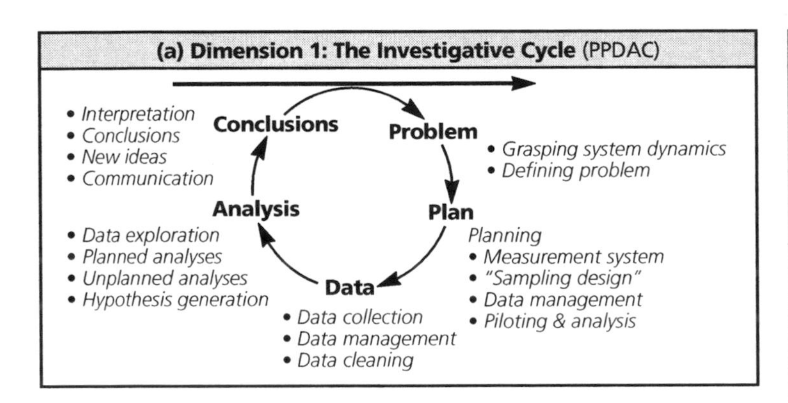

(b) Dimension 2: Types of Thinking

General Types

- **Strategic**
  - — planning, anticipating problems
  - — awareness of practical constraints
- **Seeking Explanations**
- **Modelling**
  - — construction followed by use
- **Applying Techniques**
  - — following precedents
  - — recognition and use of archetypes
  - — use of problem solving tools

Types Fundamental to *Statistical* Thinking (Foundations)

- **Recognition of need for data**
- **Transnumeration**
  (Changing representations to engender understanding)
  - — capturing "measures" from real system
  - — changing data representations
  - — communicating messages in data
- **Consideration of variation**
  - — noticing and acknowledging
  - — measuring and modelling for the purposes of prediction, explanation, or control
  - — explaining and dealing with
  - — investigative strategies
- **Reasoning with statistical models**
  - — aggregate-based reasoning
- **Integrating the statistical and contextual**
  - — information, knowledge, conceptions

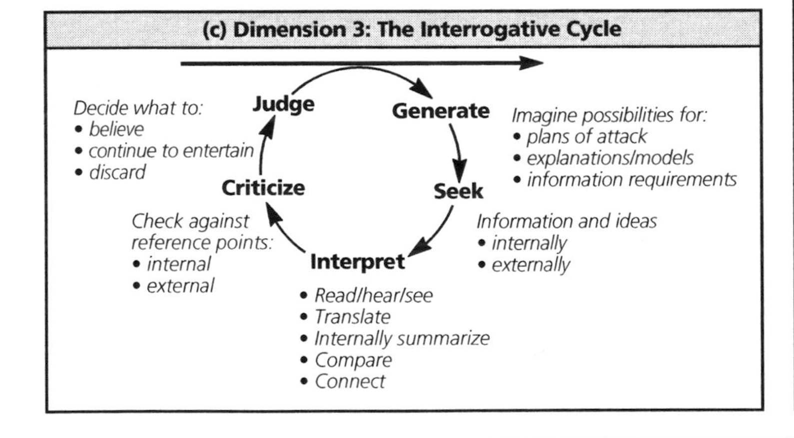

(d) Dimension 4: Dispositions

- **Scepticism**
- **Imagination**
- **Curiosity and awareness**
  - — observant, noticing
- **Openness**
  - — to ideas that challenge preconceptions
- **A propensity to seek deeper meaning**
- **Being Logical**
- **Engagement**
- **Perseverance**

Figure 21.1 A 4-dimensional framework for statistical thinking in empirical enquiry (From "Statistical Thinking in Empirical Enquiry," by C. J. Wild and M. Pfannkuch, 1999, *International Statistical Review*, 67, p. 226. Copyright 1999 by *International Statistical Institute*. Reprinted with permission)

their model of statistical thinking can also be used to analyze student thinking, not just a statistician's thinking, and that such analyses can inform both teaching and curriculum development in statistics. The initial building blocks for their model came from three data sources: students' work on statistical tasks; interviews with student team leaders of statistical projects; and, perhaps most importantly, interviews with six statisticians working in various settings (business, marketing, medicine, etc.). In developing their model Wild and Pfannkuch tapped and expanded upon the writings of previous statistics educators, such as Moore (1990, 1997), who discussed the omnipresence of variability, the need for data and data production strategies, and the measuring and modeling of variability. Statistics educators are likely to resonate with Wild and Pfannkuch's model, whereas mathematics educators who have not delved much into the area of statistics may encounter some thought-provoking and unfamiliar issues. Cobb and Moore (1997) have written eloquently and persuasively about the differences between the disciplines of statistics and mathematics. Some of the differences between solving mathematical problems and solving statistical problems surface in the Wild and Pfannkuch model.

The Investigative cycle, tabbed PPDAC (Problem, Plan, Data, Analysis, Conclusion), is really the mantra of all statistical investigations and is reminiscent of Polya's (1945) seemingly timeless four-step model of mathematical problem solving (Understand, Plan, Execute, Review). However, there are some crucial differences between Polya's model and PPDAC, chiefly lying in the two Ps and the D. Statistical problems are often ill posed at first, as they arise out of messy contexts. Often there is a need for several iterative cycles just between the Problem ↔ Plan phases of the Investigative cycle, in order to adequately formulate a statistical problem from a murky prestatistical situation. The D, for data in the PPDAC cycle of a statistics problem, is very different from the type of data that one might encounter in a mathematical problem. Data for statistics problems are accompanied by excess baggage such as bias, uncontrollable sources of variation, context issues, and so on. Most of the current statistics education in the United States places a heavy emphasis on the DAC parts of the Investigative cycle, but precious little time is devoted in classrooms to the PP parts. If students are given only prepackaged statistics problems, in which the tough decisions of problem formulation, design, and data production have already been made for them, they will encounter an impoverished, three-phase, investigative cycle and will be ill equipped to deal with statistics problems in their early formulation stages.

The dimension on Types of thinking is two-pronged in the Wild and Pfannkuch model. Although some thinking is inherently of a statistical nature, there are also more general types of strategic thinking that remind one of Polya's problem-solving heuristics. The inherently statistical types of thinking highlighted by Wild and Pfannkuch include (a) the need for *data*; (b) attention to *variation*; (c) the use of historical, statistical, and probabilistic *models* (such as those used in inference); (d) the critical importance of *context* knowledge; and (e) *transnumeration,* a word coined by Wild and Pfannkuch. Except for the modeling tools aspect, Wild and Pfannkuch's list is quite different than a typical list of the hallmarks of mathematical thinking, which include processes such as looking for patterns, abstracting, generalizing, specializing, and generating and applying algorithms. In particular, the invented word *transnumeration* needs some further explication.

Wild and Pfannkuch created the word transnumeration because sometimes in the data organization and analysis phase, a particular representation of the data can reveal entirely new or different features that were previously hidden. These hidden features may have a major impact on how one interprets the data in that particular context. Wild and Pfannkuch needed a word that went beyond a mere transformation or re-representation of the data, so as to identify instances in which striking features of a context are suddenly revealed. An analogy might be the sudden insight *Eureka*! experience that mathematical problem solvers often speak about.

An example of transnumeration occurred in the reanalysis of the O-ring data for the Space Shuttle after the *Challenger* incident (Dalal, Fowlkes, & Hoadley, 1989). It was hypothesized that there might be a relationship between the air temperature at launch time and O-ring failures on the shuttle. The original data set that was being analyzed prior to the launch of the *Challenger* included only information on the launch temperature and number of O-ring failures per launch. No significant trends could be found. However, when data on launch temperatures and the number of O-ring *successes* was included, the augmented data set revealed a cut-off temperature above which O-ring failure had never occurred on the shuttle. The fact that it was probably too cold to launch the *Challenger* was not clearly evident until after this *transnumeration* process.

Although transnumeration is more likely to be encountered by statisticians as they investigate a statistical situation, it can also occur during teaching episodes with students. For example, while watching students work with a data set on the wait time between blasts for the Old Faithful Geyser, Shaughnessy and Pfannkuch (2002) found that when students moved from representing the data in box plots or histograms to creating plots over time, a whole new vista on the data was opened up to them. (The geyser data are discussed in more detail later in this chapter. See for example Figures 2 and 3). Plots over time revealed a short-long cycle of wait times, whereas box plots and other graphical representations of the geyser data masked the nature of the variability. A teaching and learning culture that encourages transnumeration might evolve if teachers and curriculum developers heed the advice of some researchers who have recommended that students have more opportunities to construct their own representations of data rather than working primarily with canned tables and graphs (Cobb, 1999; Lehrer and Romberg, 1996). In this regard, the *Tinkerplots* software (Key Curriculum, 2005a) allows students considerable flexibility in constructing their own visual representations of data and in generating their own hypotheses and conjectures, and thus it encourages the transnumeration process.

The Interrogative cycle in Wild and Pfannkuch's model involves explicit metacognitive activity. There is an intense reflective component in the Interrogative cycle, as statistical problem solvers must always deal with beliefs, emotions, and the danger of their own, or their client's, narrow perspectives. The Interrogative cycle echoes the claim by Shaughnessy and Pfannkuch (2002) that statistical problem solving needs to be done by "data detectives" who continually question and reflect upon the processes of data production and data analysis.

The Dispositions dimension of Wild and Pfannkuch's model has much in common with not only mathematical problem solving, but also problem solving in any arena. All problem solvers need to be curious, be aware, have imagination, be skeptical, be open to alternative interpretations, and seek deeper meaning, as in the look back stage of Polya's model.

Models Focused on Statistical Literacy

"Statistical thinking will one day be as necessary for efficient citizenship as the ability to read and write." This quote from H.G. Wells appeared in Darrell Huff's famous little book *How to Lie with Statistics* (1954). Since then numerous authors have addressed the importance of statistics for an educated and competent citizenry. Consider: "The need for statistical thinking in social decision-making is exemplified every day in the news media" (Watson, 1997, p.107); or, to be statistically literate, one must have ". . . the ability to interpret, critically evaluate, and express one's own opinions about statistical information and data-based messages" (Gal, 2003, p.16). There is a clear, overarching importance for students and adults alike to be able to critically read and evaluate information in tables, graphs, and media reports, and to adopt a healthy questioning attitude towards what is presented by sellers and buyers, by scientists and by the government, by politicians and by the news media. Perhaps someday quoting numbers without an adequate basis in fact will be outlawed. Meanwhile everyone is susceptible to those who would stretch the truth, or only tell that part of a data story that suits their purposes. Better that we pay heed to quotes such as those above, than that we fall prey to statements such as this one made by a mid-western congressman several years ago while on the campaign trail: "We will legislate that all children in our state schools will be above average." Although a thorough trek through the literature on statistical literacy is well beyond the scope or intention of this chapter, work in statistical literacy has provided several frameworks that are useful lenses to consider as we examine research in statistics education.

Among the abilities subsumed under statistical literacy are Document Literacy, Prose Literacy, and Quantitative Literacy that comprise the three facets of adult literacy identified by Kirsch, Jungeblut, and Mosenthal (1998). Kirsch et al. presented a five-category framework—*locating, cycling, integrating, generating, and making inferences*—that adults use when they read information from tables and graphs. Mosenthal and Kirsch (1998) illustrated the power of using this framework for understanding the types of skills that adults need in order to read documents by providing a measure of document complexity. This framework for analyzing the graph- and table-reading skills of adults has been used in publications of several large-scale surveys, such as The National Adult Literacy Survey (NALS) and the International Adult Literacy Survey (IALS). The categories in Kirsch et al.'s adult literacy framework are somewhat similar to Curcio's (1989) three levels of reading graphs, "Reading the Graph," "Reading Within the Graph," and "Reading Beyond the Graph."

Reading the Graph corresponds to *locating* in the Kirsch et al. model. *Integrating and generating* are suggestive of Curcio's Reading Within the Graph, and Reading Beyond the Graph includes such activity as *making inferences* in the Kirsch model. The resonance between these two frameworks for graph and table literacy, Kirsch et al.'s and Curcio's, is indicative of what seems to be occurring more and more in research in statistics education: similar statistical reasoning phenomena have been identified independently by separate researchers who are pursuing related but not necessarily identical questions. Although the language is different, the types of categories that Curcio and Kirsch et al. developed point to a similar complexity on a scale of graphical literacy. When interpretive frameworks align like this, they cross validate one another in powerful ways.

Gal has been one of the champions in researching and promoting statistical literacy, mainly as part of his interest in general adult literacy (Gal, 2002, 2004). In an article in the *International Statistical Review*[1] (Gal, 2002), Gal claimed that statistically literate behavior by adults depends on their ability to access five knowledge bases: *general literacy* knowledge (like the previously mentioned document, prose and quantitative types of literacy), *statistical* knowledge, *mathematical* knowledge, knowledge of the *context*, and knowledge of how to be *critical* and question claims. Gal also examined how these five knowledge bases can interact with a person's dispositions, beliefs, and attitudes towards data and statistics in general. There is some overlap between Gal's model of statistical literacy and Wild and Pfannkuch's model of statistical thinking discussed above, although they are focused on different constructs: what adults need to do to be statistically literate versus what statisticians do on their job.

In another study Gal (2003) examined six web sites with large databases to analyze the types of knowledge and skills needed to access, process and manipulate the information on those sites. The web sites Gal examined contained information about national statistics, health issues, international education and international economic development. Gal noted that there were press releases, reports, executive summaries, and aggregate data sets available on each of these web sites but they were often in a form primarily intended for policy makers. A considerable amount of context and background knowledge was necessary for someone from the general public to wade through these websites and to make sense of the information that was provided. Gal concluded that the abilities to access, define, locate, extract, and filter information were additional critical skills needed for statistical literacy.

Clearly the type of statistical literacy that Gal described is a different kind of statistical literacy than just reading and evaluating data and graphs. There are many levels and contexts for literacy in Gal's work. He implied that it is important for adults, particularly in their roles as consumers and voters, to become "web-data" literate. However, a person does not just automatically step in and start interpreting and critiquing data and document information on the web. It is challenging to unpack data that appear in reduced formats on the web, especially when the data come from contexts that are unfamiliar. Gal's reflections on adult literacy should give both researchers and curriculum developers some pause as they think about what pre-cursor skills are necessary for adults to develop this type of web literacy. This type of literacy also should start with students when they are in school. Perhaps students need to begin with digestible chunks from large databases that will help them scaffold from a basic level of statistical literacy (say, reading and interpreting tables and graphs) to the type of literacy that Gal is suggesting. In this regard, the Watson (1997) framework for statistical literacy may provide a roadmap to develop students' statistical literacy skills.

While investigating students' developmental progression in statistical literacy, Watson (1997) identified three Tiers of literacy that arose as students worked on statistical tasks involving the media, newspapers, magazines, news reports, and so on. Tier 1 is an understanding of basic statistical terminology; Tier 2 involves considering and embedding statistical terms within a real context; and Tier 3 encompasses a critical reasoning component. People at Tier 3 are able to question statistical claims and critique media items that involve statistics using their understanding of the basic statistical concepts and their understanding of the context (Tiers 1 and 2).

Watson and her colleagues began to use these Tiers in their research studies to help analyze student thinking on statistical tasks (Watson & Moritz, 1997a, 2000a, 2000b). Much of Watson and her colleagues' subsequent work on statistical literacy arose from the planning, execution, and multiple analyses of the results of a large longitudinal study of the development of school students' knowledge of probability and statistics. About a thousand students in Australia were initially tested on a large set of survey items on statistics and probability, and subsequently many of these students were retested 2 years later and some even 4 years later. Students' understanding of such concepts as means and middles, chance, graphs, variability, sampling, comparing data sets, and their thinking on items from the media involving data and graphs were examined. As a result, over the past decade, Watson has been one of the most active and prolific researchers in statistics education in the world. Her work on students' understandings of particular statistical concepts will be addressed throughout this chapter.

In a reanalysis of data from survey tasks given to students in Grades 3 to 9 over a 7-year period (1993–2000), Watson and Callingham (2003) used Rasch model techniques and combined results on statistical tasks from

[1] *International Statistical Review* is a publication of the International Statistics Institute (ISI).

multiple student surveys from over 3000 students. They found quantitative support for the identification of six levels of development along a uni-dimensional statistical literacy scale, ranging from idiosyncratic to critical mathematical. These six levels are closely associated with Watson's (1997) three Tiers of statistical literacy that initially arose through qualitative analyses of students' responses to the tasks. Watson has thus provided substantial documentation in multiple research studies for the developmental progression of a complex construct called statistical literacy, as well as some of its components (Watson, 1997; Watson & Callingham, 2003; Watson, Kelly, Callingham, & Shaughnessy, 2003). In addition to the research contributions Watson has made in the area of statistical literacy, her three Tier framework—statistical terms, considering terms in context, and justifying statistical claims—provides a useful introductory framework for scaffolding instruction in statistical literacy.

Both Gal and Watson have made it clear that any model of statistical literacy *must* accord a major role to context, as well as the use of statistical terms, tools, and techniques. Furthermore, they both have noted the importance of being able to communicate reactions to statistical information and to critique it (Watson 1997; Gal, 2002). According to Watson and Moritz (1997a) "Judging statistical claims from the media is fundamental to being statistically literate" (p. 129). The research into statistical literacy has unveiled a very deep construct involving a myriad of types of skills and cognitive processes. It is therefore important for researchers to be quite clear about what they mean by statistical literacy when they use the term, as it has many levels of meaning.

Models That Capture Statistical Reasoning

Over the last 15 years a number of researchers have conducted studies that focus on students' reasoning about particular statistics concepts or processes, such as students' notions of average, variation, sampling, comparing data sets, graph sense, and data representations. In most of these studies an explicit, or implicit, conceptual analysis occurs. Researchers attempt to distill and categorize the ways students reason with and think about the concepts, sometimes resulting in the identification of levels of reasoning. In this section I'll first discuss a model that has often been used to identify and assess levels of student reasoning in mathematics and statistics and then point to several emerging models for student reasoning.

The SOLO Model

The SOLO Model is a neo-Piagetian model created to analyze the complexity of student responses to tasks. SOLO, which stands for Structure of Observed Learning Outcomes, was first developed by Biggs and Collis (1982) as a general model for evaluating learning in any context or environment. It has been heavily used in the past decade particularly by Australian researchers, and often to help code and analyze collections of responses to tasks that arise in clinical interviews. I discuss SOLO here briefly because it has become a prevalent analytical tool in some research circles in statistics education.

SOLO posits five modes of reasoning: sensori motor, ikonic (images), concrete symbolic, formal, and post formal, adding "post formal" to Piaget's original four modes. SOLO also postulates U-M-R (uni-structural, multi-structural, relational) cycles within each mode. Within a given mode, for a particular task, there may be several such U-M-R cycles. These cycles represent increasing orders of complexity while a person functions within a particular mode. Uni-structural responses suggest attention to only one relevant aspect of a task, whereas multi-structural responses involve several disjoint but relevant aspects, and relational responses suggest an understanding that integrates several aspects of a task within a mode. Analyzing student verbal responses is a tricky business, as anyone who has done it is aware. In studies that employ the SOLO methodology, boundaries between the U-M-R levels are often blurred, but that does not necessarily prevent the model from assisting a researcher in identifying a spectrum of complexity in student responses.

A good example of the potential advantages of the SOLO framework can be found in Watson, Collis, Callingham, and Moritz (1995). They introduced the SOLO model and then applied it at the concrete symbolic mode to analyze student responses to a sorting task. Students were given a set of data cards containing information about a set of people, including their eating habits, their TV watching habits, and their weight. The authors originally hypothesized that there would be two U-M-R cycles within the concrete symbolic mode on the data cards task given to students in Grades 6 and 9. Students were first interviewed individually while working on the task, and then groups of students worked together with the data cards and presented a report. In their analysis, the authors outlined a U-M-R cycle with the data cards as follows:

U—Students form images of individual people and invent stories about what the people must be like. For example, they might describe an individual person, who watches a lot of TV, eats a lot of fast food, and seems of higher weight, but no real sorting occurs.

M—Students begin to group cards by aspects, one at a time—perhaps all the high and low TV watchers are grouped in separate piles.

R—Students are sorting by multiple aspects, and making conjectures. For example, people who were high in both TV watching and fast food consumption might be grouped, and conjectures might be made about their weight.

The second cycle that Watson et al. (1995) hypothesized was not found in individual responses but was found when groups of students worked on the card sorting task. This second cycle involved graphical representations of the data (U2, M2), and conjecturing and defending conjectures about the graphs (R2). Groups of students made both histograms and scatterplots of the card data.

SOLO may be a useful research tool for statistics educators, as its framework is designed to assess responses to open-ended complex tasks that elicit a hierarchy of student reasoning. The SOLO model also helps researchers to identify partial success on tasks, so that a spectrum of student reasoning can be observed, rather than an all-or- nothing, right-or-wrong approach. For example, in an investigation of elementary school students' reasoning about data sets, Jones, Thornton, Langrall, Mooney, Perry and Putt (2000) modeled a spectrum of student reasoning using an approach quite similar to SOLO to describe four stages of reasoning: idiosyncratic, transitional, quantitative, and analytical. This four-stage model was subsequently employed by Mooney (2002) to characterize middle school students' statistical reasoning.

Some Emerging Models

The recent development of cognitive and developmental models to interpret student reasoning in statistics is a healthy sign of growing research maturity in the field. A detailed review and analysis of the use of models in statistics education research can be found in Jones, Langrall, Mooney, and Thornton (2004). Among the models they discussed are a series of four-stage, SOLO-like models that they have developed for interpreting and analyzing primary and middle school students' responses to statistics and probability tasks (Jones, Langrall, Thornton, & Mogill, 1999; Jones et al., 2000; Mooney, 2002). In the context of reading and interpreting data in graphs, Jones et al. (2000) superimposed their four stage model of student reasoning (idiosyncratic, transitional, quantitative, and analytical) on Curcio's (1989) three stages of graph sense: reading the data, reading between the data, reading beyond the data. They provided rich descriptions of student reasoning within each cell of the resulting 3 × 4 (Jones × Curcio) matrix.

In addition to the models that have emerged from SOLO theory (e.g., Jones et al., 1999, 2000), several groups of researchers, initially working quite independently, have identified and begun to use similar terms to describe types of student thinking. For example, Saldanha and Thompson (2003) referred to "additive" and "multiplicative" thinking when they describe students' work with sampling distributions. Shaughnessy, Ciancetta, & Canada (2004a) identified three levels of student reasoning about variability in a repeated samples environment, "additive, proportional, and distributional." Watson (2002) referred to "additive" and "distributional" reasoning when students are comparing data sets, and Cobb (1999) to "additive" and "multiplicative" thinking when he discussed students' thinking while comparing data sets. The "additive/multiplicative" terminology has been borrowed from other areas of research in mathematics education (Vergnaud, 1983; Doerr, 2000; Thompson & Saldanha, 2003). The terms *proportional* and *distributional* are particularly important for statistics and can have specialized meanings. Examples of the importance of proportional reasoning for handling statistical tasks are presented in Watson and Shaughnessy (2004). Ways of characterizing distributional reasoning are being developed by several groups of researchers, for example, Bakker and Gravemeijer (2004), Shaughnessy et al. (2004a), Shaughnessy, Ciancetta, Best, & Noll (2005). Further discussion of elements of these emerging models is provided in other sections of this chapter.

In this section I have presented some models of statistical thinking, statistical literacy, and statistical reasoning—interpretive and descriptive frameworks that researchers have used when analyzing student responses to statistical tasks. The spectrum of the models and frameworks discussed includes both the particular and the general, from Wild and Pfannkuch's model of statistical thinking that is specifically tailored to statistics, to the SOLO model that is a general framework for analyzing student responses in any content area or discipline. In the next sections I discuss some of the research that has focused on students' understanding of particular statistical concepts and processes, such as centers and average, variability, information from samples, comparison of data sets, and graph sense.

RESEARCH ON STUDENTS' UNDERSTANDINGS OF SOME STATISTICAL CONCEPTS

The research on students' conceptual understanding of statistics has been conducted with a wide range of students from primary, to middle school, to secondary, to tertiary. In this regard, there has been a change from the research reported in the first *Handbook of Research in Mathematics Education* (Shaughnessy, 1992) in two ways. First, much of the research reported in the first *Handbook* was conducted with college level students. Second, the stochastics research at the time of the first *Handbook* had concentrated more on probability concepts, or probability distributions, than it did on statistical concepts. This time around a majority of the research in stochastics has been conducted on students' understanding of specific statistics concepts.

Perhaps the overarching goal of statistics education is to enable students (of any age) to read, analyze, critique, and make inferences from distributions of data. The concept of a distribution in statistics is very complicated, and the word is used in different ways. Statisticians talk about *distributions of data,* but they also talk about *sampling distributions* and *probability distributions.* Distributions of data sometimes have an underlying probability distribution, such as the normal distribution or the binomial distribution, or, there may be no apparent normative probability distribution for a data set. The word *sampling distribution* often refers to a finite frequency distribution of repeated observations of some statistic—such as a distribution of means, or sample proportions, or standard deviations—that have been calculated for samples drawn from some population. On the other hand, the word *sampling distribution* may refer to the "distribution of all possible such statistics" for a given population, an infinite, theoretical construct. A more detailed discussion of these various uses of the word *distribution* can be found in Shaughnessy and Chance (2005).

For the purposes of this chapter, an understanding of students' reasoning about distributions is the (oft unstated) goal behind much of the research that has been conducted on students' understandings of particular statistical concepts. The notion of distribution includes concepts like center, shape, and spread. When researchers investigate student thinking about means and middles or about variation, or when they investigate students' thinking while comparing data sets or making decisions or inferences from graphs, they are researching students' understandings of aspects of distributions of data.

Bakker and Gravemeijer (2004) discussed examples of both upwards (from particular data points to entire distributions of data) and downwards (from entire distributions back to particular data points) reasoning by students as they learn to reason about distributions of data. Shaughnessy, Ciancetta, Best, and Noll (2005) presented examples of students' thinking about centers, spread, and shapes as they reasoned about distributions of data.

In the next sections of this chapter, as the research lens focuses on students' understanding of various statistical concepts such as centers, or variation, or information from samples, or on students' graph sense, keep in mind that this division into subsections is one of convenience. The aspects of a distribution—center, shape, and spread—are quite interrelated. Furthermore, distributions are often represented in graphical form so that reasoning about distributions is also confounded with reasoning about the graphs themselves. Though all these concepts and constructs are connected within the field of statistics, research that focuses on a particular concept in statistics can sometimes reveal aspects of student thinking that help to inform the teaching of statistics.

Research on Students' Understanding of Average

Concepts of average—middles and means in particular—are very powerful in statistics because means and other measures of center are used to help summarize information about an entire data set. Furthermore, if the data set is a sample that has been appropriately drawn from a parent population, the sample should mirror aspects of the parent population, and the sample mean should provide an estimate for the mean of the parent population from which the sample was drawn. On a more basic level the mean of a data set or a sample is representative, or in some sense typical, of that sample or data set. "Typical" values of different data sets can provide an efficient, albeit at times potentially misleading, mechanism to compare and contrast those whole data sets. Indeed the mean is a very powerful idea in statistics, and it is fundamental for students' understanding of summary statistics and statistical tests. Thus, there is plenty of motivation for researchers to study students' conceptions of average. It is one of the most important concepts in all of the mathematical sciences. And yet, students' school experiences with concepts of average are often reduced to a computational shell.

What do students do if they are given an opportunity to think intuitively or conceptually about averages? What notions of average, if any, do they employ when making decisions? Are students' notions of average the common statistical notions, such as mean or middles, or do they have other conceptions of what *average* means?

Early Work on Average

Much of the early research on students' understanding of averages was done with tertiary students (e.g., Pollatsek, Lima, & Well, 1981; Mevarech, 1983;

Pollatsek, Konold, Well, & Lima, 1984). Pollatsek et al. (1981) found that when college students were given the population mean (500) for SAT scores and a rather extreme data value from a sample of SAT scores, the students did not take the extreme value into account or adjust their prediction for a sample mean away from 500. Rather, they just predicted that the mean of the sample would also be 500. Students often believe that the mean is always their best bet for any prediction. Sometimes they even believe the mean is the most likely result to occur in a sample, even if the mean itself is not a possible data point. The Pollatsek studies also found that when tertiary students were given means for two unequal-sized samples and asked to find the mean of the combined sample, that the students tended to weight the samples equally and find the midpoint of the two sample means (average of averages—a common mistake by students). Mevarech (1983) referred to this as the "closure misconception," as she hypothesized that students have a "group structure" in the back of their mind when operating with means or variances in statistics. These early studies on students' understanding of the mean indicated that many tertiary students have at best a very procedural understanding of the mean, as something to be calculated.

Strauss and Bichler (1988) conducted one of the first studies reported on younger students' (Age 8–14) conceptions of average. They asked students a series of structured questions in a one-on-one setting to probe students' understanding of "statistical" or "abstract" properties of average, as well as whether students realized that an average was representative of a set of values. Strauss and Bichler concentrated on computational and measurement properties of the mean, rather than on conceptual properties. For example, they tested to see if students realized that the mean had to be located between the extreme values in a data set. They tested to see if students were aware that the mean is influenced by particular values in a data set. They tested to see if students realized that the average itself did not have to be one of the values in the set, or if students knew that the sum of deviations of data points from the average was zero. They also tested to see if students realized that the mean is in some sense closest to all the values in a data set. (Formally, this would mean that the mean is the value that minimizes the sum of the squared deviations from data points, a fairly advanced concept for students Ages 8 to 14).

In their analysis Strauss and Bichler identified two very different levels of difficulty in their tasks. On the one hand, students were quite aware that the mean was between extremes, and that particular data values can influence the mean. However, more sophisticated measurement ideas like minimizing deviations, or that a zero data value must also be included and accounted for when calculating the mean, proved extremely difficult for their students. The process of minimizing deviations from the mean normally surfaces in regression analysis in high school or college, and it can be challenging for students at that level, so it is no wonder these younger students had trouble with it. Strauss and Bichler found that children do not think about the concept of the mean in the same ways that statistically mature adults do.

Conceptual Models of Students' Thinking About Averages

Mokros and Russell (1995) conducted one of the first studies that investigated young students' conceptual understanding of averages (see also Russell & Mokros, 1996). They interviewed Grade 4, 6, and 8 students in what they called "messy data" situations, using contexts like allowance money and food prices that were familiar to students. The tasks went beyond straightforward algorithmic computations and tried to elicit students' own developing constructs of average. All of their students had been taught the procedure for finding the arithmetic average, so they had some familiarity with computing means.

The types of tasks Mokros and Russell used tended to ask students to work backwards from a mean to possibilities for a data set that could have that mean. For example, in the Potato Chips problem, students were told that the mean cost of a bag of potato chips was $1.35, and then they were asked to construct a collection of bag prices that had that mean of $1.35. The Allowance problem was similar except that students were given several existing data points along with the mean, were asked to create the rest of the collection of data points that would have that mean, *and* were told that they could not use the mean itself as a data point. Mokros and Russell were searching for students' own preferred strategies when dealing with averages. Their analysis of students' thinking resulted in the identification of five different mental constructs that students had of average: average as *mode*, average as *algorithm*, average as *reasonable*, average as *midpoint*, and average as *point of balance*.

Mokros and Russell found that students who focused on *modes*, or "mosts," in data sets had difficulty working backwards from the mean to construct a data set if they were not allowed to use the actual average value itself as a data value. They concluded that modal thinking students do not see the whole data set, the distribution, as an entity in itself. They see only individual data values. About the same time as the Mokros and Russell study, research by Cai (1995) found that although most students could calculate a mean when given all the data, they had great difficulty working backwards, filling in missing values when given the mean.

Mokros and Russell also discovered that students who had a purely *algorithmic* conception of average were unable to make connections from their computational procedures back to the actual context. For example, in the Potato Chips problem, some students multiplied $1.35 by 9, and then just divided by 9 again, thus generating a data set that had $1.35 for an average, but every single value in the data set was also $1.35. For such students average is something that you *do* with numbers, they have no rich conceptual understanding of average. On the basis of responses gathered during student interviews Mokros and Russell suggested that students who preferred a rule or algorithm for averages may actually have had their own intuitive thinking about average as "typical" interfered with during their schooling.

Students who thought of average as *reasonable* tended to refer to information from their own lives. Perhaps they thought of average as a mathematically reasonable, but not necessarily precise, approximation for a set of numbers. These students thought that no precise answer existed for problems like the Elevator problem, which asked "If there are 6 women who average 120 pounds, and 2 men who average 150 pounds in an elevator, what is the average weight of everyone in the elevator?" Students would say, "But, you don't know everyone's exact weight." For such students, the mean is somewhat representative of a situation, but not always calculable.

Even though students might not formally know what a *median* is, they have a good sense of average as midpoint. Mokros and Russell found that some students worked backwards to a distribution by symmetrically choosing values above and below the average, for example, above and below the $1.35 value for the Potato Chips problem. Similar to the students who focused on modes, these students also had some trouble when they were not allowed to use the average itself as one of the data points.

Mokros and Russell provided one of the first attempts to build a developmental framework for characterizing students' thinking about average. In their research they were disappointed to not find more students who had richer conceptions of average, such as average as *balance point*, which might be adapted more naturally to computational algorithms. They concluded that the development of higher-level conceptions of average may need to be scaffolded for students via instructional interactions. They also lamented the fact that some students' preferred intuitive notions of average, such as average as *modal* or average as *reasonable*, conceptions that do not provide good underpinnings for connecting to computational algorithms for average.

Foreman and Bennett (1995) make the case that a related intuitive concept of average, average as *fair share* does in fact translate well to computational algorithms. Students represent data in columns of blocks, and then "level" the blocks to produce a visual display of fair shares. This approach starts with whole numbers but can generalize to rational numbers (as stacks of blocks can be sketched and "cut" when necessary). The fair-share approach allows students to generate their own computational algorithms that reflect the physical leveling of stacks of blocks. This ''leveling" approach can also lead to the mean as the point of *balance* for a distribution, as it highlights the value of the mean as a representative value of an entire data set. Friel (1998) discussed connections to computational algorithms for both the leveling and the balancing models.

In a longitudinal study of the development of students' concepts of average in Grades 3 to 9, Watson and Moritz (2000c, 2000d) built upon the work of Mokros and Russell, as well as work of other researchers. Watson and Moritz included questions such as "Have you heard of the word *average*? What does it mean?" and "How do you think they got the average of 3 hours a day for watching TV?" taken from a media report. In another question they told students "On average, Australian families have 2.3 children. What can you tell from this?" In addition to probing for initial understandings of average in media or everyday contexts, Watson and Moritz asked students to work backwards and fill in missing data values when given the mean, and to find the average in weighted mean situations, similar to the tasks of Mokros and Russell.

Watson and Moritz (2000c, 2000d) used the SOLO taxonomy to model student responses on tasks about average and they described six levels of student understanding of average. For example, a response by a student who invented a story about a secret camera that was used to keep an eye on students to see how much television they were viewing instead of doing homework was classified as *pre-structural.* Responses involving colloquial terms for average like "normal" or "alright" were coded *uni-structural.* A reference to a computational algorithm without strong conceptual connections was coded as *multi-structural.* If students' thinking suggested that they considered the average as a representative for an entire data set, their response would be considered *relational.* Watson and Moritz also later identified two response levels beyond relational with a subset of these same students who successfully dealt with applications of the mean in multiple contexts.

Watson and Moritz have provided strong evidence from a large sample of student interviews (over 90) that students' conceptions of average follow a developmental path that proceeds from "stories" then to "mosts and middles" and finally to the mean as "representative" of a data set. At Grade 3, 71% of the students' responses were either pre-structural or uni-structural. At Grade 5, about 50% of the responses were above the uni-structural level, and by Grade 7 all the students in their study were at least at the multi-structural level. A strong positive association was found between the SOLO levels and student responses by grade level. A partial explanation for this might be that by Grade 7 the three measures of center—mode, median and mean—were to have been covered in the Australian national curriculum recommendations. However, there is also a purely developmental component to the students' understanding of average that is very robust. Watson and Moritz (2000c) conducted follow-up interviews 3 to 4 years later with these same students and found that they had maintained the level of their thinking about average. Once students had developed powerful and flexible conceptions of the mean, including the ability to calculate weighted averages in real contexts, they maintained those conceptions over time.

Watson and Moritz's results suggest that it takes many years for students to develop their concept of the mean to the point where it is a representative of a data set. Whereas Mokros and Russell identified each student's dominant conception of average, Watson and Moritz have documented and analyzed the great variety of conceptions of average that a given student can have. They claimed that, "It would appear that many students hold eclectic ideas associated with average, even when they have not been taught formally" (Watson & Moritz, 2000c, p. 46). They strongly recommended that the teaching of the concept of average should build up from students' initial preferences for "middles" and "mosts" to the more normative conception of mean as representative of a data set. They also recommend delaying the formal introduction of the mean until students' developmental stages have had an opportunity to unfold. Bakker and Gravemeijer (2004) made similar recommendations for delaying the introduction of some of the more formal concepts in statistics, like the mean. They found a wealth of student ideas to build upon, including informal conceptions of center and shape, when students were asked to compare distributions of data, or to generate their own hypothetical distributions of data.

Konold and Pollatsek (2002) added to the literature on the complexity of the *average* concept with a theoretical reflection that extends Mokros and Russell's earlier work. They postulated four conceptual perspectives for the mean: mean as *typical value*, mean as *fair share*, mean as *data reducer*, and mean as *signal amid noise*. They argued that from a statistical point of view that the mean as "signal amid noise" is the most important and most useful conception of the mean, because they feel that this conception of the mean is the most helpful for comparing two data sets. Furthermore, they recommended that the mean should be introduced to students in the context of comparing data sets. They also argued that inasmuch as conceptions of average like typical or fair share are not as powerful for comparing groups, they shouldn't be emphasized with students. Konold and Pollatsek's conceptions of mean as typical or mean as fair share are more closely tied to a data analysis perspective on statistics, while their conceptions of mean as a data reducer and or mean as a signal are more closely connected to decision-making in statistics. Data reduction is necessary in decision-making in order to locate an informative signal amid the noise of variability.

From a normative point of view (what statisticians are looking for) Konold and Pollatsek may have a good argument. However, on the basis of the work of Mokros and Russell and of Watson and Moritz, mean as fair-share, and subsequently mean as typical value, are perhaps better first introductions to the notion of measures of center, because they build on students' primary intuitions. The mean as data reducer requires more sophistication from students, and a willingness on their part to let go of some pieces of information. This can be difficult for students to do, especially in those instances when *their own* information is included in a data set, for they feel their information will get lost in the data reduction process. The mean as signal amid noise involves yet another level of complexity. It takes considerable experience with data sets to realize that there even is such a thing as "noise" in data. Noise could appear in data just from random variability in samples or in gathering data from probability experiments. Noise could be the result of measurement error, either systematic error or careless error. Noise could be introduced by poor data-production techniques, or biases in sampling procedures. Teachers and students must spend some time focusing on the noise itself before getting too carried away with determining a signal in the data. I am convinced that it is just as important, if not more so, to analyze the variability in data, and to look for both special-cause and common-cause variation in data as it is to compare data sets by comparing their means. Wild and Pfannkuch (1999) agree, and identify variability as the critical component in the development of students' understanding of distributions of data.

Research on Students' Understanding of Variability

Like average, variability is itself a very complex construct. Researchers have often tended to use the terms *variability* and *variation* somewhat interchangeably. I prefer the distinction pointed out in Reading and Shaughnessy (2004), in which variability is the propensity for something to change, and variation is a description of or a measurement of that change. "A survey of various dictionaries demonstrated that *variation* is a noun used to describe the act of varying or changing condition, and *variability* is a noun form of the adjective *variable,* meaning that something is apt or liable to vary or change. The term *variability* will be taken to mean the characteristic of the entity that is observable, and the term *variation* to mean the describing or measuring of that characteristic." (Shaughnessy & Reading, 2004, pp. 201–202). However, I will attempt to remain faithful to the way that particular researchers refer to the concept, variability or variation, when discussing their research.

Within the field of statistics variability arises everywhere. Even in formal statistics variability is a slippery concept. Data vary, samples vary, and distributions vary. Furthermore, variation occurs both within samples and distributions as well as across samples and distributions. A large part of statistical analysis often involves parsing out the relative contributions and locations of sources of variation. The objects within which variability occurs in statistics—in data, in samples, and in distributions—are clearly not independent of one another. Statisticians may be interested in the data in a particular sample, or the distribution of a sample, or the sampling distribution of a statistic, or even variation across several distributions of data.

Similarly, research on students' thinking and understanding about variability could focus on variation in data, or on students' conceptions of variability in samples, or on the variability across several distributions of data that are being compared. Thus, variability occurs within many levels of statistical objects, and our students need to develop their intuition for what is a reasonable or an unreasonable amount of variability in these objects.

It is a bit surprising then, given the overarching importance of variability in statistics (Moore, 1997), that practically no research on students' conceptions of variability was reported prior to 1999. A number of researchers had called for research on the teaching and learning of variability and had raised concerns about the lack of explicit attention to the concept of variability in introductory statistics texts or in school curriculum materials (Green, 1993; Batanero, Godino, Vallecillos, Green & Holmes, 1994; Loosen, Lioen, & Lacante, 1995; Shaughnessy, 1997). It wasn't until the turn of the millennium that research on and about variability began to appear at professional meetings or in scholarly journals.

Several research efforts have identified a variety of components of variability that indicate the potential depth of students' thinking and understanding about variability. For example, Wild and Pfannkuch (1999) included a number of aspects of variation in their model of statistical thinking, such as *acknowledging, measuring, explaining,* and *controlling* variation. Reading and Shaughnessy (2004) extended Wild and Pfannkuch's list of aspects of variation to include *describing* and *representing* variability, and Canada (2004) provided a detailed framework for analyzing students' thinking about *noticing, describing,* and *attributing causes of* variation. Reading and Shaughnessy also provided evidence from student work on several statistical tasks for both a *description hierarchy* and a *causation hierarchy* for variability. For example, in the description hierarchy lower level responses might be concerned only with outliers or only with middles (uni-structural), whereas a higher level response might mention both middles and extremes (multistructural). At an even higher level a student response might discuss the deviations of data from some fixed value like the mean, thus making connections between the concepts of center and variability (relational).

Some questions come to mind in considering student thinking about variability. Do students acknowledge variability? If so, how do they describe it or talk about it? Are students' conceptions of variability influenced by context? Do they recognize potential sources of variability? Will they attempt to control aspects of an experiment in order to minimize variability? Do students think about variability in a variety of ways, similar to the different conceptions of average that were discussed in the previous section? The next section reviews research that has investigated some of these questions and discusses students' thinking about variability across the spectrum of statistical objects: in data, among samples, and across distributions.

Researchers have now begun to study students' thinking about variability in a number of statistical contexts, such as when reasoning about distributions, dealing with samples and sampling, reasoning about the outcomes of a probability experiment, and comparing data sets (Shaughnessy, Watson, Moritz, & Reading, 1999; Melitou, 2000, 2002; Reading & Shaughnessy, 2000, 2004; Torok, 2000; Torok & Watson, 2000; Watson & Kelly, 2002; Reading, 2003; Ciancetta, Shaughnessy, & Canada, 2003; Shaughnessy, Ciancetta, & Canada, 2003; Watson, Kelly,

Callingham, & Shaughnessy, 2003; Canada, 2004; Shaughnessy, Ciancetta, Best, & Canada, 2004b; Watson & Kelly, 2004). Melitou (2002) presented one of the first reviews of the literature on research on the teaching and learning of variability and described a number of the ways that researchers were beginning to investigate student thinking about the construct. As this chapter was being written, an issue of the *Statistics Education Research Journal* (Vol. 4, No. 1) with a special section on reasoning about variation was in process (J. Garfield & D. Ben-zvi, Eds.). In the next several sections I discuss research on students' thinking about variability, in data, in samples, and in distributions. While this is a convenient approach to discuss student thinking about variability, it is also a somewhat artificial categorization because data, samples of data, and distributions of data all interconnect with one another.

Variability in Data

Data sets tell stories, and the heart of any statistical story is usually contained in the variability in the data. When analyzing data, the role of a student or a statistician is to be a "data detective," to uncover the stories that are hidden in the data. From a data detective point of view, there are important *signals in the variability* as well as in measures of center. In fact, premature attention to measures of center can result in missing the important trends in the variability in the data. For example, consider the data for a series of wait times for eruptions of the Old Faithful geyser in Yellowstone National Park (Figure 21.2), and the accompanying student graphs of the geyser data (Figure 21.3). This example, taken from Shaughnessy and Pfannkuch (2002), presents data for approximately 3 consecutive days of wait times (in minutes) between eruptions of Old Faithful. Students were asked to work in groups, to analyze the data, and to make a decision on how long they would expect to wait for an eruption of Old Faithful.

Typically, many beginning students first just calculate a mean or determine a median for a day's worth of Old Faithful data and then base their initial prediction on a measure of central tendency. (The mean of First day is 70.1, the median is 70.5; mean of Second day is 79.9, median is 77, and so forth). While the mean does give a one number summary of the data set, it can also mask important features in the distribution of the data. When these students were subsequently asked to graph the data set, some of their representations uncovered a pattern in the data that was highly bimodal. More importantly, students discovered an alternating short-long pattern in the Old Faithful blasts when they created plots over time, or dot plots or bar graphs (See Figure 21.3).

This oscillating pattern can be completely missed if one just calculates a mean or draws a box-plot for the data. *The signal in the variability is much stronger than the signal in the center in the Old Faithful example.* The variation in the Old Faithful data is not random. There are likely to be some underlying geological causes or relationships for the variation. Shaughnessy and Pfannkuch found that students who attended to the variability in the data were much more likely to predict a range of outcomes or an interval for the wait time for old Faithful, such as "Most of the time you'll wait from 60 to 85 minutes," than to predict a single number (such as 77 minutes) for the wait time. Means mask what is going on in the Old Faithful data set. The definitive long-short pattern is lost in an average. Means may be critical in making statistical inferences when comparing groups, but variation is often even more important to the data detective.

In the first edition of this *Handbook* I discussed some differences between the points of view of psychologists and of mathematics educators on research in the teaching and learning of statistics and probability. Konold and Pollatsek's (2002) argument on the importance of the mean as the signal in data sets may have arisen from their frustration with students lack of attention to the mean (Konold et al., 1997) as a great tool for making comparisons between data sets, something that psychologists often wish to do in their research work. However, mathematic educators can be equally frustrated when students rush to compute a mean without even considering what the variability in a data set might reveal about the context (Shaughnessy & Pfannkuch, 2002). Students also need to recognize and investigate potential sources of variation within the data and not just rush to looking at centers.

Day 1:	51	82	58	81	49	92	50	88	62	93	56	89	51	79	58	82	52	88
Day 2:	86	78	71	77	76	94	75	50	83	82	72	77	75	65	79	72	78	77
Day 3:	65	89	49	88	51	78	85	65	75	77	69	92	68	87	61	81	55	93

Figure 21.2 Wait times in minutes for the eruption of the Old Faithful Geyser.

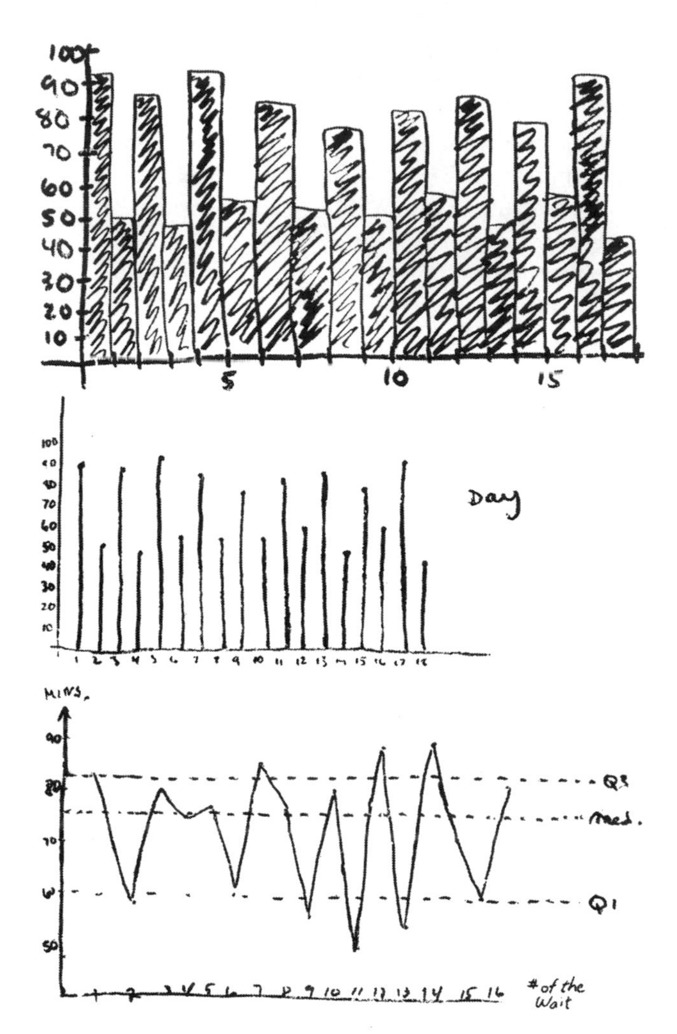

Figure 21.3 Student work samples of graphs of the Old Faithful data.

Variability Among Samples: From Data to Samples

The Old Faithful Geyser data in the previous section provided an example of the importance for students to understand the role of variability in the data itself. Do students acknowledge the variability that can occur from one entire sample to another sample?

A series of studies on students' thinking about how data vary in a sampling environment was precipitated by the Gumball Task in the 1996 NAEP (Figure 21.4).

In a sample of several hundred responses that was pulled from over a 1,000 responses to this item, Zawojewski and Shaughnessy (2000) found that only one student predicted an actual range, from 4–6 reds, rather than a specific number of reds, like 5, or 3, or 10 reds. This may have occurred because students are very accustomed to being asked questions in data and chance that prompt them to respond with single point value answers. For example: What is the Probability that . . . ? How many . . . would you expect? What is the average of . . . ? The Gumball problem was worded in such a way that it was somewhat natural for students to answer with just one number. Several research issues arose from the Gumball problem. First, what would students do if they were explicitly encouraged to think about a range of possibilities for the data in the Gumball problem? Second, the gumball task was administered only to Grade 4 students in the 1996 NAEP. How would older students respond to the gumball task?

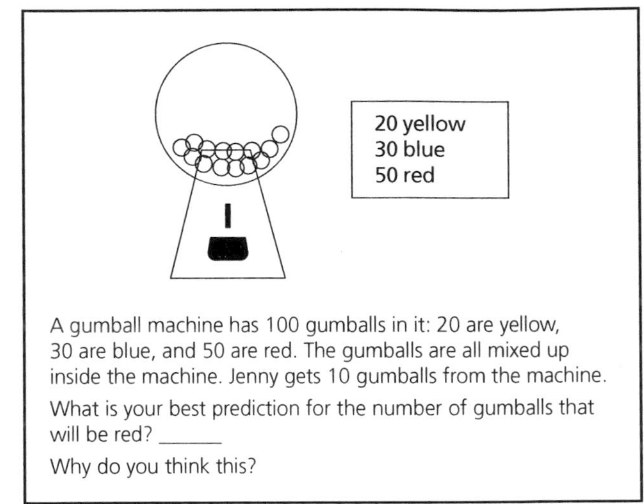

20 yellow
30 blue
50 red
A gumball machine has 100 gumballs in it: 20 are yellow, 30 are blue, and 50 are red. The gumballs are all mixed up inside the machine. Jenny gets 10 gumballs from the machine.
What is your best prediction for the number of gumballs that will be red? ______
Why do you think this?

Figure 21.4 The NAEP Gumball Problem (From Zawojewski & Shaughnessy, 2000, p. 259).

In order to explore students' thinking about the variability of the data in a sampling situation, several reformulations of the Gumball problem, called the Lollie[2] problems, were administered in survey form to over 300 students in Grades 4–6, 9, and 12 in the United States, Australia, and New Zealand (Shaughnessy et al., 1999). Three different versions of the Lollie task (List, Choice, and Range) were given to all the students (See Figure 21.5). In each case, students were asked to supply reasons for their answer.

The Lollie tasks have subsequently been administered to thousands of students in grades 3–12, primarily in Australia and the United States. (Reading & Shaughnessy, 2000; Torok & Watson, 2000; Kelly & Watson, 2002; Shaughnessy et al., 2004a). Responses to the Lollie tasks indicated that there are differences among students on how they acknowledge variability in samples. The types of responses (*high, low, wide, narrow, reasonable*) and the reasons students provide have remained very robust across all these studies. For

[2] In Australia the word *lollies* is used for hard candies that are wrapped in cellophane or paper.

The Lollie Task

A bowl has 100 wrapped lollies in it. 20 are yellow, 50 are red, and 30 are blue. They are well mixed up in the bowl.

Jenny pulls out a handful of 10 lollies, counts the number of reds, and records it on the board. Then, Jenny puts the lollies back into the bowl, and mixes them all up again.

Four of Jenny's classmates, Jack, Julie, Jason, and Jerry do the same thing. One at a time they pull ten lollies, count the reds, and write down the number of reds, and put the lollies back in the bowl and mix them up again.

What do you think? (List Version)

1. I think the numbers of reds the students pulled were

________ ________ ________

________ ________

I think this because:

2. I think the list for the number of reds is most likely to be (circle one)

A) 8,9,7,10,9 D) 2,4,3,4,3 (Choice Version)
B) 3,7,5,8,5 E) 3,0,9,2,8
C) 5,5,5,5,5

I think this because:

3. I think the numbers of reds went from (a low of) ______ to a high of ______.

I think this because: (Range Version)

Figure 21.5 The Lollie Task.

example, some students predicted all *high* numbers of reds, like 6, 7, 5, 8, 9, mostly numbers above the expected value of 5 for samples of 10 lollies from this 50% red mixture. These students usually reasoned that there were "a lot of red in there, so it (the red ones) will happen a lot." Other students, mostly among the Grade 4 students, predicted all *low* numbers (all numbers ≤ 5) and said that there were a lot of "non-reds" in the mixture that would prevent the reds from being pulled very often. Still other students predicted a *wide* list of outcomes, for example, 1, 5, 7, 9, 2 (range ≥ 8) because "any result could occur, you never know" suggesting that they may be using "outcome approach" reasoning (Konold, 1989) or an equi-probability conception (LeCoutre, 1992). Still other students predicted a very *narrow* list for the numbers of reds, for example, 5, 5, 5, 5, 5, or 5, 6, 5, 5, 6 (range ≤ 1). The "narrow" responses, especially all 5s, occurred more frequently among the older students, Grade 12 in particular. Students who predicted *narrow* reasoned that, "5 is the most likely outcome" or "5 is what you are supposed to get." Reading and Shaughnessy (2000) found that the *narrow* predictors were reticent to change their answers, even after they acknowledged that the repeated samples were unlikely to all be identical. Finally, some students' lists were *reasonable* in that they were distributed in a more normative way around 5, such as 3, 7, 5, 6, 5, centered around the expected value within a reasonable range.

There is some evidence that when students are given a chance to actually draw their own samples from a lollie type of mixture, a higher percentage of them will subsequently give *reasonable* predictions for the Lollie task. Shaughnessy et al. (1999) reported an increase from 17% to 55% in *reasonable* responses when a sample of 94 middle school students actually conducted a simulation of the lollie problem with colored cubes in a box. On the other hand, in a study using student interviews, Kelly and Watson (2002) did not find such changes in students' predictions, though they admit there may have been too few trials in their experiment to change students' predictions. The tenacity of beliefs and intuitions about probabilistic and statistical phenomena has been well documented over the years (Kahneman & Tversky, 1972, 1973a, 1973b; Tversky & Kahneman, 1974, 1983; Shaughnesssy & Dick, 1991; Shaughnessy & Bergman, 1993). Beliefs and conceptions about data and chance are very difficult to change, and research has suggested that empirical experiments and simulations must be systematically built into instruction over a longer period of time in order to change the patterns of students' intuitive conceptions (Shaughnessy, 1977; Fischbein & Schnark, 1997;).

The emerging conceptual model introduced previously in this chapter has been used by several groups of researchers to describe the progression of student reasoning about variation on the Lollie tasks, from ikonic, to additive, to proportional, and finally to distributional. According to Kelly and Watson (2002), some students, particularly younger students, reason "ikonically," by using physical circumstances or personal stories when they predict sample proportions for the lollie task. Students who are reasoning ikonically may say, "They might get more reds because their hand could find them," or "Maybe they are lucky and will get all reds," without any reference to the actual contents or the proportion of colors in the bowl. In a sample of 272 students in Grades 6–12, Shaughnessy et al. (2004a) classified students' reasoning on the Lollie task as predominantly additive, proportional, or distributional. Additive reasoners focus on frequencies, rather than on relative frequencies. Proportional reasoners tend to predict "around 5" for the Lollie problem and defend it with statements such as, "There are 50 red," or "I'd expect 5 red out of the 10 candies." Proportional reasoners explicitly discuss the connections between sample proportions and population proportions. Distributional reasoners combine both centers and spreads in their reasoning about the Lollie problem. They make comparisons between the sample proportions and the population proportion, and they also *explicitly mention variation about the expected value*. Shaughnessy et al. also discussed transitional phases in students' thinking that occur between proportional reasoning and distributional reasoning. Attention to just one of the aspects of a distribution (e.g., center, shape, or spread) does not necessarily guarantee attention to the other aspects. It takes some time for students to be able to integrate these various aspects of a distribution.

Variability: From Samples to Sampling Distributions

Distributional reasoning involves making connections from populations to samples, and back again. In order to adequately reason about samples pulled from populations, students must have a strong concept of population proportion, what Kahneman and Tversky (1972) have called *the base rate*. The propensity for subjects to ignore the base rate was so prevalent across Kahneman and Tversky's research that recognition of population proportions is clearly an important idea for instructors to emphasize when they first introduce sampling to students. The complexity of the levels of statistical objects involved in sampling, and the proportional nature of the relations among those objects, is highlighted in work by Saldanha and Thompson (2003). They designed a teaching experiment to develop secondary students' understandings of the concept of sampling distribution. In the experiment the students wrestled with a multitude of statistical objects: individual *data values*; *collections of data* values (samples); *statistics* for a sample (like the mean of a sample); and finally, *collections of statistics* for many samples (for example, a distribution of sample means). The concept of population proportion was critical to Saldanha and Thompson's teaching experiment, as was the relationship between sample proportion and population proportion. Saldanha and Thompson emphasized the differences among data, samples, and populations with the students in their teaching experiment. However, they expressed some frustration that even with this explicit emphasis, their students had great difficulty conceiving of and distinguishing among the various levels of statistical objects while working on sampling tasks. Their students "did not have a sense of variability that extended to ideas of distribution" (Saldanha & Thompson, 2003, p. 264).

Saldanha and Thompson's students interacted primarily with a computer environment, generating samples and sampling distributions. One wonders if Saldanha and Thompson's students would have had more success if they had spent more time experiencing a hands-on approach, doing simulations with objects to generate sampling distributions, rather than in the computer environment. Studies that have involved substantial hands-on simulation activity have shown some success in influencing student thinking in stochastic situations (Shaughnessy, 1977; Shaughnessy et al., 1999). In a second teaching experiment Saldanha (2003) found that the addition of physical, hands-on sampling provided better support for student understanding of sampling distributions than the computer environment alone.

Rubin, Bruce, & Tenney (1991) pointed out that peoples' understanding of a what a sample tells us ranges along a spectrum from, "knowing everything" to "knowing nothing." Knowing everything is the result of overconfidence in how well the sample represents the population from which it is drawn, reminiscent Kahneman and Tversky's (1972) representativeness heuristic. At the other extreme, knowing nothing reflects a belief that a sample is just chance, because absolutely anything can happen in a sample. This other extreme of Rubin's spectrum reflects reasoning similar to Konold's (1989) outcome approach. Another way to think about these two extremes is that at one end people are too fixated on centers—they feel all samples should be perfectly representative—and at the other end they are too fixated on variability, and all the possibilities for individual outcomes. Students need to have a balance between representativeness and variability, between expectation and variation (Watson and Kelly, 2004).

Rubin et al. created several sampling scenarios to investigate how students might attempt to balance sample representativeness with sample variability. For example, in one scenario students were told that packets containing six Gummy Bears were handed out to all the children attending a parade. The packets were assembled from a huge vat of 1 million red and 2 million green Bears that were all mixed up. Then the students were asked how many children out of 100 they thought got packets with exactly 4 green and 2 red bears? Half the students said over 75% would get packets with 4 greens. All but one student said that over 50% of the children would get 4 greens. The wording of this problem focused the students' attention on the population proportion of two-thirds green, and likely tipped the students' thinking towards the representativeness end of Rubin's spectrum. However, when a different version of the problem was given to the students, asking "out of 100 packets, how many children would get 0, 1, 2, . . . , 6 green Bears," they tended to spread their predictions out across all the possible outcomes, tipping their responses toward the variability end of Rubin's spectrum. Rubin et al. found that students had little or no understanding of the interaction between representativeness and variability on these tasks, and that students had very little intuition for the shape of the sampling distribution for the Gummy Bears packets.

Other researchers have found results similar to Rubin et al. Shaughnessy et al. (2004a) gave a lollies type question to secondary students that asked them to predict a sampling distribution for repeated samples drawn from a lollie jar. They found that many students focused heavily on aspects of variability, predicting overly wide ranges for the outcomes in a sampling distribution that too many extreme results would occur. On the other hand, Saldanha and Thompson (2003) found that students' concept of sampling did not entail much of a sense of variability at all. Their secondary students tended to judge samples purely on representativeness, perhaps relying too heavily on the underlying population proportion.

This series of studies (Rubin et al., 1991; Saldanha & Thompson, 2003; Shaughnessy et al., 2004a, Watson & Kelly, 2004) clearly shows that there is always a tension between representativeness and variability in sampling situations. Furthermore, the framing of the task itself can tip students' thinking in the direction of either an over reliance on the population proportion, or an over reliance on variability. The goal for teachers and curriculum developers is to help students to grow a middle ground between representativeness and variability, between knowing all and knowing nothing, to "knowing something," so that both population proportion and variability are taken into account together. This balance is necessary so that students understand that one can actually know something with a reasonable likelihood in sampling situations. One approach to sampling that may promote a balance for students between expectation and variation involves student-generated confidence intervals (Landwehr, Watkins, & Swift, 1987).

Variability across Distributions: From Informal to Formal Inference

Some research has focused on student thinking about variability across distributions, using tasks in which students are asked to compare several distributions or to make decisions based on a collection of distributions. Gal, Rothschild and Wagner (1989, 1990) found that middle grade students normally did not use the mean when they were asked to compare two data sets. Instead Gal et al. (1989) found that students used statistical strategies, quantitative data summaries, or proto-statistical strategies that focused on incomplete features of the data sets. Some students invented their own stories to compare data sets. Gal et al. (1989) were the among the first researchers to point out the essential role that proportional reasoning can play when students are asked to compare two data sets. Konold, Pollatsek, Well, and Gagnon (1997) reported similar results with some upper secondary students, who also neglected to use the mean as a comparative measure. Their students often focused on particular data points or individual features of the data rather than on making global comparisons of the distributions. Evidently the power of the mean as a representative measure for making comparisons of data sets is not a primary intuition (Fischbein, 1987), and must be carefully developed within a teaching-learning setting.

In an 8-week design experiment with a fourth grade class Petresino, Lehrer, & Schauble (2003) introduced distributions of measurement data as objects for comparison and decision-making. Students gathered data on the heights of model rockets, some with pointed noses and others with rounded noses and then were asked to decide which was the best design. The students found the median height for both types of rockets, but this did not entirely satisfy them as a comparison basis, because the data for one type of rocket was more inconsistent than the other. They decided to find a way to measure the inconsistency, that is, the variability. This class devised a way to compare the variation for the two rockets by computing differences from rocket height to median rocket height for each type of rocket. Then they created two new distributions from the original distributions, the distributions of height differences from the medians. These fourth graders were thus comparing distributions of residuals—a rather sophisticated concept, but one that arose naturally in this particular design experiment.

The study by Petresino et al. provides a glimpse into classroom dynamics and into the social-conceptual growth of a class of students who attempted to use both centers and variability in their decision making process. It is interesting that the students in the Petresino et al. study looked for a *signal from the variation* when they compared the distributions, rather than a signal from the centers. They felt that the center values for the two types of rockets could be misleading, as one rocket design was more inconsistent than the other in the heights it attained. In this case, it was not just the mean or the median that became the signal to inform a decision (see Konold & Pollatsek, 2002). The variability was the primary *signal* that aided the students' decision. Petresino et al. Also reported that their fourth graders performed very well on the 1996 NAEP Grade 4 statistics tasks that were administered to the class after the 8-week teaching experiment. This suggests that there may be long-term retention benefits when students invent their own approaches to data analysis and the comparison of data sets.

In a series of articles, Watson reported on Grade 3 to 9 students' types of strategies when comparing pairs of data sets (Watson and Moritz, 1999; Watson, 2001a, 2001b). A version of the data sets used for these studies is presented in Figure 21.6.

In interview settings, Watson and Moritz presented students with the pairs of the data sets from Figure 21.6 (e.g., Yellow-Brown, Pink-Black) and told them that the graphs showed the results of students' test scores in two different classes. They then asked students which class did better in each pair, and why.

In the first study, Watson and Moritz (1999) report that students' strategies fell into two SOLO cycles. The first cycle dealt with equal sized data sets (e.g., Yellow-Brown). Student explanations ranged from a comparison of individual data values, to calculating the total class values (adding up all the values in the data set), to visual comparison strategies ("Yellow goes higher"), and to combinations of these explanations. In the second SOLO cycle, data sets of different sizes were compared (e.g., the Pink-Black data sets in Figure 21.6) and proportional reasoning played a more prominent role. Students' reasoning strategies again included totaling up the scores and visual strategies, but students also calculated the means to compare the two groups. Watson and Moritz thus identified both additive and proportional reasoning strategies among their students. They concluded that students use "bottom line" strategies, such as attending to certain features of the graphs or reasoning about modal clumps, when comparing data sets. Watson and Moritz argued that these informal reasoning strategies can provide opportunities for teachers to introduce the comparison of two data sets prior to calculating means. Bakker and Gravemeijer (2004) came to the same conclusion: Students can, and should, compare data sets using their own intuitive strategies prior to formal statistics. In a follow-up study conducted 3 years later with some of the same students Watson (2001a) found considerable improvement in the level and complexity of students' responses on these tasks.

In a third study Watson (2002) used an interesting methodology to set up cognitive conflict among students and to challenge their thinking. After they had responded to the comparing-data-sets tasks in Figure 21.6, Watson showed students video clips of other students who had compared the data sets using a different strategy. For example, if a student reasoned that, "The Pink Class did better, because they have more students who scored higher," Watson might show this student a clip of a student who computed the means of the Pink and Black class and claimed that Black was better. Or, if a student said that "Yellow did better because they got high numbers" (referring to the height of the 5 column in the yellow data set) she might show a clip of a student who said that there was no difference, because both classes totaled 45, or there was no

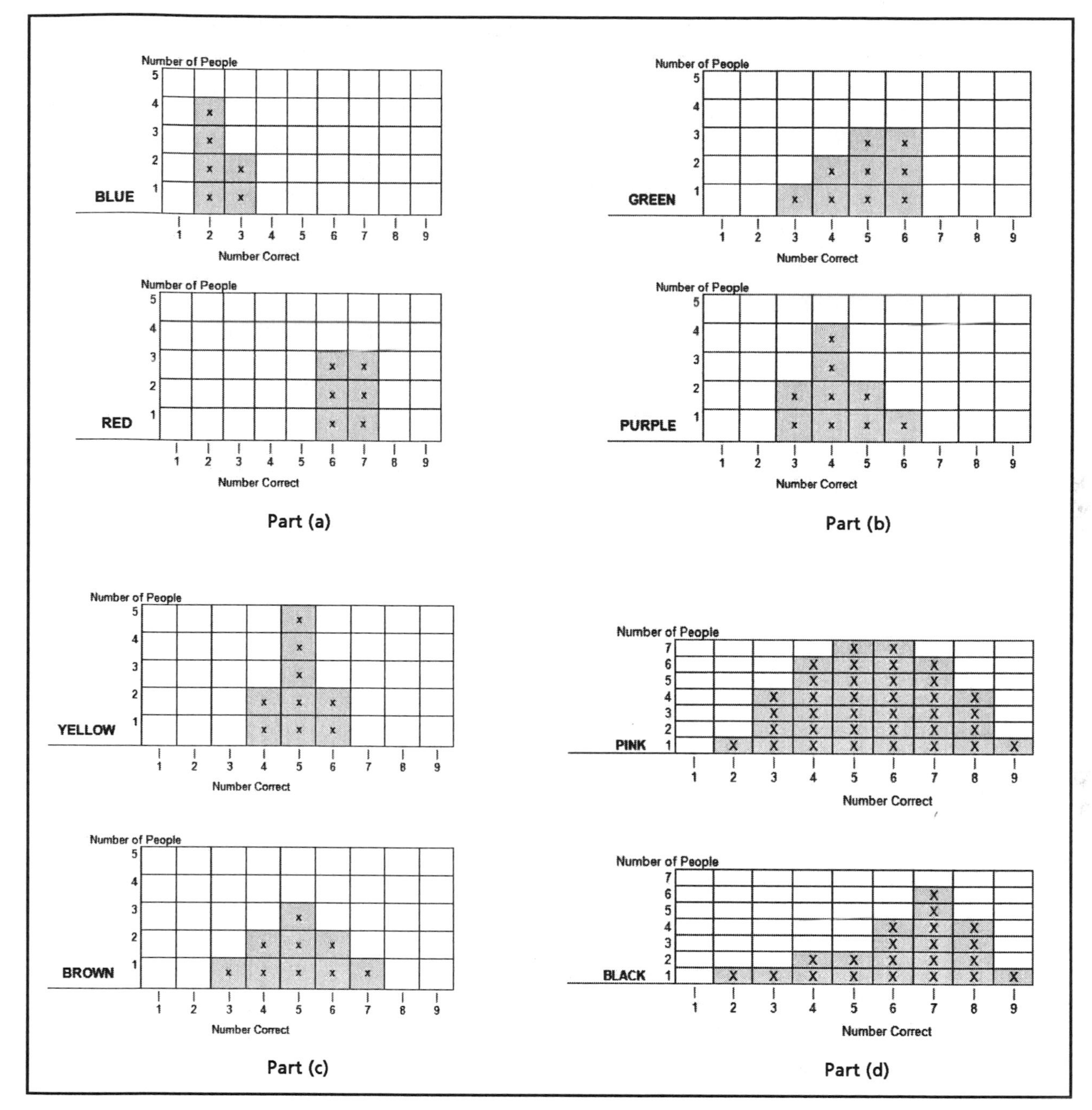

Figure 21.6 Paired data set comparisons. (From Watson & Shaughnessy, 2004. Proportional reasoning: Lessons from research in data and chance. *Mathematics Teaching in the Middle School,* p. 105)

difference because both classes had an average of 15, or that the Brown class did better because it had a 7.

Of those students whose reasoning could improve the second time around (that is, students who didn't reason at the top level the first time), 57% improved on the Yellow–Brown comparison, but only 30% improved on the Black–Pink comparison, after being shown a clip of another student. Students need to reason proportionally to get the Pink–Black comparison correct. Students who only reason additively will usually pick Pink (because it has higher totals), and it is difficult to change their minds if they are not proportional reasoners. Some students reasoned proportionally on this problem without actually calculating the means. For example, they said, "a higher percentage of the Black class got higher scores." This sort of reasoning is proportional, because students are comparing relative frequencies, not just absolute frequencies. Watson (2002) noted that the improvement rate after the cognitive conflict was nearly identical to improvement rate over 4 years in the (2001a) longitudinal study. Thus cognitive conflict might help accelerate students in their ability to analyze and compare data sets. A question that comes to mind, one acknowledged by Watson, is how stable are the students' responses after the cognitive conflict? What would they have said if asked again 3 or 4 years later?

Shaughnessy (2003b) has used three of these comparing-groups tasks from Figure 21.6 in a survey form with middle and secondary school students. Percentages of types of reasoning, including additive, proportional, and distributional reasoning, on the Brown–Yellow and Pink–Black tasks are shown in Tables 21.1 and 21.2. Although there was some growth across grade levels on the Pink–Black task, comparing unequal sized groups remained a challenge for many students. Nearly three-fourths of the students in the survey said that the Pink class did better. More work with students at comparing unequal sized groups is clearly an important target for teaching and assessment.

Most of the research in this section has dealt with students' informal inference strategies as they compare distributions. Pfannkuch (2005) reviewed the literature on probability and statistical inference, and suggested some approaches to help teachers move from informal inference towards the use of more formal inference tools, such as using simulations in a re-sampling approach to statistical inference with students as early as Grade 12. However, Lipson (2002) found that software capabilities disrupted the transition between empirical and theoretical approaches to inference. Also, delMas, Garfield, and Chance (1999) found no substantial evidence that simulations improved their students' conceptual understanding of sampling distributions. Despite these discouraging results Pfannkuch remains optimistic that resampling techniques and simulations can provide a bridge to help students transition from informal inference strategies to a more formal (normative) approach to inference. The concepts surrounding

Table 21.1 Frequencies (%) of Responses for Codes for Yellow–Brown Group Comparisons (Survey Results from NSF Grant No. REC-0207842)

Response	Code	MS (*N* = 84)	HS (*N* =188)	Total (*N* = 272)
Unclear; Omit	0	19 (23)	56 (30)	75 (28)
Sums	1	53 (63)	76 (40)	129 (47)
Average	2	9 (10)	44 (23)	53 (19)
Variation	3	3 (4)	8 (4)	11 (4)
Variation & Centers	4	0 (0)	4 (2)	4 (1)

Codes: 0—No response, misread, unclear reasoning "they look the same"
1—Using Sums; individual graph characteristics e.g., "there's more 5's in yellow," or "there's a 7 in brown"
2—Using Average (if computed)
3—Variation explanation; spread or tightness of distributions
4—Sophisticated use of Average and Variation in combination

Table 21.2 Frequencies (%) of Responses for Codes for Pink–Black Group Comparisons (Survey Results from NSF Grant No. REC-0207842)

Response	Code	MS (*N* = 84)	HS (*N* = 188)	Total (*N* = 272)
Pink (mostly)	0	67 (80)	135 (72)	202 (74)
Black—graph characteristics	1	7 (8)	12 (6)	19 (7)
Black—averages	2	10 (12)	31 (16)	41 (15)
Black—Distribution characteristics	3	0 (0)	10 (5)	10 (4)

Codes: 0—No response, or "Pink because of more scores or higher total"; or, black or equal with bad reasoning
1—Black, use of graph characteristics
2—Black, use of averages
3—Black, use of characteristics of the distribution, i.e., "higher proportion of scores are . . ."

statistical inference are very complex, and the transition for students to formal inference is likely to be in process for several years. There is no quick fix to understanding these concepts, anymore than there is a quick fix to understanding the formal concept of a limit in mathematics.

Variability and Random Outcomes in Probability Experiments

Probability is the focal point of another chapter in this volume (Jones, Langrall, & Mooney, this volume), but there are important connections between probability and statistics, particularly when repeated trials of probability experiments generate a distribution of possible outcomes.

Truran (1994) wanted to push students to investigate how far experimental probabilities would have to deviate from expectations before students would consider revising their predictions. Sampling with replacement, Truran asked 32 students in Years 4, 6, 8, and 10 what they would expect in samples of size 9, and samples of size 50 drawn from an urn with 2 green balls and 1 blue ball. Extremes were found to be surprising to most of these students, and some of them thought that their results would tend toward 50–50 if they "did it enough times." Several students offered personal confidence intervals from 4–8 greens for the small sample, and from 27–40 greens for the large sample. Many of Truran's students gave responses that involved personal theories about the balls or the urn that did not relate to the actual mixture proportions.

The probability task in Figure 21.7 is an extended constructed response that was given to twelfth graders on the 1996 NAEP. Zawojewski and Shaughnessy (2000) discovered some startling responses to this task.

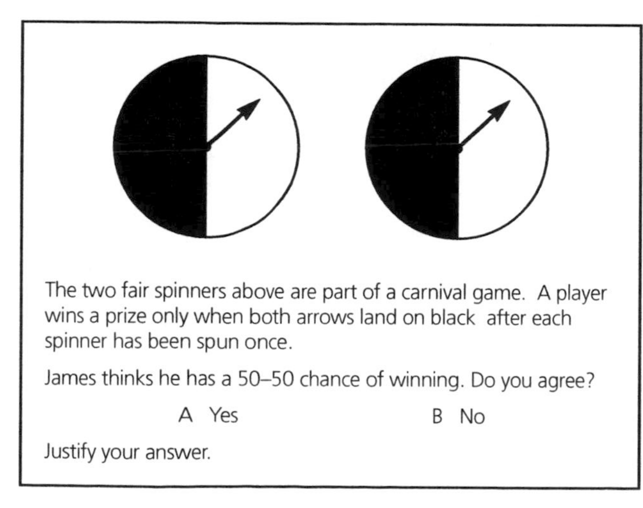

The two fair spinners above are part of a carnival game. A player wins a prize only when both arrows land on black after each spinner has been spun once.
James thinks he has a 50–50 chance of winning. Do you agree?
A Yes
B No
Justify your answer.

Figure 21.7 The 1996 NAEP Spinner Task (From Zawojewski & Shaughnessy, 2000, p. 263)

Nearly 50% of the 1270 Grade 12 students tested agreed with James, and said that there was a 50–50 chance that both spinners would land black. In a convenience sample of responses (N = 306) the reason given by a third of the students in support of a 50–50 chance of both spinners landing black was that since the spinners each were half white and half black there was a 50–50 chance that both of them would land on black. Only 8% of the students in the NAEP sample claimed that the probability was 1/4 *and* gave an adequate justification for their answer.

Responses to this item speak volumes about the state of probability instruction in the United States. Simple compound probability problems like this spinner problem are mentioned among the important probability concepts in the *Principles and Standards for School Mathematics* (NCTM, 2000). This spinner problem is not a particularly difficult compound probability problem. In fact, isomorphic versions of this problem appear in many middle school mathematics programs. Shaughnessy and Zawojewski (1999) outlined ways that teachers could use this spinner task, and similar ones, as teaching and assessment tools to explore students' understanding of probability.

Shaughnessy and Ciancetta (2002) found somewhat more promising results for the spinner task when they surveyed a sample of 652 students in Grades 6–12 and had a subset of students gather actual data. Correct responses initially ranged from 20% in Grades 6–7 to near 90% in classes of students who were studying more advanced mathematics such as precalculus, AP calculus, or AP statistics. Follow-up interviews with 28 of the students from Grade 8–12 were conducted in a setting where students first predicted the results for 10 repeated trials of the spinners tasks and then gathered their own spinner data for 10 actual trials. After conducting their own experimental trials, only 5 students persisted in believing that the chance of both spinners landing black was 50%. As the students experienced the variability in results across several sets of 10 trials and witnessed the actual oscillation in the number of times both spinners landed black, most students realized that the true expectation of winning the spinner game was not 50%, but something much smaller.

Watson and Kelly (2004) assigned SOLO levels to student responses on the 50–50 spinners task. Typical responses in the main three SOLO levels were, "it's 50–50 because it's half white" (uni-structural); "the chance that both land black is less than 50% because only one of the three outcomes has both black (multi-structural); and "there are four outcomes, BB, BW, WB, and WW, so the chance is 25% both land black" (relational). Zawojewski and Shaughnessy (2000 found that these same three response categories accounted for most of the student responses to this task on the 1996 NAEP. Watson and Kelly concluded that the concept of independence of the two spinners was not intuitive for their students, and it was especially difficult for the middle school students in their study. This points to a major challenge for teaching and curriculum development in statistics—how do we help students get a better intuitive feel for what a reasonable spread is in experimental probability situations?

Students' recognition of the appropriate variability across the outcomes of a probability experiment is often very dependent on the task. Shaughnessy, Ciancetta, and Canada (2003) compared 84 middle school students' predictions about variability on repeated trials across three task environments: the Lollie task, a task that involved repeated sets of 20 tosses of a regular six sided die, and a single spinner task. The percentage of students' responses that fell within "reasonable" variation guidelines established for these three tasks were 70% for the lollies, 30% for the die, and 49%, for the spinner task. These results suggest that in the mental tug of war that students face between variability (spread) and expectation (centers), that the Lollie problem pulls students more in the direction of acknowledging variability, whereas the die problem pulls students more in the direction of predicting expectation. The spinner problem split the students down the middle into about 50% reasonable and 50% unreasonable predictions for variability across repeated sets of trials. There seems to be an influence from probability on these types repeated trials tasks, especially in the die problem but also somewhat in the spinner problem. Probability instruction might interfere with student thinking about variability, since students may tend to predict 'what should happen, theoretically. In order for students to reason distributionally, in order for them to grow beyond a mere focus on expectation, they must develop their intuition for a reasonable amount of variation around an expected value, not just the expected value itself.

Student Thinking About Association, Covariation, and Correlation

Most of the research studies in the previous sections have dealt with univariate variability. However the most interesting problems in statistics usually involve multivariate situations in which association, dependence, and possible causes of variation lurk, all waiting to be unearthed by a persistent data detective.

In an attempt to see if students would raise questions about a suggested association between two variables, Watson and Moritz (1997a) asked students in Grades 6 and 11 to produce graphical representations of a nearly perfect relationship between an increase in heart deaths and an increase in motor vehicle use as described in a newspaper article. They also asked students what questions they would raise about such a claim. Using their three-tiered framework for analyzing statistical claims, Watson and Moritz found some students who were reasoning at the third tier because they questioned whether there really was a cause-and-effect relationship between the incidence of heart deaths and motor vehicle use. However, most of their students just accepted the claim about an association between driving a car and having a heart attack without questioning it. Students are often too willing to accept anything that they read in print, and are probably inexperienced in critiquing statistical claims about relationships between variables.

Moritz (2000, 2004) provided an overview and analysis of the research on students' reasoning association and about covariation, and presented bivariate tasks to students in Grades 3 to 9 that involved both positive and negative association. In one task, Moritz gave students the description and graph in Figure 21.8 that portrays a negative association between the number of people in a classroom and the noise level in that classroom. Then, he asked students to explain the graph to someone who could not actually see it.

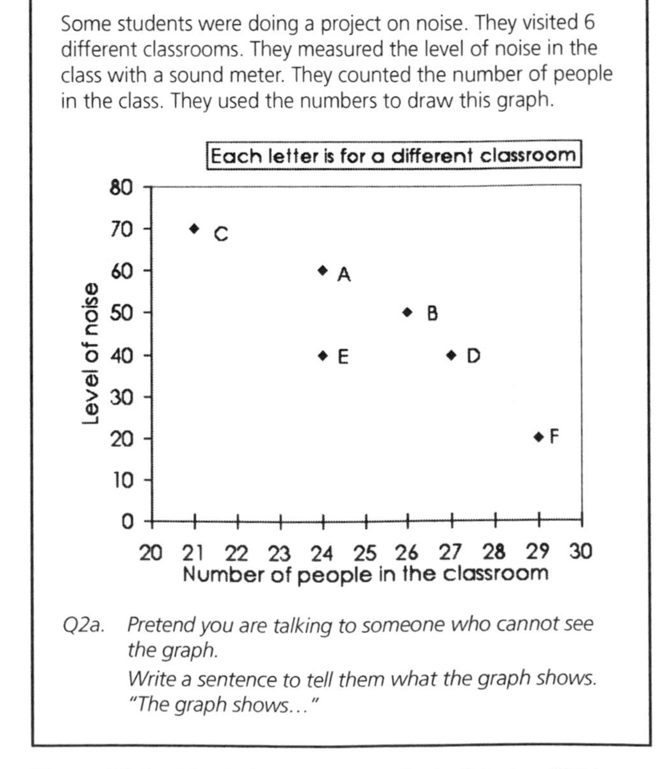

Figure 21.8 Moritz's Covariation Task (Moritz, 2004, p. 238)

Moritz identified four levels in responses to this task: nonstatistical, single aspect, inadequate covariation, and appropriate covariation. Nonstatistical responses did not address anything about covariation, for example "It's a graph of noise of students." Single-aspect responders tended to pick out a few data points, perhaps extremes or outliers, and talk about them but they used only a very few data points. If students only talked about one of the two variables, or if they compartmentalized their responses on the two variables and did not talk about how the two were related *to* one another, their responses were coded as inadequate covariation. Finally, responses that related the changes in the two variables to one another were coded as appropriate covariation.

Moritz (2004) reflected on approaches that might help students to transition through several stages of statistical reasoning, from single data points to the variation within individual variables, to a consideration of how both variables change simultaneously. He agreed with Nemirovsky (1996) who in his work in algebraic thinking suggested that covariation might best be introduced to students by using time as the independent variable. Shaughnessy (2003b) also used graphs over time such as the one in Figure 21.9 in interviews and in classroom teaching episodes with middle and secondary level students to investigate whether students would look for potential causes of variability in food consumption over time.

Most students did make some conjectures about the variability over time in such per capita food consumption graphs (e.g., per capita milk, coffee, soft drink, & bottled water consumption). Students will even make up something up if they cannot provide a solid contextual explanation for the humps and dips in these food consumption graphs over time. Students have said, "maybe it was inflation," or "maybe the economy was bad." They rarely attribute the big jumps or big dips in the food consumption graphs to random variation. Rather, students tend to look for what Wild and Pfannkuch call "special cause" variation. Shaughnessy (2003b) found students who attributed the swings in these food consumption graphs to the baby boom, to improved production and distribution of food, to the Depression, to World War II, to the war in Vietnam, and finally, if all else failed, some students claimed, "It must have been the hippies." Students do try to make contextual conjectures for why such graphs vary. Nemirovsky (1996) had a point, covariation might be best introduced with time as one of the variables, because students are interested in trends over time, and this type of data connects naturally to topics that are of interest to them.

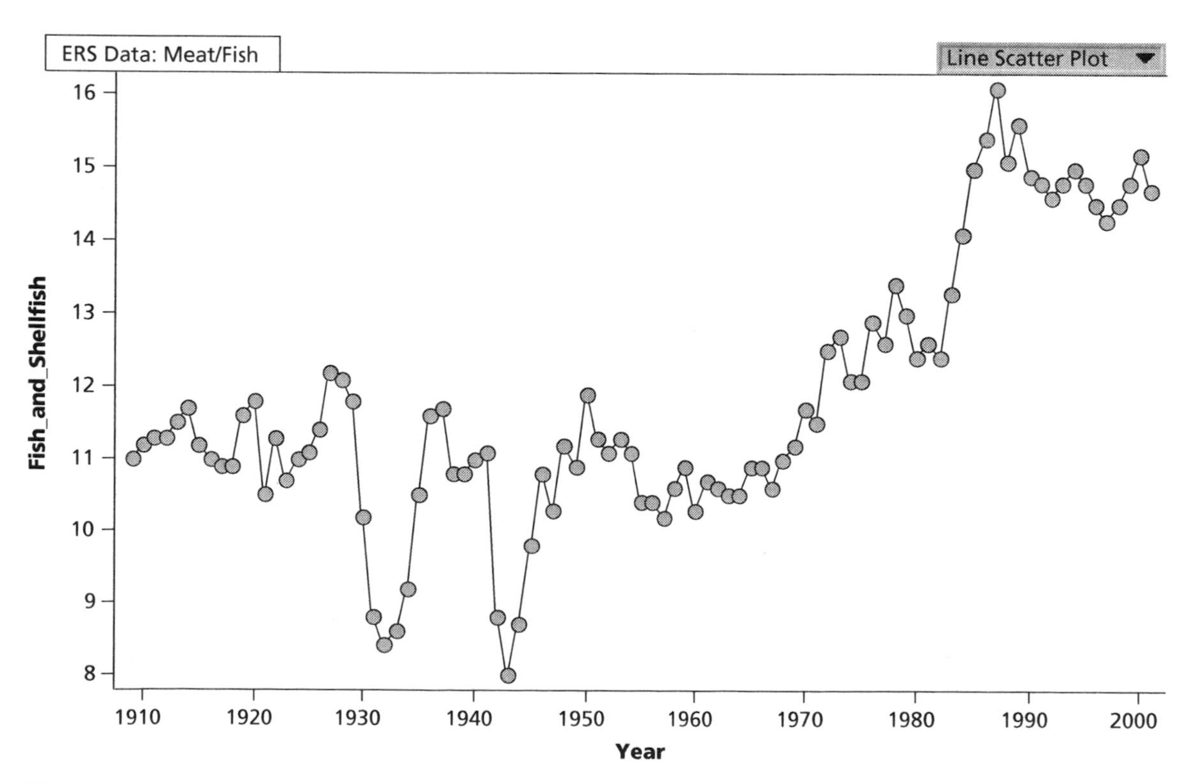

Figure 21.9 Shellfish consumption over time (USDA, 2004).

Whereas Moritz looked at younger student's understanding of covariation, Batanero, Estepa, Godino, & Green (1996) researched beginning tertiary students' reasoning about association of variables in 2×2 and 3×3 contingency tables. Among their findings was the tendency for tertiary students to make decisions about association between two variables on the basis of the frequency in *only one cell* of a contingency table. When reasoning with contingency tables the critical information is usually in the row and column proportions rather than in the cell frequencies. For example the data in Table 21.3 suggest that there may be a relationship between travel class and death rates on the Titanic. The death rate was 38%, 64%, and 76%, respectively, for first, second, and third class travel accommodations (Takis, 1999). The proportion of those who died in each class level is the critical issue, not the absolute frequencies.

Table 21.3 Deaths by Travel Class on the Titanic (From Takis, 1999, p. 483)

Travel Class	Died	Survived	Total
First	122	197	319
Second	167	94	261
Third	476	151	627
Total	765	442	1207

Batanero et al. (1996) found that many students had correct or partially correct strategies when reasoning about association of variables in contingency tables, but they also found three main incorrect reasoning patterns. Some students expected a perfect correspondence between variables, that there should be no exceptions in the data. Batanero et al. called this a *deterministic approach* to association. Other students thought that association could only be in the positive direction and ignored situations in which variables were inversely related, and this they called the *uni-directional* misconception. A third incorrect strategy, called the *localist* approach, students considered only part of the information, such as just one cell, or perhaps just the main diagonal. Batanero et al. concluded that their tertiary students lacked the necessary proportional reasoning skills to reason about contingency tables.

In another study with senior secondary students, Estepa, Batanero, & Sanchez (1999) explored students' ability to make associations between two variables when data sets were presented in two-way tables. Some students used statistical approaches, such as means, totals, percentages, or attempted to compare the whole distributions. Other students used what Estepa et al. called deterministic strategies, like comparing lowest and highest values, comparing ranges, looking at coincidences, or arguing from their own personal beliefs. Gal (1998) discussed a variety of levels of questions to assess students' understanding of data in two-way tables. Gal noted that percentages are needed to back up opinions or to defend claims that are made about data in two-way tables, because individual cell frequencies are inadequate. Gal encouraged teachers to pose more open-ended, less directive types of questions about two-way tables to promote a higher level of statistical reasoning by students.

Types of Conceptions of Variability

One of the strengths of the research on average noted above has been the spectrum of conceptions of average that researchers have identified, both in investigations of students' thinking and in theoretical conceptual analyses. For example, Mokros and Russell (1995) pointed to average as middle, average as most, average as balance, and average as reasonable. Konold and Pollatsek (2002) mentioned mean as typical, mean as fair share, mean as data reducer, and mean as signal. In a similar way, research efforts on students' understanding of variability are beginning to identify a variety of conceptions of variability, in part because of the variety of statistical objects that can vary, such as data, samples, and distributions. The types of student conceptions of variability identified by recent research include the following:

1. Variability in *particular values*, including *extremes* or *outliers*. In this conception of variability, students focus their attention on particular data values as pointers, often on very large or very small values, or very strange individual values in a graph or in a data set (Konold and Pollatsek, 2002; Konold, Higgens, Russell, & Khalil, unpublished manuscript).
2. Variability as *change over time*. As discussed, this conception of variability may be a good starting point to introduce covariation (Nemirovsky, 1996; Moritz, 2004).
3. Variability as *whole range*—the spread of all possible values. This conception of variability involves the spread of an entire data set or distribution and is closely related to the concept of sample space in probability, the set of *all* the possible outcomes. In this conception students have begun to move away from seeing data only as individual values that

vary, to recognizing that entire samples of data can also vary (Shaughnessy et al., 1999).

4. Variability as the *likely range* of a sample. This conception of variability arises in tasks like the Lollie problem (Reading & Shaughnessy, 2004) or in repeated trials of probability experiments (Shaughnessy et al., 2003). It can lead to statistical tools for representing variability within or across samples, such as box plots or frequency distributions. This conception of variability requires the concept of relative frequency and thus relies on proportional reasoning. It can also lead to the concept of a sampling distribution when applied to the likely range of a distribution of means, or distributions of other sample statistics (Saldanha & Thompson, 2003).
5. Variability as *distance or difference from some fixed point*. This concept involves an actual or a visual measurement, either from an endpoint value (as in a geometric distribution) or from some measure of center (usually the mean or median). Here students are predominantly concerned with the variability of one data point at a time from a center rather than the variability of an entire distribution of data from a center (Moritz, 2004).
6. Variability as the *sum of residuals*. This is a measure of the collective amount a distribution is "off" from some fixed value and provides a measure of the total variability of an entire distribution of data. This is the notion of variability that arose in the design experiment of the Grade 4 class investigated by Petrosino et al. (2003). It provides the foundation for such concepts as standard deviation and regression analysis.
7. Variation as *covariation or association*. This conception of variability involves the interaction of several variables, and how changes in one may correspond to (though not necessarily cause) changes in another. Covariation raises issues about the strength of relationships among variables and poses challenges to parsing out what part of variation is due to chance, and what part may actually be due to cause and effect (Batanero et al., 1996; Moritz, 2004).
8. Variation as *distribution*. Distributions themselves can vary. When the variation between or among a set of distributions is compared, the specter of statistical significance arises. Theoretical probability distributions emerge to help make decisions about distributions of data, or about sampling distributions (Bakker and Gravemeijer, 2004; Shaughnessy, Ciancetta, Best, & Noll, 2005).

Student Thinking About Information Obtained from Samples and Surveys

In her work with Grades 4 and 5 students, Jacobs (1997, 1999) found that children's evaluation of survey methods fell into four main categories: *potential for bias*, f*airness*, *practical issues*, and *results*. Some students noted the potential for bias in certain survey methods, but others were more concerned with *fairness* issues. For many students a survey is not fair unless it has representation from all possible subgroups in the survey population. For example, a school survey would have to have a boy and a girl from each class in the school, in order to be "fair." On the surface, this looks like a sophisticated stratified sampling scheme, but in practice those who favor the fair-sample approach would reject any part of randomization. Jacobs also found that students who were concerned with *results* of surveys disagreed with them if they did not match their own preconceived notions of what should happen. Students also were concerned about how decisive the results of a survey really were. In their minds it was supposed to "be right."

Several of Jacob's tasks asked students to evaluate three different survey techniques that she classified as restricted, self-selection, and random. In one task, Jacobs presented students with six different survey scenarios in a school on whether students were interested in conducting a raffle in a school to raise money. Watson et al. (2003) used this task in Grades 3 to 9 as part of their study to measure students' understanding of sources of variation. Shaughnessy (2003b) used a version of Jacob's school survey task with secondary school students to see if they could identify important aspects of sampling. The task is presented in Figure 21.10, accompanied by the results from a survey of students in Grades 7–12 in Table 21.4.

Responses to Part 1 of the task were scored 0 to 4, based on the sampling methods described by the students and whether their methods did or did not include explicit reference to (a) sampling, (b) appropriate size (30 to 120 was rated appropriate), (c) stratification, and (d) randomness. Only about a third of the students scored a 3 or 4 on this task, indicating that they had a statistically appropriate sampling plan. Many of the students wanted to survey most, if not all, of the students in the entire school.

The percentage of students who rated each of the three different sampling methods (second part of the

Part 1. A class wanted to raise money for their school trip to Disney World. They could raise money by selling raffle tickets for a Nintendo Game system. But before they decided to have a raffle they wanted to estimate how many students in their whole school would buy a ticket.

So they decided to do a *survey* to find out first. The school has 600 students in grades 7–12 with 100 students in each grade.

How many students would *you* survey and how would *you* choose them? Explain why?

Part 2. *Three students in the school suggested different methods to surveys the students in the school about buying the raffle tickets.*

a) **Shannon** got the names of all 600 children in the school and put them in a hat, and then pulled out 60 of them. What do you think of Shannon's survey?

b) **Raffi** surveyed 60 of his friends. What do you think of Raffi's survey?

c) **Claire** set up a booth outside of the cafeteria. Anyone who wanted to stop and fill out a survey could. She stopped collecting surveys when she got 60 kids to complete them. What do you think of Claire's survey?

(After each of these sampling methods, students were asked to rate the method, and to give a reason for their rating).

❑ GOOD ❑ BAD ❑ NOT SURE

Why?

d) Who do you think has the best survey method? Why?

Figure 21.10 A version of Jacob's Task on Sampling Procedures.

Table 21.4 Percentages Favoring Three Sampling Methods (Survey Results from NSF Grant No. REC-0207842)

Rating/ Method	Shannon (Random)	Raffi (Friends)	Claire (Cafeteria)
Good	24[a]	37	59
Not sure	9	10	29
Bad	67	53[a]	12[a]
Best Method	31[a]	5	64

[a] Indicates the expected response

task) as *good, not sure,* or *bad,* is listed in Table 21.4. Shaughnessy's sample of students heavily favored the self-selection approach outside the cafeteria in this task. This is consistent with results found by Jacobs (1997) and Watson and Moritz (2000a). A third of these students felt that asking friends (Raffi's way) was a good way to get a sense of the opinion in the school. Their reasons indicated that they wanted to predetermine the survey results and this was a way to make it happen. For teachers of statistics who promote the importance of random sampling this is a troubling result. So much for democracy!

Another striking result was that only 12% of the students recognized the potential for bias in the self-selection method outside the cafeteria. For example, some groups of students might not eat lunch in the cafeteria. Only about a third of the students surveyed preferred the random sampling approach (Shannon's way), whereas over half felt that Claire's self selection survey approach was the best, because "everyone has the same chance this way." The fairness criterion for sampling that Jacobs (1997) found so prevalent among younger students is still quite robust among older students.

Watson and Moritz reported on a series of studies to investigate whether students knew what a sample was, whether they would be sensitive to sample size, and whether they would recognize the possibility for bias in real sampling situations from the media (Watson & Moritz, 2000a, 2000b; Watson, 2004). In one longitudinal study they examined students over a 3 to 4-year period, administering the same four tasks three times to students who were in Grades 3 through 11. Students were asked the same questions 2 years later and again 4 years later to see if they had grown in their understanding of a sample as representative of the population from which it was drawn (Watson & Moritz, 2000b). The first question asked was "If you were given a sample, what would you have?" Some students gave examples of samples—samples of food, blood sample, a free sample in the mail, when asked what they would have if they had a sample. Others responded "a little bit," or "a small portion." Still others mentioned a "bit of something" or "a part of a group." Overall, students' thinking ranged from personal examples, to the notion of a piece, to the idea that a sample should be a representative piece of something larger.

The second question asked students whether they would put more faith in a friend's recommendation for a car purchase or in the recommendation of Consumer Reports magazine, or if it did not matter one way or the other. Watson & Moritz's other two questions were based on articles from a newspaper. One article claimed that over 90% of those who phoned in on a survey were in favor of legalizing marijuana, and another article generalized a claim 'that 6 of every 10 students from a sample in Chicago could easily bring a handgun to school' to all of the United States. Thus all three of these contexts had the clear potential for bias.

Watson & Moritz describe a progression in thinking about samples in which students (a) do not distinguish between sample and population, then (b) recognize the difference between sample and population but really wanted to sample *everyone*, and finally (c) realize that samples can be used to represent the population and to estimate population parameters. Watson and Moritz found considerable growth in their longitudinal study—50% of the students improved on the four questions after 2 years, and 75% had improved from their initial responses after 4 years. The cumulative school experience and outside world experience seems to have improved these students' understanding of the concept of a sample.

In another study Watson and Moritz (2000a) interviewed 62 students who had completed the four survey questions in their longitudinal study. In the interviews Watson and Moritz included some more general questions on sampling and sample size such as a "hospital-type" problem in order to investigate whether students would be sensitive to issues of sample size and variability. (The prototype of this task involved the gender percentages of babies born in large or small hospitals. See for example, Kahneman and Tversky, 1972; Shaughnessy, 1977). The hospital-type task went as follows: A sample of 50 students was taken from a large urban school and a sample of 20 students was taken from a small rural school. One of these samples was strange in that it had 80% boys. Which do you think is more likely? (a) The sample is from the small school, (b) it is from the large school, or (c) it could be from the large school or the small school? As part of this study, Watson and Moritz (2000a) identified six different levels of understanding sampling, using *selection criteria* and *sample size* as coding variables. They determined these levels by applying both the SOLO model and their three-tiered model of statistical literacy—knowing terms, applying terms in context, and asking questions and critiquing the media (Watson, 1997). Their six levels of understanding samples ranged from small samplers with no sampling method to large samplers using random sampling methods that were sensitive to bias. Small samplers were those students who were content to ask just a few people, or to listen to their friend rather than Consumer Reports about the car purchase question. Large samplers recognized the increased power of information when it represented a larger fraction of the population. Students' own preferred sampling methods ranged from pre-selected or distributed criteria so that "things were fair," like in the Jacobs (1997) study, to large samplers who wanted to control for bias by pulling random samples.

Watson and Moritz were troubled that so many students failed to recognize the bias in the handgun newspaper article that made a claim about the entire United States on the basis of a survey of just one high school in Chicago. Some students did question the validity of the claims of the Handgun newspaper article, exhibiting Tier 3 reasoning on the literacy scale, but such students were rare. From 55–65% of the students surveyed in each grade level were quite content to believe that those reported interviewed in the Handgun question were representative of everyone in the United States. As for the sample size task, only 9 of 41 students interviewed said that the strange result of 80% boys in the school was more likely to occur at the rural school, because there is more relative variability in small samples. Many of the students who got the School Problem wrong did see the need for and benefits of larger samples on the other questions. They just could not apply what they knew about large/small samples to the School problem. This suggests teachers have their work cut out for them to help students become aware of potential sources of bias in sampling, and to reach the level of critical thinkers at Tier 3 on Watson's statistical literacy hierarchy. Students can recognize the importance of getting a larger sample as a means to "avoid getting all the same," or as a means to "avoid too many extreme outcomes" or as a means to "balance" the input more fairly. But they may still not see that larger samples provide a method of tightening relative sample variability. Watson and Moritz sum up what they found about the development of students' understanding of samples as follows:

> Students initially build a concept of sample from experiences with sample products in medical and science related contexts, perhaps associating the term *random* with sampling. As students begin to acknowledge variation in the population, they recognize the

> importance of sample selection, at first attempting to ensure representation by predetermined selection but subsequently by realizing that adequate sample size coupled with random or stratified selection is a valid method to obtain samples representing the whole population. (Watson and Moritz, 2000a, p. 63)

Research on Students' Understanding of Graphs

Graphs are critical for data representation, data reduction, and data analysis in statistical thinking and reasoning. A complete review of all the research on reasoning about graphs is well beyond the scope of this chapter. A thorough review of the literature concerning sense-making in graphs is available in Friel, Curcio, & Bright (2001). They define graph comprehension as "the ability of graph readers to derive meaning from graphs created by others or by themselves (p.132)." They discussed the influences of visual perception, the characteristics of graph readers, and the effect that experience with statistics all have on the ability of people to make sense of graphs. Friel et al. (2001) were particularly interested in students' ability to comprehend statistical graphs, as contrasted with graphs of functions in algebra or calculus. It is interesting to compare the references in the Friel et al. review with another article by Roth and Bowen (2001) on reading graphs that appeared in the same issue of that journal. Of the hundreds of references in these two articles, only four or five references appear in both. The overall scope of research on understanding graphs has become enormous. For my purposes, I want to concentrate on a few areas of research on graph sense that are of particular importance to statistical reasoning.

Research on Understanding Particular Types of Statistical Graphs

An analysis of the student results on graph items from the 1996 NAEP indicated that although students performed well when reading information represented in pictographs and stem-and-leaf plots, Grade 12 students ability to read and interpret histograms or box plots lagged behind their performance with other graphical representations (Zawojewski & Shaughnessy, 2000). Performance on histograms, which requires some proportional reasoning, was especially low. Students tested on 1996, 2000 and 2003 NAEP administrations could read graphs fairly well but had trouble interpreting graphs, and even more trouble making predictions based on graphical information. (Zawojewski & Shaughnessy, 2000; Tarr & Shaughnessy, in press). On one extended constructed response task students were given data for railroad ticket sales and asked which of two different graphical representations of the same data they would use in a presentation on how sales had grown. Shaughnessy and Zawojewski (1999) reported that only 2% of Grade 8 students gave a reason for their choice of graphs that merited the highest score on the NAEP rubric for the rail task. Even though an additional 18% of students gave partially correct answers, the overall performance, especially the number of omits on this NAEP task, suggests that students' graph interpretation skills are weak.

Other research has confirmed some difficulties that students have when reading and interpreting particular types of graphs. Researchers have examined student thinking on bar graphs (Pereira-Mendoza & Mellor, 1991), line graphs (Aberg-Bengtsson & Ottoson, 1995), stem-and-leaf plots (Pereira-Mendoza & Dunkels, 1989; Dunkels, 1994), box-plots (Carr & Begg, 1994), scatter-plots (Estepa & Batanero, 1994), pictographs (Watson & Moritz, 2001), and histograms (Melitiou & Lee, 2002). Pereira-Mendoza (1995) suggested that children should:

1. Explore the assumptions underlying the classification of data and interpretation of the meaning of data.
2. Discuss and explore the possibility of alternative representations.
3. Predict from the data. (Pereira-Mendoza, 1995, p. 6).

His point is that by directing students' attention to alternative representations, teachers can help move students beyond mere drawing and tabulating of data to more critical elements in graph sense.

Carr and Begg (1994) introduced box-plots to 11- and 12-year-old students in order to investigate whether such graphs were appropriate for elementary school students. Their informal observational study of 8 students included a brief instructional component on constructing box-plots, followed by unstructured student interviews to determine if students understood the ideas of center and spread. They concluded that box-plots are an appropriate topic for students in this age group provided that teachers emphasize the understanding and interpretation of the plots, and not just the construction of them. However, not all researchers are in agreement about early introduction of box-plots. Bakker et al (2004) pleaded for delaying the introduction of box-plots due to the difficulties that middle school students have with the proportional reasoning needed to construct and interpret box-plots.

The research on students' understanding of particular graph types suggests that teachers should (a) include a variety of graphical representations and (b) go beyond mere graph construction to discuss the meanings and interpretations of the graphs. Lehrer and Romberg (1996) note that most examples of graphs in textbooks are preprocessed, and they recommend more attention be given to having students construct and share their own graphical representations of data.

A Closer Look at "Bar" Graphs

Given the research on students' understanding of particular graphs, consider the complexity of just one type of graph, the bar graph. Bar graphs are first introduced in the elementary grades to represent frequency counts of categorical or integer data. They follow in a sort of natural way from stacked dot plots, allowing data representation to move from discrete to continuous displays. But this can be a big jump for students, as the exact count that was visible in a dot plot is no longer obvious in a bar and the vertical axis may no longer be integer valued. Bar graphs are first used to represent frequencies, but later on they are used with the bar heights that represent relative frequencies. Still later, when histograms are introduced to represent the frequency or relative frequency of continuous data within continuous intervals, the horizontal axis is no longer categorical or integer valued, it also becomes a continuous scale. Relative frequency distributions are a critical tool for normalizing data in order to compare unequal-sized data sets—but research has shown that this is a very challenging issue for students to grasp.

In their analysis of *document literacy* among adults (discussed previously in this chapter), Kirsch, Jungeblatt, & Mosenthal (1998) and Mosenthal and Kirsch (1998) have done large-scale studies that point out the difficulties that adults have with bar graphs or line graphs that require more than simply reading the data. When integration of information in bar graphs is required, or when adults are asked to summarize in writing the conclusions from a graph, they perform much more poorly than on simple reading-the-graph tasks. These results are consistent with reports about NAEP tasks (Zawojewski & Shaughnessy, 2000; Tarr & Shaughnessy, in press) and with what Gal (2002) discussed about the Grade 12 TIMMS results on graph sense.

The conceptual complexity of bar-type graphs continually increases throughout the teaching and learning process. The values and scales on the horizontal and vertical axes morph from counts, to percentages, and then to percentages over an interval. Watson and Moritz (1997b) discussed the complexity in this transition among graphical representations (see also Moritz & Watson, 1997). The cognitive transition in bar graphs mirrors the transition from additive to proportional to distributional thinking previously discussed in this chapter, with a heavy reliance on proportional thinking. Without proportional thinking, students see only counts and miss the power of graphs to reduce data and show trends. Proportional thinking is the lynchpin in making sense of statistical graphs.

Research on Components of Graph Sense

Curcio (1987) introduced a framework for graph sense by building on research in the literature on reading. She identified three elements—form, content, and topic—which contributed to the ability of students to construct their own graph-reading schemata. In later work she characterized three levels of reading of graphs: *reading* the graph, reading *within* the graph, and reading *beyond* the graph (Curcio, 1989). These categories help to shed light on the complex nature of understanding graphs. For example consider the graph of fruit juice consumption over time in Figure 21.11. This graph is based on data collected by the United States Department of Agriculture (USDA, 2004) and posted on their website.

In order to be *Reading* the graph, students have to at least understand the scale and the measurement units. This is a graph over time, from 1970 to 2000, and the vertical axis is in gallons per person per year. It is also important for students to realize and understand that "per capita" data are rate data. Reading *within* the graph is especially important for graphs over time, in order to be able to discuss general trends. For example, the timeline shows a fairly steady increase in consumption, with a couple of sudden dips. Why is there a general increase, and why did those dips occur? Reading *beyond* the graph, Curcio's highest level of graph comprehension, includes such skills as projecting into the future and asking questions about the data. For example, one might suggest that fruit juice consumption appears to be dropping at the end of the graph, and that it may level off at some point in the future. Or, one might ask where these data came from, how they were collected, and how the USDA managed to calculate or estimate "gallons of fruit juice per capita" for each of the years.

I have previously argued (Shaughnessy, Garfield, & Greer, 1996) for another level of graph comprehension beyond Curcio's three levels, that is, reading *behind* the data or graph. This involves more than just reading *beyond* the graph. Statistics are within a context. As Wild and Pfannkuch (1999) have noted, one must always search for special causes of variation in the data. The special role of statistics as a scientific dis-

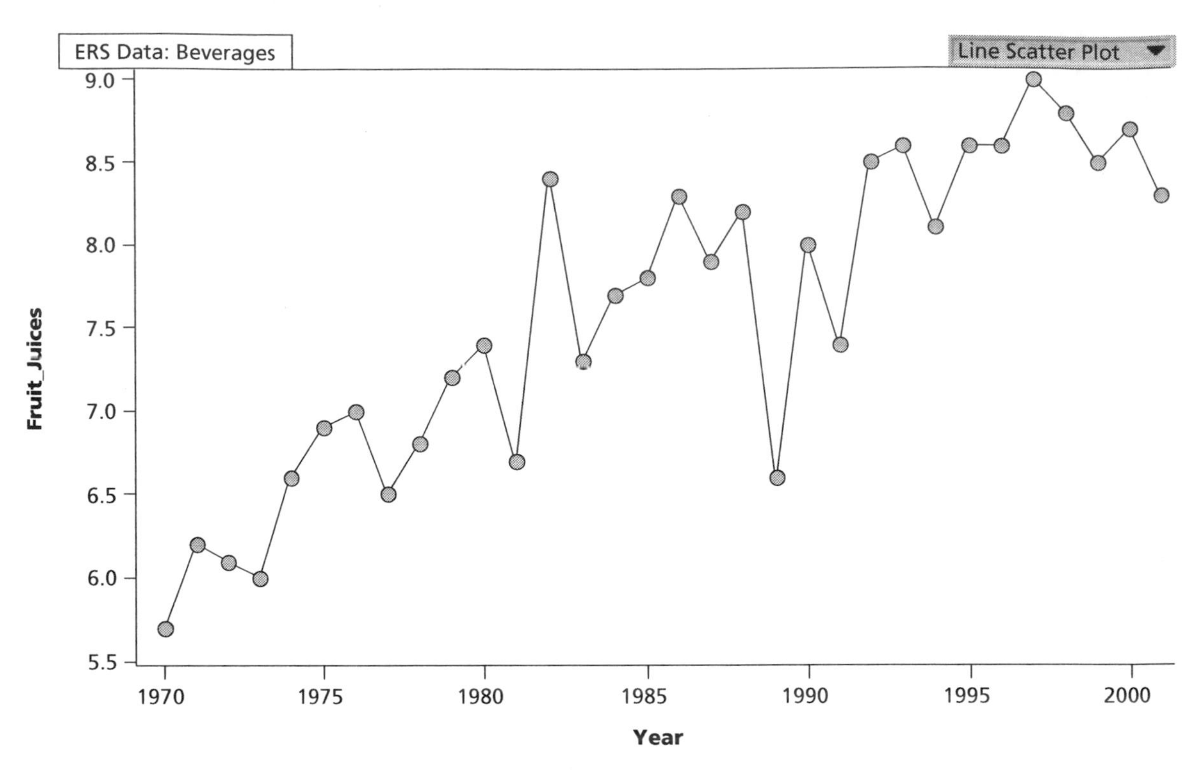

Figure 21.11 Fruit Juice Consumption, Gallons per capita, over time (USDA, 2004).

cipline lies in making connections between the context and the graph, that is, what lies *behind* the graph. For example, in the juice consumption graph one might look at historical, economic, or demographic influences that may have affected juice consumption over the last 40 years. In reading *behind* this graph, one might conjecture that in the mid 1980s, better production of juice came about. Perhaps a switch occurred from smaller orchards to larger centralized locations for collecting and processing fruit for juice. As time went on, a global economy arose so that fruit juice could be produced and shipped more easily all year round from somewhere in the world. This greater availability could contribute to a more rapid growth in fruit juice consumption. Perhaps a killing frost explains the two years, 1981 and 1989, that show a drastic drop in fruit juice consumption. Whatever the cause of the drop *behind* the data in those years, a traumatic event affected juice consumption but was followed by a quick rebound to the status quo in subsequent years. The drop-off during more recent years might be due to increased consumption of other beverages, such as soft drinks or bottled water. These are just a few of the possible special causes that might affect fruit juice consumption over time, which involve looking *behind* the graph, and using some data detective skills to analyze graphical information.

Watson and Moritz (1997a) applied their three tiers of statistical literacy—(a) knowledge of basic statistical terminology, (b) applying statistics in context, and (c) challenging statistical claims—to analyze students' responses to some graphs taken from newspapers. The tiers of statistical literacy match up fairly well with Curcio's three levels of graph sense when one considers statistical literacy about graphs. Watson and Moritz found a lack of attention to scaling issues in the data and graphs they gave their students. Furthermore, only 10% of their students were aware of the complexity of or questioned some of the features of one particularly misleading graph. When students were asked to make some calculations based on information in the graph, many of them ignored the graph altogether and reverted to personal experiences. Watson and Moritz discovered gaps in students' graphical literacy within all three of their levels of statistical literacy.

In another series of studies that investigated students' thinking about pictographs Watson and Moritz (2001) identified three levels of student reasoning. In follow-up studies 3 years later they documented growth in students' thinking about pictographs. They also found that cognitive conflict could improve students' initial reasoning. When a sample of Grade 3 students was shown responses from other children about the pictographs, their reasoning about the graphs tended to improve.

Investigations of students' graph comprehension and graph interpretation skills have been one approach to exploring the components necessary to develop good graph sense. Another approach has involved analyzing student-generated graphs. Following on earlier work in relation to the SOLO model (Watson et al., 1995), Chick and Watson (2001, 2002) used the data card tasks that included information about students' weight, age, favorite activity, and number of fast food meals eaten per week. Students in Grades 5 and 6 were put into groups of three and asked to generate some hypotheses and to construct graphical representations of the information on the case cards. Chick and Watson's students produced three levels of representations of information in the data cards. Some students depicted individual aspects of the data set, such as a table of single values of one variable with no interpretation offered. Other students produced a graph of a single variable, and still others produced a bi-variate representation, such as a scatter graph, indicating a possible relationship between two of the variables. Students' reasoning on the task followed a pattern similar to their graphical representations: looking at individual data points, considering the entire range of values for one variable, or conjecturing cause-and-effect relationships among several variables. Nearly half the students could interpret the data at a higher level than they could represent it graphically, suggesting that students' graphical sense may lag behind their data interpretation skills.

The attention to individual data points exhibited by Chick and Watson's students has been pointed out by other researchers, notably Cobb (1999) and Konold and Higgens (2002, 2003). These researchers have suggested that case-value representations of data are a good starting point to help students develop graph sense and begin making connections to the context. Konold and Higgens (2003) presented case value data in a *value bar* representation and demonstrated a progression of representations from value bars, to end points of the value bars, to stacked dot-plots. This transition path could aid students and teachers to move beyond students' initial preferences for individual case value plots to stacked dot plots that provide a visual representation of the entire distribution of the data, including information about shape, center, and spread.

A summary analysis of the literature on students' understanding of graphs by Friel et al. (2001) identified six behaviors that they considered to be closely associated with graph sense. Each of these six behaviors seems to fit nicely with one of Curcio's three levels of graph reading (read the data, read within the data, read beyond the data). I also suggest two additional behaviors that fall under the level of reading *behind* the data.

1. Recognizing components of graphs (*Reading* the data).
2. Speaking the language of graphs (*Reading* the data).
3. Understanding relationships among tables, graphs, and data (Reading *within* the data).
4. Making sense of a graph, but avoiding personalization and maintaining an objective stance while talking about the graphs. (Reading *within* the data).
5. Interpreting information in a graph and answering questions about it (Reading *beyond* the data).
6. Recognizing appropriate graphs for a given data set and its context (Reading *beyond* the data).

In addition I would add

7. Looking for possible causes of variation (Reading *behind* the data).
8. Looking for relationships among variables in the data (Reading *behind* the data).

The cumulative results from a number of researchers on graph sense indicate that students have poor graphical interpretation skills and are often unable to reason beyond graphs. Also unless the graph is very straightforward, students may even have trouble just reading a graph at all. In the special case of statistical graphs, reading *behind* the data is critical to making connections between the context and the data. Basic graph sense such as reading, reading within, and reading beyond graphs is critical to statistical thinking, reasoning, and literacy.

Technology and Research on Learning Statistics

My principal research interest in the use of technology in teaching statistics is the development of students' statistical thinking in the conceptual areas discussed in the previous section: averages, variability, information from samples and surveys, and graph sense. The availability of powerful computing tools has led to improved methods of analyzing data and exploring data graphically (Hawkins, Jolliffe, & Glickman, 1992). What sorts of technological environments can enhance student learning of the major concepts in statistics?

Biehler (1993, 1994a, 1994b, 1997) has written extensively on the type of software tools that are necessary to enhance the teaching and learning of statistics. Over a decade ago Biehler described aspects of technology needed to empower students to do interactive exploratory data analysis, using visualization and simulations tools to understand statistical concepts and methods. He proposed the following components for ideal software:

- Student tools for data analysis, for method construction and evaluation, for modeling and for visualization, that can grow and expand along several paths into a professional version, instead of mere technically reduced student versions of professional systems
- A system of coordinated computer experiments, learning environments, and major visualizations that can be adapted to students' and teachers' needs. (Biehler, 1994b, p. 3).

In another early analysis of and report on the use of technology for teaching statistics, Garfield (1990) led a Working Group on Technology and Data of the American Statistical Association (ASA). This Working Group outlined attributes of technological environments that can facilitate the learning of data handling, including:

Direct access, which allows students to view and explore data in different forms, including subsets of data and different visual representations.

Flexibility, which allows students to experiment with and alter displays of data, change intervals on a graph, and explore different models that may fit the data.

Connectedness, so that students are able to access resources on the Internet, as well as to obtain software or data used in the study of other disciplines.

Representations, including dynamic ones from which students may choose among different graphs in order to select the best way to interpret and display a data set.

Both Biehler and Garfield have continually recommended that in order to truly enhance statistics education, technology for learning statistics should go way beyond mere statistical packages that carry out procedures. They advocate for technology that puts the design and representation of data structures in the hands of the student, rather than merely allowing selection of procedures from an already existing repertoire. In an analysis of existing and desirable statistical software, Biehler (1997) distinguished among *tools, resources,* and *microworlds* for the teaching and learning of statistics with technology. By *tools,* Biehler means the type of software and hardware support that professional statisticians use to practice their trade. For example, *Minitab* (2005) is a software package that consists primarily of statistical *tools* for data analysis. *Resources* would include data sets with references, and background and context information. These types of data sets can often be found at Internet sites. The ERS Web site maintained by the Department of Agriculture (2005) from which the food consumption data (see Figures 9 & 11) were downloaded is one example of what Biehler means by resources. *Microworlds* provide a creative, exploratory environment for students to represent data, and to carry out simulations. Both *Tinkerplots* (Key Curriculum Press, 2005a) and *Fathom* (Key Curriculum Press, 2005b) are examples of microworlds that are flexible and that put a lot of the power in the hands of the students to create a variety of types of graphs and data representations.

Ben-Zvi (2000) presented examples of types of technological tools for learning statistics that are similar to Biehler's categories, such as statistical software packages, microworlds, tutorials, and resources on the Internet. According to Ben-Zvi, tools for statistical learning have been developed to support these areas:

1. Students' active knowledge construction, by "doing" and "seeing" statistics.
2. Opportunities for students to reflect on observed phenomena.
3. The development of students' metacognitive capabilities, that is, knowledge about their own thought processes, self-regulation, and control.
4. The renewal of statistics instruction and curriculum on the basis of strong synergies among content, pedagogy, and technology. (Ben-Zvi, 2000, p. 128)

Ben-Zvi shared a detailed example of some middle school students as they worked with spreadsheets and constructed graphical representations of some data on Olympic records for the 100-meter dash. He documented how even a simple tool like a spreadsheet can be very powerful in the hands of students who are directing their own learning and how a spreadsheet can provide support for the four areas he identified.

Bakker (2002) referred to *Fathom* and *Tinkerplots* as "landscape-type" tools and contrasted them with "route-type" tools with which students have far fewer choices at their disposal. Examples of route-type tools include the *Mini-tools* discussed by Cobb et al. (1999) and the tools created by the Chance–Plus project such as *Prob-Sim* and *Data-Scope* (Konold & Miller, 1994). Landscape-type tools put students in very open situations, with a variety of powerful choices, including sorting and arranging data into a variety of visual display formats. Bakker expressed concern about starting students with these landscape-type tools, as they might overwhelm students with too many representational choices. He argued that it is better to first introduce students to some more limited route-type tools, such as *Prob-Sim or Minitools,* that can provides students with focused opportunities to explore particular concepts. Friel (2007) has provided a detailed review of research on the interaction of technology with the teaching and learning of statistics, including an analysis of the strengths and weaknesses of a variety of types of technological tools for teaching statistics, including both route-type and landscape-type tools.

Research on what students learn about statistics or how their statistical thinking develops while using new statistical tools like *Tinkerplots* and *Fathom* is relatively sparse as of this writing, especially when compared to the general explosion of research in other areas of statistics education. During the development phases of statistical software most of the developers' energy is focused on the tools themselves, or on professional development work with teachers to help them use the new software. Normally very little research is done on student learning or on student conceptual growth in statistics during the developmental phase of software. Finzer (2002) made a good case for the simultaneous integration of research on student learning with software development, so that students' learning issues can inform the software development team in mid-stream, and vice-versa:

> In some ideal world the boundary between development of software for use in research on learning and development of software for classroom use would be truly porous so that researchers could easily adapt classroom software for research purposes and classrooms would reap the benefits of educational research. (Finzer, 2002, p. 1)

General overall access for all students and teachers to statistical software is still a problem. Too few teachers and schools are currently using data analysis tools like *Fathom* or *Tinkerplots.* Even AP statistics teachers are more likely to use graphing calculator statistics packages or some of the more traditional data-crunching packages such as *Minitab.* To argue the flip side of Bakker's concern about landscape-type software being too open, in the current teaching of statistics there may be too much route-type and not enough landscape-type use of technology in the teaching of statistics. The landscape-type software is more likely to help students to become good data detectives. Open-ended micro-worlds give students a lot of autonomy and power in data exploration and analysis, whereas some route-type software might be a bit confining and sometimes overly procedurally oriented.

Some Research on Students' Statistical Thinking with Technology

Bakker and Gravemeijer (2004) presented a detailed analysis of a teaching experiment with seventh Graders that was designed to get students to reason about distributions in informal ways. They noted four critical components of distributions: centers, spreads, density or clustering, and skewness. Their teaching experiment was designed to have students encounter these four components and to move students along from looking at data as individual case values to reasoning about entire distributions of data. Using two computer Mini-tools, a value bar tool and a stacked dot-plot tool, Bakker and Gravemeijer had students investigate two data sets on battery life, data sets that had also been used by other researchers (Cobb, 1999; Cobb, McClain, & Gravemeijer, 2003).

The students first compared the two battery data sets using the value bar tool. During this exploration, students began to use terms like *majority, outliers, reliability,* and *spread out.* Bakker and Gravemejier claimed that the value bar tool helped to provide a visual representation of the mean for students. Subsequently the stacked dot-plot tool allowed students to develop qualitative notions of more advanced aspects of distributions such as frequency, classes, spread, quartiles, median, and density (Bakker & Gravemeijer, 2004). The dot plot tool also allowed students to partition data into 2, 4, or more equal-sized groups. This provided an underlying structure for development of concepts such as median, box-plot, density, and eventually histogram. Students were able to reason about clumps of data and compare clumps across the two battery-life distributions, so that the underpinnings of Konold's "signal amid the noise" conception of the mean were available. Cobb (1999) described a similar learning trajectory for a group of middle school students who were exploring the battery data. Cobb's principal lens for the study was classroom discourse and social interaction, which he claims is a necessary component for the type of growth and learning witnessed in his work.

On the basis of the statistical development that they witnessed among the students in their studies, Bakker and Gravemeijer, and Cobb, concluded that it is important to provide opportunities for students to contribute their own ideas to the statistical learning process and that teachers need to provide a lot of time for discussion and interaction during class explorations of data. They also believe that formal measures to describe distributions, such as median and quartiles, should not be introduced until after students have had opportunities develop their own intuitive notions about distributions.

In reading the Bakker and Gravemeijer study, it is striking how natural the integration of the Mini-tool environment was in the teaching experiment. Furthermore, the process of students' growth from first looking at data as individual data cases, then considering clumps of data, and finally making comparisons and decisions based on clumps and spreads, was clearly supported and enhanced by the Mini-tools technology. Even though these two Mini-tools are more in Bakker and Gravemeijer's route-type software category, the students were able to grow considerably despite the software's limited capabilities. This study was enhanced by the strong alignment between the task environment and the Mini-tool's capability to explore that task environment. Based on these studies on teaching statistical concepts with technology, the need for a close match between the statistical questions that are being investigated and the capability of the software to directly tackle those questions seems paramount. These studies also suggest that the type of classroom discourse that takes place during a statistical exploration also has a major impact on students' conceptual growth in statistics.

Friel (2007) shared a number of visual representations of data from two different types of roller coasters using *Tinkerplots*. Stacked dot plots allow students to partition the two data sets in different ways, to compare spreads, middles, and clumps. Gradually the shape of the two distributions becomes apparent. The roller coaster comparison task is rich in student exploration possibilities, similar to the battery-life task discussed above. The combination of interesting data sets with versatile software tools has made these two environments, the roller coasters and the batteries, excellent candidates for future teaching and research projects.

In reviewing a series of efforts in which researchers have asked students to explore and compare data sets within rich technological environments, Konold et al. (unpublished manuscript) summarized a number of different ways to consider data. Among their research sources were cases of primary teachers who wrote about how their students dealt with data (Russell, Shifter, & Bastable, 2002), some research by Cobb et al. (2003) on how students describe co-variation in scatter-plots, and some of their own work with secondary school students who explored a large multivariate data set containing information on a number of variables about high school students (Konold, Pollatsek, Well, & Gagnon, 1997). According to Konold et al. (unpublished manuscript) data can be viewed as *pointers, case values, classifiers,* or *aggregates.* A major issue in statistics education is to find ways to help students' thinking about data to evolve along this continuum. Although Konold has written about the importance of the mean as the "signal amid the noise," indicating that he considers data as aggregate to be the most important way for students view data, in this working paper he has also acknowledged the potential benefits of viewing data in other ways. For example, case values can help to highlight the variability in data, a concept that gets short shrift in school mathematics (Shaughnessy et al., 1999). The types of technological tools that Konold et al. include in their paper support student conceptual growth on centers, as well as on variability.

As noted earlier in this chapter, students have a very difficult time learning about sampling distributions and the accompanying statistical ideas such as the law of large numbers (Saldanha and Thompson, 2003). Chance et al. (2004) shared research on a series of experiments on sampling distributions conducted with college students in introductory statistics courses. This work built upon earlier explorations of tertiary students' understanding of sampling distributions (del Mas, Garfield, & Chance, 1998; Garfield, delMas, & Chance, 1999), and used a software tool developed by del Mas (2001) to run simulations that generate sampling distributions.

In their research Chance et al. identified four prerequisite concepts that students need in order to understand sampling distributions: variability, distribution, sampling, and the concept of the normal distribution. Research reviewed in previous sections of this chapter has already noted the complexity and diversity of student thinking about variability and about samples and sampling, and has noted issues that students deal with when comparing distributions. With the addition of the normal distribution as a prerequisite concept, it is no wonder that Saldanha and Thompson (2003) encountered so much difficulty in their teaching experiment on sampling distributions with secondary students. Chance et al. used the developmental model of statistical thinking that Jones et al. (2000) had introduced when working with elementary and middle school students. Chance et al. uncovered and subsequently validated five different levels of reasoning about sampling distribution among their college students, namely, idiosyncratic, verbal, transitional, procedural, and integrated types of reasoning.

Even though Chance et al. spent considerable time targeting tasks in which their students ran software simulations to explore sampling distributions, they found only a few students who actually managed to integrate all the statistical concepts involved as they explored properties of sampling distributions.

> Part of the problem in developing a complete understanding of sampling distributions appears to be due to students' less than complete understanding of related concepts, such as distribution and standard deviation. We have found our own research progressing backward, studying the instruction of topics earlier in the course and the subsequent effects on students' ability to develop an understanding of sampling distributions. For example, initially we explored student understanding of the effect of sample size on the shape and variability of distributions of sample means. We learned, however, that many students did not fully understand the meanings of *distribution* and *variability*. Thus, we were not able to help them integrate and build on these ideas in the context of sampling distributions until they better understood the earlier terminology and concepts. (Chance et al., 2004, p. 312)

Like Saldanha and Thomspon, they go on to conclude that the "concept of sampling distribution is a difficult concept."

> The few pages given in most textbooks, a definition of the Central Limit Theorem, and static demonstrations of sampling distributions are not sufficient to help students develop an integrated understanding of the processes involved, nor to correct the persistent misconceptions many students bring to or develop during a first statistics course. Our research suggests that it is vital for teachers to spend substantial time in their course on concepts related to sampling distributions." (Chance et al., 2004, p. 312)

If the results of the research on student understandings of average, variability, and distribution that were discussed in earlier sections of this chapter can be effective in promoting better and earlier teaching and curriculum development in K–12 statistics, perhaps Chance and others will eventually have a stronger statistical foundation to build upon when they work with tertiary students. In the meantime, tertiary statistics instructors need to heed Chance et al.'s advice and start teaching from where their students actually are in their knowledge of statistics. The research reviewed in this section suggests that technological tools are very important for helping students to transition from those naive conceptions to richer, more powerful understandings of statistical concepts.

RESEARCH ON AND DEVELOPMENT OF TEACHERS' UNDERSTANDING OF STATISTICS

Much of the research on teachers' understanding of statistical concepts has been derived from professional development work that is intended to extend teachers' knowledge and competence in statistics. This is partly a matter of convenience, combining research with professional development. It is also partly because it is difficult to obtain direct information about teachers' knowledge about statistical concepts. Most K–12 mathematics teachers in the United States have very little background in statistics. The exceptions are those teachers who may have had a concentration in statistics during their masters program for secondary teachers, or middle school teachers who completed one of the few special programs that exist in the United States for middle school mathematics teachers. Teachers know that their statistical understanding is shaky, and so attempts to obtain direct information from them using surveys or interviews can be embarrassing for them. As a result, a good deal of what is known about how teachers think and reason about statistics has tended to be somewhat anecdotal. Sometimes field notes can be taken during observations of classroom statistics activity, or retrospective statements can be obtained from teachers while debriefing trial lessons of new statistics materials.

Research on Teachers' Understanding of Statistics Within Professional Development

Over the course of several curriculum development and professional development projects, Rubin used field notes and reflections after interviews to gather information on teachers' thinking about statistical concepts. In one project several teachers volunteered to trial the Stretchy Histograms and Shifty Lines software from the Elastic project with their students (Rubin & Rosebery, 1988). These two tools were among the very first tools to allow users to click and drag elements of graphs, like bars in histograms, or lines in scatter-plots, and to observe changes in the mean and median of distributions or changes in the equations of regression lines. Rubin and Rosebery found that when teachers asked their students to alter histograms by adding values to the original graph, the teachers became puzzled. For example, they wondered why did the median not change, when the mean did change? The teachers were unaware that there might be multiple occurrences of the median value in the data set, or that the median remains constant if the same number of data values are added in on either side of it. As a result of Rubin and Rosebery's work, recent software packages like *Tinkerplots* prefer to use stacked dot-plots as the default representation mode rather than histograms. Dot plots preserve every data point, and the median value can easily be highlighted in a dot plot.

In another project, Hammerman and Rubin (2003) worked with middle school and secondary school teachers as they used *Tinkerplots* to clump data into bins, thereby reducing data complexity. This helped the teachers to find ways to make comparisons between groups of unequal sizes. Research discussed earlier in this chapter documented the difficulty that students have when comparing groups of unequal size (e.g., Watson, 2002), and one would expect teachers to have some of those same difficulties. One data set that Hammerman and Rubin gave teachers consisted of T-cell counts for two groups infected with the HIV virus, a control group with n =186 and an experimental group with n = 46. Gender information within the groups was also of unequal size. (The T-cell data set had previously been used by Cobb, Gravemeijer, Doorman, and Bowers, 1999.) *Tinkerplots* allowed the teachers to create bins by T-cell count, then to count the total number of people in each bin, and finally to create pie-graphs of the percentage of men and women in each bin. The pie-graph percentages can disguise the actual bin counts, and decisions made on percentages alone may be misleading because of small sample sizes within some bins. Conversely, actual bin counts can also be misleading if relative sizes are neglected. However, these teachers were simultaneously able to compare both frequencies *and* relative frequencies for T-cell count by gender, and to use the information to make some conjectures about the relative success of a treatment for HIV for each gender. Hammerman and Rubin (2003) argued that when faced with a lot of complex data, their teachers sought ways to reduce or manage variability, so they could make better judgments.

Several things are striking in the work of Hammerman and Rubin. First, the teachers found ways to use the software to create proportions within bin cells, and to reason proportionally when comparing data sets, rather than purely additively with frequencies. Even though proportional reasoning had its potential pitfalls in this case, the fact that the teachers used proportional comparisons of unequal-sized data sets is promising. Second, the types of graphical representations that are now available to compare bivariate data (gender × treatment in this case) have the potential to revolutionize data analysis in classrooms in the future. At the present a typical AP statistics class approach to the T-cell problem might be to put the data into a 2 × 2 contingency table, sex × treatment, and examine cell values and margin proportions. Imagine in a few years, if middle school students have already explored bivariate data using powerful visual data-representation tools, the standard numerical statistical procedures for analyzing data might not be their first choice to approach bivariate problems. More of the decision-making power for how the analysis is to proceed could be put directly into the students' hands. The work of Rubin and her colleagues has important implications for the statistical education of teachers.

> We believe that one powerful way for teachers to gain expertise in statistical reasoning is to have more experience in "being statisticians" themselves. In the area of exploratory data analysis, teacher education should consist largely of teachers investigating statistical problems that interest them, collecting data, analyzing it, and drawing conclusions in the same way that statisticians would. (Rubin and Rosebery, 1988, p. 17)

In fact, this vision for the statistical education of teachers should be the norm for the statistical education of all of our students.

Makar and Confrey (2002) have also conducted professional development projects in statistics with an imbedded research component. They cleverly merged their teachers' concerns about their students' performance on large-scale state testing in mathematics with an upgrade of the teachers' own understanding of statistics. Makar and Confrey introduced teachers to data analysis with *Fathom* using data from students' performance on the state mathematics test. Teachers were particularly interested in comparing different groups of students on the state test, so Makar and Confrey took advantage of that context. They identified four constructs they wished their teachers to consider as they decided whether differences between groups of students' test scores were really meaningful differences: measurable conjectures, tolerance for variability, context, and an ability to draw conclusions. For example, they pointed out that *tolerance for variability* requires a very different mindset than the deterministic foundations of traditional mathematics courses. Statistics requires decision making under uncertainty. Their teachers discovered a new appreciation for variability within the context of student scores on the state test. What began as a surface-level analysis of just the bare numbers transitioned to a rich discussion related to context when the teachers were told that the numbers in their data sets were actual state test scores for a set of different classes. In analyzing the teachers' work Makar and Confrey created a framework of five levels of reasoning used by the teachers when they compared data sets: pre-descriptive, descriptive, emerging distributional, transitional, and emerging statistical. Makar and Confrey's levels describe the growth in the teachers' understanding of the role and importance of variability when comparing data sets, and the difficulty of making conclusions about measurable conjectures.

In another study Makar and Confrey (2005) investigated preservice mathematics and science teachers' understanding of variation and distribution during a teacher education course. The teachers were introduced to the *Fathom* software and then given some assessment data to analyze. Makar and Confrey were interested to see if their participants would compare groups just using means, or if they would look at variation and spread in their comparisons. In interviews these preservice teachers made comparisons between groups of students using conventional statistical terms, such as means, measures of spread, discussions of shape, and the proportion of students who improved on the assessment. However, Makar and Confrey also encountered a good deal of non-statistical language among their preservice teachers, such as "clustered" or "spread out" or "bulk of" or "majority" or "clumps" or "big chunk." In their analysis of the responses Makar and Confrey concentrated on "clumps" and "chunks." A *clump* means exactly what it sounds like, including modal clumps of data. A *chunk* is a contiguous subset of a distribution of data, not necessarily a clump. This language is reminiscent of the "bumps" and "gaps" language used by Friel, Mokros, & Russell (1992) in their work with elementary children in the *Used Numbers Project*. It is an intuitive, natural language used to describe a collection of data. Makar and Confrey were capturing the primary intuitions of the student teachers (Fischbein, 1987), the raw material upon which secondary intuitions can be developed through teaching environments. Makar and Confrey warned that the nonstandard terminology that students use should not be ignored, as it might be a rich source of student understanding that could be overlooked.

Heaton and Mickelson have also explored K–6 teachers' understanding of some statistical concepts (Heaton & Mickelson, 2002; Mickelson & Heaton, 2003, 2004). Several of their research efforts have involved a single case study of one third-grade teacher as she integrated statistics investigations into other parts of her curriculum, such as science, literature, or social studies investigations. One report focused on a teacher's understanding of data and distribution (Mickelson & Heaton, 2004), and another on manifestations of that same teacher's understanding of variability (Mickelson & Heaton, 2003). In some contexts, the teacher exhibited strong statistical reasoning skills, and in other contexts, only much more naive statistical skills. This teacher was acting as both a learner and a teacher in their research project, a duel role that Mickelson and Heaton urge more researchers and professional development leaders to consider, especially with elementary teachers who are trying new ideas out in statistics and data analysis in their classrooms. Although this third-grade teacher had had considerable professional development work in data analysis herself, at times she had difficulty finding an appropriate vehicle to transfer what she herself had learned to the students in her own classroom. Teachers need opportunities to create statistical activities and investigations themselves and to try them out while supported in their classroom by a statistics educator. The teaming of teachers with researchers in the classroom may help to facilitate teachers' abilities to transfer their own insights in statistics to experiences for their students.

Larger Scale Research on Teachers' Understanding of Statistics

Professional development cases, such as those described by Hammerman and Rubin and by Makar and Confrey in the previous section, tend to be with small groups of teachers who are special volunteers. In several studies that involved a larger number of teachers, Bright and Friel (1993) asked primary teachers to create concept maps for probability and statistics. They found that the maps were disconnected and rather sparse in statistical content. Greer and Ritson (1993) surveyed teachers at all levels in Northern Ireland and found a universal need for in-service teacher training in probability and statistics across all grade levels.

Watson (2001b) found a rather clever way to amass a considerable amount of information from teachers on their perceptions of statistics and on their own strengths and weaknesses in the data and chance curriculum in Australia. Using an information profile to assess the need for professional development, Watson asked teachers to rate the importance of certain statistical concepts, to describe their own lesson practices, to estimate their confidence in teaching certain statistical topics, and to suggest two possible student responses to some statistical tasks, one appropriate and the other inappropriate. This proved to be a less threatening methodology for obtaining information on teacher knowledge of statistics than testing them directly. Watson used Shulman's (1987) knowledge typology to guide the categories of questions on the teacher profile, thereby obtaining information about teachers' content knowledge, knowledge of teaching, pedagogical content knowledge, and other knowledge categories identified by Shulman. Watson was able to administer the profile to over 40 teachers, some in individual interviews, some via written survey, and some via an Internet survey. Of note in Watson's findings were the teachers' lack of confidence in statistics and their lack of knowledge about sampling. Also of note was the need expressed by many of her teachers to further their own professional development in data and chance, as well as their frustration with the lack of support from local authorities to help them in teaching statistical topics.

Watson's profile was designed to provide information about teachers' content knowledge in statistics while assessing their professional development needs. A next step would be to administer such a profile anonymously, directly to teachers, to get a better understanding of their content knowledge. Inasmuch as this is a potentially risky and threatening venture for classroom teachers, it might be easier to investigate the statistical content knowledge of preservice teachers. Canada (2004) conducted a study of preservice elementary teachers' conceptions of variability in three contexts, data and graphs, sampling, and probability. Building upon the aspects of variability identified by Wild and Pfannkuch (1999), Canada began by documenting a variety of types of preservice elementary teachers' thinking about variability. From his analysis of students' responses to survey tasks, an emerging framework evolved consisting of three main aspects: expecting variation, displaying variation, and interpreting variation. Canada subsequently used his framework to compare and contrast the thinking of 6 case study students before and after he imbedded teaching episodes on statistics in a mathematics course for preservice elementary teachers. Canada found growth in all three aspects among the case-study students, including richer conceptions involving expectation of variation in data, more flexibility with displays of variation, and stronger interpretations of variation. Canada's evidence suggests that class interventions can help to strengthen preservice teachers' conceptions of variation.

In response to the tremendous need for professional development in data and chance that a number of researchers have identified, Friel and her colleagues developed some excellent professional development and curricular activities for primary teachers to help strengthen teachers' statistical content knowledge and pedagogical content knowledge (Friel & Bright, 1998; Teach STAT, 1996a, 1996b;). In project *Teach-STAT* an intensive 3-week in-service course in data analysis for teachers in Grades 1–6 led to the creation of a professional development manual for statistics educators. The in-service materials for teachers were based on the *Used Numbers* curriculum materials that were written for elementary students (Russell & Corwin 1989; Corwin & Friel, 1990; Friel, Mokros, & Corwin 1992).

In Australia, several trials to deliver professional development in statistics for teachers from a wide geographical area were conducted by the Australian Association of Mathematics Teachers (AAMT) (Watson, 1998). The professional development was delivered by regional satellite television, by video-conferencing across the nation, by the preparation of a CD-ROM for teachers, and by the development of a Website with the help of a capital city newspaper. The material delivered was based on workshops and other material to promote data and chance as suggested by the national mathematics statement (Australian Educational Council, 1991). Evaluation by consultants and participants was quite positive (Watson & Moritz, 1997b). However, as with many such technology-based innovations, when the government funding was exhausted the ability to reach teachers was greatly diminished.

A number of other efforts have addressed the professional development needs of teachers in statistics, such as books by Hawkins (1990) and Hawkins et al. (1992) in the United Kingdom, the six books on data and chance of the *Navigations* (2004a, 2004b) series of the National Council of Teachers of Mathematics (see e.g., Burrill, Franklin, Godbold, & Young, 2004; Shaughnessy, Barrett, Billstein, Kranendonk, & Peck, 2004c), and the book *Statistical Questions from the Classroom* (Shaughnessy & Chance, 2005), just to name a few. As documented in a number of studies discussed above, professional development materials for teachers can lead to windows of opportunity for researchers to explore teachers' own understandings of statistical concepts.

SOME RECOMMENDATIONS FOR FUTURE RESEARCH

At the end of my chapter in the first edition of this *Handbook* I put forth some recommendations for future research in the teaching and learning of probability and statistics (Shaughnessy, 1992). One wonders if anyone ever really pays attention to such recommendations, but editors of research handbooks and readers who search through the chapters of such handbooks usually expect a look back over where the research has been, and some suggestions for where the research could (should?) go next. As an author, it would be fun to have a conversation with readers at this point in the chapter to see what recommendations the readers themselves would make for future research after plowing through this lengthy tome. I do have some thoughts for future research, but first I will take a peek back at my recommendations in the first edition of this *Handbook*. How have they fared?

Research Recommended in the First Edition of the Handbook

As I mentioned in the introduction to this chapter, research in probability and statistics was just beginning to take hold in a number of countries during the decade prior to the arrival of the first *Handbook of Research*. At that time the field was rather naive, and the suggestions I made for future research at that time were probably equally naive. That wish list included: The need for assessment instruments; a request to investigate secondary students' and classroom teachers' conceptions of probability and statistics; a methodological recommendation for more teaching experiments in probability and statistics; and a request to investigate the effects of technology on student learning of statistical concepts. There were also recommendations for cross-cultural studies comparing students' statistical thinking, and a suggestion to investigate the role of metacognition in solving statistical problems, but to my knowledge no one took those two recommendations very seriously.

Progress has been made on some of the recommendations, although in some cases not as much as I would like to have seen. Researchers have investigated secondary students' conceptions of statistics, for example, Saldanha and Thompson (2003) and Shaughnessy et al., (2004a, 2004b, 2005). In addition, many of the studies by Watson and her colleagues reviewed in this chapter have been conducted with students from a range of grade levels, including Grades 9 and 11. However, much more of the research on students' understanding of statistics has been conducted with students from Grades 3 to 8. Perhaps this is because students in these grade levels are actually *doing* some statistics, whereas there is still no guarantee that secondary students, at least in the United States, will study any statistics. Statistics is growing in the United States secondary schools but it is still treated as an optional content area in many schools. It should not be optional.

The research on teachers' understanding of statistics has also begun to blossom with work like that of Hammerman and Rubin (2003), Confrey and Makar (2002), Watson (2001b), and Heaton and Mickelson (2004). And, there are a few teaching experiments that have been conducted (Saldanha & Thompson, 2003; Saldanha, 2003; Cobb et al., 2003) as well as instances of the more modern version of a teaching experiment, the "design experiment" (Petrosino et. al., 2003), but teaching experiments are still not a frequent occurrence in statistics education. They yield a wealth of information, but they are hard to conduct, very intense, and very time-consuming if done properly. As for research on the influence of technology, even though some interesting "landscape-type" software packages are now available like *Tinkerplots* or *Fathom*, very little research has been conducted on how or what students learn about statistics with these powerful statistical tools. Student reasoning with technology has more often been researched using "route-type" tools (Bakker & Gravemeijer, 2004; Ben-Zvi, 2000).

Concerning the recommendation to develop assessment tools in statistics in the first edition, a good deal of progress has been made. The book *The Assessment Challenge in Statistics Education* (Gal & Garfield, 1997) contains a wealth of assessment information. In one chapter a framework for statistical assessment is presented by Friel et al. (1997). Among the authors who contributed to the development of assessment tools in statistics education are Schau and Mattern (1997) on the use of concept maps, Watson (1997) on using the media, Curcio and Artzt (1997) on small group work, Starkings (1997) and Holmes (1997) on statistical projects, Keeler (1997) on portfolios, Lajoie on using technology in assessment, Lesh, Amit, & Schorr (1997) on using real-life problems, and Jolliffe (1997) on instrument construction. More recently, Garfield, delMas, & Chance (2005) have led a project which has created a website of assessment items for improving statistical thinking.

Recommendations for Future Research: Take Two

This time around my recommendations for future research directions fall under three broad categories: (a) research on conceptual issues in statistics, (b) research on teaching issues in statistics, and (c) some methodological issues for research in statistics.

Conceptual Issues

The notion of distribution needs to be clarified and students' conceptions of the interrelationships of the aspects of a distribution deserve more research attention. Statisticians and statistics educators use the word distribution all the time. The word is used to refer to a single distribution of data, to a sampling distribution of statistics, to a probability distribution. We all sort of know what we mean, some kind of collection of observations or numbers that is represented in a table or a graph, but it is not very well defined. Probability distributions can be defined in terms of the values that are taken on by random variables (which are themselves a type of function), so clearly the word distribution involves a "range of meanings." It would be beneficial for statistics educators, present company included, to pin down what we mean when we use the term distribution. Researchers are beginning to talk about "distributional reasoning." For some researchers, distributional reasoning in data analysis involves the explicit integration of multiple aspects of a distribution, such as centers, shape, and variability (see for example, Shaughnessy et al. 2004a, 2005). A recent meeting of the International Forum on Statistical Reasoning, Thinking, and Literacy held in Auckland, New Zealand (SRTL IV, August, 2005)[3] was even entitled "Reasoning about Distributions." A number of researchers are beginning to articulate their own thinking about distributions and are sharing their research on students' understanding of distributions in a more prominent way.

More research is needed on tertiary students' conceptions of statistics. At the time of the first edition of the *Handbook* much of the research had investigated tertiary students' conceptions of probability and statistics, and very little research had been done with K–12 students. This time it is the other way around. There has been a major influx of studies of K–12 students' statistical thinking in school mathematics settings, and not so much has been done with tertiary students. What are tertiary students' conceptions of centers and average? What are their conceptions of variability? How do they reason when comparing data sets? Do tertiary students show any noticeable improvement in reasoning over their K–12 counterparts, or are they stuck at the same levels on the same issues as elementary and secondary students are? Statistics is now required in almost all collegiate majors. Are students learning anything beyond procedures in those courses? Are they learning to be data detectives, or just data crunchers?

The use of student-generated graphs is a very promising area for mining student thinking. Student-generated graphs, an instance of what been called student *inscriptions*, are very powerful tools to investigate student thinking. Student graphs not only provide information about students' graphical skills, but also information on the level of their thinking about data—cases, clumps, aggregates, and so forth. Inscriptions can also inform us on how students do or do not think about trends over time, spreads, centers, and shape. Bakker and Gravemeijer (2004), Moritz (2004), and Kelly and Watson (2002) have provided good examples of research that has uncovered a wealth of information from student-generated graphs.

More research is needed on students' conceptual growth in statistics when they work in technology-rich environments. Too little research has been conducted on the effects of statistical software packages on students' conceptual growth and thinking in statistics. Studies by Bakker and Gravemeijer (2004), Saldanha and Thompson (2003), Ben-Zvi (2000), and Chance et al. (2004) are among the welcomed exceptions to the dearth of research on the effects of technology on statistical thinking. Currently there is very little research on the effects of "landscape-type" data analysis tools such as *Tinkerplots* or *Fathom.*

More research is needed on teachers' conceptions of statistics. Teachers have the same difficulties with statistical concepts as the students they teach. Research needs to find ways to help teachers develop in their statistical knowledge and thinking, especially now that statistics has a more prominent role in the K–12 curricula. Some promising inroads into teachers' understandings of statistics have been made within professional development programs (Watson, 2001b; Heaton & Mickelson, 2002; Hammerman & Rubin, 2003; Makar & Confrey, 2004). Student work samples or the results of student test scores have been used by researchers to catalyze teachers' reflections and reasoning about statistical issues. This technique is a promising indirect, non-threatening approach to finding out more about what teachers know and do not know about statistics.

Teaching Issues

What is the statistical knowledge necessary for teaching? Ball and Bass (2000) and Ball, Lubienski, and Mewborne (2001) have argued that a special type of knowledge is needed to teach mathematics, that this knowledge is different than just more mathematical content, and that it is more than just what Shulman (1987) called pedagogical content knowledge. For example, a special knowledge is needed for teaching algebra that involves knowing the types of symbolic mistakes and misconceptions that students will have, and how to address them. A special knowledge is needed for geometry that involves the role of examples and counterexamples, and reasoning with proof. Within each content area of mathematics there are special types of knowledge that are critical to effective teaching of that content. What is needed to success-

[3] A forthcoming issue of the Statistics Education Research Journal (SERJ) will include edited papers about research on students' thinking about distribution from the SRTL IV conference (Vol. 5, No. 2, November 2006). There is also a CD of the proceedings available (Makar, 2005).

fully teach statistics? What are the particularly tough concepts and areas for potential misunderstandings in statistics? Examples of statistical knowledge for teaching are addressed in Shaughnessy and Chance (2005) in their discussions of questions that arise from teachers and students in the classroom.

Research is needed on classroom discourse in statistics. Are students being asked to analyze data at a high level? Are statistical tasks posed and discussed in classrooms in ways that promote high level thinking, critical analysis, multiple representations, and thoughtful communication of results and solutions? This type of analysis of classroom processes has been undertaken in research on the teaching of mathematics, but so far no such research has been reported for the teaching of statistics. Analyzing, critiquing, communicating, and representing are critical skills that have been identified by Wild and Pfannkuch (1999) and by Watson (1997) in their models of statistical thinking and statistical literacy, respectively. Are we doing enough to identify and promote the types of teaching that will enhance our students' discourse skills in statistics? How well is statistics taught by mathematics teachers? Statistics is not the same as mathematics (Cobb & Moore, 1997). Are we balancing our classroom discourse between exploratory data analysis (data detective work) and the teaching of statistical concepts and procedures? Discourse analysis in statistics education is a wide-open area for future research.

There has been very little research into students' and teachers' beliefs and attitudes towards statistics. This is another area of research that is wide open for statistics education. There has been research into teacher affect and teachers' beliefs about mathematics (Thompson, 1992) as well as research into student attitudes and beliefs about mathematics (McCleod, 1992), but very little work has been done on students' or teachers' attitudes and beliefs specifically about statistics.

Methodological Issues and Recommendations

The use of multiple research methodologies in research on students' conceptions of statistics has powerful payoffs. There has been a growing trend for researchers to use both quantitative and qualitative methodologies in their research on students' reasoning on statistical tasks. In many studies researchers have gathered and quantified results of surveys on statistical tasks administered to large numbers of students but have also conducted clinical interviews with smaller numbers of students, either in conjunction with survey administration or as follow-up work to the surveys. Hypotheses generated about why students are answering survey questions in particular ways can be validated in detailed clinical interviews. Also, interviews often reveal lines of thought that were in the survey data all the time but that were initially missed by the researchers. I applaud the use of multiple research methodologies and urge my fellow researchers to continue this practice.

The statistics education research community would benefit from a thoughtful discussion and debate about the strengths and limitations of using the SOLO model. The SOLO model has been used by many researchers, particularly in Australia where it was born, to identify levels of student reasoning on a number of constructs. Many research studies reviewed in this chapter employed the SOLO model in an analysis of student responses to statistical tasks. The SOLO model is based on the assumption that development can be represented in hierarchical structures. Is that assumption warranted? One of the criticisms leveled against the SOLO model is that it is not falsifiable, so the validity of any conclusions reached via a SOLO approach cannot be easily challenged. On the other hand, the SOLO model has been genuinely useful in helping to describe student reasoning on a number of concepts in statistics like average, variation, comparison of data sets, and so on. I recommend that the statistics education community engage in healthy debate on the merits and demerits of the SOLO model, and that the debate be published in a public forum.

The statistics education research community would benefit from a thoughtful discussion and debate about the strengths and limitations of using the Rasch model. Recently several statistics educators have begun to use the Rasch model to quantify students' statistical literacy, including tasks involving such concepts as average and variability (see for example, Callingham & Watson, 2005; Watson & Callingham, 2003; Watson et al., 2003). In several instances Rasch measurement has been used in conjunction with the SOLO model in order to scale students' responses to statistical tasks along a SOLO type hierarchy. The results of the Rasch analysis in these studies appear to be very robust, and Rasch provides a way to quantify and measure student constructed responses on surveys and interviews. It is an underpinning assumption of Rasch measurement that the target construct is unidimendional. One wonders if this is really the case with such constructs as average and variability, for which an eclectic variety of types of conceptions have been identified in the literature. Are linear models of complex concepts sufficient for valid measurements? This is a question for debate and discussion within the statistics education community.

Some Implications From Research for the Teaching of Statistics

There are so many recommendations that could be made from research for the teaching and learning of statistics that this section could be a separate chapter just in itself. I will contain myself to just a few.

- *Emphasize variability as one of the primary issues in statistical thinking and statistical analysis.* In the past there has been a tendency to overemphasize centers as the principal concept in statistics, and the important role of variability has been neglected. Students need to integrate the concepts of centers and variability when they investigate data so that they can reason about properties of aggregates of data.
- *Introduce comparison of data sets much earlier on with students, prior to formal statistics.* Bakker and Gravemeijer (2004), Konold and Higgens (2003), and Watson and Moritz (1999) have all found that students can develop their own powerful, intuitive ways to compare data sets prior to the introduction of formal concepts like mean, median, variation, or standard deviation. Have students compare data sets right from the start of their statistical education.
- *Build on students' intuitive notions of center and variability.* Research has uncovered a spectrum of student conceptions about these two important concepts. Students will be in transition from their own colloquial understandings of centers and variability—such as average as "typical" or variability as "things that change over time"—to more statistical understandings of these concepts, such as means or likely ranges. We should start with what students bring to the table on these concepts and build from there.
- *Make the role of proportional reasoning in the connections between populations and samples more explicit.* Although it might seem obvious that carefully chosen samples should be representative of whole populations, a long history of research evidence suggests that people ignore base rates when making inferences from samples or predictions to populations (see for example, Kahneman & Tversky 1972, 1973; Tversky & Kahneman, 1974). Students should have repeated opportunities to actually choose samples themselves, so they have chances to *see* the proportional relationship firsthand.
- *Remember that there is a difference between statistics and mathematics!* Wild and Pfannkuch's work (1999) and the writings of David Moore (1990, 1997) are signals to everyone who teaches mathematics that there are ways of thinking and analytical tools that are specific to statistics. In particular, statistics is fraught with contextual issues, which is the nature of the discipline, whereas often mathematics strips off the context in order to abstract and generalize.

Concluding Remarks

I had to make choices in developing this chapter so I tried to take to heart the old adage for authors, "Write what you know." Or, in this case perhaps it is, "Write what you think you know." There are many opportunities for thoughtful and knowledgeable readers to fill in the many gaps that I left behind. Hopefully the chapter will provide a sufficient network of references so that an interested reader can pursue a particular topic by going directly to the original works and tracing back further among other references. I want to remind the reader that in this chapter you have encountered statistical research through an indirect mode, through the lens of my own personal filter. There is some efficiency in this approach, and that is why people are asked to write reviews and syntheses of the literature. However, this is never any substitute for reading the original sources. I urge all interested readers to do just that, especially in those instances that peak your curiosity. Do not just take my word for it, go and find out for yourself!

REFERENCES

Aberg-Bengtsson, L., & Ottoson, T. (1995). *Children's understanding of graphically represented quantitative information.* Paper presented at the 6th Conference of the European Association for Research on Learning and Instruction. Nijmegen, The Netherlands.

Australian Educational Council. (1991). *A National Statement on Mathematics for Australian Schools.* Canberra, Australia: Author.

Bakker, A. (2002). Route-type and landscape-type software for learning statistical data analysis. In B. Phillips (Ed.), *Developing a Statistically Literate Society: Proceedings of the Sixth International Conference on Teaching Statistics.* Voorburg, The Netherlands: International Statistical Institute.

Bakker, A., Biehler, R., & Konold, C. (2004). Should young students learn about box plots? In G. Burrill & M. Camden (Eds.), *Curricular Development in Statistics Education. International Association for Statistical Education (IASE) Roundtable, Lund, Sweden,* (pp. 163–173). Voorburg, The Netherlands: International Statistical Institute.

Bakkar, A., & Gravemeijer, K. P. E. (2004). Learning to reason about distribution. In J. Garfield & D. Ben Zvi (Eds.), *The challenge of developing statistical literacy, reasoning and thinking* (pp. 147–168). Dordrecht, The Netherlands: Kluwer.

Ball, D. L., & Bass, H. (2000). Interweaving content and pedagogy in teaching and learning to teach: Knowing and using mathematics. In J. Boaler (Ed.), *Multiple perspectives on the teaching and learning of mathematics* (pp. 83–104). Westport, CT: Ablex.

Ball, D. L., Lubienski, S. T., & Mewborne, D. S. (2001). Research on teaching mathematics: The unsolved problem of teachers' mathematical knowledge. In V. Richardson (Ed.), *Handbook of research on teaching* (4th ed., pp. 433–457). New York: Macmillan.

Batanero, C., Estepa, A., Godino, J. D., & Green, D. R. (1996). Intuitive strategies and preconceptions about association in contingency tables. *Journal for Research in Mathematics Education, 27,* 151–169.

Batanero, C., Godino, J. D., Vallecillos, A., Green, D. R., & Holmes, P. (1994). Errors and difficulties in understanding elementary statistical concepts. *International Journal of Mathematics Education in Science and Technology, 25,* 527–547.

Ben-Zvi, D. (2000). Toward understanding the role of technological tools in statistics learning. *Mathematical Thinking and Learning, 2,* 127–155.

Biehler, R. (1993). Software tools and mathematics education: The case for statistics. In C. Keitel & K. Ruthven (Eds.), *Learning from computers: Mathematics education and technology* (pp. 68–100). NATO ASI Series F, Computers and Systems Sciences. Berlin, Germany: Springer-Verlag.

Biehler, R. (1994a, July). Cognitive technologies for statistics education: Relating the perspective of tools for learning and of tools for doing statistics. In L. Brunelli & G. Cicchitelli (Eds.), *Proceedings of the First Scientific Meeting of IASE* (pp. 173–190). Perguia, Italy: University of Perugia Press

Biehler, R. (1994b, July). *Requirements for an ideal software tool in order to support learning and doing statistics.* Paper presented at the Fourth International Conference on Teaching Statistics, Marrakech, Morocco.

Biehler, R. (1997). Software for learning and for doing statistics. *International Statistical Review, 65,* 167–189.

Biggs J. B., & Collis, K. F. (1982). *Evaluating the quality of learning: The SOLO taxonomy.* New York: Academic Press.

Bright, G. W., & Friel, S. (1993, April). *Elementary teachers' representations of relationships among statistics concepts.* Paper presented at the Annual Meeting of the American Educational Research Association, Atlanta, GA.

Burrill, G., Franklin, C., Godbold, L., & Young, L. (2004). *Navigating through data analysis, grades 9–12.* Reston, VA: National Council of Teachers of Mathematics.

Cai, J. (1995). Beyond the computational algorithm: Students' understanding of the arithmetic average concept. In L. Meira & D. Carraher (Eds.), *Proceedings of the 19th Psychology of Mathematics Education Conference* (Vol. 3. pp. 144–151). Sao Paulo, Brazil: PME Program Committee.

Canada, D. (2004). *Pre-service elementary teachers conceptions of variability.* Unpublished doctoral dissertation, Portland State University, Portland, OR.

Carr, J., & Begg, A.(1994). Introducing box and whisker plots. In J. Garfield (Ed.), *Research Papers from the Fourth International Conference on Teaching Statistics.* Minneapolis, MN.

Chance, B., delMas, R., & Garfield, J. (2004). Reasoning about sampling distributions. Data driven mathematics. In J. Garfield & D. Ben-Zvi (Eds.), *The challenge of developing statistical literacy, reasoning and thinking* (pp. 295–324). Dordrecht, The Netherlands: Kluwer.

Callinghan, R. A., & Watson, J. M. (2005). Measuring statistical literacy. *Journal of Applied Measuremment, 6,* 19–47.

Chick, H. L., & Watson, J. M. (2001). Data representations and interpretation by primary school students working in groups. *Mathematics Education Research Journal, 13,* 91–111.

Chick, H. L., & Watson, J. M. (2002). Collaborative influences on emergent statistical thinking—A case study. *Journal of Mathematical Behavior, 21,* 317–400.

Ciancetta, M., Shaughnessy, J. M., & Canada, D. (2003, July). Middle school students' emerging definitions of variability. In N. Pateman, B. Dougherty, & J. Zilliox (Eds.), Poster Session in the *Proceedings of the 27th Conference of the International Group for the Psychology of Mathematics Education* (Vol. 4, p. 481). Honolulu: University of Hawaii.

Cobb, G. W., & Moore, D. S. (1997). Mathematics, statistics, and teaching. *American Mathematical Monthly, 104,* 801–823.

Cobb, P. A. (1999). Individual and collective mathematics development: The case of statistical data analysis. *Mathematical Thinking and Learning, 1,* 5–44.

Cobb, P., Gravemeijer, K. P. E., Doorman, M., & Bowers, J. (1999). Computer *Mini-tools* for exploratory data analysis (Version Prototype). Nashville, TN: Vanderbilt University.

Cobb, P., McClain, K., & Gravemeijer, K. P. E. (2003). Learning about statistical co-variation. *Cognition and Instruction, 21*(1), 1–78.

Connected Mathematics. (1998). Palo Alto, CA: Dale Seymour Publications.

Core-Plus Mathematics Project. (1997). *Contemporary mathematics in context: A unified approach.* Dedham, MA: Janson Publications.

Corwin, R. B., & Friel, S. N. (1990) *Used numbers: Prediction and sampling.* Palo Alto, CA: Dale Seymour.

Curcio, F. R. (1987). Comprehension of mathematical relationships experienced in graphs. *Journal for Research in Mathematics Education, 18,* 382–393.

Curcio, F. R. (1989). *Developing graph comprehension.* Reston, VA: National Council of Teachers of Mathematics.

Curcio, F. R., & Artz, A. F. (1997). Assessing students' statistical problem solving behaviors in a small group setting. In I. Gal & J. Garfield (Eds.), *The assessment challenge in statistics education* (pp. 107–122). Amsterdam: IOS Press.

Dalal, S. R., Fowlkes, E. B., & Hoadley, B. (1989). Risk analysis of the space shuttle: Pre-Challenger prediction of failure. *Journal of the American Statistical Association, 84,* 945–951.

Data Driven Mathematics (1999). White Plains, NY: Dale Seymour.

delMas, R. (2001). *Sampling SIM* (Version 5). Retrieved April 23, 2003, from http://www.gen.umn.edu/faculty_staff/delMas/stat_tools.

delMas, R., Garfield, J., & Chance, B. (1998). Assessing the effects of a computer microworld on statistical reasoning. In L. Pereira-Mendoza, L. S. Kea, T. W. Kee, & W. Wong (Eds.), *Proceedings of the Fifth International Conference on Teaching Statistics* (pp. 1083–1089), Nanyang Technological University. Singapore: International Statistical Institute.

delMas, R., Garfield, J., & Chance, B. (1999). A model of classroom research in action: Developing simulation activities to improve students' statistical reasoning. *Journal of Statistics Education, 7,* 3. Retrieved July 10, 2005, from www.amstat.org/publications/jse/v7n3.

Doerr, H. M. (2000). How can I find a pattern in this random data? The convergence of multiplicative and probabilistic reasoning. *Journal of Mathematical Behavior, 18,* 431–454.

Dunkels, A. (1994). Interweaving numbers, shapes, statistics, and the real world in primary school and primary teacher education. In D. F. Robitaille, D. H. Wheeler, & C. Kieran (Eds.), *Selected Lectures from the 7th International Congress on Mathematical Education* (pp. 123–135). Sainte-Foy, Quebec, Canada: Laval University Press.

Estepa, A., & Batanero, C. (1994, July). *Judgments of association in scatter-plots: An empirical study of students' strategies and preconceptions.* Paper presented at the Fourth International Conference on Teaching Statistics: Marrakech, Morocco.

Estepa, A., Batanero, C., & Sanchez, F. T. (1999). Students' intuitive strategies in judging association when comparing two samples. *Hiroshima Journal of Mathematics Education, 7,* 17–30.

Finzer, W. (2002). The Fathom experience—is research-based development of a commercial statistics learning environment possible? In B. Phillips (Ed.), *Developing a statistically literate society: Proceedings of the Sixth International Conference on Teaching Statistics.* [CD-ROM]. Voorburg, The Netherlands: International Statistical Institute.

Fischbein, E. (1987). *Intuition in science and mathematics.* Dordrecht, The Netherlands: D. Reidel.

Fischbein, E., & Schnarck, D. (1997). The evolution with age of probabilistic intuitively based misconceptions. *Journal for Research in Mathematics Education, 28,* 96–105.

Friel, S. N. (1998). Teaching statistics: What's average? In L. J. Morrow (Ed.), *The teaching and learning of algorithms in school mathematics* (pp. 208–217). Reston, VA: National Council of Teachers of Mathematics.

Friel, S. N. (2007). The research frontier: Where technology interacts with the teaching and learning of data analysis and statistics. In G. Blume & K. Heid (Eds.), *Research on technology and the teaching and learning of mathematics: Cases and perspectives* (Vol. 1, pp. 279–331). Greenwich, CT: Information Age.

Friel, S. N., & Bright, G. W. (1998). Teach-Stat: A model for professional development and data analysis for teachers K–6. In S. Lajoie (Ed.), *Reflections on statistics: Learning, teaching, and assessment in grades K–12* (pp. 89–117). Mahwah, NJ: Erlbaum.

Friel, S. N., Mokros, J. R., & Russell, S. J. (1992). *Used numbers: Middles, means, and in-betweens.* Palo Alto, CA: Dale Seymour.

Friel, S. N., Bright, G. W., Frierson, D., & Kader, G. D. (1997). A framework for assessing knowledge and learning in statistics (K–8). In I. Gal and J. Garfield (Eds.), *The assessment challenge in statistics education* (pp. 55–64). Amsterdam: IOS Press.

Friel, S. N., Curcio, F. R., & Bright, G. W. (2001). Making sense of graphs: Critical factors influencing comprehension and instructional implications. *Journal for Research in Mathematics Education, 32,* 12–158.

Foreman, L. C., & Bennett, A. B. (1995). *Math alive, Course I.* Salem, OR: The Math Learning Center.

Gal, I. (1998). Assessing statistical knowledge as it relates to students' interpretation of data. In S. Lajoie (Ed.), *Reflections on statistics: Learning, teaching, and assessment in grades K–12* (pp. 275–295). Mahwah, NJ: Erlbaum.

Gal, I. (2002). Adults' statistical literacy: Meanings, components, responsibilities. *International Statistical Review, 70,* 1–25.

Gal, I. (2003). Expanding conceptions of statistical literacy: An analysis of products from statistics agencies. *Statistics Education Research Journal, 2,* 3–21. Retrieved March 15, 2005, from http://fehps.une.edu.au/serj.

Gal, I. (2004). Statistical literacy: Meanings, components, responsibilities. In J. Garfield & D. Ben-Zvi (Eds.), *The challenge of developing statistical literacy, reasoning and thinking* (pp. 47–78). Dordrecht, The Netherlands: Kluwer.

Gal, I., & Garfield, J. (1997). *The assessment challenge in statistics education.* Amsterdam: IOS Press.

Gal, I., Rothschild, K., & Wagner, D. A. (1989, April). *Which group is better? The development of statistical reasoning in school children.* Paper presented at the meeting of the Society for Research in Child Development, Kansas City, KS.

Gal, I., Rothschild, K., & Wagner, D. A. (1990, April). *Statistical concepts and statistical reasoning in children: Convergence of divergence?* Paper presented at the meeting of the American Educational Research Association, Boston, MA.

Garfield, J. (1990). *Technology and data: Models and analysis.* Report of the Working Group on Technology and Statistics. Madison, WI: NCRMSE.

Garfield, J. (2002). The challenge of developing statistical reasoning. *Journal of Statistics Education, 10*(3). Retrieved April 23, 2003, from http://www.amstat.org/publications/jse/.

Garfield, J., & Ben-Zvi, D. (2004). *The challenge of developing statistical literacy, reasoning and thinking.* Dordrecht, The Netherlands: Kluwer.

Garfield, J., delMas, R., & Chance, B. (1999, August). *Developing statistical reasoning about sampling distributions.* Presented at the First International Research Forum on Statistical Reasoning, Thinking, and Literacy (SRTL I). Kibbutz Be'eri, Israel.

Garfield, J., delMas, B., & Chance, B. (2005). *ARTIST*: Assessment resource tools for improving statistical thinking. Retrieved March 20, 2005, from http://data.gen.umn.edu/artist/index.html.

Green, D. (1993). Data analysis: What research do we need? In L. Pereira-Mendoza (Ed.), *Introducing data analysis in the schools: Who should teach it?* (pp. 219–239). Voorburg, The Netherlands: International Statistics Institute.

Greer, B., & Ritson, R. (1993). *Teaching data handling with the Northern Ireland Mathematics Curriculum: Report on survey in schools.* Belfast, Ireland: Queen's University.

Hammerman, J. K., & Rubin, A. (2003). Reasoning in the presence of variability. In C. Lee (Ed.), *Proceedings of the Third International Research Forum on Statistical Reasoning, Thinking, and Literacy (SRTL-3, CD-ROM).* Mt. Pleasant: Central Michigan University.

Hawkins, A. (Ed.). (1990). *Teaching teachers to teach statistics.* Voorburg, The Netherlands: International Statistics Institute.

Hawkins, A., Jolliffe, F., & Glickman, L. (1992). *Teaching statistical concepts.* London: Longman Publishing.

Heaton, R., & Mickelson, W. (2002). The learning and teaching of statistical investigation in teaching and teacher education. *Journal of Mathematics Teachers Education, 5,* 35–59.

Holmes, P. (1997). Assessing project work by external reviewers. In I. Gal & J. Garfield (Eds.), *The assessment challenge in statistics education* (pp. 153–164). Amsterdam: IOS Press.

Huff, D. (1954). *How to lie with statistics.* New York: W.W. Norton Publishing.

Investigations into Number, Data, and Space. (1998). White Plains, NY: Dale Seymour Publications.

Jacobs, V. R. (1997, April). *Children's understanding of sampling in surveys.* Paper presented at the annual meeting of the American Educational Research Association, Chicago.

Jacobs, V. R. (1999). How do students think about statistical sampling before instruction? *Mathematics Teaching in the Middle School, 5,* 240–246, 263.

Jones, G. A., Langrall, C. W., Thornton, C. A., & Mogill, A. T. (1999). Students' probabilistic thinking in instruction. *Journal for Research in Mathematics Education, 30,* 487–519.

Jones, G. A., Thornton, C. A., Langrall, C. W., Mooney, E. S., Perry, B., & Putt, I. J. (2000). A framework for characterizing students' statistical thinking. *Mathematics Thinking and Learning, 2,* 269–307.

Jones, G. A., Langrall, C. W., Mooney, E. S., & Thornton, C. A. (2004). Models of development in statistical reasoning. In J. Garfield & D. Ben Zvi (Eds.), *The Challenge of Developing Statistical Literacy, Reasoning and Thinking* (pp. 97–118). Dordrecht, The Netherlands: Kluwer.

Jones, G. A., Langrall, C. W, & Mooney, E. S. (this volume). Research in Probability: Responding to Classroom Realities. In F. Lester, (Ed.), *Handbook of research on the teaching and learning of mathematics* (2nd ed.). Reston, VA: National Council of Teachers of Mathematics.

Jolliffe, F. (1997). Issues in constructing assessment instruments for the classroom. In I. Gal & J. Garfield (Eds.), *The assessment challenge in statistics education* (pp. 191–204). Amsterdam: The International Statistics Institute.

Kahneman, D., & Tversky, A. (1972). Subjective probability: A judgment of representativeness. *Cognitive Psychology, 3,* 430–454.

Kahneman, D., & Tversky, A. (1973a). On the psychology of prediction. *Psychological Review, 80,* 237–251.

Kahneman, D., & Tversky, A. (1973b). Availability: A heuristic for judging frequency and probability. *Cognitive Psychology, 5,* 207–232.

Kelly, B. A., & Watson, J. M. (2002) Variation in a chance sampling setting: The lollies task. In B. Barton, K. C. Irvin, M. Pfannkuch, & M. J. Thomas (Eds.), *Proceedings of the 25th annual conference of the Mathematics Education Research Group of Australasia: Mathematics education in the South Pacific, Auckland* (Vol. 2, pp.366–373). Sydney, Australia: MERGA.

Keeler, C. M. (1997). Portfolio assessment in graduate level statistics courses. In I. Gal and J. Garfield (Eds.), *The Assessment Challenge in Statistics Education* (pp. 165–178). Amsterdam: IOS Press.

Key Curriculum Press. (2005a). Tinkerplots Dynamic Data Exploration (Version 1.0) [Computer software]. Emeryville, CA: Author.

Key Curriculum Press. (2005b). Fathom Dynamic Data Software (Version 2.0) [Computer software]. Emeryville, CA: Author.

Kirsch, I. S., Jungeblut, S. S., & Mosenthal, P. M. (1998). The measurement of adult literacy. In S.T. Murray, I.S. Kirsch, & L. B. Jenkins (Eds.), *Adult literacy in OECD countries: Technical report on the first International Adult Literacy Survey* (pp. 105–134). Washington, DC: National Center for Education Statistics, U.S. Department of Education.

Konold, C. (1989). Informal conceptions of probability. *Cognition and Instruction, 6,* 59–98.

Konold, C., & Higgens, T. (2002). Working with data: Highlights related to research. In S. J. Russell, D. Schifter, & V. Bastabel (Eds.), *Developing mathematical ideas: Collecting, representing, and analyzing data* (pp. 165–201). Mahwah, NJ: Erlbaum.

Konold, C., & Higgens, T. (2003). Reasoning about data. In. J. Kilpatrick, W. G. Martin, & D. Schifter (Eds.), *A research companion to principles and standards for school mathematics* (pp. 193–215). Reston, VA: National Council of Teachers of Mathematics.

Konold, C., & Miller, C. (1994). Data scope and prob-sim. [Computer software]. Amherst: University of Massachusetts, SRRI.

Konold, C., & Pollatsek, A. (2002) Data analysis as a search for signals in noisy processes. *Journal for Research in Mathematics Education, 33,* 259–289.

Konold, C., Pollatsek, A., Well, A., Lohmeier, J., & Lipson, A. (1993). Inconsistencies in students' reasoning about probability. *Journal for Research in Mathematics Education, 24,* 392–414.

Konold, C., Pollatsek, A., Well, A., & Gagnon, A. (1997). Students analyzing data: Research of critical barriers. In J. B. Garfield & G. Burrill (Eds.), *Research on the role of technology in teaching and learning statistics.* Voorburg, The Netherlands: International Statistical Institute.

Konold, C., Higgens, T., Russell, S. J., & Khalil, K. (February, 2004). Data seen through different lenses. Unpublished manuscript, Amherst, MA: University of Massachusetts.

Landwehr, J. M., Watkins, A.E., & Swift, J. (1987). *Exploring surveys: Information from samples.* Palo Alto, CA: Dale Seymour.

Lajoie, S. P. (1997). Technologies for assessing and extending statistical learning. In I. Gal & J. Garfield (Eds.), *The assessment challenge in statistics education* (pp. 179–190). Amsterdam: IOS Press.

LeCoutre, V.P. (1992). Cognitive models and problem spaces in "purely random" situations. *Educational Studies in Mathematics, 23,* 557–568.

Lehrer, R. & Romberg, T. (1996). Exploring children's data modeling. *Cognition and Instruction, 14,* 69–108.

Lesh, R., Amit, M., & Schorr, R. Y. (1997). Using "real-life" problems to prompt students to construct conceptual models. In I. Gal & J. Garfield (Eds.), *The assessment challenge in statistics education* (pp. 65–84). Amsterdam: IOS Press.

Lipson, K. (2002). The role of computer based technology in developing understanding of the concept of sampling distribution. In B. Phillips (Ed.), *Proceedings of the Sixth International Conference on Teaching Statistics: Developing a statistically literate society, Cape Town, South Africa.* [CD-ROM]. Voorburg, The Netherlands: International Statistical Institute.

Loosen, F., Lioen, M., & Lacante, M. (1985). The standard deviation: Some drawbacks of an intuitive approach. *Teaching Statistics, 7,* 29–39.

Makar, K. (2005). Reasoning about distributions: A collection of recent research studies. Proceedings of the Fourth International Research Forum for Statistical Reasoning, Thinking, and Literacy, Auckland, NZ. Brisbane Australia: University of Queensland.

Makar, K., & Confrey, J. (2002, August). *Comparing two distributions: Investigating secondary teachers' statistical thinking.* Paper presented at the Sixth International Conference on Teaching Statistics. Cape Town, South Africa.

Makar, K., & Confrey, J. (2005). "Variation Talk": Articulating meaning in statistics. *Statistical Education Research Journal, 4*(1), 27–54. Retrieved October 15, 2006, from http://www.stat.auckland.ac.nz/~iase/publications.php?show=serjarchive.

Makar, K., & Confrey, J. (2004). Secondary teachers' reasoning about comparing two groups. In D. Ben-Zvi & J. Garfield (Eds.), *The challenges of developing statistical literacy, reasoning, and thinking* (pp. 353–374). Dordrecht, The Netherlands: Kluwer.

Mathematics in context. (1994). New York: Rand McNally.

McCleod, D. (1992). Research on affect in mathematics education: A reconceptualization. In D. Grouws (Ed.), *Handbook of Research on Mathematics Teaching and Learning* (pp. 575–596). Reston, VA: National Council of Teachers of Mathematics.

Meletiou, M. (2000). *Developing students' conceptions of variation: An untapped well in statistical reasoning.* Unpublished doctoral dissertation, The University of Texas at Austin.

Meletiou, M. (2002). Conceptions of variation: A literature review. *Statistics Education Research Journal, 1*(1), 46–52.

Meletiou, M. & Lee, C. (2002). Student understanding of histograms: A stumbling stone to the development of intuitions about variation. In B. Phillips (Ed.), *Proceedings of the Sixth International Conference on Teaching Statistics: Developing a statistically literate society, Cape Town, South Africa.* [CD-ROM]. Voorburg, The Netherlands: International Statistical Institute.

Mevarech, Z. (1983). A deep structure model of students' statistical misconceptions. *Educational Studies in Mathematics, 14,* 415–429.

Mickelson, W., & Heaton, R. (2003). Purposeful statistical investigation merged with K–6 content: Variability, learning, and teacher knowledge use in teaching. In C. Lee (Ed.), *Proceedings of the Third International Research forum on statistical reasoning, thinking, & literacy* (SRTL-3, CD-ROM). Mt. Pleasant: Central Michigan University.

Mickelson, W., & Heaton, R. (2004). Primary teachers' statistical reasoning about data. In D. Ben-Zvi & J. Garfield (Eds.), *The challenge of developing statistical literacy, reasoning, and thinking* (pp. 327–352). Dordrecht, The Netherlands: Kluwer.

Minitab Statistical Software. (2005). State College, PA: Pennsylvania State University Press.

Mokros, J., & Russell, S. J. (1995). Children's concepts of average and representativeness. *Journal for Research in Mathematics Education, 26,* 20–39.

Mooney, E. S. (2002). A framework for characterizing middle school students' statistical thinking. *Mathematical Thinking and Learning, 4,* 23–63.

Moore, D. (1990). Uncertainty. In L. Steen (Ed.), *On the shoulders of giants: New approaches to numeracy* (pp. 95–137). Washington, DC: National Academy Press.

Moore, D. (1997). New pedagogy and new content: The case of statistics. *International Statistics Review, 65,* 123–165.

Moritz, J. B., & Watson, J. M. (1997). Graphs: Communication lines to students. In F. Biddulph & K. Carr (Eds.), *People in Mathematics Education* (Vol. 2, pp. 344–351). *Proceedings of the Twentieth Annual Meeting of the Mathematics Education Research Group of Australasia.* Waikato, New Zealand: The University of Waikato Printery.

Moritz, J. B. (2000). Graphical representations of statistical associations by upper primary students. In J. Bana & A. Chapman (Eds.), *Mathematics education beyond 2000. Proceedings of the 23rd Annual Conference of the Mathematics Education Research Group of Australasia, 2* (pp. 440–447). Perth, Australia: Mathematics Education Research Group of Australasia.

Moritz, J. B. (2004). Reasoning about co-variation. In J. Garfield & D. Ben-Zvi (Eds.), *The challenge of developing statistical literacy, reasoning and thinking* (pp. 227–256). Dordrecht, The Netherlands: Kluwer.

Mosenthal, P. M., & Kirsch, I. S. (1998). A new measure for assessing document complexity: The PMOSE/IKIRSZCH document readability formula. *Journal of Adolescent and Adult Literacy, 41*(8), 638–657.

National Assessment Governing Board. (1994). *Mathematics Framework for the 1996 National Assessment of Educational Progress.* Washington, D.C.: Author.

National Council of Teachers of Mathematics. (1989). *Curriculum and evaluation standards for K–12 mathematics.* Reston, VA: Author.

National Council of Teachers of Mathematics. (2000). *Principles and standards for school mathematics.* Reston, VA: Author.

National Council of Teachers of Mathematics. (2004a). *Navigating through data.* Reston, VA: Author.

National Council of Teachers of Mathematics. (2004b). *Navigating through probability.* Reston, VA: Author.

Nemirovsky, R. (1996). Mathematical narratives, modeling, and algebra. In N. Bednarz, C. Kieran, & L. Lee (Eds.), *Approaches to algebra: Perspectives for research and teaching* (pp. 197–220). Dordrecht, The Netherlands: Kluwer.

Petrosino, A. J., Lehrer, R., & Schauble, L (2003). Structuring error and experimental variation as distribution in 4th grade. *Mathematics Thinking and Learning, 5,* 131–136.

Pereira-Mendoza, L. (1995). Graphing in the primary school: Algorithm versus comprehension. *Teaching Statistics, 17,* 2–6.

Pereira-Mendoza L., & Dunkels, A. (1989). Stem-and-leaf plots in the primary grades. *Teaching Statistics, 11,* 34–37.

Pereira-Mendoza, L. & Mellor, J. (1991). Students' concepts of bar graphs—some preliminary findings. In D. Vere-Jones (Ed.), *Proceedings of the Third International Conference on Teaching Statistics. Vol. I: School and General Issues* (pp. 150–157). Voorburg, The Netherlands: International Statistics Institute.

Pfannkuch, M. (2005). Probability and statistical inference: How can teachers enable learners to make the connection? In G. Jones (Ed.), *Exploring probability in school: Challenges for teaching and learning* (pp. 1–32). Dordrecht, The Netherlands: Kluwer.

Pfannkuch, M., & Watson, J. (2005). Statistics education. In B. Perry, G. Anthony, & C. Diezmann (Eds.), *Research in mathematics education in Australasia 2000–2003* (pp. 265–289). Brisbane, Australia: PostPressed.

Pfannkuch, M., & Wild, C. J. (2000). Statistical thinking and statistical practice: Themes gleaned from professional statisticians. *Statistical Science, 15,* 132–152.

Pfannkuch, M., & Wild, C. J. (2004). Towards an understanding of statistical thinking. In J. Garfield & D. Ben-Zvi (Eds.), *The challenge of developing statistical literacy, reasoning and thinking* (pp. 17–46). Dordrecht, The Netherlands: Kluwer.

Pollatsek, A., Lima, S., & Well, A. D. (1981). Concept or computation: Students' understanding of the mean. *Educational Studies in Mathematics, 12,* 191–204.

Pollatsek, A., Konold, C., Well, A., & Lima, S. (1984). Beliefs underlying random sampling. *Cognition and Instruction, 12,* 395–401.

Polya, G. (1945). *How to Solve it.* Princeton, NJ: Princeton University Press.

Reading, C. (2003, July). *Student perceptions of variation in a real world context.* Paper presented at the Third International Research Forum on Statistics Reasoning, Thinking, and Literacy (SRTL 3). Lincoln, NE.

Reading, C., & Shaughnessy, J. M. (2000). Student perceptions of variation in a sampling situation. In T. Nakahara & M. Koyama (Eds.), *Proceedings of the 24th Conference of the Interna-*

tional Group for the Psychology of Mathematics Education (Vol. 4, pp. 89–96). Hiroshima, Japan: Hiroshima University.

Reading, C., & Shaughnessy, J. M. (2004). Reasoning about variation. In J. Garfield & D. Ben-Zvi (Eds.), *The challenge of developing statistical literacy, reasoning and thinking* (pp. 201–226). Dordrecht, The Netherlands: Kluwer.

Roth, W. -M & Bowen, G. M. (2001). Professionals read graphs: A semiotic analysis. *Journal for Research in Mathematics Education, 32,* 159–193.

Rubin, A., & Rosebery, A. S. (1988, August). *Teachers' misunderstandings in statistical reasoning: Evidence from a field test of innovative materials.* Paper presented at the International Statistics Institute Round Table Conference, Budapest, Hungary. Voorburg, The Netherlands: International Statistics Institute.

Rubin, A., Bruce, B., & Tenney, Y. (1991). Learning about sampling: Trouble at the core of statistics. In D. Vere-Jones (Ed.), *Proceedings of the Third International Conference on Teaching Statistics* (Vol. 1, pp. 314–319). Voorburg, The Netherlands: International Statistical Institute.

Russell, S. J., & Corwin, R. B. (1989). *Used numbers: The shape of the data.* Palo Alto, CA: Dale Seymour.

Russell, S. J., & Mokros, J. (1996). What do children understand about average? *Teaching Children Mathematics, 2,* 360–364.

Russell, S. J., Schifter, D., & Bastable, V. (2002). *Developing mathematical ideas: Working with data.* Parsippany, NJ: Dale Seymour.

Saldanha, L. (2003). *Is this sample unusual? An investigation of students exploring connections between sampling distributions and statistical inference.* Unpublished doctoral dissertation, Vanderbilt University, Tennessee.

Saldanha, L., & Thompson, P. (2003). Conceptions of sample and their relationship to statistical inference. *Educational Studies in Mathematics, 51,* 257–270.

Schau, C., & Mattern, N. (1997). Assessing students' connected understanding of statistical relationships. In I. Gal & J. Garfield (Eds.), *The assessment challenge in statistics education* (pp. 91–106). Amsterdam: IOS Press

Shaughnessy, J. M. (1977). Misconceptions of probability: An experiment with a small-group, activity-based, model building approach to introductory probability at the college level. *Educational Studies in Mathematics, 8,* 285–316.

Shaughnessy, J. M. (1992). Research on probability and statistics: Reflections and Directions. In D. Grouws (Ed.), *Handbook of research on mathematics teaching and learning* (pp. 465–494). Reston, VA: National Council of Teachers of Mathematics.

Shaughnessy, J. M. (1997) Missed opportunities in research on the teaching and learning of data and chance. In F. Biddulph & K. Carr (Eds.), *People in mathematics education* (Vol. 1, pp. 6–22). Proceedings of the Twentieth annual meeting of the Mathematics Education Research Group of Australasia. Waikato, New Zealand: The University of Waikato Printery.

Shaughnessy, J. M. (2003a). Research on students' understanding of probability. In J. Kilpatrick, W. G. Martin, & D. Schifter (Eds.), *A research companion to principles and standards for school mathematics* (pp. 216–226). Reston, VA: National Council of Teachers of Mathematics.

Shaughnessy, J. M. (2003b). *The development of secondary students' conceptions of variability.* (Annual report year 1. NSF Grant No. REC 0207842). Portland, OR: Portland State University.

Shaughnessy J. M., & Bergman, B. (1993). Thinking about uncertainty: Probability and statistics. In P. Wilson (Ed.), *Research ideas for the classroom: High school mathematics* (pp. 177–197). Reston, VA: National Council of Teachers of Mathematics.

Shaughnessy, J. M., & Chance, B. (2005). *Statistical questions from the classroom.* Reston, VA: National Council of Teachers of Mathematics.

Shaughnessy, J. M., & Ciancetta, M. (2002). Students' understanding of variability in a probability environment. In B. Phillips (Ed.), *CD of the Proceedings of the Sixth International Conference on Teaching Statistics: Developing a statistically literate society, Cape Town, South Africa.* Voorburg, The Netherlands: International Statistics Institute.

Shaughnessy, J. M., & Dick, T. P. (1991). Monty's Dilemma: Should you stick or switch? *The Mathematics Teacher, 84,* 252–256.

Shaughnessy, J. M., & Pfannkuch, M. (2002). How faithful is Old Faithful? Statistical thinking: A story of variation and prediction. *The Mathematics Teacher, 95,* 252–259.

Shaughnessy, J. M., & Zawojewski, J. S. (1999). Secondary students' performance on data and chance in the 1996 NAEP. *The Mathematics Teacher, 92,* 713–718.

Shaughnessy, J. M., Garfield, J., & Greer, B. (1996). Data Handling. In A. J. Bishop, K. Clements, C. Keitel, J. Kilpatrick, & C. Laborde (Eds.), *International Handbook of Mathematics Education* (pp. 205–237. Dordrecht, The Netherlands: Kluwer.

Shaughnessy, J. M., Watson, J. M, Moritz, J. B., & Reading, C. (1999, April). *School mathematics students' acknowledgement of statistical variation: There's more to life than centers.* Paper presented at the Research Pre-session of the 77th Annual meeting of the National Council of Teachers of Mathematics, San Francisco, CA.

Shaughnessy, J. M., Ciancetta, M., & Canada, D. (2003). Middle school students' thinking about variability in repeated trials: A cross-task comparison. In N. Pateman, B. Dougherty, & J. Zillah (Eds.). *Proceedings of the 27th Conference of the International Group for the Psychology of Mathematics Education* (Vol. 4, pp. 159–166). Honolulu, HI: University of Hawaii.

Shaughnessy, J. M., Ciancetta, M., & Canada, D. (2004a). Types of student reasoning on sampling tasks. In M. Johnsen Hoines & A. Berit Fuglestad (Eds.), *Proceedings of the 28th meeting of the International Group for Psychology and Mathematics Education* (Vol. 4, pp. 177–184). Bergen, Norway: Bergen University College Press.

Shaughnessy, J. M., Ciancetta, M., Best, K, & Canada, D. (2004b, April). *Students' attention to variability when comparing distributions.* Paper presented at the Research Pre-session of the 82nd annual meeting of the National Council of Teachers of Mathematics, Philadelphia, PA.

Shaughnessy, J. M., Barrett, G., Billstein, R., Kranendonk, H. A., & Peck, R. (2004c). *Navigating through probability, grades 9–12.* Reston, VA: National Council of Teachers of Mathematics.

Shaughnessy, J. M., Ciancetta, M., Best, K., & Noll, J. (2005, April). *Secondary and middle school students' attention to variability when comparing data sets.* Paper presented at the Research Pre-session of the 83rd annual meeting of the National Council of Teachers of Mathematics, Anaheim, CA.

Shulman, L. S. (1987). Knowledge and teaching: Foundations of the new reform. *Harvard Educational Review, 57,* 1–22.

Starkings, S. (1997). Assessing statistical projects. In I. Gal & J. Garfield (Eds.), *The assessment challenge in statistics education* (pp. 139–151). Amsterdam: IOS Press.

Strauss, S., & Bichler, E. (1988). The development of children's concepts of the arithmetic average. *Journal for Research in Mathematics Education, 19,* 64–80.

Tarr, J., & Shaughnessy, J. M. (in press). Statistics and probability. In P. Kloosterman & F. Lester (Eds,), *Results from the Eighth & Ninth Mathematics Assessment of the National Assessment of Educational Progress.* Reston, VA: National Council of Teachers of Mathematics.

Takis, S. L. (1999). Titanic: A statistical exploration. *The Mathematics Teacher, 92,* 660–664.

Teach-STAT. (1996a). *Teach-STAT: Teaching statistics grades 1–6: A Key for Better Mathematics.* The University of North Carolina Mathematics and Science Education Network. Palo Alto, CA: Dale.

Teach-STAT. (1996b). *Teach-STAT for statistics educators.* The University of North Carolina Mathematics and Science Education Network. Palo Alto, CA: Dale Seymour.

Thompson, A. (1992). Teachers' beliefs and conceptions: A synthesis of the research. In D. Grouws (Ed.), *Handbook of research on mathematics teaching and learning* (pp. 127–146). Reston, VA: National Council of Teachers of Mathematics.

Thompson, P. W., & Saldanha, L. A. (2003). Fractions and multiplicative reasoning. In. J. Kilpatrick, W. G. Martin, & D. Schifter (Eds.), *A research companion to principles and standards for school mathematics* (pp. 95–113). Reston, VA: National Council of Teachers of Mathematics.

Torok, R. (2000). Putting the variation into chance and data. *Australian Mathematics Teacher, 56,* 25–31.

Torok, R., & Watson, J. (2000). Development of the concept of statistical variation: An exploratory study. *Mathematics Education Research Journal, 9,* 60–82.

Truran, J. (1994). Children's intuitive understanding of variance. In J. Garfield (Ed.), *Research papers from the Fourth International Conference on Teaching Statistics.* Minneapolis, MN: The International Study Group for Research on Learning Probability and Statistics.

Tversky, A., & Kahneman, D. (1974). Judgment under uncertainty: Heuristics and biases. *Science, 185,* 1124–1131.

Tversky, A., & Kahneman, D. (1983). Extensional versus intuitive reasoning: The conjunction fallacy in probability judgment. *Psychological Review, 90,* 293–315.

United States Department of Agriculture (2004). *Economic Research Service.* Retrieved October 13, 2005, from www.ers.usda.gov/data/foodconsumption/foodavailspreads.

Vergnaud, G. (1983). Multiplicative structures. In R. Lesh & M. Landau (Eds.), *Acquisition of mathematics concepts and processes* (pp. 127–174). New York: Academic Press.

Watson, J. M. (1997). Assessing statistical thinking using the media. In I. Gal & J. Garfield (Eds.), *The assessment challenge in statistics education* (pp. 107–121). Amsterdam: IOS Press.

Watson, J. M. (1998). Professional development for teachers of probability and statistics: Into an era of technology. *International Statistical Review, 66,* 271–289.

Watson, J. M. (2001a). Longitudinal development of inferential reasoning by school students. *Educational Studies in Mathematics, 47,* 337–372.

Watson, J. M. (2001b). Profiling teachers' competence and confidence to teach particular mathematics topics: The case of data and chance. *Journal of Mathematics Teacher Education, 4,* 305–337.

Watson, J. M. (2002). Inferential reasoning and the influence of cognitive conflict. *Educational Studies in Mathematics, 51,* 225–256.

Watson, J. M. (2004). Developing reasoning about samples. In J. Garfield & D. Ben-Zvi (Eds.), *The challenge of developing statistical literacy, reasoning and thinking* (pp. 277–294). Dordrecht, The Netherlands: Kluwer.

Watson, J. M., & Callingham, R. (2003). Statistical literacy: A complex hierarchical construct. *Statistics Education Research Journal, 2,* 3–46.

Watson, J. M., & Kelly, B. A. (2002). Can grade 3 students learn about variation? In B. Phillips (Ed.), *Proceedings of the Sixth International Conference on Teaching Statistics: Developing a statistically literate society, Cape Town, South Africa.* [CD-ROM].Voorburg, The Netherlands: International Statistics Institute.

Watson, J. M., & Kelly, B. A. (2004). Expectation versus variation: Students' decision making in a chance environment. *Canadian Journal of Science, Mathematics, and Technology Education, 4,* 371–396.

Watson, J. M., & Moritz, J. B. (1997a). Student analysis of variables in a media context. In B. Phillips (Ed.), *Papers on Statistical Education Presented at ICME 8* (pp.129–147). Hawthorn, Australia: Swinburne Press.

Watson, J. M., & Moritz, J. B. (1997b). The C&D PD CD: Professional development in chance and data in the technological age. In N. Scott & H. Hollingsworth (Eds.), *Mathematics creating the future. Proceedings of the 16th Biennial Conference of the Australian Association of Mathematics Teachers*(pp. 442–450). Adelaide, South Australia: AMMT.

Watson, J. M., & Moritz, J. B. (1999). The beginning of statistical inference: Comparing two data sets. *Educational Studies in Mathematics, 37,*145–168.

Watson, J. M., & Moritz, J. B. (2000a). Developing concepts of sampling. *Journal for Research in Mathematics Education, 31,* 44–70.

Watson, J. M., & Moritz, J. B. (2000b). Development of understanding of sampling for statistical literacy. *Journal of Mathematical Behavior, 19,* 109–136.

Watson, J. M., & Moritz, J. B. (2000c). The longitudinal development of understanding of average. *Mathematical Thinking and Learning, 2,* 11–50.

Watson, J. M. & Moritz, J. B. (2000d). The development of concepts of average. *Focus on Learning Problems in Mathematics, 21,* 15–39.

Watson, J. M., & Moritz, J. B. (2001). Development of reasoning associated with pictographs: Representing, interpreting, and predicting. *Educational Studies in Mathematics, 48,* 47–81.

Watson, J. M., & Shaughnessy, J. M. (2004). Proportional reasoning: Lessons from research in data and chance. *Mathematics Teaching in the Middle School, 10,* 104–109.

Watson, J. M., Collis, K. F., Callingham, R. A., & Moritz, J. B. (1995). A model for assessing higher order thinking in statistics. *Educational Research and Evaluation, 1,* 247–275.

Watson, J. M., Kelly, B.A., Callingham, R.A., & Shaughnessy, J. M. (2003). The measurement of school students' understanding of statistical variation. *International Journal of Mathematical Education in Science and Technology, 34,* 1–29.

Wild, C. J., & Pfannkuch, M. (1999). Statistical thinking in empirical enquiry. *International Statistical Review, 67,* 223–265.

Zawojewski, J. S., & Shaughnessy, J. M. (2000). Data and chance. In E. A. Silver & P. A. Kenney (Eds.), *Results from the Seventh Mathematics Assessment of the National Assessment of Education-*

al Progress (pp. 235–268). Reston, VA: National Council of Teachers of Mathematics.

AUTHOR NOTE

I would like to express my heartfelt thanks to Iddo Gal, Jane Watson, and Joan Garfield for their insightful suggestions in reviewing two earlier versions of this chapter. A chapter of this magnitude is often a community effort in a field of inquiry, and that was truly the case for this chapter. Iddo, Jane, and Joan helped me to more clearly articulate the big themes in our research in statistics education, and pointed me in directions that I had not previously been aware of. I am very fortunate to have such thoughtful and professional colleagues who helped make the chapter stronger.

22

MATHEMATICS THINKING AND LEARNING AT POST-SECONDARY LEVEL

Michèle Artigue
UNIVERSITÉ PARIS 7

Carmen Batanero
UNIVERSIDAD DE GRANADA

Phillip Kent
UNIVERSITY OF LONDON

INTRODUCTION

This chapter deals with research carried out on mathematics thinking and learning at post-secondary level. It tries to point out the evolution of research in this area since the first *Handbook* was published in 1992, its most important advances, their potential and limit for understanding and improving teaching and learning processes at this advanced level.

Synthesizing research advances in a particular area of mathematics education has always been difficult, due to the diversity of educational structures and cultures, and to the diversity of research paradigms. Regarding post-secondary education, the exercise of looking back 10 years could nevertheless seem easier than in other domains of mathematics education. Thanks to the existence of the International Group for the Psychology of Mathematics Education (PME), and its annual conferences, researchers from different origins reflecting on these issues established a pattern of regular exchanges and collaborative work. In 1991 this work led to the book edited by Tall: *Advanced Mathematical Thinking*, a good representation of the state of the art[1] at that time. Both the structure and the content of this book show that the predominant concerns then were cognitive ones: identifying cognitive processes underlying the learning of mathematics at advanced levels, investigating the relationships of these processes with respect to those at play at more elementary levels, and understanding students' difficulties with advanced mathematical concepts. Different theoretical constructs supported this research, such as the notions of *concept definition* and *concept image* (Tall & Vinner, 1981), the *process-object duality* (Dubinsky, 1991; Sfard, 1991), or the notion of *epistemological obstacle* due to Bachelard and developed in the didactic field by Brousseau (Brousseau, 1983). However, the book shows an evident common interest in visions of knowledge growth focusing on

[1] There were some significant exceptions: For instance, research carried out in this group did not consider the stochastic domain.

misunderstandings, cognitive conflicts, discontinuities, and hierarchies. Moreover, the research tended to concentrate on just a few mathematical domains: calculus and associated concepts, mathematical rationality, and proof. This vision, unsurprisingly, is reflected in the chapter that refers most directly to post-secondary level mathematics in the 1992 *Handbook*, "The Transition to Advanced Mathematical Thinking: Functions, Limits, Infinity, and Proof" (Tall, 1992).

The situation today is far from being the same, for a lot of reasons, linked both to the internal evolution of mathematics education as a scientific field and to external changes affecting post-secondary education. Research at post-secondary level was firstly influenced by the evolution of the dominant research paradigms from constructivist cognition and cognitive development towards sociocultural and anthropological ones (Lerman & Sierpinska, 1996). Sociocultural and anthropological paradigms were considered in the 1992 *Handbook*–see for instance chapter 20 by Schoenfeld who writes, "This cultural perspective is well grounded anthropologically but it is relatively new to the mathematics education literature" (p. 340). From that time, sociocultural and anthropological approaches have taken an increasing importance, and references to Vygotsky have tended to supplant those to Piaget. Research on mathematical learning at post-secondary level could neither escape this general influence nor ignore the central role these approaches give to the analysis of social and institutional practices in the understanding of knowledge growth. Research was also influenced by the emphasis these approaches put on semiotic mediations, and thus on the semiotic tools of mathematical activity. One can easily understand that such an evolution has a particular resonance in research on advanced topics: in advanced mathematics, students' relationship with symbolism becomes an essential feature of their relationship with mathematics. More recently, research has also been influenced by the increasing interest in the neuroscientific study of cognition and embodied cognition.

At the same time as these internal changes, the field was affected by a number of external changes, which have been analysed in the ICMI Study on the Teaching and Learning of Mathematics at University Level (Holton, 2001). For example, the following quotation comes from the ICMI discussion document that launched this Study and explained its rationale[2]:

> A number of changes have taken place in recent years which have profoundly affected the teaching of mathematics at university level. Five changes which are still having considerable influences are (i) the increase in the number of students who are now attending tertiary institutions; (ii) major pedagogical and curriculum changes that have taken place at pre-university level; (iii) the increasing differences between secondary and tertiary mathematics education regarding the purposes, goals, teaching approaches and methods; (iv) the rapid development of technology; and (v) demands on universities to be publicly accountable. Of course, all of these changes are general and have had their influence on other disciplines. However, because of its pivotal position in education generally, and its compulsory nature for many students, it could be argued that these changes have had a greater influence on mathematics than perhaps on any other discipline. . . . As a result of the changing world scene, ICMI feels that there is a need to examine both the current and future states of the teaching and learning of mathematics at university level. The primary aim of this ICMI Study is therefore to pave the way for improvements in the teaching and learning of mathematics at university level for all students. (ICMI, 1997).

All these evolutions have contributed to make the field of research on mathematics thinking and learning in post-secondary education much more diverse today than was the case 10 years ago. Consequently, trying to present a systematic survey of the results obtained in the last 10 years and of the different existing trends could result in a pointillist painting whose driving forces would remain invisible to the reader. Trying to avoid this trap while taking into account the state of the field obliged us to make some choices, and we acknowledge that the vision we present is a personal vision. Our main choice has been to reflect and question some major evolutions we perceive in the field rather than to give a comprehensive view of it. We also decided to approach these evolutions in two different ways: on the one hand by showing how classical research topics such as calculus or linear algebra have been partially renewed in the last 10 years and on the other hand by addressing two emerging research themes: mathematics in engineering courses, and stochastics.

Because this *Handbook* has no chapter devoted to technology, we would additionally like to present our own reflections on technology in this chapter. Consistent with the spirit of the *Handbook*, our choice is not to have a specific section dealing with technology, but rather to integrate the discussion of technology throughout the chapter.

[2] This discussion document was disseminated through different channels and can be found in the *ICMI Bulletin* number 43 (December 1997) which is accessible on the ICMI website (www.mathunion.org/Organization/ICMI).

The chapter is structured into four main parts. In the first part, we briefly recall the state of the art in the early nineties. In the second part, we consider evolutions that, in our opinion, exemplify how the internal evolution of the theoretical frames in mathematics education has influenced and is influencing research at post-secondary level. For this purpose, we mainly consider research in traditional domains such as calculus and linear algebra. In the third part, we consider evolutions more linked to external factors such as the evolution of the context of post-secondary mathematics education in its institutional, social, cultural, and technological dimensions and the evolution in the relative importance of the different mathematical domains. For this part, which did not have its exact counterpart in the previous *Handbook*, we focus on research carried out about the teaching and learning of mathematics in engineering courses, and research on probability and statistical learning. We hope that this choice helps us question theories and positions as regard mathematical learning and thinking that have generally been established having in mind more or less explicitly the mathematical education of "pure mathematicians" or mathematics teachers and their particular needs, and also that have focused on few and classical mathematical domains. This choice also obliges us to consider the role of technology in learning in a rather different way. Indeed, what is often at stake in these domains is not simply the use of technology for developing usual mathematical knowledge but the way in which technology is changing mathematical activity and understanding, including problem solving, proving, reasoning, modeling, and symbolizing. Mathematical and technological expertise and needs are there much more tightly intertwined.

Finally, in the last part of the chapter, we come back to more general issues, pointing out both the potential and limits of existing research for understanding and for helping improve the current situation of thinking and learning at post-secondary level, pointing out also some evident research needs that are apparent from the literature reviewed in the chapter.

THE EARLY NINETIES

As was mentioned in the introduction, the state of research on mathematical thinking and learning at post-secondary level in the early nineties is rather well represented in the book *Advanced Mathematical Thinking* published in 1991. A main theme of interest for researchers, at that time, was to characterize advanced mathematical thinking (AMT in the following) with respect to more elementary forms of mathematical thinking. It was also to clarify the mental processes that allow students to enter into AMT, and the difficulties students meet in developing such mental processes.

The Nature of AMT and AMT Processes

As pointed out by Dreyfus (1991) in his contribution to the book, characterizing AMT is not something as easy as "there is no sharp distinction between many of the processes of elementary and advanced mathematical thinking" (p. 26). As with more elementary levels, according to Dreyfus, the processes of AMT can be described in terms of representing, visualizing, generalizing, classifying, conjecturing, inducing, analyzing, synthesizing, abstracting, or formalizing. Referring to famous texts by Hadamard (1945) and Poincaré (1913), Dreyfus also pointed out the diversity of mathematical thinking modes, but differentiated essentially two main mathematical styles, according to the relative importance given to visualization and intuition, or to symbolic and analytic approaches.

Beyond these general considerations about AMT, the crucial point, as stressed by Tall in the final chapter of the book, was certainly the acknowledgement of the "thorny nature of the full path of mathematical thinking, so much more demanding and rewarding than the undoubted aesthetic beauty of the final edifice of formal definition, theorem and proof" (p. 251). For the authors, the gap between the logic of the mathematical edifice and the logic of cognitive processes explained the observed inefficiency of university teaching strategies based on the former, for the majority of students. This was evidenced at the time by the high rates of failure in fundamental courses, such as calculus, and also by the limited ability demonstrated later by those students who had passed the fundamental courses. The study of Selden, Mason, and Selden (1989) is from this point of view especially illustrative. These researchers presented students with problems[3] that could easily be solved with the techniques at their disposal but were not presented in the usual way. Not one student solved an entire problem correctly, and most of them could not do anything.

The gap mentioned above has been the motivation for several theoretical constructs, and we briefly present the most important ones in the following section.

[3] One of these problems was the following: Find at least one solution to the equation $4x^3 - x^4 = 0$ or explain why no such solution exists.

Concept Definition and Concept Image

These notions were introduced by Tall and Vinner (1981) and defined in the following way:

> We shall use the term of concept image to describe the total cognitive structure that is associated with the concept, which includes all the mental pictures and associated properties and processes. It is built over the years through experiences of all kinds, changing as the individual meets new stimuli and matures. . . . As the concept image develops it need not be coherent at all times. The brain does not work that way. Sensory input excites certain neuronal pathways and inhibits others. In this way different stimuli can activate different parts of the concept image, developing them in a way which need not make a coherent whole. (p. 152)

One example of this lack of global coherence, often quoted in the literature, is the following: Students are asked first to compare 0.999. . . and 1, then to calculate the sum of the series: $\Sigma 9/10^n$. Many students answer that 0.999. . . < 1 to the first question whereas they correctly answer that the sum is 1 to the second. Several reasons have been given for explaining the first answer. They often rely on the process/object duality we will evoke later on: Students are bound to a process view of the symbolic notation 0.999. . . , and this view prevents them from seeing beyond the infinite process whose terms are all less than 1, the object "number 1" that results from it.[4] When asked the second question, they recognize a geometric series, and activating the formula for its sum they get the correct answer. Notice that this answer generally does not lead students to reconsider their answer to the first question, which is seen as a different one.

The concept image is generally at variance with the concept definition, and this was well evidenced at that time by research carried out on functions. Convergent results showed that many students, even when able to give a correct set-based definition of the notion of function, when asked to say if such or such object given by a discursive, tabular, symbolic, or graphic representation was or was not a function, gave answers at variance with their definition, and answers that could be different for the same object according to the semiotic representation used.[5]

Epistemological Obstacles

As mentioned in the introduction, other approaches had been developed in the eighties in order to approach the complexity of learning processes at post-secondary level and the distance between these and the current logical organization of mathematical knowledge. One of these approaches, also well represented in the AMT book, relied on the notion of epistemological obstacle, initially due to the philosopher Bachelard, and imported by Brousseau into the educational field. According to Bachelard (1938), scientific knowledge supposes the rejection of common knowledge. In the educational field, this notion has been introduced in order to better understand the status of students' errors and to acknowledge that some of these, generally the most resistant ones, result not from a lack of knowledge but from knowledge that has stabilized because of its efficiency. Let us stress that this knowledge can be social or cultural but that it can also result from school apprenticeship. Some researchers working at post-secondary level soon appropriated this notion in order to understand students' difficulties with advanced mathematical concepts such as the concepts of limit, derivative, and integral. Regarding the limit concept for instance, Cornu and Sierpinska first (Cornu, 1983, 1991; Sierpinska, 1985, 1987) and then Schneider (1991) evidenced the existence of epistemological obstacles related to

> The everyday meaning of the word limit, which induces resistant conceptions of the limit as a barrier or as the last term of a process, or tend to restrict convergence to monotonic convergence;
>
> The overgeneralization of properties of finite processes to infinite processes, following the continuity principle stated by Leibniz;
>
> The strength of a geometry of forms which prevents students from clearly identifying the objects involved in the limit process and their underlying topology. This makes it difficult for students to appreciate the subtle interaction between the numerical and geometrical settings in the limit process. (Artigue, 2001, p. 211)

Overgeneralization leads for instance to a belief that the limit of a sequence of strictly positive numbers is strictly positive, or that the limit of a sequence of continuous (resp. differentiable, integrable) functions is contin-

---

[4] Edwards (1997) proposed for instance an alternative interpretation based on the difference regularly observed in the treatment of the two equalities : 0.333. . . = 1/3 and 0.999. . . = 1. This would show that, when manipulating infinite decimal expansions, students do not refer to the definition they have been given but to their practice of division. This practice supports the first equality but cannot make sense of the second one.

[5] For instance, recognizing a constant function as a function if given by a graphical representation but refusing it this status if given by an algebraic expression without an explicit variable.

uous (resp. differentiable, integrable). The geometrical obstacle leads to a belief that whenever a sequence (F_n) of geometrical objects has, as a limit, an object A, all the magnitudes attached to the F_n have as a limit the corresponding magnitudes for A. As shown by Schneider, this geometrical obstacle often combines with an obstacle she labels the *heterogeneity of dimensions obstacle*, which underlies for instance the following reasoning concerning the computation of the area under a curve: When the size of the subdivision tends towards 0, each of the rectangles tends towards a segment. Thus at the limit, the total area under the curve which is the sum of the areas of the segments is necessarily 0, which is impossible![6]

Process/Object Duality

Another approach, which was to take increasing importance in the research community, was also already present: that based on the process-object duality. In the early nineties, this approach was associated with some particular names, especially those of Sfard and Dubinsky. In the AMT book, it was especially visible in the chapter written by Dubinsky where he introduced *APOS theory*. According to APOS, conceptualization begins with manipulating previously constructed mental or physical objects to form *actions*; actions are then interiorized to form *processes* that are then encapsulated to form *objects*. Finally, actions, processes and objects are organized more or less coherently in *schemas*, and "a subject's tendency to invoke a schema in order to understand, deal with, organize, or make sense out of a perceived problem situation is her or his knowledge of an individual concept in mathematics" (Dubinsky, 1991, p. 102).

Dubinsky's ambition was to isolate small portions in the complex structures of a subject's schemas and to give an explicit description of these, and especially to develop genetic decompositions of particular concepts such as induction and function, two examples he presented in the same chapter. What was also stressed by Dubinsky is the fact that a learner cannot successfully engage in AMT without developing an object view of the mathematical ideas at stake.

The 1992 *Handbook*

As mentioned earlier, Tall's chapter of the *Handbook* closely reflects the AMT book. Tall introduced the different constructs mentioned above together with illustrative examples, and we just refer here to the conclusion of the chapter where he gave his own characterization of AMT and proposed an agenda for research. As regards AMT, the crucial point, according to him, is the change in the relationship one develops with mathematical concepts: "To move to more advanced mathematical thinking involves a difficult transition, from a position where concepts have an intuitive basis founded on experience, to one where they are specified by formal definitions and their properties reconstructed through logical deductions" (p. 495), but he also stressed that "In taking students to the transition to advanced mathematical thinking, we should realize that the formalizing and systematizing is the final stage of mathematical thinking, not the total activity."

According to him, research has thus to investigate and understand the difficult cognitive changes and restructurings that this transition involves. Research evidences the existence of conflicts between students' intuitive views and formal mathematics; these have to be clarified through clinical interviews, and at the same time formal mathematics has to be itself placed into perspective as a human activity that attempts to organize the complexities of human thought into a logical system. This is the research agenda he proposed.

Beyond the work done under the AMT umbrella, post-secondary research is also present in the first *Handbook* through Schoenfeld's chapter "Learning to Think Mathematically: Problem Solving, Metacognition and Sense Making in Mathematics". Synthesizing different approaches to these issues coming from mathematics education and related fields, and the personal work he has carried out in universities on problem solving, reporting on differences between students' and experts' problem-solving behaviors, Schoenfeld introduced a framework for the analysis of mathematical cognition. This framework is organized around five dimensions: the knowledge base, problem-solving strategies, monitoring and control, beliefs and affects, practices. We would like to stress here the attention paid by Schoenfeld to the role played by the two last categories: beliefs and practices, and the links he establishes between these, as this attention shows the emergence of the more global and cultural views on cognition whose influence was to increase in the next decade.

Taking into account the specific theme of this chapter, we have focused in this brief summary on research on learning processes and mathematical thinking. Nevertheless we would like to mention that in the early nineties research on post-secondary mathemat-

[6] Note that recently Gonzalez-Martin and Camacho (2004) evidenced the existence in university students of a variant of this obstacle, when working on students' conceptions of generalized integrals. It is expressed in the conviction held by students that, for a positive function the integrals $\int_a^{+\infty} f(x)dx$ and $\int_a^{+\infty} f^2(x)dx$ have necessarily the same nature because to a finite area (resp. infinite area) there necessarily corresponds by rotation around an axis (here Ox) a finite volume (resp. infinite volume).

ics education was not restricted to this aspect. Teaching experiments were also developed consistent with these perspectives on learning. For instance, Tall used the notion of generic organizer[7] to build the software Graphic Calculus for introducing students to calculus concepts; Dubinsky and his colleagues developed the language ISETL and began to use it for teaching functions and algebraic structures; French researchers such as Artigue, Legrand, and Rogalski used the theory of didactic situations and the notion of scientific debate to develop (jointly with physicists) "didactic engineering" for the teaching of differentiation, integration, and differential equations (Artigue, 1991). Many of these developments incorporated technology as a generic organizer, as support for visualization and coordination between semiotic registers, or involved a programming language to support the interiorization and encapsulation of processes.

Thus, if we consider the field of research on mathematics learning at post-secondary level at the beginning of the nineties, there is no doubt that different theoretical constructs had been elaborated and worked out. These structured the didactic reflections of researchers about students' learning processes and difficulties, allowing researchers to better understand the failure of ordinary educational practices for the majority of students, to design alternatives to these and to test them. Nevertheless, the results obtained were limited to small areas of post-secondary education, both in terms of categories of students and mathematical subjects; the theoretical constructs were not generally integrated into more global didactic structures connecting both learning and teaching phenomena; they remained essentially cognitive constructs and, within this cognitive perspective, relied on the dominant constructivist epistemology. Both the evolution of the field and the evolution of the educational context have produced changes to this situation, and these we present and discuss in the following parts.

EVOLUTIONS DEALING WITH ALREADY-DEVELOPED RESEARCH AREAS AND PERSPECTIVES

In this part we present and discuss evolutions, focusing on those that can be easily connected to research areas and perspectives already mentioned above. First we consider the evolution of ideas about AMT itself. Second we consider the reinforcement and extension of existing approaches. We discuss the evolution of the process-object oriented approaches through the development of APOS theory, and the development of the proceptual approach. We also consider approaches that, starting from epistemological and/or historical analysis of mathematical knowledge, propose alternative categorizations for approaching the analyses of conceptualization. Third, we discuss the impact of some newer cognitive approaches, focusing mainly on the increasing influence of embodied cognition and of the linguistic approach developed by Lakoff and Nuñez. Fourth, we come to another dimension of the evolution that we see as a consequence of the increasing influence of anthropological and sociocultural approaches in the educational field. Cognition is there seen as something emerging from institutional practices, and understanding learning processes cannot be achieved without analyzing these institutional practices and identifying the norms and values underlying them. We show how these approaches complement the preceding ones for understanding the interaction between the individual and the collective in learning processes and supporting didactic design, and also for understanding issues that have become more and more crucial, such as the secondary/tertiary transition, or the relationships that new generations of students develop regarding mathematics and mathematical activity, and how these affect their learning of mathematics.

Evolution of Ideas About the Nature of Advanced Mathematical Thinking

The publication of the AMT book did not close the discussion about the nature of advanced mathematical thinking and its development. Critical reviews argued about what was referred to by the term *advanced* in the AMT book: mathematics? thinking? both? This was unclear, and the criteria used in the distinction between elementary and advanced mathematical thinking were not really convincing, according to the critics. From that time, the definition of AMT has been a question regularly addressed but unresolved. In the following, we have chosen to illustrate some of the associated discussions by referring to the work carried out since 1998 by a working group of PME-NA titled "The Role of Advanced Mathematical Thinking in Mathematics Education Reform" and the special issue of the journal *Mathematical Thinking and Learning* that

[7] Tall (1991) defined a *generic organizer* as "an environment that provides the user the facilities of manipulating examples (and, where possible, non-examples) of a concept. The word 'generic' means that the learner's attention is directed to certain aspects of the examples which embody the more abstract concept" (p. 187).

recently emerged from this work (Selden & Selden, 2005). As the editors explained in the introduction to this special issue, the working group "began by discussing such questions as what kinds of earlier experiences might help students make the transition to the kinds of AMT that post-secondary students are often asked to engage" and its interest "rather naturally metamorphosed into efforts at characterizing AMT and looking for seeds thereof that are, or could be, planted early in students' mathematical careers" (p. 3). The special issue evidenced the diversity of the answers offered within the group. For instance, starting from the fact that manipulating advanced concepts such as the concept of limit does not necessarily require advanced modes of thinking, and that students have difficulty finding referents for these abstract mathematical concepts in their familiar world, Edwards, Dubinsky, and McDonald defined AMT in the following way: AMT is "thinking that requires deductive and rigorous reasoning about mathematical notions that are not entirely accessible to us through our five senses" (2005, pp. 17–18). Harel and Sowder, for their part, defined AMT by referring to the notion of epistemological obstacle. Considering the three conditions imposed by Duroux and Brousseau (Brousseau, 1997) for epistemological obstacles: (a) to have traces in the history of mathematics; (b) to be a piece of knowledge producing valid answers in particular contexts, and invalid responses outside this context; (c) to "withstand both occasional contradictions and the establishment of a better piece of knowledge" (p. 34). Harel and Sowder claimed that "mathematical thinking is advanced, if its development involves at least one of the above three conditions" (2005, pp. 17–18).

These five authors proposed various examples in order to show the pertinence of the perspectives they developed but these examples, in spite of their interest, are insufficient to make the proposed definitions fully convincing. For instance, the first definition relies on the claim that some mathematical objects are directly accessible to our senses, an assertion that can be seriously discussed from an epistemological point of view. This definition also opposes rigorous and deductive reasoning to automatic reasoning or routine, which does not pay attention to the fact that most reasoning processes subtly intertwine routine and rigorous reflection. The separation between what is routine and what needs reflection changes for each learner according to the development of the learning process and even the academic institution where the learning takes place. Thus the operationality of this definition is questionable. As regard the second definition, the separation it artificially introduces between the three complementary conditions imposed by Duroux on the notion of epistemological obstacle, and the fact that these three conditions are given an equivalent role for qualifying AMT, seems difficult to sustain.

More than the attempts at defining AMT, which we do not consider as really successful, what we take from this special issue is the hypothesis made by most of the contributors, and especially Rasmussen, Zandieh, King, and Teppo, that AMT is not specific to a particular level of schooling. As these authors did, we would prefer to speak of *advanced mathematical practices*. Underestimating the relative nature of these, their dependence on institutional context, and linking them too much to formal mathematics can turn an idea that is a priori pertinent for analyzing learning processes into an obstacle. Our ambition in this chapter is to investigate how research carried out up to now at post-secondary level helps us to understand the ways students learn or do not learn, and could better learn mathematics, in the diversity of existing post-secondary institutions, without establishing a priori hierarchies of values among the different mathematics cultures these institutions offer. We need thus to be sensitive to the implicit values carried by the research constructs we use, and our review of AMT research leads us to be very critical in this regard about the construct of advanced mathematical thinking.

The Reinforcement and Development of Existing Approaches

Process–Object Duality

Expanding on the idea of process/object duality introduced above, we would like to focus on two trends: (a) the development of the APOS theory whose influence in research at post-secondary level has strongly increased in the last 10 years, thanks to the collaborative work of several groups of researchers and the existence of institutions such as RUMEC,[8] and (b) the development of the proceptual approach.[9]

The fundamental base of APOS remains the same as what was extensively presented by Dubinsky (1991). But, since that time, this approach has been used for building genetic decompositions of many different concepts taught at university level, for extending

[8] Research in Undergraduate Mathematics Education Community.

[9] We have chosen to focus on these two examples because they nicely illustrate what can be the long-term development of approaches based on the process-object duality, as far as these are widely used. A similar analysis could have been carried out starting from the theory of reification initiated by Sfard.

research based on APOS towards secondary mathematics education, and for elaborating and testing teaching designs based on the theory. APOS theory is a cognitive theory, and this obliges researchers to rely for teaching design on educational approaches complementing what APOS can offer. Thus the idea of cooperative learning has been linked to APOS, leading to a number of different projects (Dubinsky & Schwingendorf, 1997).

The fundamental basis of APOS has not changed, but one evolution is worth mentioning. It resulted from difficulties met by the researchers in satisfactorily explaining all the data they had collected. What was mainly at stake was the schema part of APOS. According to APOS (Dubinsky & McDonald, 2001, p. 277),

> a schema for a certain mathematical concept is an individual's collection of actions, processes, objects and other schemas which are linked by some general principles to form a framework in the individual's mind that may be brought to bear upon a problem situation involving that concept. This framework must be coherent in the sense that it gives, explicitly or implicitly, means of determining which phenomena are in the scope of the schema and which are not.

Data obtained in research concerning the chain rule, and then the properties linking the graph of a function and its derivatives, led to a reconsideration of this idea of schema, and to the incorporation of the triad introduced by Piaget and Garcia (1989) in order to better explain the construction of schemas. Incorporating the triad led to the introduction of three different stages in the construction: the Intra, Inter, and Trans stages. As explained in Dubinsky and McDonald (2001, p. 280),

> The Intra stage of schema development is characterized by a focus on individual actions, processes, and objects in isolation of other cognitive items of a similar nature. . . . The Inter stage is characterized by the construction of relationships and transformations among these cognitive entities. . . . Finally, at the Trans stage the individual constructs an implicit or explicit underlying structure through which the relationships developed in the Inter stage are understood and which gives the schema a coherence by which the individual can decide what is in the scope of the schema and what is not.

Taking as an example the function concept, one could say that at the Intra level, an individual tends to focus on functions seen as isolated objects and on the activities they perform on these; at the Inter level, he or she begins to make connections between functional objects, and to see how new functional objects can be created through these connections, to give sense to the idea of transformation of functional objects; at the Trans level, he or she can consider systems of transformations and the mathematical structures that emerge from these.

The second trend in process-object duality we evoke here is that developed by Tall. According to APOS, mathematical learning is achieved through the construction of mental actions, processes and objects, and their organization into schemas. Over the years, the distance between Tall's vision and APOS has progressively increased, and the model he proposes today is something quite different. According to him, cognitive growth of mathematical knowledge presents three main different paths corresponding to three different mathematical worlds (Tall, 2004):

> The first grows out of our perceptions of the world and consists of our thinking about things that we perceive and sense, not only in the physical world, but in our own mental world of meaning. By reflection and by the use of increasingly sophisticated language, we can focus on aspects of our sensory experience that enable us to envisage conceptions that no longer exist in the world outside, such as a "line" that is "perfectly straight." (p. 285)

This first world, which Tall names the *embodied world*, applies to the developmental path of Euclidean geometry from our perception of spaces and forms. The second path leads from enactive experiences with quantity and change to numbers, algebra, and calculus. Actions are there "encapsulated as concepts by using symbols that allow us to switch effortlessly from processes to do mathematics to concepts to think about" (p. 285), and Tall names the associated mathematical world the *procept world*. The third world is the *formal world*. Objects are there expressed in terms of formal definitions, and their properties deduced by formal proofs. They belong to mathematical structures defined through axiomatic systems. Geometry becomes axiomatic geometry; calculus becomes formal Analysis. Tall (2001) identified two different ways for the cognitive development of this formal world: natural thinking, which builds from concept imagery towards formalism on the one hand, and formal thinking, which builds from the concept definition, marginalizing imagery and focusing on logical deduction. According to him, individuals generally, according to the context, use one or the other way: "Natural thinking is appropriate for thought experiments that suggest possible theorems; formal thinking is appropriate for establishing formal proofs" (p. 235). But he also stressed that "formal thinking may also lead to 'structure theorems,' whose properties may be used to develop more subtle visual imagery, now enhanced by the network of formal relationships" (p. 235). Such distinctions are illustrated (Tall, 2001) using examples based on real numbers and infinitesimals. One can understand for instance the transitive property of the real order by relying on the image of the number line; and one can understand it as a consequence of the existence of a subset of positive elements P in a field satisfying the following axioms that make it an ordered field for the relation $x < y$ if $y - x$ belongs to P:

1) if x and y belong to P, $x + y$ and xy belong also to P,
2) Any x in the field satisfies one and only one of these properties: x belongs to P, $-x$ belongs to P, $x = 0$.

These two modes of understanding the same property formally expressed are obviously different.

In Tall's approach, relationships between processes and objects are seen in a much more dialectical way than in APOS, and symbolism contributes in an essential way to this dialectic through the notion of procept. This notion, first introduced by Gray and Tall (1994), expresses the fact that the same symbol can evoke both a process and the concept (i.e., object for Dubinsky) produced by this process; hence the name pro-cept, the condensation of process and concept. According to Tall (1996), "Function, derivative, integral and the fundamental limit notion are all examples of procepts. The theory of functions and Calculus can be summarized in outline as the study of the "doing" and 'undoing' of the processes involved" (p. 293).

Learning in this viewpoint means developing a flexible proceptual view. What this exactly covers is not something uniform, and Tall distinguishes between three categories of procepts according to their operational characteristics. Procepts in elementary arithmetic give direct access to the object they represent: The procept "4 + 8" represents both the process of adding the number 4 and the number 8, and the number 12 that results from this addition. Algebra is the domain of *template processes*: "$2x + 3$" can be seen as a process and as an object that I can for instance substitute for y in a procept such as "$2y^2 - 1$". As a process, it is not directly workable: it only tells how to get a numerical value each time a numerical value is given to x; hence the name template process. Procepts involved in calculus are generally of a third nature as they are associated to symbolizations that do not have the same algorithmic power, even if they can support some operational work, as for instance change of variables in definite integrals. Interpreting the symbol $\Sigma 1/n^2$ both as a symbol for a process and as a concept does not give means for practically operating with it, and for solving the mathematical problems that can be attached to it. In fact, developing a flexible ability for doing and undoing the processes involved in elementary calculus is certainly a rather complex and long-term construction not only dependent on the development of symbolic techniques and abilities. Tall insists in his most recent work on the role that can be played by enactive experiences and visualization, as mentioned above. The fact that enactive experiences can help students and even young students develop an intuitive sense of calculus concepts such as those of velocity and acceleration has been evidenced by different researchers for a long time (see for instance Kaput, 1992). As regards visualization, the software Graphic Calculus, developed by Tall in the 1980s (Tall, 1986), has provided a paradigmatic example ever since of what can be achieved for developing a first approach to calculus concepts without relying on formal definitions and proofs. It exploits the fact that fundamental notions in calculus such as those of continuity and derivative can be expressed in terms of *local constancy* and *local straightness*, and that the characteristics of computer visualization make it possible to "access" local properties through a finite number of zooms, and to "escape" in some sense the underlying limit process.[10] By working on these visual representations, one can thus explore the properties of mathematical objects that in fact result from an infinite limit process. Connections can be made with symbolic representations, and the interaction between these two different semiotic registers of representation can be used to support conceptualization and the development of procepts.

A major point of interest when thinking about post-secondary mathematics learning is obviously the transition towards formal thinking. Tall has often insisted on the radical change in perspective that this transition requires, and on its difficulty. According to him, only a few students truly enter this world but what is important to stress, and we will have the opportunity to come back to this point later on in the chapter, is that a lot can be achieved through mathematical work in the first two worlds mentioned above, the embodied and the proceptual.

We have summarized above some evolutions of approaches that, in the beginning, focused on the process/object distinction. The distinction between process and object levels of conceptualization is in-

[10] A comparison can be made here with research in stochastics in which simulations serve to create "microworlds" to explore abstracts concepts, such as sampling distributions or stochastic processes.

teresting for post-secondary learning because what students are asked to learn in university courses often requires an object level, whereas what they have learnt previously, even if they have developed some familiarity with the notions, does not ensure this object level, except for some limited set of instances in the best cases. In APOS, this initial distinction remained the core of the theory. It aims at representing some *vertical organization* of knowledge whereas the notion of schema takes care of the *horizontal organization* and of the underlying interconnections. But starting from the principle that any form of conceptualization obeys this pattern or, said in another way, that through the process/object duality we approach the epistemological core of knowledge growth in any mathematical domain is a very strong hypothesis that is open to discussion. We will have the opportunity to come back to this point in the last two parts of this chapter. In the research literature moreover, more evidence is emerging that establishing connections plays a fundamental role in knowledge development. In APOS, this dimension is accounted for by the schema (S) part of the model. But this part remains rather undeveloped compared with the APO parts. In our opinion, it is not by chance that researchers met problems with this schema part, which they tried to solve through incorporating the triad of Piaget and Garcia. A new hierarchy has been introduced, but the fundamental problem of the "extension of the concepts" cannot be solved just by introducing a new hierarchy. A fundamental question seems indeed to us the following: What does it mean to have built the concept of function as an object? What does it exactly mean to have an Inter, Intra, or Trans level of the function schema for students, and even for experienced mathematicians? Can we researchers give answers that are independent of the kind of functional objects we deal with, on the kind of operations we are asked to perform on these, on the kind of structures in which these objects are embedded?

The model proposed by Tall does not expose itself exactly to the same criticisms, but some points that are especially important when one thinks about post-secondary education seem to deserve much more research. The notion of procept recognizes the essential symbolic part of mathematical activity. Nevertheless saying that a symbol can flexibly refer to both a process and a concept is not enough for understanding the role played by symbols, and the mathematical work involving them, in conceptualization. This is especially the case in calculus and analysis because, as mentioned above, the operational value of symbols is not necessarily very high. Symbols are involved in techniques that can be seen as components of mathematical practices. The term *technique* is understood here with the anthropological meaning developed by Chevallard (1992): as a way of doing things, not necessarily something algorithmic. Techniques can be attributed both a pragmatic value linked to their effectiveness in producing results and an epistemic value linked to the way they contribute to the understanding of the mathematical objects they involve. The same can be said of the symbols that support techniques, and thus of procepts. Researchers need certainly to know more about the ways in which work with symbols can help develop the epistemic value of procepts. The second point we would like to make is about the transition towards formal thinking and the reconstructions this requires. More certainly needs to be known about this, and we think that complementary insights can be offered by other approaches, such as those we will describe in the next paragraph.

Epistemological Approaches

The perspectives described in this paragraph can be seen as complementary to the preceding ones. Even if the motivation of research is, as usual, the desire to understand students' difficulties and to develop didactic strategies helping them to overcome these difficulties, the base of the thinking is here more historical-epistemological than cognitive. This trend seems to us well illustrated by the book synthesizing educational research in linear algebra edited by Dorier (2000), and we will use this example in order to give an account of this position. We will come back then to what can be offered by such approaches to areas other than linear algebra, for instance calculus. As pointed out by Dorier in his epistemological work tracing the origin and evolution of the main concepts of linear algebra such as vector, vector space, linear independence, set of generators, basis, rank, dimension and duality, and linear transformations, the modern theory called *linear algebra* results from a long historical process covering several centuries, from the initial work on linear systems carried out by mathematicians such as Euler and Cramer around 1750 to the axiomatic theory of vector spaces, firstly formulated by Peano in 1880, which became universally accepted only 50 years later. In this sense,"the concept of vector space encapsulates, in a very elaborate product, the result of a long and complex process of generalization and unification" (p. 59).

The importance of this concept did not come just from the fact that it allowed the solution of new problems[11] but more from the fact that it opened new perspectives on old problems and productive connections between problems set up in different mathematical domains once these were recognized as instances of linear problems in adequate vectorial spaces. According to Dorier and his colleagues, the fundamental values of generalization and unification at the core of linear algebra are not easily understood, and the apparent simplicity of axiomatic structures is misleading. Understanding these values requires having already gained sufficient mathematical experience of linear problems in different contexts (geometry, systems of linear equations, sequences, differential equations . . .) and being able to adopt a reflective view in order to connect these different mathematical experiences. When introduced abruptly to the axiomatic structure of vectorial spaces, many students feel understandably overwhelmed with new definitions and vocabulary, immersed in a formalism they cannot really make sense of. Robert and her colleagues termed this the obstacle of formalism (Robert et al., 2000). Making sense of this formalism and understanding abstract linear algebra as more than a pure formal game requires one to establish reflective links with one's previous linear experience.[12] This contributes to make linear algebra, as pointed out by Dorier (2000),

> an "explosive compound" of languages, settings and systems of representation. There is the geometric language of lines and planes, the algebraic language of linear equations, n-tuples and matrices, the abstract language of vector spaces and linear transformations. There are the settings of geometry, of algebra, but also of graphical representations which allow a metaphoric use of geometry in higher dimensional spaces. There are the "graphical," the "tabular" and the "symbolic" registers of the languages of linear algebra. (p. 274)

As evidenced by research, teachers and texts constantly jump among these languages, settings, and semiotic systems as if conversions among these were obvious. To this complexity one must add the diversity of the associated reasoning modes. Sierpinska (2000) for instance distinguished three different reasoning modes that strongly intertwine in linear algebra: the synthetic and geometric in which mathematical objects are, in some way, directly given to the mind, which tries to grasp and describe them; the analytic-arithmetic mode in which objects are given indirectly by formulas or equations that make calculations with them possible; and the analytic-structural mode in which objects are also given indirectly, but this time through a set of properties. Again, Sierpinska showed that university teachers jump regularly without any precaution from one mode to another, leaving the responsibility for making connections to the student.

This is also an issue of the duality between the Cartesian and parametric points of view. Alvez Diaz (1998) has shown that, even if the conversion between parametric and Cartesian representations of subspaces can be, a priori, easily achieved thanks to algorithmic techniques in finite dimensions—those attached to the solving of linear systems—a flexible connection between these two points of view is hardly ever achieved by advanced students. Once more, looking back at history shows that the apparent easiness of such conversions is misleading. Understanding and mastering these took a very long time and was tightly linked to the long-term development of the concept of duality. Once more, an analysis of textbooks and teaching practices evidences the poor sensitivity of university educators to these difficulties.

The research on linear algebra offers views on learning processes that are complementary to the process/object research, emphasizing the specific epistemological status of concepts that have a particular generalizing and unifying role in the mathematical universe and suggesting that specific strategies have to be developed if one wants to have this epistemological status accessible to students. Of course, a more practical initiation to linear algebra can be organized, and this corresponds to an increasing tendency in post-secondary systems facing students' difficulties with abstract linear algebra. Powerful algorithmic techniques in linear algebra can give access to many conceptual ideas, as for instance shown by Uhlig (2002), and one can consider that this level of conceptualization is enough for many students. But research also shows that such an approach necessarily limits the modes of control that students can have on their mathematical work in linear algebra, and that when students are deprived of means of theoretical control on the

[11] When students are introduced to calculus concepts, for instance the concept of derivative, they quickly can understand what mathematical power they gain for solving a lot of problems linked to variation and optimization. Finding problems playing a similar role for showing the power gained through introducing linear algebra ideas is not so easy. As shown by the historical development of linear algebra (Dorier, 2000), axiomatic theory took a long time to be recognized as something essential, and this was achieved only when mathematicians used this theory for dealing with infinite nondenumerable spaces in functional analysis, something out of the range of university beginners.

[12] This is the rationale for different didactic strategies presented in the book.

techniques they use, they are easily trapped by "formal skids" as shown by Alvez Diaz (1998).[13]

The research also shows the essential status of the "connective dimension" in learning processes and highlights the diversity of the connections at stake: connections between settings and contexts, between points of view, between languages, between semiotic registers, and between reasoning modes. It thus confirms that cognitive flexibility is a crucial dimension of learning at these advanced levels and shows that this cognitive flexibility requires, to be adequately approached and understood, to go beyond the idea of conversion between different types of semiotic representation—the way in which it has often been analyzed in the research literature.

Coming back to the domain of calculus, one can identify similar phenomena in the transition towards formal analysis. When the fundamental concept of limit was formally defined in the 19th century, the main ambition was to establish the field of differential and integral calculus with solid foundations. This formal concept of limit can be given the status of "proof generated concept" according to the categories introduced by Lakatos (1976). Being sensitive to such mathematical concerns requires some particular mathematical culture, and developing this sensitivity in students certainly requires specific didactic strategies, as is the case for the notion of algebraic structure.

Whatever the way we express it, we touch here on the fact that different forms of conceptualizations can be a priori attached to the same concept, each of these being a coherent whole characterized by the experiences and problems that can be approached and made sense of, the techniques it offers for solving these problems, and a specific vision of mathematical rationality, and based on a specific level of formalization and symbolic manipulation. Conceptualization in any of these forms can a priori reach a high level of sophistication, as evidenced by history, and thus placing these too strictly in a hierarchy is not necessarily the most appropriate way of understanding mathematical thinking and learning, even if this is the general tendency. The idea of hierarchy leads us to see these different forms as corresponding to different cognitive levels and to think of cognitive growth as a transition process between these different levels. This is not necessarily the most appropriate metaphor for cognitive growth, as evidenced by research that shows that in expert practices different forms coexist, that university teachers for example jump so naturally from one form to another. If we do not adopt a strict hierarchical perspective, nevertheless other issues immediately emerge: issues of connection, relative importance, and respective role. These are different but not easier to solve, and certainly do not have uniform or universal answers. What researchers can nevertheless infer from both epistemological and educational research carried out up to now is that the connection between different forms of conceptualization is cognitively costly as it obliges to connect different ways of doing and thinking mathematically in a particular area, and to give each of these a particular role, and an importance that of course may vary according to the context. Research in geometry such as that carried out by Houdement and Kuzniak (1999), inspired by Gonseth's epistemological perspectives, can be considered as a step in this direction. Following Gonseth, they distinguish three main geometrical worlds: the worlds of practical geometry, natural geometry, and axiomatic structures, and they try to describe the characteristics that mathematical work presents in these different worlds. They show that each of these develops its own approach to geometry and rationality, and each has led to the development of sophisticated techniques and results, which are incommensurable between them. They also show that textbooks and elementary courses for preservice teachers jump without any precaution from one world to another (essentially between the first two), and that this fact makes it especially difficult for elementary teachers to understand the relationships between these different geometries, and the consequences for pupils' difficulties and curricular goals.

The Integration of New Cognitive Approaches

As evidenced by the research literature, for instance by the evolution of the AMT group of PME, educational thinking on learning processes is now influenced by different cognitive approaches. In this part, we consider the influence of neurosciences and theories of embodied cognition. Neurosciences have now informed the investigation of cognition in mathematics with research on perception and geometry (Berthoz, 1998; Longo, 2003) and on numbers and arithmetic, such as those by Dehaene (1996). Thanks to visual brain imaging, researchers can now localize those brain areas actuated in the solving of different

[13] For instance, Alvez Diaz shows that students customarily manipulate linear objects such as sets of vectors, matrices, linear transformations, linear systems, and determinants through numerical tables. These tables are, a priori, efficient semiotic instruments, but students tend to operate on the lines and rows of these tables automatically, without taking into account the mathematical meaning these operations have or not according to the object the table represents. This behavior generates "formal skids" leading to absurd results or contradictions that students generally do not notice.

mathematical tasks and generate a dynamic image of brain activity. The results obtained show some very interesting facts, for instance that the human brain seems to be genetically equipped with a sort of analogic counter that allows humans to globally compare quantities, and moreover that the brain areas involved in this are not the same as those used for counting and exact calculations. Research carried out with people who have brain damage shows also up to what point our numerical abilities are, at brain level, decomposed into a complex of localized processes dependent not only on the task but also on the precise context evoked in the task.

With some exceptions, these results have not concerned up to now advanced learning processes such as those considered in this chapter, but they do renew our attention to the dependence of learning processes on the biological condition of human beings. The increasing influence taken, in the educational world, by theories of embodied cognition reflects this attention. Embodied cognition, which began to develop in cognitive sciences in the eighties (Varela et al., 1991), investigates cognition as a physically embodied phenomenon realized via a process of codetermination between the organism and the medium in which it exists. Within this perspective, abstract understandings are grounded in bodily experience; an example frequently quoted is that of the concept of balance (Johnson, 1987) whose genuine meaning stems from our physical experience of bodily equilibrium and loss of equilibrium. Such bodily experience, according to the theory, gives way to perceptual-conceptual primitives called *image-schemata* that function as patterns and allow for the organization of experience. Their extension to different domains such as art (the "balancing" of colors in a painting), accountancy ("balancing" a budget), or mathematics ("balancing" an equation), which is generally a cultural phenomenon, leads to embodied concepts that are all grounded in this primitive physical experience. These extensions occur through conceptual mappings involving conceptual metaphors, hence the idea of mathematical knowledge as something whose source is metaphorical thinking (Lakoff & Nuñez, 2000).

The article published by Nuñez, Edwards, and Matos in the special issue of *ESM* "Teaching and Learning Mathematics in Context" (Boero, 1999) illustrates quite well how this perspective can be used to reflect on the learning of advanced mathematical concepts, such as the concept of continuity. Starting from the classical distinction between a natural idea of continuity characterizing a process without gaps (the vision of Euler of a continuous curve as a curve that can be drawn without lifting the pen) and the formalized idea of continuity due to Weierstrass, the authors showed that these two mathematical ideas are grounded in two very different conceptual metaphors. The first one grounds in what they call "the fictive motion metaphor," which they summarize by the following sentence: "A line is the motion of a traveller tracing that line" (p. 56). According to this cinematic vision, the curve is not a set of points but points can be put on it like milestones along a road, and points can move on it as travellers on a road. Students' and mathematicians' language is full of references to this cinematic vision of continuity, as also in everyday discourse. Metaphors can also be associated with the Cauchy-Weierstrass notion but these are radically different: "A line is a set of points"; "Continuity is gaplessness"; "Approaching a limit is preservation of closeness near a point." Points are in this case constituents of the line, and its continuous character that results from gaplessness is no longer evident. Being continuous for a function means that it preserves gaplessness: The image of a continuous set is a continuous set. The authors use this analysis for criticizing the usual teaching approach that establishes a clear hierarchy between these two conceptions of continuity and presents the $\varepsilon\,\delta$ definition as the one that captures the essence of the mathematical idea, which is in fact multiform. These authors reject a vision of learning about continuity that would see it as involving the rejection of the natural and cinematic conception and its replacement by the set-theoretical one, and the teaching strategies that more or less explicitly rely on such a vision. Of course, this does not contradict the fact that the set-theoretic conception, through the mathematical tools it provides, allows the solution of problems inaccessible to the cinematic conception. Even if expressed through another theoretical frame, it is interesting to point out here concerns very close to, for example, those expressed above concerning the different geometrical paradigms.

The influence of embodied cognition is also visible in the way technological issues are addressed today at post-secondary level.[14] In the early nineties, two main approaches in educational research featured at this educational level: the programming approach and the visual and multirepresentational approach. Educational research began to be sensitive to what could be offered by technological devices that simulate movement, thanks to the pioneering work by Kaput

[14] This is for instance visible in the way that Tall (2004) analyzes today the cognitive role of technological tools that he began to develop a long time ago, within other theoretical frames (as mentioned above).

and Nemirovsky with "Math Car" (Kaput, 1992), but this research concerned young students with limited mathematical knowledge. The impressive amount of research work carried out on representations (see the two special issues of the *Journal of Mathematical Behavior* edited by Goldin and Janvier, 1998, for a synthesis) and on the use of multiple representations for learning concepts such as the function concept, leading to what Confrey and Smith (1994) termed an *epistemology of multiple representations*, developed rather independently. As expressed by Borba and Scheffer in a special issue of *ESM* (2004), thanks to the development of technologies like CBR (Calculator-Based Ranger) the context today is changing and new perspectives are being offered to research. Body motion that was only peripheral to the multiple representation discussion becomes an essential component, and research on multiple representations takes another dimension through research concerning embodied cognition.

This connection, and the way it can influence research at post-secondary level is well illustrated in the *ESM* special issue in a paper by Rasmussen et al. (2004). These authors analyzed how three undergraduate students having already completed 3 semesters in calculus and taking a course on differential equations progressively made sense of an unfamiliar tool called the water wheel. Through this analysis, they tried to characterize how bodily activity and tool use can combine in mathematical learning, and how this combination can suggest alternative characteristics of knowing. The water wheel is a complex object. It consists of a clear, circular acrylic disc holding 32 plastic tubes around its perimeter. The disc is mounted on an axle and is free to rotate a full 360° and tilt between 0 and approximately 45°. In the center of the disc are two concentric clear plastic cylinders that contain a variable amount of oil that acts as damping for the system. Water from a bucket with a submersible pump (with adjustable flow rate) falls into several contiguous tubes on the "higher side" of the wheel. Each tube has a small hole at the bottom that allows water to drain out and be directed back into the bucket containing the submersible pump for a continual flow of water into and out of the tubes. When the wheel is tilted, gravity causes the wheel to rotate. When connected to a computer, an optical sensor collects data with real-time displays of angular velocity versus time, angular acceleration versus time, and angular acceleration versus angular velocity. Relying on the work developed by Nemirovsky, Tierney, and Wright (1998) on students' developing expertise with a motion detector, the authors' analysis takes into account the evolution of the students' relationship with the water wheel along several dimensions: becoming aware of how the tool may respond to certain aspects of one's kinesthesic engagement and not to others, acquiring an emerging awareness of what the tool does on its own, independent of one's action; gradually distinguishing which actions may affect the tool and the conditions that need to be in place so that it continues to remain sensitive to the actions; and developing a sense for what outcomes or results are possible, impossible, or difficult to obtain. Analysis along these dimensions helps the authors understand what kind of knowledge can be built through physically interacting with the tool, and how. This leads them to elaborate on a particular type of knowing called *knowing-with*, initially introduced by Broudy (1977). Knowing-with differs from commonly-identified forms of knowing such as knowing-that or knowing-how in that it involves at once the subjective feel and the objective sides of the experience and thus seems to the authors especially appropriate for describing the kind of cognitive construct that is at stake in the interaction with the tool. The language used by the students in this study shows indeed a very strong personal connection with the tool as if they were themselves becoming the tool, and this shapes the knowledge they develop about concepts such as angular velocity and acceleration in this environment.

Without any doubt, research like this attracts attention to learning processes that have been not sufficiently taken into account in post-secondary educational research up to now. The process raises interesting questions, shows intriguing phenomena, but is still in an exploratory phase. What kind of knowledge exactly is built in the sessions with the water wheel? How can students' utterances that are mainly of a qualitative nature, and often fuzzy, give way to something more analytical? How does knowing-with connect with the other forms of knowledge that develop at post-secondary level? How can it serve to understand and work with other mechanisms, other contexts? These seem today to be mainly open questions.

Investigating how qualitative knowledge that emerges from physical interaction with a calculator can be connected to more analytical forms of knowledge that are aimed at by educational institutions was a focus in the recent doctoral thesis by Maschietto (2002). This thesis addressed the transition between algebra and calculus, and the emergence of the dialectic interplay between, on the one hand, the point-wise or global perspective that characterizes the relationship that students develop with functions in algebra courses, and on the other hand, the local perspective that is attached to calculus and analysis. From a technical point of view, this emergence requires a reconstruction of the relationships with algebraic computations: The different terms of an algebraic expression are no longer given the same weight; their treatment depends on their respective orders of magnitude and relies on the fundamental idea of relative (or absolute, in nonstandard analysis) negligibility. The mathematical world of algebraic computations is thus deeply affected, at technical level and also in terms of strategies and control. Maschietto explored how these fundamental ideas can develop through adequate interaction with calculators, and the corresponding reconstructions may be promoted by appropriate didactic engineering, without entering the field of formal analysis. Maschietto relies on the potential offered by graphic calculators for visualizing local linearity (a potential that is well recognized today). The influence of embodied cognition is visible in the sensitivity she develops about the students' gestures and discourse as they zoom in and zoom out with the calculator, and about the personal and collective development of the metaphor of microlinearity. But what is added to these perspectives is a careful attention to the mathematical limits of visualization (it shows closeness, but not the order of the approximation that is essential here), a careful attention to the way the microlinearity metaphor can become an operational tool, and to the difficulties met by students in that operationalization. Students are very soon sensitive to the phenomenon of microlinearity that they recognize as an invariant. Moreover the distinction that they make between the two perceptive categories of straight and curved leads them to think that, beyond what they see, there is a more complex process involving infinity (something that is curved cannot become straight through a finite process). But mathematizing the situation in an operational way is not obvious and cannot be left to the students' responsibility. If not carefully dealt with, the microlinearity metaphor easily loses some of its essential attributes and only supports fuzzy discourse and knowledge; the necessary reconstruction of algebraic techniques takes time. Results show that fundamental and deep ideas and techniques of analysis can be developed in such a context rather early, but also that this requires a subtle design of the whole didactic engineering (Artigue, 1988), an adequate sharing of the mathematical responsibilities between the students and the teacher, and evident mathematical and didactic expertise from the teacher.

This research influenced by embodied cognition, by the attention it pays to cognitive processes and the situations in which they develop, and the possible "ecology" of these, leads us into the final section of this part devoted to more global research approaches, inspired by the increasing influence in the educational field of sociocultural and anthropological approaches.

The Integration of More Global Approaches

In this part, we consider approaches to mathematical thinking and learning at post-secondary level in which the consideration of sociocultural and institutional practices plays an essential role. According to the research culture in which it has developed or is developing, this consideration takes different forms and relies on different theoretical frames. These frames are often drawn from outside the field of mathematics education itself, as attested by the increasing number of references to Vygotsky, to Activity Theory, and to frames developed for understanding enculturation processes in various kinds of contexts. It relies also on frames developed inside the educational field. This is the case with the emergent approach developed by Cobb and Yackel (1996) in the USA that aims at connecting psychological constructivist and sociocultural perspectives for the analysis of classroom processes and the design of classroom experiments. This is also the case with the anthropological approach initiated by Chevallard from the early nineties (Chevallard, 1992; Chevallard & Bosch, 1999), which is clearly distinct from constructivist perspectives. The theoretical constructs developed by Godino and his colleagues in Spain (Godino, 2002; Godino & Batanero, 1998; Godino, Batanero & Roa, 2005) or by Cantoral, Farfan, and their colleagues in Mexico (Cantoral & Farfan, 2003) play for other researchers a similar role. These approaches all have their differences in detail, and the distance taken from constructivist and socioconstructivist perspectives may vary, but they share the common view that mathematical objects emerge from human practices, and that these practices are institutional and sociocultural practices. The term *institution* here has a very wide sense and includes any kind of formal or informal structure that organizes or conditions our social and cultural activities. The approaches also share the view that institutions develop with regard to any object that is recognized as a mathematical object a specific idea of what it means to know that object, thus defining a set of institutional norms for knowledge. So, the personal relationship one develops with mathematical objects emerges from the institutional practices and norms one has experienced in the different institutions where this object has been met. Within this perspective, understanding institutional practices and norms is sine qua non for the understanding of learning processes. Social and cultural mediations thus play a crucial role, and not by chance do the frames we are discussing pay so much attention to the semiotic dimension of mathematical activity, semiotic tools being an essential channel for these sociocultural mediations. In this part of the chapter, we would like to show the complementary insights these approaches offer to the understanding of learning processes and students' difficulties. Different examples will help us to illustrate what is offered and to show the diversity of the research developed within these perspectives.

In all the perspectives considered here, individual knowledge emerges from sociocultural practices. Both individual and collective knowledge develop in the classroom, in a subtle intertwining, and one cannot understand individual cognitive development without considering its collective counterpart. Our first example illustrates this dimension, showing one perspective developed within the emergent approach for analyzing the collective construction of knowledge.[15] The research (Stephan & Rasmussen, 2002; Yackel, Rasmussen, & King, 2000) deals with the teaching of differential equations and was carried out in the frame of a 15-week reformed course that tried to benefit from technology for emphasizing graphical and numerical approaches to the topic. From a theoretical point of view, it relies, as mentioned above, on the emergent approach, which considers that learning is both an individual and social accomplishment with neither taking primacy over the other. The design of the instructional sequence was inspired by the theory of realistic mathematics education, initiated by Freudenthal (1991) and further developed by the Freudenthal Institute (Gravemeijer, 1999). According to this, learning trajectories are built, giving students the opportunity to create meaningful mathematical ideas as they engage in challenging mathematical tasks. The development of knowledge is organized around cycles of horizontal and vertical processes of mathematization (Treffers, 1987). Horizontal mathematization refers to "formulating a problem situation in such a way that it is amenable to further mathematical analysis" whereas vertical mathematization "consists of those activities that are grounded in and built on horizontal activity" such as "reasoning about abstract structures, generalizing and formalizing" (Rasmussen et al., 2005, p. 54). Further, in this research the collective development of knowledge is approached through the analysis of argumentation, by using Toulmin's (1969) model of argumentation. According to this model, the core of an argument consists of three parts: the data, claim, and warrant. Briefly speaking, the data provide evidence for the claim, and the warrant explains how the data leads to the claim. Moreover, in case the validity of the warrant is challenged, the presenter must provide a backing to justify why the warrant, and therefore the core of the argument, is valid. Progression of collective knowledge is associated to the development of taken-as-shared knowledge, and this is achieved through a detailed analysis of the evolution of collective argumentation during the course. The researchers consider that mathematical ideas become taken-as-shared in the classroom community when

> either (1) the backing and/or warrants for an argumentation no longer appear in students' explanations and therefore the mathematical idea expressed in the core of the arguments stands as self-evident, or (2) any of the four parts of an argument (data, warrant, claim, backing) shift position (i.e., function) within subsequent arguments and are unchallenged. (Stephan & Rasmussen, 2002, p. 462)

This allows the researchers to develop an interesting analysis of the progression of the classroom knowledge, organized around the emergence and development of six mathematical practices: predicting individual solution functions, refining and comparing individual predictions, creating and structuring a slope field as it relates to prediction, reasoning about the function P (in $P'(t) = f(P)$) as both a variable and a function, creating and organizing collections of solution functions, and reasoning with spaces of solution functions. The researchers stress that the analysis shows a nonsequential development over time of these different practices. For instance, the idea that, for an autonomous differential equation,[16] the phase portrait is invariant by horizontal translation began to emerge in the second session (not of course articulated in the expert language we use here) and went on being discussed until the 11th session. These results are of course linked to the choices made in the design of the instructional sequence regarding the learning trajectory, the precise tasks proposed to students along this trajectory, and the sharing of mathematical responsibilities between teachers and students. The mathematical practices developed here contrast with the practices usually adopted in differential equation courses and offer evidence that teaching designs that are more respectful of the epistemology of the field, both meaningful for students and reasonably ambitious[17] from a mathematical point of view, can exist. The emergent approach, by the attention it pays to the

---

[15] Another interesting dimension of this research we just mention here deals with the progressive development of sociomathematical norms in this reformed course.

[16] A differential equation $y' = f(y,t)$ is said autonomous if the function f does not depend on t. This reform course only dealt with autonomous differential equations.

[17] The course for instance approaches the dynamics of differential equations depending on parameters and introduces and discusses bifurcation diagrams.

connection between the individual and the collective, helps to understand the essential role played by the classroom community in the individual development of knowledge. By the attention it pays to social norms and sociomathematical norms, it also makes clear that new practices cannot be developed without removing some of the constraints induced by usual norms, and thus raises the issue of the institutional viability of innovative practices.

The second example we present deals with a crucial issue: that of the transition between secondary and tertiary education. This transition is widely known as a serious problem, and the difference between the mathematical cultures of secondary school and college/university is generally admitted as an essential source of it. These two cultures are moreover nowadays generally regarded as two cultures moving on roads increasingly distant, which can only make the transition harder and harder. We are thus interested in looking at this problem from an anthropological or sociocultural perspective. This was the perspective adopted by Praslon (2000) in France, in his thesis devoted to the notion of derivative at the transition between high school and university. Praslon used the anthropological approach of Chevallard and the associated conceptual tools in order to explore the characteristics of the two cultures (limited nevertheless to the case of science-focused students both at high school and university). In this approach, as stated above, knowledge emerges from practices. Practices are analyzed through the notion of *praxeology*, a complex that includes types of tasks, ways for solving tasks called techniques (not necessarily algorithmic techniques in the usual sense), a discourse that explains and justifies techniques called technology, and theories that organize and structure the technological discourse. This led Praslon to characterize the praxeologies involving the notion of derivative in the two different institutions that he considered. This characterization led to interesting results. He showed that, contrary to what is often said, the secondary-tertiary transition in this area is not a transition between informal and formal mathematics, between intuitive and rigorous approaches. Rather the transition is more an accumulation of "micro-breaches," which affect the balance between the tool and object dimensions of the derivative, the balance between the study of particular objects and objects defined by general conditions, and the balance between algorithmic techniques and techniques that have more the status of general methods to be adapted to each particular case. These shifts also result from the increased autonomy given to students: as regards the choice of appropriate settings, appropriate semiotic registers, and the connections and changes made between these during the solving process, and more globally as regards the overall development of the solving process (similar tasks often appear in the two institutions, but the number of intermediate questions in a problem is far from being the same). They also result from the incredible diversification of tasks that occurs. From a culture organized around the mastery of a restricted number of tasks that can become reasonably familiar, and in which the progression in complexity is carefully managed, students pass into a culture in which diversity is the norm and the greater number of new notions to be covered in the same time period makes routinization of practices much more difficult. Once more it is evident that cognitive flexibility is an essential criterion for success. In order to make both students and university teachers sensitive to the differences between the two cultures, Praslon has created a set of tasks that, according to his results, are situated today in the "gap" between the two cultures and reveal the main facets of the differences between these. Purposely, none of the tasks requires solution by formal analysis. They are designed to be proposed to the students before their entrance to university or at the beginning of the academic year, and the students' work discussed with university teachers.

The anthropological framework is also involved in the *instrumental* approach initiated by French researchers for addressing issues linked to the integration of Computer Algebra Systems (CAS) in mathematics education (Guin, Ruthven, & Trouche, 2004). Combining Chevallard's anthropology with perspectives coming from cognitive ergonomy (Rabardel, 1995), the instrumental approach analyzes how mathematical praxeologies are affected by the use of CAS in their technical and technological components. This approach leads to specific attention on what is called *instrumental genesis*, which is the process in which there is a transformation of a tool (here a CAS) into a mathematical instrument, either for an individual or for an institution. The results obtained (Guin et al., 2004) show the complexity of this process, which for a long time has been underestimated in educational research dealing with technology, and the deep extent to which learning processes intertwine mathematical knowledge and knowledge about the tool itself. The sensitivity to institutional aspects allowed by the anthropological frame leads also to a better understanding of how a teacher's professional work is affected by the integration of such tools and offers a vision of these changes quite different from what is generally offered by the literature. We will not enter into more details here regarding this approach, whose constructs have been elaborated and used in the context of secondary education up to now (see Guin et al., 2004, for such details). However, the results obtained on mathematical topics such as calculus that are generally taught primarily at post-secondary level lead us to think that it can be of interest in the future for research involving technology, and especially professional technology[18] as is the case for CAS, at post-secondary level. Its constructs could certainly also be usefully connected to those developed by Nemirovsky and his colleagues mentioned above, as the two approaches share the same fundamental idea: technology *cannot* be considered only as a kind of educational assistant. It deeply shapes what we learn and the way we learn it, and an efficient integration of technology in mathematics education has to take this seriously into consideration, in terms of institutional practices, values, and norms.

In the preceding, we have approached the idea of culture in terms of practices and praxeologies. Other categorizations can be used. Those introduced by Hall (1981) have been used for instance by Sierpinska (1989, 1994) and more recently Nardi (1996). Hall recognizes three types of consciousness, three types of emotional relations to things: the formal, the informal, and the technical. In the context of mathematical culture, the formal level corresponds to beliefs around what is mathematics about, what are the legitimate tools and methods for mathematical work, and so on; the informal level corresponds to schemes of action and thought, unspoken ways of doing things or thinking that result from experience and practice, what is also often called "tacit knowledge"; and the technical level corresponds to the explicit part of knowledge, to the techniques and theories. Approaching learning as an enculturation process leads researchers to try to understand how these three levels and the interactions between these can develop in a given culture, and why. Sierpinska for instance first used this approach in reflecting on the long-term program of research she had developed about the learning of the concept of limit. In this program, the determination of epistemological obstacles, further complemented by the dual notion of "act of understanding," had played an essential role. Thinking in terms of Hall's categories, Sierpinska inclined to ask herself what the cultural situation of epistemological obstacles can be. According to her, these can be situated both at formal and informal levels, but not at the technical level.

Nardi (1996) used Hall's categories in order to understand the tensions between the mathematical culture of students entering the university and that of the university. She also looked at the ways enculturation progresses, taking into account not only the technical level as is usually done but also the formal and especially the informal level, through the accumulation of experience shared with the expert (here a tutor) and in the process of appropriation by the internalising imitation of the expert's cultural practices. The data she collected led her to focus on concept image construction and on formalization in order to analyze the enculturation/cognitive processes at stake. She pointed out the tension existing between the informal-intuitive-and-verbal mode of thinking on the one hand, and the formal-abstract-and-symbolic mode on the other, and the difficulties students meet with the mechanics of formal reasoning. These tensions are especially visible in calculus, but Nardi showed that the exercises proposed to students, instead of helping them overcome the epistemological obstacles they face, can reinforce them: for instance when exercises constantly expose students to infinite sums that can be broken up and rearranged as finite sums, reinforcing their belief that infinite sums can simply be treated in the same ways as finite sums.

The two examples just presented focus on the understanding of students' learning of mathematical concepts. But students' learning is also affected more generally by the cultural practices they are part of, and by the visions of mathematical activity and mathematical learning these practices induce. As we noted above, such concerns were already present in the first *Handbook*, in the chapter written by Schoenfeld. In our opinion, even if research has remained sensitive to these issues, knowledge in this area did not substantially improve during the last decade. Today, nevertheless, one can think that changes in the context of university teaching—massification of tertiary education and the resulting diversity of the student population, increasing student disaffection for mathematics and science—can stimulate research. Moreover, the development of cultural approaches over the last decade better equips researchers for approaching the difficult questions at stake. From this point of view, the article by Zevenbergen (2001) in the ICMI Study on teaching and learning mathematics at university level is an interesting example. Zevenbergen's particular concern is about equity and the characteristics of mathematical practices at university that can be seen as excluding some groups of students, preventing them from learning and succeeding. His interest for this issue seems linked to the expansion of the higher education sector and to the resulting increased diversity of students. His approach is cultural and especially relies on sociolinguistics (Halliday, 1988). Language is seen as a form of cultural capital, and the role of language in

[18] That is, a CAS is software created to satisfy the needs of professional users of mathematics, rather than software created for the needs of mathematics learners.

the access to mathematics is seen as something essential. Hence the interest in the analysis of the linguistic practices in mathematics and in the understanding of their potentially excluding role for new categories of students entering the university: those from the working class and those who are not native English speakers. As pointed out by this author, research within these perspectives has been mainly concentrated up to now on elementary and secondary school. Very little has been done at post-secondary level.

Another piece of research that illustrates how the evolution of theoretical approaches can serve to renew research approaches is the recent article by Castela (2004). This research analyzes how tertiary institutions influence students' personal study, and thus learning, by comparing two different types of tertiary institutions that exist in France: specific courses (known as CPGE) that prepare students for entry to the most prestigious business and engineering schools, and university courses. The stimulus for her research was the observed discrepancy between the respective rates of success of students coming from these two types of institutions at the national competition for secondary teachers (CAPES), after 1 year of common preparation. From this, she investigated the characteristics of the mathematical cultures in the two institutions and how these influenced the students' vision of mathematics learning and personal achievement in mathematics, and how this achievement is more or less sufficient for the purposes of entering the CAPES competition. She showed that the type of task proposed at the CAPES (long problems covering extensive mathematical domains) requires specific competences in terms of strategy and structuration of problem solving that are not really taught. Then she tried to match these requirements with the ways that the two categories of students usually work. She for instance showed that CPGE students have a rather operational vision of mathematics and are sensitive to the need for competence in strategy and structuration. University students however perceive mathematics more in terms of content and put more emphasis in their work on lectures. She identified three different approaches to problem-solving in students' practices: a drill-and-practice approach, a reproductive approach, and a transferential approach. The first approach is more representative amongst university students, the last more amongst CPGE students. By deploying the tools provided by didactic anthropology, she then showed how these different attitudes can emerge from the adaptation of students to the institutions in which they study.

We find it interesting to connect these results with those obtained by Lithner, in his observation of students' functioning with textbook exercises (Lithner, 2003). Lithner used three categories in order to qualify the forms of reasoning used by students in task solving:

> *plausible reasoning* (PR) if the argumentation developed
>
> i) "is founded on intrinsic mathematical properties of the components involved in the reasoning, and
> ii) is meant to guide towards what probably is the truth, without necessarily having to be complete or correct" (p. 32),
>
> *established experience* (EE) if argumentation
>
> i) "is founded on notions and procedures established on the basis of the individual's previous experiences from the learning environment,"
> ii) [the same as for plausible reasoning] (p. 34),
>
> *identification of similarities* (IS), if
>
> i) "the strategy choice is founded on identifying similar surface properties in an example, theorem, rule, or some other situation described earlier in the text,
> ii) the strategy implementation is carried out through mimicking the procedure from the identified situation." (p. 35)

Even if the academic results of the three categories of students observed by Lithner are quite different, clear similarities appear in their functioning. Almost all of the students' homework time seems to be spent on exercises. IS reasoning dominates strongly, and the evaluation is done by comparing with the textbook solutions section. He also showed that, due to the organization of the textbooks used and to the exercises they propose, most exercises can be solved by IS reasoning. In Lithner's experiment, more elaborated forms of reasoning became necessary when students made errors in the implementation of their IS strategies, but, even in this case, PR remained very limited. What is observed suggests that the forms of work based on memorizing and mimicking can be considered by the students as efficient for passing exams, and he sees these as "unintended by-products of their mathematical instruction" (p. 54).

In this part of the chapter, we have focused on what we perceived as major evolutions of different theoretical approaches linked to both internal and external factors in the field of mathematics education. The choices we made for our presentation helped us structure it, but perhaps we have too strongly opposed the sociocultural and anthropological approaches to the preceding ones. The reality of research is much more complex. Complementarity is certainly a more appropriate term than opposition. Many researchers combine several frames in the same research project.

For instance, Nardi (1996) combined a cultural approach and the use of process/object duality, whilst Praslon (2000) combined an anthropological general approach with different cognitive perspectives used in educational research about calculus. Favoring certain approaches nevertheless shapes the problematics and methodology of the research, and through these the kind of results that one can access and the way they will be expressed. Hence there is an explosion of notions and terms that is not easy to make sense of, but links and partial translations are often possible, as we have also tried to show. In the next part of the chapter, we broaden our discussion to new ideas and new coherences, serving to challenge and complement those already presented, which will provide further evidence of the complexity of the issues at stake, and the fact that any single perspective can only approach this complexity very partially.

EVOLUTION THROUGH THE DEVELOPMENT OF NEW RESEARCH AREAS

In this part, we turn our attention to two areas of research that lie outside the core mathematical domain generally considered by advanced mathematics education: research on the teaching and learning of mathematics in engineering courses, and research on learning probability and statistics. In these two areas we hope to illustrate the effects of the five influences alluded to by the ICMI Study (Holton, 2001):

> (1) the increase in the number of students who are now attending tertiary institutions; (2) major pedagogical and curriculum changes that have taken place at pre-university level; (3) the increasing differences between secondary and tertiary mathematics education regarding the purposes, goals, teaching approaches and methods; (4) the rapid development of technology; and (5) demands on universities to be publicly accountable.

Although public accountability is a generally political issue that may seem to lie beyond the interest of researchers, nevertheless it impacts quite directly in several ways, largely due to influence (1), the growing proportion of young people in tertiary education. Public scrutiny of "teaching quality" in universities has become rather common in the last 15 years, as increasing amounts of public money are spent on greater numbers of students. But also there is the case that as student numbers grow then the funding mechanisms have changed in many countries, with students themselves paying growing proportions of their tuition fees. Thus there is another accountability to be met as students come to expect a good "service" for their money, and perhaps hold stronger opinions than before about the education they expect, setting up a tension with the education that their teachers believe is appropriate for them.

Statistics is one of the most widely taught topics at university level, where many service course students meet advanced stochastic thinking without any prior or concurrent experience of advanced algebra or calculus. At the same time, statistics is separating from mathematics as an academic discipline (e.g., in the training of mathematicians and statisticians; research journals and associations) and is taught by teachers with a variety of backgrounds, rarely by pure mathematicians. As regards secondary education, stochastics is receiving increasing attention in recent curricula (e.g., National Council of Teachers of Mathematics [NCTM], 2000), where it is considered as an essential component of mathematics education, both due to its instrumental role to understand other disciplines and as a vehicle to develop critical reasoning and democratic values.

In the development of mathematics courses for engineering, the main trend comes from the technological influence, particularly stimulated by changes within professional engineering practice. The traditional teaching approach, based on having students spend much time developing fluency with pen-and-paper mathematical techniques, is giving way to students learning how to use computer software (such as spreadsheets or computer algebra systems) to carry out mathematical calculations. On the one hand, this is liberating and empowering, because students can more quickly become engaged in solving realistic engineering problems. On the other hand, mathematicians, and also some engineers, have legitimate concerns that a fundamental understanding of mathematics, as it applies to engineering, could be lost in these changes.

It is, in our opinion, especially insightful to try to understand how research carried out in other areas can contribute to the consideration of mathematical thinking and learning at post-secondary level. It can help us question theories and positions as regard mathematical learning and thinking that have generally been established having in mind more or less explicitly the mathematical education of "pure mathematicians" or mathematics teachers and their particular needs, and also that have focused on a few classical mathematical domains. It also obliges us to consider the role of technology in learning in a rather different way. Indeed, what is often at stake in these domains is not simply the use of technology for developing usual mathematical knowledge but the way in which technology is changing mathematical activity and understanding, including problem solving, proving, reasoning, modeling, and symbolizing. Mathematical and technological expertise and needs are in these domains much more tightly intertwined.

Learning Advanced Mathematics: The Case of Engineering Courses

This section provides a commentary on recent trends in mathematics education from the viewpoint of "service mathematics," that is, mathematics as taught to non-mathematics specialists and students studying science, engineering, and other technical subjects. We focus on trends in the United Kingdom and with respect to engineering students (Kent & Noss, 2003). We also consider the development of technology for teaching and learning in service mathematics.

In this section, we move away from a concern with what might be called intra-mathematical structure (and cognition), such as debates on the role of proof in mathematics, or advanced mathematical thinking as discussed earlier, and we will be more concerned with questions of *modeling*—the relationships between mathematics and other knowledge domains, and the ways in which computer technologies are shaping these relationships.

We begin with a sketch of the area of nonspecialist mathematics practice and research. There is in fact rather little tradition of educational research in the nonspecialist area, although arguably this area of research could bring many benefits, given that in terms of the numbers of students involved there are far more than the number of specialist mathematics students, and it is also the area of mathematics education with perhaps the most pressing need for changes in curriculum and teaching approaches, as the "customers" (engineering students and engineering academics) are by many accounts increasingly dissatisfied with the traditional curricula and approaches offered by the mathematicians who teach nonspecialist courses. Up to now, there have been very few studies of the different ways that mathematics and engineering students think about mathematics (see Bingolbali, 2004; Maull & Berry, 2000). One of these few has been Magajna's interesting and very detailed study of students and their mathematical thinking while studying at vocational college to become engineering technicians (Magajna, 2001; Magajna & Monaghan, 2002).

Innovation in engineering mathematics teaching tends to be practitioner-led, that is, it is carried out by university lecturers themselves, who (in most countries) divide their time between teaching and research. Educational research papers, if written at all, tend to take the form of descriptive reports, not much connected with the research literature of the world of mathematics education. Where a distinct research methodology is followed, the pre-test/intervention/posttest approach is still quite common, which is nowadays out of favor amongst socioculturally influenced mathematics educators. It remains very much an open question what kinds of research can particularly contribute to promoting change amongst higher education teachers (including connecting them to educational research), who have traditionally asserted a high degree of autonomy in their professional lives, including matters of teaching, and often harbor a somewhat negative attitude towards mathematics education (the "math wars" that broke out in the 1990s in the USA illustrate some of the tensions involved here—see Ralston, 2004). There have been some attempts to build a research culture for educational innovation in university teaching. In the UK, for example, the Higher Education Academy Subject Network is a government-funded initiative that aims to support lecturers in all major subject disciplines, including the branches of engineering and science, to carry out educational projects, share their results, and generally network ideas about learning and teaching; one of the significant features of the Network is that it is based on subject-specific support centres, directed and staffed mainly by lecturers from that subject area—the implication being that lecturers tend to feel closer in thinking to their fellow subject professionals than to educational specialists.[19]

The last 10 years have been a very interesting period of change for nonspecialist mathematics teaching—for a range of accounts of these developments see the outcomes of the ICMI Study on the Teaching and Learning of Mathematics at University Level, which took place in 1999 and was published in two volumes of papers (Holton, 2000, 2001); see also the proceedings of a regular conference on developments in engineering mathematics (Hibberd & Mustoe, 1997, 2000; Mustoe & Hibberd, 1995). After many decades of stable curricula and teaching methods—typically, substantial lecture courses of up to 100 hours per academic year delivered by mathematics lecturers to large groups of nonspecialist students (and the notion of delivery very much underlines the most common

[19] See, for example, the Mathematics, Statistics and OR Network [http://www.mathstore.ac.uk] and the Engineering Subject Centre [www.engsc.ac.uk].

approach, based on transmission of mathematical content, accompanied by students doing extensive exercises in techniques)—recent years have seen some major changes. The impetus for change in mathematics teaching has come from several directions. From the direction of professional engineering comes the enormous change throughout professional practice with the arrival of cheap and ubiquitous computer technology. In many cases (in civil engineering, for example), this has fundamentally changed the technical nature of work, and this is now being reflected in a deep-rooted re-appraisal of undergraduate engineering education. Indeed, new attitudes to "knowledge" are developing in engineering education (Kent & Noss, 2003). More and more knowledge is becoming accessible and potentially relevant to the practicing engineer—not only mathematics and physical science, but knowledge concerned with materials, construction techniques, design methodology, project finance, law, environmental issues, and so on. Broad agreement is emerging amongst engineers that the way to deal with this "knowledge explosion" is to implement a shift in emphasis from teaching built around subject knowledge (i.e., the topics of engineering theory and science, and issues of professional practice) toward teaching about the process of engineering (how knowledge is operationalized), using engineering design as an organizing and motivating principle of an engineering degree.

The other direction of impetus comes from school mathematics: During the 1990s, in the UK (where expansion of university-level education was and remains a key government education policy) and in some other countries, a growing proportion of young people went to university. At the same time, this was accompanied by a widely perceived (but usually officially denied) perception that the level and overall quality of mathematical preparation in schools has declined (Smith, 2004, reports on a major investigation of this phenomenon). To what extent the decline is true or not, certainly whereas in the past a relatively elite group of students would enter university to take degrees in engineering and science, today's students come from a far more diverse background and have a far broader range of mathematical competence (cf. SEFI, 2002). School mathematics in many countries has also changed radically during the last 20 or so years—a common significant change is the "democratization" of mathematics in schools, a movement that has seen a decline in formal mathematical activities by students (exercises in geometry, algebra, calculus, etc.—mathematics regarded as a body of knowledge to be acquired) and an increase in student-generated investigative activity (projects, etc.—mathematics regarded as a process in which students can participate).

From the perspective of service mathematics, university teachers ironically seem to be unhappy with a style of mathematics curriculum in schools that emphasizes a way of working that can help to develop students' abilities in problem solving and applying mathematical ideas—of course, this very much depends on how, and how well, this is done. Yet one can sense a missed opportunity in this situation: Where there seems to be a possibility for making connections, the debate in universities has focused on the emergence of a gap between the expectations of universities about new students and the students' actual mathematical capabilities; this has been termed the *mathematics problem* in the UK. To a significant extent, however, arguably the notion of "problem" is more a reflection of a dominant conservative viewpoint towards mathematics teaching than a problematic reality. As Steen (2001) describes, through the particular influence of the computer mathematics is becoming more and more relevant to more and more professional workers, including engineers, and therefore a rich *opportunity* is opening up for mathematics education, but this will require mathematicians to be more open to consider changes in the content and approach of mathematics teaching—either that, or to see more and more mathematics teaching being undertaken by the users of mathematics, such as engineering departments, themselves. Here, for example, is a comment recorded (by PK) from a mathematician specializing in engineering mathematics teaching at a large technical university in the UK: "We're pretty traditional—you would not see much difference between what we do now and 20 years ago, except that the level is lower now."

The Influence of Technology

Implementation of technology has been a big concern generally in university teaching since the early 1990s, when personal computers became accessible in significant numbers to students in higher education. One of the striking things about this has been the tendency for researchers and especially implementers to regard the higher education setting as distinct from the school setting, and many lessons learnt from the implementation of IT in school level education have noticeably *not* been learnt by those in higher education. For example, technology enthusiasm was particularly rife in the early to mid-1990s, much as in schools about a decade earlier, often in the form of expensive projects that generally proved unsustainable after the special project funding was taken away.

> The strength of enthusiasts is that they innovate, they try new ideas, they explore where more cautious colleagues might wait and see. Their weakness is that the outcomes of their enthusiastic endeavours are often poorly contextualised, theorised or evaluated. (Burton et al., 2004, p. 221)

It is pretty much recognized, by mathematics educators at least, that for technology to have a sustainable impact on teaching and curriculum, then there has to be some deep-rooted thinking about the aims and methods of teaching—as the anthropological approach in mathematics education illustrates (see above). Of course, to do this in the complex political atmosphere that surrounds a typical university mathematics programme is no easy task, and even today one is inclined to think, pessimistically, that "university mathematics curricula are virtually immune to change; in mathematical terms, we might say that the mathematics curriculum is invariant under intellectual revolutions" (Steen, 1989).

For all the difficulties inherent in technological innovation in mathematics education, it is evident that those involved in teaching mathematics to nonspecialist students cannot ignore it, inasmuch as computer software is ubiquitous in the professional practice of engineers and is becoming a standard tool for engineering students, for example in the form of specialist Computer-Aided Design packages for the different engineering professions, or structural analysis packages in civil engineering.

The case of structural analysis software is a good illustration of the trends. Structural analysis is essentially a deeply mathematical subject and is central to the professional practice of a civil engineer, and thus also central to his or her education. As part of the design process, the engineer needs to predict the behavior of a proposed structure (the walls of a building, a roof, a bridge, etc.). Before the advent of computers, the working life of an engineer, especially in the early part of his or her career, would be dominated by actually doing structural calculations using pen-and-paper, and a large part of the civil engineering degree was therefore dedicated to giving students an understanding and fluency in a variety of calculational techniques. For the majority of engineers today, all such calculations will be done in practice using computer software. The priority now for education is that students may "do" little mathematics themselves in the form of explicit calculation, yet they are likely to *use* more mathematics than ever before, implicit within computer software:

> No longer do we have to grind through long calculations—the computer will do it for us. The challenge has changed from the ability to do this to the ability to interpret the meaning of mathematics to engineering and herein lies the challenge and change of emphasis. (Blockley & Woodman, 2002, p. 14)

A typical structural analysis curriculum of 10 years ago would begin with a solid dose of analytical theory—using matrix algebra—followed later (perhaps in the 2nd or 3rd year) by working with computer software. Nowadays, work with computer software is likely to come much earlier, and the place of learning matrix algebra as a preparation for working with software is becoming open to question.

What is the Purpose of Mathematics to an Engineer?

Structural analysis is just one of the analytically based areas of engineering where the curriculum is currently being debated and is evolving into new forms (see, for example, Allen, 2000). In general, one can see a tension between the traditional notion of "pushing" mathematical ideas into the engineering curriculum and of "pulling" ideas from mathematics—that is, the need for a particular piece of mathematics can emerge where the engineering curriculum requires it, rather than being pushed into the student prior to having a meaningful context for it. Of course, this has to be done in a systematic way—mathematics cannot in general be understood in a "just in time" fashion.

In university mathematics education, a strong tendency remains to think that *understanding* of a mathematical technique *must* precede its *application* to the engineering context. In the (precomputer) past, a valid objection to any sort of pull-based mathematics was the uncomfortable notion of anyone trying to use a mathematical idea before knowing the techniques of its application. Mathematicians have tended to regard knowledge of techniques as an essential part of what it means to understand an idea and how to apply it. What users of mathematics such as engineers are wanting however, and it seems that mathematicians will increasingly need to provide, is a form of pull-based mathematics in which the use of mathematical software makes mathematical ideas usable, as in the example of structural analysis above. Carefully designed use of IT can make it possible for students to use mathematical ideas *before* understanding techniques, and to make this part of a genuinely rounded mathematical learning experience (a few examples are given in Kent & Noss, 2003).

It is worth considering at this point the whole question of the purpose of mathematics to a (student) engineer. Kent & Noss (2003) in a UK-based study found widespread agreement across the civil engineering profession about what is the desirable mathematical competence of a graduate: "Mathematics should instil disciplined thinking and rigour in the development of arguments based on assumption and simplification in modelling, should teach the importance of controlled approximation, and, above all, impress upon students its value as a tool to be invoked when quantitative evidence is needed to underpin assertion, hypothesis, or sheer physical intuition" (Nethercot & Lloyd-Smith, quoted in Kent & Noss, 2003, p. 8). Whilst few would dissent from such a description, there are certainly significant questions about what meanings people attach to the different terms, even, what is "mathematics"? It is notable in talking to experienced engineers that the mathematics that is useful and relevant has in many cases long since come to be thought of as part of engineering, whereas mathematics that is *not* used is regarded as "mathematics."

Again according to Kent and Noss's (2003) survey, practicing engineers do not tend to regard mathematics as a problem area—contrary to the sometimes obsessive concern of academics about the "mathematics problem" (see above). Practitioners regard confidence with mathematics as crucial for the majority of engineers, and above all they require a balance of skills across engineering teams. Different employers require different balances, and even specialist analytical design consultancies in civil engineering reported that they only require 10–20% of their engineers to have specialist skills in analysis.

In the past, engineers had to learn a lot of mathematics for practical purposes. At the same time, they could be expected to absorb some understanding of mathematics as a logical way of thinking, and its importance as part of a practicing engineer's expertise. The availability of computer software for calculations has undone this relationship between practical and theoretical aspects. The teaching of practical mathematics is becoming much more focused on the process of modeling of engineering systems—this results in a decrease in the teaching of calculation techniques, but it does not mean that all manual work with mathematics could be replaced: The right balance must be found. To teach mathematics as a way of thinking, a traditional model exists in the use of formal mathematics—older engineers do look back on courses in Euclidean geometry in their schooldays as having significantly shaped their development. It seems unlikely that formal geometry can re-occupy such a dominant place in the school mathematics curriculum (for one thing, there are too many competing demands on the school curriculum), and new models for this aspect of mathematics learning are not yet widely agreed. Indeed, is a logical way of thinking *only* to be gained through studying "pure" mathematics? Could the same kinds of analytical problem solving also be developed through, for example, requiring all students to develop substantial skills in (mathematical) computer programming? Certainly in the school context this potential for computer programming has long been advocated, most famously in the case of the language Logo (see Cuoco et al., 1996; Noss & Hoyles, 1996).

Modeling and the Development of "Problem Solving" Curricula

Another emergent factor in engineering mathematics is the role of modeling. However, this should be understood more broadly than it is sometimes interpreted in undergraduate mathematics. Often, the translation between the physical situation and the mathematical analysis is regarded as a small part of the task, whose focus is on the applied mathematics of solving the equations in the model (in the worst cases, the physical situation is but an excuse for students to do a routine exercise in applied mathematics). Yet the process of translation, which involves on the one hand developing a mathematical model of the situation and on the other hand interpreting the mathematical analysis back into the context, is in general a rather complex task, and is intimately related to the issue of learning *how to use* (rather than how to do) mathematics, which we suggested above is becoming the focus of engineering mathematics education. See Bissell and Dillon (2000) for an extremely perceptive engineers' view on the nature of modeling.

As discussed above, an analogous situation to this lies in some aspects of applied mathematics teaching, such as differential equations (see, for example, Stephan & Rasmussen, 2002). As the use of differential equations has been transformed by the availability of mathematical software, both numerical and symbolic (computer algebra systems), both a possibility and a need has opened up to teach about equations differently, less focused on doing techniques of solving differential equations, more focused on the meaning of differential equations, in terms of their geometrical structures and their domains of application.

One of the major trends of recent years in engineering education generally that illustrates the issue of modeling is a growing number of experiments with various forms of *problem-based learning* (PBL). (A few exceptional universities—such as Roskilde University in Denmark—embraced this mode of learning a long time ago, cf. Niss, 2001). The idea behind this is that lecture-based teaching is largely replaced by students working on projects, usually in teams, and being assessed through group-work and continuous assessment. The projects develop increasing complexity as students progress from year to year. Problem-based learning originated in the 1960s in medical education, where it has since become increasingly common. The appropriateness of PBL has been criticized for engineering studies because of its origins in medical education: The argument goes that PBL is good for fact-based knowledge, such as medicine, but is less appropriate for an analytically based subject like engineering (cf. Spencer-Chapman, 2000). Yet making abstract analysis meaningful depends crucially on the student making connections between engineering and mathematics through solving problems (Kent & Noss, 2003); just as in the case of differential equations mentioned above, meaning develops by providing activities for students to develop connections within the subject (e.g., the geometrical structure of differential equations), as well as making connections to other subjects.

Statistics and Probability at Post-Secondary Level

In this section we summarize the research carried out in advanced stochastics, with particular emphasis in the recent emergence of a statistics education research community, which has some links but is almost independent from the advanced mathematics education research community. We reflect on the specific characteristics of advanced stochastics and the potential research challenges that this area sets to educational research.

Many features described for the teaching of "service mathematics" to engineers also apply to the case of statistics. As mentioned above, statistics is one of the most widely taught topics at university level, where it is mainly studied as a tool to solve problems in other fields. However, the variety of previous knowledge and interests of the students involved is even larger than in the case of service calculus, because statistics is taught with almost no exception in all the university majors, from engineering to education, geography, psychology, biology, sociology, journalism, economics, linguistics, and so on. There is also a general dissatisfaction with the mathematically oriented approach to teaching in traditional courses, which often emphasize the teaching of formulas for calculating statistics (e.g., correlation coefficients or confidence intervals) without much concern towards the data context and interpretative activities or simulations that could help students improve their stochastic intuitions. In many cases, the courses are over-mathematized, which involves meeting concepts of advanced stochastic thinking for students who do not have any prior or concurrent experience of advanced algebra or calculus.

Technology has also had a major impact in the practice and development of statistics in expanding the range of processes that statisticians and users of statistics can employ to collect, analyze, and interpret data and the amount and type of data they can analyze. This has resulted in the development of new statistical methods, such as exploratory or graphical data analysis, resampling techniques, or data mining and has facilitated the implementation of some techniques that were previously difficult to make use of, such as multivariate or Bayesian analysis. This no doubt has also influenced the increasing demand for statistics education, inasmuch as many people can now apply a method with scarce knowledge of the complex mathematical calculation behind the user-friendly software that is apparently easy to learn. For example, an education or psychology student can easily enter data in a computer, use statistical software to perform a factor analysis in a few minutes, and give an intuitive interpretation of the factors retained in terms of the problem without knowing what an eigenvalue is, how eigenvalues are extracted, or how the different rotations of the components have been obtained.

Modeling and problem solving are also central to the teaching of statistics to professionals. However, statistical problems are often open-ended or ill-defined and have multiple possible solutions. Although model abstraction is common to mathematics and statistics, the context plays a fundamental role in guiding the selection of a statistical model (delMas, 2004), and selecting the model is frequently harder than subsequent mathematical reasoning in working with the model. Algebraic work on the model, for example, is very limited, and although the processes of generalizing, classifying, conjecturing, inducing, analyzing, synthesizing, and abstracting described by Dreyfus (1991) for advanced mathematics still apply to statistics, the meaning of representing, visualizing, or formalizing is quite different. As we will argue later, representing and visualizing apply mainly to the data and are used to create meaning from these data, for example helping to find an adequate model or identifying unexpected sources of variation. Formalizing in statistics applies not just to the mathematical work with models, but to the definitions of units, variables, and categories for data analysis, or designing instruments to measure them, which again are rooted in the context of application. Whereas mathematical practice can be removed from real-world context, statistics practice is highly dependent on the problem context, and this dependence may lead to reasoning errors that are hard to overcome.[20]

There is an interest and tradition for educational innovation in the teaching of statistics, and this topic is also taught at advanced level by lecturers with a variety of backgrounds, mostly statisticians, but also economists, health care professionals, engineers, psychologists, or educators. Contrary to the case of engineering, mathematicians or mathematics educators teaching statistics are only a minority. This explains the fact that, until very recently, research in advanced stochastic teaching and learning has not attracted mathematics educators and thus has had a small presence at, for example, the annual Psychology of Mathematics Education Conferences. Of course, this does not mean that the teaching of advanced statistics is without problems or that existing related research on this theme is lacking.

A main difference compared with the teaching of mathematics to engineers is the long tradition of educational research related to the teaching and learning of statistics, and a substantial part of it has been carried out outside the mathematics education community. Research in stochastics thinking, teaching, and learning started during the 1950s with the pioneering work by Piaget and Inhelder (1951) and has always had an interdisciplinary character. As stated by Shaughnessy (1992, p. 465): *"The cross fertilization of research traditions and methodologies in probability and statistics makes it one of the theoretically richest branches in mathematics education."* Different fields have contributed with various research paradigms and theoretical frameworks, which have been analyzed in different surveys of research, such as Scholz (1991), Shaughnessy (1992), Jones and Thornton (2005), and other chapters in this *Handbook*. Whilst research was still relatively scarce when the first *Handbook* was published, compared with other areas of mathematics it has experienced a notable growth in the past 10 years. However, this has mainly concentrated in elementary and secondary school levels, perhaps unsurprising given the much greater emphasis that recent new curricula for these levels have given to statistics and probability.

Research into advanced stochastic reasoning has interested psychologists for decades. Because psychology is an experimental science that heavily relies on statistics, the efforts to justify the scientific character of this field have led psychologists to examine the validity of their research paradigms, including the use of statistics in empirical research. An amazing observation is that statistical inference and particularly significance tests were found to be misunderstood and misused by psychologists and experimental researchers at large over 30 years ago, and that the situation still persists in spite of strong debates ever since (Harlow, Mulaik, & Steiger, 1997; Morrison & Henkel, 1970). Moreover, researchers in the field of reasoning under uncertainty have suggested that, even after statistical instruction, students and professionals tend to continue to make erroneous stochastic judgements and decisions (Kahneman, Slovic, & Tversky, 1982). The diffusion of these research results and the increasingly easy access to powerful and user-friendly computers and statistical software, which save teaching time previously devoted to laborious calculations and allow a more intuitive approach to statistics (more real data, active learning, problem solving, and use of technology to illustrate abstract concepts through simulation) have led statistics lecturers to increase their concern towards didactical problems (see, e.g., Moore, 1997). Consequently, a general effort exists to create curricular materials and to evaluate teaching and learning at university level. The influence of the International Association for Statistical Education (IASE) created in 1991 serves to establish links among the different communities interested in statistics education and to support a more systematic research program. Below we summarize these contributions.

Psychological Research on Advanced Stochastic Thinking

Different theoretical approaches in the field of psychology have tried to explain people's poor performance in probability and statistical tasks, which have been widely documented in relation to different concepts that can be considered as part of advanced stochastics, such as randomness, compound probability, association in contingency tables, correlation, conditional probabilities, Bayes problems, sampling, and the test of hypotheses. Traditionally, decisions under uncertainty are defined by incomplete information about the situation, that is, the possible alternatives or their outcomes are only known (at best) in term of probabilities, a feature of many professional tasks (e.g., medical diagnosis, jury verdict, educational assessment). A *fallacy* is the result of a cognitive process that leads to an incorrect conclusion and may consist

---

[20] For example, a very simple rule is computing the probability of the conjunction of two events in terms of simple and conditional probabilities: $P(A \cap B) = P(A) \cdot P(B|A) = P(B) \cdot P(A|B)$. A consequence of this rule is that the probability $P(A \cap B)$ cannot be higher than the probabilities of either of the two individual events. However, when $P(A)$ is very high and $P(B)$ small, people forget the rule and consider $P(A \cap B) > P(B)$. This behavior was first described by Tversky and Kahneman (1983) and termed the *conjunction fallacy*. So, for example students who are asked whether it is more likely that the government will increase the number of grants or that the government will increase the number of grants and the salary of members of parliament will consider more likely the second event.

of the application of an inadequate model or incorrect application of intuitive rules of inference. In trying to explain the mechanisms leading to a fallacy, Kahneman and his collaborators developed the *heuristics and biases* program (Kahneman, Slovic, & Tversky, 1982), which was the dominant paradigm in the early eighties and is still very influential. This assumes that people do not follow the normative mathematical rules that guide formal scientific inference when they make a decision under uncertainty and that, instead, they use simpler judgmental heuristics. Heuristics reduce the complex tasks of assessing probabilities and predicting values to simpler judgmental operations and are in general useful; however they sometimes cause serious and systematic errors and are resistant to change. For example in the *representativeness* heuristics, people tend to estimate the likelihood for an event on the basis of how well it represents some aspects of the parent population. An associated fallacy that has been termed *belief in the law of small numbers* is the belief that even small samples should exactly reflect all the characteristics in the population distribution. Then for example a mother of two boys may incorrectly assume that the likelihood of having a third baby who is a girl is greater than the likelihood of having a baby boy, because a family with 2 boys and a girl is more representative of the general population.

A huge amount of research in stochastics education at both undergraduate and advanced level has been based on the idea of heuristics (see a summary in Shaughnessy, 1992). Large-scale studies carried out by psychologists showed misconceptions related to these heuristics, in which participants assumed that sampling distributions (distribution of the value of the mean and other parameters in repeated samples from the same population) were independent of the sample size, whereas in mathematical theory the spread of these distributions should be inversely related to the sample size. Direct consequences of this for the practice of statistics are that users have an overconfidence in the results of statistical tests, underestimate the width of confidence intervals, and expect a very close result in replication of the experiment, even with a small sample. Recent research on heuristics at advanced level includes studies on: assessment (Garfield, 2003; Hirsch & O'Donnell, 2001), the effect of teaching experiments on the use of heuristics (e.g., Barragués, 2002; Pfannkuch & Brown, 1996), and misconceptions related to concepts such as randomness (Falk & Konold, 1997) in terms of heuristics or providing alternative explanations for incorrect responses to tasks proposed in research about heuristics. For example, Konold (1991) suggested that students might confuse frequentist probability, which is objective and refers to the frequency of occurrence of an event in relation to a population, with epistemic (subjective) probability referred only to an isolated event. Then they might confuse the task of assigning a frequentist probability with the task of predicting the next outcome (the *outcome approach*) in a random experiment. This also involves a deterministic view of an uncertain situation, because in the outcome approach, participants tend to rely more on causal explanations rather than accept explanations due to chance and variability.

A different theoretical model in the psychology of decision is the *abstract–rules* framework (Nisbett & Ross, 1980) in which people are assumed to acquire a form of correct statistical reasoning and develop intuitive versions of abstract statistical rules, such as the law of large numbers, which are well adapted to deal with a wide range of problems in everyday life, for example, estimating the average time spent to perform a repetitive task or the approximate cost of the shopping in a supermarket, but the rules fail when applied beyond that range. Such rules are used to solve statistical problems, in which one recognizes some cues in the problematic situation. With respect to training in statistical reasoning the suggestion is that the quality of human inferences and judgments can be improved via statistical instruction: "Many of the inferential principles central to the education we are proposing can be appreciated fully only if one has been exposed to some elementary statistics and probability theory" (Nisbett & Ross, 1980, p. 281), although the suggestion is that statistics teaching should take into account students' intuitive strategies and errors. Moreover, for these authors, even brief formal training in inferential rules may enhance their use for reasoning about chance events; either teaching statistical rules or teaching by having students solve example problems would work. As regards training in conditional logic or conditional probability students need to be trained simultaneously in abstract logical rules and in problem-solving strategies. Problem solving is improved, according to these authors, when the sample space is clearly defined, people recognize the role of chance in the experiment, and the context forces the person to think statistically.

A more recent theoretical framework is the *adaptive algorithms* approach (Cosmides & Tooby, 1996; Gigerenzer, 1994), which supposes that people possess evolutionarily acquired cognitive algorithms that serve to solve complex probability problems, such as conditional probability or problems involving Bayes theorem. Adaptive algorithms serve to solve adaptive problems (such as finding food, avoiding predation, or communicating) and take a long time to be shaped, due to natural selection. Because adaptive algorithms are shaped by natural environments they are more effective when the tasks are presented in a format close to how data are perceived and remembered in ordinary life. According to this theory people should have little difficulty in solving statistical tasks if the probability data are presented in a natural format of absolute frequencies instead of using rates or percentages (Sedlmeier, 1999). Although the percentage, rate, or frequency representations of probability values in the problem are mathematically equivalent, inasmuch as they can be mapped onto each other, they might be not psychologically equivalent. Gigerenzer (1996) argued that the frequency representation makes some cognitive illusions in statistics disappear and that the cognitive algorithm used to solve the same problem changes with the information representation.[21]

Sedlmeier (1999) analyzed and summarized recent teaching experiments carried out by psychologists that follow either the abstract-rules or adaptive-algorithms theories and involve the use of computers. The results of these experiments favor the adaptive-algorithms approach as an explanation for people's performance in stochastic tasks and suggest that statistical training is effective if students are taught to translate statistical tasks to an adequate format, including tree diagrams and absolute frequencies. These experiments have concerned areas such as compound probabilities, conditional probability, the Bayes theorem, and the impact of sample size on the sampling distributions. However, learning is assessed through participants' performance to tasks that are very close to those used in the training, so evaluating the extent to which students could transfer this knowledge to wider types of problems is difficult. Although this research provides empirical information and potential theoretical explanations for students' difficulties in advanced stochastics, a large amount of work still must be done by mathematics and statistics educators to integrate these results and to use them to design and evaluate teaching sequences in natural settings where students are expected to meet wider curricular requirements.

The Mathematics Education Approach

The approach of mathematics educators is rather different from that taken by psychologists and has followed the research paradigms and theoretical frameworks in mathematics education. The mathematical and epistemological analyses reveal that the complexity of concepts, tasks, and students' responses investigated by psychologists is often greater than has been assumed in psychological research and suggests the need for reanalyzing them from a mathematical perspective; for example, centering on isolated types of tasks does not always reveal in depth the students' understanding of a concept, inasmuch as responses are sometimes very dependent on the task variables. Moreover, theoretical constructs taken from mathematics education can contribute to a different perspective for the same phenomena. Take, for example, the concept of correlation, a field that has been extensively studied by psychologists, including Inhelder and Piaget (1955) who considered that the evolutionary development of the concepts of correlation and probability are related and that understanding correlation requires prior comprehension of proportionality, probability, and combinatorics. In these experiments participants are presented two-way tables (two rows and two columns) in which a sample subjects are classified according to the presence-absence of two qualitative attributes, such as eye and hair color (dark or fair hair, brown or blue eyes). Participants are asked to judge if there is a relationship between the two attributes (judgment of correlation) in these 2 × 2 contingency tables and to justify the procedure to reach their conclusion. In Table 22.1 we describe the data in this type of problem, where a, b, c and d represent absolute frequencies.

Table 22.1 Typical Format for a 2 × 2 Contingency Table

	B	Not B	Total
A	a	b	a + b
Not A	c	d	c + d
Total	a + c	b + d	a + b + c + d

Piaget and Inhelder found that some adolescents who are able to compute single probabilities only ana-

[21] For example, in the following Problem 1 (Eddy, 1982) probability data are given in percentages and should be solved with the help of Bayes theorem. Giving the data in a format of absolute frequencies (Problem 2), it transforms to a simpler problem that can be solved by applying the Laplace rule: It is easy to see that out of 107 women with positive test results (8 + 99) only 8 of them will have breast cancer.

Problem 1. The probability of breast cancer is 1% for a woman at age 40 who participates in routine screening. If a woman has breast cancer, the probability is 80% that she will have a positive mammogram. If a woman does not have breast cancer, the probability is 10% that she will still have a positive mammogram. Imagine a woman from this age group with a positive mammogram. What is the probability that she actually has breast cancer?

Problem 2. Ten out of every 1,000 women at age 40 who participate in routine screening have breast cancer. Of these 10 women with breast cancer, 8 will have a positive mammogram. Of the remaining 990 women without breast cancer, 99 will still have a positive mammogram. Imagine a woman from this age group with a positive mammogram. What is the probability that she actually has breast cancer?

lyze the cases in cell [a] in Table 22.1 (presence-presence of the two characters A and B). When they admit that the cases in cell [d] (absence-absence) are also related to the existence of correlation, they do not understand that cells [a] and [d] have the same meaning concerning the association, and they only compare [a] with [b] or [c] with [d] instead. Understanding correlation requires considering quantities (a + d) as favorable to the association and (b + c) as opposed to it. The correct strategy should use the comparison of two probabilities *P(B/A)* and *P(B/NotA)*, that is

$$\frac{a}{a+b} \text{ with } \frac{c}{c+d}.$$

According to Piaget and Inhelder, recognition of this fact only happens at 15 years of age. Following Piaget and Inhelder, several psychologists have studied the judgment of association in 2 × 2 contingency tables in adults, using various kinds of tasks. Other psychologists used instead two numerical variables and different numerical, verbal, or graphical representation of data. As a consequence of all this research, it has been noted that participants perform poorly when establishing correlation (see Beyth-Marom, 1982, for a survey) and people frequently use their previous theories about the context of a problem when judging association. The general conclusion is that when data do not coincide with these expectations a cognitive conflict affects accuracy in the perception of correlation (Jennings, Amabile, & Ross, 1982).

Estepa and his collaborators began from these results and, after a mathematical and epistemological analysis of the concept of correlation, designed a questionnaire including a wider range of open-response tasks to give an alternative explanation of pre-university students' strategies and judgments in terms of *misconceptions* regarding correlation (Batanero, Estepa, Godino, & Green, 1996; Estepa, 1993). In the *determinist* conception of association, some students expect that correlated variables should be linked by a mathematical function; students with a *unidirectional* conception of association only perceive direct association (positive sign of the correlation coefficient) and interpret inverse association (negative sign) as independence; *causal* conception consists of assuming that correlation always involves a cause-and-effect relationship between the variables, and in the *local* conception students base their judgment on only part of the data (e.g., they only use one cell in 2 × 2 contingency tables). These types of conceptions have been confirmed in later research (Morris, 1997, 1999). Another finding was that some mathematical concepts might constitute an *obstacle* when learning correlation, in which case they receive different interpretations. For example, some students believe a correlation coefficient of –0.7 indicates a smaller degree of dependence than a correlation coefficient of 0.1 because –0.7 < 0.1 in the usual ordering of real numbers, whereas in fact –0.7 indicates a stronger dependency than 0.1.

Estepa and his collaborators organized two teaching experiments based on *computer learning environments* (in the sense of Biehler, 1994, 1997), that is, integrated instructional settings that allow the teacher and a group of students to work with statistical software, data sets, and related problems as well as with a selection of statistical concepts and procedures. The researchers found general improvement in students' strategies and conceptions after teaching, although the unidirectional and causal misconceptions concerning statistical association were harder to eradicate. Using qualitative methods, such as observation, interviews, and the analysis of the students' interaction with the computer, they documented specific *acts of understanding* (following Sierpinska, 1994) of correlation (Batanero, Godino, & Estepa, 1998; Estepa, 1993). For example, students must realize that *the study of the association between two variables has to be made in terms of relative frequencies;* however, in the first teaching session the students tried to solve the problems in terms of absolute frequencies. Although the lecturer pointed out this mistake to them at the end of that session, the same incorrect procedure appeared recurrently for the same students for several sessions, until the students overcame this difficulty. Another example of an act of understanding was realizing that *from the same absolute frequency in a contingency table cell one can compute two different relative conditional frequencies (conditioning by row or column), and the role of the two events in the conditional relative frequency is not interchangeable.* Students' difficulties in discriminating between the probabilities P(A/B) and P(B/A) have previously been described (e.g., Falk, 1986). Again, this was a recurrent difficulty for the students throughout the teaching experiment.

Other researchers have taken mathematics education frameworks to organize research on topics that have scarcely attracted psychologists in complex experimental settings. For example Batanero, Tauber, and Sánchez (2004) used a theoretical framework that incorporates ontological, anthropological, and semiotic ideas (Godino, 2002; Godino & Batanero, 1998) to describe the evolution of students' understanding of the normal distribution in a statistics course based on intensive use of computers. In the theoretical model used, the meaning (understanding) of any mathematical object is conceived as a complex system, that contains different types of interrelated elements: a) problems and situations from which the object emerges, such as fitting a curve to a histogram for empirical data distributions or approximating the binomial or Poisson distributions; b) representations of data and concepts, for example, histograms or density curves, verbal and algebraic representations of the normal distribution; c) procedures and strategies to solve the problem, such as computing probabilities under the curve, computing standard scores, and critical values; d) definitions and properties, for example, symmetry of the normal distribution, horizontal asymptote, areas above and below the mean, meanings of parameters; and e) arguments and proofs, including deductive and informal arguments. Godino and Batanero (1998) argued that the understanding of a concept emerges from a person's meaningful practices linked to repeated solution of problems that are specific to that concept. Mathematical problems promote and contextualize mathematical activity and, together with actions, constitute the praxemic or phenomenological component of mathematics (*praxis*) as proposed by Chevallard (1997). The three remaining components (concept-definitions, properties, arguments) are produced by reflective practice and constitute the theoretical or discursive component (*logos*). Godino and Batanero also described different dual facets or dimensions of mathematical knowledge, in particular distinguishing between the *personal* and the *institutional* meaning to differentiate between the meaning that for a given concept has been proposed or fixed in a specific institution and the meaning given to the concept by a particular person in the institution.

This theoretical framework was specially suited to describe the complexity of understanding the normal distribution in Batanero, Tauber, and Sánchez's (2004) research, because the understanding of the normal distribution in their experiment was based on the students' previous knowledge about many interrelated concepts such as statistic and random variables, frequency and probability distribution, parameter and statistics, center and spread, symmetry and kurtosis, histogram and density curve, areas under the normal curve, mathematical model and empirical data, their different representations, procedures, and properties. The teaching was based on intensive solving of real problems in which students analyzed different real data sets taken from fields in their areas of interest, which served to gradually suggest to them to reflect about the different elements of meaning of the normal distribution. The idea of different institutional meanings also describes well the changes in meaning and understanding implied by the use of computers: different types of problems (starting from real, multivariate data sets related to a problematic situation that the students should model), representations (wider and quicker availability of interactive dynamic graphs, tables, and numerical summaries), procedures (knowledge of software options and interpretative abilities substituted for classical probability computations); graphical and iconic language replacing algebraic manipulation, and types of proof (simulation and visualization, instead of deductive proof).

Finally the framework served to picture the general tendencies in the students' *personal meaning* for the normal distribution after the instruction, as well as the variety of these meanings, and to identify main agreements and differences with the intended *institutional meaning*. At the end of the teaching students were given written questionnaires to evaluate specific understanding about particular properties, representations, and procedures and an open-ended task, to be solved with the computer that referred to a data file the students had not seen before. The students were to chose a variable that could be well fitted by a normal distribution, then explain and justify their responses in detail (the file contained data from a real context with 10 different qualitative and quantitative variables, only 2 of which were acceptable solutions to the problem). The students were given freedom to use any resource of the software (statistical graphs or analysis) to support their choice. Quantitative analysis of responses to questionnaires and semiotic analysis of the students' written protocols in the open tasks as well as interviews with a small number of students were used to describe the students' correct and incorrect reasoning about the normal distribution. For example, many students could relate the idea of symmetry to skewness coefficient or to relative position of mean, median, and mode; they were able to compare the empirical histogram and density curve shapes to the theoretical pattern in a normal curve and relate the various graphic representations and data summaries to the geometrical properties of normal distribution. Even when the majority of students learnt to use the software, some of them had difficulties in interpreting areas in frequency histograms produced by the software, in computing areas under the normal curve, or in discriminating between the empirical data and the mathematical model (normal curve).

Influence from Statistics

Another strong influence on research comes from the field of statistics, where interest in education arose especially since the creation in 1949 of an Education Committee by the International Statistical Institute (ISI), through which the ISI promotes the training of official statisticians in developing countries and has organized a series of Round Table Conferences on specific educational problems since 1973 (Vere-Jones, 1997). The International Conferences on Teaching

Statistics (ICOTS) were started in 1982 by the ISI and have continued every 4 years. In 1991 the IASE was created as a separate section of the ISI and took over the organization of these conferences. The journals *Teaching Statistics,* first published in 1979, and *Journal of Statistics Education,* started in 1993, soon became important tools to improve statistics education all over the world. This activity complemented the educational sessions at international and national statistical conferences (e.g., in the biannual Sessions of the International Statistical Institute or in the American Statistical Education Conferences) as well as the Stochastic Thinking Reasoning and Literacy Research Forum started in 1999. With regard to the teaching, learning, and understanding of stochastics at post-secondary level these conference activities promoted a large and varied number of teaching experiences, and many suggestions were published in proceedings and journals. At the same time, some systematic research has been progressively emerging under the influence of the IASE and of several funded projects and initiatives by the American Statistical Association (e.g., the Undergraduate Statistics Education Initiative, USEI,[22] and the Consortium for the Advancement of Undergraduate Statistics Education, CAUSE[23]).

One example is the research carried out by Garfield, del Mas, and Chance who developed a piece of didactic software (*Sampling Distribution*) and complementary instructional materials (pretest and posttest, instructional objectives, activities) to assess students' previous misconceptions, support discovery and exploration of inferential concepts, and assess change after instruction (Garfield, del Mas, & Chance, 1997). The software allows student to choose among a variety of possible distributions for a continuous variable in a population, including nonstandard models of distributions. In addition it provides different windows where the student can simulate the drawing of samples (for any size and number of samples) from the specified population and visualize the sample outcomes as well as the values of some statistics (e.g., mean, median, variance) in the different samples (sample statistics). From these, there is the possibility to graph the set of different values for these sample statistics to get an empirical sampling distribution for the statistics. Students can add new samples to their previous results little by little, in order to discover the long-run patterns in the empirical sampling distribution and understand the features in the theoretical sampling distribution (distribution of all the possible values of the statistics for a given population and sample size). For example, they can "discover" for themselves the central limit theorem, according to which the sampling distribution of mean and other statistics will approach to a normal distribution for moderate sample sizes ($n > 30$) even for nonsymmetrical population distributions. They can see how the approximation improves with the sample size, and that as the spread in the sampling distribution decreases the average (value of the sample statistics) approaches to the value of the population parameter.

The same authors carried out a teaching experiment with three groups of students at the university introductory level (delMas, Garfield, & Chance, 1999). The students were supposed to have read a chapter on sampling distribution and the central limit theorem before engaging with hands-on simulation using the software. Class discussions served to compare the shape, center, and spread of empirical sampling distributions for different sample sizes and population distributions and whether empirical results agreed with what was expected in the central limit theorem. The authors were surprised that after the first series of experiments many of their students still showed serious misconceptions. For example, the variability the students expected in the distribution was not consistent with the sample size, and they did not understand that the sampling distribution would resemble a normal distribution with increasing sample sizes. Much better results were obtained when the authors followed a model of learning through conceptual change in which students were asked to test their predictions and confront their misconceptions, following a constructivist model of learning. The authors here concluded that students with misconceptions have to experience contradictory evidence and reflect on this contradiction with their previous expectations before they change their views about random phenomena.

To gather more detailed information about students' reasoning they carried out several guided interviews that suggested a model of developmental stages in students' statistical reasoning: idiosyncratic, verbal, transitional, procedural, and integrated process reasoning (Chance, delMas, & Garfield, 2004), based on work by neo-Piagetian cognitive researchers (e.g., Biggs & Collis, 1991) who have examined the process of cognitive development in everyday and school contexts. In this model students progress from knowing words and symbols without understanding their meaning (idiosyncratic level) to a substantial understanding of the process of sampling and sampling distributions (process level) in a series of stages. Similar mod-

[22] http://www.amstat.org/education/index.cfm?fuseaction=usei
[23] http://www.causeweb.org/

els of developmental stages have also been proposed by some mathematics educators for some elementary stochastics concepts, in research carried out with primary and secondary school students (Jones, Langrall, Mooney, & Thornton, 2004). Finally, we should note this kind of research focusing on the identification on misconceptions and developmental stages is rather classical from a mathematics education point of view.

Although many different theoretical concepts in research carried out by statisticians are taken from psychology or mathematics education, some statisticians are trying to develop specific frameworks to describe statistical thinking as a specific type of thinking that recognizes the variation around us and includes a series of interconnected processes, aimed at identifying, analyzing, quantifying, controlling, and reducing this variation in order to improve or inform decisions and actions in many different fields (Snee, 1990). A notable example is the theoretical framework developed by Wild and Pfannkuch (1999) and Pfannkuch and Wild (2004), who conducted qualitative research on the activities carried out by students and professionals when engaged in statistical investigations from which the authors developed a complex four-dimensional framework including four components. The first component is termed the *statistical investigation cycle PPDAC (Problem, Planning, Data, Analysis, Conclusion)* and describes the activities carried out when identifying and solving a statistical problem that is embedded in a wider contextual problem. This cycle includes identifying and setting of the problem, planning for its solution, collecting data, data analysis, and reaching conclusions. Secondly there is an *interrogative cycle*, which is a generic process in statistical problem solving consisting in generating possibilities for causes or explanations, seeking or recalling information, interpreting, translating or comparing information, criticizing this information from both internal and external points of view, and judging the reliability or usefulness of the information. The third component in the model describes the *types of thinking* present in statistical problem solving. In addition to strategic thinking the authors define fundamental modes of stochastic thinking, which include recognizing the need for data, transnumeration (numerical transformation to facilitate understanding, for example, classifying, coding, or representing data), identifying, explaining controlling and measuring variation, use of mathematical models, and synthesis of statistical and contextual knowledge. Finally the model describes a series of *dispositions*, such as curiosity, imagination, or skepticism. Following the publication of the Wild and Pfannkuch paper a number of research projects have focused on teaching or assessing statistical reasoning and its several components as a whole. For example Pfannkuch, Rubick, and Yoon (2002) described how students use transnumeration or perception of variation to build up recognition and understanding of relationships of variables in small data sets. Another example is the research by Biehler (2005) who analyzed about 60 students' statistical projects and used Wild and Pfannkuch's model to develop an assessment scheme and a project guide to improve the quality of these projects.

The Challenges of Advanced Stochastics

The above summary of research suggests that probability and statistics pose a number of important challenges to research on mathematics thinking and learning at post-secondary level. First, the existing research has been carried out by different scientific communities, and not just by mathematics educators, and for this reason the sources of information are widespread and not always easily available. At the same time, the diversity of research problems and approaches is very wide, something that is also common for undergraduate stochastics. Some research agendas posed by Shaughnessy (1992) and Shaughnessy, Garfield, and Greer (1996), as well as the creation of the *Statistics Education Research Journal* with the specific purpose to promote research show a tendency to unification and linkage of the isolated pieces of research towards developing a more general knowledge about statistical education. However, the influence from philosophical views about randomness, probability, and statistical inference that still continue today is reflected in stochastics teaching and research: *"in the field of probability there continues an ongoing fierce debate on the foundations, even though for pragmatic reasons only, this debate has calmed down in recent times"* (Borovcnik & Peard, 1996, p. 239). These different views influence the role given to probability in the curriculum, from being the central core, to trying to teach statistics without resort to probability (focus on exploratory data analysis only) or favoring classical, Bayesian, or mathematical-abstract approaches to inference.

Secondly, in the case of statistics, the distinction between advanced and elementary topics remains very fuzzy, inasmuch as current curricula in many countries include ideas about association and inference in secondary school alongside situations that require mature stochastic thinking for correct interpretation, such as voting, investment, research planning, or quality control, topics that increasingly form part of the information to which many citizens and professionals are exposed. So, whereas advanced mathematical thinking tends to be used only formally and after some systematic training, advanced stochastic thinking is now being formally or informally used by many people with little formal training in either mathematics or stochastics. Moreover some apparently simple concepts, such as randomness, that are taken for granted in elementary courses are in fact very complex. For example, judging whether a sequence of outcomes is likely to have been randomly generated involves being able to measure a number of parameters including relative frequency, length and distribution of runs, variation found in subsets, estimating probabilities, and so on. These skills are frequently taught in school; however research in both psychology and education has shown frequent errors and misconceptions related to such simple ideas as probability, average, and distribution. Furthermore, because no random sequence of outcomes exactly fits the expected patterns suggested by probability theory, judging randomness in a particular situation also requires an understanding of hypothesis-testing logic and sampling distribution features. These are both advanced stochastic ideas with which students have many difficulties. Moreover randomness is assumed in building a sampling distribution or in hypothesis testing, so a sound understanding of these two concepts in turn actually rests on the idea of randomness (thus creating a circular situation).

Stochastics is a field in which the need to present advanced concepts to a wide audience of students with varied backgrounds, interests, and capacities is urgent, given the implications of poor stochastic thinking in many fields of human activity. For example, recommendations to substitute or complement statistical tests with confidence intervals (e.g., Wilkinson, 1999) do not take into account the fact that their appealing feature is based on a fundamental misunderstanding (Lecoutre, 1998), which consists of thinking of the parameters as random variables and assuming that confidence intervals contain the parameters with a specified probability. Such interpretation is incorrect in a classical inferential framework, although it is acceptable in Bayesian statistics. For this reason some researchers (e.g., Lecoutre & Lecoutre, 2001) are suggesting to change the practice of statistics towards Bayesian methods and suggest that Bayes's thinking is more intuitive than frequentist probability for students and better reflects students' everyday thinking about uncertainty. Reported results from research focused on teaching Bayesian statistics are limited to a few cases (Albert, 2000), and the decision to change from classical to Bayesian approach is also dependent on researchers' own objective or subjective views of probability. The wide research done on students' and professionals' misunderstanding and misuse of advanced statistics should be complemented by a similar effort in designing and evaluating teaching experiments oriented to help students and researchers overcome these difficulties. In this sense, statistics and probability can be a paradigm for finding ways and approaches to introduce advanced mathematical ideas to wide audiences and to rethink the very meaning of what is advanced mathematical thinking.

CONCLUSION

In this chapter, we have tried to synthesize the evolution of research on mathematics thinking and learning at post-secondary level since the first *Handbook* was published in 1992, and to analyze its most important advances, their potential, and limit for understanding and improving teaching and learning processes at this advanced level. This evolution, which has been fostered by both internal and external factors, has not obeyed a simple dynamic, and the multiplicity of its facets reflects both the intrinsic diversity of educational research and the diversity of the changes that have affected post-secondary education during the last decade. Some of these evolutions were highly predictable, such as the development of the several theories of reification focusing on process-object duality; the increasing attention paid to the semiotic dimension of mathematical activity and to the essential role played by connections between representations, settings, and perspectives in mathematical thinking and learning; and the increasing influence taken by sociocultural and anthropological approaches towards learning processes. Others were less predictable, such as the increasing theoretical influence of current developments in cognitive sciences and embodied cognition, the rapid changes and growth in mathematical practice fostered by information and communication technologies, the reflection of these changes in post-secondary education mathematical curricula that today oblige educational researchers to expand their privileged fields of investigation, the increasing demand of advanced mathematics learning by varied types of students, and the interest in research on teaching and learning mathematics from a variety of disciplines (not just in mathematics education). All these evolutions make the landscape of research on mathematics thinking and learning at post-secondary level today something much more diverse and richer than was the case about 10 years ago.

Looking at what has been achieved, we have the feeling that the research initiated in the Advanced Mathematical Thinking working group of PME in this area, in spite of the diversity of its developments, has avoided a fractionalization of its perspectives and been able to integrate its previous achievements into complementary and coherent constructs, as we have tried to show in the first parts of this chapter. This certainly provides researchers with a strong and mature basis for addressing the important challenges that research has to face today.

More and more, however, the evolution of post-secondary education obliges us to reconsider the answers we have given to fundamental epistemological issues about the nature of mathematics, and the nature of mathematical learning and thinking. These fundamental issues cannot be discussed without taking into account the current reality of mathematical practices, or by considering only the practice of pure mathematicians working in traditional fields. Post-secondary educational research has from this point of view a specific epistemological role to play in educational research thanks to its proximity with the professional world of mathematics. The increasing importance taken in post-secondary mathematics education by service courses faces us with the necessity of taking a wider perspective. As has been shown in the fourth part of this chapter, we are pushed and questioned by the evolution of mathematical practices in professional fields such as engineering. In the case of statistics, we observe a progressive separation of mathematics from the applications of statistics that originates in the fact that statistics has changed much more rapidly than mathematics, is much more dependent on the context and on information technology, and is usually taught at post-secondary level by lecturers who are seldom mathematicians. At the same time, an increasing amount of research is being carried out in advanced stochastics outside the mathematics education community.

Technology has deeply changed professional practices and mathematical needs and as a consequence changed what has to be learnt and how it can be learnt. As has been argued for the case of stochastics and engineering, these changes oblige us to see in technology more than an educational help, but something constitutive of mathematical practices, and moreover having the effect of changing mathematical practices as well as changing the meaning of mathematical objects. This has an impact on the learning processes and obliges us also to consider forms of and progressions in learning that are different from the usual ones. At the same time these changed practices should support ways for students with a limited mathematical background to make reasonable sense and use of the very sophisticated mathematics that are implemented in the technology they use, thus producing new, more intuitive meanings for these mathematical objects.

From the theoretical point of view, researchers in these new research fields have taken into account only a very limited number of the theoretical concepts and paradigms developed by the post-secondary mathematics education community. Therefore the study of the extent to which these frameworks can describe the learning of service mathematics or the learning of advanced stochastics or how these frameworks should be complemented with other constructs specific to these fields is still an open challenge for researchers.

Thanks to the global evolution of theoretical frames towards sociocultural and anthropological perspectives, researchers are certainly today better equipped for addressing this issue of mathematical practices, taking into account their different components, both explicit and tacit, in order to analyze their potential cognitive effects. This evolution has also a corollary that any consideration of learning processes is necessarily relative. Knowledge emerges from practices, and the learning processes one can access are those that are made possible by the existing practices. We have thus to be aware that the answers we can provide to the questions at stake are not absolute; they depend on the current state of educational practice. Research has certainly to pay more attention to these dependences than it has done in the past. This makes it necessary to explore more systematically than has been done before what is learnt and how in educational designs different from the traditional ones, as for instance those mentioned in the fourth part of the chapter where the learning of mathematics emerges from the realization of projects and activities in which modeling and technology are given an essential role.

Even if researchers seem today better equipped for approaching the relationships between learning and practices, and sensitive to the cognitive diversity that emerges from their diversity, benefiting today from the different advances is not at all an easy task. We met this difficulty when writing this chapter. For instance, we found it essential to open the chapter to challenging areas such as stochastics or engineering, and our initial project was to end this chapter by some kind of integrated perspective on learning and thinking at post-secondary level. But connecting research in stochastics and research on AMT, two areas of research that have developed in isolated ways and pushed by different logics, turned out to be too difficult. Much more work than what we were able to do within the time devoted to the elaboration of this chapter would have been necessary in order to succeed. What is certainly expressed in this chapter is the specific coherence underlying each of these approaches, and the ways it tends to question the other one, but no more.

We end this chapter with the feeling that research on mathematical learning and thinking at post-secondary level is entering now a new and fascinating phase, with difficult and new challenges to face, challenges that will require to be solved to extend interactions and collaborations beyond the traditional community of research in mathematics education at advanced level.

REFERENCES

Albert, J. (2000). Using a sample survey project to assess the teaching of statistical inference, *Journal of Statistical Education, 8.* Retrieved April 13, 2006, from http://www.amstat.org/publications/jse/secure/v8n1/albert.cfm.

Allen, H. G. (Ed.). (2000). *Proceedings of a conference on Civil and Structural Engineering Education in the 21st Century (Southampton, April 2000).* Two volumes. Southampton, UK: Department of Civil and Environmental Engineering, University of Southampton.

Alvez Diaz M. (1998). *Problèmes d'articulation entre points de vue "cartésien" et "paramétrique" dans l'enseignement de l'algèbre linéaire* [Problems of articulation between Cartesian and parametric "point of view" in the teaching of linear algebra]. Doctoral thesis, University Paris 7. Paris: IREM Paris 7 (Institut de Recherche sur l'Enseignement des Mathématiques).

Artigue, M. (1988). Ingénierie didactique [Didactic engineering]. *Recherches en Didactique des Mathématiques, 9*(3), 281–308.

Artigue, M. (1991). Analysis. In D. Tall (Ed.), *Advanced mathematical thinking* (167–198). Dordrecht, The Netherlands: Kluwer.

Artigue, M. (2001). What can we learn from educational research at the university level? In D. Holton (Ed.), *The teaching and learning of mathematics at university level. An ICMI Study* (pp. 207–220). Dordrecht, The Netherlands: Kluwer.

Asalia, M., Brown, A., DeVries, D., Dubinsky, E., Mathews, D., & Thomas, K. (1996). A framework for research and curriculum development in undergraduate mathematics education. *CBMS Issues in Mathematics Education,* vol. 6, 1–32.

Bachelard, G. (1938). *La formation de l'esprit scientifique* [The formation of the scientific spirit]. Paris: J. Vrin.

Batanero, C., Godino, J., & Estepa, A. (1998). Building the meaning of statistical association through data analysis activities. In A. Olivier, & K. Newstead (Eds.), *Proceedings of the 22nd Conference of the International Group for the Psychology of Mathematics Education* (Vol. 1, pp. 221–236). Stellenbosh, South Africa: University of Stellenbosh.

Batanero, C., Estepa, A., Godino, J. D., & Green, D. R. (1996). Intuitive strategies and preconceptions about association in contingency tables. *Journal for Research in Mathematics Education, 27*(2), 151–169.

Batanero, C., Tauber, L., & Sánchez, L. (2004). Students' reasoning about the normal distribution. In J. B. Garfield & D. Ben-Zvi (Eds.), *The challenge of developing statistical literacy, reasoning and thinking* (pp. 257–276). Dordrecht, The Netherlands: Kluwer.

Barragués, J. I. (2002). *La enseñanza de la probabilidad en primer ciclo de universidad. Análisis de dificultades y propuesta alternativa de orientación constructivista* [Teaching probability at introductory university level. Analysis of difficulties and alternative constructivist proposal]. Unpublished dissertation, University of the Basque Country, San Sebastian, Spain.

Berthoz, A. (1998). *Le sens du mouvement* [The direction of the movement]. Paris: Odile Jacob.

Beyth-Marom, R. (1982). Perception of correlation reexamined. *Memory and Cognition, 10*(6), 511–519.

Biehler, R. (1994). Software tools and mathematics education: The case of statistics. In C. Keitel & K. Ruthven (Eds.), *Learning from computers: Mathematics education & technology* (pp. 68–100). Berlin: Springer Verlag.

Biehler, R. (1997). Software for learning and for doing statistics. *International Statistical Review, 65*(2), 167–190.

Biehler, R. (2005, February). Strengths and weaknesses in students' project work in exploratory data analysis. Paper presented at the Fourth Congress of the European Society for Research in Mathematics Education.CERME 4, Sant Feliu de Guissols, Spain.

Biggs, J. B., & Collis, K. F. (1991). Multimodal learning and intelligence behavior. In H. Rowe (Ed.), *Intelligence: Reconceptualization and measurement* (pp. 57–75). Hillsdale, NJ: Erlbaum.

Bingolbali, E. (2004). The calculus of engineering and mathematics undergraduates. Paper presented at the 10th International Congress on Mathematics Education, Copenhagen (Topic Study Group 12). Retrieved April 13, 2006, from www.icme-organisers.dk/tsg12.

Bissell, C., & Dillon, C. (2000). Telling tales: Models, stories and meanings. *For the Learning of Mathematics, 20*(3), 3–11.

Blockley, D., & Woodman, N. (2002). Civil/structural engineers and maths: The changing relationship. *The Structural Engineer,* April 2, pp. 14–15.

Boero, P. (Ed) (1999). Teaching and learning mathematics in context. *Educational Studies in Mathematics, 39*(1–3), vii–x.

Borba, M., & Scheffer, N. F. (2004). Coordination of multiple representations and body awareness. *Educational Studies in Mathematics, 57*(3), 303–321.

Borovcnik, M., & Peard, R. (1996). Probability. In A. J. Bishop, K. Clements, C. Keitel, J. Kilpatrick, & C. Laborde (Eds.), *International handbook of mathematics education* (Part 1, pp. 239–288). Dordrecht, The Netherlands: Kluwer.

Burton, L., Falk, L., & Jarner, S. (2004). Too much, too seldom. *International Journal of Mathematical Education in Science and Technology, 35*(2), 219–226.

Broudy, H. S. (1977). Types and knowledge and purposes of education. In R. C. Anderson & R. J. Spiro (Eds.), *Schooling and the acquisition of knowledge* (pp. 1–17). Hillsdale, NJ: Erlbaum.

Brousseau, G. (1983). Les obstacles épistémologiques et les problèmes en mathématiques [Epistemological obstacles and problems in mathematics]. *Recherches en Didactique des Mathématiques. 4*(2), 164–198.

Brousseau, G. (1997). *The theory of didactical situations in mathematics.* 1970–1990. Dordrecht, The Netherlands: Kluwer.

Cantoral, R., & Farfan, R.M. (2003) Mathematics education: A vision of its evolution. *Educational Studies in Mathematics, 53*(3) 255–270.

Castela, C. (2004). Institutions influencing mathematics students' private work: A factor of academic achievement. *Educational Studies in Mathematics, 57*(1), 33–63.

Chance, B., delMas, R. & Garfield, J. F. (2004). Reasoning about sampling distributions. In D. Ben Zvi & J. B. Garfield (Eds.). *The challenge of developing statistical literacy, reasoning and thinking* (pp. 295–323). Dordrecht, The Netherlands: Kluwer.

Chevallard, Y. (1992). Concepts fondamentaux de la didactique: perspectives apportées par une approche anthropologique [Basic concepts of education: Perspectives from an anthropological approach]. *Recherches en Didactique des Mathématiques, 12*(1), 77–111.

Chevallard, Y. (1997). Familière et problématique, la figure du professeur [Familiar and problematic, the face of the professor]. *Recherche en Didactique des Mathématiques, 17*(3), 17–54.

Chevallard, Y., & Bosch, M. (1999). La sensibilité de l'activité mathématique aux ostensifs. Objet d'étude et problématique [Sensitivity of mathematical activity to demonstration. Instances and problems]. *Recherches en Didactique des Mathématiques, 19*(1), 77–124.

Cobb, P., & Yackel, E. (1996). Constructivist, emergent and socio-cultural perspectives in the context of developmental research. *Educational Psychologist, 31*, 175–190.

Confrey, J., & Smith, E. (1994). Comments on James Kaput's chapter. In A. H. Schoenfeld (Ed.), *Mathematical thinking and problem solving* (pp. 172–192). Mahway, NJ: Erlbaum.

Cornu, B. (1983). Apprentissage de la notion de limite: conceptions et obstacles [Apprenticeship of the limit notion: Conceptions and obstacles]. Doctoral thesis, University of Grenoble I.

Cornu, B. (1991). Limits. In D. Tall (Ed.), *Advanced Mathematical Thinking* (153–166). Dordrecht, The Netherlands: Kluwer.

Cosmides, L., & Tooby, J. (1996). Are humans good intuitive statisticians after all? Rethinking some conclusions from the literature on judgment under uncertainty. *Cognition, 58*, 1–73.

Cuoco, A. A., Goldenberg, E. P., & Mark, J. (1996). Habits of mind: An organizing principle for mathematics curriculum. *Journal of Mathematical Behavior, 15*(4), 375–402.

Dehaene, S. (1996). *La Bosse des Maths.* Paris: Odile Jacob. [English translation : *The Number Sense: How the mind creates mathematics.* New York: Oxford University Press, 1997.]

DelMas, R. C. (2004). A comparison of mathematical and statistical reasoning. In D. Ben Zvi & J. B. Garfield (Eds.). *The challenge of developing statistical literacy, reasoning and thinking* (pp. 79–96). Dordrecht, The Netherlands: Kluwer.

DelMas, R. C., Garfield, J. B., & Chance, B. (1999). A model of classroom research in action: Developing simulation activities to improve students' statistical thinking. *Journal of Statistics Education, 7*(3). Retrieved April 13, 2006, from http://www.amstat.org/publications/jse/secure/v7n3/delmas.cfm.

Dorier, J. L. (Ed.). (2000). *On the teaching of linear algebra.* Dordrecht, The Netherlands: Kluwer.

Dreyfus, T. (1991). Advanced mathematical thinking processes, In D. Tall (Ed.), *Advanced mathematical thinking* (pp. 25–41). Dordrecht, The Netherlands: Kluwer.

Dubinsky, E. (1991). Reflective abstraction in advanced mathematical thinking. In D. Tall (Ed.), *Advanced mathematical thinking* (pp. 95–123). Dordrecht, The Netherlands: Kluwer.

Dubinsky, E., & McDonald, M. (2001). APOS: A constructive theory of learning in undergraduate mathematics education research. In D. Holton (Ed.), *The teaching and learning of mathematics at university level. An ICMI Study* (pp. 275–282). Dordrecht, The Netherlands: Kluwer.

Dubinsky, E., & Schwingendorf, K. (1997). Constructing calculus concepts: Cooperation in a computer laboratory. In E. Dubinsky, D. Mathews, & B. E. Reynolds (Eds.), *Cooperative learning for undergraduate mathematics.* MAA Notes No. 44 (pp. 241–272). Washington, DC: The Mathematical Association of America.

Edwards, B. (1997). An undergraduate student's understanding and use of mathematical definitions in real analysis. In J. A. Dossey, J. O. Swafford, M. Parmantie, & A. E. Dossey (Eds.), *Proceedings of the 19th meeting of the North American Chapter of the International Group for the Psychology of Mathematics Education, Vol. 1,* (pp. 17–22). Colombus, OH: The ERIC Clearinghouse for Sciences, Mathematics, and Environmental Education.

Edwards, B. E., Dubinsky, E., & McDonald, M.A. (2005). Advanced mathematical thinking. *Mathematical Thinking and Learning, 7*(1), 15–25.

Eddy, D. M. (1982). Probabilistic reasoning in clinical medicine: Problems and opportunities. In D. Kahneman, P. Slovic, & A. Tversky (Eds.), *Judgment under uncertainty: Heuristics and biases* (pp. 249–267). Cambridge: Cambridge University Press.

Estepa, A. (1993). Concepciones iniciales sobre la asociación estadística y su evolución como consecuencia de una enseñanza basada en el uso de ordenadores [Students' preconceptions on statistics association and its evolution after a teaching experiment based on computers]. Unpublished Ph.D. thesis, University of Granada, Spain.

Falk, R. (1986). Conditional probabilities: Insights and difficulties. In R. Davidson & J. Swift (Eds.), *Proceedings of the Second International Conference on Teaching Statistics.* (pp. 292–297). Victoria, Canada: International Statistical Institute.

Falk, F., & Konold, C. (1997). Making sense of randomness. Implicit encoding as a basis for judgment. *Psychological Review, 104*, 310–318.

Freudenthal, H. (1991). *Revisiting mathematics education.* Dordrecht, The Netherlands: Kluwer.

Garfield, J. B. (2003). Assessing statistical reasoning. *Statistics Education Research Journal,* 2(1), 22–38. Retrieved April 13, 2006, from http://www.stat.auckland.ac.nz/~iase/serj/SERJ2(1).pdf.

Garfield, J. B., del Mas, B., & Chance, B. (1997). Assessing the effects of a computer microworld on statistical reasoning. Paper presented at the *Join Meetings of the American Statistical Association,* Anaheim, CA, August, 1997. Retrieved April 30, 2005, from http://www.gen.umn.edu/research/stat_tools/asa_1997_paper/asa_paper.htm.

Gigerenzer, G. (1994). Why the distinction between single-event probabilities and frequencies is important for psychology (and vice-versa). In G. Wright & P. Ayton (Eds.), *Subjective probability* (pp. 129–161). Chichester, UK: John Wiley.

Gigerenzer, G. (1996). Why do frequency formats improve Bayesian reasoning? Cognitive algorithms work on information, which needs representation. *Behavioral and Brain Sciences, 19*(1), 23–24.

Godino, J. D. (2002). Un enfoque ontológico y semiótico de la cognición matemática [An ontological and semiotic approach to mathematical cognition]. *Recherches en Didactique des Mathématiques, 22*(2/3), 237–284.

Godino, J. D., & Batanero, C. (1998). Clarifying the meaning of mathematical objects as a priority area of research in mathematics education. In A. Sierpinska & J. Kilpatrick (Eds.), *Mathematics education as a research domain: A search for identity* (pp. 177–195). Dordrecht, The Netherlands: Kluwer.

Godino, J. D., Batanero, C., & Roa, R. (2005). A semiotic analysis of combinatorial problems and its resolution by university students. *Educational Studies in Mathematics, 60*(1), 3–36.

Goldin, G. A., & Janvier, C. (Eds.). (1998). Representations and the psychology of mathematics education. *The Journal of Mathematical Behavior, 17*(1/2) (Special Issue).

González-Martín, A. S. & Camacho, M. (2004). What is first-year mathematics students' actual knowledge about improper integrals?. *International Journal of Mathematical Education in Science and Technology, 35*(1), 73–89.

Gravemeijer, K. (1999). How emergent models may foster the constitution of formal mathematics. *Mathematical Thinking and Learning, 1*, 155–177.

Gray, E., & Tall, D. (1994). Duality, ambiguity and flexibility: A proceptual view of simple arithmetic. *Journal for Research in Mathematics Education 26*, 115–141.

Guin, D., Ruthven, K., & Trouche, L. (Eds.) (2004). *The didactical challenge of symbolic calculators: Turning a computational device into a mathematical instrument.* New York: Springer (Mathematics Education Library).

Hadamard, J. (1945). *La psychologie de l'invention mathématique.* [The Psychology of invention in the mathematical field, New York: Dover Books, 1954]

Hall, E. T. (1981/1959). *The silent language.* New York: Anchor Press.

Halliday, M. A. K. (1988). *Spoken and written language.* Geelong, VIC: Deakin University Press.

Harel, G., & Sowder, L. (2005). Advanced mathematical thinking at any age: Its nature and its development. *Mathematical Thinking and Learning, 7*(1), 27–50.

Harlow, L. L., Mulaik, S. A., & Steiger, J. H. (1997). *What if there were no significance tests?* Mahwah, NJ: Erlbaum.

Hibberd, S., & Mustoe, L. (1997). *Mathematical education of engineers II: Proceedings of the second conference, April 1997.* Southend, UK: Institute of Mathematics and its Applications.

Hibberd, S., & Mustoe, L. (2000). *Mathematical education of engineers III: Proceedings of the third conference, April 2000.* Southend, UK: Institute of Mathematics and its Applications.

Hirsch, L., & O'Donnell, A. M. (2001). Representativeness in statistical reasoning: Identifying and assessing misconceptions. *Journal of Statistics Education, 9*(2). Retrieved April 13, 2006, from http://www.amstat.org/publications/jse/v9n2/hirsch.html.

Holton, D. (Ed.) (2000). Special issue for the ICMI Study on the Teaching and Learning of Mathematics at University Level. *International Journal of Mathematical Education in Science and Technology, 31*(1).

Holton, D. (Ed.) (2001). *The teaching and learning of mathematics at university level. An ICMI Study.* Dordrecht, The Netherlands: Kluwer.

Houdement C., & Kuzniak A. (1999). Un exemple de cadre conceptuel pour l'étude de l'enseignement de la géométrie en formation des maîtres [A conceptual framework example for the study of the teaching of geometry in training of the masters]. *Educational Studies in Mathematics, 40*(3), 283–312.

Inhelder, B., & Piaget, J. (1955). *De la logique de l'enfant á la logique de l'adolescent* [The growth of logical thinking from childhood to adolescence: An essay on the construction of formal operational structures]. Paris: Presses Universitaires de France.

Jennings, D. L., Amabile, M. T., & Ross, L. (1982). Informal covariation assessment: Data-based versus theory-based judgements. In D. Kahneman, P. Slovic, & A. Tversky (Eds.), *Judgement under uncertainty: Heuristics and biases* (pp. 211–230). New York: Cambridge University Press.

Johnson, M. (1987). *The body in the mind: The bodily basis of meaning, imagination, and reason.* Chicago: University of Chicago.

Jones, G. A., Langrall, C. W., Mooney, E. S., & Thornton, C. A. (2004). Models of development in statistical reasoning. In D. Ben Zvi & J. B. Garfield (Eds.), *The challenge of developing statistical literacy, reasoning and thinking* (pp. 97–117). Dordrecht, The Netherlands: Kluwer.

Jones, G. A., & Thornton, C. A. (2005). An overview of research into the teaching and learning of probability. In G. A. Jones (Ed.), *Exploring probability in school: Challenges for teaching and learning* (pp. 65–92). Dordrecht, The Netherlands: Kluwer.

Kahneman, D., Slovic, P., & Tversky, A. (1982). *Judgment under uncertainty: Heuristics and biases.* New York: Cambridge University Press.

Kaput, J. (1992). Technology in mathematics education. In D. A. Grouws (Ed.), *Handbook of research on mathematics teaching and learning* (pp. 515–552). New York: Macmillan.

Kent, P. & Noss, R. (2003). *Mathematics in the university education of engineers (A report to The Ove Arup Foundation).* London: The Ove Arup Foundation. Retrieved April 13, 2006, from http://www.theovearupfoundation.org/arupfoundation/pages/download25.pdf

Konold, C. (1991). Understanding students' beliefs about probability. In E. von Glasersfeld (Ed.), *Radical constructivism in mathematics education* (pp. 139–156). Dordrecht, The Netherlands: Kluwer.

Lakatos, I. (1976). *Proofs and refutations.* Cambridge: Cambridge University Press.

Lakoff, G., & Nunez, R. (2000). *Where mathematics comes from: How the embodied mind brings the mathematics into being.* New York: Basic Books.

Lecoutre, B. (1998). Teaching bayesian methods for experimental data analysis. In L. Pereira-Mendoza, L. Seu Kea, T. Wee Kee, & W. K. Wong (Eds.), *Proceedings of the Fifth international Conference on Teaching of Statistics* (Vol. 1, pp. 239–244). Singapore: International Statistical Institute.

Lecoutre, B., & Lecoutre, M. P. (2001). Uses, abuses and misuses of significance tests in the scientific community: Won't the Bayesian choice be unavoidable? *International Statistical Review, 69*(3), 399–417.

Lerman, S., & Sierpinska, A. (1996). Epistemologies of mathematics and of mathematics education. In A. Bishop, K. Clements, C. Keitel, J. Kilpatrick, & C. Laborde (Eds.), *International handbook of mathematics education* (pp. 827–876). Dordrecht, The Netherlands: Kluwer.

Lithner, J. (2003). Students' mathematical reasoning in university textbook exercises. *Educational Studies in Mathematics, 52*(1), 29–55.

Longo, G. (Ed.). (2004). *Géométrie et cognition* [Geometry and cognition]. Paris: Albin Michel. Special Issue of the Revue de Synthèse, Vol. 124.

Magajna, Z. (2001). *Geometric thinking in out of school contexts.* Unpublished dissertation, University of Leeds.

Magajna, Z., & Monaghan, J. (2002). Advanced mathematical thinking in a technological workplace. *Educational Studies in Mathematics, 52*(2), 101–122.

Maschietto, M. (2002). *L'enseignement de l'analyse au lycée: les débuts du jeu local/global dans l'environnement de calculatrices* [The teaching of analysis to the high school: Beginnings of the global/local game in the environment of computers]. Doctoral thesis, University Paris 7. Paris: IREM Paris 7 (Institut de Recherche sur l'Enseignement des Mathématiques).

Maull, W. M., & Berry, J. S. (2000). A questionnaire to elicit the mathematical concept images of engineering students. *International Journal of Mathematical Education in Science and Technology, 31*(6), 899–917.

Morris, E. J. (1997). *An investigation of students' conceptions and procedural skills in the statistics topics correlation.* Technical report n. 230. Milton Keynes, U.K.: Centre for Information Technology, The Open University.

Morris, E. J. (1999). *Another look at psychology students' understanding of correlation.* Technical report n. 246. Milton Keynes, U.K.: Centre for Information Technology, The Open University.

Morrison, D. E., & Henkel, R. E. (Eds.). (1970). *The significance tests controversy. A reader.* Chicago: Aldine.

Moore, D. S. (1997). New pedagogy and new content: The case of statistics. *International Statistical Review, 635*, 123–165.

Mustoe, L., & Hibberd, S. (1995). *Mathematical education of engineers: Proceedings of the first conference, 1994.* Oxford, UK: Oxford University Press.

Nardi, E. (1996). *The novice mathematician's encounter with mathematical abstraction. Tensions in concept—image formation and formalisation.* Doctoral thesis, University of Oxford.

National Council of Teachers of Mathematics. (2000). *Principles and standards for school mathematics.* Reston, VA: Author. Retrieved April 13, 2006, from http://standards.nctm.org/.

Nemirovsky, R, Tierney, C., & Wright, T. (1998). Body motion and graphing. *Cognition and Instruction, 16*, 119–172.

Nisbett, R., & Ross, L. (1980). *Human inference: Strategies and shortcomings of social judgments.* Englewood Cliffs, NJ: Prentice Hall.

Niss, M. (2001). University mathematics based on problem-oriented student projects: 25 years of experience with the Roskilde model. In D. Holton (Ed.), *The teaching and learning of mathematics at university level: An ICMI Study.* Dordrecht, The Netherlands: Kluwer.

Noss, R., & Hoyles, C. (1996). *Windows on mathematical meanings.* Dordrecht, The Netherlands: Kluwer.

Nuñez, R., Edwards, L. D., & Matos, J. P. (1999). Embodied cognition as grounding for situatedness and context in mathematics education. *Educational Studies in Mathematics, 39*, 45–65.

Pfannkuch, M., & Brown, C. (1996). Building and challenging students' intuitions about probability: Can we improve undergraduate learning? *Journal of Statistics Education, 4*(1). Retrieved April 13, 2006, from http://www.amstat.org/publications/jse/v4n1/pfannkuch.html.

Pfannkuch, M., & Wild, C., (2004). Towards an understanding of statistical thinking. In D. Ben Zvi and J. B. Garfield (Eds.), *The challenge of developing statistical literacy, reasoning and thinking* (pp. 17–46). Dordrecht, The Netherlands: Kluwer.

Pfannkuch, M., Rubick, A., & Yoon, C. (2002). Statistical thinking: An exploration into students' variation-type thinking. *New England Mathematics Journal, 34*(2), 82–98.

Piaget, J., & Inhelder, B. (1951), *La genèse de l'idée de hasard chez l'enfant* [The origin of the idea of chance in children]. Paris: Presses Universitaires de France.

Piaget, J., & Garcia, R. (1989). *Psychogenesis and the history of science.* New York: Colombia University Press.

Poincaré, H. (1913). *Les fondations de la science.* [The foundations of Science. New York: The Science Press.]

Praslon, F. (2000). *Continuités et ruptures dans la transition terminale S/DEUG Sciences en analyse. Le cas de la notion de dérivée et son environnement* [Continuity and breaks in the final transition S/DEUG Sciences in analysis. The case of the derivative notion and its context]. Doctoral thesis, University Paris 7. Paris: IREM Paris 7 (Institut de Recherche sur l'Enseignement des Mathématiques).

Rabardel, P. (1995). *Les hommes et les technologies. Approche psychologique des instruments contemporains.* Paris: Armand Colin. [English translation: Rabardel, P. (2002) *People and technology: A cognitive approach to contemporary instruments.* Retrieved April 13, 2006, from: http://ergoserv.psy.univ-paris8.fr]

Ralston, A. (2004). Research mathematicians and mathematics education: A critique. *Notices of the AMS, 51*(4), 403–411.

Rasmussen, C., Zandieh, M., King, K., & Teppo, A. (2005). Advancing mathematical activity: A practice-oriented view of advanced mathematical thinking. *Mathematical Thinking and Learning, 7*(1), 51–73.

Rasmussen, C., Nemirovsky, R., Olszervski, J., Dost, K. & Johnson, J. L. (2004). On forms of knowing: The role of bodily activity and tools in mathematical learning. *Educational Studies in Mathematics, 57*(3), 303–321.

Robert, A., Dorier, J. L., Robinet, J., & Rogalski, M. (2000). The obstacle of formalism in linear algebra. In J. L. Dorier (Ed.), *On the teaching of linear algebra* (pp. 85–122). Dordrecht, The Netherlands: Kluwer.

Schneider, M. (1991). Un obstacle épistémologique soulevé par des découpages infinis des surfaces et des solides [An epistemological obstacle raised by cutting infi or of the surfaces and solids]. *Recherches en Didactique des Mathématiques, 11*(2/3), 241–294.

Schoenfeld, A. (1992). Learning to think mathematically: Problem solving, metacognition and sense making in mathematics. In D. A. Grouws (Ed.), *Handbook of research on mathematics teaching and learning* (pp. 334–370). New York: Macmillan.

Scholz, R. W. (1991). Psychological research in probabilistic understanding. In R. Acadia, & M. Borovcnik (Eds.), *Chance encounters: Probability in education* (pp. 213–249). Dordrecht, The Netherlands: Kluwer.

Sedlmeier, P. (1999). *Improving statistical reasoning. Theoretical models and practical implications.* Mahwah, NJ: Erlbaum.

SEFI. (2002). *Mathematics for the European engineer: A curriculum for the twenty-first century*, Report by the SEFI Mathematics Working Group. Brussels: Société Européenne pour la Formation des Ingénieurs. Retrieved April 13, 2006, from http://learn.lboro.ac.uk/mwg/core/latest/sefimarch2002.pdf.

Selden, J., Mason, A., & Selden, A. (1989). Can average calculus students solve non-routine problems? *Journal of Mathematical Behavior, 8*(2), 45–50.

Selden, A. & Selden, J. (2005). Perspectives on advanced mathematical thinking. *Mathematical Thinking and Learning, 7*(1), 1–13.

Sfard, A. (1991). On the dual nature of mathematical conceptions: Reflections on processes and objects as different sides of the same coin. *Educational Studies in Mathematics, 22*, 1–36.

Shaughnessy, J. M. (1992). Research in probability and statistics: Reflections and directions. In D. A. Grouws (Ed.), *Handbook of research on mathematics teaching and learning* (pp. 465–494). New York: Macmillan.

Shaughnessy, J. M., Garfield, J., & Greer, B. (1996). Data handling. In A. J. Bishop, K. Clements, C. Keitel, J. Kilpatrick, & C. Laborde (Eds.), *International handbook of mathematics education* (Vol. 1, pp. 205–237). Dordrecht, Netherlands: Kluwer.

Sierpinska, A. (1985). Obstacles épistémologiques relatifs à la notion de limite [Episyemological obstacles relating to the concept of limit]. *Recherches en Didactique des Mathématiques. 6*(1), 164–198.

Sierpinska, A. (1987). Humanities students and epistemological obstacles related to limits. *Educational Studies in Mathematics, 18*(4), 371–387.

Sierpinska, A. (1989). Sur un programme de recherche lié à la notion d'obstacle épistémologique [A research program related to the concept of epistemological obstacles]. In N. Bednarz & C. Garnier (Eds.), *Construction des savoirs: Obstacles et conflits.* Montreal: Agene dark Editions.

Sierpinska, A. (1994). *Understanding in Mathematics.* London: The Flamer Press.

Sierpinska, A. (2000). On some aspects of students' thinking in linear algebra. In J.L. Dorier (Ed.), *On the teaching of linear algebra,* 209–246. Dordrecht, The Netherlands: Kluwer.

Smith, A. (2004). *Making mathematics count: The report of Professor Adrian Smith's inquiry into post-14 mathematics education.* London: Department for Education and Skills. Retrieved April 13, 2006, from: www.dfes.gov.uk/mathsinquiry.

Snee, R. (1990). Statistical thinking and its contribution to quality. *The American Statistician, 44,* 116–121.

Spencer-Chapman, N. F. (2000). Problem-based learning in civil engineering education: A bridge too far? In H. G. Allen (Ed.), *Proceedings of a Conference on Civil and Structural Engineering Education in the 21st Century (Southampton, April 2000)* (pp. 229–240). Southampton, UK: Department of Civil and Environmental Engineering, University of Southampton.

Steen, L. (1989). Mathematics for a new century. *Notices of the American Mathematical Society 36,* 133–138.

Steen, L. (2001). Revolution by stealth: Redefining university mathematics. In D. Holton (Ed.), *The teaching and learning of mathematics at university level: An ICMI study* (pp. 303–312). Dordrecht, The Netherlands: Kluwer.

Stephan, M., & Rasmussen, C. (2002). Classroom mathematical practices in differential equations. *Journal of Mathematical Behavior, 21,* 459–490.

Tall, D. O., & Vinner, S. (1981). Concept image and concept definition in mathematics with particular reference to limits and continuity. *Educational Studies in Mathematics, 12*(2), 151–169.

Tall, D. O. (1986). *Graphic calculus I, II, III* [computer software]. London: Glen top Press.

Tall, D. (Ed.). (1991). *Advanced mathematical thinking.* Dordrecht, The Netherlands: Kluwer.

Tall, D. (1992). The transition to advanced mathematical thinking: Functions, limits, infinity and proof. In D.A. Grouws (Ed.), *Handbook of research on mathematics teaching and learning* (pp. 495–511). New York: Macmillan.

Tall, D. (1996). Functions and calculus. In A. Bishop, K. Clements, C. Keitel, J. Kilpatrick, & C. Laborde (Eds), *International handbook of mathematics education* (pp. 289–325). Dordrecht, The Netherlands: Kluwer.

Tall, D. (2001). Natural and formal infinities. *Educational Studies in Mathematics, 48*(2/3), 199–238.

Tall, D. (2004). Thinking through three worlds of mathematics. In M. Holmes & A. Fuglestead (Eds.), *Proceedings of the 28th Conference of the International group for the Psychology of Mathematics Education* (Vol. 4, pp. 281–288). Bergen, Norway: Bergen University College.

Toulmin, S. (1969). *The uses of argument.* Cambridge: Cambridge University Press.

Treffers, A. (1987). *Three dimensions. A model of goal and theory description in mathematics education: The Isobars project.* Dordrecht, The Netherlands: Kluwer.

Tversky, A. & Kahneman, D. (1983). Extension versus intuitive reasoning: The conjunction fallacy in probability judgment. *Psychological Review, 90,* 293–315.

Uhlig, F. (2002). The role of proof in comprehending and teaching linear algebra. *Educational Studies in Mathematics, 50*(3), 335–346.

Varela, F., Thompson, E., & Rosh, E. (1991). *The embodied mind: Cognitive science and human experience.* Cambridge: MIT Press.

Vere-Jones, D. (1997). The coming of age of statistical education. *International Statistical Review, 63*(1), 3–23.

Wild, C., & Pfannkuch, M. (1999). Statistical thinking in empirical enquiry (with discussion). *International Statistical Review, 67*(3), 223–265.

Wilkinson, L., & Task Force on Statistical Inference. (1999). Statistical methods in psychology journals: Guidelines and explanations. *American Psychologist, 54,* 594–604.

Yackel, E., Rasmussen, C. & King, K. (2000). Social and sociomathematical norms in an advanced undergraduate mathematics course. *Journal of Mathematical Behavior, 19,* 275–287.

Zevenbergen R. (2001). Changing contexts in tertiary mathematics. In, D. Holton (Ed.), *The teaching and learning of mathematics at university level. An ICMI Study* (pp. 13–26). Dordrecht, The Netherlands: Kluwer.

AUTHOR NOTE

We are grateful to Chris Rasmussen for his insightful comments on an earlier version of this chapter.

PART V

Assessment

23

KEEPING LEARNING ON TRACK

Classroom Assessment and the Regulation of Learning

Dylan Wiliam

INSTITUTE OF EDUCATION, UNIVERSITY OF LONDON

INTRODUCTION

When teachers are asked how they assess their students, they are likely to cite tests, quizzes, portfolios, projects, and other more or less formal methods. When, instead, teachers are asked how they know whether their students have learned something, the responses are typically very different (Dorr-Bremme & Herman, 1986). They mention classroom questions, group activities, discussions, posters, concept maps, and even the expressions on the faces of their students. In fact, the origin of the word *assessment* (Latin assidere, literally "to sit beside") is much closer to this more informal meaning. However, the emphasis on assessment as a formal process is pervasive, and mathematics education is no exception.

To a first approximation, then, the research literature on assessment, both generally and in mathematics education, is almost entirely about the formal methods of assessment, particularly tests and examinations. To make matters worse, even when less formal methods of assessments, such as teacher-made tests, are discussed, the purpose of an assessment is far more likely to be that of making a determination of a student's existing state of knowledge. Glaser and Silver (1994) observed that, "Aside from teacher-made classroom tests, the integration of assessment and learning as an interacting system has been too little explored" (p. 403). Thus, even when classroom assessment has been studied, the emphasis has tended to be on the concordance of such measures with external measures—in other words, on classroom assessment as an alternative to external assessments. As Kilpatrick, Swafford, and Findell (2001) noted in their survey *Adding It Up: Helping Children Learn Mathematics*, "Even less attention appears to have been paid to how teachers' assessments might help improve mathematics learning" (p. 40).

The intent of this chapter is to redress the balance by focusing on the role that assessment can play in supporting learning, rather than just measuring it—sometimes described as a distinction between assessment *for* learning and assessment *of* learning[1]—but two further qualifications are needed here. The first is that assessment can support learning in a vari-

[1] In the United States, the term *assessment for learning* is often mistakenly attributed to Rick Stiggins (2002), although Stiggins himself has always attributed the term to authors in the United Kingdom. In fact, the earliest use of this term in this sense appears to be a paper given by Mary James at the annual conference of the Association for Supervision and Curriculum Development in New Orleans (James, 1992). Three years later, the phrase was used by Ruth Sutton as the title of a book (Sutton, 1995). However, the first use of the *prepositional permutation* appears to be the third edition of a book entitled *Assessment: A Teacher's Guide to the Issues* by Caroline Gipps and Gordon Stobart, where the first chapter is entitled *Assessment of Learning* and the second *Assessment for Learning* (Gipps & Stobart, 1997). The distinction was brought to a wider audience by the Assessment Reform Group in 1999 in a guide for policymakers (Broadfoot et al., 1999).

ety of ways such as, for example, when students actually learn something while completing an assessment, as emphasized by Sternberg and Williams (1998, p. 10)—a process that might be termed assessment *as* learning. As Shavelson, Baxter, and Pine (1992) noted, "a good assessment makes a good teaching activity, and a good teaching activity makes a good assessment" (p. 22). Pellegrino, Chudowsky, and Glaser (2001) and Shepard et al. (2005) summarized the research on the design of such formal assessments, and Lester, Lambdin, and Preston (1997) dealt specifically with the possibilities for mathematics.

Accordingly the specific focus of this chapter is not on how teachers and students can use assessment activities to promote learning. The chapter is about assessment as an essentially interactive process, in which the teacher can find out whether what has been taught has been learned, and if not, to do something about it. It is therefore about assessment functioning as a bridge between teaching and learning, helping teachers collect evidence about student achievement in order to adjust instruction to better meet student learning needs, in real time.

The second qualification is that the focus of this chapter is firmly on the learning of mathematics in the mathematics classroom. Effective implementation of the kinds of exemplary practices identified in this chapter will entail careful consideration of a whole range of issues related to classroom management (see, for example, Brookhart, 2004), teacher professional development (Wiliam & Thompson, in press) and educational policy (Looney, 2005). Although these issues are clearly important, they are beyond the scope of this chapter.

The importance of the role of assessment in instruction was explicitly recognized in the National Council of Teachers of Mathematics' six *Assessment Standards for School Mathematics* (NCTM, 1995), the second of which states that "Assessment should enhance mathematics learning" (p. 13). The NCTM's *Assessment Standards* also make clear that assessment has a role to play not just in making determinations about whether particular teaching activities were successful, but also in teachers' moment-by-moment decision making (p. 46)—in other words, that teachers should use assessment to "keep learning on track."

This distinction about the *purpose* of assessment is quite different from the distinction between classroom assessment and external assessments, which is more concerned with where the assessments take place, who sets them, and who scores them (Black & Wiliam, 2004a). This is an important qualification, particularly in the United States, where the term *classroom assessment* is used primarily to mean classroom *summative* assessment.

The next section lays out the reasons for two key assumptions that have been made in the writing of this chapter, namely that assessment is independent both of curriculum and of any particular stance in psychology—in other words, that one can talk about the principles of good assessment without subscribing to a particular view of what should be in the mathematics curriculum, or even to a particular view about what happens when learning takes place. The reader who is prepared to accept these two assertions may comfortably skip the next section without loss of continuity. In subsequent sections, the research on the impact of assessment on learning is reviewed, and I suggest that the effective use of assessment for learning consists of five key strategies:

(a) Clarifying and sharing learning intentions and criteria for success;
(b) Engineering effective classroom discussions, questions, and learning tasks that elicit evidence of learning;
(c) Providing feedback that moves learners forward;
(d) Activating students as instructional resources for one another; and
(e) Activating students as the owners of their own learning.

In the final section of the chapter, these five strategies are subsumed within a broader theoretical framework, namely the regulation of learning processes, which allows assessment to be integrated with principles of instructional design.

KEY ASSUMPTIONS

There are two major challenges in synthesizing the research on assessment *for* learning (or *formative assessment* as it is sometimes called). The first is to what extent does an adequate account of assessment for learning require agreement about what mathematics students should learn? The second is to what extent does an adequate account of assessment for learning require agreement about what happens when learning takes place? Each of these is discussed in turn below.

What students should learn in mathematics is a highly contested domain. For example, in discussing the widespread preconception that mathematics is about learning to compute, Fuson, Kalchman and Bransford (2005) illustrated their discussion with the following question:

> What, approximately, is the sum of 8/9 plus 12/13?

They pointed out that some people will, sensibly, conclude that the answer is a little less than 2, just by observing that the two numbers to be added are each a little less than 1. Others, however, will attempt to find the smallest common multiple of the denominators of the two fractions. They commented:

> The point of this example is not that computation should not be taught or is unimportant; indeed, it is often critical to efficient problem solving. But if one believes that mathematics is about problem solving and that computation is a tool for use to that end when it is helpful, then the above problem is viewed not as a "request for computation," but as a problem to be solved that may or may not require computation—and in this case, it does not. (Fuson et al., p. 220)

For many people involved in mathematics education, estimation skills are at least as important as, and perhaps more important than, computation skills. Almost 30 years ago, Michael Girling defined *numeracy* as "the ability to use a four-function calculator sensibly" (Girling, 1977, p. 4), suggesting that the skill of being able to assess the reasonableness of an answer was much more important than being able to compute it accurately. Others feel that estimation is much less important than a solid grounding in basic mathematical knowledge and skills (Klein, 2003).

These differences become even clearer in more advanced mathematics. For example, many people need to use the standard formula for solving quadratic equations:

$$x = \frac{-b \pm \sqrt{b^2 - 4ac}}{2a}$$

Two points seem important here. The first is that, obviously, this formula needs to be memorized exactly if it is to be of any use. The second is that although some of the people who know this formula would be able to re-create it from scratch by the process of "completing the square," most would not. Professionals involved in mathematics education would probably agree that the person who could derive this formula from scratch has a deeper understanding than someone who can merely reproduce the formula from memory, but that is not to say that simply knowing this formula is not useful. Indeed, anyone who knows and can apply this formula accurately has what Richard Skemp called *instrumental understanding*. This is a different kind of understanding from knowing where the formula comes from and how to derive it (what Skemp called *relational understanding*) but it is still a form of understanding (Skemp, 1977). The crucial point here is that different users of mathematics have different needs, and that a well-grounded account of the role of assessment in mathematics education should serve them all.

Building on the work of Raymond Williams (1961), Paul Ernest (1991) identified five broad purposes for mathematics education:

> Acquiring basic mathematical skills, numeracy and social training in obedience;
>
> Learning basic skills and learning to solve practical problems with mathematics and information technology;
>
> Achieving understanding and capability in advanced mathematics, with some appreciation of mathematics;
>
> Gaining confidence, creativity, and self-expression through mathematics;
>
> Empowerment of learners as critical and mathematically literate citizens in society.

The obvious corollary of the fact that there are differences in people's perceptions of the purpose of mathematics is that those with different aims will emphasize different aspects of mathematics. For those who see the first of these five broad aims as the major purpose of mathematics education, they will value mathematics curricula that emphasize this purpose. For those who regard the last of the five as most important, they will value very different curricula. The important point here is that although competing groups can construct arguments to justify their claims (Niss, 1993), these are essentially *value* arguments. In particular, there is no way for adherents of one particular view of the purpose of mathematics education to show dissenters that they are wrong. This is why an adequate account of classroom assessment must support any and all of these conflicting views of mathematics education, rather than impose a certain set of views.

Similar arguments apply to the psychology of mathematics education. For the first half of the 20th century, the dominant view about what happens when learning takes place was that the individual creates associations between stimuli and responses. These *associationist* views of learning included the behaviorism of Skinner (for example), as well as a range of other views. Associationist models of learning explained some aspects of mathematics learning reasonably well but were unable to explain other aspects. For example, an associationist analysis of students' errors in learning multiplication facts would indicate that students' errors should be random—the result of insufficient reinforcement of particular links in chains of stimulus and response—which accords reasonably well with what one observes in practice. However, in other areas of mathematics learning, students' errors are clearly not random; in fact they are highly predictable (see examples in the section on *Eliciting Evidence*). The mounting evidence about the nonrandom nature of students' errors in mathematics led to the development of *constructivist* approaches to the study of learning, where it is acknowledged that students are active rather than passive in the development of their conceptions (see, for example, von Glasersfeld, 1991). Such theories were much better able to account for the systematicity in student errors, but there were some aspects of learning that constructivist views were unable to explain. For example, it was observed that some adults were able to perform calculations in some contexts (e.g., in supermarkets) but not others, such as classrooms (Lave, Murtaugh, & de la Roche, 1984). This idea that learning is often tied to the context in which learning takes place has proved to be particularly powerful in mathematics education. For example, Boaler (2002) found that in some mathematics classrooms, students were able to use the mathematics they had learned outside school, while in others, students were not. This suggests that theories in psychology are not like theories in, say, physics, where new theories include previous theories as special cases (for example in the way that Einstein's relativistic mechanics includes Newton's nonrelativistic mechanics as a special case).

Rather, in psychology, the tendency is for each new theory to be very good at explaining what previous theories did not, but generally not so good at explaining what the previous theories explained well. Many kinds of rote learning are explained well by associationist theories, whereas the regularities in students' errors are better explained by constructivist or information-processing theories, and situated theories explain transfer, or its absence, well. In this sense, each new theory does not replace the preceding theories but rather complements them. For different views of what mathematics students should learn, there may be different views of what happens when learning takes place, although it is important to note that no one view of learning will suffice for even the most specific learning (Sfard, 1998). For example, it might appear at first sight that learning multiplication facts would be primarily a matter of strengthening associations between stimuli and responses, but research has shown that many students "repair" gaps in their knowledge by using their knowledge of the processes of arithmetic to assemble and combine facts and routines that they can recall (VanLehn, 1990).

For the purpose of this chapter, the important point is that if it is to be useful, an adequate account of classroom assessment cannot dictate what mathematics students should learn nor should it be tied to a single view of what happens when learning takes place.

This chapter will, therefore, as far as possible, avoid putting the assessment cart before either the curriculum or psychology horse. The stance being taken is that assessment is a powerful servant but a bad master. As soon as assessment considerations are allowed to influence either what is to be learnt or what it means to learn, educators are likely to slip from making the important assessable to making the assessable important.

THE PURPOSES OF ASSESSMENT

Educational assessments are conducted in a variety of ways and their outcomes can be used for a variety of purposes. Differences exist in who decides what is to be assessed, who carries out the assessment, where the assessment takes place, how the resulting responses made by students are scored and interpreted, and what happens as a result (Black & Wiliam, 2004a). In particular, each of these can be the responsibility of those who teach the students, or, at the other extreme, all can be carried out by an external agency. Cutting across these differences, there are also differences in the *purposes* that assessments serve. Broadly, educational assessments serve three functions:

- supporting learning (formative)
- certifying the achievements or potential of individuals (summative)
- evaluating the quality of educational programs or institutions (evaluative)

Through a series of historical contingencies, a situation has developed in many countries in which the *circumstances* of the assessments have become conflated with the *purposes* of the assessment (Black & Wiliam, 2004a). So, for example, it is often widely assumed that the role of classroom assessment should be limited to supporting learning and that all assessments with which educational institutions can be held to account must be conducted by an external agency, even though in some countries, this is not the case (Black & Wiliam, 2005a).

In broad terms, moving from formative through summative to evaluative functions of assessment requires data at increasing levels of aggregation, from the individual to the institution and from specifics of particular skills and weaknesses to generalities about overall levels of performance (although of course evaluative data may still be disaggregated in order to identify specific subgroups in the population that are not making progress, or to identify particular weaknesses in students' performance in specific areas, as is the case in France—see Black & Wiliam, 2005a). However, the different functions that assessments may serve are also clearly in tension. The use of data from assessments to hold schools accountable has, in many cases, because of "teaching to the test," rendered the data almost useless for attesting to the qualities of individual students (apart, of course, from those qualities that are tested) or for supporting learning.

For similar reasons, some have argued that the uses of assessment to support learning and to certify the achievements of individuals are so fundamentally in tension that the same assessments cannot serve both functions adequately (Torrance, 1993). On the other hand, others have argued that ways must be found to integrate the two (e.g. Shavelson, Black, Wiliam, & Coffey, 2003). For the purposes of this chapter, the crucial point is that the use of assessment should support instruction in *any* assessment regime. Whether the assessment is for purposes of selection and certification, or for evaluation, whether it is conducted through teacher judgment, external assessments, or some combination of the two, classroom assessment must first be designed to support learning (see Black & Wiliam, 2004b, for a more detailed argument on this point). The remainder of this chapter considers further how this might be done.

FORMATIVE ASSESSMENT: ORIGINS AND EXAMPLES

In 1967, Michael Scriven proposed the use of the terms *formative* and *summative* to distinguish between different roles that evaluation[2] might play. On the one hand, he pointed out that evaluation "may have a role in the on-going improvement of the curriculum" (Scriven, 1967, p. 41), while in another role, evaluation "may serve to enable administrators to decide whether the entire finished curriculum, refined by use of the evaluation process in its first role, represents a sufficiently significant advance on the available alternatives to justify the expense of adoption by a school system" (pp. 41–42). He then proposed "to use the terms 'formative' and 'summative' evaluation to qualify evaluation in these roles" (p. 43).

Two years later, Bloom (1969, p. 48) applied the same distinction to classroom tests:

> Quite in contrast is the use of "formative evaluation" to provide feedback and correctives at each stage in the teaching-learning process. By formative evaluation we mean evaluation by brief tests used by teachers and students as aids in the learning process. While such tests may be graded and used as part of the judging and classificatory function of evaluation, we see much more effective use of formative evaluation if it is separated from the grading process and used primarily as an aid to teaching.

However, despite Bloom's extension of the term *formative* to apply to the evaluation of individual students (what in this chapter is termed *assessment*) as well as the evaluation of programs or institutions, for the next 30 years, the term *formative* was used almost exclusively in the context of program evaluation. Indeed, the index of the previous NCTM *Handbook of Research on Mathematics Education* lists only one mention of the term *formative* and that is in the section on "Evaluation" in the chapter on Research Methods (Romberg, 1992, p. 58).

Nevertheless, although the term *formative* was rarely used to describe teachers' assessment practices, a number of studies investigated the integration of assessment with instruction, the best known of which is probably Cognitively Guided Instruction (CGI).

In the original CGI project, a group of 21 elementary school teachers participated in a series of workshops over a 4 year period (an introductory 2½-hour workshop and a 2-day workshop before the beginning of the 1st school year, fourteen 3-hour workshops during the first year, four 2½-hour workshops and a 2-day reflection workshop in the 2nd year, and four 3-hour workshops and two 2½-hour review workshops in the 3rd year). During the workshops, the teachers were shown extracts of videotapes selected to illustrate critical aspects of children's thinking. Teachers were then prompted to reflect on what they had seen, by, for example, being challenged to relate the way a child had solved one problem to how they had solved, or might solve, other problems (Fennema et al., 1996,

[2] Here I am following the convention in American and British English that the term *assessment* applies to individuals, whereas *evaluation* applies to institutions or programs.

p. 407). Throughout the project, the teachers were encouraged to make use of the evidence they had collected about the achievement of their students to adjust their instruction to better meet their students' learning needs.

The teachers in the CGI program taught problem solving significantly more and number facts significantly less than did controls. They also knew more about individual students' problem-solving processes, and their students did better in number fact knowledge, understanding, problem solving, and confidence (Carpenter, Fennema, Peterson, Chiang, & Loef, 1989). More importantly, 4 years after the end of the program, the participating teachers were still implementing the principles of the program (Franke, Carpenter, Levi, & Fennema, 2001).

Another study that showed the substantial benefits of adapting instruction to meet student learning needs was that conducted by Bergan, Sladeczek, Schwarz, and Smith (1991). The performance of 428 Kindergarten students taught by 29 teachers implementing a measurement-and-planning system (MAPS) was compared with that of 410 students taught by 27 teachers who taught their classes as usual. In the MAPS program, teachers, together with an aide and a site manager, assessed their students' readiness for learning in reading and mathematics in the fall, and again in the spring (children were allowed as much time as they needed to complete the items). Teachers in the experimental group were trained on how to interpret the test results and provided with the *Classroom Activity Library*—a series of activities typical of early-grades instruction but keyed specifically to empirically validated developmental progressions—which they could use to individualize instruction, depending on the students' performance in the assessments. At the end of the year, 111 of the students in the control group (27%) were referred for placement in a special education program for the following year, and 80 (20%) were actually placed in special education programs. In the experimental group, only 25 students (6%) were referred, and only 6 students (1.4%) were placed in special education programs. In other words, students in the control group were 4.5 times more likely to be referred for placement in special education, and 14 times more likely actually to be placed in special education programs than students taught by teachers using the MAPS scheme. What is perhaps even more remarkable about this study is that all the schools in the study served districts with considerable socioeconomic disadvantage, and the socioeconomic status of the students in the experimental group was actually lower than that of the control group.

A third example of the use of assessment to improve student learning was a project involving a group of 24 (later expanded to 36) secondary school mathematics and science teachers in six schools in two districts in England (Black, Harrison, Lee, Marshall, & Wiliam, 2003). The work with teachers had two main components. The first was a series of eight workshops over an 18-month period (from February 1999 to June 2000). Seven of the workshops were of 5 hours duration and one was of 3 hours duration. During the workshops, the teachers were introduced to the research basis underlying how assessment can support learning (derived from Black & Wiliam, 1998a, 1998b), had the opportunity to develop their own plans, and, at later meetings, discussed with colleagues the changes they had attempted to make in their practice. The second component of the intervention with the teachers was a series of visits by researchers to the teachers' classrooms, so that the teachers could be observed implementing some of the ideas they had discussed in the workshops, had an opportunity to discuss their ideas, and could plan how they could be put into practice more effectively.

A key feature of the workshops was the development of action plans. As Perrenoud (1998) has pointed out, changing pedagogy requires teachers to re-negotiate the "learning contract" (cf. Brousseau, 1984) that they have evolved with their students, suggesting that radical changes are best effected at the beginning of a new school year. For the first 6 months of the project, therefore, the teachers were encouraged to experiment with some of the strategies and techniques suggested by the research, such as rich questioning, providing feedback to students in the form of comments rather than scores or grades, sharing learning intentions and success criteria with learners, and student peer- and self-assessment (see below). Each teacher was then asked to draw up, and later to refine, an action plan specifying which aspects of formative assessment they wished to develop in their practice and to identify a focal class with whom these strategies would be introduced in September 1999. Most of the teachers' plans contained reference to two or three important areas in their teaching where they were seeking to increase their use of formative assessment, generally followed by details of techniques that would be used to make this happen. In almost all cases the plan was given in some detail, although many teachers used phrases with meanings that differed from teacher to teacher (even within the same school).

Almost every teacher's plan contained some reference to focusing on or improving the teacher's own questioning techniques although only 11 gave details on how they were going to do this (for example using more open questions, allowing students more time to think of answers, or starting the lesson with a focal question). Others were less precise (for example stating that they intended using more sustained questioning of individuals, or improving questioning techniques in general). Some teachers mentioned planning and recording their questions. Many teachers also mentioned involving students more in setting questions (for homework, or for each other in class). Some teachers also saw existing standardized tests as a source of good questions.

Nearly half the teachers mentioned providing feedback in the form of comments rather than scores or grades, although only 6 of the teachers included it as a specific element in their action plans. Some of the teachers wanted to reduce the use of scores and grades but foresaw problems with this, given school policies on assessment. Four teachers planned for a module test to be taken before the end of the module, thus providing time for remediation.

Sharing the objectives of lessons or topics was mentioned by most of the teachers, through a variety of techniques (using a question that the students should be able to answer at the end of the lesson, stating the objectives clearly at the start of the lesson, getting the students to round up the lesson with an account of what they had learned). About half the plans included references to helping the students understand the rubrics used for investigative or exploratory work, generally using exemplars from the work of students from previous years. Exemplar material was mentioned in other contexts such as having work on display and asking students to assess work using rubrics provided by the teacher.

Almost all the teachers mentioned some form of self-assessment in their plans, ranging from using red, yellow, or green "traffic lights" to indicate the student's perception of the extent to which a topic or lesson had been understood, to strategies that encouraged self-assessment via targets that placed responsibility on students (e.g., "1 of these 20 answers is wrong: Find it and fix it!"). Traffic lights were mentioned in about half of the plans, and in practically all cases their use was combined with strategies to follow up the cases where the students signaled incomplete understanding. Several teachers mentioned their conviction that group work provided important reinforcement for students, as well as providing the teacher with insights into their students' understanding of the work.

The other component of the intervention, the visits to the schools, provided an opportunity for researchers to discuss with the teachers what they were doing, and how this related to their efforts to put their action plans into practice. The interactions were intended to be supportive rather than directive, but because researchers were frequently seen as "experts" in either mathematics or science education, there was a tendency sometimes for teachers to invest questions from a member of the project team with a particular significance, and for this reason, these discussions were often more effective when science teachers were observed by mathematics specialists, and vice-versa.

A detailed description of the qualitative changes in teachers' practices is beyond the scope of this chapter (see Black et al., 2003, for a full account). In quantitative terms, students taught by the teachers developing the use of assessment for learning outscored comparable students in the same schools by approximately 0.3 standard deviations, both on teacher-produced and external state-mandated tests (Wiliam, Lee, Harrison, & Black, 2004). Since one year's growth in mathematics as measured in the Trends in Mathematics and Science Study (TIMSS) is 0.36 standard deviations (Rodriguez, 2004, p.18), the effect of the intervention can be seen to almost double the rate of student learning.

FORMATIVE ASSESSMENT: PREVALENCE AND IMPACT

These studies show that integrating assessment with instruction is both possible and desirable. It is also valued. A sample of 580 principals ranked "Determining what needs to be re-taught after tests" as the most important in a list of 26 assessment competencies desired in their teachers (Marso & Pigge, 1993, pp. 137–138), although teachers rated competence in grading as the most important. However, there is little evidence of such data-driven practices being regularly enacted in classrooms. In a survey first published in 1980, Salmon-Cox stated that "student scores on standardized tests are not very useful to the classroom teacher" and concluded that teachers prefer to rely on their own judgment about student weaknesses and areas of needed help (Salmon-Cox, 1981 p. 631). Stiggins and Bridgeford (1985) found that although many teachers created their own assessments, "in at least a third of the structured performance assessments created by these teachers, important assessment procedures appeared not to be followed" (p. 282) and "in an average of 40% of the structured performance assessments, teachers rely on mental record-keeping" (p. 283).

McMorris and Boothroyd (1993) analyzed the quality of tests developed by seventh- and eighth-grade mathematics and science teachers and found that science teachers made greater use of multiple-choice items whereas mathematics teachers tended to set more computation items. However, for both mathematics and science teachers, the tests were of variable quality, with a significant correlation between test quality and the amount of training in educational measurement the teachers had received.

More recently Senk, Beckman, and Thompson (1997) conducted a survey of assessment practices in 19 mathematics classes in 5 high schools in 3 states. They found that formal tests and quizzes were the most frequently used assessment tools, with a nominal weight across the sample of 77%, whereas written projects and interviews accounted for a further 7%. The tests and quizzes used focused largely on "low-level" aspects of the domains assessed, and the "grading" function of assessment dominated the "assessment" function (58% of all assessments were reported in terms of a "brute grade"). The survey also found that teachers tended to ignore the results of standardized tests in arriving at terminal grades.

In a national survey, Dorr-Bremme and Herman (1986) found that students spent around 12 hours each year taking mathematics tests in elementary school (4th to 6th grade), and approximately twice that in 10th grade, although only about 6 hours was required by the state or district (pp. 16–17), and two substantial review articles, one by Natriello (1987) and the other by Crooks (1988), provided clear evidence that classroom assessment practices had substantial impact on students and their learning, although the impact was rarely beneficial. Natriello's review used a model of the assessment cycle, beginning with purposes; and moving on to the setting of tasks, criteria, and standards; evaluating performance and providing feedback; and then discussing the impact of these evaluation processes on students. His most significant point was that the vast majority of the research he cited was largely irrelevant because of weak theorization, which resulted in the conflation of key distinctions (e.g., the quality and quantity of feedback). Crooks's article had a narrower focus—the impact of assessment practices on students. He concluded that the summative function of assessment has been too dominant and that more emphasis should be given to the potential of classroom assessments to assist learning. Most importantly, assessments should emphasize the skills, knowledge, and attitudes regarded as most important, not just those that are easy to assess.

Bangert-Drowns, Kulik, Kulik, and Morgan (1991) reported the results of a meta-analysis of 40 research reports on the effects of feedback in what they called "test-like" events (e.g., evaluation questions in programmed learning materials, review tests at the end of a block of teaching). They found that providing feedback in the form of answers to the review questions was effective only when students could not look ahead to the answers before they had attempted the questions themselves (what they called "control for presearch availability," p. 213). Furthermore feedback was more effective when the feedback gave details of the correct answer, rather than simply indicating whether the student's answer was correct or incorrect. In the studies in which the students could not look ahead for the answers, and the feedback gave details of the correct answer, the mean effect size was 0.58 standard deviations. Reviews by Dempster (1991, 1992) confirm these findings, as does a review by Elshout-Mohr (1994) that reports many findings not available in English.

The difficulty of reviewing relevant research in this area was highlighted by Black and Wiliam (1998a) in their synthesis of research published since the reviews by Natriello and Crooks. Those earlier two articles had cited 91 and 241 references respectively, and yet only 9 references were common to both articles. In their own research, Black and Wiliam found that electronic searches based on keywords either generated far too many irrelevant sources or omitted key papers. In the end, they resorted to manual searches of each issue between 1987 and 1997 of 76 of the journals considered most likely to contain relevant research. Black and Wiliam's review (which cited 250 studies) found that effective use of classroom assessment yielded improvements in student achievement between 0.4 and 0.7 standard deviations.

Thirty-five years ago, Bloom suggested that

> evaluation in relation to the process of learning and teaching can have strong positive effects on the actual learning of students as well as on their motivation for the learning and their self-concept in relation to school learning. . . . [E]valuation which is directly related to the teaching-learning process as it unfolds can have highly beneficial effects on the learning of students, the instructional process of teachers, and the use of instructional materials by teachers and learners. (Bloom, 1969, p. 50)

At the time, Bloom cited no evidence in support of this claim, but it is probably safe to conclude that the question has now been settled: Attention to formative classroom assessment practices can indeed have a substantial impact on student achievement.

What is less clear is what exactly constitutes *effective* classroom assessment. Although the studies cited above indicate that assessment for learning can improve learning, several studies have found conflicting results. For example, in a study of 32 fifth-grade teachers in Germany, Helmke and Schrader (1987) found that teachers who had an accurate knowledge of their students (as measured by the teachers' ability to predict achievement test scores) were associated with higher levels of achievement *only* when the teachers also showed a high range of instructional techniques. Students taught by teachers who had a high knowledge of their students' achievement but lacked a range of instructional techniques actually performed worse than students taught by teachers who did not know their students' achievement. This study seems to indicate that collecting data if one cannot do anything with it is counterproductive.

Furthermore, even when teachers do manage to use information about student achievement to adjust or individualize their instruction, teachers may lack the ability to do so effectively. For example, in a 20-week study of 33 teachers in elementary and middle schools, Fuchs, Fuchs, Hamlett and Stecker (1991) found that teachers who received feedback on the achievement of students with learning difficulties in their classes made more adjustments to their teaching programs than teachers not given this information. However, the achievement of these students was improved *only* when this feedback was accompanied by advice from a computerized "expert system", because the teachers not given the feedback from the expert system tended to re-explain how to do problems with the same algorithms that had led to previous failure.

Therefore, if educators are to maximize the potential benefits of formative assessment or assessment for learning, there is a need to understand what, exactly, constitutes effective formative assessment, and this is the focus of the remainder of this chapter.

FORMATIVE ASSESSMENT: THEORETICAL CONSIDERATIONS

In education, the term *feedback* is routinely applied to any information that a student is given about their performance, and the generality of the term obscures that this is, in reality, a frozen metaphor derived from systems engineering. One of the earliest uses of the term was by Norbert Wiener. In 1940, Wiener and his colleagues had been working on automatic range-finders for antiaircraft guns, which involved mechanisms for predicting the path of airplanes, and realized that the control mechanisms needed for the range-finders were similar to control mechanisms in animals. He realized that purposeful action required the existence of a closed loop allowing the evaluation of the effects of one's actions and the adaptation of future conduct based on past performances (Wiener, 1948). He also realized that there were two kinds of loops: those involving positive feedback and those involving negative feedback. In both kinds of loops, there are inputs to the system and, some time later, outputs from the system. The crucial feature of the loop is that information about the output is fed back to the input side of the system. In positive feedback loops, the feedback serves to drive the system further in the direction it is already going, whereas in negative feedback loops, the feedback acts to oppose the current direction of the system. All positive feedback systems are unstable, driving the system towards either explosion or collapse. Examples of the explosive kind of positive feedback loops are simple population growth in the presence of plentiful supplies of food and in the absence of predators, and inflationary price/wage spirals in economics. Examples of collapse are economic depression, food hoarding in times of shortage, and the loss of tax revenue in urban areas as a result of "middle-class flight" (note that whether the effects are explosion or collapse, both are examples of positive feedback).

In contrast, negative feedback systems tend to produce stability, because they are inherently "self-correcting." One example of negative feedback is population growth with limited food supply, in which the lack of food causes a slowdown in population growth, which in turn, depending on the conditions, produces either an asymptotic approach towards or a damped oscillation about the carrying capacity of the environment. Another example is the domestic thermostat. When the temperature of the room drops below the setting on the thermostat, a signal is sent to turn on the furnace. When the room heats up above the setting on the thermostat, a signal is sent to turn off the furnace.

The foregoing discussion clearly shows that the current uses of the term feedback in education are very different from those in engineering, but, more importantly, the simplistic engineering metaphor may just not be helpful. In systems engineering, negative feedback is good, because it keeps a system under control whereas positive feedback is bad because it leads to explosion or collapse. In educational settings, things are not so clear-cut. Negative feedback may be helpful for correcting learning when it is off-course, but feedback that reinforces learning that is on track is also powerful. And the thermostat does not care how often it is told that the temperature in the room is "wrong" and needs to be corrected, whereas humans are often adversely affected by such information. This is not to say that the insights of systems engineering are irrelevant, but educators should exercise considerable caution in adopting what are essentially metaphors in complex areas like human learning, especially in terms of the relationship between *feedback* and *formative assessment*.

In the United States, the term *formative assessment* is often used to describe assessments that are used to provide information on the likely performance of students on state-mandated tests—a usage that might be described better as "early-warning summative." In other contexts, the term is used to describe any feedback given to students, no matter what use is made of it, such as telling students which items they got correct and incorrect (sometimes termed *knowledge of results*). These kinds of usages suggest that the distinction between *formative* and *summative* applies to the assessments themselves, but inasmuch as the same assessment can be used both formatively and summatively, these terms are more usefully applied to the *use* to which the information arising from assessments is put.

As Ramaprasad (1983) noted, the defining feature of feedback is that the information generated within the system must have some effect. Information that does not have the capability to change the performance of the system is not feedback: "Feedback is information about the gap between the actual level and the reference level of a system parameter which is used to alter the gap in some way" (Ramaprasad, 1983, p. 4).

Commenting on this, Sadler (1989) noted

> An important feature of Ramaprasad's definition is that information about the gap between actual and reference levels is considered as feedback *only when it is used to alter the gap*. If the information is simply recorded, passed to a third party who lacks either the knowledge or the power to change the outcome, or is too deeply coded (for example, as a summary grade given by the teacher) to lead to appropriate action, the control loop cannot be closed, and "dangling data" substituted for effective feedback. (p. 121)

In this view, formative assessments (or feedback, in Ramaprasad's terminology) cannot be separated from their instructional consequences, and assessments are formative only to the extent that they impact learning (for an extended discussion on consequences as the key part of the validity of formative assessments, see Wiliam & Black, 1996). Therefore what is important is not the intent behind the assessment, but the function it actually serves, and this provides a useful point of difference between the terms *assessment for learning* and *formative assessment.* Black, Harrison, Lee Marshall, and Wiliam (2004, p. 8) distinguished between the two as follows:

> Assessment for learning is any assessment for which the first priority in its design and practice is to serve the purpose of promoting pupils' learning. It thus differs from assessment designed primarily to serve the purposes of accountability, or of ranking, or of certifying competence. An assessment activity can help learning if it provides information to be used as feedback, by teachers, and by their pupils, in assessing themselves and each other, to modify the teaching and learning activities in which they are engaged. Such assessment becomes "formative assessment" when the evidence is actually used to adapt the teaching work to meet learning needs.

For the purpose of this chapter, then, the qualifier *formative* will refer not to an assessment, nor even to the purpose of an assessment, but the function it actually serves. An assessment is formative to the extent that information from the assessment is fed back within the system and actually used to improve the performance of the system in some way (i.e. that the assessment *forms* the direction of the improvement).

So, for example, if a student is told that she needs to work harder, and does work harder as a result, and consequently does indeed make improvements in her performance, this would *not* be formative. The feedback would be *causal,* in that it did trigger the improvement in performance, but not *formative,* because decisions about *how* to "work harder" were left to the student. Telling students to "give more detail" might be formative, but only if the students knew what "giving more detail" meant (which is unlikely, because if they knew what detail was required, they would probably have provided it on the first occasion). Similarly, a "formative assessment" that predicts which students are likely to fail the forthcoming state-mandated test is not formative unless the information from the test can be used to improve the quality of the learning within the system. To be formative, feedback needs to contain an implicit or explicit recipe for future action.

Another way of thinking about the distinction being made here is in terms of monitoring assessments, diagnostic assessments, and formative assessments. An assessment *monitors* learning to the extent that it provides information about whether the student, class, school, or system is learning or not; it is *diagnostic* to the extent that it provides information about what is going wrong; and it is *formative* to the extent that it provides information about what to do about it. A sporting metaphor may be helpful here. Consider a young fast-pitch softballer who has an earned-run-average of 10 (for readers who know nothing about softball, that is not good). This is the *monitoring* assessment. Analysis of what she is doing shows that she is trying to pitch a rising fastball (i.e., one that actually rises as it gets near the plate, due to the back-spin applied), but that this ball is not rising and therefore becomes an ordinary fastball in the middle of the strike zone, which is very easy for the batter to hit. This is the *diagnostic* assessment, but it is of little help to the pitcher, because she already knows that her rising fastball is not rising, which is why she is giving up a lot of runs. However, if a pitching coach is able to see that she is not dropping her shoulder sufficiently to allow her to deliver the pitch from below the knee, then this assessment has the potential to be not just diagnostic, but *formative*. If the athlete is able to use the advice about delivering the pitch from below the knee to make her rising fastball rise, then the feedback given by the coach will, indeed, have been formative. This use of *formative* recalls the original meaning of the term. In the same way that one's formative experiences are the experiences that shape the individual, formative assessments are those that shape learning.

The important point here is that not all diagnoses are *instructionally tractable*—an assessment can accurately diagnose what needs attention without indicating what needs to be done to address the issue. To promote learning, one must collect the correct data in the first place. This is discussed further in the section on *Clarifying and Sharing Learning Intentions* below.

In the examples given above, the action follows quite quickly on from the elicitation of the evidence about student achievement, but the definition of formative assessment given above allows cycles of elicitation, interpretation and action of any length, provided the information is used to form the direction of future learning. For example, in March, a mathematics supervisor may be planning the workshops she will make available to teachers in the summer. By looking at the scores obtained by the students in the district on last year's state-mandated tests, she might discover that, compared to other students in the state, the students are performing relatively poorly on items that assess geometry. As a result, she might plan a series of workshops on teaching geometry for the summer, and, if these workshops are successful, then the students' performance on geometry items should improve. This cycle would be over 2 years in length, given that the supervisor would be using data from tests taken the previous March, and the impact would not be felt until the students took the test the following March (and the results might not be available until July). Furthermore, the data that were used to improve learning in the district were not collected from the students who benefited, but those who were in that grade 2 years earlier. Nevertheless, data from the state tests functioned formatively in this example, since information about student performance was used to make adjustments to instruction (in this case improving the teaching of geometry) that improved the learning of mathematics in the district.

It could be argued that this kind of usage is *evaluation*, rather than *assessment*, because it is the program that is being improved through the analysis of student data, but there is no clear boundary between the two. A teacher might look through the responses of her students to a trial run of a state test and replan the topics that she is going to teach in the time remaining until the test. Such a test could still be useful as little as a week or two before the state-mandated test, as long as there is time to use the information to redirect the teaching. Again this assessment would be formative as long as the information from the test was actually used to adapt the teaching, and in particular, not only telling the teacher which topics need to be retaught, but also suggesting what kinds of reteaching might produce better results. At this level, both the program and the learning of individual students on whom the data were collected is being impacted by the assessment outcomes, so this example could be thought of as either assessment or evaluation. The crucial point is that the evidence is used to make decisions that could not be made, or at least could not be made as well, without that evidence, with the result that learning is enhanced.

The building-in of time to make use of assessment data is a central feature of much elementary and middle school teaching in Japan. A teaching unit is typically allocated 14 lessons, but the content usually occupies only 10 or 11 of the lessons, allowing time for a short test to be given in the 12th lesson, and for the teacher to use Lessons 13 and 14 to reteach aspects of the unit that were not well understood.

Another example, on an even shorter time-scale, is the use of *exit passes* from a lesson. The idea here is that before leaving a classroom, each student must compose an answer to a question that goes to the heart of the concept being taught at the end of the lesson. On a lesson on probability for example, such a question might be, "Why can't a probability be greater than 1?" Once the students have left, the teacher can look at the students' responses and make appropriate adjustments in the plan for the next period of instruction. The shortest feedback loops are those involved in the day-to-day classroom practices of teachers, where teachers adjust their teaching in light of students' responses to questions or other prompts in real time. The key point in all this is that the length of the feedback loop should be tailored according to the ability of the system to react to the feedback.

All this suggests that the conflicting uses of the term *formative assessment* can be reconciled by recognizing that almost any assessment can be formative provided it is used to make instructional adjustments, but a crucial difference between different assessments is the length of the adjustment cycle. Table 23.1 provides a terminology for the different lengths of cycles.

Table 23.1 Cycle Lengths for Formative Assessment

Type	Focus	Length
Long-cycle	across marking periods, quarters, semesters, years	4 weeks to 1 year
Medium-cycle	within and between instructional units	1 to 4 weeks
Short-cycle	within and between lessons	
day-by-day		24 to 48 hours
minute-by-minute		5 seconds to 2 hours

The foregoing discussion establishes that any assessment can be formative, and that an assessment is formative to the extent that information from the assessment is used to adjust instruction to better meet student learning needs. The adjustment can take place immediately, in time for the next instructional episode, between units, or even between years, and the beneficiaries of the adjustments may or may not be the students on whom information was collected. However, although all these different uses of assessment information may be formative, in that the information is used to adapt instruction to better meet student learning needs, the research evidence cited above suggests that not all are equally effective. To establish what, exactly, constitutes the most effective form of formative assessment requires a deeper look at the research evidence.

An important feature of Ramaprasad's definition of feedback (here *formative assessment*) is that it draws attention to three key instructional processes:

- establishing where the learners are in their learning,
- establishing where they are going,
- establishing what needs to be done to get them there.

Traditionally, this may have been seen as primarily the teacher's role, but one must also take account of the role that the learners themselves, and their peers, play in these processes. Crossing the three key instructional processes listed above with the three agents involved in the classroom produces the framework shown in Figure 23.1, which provides a way of thinking about the key strategies involved in formative assessment. The subject classroom that is the focus of Figure 23.1 is, of course, itself nested within a school, which in turn is located in a community, and so on. Any adequate account of formative assessment will have to acknowledge these multiple contexts, but they are beyond the scope of this chapter. Furthermore, given that the stance taken in this chapter is that, ultimately, assessment must feed into actions in the classroom in order to affect learning, this simplification seems reasonable, at least as a first-order approximation. For examples of sociocultural approaches to the implementation of formative assessment, see Black and Wiliam (2005b) and Pryor and Crossouard (2005).

This framework suggests that assessment for learning can be conceptualized as consisting of five key strategies and one "big idea" (Wiliam & Thompson, in press). The five key strategies are (a) engineering effective classroom discussions, questions, and learning tasks that elicit evidence of learning; (b) providing feedback that moves learners forward; (c) clarifying and sharing learning intentions and criteria for suc-

	Where the learner is going	Where the learner is right now	How to get there
Teacher	Clarifying learning intentions and sharing and criteria for success	Engineering effective classroom discussions and tasks that elicit evidence of learning	Providing feedback that moves learners forward
Peer	Understanding and sharing learning intentions and criteria for success	Activating students as instructional resources for one another	
Learner	Understanding learning intentions and criteria for success	Activating students as the owners of their own learning	

Figure 23.1 Aspects of assessment for learning.

cess; (d) activating students as the owners of their own learning; and (e) activating students as instructional resources for one another.

The big idea is that evidence about student learning is used to adjust instruction to better meet student needs—in other words that teaching is *adaptive* to the students' learning needs. The following sections describe in more detail these five key strategies and how they fit together within the more general idea of the "regulation of learning processes."

ENGINEERING EFFECTIVE CLASSROOM DISCUSSIONS, QUESTIONS, AND LEARNING TASKS THAT ELICIT EVIDENCE OF LEARNING

Teachers can use a vast range of strategies to elicit evidence of student learning, from formal testing occasions, through the activities that students routinely undertake in mathematics classrooms, to classroom discussions and informal exchanges with students. In the previous version of this *Handbook*, Webb (1992) outlined a set of principles for assessing mathematics. The chapters by Jan de Lange and Linda Dager Wilson in this volume cover large-scale and high-stakes assessment, respectively, and Clarke (1996) provides an international perspective on contemporary mathematics assessment. In addition, van den Heuvel-Panhuizen and Becker (2003) made proposals for a didactic model of assessment design in mathematics, and an excellent overview of the characteristics of and general principles for the design of "thought-revealing activities" can be found in Lesh, Hoover, Hole, Kelly, and Post (2001). The focus of this section is on classroom discussions and tasks that elicit evidence of learning.

Classroom Discussions

Teacher-led classroom discussion, along with individual seatwork, is one of the staples of mathematics instruction in the United States (and indeed most other countries). In the traditional model of classroom transactions—termed *Initiation-Response-Evaluation* or *I-R-E* by Mehan (1979)—the teacher asks a question, chooses a student to answer the question, and then makes some response to the student's answer. Within this broad structure one can say rather more.

First, it is important to note that the teacher almost invariably dominates such classroom discussions, even in the United States, where it has been shown that American teachers actually talk less than teachers in countries that are more successful in mathematics, at least as measured in international comparisons. For example, the 1999 TIMSS video study found that in U.S. classrooms there were 8 teacher words for every student word, whereas in Japan, there were 13, and in Hong Kong 16 teacher words for every student word (Hiebert et al., 2003). Of course, this means that even in a U.S. classroom of 25 students, the teacher speaks 200 times as much as any one student. It also makes clear that the quality of what the teacher is saying is much more important than the quantity, and here the evidence from the TIMSS video studies is compelling.

Although the 1999 TIMSS video study focused on eighth grade, the findings were similar to those of earlier studies (e.g., Weiss, Pasley, Smith, Banilower & Heck, 2003; Rowan, Harrison, & Hayes, 2004) and showed that mathematics teaching in the U.S. "is characterized by frequent review of relatively unchallenging, procedurally oriented mathematics" (Hiebert et al., 2005, p. 125). Why this should be the case is unclear, but the consequences are profound. To see why, it is necessary to discuss briefly some recent work on the nature and development of human abilities.

The term *intelligence* has been used in a variety of ways for at least 100 years, and there is no consensus on its meaning today. The word has been tarnished by its use by the eugenics movement in the first half of the 20th century, especially in the United States (see Selden, 1999, for an extended discussion), and more recently, the term has been used in essentially racist projects such as *The Bell Curve* by Herrnstein and Murray (1994). There is ample evidence that the conclusions reached by such authors is fundamentally flawed (see, e.g., Fish, 2001; Montague, 1999) but the intensity and the politicization of the debate makes it difficult to separate the science from the myth. One particularly unfortunate effect of this debate is that it has led many people to adopt rigid and extreme positions on the issue of intelligence.

At one extreme are those who believe that

- intelligence is determined entirely by one's genes and is fixed for life;
- intelligence tests measure the most important aspects of human thinking;
- intelligence is the most important predictor of success in the workplace; and
- other kinds of ability do not really matter.

At the other extreme, partly in response to the political motives of those who hold the extreme views listed above, are those who reject the concept of intelligence entirely or deny its relevance except in the most limited laboratory studies. For these proponents

- intelligence is determined by the environment and not one's genes;

- intelligence tests measure only the ability to take intelligence tests;
- intelligence does not matter in the real world; and
- there are several different kinds of intelligence, all independent of one another

It turns out that there *is* a high degree of consensus amongst psychologists on the science underlying intelligence, and, predictably, it is somewhere between the two extreme positions described above:

- intelligence is determined by both environment and genetics, and the genetic influence is substantial (Plomin & Petrill, 1997);
- intelligence tests correlate strongly with a range of other measurements of mental capability (Mackintosh, 2000);
- intelligence is strongly associated with success in a wide range of real world activities (Bertua, Anderson, & Salgado, 2006);
- there are several different aspects of intelligence, but most of them are strongly inter-related (Deary, 2000).

In this sense, intelligence seems to be like physical height. Physical height is inherited—tall parents tend to have tall children—but there is also a strong environmental component. The better the standard of nutrition, the taller the child grows. The same is clearly true for intelligence, most clearly evidenced by the *Flynn effect*. Named after James Flynn, this refers to the rise of about one standard deviation that has been observed in IQ scores throughout the developed world over the last 60 years (Flynn, 1984, 1987). Although researchers disagree about the causes of these gains (see, e.g., Neisser, 1998), students' educational experiences are clearly an important element.

Dickens and Flynn (2001) have shown that what is observed about intelligence is best accounted for by the idea that people select, or have selected for them, environments that match their intelligence. People with high intelligence, for example, engage in more of the activities that enhance intelligence, and so become more intelligent whereas people with lower intelligence opt out of or are denied these intelligence-enhancing activities and so lose the opportunities to enhance their intelligence. This suggests that intelligence and environment are mutually constitutive of each other: Environment causes intelligence and intelligence causes environment. However, the model proposed by Dickens and Flynn also suggests that the impact of transient improvements in environment are themselves transient. This explains why compensatory preschool programs have significant effects on intelligence while students are in the program, but the effects diminish when students leave (Barnett & Camilli, 2001), although it should be noted that the improvements in student *achievement* produced by such programs are lasting.

One concrete demonstration on the power of classroom environments to improve student achievement is a study of 191 students in 7 fourth grade classrooms following the "Thinking Together" program (Dawes, Mercer, & Wegerif, 2000) as they learned science. These students outperformed controls in similar schools by 0.74 standard deviations on tests of concept mapping and 0.29 standard deviations on a standardized science achievement test (Mercer, Dawes, Wegerif, & Sams, 2004). Perhaps even more surprisingly, these students outperformed controls by 0.27 standard deviations on a purely spatial intelligence test—Raven's Progressive Matrices (Raven, 1960). Just by increasing the amount of structured talk in the classroom, students improved at performing purely spatial tasks.

The implications of these findings for mathematics classrooms are profound. To maximize mathematics achievement, classrooms must be places where every student is required to engage cognitively with the mathematics she or he is learning, and this appears to be the hallmark of practice both in countries that are successful and exemplary classrooms in the United States (see, e.g., Boaler & Humphreys, 2005). Shulman (2005) has described such cognitively rich pedagogies as *pedagogies of engagement*.

The problem is that, as is clear from the TIMSS video studies (Hiebert et al., 2003), few mathematics classrooms in the United States rigorously employ pedagogies of engagement. In most American mathematics classrooms, the choice about whether to engage is left to the student. In classroom dialogue, the teacher asks a question and then selects a respondent from those who have signaled that they have an answer by raising their hands. Some teachers do try to counter this by occasionally calling on students who have not raised their hands, but this is frequently seen as breaching the terms of the *didactical contract* (Brousseau, 1984). One teacher summed up his predicament thus:

> I'd become dissatisfied with the closed Q&A style that my unthinking teaching had fallen into, and I would frequently be lazy in my acceptance of right answers and sometimes even tacit complicity with a class to make sure none of us had to work too hard. . . . They and I knew that if the Q&A wasn't going smoothly, I'd change the question, answer it myself or only seek answers from the "brighter students." There must have been times (still are?) where an outside observer would see my lessons as a small discussion group

> surrounded by many sleepy onlookers. (Black et al., 2004, p. 11)

The consequence of this is that some students are deeply engaged in the lesson and, as such, are increasing their capabilities, while others are avoiding engagement and thus forgoing the opportunities to increase their ability.

In other classrooms, participation is not voluntary. Magdalene Lampert (2001) described a lesson she taught in which she made a point of calling on a student who had not raised his hand even though many others had done so. She gave her reasons as follows: "I called on Richard because I wanted to teach him and others in the class that everyone would indeed be asked to explain their thinking publicly. I also wanted to teach everyone that what they said would be expected to be an effort to make mathematical sense" (p. 146).

Leahy, Lyon, Thompson, and Wiliam (2005) described mathematics classrooms in which teachers have gone further and instituted a rule of "no hands up, except to *ask* a question." After posing a question, the teacher decides which student should respond by the use of some randomizing device, such as name cards (see Webb, 2004, p. 175) or a beaker of Popsicle sticks on which the students' names are written. The important point about such classrooms is that mental participation is not optional.

Such a radical change in the "classroom contract" (Brousseau, 1984) may be unwelcome for many students who are accustomed to classrooms where participation is optional, but there is evidence that students' participation practices in mathematics classrooms are malleable (Turner & Patrick, 2004). In particular, there are many strategies that teachers can use to engage students in classroom participation. Where students reply "I don't know," Ellin Keene suggests that teachers can ask, "OK, but if you did know what would you say?" (Carol, 2006), or the teacher can solicit answers from other students and then return to the original student and ask them to select from amongst the answers they have heard. Other possibilities are allowing students to "phone a friend," or, for multiple-choice items, they can "ask the audience" or ask to go "50–50" where two incorrect responses are removed. All these strategies derive their power from the fact that classroom participation is not optional, and even when the student resists, the teacher looks for ways to maintain the student's engagement.

How much time a teacher allows a student to respond before evaluating the response is also important. It is well known that teachers do not allow students much time to answer questions (Rowe, 1974), and if they do not receive a response quickly, they will "help" the student by providing a clue or weakening the question in some way, or even moving on to another student. However, what is not widely appreciated is that the amount of time between the student providing an answer and the teacher's evaluation of that answer is just as, if not more, important. Of course, where the question is a simple matter of factual recall, then allowing a student time to reflect and expand upon the answer is unlikely to help much. But where the question requires thought, then increasing the time between the end of the student's answer and the teacher's evaluation from the average wait-time of less than a second to 3 seconds produces measurable increases in learning, although according to Tobin (1987) increases beyond 3 seconds have little effect, and may cause lessons to lose pace.

In fact, questions need not always come from the teacher. There is substantial evidence that getting students to generate their own questions enhances their learning. Rosenshine, Meister, and Chapman (1996) found that training students to generate questions while reading increased performance by 0.36 standard deviations on standardized tests, and by 0.86 standard deviations on tests developed by the experimenters. Foos, Mora, and Tkacz (1994) compared four strategies for helping students prepare for a test or examination: (a) tell the students to review the material on which the test is based, (b) provide the students with study methods and materials, (c) tell the students to generate their own study outlines, and (d) tell the students to generate their own study questions, with answers. They found that strategy (d) generated the highest performance, followed, in turn, by (c), (b), and (a). Such student-produced assessments can also be useful to the teacher, because they provide information about what the students think they have been learning, which may not be the same as what the teacher thinks the students have been learning.

Some researchers have gone even further and shown that questions can limit classroom discourse, because they tend to demand a simple answer. There is a substantial body of evidence that classroom learning is enhanced considerably by shifting from asking questions to making statements (Dillon, 1988). For example, instead of asking, "Are all squares rectangles?" which seems to require a simple yes/no answer, the level of classroom discourse (and student learning) is improved considerably by framing the same question as a statement—"All squares are rectangles"—and asking students to discuss this in small groups before presenting a reasoned assessment of the truth of this statement to the class.

Another key feature of classroom questioning is the way that teachers listen to student answers. As many authors have pointed out, when teachers listen to student responses, they attend more to the correctness of the answers rather than what they can learn about the student's understanding (Even & Tirosh, 1995, 2002; Heid, Blume, Zbiek, & Edwards, 1999). In a detailed study, Davis (1997) followed the changes in the practice of one middle school mathematics teacher, focusing in particular on how the teacher responded to student answers. Initially, the teacher's reactions tended to focus on the extent to which the student responses accorded with the teacher's expectations. After sustained reflection and discussion with the researcher over a period of several months, the teacher's reaction placed increasing emphasis on "information-seeking" as opposed to the "response-seeking" that characterized the earlier lessons. Davis termed these two kinds of listening *evaluative listening* and *interpretive listening* respectively. Towards the end of the 2-year period, there was a further shift in the teacher's practice, with a marked move away from clear lesson structures and prespecified learning outcomes, and towards the exploration of potentially rich mathematical situations, in which the teacher is a coparticipant. Most notably, in this third phase, the teacher's own views of the subject matter being taught developed and altered along with that of the students (what Davis termed *hermeneutic listening*). Similar trajectories of change from evaluative to interpretive listening have been observed in other mathematics teachers (English & Doerr, 2004) and in pre-service teachers (Crespo, 2000), as eloquently summarized by a girl in the seventh grade: "When Miss used to ask a question, she used to be interested in the right answer. Now she's interested in what we think" (Hodgen & Wiliam, 2006, p. 16).

What Makes a Good Question?

Careful analysis of students' incorrect responses to standard classroom items can reveal important insights into students' conceptions. For example, De-Corte and Verschaffel (2006) point out that when students are asked to respond to the following item:

$$\underline{\qquad} - 12 = 7$$

many students write 18 or 5 in the blank space. Both responses are, of course, incorrect, but in completely different ways, the first probably indicating an arithmetical slip, and the second more likely indicating a lack of understanding of the meaning of the equal sign.

However, understanding the thinking behind students' responses is more straightforward when the questions asked of students have been planned carefully ahead of time expressly for this purpose. Two items used in TIMSS, shown in Figure 23.2 below, illustrate one important feature of the design of good questions. Although apparently quite similar, the success rates on the two items were very different. For example, in Israel, 88% of the students answered the first item correctly, whereas only 46% answered the second correctly, with 39% choosing response (b) (Vinner, 1997). Vinner suggested that the reason for this is that many students, in learning about fractions, develop the naive conception that the largest fraction is the one with the smallest denominator, and the smallest fraction is the one with the largest denominator. This approach leads to the correct answer for the first item but leads to an incorrect response to the second. Further evidence for this interpretation is given by noting that 46% plus 39% is very close to 88%, suggesting that almost half of the students who answered the first item correctly did so with an incorrect strategy. In this sense, the first item is a much weaker item than the second, because students can get it right for the wrong reasons.

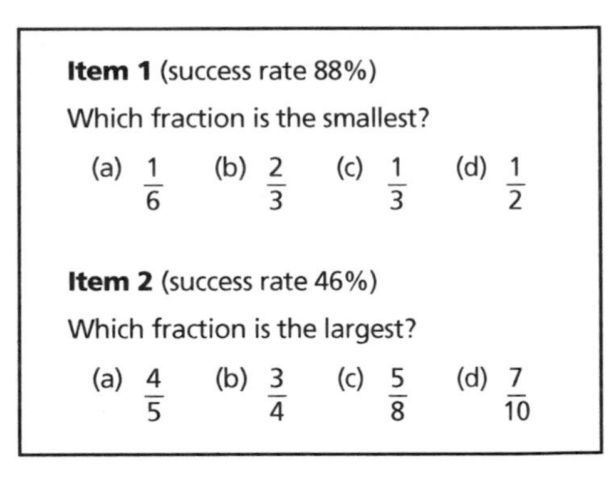

Figure 23.2 Two items from the Third International Mathematics and Science Study.

This example illustrates a very general principle in teachers' classroom questioning. By asking questions of students, teachers try to establish whether students have understood what they are meant to be learning. In other words, the teacher is trying to construct a model of the student's thinking. As von Glasersfeld (1987) noted:

> Inevitably, that model will be constructed, not out of the child's conceptual elements, but out of the conceptual elements that are the interviewer's own. It is in this context that the epistemological principle of *fit*, rather than *match* is of crucial importance. Just as cognitive organisms can never compare their conceptual organisations of experience with the structure of an independent objective reality, so the interviewer, experimenter, or teacher can never compare the model he or she has constructed of a child's conceptualisations with what actually goes on in the child's head. In the one case as in the other, the best that can

be achieved is a model that remains viable within the range of available experience. (p. 13)

The key phrase here is "within the range of available experience." If the teacher asks questions that are more like the first TIMSS item above than the second, the range of available experience is narrow, and the range of models that fit, correspondingly large. As Gay and Thomas (1993) pointedly asked, "Just because they got it right, does that mean they know it?" (p. 130).

For teacher questioning to be effective, teachers need to know what kinds of conceptualizations students are likely to have, and need tools to identify them. Consider, for example, the following pair of simultaneous equations:

$$3a = 24$$

$$a + b = 16$$

Many students find this difficult, saying that it cannot be done. The teacher might conclude that they need some more help with equations of this sort, but with this particular pair of equations, a more likely reason for the difficulty is not with mathematical skills but with the student's *beliefs* (see also Schoenfeld, 1985). If the students are encouraged to talk about their difficulty, they often say things like, "I keep on getting b is 8, but it can't be because a is." The reason that many students have developed such a belief is, of course, that before they were introduced to solving equations, they had been practicing substitution of numbers into algebraic formulas, where each letter always did stand for a different number. Although the students will not have been taught that each letter must stand for a different number, they have generalized implicit rules from their previous experience, just as because they are always shown triangles where the lowest side is horizontal, they talk of "upside-down triangles" (Askew & Wiliam, 1995 p. 14).

The important point here is that one would not have known about these unintended conceptions if the second equation had been $a + b = 17$ instead of $a + b = 16$. Items that reveal unintended conceptions—in other words that provide a "window into thinking"—are not easy to generate, but they are crucially important to improve the quality of students' mathematical learning.

Some people have argued that these unintended conceptions are the result of poor teaching. If only the teacher had phrased the explanation more carefully, had ensured that no unintended features were learned alongside the intended features, then these misconceptions would not arise, but this argument fails to acknowledge two important points. The first is that this kind of overgeneralization is a fundamental feature of human thinking. When young children say things like "I spended all my money," they are demonstrating a remarkable feat of generalization. From the huge messiness of the language that they hear around them, they have learned that to create the past tense of a verb, one adds *d* or *ed*. In the same way, if one asks young children what causes the wind, a common answer is "trees." They have not been taught this but have observed that trees are swaying when the wind is blowing and have inferred (incorrectly in this case) a causal relationship from a correlation.

The second point is that even if one wanted to, one cannot control the student's environment to the extent necessary for unintended conceptions not to arise. For example, it is well known that many students believe that the result of multiplying 2.3 by 10 is 2.30. It is highly unlikely that they have been taught this. Rather this belief arises as a result of observing regularities in what they see around them. The result of multiplying whole numbers by 10 is just to add a zero, so why should that not work for all numbers? The only way to prevent students from acquiring this (mis)conception would be to introduce decimals before one introduces multiplying single-digit numbers by 10, which is clearly absurd. The important point is that one must acknowledge that what students learn is not necessarily what the teacher intended, and it is essential that teachers explore students' thinking before assuming that students have "understood" something.

Questions that provide this "window into thinking" are hard to find, but within any school there will be good selection of rich questions in use—the trouble is that each teacher will have her or his stock of good questions, but these questions are not shared within the school and are certainly not seen as central to good teaching. In most Anglophone countries, teachers spend the majority of their lesson preparation time in grading students' notebooks or assignments, almost invariably doing so alone. In some other countries, the majority of lesson preparation time is spent planning how new topics can be introduced, which contexts and examples will be used, and so on. This is sometimes done individually or with groups of teachers working together. In Japan, however, teachers spend a substantial proportion of their lesson preparation time working together to devise questions to use in order to find out whether their teaching has been successful, in particular through the process known as *lesson study* (Fernandez & Makoto, 2004).

In generating questions, the traditional concerns of reliability and validity do not provide sound guidance as to what makes a good question. For example, many teachers think that the following question, taken from the Chelsea Diagnostic Test for Algebra (Hart, Brown, Kerslake, Küchemann, & Ruddock, 1985), is "unfair":

Simplify (if possible): 2a + 5b

This item is deemed unfair because students "know" that in answering test questions, one must do some work, so it must be possible to simplify this expression, otherwise the teacher would not have asked the question. To use this item in a test or an examination where the goal is to determine a student's achievement would probably not be a good idea. But to find out whether students understand algebra, it is a very good item indeed. If in the context of classroom work, rather than a formal test or exam, a student can be tempted to simplify 2a + 5b then the teacher should want to know that, because it means that the student has not yet developed a real sense of what algebra is about.

Asking students which of the following two fractions is larger raises similar issues:

$$\frac{3}{7} \quad \frac{3}{11}$$

In some senses this is a "trick question." There is no doubt that this is a very hard item, with typically only around one 14-year-old in six able to give the correct answer, compared with around 75% of 14-year-olds being able to select correctly the larger of two "typical" fractions—that is where the numerators and denominators are different, and each less than 12 (Hart, 1981). It may not, therefore, be a very good item to use in a test of students' achievement. But it is very important for the teacher to know if her students think that __ is larger than __. The fact that this item is seen as a trick question shows how deeply the summative function of assessment is ingrained into the practice of most teachers.

A third example, which caused considerable disquiet amongst teachers when it was used in a national test in England in the 1990s, is based on the following item, again taken from one of the Chelsea Diagnostic Tests:

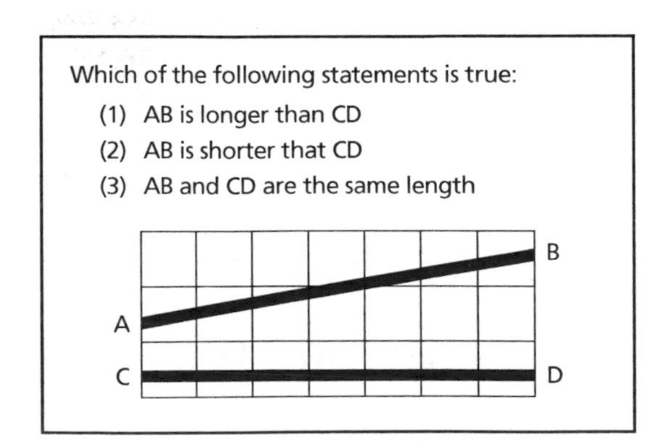

Figure 23.3 Item adapted from the CSMS tests (Hart et al., 1985).

Again, viewed in terms of formal tests and examinations, this may be an unfair item, but in terms of a teacher's need to establish secure foundations for future learning, it would seem to be entirely appropriate.

All of the questions discussed so far have the potential to elicit evidence of active student conceptions that may hinder learning, and as such, have the potential to support teachers' instructional decision-making. However, the questions discussed above, and most of the questions used by teachers, are really valuable only when the teacher can ask students to explain or elaborate their answers. Such questions are therefore good as *discussion questions* but less good as *diagnostic questions* (Ciofalo & Wylie, 2006). If it is necessary for each student to explain the reason for her or his answer, then it will take a great deal of time to get around the whole class. Furthermore, those students answering last are more likely to be choosing their justifications from amongst those already given by their peers than thinking hard about their own answers.

Consider the item shown in Figure 23.4. This item has some merit as a discussion item, because it is likely to elicit very quickly the fact that some students in a class think that $a^2 + b^2 = c^2$ for this triangle because it is true for all right triangles. However, as soon as students hear a peer mention that $a^2 + b^2 = c^2$ is true only when c is the hypotenuse, then they may well give the same response.

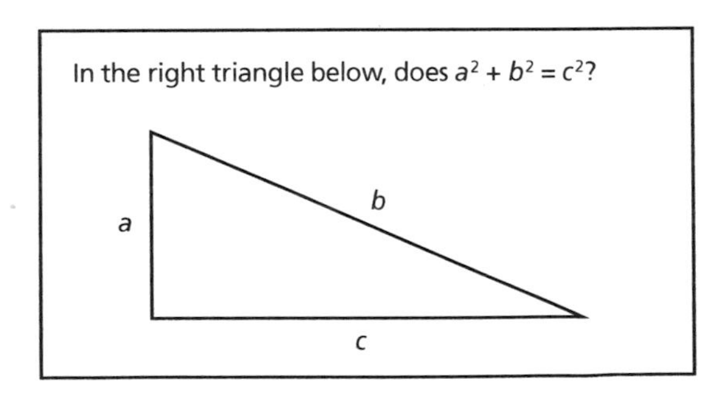

In the right triangle below, does $a^2 + b^2 = c^2$?
a
b
c

Figure 23.4 Item on Pythagoras's rule.

As an alternative, consider the item shown in Figure 23.5. For this item, there are 64 possible responses (the statement might be true for each of the six triangles giving a total of 2^6 possibilities), so the chance of a student getting this correct by guessing is less than 2%. Moreover, the item can be used to get a response from every single student in the class at the same time, by giving students a set of six cards labeled A, B, C, D, E, and F. This use of questions with an *all-student response system* (Leahy et al., 2005) provides teachers with much richer data on the level of understanding in a class than is possible with a *single-student response*

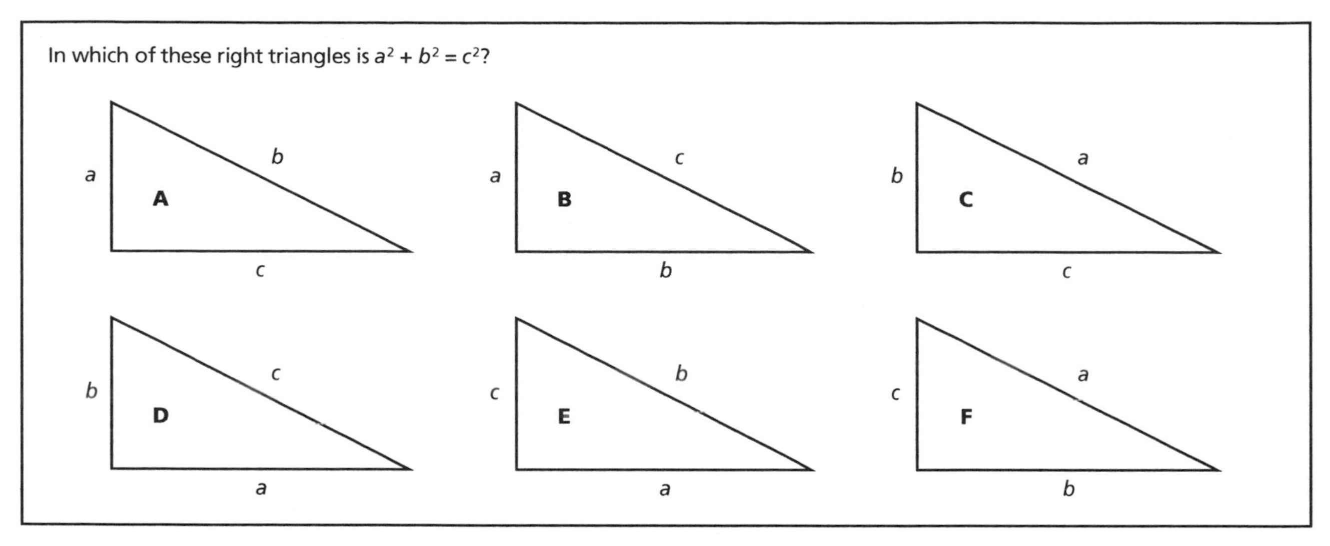

Figure 23.5 Item on Pythagoras's rule for use as an all-student response item.

system such as asking students to raise their hands, or selecting a student at random to answer.

By moving from single-student response systems to all-student response systems, the teacher is able to make more effective instructional decisions in real time by creating hinge-points in the lesson (Leahy et al., 2005). If all students answer correctly, then she can move on. If not, then the teacher has created a "teachable moment." If no one answered correctly, then the teacher might choose to reteach the material to the whole class using a different approach from the one that she used originally. However, if some of the students answer correctly and some answer incorrectly, this provides an opportunity for a class discussion of the issue. Not only does the teacher know that some students have not understood, but she is able to use the information about the responses of each student to engineer a more effective discussion, by, for example, calling on all those with a particular response to supply their reasons for that response before considering other responses. Of course the aggregation of information from student responses is greatly enhanced by the use of electronic "clickers" (Roschelle, Abrahamson, & Penuel, 2004) but note that most such systems can currently cope with only a single response, so that items such as that in Figure 23.5 would be much more difficult to use. Incorrect responses to these items may reveal "facets" of student thinking (diSessa & Minstrell, 1998) and thus provide important evidence about student learning. In this sense, it is useful if the incorrect responses made by students are interpretable. However, if assuming that students do understand something when they don't is more damaging than assuming they do not understand something when they do, then a more important requirement for teaching is that that the *correct* response is interpretable. In other words, the teacher needs to be sure that the student got the answer right for the right reasons.

Sets of lettered cards work well with multiple-choice items, but not all items can usefully be presented in such a format. For this reason, many teachers have adopted the use of "slates" or "mini white-boards" on which students can write and display an answer for the teacher to see. Such a technique is particularly useful where students' responses are difficult to predict, such as asking students to write down a fraction between $\frac{1}{6}$ and $\frac{1}{7}$ if they can. Some students will say that it can't be done, others will generate incorrect solutions, and others will generate correct solutions. The powerful feature of such unstructured response systems is that it allows students to generate substantially correct solutions in nonstandard form, such as

$$\frac{1}{6\frac{1}{2}}$$

Perhaps the most important point in all this is that questions worth asking are not likely to be generated spontaneously in the middle of a lesson. They need careful planning, preferably in collaboration with other teachers, as is practiced in Japanese lesson study (Fernandez & Makoto, 2004).

PROVIDING FEEDBACK THAT MOVES LEARNERS FORWARD

From the reviews of research conducted by Natriello (1987), Crooks (1988), Bangert-Drowns et al. (1991), and Black and Wiliam (1998a) cited above it is clear that not all kinds of feedback to students about their work are equally effective. For example, Meisels, Atkins-Burnett, Xue, Bickel, and Son (2003) explored the impact of the Work Sample System (WSS)—a system of curriculum-embedded performance assessments—on the achievement of 96 third-grade urban students in reading and mathematics, as measured by the Iowa Test of Basic Skills (ITBS). When compared with a sample of 116 third graders in matched schools, and with students in the remainder of the school district (Pittsburgh, PA), the achievement of WSS students was significantly and substantially higher in reading, but in mathematics, the differences were much smaller and failed to reach statistical significance. The details of the system in use, how it is implemented, and the nature of the feedback provided to students seem to be crucial variables, with small changes often producing large impacts on the effectiveness of the system. Some insight into the characteristics of effective feedback is provided by a study by Nyquist (2003).

The reviews by Natriello, Crooks, and Black and Wiliam, had focused on K–12 education, but Nyquist reported that the findings about the importance of feedback generalize to higher education too. In reviewing the research on the effects of feedback in 185 studies in higher education, he developed the following typology of different kinds of formative assessment:

Weaker feedback only: students are given only the knowledge of their own score or grade, often described as "knowledge of results."

Feedback only: students are given their own score or grade, together with either clear goals to work towards or feedback on the correct answers to the questions they attempt, often described as "knowledge of correct results."

Weak formative assessment: students are given information about the correct results, together with some explanation.

Moderate formative assessment: students are given information about the correct results, some explanation, and some specific suggestions for improvement.

Strong formative assessment: students are given information about the correct results, some explanation, and specific activities to undertake in order to improve.

He then calculated the average standardized effect size for the studies for each type of intervention, and the results are given in Table 23.2.

Table 23.2 Standardized Effect Sizes for Different Kinds of Feedback Interventions

	N	Effect
Weaker feedback only	31	0.16
Feedback only	48	0.23
Weaker formative assessment	49	0.30
Moderate formative assessment	41	0.33
Strong formative assessment	16	0.51
Total	185	

Nyquist's results echo the findings of Bangert-Drowns et al. discussed above. Just giving students feedback about current achievement produces very little benefit, but where feedback engages students in mindful activity, the effects on learning can be profound.

In one study (Elawar & Corno, 1985), 18 sixth-grade teachers in three schools in Venezuela received 7 hours of training on how to give constructive written feedback on the mathematics homework produced by their students (specific comments on errors, suggestions about how to improve, and at least one positive remark). Another group of teachers graded homework as normal (i.e., just scores), and a third group gave constructive feedback to half their classes and just scores to the other half. The students receiving the constructive feedback learned twice as fast as the control-group students (in other words, they learned in 1 week what the others would have taken 2 weeks to learn). Moreover, in classes receiving the constructive feedback, the achievement gap between male and female students was reduced, and attitudes towards mathematics were more positive.

The negative impact of grades was also established in a study of Israeli students conducted by Butler and Nisan (1986). In this study, one group of students received nonthreatening task-related evaluations of their work, one group received normative grades, and one group received no feedback. The level of intrinsic motivation of students given grades was lower than those given the task-related evaluations, and this was also reflected in the level of achievement. Interestingly, those given no feedback also showed lower levels of intrinsic motivation. This suggests that task-related feedback has an important role to play in building intrinsic motivation (see section on *Activating Students as Owners of Their Own Learning* below) and that lack of feedback can, under some circumstances, be as deleterious as the wrong sort of feedback, but the relationship is clearly complex, inasmuch as other studies have not found lack of feedback to be consistently associated with lower levels of learning. For example, Grolnick and Ryan (1987) reported a study of 91 fifth-grade children that assessed the effects of different conditions on emotional experience and performance on a learning task. Two directed-learning conditions, one controlling (DLC) and one noncontrolling (DLN), were contrasted with each other and with a third nondirected, spontaneous-learning context (NLC). In both the DLC and DLN conditions, what was to be learned was specified and students knew that there would be an assessment, but in the DLC condition, students were told they would be given grades as they worked, whereas those in the DLN condition were told there would be no grades and that the purpose of the activity was to see what they could learn. In the NLC condition, students were exposed to material without specifications and had no knowledge of the subsequent assessment. On measures of rote learning, both the DLC and DLN groups scored higher than the NLC, but the level of interest and the amount of conceptual learning was lower in the DLC group than in the other two. Furthermore, children in the DLC condition evidenced a greater deterioration in rote learning in a follow-up assessment approximately 1 week later.

In a subsequent study, Butler (1988) investigated the effectiveness of different kinds of feedback in 12 sixth-grade classes in four Israeli schools. For the first lesson, the students in each class were given a booklet containing a range of divergent thinking tasks. At the end of the lesson, their work was collected. Independent scorers then assessed this work. At the beginning of the next lesson, two days later, the students were given feedback on the work they had done in the first lesson. In 2 of the classes students were given scores (which were scaled so as to range from 40 to 99) while in another 4 of the classes, students were given comments, such as "You thought of quite a few interesting ideas; maybe you could think of more ideas." In the other 4 classes, both scores and comments were written in the students' notebooks. Then, the students were asked to attempt some similar tasks and told that they would get the same sort of feedback as they had received for the first lesson's work. Again, the work was collected and assessed.

In order to explore the differences in impact of these treatments on students, the achievement and attitudes toward the subject under study of the highest performing students in each class were compared with the lowest achievers in each class. Those given only scores made no gain from the first lesson to the second. Those who had received high scores in the tests were interested in the work, but those who had received low scores were not. The students given only comments scored, on average, 30% higher on the work done in the second lesson than on the first, and the interest of all the students in the work was high. However, those given both scores and comments made *no gain* from the first lesson to the second, and those who had received high scores showed high interest whereas those who received low scores did not. The results are summarized in Table 23.3.

Table 23.3 Impact of Feedback Interventions on Achievement and Attitude (Butler, 1988)

	Achievement gain	Attitude toward subject
Scores only	none	High achievers: positive Low achievers: negative
Comments only	30%	All positive
Scores and comments	none	High achievers: positive Low achievers: negative

Far from producing the best effects of both kinds of feedback, giving scores alongside the comments completely washed out the beneficial effects of the comments. When given both scores and comments, most students looked at the score first. The next thing most of these students looked at was the score of their nearest neighbor. In other words, if teachers write careful diagnostic comments on a student's work and then put a score or grade on it, they are likely wasting the time they spend writing comments. The students who get the high scores do not need to read the comments, and the students who get the low scores do not want to. And yet the use of both scores and comments is probably the most widespread form of feedback in the United States, despite the fact that this study (and others like it—see below) shows that it is no more effective than scores alone. The teacher would be better off just giving a score. The students will not learn anything from this, but the teacher will save herself a great deal of time.

A clear indication of the role that ego plays in learning is given by another study by Ruth Butler (1987). In this study, 200 fifth- and sixth-grade students in eight classes spent a lesson working on a variety of divergent thinking tasks. Again, the work was collected and the students were given one of four kinds of feedback on this work at the beginning of the second lesson (again 2 days later):

In two of the classes, students were given written comments;

In two of the classes, students were given grades;

In two of the classes, students were given written praise; and

In two of the classes, students were given no feedback at all.

The quality of the work done in the second lesson was compared to that done in the first. The quality of work of those given comments had improved substantially compared to the first lesson, but those given grades and praise had made no more progress than those given no feedback. At the end of the second lesson, the students completed a questionnaire about the factors that influenced their work that was designed to elicit from the students the reasons for their level of motivation. In particular the questionnaire sought to establish whether the students' effort investment was due to ego-related concerns (such as class rank) or task-related concerns (such as learning the material). The students who had received comments during their work on the topic had high levels of task involvement, but their levels of ego involvement were the same as those given no feedback. However, although those given grades and those given written praise had comparable levels of task involvement to the control group, their levels of ego involvement were substantially higher. The only effect of the grades and the praise, therefore, was to increase the sense of ego involvement without increasing achievement. These findings are consistent with those of Cameron and Pierce (1994), who found that although verbal praise and supportive feedback did increase students' interest in and attitude towards a task, such feedback had little if any effect on performance.

These findings are consistent with the research on praise carried out in the 1970s that showed clearly that praise was not necessarily "a good thing"—in fact the best teachers appear to praise slightly less than average (Good & Grouws, 1975). It is the quality rather than the quantity of praise that is important. In particular, teacher feedback is far more effective when it is infrequent, credible, contingent, specific, and genuine (Brophy, 1981) and focuses on features of the work that are within the students' control (Dweck, 2000; Siero & van Oudenhoven, 1995).

From a young age, students' attitudes to learning are clearly shaped by the feedback they receive. In a year-long study of eight kindergarten and first-grade classrooms in six schools, Tunstall and Gipps (1996a, 1996b) identified a range of roles that feedback played in classrooms in England. Like Torrance and Pryor (1998), they found that a substantial proportion of the feedback given to students focused on socialization: "I'm only helping people who are sitting down with their hands up" (p. 395). Beyond this socialization role, they identified four types of feedback on academic work, which they labeled A, B, C and D as shown in Table 23.4. In the first type was placed feedback that rewarded or punished the students for their work, for example when students were allowed to leave the classroom for lunch early if they had done good work, or where students were told that if they had not completed assigned tasks, they would have to stay in during lunch to finish their work. Type B feedback was, like Type A, evaluative, but rather than focusing on rewards and sanctions, the feedback indicated the teacher's approval or disapproval (e.g., "I'm very pleased with you" versus "I'm very disappointed in you today").

In addition to the kinds of *evaluative* feedback described above, they identified two types of *descriptive* feedback. Type C feedback focused on academic work as product and emphasized the adequacy of the work with respect to the teacher's criteria for success. At one end of the continuum, feedback focused on the extent to which the work already satisfied the criteria, while at the other end, the feedback focused on what the student needed to do to improve the work. An example of the former is, "This is extremely well explained," pointing out the way that the student had satisfied the teacher's criteria. An example of the latter is, "I want you to go over all of them and write your equals sign in each one"—here the emphasis is on what needs to be done to improve the work.

In contrast to the idea of work as product, Type D feedback focused on the process aspects of work, with the teacher cast in the role of facilitator, rather than judge or coach. As Tunstall and Gipps (1996a) explain, teachers engaged in this kind of feedback "conveyed a sense of work in progress, heightening awareness of what was being undertaken and reflecting on it" (p. 399).

Table 23.4 Typology of Teacher Feedback (Tunstall & Gipps, 1996a)

Evaluative feedback	**Type A**	**Type B**
Positive	Rewarding	Approving
Negative	Punishing	Disapproving
Descriptive feedback	**Type C**	**Type D**
Achievement feedback	Specifying attainment	Constructing achievement
Improvement feedback	Specifying improvement	Constructing the way forward

The timing of feedback is also crucial. If it is given too early, before students have had a chance to work on a problem, then they will learn less. Most of this research has been done in the United States, where it goes under the name of "peekability research," because the important question is whether students are able to "peek" at the answers before they have tried to answer the question. However, a British study undertaken by Simmons and Cope (1993) found similar results. Pairs of students aged between 9 and 11 worked on angle and rotation problems. Some of these worked on the problems using the computer language Logo and some worked on the problems using pencil and paper. The students working in Logo were able to use a "trial and improvement" strategy that enabled them to get a solution with little mental effort. However, for those working with pencil and paper, working out the effect of a single rotation was much more time consuming, and thus the students had an incentive to think carefully, and this greater "mindfulness" led to deeper learning.

The effects of feedback highlighted above might suggest that the more feedback, the better, but this is not necessarily the case. Day and Cordón (1993) looked at the learning of a group of 64 third-grade students on reasoning tasks. Half of the students were given a "scaffolded" response when they got stuck—in other words they were given only as much help as they needed to make progress, whereas the other half were given a complete solution as soon as they got stuck, and then given a new problem to work on. Those given the scaffolded response learned more and retained their learning longer than those given full solutions. In a sense, this is hardly surprising, inasmuch as those given the complete solutions had the opportunity for learning taken away from them. As well as saving time for the teacher, developing skills of "minimal intervention" promotes better learning.

A good example of such feedback is given by Saphier (2005, p. 92):

Teacher: "What part don't you understand?"
Student: "I just don't get it."
Teacher: "Well, the first part is just like the last problem you did. Then we add one more variable. See if you can find out what it is, and I'll come back in a few minutes."

Sometimes, the help need not even be related to the subject matter. Often, when a student is given a new task, the student asks for help immediately. When the teacher asks, "What can't you do?" it is common to hear the reply, "I can't do any of it." In such circumstances, the student's reaction may be caused by anxiety about the unfamiliar nature of the task, and it is frequently possible to support the student by saying something like "Copy out that table, and I'll be back in five minutes to help you fill it in." This is often all the support the student needs. Copying out the table forces the student to look in detail at how the table is laid out, and this "busy work" can provide time for the student to make sense of the task herself (see discussion of performance versus mastery goals below).

The consistency of these messages from research on the effects of feedback extends well beyond school and other educational settings. A review by Kluger and DeNisi (1996) of 131 well-designed studies in educational and workplace settings found that, on average, feedback did improve performance, but this average effect disguised substantial differences between studies. Perhaps most surprisingly, in 38% of the studies, giving feedback had a negative impact on performance. In other words, in 2 out of every 5 carefully controlled scientific studies, giving people feedback on their performance made their performance worse than if they were given no feedback on their performance at all! On further investigation, the researchers found that feedback made performance worse when it was focused on the self-esteem or self-image (as is the case with grades and praise). The use of praise can increase motivation, but then it becomes necessary to use praise all the time to maintain the motivation. In this situation, it is very difficult to maintain praise as genuine and sincere. In contrast, the use of feedback improves performance when it is focused on what needs to be done to improve, and particularly when it gives specific details about *how* to improve (see section on *Activating Students as Owners of Their Own Learning* below).

This suggests that feedback, as the term is used currently in education, is not the same as formative assessment. Feedback is a necessary first step, but feedback is formative *only if the information fed back to the learner is used by the learner in improving performance.* If the information fed back to the learner is intended to be helpful but cannot be used by the learner in improving her own performance, it is not formative. It is rather like telling an unsuccessful comedian to "be funnier."

CLARIFYING AND SHARING LEARNING INTENTIONS AND SUCCESS CRITERIA WITH LEARNERS

In a chapter entitled "The View From the Student's Desk" Mary Alice White (1971) suggested that students often had no real idea where they were going in their learning.

> The analogy that might make the student's view more comprehensible to adults is to imagine oneself on a

> ship sailing across an unknown sea, to an unknown destination. An adult would be desperate to know where he [*sic*] is going. But a child only knows he is going to school. . . . The chart is neither available nor understandable to him. . . . Very quickly, the daily life on board ship becomes all important. . . . The daily chores, the demands, the inspections, become the reality, not the voyage, nor the destination. (White, 1971, p. 340)

Certainly many authors, notably Bernstein (1975), have shown that not all students share the teacher's understanding of what they are meant to be doing in classrooms. As a simple example, consider the task of identifying the "odd one out" in the following list of objects: knife, fork, hammer, bottle of tomato ketchup. Some students believe that the bottle of tomato ketchup is the odd one out, because the others are all metal tools, whereas other students believe that the hammer is the odd one out because the others appear on the table at meal-times. Neither answer is wrong or right, of course, but as Keddie (1971) pointed out over 35 years ago, schools value some kinds of knowledge more than others, although what precisely is desired or valued is not always made clear.

For example, in a study of 72 students between the ages of 7 and 13, Gray and Tall (1994) found that the reasoning of the higher achieving students was qualitatively different from that of the lower achieving students. This, of course, is not surprising, as the research on expertise (Berliner, 1994) shows that this is a characteristic of expertise. What was surprising in Gray and Tall's study was in the nature of the difference. In particular, the higher achieving students were able to work with unresolved ambiguities about whether mathematical entities are concepts or procedures. In contrast, lower attaining students, by refusing to accept the ambiguities inherent in mathematics are, in fact, attempting a far more difficult form of mathematics, with a far greater cognitive demand.

A simple example may be illustrative here. Consider the number $6\frac{1}{2}$. The mathematical operation between the 6 and the $\frac{1}{2}$ is actually addition. $6\frac{1}{2}$ is a shorthand expression for $6+\frac{1}{2}$, but when writing an expression in algebra by concatenating two terms, such as $6x$, the implied operation between the 6 and the x is multiplication, not addition. The relationship between the 6 and the 1 in 61 is different again. And yet, very few people who are successful in mathematics are aware of these inconsistencies or differences in mathematical notation. In a very real sense, being successful in mathematics requires knowing what to worry about and what not to worry about. Students who do not understand what is important and what is not important will be at a very real disadvantage.

Quite how big a difference this can make is brought out in a study of seventh-grade science classes conducted by White and Frederiksen (1998). The study involved three teachers, each of whom taught 4 parallel seventh grade classes in two U.S. schools. The average size of the classes was 31 students. In order to assess the representativeness of the sample, all the students in the study were given a basic skills test, and their scores were close to the national average. All 12 classes followed a novel curriculum (called ThinkerTools) for 14 weeks. The curriculum had been designed to promote thinking in the science classroom through a focus on a series of seven scientific investigations (approximately 2 weeks each). Each investigation incorporated a series of evaluation activities. In two of each teacher's four classes, these evaluation episodes took the form of a discussion about what they liked and disliked about the topic. For the other two classes, they engaged in a process of "reflective assessment." Through a series of small-group and individual activities, the students were introduced to the nine assessment criteria (each of which was assessed on a 5-point scale) that the teacher would use in evaluating their work. At the end of each episode within an investigation, the students were asked to assess their performance against two of the criteria. At the end of the investigation, students had to assess their performance against all nine. Whenever they assessed themselves, they had to write a brief statement showing which aspects of their work formed the basis for their rating. At the end of each investigation, students presented their work to the class, and the students used the criteria to give one another feedback.

As well as the students' self-evaluations, the teachers assessed each investigation, scoring both the quality of the presentation and the quality of the written report, each being scored on a 1 to 5 scale. The possible score on each of the seven investigations therefore ranged from 2 to 10. The mean project scores achieved by the students in the two groups over the seven investigations are summarized in Table 23.5, classified according to their score on the basic skills test.

Table 23.5 Mean Project Scores for Students

	Score on basic skills test		
Group	**Low**	**Intermediate**	**High**
Likes and dislikes (control group)	4.6	5.9	6.6
Reflective assessment (treatment group)	6.7	7.2	7.4

Note. The 95% confidence interval for each of these means is approximately 0.5 either side of the mean.

Two features are immediately apparent in these data. The first is that the mean scores are higher for the students doing reflective assessment, when compared with the likes-and-dislikes group—in other words, all students improved their scores when they thought about what it was that counted as good work. However, much more significantly, the difference between the "likes and dislikes" group and the "assessment" group was much greater for students with weak basic skills. This suggests that, at least in part, low achievement in mathematics is exacerbated by students' not understanding what it is they are meant to be doing.

Although students clearly need to understand the standards against which their work will be assessed, the study by White and Frederiksen shows that the criteria themselves are only the starting point. At the beginning, the words do not have the meaning for the student that they have for the teacher. Just giving "quality criteria" or "success criteria" to students will not work, unless students have a chance to see what this might mean in the context of their own work. Furthermore, as Stiggins (2001, pp. 314–322) pointed out, there is a very real tension between task-specific and generic rubrics. Task-specific rubrics can be written in very clear language and thus can communicate accurately to students what is required, but students need to come to grips with a new rubric for each task. Generic rubrics, on the other hand, can span a number of tasks but are unlikely to provide clear guidance to a student about how to improve her or his work. Another weakness of task-specific rubrics is that, by definition, they focus only on the qualities needed for the specific task and therefore may not contribute to more generalizable skills. Too much attention to one's performance relative to a detailed rubric may, in fact, be counterproductive (Kohn, 2006). For this reason, Arter and McTighe (2001) suggested that task-specific rubrics are more appropriate for summative assessment, where consistency of scoring is paramount, but that generic rubrics are more appropriate for formative assessment, inasmuch they focus on qualities that transcend the immediate task (for more on the characteristics of good rubrics, see Relearning by Design, 2000).

All the features of good rubrics discussed above apply to the use of rubrics for both summative and formative purposes, but if rubrics are to function formatively, additional considerations apply. Effective summative assessment requires those engaged in assessment to share the same construct of quality, and rubrics can play an important role in promulgating common standards (Sadler, 1987). Effective formative assessment requires the student to share the same construct:

> The indispensable conditions for improvement are that the student comes to hold a concept of quality roughly similar to that held by the teacher, is continuously able to monitor the quality of what is being produced during the act of production itself, and has a repertoire of alternative moves or strategies from which to draw at any given point. (Sadler, 1989, p. 121)

However, it is possible for a rubric to identify accurately different degrees of quality in a piece of work without explicitly identifying the relationship between the levels. Both a teacher and a student may be able to identify that a particular piece of work merits a 4 on a six-point rubric, but be unclear about what needs to be done to move the piece of work forward to a 5 or a 6, much in the same way as one can tell a softball pitcher that she needs to lower her ERA or get her rising fast-ball to rise without having any idea how to accomplish this. For this reason, it is particularly helpful for formative purposes if rubrics are longitudinal or developmental, so that they are not just descriptions of quality, but effectively *anatomies* of quality—identifying the immediate steps that can be taken to begin improvement.

In clarifying and sharing learning intentions and success criteria with learners, it is also important to be clear about whether the particular aspect of quality sought can be communicated with a rubric at all. Some aspects of quality, such as whether calculations are correct or whether diagrams are drawn in pencil and labeled, can easily be set down in a rubric. On the other hand, some features, such as whether an approach to a mathematical investigation is systematic or not, cannot be so easily described. Using a term like "Adopts a systematic approach" in a rubric may look like a definition of quality, but it is unlikely to be meaningful to those who do not already know what it means to be systematic. In Michael Polanyi's terms, it is not a rule that can be easily put into effect but is rather a *maxim*:

> Maxims cannot be understood, still less applied by anyone not already possessing a good practical knowledge of the art. They derive their interest from our appreciation of the art and cannot themselves either replace or establish that appreciation. (Polanyi, 1958, p. 50)

In such situations, the best we can do is to help students develop what Guy Claxton called a "nose for quality" (Claxton, 1995). Rubrics may have a role to play in this process, as they did in the study by White and Frederiksen. Rubrics were shared with students, but the students were given time to think through, in discussion with others, what this might mean in practice, applied to their own work. One should not assume that the students will understand these right away, but the criteria will provide a focus for negotiating with students about what counts as quality in the mathematics classroom.

Another way of helping students understand the criteria for success is, before asking the students to embark on (say) an investigation, to get them to look at the work of other students (suitably anonymized) on similar (although not, of course, the same) investigations. In small groups, they can then be asked to decide which pieces of students' work are good investigations, and why. It is not necessary, or even desirable, for the students to come to firm conclusions and can generate a definition of quality—what is crucial is that they have an opportunity to explore notions of *quality* for themselves. Spending time looking at other students' work, rather than producing their own work, may seem like time off-task, but the evidence is that it is a considerable benefit, particularly for students who do not find mathematics easy to learn.

ACTIVATING STUDENTS AS OWNERS OF THEIR OWN LEARNING

The power of getting students to take some ownership of their own work is shown very clearly in an experiment by Fontana and Fernandez (1994). A group of 25 Portuguese primary school teachers met for 2 hours each week over a 20-week period during which they were trained in the use of a structured approach to student self-assessment. The approach to self-assessment involved an exploratory component and a prescriptive component. In the exploratory component, each day, at a set time, students organized and carried out individual plans of work, choosing tasks from a range offered to them by the teacher, and had to evaluate their performance against their plans once each week. The progression within the exploratory component had two strands—over the 20 weeks, the tasks and areas in which the students worked were to take on the students' own ideas more and more, and secondly, the criteria that the students used to assess themselves were to become more objective and precise.

The prescriptive component took the form of a series of activities, organized hierarchically, with the choice of activity made by the teacher on the basis of diagnostic assessments of the students. During the first 2 weeks, children chose from a set of carefully structured tasks and were then asked to assess themselves. For the next 4 weeks, students constructed their own mathematical problems following the patterns of those used in weeks 1 and 2 and evaluated them as before but were required to identify any problems they had, and whether they had sought appropriate help from the teacher. Over the next 4 weeks, students were given further sets of learning objectives by the teacher and again had to devise problems, but now they were not given examples by the teacher. Finally, in the last 10 weeks, students were allowed to set their own learning objectives, to construct relevant mathematical problems, to select appropriate manipulatives, and to identify suitable self-assessments.

Another 20 teachers, matched in terms of age, qualifications, and experience, using the same curriculum scheme, for the same amount of time, and receiving the same amount of professional development, acted as a control group. The 354 students being taught by the 25 teachers using self-assessment and the 313 students being taught by the 20 teachers acting as a control group were each given the same mathematics test at the beginning of the project, and again at the end of the project 20 weeks later. Over the course of the experiment, the scores of the students taught by the control-group teachers improved by 7.8 points. The scores of the students taught by the teachers developing self-assessment improved by 15 points—almost twice as big an improvement. In other words, the students participating in the self-assessment learned in 20 weeks what the students in the control group classrooms would take 40 weeks to learn—a doubling of the rate of learning.

Whether the students in this study were able to assess their own performance objectively is unclear, and in general, such questions are a matter of heated debate, but very often the debate takes place at cross-purposes. Opponents of self-assessment say that students cannot possibly assess their own performance objectively, but this is an argument about the *summative* function of self-assessment, which is beyond the scope of this chapter. The focus here is whether activating students as owners of their own learning does, indeed, enhance learning, and in this regard, concordance with other, external judgments is a secondary concern. What matters is whether enhancing students' engagement with and ownership of their own learning enhances learning. For this to happen students have to be motivated (literally, moved) to do so, but they also have to possess the necessary cognitive resources.

The key concept here is *self-regulation* (Boekaerts, Maes, & Karoly, 2005), defined as "a multilevel, multicomponent process that targets affect, cognitions, & actions, as well as features of the environment for modulation in the service of one's goals" (Boekaerts, 2006, p. 347). Put simply, students who lack self-regulation skills are unlikely to be able to take control of and guide their own learning (DeCorte, Verschaffel, & Op't Eynde, 2000). Inevitably, with such a broad area, different researchers have emphasized different aspects of self-regulation.

Winne (1996) emphasized the cognitive aspects of this process, defining self-regulated learning as a "metacognitively governed behavior wherein learners adaptively regulate their use of cognitive tactics and strategies in tasks" (p. 327). Others have observed that many students appear to possess the necessary skills of self-regulation but do not use them in classrooms (and especially in mathematics classrooms), which suggests that the problem in not a lack of skill, but rather a lack of motivation or volition (Corno, 2001). Still others have argued for the integration of different perspectives within sociocultural (Hickey & McCaslin, 2001; McCaslin & Hickey, 2001) or social constructivist (Op't Eynde, DeCorte, & Verschaffel, 2001) perspectives.

These two broad areas—cognition and motivation—have been extremely fertile areas of inquiry over the last quarter century or so, but unfortunately most of the research undertaken has been firmly in one area or the other (Wigfield, Eccles, & Rodriguez, 1998). As Sorrentino and Higgins (1986) pointed out there has been an "almost total exclusion of motivation in research on cognition" (p. 3); Nisbett and Ross (1980), in their influential book on human reasoning, lamented psychology's "inability to bridge the gap between cognition and behavior" (p. 11). Nevertheless, to realize the potential for self-regulation to improve learning, ways of bridging this gap must be found, because, as Boekaerts (2006) showed, self-regulated learning is both metacognitively governed *and* affectively charged (p. 348).

In the next two sections, the research on cognition and motivation will be briefly reviewed, focusing on the aspects that are most relevant to the idea of activating students as owners of their own learning. The implications of this research for using assessment to activate students as owners of their own learning will then be explored through the use of the dual-processing model of self-regulation proposed by Boekaerts (2006).

Metacognition

It is perhaps not too much of an overstatement to say that everyone agrees that metacognition is important, but no one can agree on what it is, at least in a sufficiently precise definition to put into practice. Even John Flavell, widely credited with inventing the term, acknowledged that it was a "fuzzy concept" (Flavell, 1981 p. 37). Literally, it is "beyond thinking" or thinking about thinking. Flavell (1976) defined the concept thus:

> "Metacognition" refers to one's knowledge concerning one's own cognitive processes and products or anything related to them, e.g., the learning-relevant properties of information and data. For example I am engaging in metacognition (metamemory, metalearning, metaattention, metalanguage, or whatever) if I notice that I am having more trouble learning A than B; if it strikes me that I should double-check C before accepting it as a fact; if it occurs to me that I had better scrutinise each and every alternative in any multiple-choice type task situation before deciding which is the best one; if I sense that I had better make a note of D because I may forget it; if I think to ask someone about E to see if I have it right. In any kind of cognitive transaction with the human or nonhuman environment, a variety of information processing activities may go on. Metacognition refers, among other things, to the active monitoring and consequent regulation and orchestration of these processes in relation to the cognitive objects or data on which they bear, usually in the service of some concrete goal or objective. (p. 232)

It is thus an "umbrella term [that encompasses all] knowledge and cognition about cognitive phenomena" (Flavell, 1979, p. 906), including knowing what one knows (*metacognitive knowledge*), what one can do (*metacognitive skills*), and what one knows about one's own cognitive abilities (*metacognitive experience*). Kluwe (1982, p. 202) broadened the term further, by including knowledge about the thinking of others as well as oneself. Brown (1987) identified "four historically separate, but obviously interlinked, problems in psychology that pertain to issues of metacognition" (p. 69):

- whether individuals can have access to their own cognitive processes
- who or what is responsible for executive control of cognition
- what is the role of self-regulation within learning
- what is the role of regulation by others in learning

although there is a continuing debate about whether metacognition has to be a conscious process (Carr, Alexander, & Folds-Bennett, 1994).

Studies have repeatedly shown that students with greater awareness of their own cognitive processes have higher achievement (DeCorte, Verschaffel, & Op't Eynde, 2000), as do students who display more conscious self-regulation (see e.g., Stillman & Galbraith, 1998). Butler and Winne (1995) summarized the situation thus: "Theoreticians seem unanimous—the most effective learners are self-regulating" (p. 245), but this leaves unresolved the question of whether improving metacognitive skills improves achievement, and this is where the lack of an agreed operational definition of metacognition is most serious.

Many studies have shown that training students in metacognitive strategies improves their performance even in the early school years (e.g., Lodico, Ghatala, Levin, Pressley, & Bell, 1983) and can also generalize metacognitive strategies to new situations (see Hacker, Dunlosky, & Graesser, 1998, for a review), but other studies have failed to find benefits for metacognitive training. For example, in a multilevel study of 444 seventh grade students in 18 heterogeneous classrooms in the Netherlands, Hoek, van den Eeden, and Terwel (1999) found that the achievement of students trained in metacognitive and social strategies simultaneously was no greater than controls just told to work together (although differences in two of the three outcome measures were found in single-level models). Without a clear operational definition of metacognition, this result is difficult to interpret. Those who advocate the benefits of metacognitive training can just claim that the unsuccessful efforts were not "real" metacognition.

What is clear is that many studies *have* found clear benefits for metacognitive interventions in real classroom settings as well as laboratories. In the previous edition of this *Handbook*, Schoenfeld (1992) reviewed the research on metacognition with a specific focus on the learning of mathematics, although most of the studies available at that time had been conducted in laboratory settings or in small-scale trials. Since then, a series of studies have shown that training in self-evaluation improves student achievement in real settings, and across extended periods of time.

For example, Ross, Hogaboam-Gray, and Rolheiser (2002) reported on a study in which 12 fifth- and sixth-grade mathematics teachers were trained in the promotion of systematic student self-evaluation, and the performance of their students ($N = 259$) was compared with students ($N = 257$) taught by teachers who had not been so trained. After a 12-week period, the students undertaking self-evaluation outscored the other students by 0.4 standard deviations on a mathematics task.

Mevarech and her colleagues in Israel have demonstrated that metacognitive training improves student achievement in a range of settings, and across even longer periods of time. Their method is termed *IMPROVE* after the initial letters of the key components:

Introducing the new concepts;

Metacognitive questioning in small groups;

Practicing, **R**eviewing, and reducing difficulties;

Obtaining mastery; and

Verification and **E**nrichment.

In one year-long study 99 seventh-grade students in three classrooms in Israel were taught using the IMPROVE method, and their performance was compared with 148 students from five other classes (Mevarech & Kramarski, 1997). Although the IMPROVE students scored slightly lower than the control group students on a pretest, they outscored controls on both an "Introduction to Algebra" test and on a test of mathematical reasoning. The standardized effect size d (Cohen, 1988) of the advantage for the IMPROVE students was 0.31 standard deviations for the "Introduction to algebra" test and 0.44 standard deviations on the test of mathematical reasoning. A second study involving 265 seventh graders found a similar result ($d = 0.38$).

Other studies found that the IMPROVE method was more effective than just training in metacognitive skills (Kramarski, Mevarech, & Lieberman, 2001) or structured collaborative work (Kramarski, Mevarech, & Arami, 2002), and the benefits extended to both traditional measures of achievement and authentic tasks. More recent work has established that achievement gains are still present after 1 year and are larger for lower achieving students (Mevarech & Kramarski, 2003).

Another research program—Cognitive Acceleration in Mathematics Education or CAME (Adhami, Johnson, & Shayer, 1998)—also aims to increase student achievement in mathematics by attention to metacognitive processes. Although it shares many features with IMPROVE, the program seems to make somewhat greater demands on teachers and is therefore more difficult to implement than the IMPROVE project. Results from 68 classrooms monitored over a 2-year period suggest that the average increase in student achievement was approximately 0.34 standard deviations (Adhami et al., 1997), although the distribution was trimodal, with one peak around 0.8 standard deviations (almost all the classes coming from one school), one peak around 0.5 standard deviations, and another at 0 (which appears to be attributable to schools that did not implement the program as intended).

Other cognitive-behavior interventions have been reviewed by Boekaerts and Corno (2005, pp. 213–221), but the lack of an adequate theorization of metacognition makes comparing different interventions difficult, if not impossible. Nevertheless, the results from IMPROVE, CAME, and similar interventions suggest that metacognitive interventions can have a substantial impact on student achievement in mathematics, even if the design of such interventions is currently more art than science. It is also no coincidence that the most effective programs for developing metacognitive skills are firmly rooted in specific domains of knowledge (Bransford, Brown, & Cocking, 2000). However, just possessing the skills of being able to monitor one's own learning is not enough for students to learn mathematics. They also have to want to do so, and, as any mathematics teacher is all too aware, this cannot be taken for granted.

Motivation and Learning

As Edward Deci (1996) noted, "Most people seem to think that the most effective motivation comes from outside the person, that it is something that one skilful person does to another" (p. 9), although there is substantial evidence that the situation is not that simple. In their explanation of Self-Determination Theory (SDT), Deci and Ryan (1985) distinguished between different types of motivation based on the different reasons or goals that give rise to a particular action. "The most basic distinction is between *intrinsic motivation*, which refers to doing something because it is inherently interesting or enjoyable, and *extrinsic motivation*, which refers to doing something because it leads to a separable outcome" (Ryan & Deci, 2000, p. 55).

Because of its power to drive quality learning, the notion of intrinsic motivation has been extensively studied, especially how the actions of teachers and parents can build or undermine it (Connell & Wellborn, 1991; Ryan & Stiller, 1991), and the basic distinction between intrinsic and extrinsic motivation has held up pretty well. However, although the benefits of intrinsic motivation are clear, it is far from clear that extrinsic motivation is antithetical to learning. For example, a meta-analysis of 61 studies that compared a group receiving rewards with a group that did not found that the effects were, on average, very small. Unanticipated verbal rewards increased both attitudes towards tasks and the time that participants chose to spend on them. Anticipated rewards had no impact on attitude but did reduce time spent on tasks, with a much stronger effect when the reward was independent of the quality of performance (Eisenberger & Cameron, 1996). As Ryan and Deci (2000) themselves pointed out, although intrinsic motivation is by definition autonomous, the fact that motivation is extrinsic does not mean that the motivation is necessarily *not* autonomous:

> For example, a student who does his homework only because he fears parental sanctions for not doing it is extrinsically motivated because he is doing the work in order to attain the separable outcome of avoiding sanctions. Similarly, a student who does the work because she personally believes it is valuable for her chosen career is also extrinsically motivated because she too is doing it for its instrumental value rather than just because she finds it interesting. (Ryan & Deci, 2000, p. 60)

In these two cases, neither student is intrinsically motivated, but the first is primarily just complying with an external control, whereas the second has embraced the task and is working because she values the goal. In other words, what controls the behavior of the individual involves both whether the motivation itself is internal or external, and whether the value system is internal or external.

When both are external, then behavior is *externally* regulated "by contingencies overtly external to the individual" (Deci & Ryan, 1994, p. 6), whereas *introjected* regulation "refers to behaviours that are motivated by internal prods and pressures such as self-esteem-relevant contingencies" (p. 6), such as the student described above doing his homework under threat of sanctions. *Identified* regulation "results when a behaviour or regulation is adopted by the self as personally important or valuable" (p. 6), as in the case of the student doing the work for her career, whereas *integrated* regulation "results from the integration of identified values and regulations into one's coherent sense of self" (p. 6). These four types of extrinsic motivation—external, introjected, identified, and integrated— form a continuum, in that their intercorrelations have been shown to conform to a simplex pattern in which the largest intercorrelations are closest to the leading diagonal (Ryan & Connell, 1989) and so, together with intrinsic motivation and amotivation (where the individual lacks any intention to act), provide a taxonomy of six kinds of motivation. Although there is undoubtedly a clear hierarchy in terms of the degree of self-determination, this is clearly not a developmental continuum: A student may start out intrinsically motivated but maintain activity only because of external sanctions or instrumental goals, whereas someone may engage in an activity because of the rewards associated with it but then maintain the activity even when rewards are withdrawn (Eisenberger & Cameron, 1996).

The research on internal and external drivers of motivation accounts well for a range of observed student behaviors but pays relatively little attention to why an individual learner might find something interesting or not. Hidi and Harackiewicz (2000) defined *interest* as "an interactive relation between an individual and certain aspects of his or her environment (e.g., objects, events ideas), and is therefore content specific" (p. 152) involving both cognitive and affective components. They pointed out that researchers have placed different emphases on different aspects of interest, with some focusing on specific stimuli that focus attention that may or may not last (generally termed *situational* interest) whereas others have focused on more stable motivational orientations or personal dispositions towards particular topics or domains (termed *individual* or *personal* interest). As Dewey (1913) noted, catching and holding student interest are two very different things. Mitchell (1993) found that it was possible to generate student interest in mathematics through the use of group work, puzzles, and computer-based activities, but that such interest waned over time. In contrast, meaningfulness and involvement in the tasks built interest that *was* sustained over time.

The approach to motivation discussed above seeks to examine motivation as a cause of engagement in activity. This has been very powerful in understanding why people and, in particular, students in mathematics classes do what they do. However, Csikszentmihalyi (1990) turned this on its head by looking at motivation as an *emergent* property of the interaction between the task in which the individual is engaged and the competences that the individual brings to the task. In this view, motivation is therefore not the cause, but the consequence of engagement. When the task demand is high and the skill levels is low, then the individual experiences anxiety, whereas if the task demand is low and the skill level is high, the individual experiences boredom. But if the task demand is just at the limit of the skill level, then the individual experiences "flow":

> A dancer describes how it fees when a performance is going well: "Your concentration is very complete. Your mind isn't wandering, you are not thinking of something else; you are totally involved in what you are doing. . . . Your energy is flowing very smoothly. You feel relaxed, comfortable and energetic."
>
> A rock climber describes how it feels when he is scaling a mountain: "You are so involved in what you are doing [that] you aren't thinking of yourself as separate from the immediate activity. . . . You don't see yourself as separate from what you are doing."
>
> A mother who enjoys the time spent with her small daughter: "Her reading is the one thing she's really into, and we read together. She reads to me and I read to her, and that's a time when I sort of lose touch with the rest of the world, I'm totally absorbed in what I'm doing."
>
> A chess player tells of playing in a tournament: ". . . the concentration is like breathing—you never think of it. The roof could fall in and, if it missed you, you would be unaware of it." (Csikszentmihalyi, 1990, pp. 53–54)

As well as the drivers of interest that Hidi and Harackiewicz identified, therefore, it is also necessary to attend to matching the level of demand to the skill of the individual.

Of course, the research reviewed above represents somewhat of a "counsel of perfection." Ideally one would want all students to be intrinsically motivated, and to experience flow in their mathematics classrooms, but this is impossible to achieve in even the best mathematics classrooms all the time. Researchers therefore need to take account of why students do what they do even if they are not intrinsically motivated.

Eccles et al. (1983) identified four components of task value: attainment value, intrinsic value, utility value and cost. *Attainment value* refers to the value to an individual of doing well on a task, particularly in terms of confirming or disconfirming particular aspects of one's identity (see also Boaler, Wiliam, & Zevenbergen, 2000). *Intrinsic value* corresponds to the ideas of intrinsic motivation and flow discussed above. *Utility value* encompasses the value of the task in terms of current and future goals, and is closely related to the identified, introjected and external forms of regulation described above. Finally, Eccles et al. (1983) emphasized the importance of a trade-off between the perceived value (whether it is attainment value, intrinsic value, or utility value) of a task and its cost. Cost includes both the opportunity cost that attempting a task might take, and also the negative consequences such as performance anxiety, risk to one's view of self if unsuccessful, and so on.

The goals that students actually pursue in mathematics classrooms will therefore depend on a complex calculus of cost and benefit. Bandura (1986) and Schunk (1990, 1991) have shown that students are more motivated to reach goals that are specific, within reach, and offer some degree of challenge, but more recently, researchers have focused on broader issues, and in particular whether the student's primary orientation is towards *mastery* or *performance*. Ames (1992) defined an *achievement goal* as follows:

> An *achievement goal* concerns the purposes of achievement behavior. It defines an integrated pattern of beliefs, attributions, and affect that produces the intentions of behavior and that is represented by different ways of approaching, engaging in, and responding to achievement activities. (p. 261)

Mastery achievement goals are those related to the acquisition of new skills or increasing the level of competence, whereas performance achievement goals are those that are related to the level of performance one is able to produce. Students with mastery goals seek to increase their level of competence, whereas those with performance goals are motivated either to show what they can do (performance approach goals) or to avoid failing (performance avoidance goals). Although mastery goals are clearly beneficial and performance-avoidance goals are clearly detrimental to learning, there is still much debate about the usefulness of performance-approach goals. In a study of a 5-week elementary school math unit, Linnenbrink (2005) found that a combined approach, emphasizing mastery goals within small groups but emphasizing performance goals between groups produced the highest achievement, provided the competition between groups was focused on relative improvement amongst the groups (see also the discussion on *Activating Students as Instructional Resources for One Another* below).

Whether students choose to pursue mastery or performance goals depends, of course, on a range of factors, including the students' beliefs about the likelihood of success, and this, in turn, is influenced by students' ideas about the causes of success in school. Through an extensive research program going back over a quarter of a century, involving questionnaires and interviews with thousands of students, Dweck (2000) has shown that there are substantial differences between students in their beliefs about the causes of success and failure in the classroom. Two dimensions appear to be particularly important. The first is *personalization*: whether success is due to *internal* factors (such as one's own performance) or *external* factors (such as getting a lenient or severe grader). The second is *stability*: whether success is due to *stable* factors (such as one's ability) or *unstable* factors (such as effort or luck).

Dweck and others have found that many if not most students attribute success and failure differently. So for example, the student who says "I got an A" but "She gave me a C" is attributing success internally but attributing failure externally. In fact many studies have found that boys are more likely to attribute their successes to internal, stable causes (such as ability) and their failures to external, unstable causes (such as bad luck or hostile teachers). This would certainly explain the high degree of confidence with which many boys approach tests or examinations for which they are completely unprepared (the important question for such students is not "Have I thoroughly reviewed the material for this test?" but "Do I feel lucky today?"). More controversially, the same research suggests that girls attribute their successes to internal unstable causes (such as effort) and their failures to internal stable causes (such as lack of ability), leading to what has been termed *learned helplessness* (Dweck, 2000). What *is* clear is that the most adaptive beliefs are that both success and failure are attributable to internal, unstable causes: "it's down to you, and you can do something about it." This is why the strategy of *Clarifying and Sharing Learning Intentions and Success Criteria with Learners* discussed above is so important. When students are clear about the criteria for success, they are more likely to attribute success and failure to internal unstable causes.

In the 1970s and 1980s, efforts to raise the confidence of students that they could achieve in mathematics focused on self-esteem as a critical factor, but Bandura (1977) has convincingly argued that self-efficacy—a generative capacity to carry one's plans through to completion—is a far more useful focus of inquiry. Students' expectations about their likelihood of success strongly influence their decisions about which tasks to attempt, how much effort to expend, and their persistence in the face of difficulties (Bandura, 1997), and increasing students' self-efficacy has been shown to increase performance in mathematics. Indeed, it appears that self-efficacy may be more important in mathematics than in any other school subject (Pajares, 1996, p. 555).

Students' decisions about whether to pursue mastery or performance goals are also influenced by teachers' practices. In a study of 1571 students in 84 mathematics classrooms from 5th to 12th-grade Deevers (2006) used hierarchical linear modeling to explore the relationship between teachers' formative assessment practices and student attitudes and beliefs about mathematics. He found that although students' self-efficacy beliefs and motivation to learn declined steadily from 5th to 12th grade, students provided with positive constructive feedback were more likely to display mastery orientation, even though teachers gave less of this kind of feedback as students got older.

Closely related to the work on self-efficacy and goal orientation are studies undertaken by Carol Dweck and her colleagues into student perceptions and beliefs about the nature of ability in school (see Dweck, 2000, for a comprehensive summary of this work). Dweck and Leggatt (1986) showed that many students believe that ability is fixed: There are smart people and not-so-smart people. Students holding this "entity" view of ability are likely to adopt a performance orientation to challenging tasks that are set for them. If they are confident in their ability to achieve what is asked of them, then they will attempt the task. However, if their confidence in their ability to carry out their task is low, then, unless they believe that the task is so hard that no one is expected to succeed, they will avoid the challenge. In short, these students are deciding that they would rather be thought lazy than dumb.

In contrast other students see ability as incremental. They see challenging tasks as chances to learn—to get smarter—and therefore in the face of failure will try harder (Mayer, Turner, & Spencer, 1997). What is perhaps most important here is that these views of ability are generally not global—the same students often believe that ability in mathematics is fixed, while at the same time they believe that ability in, for example, athletics is incremental, in that the more one trains, the more one's ability increases. The crucial thing, therefore, is that teachers inculcate in their students a belief that ability is incremental rather than fixed. Doing so would be advantageous even if it were untrue, because the consequences would be so beneficial, but, as was discussed in the section on *Eliciting Evidence* above, there is overwhelming evidence that "smart is not something you are; it's something you get."

Integrating Motivational and Cognitive Perspectives

A number of ways of bringing together the motivational and cognitive perspectives on self-regulation have been proposed. One of the most promising is the *dual processing* theory developed by Boekaerts (1993). In the model,

> It is assumed that students who are invited to participate in a learning activity use three sources of information to form a mental representation of the task-in-context and to appraise it: (1) current perceptions of the task and the physical, social, and instructional context within which it is embedded; (2) activated domain-specific knowledge and (meta)cognitive strategies related to the task; and (3) motivational beliefs, including domain-specific capacity, interest and effort beliefs. (Boekaerts, 2006, p. 349)

As a result of the appraisal of the task, the student begins to act along one of two pathways. If the task appraisal is positive, the student "crosses the Rubicon" (Winne, 2005, pp. 234–236) and begins activity along the "growth pathway" where the goal is to increase competence. In this case, the self-regulation is *top-down* in that the flow of energy is directed by the student. If, on the other hand, the task appraisal is negative, attention shifts away from the learning task and towards the well-being pathway. The student then becomes focused on self-appraisal rather than task appraisal, concentrating on preventing threat, harm or loss. This form of self-regulation is termed *bottom-up* by Boekaerts because it is triggered by cues in the environment, rather than by learning goals. Such bottom-up regulation is not necessarily negative, because by attending to the well-being pathway, the student may find a way to restore well-being, thus allowing a shift of energy and attention back to the growth pathway.

An important feature of the dual regulation model is that it hypothesizes that the balance between top-down and bottom-up pathways of regulation is dynamic, rather than being a dispositional feature of an individual student. Experimental support for this hypothesis was provided by a series of studies summarized in Boekaerts (2001, pp. 24–26) in which it was found that there was no direct link between domain-specific motivational beliefs and learning intention in any of the mathematics classrooms they studied. Rather, students' decisions about whether to invest effort in a mathematics assignment depended largely on their appraisal of the specific mathematics task in front of them. There is also evidence that interpretations of assessment outcomes from friends and parents mediate the relationship between the assessment outcomes themselves and how they impact students beliefs (Ross, Rolheiser, & Hogaboam-Gray, 2002).

The dual-processing model allows one to relate the various perspectives on cognition and motivation and learning described above. To simplify somewhat, students who are *personally*, as opposed to *situationally*, interested in a task are likely to engage in activity along the growth pathway, although for students who are not personally interested in a task, a range of factors related to the task-in-context may spark situational interest, thus also triggering activity along the growth pathway. Where personal interest is not the main driver of attention, more concern will be given to considerations of task *value* versus *cost*. The *integrated* and *identified* forms of regulation defined by Deci and Ryan are related to activity along the growth pathway, whereas *external* or *introjected* forms of regulation are related to activity along the well-being pathway. Students who display *mastery* orientation are activating the growth pathway, and those displaying *performance* orientation are activating the well-being pathway.

Self-efficacy beliefs can drive progress along either pathway. Along the growth pathway, self-efficacy drives adaptive (meta)cognitive strategy use, whereas along the well-being pathway, self-efficacy beliefs are likely to steer the learner away from *performance-avoidance goals* and towards *performance-approach goals*. Similarly views of ability as *incremental* help the learner stay on the growth pathway, whereas *entity* views of ability direct activity towards the well-being pathway, where details of the task-in-context, appraised in the light of views of personal capability, will influence decisions about whether to engage in the task.

This extended discussion of the research on metacognition, motivation, and beliefs may seem distant from the central theme of this chapter, but an understanding of these issues is essential in developing an understanding of the role of assessment in learning. Adaptive adjustments to instruction are necessary to maximize learning, so assessment is an essential part of learning, but assessment processes themselves impact the learner's willingness, desire, and capacity to learn (Harlen & Deakin-Crick, 2002). The research reviewed briefly above shows clearly that practices ostensibly designed to support learning actually prevent it from taking place (recall that the research reviewed by Kluger and DeNisi, 1996, discussed above showed that in 38% of rigorously designed and conducted studies, feedback designed to improve performance actually had the opposite effect).

Although many issues are still unresolved, and a full theoretical synthesis of all the research perspectives is some way off if it is even possible, the existing research on cognition and motivation paints a reasonably coherent picture and provides strong guidance on the design of classroom assessment environments (Brookhart, 1997) that can activate students as owners of their own learning. Feedback to learners should focus on what they need to do to improve, rather than on how well they have done, and should avoid comparison with others. Students who are used to having every piece of work scored or graded will resist this (see, e.g., Smith and Gorard, 2005), wanting to know whether a particular piece of work is good or not, and in some cases, depending on the situation, the teacher may need to go along with this. In the long term, however, it seems that teachers should aim to reduce the amount of ego-involving feedback given to learners (and with new entrants to the school, perhaps not begin the process at all) and focus on the student's learning needs (Kohn, 1999). Furthermore, feedback should not just tell students to work harder or be "more systematic"—the feedback should contain a recipe for future action, for otherwise it is not formative. Finally, feedback should be designed so as to lead all students to believe that ability—even in mathematics—is incremental. In other words the more students "train" at mathematics, the smarter they get.

Although there is a clear set of priorities for the development of feedback, there is no "one right way" to do this. The feedback routines in each class will need to be thoroughly integrated into the daily work of the class, and so they will look slightly different in every classroom. This means that no one can tell teachers how this should be done—it will be a matter for each teacher to work out how to incorporate some of these ideas into her or his own practice.

ACTIVATING STUDENTS AS INSTRUCTIONAL RESOURCES FOR ONE ANOTHER

As Slavin, Hurley, and Chamberlain (2003) noted, "Research on cooperative learning is one of the greatest success stories in the history of educational research" (p. 177). However, despite the vast amount of research that has been done in this area and the consensus that cooperative learning does increase student achievement, there is much less agreement on why cooperative learning is effective. Slavin et al. (2003) suggested that there are four major theoretical perspectives on cooperative learning and its effects on achievement.

Two of the perspectives—the *motivational* perspective and the *social cohesion* perspective—focus on the role of motivation. Adherents of the motivational perspective hold that students help their peers to learn because it is in their own interests to do so; the only way they can achieve their personal goals is to ensure that all the individuals in the group are successful. In contrast, theorists within the social cohesion perspective propose that students help their peers because they care about the group. In both of these perspectives, increased student achievement comes primarily from increased motivation and effort.

The other two perspectives—the *developmental perspective* and the *cognitive elaboration* perspective—emphasize the idea that increased learning in cooperative groups stems from the increased cognitive engagement produced in small groups. Empirical work in these two perspectives tends not to emphasize the importance of group goals, which are the hallmark of the motivational perspective, nor the building of team spirit, as required in the social cohesion perspective. Indeed, many writers in this tradition, such as Damon (1984), have explicitly rejected the idea that extrinsic incentives are important in group learning situations (p. 337).

The developmental perspective draws strongly on the work of Piaget and Vygotsky, and in particular the latter, although as Shayer (2003) noted, the differences between these two are much less than is often assumed. The starting point for much of this work is Vygotsky's definition of the *zone of proximal development* (zpd) as "the distance between the actual developmental level as determined by independent problem solving and the level of potential development as determined through problem solving under adult guidance or in collaboration with more capable peers" (Vygotsky, 1978, p. 86).[3] Research in the developmental perspective emphasizes the role of more advanced peers in guiding the learning of others.

The fourth perspective—the cognitive elaboration perspective—like the developmental perspective, locates the benefit of cooperative learning in the in-

[3] As Chaiklin (2003) made clear, in reading Vygotsky, it is important to understand the cultural and historical context of his work, and in particular, Vygotsky's attempt to distinguish between learning and development. For Vygotsky, development requires changes in the psychological functions that the child can deploy, whereas learning does not. The zone of proximal development (zpd) is not, therefore, just a way of describing what a student can do with support—that could just be learning. Rather the zpd is a description of the *maturing* psychological functions rather than those that already exist, and a focus in instruction on the maturing psychological functions is most likely to produce a transition to the next developmental level.

teraction between students rather than in the motivation for engagement. However, whereas the developmental perspective predicts benefits for the recipient of the support, the cognitive elaboration perspective emphasizes that providers of support also benefit.

Of course, none of these perspectives is contradictory. Slavin et al. (2003) suggested a *logic model* in which group goals lead to both an increased personal motivation to learn as well as a motivation to ensure one's group-mates also learn. This in turn yields peer assessment and correction, with elaborated explanations and peer modeling, resulting in enhanced learning. However, although the four perspectives may be complementary, there are marked differences in the degree of empirical support for each.

Slavin (1995) reviewed studies of cooperative learning in elementary and secondary schools and found 99 that involved interventions of at least 4 weeks' duration in which the performance of students involved in cooperative learning was compared with students who were not. Of the 64 studies that provided *group* rewards that were contingent on the aggregate of the learning of *individual* members, more than 75% (50 out of 64) found significant positive effects on learning, and in none of the 64 studies was the effect of cooperative learning negative. The median effect size in these studies was 0.32 standard deviations, compared with a median effect size of 0.07 in the studies where there were no group rewards or where the group rewards were based on a single group product.

In reviewing the available research on interventions designed to increase social cohesion, Slavin (1995) concluded that interventions that are designed to increase social cohesion but do not provide rewards based on the learning of all group members are no more effective than traditional instruction.

A review of 17 studies looking at the effects of peer interaction on the learning of mathematics, ranging from 2nd grade to 11th grade (Webb, 1991), found that those who provided help benefited little when this help was in the form of answers, procedural information, or managerial information. However, when the feedback was in the form of elaborated explanations, those providing help benefited substantially. The median partial correlation between giving help in the form of elaborated explanations and achievement in mathematics, across the 10 studies on which the relevant information was available, was 0.27. In other words, a 1 standard deviation increase in giving elaborated help resulted in a 0.27 increase in achievement for the help-giver even after controlling for ability. The same study also found a strong negative effect of approximately the same magnitude for students who asked for help but were given only the answer. The fact that these results were similar across topics and grades suggests that they are likely to generalize to a variety of mathematics classrooms.

Perhaps surprisingly, the effect of peer tutoring can be almost as strong as one-to-one instruction from a teacher and can be stronger than small-group instruction from a teacher, although some studies have found a strong correlation between the mathematical ability of the student giving help, the quality of explanation given, and the amount of learning that results (Fuchs et al., 1996). In a study by Shacter (2000) 109 students drawn from one fourth-grade classroom, one-fifth grade classroom and three combined fifth- and sixth-grade classrooms were randomly assigned to one of four treatments: teacher-student dyad, student-student dyad, same-age peer group with a teacher, and same-age peer group without a teacher. Compared with student-student pairs, the highest achievement was found in teacher-student pairs (standardized effect size of 0.53, 0.86, and 0.55 on concept mapping, essay, and declarative knowledge respectively), but the achievement of students in student-led groups was almost as high (effect sizes 0.50, 0.57, and 0.30), and on average higher than the achievement in teacher-led groups (0.33, 0.67, 0.14). These effects may appear to be larger than some of the results reported above, but the measures in this study were much more sensitive to instruction than is typical for standardized tests.

In fact studies that have explored the effects of group rewards in combination with other interventions have consistently concluded that the "active ingredient" is the establishment of group goals and individual accountability. Fantuzzo, King, and Heller (1992) compared the effects of reward and structure in the context of reciprocal peer teaching in mathematics. In one group, pairs of students were rewarded with a choice of activities if the sum of their scores on a test exceeded a threshold set by the teacher (reward only). In a second group students were trained in a structured method for tutoring each other, taking turns to be tutor and tutee (structure only). A third group of students received both interventions (reward + structure) while a fourth received neither (control). The results showed that reward had a greater effect on mathematics achievement than structure, but the combined effect was greater still. In another study (Fantuzzo, Davis, & Ginsburg, 1995), 72 fourth- and fifth-grade students in urban schools evidencing difficulties in mathematics were assigned randomly to one of three treatments. One-third of the students received a home-based parental involvement program, one third of the students received the parental involvement together with a reciprocal peer-tutoring intervention, and one third of the students were assigned to a control group where they just practiced work on which they were having difficulties. Students engaged in both reciprocal peer teaching and parental involvement were more confident and had higher mathematics achievement (on both classroom assessments and standardized tests) than either of the other two groups.

However, despite the clarity of the research evidence about the importance of group goals and individual accountability, it appears that although teachers acknowledge the importance of collaborative learning, few implement it in a way that the research suggests would be necessary for it to be effective. Antil, Jenkins, Wayne, and Vadasy (1998) surveyed 85 elementary school teachers in two districts, and although 93% of the teachers said they employed collaborative learning, follow-up interviews with 21 of the teachers showed that only 5 teachers implemented collaborative learning in such a way as to create group goals together with individual accountability. Furthermore, only 1 of the 21 teachers implemented collaborative learning that satisfied the more complex criteria proposed by Cohen (1994): open-ended tasks that emphasize higher order thinking, group tasks that require input from other members, multiple tasks related to a central intellectual theme, and roles assigned to different group members. Whatever the research says about the benefits of peer collaboration, it seems to be more difficult for teachers to implement than would at first appear.

Whether cooperative learning is more effective for some subgroups of students than others is still a matter of some debate. Although some studies have found greater effects for higher attaining students (e.g., Stevens & Slavin, 1995), others have found more significant effects for lower-attaining students (e.g., Boaler, 2002). Slavin et al. (2003) concluded that most interventions appear to benefit high, medium, and low achievers equally. There is, however, evidence that cooperative learning interventions are particularly beneficial for students of color. A meta-analysis of 37 studies on the effects of small-group collaborative learning on achievement in postsecondary science, mathematics, engineering, and technology (Springer, Stanne, & Donovan, 1999) found a mean effect size of 0.51, with similar impact on persistence (d = 0.46) and attitudes (d = 0.55). The effects were larger for groups that were held outside class time (d = 0.65) than for those held inside class time (d = 0.44), and larger in 4-year colleges (d = 0.54) than 2-year colleges (d = 0.21). The effect sizes on achievement were also larger for groups of African-American and Latina/o students (d = 0.76) than for white (d = 0.46) or heterogeneous groups (d = 0.42). The particular value of collaborative learning environments for African-American students has also been emphasized by Boykin and his colleagues (Boykin, Coleman, Lilja, & Tyler, 2004; Boykin, Lilja, & Tyler, 2004).

THE REGULATION OF LEARNING

The preceding sections have discussed in some detail the five key strategies of assessment for learning:

- Clarifying and sharing learning intentions and criteria for success;
- Engineering effective classroom discussions, questions, and learning tasks that elicit evidence of learning;
- Providing feedback that moves learners forward;
- Activating students as instructional resources for one another; and
- Activating students as owners of their own learning.

These five strategies can be integrated within a more general theoretical framework of the *regulation of learning processes* as suggested by Perrenoud (1991, 1998).[4] Within such a framework, the actions of the teacher, the learners, and the context of the classroom are all evaluated with respect to the extent to which they contribute to guiding the learning towards the intended goal. In some environments, the responsibility for regulating learning, or "keeping learning on track" is with the teacher, in some it is with the student, and in some it is shared (see Vermunt, 2003, for a classification of different kinds of learning environments in terms of the responsibility for regulation).

From this perspective, the task of the teacher is not necessarily to teach, but to create situations in which students learn. This focus emphasizes what it is that students learn, rather than what teachers do, although in an accountability-driven culture, it is hard to maintain this focus. One rather startling example of this is provided by a study of teachers being trained in teaching problem-solving skills to their students that was undertaken by Deci, Speigel, Ryan, Koestner, and Kauffman (1982). Participating teachers were

[4] In English, the noun *regulation* has two meanings; one refers to the act of regulating and the other to a rule or law to govern conduct, and so, although the former sense is intended here, the word has the unfortunate connotation of the second. In French, the two senses have separate terms (*régulation* and *règlement*) and so the problem does not arise.

randomly allocated to one of two groups, and both groups were introduced to the important ideas about teaching problem-solving. In both groups teachers were given time to practice the problems and were given both a list of useful hints and the actual solutions to all the problems. The only difference between the two groups was that at the end of the training session, one additional statement was made to the teachers in one group: "Remember, it is your responsibility as a teacher to make sure your students perform up to high standards." Subsequent analysis of recordings of the lessons of the teachers who had been told about the importance of performing up to high standards spent twice as much time talking during the teaching session as the other teachers and made three times as many directives and three times as many controlling statements (e.g., words like *should* and *must*).

Obviously, regulating or controlling the activities in which students engage is only indirectly related to the learning that results (Clarke, 2001) and yet teachers appear to find it difficult to shift from planning activities to planning learning (Franke, Fennema, & Carpenter, 1997). This is especially evident in interviews before lessons in which teachers focus much more on the planned activities than on the resulting learning (e.g., "I'm going to have them do X"). In a way, this is inevitable, because only the activities can be manipulated directly. Nevertheless, in teachers who have developed their formative assessment practices, there is clearly a strong shift in emphasis away from regulating the activities in which students engage, and towards the learning that results (Black et al., 2003). Indeed, from such a perspective, even to describe the task of the teacher as teaching is misleading, since it is rather to "engineer" situations in which student learn.

However, in this context, it is important to note that the *engineering of learning environments* does not guarantee that the learning is proceeding in fruitful ways. Many visual-arts classrooms are *productive*, in that they do lead to significant learning on the part of students, but what any given student might learn is impossible to predict. An emphasis on the regulation of learning processes entails ensuring that the learning that is taking place is as intended.

When the learning environment is well regulated, much of the regulation is proactive, through the setting up of *didactical situations* (Brousseau, 1997). The regulation can be unmediated within such didactical situations when, for example, a teacher "does not intervene in person, but puts in place a 'metacognitive culture', mutual forms of teaching and the organization of regulation of learning processes run by technologies or incorporated into classroom organization and management" (Perrenoud, 1998, p. 100). For example, a teacher's decision to use realistic contexts in the mathematics classroom can provide a source of proactive regulation, because then students can determine the reasonableness of their answers. If students calculate that the average cost per slice of pizza (say) is $200, provided they are genuinely engaged in the activity, they will know that this solution is unreasonable, and so the use of realistic settings provides a self-checking mechanism. They are able to keep their learning on track themselves, rather than requiring a teacher's intervention, because they are "metacognitively, motivationally, and behaviorally active participants in their own learning process" (Zimmerman, 1986, p. 308).

On the other hand, the didactical situation may be set up so that the regulation is achieved through the mediation of the teacher or peers. McCaslin and Good (1996) proposed the term *co-regulation* to describe

> the process by which the social instructional environment supports or scaffolds the individual via her relationships within the classroom, relationships with teachers and peers, objects and setting, and ultimately the self. Internalization of these supportive relationships empowers the individual to seek new challenges within co-regulated support. (p. 660)

The teacher, in planning the lesson, can create questions, prompts or activities that evoke responses from the students that the teacher can use to determine the progress of the learning (such as the "hinge-point" questions described above) and, if necessary, to make adjustments to the instruction. Examples of such questions are, "Is calculus exact or approximate?" or "Would your mass be the same on the moon?" (In this context it is worth noting that each of these questions is "closed" in that there is only one correct response—their value is that although they are closed, each question is focused on a specific conception.) In classroom discussion, through careful scaffolding of student discussion, teachers can help their students develop the skills of self-regulation (Mayer & Turner, 2002), especially through the establishment and maintenance of sociomathematical norms (Yackel & Cobb, 1996) allowing students to support one another's learning in appropriate ways (McClain & Cobb, 2001; Ross, 1995).

Thoughtful *upstream* planning (before the lesson) therefore creates, *downstream* (during the lesson), the possibility that the learning activities may change course in the light of the students' responses. These *moments of contingency*—points in the instructional sequence when the instruction can proceed in different directions according to the responses of the student—are at the heart of the regulation of learning.

These moments arise continuously in whole-class teaching, where teachers constantly have to make sense of students' responses, interpreting them in terms of learning needs and making appropriate responses. But they also arise when the teacher circulates around the classroom, looking at individual students' work, observing the extent to which the students are on track. In most teaching of mathematics, the regulation of learning will be relatively tight, so that the teacher will attempt to "bring into line" all learners who are not heading towards the particular goal sought by the teacher—in these topics, the goal of learning is generally both highly specific and common to all the students in a class. In contrast, when the class is doing an investigation, the regulation will be much looser. Rather than a single goal, there is likely to be a broad *horizon* of appropriate goals (Marshall, 2004), all of which are acceptable, and the teacher will intervene to bring the learners into line only when the trajectory of the learner is radically different from that intended by the teacher. In this context, it is worth noting that there are significant cultural differences in how to use this information. In the United States, the teacher will typically intervene with individual students where they appear not to be on track whereas in Japan, the teacher is far more likely to observe all the students carefully, while walking round the class and then will select some major issues for discussion with the whole class. This is consistent with what Bromme and Steinbring (1994) discovered in their expert-novice analysis of two mathematics teachers: The novice teacher tended to treat students' questions as being from individual learners, whereas the expert teacher's responses tended to be directed more to a "collective student."

One of the features that makes a lesson *formative*, then, is that the lesson can change course in the light of evidence about the progress of learning. This is in stark contrast to the traditional pattern of classroom interaction, exemplified by the following extract:

> "Yesterday we talked about triangles, and we had a special name for triangles with three sides the same. Anyone remember what it was? . . . Begins with E . . . equi- . . ."

In terms of formative assessment, there are two salient points about such an exchange. First, little is contingent on the responses of the students, except how long it takes to get on to the next part of the teacher's script, so there is little scope for downstream regulation. The teacher is interested only in getting to the word *equilateral* in order that she can move on, and so all incorrect answers are treated as equivalent. The only information that the teacher extracts from the students' responses is whether they can recall the word equilateral.

The second point is that the situation that the teacher set up in the first place—the question she chose to ask—has little potential for providing the teacher with useful information about the students' thinking, except, possibly, whether the students can recall the word *equilateral.* This is typical in situations where the questions that the teacher uses in whole-class interaction have not been prepared in advance (in other words, when there is little or no proactive or upstream regulation).

In contrast, the vignettes of mathematics teaching given by Kilpatrick et al. (2001) show how exemplary teachers actually design these teachable moments into their lessons: "Mr. Hernandez and Ms. Kaye have each designed the lesson to afford them critical information about their students' progress. The tasks they frame create a strategic space for students' work and for gaining insights into students' thinking" (p. 349). Similar considerations apply when the teacher collects the students' notebooks and attempts to give helpful feedback to the students in the form of comments on how to improve rather than grades or percentage scores. If sufficient attention has not been given upstream to the design of the tasks given to the students, then the teacher may find that she has nothing useful to say to the students. Ideally, from examining the students' responses to the task, the teacher would be able to judge (a) how to help the learners learn better and (b) what she might do to improve the teaching of this topic. In this way, the assessment could be formative for the students, through the feedback she provides, and formative for the teacher herself, in that appropriate analysis of the students' responses might suggest how the lesson could be improved.

As a concrete illustration of these ideas in action, consider the following account from a research project (Lyon, Leahy, Morris, & Thompson, 2005) of a teacher who was working through a task entitled "Up, Up and Away" (Marquez & Boxley, 2003). The task requires students to complete a table and a graph for a weather balloon that rises at a rate of 8 feet per second. The students are then given a graph of a flare that was set off 2 seconds after the weather balloon and asked three questions:

1. What is the maximum height, in feet, that the flare reaches?
2. From the time the flare is set off, how many seconds will the flare take to reach its maximum height?
3. What is the average speed of the flare, in feet per second, from the ground to its highest point from the ground?

This particular class began to work in groups when the graph of the flare was introduced. The teacher said to the class, *"I'm hearing some problems, especially for the third part. Write your answers to the first part on the dry erase."* All the answers held up by the students were correct. The teacher then asked the students to hold up their answers to the second question. She said, *"We have 3 different answers: 3, 4, and 5 seconds. It hits the maximum at what time?"* One student answered, *"5 seconds."* The teacher asked, *"And starts at?"* A student said, *"2 seconds."*

The teacher asked the students to hold up their answers for the third question and saw that there were still different answers. She said, *"We have different answers. We are going to look at a different problem"*. She then took the students through a similar activity, again looking at average speed, but this time in the context of car journeys where speed was measured in miles per hour. After some discussion, students showed, again by displaying their answers on dry-erase boards, that they understood how to calculate from a distance-time graph the average speed over a specified time interval of a body moving at nonconstant speed.

The students then resumed work on the "Up, Up, and Away" task. After a few moments the teacher asked the groups to hold up their new answers to the third question. Almost all the posted answers were correct, so the teacher asked *"If the average speed is 48 feet per second does that mean that the flare is always traveling at that speed?"* The students answered, *"To have a constant speed the graph would be a straight line.* In the post-interview the teacher was asked if this was an on-the fly change to the lesson plan or if the additional problem had been part of her lesson design. The teacher had anticipated that the students might have a problem with the average speed and therefore designed the problem in case there was difficulty during the lesson. She only planned on using the additional problem if it were needed.

This brief extract illustrates several features of the effective regulation of learning. Before the lesson began, the teacher had planned the questions she was going to ask and had provided the students with dry-erase boards, so that she could require responses from all students. She had also developed, within the students, a willingness to be open about their answers, as a step towards the "metacognitive culture" mentioned by Perrenoud above. These are examples of the proactive regulation of learning. As a result of teaching this material previously, she had also identified that the third question might pose particular difficulties—in other words, that this was a hinge point in the lesson—and after some reflection on why this might have occurred, she planned the backup activity. This use of reflection on past teaching to modify future teaching is an example of *retroactive* regulation of learning. During the lesson, there was also a significant amount of the *interactive* regulation of learning. She listened carefully to the students' discussions and looked at their work as she walked around the classroom. It was as a result of this that she said, "I'm hearing some problems" and decided to collect evidence more systematically about the understanding of her students by asking them to display their responses on dry-erase boards. Because she had evidence from all students, she was in a much better position to make the on-the-fly decision to use the backup activity she had planned. It is also worth noting that the teacher used all-student response systems selectively. At times, she relied on her intuitive perceptions of the class, at times she decided to collect information systematically, and at other times, she allowed students to volunteer contributions. Shulman (2005) describes practices such as these as *pedagogies of uncertainty*. This term accurately reflects the fact that it is impossible to be certain about what students will learn as a result of a particular episode of instruction but the term *uncertainty* suggests that it is impossible to predict anything in advance. As the vignette above shows, in many cases, one might not know what will happen, but one can often reduce the uncertainty by careful planning. In this case, the teacher reduced the possibilities to a simple decision about whether to use the supplementary activity she had prepared. In other cases, the teacher may plan for a more complex range of possibilities, as in the case with the item that asked students for a fraction between $\frac{1}{6}$ and $\frac{1}{7}$, but again, careful consideration of the kinds of responses that students may produce allows teachers to "prefilter" the students' responses. By anticipating the students' responses the teacher can simplify the task she is faced with in the classroom. Rather than pedagogies of uncertainty, therefore, it seems more appropriate to describe such pedagogies as *pedagogies of contingency*. The essence of formative assessment is that instruction is contingent on what it is that the students have learned. Teachers can regulate learning proactively by creating moments of contingency, for example by identifying a hinge point in a sequence of instruction as discussed above, and designing a hinge-point question to be used at that time. Teachers can also regulate learning interactively by capitalizing on moments of contingency that may arise spontaneously in their teaching. Through careful reflection after the lesson, they can plan better for the future, both in terms of how to create more moments of contingency, and in terms of how to recognize and take advantage of such moments when they arise.

SUMMARY

In this chapter, I have outlined some of the research that suggests that focusing on the use of day-to-day formative assessment is one of the most powerful ways of improving learning in the mathematics classroom. In other words, even if teachers do not care about deep understanding and instead wish only to increase their students' test scores, then attention to formative assessment appears to be one of, if not the, most powerful way to do this.

To be effective, these strategies must be embedded into the day-to-day life of the classroom and must be integrated into whatever curriculum scheme is being used. That is why there can be no recipe that will work for everyone. Each teacher will have to find a way of incorporating these ideas into her or his own practice, and effective formative assessment will look very different in different classrooms. It will, however, have some distinguishing features. Students will be thinking more often than they are trying to remember something, they will believe that by working hard they get smarter, they will understand what they are working towards, and they will know how they are progressing. The teachers will ensure that students understand what it is that they are meant to be learning, they will be collecting evidence frequently about the extent of students' progress towards the goal, and they will be making frequent adjustments to the instruction to better meet the learning needs of the students.

In some ways, this is an old-fashioned message—indeed, none of the techniques that teachers have used to put these principles into practice in their classrooms is new. What is new is that researchers now have hard empirical evidence that quality learning does lead to higher achievement, even when performance is measured through externally mandated tests. What is also new is the broad theoretical framework of the regulation of learning, which may help teachers to understand how these ideas can be implemented effectively, so that teachers and students can, together, keep the learning of mathematics "on track."

REFERENCES

Adhami, M., Johnson, D. C., & Shayer, M. (1997, November). *Does CAME work? Summary report on phase 2 of the Cognitive Acceleration in Mathematics Education (CAME) project.* Paper presented at the Day Conference of the British Society for Research into Learning Mathematics, Bristol, UK. Retrieved May 14, 2006, from http://www.bsrlm.org.uk/IPs/ip17-3/BSRLM-IP-17-3-Full.pdf.

Adhami, M., Johnson, D. C., & Shayer, M. (1998). *Thinking maths: Accelerated learning in mathematics (KS3 and S1/S2).* Oxford, UK: Heinemann Educational.

Ames, C. (1992). Classrooms: Goals, structures, and student motivation. *Journal of Educational Psychology, 84*(3), 261–271.

Antil, L. R., Jenkins, J. R., Wayne, S. K., & Vadasy, P. F. (1998). Cooperative learning: Prevalence, conceptualization and the relation between research and practice. *American Educational Research Journal, 35*(3), 419–454.

Arter, J. A., & McTighe, J. (2001). *Scoring rubrics in the classroom.* Thousand Oaks, CA: Corwin Press.

Askew, M., & Wiliam, D. (1995). *Recent research in mathematics education 5–16.* London: Her Majesty's Stationery Office.

Bandura, A. (1977). Self-efficacy: Towards a unifying theory of behavioral change. *Psychological Review, 84*(2), 191–215.

Bandura, A. (1986). *Social foundations of thought and action: A social cognitive theory.* Englewood Cliffs, NJ: Prentice Hall.

Bandura, A. (1997). *Self-efficacy: The exercise of control.* New York: W. H. Freeman.

Bangert-Drowns, R. L., Kulik, C.-L. C., Kulik, J. A., & Morgan, M. (1991). The instructional effect of feedback in test-like events. *Review of Educational Research, 61*(2), 213–238.

Barnett, W. S., & Camilli, G. (2001). Compensatory preschool education, cognitive development and "race". In J. M. Fish (Ed.), *Race and intelligence: Separating science from myth* (pp. 369–406). Mahwah, NJ: Erlbaum.

Bergan, J.R., Sladeczek, I. E., Schwarz, R. D., & Smith, A. N. (1991). Effects of a measurement and planning system on kindergartners' cognitive development and educational programming. *American Educational Research Journal, 28*(3), 683–714.

Berliner, D. C. (1994). Expertise: The wonder of exemplary performances. In J. N. Mangieri & C. C. Block (Eds.), *Creating powerful thinking in teachers and students: Diverse perspectives* (pp. 161–186). Fort Worth, TX: Harcourt Brace College.

Bernstein, B. (Ed.). (1975). *Class, codes and control volume 3: Towards a theory of educational transmissions.* London: Routledge and Kegan Paul.

Bertua, C., Anderson, N., & Salgado, J. F. (2006). The predictive validity of cognitive ability tests: A UK meta-analysis. *Journal of Occupational and Organisational Psychology, 78*(3), 387–409.

Black, P. J., & Wiliam, D. (1998a). Assessment and classroom learning. *Assessment in Education: Principles policy and practice, 5*(1), 7–73.

Black, P. J., & Wiliam, D. (1998b). Inside the black box: Raising standards through classroom assessment. *Phi Delta Kappan, 80*(2), 139–148.

Black, P. J., & Wiliam, D. (2004a). Classroom assessment is not (necessarily) formative assessment (and vice-versa). In M. Wilson (Ed.), *Towards coherence between classroom assessment and accountability: 103rd Yearbook of the National Society for the Study of Education* (Pt. 2, pp. 183–188). Chicago: University of Chicago Press.

Black, P. J., & Wiliam, D. (2004b). The formative purpose: Assessment must first promote learning. In M. Wilson (Ed.), *Towards coherence between classroom assessment and accountability: 103rd Yearbook of the National Society for the Study of Education* (Pt. 2, pp. 20–50). Chicago: University of Chicago Press.

Black, P., & Wiliam, D. (2005a). Lessons from around the world: How policies, politics and cultures constrain and afford assessment practices. *Curriculum Journal, 16*(2), 249–261.

Black, P., & Wiliam, D. (2005b). Developing a theory of formative assessment. In J. Gardner (Ed.), *Assessment and learning* (pp. 81–100). London: Sage.

Black, P., Harrison, C., Lee, C., Marshall, B., & Wiliam, D. (2003). *Assessment for learning: Putting it into practice.* Buckingham, UK: Open University Press.

Black, P., Harrison, C., Lee, C., Marshall, B., & Wiliam, D. (2004). Working inside the black box: Assessment for learning in the classroom. *Phi Delta Kappan, 86*(1), 8–21.

Bloom, B. S. (1969). Some theoretical issues relating to educational evaluation. In R. W. Tyler (Ed.), *Educational evaluation: New roles, new means: The 68th yearbook of the National Society for the Study of Education* (Pt. 2, pp. 26–50). Chicago: University of Chicago Press.

Boaler, J. (2002). *Experiencing school mathematics: Traditional and reform approaches to teaching and their impact on student learning.* Mahwah, NJ: Erlbaum.

Boaler, J., & Humphreys, C. (2005). *Connecting mathematical ideas: Middle school video cases to support teaching and learning.* Portmsouth, NH: Heinemann.

Boaler, J., Wiliam, D., & Zevenbergen, R. (2000). The construction of identity in secondary mathematics education. In J. F. Matos & M. Santos (Eds.), *Mathematics education and society* (pp. 192–202). Montechoro, Portugal: Centro de Investigação em Educação da Faculdade de Ciências Universidade de Lisboa.

Boekaerts, M. (1993). Being concerned with well being and with learning. *Educational Psychologist, 28*(2), 149–167.

Boekaerts, M. (2001). Context sensitivity: Activated motivational beliefs, current concerns and emotional arousal. In S. Volet & S. Järvelä (Eds.), *Motivation in learning contexts: Theoretical advances and methodological implications* (pp. 17–31). Oxford, UK: Pergamon.

Boekaerts, M. (2006). Self-regulation and effort investment. In K. A. Renninger & I. E. Sigel (Eds.), *Handbook of child psychology: Vol. 4. Child psychology in practice* (6th ed., pp. 345–377). New York: Wiley.

Boekaerts, M., & Corno, L. (2005). Self-regulation in the classroom: A perspective on assessment and intervention. *Applied Psychology: An International Review, 54*(2), 199–231.

Boekaerts, M., Maes, S., & Karoly, P. (2005). Self-regulation across domains of applied psychology: Is there an emerging consensus? *Applied Psychology: An International Review, 54*(2), 149–154.

Boykin, A. W., Coleman, S. T., Lilja, A., & Tyler, K. M. (2004). *Building on children's cultural assets in simulated classroom performance environments: research vistas in the communal learning paradigm* (Rep. No. 68). Washington, DC: Howard University.

Boykin, A. W., Lilja, A., & Tyler, K. M. (2004). The influence of communal vs. individual learning context on the academic performance in social studies of grade 4–5 African-Americans. *Learning Environments Research, 7*(3), 227–244.

Bransford, J. D., Brown, A. L., & Cocking, R. R. (Eds.). (2000). *How people learn: Brain, mind, experience, and school.* Washington, DC: National Academy Press.

Broadfoot, P. M., Daugherty, R., Gardner, J., Gipps, C. V., Harlen, W., James, M., et al. (1999). *Assessment for learning: Beyond the black box.* Cambridge, UK: University of Cambridge School of Education.

Bromme, R., & Steinbring, H. (1994). Interactive development of subject matter in the mathematics classroom. *Educational Studies in Mathematics, 27*(3), 217–248.

Brookhart, S. M. (1997). A theoretical framework for the role of classroom assessment in motivating student effort and achievement. *Applied Measurement in Education, 10*(2), 161–180.

Brookhart, S. M. (2004). Classroom assessment: Tensions and intersections in theory and practice. *Teachers College Record, 106*(3), 429–458.

Brophy, J. (1981) Teacher praise: A functional analysis. *Review of Educational Research, 51*(1) 5–32.

Brousseau, G. (1984). The crucial role of the didactical contract in the analysis and construction of situations in teaching and learning mathematics (G. Seib, Trans.). In H.-G. Steiner (Ed.), *Theory of mathematics education: ICME 5 topic area and miniconference* (Vol. 54, pp. 110–119). Bielefeld, Germany: Institut für Didaktik der Mathematik der Universität Bielefeld.

Brousseau, G. (1997). *Theory of didactical situations in mathematics* (N. Balacheff, M. Cooper, R. Sutherland, & V. Warfield, Trans.). Dordrecht, The Netherlands: Kluwer.

Brown, A. (1987). Metacognition, executive control, self-regulation and other more mysterious mechanisms. In F. E. Weinert & R. H. Kluwe (Eds.), *Metacognition, motivation and understanding* (pp. 65–116). HIllsdale, NJ: Erlbaum.

Butler, D. L., & Winne, P. H. (1995). Feedback and self-regulated learning: A theoretical synthesis. *Review of Educational Research, 65*(3), 245–281.

Butler, R. (1987) Task-involving and ego-involving properties of evaluation: Effects of different feedback conditions on motivational perceptions, interest and performance. *Journal of Educational Psychology, 79*(4), 474–482.

Butler, R. (1988) Enhancing and undermining intrinsic motivation: The effects of task-involving and ego-involving evaluation on interest and performance. *British Journal of Educational Psychology, 58*, 1–14.

Butler, R., & Nisan, M. (1986). Effects of no feedback, task-related comments, and grades on intrinsic motivation and performance. *Journal of Educational Psychology, 78*(3), 210–216.

Cameron, J. A., & Pierce, W. D. (1994). Reinforcement, reward, and intrinsic motivation: A meta-analysis. *Review of Educational Research, 64*(3), 363–423.

Carol. (2006, February 22). *Conversations with children and literacy.* Retrieved May 19, 2006, from feed://themediansib.com/category/this-n-that/feed/.

Carpenter, T. P., Fennema, E., Peterson, P. L., Chiang, C. P., & Loef, M. (1989). Using knowledge of children's mathematics thinking in classroom teaching: an experimental study. *American Educational Research Journal, 26*(4), 499–531.

Carr, M. M., Alexander, J., & Folds-Bennett, T. (1994). Metacognition and mathematics strategy use. *Applied Cognitive Psychology, 8*, 583–595.

Chaiklin, S. (2003). The zone of proximal development in Vygotsky's analysis of learning and instruction. In A. Kozulin, B. Gindis, V. S. Ageyev & S. M. Miller (Eds.), *Vygotsky's educational theory and practice in cultural context* (pp. 39–64). Cambridge, UK: Cambridge University Press.

Ciofalo, J., & Wylie, E. C. (2006, January 10). Using diagnostic classroom assessment: One question at a time. *Teachers College Record.* Retrieved January 18, 2006, from http://www.tcrecord.org/content.asp?contentid=12285.

Clarke, D. (1996). Assessment. In A. J. Bishop, K. Clements, C. Keitel, J. Kilpatrick, & C. Laborde (Eds.), *International handbook of mathematics education* (pp. 327–370). Dordrecht, The Netherlands: Kluwer.

Clarke, S. (2001). *Unlocking formative assessment.* London: Hodder & Stoughton.

Claxton, G. L. (1995). What kind of learning does self-assessment drive? Developing a 'nose' for quality: Comments on Klenowski. *Assessment in Education: Principles, policy and practice, 2*(3), 339–343.

Cohen, E. G. (1994). Restructuring the classroom: Conditions for productive small groups. *Review of Educational Research, 64*(1), 1–35.

Cohen, J. (1988). *Statistical power analysis for the behavioural sciences.* Hillsdale, NJ: Erlbaum.

Connell, J. P., & Wellborn, J. G. (1991). Competence, autonomy, and relatedness: A motivational analysis of self-system processes. In M. R. Gunnar, & L. A. Sroufe (Eds.), *Minnesota symposia on child psychology* (Vol. 23, pp. 43–77). Hillsdale, NJ: Erlbaum.

Corno, L. (2001). Volitional aspects of self-regulated learning. In B. J. Zimmerman, & D. H. Schunk (Eds.), *Self-regulated leaning and academic achievement: Theoretical perspectives* (2nd ed., pp. 191–225). Hillsdale, NJ: Erlbaum.

Crespo, S. (2000). Seeing more than right and wrong answers: Prospective teachers' interpretations of students' mathematical work. *Journal of Mathematics Teacher Education, 3,* 155–181.

Crooks, T. J. (1988). The impact of classroom evaluation practices on students. *Review of Educational Research, 58*(4), 438–481.

Csikszentmihalyi, M. (1990). *Flow: The psychology of optimal experience.* New York: Harper & Row.

Damon, W. (1984). Peer education: the untapped potential. *Journal of Applied Developmental Psychology,* **5**, 331–343.

Davis, B. (1997). Listening for differences: An evolving conception of mathematics teaching. *Journal for Research in Mathematics Education, 28*(3), 355–376.

Dawes, L., Mercer, N., & Wegerif, R. (2000). *Thinking Together: A programme of activities for developing thinking skills at KS2.* Birmingham, UK: Questions Publishing Company.

Day, J. D., & Cordón, L. A. (1993). Static and dynamic measures of ability: An experimental comparison. *Journal of Educational Psychology, 85*(1), 76–82.

Deary, I. (2000). *Looking down on human intelligence* (Vol. 34). Oxford, UK: Oxford University Press.

Deci, E. L. (1996). *Why we do what we do.* New York: Penguin.

Deci, E. L., & Ryan, R. M. (1985). *Intrinsic motivation and self-determination in human behavior.* New York: Plenum.

Deci, E. L., & Ryan, R. M. (1994). Promoting self-determined education. *Scandinavian Journal of Educational Research, 38*(1), 3–14.

Deci, E. L., Speigel, N. H., Ryan, R. M., Koestner, R., & Kauffman, M. (1982). The effects of performance standards on teaching styles: The behavior of controlling teachers. *Journal of Educational Psychology, 74,* 852–859.

DeCorte, E., & Verschaffel, L. (2006). Mathematical thinking and learning. In K. A. Renninger & I. E. Sigel (Eds.), *Handbook of child psychology: Vol. 4 . Child psychology in practice* (6th ed., pp. 103–152). New York: Wiley.

DeCorte, E., Verschaffel, L., & Op't Eynde, P. (2000). Self-regulation: A characteristic and a goal of mathematics education. In M. Boekaerts, P. R. Pintrich, & M. Zeidner (Eds.), *Handbook of self-regulation* (pp. 687–726). San Diego, CA: Academic Press.

Deevers, M. (2006, April). *Linking classroom assessment practices with student motivation in mathematics* Paper presented at Annual meeting of the American Educational Research Association. San Francisco, CA.

Dempster, F. N. (1991). Synthesis of research on reviews and tests. *Educational Leadership, 48*(7), 71–76.

Dempster, F. N. (1992). Using tests to promote learning: A neglected classroom resource. *Journal of Research and Development in Education, 25*(4), 213–217.

Dewey, J. (1913). *Interest and effort in education.* Boston: Riverside Press.

Dickens, W. T., & Flynn, J. R. (2001). Heritability estimates versus large environmental effects: The IQ paradox resolved. *Psychological Review, 108,* 346–369.

Dillon, J. T. (1988). *Questioning and teaching: A manual of practice.* London: Croom Helm.

diSessa, A. A., & Minstrell, J. (1998). Cultivating conceptual change with benchmark lessons. In J. G. Greeno & S. V. Goldman (Eds.), *Thinking practices in mathematical and science learning* (pp. 155–187). Mahwah, NJ: Erlbaum.

Dorr-Bremme, D. W., & Herman, J. L. (1986). *Assessing student achievement: A profile of classroom practices* (Vol. 11). Los Angeles: University of California Los Angeles Center for the Study of Evaluation.

Dweck, C. S., & Leggett, E. L. (1986). Motivational processes affecting learning. *American Psychologist, 41*(10), 1040–1048.

Dweck, C. S. (2000). *Self-theories: Their role in motivation, personality and development.* Philadelphia: Psychology Press.

Eccles, J. S., Adler, T. F., Futterman, R., Goff, S. B., Kaczala, C. M., Meece, J. L., et al. (1983). Expectancies, values, and academic behaviors. In J. T. Spence (Ed.), *Achievement and achievement motivation* (pp. 75–146). San Francisco: W. H. Freeman.

Eisenberger, R., & Cameron, J. A. (1996). Detrimental effects of reward: Reality or myth? *American Psychologist, 51*(11), 1153–1166.

Elawar, M. C., & Corno, L. (1985). A Factorial experiment in teachers' written feedback on student homework: Changing teacher behavior a little rather than a lot. *Journal of Educational Psychology, 77*(2), 162–173.

Elshout-Mohr, M. (1994). Feedback in self-instruction. *European Education, 26*(2), 58–73.

English, L. D., & Doerr, H. M. (2004). *Listening and responding to students' ways of thinking.* Paper presented at the 27th annual conference of the Mathematics Education Research Group of Australasia: Mathematics education for the third millennium: towards 2010. Townsville, Queensland, Australia.

Ernest, P. (1991). *The philosophy of mathematics education* (Vol. 1). London: Falmer Press.

Even, R., & Tirosh, D. (1995). Subject-matter knowledge and the knowledge about students as sources of teacher presentations of the subject-matter. *Educational Studies in Mathematics, 29*(1), 1–20.

Even, R., & Tirosh, D. (2002). Teacher knowledge and understanding of students' mathematical learning. In L. D. English (Ed.), *Handbook of international research in mathematics education* (pp. 219–240). Mahwah, NJ: Erlbaum.

Fantuzzo, J. W., Davis, G. Y., & Ginsburg, M. D. (1995). Effects of parent involvement in isolation or in combination with peer tutoring on student self-concept and mathematics achievement. *Journal of Educational Psychology, 87*(2), 272–281.

Fantuzzo, J. W., King, J., & Heller, L. R. (1992). Effects of reciprocal peer tutoring on mathematics and school adjustment: A component analysis. *Journal of Educational Psychology, 84*, 331–339.

Fennema, E., Carpenter, T. P., Franke, M. L., Levi, L., Jacobs, V. R., & Empson, S. B. (1996). A longitudinal study of learning to use children's thinking in mathematics instruction. *Journal for Research in Mathematics Education, 27*(4), 403–434.

Fernandez, C., & Makoto, Y. (2004). *Lesson study: A Japanese approach to improving mathematics teaching and learning.* Mahwah, NJ: Erlbaum.

Fish, J. M. (Ed.). (2001). *Race and intelligence: Separating science from myth.* Mahwah, NJ: Erlbaum.

Flavell, J. H. (1976). Metacognitive aspects of problem solving. In L. B. Resnick (Ed.), *The nature of intelligence* (pp. 231–235). Hillsdale, NJ: Erlbaum.

Flavell, J. H. (1979). Metacognition and cognitive monitoring: A new area of cognitive-developmental inquiry. *American Psychologist, 34*, 906–911.

Flavell, J. H. (1981). Cognitive monitoring. In W. P. Dickson (Ed.), *Children's oral communication skills* (pp. 35–60). New York: Academic Press.

Flynn, J. R. (1984). The mean IQ of Americans: Massive gains 1932 to 1978. *Psychological Bulletin, 95*(1), 58–64.

Flynn, J. R. (1987). Massive IQ gains in 14 nations: What IQ tests really measure. *Psychological bulletin, 101*(2), 171–191.

Fontana, D. & Fernandes, M. (1994). Improvements in mathematics performance as a consequence of self-assessment in Portuguese primary school pupils. *British Journal of Educational Psychology, 64*, 407–417.

Foos, P. W., Mora, J. & Tkacz, S. (1994). Student study techniques and the generation effect. *Journal of Educational Psychology, 86*(4), 567–576.

Franke, M. L., Carpenter, T. P., Levi, L., & Fennema, E. (2001). Capturing teachers' generative change: A follow-up study of professional development in mathematics. *American Educational Research Journal, 38*(3), 653–689.

Franke, M. L., Fennema, E., & Carpenter, T. P. (1997). Teachers creating change: Examining evolving beliefs and classroom practices. In E. Fennema & B. S. Nelson (Eds.), *Mathematics teachers in transition* (pp. 255–282). Mahwah, NJ: Erlbaum.

Fuchs, L. S., Fuchs, D., Hamlett, C. L., & Stecker, P. M. (1991). Effects of curriculum-based measurement and consultation on teacher planning and student achievement in mathematics operations. *American Educational Research Journal, 28*(3), 617–641.

Fuchs, L. S., Fuchs, D., Karns, K., Hamlett, C. L., Dutka, S., & Katzaroff, M. (1996). The relation between student ability and the quality and effectiveness of explanations. *American Educational Research Journal, 33*(3), 631–664.

Fuson, K. C., Kalchman, M., & Bransford, J. D. (2005). Mathematical understanding: An introduction. In M. S. Donovan, & J. Bransford (Eds.), *How students learn: History, mathematics and science in the classroom* (pp. 217–256). Washington, DC: National Academies Press.

Gay, S., & Thomas, M. (1993). Just because they got it right, does it mean they know it? In N. L. Webb & A. F. Coxford (Eds.), *Assessment in the mathematics classroom: 1993 yearbook of the National Council of Teachers of Mathematics* (pp. 130–134). Reston, VA: National Council of Teachers of Mathematics.

Gipps, C. V., & Stobart, G. (1997). *Assessment: A teacher's guide to the issues* (3rd ed.). London: Hodder and Stoughton.

Girling, M. (1977). Towards a definition of basic numeracy. *Mathematics Teaching, 81*, 4–5.

Glaser, R., & Silver, E. A. (1994). Assessment, testing and instruction: Retrospect and prospect. In L. Darling-Hammond (Ed.), *Review of research in education* (Vol. 20, pp. 393–419). Washington, DC: American Educational Research Association.

Good, T. L., & Grouws, D. A. (1975). *Process-product relationships in fourth grade mathematics classrooms.* (Rep. No. NE-G-00-0-0123) National Institute of Education, Columbia: University of Missouri.

Gray, E. M., & Tall, D. O. (1994). Duality, ambiguity and flexibility: A 'proceptual' view of simple arithmetic. *Journal for Research in Mathematics Education, 25*(2), 116–140.

Grolnick, W. S., & Ryan, R. M. (1987). Autonomy in children's learning: An experimental and individual difference investigation. *Journal of Personality and Social Psychology, 52*(5), 890–898.

Hacker, D. J., Dunlosky, J., & Graesser, A. C. (Eds.). (1998). *Metacognition in educational theory and practice.* Mahwah, NJ: Erlbaum.

Harlen, W., & Deakin-Crick, R. (2002). A systematic review of the impact of summative assessment and tests on students' motivation for learning (version 1.1). *Research Evidence in Education Library.* London: University of London Institute of Education Social Science Research Unit. Retrieved May 28, 2006, from http://eppi.ioe.ac.uk/EPPIWebContent/reel/review_groups/assessment/ass_rv1/ass_rv1.pdf.

Hart, K. M. (Ed.). (1981). *Children's understanding of mathematics: 11–16.* London: John Murray.

Hart, K. M., Brown, M. L., Kerslake, D., Küchemann, D., & Ruddock, G. (1985). *Chelsea diagnostic mathematics tests.* Windsor, UK: NFER-Nelson.

Heid, M. K., Blume, G. W., Zbiek, R. M., & Edwards, B. S. (1999). Factors that influence teachers learning to do interviews to understand students' mathematical understandings. *Educational Studies in Mathematics, 37*(3), 223–249.

Helmke, A., & Schrader, F. W. (1987). Interactional effects of instructional quality and teacher judgement accuracy on achievement. *Teaching and Teacher Education, 3*(2), 91–98.

Herrnstein, R. J., & Murray, C. (1994). *The bell curve.* New York: Free Press.

Hickey, D. T., & McCaslin, M. (2001). A comparative, sociocultural analysis of context and motivation. In S. Volet, & S. Järvelä (Eds.), *Motivation in learning contexts: Theoretical advances and methodological implications* (pp. 33–55). Oxford, UK: Pergamon.

Hidi, S., & Harackiewicz, J. M. (2000). Motivating the academically unmotivated: A critical issue for the 21st century. *Review of Educational Research, 70*(2), 151–179.

Hiebert, J., Gallimore, R., Garnier, H., Givvin, K. B., Hollingsworth, H., Jacobs, J. K., et al. (2003). *Teaching mathematics in seven countries: Results from the TIMSS 1999 Video Study.* Washington, DC: National Center for Education Statistics.

Hiebert, J., Stigler, J. W., Jacobs, J. K., Givvin, K. B., Garnier, H., Smith, M., et al. (2005). Mathematics teaching in the United States today (and tomorrow): Results from the TIMSS 1999 Video Study. *Educational Evaluation and Policy Analysis, 27*(2), 111–132.

Hodgen, J., & Wiliam, D. (2006). *Mathematics inside the black box: Assessment for learning in the mathematics classroom.* London: NFER-Nelson.

Hoek, D., van den Eeden, P., & Terwel, J. (1999). The effects of integrated social and cognitive strategy instruction on the mathematics achievement in secondary education. *Learning and Instruction, 9*(5), 427–448.

James, M. (1992). *Assessment for learning.* Annual Conference of the Association for Supervision and Curriculum Development (Assembly session on Critique of Reforms in Assessment and Testing in Britain). New Orleans, LA.

Keddie, N. (1971). Classroom knowledge. In M. F. D. Young (Ed.), *Knowledge and control.* London: Collier-Macmillan.

Kilpatrick, J., Swafford, J. O., & Findell, B. (Eds.). (2001). *Adding it up: Helping children learn mathematics.* Washington, DC: National Academy Press.

Klein, D. (2003). A brief history of American K–12 mathematics education in the 20th century. In J. M. Royer (Ed.), *Mathematical cognition.* Greenwich, CT: Information Age Publishing.

Kluger, A. N., & DeNisi, A. (1996). The effects of feedback interventions on performance: A historical review, a meta-analysis, and a preliminary feedback intervention theory. *Psychological Bulletin, 119*(2), 254–284.

Kluwe, R. H. (1982). Cognitive knowledge and executive control: Metacognition. In D. R. Griffin (Ed.), *Animal mind—human mind* (pp. 201–224). New York: Springer-Verlag.

Kohn, A. (1999). *Punished by rewards: The trouble with gold stars, incentive plans, A's, praise and other bribes* (2nd ed.). Boston: Houghton-Mifflin.

Kohn, A. (2006). The trouble with rubrics. *English Journal, 95*(4), 12–15.

Kramarski, B., Mevarech, Z. R., & Arami, M. (2002). The effects of metacognitive instruction on solving mathematical authentic tasks. *Educational Studies in Mathematics, 49*(2), 225–250.

Kramarski, B., Mevarech, Z. R., & Lieberman, A. (2001). Effects of multilevel versus unilevel metacognitive training on mathematical reasoning. *Journal of Educational Research, 94,* 292–300.

Lampert, M. (2001). *Teaching problems and the problems of teaching.* New Haven, CT: Yale University Press.

Lave, J., Murtaugh, M., & de la Roche, O. (1984). The dialectic of arithmetic in grocery shopping. In B. Rogoff, & J. Lave (Eds.), *Everyday cognition: Its development in social context* (pp. 67–94). Cambridge, MA: Harvard University Press.

Leahy, S., Lyon, C., Thompson, M., & Wiliam, D. (2005). Classroom assessment: Minute-by-minute and day-by-day. *Educational Leadership, 63*(3), 18–24.

Lesh, R., Hoover, M., Hole, B., Kelly, A. E., & Post, T. (2001). Principles for developing thought-revealing activities for students and teachers. In A. E. Kelly, & R. A. Lesh (Eds.), *Handbook of research design in mathematics and science education* (pp. 591–646). Mahwah, NJ: Erlbaum.

Lester Jr., F. K., Lambdin, D. V., & Preston, R. V. (1997). A new vision of the nature and purposes of assessment in the mathematics classroom. In G. D. Phye (Ed.), *Handbook of classroom assessment: Learning, adjustment and achievement* (pp. 287–319). San Diego, CA: Academic Press.

Linnenbrink, E. A. (2005). The dilemma of performance-approach goals: The use of multiple goal contexts to promote students' motivation and learning. *Journal of Educational Psychology, 97*(2), 197–213.

Lodico, M. G., Ghatala, E. S., Levin, J. R., Pressley, M., & Bell, J. A. (1983). The effects of strategy-monitoring training on children's selection of effective memory strategies. *Journal of Experimental Child Psychology, 35*(2), 263–277.

Looney, J. (Ed.). (2005). *Formative assessment: Improving learning in secondary classrooms.* Paris: Organisation for Economic Cooperation and Development.

Lyon, C., Leahy, S., Morris, T., & Thompson, M. (2005). *Lessons learned (the hard way) in the Evidence-Centered Teaching in Algebra project.* [Research Memorandum]. Princeton, NJ: Educational Testing Service.

Mackintosh, N. J. (2000). *IQ and human intelligence.* Oxford, UK: Oxford University Press.

Marquez, E., & Boxley, B. (2003). *Teacher assistance package guide 3: Using mathematical models to represent and understand quantitative relationships.* Princeton, NJ: Educational Testing Service.

Marshall, B. (2004). Goals or horizons—the conundrum of progression in English: Or a possible way of understanding formative assessment in English. *Curriculum Journal, 15*(2), 101–113.

Marso, R. N., & Pigge, F. L. (1993). Teachers' testing knowledge, skills, and practices. In S. L. Wise (Ed.), *Teacher training in measurement and assessment skills* (pp. 129–185). Lincoln, NE: Buros Institute of Mental Measurements.

Mayer, D. K., & Turner, J. C. (2002). Using instructional discourse analysis to study the scaffolding of student self-regulation. *Educational Psychologist, 37*(1), 17–25.

Mayer, D. K., Turner, J. C., & Spencer, C. A. (1997). Challenge in a mathematics classroom: Students' motivation and strategies in project-based learning. *Elementary School Journal, 97*(5), 501–521.

McCaslin, M., & Good, T. L. (1996). The informal curriculum. In D. C. Berliner, & R. C. Calfee (Eds.), *Handbook of educational psychology* (pp. 622–670). New York: Macmillan.

McCaslin, M., & Hickey, D. T. (2001). Educational psychology, social constructivism, and educational practice: A case of emergent identity. *Educational Psychologist, 36*(2), 133–140.

McClain, K., & Cobb, P. (2001). An analysis of development of sociomathematical norms in one first-grade classroom. *Journal for Research in Mathematics Education, 32*(3), 236–266.

McMorris, R. F., & Boothroyd, R. A. (1993). Tests that teachers build: An analysis of classroom tests in science and mathematics. *Applied Measurement in Education, 6,* 321–342.

Mehan, H. (1979). *Learning lessons: Social organization in the classroom.* Cambridge, MA: Harvard University Press.

Meisels, S. J., Atkins-Burnett, S., Xue, Y., Bickel, D. D., & Son, S.-H. (2003). Creating a system of accountability: The impact of instructional assessment on elementary children's achievement test scores. *Education Policy Analysis Archives, 11*(9). Retrieved December 12, 2005, from http://epaa.asu.edu/epaa/v11n9/.

Mercer, N., Dawes, L., Wegerif, R., & Sams, C. (2004). Reasoning as a scientist: Ways of helping children to use language to learn science. *British Educational Research Journal, 30*(3), 359–377.

Mevarech, Z. R., & Kramarski, B. (1997). IMPROVE: A multidimensional method for teaching mathematics in heterogeneous classrooms. *American Educational Research Journal, 34*(2), 365–394.

Mevarech, Z. R., & Kramarski, B. (2003). The effects of metacognitive training versus worked-out examples on students' mathematical reasoning. *British Journal of Educational Psychology, 73*(4), 449–471.

Mitchell, M. (1993). Situational interest: Its multifaceted structure in the secondary school mathematics classroom. *Journal of Educational Psychology, 85*, 424–436.

Montague, A. (Ed.). (1999). *Race and IQ*. New York: Oxford University Press.

National Council of Teachers of Mathematics. (1995). *Assessment standards for school mathematics*. Reston, VA: Author.

Natriello, G. (1987). The impact of evaluation processes on students. *Educational Psychologist, 22*(2), 155–175.

Neisser, U. (Ed.). (1998). *The rising curve: Long-term gains in IQ and related measures*. Washington, DC: American Psychological Association.

Nisbett, R. E., & Ross, L. D. (1980). *Human inference: Strategies and shortcomings of social judgment*. Englewood Cliffs, NJ: Prentice Hall.

Niss, M. (1993). Assessment in mathematics education and its effects. In M. Niss (Ed.), *Investigations into assessment in mathematics education: An ICMI study* (pp. 1–30). Dordrecht, The Netherlands: Kluwer.

Nyquist, J. B. (2003). *The benefits of reconstruing feedback as a larger system of formative assessment: a meta-analysis*. Unpublished master's thesis, Vanderbilt University, Nashville, Tennessee.

Op't Eynde, P., DeCorte, E., & Verschaffel, L. (2001). "What to learn from what we feel?" The role of students' emotions in the mathematics classroom. In S. Volet, & S. Järvelä (Eds.), *Motivation in learning contexts: Theoretical advances and methodological implications* (pp. 149–167). Oxford, UK: Pergamon.

Pajares, F. (1996). Self-efficacy beliefs in academic settings. *Review of Educational Research, 66*(4), 543–578.

Pellegrino, J. W., Chudowsky, N., & Glaser, R. (Eds.). (2001). *Knowing what students know: The science and design of educational assessment*. Washington, DC: National Academy Press.

Perrenoud, P. (1991). Towards a pragmatic approach to formative evaluation. In P. Weston (Ed.), *Assessment of pupil achievement* (pp. 79–101). Amsterdam, The Netherlands: Swets & Zeitlinger.

Perrenoud, P. (1998). From formative evaluation to a controlled regulation of learning. Towards a wider conceptual field. *Assessment in Education: Principles Policy and Practice, 5*(1), 85–102.

Plomin, R., & Petrill, S. A. (1997). Genetics and intelligence: What's new? *Intelligence, 24*, 53–77.

Polanyi, M. (1958). *Personal knowledge*. Chicago, IL: University of Chicago Press.

Pryor, J., & Crossouard, B. (2005). *A sociocultural theorization of formative assessment*. Paper presented at Sociocultural Theory in Educational Research and Practice Conference. University of Manchester, Manchester, UK..

Ramaprasad, A. (1983). On the definition of feedback. *Behavioural Science, 28*(1), 4–13.

Raven, J. (1960). *Guide to the standard progressive matrices: Sets A, B, C and D*. London: Lewis.

Relearning by Design. (2000). What is a rubric? Retrieved May 1, 2006, from http://www.relearning.org/resources/PDF/rubric_sampler.pdf.

Rodriguez, M. C. (2004). The role of classroom assessment in student performance on TIMSS. *Applied Measurement in Education, 17*(1), 1–24.

Romberg, T. A. (1992). Perspectives on scholarship and research methods. In D. A. Grouws (Ed.), *Handbook of research on mathematics teaching and learning* (pp. 49–64). New York: Macmillan.

Roschelle, J., Abrahamson, L., & Penuel, W. R. (2004, April). *Integrating classroom network technology and learning theory to improve classroom science learning: a literature synthesis*. Paper presented at the annual meeting of the American Educational Research Association, San Diego, CA.

Rosenshine, B. V., Meister, C., & Chapman, S. (1996). Teaching students to generate questions: A review of intervention studies. *Review of Educational Research, 66*(2), 181–221.

Ross, J. A. (1995). Effects of feedback on student behavior in cooperative learning groups in a grade-7 math class. *Elementary School Journal, 96*(2), 125–143.

Ross, J. A., Hogaboam-Gray, A., & Rolheiser, C. (2002). Student self-evaluation in grade 5–6 mathematics: Effects on problem solving achievement. *Educational Assessment, 8*(1), 43–58.

Ross, J. A., Rolheiser, C., & Hogaboam-Gray, A. (2002). Influences on student cognitions about evaluation. *Assessment in Education: Principles Policy and Practice, 9*(1), 81–95.

Rowan, B., Harrison, D. M., & Hayes, A. (2004). Using instructional logs to study mathematics curriculum and teaching in the early grades. *Elementary School Journal, 105*, 103–127.

Rowe, M. B. (1974). Wait time and rewards as instructional variables, their influence on language, learning and fate control. *Journal of Research in Science Teaching, 11*, 81–94.

Ryan, R. M., & Connell, J. P. (1989). Perceived locus of causality and internalization: Examining reasons for acting in two domains. *Journal of Personality and Social Psychology, 57*, 749–761.

Ryan, R. M., & Deci, E. L. (2000). Intrinsic and extrinsic motivations: Classic definitions and new directions. *Contemporary Educational Psychology, 25*, 54–67.

Ryan, R. M., & Stiller, J. (1991). The social contexts of internalization: Parent and teacher influences on autonomy, motivation, and learning. In M. L. Maehr, & P. R. Pintrich (Eds.), *Advances in motivation and achievement* (Vol. 7, pp. 115–149). Greenwich, CT: JAI Press.

Sadler, D. R. (1987). Specifying and promulgating achievement standards. *Oxford Review of Education, 13*, 191–209.

Sadler, D. R. (1989). Formative assessment and the design of instructional systems. *Instructional Science, 18*, 119–144.

Salmon-Cox, L. (1981). Teachers and standardized achievement tests: What's really happening? *Phi Delta Kappan, 62*(5), 631–634.

Saphier, J. (2005). Masters of motivation. In R. DuFour, R. Eaker, & R. DuFour (Eds.), *On common ground: The power of professional learning communities* (pp. 85–113). Bloomington, IL: National Education Service.

Schoenfeld, A. H. (1985). *Mathematical problem-solving*. New York: Academic Press.

Schoenfeld, A. H. (1992). Learning to think mathematically: Problem solving, metacognition and sense making in mathematics. In D. A. Grouws (Ed.), *Handbook of research on mathematics teaching and learning* (pp. 334–370). New York: Macmillan.

Schunk, D. H. (1990). Goal-setting and self-efficacy during self-regulated learning. *Educational Psychologist, 25*, 71–86.

Schunk, D. H. (1991). Self-efficacy and academic motivation. *Educational Psychologist, 26*, 207–231.

Scriven, M. (1967). The methodology of evaluation. In R. W. Tyler, R. M. Gagné & M. Scriven (Eds.), *Perspectives of curriculum evaluation* (Vol. 1, pp. 39–83). Chicago: Rand McNally.

Selden, S. (1999). *Inheriting shame: The story of eugenics and racism in America.* New York: Teachers College Press.

Senk, S. L., Beckman, C. E., & Thompson, D. R. (1997). Assessment and grading in high school mathematics classrooms. *Journal for Research in Mathematics Education, 28*(2), 187–215.

Sfard, A. (1998). On two metaphors for learning and on the dangers of choosing just one. *Educational Researcher, 27*(2), 4–13.

Shacter, J. (2000). Does individual tutoring produce optimal learning? *American Educational Research Journal, 37*(3), 801–829.

Shavelson, R. J., Baxter, G. P., & Pine, J. (1992). Performance assessments: Political rhetoric and measurement reality. *Educational Researcher, 21*(4), 22–27.

Shavelson, R. J., Black, P. J., Wiliam, D., & Coffey, J. (2003). *On aligning formative and summative functions in the design of large-scale assessment systems.* Paper presented at National Research Council workshop on Assessment In Support of Instruction and Learning: Bridging the Gap Between Large-Scale and Classroom Assessment, Washington, DC.

Shayer, M. (2003). Not just Piaget, not just Vygotsky, and certainly not Vygotsky as alternative to Piaget. *Learning and Instruction, 13*(5), 465–485.

Shepard, L. A., Hammerness, K., Darling-Hammond, L., Rust, F., Snowden, J. B., Gordon, E., et al. (2005). Assessment. In L. Darling-Hammond, & J. Bransford (Eds.), *Preparing teachers for a changing world: What teachers should learn and be able to do* (pp. 275–326). San Francisco: Jossey-Bass.

Shulman, L. S. (2005). *The signature pedagogies of the professions of law, medicine, engineering, and the clergy: Potential lessons for the education of teachers* Paper presented at National Science Foundation Mathematics and Science Partnerships Workshop: Teacher Education for Effective Teaching and Learning held at National Research Council Center for Education, Irvine, CA.

Siero, F., & van Oudenhoven, J. P. (1995). The effects of contingent feedback on perceived control and performance. *European Journal of Psychology of Education, 10*(1), 13–24.

Simmons, M., & Cope, P. (1993). Angle and rotation: Effects of differing types of feedback on the quality of response. *Educational Studies in Mathematics, 24*(2), 163–176.

Skemp, R. R. (1977). Relational understanding and instrumental understanding. *Mathematics teaching, 77*, 20–26.

Slavin, R. E. (1995). *Cooperative learning: Theory, research and practice* (2nd ed.). Boston: Allyn & Bacon.

Slavin, R. E., Hurley, E. A., & Chamberlain, A. M. (2003). Cooperative learning and achievement. In W. M. Reynolds, & G. J. Miller (Eds.), *Handbook of psychology: Vol. 7. Educational psychology* (pp. 177–198). Hoboken, NJ: Wiley.

Smith, E., & Gorard, S. (2005). "They don't give us our marks": The role of formative feedback in student progress. *Assessment in Education: Principles Policy and Practice, 12*(1), 21–28.

Sorrentino, R. M., & Higgins, E. T. (1986). Motivation and synergism: Warming up to synergism. In E. T. Higgins, & R. M. Sorrentino (Eds.), *Handbook of motivation and cognition: Foundations of social behavior* (pp. 3–19). New York: Guildford Press.

Springer, L., Stanne, M. E., & Donovan, S. S. (1999). Effects of small-group learning on undergraduates in science, mathematics, engineering and technology: A meta-analysis. *Review of Educational Research, 69*(1), 21–51.

Sternberg, R. J., & Williams, W. (1998). Applying the triarchic theory of human intelligence in the classroom. In R. J. Sternberg, & W. Williams (Eds.), *Intelligence, instruction and assessment: Theory into practice* (pp. 1–15). Mahwah, NJ: Erlbaum.

Stevens, R. J., & Slavin, R. E. (1995). Effects of a cooperative learning approach in reading and writing on academically handicapped and nonhandicapped students. *Elementary School Journal, 95*(3), 241–262.

Stiggins, R. J. (2001). *Student-involved classroom assessment* (3rd ed.). Upper Saddle River, NJ: Prentice-Hall.

Stiggins, R. J. (2002). Assessment crisis: The absence of assessment for learning. *Phi Delta Kappan, 83*(10), 758–765.

Stiggins, R. J., & Bridgeford, N. J. (1985). The ecology of classroom assessment. *Journal of Educational Measurement, 22*(4), 271–286.

Stillman, G. A., & Galbraith, P. L. (1998). Applying mathematics with real world connections: Metacognitive characteristics of secondary students. *Educational Studies in Mathematics, 36*(2), 157–194.

Sutton, R. (1995). *Assessment for learning.* Salford, UK: RS Publications.

Tobin, K. (1987). The role of wait time in higher cognitive level learning. *Review of Educational Research, 57*(1), 69–95.

Torrance, H. (1993). Formative assessment: Some theoretical problems and empirical questions. *Cambridge Journal of Education, 23*(3), 333–343.

Torrance, H., & Pryor, J. (1998). *Investigating formative assessment.* Buckingham, UK: Open University Press.

Tunstall, P., & Gipps, C. V. (1996a). Teacher feedback to young children in formative assessment: A typology. *British Educational Research Journal, 22*(4), 389–404.

Tunstall, P., & Gipps, C. V. (1996b). "How does your teacher help you to make your work better?" Children's understanding of formative assessment. *The Curriculum Journal, 7*(2), 185–203.

Turner, J. C., & Patrick, H. (2004). Motivational influences on student participation in classroom learning activities. *Teachers College Record, 106*(9), 1759–1785.

van den Heuvel-Panhuizen, M., & Becker, J. (2003). Towards a didactic model for assessment design in mathematics education. In A. Bishop, M. A. Clements, C. Keitel, J. Kilpatrick, & F. K. S. Leung (Eds.), *Second international handbook of mathematics education* (pp. 689–716). Dordrecht, The Netherlands: Kluwer.

VanLehn, K. (1990). *Mind bugs: The origins of procedural misconceptions.* Cambridge, MA: MIT Press.

Vermunt, J. D. (2003). The power of learning environments and the quality of student learning. In E. DeCorte, L. Verschaffel, N. Entwistle, & J. van Merriënboer (Eds.), *Powerful learning environments: Unravelling basic components and dimensions* (pp. 109–124). Oxford, UK: Pergamon.

Vinner, S. (1997). From intuition to inhibition—mathematics, education and other endangered species. In E. Pehkonen (Ed.), *Proceedings of the 21st conference of the International Group for the Psychology of Mathematics Education* (Vol. 1, pp. 63–78). Lahti, Finland: University of Helsinki Lahti Research and Training Centre.

von Glasersfeld, E. (1987). Learning as a constructive activity. In C. Janvier (Ed.), *Problems of representation in the teaching and learning of mathematics* (pp. 3–17). Hillsdale, NJ: Erlbaum.

von Glasersfeld, E. (Ed.). (1991). *Radical constructivism in mathematics education.* Dordrecht, The Netherlands: Kluwer.

Vygotsky, L. (1978). *The development of higher psychological processes.* Cambridge, MA: Harvard University Press.

Webb, D. C. (2004). Enriching classroom assessment opportunities through discourse. In T. A. Romberg (Ed.), *Standards-based mathematics assessment in middle school* (pp. 168–187). New York: Teachers College Press.

Webb, N. L. (1992). Assessment of students' knowledge of mathematics: Steps towards a theory. In D. A. Grouws (Ed.), *Handbook of research on mathematics teaching and learning* (pp. 661–683). New York: Macmillan.

Webb, N. M. (1991). Task-related verbal interaction and mathematics learning in small groups. *Journal for Research in Mathematics Education, 22*(5), 366–389.

Weiss, I. R., Pasley, J. D., Smith, P. S., Banilower, E. R., & Heck, D. J. (2003). *Looking inside the classroom: A study of K–12 mathematics and science education in the United States.* Chapel Hill, NC: Horizon Research.

White, B. Y., & Frederiksen, J. R. (1998). Inquiry, modeling, and metacognition. Making science accessible to all students. *Cognition and Instruction, 16*(1), 3–118.

White, M. A. (1971). The view from the student's desk. In M. L. Silberman (Ed.), *The experience of schooling* (pp. 337–345). New York: Rinehart and Winston.

Wiener, N. (1948). *Cybernetics, or control and communication in the animal and the machine.* New York: John Wiley.

Wigfield, A., Eccles, J. S., & Rodriguez, D. (1998). The development of children's motivation in school contexts. In P. D. Pearson, & A. Iran-Nejad (Eds.), *Review of research in education* (Vol. 23, pp. 73–118). Washington, DC: American Educational Research Association.

Wiliam, D., & Black, P. J. (1996). Meanings and consequences: A basis for distinguishing formative and summative functions of assessment? *British Educational Research Journal, 22*(5), 537–548.

Wiliam, D., Lee, C., Harrison, C., & Black, P. J. (2004). Teachers developing assessment for learning: Impact on student achievement. *Assessment in Education: Principles Policy and Practice, 11*(1), 49–65.

Wiliam, D., & Thompson, M. (in press). Integrating assessment with instruction: What will it take to make it work? In C. A. Dwyer (Ed.), *The future of assessment: Shaping teaching and learning.* Mahwah, NJ: Erlbaum.

Williams, R. (1961). *The long revolution.* London: Chatto & Windus.

Winne, P. H. (1996). A metacognitive view of individual differences in self-regulated learning. *Learning and Individual Differences, 8,* 327–353.

Winne, P. H. (2005). Key issues in modeling and applying research on self-regulated learning. *Applied Psychology: An International Review, 54*(2), 232–238.

Yackel, E., & Cobb, P. (1996). Sociomathematical norms, argumentation, and autonomy in mathematics. *Journal for Research in Mathematics Education, 27*(4), 458–477.

Zimmerman, B. J. (1986). Becoming a self-regulated learner: Which are the key subprocesses? *Contemporary Educational Psychology, 11,* 307–313.

AUTHOR NOTE

At various times in the preparation of this chapter, I received helpful comments from Susan M. Brookhart, William S. Bush, Edith Aurora Graf, Siobhan Leahy, and Marnie Thompson, and as a result, the chapter changed considerably from the drafts on which they had commented. Consequently, I am solely responsible for the final version and any errors or omissions that remain, or indeed, were reintroduced.

24

HIGH-STAKES TESTING IN MATHEMATICS

Linda Dager Wilson

Test results in schools are often used to make decisions that have serious consequences. The term *high stakes* is used for such tests.[1] For whom are the stakes high? Students are often the constituents who reap both the rewards and the punishments of the tests they take. For example, students may be required to pass a test or a series of tests to earn a high school diploma, or to be promoted to the next grade. Tests might also be used to place students in different courses, or to move them from one stream of the curriculum to another. Students may earn college credit by high scores on advanced placement tests or earn an International Baccalaureate diploma through testing. In Japan, Germany, and France students must pass entrance exams for admission to universities. Thus the stakes associated with tests may be high for students, when either rewards or sanctions are attached to the outcomes.

Other constituents are also affected by the consequences of high-stakes tests. Teachers may be held accountable for their students' test scores, but they may also be required to pass tests themselves in order to earn or maintain their certification. Schools, districts, and states may have accountability measures in place that use the results of student tests to make funding decisions. The No Child Left Behind (NCLB) legislation from the Bush administration is an example of federal funding that is linked, at least in part, to the results of standardized achievement tests at the state level. Every state has now instituted some sort of state-wide testing program, partially in response to this legislation.

Mathematics as a content area is prevalent in high-stakes testing programs. The NCLB legislation calls for annual testing in reading and mathematics at Grades 3–8 and once in Grades 10–12. Advanced placement tests, such as those administered by the College Board, are offered in statistics and at two levels of calculus. The National Board of Professional Teaching Standards offers national certification for teachers in mathematics, among other content areas. End-of-course exams, required for a high school diploma, are offered in North Carolina, for example, in Algebra 1, Algebra 2, and Geometry. So the teaching and learning of mathematics has the potential to be greatly affected by high-stakes testing.

International high-stakes testing. Although the current political atmosphere in the U.S. seems to call for more and more high-stakes testing in mathematics, as well as other major content areas, some researchers have questioned whether this trend is operating in the reverse in other parts of the world (U.S. Congress, 1992; Kellaghan, Madaus, & Raczek, 1996). Apparently, this is not so. Phelps (2000) looked at data from 31 countries and provinces and concluded that there was, in fact, an overall net increase internationally in all testing from 1974–1999. This included an increase in high-stakes testing during those years as

[1] The term *tests* in this chapter refers to standardized tests. This is not to be confused with multiple-choice, machine-scored, or any other test format, item type, or scoring procedure. Rather, standardized tests are tests that are given under uniform conditions, in which all examinees are asked to respond to the same, or nearly the same, test questions.

well. In the case of large-scale, external testing that was added during those years, most were assessments used for diagnostic or monitoring purposes, and thus not high-stakes. However, the second most frequently added type of test was upper secondary exit exams, followed by university entrance exams and subject-area end-of-course exams. One would expect these three categories to all include mathematics. Phelps also refuted the claim that there is little or no testing in other countries at ages below 16, citing the prevalence of lower secondary exams in such countries as France, Italy, Iceland, Ireland, Denmark, New Zealand, Switzerland, Japan, and Korea (p. 18).

Both commonalities and differences are apparent in the use of high-stakes testing in mathematics in the U.S. and around the world. On the one hand, the use of mathematics exams for promotion (such as a secondary diploma) seems to be similar, as is the use of mathematics exams for entrance to the university. Most U.S. colleges and universities require the mathematics portion of either the Scholastic Achievement Test (SAT) or the American College Test (ACT) for entrance. On the other hand, U. S. students appear to be much more likely to be tested in mathematics in the primary grades.

HISTORICAL CONTEXT OF HIGH-STAKES TESTING IN MATHEMATICS

Much of the origin and development of school testing began in the U.S., starting with the first school examinations in the Boston schools in 1845, which were championed by Horace Mann as a superior way to examine all pupils. Thorndike's (1904) book on mental, social, and educational measurements stimulated the development of the first standardized achievement tests, beginning with arithmetic and English composition, spelling, drawing, and handwriting. Testing as an industry grew rapidly with the use of intelligence testing during World War I. The primary function of such tests was to sort and classify students according to "innate" ability, an illustration of some of the earliest uses of large-scale tests for policy decisions. Psychometrics as a field of study developed in the 1920s as statistical procedures were applied to standardized tests. From their earliest uses, standardized achievement tests had some common features. First, the assumption was made that a single measure or index could be developed to compare individuals on a general, fixed, unidimensional trait (Romberg, 1992). Second, the tests were administered under uniform conditions and constraints. Third, the tests comprised a collection of items, each having one unambiguous answer. Since those early beginnings, the testing industry in the U.S. has grown into a multi-million dollar business, and the number of tests produced has expanded exponentially.

Today, the use of high-stakes standardized tests in mathematics has the following profile in U.S. schools:

- 26 states currently have or will implement exit exams in mathematics by 2008
- 50 states require mathematics testing in Grades 3–8 and at some year of high school
- 96% of public colleges and universities require an entrance exam in mathematics
- 91% of private colleges and universities require an entrance exam in mathematics

(Breland et al., 2002; Gayler et al., 2003; Potts et al., 2002; U.S. Department of Education NCLB website).

From Grade 3 on, a student in a U.S. public school can expect to be tested at least once each year on a high-stakes test in mathematics. During the high school years the student may be subjected to numerous mathematics tests for a variety of purposes, including state-wide end-of-course exams, an exam for a diploma requirement, and a series of college-entrance and placement exams. Meanwhile that student's teachers have likely been required to take mathematics exams as a certification requirement.

The story of the content and format of large-scale tests in mathematics has not been linear. The early tests established models for the types of instruments that were used in schools throughout the decades, up until the "standards movement" that began in the late 1980s. The first content standards were published by the National Council of Teachers of Mathematics (NCTM) in 1989, and this unleashed efforts in other content areas to do the same (American Association for the Advancement of Science (AAAS), 1993; National Committee on Science Education Standards and Assessment, 1996). The *Curriculum and Evaluation Standards for School Mathematics* (NCTM, 1989) were not only influential in advocating the setting of standards to define specific goals for what students should know and be able to do in mathematics, but the document also called for reform in the way students are assessed. This effort peaked in 1995 with the publication of the *Assessment Standards for School Mathematics* (NCTM, 1995). Both standards documents called for changes in external testing, moving away from the discrete objectives and the sole use of "objective" items and toward the use of more contextual, constructed-response items that would require human scorers. The middle of the decade of the 1990s saw changes in state testing systems in response to these and other calls for reform. For example, Kentucky instituted an innovative assessment system as part of the Kentucky Education Reform Act of 1990, and Vermont took the bold step towards a portfolio system of assessment in mathematics. Many other states began to supplement their machine-scored, mainly multiple-choice standardized tests with an increased number of constructed-response items. At the same time, nearly every state was developing its own content standards in mathematics, as well as other content areas.

As content standards were developed, interest increased in examining the alignment of standards with assessments (e.g., Romberg, 1995; Romberg & Wilson, 1992). The more simplistic notions of alignment consisted of a simple content match between a test item and a given statement in a set of standards, to determine by examination whether the test items seemed to be measuring the same mathematics called for in the standard statement. In the last decade, alignment analyses have grown in depth, complexity, and scope (AAAS, 2002; Webb, 1997).

In addition to the evolution of content standards and alignment studies, another contextual factor that has affected the nature of high-stakes testing is the growth of technological tools in the teaching and learning of mathematics. Calculators, once nonexistent in the classroom, are now prevalent throughout the grades (10% of fourth-graders, 48% of eighth-graders and 69% of twelfth-graders report using a calculator almost every day to do mathematics in school (National Center for Education Statistics, 2005)). In addition, the type of calculators being used is more sophisticated and their use is at ever earlier grades than before. In addition, the use of computer-assisted technologies has increased (Parsad & Jones, 2005). Standardized tests, in turn, have responded by developing more test items that incorporate the use of calculators, and testing companies are experimenting with ways to situate test questions in a computer setting, rather than with the use of paper and pencil alone (Bennett, 2002; Russell & Haney, 2000).

Currently 24 states have exit exams in mathematics. In all of those states the passing of the exit exam in mathematics either is a diploma requirement or is designed to be a diploma requirement at some future date. Some of these exit exams are described as general mathematics. Some states, such as North Carolina, base the exam on eighth-grade mathematics standards. Others specify exams that are more directly tied to course content, such as Mississippi's Algebra exam (Center on Education Policy, 2004).

In January, 2002 President Bush signed into law the No Child Left Behind (NCLB) Act, which affects every program authorized under the federal Elementary and Secondary Education Act, first enacted in 1965. Although the federal government allocates only about 7% of its budget to education, which is predominantly the purview of the states, the NCLB Act is having far-reaching consequences for high-stakes testing in mathematics across the country. The law requires that states develop a set of content standards in mathematics, create annual assessments that are aligned with those standards, and administer those tests to nearly all students annually in Grades 3–8, and at least once in Grades 10–12. Further, states must show that all groups of students, disaggregated by poverty, race and ethnicity, disability, and limited English proficiency, reach proficiency in mathematics by 2013–2014. Schools must show evidence of "adequate yearly progress" toward specific annual measurable objectives, and the sanctions associated with not meeting these goals range from mandated changes in instruction to "fundamental restructuring" of "any school that fails to improve over an extended period of time" (U.S. Department of Education, p. 8). The NCLB Act, in effect, legislates that all states will have content standards, will have a set of tests that are aligned with those standards, and will show progress for all students on those instruments, with the threat of serious sanctions if they do not participate. The stakes are high at the school, district, and state levels.

Another piece of the NCLB Act that is having an immediate impact in mathematics assessment is the role of the National Assessment of Educational Progress (NAEP). Since 1972 NAEP has been used as a measure of student achievement in mathematics. Historically, there have been three types of NAEP measures: one for long-term trends, one at the state level, and one at the national level. Since 1973 there have been 9 administrations of the long-term trend instrument in mathematics. These tests are the only consistent gauges of student achievement given over time to a nationally representative sample in the U.S. They are given to students of ages 9, 13, and 17. Although NAEP is not a high-stakes assessment, it is interesting to note here that the average scores in 1999 at all three ages were significantly higher than in 1973 (Campbell, Hombo, & Mazzeo, 2000). The national or "main" NAEP has been administered approximately every 4 years to a nationally representative sample of students in Grades 4, 8 and 12. The instruments and scales for "main" NAEP are not comparable to the long-term trend NAEP, and in mathematics they have been based on different frameworks over time. The instruments administered in 1990–2003 made use of the 1990 framework (NAEP, 1988), whereas those administered beginning in 2005 will be based on the 2005 framework. In addition to the national results, beginning in 1990 states were given the opportunity to voluntarily participate in main NAEP, in order to be able to see state-by-state comparisons of the results.

The advent of the NCLB legislation has changed the voluntary nature of participation in main NAEP at the state level. Under the law, NAEP would be used as a benchmark against which state results can be compared. To participate in NCLB a state is required to participate in the biennial administrations of NAEP at Grades 4 and 8. States are also required to set proficiency standards in reading and mathematics, and many states have chosen to use the NAEP achievement levels (Below Basic, Basic, Proficient, and Advanced) as the model for those levels. The stated goal of NCLB is for 100% of students in any participating state to attain the level of Proficient in both reading and mathematics by the year 2014. As Linn (2003) has pointed out, if the NAEP achievement levels are used as a guide, this is probably an impossible goal.

This change in the role of NAEP, from a voluntary measure of student achievement to a mandatory benchmark for accountability, was predicted in a congressionally mandated evaluative study of NAEP conducted by the National Research Council (Pellegrino et al., 1999). That committee referred to the increased pressures on NAEP to do "more and more beyond its established purposes" (p. 2). In response to these changing demands, the committee endorsed a second redesign of NAEP.

Ever since the first NAEP mathematics assessment in 1972, the National Council of Teachers of Mathematics has produced "interpretive reports" of the NAEP results (see Kloosterman & Lester, 2004). These reports have examined the results in more depth than the more general reports published by the National Center for Education Statistics (Braswell et. al., 2000). In each report, mathematics educators have looked within specific content areas, such as Number or Geometry, and analyzed the achievement results, often including disaggregated results by student subgroups. When possible, they have also analyzed samples of student work on open-ended items. The reports also include an analysis of survey data on attitudes, beliefs, and instructional context that have been collected from students and classroom teachers in conjunction with the assessment. The audience for these reports is usually classroom teachers, though they are also intended to be accessible to a more general audience. Examples of statements from the interpretive reports include the following from the Sixth NAEP Mathematics Assessment administered in 1992:

> "Performance and participation differences between males and females are disappearing." (Silver & Kenney, 1997, p. 57)
>
> "Interpreting fractions and decimals as locations on a number line was more difficult for students than interpreting fractions as part of geometric regions." (p. 115)

As NAEP begins the cycle of biennial testing in the service of No Child Left Behind, the availability of data for the interpretive reports will likely suffer. Beginning with the 2000 administration, researchers for those reports were no longer given access to secure items. So long as NAEP is given this high-stakes role, such a trend will likely continue. Fewer items will be released, researchers will likely be denied access to secure items, and student work will probably not be accessible. In terms of more general usefulness as a data base for the study of student achievement, Johnson argued (2000):

> in order to use the data for school improvement or theory building there would need to be more complex NAEP data gathered at the level of the student, the family, the teacher, and school and the state. This emphasis, however, runs counter to the current push to simplify and speed up the data collection and reporting process. (p. 99)

The student achievement data, not to mention the classroom practice data, which have been collected by NAEP over the past 35 years are invaluable to the mathematics education research community. Under the current political pressures to use NAEP as a high-stakes tool with biennial testing, this role for NAEP may be greatly diminished, at the very least.

TECHNICAL ISSUES IN HIGH-STAKES TESTING

When tests are used to make decisions with critical consequences for various constituencies, the technical aspects of the test must be sound. Inasmuch as tests are essentially measures, and all measurement involves error, anyone who is in the position of developing, using, or interpreting a test of student achievement must attend to the potential for error and be mindful of the need to minimize error. In this section

I review recent literature on the possible sources of error in measuring student achievement and how it might be avoided.

Validity

All types of mathematics assessment, whether high-stakes or not, are samples of what students know and can do. This sample data is used to make inferences about certain constructs, such as knowledge, attitudes, or achievement. *Validity* refers to the degree to which the inferences made on the basis of the assessment are meaningful, useful, and appropriate (Brualdi, 2002). Traditional concepts of validity (content-related, criterion-related, and construct-related) have been expanded in recent years, largely through the work of Messick (1989, 1996). He argued that validity should take into account information about actual as well as potential consequences of score interpretations. Validity, according to this view, is not a property of the test, but instead a property of inferences or interpretations made from test scores and is thus essential for all types of assessments, but particularly for high-stakes tests. (See Messick, 1996, for a discussion of the six interdependent aspects of validity that form a general theory).

Two major threats to validity in high-stakes tests are construct underrepresentation and construct-irrelevant variance (Brualdi, 2002). The former refers to a situation in which the concepts or skills that are measured in the assessment fail to include important dimensions or facets of the intended construct (presumably, some content domain in school mathematics). In such a case, the test results are unlikely to reveal a student's true ability regarding that construct. For example, a test of measurement that included only items using the metric system would not be a valid measure of students' ability to solve problems involving nonmetric systems of measure. Construct-irrelevant variance can take two forms. In the first, extraneous clues in the item or the task format may permit some students to respond correctly or appropriately in ways that are irrelevant to the construct being assessed. For example, a student might use "test-wiseness" to ferret out the correct answer without actually knowing or being able to do the mathematics being tested. The other possibility is that extraneous aspects of the task make it irrelevantly difficult for some individuals or groups, such as in the case of a context that is so unfamiliar as to limit access for some students to the mathematics in the task. In either case, the resultant score would not be one upon which to make valid inferences about the student's achievement.

When high-stakes tests are used for accountability purposes, as is proposed in the NCLB legislation, validity issues also arise in the setting of performance standards for the test. Such standards, or levels of achievement, serve the purpose of classifying the test takers into categories, such as "basic," "proficient," "advanced," etc. Performance standard-setting is a judgmental process, often carried out by committees of "experts." To achieve the highest possible level of validity, the process must be carefully planned and executed (see Jaeger, Mullis, Bourque & Shakrani, 1996 for a discussion of the challenges and issues of standard setting, especially for performance assessments). Other aspects of validity are that the distinctions between the levels should be "meaningful and useful" (Stecher, Hamilton, & Gonzalez, 2003, p. 21) for the purposes of the test.

In an analysis of the content validity of selected standardized tests of mathematics achievement that are often used to make decisions about students with disabilities, Parmar et al. (1996) cited numerous technical flaws in the tests. These include inadequate representation of content domains, inappropriate sequencing and placement of items, inappropriate use of age and grade-equivalent scores, and incorrect descriptors assigned to items. One of the conclusions of the study is that "the majority of instruments and procedures used for assessment of students in special education do not reflect important mathematics, nor do they provide information that will drive or guide curriculum or instructional decisions" (p. 128).

Because the content covered in standardized tests is nearly always too narrow to reflect the complexities of teaching and learning, Stake (1995) argued that they are a "flimsy indicator" of the mathematics that students have learned (p. 174). He makes the point that much of the evidence for the learning of mathematics comes from reflective interaction with individual students and with classes of students, so that the complexities of learning can be fully captured. Standardized tests often rely on a single-dimension concept of achievement and thus at best should only be considered as one component in an array of data.

Reliability

No measurement is error-free, and tests are no exception. The reliability of a test refers to the extent to which the measurement is free of random errors, or errors that are not systematic. Reliability is a characteristic of both the population of test takers and the instrument itself, because of the way reliability is computed. Thus a test will show different reliability results for different populations of test takers.

Three potential sources of random error exist in all tests, not just high-stakes tests. Those sources are the test itself, the test takers, and the scoring process. The test may contain random errors that stem from inadequate sampling of the content domain, the ineffectiveness of the distractors (the incorrect options) in multiple-choice items, partially correct distractors, more than one correct distractor, and the difficulty of the items relative to the student's ability, in that a test that is too difficult for a student will have low reliability for that student or population of students. Test takers introduce error by their human nature. Many factors, both external and internal, can affect the consistency of a student's score in a given testing situation, including motivation or illness. An overtired student, for example, may misread the test directions or inadvertently omit sections of the test. Scoring, especially for open–response items, is a third source of random error. Even machine-scored tests can have error if the machine does not correctly read the answers the student intended (an interaction between student error and scoring error). Open-response items introduce more error if scoring rubrics are unclear or if scorers are not highly trained and monitored for high interrater reliability.

Another useful aspect of reliability for high-stakes tests is described by Parmar et al. (1996) as *user reliability*, which they define as the extent to which different users would come to the same diagnostic conclusions when faced with the same results, and the extent to which they would make similar curriculum and instruction recommendations based on that diagnosis. Their analysis of a selection of standardized mathematics tests used to make decisions about students with disabilities cited numerous problems with user reliability of those tests.

Equity

When the results of achievement tests are disaggregated for various subpopulations, one can measure the gaps in achievement among different groups. These achievement gaps among different genders, races, or groups with differing socioeconomic status have changed over the years in mathematics, both for better and for worse. The NAEP data, although not a high-stakes test prior to 2003, shows these differences. For example, in the five NAEP mathematics tests given between 1990 and 2003 (U.S. Department of Education, 2004), the gender gap at Grade 4 has remained small but unchanged (in a meta-analysis of 100 studies on gender differences in mathematics performance, Hyde et. al., 1990, found females outperforming males by a negligible amount). However, the gap between Black and White students, as reported in school records of race/ethnicity, indicates a steady narrowing of the gap over those years. The White/Hispanic gap narrowed from 2000 to 2003, after remaining steady from 1990 to 2000. Finally, the gap between those eligible and not eligible for free or reduced-price lunch also narrowed from 2000 to 2003, when it had not narrowed from 1996 to 2003. Although the news looks more promising in recent years, there is still evidence of disproportionate numbers of minority, low-income, and special-needs students who are failing tests for promotion and graduation (Catalyst Chicago, 2004; Darling-Hammond, 2003).

Aside from the data from high-stakes tests that track gaps among different subpopulations, another issue is the interactions among those groups and the tests themselves. In the literature on testing, the notion of a "fair test" is one that yields comparably valid inferences from person to person and from group to group (National Research Council (NRC), 1999; Pellegrino et al., 2001). In order to avoid bias, test developers must ensure that no task has characteristics that would have different meanings for different subgroups of test takers. This might mean, for example, avoiding contexts or language associated with one particular culture, so that any inferences drawn from those test scores would be comparable across all subgroups. On large-scale tests expert review panels are often set up to review assessment items for this type of bias. Psychometricians might also examine the results of statistical tests for differential item functioning (DIF), which identifies items that produce differing results for members of particular groups after the groups have been matched with regard to the attribute being measured. Some researchers have begun to follow up DIF analyses with cognitive analyses, to try to uncover why certain items function differently across groups. The purpose of such studies is to learn more about how students think about and approach the problems (Lane, Wang, & Magone, 1996).

Many high-stakes tests provide accommodations for certain students who demonstrate special needs. NAEP, for example, began allowing accommodations on the mathematics tests in 1996. These include extra time and bilingual Spanish editions. Accommodations generally fall into four categories: presentation, response, setting, or timing/scheduling (Thompson et al., 2002). Test presentation accommodations may include oral presentations, changes in test content or test format. Response accommodations include allowing students to dictate their answers or write their answers directly into their test booklets. Setting accommodations often involve testing students individually or in a separate room. Finally, timing or scheduling accommodations include allowing extra time or administering smaller pieces of the test in separate sessions. In a meta-analysis of 150 studies of test accommodations, Sireci et al. (2004) critiqued the studies with respect to an *interaction hypothesis*. The hypothesis states that test accommodations should improve the test scores for targeted groups but not improve the test scores for examinees for whom the accommodations are not intended. Among their conclusions they found that "many accommodations are justified and effective for reducing construct-irrelevant barriers to students' test performance" (p. 61). In particular, they found that extended time improved the performance of students with disabilities more than it improved the performance of students without disabilities. The authors also argued for the need for future tests to be "constructed and administered more flexibly so that accommodations become unnecessary" (p. 67).

In terms of high-stakes tests, such as those used under the NCLB legislation, Abedi (2004) argued that major issues are involved with measuring adequate yearly progress for students with limited English proficiency (LEP). Among the issues cited are inconsistent LEP classification, the sparse population of LEP students in many states that threatens validity, and the LEP subgroup's lack of stability. Abedi concluded that NCLB's mandates may unintentionally place undue pressure on schools with high numbers of LEP students.

Large-scale standardized tests, such as are used in most high-stakes testing situations, are prone to error. In addition to random measurement error, as described above under Validity and Reliability, non-random human errors are associated with tests. Rhoades and Madaus (2003) divided human errors in testing into two types: active and latent. Active errors are those that are committed by specific individuals and are often easy to identify. Examples of active errors would include incorrect scoring of a test or a computer programming error that produces incorrect percentile rankings. Latent errors are systemic in nature and often stem from poor administrative decisions or a broad constellation of circumstances that provide fertile ground for human mistakes. Examples of latent error are state Departments of Education that rely on a single test score to make high-stakes decisions, or an overly demanding testing schedule that puts undue demands on an understaffed or underqualified team to develop, administer, score and report tests. Error can be found in all phases of the testing process, from the development of items to printing of test booklets to the setting of a passing score. Of particular concern is the use of tests to measure change in achievement over time, or trends. A common error is to analyze results from one year to the next of administrations of the same test, in that the test may have undergone changes that can alter the results. As Beaton (1990) pointed out, "If you want to measure change, do not change the measure" (Beaton et al., 1990, p. 165).

Another issue of particular concern with respect to the NCLB legislation is the determination of adequate yearly progress (AYP). Under this requirement states establish goals for what percentages of students in various subgroups (e.g., low income, minority, limited English proficiency) will meet or exceed proficiency levels on the state's assessments each year. In order to confirm gains from one measure to the next, or across grade levels, test scores must be comparable from test to test and year to year. *Scaling* is a process used to accomplish this, in which raw scores are converted by means of a mathematical function (which might be either linear or nonlinear) to a common scale. *Vertical scaling* is the process of equating scores of students at different levels. Several assumptions that underlie the process of vertical scaling and equation may be problematic. For example, the tests are assumed to measure the same general content, which is an assumption not easily met by tests that span one or more grade levels. In addition, both the nature of the items and the assessment process may change over grades, in which case vertical equating confounds content changes with method changes. Lissitz and Huynh (2003) have alternative approaches that may resolve some of these issues.

Many testing programs use cut scores to determine the predetermined level at which a student reaches a certain level of achievement. That level might denote the difference between passing and failing, or it might determine categories of achievement, such as "basic," "proficient," "advanced," and so on. The setting of cut scores is used to separate students into different classes of achievement, and it is based upon predetermined concepts of student achievement. Regardless of which method of setting cut scores is used, human judgment is embedded in the process, and therefore the process cannot be regarded as objective. (See Horn et al., 2000, for further discussion of cut scores).

Teachers, as well as students, may be subjected to high-stakes testing, and often the result is not exemplary. In 2004, two thirds of the middle school mathematics teachers in the Philadelphia school district failed the certification test that is used to determine if they meet the qualifications for "highly qualified" under the federal No Child Left Behind Act. These tests are no less immune to error than any other standardized tests.

Many researchers have attempted to find causative links between teacher quality and student achievement. Writing about testing for teacher quality (as in whether teachers' test scores or observers' ratings are good predictors of professional effectiveness), Ferguson and Brown (2000) described potential sources of error thus:

> teacher quality, including teachers' test scores, are measured with error; student quality in a school or district can affect which teachers choose to apply there, creating reverse causation from student performance to teacher quality; particular measures of teacher quality may matter more or less depending on other variables, such as class size, so simple linear models that ignore interactions may produce misleading results; correlations between teacher quality and other inputs such as parental effectiveness can produce biased estimates for the effect of teacher quality if parental variables are omitted from the analysis or measured with considerable error. Further, most studies lack the type of data necessary for sorting out the issues that the advocates of multilevel estimation emphasize. (p. 149)

Goldhaber et al. (2004) studied 3 years of teacher and student data from North Carolina and concluded that teachers who received national certification through the National Board for Professional Teaching Standards were more effective at raising student achievement than teachers who pursued, but failed to obtain, certification. On the one hand, this study looks at one outcome score for students; on the other hand, the national certification process entails an extensive performance-based evaluation, including written exams, lesson portfolios, and classroom videos.

EFFECTS OF HIGH-STAKES TESTING IN MATHEMATICS

When the stakes change from low to high, there are often consequences. DeMars (2000) studied the interactions between test stakes and item format interactions on math and science sections of a high school diploma endorsement test. The findings indicate an interaction between response format and test consequences. Under both multiple-choice and constructed-response formats, students performed better under high stakes (diploma endorsement) than under low stakes (pilot test), but the difference was larger for constructed-response items. Gender and ethnicity did not interact with test stakes; the means of all groups increased when the test had high stakes. Gender interacted with format; boys scored higher than girls on multiple-choice items, girls scored higher than boys on constructed-response items.

Test format (open-ended, multiple-choice, etc) on high-stakes tests also has an influence on instruction. During the school reform movements of the 1990s, much effort was expended on building tests worthy of "teaching to the test" (Resnick & Resnick, 1992). Others questioned at the time whether any one test can be so representative of a desired curriculum that teaching to such a test would equal good instruction (Linn, Baker, & Dunbar, 1991). Under the pressures of No Child Left Behind and other large-scale accountability measures, many of the efforts toward more open, performance-based testing, such as those in Kentucky and Vermont, have been replaced with more traditional tests. When tests use more open-ended items that require writing, teachers place more emphasis on writing in mathematics classes (Taylor et al., 2003). In general, teachers tend to tailor their instructional practices to the format of the tests (Abrams & Madaus, 2003; Pedulla et al., 2003). The format of the test can also have a substantial impact on test results. In a district reporting gains on a high-stakes multiple-choice test, Koretz et al. (1991) gave students a similar test with more open-ended questions and found that student performance dropped off as much as half a standard deviation.

Researchers have documented effects of high-stakes testing on the mathematics curriculum. Taylor et al. (2003) found that teachers in Colorado added new content, such as probability and geometric topics, and placed more emphasis on problem solving and writing in mathematics classes. At the same time, they found that teachers decreased the amount of time spent on other curricular areas, such as science and social studies. Clarke et al. (2003) found that educators in Massachusetts, Kansas and Michigan were more likely to remove content topics than to add them. In those three states, about 10% of educators interviewed thought that the state test had no impact on what was taught. For high school exit exams, Gayler et al. (2004) reviewed existing literature on impact and found that exit tests resulted in broader coverage of state standards, alignment of curriculum with those standards, and targeted course offerings for at-risk students. At the same time, these tests can also narrow the curriculum and force a sequence and pace of instruction that is inappropriate for some students. In a national survey of teachers, Pedulla et al. (2003) found a strong interaction between the level of the stakes in the test and the degree to which it impacted the curriculum; in particular, far more teachers in high-stakes states reported that their own tests reflected the format of the state test than did teachers in low-stakes states. In addition, a substantial majority of teachers at each grade level, but especially at the elementary level, reported that testing programs in their state have led them to teach in ways that contradicted their ideas of sound instructional practices.

The effects of high-stakes testing are also linked to the grades at which testing is conducted, as documented in a study by Stecher and Barron (2001). Grades in which testing occurs show more profound effects on curriculum and instruction than those grades that are not included in the testing. A more insidious consequence of grade-specific effects is found in a significant nationwide bulge in students enrolled in 9th grade and a tripling of the attrition rate between 9th and 10th grades over the last 30 years (Haney et al., 2004). The change is due, according to their analysis, to the increased practice of holding students back from the pivotal high-stakes exit exams in 10th and 11th grades, as well as to increased course requirements. Frequently, such students drop out of school. These findings are in line with those of Jacob (2001), who reported that high school graduation tests, while they did not have an effect on student achievement in mathematics or reading, did increase the probability of dropping out among the lowest ability students.

STANDARDS FOR APPROPRIATE USE OF HIGH-STAKES TESTS

In light of the effects of high-stakes testing, many educators have called for reforms in the uses to which such tests are put. Many have found fault with the NCLB legislation's emphasis on testing and offered advice on reforms (Abrams & Madaus, 2003; Elmore, 2003; Harvey, 2003; Neill et al., 2004; Stecher et al., 2003). In a broader reaction to the current climate in the U.S. of widespread high-stakes tests for educational policy purposes, the National Commission on Testing and Public Policy (2002) issued a statement that calls for, among other things, less reliance on multiple-choice tests, greater use of multiple sources of evidence for decision-making, critical evaluations of tests for fairness and accuracy, and a separation of testing that is used for instructional purposes and that used for accountability purposes. For more general application of educational testing, the *Standards for Educational and Psychological Testing* (1999) delineates standards for practice in the use of educational tests. It was a joint venture of the American Educational Research Association, the American Psychological Association, and the National Council on Measurement in Education. Among other standards these groups emphasize that when stakes are high, it is particularly important that the inferences drawn from an assessment be valid, fair, and reliable.

As the research has shown, high-stakes testing in mathematics can have a powerful influence on educational practice, whether for good or ill. It can affect teaching practices, curricular decisions, and the decisions made about schools or individual students. The potential power of testing makes it an alluring policy tool, but a tool that is fraught with tensions. As the National Research Council (1999) pointed out, one of the dilemmas at the heart of this tension is that "policy and public expectations of testing generally exceed the technical capacity of the tests themselves" (p. 30). The primary reason for this is that tests are too often used for purposes other than those for which the tests were designed and validated. It is not uncommon for tests that were developed to produce valid measures of student learning at the aggregate level (e.g., school, state, national) to be used to make decisions about teachers (classroom-level data) or even individual students. Such practices may result in invalid inferences, and thus unfair decisions, being made about those stakeholders.

RESEARCH NEEDS

There are many potential arenas for research in high-stakes testing in mathematics. Many of these were described in the report of the Assessment Working Group of the 2003 Research Catalyst Conference sponsored by the National Council of Teachers of Mathematics (Lester & Ferrini-Mundy, 2003). Though the focus of the conference was on the impact of standards, the assessment group's research agenda covered many components that involve high-stakes tests. The group called for descriptive studies on the evolution of large-scale assessments over the last decade as well as on college entrance exams. They also called for investigations into the construction of assessment instruments that are valid and fair for all students, particularly how test items can be designed to allow greater access to all groups of students. In addition, the agenda designates for high priority studies that investigate how high-stakes tests can be used to improve classroom learning.

The NRC (2002) has noted the need for research on the impact of standards on assessment in mathematics, science, and technology. Among other issues, they emphasize the need for more research into the extent to which teachers have modified their classroom assessment practices to respond to calls for change in standards documents, and to what extent teachers use assessment data to adjust instruction.

They also call for research into college admission tests and the extent to which they are aligned with nationally developed standards.

Because of the important social and political impact of high-stakes testing in mathematics, further research is urgently needed. At the same time, there are potential barriers to the undertaking of research in this area, many of which stem from the nature of high-stakes testing. For example, high-stakes tests are often held in secret, making access to item pools or student data difficult, if not impossible. Further, random assignment and experimental intervention is problematic when students and teachers face serious consequences from the assessment. Yet the very fact that the stakes are high should underscore the critical nature of this research agenda.

REFERENCES

Abedi, J. (January/February 2004). The no child left behind act and English language learners: assessment and accountability issues. *Educational Researcher, 33*(1), 4–14.

Abrams, L., & Madaus, G. (2003, November). The lessons of high-stakes testing. *Educational Leadership 61*(3), 31–35.

American Association for the Advancement of Science, Project 2061 (1993). *Benchmarks for science literacy.* New York: Oxford University Press.

American Association for the Advancement of Science, Project 2061 (March, 2002 draft). Analysis of science and mathematics assessment tasks. Author.

American Educational Research Association, American Psychological Association and National Council on Measurement in Education (1999). *Standards for educational and psychological testing.* Washington, DC: American Educational Research Association.

Beaton, A., Zwick R., Yamamoto, K., Mislevy, R., Johnson, E., & Rust, K. (1990). *The effect of changes in the National Assessment: Disentangling the NAEP 1985–1986 reading anomaly* (No. 17-TR-21). Princeton, NJ: National Assessment of Educational Progress, Educational Testing Service.

Bennett, R. (2002, June). Inexorable and inevitable: The continuing story of technology and assessment. *The Journal of Technology, Learning, and Assessment, 1*(1), www.jtla.org.

Braswell, J., Lutkus, A., Grigg, W., Santapau, S., Tay-Lim, B., & Johnson, M. (2001, August). *The nation's report card: Mathematics 2000.* Washington, DC: National Center for Education Statistics, U.S. Department of Education.

Breland, H., Maxey, J., Gernand, R., Cumming, T., & Trapani, C. (March, 2002). *Trends in college admission 2000: A report of a national survey of undergraduate admission policies, practices, and procedures.* Sponsoring organizations: ACT, Inc., Association for Institutional Research, The College Board, Educational Testing Service, The National Association for College Admission Counseling.

Brualdi, A. (2002). Traditional and modern concepts of validity. In L. Rudner & W. Schafer (Eds.). *What teachers need to know about assessment.* Washington, DC: National Education Association of the United States.

Campbell, J., Hombo, C., & Mazzeo, J. (2000, August). *NAEP 1999 trends in academic progress: Three decades of student performance.* Washington, DC: National Center for Education Statistics, U.S. Department of Education.

Catalyst Chicago website. Retrieved November 2004, from http://www.catalyst-chicago.org/crsinfo.htm.

Center on Education Policy. (2004, August). *State high school exit exams: Put to the test.* Retrieved from www.ctredpol.org/highschoolexit/1/exitexam4.pdf.

Clarke, M., Shore, A., Rhoades, K., Abrams, L., Miao, J., & Li, J. (2003, January). *Perceived effects of state-mandated testing programs on teaching and learning: Findings from interviews with educators in low-, medium-, and high-stakes states.* Boston: Lynch School of Education, National Board of Educational Testing and Public Policy.

Darling-Hammond, L. (n.d.). Standards and assessments. Where we are and what we need. *Teachers College Record.* Retrieved February 28, 2003, from http://www.tcrecord.org/Content.asp?ContentID=11109.

DeMars, C. (2000). Test stakes and item format interactions. *Applied Measurement in Education, 13*(1), 55–77.

Elmore, R. (2003, November). A plea for strong practice. *Educational Leadership, 61*(3), 6–10.

Ferguson, R., & Brown, J. (2000). Certification test scores, teacher quality, and student achievement. In D.Grissmer & J. Ross (Eds). *Analytic issues in the assessment of student achievement.* Proceedings from a research seminar jointly sponsored by National Center for Education Statistics, National Institute on Student Achievement, Curriculum, and Assessment, RAND.

Gayler, K., Chudowsky, N., Kober, N., & Hamilton, M. (2003). *State high school exit exams: Put to the test.* Washington, DC: Center on Education Policy.

Goldhaber, D., Perry, D., & Anthony, E. (2004). *National Board Certification: Who applies and what factors are associated with success?* The Urban Institute, Education Policy Center, University of Washington, working paper.

Haney, W., Madaus, G., Abrams, L., Wheelock, A., Miao, J., & Gruia, I. (2004, January). *The education pipeline in the United States 1970–2000.* Boston: Lynch School of Education, National Board of Educational Testing and Public Policy.

Harvey, J. (2003, November). The matrix reloaded. *Educational Leadership, 61*(3), 18–21.

Horn, C., Ramos, M., Blumer, I., & Madaus, G. (April, 2000). *Cut scores: Results may vary.* Volume 1, number 1, monograph from The National Board on Educational Testing and Public Policy.

Hyde, J. S., Fennema, E., & Lamon, S. (1990). Gender differences in mathematics performance. *Psychological Bulletin, 107*(2), 139–155.

Jacob, B. (2001, Spring). Getting tough? The impact of high school graduation exams. *Educational Evaluation and Policy Analysis 23*(2), 99–121.

Jaeger, R., Mullis, I., Bourque, M.L., Shakrani, S. (1996). Setting performance standards for performance assessments: Some fundamental issues, current practice, and technical dilemmas. In G. Phillips (Ed.), *Technical issues in large-scale performance assessment* (pp. 79–116). Washington, DC: National Center for Education Statistics and U.S. Department of Education.

Johnson, S. (July, 2000). Response: Guidance for future directions in improving and use of NAEP data. In D. Grissmer & J. Ross (Eds.), *Analytic issues in the assessment of student achievement.* Proceedings from a research seminar jointly sponsored by National Center for Education Statistics, National Institute on Student Achievement, Curriculum, and Assessment, RAND.

Kellaghan, T., & Madaus, G. F. (1995). National curricula in European countries. In E. W. Eisner (Ed.), *The hidden consequences of a national curriculum* (pp. 79–118). Washington, DC: American Educational Research Assocation.

Kenney, P. A., & Silver, E. A. (Eds.). (1997). *Results from the sixth mathematics assessment of the National Assessment of Educational Progress.* Reston, VA: National Council of Teachers of Mathematics.

Kloosterman, P., & Lester, F. (Eds.). (2004). *Results and interpretations of the 1990 through 2000 mathematics assessments of the National Assessment of Educational Progress.* Reston, VA: National Council of Teachers of Mathematics.

Koretz, D., Linn, R.L., Dunbar, S.B., & Shepard, L. (1991, April). *The effects of high-stakes testing on achievement: Preliminar findings about generalization across tests.* Paper presented at American Educational Research Association, Chicago.

Lane, S., Wang, N., & Magone, M. (1996). Gender-related differential item functioning on a middle-school mathematics performance assessment. *Educational Measurement: Issues and Practices, 12*(2), 16–23.

Linn, R. (October, 2003). Accountability: Responsibility and reasonable expectations. *Educational Researcher 32*(7), 3–13.

Linn, R.L., Baker, E.L., & Dunbar, S.B. (1991). Complex, performance-based assessment: Expectations and validation criteria, *Educational Researcher 20*(8), 15–21.

Lindquist, M.M. (Ed.) (1989). *Results from the fourth mathematics assessment of the National Assessment of Educational Progress.* Reston, VA: National Council of Teachers of Mathematics.

Lissitz, R., & Huynh, H. (2003). Vertical equating for state assessments: issues and solutions in determination of adequate yearly progress and school accountability. *Practical Assessment, Research & Evaluation, 8*(10). Retrieved February 21, 2005, from http://PAREonline.net/getvn.asp?v=8&n=10.

Messick, S. (1989). Validity. In R. L. Linn (Ed.) *Educational Measurement* (3rd ed., pp. 13–103). New York: Macmillan.

Messick, S. (1996). *Standards-based score interpretation: Establishing valid grounds for valid inferences.* Proceedings of the joint conference on standard setting for large scale assessments. National Assessment Governing Board and the National Center for Education Statistics. Washington, DC: Government Printing Office.

National Center for Education Statistics (2005). Retrieved February 28, 2005, from http://nces.ed.gov/nationsreportcard/NAEPdate/getdata.asp.

National Commission on Testing and Public Policy (2002). *From gatekeeper to gateway: Transforming testing in America.* Boston: National Board on Educational Testing and Public Policy, Lynch School of Education.

National Committee on Science Education Standards and Assessment(1996) *National Science Education Standards.* Washington, DC: National Academies Press.

National Council of Teachers of Mathematics (1989). *Curriculum and evaluation standards for school mathematics.* Reston, VA: Author.

National Council of Teachers of Mathematics (1995). *Assessment standards for school mathematics.* Reston, VA: Author.

National Research Council (2002). *Investigating the influence of standards: A framework for research in mathematics, science, and technology education.* I. Weiss, M. Knapp, K. Hollweg, & G. Burrill. Committee on understanding the influence of standards in K–12 science, mathematics, and technology education, Center for Education, Division of Behavioral and Social Sciences and Education. Washington, DC: National Academy Press..

Parmar, R., Frazita, R., & Cawley, J. (Spring, 1996). Mathematics assessment for students with mild disabilities : An exploration of content validity. *Learning Disability Quarterly, 19,* 127–136.

Parsad, B., & Jones, J. (2005). *Internet access in U.S. public schools and classrooms: 1994–2003 (NCES 2005-015).* U.S. Department of Education. Washington, DC: National Center for Education Statistics.

Pedulla, J., Abrams, L., Madaus, G., Russell, M., Ramos, M., & Miao, J. (2003, March). *Perceived effects of state-mandated testing programs on teaching and learning: Findings from a national survey of teachers.* Boston: Lynch School of Education, National Board on Educational Testing and Public Policy.

Pellegrino, J., Jones, L., & Mitchell, K. (Eds.). (1999). *Grading the nation's report card: Evaluating NAEP and transforming the assessment of educational progress.* Washington, DC: National Academy Press.

Pellegrino, J., Chudowsky, N., & Glaser, R. (Eds.) (2001). *Knowing what students know: The science and design of educational assessment.* Washington, DC: National Academy Press.

Phelps, R. (Spring, 2000). Trends in large-scale testing outside the United States. *Educational Measurement: Issues and Practice, 19*(1) 11–21.

Potts, A., Blank, R., & Williams, A. (2002). *Key state education policies on K–12 education.* Washington, DC: Council of Chief State School Officers.

Resnick, L. (February, 1999). *Reflections on the future of NAEP: Instrument for monitoring or for accountability?* CSE Technical Report 499. Center for the Study of Evaluation, National Center for Research on Evaluation, Standards, and Student Testing, UCLA.

Resnick, L. B., & Resnick, D. P. (1992) Assessing the thinking curriculum: New tools for educational reform. In B. R. Gifford & M. C. O'Connor (Eds.). *Changing assessments: Alternative views of aptitude, achievement, and instruction.* Boston: Kluwer.

Rhoades, K., & Madaus, G. (May, 2003). *Errors in standardized tests: A systemic problem.* National Board on Educational Testing and Public Policy, Lynch School of Education, Boston College.

Romberg, T. A. (1992) Evaluation: A coat of many colors. In T. A. Romberg (Ed.), *Mathematics assessment and evaluation: imperatives for mathematics educators* (pp. 10–36). Albany, NY: SUNY Press.

Romberg, T. A. (1995). *Reform in school mathematics and authentic assessment.* Albany, NY: SUNY Press.

Romberg, T. A., & Wilson, L. (1992). Alignment of tests with the *Standards. Arithmetic Teacher, 40*(1), pp. 18–24.

Russell, M., & Haney, W. (2000, January). The gap between testing and technology in schools. *National Board on Educational Testing and Public Policy statements, 1*(2).

Sireci, A., Li, S., & Scarpati, S. (2004). *The effects of test accommodation on test performance: A review of the literature.* Amherst, MA: School of Education, University of Massachusetts Amherst. Center for Educational Assessment Research Report no. 485.

Snyder, S., & Mezzacappa, D. (March 23, 2004). Teachers come up short in testing. *Philadelphia Inquirer,* Retrieved from http://www.philly.com/mld/inquirer/news/local/8252143.htm?1c.

Stake, R. (1995). Standardized testing for measuring mathematics achievement. In T. Romberg, (Ed.), *Reform in school mathematics and authentic assessment* (pp. 173–235). Albany, NY: SUNY Press.

Stecher, B., Hamilton, L., & Gonzalez, G. (2003). *Working smarter to leave no child behind: Practical insights for school leaders.* White paper of the Rand Corporation. Retrieved from http://www.rand.org/publications.

Taylor, G., Shepard, L., Kinner, F., & Rosenthal, J. (2003). *A survey of teachers' perspectives on high-stakes testing in Colorado: What gets taught, what gets lost* (CSE Technical Report 588). Los Angeles: CRESST.

Thompson, S., Blount, A., & Thurlow, M. (2002). *A summary of research on the effects of test accommodations: 1999 through 2001* (Technical Report 34). Minneapolis, MN: University of Minnesota, National Center on Educational Outcomes. Retrieved June 2003, from http://education.umn.edu/NCDO/OnlinePubs/Technical34.htm.

U.S. Congress, Office of Technology Assessment. (1992). *Testing in American schools: Asking the right questions* (OTA-SET-419). Washington, DC: U.S. Government Printing Office.

U.S. Department of Education website for NCLB Retrieved September 2004.

Webb, N. L. (1997). *Criteria for alignment of expectations and assessments in mathematics and science education.* Council of Chief State School Officers and National Institute for Science Education Research Monograph No. 6. Madison: University of Wisconsin, Wisconsin Center for Education Research.

AUTHOR NOTE

I'd like to thank John Dossey and Pat Kenney for their reviews of earlier drafts of this manuscript, Carol Boston and Peter Kloosterman for their assistance, and Frank Lester for masterful editing.

25

LARGE-SCALE ASSESSMENT AND MATHEMATICS EDUCATION

Jan de Lange

FREUDENTHAL INSTITUTE

INTRODUCTION: IMPORTANCE AND IMPACT

Public View

The science of mathematics education is not a very sexy science if publications in the popular media are taken as an indicator. In many countries one could also use the amount of money spent on this science as a good measure. But sometimes the science of mathematics education hits the media in a big way or becomes even a hot issue that will not disappear from the headlines for years to come.

Recently Germany delivered a prime example when the results of the PISA 2000 study were published. The date of the data release was December 5, 2001, and the cover of *Der Spiegel* (a magazine comparable to the American publications *TIME* and *Newsweek*) of 10 December 2001 tells it all: *PISA-Study: The New Education Catastrophe.* The 16-page-long article in small print makes the message more than clear: Compared internationally the German schools fail. The students are poor readers, are insufficient in arithmetic, and find problem solving completely beyond their possibilities. For Germans Pisa is not an Italian city with a leaning tower; Germans are suffering from the PISA "shock."

The words and adjectives used in the article tell a story of their own: disaster, catastrophe, shock, hopeless, emergency plan. What makes the story extra difficult to accept for Germans is the fact that all neighbors of Germany score significantly better, and that the Germans are at the same level as the U.S.A. Embarrassing for the Germans, but the Americans do not see any reflection on the PISA results in their *TIME* and *Newsweek*: No media attention at all is probably the fair picture of how the results of PISA 2000 were received in the U.S.A. No news is good news seemed to be prevailing judgment here.

It is worth noting that, according to *Der Spiegel* and many others, PISA had a winner: Finland. The horse race was back, and whatever the organizers of international comparative studies have in mind, it seems inevitable that the horse race component always reaches the front pages. Finland has known it: Flocks of experts went to Finland to see the educational miracle in action. According to *Der Spiegel* the Finns are proud of their comprehensive system. Policy issues are quite often the focus of the interest: the comprehensive system in Finland, the brilliance of the Japanese "average" student in an otherwise miserable system.

There seems to be no interest in the content of the study: There was no discussion of the items that were used, the competencies needed, the quality of the instruments, or the relation to the curricula. And there certainly was no discussion about the whole study being comparative, or norm-referenced, and the fact that it does not indicate any "absolute" quality. It is definitely not criterion referenced. This often-ignored fact can cause serious interpretation problems, especially when considering policy measures. According to the national Dutch report on PISA, the Netherlands was first in Mathematical Literacy ("The Winner"). In the official Organisation for Economic Co-operation and Development (OECD) publication one cannot find the Netherlands because the country did not meet the statistical standards. The conclusion of the Dutch Ministry of Education was one of being proud of the system, the schools, and the teachers. The system was above all "effective" because the Netherlands is "underspending" on education in the OECD. At the same time math education experts (not excluding the present author) were trying to make a point about the very low level of math education in the Netherlands.

This paradox also runs among experts: In 1997 Lyle Jones argued in his William H. Angoff Memorial Lecture that NAEP (National Assessment of Educational Progress) has raised a high level of controversy because the NAEP report of 1996 shows only 18% of U.S. fourth graders "proficient" and only 2% "advanced" in math (Reese, Miller, Mazzeo,& Dossey, 1997) whereas recent results from TIMSS show that the average math performance for U.S. fourth graders is significantly above the international average (Mullis et al., 1997). Jones asked his audience, "When U.S. fourth-graders perform reasonably well in an international comparison, isn't it unreasonable that only 20 percent are reported to be 'proficient'?" Of course the answer is not the seemingly suggested "it is unreasonable," but there can be very good reasons that this is completely normal—although not always easy to explain to policymakers or journalists.

Mathematics Education View

Not only are international comparative studies popular with the media and policymakers, they are also a subject for much controversy within the math education community itself. One of the key issues is about the preoccupation with competition (the horse race). Another is the assumptions on which comparative studies in mathematics are predicated. Keitel and Kilpatrick (1999) questioned in particular the treatment of the mathematics curriculum as unproblematic. The next is the assumption that a single test can give comparable measures of curriculum effects across countries. Jablonka (2003) discussed the fundamentally situated nature of mathematical literacy (like in PISA, but also in any assessment in context). She argued that contexts will be familiar to some students and not others. And thus any attempt to use a single instrument to assess mathematical literacy beyond the most local context would appear to be self-defeating. Cultural differences exist between countries and will play a major role. Freudenthal found this argument the most important one in discarding international comparative studies. Another problem is the fact that the single scale instrument does not mean that when two countries have (more or less) the same score, the consequences for policymakers are the same. As an example we analyzed the situation in two neighboring countries, the Netherlands and Belgium, who score almost the same on TIMSS but show big differences in curricula, didactics, culture, and policies (de Lange, 1997a). So if a policymaker wants to learn from these relatively high-scoring countries, where should she go? Probably to Finland.

Politics

One of the concerns that comes up after every large-scale assessment is whether or not such tests are really about measurement. Especially people that look more at the policy side of such studies argue that performance measures are not really about measurement, but about political communication. Political scientists have expressed this view on many occasions, and there is certainly reason to take this view seriously.

Clearly politicians make up their own interpretation of the facts, to their own liking. This can be observed in most countries, but a precondition to make this observation is knowing the facts in some detail. This is most often not the case and beyond what journalists are interested in. If one looks at the large number of national reports that tend to come out parallel with the international report, it is amazing to see how the data were selected. It takes some effort, but it is worth it: large-scale international assessments are indeed about politics, about policy, about political communication.

In the discussions, whether scientific or more public, these politics play a large role. The chain of reasoning is quite often influenced by political views. This may lead to chains of reasoning in the arguments like the ones mentioned earlier that are open to some kind of criticism because of the weak reasoning, not being based on the actual data, or taking the data too seriously. Quite often the points made are mere opinions, not seldom without proper knowledge of the methodology of the study involved. Where appropriate I will deal with a number of controversial issues, hopefully without falling into the trap of weak causal reasoning and mere opinion.

NATURE OF LARGE-SCALE ASSESSMENTS

Large-Scale Assessment As Information

To identify "large-scale assessment" as a subject worthy of inclusion in this book is one thing, to find out what it really is, is quite another. Remarkably, many of my colleagues seem to agree: They shy away from giving a definition at all, and those who even try come up with rather unsuccessful propositions. In the publication *Reinventing Assessment* (1998) Randy Elliott Bennett proposed the following:

> Large-scale educational assessment consists of those tests administered to sizable numbers of people for such purposes as placement, course credit, graduation, educational admissions, and school accountability. It includes group-administered, standardized tests used most often in the secondary through post secondary years. (p. 1)

Bennett is clever enough to use words as "for such purposes as," leaving room for other purposes. The second sentence seems a bit redundant. But there are few definitions around, so I start with this one and look to improve on this definition, if only for this chapter of the book.

The National Research Council (NRC) publication *Knowing What Students Know* (2001) saw two components in educational assessment: The first context in which educational assessment occurs is the classroom. Here assessment is used by teachers and students mainly to assist learning, but also to gauge students' summative achievement over the longer term. Second is large-scale assessment, used by policymakers and educational leaders to evaluate programs or obtain information about whether individual students have met learning goals.

The committee adds that large-scale, standardized assessments can communicate across time and place, but by so constraining the content and timelines of the message, they often have limited utility in the classroom. Thus the contrast between classroom and large-scale assessments arises from the different purposes they serve and contexts in which they are used. And large-scale assessments are further removed from instruction but can still benefit learning if well designed and properly used. Finally, large-scale assessments not only serve as a means for reporting on student's achievement but also reflect aspects of academic competence societies consider worthy of recognition and reward. Thus large-scale assessments can provide worthwhile targets for educators and students to pursue.

The problem with the definition and description just given is that it seems to fit seamlessly to the American situation but ignores the realities in many other countries. International comparative studies may be blooming; international cooperation and recognition still has a long way to go.

Paul Barton made the point of the relatively isolated position in the international large-scale assessment landscape clear in his 1999 publication *Too Much Testing*. In one area there is very little (large-scale) testing whatsoever in the United States. and a lot in other developed countries: extensive examinations at the exit point for secondary education. In effect, in many countries large-scale testing is quite often identified with these exit-examinations, which are often very closely related to the curriculum. This has a very desirable side-effect that the distinction between classroom testing and large-scale testing, as clearly identified by the committee that is responsible for *Knowing What Students Know*, becomes a fluent connection, especially in those countries where a classroom assessment is part of the large-scale examination. If asked for a definition of large-scale assessments, these countries would probably suggest quite different definitions than the ones above

Large-Scale Assessment as Entrance/Exit

The isolation of the United States in this area has been properly phrased by Eckstein and Noah (1993):

> The United States is unique among the countries we have studied (U.S., China, Japan, Germany, England, Wales, France, Sweden, and the former Soviet Union) in having no coordinated, public, national system for assessing student achievement at the end of secondary school. (p. 238)

A more recent study (Stevenson & Lee, 1997) focused on entrance and exit examinations in Japan, Germany, France, and the United Kingdom. The authors observed:

> Entrance and exit examinations in these countries are based on a curriculum established by ministries of education at the local, regional or national level. Rather than imposing some arbitrarily defined standards of achievement, the examinations are closely tied to what students have studied in high school. Because teachers are aware of what students are expected to know in examinations, it becomes their responsibility to equip students with the information and skills needed to pass the examination. (p. 47)

This last observation is important because it points to the fact that large-scale assessments need not to be completely different from classroom assessment, as a matter of principle. On the contrary: For students in many countries the exit examination fits seamlessly with their classroom assessment. This fact surely complicates the task to come up with a fitting definition of large-scale assessment. Especially because of the fact that these exit examinations also do not always fit the pencil-and-paper format that people tend to link to large-scale assessments, not to mention the emphasis on the multiple-choice format.

As Stevenson and Lee observed, the exit examinations typically include open-ended questions that require organization and application of knowledge, and oral examinations that require students to express themselves verbally. Thus exist large-scale assessments that do have more than one measure, use different formats, and try to measure different competencies.

Large-Scale Assessment As Proficiency

Finally I need to mention the developments in large-scale assessments that are the results of information technology. Several examples in existence now illustrate how technology can help infuse ongoing formative assessment into the learning process. The intelligent tutors form a category that makes a difference. *Knowing What Students Know* observes that these intelligent tutoring systems are powerful examples of the use of cognitively based classroom assessment tools blended with instruction. The systems have in common that when the student makes a mistake the system offers effective remediation.

In relation to large-scale assessment caution is necessary when such intelligent tutoring systems are being used nationwide or even made compulsory—as is being considered in the Netherlands. If all schools in a country or state are using a (diagnostic) student tutoring system that also keeps track of student progress over time (maybe over a somewhat narrow band of competencies), can one then conclude that this is large-scale assessment, or should one follow *Knowing What Students Know* and classify this as formative or classroom assessment, which would make it distinct from large-scale assessment?

Definition

Clearly trying to formulate a definition of large-scale assessment, even if only for this paper, is not a simple task. For the scope of this article I will use the following formulation:

> Assessment is a process by which educators use students' responses to specially created or naturally occurring stimuli to draw inferences about the students' knowledge and skills (Popham, 2000). Large-scale assessments are assessment instruments intended to be administered to a large number of students for a wide variety of reasons. Three broad purposes can be identified: to assist learning, to measure individual achievement, and to evaluate programs (NRC, 2001). They can be delivered by different media, be individual or teamwork, be formative or summative, use a variety of formats, can be aligned with a curriculum or not, can be standards based or not, may operationalize competencies (in contrast with a content-oriented curriculum), and can be part of a comprehensive balanced assessment system or not.

STRUCTURING THE LARGE-SCALE ASSESSMENTS DISCUSSION

The definition gives way to a variety of ways to structure the discussion. One class of assessments seems a very natural one, and it is also a very visible one: the class of international comparative studies. In mathematics this class is usually restricted to the IEA (International Association for the Evaluation of Educational Achievement) studies like SIMSS, TIMSS, TIMMS-R and the recent TIMSS—with the meaning of the *T* changed from *Third* to *Trends*—and the OECD studies like PISA 2000, PISA 2003, and so on. But there are numerous other studies, especially in the area of adult mathematical literacy. I will discuss the TIMSS and PISA studies in some detail and make a relation with some of the other studies. All these studies have as a common denominator the international comparative component.

At the national level we run already into problems because of the large spread in formats and uses of large-scale assessments. In many countries exit examinations would be the first assessment that people think of. In others, most notably the United States, the one that comes closest to a national test is the NAEP project. This is a national survey intended to provide policymakers and the public with information about academic achievement of students across the nation. Its goal is of a similar nature as the IEA and OECD studies, albeit on a national level. Trend studies at national level include, for instance, the PPON (Periodieke Peiling van het Onderwijsniveau, Periodic Sounding of the Education Level) study in the Netherlands at elementary level. Of course aside from the NAEP study that has a reputation in regard to the use of standardized tests and the industry that comes with them.

I will discuss in some detail at the national level both national examinations and national trend studies, and to some extent, standardized tests both more traditional and more innovative.

In some countries it is relevant to go down one more level: to that of the states. Of course this is especially relevant in the United States, but other countries have state tests as well, for instance Australia, where Victoria has been very visible when introducing their quite innovative and balanced assessment system. I will not discuss some consequences of the No Child Left Behind Act in the United States as this issue is addressed in Wilson's chapter (this volume).

It seems reasonable to devote a discussion on the future of large-scale assessment in light of evolving information technology. The trend towards more problem-solving skills and more complex competencies like knowledge organization, problem representation, strategy use, and metacognition in mathematics education seems very challenging, as the underlying assessment principles in these systems are traditional.

The same trend towards more complex competencies in mathematics education also leads to renewed interest in teamwork in real-world problem solving. I will mention some examples that have been in place for quite some time and certainly can be regarded as large-scale assessments.

This leaves a structure with three key tenets: international, national, and state; technology; and assessing real-world problem solving. Of course this is no real taxonomy by any measure but seems a way of clustering that might help the reader to grasp this complex subject. Within each of the five clusters I will discuss some examples, problems and benefits, influence or results, and some suggestions for the future.

REVIEW OF ISSUES IN LARGE-SCALE ASSESSMENTS

International Comparative Studies

TIMSS and PISA

The IEA has conducted studies of cross-national studies since 1959. TIMSS, originally named the Third International Mathematics and Science Study, from 2003 onwards Trends in International Mathematics and Science Study, is the most recent one of the IEA series to measure trends in students' mathematics and science achievement. Additionally the countries that participate are provided with a resource for interpreting the achievement results and to track changes in instructional practices. TIMSS asks students, their teachers, and their school principals to complete questionnaires about the contexts for learning mathematics and science. TIMSS 2003 assesses the mathematics and science achievement of children in two target populations. One target population, sometimes referred to as Population 1, includes children Ages 9 and 10. It is defined as "the upper of the two adjacent grades with the most 9-year-olds." In most countries this is the fourth grade. Population 2 includes children Ages 13 and 14 and is defined as "the upper of the two adjacent grades with the most 13-year-olds." In most countries this is eighth grade (Mullis et al., 2001b). By assessing these grades using the same target populations as in 1995 and 1999, TIMSS 2003 will provide trend data at three points over an 8-year period. In addition, TIMSS data will complement IEA's Progress in International Reading Literacy Study (PIRLS) being conducted at the fourth grade. According to the authors of the *TIMSS Assessment Frameworks and Specifications 2003* (Mullis et al., 2001b), countries

> participating in PIRLS and TIMSS will have information at regular intervals about how well their students read and what they know and can do in mathematics and science. TIMSS also complements another international study of student achievement, the OECD's Programme for International Student Achievement (PISA), which assesses the mathematics and science literacy of 15-year olds. (p. 6)

According to the OECD (2000):

> The OECD's Programme for International Student Assessment (PISA) is a collaborative effort among the member countries of the OECD to measure how well young adults, at age 15 and therefore approaching the end of compulsory schooling, are prepared to meet the challenges of today's knowledge societies. The assessment is forward looking, focusing on young people's ability to use their knowledge and skills to meet real-life challenges, rather than on the extent to which they have mastered a specific school curriculum. This orientation reflects a change in the goals and objectives of curricula themselves, which are increasingly concerned with what students can do with what they learn at school, and not merely whether they have learned it. The term 'literacy' is used to encapsulate this broader conception of knowledge and skills. (p. 12)

The first PISA survey was carried out in 2000 in 32 countries, including 28 OECD member countries. Another 13 countries completed PISA 2000 in 2002, and from PISA 2003 onwards more than 45 countries will participate "representing more than one third of the world population" (p. 22). PISA 2000 surveyed reading literacy, mathematical literacy, and scientific literacy, with the primary focus on reading. In 2003 the main focus was on mathematical literacy (published in 2004), and in 2006 scientific literacy will be highlighted.

It will be clear that both studies have a lot of similarities resulting in improper identification of the two series of studies in the media, which is undesirable and confusing. But the descriptions of the organizations that are responsible show that they both claim similar relevance for the studies. It seems safe to say that, for both TIMSS and

PISA, countries participating in their studies will be given information at regular intervals about how well their students read and what they know and can do in mathematics and science. Both studies provide such information and they provide it, methodologically speaking, in a very similar way (based on Item Response Theory, IRT). Even the reporting tables in the respective reports look very similar. We refer to the Technical Aspects Reports for both studies for the methodological aspects (OECD, 2001; Yamamoto & Kulick, 2000).

If there is a problem that both studies share it is the design of the measuring instrument in relation to the validity of the outcomes. Traditionally, validity concerns associated with tests have centered about test content, meaning how the subject domain has been sampled. Typically evidence is collected through expert appraisal of alignment between the content of the assessment tasks and the curriculum standards (in case of TIMSS) and "subject matter" assessment framework (PISA). Nowadays, empirical data are often used before an item is included in a test.

Traditionally validation emphasized consistency with other measures, as well as the search for indirect indicators that can show this consistency statistically. More recently is the recognition that these data should be supplemented with evidence of the cognitive or substantive aspect of validity (Linn, Baker, & Dunbar, 1991; Messick, 1993). Or as *Knowing What Students Know* summarized: "The trustworthiness of the interpretation of test scores should rest in part on empirical evidence that *the assessment tasks actually tap the intended cognitive process.*" (p. 7). One method to do this is a protocol analysis in which students are asked to think aloud as they solve problems; another is an analysis of reasons in which students are asked to provide rationales for their responses; and a third method is an analysis of errors in which one draws inferences about processes from incorrect procedures, concepts, or representations of problems. Although some of these methods are applied only after the test is administered, there is a trend that large-scale assessments like TIMSS and PISA use these methods as well. The use of cognitive laboratories to gauge whether students respond to the items in ways the developers intended has become a new instrument in the developmental process. The use of double-digit coding is another sign of interest in the process of problem solving instead of just judging whether an answer is incorrect or correct. A "correct" or "partly correct" score given not only to each work of the student, but also to which strategy was used or where in the process the students "lost track."

A point of critique about the validity that remains to be discussed is the way of assessing, and more specifically the choice of item formats. Clearly tension exists between the frameworks (what educators want to measure) and the choice of item format (what educators can afford in a practical and economical way).

Frameworks

The original *Curriculum Frameworks for Mathematics and Science* of TIMSS did not mention item formats (Robitaille et al., 1993). In practice TIMSS relied heavily on the multiple-choice format. That the alignment between the framework and test items leaves some room for improvement may be illustrated by the example shown in Figure 25.1.

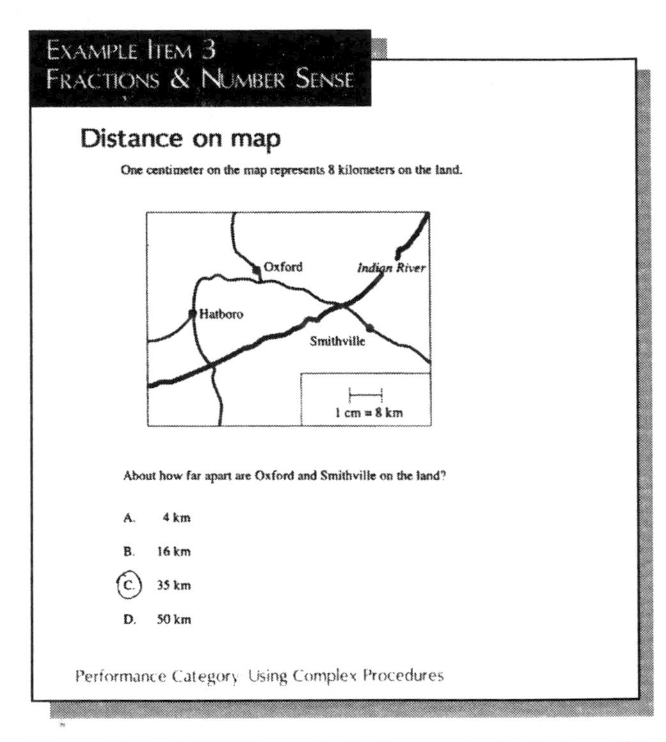

Figure 25.1 TIMSS-item, 1996 (Beaton et al., 1996, p. 63).

This item is from the Population 2 cycle of 1996 (Beaton et al., 1996, p. 63). This rather straightforward problem, in multiple-choice format, is intended to show the use of "complex (mathematical) procedures." If this is true it shows clearly that TIMSS is not criterion referenced but very normative and that the alignment between framework and test items leaves some room for improvement, especially given the ambitious performance expectations.

The more recent *TIMSS Assessment Frameworks and Specifications 2003* (Mullis et al., 2001b) made major improvements by including a "Question Types and Scoring Procedures" chapter. This made clear that the instrument heavily relies on the multiple-choice formats. The Framework left it to the reader to do the arithmetic to find out how many questions still are multiple-choice:

> Two question formats will be used—multiple-choice and constructed response. Each multiple-choice question will be worth one point. Constructed-response questions generally will be worth one, two or three points. However, extended-problem solving and inquiry-items may be worth up to five points. Up to two-thirds of the total number of points will come from multiple choice items. (p. 88)

The wording in the PISA Frameworks is different, and this study seems to be more open to open questions. The PISA 2000 Framework (OECD, 1999) explicitly mentioned the somewhat problematic status of multiple-choice items:

> Travers and Westbury (1989) state when discussing the second IEA mathematics study (SIMSS) that: "The construction and selection of multiple choice items was not difficult for the lower levels of cognitive behavior—computation and comprehension." But they continue: "difficulties were presented at the higher levels." There is place for the use of multiple-choice formats but only to a limited extend and only for the lowest goals and learning outcomes. For any higher-order goals and more complex processes, other test formats should be preferred, the simplest being open questions. (p. 54)

In practice, PISA 2000 ended up with around 50% of the items being multiple-choice. What is interesting in the 2000 Framework is appendix 2: "Considerations for future survey cycles of OECD/PISA." In fact it is a plea for a balanced assessment instrument in which students can better show what they can do in a more constructive way: extended-response essay tasks, oral tasks, two-stage tasks, and production items.

For the PISA 2003 Framework, the test instrument consisted of "a combination of items with open-constructed response types, closed constructed-response types and multiple-choice types. About equal numbers of each of these types will be used in constructing the test instruments for OECD/PISA 2003." (p. 50).

In practice the amount of multiple-choice items is higher than indicated in the Framework. None of the recommendations of the appendix of 2000 were realized.

The validity of the test instrument remains a complex issue. It goes without saying that there is an inherent tension between the traditional choice of item formats, usually with very restricted time (1–2 minutes per item), and the rather ambitious definitions of what the instrument is intended to measure. As Barton (1999) stated:

> In the 1990s, there have been constructive attempts to improve the testing enterprise. Serious efforts have been made to broaden tests beyond multiple-choice questions, and to include open-ended questions, performance assessments, and portfolios. However, the assessment reform movement has been slowed over issues of reliability and measurement error. (p. 4)

But not only the concern about "errors" plays an important role in relying so much on multiple-choice, it is also an economic issue: Many countries participating in these large cooperative studies are unwilling or unable to fund much more expensive multiple marker studies, even if such studies have demonstrated their efficacy.

PISA 2003 also had a problem-solving component. Many of the items (Figure 25.2) would fit the

HOLIDAY

This problem is about planning the best route for a holiday.
Figures 1 and 2 show a map of the area and the distances between towns.

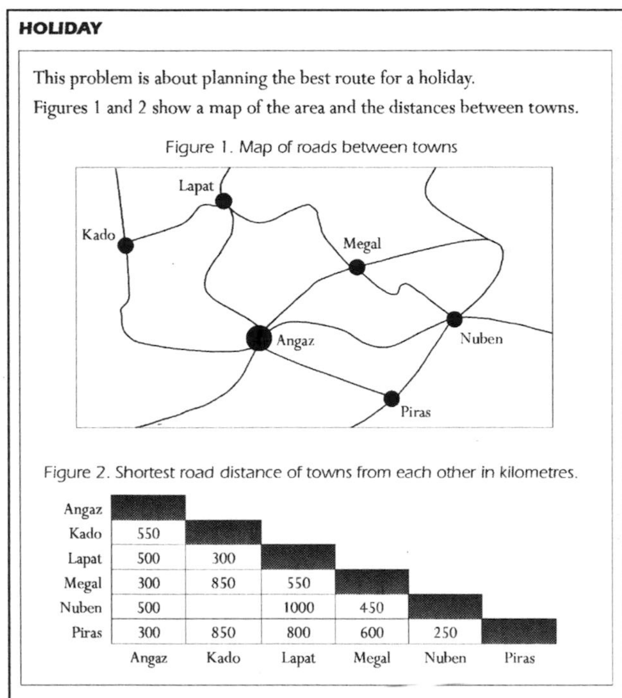

Angaz						
Kado	550					
Lapat	500	300				
Megal	300	850	550			
Nuben	500		1000	450		
Piras	300	850	800	600	250	
	Angaz	Kado	Lapat	Megal	Nuben	Piras

HOLIDAY – Question 1

Calculate the shortest distance by road between Nuben and Kado.

Distance: .... kilometres.

HOLIDAY – Question 2

Zoe lives in Angaz. She wants to visit Kado and Lapat. She can only travel **up to 300 kilometres** in any one day, but can break her journey by camping overnight anywhere between towns.

Zoe will stay for **two nights** in each town, so that she can spend one whole day sightseeing in each town.

Show Zoe's itinerary by completing the following table to indicate where she stays each night.

Day	Overnight Stay
1	Camp-site between Angaz and Kado.
2	
3	
4	
5	
6	
7	Angaz

Figure 25.2 PISA-item 2003 (OECD, 2004, pp. 70–72).

mathematics Framework, and given the fact that the instrument for problem solving (PS) had much more open "constructive" items, a study relating the math items and the PS items could be very helpful in advancing the discussion on item instruments and their restrictions in large-scale international studies.

This example (see Figure 25.2) from PISA Problem Solving shows not only the relation to mathematics, but also that students are given the opportunity to engage in a more open problem, like question 2. According to the PISA report on problem solving

> [T]he items for problem solving give a first glimpse of what students can do when asked to use their total accumulated knowledge and skills to solve problems in authentic situations that are not associated with a single part of the school curriculum. (p. 18)

One can easily argue that this is always the case in a curriculum: For mathematical literacy, mathematics as taught at school will not suffice. Students need to read, need to interpret tables and graphs (seen by many as belonging to reading literacy), and, indeed, need problem-solving strategies. But seen from the perspective of promising developments on item formats and item quality, the problem-solving component of PISA is interesting, at least. And if TIMSS implements their intent to "place more emphasis on questions and tasks that offer better insights into students' analytical, problem-solving, and inquiry skills and capabilities," innovation in large-scale assessments could materialize.

PISA versus TIMSS

The main differences between TIMSS and PISA seem to be the following:

- curricular emphasis for TIMSS versus functional aspect (literacy) for PISA;
- grade-specific structure of TIMSS versus age-specific structure of PISA.

TIMSS uses the curriculum as the major organizational aspect. The TIMSS curriculum model (see Figure 25.3) has three aspects: the intended curriculum, the implemented curriculum, and the attained curriculum. These represent, respectively, the mathematics and science intended for students to learn, and how the education system should be organized to facilitate this learning: what is actually taught in the classrooms, who teaches it, and how is it taught; and finally, what it is that students have learned, and what they think about those subjects.

International curricular diversity was a serious point of concern to the TIMSS study. The goal was

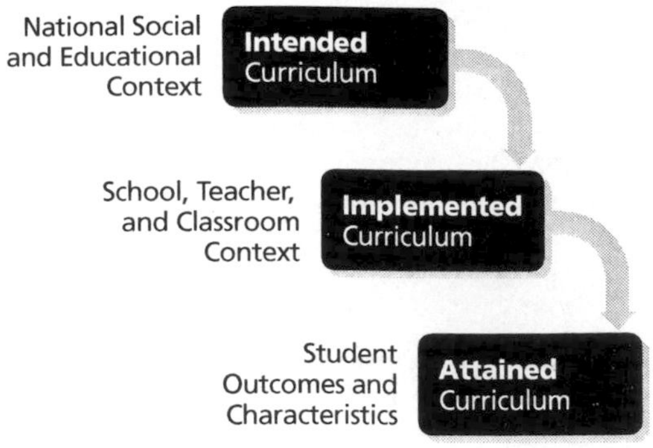

Figure 25.3 The TIMSS curriculum model (Mullis et al., 2001b, p. 3).

to develop an international test that would be equally fair to all participating countries. Therefore subject-matter specialists from all countries were consulted and asked to contribute to the process of test development (Garden & Orpwood, 1996). Most countries participating in TIMSS had an intended mathematics curriculum that matched with more than 90% of the items (Beaton et al., 1996; Kuiper, Bos, & Plomp, 1997). The outliers were the United States and Hungary with 100% matching, and the Netherlands, with 71% matching.

Insiders have discussed the procedure and its validity of this equally unfair analysis. The question not satisfactorily answered is how the mathematics education communities in the different countries were involved, and how representative they were. But if these numbers are accepted, in this context it is worth looking at the minimal matching result of the Netherlands.

As Vos (2002) observed, it was expected that students of other countries would outperform Dutch students. However, contrary to expectations, in 1995 Dutch Grade 8 students performed well on the TIMSS test. Their score was significantly above the international average, just below the four Asian top-scoring countries (Beaton et al., 1996). After some additional research by Beaton it was concluded that somehow the Dutch students were knowledgeable about the 29% of test items that were *remote* from their intended curriculum. In the end it was concluded that the students had the abilities for transfer of their knowledge and skills to items that did not match with their intended curriculum (Kuiper, Bos, & Plomp, 2000). As Heubert and Hauser (1999) explained it can be very appropriate to test students on material they have not been taught, if the test is used to find out whether the schools are doing their job.

PISA takes this point even further: It is based on a dynamic model of lifelong learning in which new knowledge and skills necessary for successful adaptation to a changing world are continuously acquired throughout life. It focuses on young people's ability to use their knowledge and skills to meet real-life challenges, rather than on the extent to which they have mastered a specific school curriculum.

Given the observation of the TIMSS researcher about the transfer skills of Dutch students, combined with goals of PISA, it may come as no surprise that Dutch students outperformed all other countries in mathematics but did not appear in the official report because "the response rate was too low" (OECD, 2001, p. 236).

The two different approaches can both be critiqued: What does it mean that the Netherlands scored so high with the minimal relation with its curriculum? What does it mean if PISA will not constrain itself to any national curricula? It is clearly *not* true, as Clarke (2003) suggested as a proposition and as Freudenthal supported in 1975 already, that international studies of student achievement may be unintentionally measuring little more than the degree of alignment between the test instrument and the curriculum. What it *does* measure is still a question open to interpretation.

Another indication that shows how difficult it is to make statements that go beyond well-intended opinions can be found in the observation of Westbury in 1992, in relation to SIMSS, when he observed that the lower achievement of the United States is the result of curricula that are not as well matched to the SIMSS test as are the curricula of Japan. But in TIMSS the match was 100% (see earlier), and still the United States did not perform very well.

The Real World

Another concern that the math education community has is about the implicit imposition of a global mathematics curriculum. Keitel and Kilpatrick (1999) have problematized the assumptions on which international comparative studies are predicated. In particular, they questioned the treatment of the mathematics curriculum as unproblematic and the associated assumption that a single test can give comparable measures of curriculum effects across nations. They further suggested that the concept of an "idealized international curriculum" lies behind even the most sophisticated research designs. The problem with this judgment is that no one knows which part of the learning process as demonstrated in the tests can be attributed to schooling and the curriculum.

Without defining what a curriculum is, it is difficult to have a proper discussion. And the question can even be stated whether the whole question about analyzing curricula is a relevant one, given the outcomes of the Netherlands in TIMSS and the PISA conception of learning in society with schools being one component of the learning process. This of course is a correct observation.

Keitel and Kilpatrick's criticism seems to focus on the underlying assumption of the international "common" curriculum. The TIMSS model seems to invite this kind of criticism, but PISA much less, if at all. However, the problem of creating such an international curriculum seems very much alive as PISA expands to about 70 countries in 2006. Many countries will look at the PISA items as something worth teaching, if alone to get better results next time. This is indeed a very serious problem as many politicians lack the knowledge and skills to realize that PISA only measures part of mathematics, more precisely: literacy. This has led to communication problems in several countries, and by many mathematicians, who tend to see PISA mathematics as representing mathematics as a discipline.

Another concern expressed in Clarke's (2003) article "International Comparative Research in Mathematics Education" is that differences in societies—like Ethiopia and the United States—automatically lead to differences in skills required for effective participation in these societies. This point seems very straightforward, but many of the competencies identified in the PISA Framework or item-collection intend to operationalize competencies that are of some relevance in any society. To state that "to assess the extent to which Ethiopian students possess the mathematical skills required for effective participation in American society seems both futile and uninformative" (p. 153) seems futile itself.

Another point that should be discussed, as it refers to an important aspect given the trend towards more "real-world" problem solving, is the link between mathematical literacy and social practice. Jablonka (2003) argued that even within a single school system there will be contexts familiar to some students and not others. Any attempt to use a single instrument to assess mathematical literacy beyond the most local of contexts would appear to be self-defeating. Again the chain of reasoning is not evidence based. The choice of context in relation to item construction and validation is a very complex one, as shown by many examples from previous experiments in designing curricular materials (e.g., Feijs, 2005). These experiments indicate that one cannot say anything firm about the relationship "context familiarity: success rate." As an example, due to the fear of experts that an item for an international study about the tides of an ocean should not be included because of the participation of land-locked countries, the item was rejected forcefully by the field trials. The influence of contexts should be studied much more systematically than is presently the case, and we researchers should refrain from strong statements that have proven to be of disputable quality until we have firmer evidence. For the present studies all countries had to agree on the items, including the role of the context.

In a working paper (Nohara, 2001) from the (U.S.) National Center for Education Statistics in which TIMSS-R(epeat), PISA, and NAEP were compared, remarks were made about the inclusion of real-life situations, defined as items not presented strictly in the language of mathematics. According to the paper, this characteristic is significant because connecting mathematics to the world outside of school is a major goal of many mathematics education reform initiatives. It is also significant because it means that students have to choose for themselves the operations and solutions most appropriate for the problem and figure out how they relate to the information provided, thereby "adding to the difficulty of the item." One may comment here that clearly such contexts require different cognitive demands from the students, but whether one can conclude that this leads to more difficult items is a matter to be studied in more detail, and especially PISA 2003 might offer some direction. The reason for that somewhat optimistic expectation can be found in the same working paper when the author stated that both TIMSS-R and PISA (and NAEP) contained many items situated in real-world contexts: 44% of items for TIMSS-R, 48% for NAEP, and 97% for PISA. The author also noted that in reviewing PISA items, panel members noted that several items in real-life situations presented students with significantly more challenges than others. These contexts were highly unique, that is, not typically encountered in mathematics instruction or textbooks, or required significantly more thought regarding how the nature of the context affects the mathematics involved in the problem. This type of item can be contrasted with the standard word problems typically used in mathematics classes, which can be described as "proxies for reality." Only a few such items can be found in TIMSS-R or NAEP.

Meaning of Measures

Via the real or perceived influence of context I now come back to the relative meaning of outcomes of large-scale international assessments. The use of context is or could be one factor contributing to the relative difficulty of the assessments. Other factors contributing to the relative difficulty are (Nohara, 2001) the number of items that use the extended-response format, the number that require multistep reasoning, and the amount of computation as "the presence of a computation requirement does not present an additional degree of difficulty as it would in an item classified in another content strand" (p. 14).

Looking only at these four factors PISA appears to be the most difficult: It has the highest percentages in all four categories.

If it is true that one can attribute item difficulty to these four factors and conclude that PISA is the most difficult large-scale international assessment, it might give some additional insight into what one can say about the *absolute* level in relation to the *relative* level. In addition to the TIMSS item presented before in this contribution it might be interesting to look at a PISA item that was considered "too difficult" to be included in the final item collection.

In my opinion, Figure 25.4 shows, as part of a wider pattern if one analyzes the items of large-scale assessments, that the "cognitive competency level" operationalized by these tests leaves lots of room for improvement. This implies that the "results" of such studies should be interpreted much more carefully than is usually the case. The horse race outcomes will be differently interpreted if policymakers understand that the metaphor itself is quite inappropriate if the horses do not have to meet a certain qualifying time over a fixed distance but merely are allowed to underperform.

Indeed, as Clarke (2003) summarized the feeling of most of the mathematics education community, it is "the current preoccupation with competition" that always seems to make it into the headlines of the media and always seems to support the present politicians and policymakers. The "Mathematics Olympics" under the banner of PISA 2000 made a media uproar in Germany, as did TIMSS in the United States in 1997.

What is missing quite often is a discussion by somewhat informed mathematics educators about what is being measured from a "content" view and what it means. The quoted working paper from NCES is a laudable attempt to shed some light on this aspect of international comparative studies. Keitel and Kilpatrick (1999) observed correctly the striking fact that almost all of the people with the primary responsibility for conducting the study (TIMSS) have been empirical researchers in education, psychometricians, or experts in data processing. The problem of the content of mathematics education, Keitel and Kilpatrick (1999) argued, has been dealt with as no more than a technical question:

> The most monumental and most accurately treated feature of the studies has been the handling of data once they have been collected. Particularly notable has been the fashion in which problems of methodological validity, reliability and quality have been re-

Mathematics Unit 3

HEARTBEAT

For health reasons people should limit their efforts, for instance during sports, in order not to exceed a certain heartbeat frequency.

For years the relationship between a person's recommended maximum heart rate and the person's age was described by the following formula:

Recommended maximum heart rate = 220 – age

Recent research showed that this formula should be modified slightly. The new formula is as follows:

Recommended maximum heart rate = 208 – (0.7 × age)

Mathematics Example 3.1

A newspaper article stated: "A result of using the new formula instead of the old one is that the recommended maximum number of heartbeats per minute for young people decreases slightly and for old people it increases slightly."

From which age onwards does the recommended maximum heart rate increase as a result of the introduction of the new formula? Show your work.

The classification of the situation depends of course on whether or not people are actually interested in data about their own health and body. One can safely argue that this item is somewhat scientific (because of the use of formulas) but many sportsmen and women (joggers, bicyclists, rowers, walkers, etc.) really do measure their heartbeat quite regularly during their exercises. More and more inexpensive instruments using micro-technology have made this aspect of human well-being much more accessible to ordinary people. This esplains the situation classification as "Public/Personal."

Because we are really dealing more with modelling and less with trivial problem solving, a classification to the *connections* cluster seems rather straightforward, as well as the overarching idea *change and relationship*.

Comparing two formulas, even though they are merely rules of thumb, that relate to a person's well-being can be an intriguing activity, especially as they are partly presented as "word" formulas. This usually makes them more accessible to students. Even without asking a question, an initial reaction of students might be to see how their own age will lead to different recommended outcomes. As the PISA students are 15 years of age, the result under the old formula is 205 hearbeats per minute (realizing that the information that the rate is *per minute* is not given), and under the new formula it is 198 (or 197). Thus they might already have found an indication that the statement in the newspaper article seems to be correct.

The example posed is somewhat more complex than this. It requires the students to find out when (at which age) the two formulas give the same result. This can be done by trial and error (a well-established strategy by many students) but the more algebraic way seems more likely: 220 – a = 208 – (0.7 × a), leading to an answer around 40.

From the viewpoint of mathematical literacy as well as from the viewpoint of more curricular-oriented mathematics, that is quite an interesting and relevant problem. We note that the field trial data indicate that 15-year-old students found this problem quite difficult.

Mathematics Example 3.2

The formula *recommended maximum heart rate* = 208 – (0.7 × *age*) is also used to determine when physical training is most effective. Research has shown that physical training is most effective when the heartbeat is at 80% of the recommended maximum heart rate.

Write down a formula for calculating the heart rate for most effective physical training, expressed in terms of age.

This example *seems* to measure exactly the same competencies as Example 3.1. The correct response rate is almost identical (during field trial). But there is a notable difference: In Example 3.1 students have to compare two formulas and to decide when they give the same result. In Example 3.2 the students are asked to "construct" a formula, something they are not frequently asked to do during their school career, in many countries. From a strictly mathematical viewpoint the question is not difficult at all: just multiply the formula by 0.8—for instance, *Heart Rate* = (208 – 0.7 × *age*) × 0.8. It would seem that even such simple manipulation of algebraic expressions, expressed in a practical and realistic context, presents a substantial challenge to many 15-year-olds.

Figure 25.4 PISA item 2003 (OECD, 2003, pp. 64–66).

> solved from a purely formal point of view. Questions of content—in all its aspects—have usually been seen as secondary. (p. 245)

I agree with this point and also note that in relation with this point several countries carried out national options, in part because content-related matters gave concern about the validity of the assessment as appropriate for the specific country. An impressive effort in relation to PISA 2000 was carried out by Germany. Although the report (Baumert et al., 2001) was in general positive about PISA 2000, the Germans opted for a National Option because of the fact that there was a need (in Germany) for more "inner-mathematical problems," for more "basic-skills problems" and "facts." This was done to make the PISA test "closer" to the German curriculum.

A similar action, but then in the opposite direction, was taken by the Netherlands in reaction to TIMSS. As I reported in 1997, the Netherlands (here meaning some influential math educators) concluded that TIMSS did not fit in a satisfying way to the intended/perceived/implemented curriculum, not on the basis of the earlier mentioned relatively low "alignment-quotient," but because of too many multiple-choice items operationalizing too many lower level thinking skills (de Lange, 1997). Therefore a National Dutch Option was carried out. The international test and the national option had an intersection of carefully selected "anchor items." Methodologically speaking, the results on the anchor items within the national option or in the international option should be the same. In reality the students did much better on the TIMSS items when they were part of the national option. This outcome supports the arguments to first think what is really being measured and with which instruments before making policy statements regarding the horse race. In fact, the TIMSS score in reality was 541 in Grade 8. But based on the national option as a more reliable measurement, the same TIMSS items would lead to a score of 585. The only explanation brought up by the national experts, including the IEA group, was that the "national" context made children feel better and more motivated. Or, to put it slightly differently, perhaps the fact that the anchor items were culturally correctly embedded made students perform better.

The content is a point of concern that receives relatively little attention, contrary to the question of the influence of the different cultures. It goes without saying that we measure "cultures." Bracey's (1997) observation that "international achievement differences become similarities if we compare the performance of students of Asian cultural affiliation, no matter what school system they attend" is true to a certain extent. One does not have to be a rocket

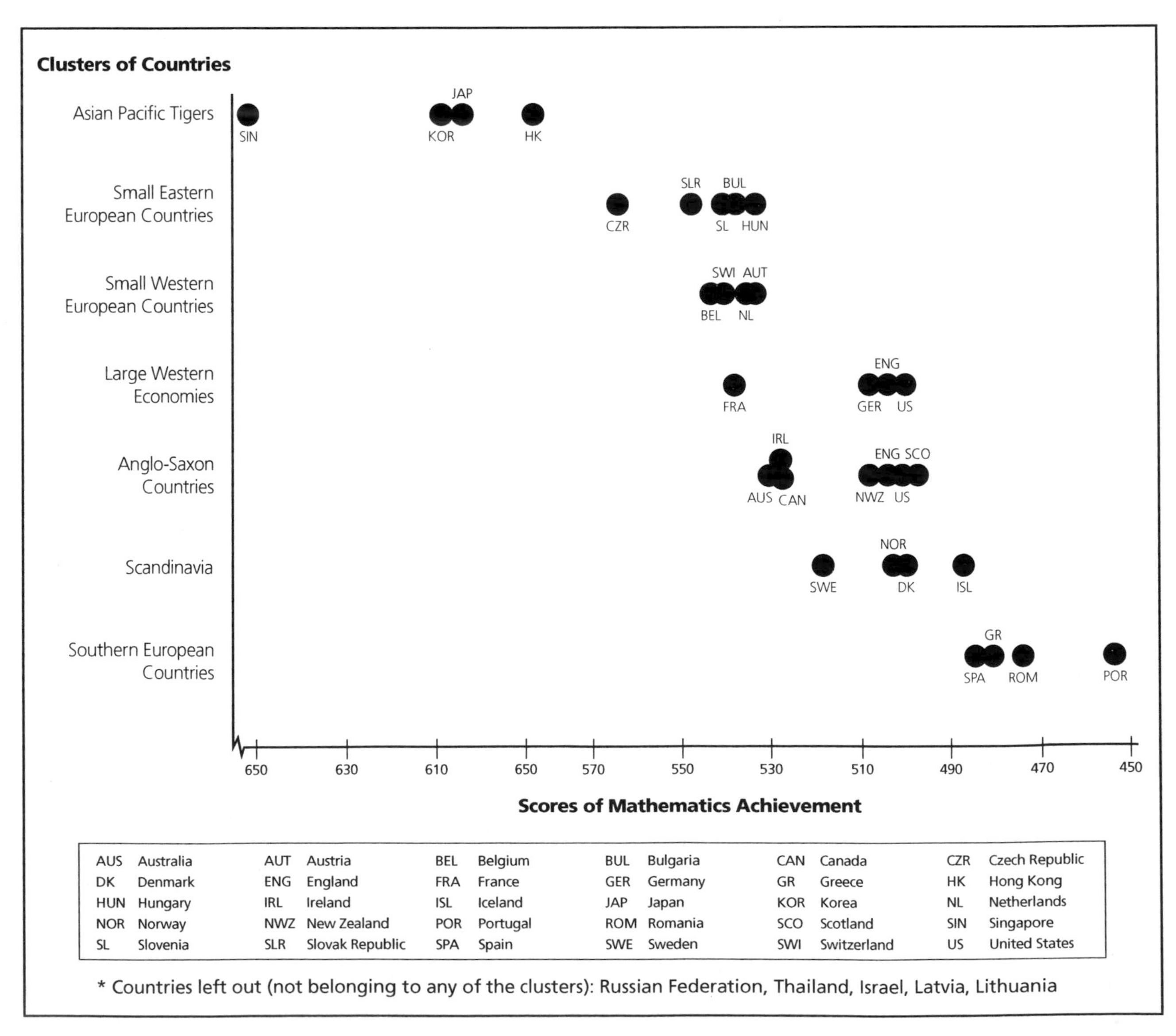

Figure 25.5 Graph showing countries participating in TIMSS 1996 (de Lange, 1997).

scientist, let alone a math education researcher, to see a certain socio-economic-cultural structure in the listing of the horses of TIMSS.

In my opinion one can see a certain clustering of countries, starting with the Pacific Rim countries (referred to as "Asian Pacific Tiger" countries in Figure 25.5), via small European countries, to France, to (former) commonwealth countries, to Germany and the United States, to Scandinavia, and finally Southern Europe (de Lange, 1997). The mathematics education community has the responsibility not only to contribute to such studies (and to influence them in a positive way), but also to critique them when necessary, to draw attention to the content part of such studies, and to inform policymakers what the study might mean beyond the horse race.

Missing Content

In a review of the PISA 2000 report Bonnet (2002) underscored the lack of interest—in content matters—in trying to understand in pedagogical terms how and why pupils gave the responses, or absence of responses, that they actually gave. Instead, he observed, the focus of the report is on trying to analyze results in the light of socioeconomic data and the organization of education systems and schools. One would have wished for more space to be devoted to trying to understand what happens in the classroom in terms of the teaching and learning process. This exactly is what the title of the earlier mentioned IEA/TIMSS promises but fails to materialize. Both studies, TIMSS and PISA, should and could do much better in this respect. It is difficult to point to the real reason behind this phenomenon. Keitel and Kilpatrick (1999) suggested looking at who conducts the study in order to better understand the results.

Taking into account that PISA is organized by the OECD, Bonnet (2002) observed that it is illuminating that chapter 3 of the PISA 2000 report ends at page 91 with tables showing the relationships between the average country performance in the three combined scales (reading, mathematics, and science) and the GDP (Gross Domestic Product) per capita. The true worth of such correlations becomes apparent with the inspiring conclusions that "countries with higher income per capita tend to perform better, on average, but some countries do better or worse than their income would predict" (p. 92). In my opinion, Bonnet was right in concluding that the bias in the report in favor of assessing education from the point of view of economic efficiency is overshadowing another conception of education, prevailing in Europe, whereby education is just as much about social cohesion, personal fulfillment, and cultural development. It is very frustrating for those who care about children if a minister of education proudly states that the Netherlands has a very efficient educational system, because we outperform most other countries, while spending well below the OECD norm. In this way international comparative studies can be very destructive to the improvement of the teaching and learning process.

Impact of Studies

Consider TIMSS. It is very tempting to use as a source the publication by Robitaille, Beaton, and Plomp for IEA/TIMSS with the title: *The Impact of TIMSS on the Teaching and Learning of Mathematics and Science* (2000). The title of the book is a bit misleading as it deals only in a minor way with the "impact"; the book reports also on national education systems, and the results and findings per country. Although the title of the book stresses—as it should do—the "learning and teaching," this aspect is difficult to find in the book. This is also to some extent the conclusion of the authors as they state:

> the results seem, at least until now, to have made a greater impact on the general public and in political circles through the mass media than they have on educators. On the other hand, it is also true that many of the reports refer to the impact of the study on the process of curriculum reform or on national evaluation programs. Those kind of processes tend to take a long time to bring to fruition, and they are then followed by an equally lengthy and perilous process of implementation. So it is probably unrealistic to expect that the results would have made more of an impact at the classroom level since only three years have elapsed since the publication of the first set of international results. (p. 168)

Without trying to be exhaustive a rather typical example of a specific country's reaction under the chapter "Impact" will resemble something like the following (based on Switzerland):

> TIMSS showed primarily where Switzerland stands in comparison to other countries with respect to achievement in mathematics . . . [and] gave a detailed picture of current levels and patterns of achievement. This is a valuable basis for discussion of needed improvements. TIMSS also gives some indications of possible reasons for such differences in achievement. . . . However, given the way TIMSS was designed, such statements of causality are only valid to a limited extent. It is not easy to deduce directly from TIMSS how improvements might be made. While TIMSS had a direct effect on debate and reflections concerned with educational policy, educational planning, pedagogy, educational science, and teacher preparation, it had only an indirect effect on corresponding practice. (p. 158)

One is tempted to comment that these observations cannot come as a big surprise. Having some experience and involvement with both TIMSS and PISA (see note at end), I find it striking to note the relatively little interest that IEA and OECD have in both content and teachers and students, the key factors in math education. As the studies are usually announced and described as an educational system analysis, and as relations are made between the performance of educational systems and the economy, it is hard to expect much interest from the "real" stakeholders in mathematics education. The introductory remarks for the "impact study" from the executive director of IEA, Hans Wagemaker, make clear where the emphasis lies:

> International comparative studies of educational performance . . . can therefore play a central role in assisting policy-makers, curriculum experts, and researchers in judging the extent to which an education system meets national expectations and the extent to which it is likely to contribute to the nation's overall economic and social well-being. (p. 7)

I now turn my attention to examples of the possible impact of the PISA studies. They are both not about system changes, but about content-oriented matters, taking teachers, children, and content as key variables. In the examples I will discuss two completely different ways the outcomes of PISA 2000 have been used—one of them being by Germany. This cannot come as a surprise after the introduction of this article focusing on the PISA shock; but Germany is also a good starting point for quite another point that I will only mention here, but that is worth further research and analysis. The Germans produced a national PISA report of 550 pages, the international OECD report was 330 pages, and the Dutch report a mere 65 pages. Most countries had something around 150 pages. It is not the statistics that are interesting here, but the message from the report and what has been selected to be included. Even a superficial analysis, which was carried out for this article with the reports mentioned and the one from the United States, makes significant differences visible. As Keitel and Kilpatrick (1999) pointed out very correctly, there is a common myth that "numbers do not lie." It is now widely accepted that data can be gathered, processed, mathematized, and interpreted in a variety of ways. So a key issue is the question of who influences this process, for what reasons, and through what means. The four studies just mentioned underscore this concern apart from the fact that even numbers can lie.

Back to the very *gründliches* German report. Not only did the German PISA *Konsortium* do an excellent and thoughtful job, it also made recommendations for immediate improvement, including ones that directly affect the content. According to Blum (2000), the changes should include

- more integration of inner- and outer-mathematical "networks";
- fewer calculations;
- more thinking activities and student mental "constructions";
- more reflection;
- more flexible use of schoolbooks.

These goals can be reached when the recommendations that were formulated after TIMSS are implemented:

- development of a different math-problems culture: more open-ended, more "real-world";
- a new teaching-and-learning culture, with a more exiting cognitive school environment (Klieme, Schümer, & Knoll, 2001);
- more and different professionalization of teachers, emphasizing teamwork.

PISA adds to these recommendations a "very different conceptualization" of mathematical concepts and emphasis of modeling and mathematization, situated in contexts. And, argued the report, the Germans have definitely not reached the optimum in using different representations as a tool to build better conceptual understanding.

Mathematics education is in a state of transition, in part because of the fact that both TIMSS and PISA were taken seriously. Surprisingly the shock and catastrophe that struck Germany as some kind of natural disaster, if one had only the popular media as a resource, has resulted in a government-supported nationwide action-plan with a very strong content part that will result in a different mathematics education culture at schools. Of course, the success of these changes will be measured by PISA 2003, 2006, 2009, and so on. At least in part.

A very different way of using the "very different conceptualization" of PISA was reported by Dekker and Feijs (2005). Classroom teachers (in Philadelphia and Milwaukee, WI) were invited to reflect on their own classroom assessment, initially in their own way using their principles. Teachers were questioned why they asked the questions they asked, why they ordered the questions as they did, what were the presumed learning goals that the items tried to operationalize, and so on.

After getting familiar with the different competency clusters (computation, connections, reflections) and discussing PISA items and other "good" items that were more of an open kind and needed more student construction, the teachers were again confronted with their own classroom assessments. This Classroom Assessment as a Tool for Teacher Change (CATCH) project was considered powerful because it not only changed the teacher's attitudes towards the teaching-learning-assessing process, but also changed their actual classroom practices—not just their classroom assessments. It seems that using an international comparative study by focusing on the concept of the study from a mathematical point of view, in combination with the available instruments, and by inviting teachers to make this instrument better and more fit for the classroom practice should be strongly encouraged as it can be a powerful tool from the bottom-up in combination with top-down approaches.

In the CATCH project a competency model was used that was quite similar to that of the PISA Framework (2003). The three different competency clusters were visualized in the Assessment Pyramid (de Lange, 1994, 1999) as competency levels. This pyramid turned out to be very effective in changing teachers' attitudes and practices in the classroom. Some of this effect can be attributed to the fact that PISA had adapted a similar model, which is seen as adding validity to the Pyramid.

As indicated before, I would like to close this "international comparative studies" part with a short discussion about other large-scale assessments in mathematics education other than the TIMSS and PISA studies that I have discussed in some detail.

Numeracy, Quantitative, and Mathematical Literacy Assessments

Studies like NAEP and TIMSS are connected to some kind of "idealized" curriculum. PISA is not, as it claims to test mathematical literacy, a term described as

> an individual's capacity to identify and understand the role that mathematics plays in the world, to make well-founded mathematical judgments and to engage in mathematics in ways that meet the needs of that individual's current and future life as a constructive, concerned and reflective citizen (OECD, 2001, p. 22).

If the TIMSS publication on the impact of the study expressed some concern about the relative lack of interest in the results, this lack of interest is even more evident in studies about numeracy, quantitative literacy, and mathematical literacy, especially if these are about adults and not, like PISA, for 15-year-olds.

An important event in this respect was the Young Adults Literacy Skills (YALS) project in the United States, in which young adults (12–15) were assessed on tasks related to three literacy scales, among them quantitative literacy. The description of quantitative literacy was somewhat limited compared to the more recent PISA definition: "The knowledge and skills required to apply arithmetic operations, either alone or sequentially, using numbers embedded in printed material (e.g. balancing a checkbook, completing an order form" (Dossey, 1997, p. 46; Gal et al., 1999, p.12). The results of this assessment led to conclusions like "performances on the scale were less than what the nation required in human talent for long-term international competitiveness" (Dossey, 1997, p. 48).

The follow-up of YALS was the National Adult Literacy Survey (NALS), carried out in 1992. Gal (1993) concluded that roughly 50% of American adults would either have major difficulty with or be fully unable to handle real world tasks such as

- using a bus schedule to determine departure time;
- identifying a trend on a simple graph about sales figures;
- using a calculator to find the difference between the regular price and the sale price;
- estimating the unit price of a grocery item;
- understanding a table summarizing a school survey;
- calculating interest charges related to a home loan.

In the technical report *Issues and Challenges in Adult Numeracy* (1993) Gal observed that there are two "extreme" views on numeracy. One extreme is to see numeracy as the lower end of mathematics, or whatever math educators attempt to achieve in the early grades (Van Groenestijn, 2002) whereas the other end is viewed as "encompassing a broad set of skills, knowledge, strategies, beliefs and dispositions that people need to autonomously engage in and effectively manage situations involving numbers, quantitative or quantifiable data, or information based on quantitative data" (Gal, 1993, p. 2).

More recently the definition has broadened even more, as in PISA, to reflect that literacy in the area of mathematics goes well beyond the strictly numerical (numeracy) or quantitative aspects (quantitative literacy). Confronted with this problem, I proposed in Madison and Steen's publication on quantitative literacy (2003) to see numeracy as part of quantitative literacy. This includes understanding of and mathematical abilities concerned with certainties (quantities), uncertainties (quantity as well as uncertainty), and relations (types of, recognition of, changes in, and reasons for those changes) and defines spatial literacy, most likely the most basic and neglected aspect of mathematical literacy, as understanding of the (three dimensional) world in which we live and move (de Lange, 2003). Shaeffer (2003) supported this view because of the indisputable fact that much more in mathematics is useful besides numbers. Indeed, many aspects of statistical thinking (uncertainty) are not about numbers as much as about concepts and habits of mind, he argued.

This discussion is relevant to the problems of large-scale assessments as it is not clear at all what role school curricula play in setting up large-scale literacy assessments, although clearly some are very much tuned to traditional curricula and others are more society oriented. An example is the International Numeracy Survey (INS) in which only a limited number of countries took place: U.K., France, the Netherlands, Sweden, Denmark, Japan, and Australia, organized by The Basic Skills Agency in London in 1997. Two typical examples (out of only 12 problems) are:

> The total of 4.25, 6 and 7.74 =
>
> Work out 15% of 700.

Van Groenestijn (2002, p. 34) observed that "though in this study 'numeracy' was seen as doing mathematical operations on school-based tasks, it does indicate that the performance on such school-related tasks is disappointing and presumably undermines numerate behavior." This observation seems somewhat questionable as it directly relates literacy to school mathematics in a restricted basic skills sense and fails to clarify how such skills "undermines" numerate behavior. I cannot but agree on the observation that the performance on the task is disappointing. What can be learned from that is quite another matter.

The NALS was followed by an international study, set up by the OECD: the International Adults Literacy Survey (IALS). This survey was carried out in 1997 in OECD countries. Apart from the United States and Canada 21 countries including Germany, Sweden, Switzerland, and the Netherlands took part. This study defined quantitative literacy quite narrowly, trying to connect traditional arithmetic with the problems of daily life: "To apply arithmetic operations to numbers embedded in printed materials, such as balancing a check-book, figuring out a tip, completing an order form, or determining the amount of interest on a loan from an advertisement"(NRC, 1989, p. 7–8).

Finally I mention the follow-up survey of IALS, the Adult Literacy and Life Skills Survey (ALL) organized by the National Center for Education Statistics (NCES) and Statistics Canada, which uses a more elaborated numeracy domain. An interesting "Policy Information Report" about this study was published in 2002 (Sum, Kirsch, & Taggart, 2002) for the United States. The report provided a good view of the role of international large-scale assessments as the goal of the report is to call attention to important findings from the IALS and *their implications in terms of educational and workforce development policies.* The report also stated that it is interesting that the results of IALS are becoming available at a time when the role of human capital in influencing the fate of individuals and nations is receiving increased attention. According to a recent report released by the OECD (2001), research shows that the development of human capital is correlated with better health, lower crime, political and community participation, and social cohesion. Some studies, they reported, even *suggest* that the social impacts of acquiring these knowledge and skills could be as large as their impacts on economic productivity. Elsewhere in this article some doubt is cast on this statement, or more precisely on the relation between achievement in education and economic performance in a causal way.

One of the more interesting results is the fact that 46% of the respondents in the United States belonged to Level 1 and 2, "possessing a very-limited to limited mathematical literacy proficiency." Also noteworthy is the fact that there is no "horse race" ranking in the report: The United States is compared only with all high-income countries, but only as a group.

From the summary I quote the following remarks, clearly identifying the purpose of the report as the title is *Summary of Findings and Policy Recommendations.*

- The U.S. educational system is clearly less productive in raising literacy skills of students per dollar spent.
- The U.S. appears to be living off its past as higher educational investments will inevitably lose ground in the coming decade.
- The nation's changing demographics will likely exacerbate the literacy skill deficit in coming years.
- Although skills influence labor market success everywhere in the world, this is especially true in the U.S. and other English-speaking countries.
- Mediocre skills and inequality in the distribution of skills may have worked reasonably well in recent years but are not likely to do so in the future.

Given these results it may hardly come as a surprise that mathematics educators pay relatively little attention to these studies, as the report gave no indication whatsoever about the instruments used and the reasons behind them. It is my opinion that more balanced reports, not only this one, showing the links between policy implications and the assessment instruments are in the interest of everyone.

In 2000 the OECD carried out the earlier discussed PISA study, also with a much wider ranging definition of mathematical literacy. It seems not very reasonable that the difference in definition of literacy can be attributed to the difference in target population: adults or 15-year-olds. Hopefully this change—by OECD—can contribute to newer insights into the importance of mathematics in society and the relevance of seeing mathematics function in the real world.

NATIONAL LARGE-SCALE ASSESSMENTS

Introduction

Two major clusters comprise large-scale assessments on a national level: a discussion of NAEP and the TIMSS Benchmarking Study for the United States at one hand and at the other side the national exit examinations as they exist in many countries around the globe. It goes without saying that, like this whole chapter, I can touch on the subjects discussed and then only in a quite selective way.

NAEP

In 1969 the Carnegie Corporation of New York appointed the Exploratory Committee on Assessing the Progress of Education (ECAPE) with Ralph W. Tyler as chair. According to Tyler (1969) a major purpose of the national assessment program was to provide the lay public with census-like data on the educational achievements of children, youth, and adults—data that furnish dependable background information about educational attainments, the progress being made, and the problems still faced in achieving educational aspirations.

This was the first description for the program that is presently known as the National Assessment of Educational Progress. NAEP is a national survey intended to provide policymakers and the public (definitely lay people) with information about the academic achievement of students across the nation. It serves as a source of information for policymakers, school administrators, and the public for evaluating the quality of their educational programs. Just like TIMSS, NAEP is not tied to a specific curriculum. Whereas TIMSS tried to find a set of instruments that best fit the existing curricula of all participating countries, NAEP is based on a set of frameworks that describe the knowledge and skills to be assessed in each subject area. The performances assessed are intended to represent the leading edge of what all students are learning.

For over 30 years, the NAEP has served as the United States' only ongoing gauge of student achievement across time. One has to understand that just like with the other large-scale assessments discussed before, the studies do not report or even intend to report on student's individual achievements. This is caused by the multiple-matrix sampling method, the design selected by the founders of NAEP. Not all students will be sampled. On the contrary, a strategically selected sample can support the targeted inferences about groups of students with virtually the same precision as the familiar approach of testing every student. Moreover, like in PISA and TIMSS, not all students are administered all items. It can hardly come as a surprise that one of the current debates is about this aspect of large-scale assessments of this kind: the impossibility to provide student-level reports, something that exit examinations do provide, sometimes in great detail. In NAEP, measurement at the level of individual students is poor, and individuals cannot be ranked, compared, or diagnosed (NRC, 2001, p. 224). Worse even, students who have similar performances on one particular item can have different competencies—something known not only at the student level, but also at country level in TIMSS or PISA.

Silver, Alacaci, and Stylianou (2000) have demonstrated some limitations of scoring methods used by NAEP for capturing the complexities of learning. For a particular item NAEP concluded that 11% of the students gave satisfactory or better responses but can give no further detail about the important differences in the quality of the reasoning demonstrated. In this particular case the reasoning varied from surface-level reasoning to analytic reasoning and to even more sophisticated reasoning.

Dossey, Jones, and Martin (2002) confirmed that indeed a "correct" answer is only part of the story. They did so in the context of the TIMSS study, which used the double-digit coding to find out more information about the strategies used by students in an effective way. In a case that involves three distinct different ways of reasoning one can code a correct answer with 1.1, 1.2, or 1.3, the 1 standing for "correct," the 1, 2, and 3 for the strategy used. In this way important information about the process or strategy of the student can be made visible to a certain extent. But the data offer more for those interested in the actual solution process, in addition to double-digit coding. This can be interesting at the national level, but also at the international level: In one country most students could be using one—often computational—technique, whereas in another country students use a variety of strategies.

A national analysis of the actual process the students have gone through is done by studying and investigating actual students' work, in combination with double-digit coding and PISA scores—for instance in the Netherlands. Indeed, it seems that stakeholders are increasingly interested in better—which often means more complex—assessments, even on a large scale and even if this means higher costs and less reliability because of intersubjective scoring. I will come back to this issue later, when discussing future developments.

Over the past thirty years (since 1973) three distinct NAEP projects have evolved: the Main NAEP, the Long-Term Trend NAEP, and the State NAEP. The Main NAEP periodically assesses students' achievement in reading, mathematics, science, writing, U.S. history, civics, geography, the arts, and other subjects at Grades 4, 8, and 12.

Trial assessments with voluntary state-by-state assessments were conducted in 1990 and 1992 and were considered so successful that the U.S. Congress authorized regular state assessments starting in 1996 (Allen, Jenkins, Kulick, & Zelenak, 1997).

The content of both the Main and State NAEP programs follows curriculum frameworks developed by the National Assessment Governing Board (College Board, 1996). Kane and Webb (2004) argued that test-item types for the Main and State NAEP assessments are consistent with the current state-of-the-art in achievement testing and have evolved over time with curriculum changes. In contrast, both the students sampling frame and the content of the Long-Term NAEP have remained essentially unchanged.

This Long-Term NAEP, or the Long-Term Trend Mathematics Assessments (LTT Mathematics), has provided a measure of students' achievement over time by administering similar assessments periodically since 1973 (Galindo, Caulfield, Mohr, & McCormick, 2004). It should be noted that the LTT has a computational focus, measuring students' knowledge of basic facts, their ability to carry out paper-and-pencil calculations, their knowledge of basic measurements formulas, and their ability to apply mathematics to daily-life situations such as those dealing with time and money. This focus reflects the school mathematics in the United States in the 1970s. The content areas of the LTT are represented in the Main NAEP. The three types of items used on Main NAEP—multiple-choice, short constructed response, and extended constructed response—are reduced to only multiple-choice and short constructed response in the LTT.

Not only in content and item-type format is the LTT different from the Main NAEP, but also the reporting scales. The Main NAEP is more comparable in many respects with the international comparative studies like TIMSS and PISA. The Main NAEP has three achievement levels: Basic, Proficient, and Advanced. The levels are based on collective judgments of experts about what students are expected to know and do at each grade level (NCES, 2001). The performance is based on a 0 to 500 scale, just like with LTT (and PISA and TIMSS), but in the Trend Study the students' performance is reported in terms of percentages of students, which corresponds to five points on the scale: 150, 200, 250, 300, and 350, with level 150 described as "Simple Arithmetic Facts" to Level 350 as "Multi-step Problem Solving and Algebra." The five LTT performance levels were arbitrarily set at 50-point increments.

Although the results of the LTT study indicate an almost continuous upward—though very slow—trend, the gains document the fact that students' ability to complete traditional mathematics tested by means of a multiple-choice or short-answer format questions has improved across the 30 years, and that the gains were much smaller than the gains on Main NAEP during the same period (Galindo et al., 2004). The "deeper" meaning of this phenomenon is somewhat unclear, as it is unknown how much influence familiarity with the test instrument has on the outcomes. It is a fact that the longer a test exists the more familiar students and other stakeholders become with the nature of the test items. Experts can already recognize a NAEP, TIMSS, or PISA item from a distance, although the items seem to converge, which may be an undesirable development. This familiarity might itself be responsible for the observed almost continuous, but slow, upward trend. Policymakers even go so far as to speak of a TIMSS curriculum, indicating that they think one can identify a framework behind these studies that is reflected in the items and item formats. A familiar request that reaches persons or organizations responsible for these studies is "Please help me to prepare my students for the next study."

Linking the LTT NAEP with the Main NAEP is no easy task because of different content, different scaling, and different item formats. Also linking State NAEP results to those administered by states or local schools is difficult, if not impossible, to achieve (Feuer, Holland, Green, Bertenthal, & Hemphill, 1999); Linn (2000) observed that comparability of state assessments with NAEP scores is sometimes compromised when purposes differ. Add to this the evidence suggested by Kenney and Silver (1997) and Linn, Koretz, Baker, and Burstein (1991) that variation in item format may result in different estimates of student knowledge and skills across content topics and content dimensions. This point was also made in an explorative study in assessing mathematics problem solving for 16–17-year-olds (de Lange, 1987).

Linking NAEP with TIMSS and PISA is also difficult: The three assessments are targeted towards slightly different student populations, place different emphases on content areas within mathematics, include questions requiring different types of responses and thinking skills, and report results in different ways.

From the content point of view the most commonly addressed NAEP mathematics Content Strand on both NAEP and TIMSS was number sense, properties, and operations, addressed by 32% of NAEP items and 46% of TIMSS-R items, compared to only 9% of PISA items (Nohara, 2001). PISA did pay a lot of attention to Data Analysis, Statistics, and Probability: 31% of the items belonged to this cluster, with 14% in NAEP and 11% in TIMSS. (Note that these data are for PISA 2000; PISA 2003 has a more balanced distribution matrix.) Probably more important than attribution of items to content strands is the fact that one fourth of the items of Data Analysis, Statistics, and Probability were related to the subcategory "read, interpret, and make decisions using tables and graphs," which may come as no surprise given the fact that PISA aims at measuring mathematical literacy. The more curriculum oriented studies, NAEP and TIMSS, have far fewer items in this subcategory: 4% and 7%, respectively.

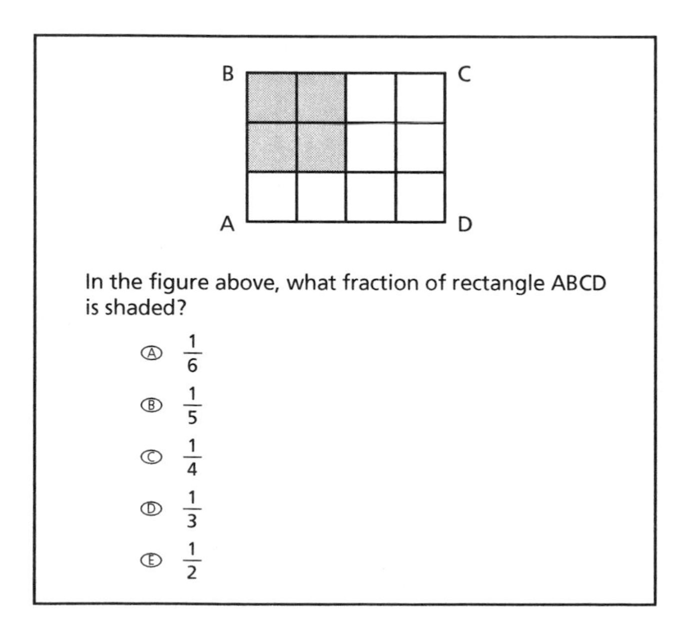

B
C
A
D
In the figure above, what fraction of rectangle ABCD is shaded?
A 1/6
B 1/5
C 1/4
D 1/3
E 1/2

Figure 25.6 Example item from NAEP (NAEP, 2004, p. 29).

Given these observations it may come as no surprise that NAEP and TIMSS address similar sets of subcategories within each of the content strands. This becomes also clear in the percentage of items that present students with real-life situations or scenarios as settings for the problem: NAEP scores 48%, TIMSS 44%, and PISA again is very different: 97%. One should keep in mind that these numbers give some indication of the emphasis placed on real-world situations but do not tell anything about the artificiality, relevance, role, mode of representation, familiarity, and so on, of the context.

A weak spot in any existing large-scale assessment is addressed by many authors and summarized very well in *Knowing What Students Know*. The authors commented on NAEP by noting that NAEP is beginning to be influenced by the call for more cognitively informed assessments of educational programs:

> Recent evaluations of NAEP (National Academy of Education, 1997; NRC, 1999) emphasize that the current survey does not adequately capitalize on advances in our understanding of how people learn particular subject matter. These study committees have strongly recommended that NAEP incorporate a broader conceptualization of school achievement to include aspects of learning that are not well specified in the existing NAEP frameworks or well measured by current survey methods. The National Academy of Education panel recommended that particular attention be given to such aspects of student cognition as problem representation, the use of strategies and self-regulatory skills, and the formulation of explanations and interpretations, contending that consideration of these aspects of students achievement is necessary for the NAEP to provide a complete and accurate assessment of achievement in a subject area. (pp. 244–245)

The NRC agreed with this position and added that NAEP should include carefully designed, targeted assessments administered to smaller samples of students that could provide in-depth descriptive information about more complex problem solving over longer periods of time. For instance, smaller data collections could involve observations of solving mathematical problems in groups.

It should be noted that the recommendations suggested in the above publications are not new: Both TIMSS and PISA designers realized this problem and have expressed intentions to address the issues of more complex processes. But the realization of these intentions is no simple task: We have already noted the intentions for more complex assessments that were mentioned in the PISA 2000 framework but have not been realized so far. One can easily guess the reasons: The required "careful design" is extremely difficult, expensive, and not easy to score even if just settling for qualitative descriptions. Time is another big issue.

As I will discuss later there are already rather large-scale assessments in existence, even on problem solving in group work, but they take at least one full day to complete. On the other hand, there are also examples of national large-scale (exit) examinations that consist of more than one paper-and-pencil session of 50 minutes or so and do address more complex processes. To discuss these national tests will be the next part of this chapter. But before going to this interesting area I would like to mention an interesting study that tries to connect an international comparative study (TIMSS) with state or local assessments: *The Mathematics Benchmarking Report TIMSS 1999* (Mullis et al., 2001a).

TIMSS 1999, also known as TIMSS-R, focused on mathematics achievement of eighth-grade students. The United States and 37 other countries participated, and important for the United States, 13 states and 14 districts or consortia participated in a voluntary Benchmarking Study. The study provided those states and consortia an opportunity to assess the comparative international standing of their students' achievement and to evaluate their mathematics programs in an international context. And, of course, compare themselves with the score of the Unites States as a whole.

The Benchmarking Study confirms the extreme importance of looking beyond the averages (horse race) to the range of performance found across the nation. It may come as no surprise that performance across the participating school districts and consortia reflected nearly the full range of achievement internationally: Although achievement was not as high as Singapore, Korea, and Chinese Taipei, the top-performing Benchmarking jurisdictions of Naperville School District and the First in the World Consortium (both Illinois) performed similarly to Hong Kong, Japan, Belgium (Flemish), and the Netherlands. The problem is, as mentioned before, that the international context consists of four very different socioeconomic cultural entities. And one wonders what to do with the finding that the Chicago public schools, the Rochester school district, and the Miami-Dade county public schools are at par with Thailand, Macedonia, and Iran.

This item from the TIMSS Benchmark study (Figure 25.7) illustrates an item from the content area Fractions and Number sense, 8th grade mathematics. Singapore scored 89% correct, the American First in the World Consortium scored 71%, England and the Russian Federation 52%, Maryland 42%, and the Dade County Public Schools scored 20% correct.

Another positive development regarding TIMSS is the inclusion of videotaping in classrooms and Case Studies. In doing this it follows the recommendations of the authors of *Knowing What Students Know* in an unprecedented way. In his overview of the Case Studies, Stevenson (1998) argued that the questionnaires, as used in the main studies of TIMSS and PISA, have both strengths and weaknesses. Their major strength—as with multiple choice—is to cheaply gather large amounts of data on a wide variety of topics from many individuals. On the other hand, and this causes serious problems, respondents may provide answers that are difficult to interpret. Deprived of the opportunity to question the respondents about the meaning of certain replies, to probe for more complete answers, or to ascertain the respondents' understanding of the questions, the investigator can only conjecture about what an answer might have meant. And the opposite is a problem too: What is the real meaning of the question for the respondent?

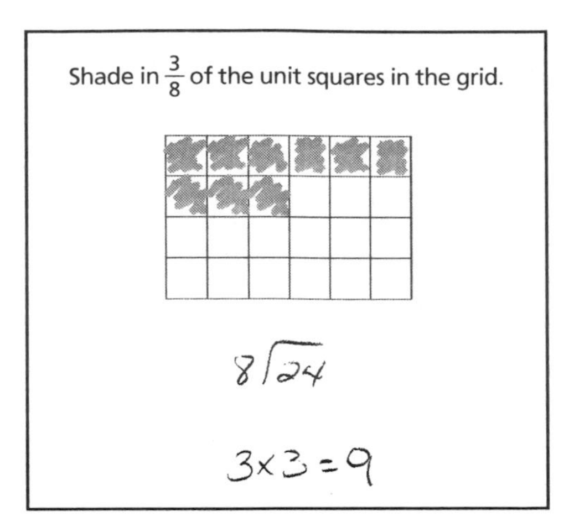

Figure 25.7 TIMSS Benchmark Study-item (Mullis et al., 2001a, p. 72).

Case studies, which allow for one-to-one interactions between investigator and respondent, can help to understand the contexts and relationships that lie behind more traditional quantitative surveys of beliefs, attitudes, and practices, especially when seen in combination with the more quantitative data.

The five-volume report from TIMSS on Japan, Germany, and the United States offers examples from standards to homework, from teacher training to response to individual differences. From these examples Stevenson (1998) concluded that the education systems of the different countries are embedded within the culture of each country. As a result of this close relationship, one cannot hope to learn how education systems can be improved or academic achievement increased without understanding the actions, beliefs, and attitudes related to education in that culture. However it is impossible to make causal conclusions about the relation between scores on a test and the culture.

National Exit Examinations, a Personal Experience

Reaching this part of this chapter I am tempted to reflect on this subject in a very personal way. As a student who underwent a national exit examination, as a teacher who was responsible to prepare students for this event, and as a researcher interested in finding out if different tests measure different competencies. As a student I was perfectly aware of the goal of our 11-year-long educational process: to pass the national examination, a national event for all students at exactly the same time, somewhere in May. The final year was merely used to prepare us for the main event: to do three 3-hour long written tests—by no means with multiple-choice but by solving large and complex problems, although not in context. To be prepared was quite easy, to pass difficult. Easy to prepare because there were numerous examples of examination problems from previous years, as all examination test problems are in the public domain after the examination. Passing the exam was difficult because students needed high technical skills and preciseness. Apart from the written tests (9 hours total) most students had to do an oral session as well, under the principle that different cognitive demands could be operationalized under different formats. Some deeper conceptual understanding should be measured in an oral session carried out by the teacher and an outside "expert," quite often a university mathematician. But if a student had very good grades for the written exams, one did not have to do the orals. And after passing the exam, one had a feeling of absolute freedom: One could enter any university to study an almost unlimited number of disciplines.

As a teacher I felt that the final national examination was a rather threatening event, the moment of truth whether I was a good teacher or not. The only good thing about the last year of schooling was that I had the undivided attention of most students, as they were all very interested in passing the exam. But nightmares for beginning teachers were common: All students failing and the teacher being fired was a dreaded scenario. Of course the teacher was also responsible for scoring the test, but regrettably a randomly appointed colleague had to check the work, and the scores were expected to be within a very narrow band. Reflecting on this practice it is quite surprising that parents completely trusted this process, even today in the twenty-first century. Nowadays the oral has been discontinued because of economical reasons—a problem that again and again overshadows good assessment practices. It was replaced by a school exam, to be designed and scored by the teacher. It accounted for 50% of the total score of the final examination. There was a large degree of freedom for the teacher resulting in many teachers' trying to copy the "real thing" at the end of the same school year. But more adventurous teachers tried something very different like essays and open-ended problem-solving items.

Although the examinations included quite extensive open-ended problems, the researcher in me was interested in trying to find other ways to operationalize the so-called higher order thinking skills, with expected but still surprising results. I used the liberty of the 50% school examination to find out differences to test out the restricted-time written test, a take-home test, and an oral test. The most striking observation I was able to measure was that the correlation between the different formats was low, indicating that indeed one test cannot measure all cognitive demands or, in this case, that different formats measure different competencies (de Lange, 1987).

Exit Examinations

School exit examinations are not a very popular object of study. One of the few studies, *Secondary School Examinations: International Perspectives on Policies and Practices* by Eckstein and Noah (1993), was announced as filling an informational and conceptual vacuum. The study observed that most (industrialized) countries require students to take extensive examinations at the end of secondary education to determine whether they have satisfactorily met national criteria and are qualified for further education. These examinations are also used to assess individual schools (and teachers, as indicated in my personal note) and a nation's educational program as a whole. It is remarkable that the United States does not have high school tests at the national level: "The United States is unique among the countries we have studied in having no coordinated, public, national system for assessing student achievement at the end of secondary school" (Eckstein & Noah, 1993, p. 238). In Barton's paper (1999) the author commented that the examinations in the countries studied by Eckstein and Noah (U.S.A., China, Japan, Germany, England and Wales, France, Sweden, and the former Soviet Union) are closely related to the curriculum. In the United States, he continued, it is hard to conceive of any *national* exam being closely related to the actual curriculum for such a high-stakes examination because of the decentralized control over the curriculum. According to Eckstein and Noah the governmental control of the school curriculum in the United States has been extraordinarily weak.

But this argument of decentralization cannot be used to not introduce decentralized exit examinations. In several countries there are no *national* exit examinations but state examinations (like in Germany and Australia).

Japan, the United Kingdom, France, and Germany were the subjects of a more recent study: *Comparisons of Entrance and Exit Examinations* (Stevenson & Lee, 1997). The authors observed also the relation between curriculum and examination, noting that:

> rather than imposing some arbitrarily defined standard of achievement, the examinations are closely tied to what the students have studied in high school. Because teachers are aware of what students are expected to know in examinations, it becomes their responsibility to equip students with the information and skills needed to pass the examination. (p. 47)

It is amazing to see how this fits with the personal experiences described before.

One of the observations in the introduction of the Eckstein and Noah study is the relative lack of systemic, cross-national comparative work, and the fact that, with only a few notable exceptions, the comparative study of examinations has been a largely neglected aspect of comparative education. Although the study is somewhat dated (1993), many observations and conclusions are still very informative if only as sketching possibilities for future development. Quite often things of the past are neglected just for that reason, without looking at the benefits from a strictly educational point of view instead of a practical or economical perspective. A nice example is the disappearance of the oral part of the mathematics examination in the Netherlands: Officially one argument was the lack of validity and reliability with two persons examining a student. In reality is was an economical and logistical argument that did away a format that quite often was highly regarded by students. It is hopeful to observe that the new rules allow for a variety of formats in the school examination part of the exam, bringing back a more balanced assessment system.

The Eckstein-Noah study observed that mathematics is a subject that all must take in some countries, with accommodation for more and less advanced studies. In others, it is differentiated according to specializations in the subject, all at an advanced level, whereas in yet other countries it may not be required at all. Substantial differences among countries in the knowledge required of candidates in mathematics exist to a surprising extent. But what is even more remarkable, especially in light of the (inter)national large-scale assessments, is the observation that format and content are inseparable, an ever-returning point in any discussion about valid and fair assessment when the designers of large-scale assessments point out that multiple-choice and short-answer questions quite often offer enough possibilities to test the students. It is commonplace that the format of examinations strongly influences both the content and style of teaching. As a consequence, the format of examinations, as much as their content, is a target of the debate over the proper objectives of instruction and over what teachers and schools are supposed to do. The United States and Japan continue to rely heavily on multiple-choice formats. In China as well, in response to the large number of candidates, the high cost of university entrance examinations, and concern about their reliability and objectivity, the examination authorities are increasingly substituting short-answer and multiple-choice questions for the traditional extended-answer format. Concern for reliability and validity thus influences the choice of formats, which in turn influences the content selected for examination, which in turn influences the content and style of classroom instruction.

These remarks are even more remarkable as they were made more than a decade ago, and one assumes that some progress has been made in this area. But this progress fails to show up in the international large-scale assessments to a convincing extent. TIMSS still has a large amount of multiple-choice and short-answer questions, and although PISA may have less than 50% multiple-choice, it still limits itself to very short restricted time-open questions. I noted before that *Knowing What Students Know* recommended strongly that NAEP should include carefully designed, targeted assessments that could provide in-depth descriptive information about more complex activities that occur over longer periods of time (longer than 50 minutes). Having participated in several discussions about the trade-off between "fair-and-valid" assessments and "practical-and-cheap" assessments I have noticed that the decision makers often tend to favor the "practical-and-cheap," which fits in nicely with the earlier observation that these stakeholders show little interest in the content domain of large-scale assessments.

"Broader" Assessment Systems

That many countries and states have shown that a broader "assessment system" can work in reality has been indicated by the study about the examinations. The variety of formats used for this form of "high-stakes testing" is almost unlimited. An example from the United States that gives indications of future types of large-scale assessments is the New Standards Project. The choice for this example is not completely at random.

VCE

One of the quite interesting new approaches to something like the New Standards Project was undertaken in the state of Victoria, Australia. One of the principles behind this examination of the Victoria Certificate Assessment Board was that teacher would like to teach to the test. So the test had four different formats, testing very different cognitive demands. I have described this daring approach earlier (de Lange, 1996), and many articles are around by people directly involved. In one of their earlier publications, Money and Stephens stated that applications and modeling are central features of the Mathematics Study Design in the Victorian Certificate of Education (VCE) in the state of Victoria, Australia (Money & Stephens, 1993). All students undertaking mathematics in the final 2 years, that is Years 11 and 12, engage in substantial applications and modeling, with content options up to 50% to provide for the widely varying abilities and interest of students. In my opinion it is fair to say that the new test requirements lead to different practices in teaching (test driven innovation) or, as Money and Stephens saw it: "The new arrangements have had a significant impact on the teaching and learning of mathematics." (p. 336). Much responsibility is given to the teachers because the Mathematics Study Design for the VCE provides a framework within which *teachers* develop courses, each normally 1 year in length.

Three work requirements, detailed by the teacher but each taking up between 20% and 60% of the course time, provide the basis for determining satisfactory completion and certificate credit for all semester units within VCE Mathematics:

- *Skills Practice and Standard Applications:* the study of aspects of the existing body of mathematical knowledge through learning and practicing mathematical algorithms, routines, and techniques, and using them to find solutions to standard problems.
- *Problem Solving and Modeling:* the creative application of mathematical knowledge and skills to solve problems in unfamiliar situations, including real-life situations.
- *Projects:* extended independent investigations involving the use of mathematics.

For teachers, these work requirements provide the framework for a shift in the balance of instructional patterns and a shift in role from dispenser of information towards learning facilitator, catalyst, and coach. For students, the work requirements provide an accessible pathway towards formal recognition of the learning processes they have undertaken.

It seems rather trivial to observe that one can only introduce a modeling and applications curriculum when the assessments do really operationalize the goals of such a curriculum (de Lange, 1987), a point repeatedly underlined by many authors (e.g., Blum & Niss, 1991; Money & Stephens, 1993). To achieve this, the Common Assessment Task (CAT) initially consisted of four parts:

- CAT 1, an Investigative Project, is based on a centrally set theme and is intended to represent 15 to 20 hours of students' work;
- CAT 2, a Challenging Problem, is chosen from four centrally set problems for each course, representing about 6 to 8 hours of students' work;
- CAT 3, a Facts and Skills Task, is a multiple-choice test of 90 minutes duration, covering the full range of core and optional content;
- CAT 4, an Analysis Task, is a test, also of 90 minutes, consisting of about four structured questions that lead from routine to nonroutine aspects of a problem.

As the reader can judge, my earlier qualification of a "daring" approach seems quite appropriate, especially for real large-scale assessments on a state level.

It cannot come as a surprise that not all CATs survived, but this does not affect the message: To design appropriate large-scale assessments, incorporating the more complex competencies, one is confronted with a daunting and challenging task.

I conclude with some examples from the *Mathematical Methods* Written Examinations 1 and 2 from November 2003. Written Examination 1 (Facts, Skills, and Applications) consists of two parts: Part I and Part II. Part I has 27 multiple-choice items, Part II has 6 short-answer questions. The 27 items of Part I are good for a score of 27 points, the 6 questions for Part II are worth 23 points. Figure 25.8 shows an example of both parts:

Another 1½ hours is given to the students to complete four questions in Written Examination 2: Analysis Task. The four questions have several subquestions (see Figure 25.9), and the total is worth 55 points.

New Standards

As the United States is often seen as "the mother of multiple-choice and other bad habits in assessment," in part because of the existence of a commercial testing

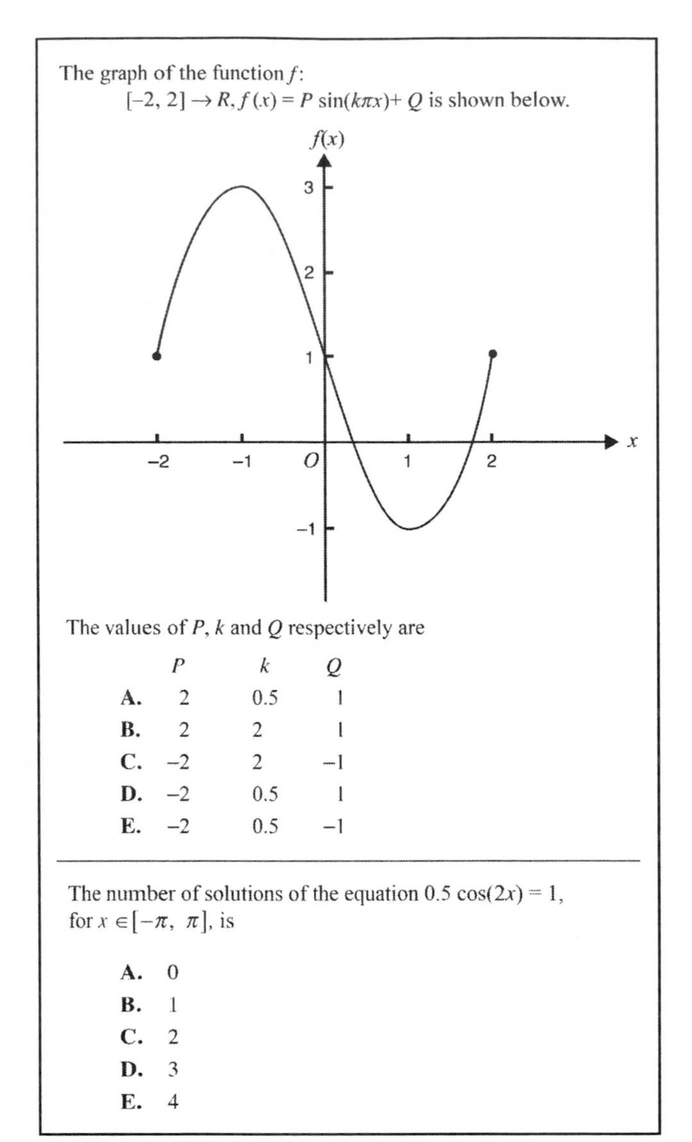

The graph of the function f:
$[-2, 2] \to R, f(x) = P \sin(k\pi x) + Q$ is shown below.

The values of P, k and Q respectively are

	P	k	Q
A.	2	0.5	1
B.	2	2	1
C.	−2	2	−1
D.	−2	0.5	1
E.	−2	0.5	−1

The number of solutions of the equation $0.5\cos(2x) = 1$, for $x \in [-\pi, \pi]$, is

A. 0
B. 1
C. 2
D. 3
E. 4

Figure 25.8 Two examples from Part I (VCA, 2003a, pp. 5–6).

Andrew is making a skateboard ramp. He draws a cross-section diagram with coordinate axes as shown below.

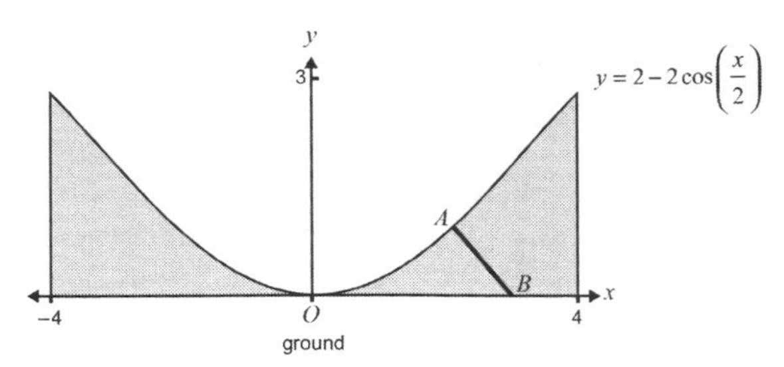

The curve has the equation $y = 2 - 2\cos\left(\frac{x}{2}\right)$, $-4 \le x \le 4$.

All measurements are in meters; the horizontal length of the structure is 8 meters.

a. How many meters above the ground is the highest point of the ramp? Give your answer correct to two decimal places.
b. Use calculus to show that the gradient of the ramp is always less than or equal to 1.
c. Use calculus to find the area of the shaded region, correct to two decimal places.

There is a supporting beam AB on the structure as shown.
A is a point on the curve one meter vertically above the x-axis.
B is a point on the x-axis such that AB is normal to the curve at A.

d. i. Find the exact value of the x-coordinate of A.
ii. Use calculus to find the exact value of the gradient of the normal to the curve at A.
iii. Find the exact value of the length of AB.

Figure 25.9 Example from Part II (VCA, 2003b, pp. 4–6).

industry and its collaborators (Kaplan Test Preparation Courses), it is remarkable that a project like the New Standards Project actually made an impact in the United States. The project illustrates a way to approach a number of issues related to large-scale assessments that differ from the earlier described international and national large-scale assessments.

The program was designed to provide clear goals for learning and assessments that are closely tied to those goals. A combination of on-demand and embedded assessments was to be used to tap a broad range of learning outcomes. Development of the program was through collaboration between the Learning Research and Development Center of the University of Pittsburgh and the National Center on Education and the Economy, in partnership with states and urban school districts. Challenging standards were established for student performances at Grades 4, 8, and 10, along with large-scale assessments designed to measure attainment of those standards.

The New Standards Project consists of three interrelated components: performance standards, a portfolio assessment system, and an on-demand exam. The performance standards describe what students should know and the ways they should demonstrate the knowledge and skills they have acquired. The performance standards include samples of student work that illustrate high-quality performances, accompanied by commentary that shows how the work sample reflects the performance standards (New Standards, 1994, 1996, 1997). According to *Knowing What Students Know* they go beyond most content standards by describing how good is good enough, thus providing clear targets to pursue (NRC, 2001, p. 251).

Balanced Assessment Project

Another influential project in the United States that clearly falls under the definition of large-scale assessment is the Balanced Assessment Project. The collaboration involves not only the U.S. partners (initially Berkeley, Harvard, Michigan State) but also the University of Nottingham, U.K. The project "takes off where multiple-choice ends" (Figure 25.10). Multiple-choice is seen as falling into the range of transforming and manipulating; what the project wants to do is performance assessment that cuts across ranges of mathematical processes, including modeling and formulating, inferring and drawing conclusions, checking, evaluating, and reporting.

T-Shirt

The aim of this assessment is to provide the opportunity for you to:

- systematically communicate about geometric shapes;
- locate shapes on a grid;
- give a clear set of directions.

The design below, including the 10 by 10 grid, is going to be used on a math team T-shirt. You accidentally took the original design home, and your friend Chris, needs it tonight. Chris has no fax machine, but has a 10 by 10 grid just like yours. You must call Chris on the telephone and tell him very precisely how to draw the design on his grid.

Prepare for the phone call by writing out your directions clearly, ready to read over the telephone.

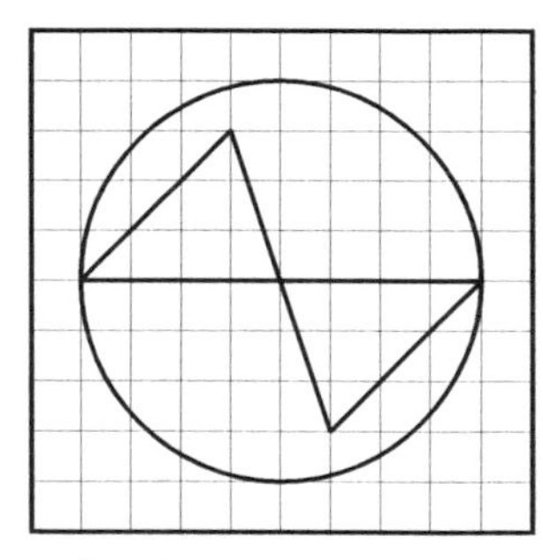

Write your directions on another sheet of paper.

This is the grid that Chris has in front of him.

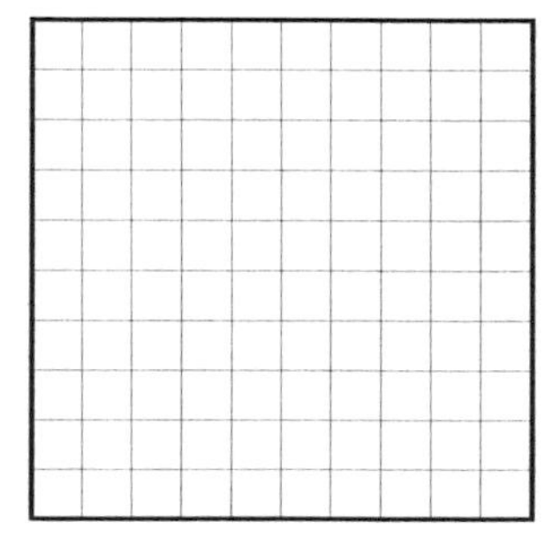

Figure 25.10 Grade 8 sample task from the Balanced Assessment Project (1995).

NCEM

An interesting point that permeates almost all discussions about exit examinations is whether examinations or other large-scale assessments that can promote and stimulate worthwhile educational goals and can function as a high-stakes assessment at the same time. Another point is the fact that in many countries the examination tool becomes a tool for selection of the "best and the brightest." In their article on reform on the national examination, Amit and Fried (2002) observed the struggle between society's ambition for high achievement and its simultaneous ambition for mathematical literacy for all. On the one hand, because of the prominent place assumed by mathematics in professional and everyday life, it has become a critical filter, via higher education, for social and economic advancement (Sells, 1978). At the same time, and for the same reason, mathematical literacy is recognized as a basic societal goal, so that good mathematics education, therefore, becomes a basic right for all members of society. In their article Amit and Fried argued that it is possible to align the role of completion examinations in promoting the high quality of advanced studies with the original role as an agent of democracy. They based this conclusion on the new model for the National Completion Examinations in Mathematics (NCEM) in Israel.

The original NCEM consisted of three levels: Ordinary, Advanced, and Highly Advanced. The progression from one level to the next can be characterized by an increasing range of mathematical subjects, but also by an increasing demand for depth in the approach to mathematical ideas. But in the early 1990s Israel's educational policymakers, mathematics educators, and researchers called for reform of the system because of the rising numbers of students not opting to take the ordinary level examination and therefore not getting the examination certificate.

The answer to these problems came in 1996 when both a new curriculum and a new ordinary level NCEM were introduced. The new examination comprised two separate questionnaires: the basic and the supplementary questionnaire. The first is oriented towards basic mathematical literacy, and the second towards a more complete mastery of mathematical concepts and techniques. The basic part can be taken and retaken independently of the supplementary part. This feature allows schools to develop special tracks geared towards students who will take only the basic questionnaire. It is this basic part that makes the new system work in merging the different objectives. Figure 25.11 gives an example of problems from the basic component (that can be taken when students are in 10th, 11th or 12th grade).

An evaluation by the Ministry of Education showed an increase in the number of students taking the basic examination, with the increase in the basic part being significantly higher than the number of students tak-

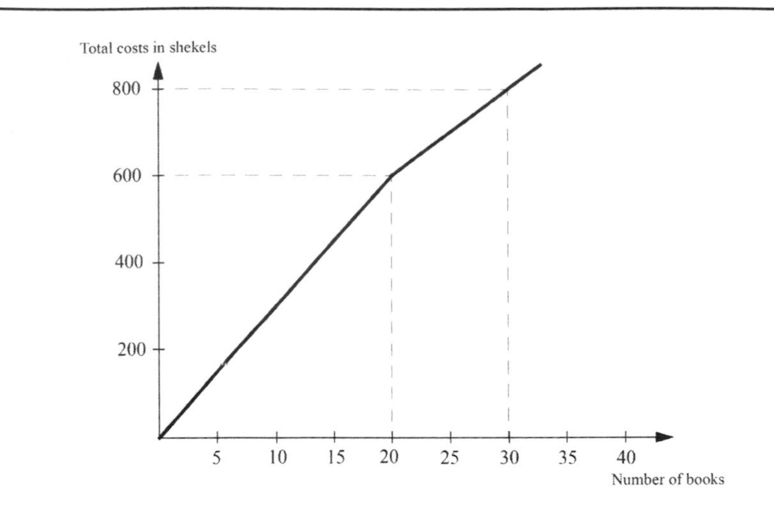

Example 1
The graph shows the price of books according to the number purchased (see figure above). Answer the questions according to the graph.
(a) How much do you pay when you purchase 25 books?
(b) How much does each book among the first 20 cost?
(c) How much does each book after the 21st book cost?

Example 2
Total 15 people where tested and the following scores where obtained: 8, 7, 7, 9, 8, 8, 9, 5, 7, 6, 6, 4, 6, 10, 8.
(a) Calculate the average score.
(b) What is the probability that the score of one of the people chosen at random will be higher than the average?

Example 3
Given a right triangle ABC in which angle D is right. The length of side AD is 40 cm. Angle BAD is 41°. C is a point on BD such that the angle CAD is 38° (see figure below).
(a) Calculate the length of the leg BD.
(b) Calculate the length of the segment BC.

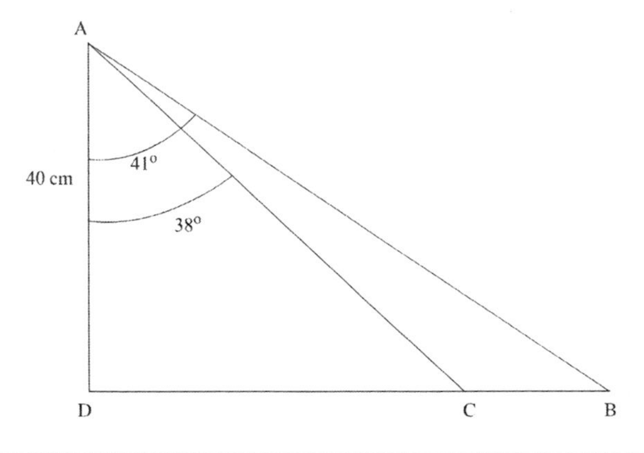

Figure 25.11 Three examples of questions in the Basic Questionnaire (Amit & Fried, 2002, p. 507).

ing the complete examination. Amit and Fried concluded that over the years the gap between the number of examinees sitting for the basic questionnaire and number of examinees sitting for the supplementary part has increased. This indicates that many students take only the basic part and do not continue to complete the requirements for the ordinary level. It is reasonable to assume, based on experience, that students would not have sat for the examination at all for mathematics if this opportunity had not existed. More students are now enabled to continue to study mathematics throughout high school due to a rather simple intervention: more attention for the functionality of mathematics in the curriculum—and as noted before Israel is certainly not alone in this development—and consequential changes in the assessment system.

Let me make a concluding remark or two before moving on to technology in large-scale assessments. Large-scale and high-stakes assessments clearly can go together in an appropriate way. Regrettably, some large-scale assessments are "misused" as certain jurisdictions have adopted policies that rely exclusively on achievement test scores to make high-stakes decisions. Many standardized tests do not meet desired criteria, or, as *Knowing What Students Know* stated, current assessment practices are the cumulative product of theories of learning and models of measurement that were developed to fulfill the social and educational needs of a different time. The wish to refashion assessments to meet current and future needs for quality information is recognized beyond doubt, in many countries, and even implemented to a certain extent in some.

LARGE-SCALE ASSESSMENTS, PROBLEM SOLVING, AND TECHNOLOGY

As pointed out before one of the key issues in assessment and also large-scale assessment is not only the need to pay more attention to the mathematical content domain, but also in realizing the idea that assessments are based on modern knowledge of cognition and its measurements. One of the major problems in large-scale assessments is that it should be seen not only as a policy evaluation, but more as a tool for practitioners. Therefore assessment instruments should relate clearly to intended learning outcomes and can be used to improve students' achievement. These points relate to two major threats to validity in high-stakes tests (Wilson, this volume). The first is *construct underrepresentation* and refers to a situation in which the concepts and skills that are measured in the assessment fail to include important dimensions or facets of the intended construct, in this case the content domain in mathematics. The second is *construct-irrelevant variance* (Brualdi, 2002). This can take the form of extraneous clues in the item or the task format that may permit some students to respond correctly or appropriately in ways that are irrelevant to the construct being assessed. Another possibility is that extraneous aspects of the task make it irrelevantly difficult for some individuals or groups.

Various technologies have been applied to bring greater efficiency, timeliness, and sophistication to multiple aspects of assessment design and implementation. Examples include technologies that generate items; immediately adapt items on the basis of the examinee's performance; analyze, score, and report assessment data; allow learners to be assessed at different times and in distant locations; enliven assessment tasks with multimedia; and add interactivity to assessment tasks. In many cases, these technology tools have been used to implement conventional theories and methods of assessment, albeit more effectively and efficiently (NRC, 2001). It can even be argued that access to modern technologies has hindered the development of more valid tasks in respect to the earlier made points. Therefore the focus of the research relating assessments and technologies should shift towards these goals and challenges. As a matter of fact it seems that a precondition to develop valid and interesting tasks is the availability of modern technologies.

Knowing What Students Know provided a number of promising new developments varying from theory-based item generation, developing concepts maps (individual or in groups), complex problem solving in a variety of ways, analysis of complex solution strategies (using complex neural network technology), facilitating formative assessment, technology-based or technology-enhanced learning environments, and more. Many developments seem to be happening at very different levels, and many of the implications are unclear. One of the dilemmas we educators are facing is that we have to reinvent the definition of learning goals in these new environments: Are we abiding with the traditional curricular-based learning goals, or do we see quite new and unexpected horizons? At one side we see successful mathematical modeling competitions in which students work in teams for an extended period of time using technology in a variety of ways on quite a large scale. This might be considered as an existence proof for the merging of large-scale assessments and complex problem solving in groups in a somewhat traditional setting (just curricular mathematics) but where the role of technology is crucial.

An example of large-scale assessment in groups, among others, is the math A-lympiad (see Figure 25.12). This is a modeling competition for students from Germany, Denmark, and the Netherlands (with guests from other countries as well) that has been in existence since 1989 (de Haan & Wijers, 2000). Thousands of students work in many hundreds of teams on an assignment—a very open-ended problem—in which mathematical problem solving and higher order thinking skills must be used to solve a real-world problem. The result is a written report, just like in most other modeling competitions.

The A-lympiad has two rounds: the preliminary round competing at school for a whole day at their own schools, and the international final for the best 16 teams. This final takes most of 2 days. The aforementioned publication has an interesting chapter on the "Assessment of the Papers." The conclusion is in line with earlier research on problem-solving assessment in mathematics: It takes time, effort and therefore is expensive, but the intersubjective scoring is quite reliable (de Lange, 1987). This is in line with the intermarker reliability found in the PISA study.

Security is an art unto itself

Introduction

A museum for modern art will organize a big exhibition with pieces of a number of great artists in the field of modern art. The organization is busy with preparations for the exposition, and meets with difficulties concerning security.

A new security system

The present security system with video cameras doesn't satisfy the demands of an exhibition this size. The cameras aren't movable enough (to show another corner), but the main objection is, that the present system cannot cover the complete space in the museum. To organize the exposition, it's necessary to get a new security system.

The new type of camera that will be used is already chosen. This camera is so fast in moving (in all directions) and focusing, that one can say that this camera really secures the complete surrounding space (see the figure below).

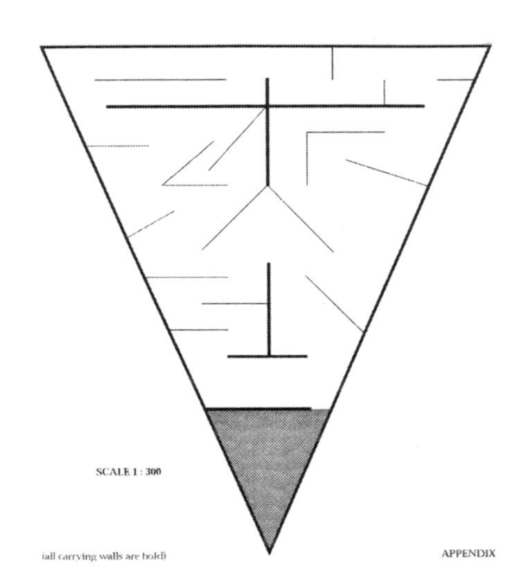

All walls in the museum go from floor to ceiling.

A disadvantage of this camera is its high price. That's why it's necessary to investigate carefully where the cameras should be placed, for you want the least possible amount of cameras.

Enclosed is a map of the museum. The museum has a triangular formed top view. The grey part contains the entrance, bathrooms, cloakroom and office rooms. It's not necessary to replace the old cameras in this part. The new security system is just needed for the part that is not grey.

Exercise 1

Think of a placement of cameras, in such a way that you use the smallest possible amount of cameras to secure the exhibition room.
Describe the system you used to come to this placement of cameras.
Mark where the cameras should be placed and show that the whole exhibition room will be secured this way.

The exhibition contains a little less than 100 paintings. To show these paintings in a proper way (with enough space in between), at least 280 meters of wall capacity is needed, but: the more wall capacity, the better.

On the map, you can see a number of bold walls. These are carrying walls; the other walls can be removed.

Because the new security system is very expensive, the organization wants to reduce the expenses. To do this, there is the possibility to reduce the number of cameras, but there is also the possibility to remove non-carrying walls. Removal of wall costs *ƒ* 500, per meter. A video camera costs *ƒ* 10.000, each.

Exercise 2

How is it possible through removing walls (keep the minimum amounts of capacity you need in mind!) to save expenses on the cameras? In other words: how is it possible, by removing as little wall capacity as possible, to make as many cameras as possible superfluous? And yet, all walls stay secured?
Make a proposal and calculate the savings that this solution will give you.

After this exhibition, the inside of the museum will be changed drastically. First, all non-carrying walls will be removed and just the carrying walls will stay. Because the new inside of the museum will be used for a longer period of time, and for several more exhibitions, an architect agency will be hired to create the new inside. The assignment will be to add 150 meters of wall capacity to the existing carrying walls. Furthermore, 6 cameras must be sufficient for security, yet, there must be an attractive partition of space.

Exercise 3

A number of architect agencies is being asked to hand in proposals in which the conditions are met. Your team works for one of the architect agencies, and your agency really wants the assignment. Of course you won't be a strong rival if you just meet the minimum demands.
Make a proposal for the museum board (including working drawing and short 'covering letter').

Figure 25.12 Assignment in preliminary round of Mathematics A-lympiad, 1996–1997 (de Haan & Wijers, 2000, pp. 59–61).

At the other side is complex problem solving in a less traditional setting: a MashpeeQuest assessment task gives students an opportunity to put complex problem-solving skills to use in a web-based environment that structures their work (Mislevy, Steinberg, Almond, Haertel, & Penuel, 2000). In this example, technology plays at least two roles. The first is conceptual as the information analysis skills to be assessed, and the behaviors that serve as evidence are embedded within the web-based environment. The second is more an operational issue: Because the actions take place in a technological environment, some observations of students' performance can be made automatically. Staying with the area of complex problem solving with technology, some interesting experiments were carried out in Germany by Klieme, Schümer, and Knoll in 2001.

Technology and Assessment: Thinking Ahead (NRC, 2002) argued that technology can help to merge the advances in cognitive sciences and measurement theories, which then should result in a significant leap forward in the science and practice of assessment. In the first paper Gitomer and Bennett (2002) illustrated how researchers at the Educational Testing Service use computer technologies to address a long-standing and very valid criticism of standardized tests. They tend to consist of certain types of traditional test items and have lost sight of the underlying constructs, or cognitive competencies. The authors argued that computer technologies can help "unmask" the constructs underlying traditional assessments, and to make the constructs more visible and explicit in the design of new assessments.

Another paper in this publication by Means and Haertel (2002) described, according to the introduction, the effort to develop computer-based, quality assessments of scientific inquiry. One of the key points they made is that in contrast to standardized tests and the more conventional paper-and-pencil tests used in most (science) classrooms, the new technology-based learning environments reflect the richness and complexity of science inquiry. It is at least interesting to make the observation that the authors and their colleagues saw very exciting opportunities in technology-based assessments in comparison with what they identify as "traditional standardized achievement test," but failed to observe that most of the constructs or competencies can be measured with pencil-and-paper, with or without the use of technology. It is my opinion that the perceived contrast between "Innovative Technology-Supported Assessments" and "Traditional Tests" (p. 15) is a construct of the assessment industry. Technology should not drive innovations in assessments, as has been the case so often in the past. The drive should come from advances in mathematical cognition, from advances in measurement theory, and should only be facilitated by the use of technology.

Eva Baker (2000) stated that

> the previous discussion has focused on the use of technology to make substantial improvements in our ability to assess and understand educational quality, principally measured by student achievement. But there is plenty of resistance to even this level of technology application. (p. 13)

Again, nothing seems new: There has always been plenty of resistance against innovations in assessments—looking back some 30 years shows that many efforts have been made to introduce more valid and fair assessments, with or without technology. Central should be what to measure and how to measure taking the kids and their competencies in the area of mathematics seriously.

Large-scale assessments have committed themselves to measure a wider variety of mathematical competencies. They have indicated that they will include more and more extended formats, and include technology use, starting with the calculator, but that they are also offering technology-based items (like Science in PISA 2006). Finally, high-stakes examinations can be done, although still on a voluntary basis, in a computer environment (Netherlands). There is mention of evaluating or measuring group work as this is regarded widely as an essential competency, and there are working examples in existence. Resistance exists, especially with policymakers, with the people who pay for the large-scale assessment studies. But some progress has been made, and with educators' combined efforts, further advances are around the corner.

REFERENCES

Allen, N. L., Jenkins, F., Kulick, E., & Zelenak, C. A. (1997). *Technical Report of the NAEP State Assessment Program in Mathematics.* [Electronic version]. Washington, DC: National Center for Education Statistics.

Amit, M., & Fried, M. N. (2002). High-stakes assessment as a tool for promoting mathematical literacy and the democratization of mathematics education. *Journal of Mathematical Behavior 21*, 499–514.

Baker, E. L. (2000). *Understanding educational quality: Where validity meets technology. William Angoff Memorial Lecture Series.* Princeton, NJ: Educational Testing Service. Retrieved April 13, 2006, from http://www.ets.org/Media/Research/pdf/PICANG5.pdf.

Balanced Assessment Project. (1995). *Grade 8 Sample Task.* Retrieved April 13, 2006, from Mathematics Assessment Resource Service website http://www.nottingham.ac.uk/education/MARS/tasks/g8_1/full.html.

Barton, P. E. (1999). *Too much testing of the wrong kind; Too little of the right kind in K–12 education.* Princeton, NJ: Educational Testing Service. Retrieved April 13, 2006, from http://www.ets.org/Media/Research/pdf/PICTOOMUCH.pdf.

Baumert, J., Klieme, E., Neubrand, M., Prenzel, M., Schiefele, U., Schneider, W., et al. (2001). *PISA 2000: Basiskompetenzen von Schülerinnen und Schülern im internationalen Vergleich* [Basic competencies of schoolgirls and schoolboys in international comparison]. Leverkussen, Germany: Leske + Budrich Verlag.

Beaton, A. E., Mullis, I. V. S., Martin, M. O., Gonzalez, E. J., Kelly, D. L., & Smith, T. A. (1996). *Mathematics achievement in the middle school years: IEA's Third International Mathematics and Science Study (TIMSS).* Chestnut Hill, MA: TIMSS International Study Center, Boston College.

Bennett, R. E. (1998). *Reinventing assessment: Speculations on the future of large scale educational testing.* Princeton, NJ: Policy and Information Center, Educational Testing Service. Retrieved April 13, 2006, from http://www.ets.org/Media/Research/pdf/PICREINVENT.pdf

Blum, W. (2000). TIMSS und PISA ? Herausforderung und Chance für Mathematikunterricht und Mathematikdidaktik [Challenges and Chances for Mathematics Education and Didactics of Mathematics]. In *Empirische Schulleistungsvergleiche—Nutzen, Risiken, Interessen* (pp. 113–128). Bonn, Germany: Sekretariat der deutschen Kultusminister-Konferenz.

Blum, W., & Niss, M. (1991). Applied mathematical problem solving, modelling, applications, and links to the other subjects—state, trends and issues in mathematics instruction. *Educational Studies in Mathematics, 22,* 37–68.

Bonnet, G. (2002). Reflections in a critical eye: On the pitfalls of international assessment. *Assessment in Education: Principles, Policy and Practice, 9*(3), 387–399.

Bracey, G. W. (1997). On comparing the incomparable: A response to Baker and Stedman. *Educational Researcher, 26*(2), 365–377.

Brualdi, A. (2002). Traditional and modern concepts of validity. In L. Rudner & W. Schafer (Eds.), *What teachers need to know about assessment.* Washington, DC: National Education Association.

Clarke, D. (2003). International comparative research in mathematics education. In A. J. Bishop, M. A. Clements, C. Keitel, J. Kilpatrick & F. K. S. Leung (Eds.), *Second international handbook of mathematics education,* (Pt. 1, pp. 143–184). Dordrecht, The Netherlands: Kluwer.

College Board. (1996). *Mathematics framework for the 1996 and 2000 National Assessment of Educational Progress.* Washington, DC: National Assessment Governing Board. Retrieved April 13, 2006, from http://www.nagb.org/pubs/96-2000math/toc.html

de Haan, D., & Wijers, M. (2000). *10 Years Math A-lymiad. The Real World Mathematics Team Competition.* Utrecht, The Netherlands: Freudenthal Institute, Utrecht University.

Dekker, T., & Feijs, E. (2005). Scaling up strategies for change. *Assessment in Education: Principles, Policy & Practice, 12*(3), 237–254.

de Lange, J. (1987). *Mathematics, insight, and meaning. Teaching, learning, and testing of mathematics for the life and social sciences.* Utrecht, The Netherlands: Vakgroep Onderzoek Wiskundeonderwijs en Onderwijscomputercentrum (OW&OC).

de Lange, J. (1994). *The three levels of mathematical thinking.* Utrecht, The Netherlands: Freudenthal Institute.

de Lange, J. (1996). Using and applying mathematics in education. In A. J. Bishop, , K. Clements, C. Keitel, J. Kilpatrick, & C. Laborde (Eds.), *International handbook of mathematics education* (pp. 49–97). Dordrecht, The Netherlands: Kluwer.

de Lange, J. (1997a). Looking through the TIMSS-mirror from a teaching angle. In National Research Council (Ed.), *Learning from TIMSS: An NRC symposium on the results of the Third International Mathematics and Science Study* (pp. 91–107). Papers Prepared for a Symposium. Washington, DC: National Academy of Sciences, National Research Council.

de Lange, J. (1997b). *Graph showing countries participating in TIMSS 1996.* Utrecht, The Netherlands: Freudenthal Institute.

de Lange, J. (1999). *A framework for classroom assessment in mathematics.* Unpublished manuscript, National Center for Improving Student Learning and Achievement in Mathematics and Science, Assessment Study Group.

de Lange, J. (2003). Mathematics for literacy. In B. L. Madison, & L. A. Steen (Eds.), *Quantitative literacy. Why numeracy matters for schools and colleges* (pp. 75–89). Princeton, NJ: The National Council on Education and the Disciplines.

Dossey, J. A. (1997). National indicators of quantitative literacy. In L. A. Steen (Ed.), *Why numbers count. Quantitative literacy for tomorrow's America.* New York: College Entrance Examination Board.

Dossey, J. A., Jones, C. O., & Martin, T. S. (2002). Analyzing student responses in mathematics using two-digit rubrics. In D. F. Robitaille & A. E. Beaton (Eds.), *Secondary analysis of the TIMSS Data* (pp. 21–45). Dordrecht, The Netherlands: Kluwer.

Eckstein, M. A., & Noah, H. J. (1993). *Secondary school examinations. International perspectives on policy and practice.* New Haven, CT: Yale University Press.

Feijs, E. (2005). Constructing a learning environment that promotes reinvention. In R. Nemirovsky, A. S. Rosebery, J. Solomon, & B. Warren (Eds.), *Everyday matters in science and mathematics: Studies of complex classroom events.* Mahwah, NJ: Erlbaum.

Feuer, M. J., Holland, P. W., Green, B. F., Bertenthal, M. W., & Hemphill, F. C. (Eds.). (1999). *Uncommon measures: Equivalence and linkage among educational tests.* Washington, DC: National Research Council, National Academy Press.

Freudenthal, H. (1975). Pupils' achievements internationally compared? The IEA. *Educational Studies in Mathematics, 6,* 127–186.

Gal, I. (1993). *Issues and challenges in adult numeracy.* (Tech. Rep. No. TR93-15). Philadelphia: National Center of Adult Literacy, University of Pennsylvania. Retrieved April 13, 2006, from http://www.literacy.org/products/ncal/pdf/TR9315.pdf.

Gal, I., Van Groenestijn, M., Manly, M., Schmitt, M.J., & Tout, D. (1999). *Numeracy framework for the international adult literacy and lifeskills survey (ALL).* Ottawa, Canada: Statistics Canada and Washington, DC: National Center for Education Statistics.

Galindo, E., Caulfield, R., Mohr, D., & McCormick, K. (2004). Long-Term NAEP: Mathematics learning over three decades. In P. Kloosterman & F. K. Lester, Jr. (Eds.), *Results and interpretations of the 1990 through 2000 mathematics assessment of the National Assessment of Educational Progress* (pp.

175–193). Reston, VA: National Council of Teachers of Mathematics.

Garden, R. A., & Orpwood, G. (1996). Development of the TIMSS achievement tests. In M. O. Martin, & D. L. Kelly (Eds.), *Third International Mathematics and Science Study Tech. Rep., Vol. 1. Design and Development* (pp. 2-1 to 2-19). Boston, MA: IEA, Boston College.

Gitomer, D. H., & Bennett, R. E. (2002). Unmasking constructs through new technology, measurement theory, and cognitive sciences. In National Research Council (Ed.), *Technology and assessment: Thinking ahead. Proceedings from a workshop* (pp. 1–11). Washington, DC: National Academy Press.

Heubert, J. P., & Hauser, R. M. (Eds.). (1999). *High stakes: Testing for tracking, promotion, and graduation* [Electronic version]. Washington, DC: National Research Council, National Academy Press.

Jablonka, E. (2003). Mathematical literacy. In A. J. Bishop, M. A. Clements, C. Keitel, J. Kilpatrick, & F. K. S. Leung (Eds.), *Second international handbook of mathematics education* (pp. 75–102). Dordrecht, The Netherlands: Kluwer.

Jones, L. V. (1997, October). *National tests and education reform: Are they compatible?* [William H. Angoff Memorial Lecture Series, October 8, 1997. Princeton, NJ].

Kane, J. K., & Webb, N. L. (2004). State NAEP and mathematics achievement. In P. Kloosterman, & F. K. Lester, Jr. (Eds.), *Results and interpretations of the 1990 through 2000 mathematics assessment of the National Assessment of Educational Progress* (pp. 193–217). Reston, VA: National Council of Teachers of Mathematics.

Keitel, C., & Kilpatrick, J. (1999). The rationality and irrationality of international comparative studies. In G. Kaiser, E. Luna, & I. Huntley (Eds.), *International comparisons in mathematics education* (pp. 241–256). London: Falmer Press.

Kenney, P. A., & Silver, E. A. (Eds.). (1997). *Results from the seventh mathematics assessment of the National Assessment of Educational Progress.* Reston, VA: National Council of Teachers of Mathematics.

Klieme, E., Schümer, G., & Knoll, S. (2001). Mathematikunterricht in der Sekundarstufe I: "Aufgabenkultur" und Unterrichtsgestaltung. Forschungsbefunde, Reforminitiativen, Praxisberichte und Video-Dokumente [Mathematics teaching in the second degree 1: "Assessment culture" and teaching representation. Research findings, reform initiatives, practice reports and video documents]. In E. Klieme, & J. Baumert (Eds.), *TIMSS – Impulse für Schule und Unterricht* [Electronic version]. Bonn, Germany: Bundesministerium für Bildung und Forschung.

Kuiper, W. A. J. M., Bos, K. Tj., & Plomp, Tj. (1997). *Wiskunde en de natuurwetenschappelijk vakken in het leerjaar 1 en 2 van het voortgezet onderwijs. Nederlands aandeel in TIMSS populatie 2* [Mathematics and the scientific courses in the first and second years of high school. The Dutch share in TIMSS population 2]. Enschede, The Netherlands: Universiteit Twente.

Kuiper, W. A. J. M., Bos, K. Tj., & Plomp, Tj. (2000). The TIMSS National Option Test. *Studies in Educational Evaluation, 26,* 43–60.

Linn, R. L. (2000). Assessments and accountability. *Educational Researcher, 29*(2), 4–16.

Linn, R. L., Baker, E. L., & Dunbar, S. B. (1991). Complex, performance-based assessment: Expectations and validation criteria. *Educational Researcher, 20*(8), 15–21.

Linn, R. L., Koretz, D., Baker, E. L., & Burstein, L. (1991). *The validity and credibility of the achievement levels for the 1990 National Assessment of Educational Progress in Mathematics.* Los Angeles: Center for Research on Evaluation, Standards, and Student Testing, University of California.

Madison, B. L., & Steen, L. A. (Eds.). (2003). *Quantitative literacy: Why numeracy matters for schools and colleges.* Princeton, NJ: The National Council on Education and the Disciplines.

Means, B., & Haertel, G. (2002). Technology supports for assessing science inquiry. In National Research Council (Ed.), *Technology and assessment: Thinking ahead. Proceedings from a Workshop* (pp. 12–25). Washington, DC: National Academy Press.

Messick, S. (1993). Validity. In R. L. Linn (Ed.), *Educational measurement* (3rd ed., pp. 13–103). Phoenix, AZ: Oryx Press.

Mislevy, R. J., Steinberg, L. S., Almond, R. G., Haertel, G. D., & Penuel, W. R. (2000). *Leverage points for improving educational assessment.* Menlo Park, CA: SRI International.

Money, R., & Stephens, M. (1993). Linking applications, modelling and assessment. In J. de Lange, I. Huntley, C. Keitel, & M. Niss (Eds.), *Innovation in mathematics education* (pp. 323–336). New York: Ellis Horwood.

Mullis, I. V. S., Martin, M. O., Beaton, A. E., Gonzalez, E. J., Kelly, D. A., & Smith, T. A. (1997). *Mathematics achievement in the primary school Years.* Chestnut Hill, MA: TIMSS International Study Center, Boston College.

Mullis, I. V. S., Martin, M. O., Gonzalez, E. J., O'Connor, K. M., Chrostowski, S. J., Gregory, K. D., et al. (2001a). *Mathematics benchmarking report TIMSS 1999—Eighth grade: Achievement for U.S. States and districts in an international context.* Chestnut Hill, MA: International Study Center, Lynch School of Education, Boston College.

Mullis, I. V. S., Martin, M. O., Smith, T. A., Garden, R. A., Gregory, K. D., Gonzalez, E. J., et al. (2001b). *TIMSS assessment frameworks and specifications 2003.* Paris: OECD Publications.

National Assessment of Educational Progress (NAEP). (2004). *Demonstration Booklet. 2004 Field tests in reading, mathematics, and science—Grades 4, 8, & 12.* Washington, DC: NAEP, NCES.

National Center for Education Statistics (NCES). (2001). *The nation's report card: Mathematics 2000.* By J. S. Braswell, A. D. Lutkus, W. S. Grigg, S. L. Santapau, B. Tay-Lim, & M. Johnson, U.S. Department of Education. Office of Educational Research and Improvement [Electronic version]. Washington, DC: National Center for Education Statistics.

National Research Council (NRC). (1989). *Everybody counts.* Washington, DC: National Academy Press.

National Research Council (NRC). (2001). *Knowing what students know. The science and design of educational assessment.* Committee on the Foundations of Assessment. By J. W. Pellegrino, N. Chudowsky, & R. Glaser (Eds.), Board on Testing and Assessment, Center for Education. Division of Behavioral and Social Sciences and Education. Washington, DC: National Academy Press.

National Research Council (NRC) (2002). *Technology and assessment: Thinking ahead. Proceedings from a workshop.* Board on Testing and Assessment, Center for Education, Division of Behavioral and Social Sciences and Education. Washington, DC: National Academies Press.

New Standards. (1994). *The 1994 mathematics reference examination: Design, administration, results, and accuracy assessment.* Oakland, CA: New Standards.

New Standards. (1996). *1996 New standards reference examination technical summary.* Oakland, CA: New Standards.

New Standards. (1997). *Performance standards* (Vols. 1–3). Rochester, NY: National Center on Education and the Economy.

Nohara, D. (2001). *A comparison of the National Assessment of Educational Progress (NAEP), the Third International Mathematics and Science Study Repeat (TIMSS-R), and the Programme for International Student Assessment (PISA)* [Working Paper No. 2001-07]. Washington, DC: National Center for Educational Statistics. Retrieved April 13, 2006, from http://nces.ed.gov/pubs2001/200107.pdf.

Organisation for Economic Co-operation and Development (OECD) & Statistics Canada. (1997). *Literacy skills for the Knowledge Society; Further results from the International Adults Literacy Skills Survey.* Paris: OECD Publications.

Organisation for Economic Co-operation and Development (OECD). (1999). *Measuring student knowledge and skills - A new framework for assessment.* Paris: OECD Publications.

Organisation for Economic Co-operation and Development (OECD). (2000). *Literacy skills for the world of tomorrow - Further results from PISA 2000.* Paris: OECD, UNESCO.

Organisation for Economic Co-operation and Development (OECD). (2001). *Knowledge and skills for life - First results from PISA 2000.* Paris: OECD Publications.

Organisation for Economic Co-operation and Development (OECD). (2003). *The PISA 2003 assessment framework - Mathematics, reading, science and problem solving knowledge and skills.* Paris: OECD Publications.

Organisation for Economic Co-operation and Development (OECD). (2004). *Problem solving for tomorrow's world. First measures of cross-curricular competencies from PISA 2003.* Paris: OECD Publications.

Popham, W. J. (2000). *Modern educational measurement: Practical guidelines for educational leaders.* Needham, MA: Allyn and Bacon.

Reese, C. M., Miller, K. E., Mazzeo, J., & Dossey, J. A. (1997). *NAEP 1996 mathematics report card for the nation and the states.* Washington, DC: National Center for Education Statistics.

Robitaille, D. F., Beaton, A. E., & Plomp, T. (Eds.). (2000). *The impact of TIMSS on the teaching & learning of mathematics & science.* Vancouver, Canada: Pacific Educational Press.

Robitaille, D. F., Schmidt, W. H., Raizen, S., Mc Knight, C., Britton, E., & Nicol, C. (1993). *TIMSS Monograph No. 1. Curriculum framework for mathematics and science.* Vancouver, BC, Canada: Pacific Educational Press.

Sells, L. W. (1978). Mathematics: A critical filter. *Science Teacher, 45,* 28–29.

Shaeffer, R. L. (2003). Statistics and quantitative literacy. In B. L. Madison & L. A. Steen (Eds.), *Quantitative literacy. Why numeracy matters for schools and colleges* (pp. 145–152). Princeton, NJ: National Council on Education and the Disciplines.

Silver, E. A., Alacaci, C., & Stylianou, D. A. (2000). Students' mathematical problem solving, reasoning and communication: Examining performance on extended constructed response tasks. In E. A. Silver & P. A. Kenney (Eds.), *Results from the Seventh Mathematics Assessment of the National Assessment of Educational Progress* (pp. 301–341). Reston, VA: The National Council of Teachers of Mathematics.

Stevenson, H. W. (1998). A study of three cultures: Germany, Japan, and the United States. An Overview of the TIMSS Case Study Project. *Phi Delta Kappan, 79*(7). Retrieved June 2005, from http://www.pdkintl.org/kappan/kste9803.htm.

Stevenson, H. W., & Lee, S. (1997). *International comparisons of entrance and exit examinations: Japan, United Kingdom, France, and Germany.* Washington, DC: U.S. Department of Education.

Sum, A. M., Kirsch, I. S., & Taggart, R. (2002). *The twin challenges of mediocrity and inequality: Literacy in the U.S. from an international perspective.* Princeton, NJ: Educational Testing Service. Retrieved April 13, 2006, from http://www.ets.org/Media/Research/pdf/PICTWIN.pdf.

Travers, K. J., & Westbury, I. (1989). *The IEA study of mathematics, Vol. 1: Analysis of mathematics curricula.* Oxford, United Kingdom: Pergamon Press.

Tyler, R. W. (1969). National Assessment—Some valuable by-products for schools. *The National Elementary Principal, 48,* 42–48.

Van Groenestijn, M.J.A. (2002). *A gateway to numeracy. A study of numeracy in adult basic education.* Utrecht, The Netherlands: CD Press.

Victorian Curriculum and Assessment Authority (VCAA). (2003a). *Mathematical Methods. Written Examination 1 (Facts, Skills and Applications). Part1—Multiple-Choice Question Book.* Victorian Certificate of Education 2003. Melbourne, Australia: Victorian Curriculum and Assessment Authority.

Victorian Curriculum and Assessment Authority (VCAA). (2003b). *Mathematical Methods. Written Examination 2 (Analysis Task). Question and Answer Book.* Victorian Certificate of Education 2003. Melbourne, Australia: Victorian Curriculum and Assessment Authority.

Vos, P. (2002). *Like an ocean liner changing course; the Grade 8 mathematics curriculum in the Netherlands 1995–2000.* Enschede, the Netherlands: University of Twente.

Westbury, I. (1992). Comparing American and Japanese Achievement: Is the United States really a low achiever? *Educational Researcher, 21*(5), 18–24.

Yamamoto, K., & Kulick, E. (2000). Scaling methodology and procedures for the TIMSS Mathematics and Science Scales. In M. O. Martin, K. D. Gregory, K. M. O'Connor, & S. E. Stemler (Eds.), *TIMSS 1999 Benchmark Technical Report.* Chestnut Hill, MA: Boston College. Retrieved April 13, 2006, from http://timss.bc.edu/timss1999b/pdf/T99B_TR_Chap13.pdf.

AUTHOR NOTE

In a chapter like this it seems important to make visible some of the author's activities as a professional in this area. I have been on the National Dutch Committee for TIMSS and member of the Subject Matter Item Replacement Committee for TIMSS.

I am Chairman of the International Mathematics Expert Group of OECD/PISA, an advisory body that feels itself philosophically responsible for the mathematics framework. Furthermore I was Principal Investigator of the CATCH-project, and founder of the Math A-lympiad.

I would like to thank my colleague Truus Dekker and the reviewers John Dossey and Ken Travers for their valuable and constructive comments on earlier versions of this article. It is good to know that there are always colleagues willing to spend their time on work of others in order to get a better product.

PART VI

Issues and Perspectives

26

ISSUES IN ACCESS AND EQUITY IN MATHEMATICS EDUCATION

Alan J. Bishop and Helen J. Forgasz

MONASH UNIVERSITY, MELBOURNE, AUSTRALIA

In the 1992 *Handbook*, two chapters dealt with issues of equity and access. Gilah Leder (1992) contributed the chapter on "Mathematics and Gender: Changing Perspectives," and Walter Secada's (1992) chapter was entitled "Race, Ethnicity, Social class, Language and Achievement in Mathematics." Other chapters alluded to an issue or two, but there was no chapter with the same brief as this current one. Both of the previous authors discussed the problems of terminology and argued for different approaches to research. In Leder's case she argued for more "in-depth, small sample" research studies, whereas Secada looked for more research within mainstream mathematics education that takes issues of student diversity as a part of the study.

In this chapter, being conscious of the fact that several chapters in this volume focus on students' learning, and on race, class, and so on, we focus on the issues and perspectives that relate to equity and access. With that in mind we disclaim any attempt to cover the whole research field of equity in mathematics education and instead intend the chapter to clarify and discuss the significant issues that need to be addressed in any current research in this field. We will of course draw on the published research literature in the course of our discussions and clarifications.

High on the list of issue areas is one that must come into this chapter, namely that of terminology. Secada's (1992) chapter dealt extensively with this topic and raised the significant point that although research on "disadvantaged" or "minority" students is to be welcomed, "it should not preclude inquiry into how the categories of ethnicity, race, social class and language are constructed and how they are maintained in our society and, by extension, in our research" (p. 623). Who is defining *disadvantage*, why, and from what perspective? Who are the 'minority' and do *they* define themselves that way? In South Africa, the Black Africans were referred to as "the minority," despite being clearly in the majority in terms of numbers.

We are reminded of the study by Becker (1963), who had an interesting perspective on the powerful strategy of labeling, based on his research into deviance in society.

> The central fact of deviance is that it is created by society. I do not mean this in the way it is ordinarily understood, in which causes of deviance are located in the social situation of the deviant, or the social factors which prompted his [sic] action. I mean, rather, that social groups create deviants by making the rules whose infraction constitutes deviance, and by applying those rules to particular people and labelling them as outsiders. From this point of view, deviance is not a quality of the act a person commits, but rather a consequence of the application by others of rules and sanctions to an 'offender'. The deviant is one to whom the label has been successfully applied. Deviant behaviour is behaviour that people so label. (pp. 8, 9)

Deviants are therefore not homogeneous. They share the label and the experiences of being so labelled, and that is all. It is the same with immigrant students, with Black, or Hispanic, or female, or blind, or working-class students. They are not homogeneous groups, but they share the labels and the experience of being so labelled, as does anyone who is perceived as an outsider to whatever is the mainstream culture. What is important for this chapter is the fact that this labelling affects and indeed controls the educational consequences for those learners so labelled. The provocative research by McDermott (1974, 1996) shows some of these hidden consequences, as can be seen from the two titles "Achieving School Failure..." and "The Acquisition of a Child by a Learning Disability."

In the case of mathematics education, equity and access have their special problems of definition and interpretation. At one level, we can confidently assert that without *access* to mathematics education there can be no equity; thus in this view *equity* is referring to an outcome whereas *access* is a means of getting there. At another level, we can argue that there are many equities: equity of predisposition, equity of teaching, equity of outcome, as well as equity of access. Equity here is being used as one criterion for judging or evaluating any educational variable, including access. And what precisely does "access to mathematics education" mean? "It is there if you want it." Or "You can always buy the books." Or "You have as much chance as anyone of studying mathematics at university!" or "You have the opportunity to learn, what you do with it is up to you." Where does the responsibility for guaranteeing access lie? And is access, like beauty, in the eyes of the beholder?

The research literature generally indicates that the following student groups have all suffered in some way from conflicts with mainstream mathematics education, largely based on what can be called *Western Mathematics* (Bishop, 1988):

- Girls in "Western" and other societies
- Ethnic minority children in Western societies
- Indigenous "minorities" in Westernized societies (including Black students in South Africa who are numerically in the majority)
- Western "ex-colonial" student groups
- Non-Judeo-Christian religious student groups
- Rural learners, particularly in developing countries
- Learners with physical and mental impairments
- Children from lower class (or lower caste) families

Thus there is no shortage of literatures to help with this chapter; indeed the problem is almost that there is too much to include. However, what this listing, and the consequent surfeit of literature, demonstrates is that equity and access are not just about students who have physical or mental impairments. Also there are many aspects of commonality between those groups and their problems, which makes an approach based on the different groups not so relevant for this chapter. Instead we shall deal with the common issues and obstacles that all disadvantaged groups face, and the attempts by mathematics educators to research ways that will help them overcome those obstacles.

Firstly there is research at the individual level, dealing with the individual learners and their experiences of inequitable mathematical treatment, which tends to be based mainly on cognitive or affective psychology (see for example, McLeod, 1992; Secada, 1991; Willis, 1992). The construct of *learning disabled* (LD), and, more particularly, of *mathematically disabled* (MD), is one with which cognitive science researchers are familiar (see, for example, Siegler, 2003; Zeleke, 2004). A more recent area of research, cultural psychology, has given further insights into the cultural conditions of learning that inevitably structure the provision of inequitable mathematical treatment and outcomes (see Carraher, Carraher, & Schliemann, 1985; Cole, 1996).

At the second level of the classroom, there are studies of access given or denied to students by virtue of the conditions of pedagogy established by the teacher and the peers (see, for example, Boaler, 1997; Khisty, 1995; Zevenbergen, 2001).

The classroom is part of a larger institution that also structures and constrains the behaviors of teachers and students, and at this institutional level the political and sociological literatures are more active (for example, Apple, 1995; Lubienski, 1997). These literatures also sensitize readers to the fact that it is society at large where the major political battles take place over access, equity, and of course funding, which underlies any decision about educational policies, provision, and resource priorities (Frankenstein, 1995; Knijnik, 1998). In the United States, for example, the influence on equity thinking and practice of the so-called No Child Left Behind Act (The Elementary and Secondary Education Act 2001) is huge and far-reaching, with its emphasis on student outcomes, and on other foci such as the insistence on the use of randomly assigned treatment groups as the only "scientific" kind of research.

Finally, at the broader cultural level are the values, history, and customs that subconsciously but nonetheless effectively shape, constrain, and foster dispositions toward educational equity and access (Keitel, Damerow, Bishop, & Gerdes, 1989). For example, it is not the case that all countries/cultures subscribe to the aspiration of "Mathematics for all" as shown by D'Ambrosio (1994, 1998).

Thus there are many ways we could have structured this chapter, but we have chosen a relatively simple structuring in that following this introduction, the next section presents some of the rich evidence about inequities in student mathematical performance. The subsequent sections focus on issues arising from studies of policy and intention, regarding equity and access, and on research on practice and policy implementation. The fourth substantive section will consider some of the most significant research issues arising from the two previous sections. We close the chapter by considering some of the research challenges for the future, as a kind of research agenda.

EQUITY IN MATHEMATICS LEARNING OUTCOMES?

One example of the kind of outcomes-based research that makes us aware of inequities in the provision of mathematics education is the mathematics achievement data of approximately 190,000 fourth graders and 153,000 eighth graders from across the United States that were found in the 2003 NAEP results for mathematics (National Center for Education Statistics [NCES], 2003). The results continue a trend in mathematics covering the years 1990, 1992, 1996, 2000, and 2003. The mathematics questions in 2003 covered five content strands: number sense, properties, and operations; measurement; geometry and spatial sense; data analysis, statistics, and probability; and algebra and functions.

According to NCES (2003), the major findings of the national results for students in Grades 4 and 8 included the following:

- for both grades, 2003 average mathematics scores were higher than in all the previous assessment years (e.g., 1990: Grade 4—213, Grade 8—263; 2003: Grade 4—235; Grade 8—278)
- among Grade 4 students in 2003, the percentages performing at or above the Basic level, at or above Proficient, and at Advanced were all higher than in all previous assessment years since 1990.

Of particular interest to us in this chapter are the findings for various subgroups of Grade 4 and Grade 8 students in 2003. The results are shown in Table 26.1 and summarized below.

Table 26.1 Mean Scores for 2003 NAEP Results for Grade 4 and Grade 8 From Various Subgroups

Subgroup	Grade 4	Grade 8
Male	236	278
Female	233	277
White	243	288
Black	216	252
Hispanic	222	259
Asian/Pacific Islander	246	291
American Indian/Alaska Native	233	263
Eligible for free/reduced price school lunch	222	259
Ineligible for free/reduced price school lunch	244	287

Gender

- At both grades, male students scored higher on average than female students.
- At both grades, the average scores for male and female students were higher than in any of the previous assessment years.
- At both grades, the apparent gap in scores was not found to be significantly different than the gap in any of the previous assessment years.

Race and Ethnicity

- At both grades, Asian/Pacific Islander students scored higher on average than did White students. Also, White students and Asian/Pacific Islander students had higher average scores than their Black, Hispanic, and American Indian/Alaska Native peers.
- At both grades, Hispanic students and American Indian/Alaska Native students scored higher on average than Black students.
- At both grades, White, Black, and Hispanic students had higher average scores than in any of the previous assessment years since 1990.
- No significant change was detected in the average score for Asian/Pacific Islander students between 2000 and 2003 at Grade 8. The average scores for Asian/Pacific Islander students were higher in 2003 than in 1990 at both Grades 4 and 8.

Socioeconomic Status (using eligibility for free/reduced-price school lunch as the criterion)

- At both grades, the average mathematics score for students who were eligible for free/reduced-price lunch was lower than for students who were not eligible.
- At both grades, average mathematics scores were higher in 2003 than in any of the previous assessment years for both groups of students.

These kinds of data are frequently found in both national and international studies such as the *Third* (and more recent *Trends in*) *International Mathematics and Science Study* (TIMSS), with gender, race, and class (or socioeconomic status) being the predominant difference variables of interest. The gender variable is arguably the least contentious of the three to be measured in quantitative studies and is thus the most represented in achievement-oriented equity research.

One of the key issues in this research area concerns the extent to which any research is actually trying to study equity and access, or whether the outcomes data happen to show differences between certain defined groups of students. In other words, the differences may be found by accident, so to speak. Thus, for example, in international studies such as the TIMSS, certain sampling decisions were taken in order to get "comparable" samples of students and teaching contexts across countries. This criterion, and those decisions, inevitably meant that any differences found *within* those countries might well have been due to the exigencies and "accidents" of the samples. A classic example is given by the use of the TIMSS tests in Australia to measure the comparative performance of indigenous students. As Lokan, Ford, and Greenwood (1996) reported,

> Only about 400 students from the 13 700 or so TIMSS participants identified themselves as Aboriginal or Torres Strait Islander students. To enable more accurate results to be reported for Indigenous students, some additional sampling of schools with relatively high Aboriginal and Torres Strait Islander enrolments was done.... Of the 608 students, 330 were in the upper year level, 278 in the lower. . . . The results for the total group are descriptive only, since the additional schools were not randomly chosen. Nevertheless, the very much lower achievement levels, and the smaller differences between lower and upper year levels of the Indigenous students in comparison with the overall Australian results presented in Chapter 3 seem likely to signal real differences. (p. 161)

Another variable, the study of which produces some salutary lessons for equity research, concerns the student's language and whether that presents any disadvantage for mathematics achievement. For example, an assumption often made is that students who work and study in their second or third languages are necessarily handicapped in their studies and likely to show underachievement. However, in English-speaking countries, one should not assume that students whose first language is not English are necessarily disadvantaged in mathematics. Among the background data collected in the TIMSS was the main language spoken at home (English or other) and students' place of birth (English- or non-English- speaking countries) (Lokan, Ford, & Greenwood, 1997). Australian data from Population 1 (9-year-olds) were analyzed by the combination of these two factors. Of the four resulting groups, the lowest achievers were those born in Australia whose families used a language other than English at home (about 4% of the cohort; Lokan et al., 1997). The highest achievers were children living in Australia but born in non-English- speaking countries, whose families had adopted English as the main language used at home.

Huang (2000) analyzed the TIMSS findings of students from five English-speaking countries: England, the United States, Canada, Australia, and New Zealand, confirming the Australian findings. Compared to students who spoke English at home, students whose home language was not English performed substantially less well; this was true whether or not they were immigrants.

Finally, the special education literature is replete with data concerning LD (learning disabled) and MD (mathematically disabled) students and their disadvantages in achieving mathematical understandings and competences, with large age/grade performance differences. As a typical example, Cawley and Miller (1989) reported that 8- and 9- year- old LD children performed at a first- grade level on tasks of computation and application. Zeleke (2004) analyzed and summarized the research and data concerning LD and MD students' performance on simple arithmetic tasks and concluded, "MD children are characterized by poor strategy choices . . . frequent use of immature strategies . . . less frequent use of retrieval strategies . . . less varied collection of usable strategies . . . and use of significantly fewer problem solving strategies." (p.12) Clearly there are many implications here, not just for special education and mainstream teachers, but also for policy decisions regarding access and equity.

POLICY ISSUES OF ACCESS AND EQUITY

Given the vast amount of evidence both national and international about systematic differences in mathematical learning outcomes, we will now explore some of the research and ideas helping to explain the data, and hopefully offering strategies to combat their causes. This section will draw on studies that focus on policy aspects, intended curricula, and other mechanisms, such as examinations, teacher training, inservice aspects, and school structures. It will also give more of the educational and mathematical context of this review.

In the late 1980s and at the turn of the 1990s three international developments had a significant influence on our international and increasingly intercultural field. The first was ICME-6, which took place in Budapest, Hungary, in 1988. At that ICME, and following widespread pressure, a whole day was devoted to the issues and possibilities of Mathematics, Education, and Society. More than 120 people contributed formally to the program that day, and the UNESCO publication that followed (Keitel et al., 1989) drew everyone's attention to the social and cultural aspects that lay behind inequities of policy and practice in many countries. It also built on the groundwork laid at ICME-5 in Adelaide, Australia, where not only was there a whole Theme Group on the topic of "Mathematics for All" (Damerow, Dunkin, Nebres, & Werry, 1984) but also one of the plenary lectures was by D'Ambrosio (1986). In this he introduced the topic of ethnomathematics, thereby changing forever mathematics educators' conceptions of mathematics, and of the mathematics curriculum (see also Ascher, 1981).

The second development took place in 1989, when the National Council of Mathematics Teachers (NCTM) of the United States published *The Curriculum and Evaluation Standards for School Mathematics*, a wide-ranging and far-reaching document that would spearhead the reform of mathematics education across the country. Not only was the aim to change mainstream mathematics education, but particularly to emphasize that this was for *all* students, because all were capable of learning. Later the six principles in the new framework were spelled out in more detail (NCTM, 2000), and one of them is the Equity Principle, which significantly was put first in the order of priorities "Excellence in mathematics education requires equity—high expectations and strong support for all students."(p.12) The NCTM also issued this caveat:

> Equity does not mean that every student should receive identical instruction; instead, it demands that reasonable and appropriate accommodations be made as needed to promote access and attainment for all students. . . . Technology can assist in achieving equity and must be accessible to all students. (NCTM, 2000, p. 12)

More reference will be made to this Principle later.

Thirdly, in 1990 UNESCO, together with UNICEF, UNDP, and the World Bank, organized the World Conference on Education for All (WCEFA) involving hundreds of representatives of governments, NGOs, educational bodies, and foundations. The Conference concluded by agreeing on two key sets of recommendations: the "framework for action to meet basic learning needs" and the "world declaration on education for all." They left the educational world in no doubt about their commitment to universal education, including numeracy for all (UNESCO, 1990).

All three of these initiatives highlighted the fact that from the 1990s forward issues of equity and access to mathematics education could not be ignored by those in policy- and decision-making positions. So what are the policy issues and challenges facing those mathematics educators who are seeking to provide an equitable education for all students, and what options does research has to offer?

The Structuring of Mathematics Education by Society, and Vice Versa

The first significant challenge to be faced is that of the valuing and structuring of mathematics by cultures and societies, which have profound effects on the issues of equity and access. Ever since the mathematization of technology in the 18th and 19th centuries, mathematical knowledge has played a strong role in the psyche of modernist and economically focused governments. As Skovsmose (1994) argued in his chapter "A Formatting Power," the formatting of society by mathematics is ever increasing, to which Woodrow (2003) added:

> Mathematics has a peculiarly special position in the social and political discourse across the world—to use a new dangerous word it holds global significance. As the 'language of science' it had long assumed power and influence as the terminology of science but during the last half-century it has permeated many of the social sciences, including not only economics but also such social areas of debate as wealth distribution (poverty and affluence) or crime and its causes and consequences. (p. 10)

In every country in the world mathematics now holds a special position, and those who excel at it or its applications also hold a significant position in their societies. The corollary for us in this chapter is that students' mathematical attainment has been used as a selective "filter" for entry to various professions, a fact that mitigates strongly against the ethic of an equitable and accessible mathematics education for all.

Who are the groups controlling this filter effect, and who becomes disadvantaged by it? Professional bodies are the obvious controllers, including those who prescribe the entry requirements for their various professions, but more frequently it is the universities and other educational institutions that act as surrogate filter mechanisms for the professions. Of course universities also want to recruit the best mathematics students for their own purposes, such as to continue and develop the mathematics research programs, and this gives universities a powerful voice in the filtering process. As Woodrow (2003) rightly pointed out, "The 'cultural capital' represented by knowing mathematics is 'gilt edged' and it has been socially and culturally advantageous to mathematicians to maintain the investment" (p. 19).

In the United States we have seen the developing phenomenon of the "Math Wars" whereby proponents of the NCTM's reform agenda for mathematics education (see the previous section) ran headlong into the brick walls of the university mathematics establishment, who perceived the reforms as diluting and lowering the high filtering standards they believed they set for their entrants (e.g., Jackson, 1997). The arguments may be cloaked in terms of mathematics content, which is the equivalent of the moral high ground for mathematicians, but the issues are clearly all about equity and access. The debates have spawned web sites (such as http://mathematicallysane.com) where evidence of various kinds is disseminated.

The United States is not the only country where these debates take place. In part the arguments in any country are fuelled by the results of international competitive testing, such as TIMSS and its analysts (see Schmidt, McKnight, Valverde, Houang, & Wiley, 1997), where the league tables of results produced by this and similar studies are used by all sides in these debates to support or destroy the arguments for equity. Equity and competition do not sit easily together, and any system that builds its mathematics education on the precepts of competition and rewarding high but narrow achievement will necessarily cast aside arguments for equitable and accessible approaches.

Once again, different countries' policies regarding mathematics education show certain important variations, all of which have equity implications, due in part to their different histories, cultures, values, and perceived needs. Robitaille (1997) edited a text documenting the various national contexts for mathematics and science in the TIMSS-participating countries, where one can find, for example, that streaming and tracking are not sanctioned everywhere, there are not always policies to integrate students with disabilities in mainstream schools, and private schools do not necessarily get financial or other assistance from governments. Mathematics is not always a compulsory school subject towards the end of formal schooling, and although there are general efforts to encourage girls to continue their mathematics education for longer, in Australia girls already do stay in schooling longer than boys. In some countries the textbooks are mandated by government whereas in others, teachers can choose what texts and teaching materials to use.

However, despite these differences, fundamentally for mathematics educators the most important value that any particular society puts on mathematical education is to a large degree reflected in how the subject is structured in any intended curriculum, and it is to that level that we now turn.

The New Curriculum—Mathematics or Numeracy or Both?

What then are the issues and challenges concerning the nature of the "intended" school mathematics curriculum, and the extent to which it could offer access to all? By saying *intended* here we are using the terminology of the IEA in its international surveys, where they distinguish the curriculum intended by the state or school district, the curriculum implemented by the schools and teachers, and the curriculum attained by the students (Robitaille & Garden, 1989). In this main section of the chapter we are concerned with the intended curriculum and the access issues surrounding its determination. In the following section on "practices" in institutions, we discuss the issues of equity as they relate to the implemented curriculum and the opportunities to learn.

In the 1990s the interpretations of *mathematics* and of *the mathematics curriculum* were debated and continue to be debated. It may seem strange to be debating the nature of mathematics, although philosophers throughout the ages have always done so. But given that mathematics educators approach this problem from the perspective of what a mathematical education is to contribute to all future citizens in a democratic society, this is not an arid philosophical discussion point. It should be at the heart of our teaching, and at the heart of our research.

In democratic societies one must always consider the notion of *democratic education*. Malloy (2002), for example, argued that democratic access to powerful mathematics for all children was a human right and, instead of universal enculturation, "students are educated in a world that concentrates on differences, which consciously or unconsciously separate the rich and poor, educated and non-educated, leaders and followers, and racial and ethnic groups" (p. 18). Povey

(2003) argued in terms of *democratic* mathematics in the context of *citizenship* education. Citizenship was defined in the British context, and the problematic nature of what constituted citizenship and democratic mathematics and what democratic mathematics classrooms might look like were discussed. There was a perceived need for mathematics education to move towards "a deep democracy in which equality and person centredness, engagement, and creativity are recognized as key to effective participation" (p. 61).

In the late 1980s, the construct of ethnomathematics focused the attention of mathematics educators on the "democratic" question of what kind of mathematics should be taught and learned in schools. Various analyses of this construct (Bishop, 1988; Carraher et al., 1985; D'Ambrosio, 1986; Gerdes, 1995; Powell & Frankenstein, 1997) have stretched and generalized it from its roots in the knowledge systems of indigenous communities in third-world countries to encompass any specific and localized forms of mathematical knowledge in any society—carpenters, gamblers, street traders, and so on—who all have their specific knowledge and practices. Bishop (1988) for example wrote about and differentiated between *Mathematics* with an upper-case *M* and *mathematics* with a lower-case *m*, where the second referred to all mathematical knowledges and the first referred to the kind of supposedly culture-neutral and "universal" mathematics that the university mathematicians in Western societies research and teach, and that forms the basis of most mathematics curricula in schools today.

This is a similar position to the argument that the concept of ethnomathematics embraces every form of mathematical knowledge, including the so-called universal mathematics of Western university mathematicians (see, for example, D'Ambrosio & Ascher, 1994), and therefore the educational argument should be about what kind of ethnomathematics to teach. Should one teach Mathematics or mathematics, including the "specific" ethnomathematical knowledges and practices relevant to the cultural and societal context of the students? And if the answer is "both" then what are the best ways of factoring these different knowledges into a school mathematics curriculum?

This enlarged construct of ethnomathematics does not appear to have been widely accepted as offering a structure for determining the appropriate mathematics curriculum in any given educational context. For example, according to Kassem (2001), despite early hopes and intentions of including cultural referents and contexts, the UK national curriculum has never had a multicultural dimension. The NCTM (2000) *Principles and Standards* document likewise makes no reference to any cultural contexts for reinforcing or contrasting mathematical practices.

A more widespread and accessible democratizing construct nowadays appears to be that of *numeracy*, despite the fact that different definitions of the term abound (see Jablonka, 2003; Steen, 1990). Gellert, Jablonka, and Keitel (2001) wrote about different conceptions of *mathematical literacy* and their relationships to mathematics, reality, and society, in the context of individuals' needs for civic participation. Frankenstein (1998) presented four goals (with examples) of a critical mathematical literacy curriculum that would enable teachers to "teach mathematics as a tool to interpret and challenge inequities in our society" (p. 180). The four goals are understanding the mathematics, understanding the mathematics of political knowledge, understanding the politics of mathematical knowledge, and understanding the politics of knowledge.

What the introduction of this term contributes to the equity discussion on school mathematics education is that, despite the different details of the definitions, it focuses attention on the mathematical knowledge deemed necessary to be a competent citizen in a democratic society (see various articles in Kelly, Johnston & Yasukawa, 2003; Steen, 2001). This is surely what equity in mathematics education is all about.

From an equity perspective we would argue that schools certainly *should* give all students access to the so-called universal mathematics knowledge (from here on we shall use the term *Mathematics* with a capital *M*) but schools *must* prepare all students to be competent citizens and should therefore be teaching numeracy to all students.

Numeracy is a widely used term with a myriad of definitions that embrace many of the "mathematics for all" notions described above. According to the Government of Canada (n.d.),

> Before the word "numeracy" came into use, there was discussion of terms such as "mathematical literacy" and "quantitative literacy," which placed the focus on calculations and the ways in which numbers and mathematical concepts were embedded in texts, but which did not take into account the wider practical uses of numbers and mathematics in the workplace and in personal life on an everyday basis.

The Australian Association of Mathematics Teachers (AAMT, 1998) defined what it means to be numerate as being able "to use mathematics effectively to meet the general demands of life at home, in paid work, and for participation in community and civic life" (p. 1). For school education, numeracy was considered to be "a fundamental component of learning, discourse and critique across all areas of the curriculum" (AAMT, 1998, p. 1) and involved

> the disposition to use, in context, a combination of: underpinning mathematical concepts and skills from across the discipline (numerical, spatial, graphical, statistical and algebraic); mathematical thinking and strategies; general thinking skills; [and] rounded appreciation of context. (AAMT, 1998, p. 1)

The relationships between numeracy and Mathematics seem logically to be of five kinds:

1. Mathematics and numeracy intersect, that is they share aspects but do not include each other.
2. Numeracy is a subset of Mathematics.
3. Mathematics is a subset of numeracy.
4. Numeracy is Mathematics.
5. Mathematics and numeracy are two very different phenomena, having no relationship.

From an equity perspective, choices 4 and 5 do not appear to offer any useful ideas for structuring the intended school curriculum, which satisfy the criteria described in the paragraph above. Nor do many argue for such positions. Only Choices 1, 2, and 3 are supported in the numeracy literature (see Jablonka, 2003; Kelly, Johnston & Yasukawa, 2003), with 2 occupying the least radical position, followed by 1, and then 3, the most radical.

Willis (1996) identified four broad perspectives on the relationships between the mathematics curriculum, disadvantage, and social justice: remedial, non-discriminatory, inclusive, and socially critical. A framework was developed that included statements completing the following sentence stems with respect to each of the four perspectives: "The mathematics curriculum is . . . ," "The problem with 'disadvantage' lies with . . . ," "The solution is to . . . ," and "The educational task is to..." (Willis, 1996, p. 47). The framework was used with teachers and curriculum writers. Willis (1996) claimed that this enabled them to understand their own positions better and provided "a starting point for developing a more consistent approach to issues of gender justice and the curriculum" (p. 48).

Various curricular approaches have attempted to embed the mathematics in realistic contexts as a way of trying to get the equitable "best of both the mathematical and numeracy worlds," so to speak. In Holland the most successful approach has been that of the Realistic Mathematics Education (RME) project, based on the educational philosophy of Hans Freudenthal (1983). Under his philosophy the learners should begin with real, or at least realistic, problem situations from which the teacher can lead them into exploring the relevant and significant mathematical concepts and techniques. From the perspective of the Mathematics in Context project, Romberg and Shafer (2003) described a design study based on RME principles of first-year algebra teaching in a U.S. high school. In this study, all the students had the opportunity to succeed and, in doing so, exceeded the researchers' expectations. There was also consistent evidence of higher-order thinking and analysis in all of the classes, and not just in the honours class.

Everything depends of course on definitions of *realistic, problem-based, numeracy,* and so on, but for determining appropriate curriculum structures to be equitable and accessible to all, the key arguments are the same: Will all students study the same curriculum? If so, how and by whom will the pedagogical and microcurricula decisions be made, in order to involve all students? If not all students will study the same curriculum, how will the choices be made? How will the different curricula be managed and structured? Will different curricular sequences run in parallel, or will the differences relate to the different timing or depth of topics taught?

Mathematics education in schools is thus seen to have a dual function: to prepare students to be mathematically functional as citizens of their societies—arguably provided equitably for all—and to prepare some students to be the future professionals in careers in which mathematics is fundamental, with no one precluded from or denied access to participation along this path. The notion of more than one purpose for mathematics education raises issues of what constitutes an equitable mathematics curriculum and raises questions about the equity implications of systemic, school-based, or classroom-based practices in mathematics education such as tracking/streaming or single-sex settings.

This then moves us into considering the institutions of mathematics education and the research that focuses on issues and challenges at that level.

INSTITUTIONAL STRUCTURING OF MATHEMATICS EDUCATION

Although institutions can be considered in various ways, the most appropriate for this chapter's focus is the idea of the Opportunity to Learn (OTL). This construct first made its appearance in the studies organized by the International Association for the Evaluation of Educational Achievement (IEA; Keeves,

1995), as the opportunity that students have to learn the content of the items in the tests. Clearly however this definition is limited and could go much wider, particularly if one is interested in equity research.

As an example of the broadening of the OTL concept, in 1994, US President Clinton signed education reform legislation known as *Goals 2000: Educate America Act*, making *opportunity to learn* standards voluntary (NCEO Policy Directions, 1995). In *Goals 2000*, OTL standards were defined as

> the criteria for, and the basis of, assessing the sufficiency or quality of the resources, practices, and conditions necessary at each level of the education system (schools, local educational agencies, and States) to provide all students with an opportunity to learn the material in voluntary national content standards or State content standards. (Goals 2000: Educate America Act, 1994, Section 3a(7))

The OTL standards addressed the following educational dimensions (NCEO Policy Directions, 1995):

- curricula, instructional materials, and technologies
- teacher capability
- continuous professional development
- alignment of curriculum, instructional practices, and assessments with content standards
- safety and security of the learning environment
- nondiscriminatory policies, curricula, and instructional practices
- other factors that help students receive a fair opportunity to achieve the knowledge and skills in the content standards

Whereas poor and minority students were recognized by proponents of OTL, NCEO Policy Directions (1995) identified the exclusion of students with disabilities.

Oakes (1995) wrote in support of the OTL standards, claiming that they would provide information enabling insights to be gained into why students in particular schools scored as they did in content-standards-related testing—the focus would no longer be exclusively on student performance.

Lynch and Baker (2005) discussed equality in education from the perspective of *condition*, which they defined as "the belief that people should be as equal as possible in relation to the central condition of their lives" (p. 132). They claimed that there were five dimensions of equality: "resources; respect and recognition; love, care and solidarity; power; and working and learning" (p. 137). The OTL characteristics and Lynch and Baker's dimensions of equality of condition resonate strongly with mathematics educators' specific attempts to grapple with definitions of equity, equality, justice, and opportunity to learn within the framework of mathematics learning (see, for example, Hart, 2003). Fennema (1993) defined equity in mathematics education as composed of three parts: "(1) equity as equal educational opportunity, (2) equity as equal educational treatment, and (3) equity as equal educational outcome" (p. 2). She contended that

> while there is a semblance of equal opportunity to learn mathematics in today's world, mere semblance is not enough. . . . Are there less obvious ways in which inequality exists, . . . Legally enforcing equity is a prerequisite to the achievement of equity; but it is not a sufficient condition. . . . Justice in mathematics will not be achieved until the goals of education are met equally by both sexes. (Fennema, 1993, p. 5)

The notions of *equity, justice,* and *opportunity to learn* were intertwined in Fennema's (1993) words. Although Fennema's definitions and discussion were set in the context of gender, they are clearly applicable to all dimensions of disadvantage and inequality with respect to mathematics education. The work of Allexsaht-Snider and Hart (2001) also extends the notion of equity in mathematics education beyond gender. They contended that "teachers' work with students in individual classrooms is at the heart of any program designed to attain equity for all students" (p. 93). They recognized that teachers could not do this alone and identified the need for collaborative effort among members of the wider educational community. They used a definition of equity based on two premises:

- that all students, irrespective of race, ethnicity, class, gender, or language proficiency, will learn and use mathematics
- that all associated with children's education be aware of social, economic, and political factors that hinder or facilitate the mathematics learning of underrepresented students.

For equity in mathematics education to be achieved, Allexsaht-Snider and Hart (2001) maintained that there were three conditions to be met:

1. equitable resource distribution: schools, students, and teachers
2. equitable instructional quality: teachers' mathematical knowledge, their preparation for mathematics teaching, and their beliefs about and skills in teaching students from diverse backgrounds;

3. equitable student outcomes: achievement, course enrollments, interest, motivation, and valuing of mathematics

Equity would be achieved, they claimed, "when differences among sub-groups of students in these three areas are decreasing or disappearing" (p. 93). Allexsaht-Snider and Hart's (2001) three conditions for the achievement of equity in mathematics education clearly resonate with Fennema's (1993) definition of equity in mathematics education. The first two common categories of equity are the organizing parameters for the discussions that follow in this section of the chapter.

Rousseau and Powell (2005) framed their study of equity in reform mathematics on the three variables which they argued shape students' opportunities to learn mathematics: time, quality, and design. They identified several factors that operated as barriers to reform for one of the two teachers in the study—large class size, high-stakes-standardized tests, high student mobility and absenteeism, and lack of a high-quality curriculum. This female teacher worked in an urban school and her practices were closer to the traditional end of the traditional-reform continuum than were those of the other teacher, a male who taught in a suburban school. According to Rousseau and Powell (2005), the identified barriers to reform noted "were more likely to be present in schools with high proportions of students of color or students living in poverty," students who were "more likely to be denied access to the kinds of curricula and teaching practices that are associated with mathematics reform" (p. 29).

According to Silver (2003), rural communities worldwide have high poverty levels and low educational attainment levels, and mathematics educational researchers focusing on equity considerations have largely ignored rural students. It is also interesting to note that learning disabilities and physical disabilities are frequently overlooked as categories of inequity in the mathematics education literature. In Australia, van Kraayenoord, Elkins, Palmer, and Rickards (2000) conducted a literature search on literacy, numeracy, and students with disabilities and concluded that there had been greater research emphasis on literacy than on numeracy and that there was a lack of information on all groups of students with disabilities.

Lubienski and Bowen (2000) reported that there were many articles on mathematics and disability published between 1982 and early 1998 to be found in the ERIC database. They claimed that about half of the articles dealt with cognition, most articles were not found in mathematics education journals, and only a handful also included other categories of equity. It is not clear from their writing, but it appears that Lubienski and Bowen were referring to articles about learning disabilities and not physical disabilities. For this chapter, some research on mathematics learning and physical disabilities, and on mild learning disabilities, have been included in the review.

Equity of Access—Resources: Schools, Students, and Teachers

Oakes (1995 argued that "an adequate and fair distribution of resources, programs, and teachers won't, by itself, guarantee that disadvantaged students will learn well." (p. 86). It was noted, however, that not much improvement could be expected from the students without sufficient funding being directed into high-quality instruction. In this section, a selection of research studies illustrating inequities associated with a range of resource issues is presented.

Fenwick (1996) reported that students in poverty (i.e., from low socioeconomic backgrounds) in the United States "are more likely to be taught by inexperienced teachers . . . and by teachers who are either without any certification or not certified to teach the subject they are teaching" (p. 4). This was exacerbated for the poor urban students (mainly Hispanic and African American). These teachers were also likely to have negative attitudes towards the students and were less likely to encourage them.

Secada, Cueto, and Andrade (2002) argued that "group-based inequalities in the distribution of social good," in their case the opportunity for students to learn mathematics, was "bad social policy from a position of socially enlightened self-interest" (p. 108). Using a wide range of data-gathering methods, their study involved an examination of the opportunities to learn mathematics of Peruvian students from Puno (in southern Peru) and the kinds of schools attended by fourth and fifth graders. The general poverty of the people was reflected in the schools' resources. Particularly in the regional areas, many schools lacked basic facilities such as water, sewage, and telephones. Good blackboards, however, were found in all schools. The recently reformed mathematics curriculum was described as having a balanced representation of mathematics, yet there had been little change in practices. Arithmetic was the focus, and it was taught by rote. Secada et al. regarded their findings as "showing the crisis in Puno's overall educational system" (p. 129).

Kitchen (2001) described Guatemala as a country in which political turmoil and student activism were commonplace, where there was inadequate government support for education, teacher professional development was nonexistent, and teachers often had to work additional jobs to make ends meet. The two teachers in Kitchen's study both worked in inner-city schools attended by an ethnic mix of poor students; one was a school for girls and the other a school for boys. According to one of the teachers, "economic class, not ethnicity, was the great equalizer in Guatemalan society" (Kitchen, 2001, p. 159). Kitchen (2001) maintained that the political and social context in which the two teachers worked was the basis for the conventional pedagogical style they adopted.

In the context of post-apartheid South Africa, Adler's (2001) findings from a 3-year study that focused on learner-centred pedagogy and "mathematical practice that moved beyond 'mathematics as procedures'" (p. 187) challenge notions that equality of resourcing will necessarily bring about equity in opportunities to learn. Classroom observations were conducted in various school settings: urban and rural, primary and secondary, and in a range of relatively well-resourced to impoverished schools. For some students, teachers' use of more resources extended their opportunities for mathematical engagement and growth. For others, particularly in the poorest schools, students' learning opportunities were narrowed. Adler (2001) maintained that the teachers' personal mathematical and professional backgrounds and the learning contexts in which they operated had shaped their changing practices. As a consequence, Adler (2001) contended, it appeared that there had been an exacerbation, rather than amelioration, of the existing inequalities.

Using Fennema's (1993) three components of equity for mathematics learning—access, treatment, and outcomes—equity issues associated with the use of technology for mathematics learning were discussed by Hennesy and Dunham (2002). Inequities with respect to computing technology and the Internet included

- the "global digital divide" between rich and poor countries in terms of numbers of computers per student and Internet access
- inequitable distribution of technology resources between schools with different racial/ethnic profiles and socioeconomic status, and in urban and rural areas
- unreliable technology more likely to be found in disadvantaged schools, and
- gender gaps in use at home and at school.

Access to hand-held technologies (calculators) in schools was considered more equitable; with sophistication and cost increases, inequities were considered likely to increase, however. The extent of technology use was found to be related to the grouping level in which students are taught when setting/streaming/tracking by "ability" is practiced. Home access to computers was reported as more likely in middle-class and suburban homes. Teachers' attitudes towards and experiences with technology, as well as their biased/stereotyped beliefs and expectations of particular groups of students, were factors that could affect the extent of technology use in their teaching. Inequities in outcomes included achievement and attitudinal differences. Hennesy and Dunham (2002) concluded that "opportunities for access are only a starting point; they may mask physical or psychological restrictions on use and may not translate into equitable use of technology in practice" (p. 158), and they claimed that a radical departure from conventional, inequitable schooling methods, particularly for disadvantaged groups, was needed.

In his book on educating the deaf, Moores (2001) discussed several issues associated with mathematics learning. He stated that "for more than two hundred years, the primary emphasis among educators of the deaf has been communication, with academic achievement receiving secondary attention." (p. 8). Thus, across the curriculum, deaf students were said to spend less time than hearing students on learning academic subjects, including mathematics. At 7 years of age, Moores (2001) claimed, their mathematics achievement scores reflected this difference. Mathematics received "relatively little attention in programs for young deaf children," the time allocated to mathematics was minimal, and "teachers of deaf students tend to underestimate the need for mathematics instruction" (Moores, 2001, p. 336). In preservice courses for teachers of the deaf, communication was central and those teaching mathematics to deaf students from kindergarten through to Grade 12 were relying on "traditional practices such as rote memorization, work sheets, and drill and practice" (Moores, 2001, p. 9). Deficiencies in the preservice and inservice education programs for mathematics teachers of the deaf with respect to mathematics content and the mathematics background of participants were identified. Moores (2001) felt that much was needed "to provide both equity and excellence to . . . deaf students" (p. 10).

Equity and Separated Settings

> Segregation has been a common institutional response to the management of differences in education. The degree of segregation varies historically and cross-culturally for different social groups, and often occurs invisibly through broader patterns of residential segregation, selection procedures and parental choice. (Lynch & Baker, 2005, pp. 145–146)

In the educational literature researchers have continued to debate the relative merits of educating particular groups of students in segregated or mixed settings, for example, coeducational or single-sex settings, regular or special classes for those with learning disabilities or physical disabilities. As noted by Lynch and Baker (2005), segregation can result indirectly through circumstance, geography, or choice. Thus, for example, segregation by socioeconomic status (e.g., public versus private education) or by ethnic or racial groupings are found in various contexts. Issues of equity are clearly associated with these issues and are discussed in the literature. Because mathematics serves as a gateway to a number of career options, researchers have shown considerable interest in exploring which educational settings optimize equitable access to, and opportunity for, mathematical learning.

In the area of physical disabilities, the example of hearing-impaired students illustrates the issues. The increasing range and definitions of placement options currently open to students with hearing disabilities in the United States, and the legislative changes that have led to them, were described by Moores (2001). The growth in these options is fairly recent and ranges from residential schools, day schools (only for the deaf), and day classes (classes for the deaf within mainstream schools), to resource rooms (where deaf students return from regular classes to receive special attention) and itinerant programs (students attend regular classes and itinerant teachers assist—schedule of assistance can vary). In Australia, another option is available; students with disabilities (including the deaf) can attend mainstream schools with personal aides (publicly funded) accompanying and assisting them in all lessons. Research on the effects of the different learning settings is sparse. Moores (2001), however, noted that

> growing numbers of deaf children are being taught in public school settings. . . . Interest in applying the results of work with hearing children to education of deaf children is growing. Improvements in academic achievement have already been documented, although the gap between scores of deaf and hearing children is still unacceptably large. (p. 337)

In the past, hearing-impaired students' difficulties with mathematics had been thought to be related to the linguistic content of problems or with the students' English-language competence. Kelly, Lang, and Pagliaro (2003) surveyed Grade 6–12 teachers of deaf children in mainstream settings (some in integrated classrooms; others from self-contained classes for the deaf) and in center schools (boarding or day schools for the deaf only). They found that, irrespective of the setting, deaf students were insufficiently engaged with challenging word problems. Interestingly, teachers in the mainstream classes were found to have higher perceptions of deaf students' problem-solving abilities than the teachers in deaf-only classes. Consistent with Moores's (2001) claim regarding the main focus in their preservice education, the teachers of deaf students attributed the students' low abilities with word problems primarily to their skills and abilities with English.

As another example, one aspect of Tracey's (2002) doctoral study focusing on children with mild intellectual disabilities was also concerned with investigating the effect of regular class placement and special class placement on the children's self-concepts, including mathematics self-concept. It was found that students placed or moved into special classes had higher academic, including mathematics, self-concepts than their counterparts in regular classes, a finding consistent with an earlier study (Burton, 1988). Tracey (2002) claimed that the findings questioned the appropriateness of the continued trend internationally towards the inclusion of students with disabilities into regular classrooms. In a similar vein, a study of students in their first year of schooling by Arnold (1994) found that parents of children with disabilities in the mainstream classes were very satisfied with their social outcomes but less satisfied with their academic progress. Achieving the best balance between those two goals is a major issue for teachers and parents regarding separated educational settings.

Moreover, particular challenges occur when the mainstream curriculum is changed. For example, according to the ERIC Development Team (2002), "many students with disabilities experience difficulties with the [U.S.] reformed math curriculum" (p. 2) including processing and distinguishing relevant information, computational skills, and a lack of reasoning and problem-solving skills. However developments in new information and communication technology can facilitate the learning situation particularly for students with disabilities. For example, in the past, blind students who were unable to read or write the symbols that comprise mathematics had to learn concepts and perform calculations mentally, thus limiting their ability to master the details of mathematics. The development of interactive software was said to help students study the Braille code of mathematics (the Nemeth Code).

Steele's (2004) answer to the dilemma about special versus mainstream teaching was to argue that as the majority of students with learning disabilities (LD) are today being educated in mainstream classrooms, more research attention should be paid to appropriate teaching strategies that mainstream teachers can use. She also argued that such changes could equally facilitate the mathematics learning of the students without disabilities. Her summary of these strategies included the following suggestions:

- Provide advance organizers to introduce the purpose of the lesson.
- Provide additional review of all prerequisites as needed.
- Prioritize, teach, and review major concepts frequently.
- Teach generalization and application to real-life situations.
- Model sequential procedures at a slow pace and with extra clues.
- Present new skills using concrete materials, then pictures, and finally abstract explanations.
- Provide additional practice in small steps with sufficient guidance.
- Be sure directions are clear before starting independent practice.
- Teach students to keep track of their progress with charts and graphs.
- Check for error patterns and related corrections when providing guidance.

Another context in which the issue of segregated or integrated mathematics education is hotly debated concerns single-sex or coeducational settings. The extent of single-sex schooling varies from country to country around the world. Within some countries, gender-segregated schooling is a religious or a cultural norm; in others, it is seen as an alternative for parents to choose for their offspring. Gill (1988) reported that the findings from comparisons of the academic performance and school-related attitudes of students attending single-sex and coeducational schools were inconsistent. In Australia (and the UK), "the distinction between single sex and coeducational schools is interwoven with the division between private and public schooling" (Gill, 1988, p. 3), which, in turn, is closely associated with social class differences. Often, too, the arguments mounted (particularly in the popular press) are shaped by the vested interests of the protagonists, the indicators of success adopted, and factors within the societal context in which the debate is held (Haag, 2000; Leder & Forgasz, 1997).

Single-sex education has been viewed as anachronistic, reflecting times when males and females had different educational needs related to their gendered future roles in society, and for some time, coeducation was considered a means for achieving gender equality (Leder, 1992). More recently, however, the equity assumptions in support of coeducation have been challenged.

A number of single-sex interventions have addressed gender inequities in mathematics learning outcomes. Leder, Forgasz, and Solar (1996) described a vast range of such interventions that have taken place in different settings, for different lengths of time, for different age groups, and with different aims. The SummerMath program (Morrow & Morrow, 1995) in the United States was one short-term out-of-school intervention that succeeded in assisting 13–18-year-old females to see that they could do mathematics confidently and that mathematics is connected to people and their lives. Several researchers have examined single-sex interventions implemented within regular coeducational secondary school settings (e.g., Forgasz & Leder, 1995; Leder & Forgasz, 1997; Rennie & Parker, 1997). The findings were mixed. Forgasz and Leder (1995), for example, concluded that although "beliefs were that females would benefit most from single-sex classes, there were signs that males derived equal, if not more, benefit from the program than the females" (p. 44). Rennie and Parker (1997) found evidence from teachers that

> single-sex classes appeared to hold the most benefit for specific groups of girls who were experiencing a great deal of harassment from boys in mixed-sex classes, and the least benefit for high-achieving girls and boys and for boys in some classes which were particularly difficult to discipline. (Parker & Rennie, 1995, p. 8)

Interestingly, Parker and Rennie (1995) found that the teachers used different strategies in the all-male and all-female classes and that they became more aware of the different needs of girls and boys that may have been missing in coeducational settings.

In the United States, Becker (2001) described the circumstances behind the establishment of public, single-gender academies on a single campus that had previously housed a coeducational high school. Due to a lack of planning and a strong philosophical undertaking for the single gender academies, Becker (2001) felt that the opportunity to experiment with curriculum and instructional approaches to enhance females' learning opportunities had been wasted, particularly in the mathematics classroom.

> The findings discussed above are consistent with the conclusion of Forgasz, Leder, and Vale (2000), who summarized several studies on single-sex mathematics interventions:
>
> Simplistic solutions, such as single sex classes per se, do not appear to have been successful in themselves

in achieving equity and there have been calls for new strategies to be explored. . . . Epistemology, pedagogy and parents' perceptions remain important factors. (p. 323)

Equity and Streaming/Tracking/Setting for Mathematics

The debate on the advantages and disadvantages of streaming/tracking/setting students generally is a long and bitter one (e.g., Edwards, 1998; Loveless, 1998). Over time, mathematics appears to be one of the few disciplines that has attracted continued support for classroom segregation by achievement level, often termed *ability grouping*. In various national and international settings, streaming/tracking/setting of students is practiced to different degrees either systemically or informally at all levels of education. Segregation can occur in several different ways: within-classroom groupings; grade levels within schools with students ranked and divided into homogenous groups, or "topped and tailed" (that is, creaming off the highest and the lowest achievers, with the rest assigned to fairly mixed groups); within-school course selection tracks, for example, academic and technical tracks; within-system tracks, for example, academic (university) bound schools and technical schools.

When streaming/tracking/setting occurs, sometimes the core mathematics curriculum offered is said to be the same, with more time for extending beyond the core for the higher achievers; at other times the curriculum is clearly differentiated with less demanding mathematics offered to the lower streams (Oakes, 1995) and less abstract, theoretical mathematics and more "practical" mathematics offered to those in the technical tracks. The equity-related questions that arise as a consequence of streaming, tracking, or setting, particularly when they occur during the compulsory years of schooling, include the following:

- Who is found in and who teaches the lower streams and technical tracks?
- Are there differences in enrollments in the lower streams and technical tracks by gender, race/ethnicity, social class, or language capabilities?
- Is a differentiated mathematics curriculum equitable?
- Have students' life opportunities been limited by the practices of streaming/tracking/setting?

According to Oakes (1995), minority students are overrepresented in low-track classes and underrepresented in advanced courses and programs for gifted and talented students. Schools also tend "to place their least qualified teachers with low-ability classes" (p. 87), and expectations of the students in these classes are low. Moody (2001) studied the social and cultural factors in the mathematics learning experiences of two successful female African American students. Both were exposed to tracking at various levels of their schooling. Through tracking, one of the students was isolated from her African American peers who were in lower tracks, many of whom "associated being in high tracks with acting White" (p. 271). Moody concluded that the girls' "experiences exemplify the attributes of tracking that maintain and perpetuate inequalities and inequities in society . . . [and that] tracking seems to *resegregate* our mathematics classrooms, our schools, and our society" (p. 272).

Citing a report from the media, Bartholomew (2002) reported that by the upper secondary years, 94% of students in Britain are taught in setted (streamed) mathematics classes and that

> the composition of the different groups is sharply polarised along social class lines, with middle class students concentrated in the higher sets and working class students in the lower sets. . . . In mathematics, it is also the case that boys are frequently over-represented in top set classes. (p. 138)

Bartholomew (2002) also noted that boys continued to outperform girls in top-set mathematics classes.

Even in *high* sets (upper streams), some students have been found to be disadvantaged (Boaler, 1997). Boaler reported that the students' lower-than-expected achievement levels and high anxiety levels, particularly among those from lower socioeconomic backgrounds, were due to the procedural teaching approaches adopted and the fast pace of lessons.

Linchevski and Kutscher (1998) reported findings from three related studies involving Israeli students' learning of mathematics in mixed-ability and same-ability settings. They found that students' mathematics achievements need not be compromised by learning in heterogeneous ability settings. Any increase in the achievement gap due to learning in homogeneous settings was said to be due to the lowered achievements of weaker students rather than the higher achievements of stronger students. There were also positive effects on teachers' attitudes towards mixed-ability mathematics teaching as a consequence of participation in the study.

Equity in Quality of Teaching and Learning

To attain equity in the mathematics classroom, Burton (1996) wrote of a *socially just pedagogy* that involved capitalizing on the heterogeneity in the classroom. Burton saw the need to "celebrate the diversity to be found in classrooms rather than attempting to turn all pupils into carbon copies" (p. 141) or matching "every pupil against a desired norm" (p. 142). Expanding on this, she maintained that "the clash of meanings, interpretations, styles of learning, methods, representations, knowledges, uses for mathematics—all of these, and others I have failed to mention—underpin a rich culture in every classroom" (Burton, 1996, p. 141). For this model of a socially just pedagogy, Burton (1996) claimed that there was a need to change one's understandings of the nature of mathematics and the conditions supporting its learning, that is a shift from an objective, context-neutral, value-free view of mathematics to one in which mathematics is accepted as a human construct with an emphasis on relationships rather than properties.

Ladson-Billings (1995) defined the notion of an *equity pedagogy* as "the opportunities that *all* children have to benefit from classroom instruction" (p. 130). She described several learning contexts in which inequities in teaching have been identified:

> In the single classroom a teacher can deny some students access to educational opportunities while providing special opportunities for others. Sorting, grouping, or tracking students into lower levels where they receive minimal or no instruction from the teacher, while the higher-track students are presented with challenging and intellectually stimulating curriculum and instruction, is an example of how such inequity is structured. (Ladson-Billings, 1995, p. 130)

What can be inferred from the arguments of both Burton and Ladson-Billings is that equity in the mathematics classroom is most likely to be attained in the heterogeneous classroom in which teachers deal successfully with all aspects of the diversity among students—gender, race/ethnicity, language capabilities, social class, special needs, and the full range of mathematics achievement levels. Some researchers exploring equity issues have worked in heterogeneous classrooms. Others have focused on one dimension of equity by examining more homogenous settings including, for example, single-sex classes (homogeneity of gender) and tracked or streamed classes (homogeneity of achievement). One could argue that the Burton and Ladson-Billings models require teachers to have sound mathematical content knowledge and an exceptional range of pedagogical skills. Classroom based research on equity has also involved a focus on pedagogical approaches. In other studies, preservice teacher education and ongoing professional education needs have been explored. Representative examples of the range of research undertaken related to mathematics classrooms and equity are described below.

In a year-long ethnographic study, Zevenbergen (2001) analyzed classroom interactions in two Australian classrooms—one in a nongovernment school with a middle-class and elite clientele; the other in a government school serving children of working-class parents. Analyses of episodes with examples of triadic dialogue—teacher asks question, student responds, teacher evaluates response—revealed that the middle-class students participated in the exchanges, whereas the working-class children seemed to resist or failed to recognize the structure of this form of interaction. Zevenbergen (2001) proposed that "one subtle and coercive way that some students are advantaged and others excluded in and through the practices of mathematics is through the dis/continuities between the linguistic habitus[1] of the students and the practices of classroom interactions" (p. 213).

Whereas Zevenbergen focused on social class, Khisty's (1995) concern in another ethnographic study was with Hispanic students whose mathematics achievements were consistently below the U.S. national average. Rather than focusing on deficit models that assume the need to change students, Khisty (1995) argued for a paradigm shift premised on school failure that involved inadequate and inappropriate instruction given to disadvantaged poor and minority students. Five fully bilingual elementary classroom teachers in schools with high Hispanic populations were observed at various times over a year; 7 to 10 hours of videotape were gathered from each and the classroom discourse analyzed. Two patterns of teacher discourse were identified. The first involved teachers' efforts in attending to the language of mathematics, and effective and ineffective techniques were noted. The second pattern that emerged concerned the minimal use of Spanish in the mathematics context, with the exception of one teacher. This had occurred even though many students either did not speak or were weak at English. The second pattern is unlikely to have surprised Barwell (2002), who argued that even in mathematics education research, the disadvantage of those who speak less favored languages—and English is often the default, favored language—can be reinforced.

But what happens when students from a range of different language backgrounds are found in the one mathematics classroom and a common language of communication is needed? In the Australian setting, where English is the dominant language, Thomas

[1] Zevenbergen (2001) defined *habitus* as "the embodiment of culture and the lens through which the world is interpreted" (p. 202).

(1997) proposed a range of school-and classroom-based strategies to address the needs of these students: providing more time through extra classes; recognizing that mathematics symbols are not universal, and that in English, mathematical language can have different forms; developing concepts before language; and teaching the four language skills of speaking, listening, reading, and writing within the mathematics classroom.

Carey, Fennema, Carpenter, and Franke (1995) claimed that the development of *diverse curricula for diverse groups* and of *one curriculum for all* were both failed curricular strategies aimed at achieving equity in mathematics education. They maintained that research findings suggested that certain universals in the ways people learn mathematics cut across cultures and that when these were taken into account learning was improved. The *Cognitively Guided Instruction* (CGI) program (see, for example, Carpenter, Fennema, Franke, Levi, & Empson, 1999), they claimed, was one that blends equity concerns with knowledge derived from mathematics education research about how people learn mathematics and contributes to the development of equitable classrooms. Its success was dependent on teachers being concerned about every child and willing to challenge and improve their teaching practices.

Berry (2003) presented findings from three studies in which mathematics classroom practices were consistent with the NCTM (2000) standards and argued that they provided evidence that such classrooms had positive effects on the mathematics achievement of African American students.

Gutstein (2003) described his own attempts, over a 2-year period, to teach mathematics for social justice to a class of Latino students. The three main goals were to help students develop their own social and political consciousness, sense of agency, and social and cultural identities. To achieve these goals, Gutstein (2003) used a curriculum based on the *Mathematics in Context* (Freudenthal Institute) and real-world projects. Gutstein (2003) concluded that although the process was complex, it was possible to teach mathematics so that students could "read the world" with mathematics, develop mathematical power, and change their attitudes towards mathematics. The main factor contributing to the students' growth, Gutstein (2003) contended, was the classroom environment engendered, in which meaningful issues of justice and equity were discussed.

In another self-study, Lubienski (2000) used a problem-centred curriculum and pedagogy with a focus on socioeconomic backgrounds to see how students reacted to learning mathematics through problem solving. The lower socioeconomic background students, particularly the females, indicated a strong preference for traditional instruction. They were the most frustrated with the openness of the problems and the teaching approach. Lubienski (2000) was careful not to generalize beyond the sample of one class of students who participated in the study. It was suggested, however, that reform-oriented pedagogy might improve low and high socioeconomic students' understanding of mathematics while, at the same time, increasing the gap in performance in favor of the high socioeconomic students.

Equity and Affective Factors

The affective domain may not have received as much attention as the cognitive domain among mathematics education researchers. Nonetheless its relevance to mathematics learning is clearly recognized, and the interaction of affect and cognition, for example, was embodied in the research studies reported by McLeod and Adams (1989). With respect to equity, affective factors were clearly identified in the past as contributors to observed gender differences favoring males in mathematics achievement and participation rates (e.g., Fennema & Leder, 1993; Leder, 1992). More recently, research on affective factors associated with mathematics teachers, and with students of mathematics, has been associated with a range of equity dimensions. Findings from some of this work are discussed below.

Zevenbergen (2002) gathered data from 50 Australian teachers on their beliefs about teaching mathematics to socially disadvantaged students. She found that "there was a heavy emphasis on deficit models and a pathologizing of students and families" (p. 148) and that the teachers had low expectations of the students. The literature on beliefs, Zevenbergen (2002) argued, would suggest that the teachers' practices would reflect their beliefs, with social differences becoming educational ones.

The beliefs of large samples of Australian and U.S. students about the gender stereotyping of mathematics were examined by Forgasz (2001). Data were also gathered on preservice teachers' views of how students would respond to the same sets of items. For some aspects of mathematics learning, the students' views were inconsistent with previous research, for example, girls were considered more likely than boys to enjoy mathematics and to be good at it. However, the preservice teachers' views on how students would respond were consistent with the predicted gendered directions of responses based on earlier findings in the field. Interestingly, in both countries, the views of the students were similar, as were the views of the preservice teachers; this was considered partially attributable to the similarities in the mainstream cultural and social dimensions in Australia and the United States. Forgasz (2001) concluded that the preservice teachers' views were "out of touch" with contemporary high school students' views on the gender stereotyping of mathematics.

Atweh, Bleicher, and Cooper (1998) examined the effects of students' socioeconomic backgrounds and gender on teachers' perceptions of their needs and abilities. The teachers and students in two Grade 9 classes participated in the study; one was an all-girls class in a low socioeconomic school, and the other an all-boys class in a high socioeconomic school. Based on interviews and classroom observations, two conclusions were reached. First, the teachers' conduct of their classes was consistent with their perceptions of the students. Second, both teachers were aware of their students' gender and socioeconomic backgrounds. Classroom interactions, it was argued, were consistent with teacher perceptions, and this would "tend to have a self-fulfilling role for teacher expectations" (p. 80).

Ortiz-Franco and Flores (2001) claimed that in the absence of adequate support, many teachers have been successful in providing quality mathematics education for Latino students. Others, however, were said to have "biased views and low expectations for their Latino students that negatively influence their teaching and consequently affect student learning" (p. 245). In their study, two intervention programs had been successful in challenging and changing teachers' attitudes, beliefs, and classroom practices with consequential positive effects on students. Findings from a 3-year study on equity and technology use in secondary mathematics learning indicated that teachers who believed there were differences in the ways boys and girls worked with computers considered girls to be less confident, competent, and interested in using computers than boys (Forgasz, in press).

In Britain, deaf children's mathematics achievements are lower than those of hearing children, and there has been no change in that situation in more than two decades (Nunes & Moreno, 2002). An intervention program focused on visual representations was undertaken with deaf pupils aged 7–11 in deaf schools or schools with deaf units. The researchers found that the students in the program had higher achievement levels than a similar group of deaf children from the same schools from the previous year who had not been exposed to the program. Cautious in their conclusions about the effects of the intervention, Nunes and Moreno (2002) attributed the outcomes to a combination of cognitive and motivational factors; the students were said to have responded positively to the intervention and to have enjoyed working with the booklets. Similar results were found in the research of Barham and Bishop (1991), which used computer-generated visual imagery for teaching key logical connectives.

Rodd (2003) highlighted the role of affect in the mathematics learning of children with "special needs," those with "sense impairments, . . . medical, mobility and developmental conditions, . . . and children who have suffered adverse social conditions" (p. I-127). In her short article, she discussed the relevance of several theoretical frameworks for the analysis of the role of affect in mathematics learning. None, she claimed, included an explanation of the relationship between learning styles and affect. Teaching special-needs students, she maintained, "demands acknowledgment of their specific learning styles otherwise frustration and possibly anger or panic arise" (Rodd, 2003, p. I-129). Although no research evidence was provided, she suggested that autistic children, impaired with respect to their grasp of social situations, "may be more comfortable accessing mathematics via pattern and logic rather than via a (social) context" (p. I-129), and that kinaesthetic activities might suit attention-deficit hyperactive students whose "need to move can be channelled into embodying mathematical relationships" (p. I-129).

ISSUES IN RESEARCHING ACCESS AND EQUITY

In this section we reflect on the research studies mentioned in the previous sections and take a critical view on them. The aim however is not to critique individual studies but to critique generally research in this area, and thereby to help plan worthwhile future studies. Many studies have relevance for the issues of access and equity in mathematics education, and our aim is not to repeat the conclusions from them, but to raise questions that apply to any research in this field, and that we feel have been relatively ignored in the literature.

Research Focus

We must first deal with the issues surrounding the research focus itself. Research is undertaken for various reasons, but in this field of equity in mathematics education, the key motivating force is usually to challenge inequitable practices, and to promote instead equity and access in all their facets. That is, the research focus is often on issues related to aspects of inequitable practices.

However much research undertaken in mathematics education has generated relevant data and also raised possible implications for questions of access and equity. For example, large-scale international as well as at the local national-level quantitative comparative studies such as TIMSS and NAEP often have data on disadvantaged students and indeed may help to identify specific populations suffering inequities. The problems however are about the focus of that research, and if that focus is not on equity itself, the data may be skewed because of sampling, the tests given, the data collecting procedures, and so on.

We maintain therefore that important questions need to be raised whenever equity is discussed in relation to a particular study:

- What is the research focus?
- What is the background to the research?
- Who is doing the research, and why?
- What is the position of the researcher(s), and what equitable (or inequitable) assumptions lie hidden beneath the study and its data?
- Who is funding the research, and what are the equity assumptions behind the funding agency's stance?

Research Questions

The quality of a research study rests first with the quality of the research questions asked. These questions can be generated from a variety of sources, but hopefully they build on the knowledge gained from previous studies, by extending the range of the earlier studies, by dealing with an aspect revealed by the previous study, or by challenging a finding from a previous study. The questions asked will clearly relate to the research focus referred to above and will themselves be subject to the kinds of assumptions underlying that focus.

Another approach to generating the research questions is to base them overtly on a theoretical model or educational position. In access and equity research that is a common approach and, as we have shown in this chapter, many theoretical positions are available to researchers. There is however the danger that competing theoretical models breed competing research studies, which produce data that are sometimes hard to reconcile or even compare. A good example of this is the difference between psychological and sociocultural models of underachievement that have given rise to very different research studies (see, for example, in gender-related research). Although this issue exists in any field of research, it seems to be particularly prevalent in equity research, where there are many competing explanations for the presence of inequitable achievements and educational practices.

A third aspect is the extent to which the research study is about access and equity in *mathematics* education. It is true that for most types of educational disadvantage, the equity issues are similar for all school subjects, but it is not true for all types of disadvantage. For example, blind and deaf students face specific challenges related to the subject of mathematics, and gender-related studies have also shown specific aspects of disadvantage because the subject of concern is mathematics. So although nonspecific subject research studies may be of value in interpreting some aspects of nonaccess and inequity within mathematics education, researchers should not necessarily rely on studies of that type to improve access and equity in mathematics education.

Thus some of the questions that need to be raised about any research study, but particularly with any equity-related mathematics research study, are the following:

- What are the research questions based on?
- What previous research has been done in the area?
- What is the theoretical position underlying the research questions?
- How are the questions related to access and equity issues in education?
- How are the questions related to access and equity in *mathematics* education?
- Who is defining these research questions?
- Who is defining equitable and inequitable practices?

Research Participants

If the research study is empirically based, there are many issues concerned with the participants of the research. The field of access and equity is plagued by issues of definition, with the categorizing and labeling of disadvantaged students being the most challenging area. Typically it is the students who are labeled, although the research context is also labeled rather than described, with the consequences of institutionalised self-fulfillment described by McDermott (1974, 1996).

This is the reason for exercising caution in the research process in this area. Whereas it is relatively easy and convenient to use low-inference categories such as *age*, *language*, *place of residence*, and *gender*, researchers must still recognize that the meaning and relevance of those terms are problematic, and that they need to be problematized in any study. They however are relatively easy to deal with compared with research that involves such high-inference categories as *sightedness*, *class*, *indigenous culture*, or even *Afro-American*.

This is also the area where ethical aspects of research studies come to the fore, involving sampling issues, the selection of participants, confidentiality, feedback, and the rights of those participants and their guardians in the research study. The significant questions to be raised in this section are

- Who is the focus of the research?
- Who else is involved but not directly?
- How are they defined, labeled or described?
- What sampling and selection procedures are followed?
- What confidentiality agreements are in place?
- What access do the participants have to the data about them?
- What benefits will accrue to them as a result of the research?

Methodology Issues and Multiple Perspectives

Issues of sampling and definition of the participants are part of the methodology of the research, but it is important to reflect on other specific issues concerning the choice of methodology in research on equity and access. For example, large-scale quantitative studies can certainly point to differences in access and achievement, but they may not reveal a great deal about the reasons for those differences. For that one must turn to more qualitative studies involving case studies and interviews.

In addition, large-scale studies rarely reveal what approaches are beneficial in overcoming inequity and improving access. Small-scale experiments can serve that purpose far better, although once again the involvement of participants in experimental procedures raises ethical issues. Multiple perspectives and approaches are to be recommended to overcome the shortcomings with any one type of research approach.

Experimental work raises another issue of importance that concerns the research site, also involving considerations of the sociocultural context in which the experimental work is done, together with any political pressures that may be applied. Of course these considerations apply equally to any experimental research, but in relation to inequity and access they are far more significant. Whenever there is consideration of unequal opportunities to learn or other inequitable practices, there are histories and contexts to these practices, and there are people who have established and who may be upholding these practices. One has only to remember the violence associated with the race issue in education in the southern United States to understand the significance of these issues when planning any experimental research into overcoming inequitable practices.

It should also be acknowledged that in some contexts and settings (albeit frequently small-scale models of successful practice) access and equity issues are minimal. Too often, researchers are concerned with identifying problems and ignore the opportunities to research and identify factors that contribute to these success stories.

Thus the significant questions to be raised here are

- What is the sociocultural context of the research?
- What are the histories of the situation or the practice?
- Who are the holders of those histories?
- Who is behind any proposals for change?
- Who are the stakeholders in any future situation?
- Who has the most to lose, or to gain from the research?
- How will the different goals of the stakeholders be balanced in the research?
- What can be learnt from models of successful practice?

FINAL WORDS

In this chapter we have explored access and equity issues in mathematics education from a range of perspectives. Terminology and definitions have been highlighted. Illustrative research findings have been presented that cover a range of issues identified as contributing to inequities including achievement differences, policy and societal factors, the intended and implemented mathematics curriculum, and institutional practices. We concluded the chapter by raising research-related issues that must be considered if research in this field is to provide directions for access and equity to be achieved, and for there to no longer be groups of learners for whom there remain impediments to the learning of mathematics.

REFERENCES

Adler, J. (2001). Resourcing practice and equity: A dual challenge for mathematics education. In B. Atweh, H. Forgasz, & B. Nebres (Eds.), *Sociocultural research on mathematics education. An international perspective* (pp. 185–200). Mahwah, NJ: Erlbaum.

Allexsaht-Snider, M., & Hart, L. E. (2001). "Mathematics for all": How do we get there? *Theory into Practice, 40*(2), 93–101.

Apple, M. (1995). Taking power seriously: New directions in equity in mathematics education and beyond. In W. G. Secada, E. Fennema & L. B. Adajian (Eds.), *New directions for equity in mathematics education* (pp. 329–349).Cambridge, UK: Cambridge University Press

Arnold, N. (1994). *Disability, mainstreaming and academic progress in prep grade: The case of reading and mathematics.* Unpublished master's dissertation, Royal Melbourne Institute of Technology, Australia

Ascher, M. (1981). *Ethnomathematics—A multi-cultural view of mathematical ideas,* Pacific Grove, CA: Brooks/Cole.

Atweh, B., Bleicher, R. E., & Cooper, T. J. (1998). The construction of the social context of mathematics classrooms: A sociolinguistic analysis. *Journal for Research in Mathematics Education, 29*(1), 63–82.

Australian Association of Mathematics Teachers. (1998). *Policy on numeracy education in school.* Adelaide, Australia: Author.

Barham, J., & Bishop, A. (1991). Mathematics and the deaf child. In K. Durkin & B. Shire (Eds.), *Language in mathematical education: Research and practice* (pp. 179–187). Philadelphia: Open University Press.

Bartholomew, H. (2002). Negotiating identity in the community of the mathematics classroom. In. P. Valero & O. Skovsmose (Eds.), *Proceedings of the Third International Mathematics Education and Society Conference* (pp. 133–143). Centre for Research in Learning Mathematics, the Danish University of Education: Roskilde, Denmark.

Barwell, R. (2002). Linguistic discrimination and mathematics education research. In P. Valero & O. Skovmose (Eds.), *Proceedings of the Third International Mathematics Education and Society Conference* (pp. 154–164). Centre for Research in Learning Mathematics, the Danish University of Education: Roskilde, Denmark.

Becker, E. (1963). *Outsiders: Studies in the sociology of deviance.* New York: The Free Press.

Becker, J. R. (2001). Single-gender schooling in the public sector in California: Promise and practice. In B. Atweh, H. Forgasz, & B. Nebres (Eds.), *Sociocultural research on mathematics education. An international perspective* (pp. 367–378). Mahwah, NJ: Erlbaum.

Berry, III, R. Q. (2003). Mathematics standards, cultural styles, and learning preferences. The plight and the promise of African American students. *The Clearing House, 76*(5), 244–249.

Bishop, A. J. (1988). *Mathematical enculturation.* Dordrecht, The Netherlands: Kluwer.

Boaler, J. (1997). When even the winners are losers: Evaluating the experiences of 'top set' students. *Journal of Curriculum Studies, 29,* 165–182.

Burton, T. (1988, September). The self-concept of MIH children: Special school placement versus mainstreaming. In *Fifth guidance conference: Proceedings,* Banyo Seminary, Brisbane, Australia.

Burton, L. (1996). A socially just pedagogy for the teaching of mathematics. In P. F. Murphy & C. V. Gipps (Eds.), *Equity in the classroom. Towards effective pedagogy for girls and boys* (pp. 136–145). London: The Falmer Press.

Carey, D. A., Fennema, E., Carpenter, T. P., & Franke, M. L. (1995). Equity and mathematics education. In W. G. Secada, E. Fennema, & L B. Adajian (Eds.), *New directions for equity in mathematics education* (pp. 93–125). Cambridge, UK: Cambridge University Press.

Carpenter, T. P., Fennema, E., Franke, M. L., Levi, L., & Empson, S. B. (1999). *Children's mathematics. Cognitively guided instruction.* Portsmouth, NH: Heinemann.

Carraher, T. N., Carraher, D. W., & Schliemann, A. D. (1985). Mathematics in the streets and in schools. *British Journal of Educational Psychology, 3,* 21–29.

Cawley, J. F., & Miller, J. H. (1989). Cross-sectional comparisons of the mathematical performance of children with learning disabilities: Are we on the right track toward comprehensive programming? *Journal of Learning Disabilities, 22*(4), 250–254, 259.

Cole, M. (1996). *Cultural psychology: A once and future discipline.* Cambridge, MA:Harvard University Press.

D'Ambrosio, U. (1986). Socio-cultural bases for mathematical education. In M. Carss (Ed.) *Proceedings of the Fifth International Congress on Mathematical Education* (pp. 1–6). Boston, MA: Birkhauser.

D'Ambrosio, U. (1994). Cultural framing of mathematics teaching and learning. In R. Biehler, R. W. Scholz, R. Strasser, & B. Winkelmann (Eds.), *Didactics of mathematics as a scientific discipline* (pp. 443–455). Dordrecht, The Netherlands: Kluwer.

D'Ambrosio, U. (1998). *Literacy, matheracy and technocracy—the new trivium for the era of technology.* Plenary paper presented at the Mathematics, Education and Society, an International Conference, Centre for the study of mathematics education, Nottingham, UK.

D'Ambrosio, U., & Ascher, M. (1994). Ethnomathematics: A dialogue. *For the Learning of Mathematics, 14*(2), 36–43

Damerow, P., Dunkin, M. E., Nebres, B. F., & Werry, B. (1984). *Mathematics for all.* Paris: UNESCO.

Edwards, B. (1998). *Tracking re-visited: Old measures for new times.* Paper presented at the annual conference of the Australian Association for Research in Education, Adelaide. Retrieved August 1, 2005, from http://www.aare.edu.au/98pap/edw98006.htm

ERIC Development Team. (2002). *Helping students with disabilities participate in Standards-based mathematics curriculum.* ERIC/OSEP Digest. (ERIC Document No. ED468579).

Fennema, E. (1993). Justice, equity and mathematics education. In E. Fennema & G. C. Leder (Eds.), *Mathematics and gender* (pp. 1–9). Brisbane, Australia: Queensland University Press.

Fennema, E., & Leder, G. C. (Eds.). (1993). *Mathematics and gender.* Brisbane, Australia: Queensland University Press.

Fenwick, L. T. (1996). *A perspective on race equity and science and math education: Toward making science and math for all.* Paper presented at the Annual Conference of the Georgia Initiative in Mathematics and Science, Atlanta. [ERIC Document No. ED402194].

Forgasz, H. J. (in press). Teachers, equity, and computers for secondary mathematics learning. *Journal for Mathematics Teacher Education.*

Forgasz, H. J. (2001, April). *Mathematics: Still a male domain? Australian findings.* Paper presented at the annual meeting of American Education Research Association [AERA] as part of the Symposium Mathematics: Still a male domain? Seattle, WA [ERIC Document No. ED452071].

Forgasz, H. J., & Leder, G. C. (1995). Single-sex mathematics classes: Who benefits? *Nordisk Matematik Didaktik (Nordic Studies in Mathematics), 3*(1), 27–46.

Forgasz, H. J., Leder, G. C., & Vale, C. (2000). Gender and mathematics: Changing perspectives. In K. Owens & J. Mousley (Eds.), *Research in mathematics education in Australasia 1996–1999* (pp. 305–340). Turramurra, NSW, Australia: Mathematics Education Research Group of Australasia Inc.

Frankenstein, M. (1995). Equity in mathematics education: Class in the world outside the class. In W. G.Secada, E. Fennema, & L. B. Adajian (Eds.), *New directions for equity in mathematics education* (pp. 165–190). Cambridge, UK: Cambridge University Press.

Frankenstein, M. (1998). Reading the world with maths: Goals for a critical mathematical literacy curriculum. In P. Gates & T. Cotton (Eds.), *Proceedings of the First International Mathematics Education and Society Conference* (pp. 180–189). Centre for the Study of Mathematics Education, Nottingham University, UK.

Freudenthal, H. (1983) *Didactical phenomenology of mathematical structures.* Dordrecht, Holland: Reidel.

Gellert, U., Jablonka, E., & Keitel, C. (2001). Mathematical literacy and common sense in mathematics education. In B. Atweh, H. Forgasz, & B. Nebres (Eds.), *Sociocultural research on mathematics education. An international perspective* (pp. 57–76). Mahwah, NJ: Erlbaum.

Gerdes, P. (1995). *Ethnomathematics and Education in Africa.* Stockholm, Sweden: Institute of International Education, Stockholm University.

Gill, J. G. (1988). *Which way to school? A review of the evidence on the single sex versus coeducation debate and an annotated bibliography of the research.* Canberra, Australia: Curriculum Development Centre.

Goals 2000: Educate America Act. (1994). Retrieved December 12, 2003, from http://www.ed.gov/legislation/GOALS2000/TheAct/index.html(Editor: this is how it was stated in the WWW)

Government of Canada. (n.d.). Numeracy roundtable Queen's University—Improving numeracy in Canada. Retrieved February 20, 2006 from: http://www.hrsdc.gc.ca/asp/gateway.asp?hr=en/hip/lld/nls/Publications/B/improve-b.shtml&hs=lxa

Gutstein, E. (2003). Teaching and learning mathematics for social justice in an urban, Latino school. *Journal for Research in Mathematics Education, 34*(1), 37–73.

Hart, L. (2003). Some directions for research on equity and justice in mathematics education. In L. Burton (Ed.), *Which way social justice in mathematics education* (pp. 27–49). Westport, CT: Praeger.

Hennesy, S., & Dunham, P. (2002). Equity issues affecting mathematics learning using ICT. In J. Winter & S. Pope (Eds.), *Research in mathematics education* (Vol. 4, pp. 145–165). London: British Society for Research into Learning Mathematics.

Huang, G. G. (2000). Mathematics achievement by immigrant children: A comparison of five English-speaking countries. *Education Policy Analysis Archives, 8*(25). Retrieved December 12, 2003, from http://epaa.asu.edu/epaa/v8n25/

Jablonka, E. (2003). Mathematical literacy. In A. J. Bishop, M. A. Clements, C. Keitel, J. Kilpatrick, & F. K. S. Leung (Eds.), *Second international handbook of mathematics education* (pp. 75–102). Dordrecht, The Netherlands: Kluwer.

Jackson, A. (1997).The math wars: California battles it out over mathematics reform. *Notices of the American Mathematical Society, 44*(6), 695–702.

Kassem, D. (2001). Ethnicity and mathematics education. In P.Gates (Ed.), *Issues in mathematics teaching* (pp. 64–76). London: Routledge Falmer.

Keeves, J. P. (1995). The case for international comparisons. In J. L. Lane (Ed.), *Ferment in education: A look abroad* (pp. 169–189). Chicago: University of Chicago Press.

Keitel, C., Damerow, P., Bishop, A. J., & Gerdes, P. (1989). *Mathematics, education, and society.* Paris: UNESCO.

Kelly, R. R., Lang, H. G., & Pagliaro, C. M. (2003). Mathematics word problem solving for deaf students: A survey of practices in grades 6–12. *Journal of Deaf Studies and Deaf Education, 8*(2), 104–119.

Kelly, S., Johnston, B., & Yasukawa, K.(Eds.). (2003). *The adult numeracy handbook.* Sydney, Australia: NSW Adult Literacy and Numeracy Australian Research Consortium.

Khisty, L. L. (1995). Making inequality: Issues of language and meanings in mathematics teaching with Hispanic students. In W. G. Secada, E. Fennema, & L B. Adajian (Eds.), *New directions for equity in mathematics education* (pp. 279–297). Cambridge, UK: Cambridge University Press.

Kitchen, R. S. (2001). The sociopolitical context of mathematics education in Guatemala through the words and practices of two teachers. In B. Atweh, H. Forgasz, & B. Nebres (Eds.), *Sociocultural research on mathematics education. An international perspective* (pp. 151–162). Mahwah, NJ: Erlbaum.

Knijnik, G. (1998). Ethnomathematics and political struggles. *Zentralblatt fur Didaktik der Mathematik, 98*(6), 188–194.

Ladson-Billings, G. (1995). Making mathematics meaningful in multicultural contexts. In W. G. Secada, E. Fennema, & L B. Adajian (Eds.), *New directions for equity in mathematics education* (pp. 126–145). Cambridge, UK: Cambridge University Press.

Leder, G. C. (1992). Mathematics and gender: Changing perspectives. In D. A. Grouws (Ed.), *Handbook of research on mathematics teaching and learning* (pp. 597–622). New York: Macmillan.

Leder, G. C., & Forgasz, H. J. (1997). Single-sex classes in a co-educational high school: Highlighting parents' perspectives. *Mathematics Education Research Journal, 9*(3), 274–291.

Leder, G. C., Forgasz, H. J., & Solar, C. (1996). Research and intervention programs in mathematics education: A gendered issue. In A. Bishop, K. Clements, C. Keitel, J. Kilpatrick, & C. Laborde (Eds.), *International handbook of mathematics education* (Pt. 2, pp. 945–985). Dordrecht, The Netherlands: Kluwer.

Linchevski, L., & Kutscher, B. (1998). Tell me with whom you're learning and I'll tell you how much you've learned: Mixed-ability versus same-ability grouping in mathematics. *Journal for Research in Mathematics Education, 29*(5), 553–554.

Lokan, J., Ford, P., & Greenwood, L. (1996). *Maths & science on the line: Australian junior secondary students' performance in the Third International Mathematics and Science Study (TIMSS Australia Monograph No.1).* Melbourne: ACER Press.

Lokan, J., Ford, P., & Greenwood, L. (1997). *Maths & science on the line: Australian middle primary students' performance in the Third International Mathematics and Science Study (TIMSS Australia Monograph No.2).* Melbourne: ACER Press.

Loveless, T. (1998). *The tracking and ability grouping debate.* Retrieved August 1, 2005, from http://www.edexcellence.net/foundation/publication/publication.cfm?id=127

Lubienski, S. T. (1997). Class matters: A preliminary excursion. In J. Trentacosta & M.J.Kenney (Eds.), *Multicultural and gender equity in the mathematics classroom: The gift of diversity* (pp. 46–59). Reston, VA: NCTM.

Lubienski, S. T. (2000). Problem solving as a means towards mathematics for all: An exploratory look through a class lens. *Journal for Research in Mathematics Education, 31*(4), 454–482.

Lubienski, S. T., & Bowen, A. (2000). Who's counting? A survey of mathematics education research, 1982–1998. *Journal for Research in Mathematics Education, 31*(5), 626–633.

Lynch, K., & Baker, J. (2005). Equality in education. An equality of condition perspective. *Theory and Research in Education, 3*(2), 131–164.

Malloy, C. (2002). Democratic access to mathematics through education: An introduction. In L. D. English (Ed.), *Handbook of international research in mathematics education* (pp. 17–26). Mahweh, NJ: Erlbaum.

McDermott, R. P. (1974). Achieving school failure: An anthropological approach to illiteracy and social stratification. In G.D.Spindler (Ed.), *Education and cultural process: Towards an anthropology of education* (pp. 82–118). New York: Holt, Rinehart and Winston.

McDermott, R. P. (1996). The acquisition of a child by a learning disability. In S.Chaiklin & J.Lave (Eds.), *Understanding practice: Perspectives on activity and context* (pp. 269–305). Cambridge, UK: Cambridge University Press.

McLeod, D. B. (1992). Research on affect in mathematics education: A reconceptualization. In D. A. Grouws (Ed.), *Handbook of research on mathematics teaching and learning* (pp. 575–596). New York: Macmillan.

McLeod, D. B., & Adams, V. M. (Eds.) (1989). *Affect and mathematical problem solving: A new perspective.* New York: Springer-Verlag.

Moody, V. R. (2001). The social constructs of the mathematical experiences of African-American students. In B. Atweh, H. Forgasz, & B. Nebres (Eds.), *Sociocultural research on mathematics education. An international perspective* (pp. 255–276). Mahwah, NJ: Erlbaum.

Moores, D. F. (2001). *Educating the deaf* (5th ed.). Boston, MA: Houghton Mifflin Company.

Morrow, C., & Morrow, J. (1995). Connecting women with mathematics. In P. Rogers & G. Kaiser (Eds.), *Equity in mathematics education. Influences of feminism and culture* (pp. 13–26). London: Falmer Press.

National Council of Teachers of Mathematics (NCTM). (1989). *Curriculum and evaluation standards for school mathematics.* Reston, VA: Author.

National Council of Teachers of Mathematics (NCTM). (2000). *Principles and standards for school mathematics.* Reston, VA: Author.

NCEO Policy Directions. (1995). Opportunity-to-learn standards. NCEO Policy Directions, No.4. Retrieved December 12, 2003, from http://education.umn.edu/nceo/OnlinePubs/Policy4.html

National Center for Education Statistics [NCES]. (2003). *The nation's report card. Mathematics highlights 2003.* U.S. Department of Education, Institute of Education Sciences, National Center for Education Statistics. Retrieved December 12, 2003, from http://nces.ed.gov/nationsreportcard/mathematics/results2003/.

Nunes, T., & Moreno, C. (2002). An intervention program for promoting deaf pupils' achievement in mathematics. *Journal of Deaf Studies and Deaf Education, 7*(2), 120–133.

Oakes, J. (1995). Opportunity to learn: Can standards-based reform be equity-based reform. In I. M. Carl (Ed.), *Seventy-five years of progress: Prospects for school mathematics* (pp. 78–98). Reston, VA: NCTM.

Ortiz-Franco, L., & Flores, W. V. (2001). Sociocultural considerations and Latino mathematics achievement. In B. Atweh, H. Forgasz, & B. Nebres (Eds.), *Sociocultural research on mathematics education. An international perspective* (pp. 233–253). Mahwah, NJ: Erlbaum.

Parker, L. J., & Rennie, L. H. (1995, November). *To mix of not to mix? An update on the single-ses/coeducation debate.* Pre-conference draft. Paper presented at the Australian Association for Research in Education annual conference, Hobart, Australia.

Povey, H. (2003). Teaching and learning mathematics: Can the concept of citizenship be reclaimed for social justice? In L. Burton (Ed.), *Which way social justice in mathematics education?* (pp. 51–64). Westport, CT: Praeger.

Powell, A. B., & Frankenstein, M. (Eds.). (1997). *Ethnomathematics.* Albany: State University of New York Press.

Rennie, L. R., & Parker, L. H. (1997). Students' and teachers' perceptions of single-sex and mixed-sex mathematics classes. *Mathematics Education Research Journal, 9*(3), 257–273.

Robitaille, D. (Ed.). (1997). National contexts for mathematics and science education: An encyclopedia of the education systems participating in TIMSS. Vancouver, BC, Canada: Pacific Educational Press.

Robitaille, D., & Garden, R. (1989). *The IEA study of mathematics II: Contexts and outcomes of school mathematics.* Boston: Pergamon.

Rodd, M. (2003). Special students feeling mathematics. In M. J. Høines, & A. B. Fuglestad. (Eds.), *Proceedings of the 28th conference of the International Group for the Psychology of Mathematics Education* (Vol. 1, pp. I-127–I-129). Bergen, Norway: Bergen University College.

Romberg, T. A., & Shafer, M. C. (2003). Mathematics in context (MiC)—Preliminary evidence about student outcomes. In S. L. Senk & D. R. Thompson (Eds.), *Standards-based school mathematics curricula: What are they? What do the students learn?* (pp. 225–250). Mahwah, NJ: Erlbaum.

Rousseau, C. K., & Powell, A. (2005). Understanding the significance of context: A framework to examine equity and reform in secondary mathematics education. *The High School Journal, 88*(4), 19–31.

Schmidt, W. H., McKnight, C. C., Valverde, G. A., Houang, R. T., & Wiley, D. E. (1997). *Many visions, many aims. Volume 1: A cross-national investigation of curricular intentions in school mathematics.* Dordrecht, The Netherlands: Kluwer.

Secada, W. G. (1991). Diversity, equity, and cognitivist research. In E. Fennema, T. Carpenter, & S. J. Lamon (Eds.), *Integrating research on teaching and learning* (pp. 17–53). Albany: State University of New York Press.

Secada, W. G. (1992). Race, ethnicity, social class, language and achievement in mathematics. In D. A. Grouws (Ed.), *Handbook of research on mathematics teaching and learning* (pp. 623–660). New York: Macmillan.

Secada, W. G., Cueto, S., & Andrade, F. (2002). Opportunity to learn mathematics among Arymara-, Quechua-, and Spanish-speaking rural and urban fourth- and fifth-graders in Puno, Peru. In L. Burton (Ed.), *Which way social justice in mathematics education* (pp. 103–132).Westport, CT: Praeger.

Siegler, R. S. (2003). Implications of cognitive science research for mathematics education. In J. Kilpatrick, W. B. Martin, & D. E. Schifter (Eds.), *A research companion to principles and standards for school mathematics* (pp. 219–233). Reston, VA: National Council of Teachers of Mathematics.

Silver, E. A. (2003). Attention deficit disorder? *Journal for Research in Mathematics Education, 34*(1), 2.

Skovsmose, O. (1994). *Towards a philosophy of critical mathematics education.* Dordrecht, The Netherlands: Kluwer.

Steele, M. M. (2004, Spring). A review of literature on mathematics instruction for elementary students with learning disabilities. *Focus on Learning Problems in Mathematics.* Retrieved November 15, 2005, from http://findarticles.com/p/articles/mi_m0NVC/is_2_26/ai_n6154494

Steen, L. A. (Ed.). (1990). *On the shoulders of giants: New approaches to numeracy.* Washington, DC: National Academy Press.

Steen, L. A. (Ed.). (2001). *Mathematics and democracy: The case for quantitative literacy.* Washington, DC: National Council on Education and the Disciplines.

Thomas, J. (1997). Teaching mathematics in a multicultural classroom. Lessons from Australia. In J. Trentacosta (Ed.), *Multicultural and gender equity in the mathematics classroom. The gift of diversity. 1997 yearbook.* Reston, VA: National Council of Teachers of Mathematics.

Tracey, D. K. (2002). *Self-concepts of pre-adolescents with mild intellectual disability.* Unpublished doctoral dissertation, University of Western Sydney, Australia.

UNESCO. (1990). *World declaration on education for all.* Bangkok, Thailand: UNESCO Regional Office for Education in Asia and the Pacific.

van Kraayenoord, C., Elkins, J., Palmer, C., & Rickards, F. (2000). *Literacy, numeracy and students with disabilities* (Vols. 1–4). Canberra, Australia: Commonwealth Department of Education, Training and Youth Affairs. Retrieved September 9, 2002, from http://www.gu.edu.au/school/cls/clearinghouse/

Willis, M. G. (1992). Learning styles of African-American children: Review of the literature and interventions. In A. K. H. Burlew, W. C. Banks, H. P. McAdoo, & D. A. Azibo (Eds.), *African-American psychology* (pp. 260–278). Newbury Park, CA: Sage.

Willis, S. (1996). Gender justice and the mathematics curriculum: Four perspectives. In L. H. Parker, L. J. Rennie, & B. J. Fraser (Eds.), *Gender, science and mathematics. Shortening the shadow* (pp. 41–51). Dordrecht, The Netherlands: Kluwer Academic.

Woodrow, D. (2003). Mathematics, mathematics education and economic conditions. In A. J. Bishop, M. A. Clements, C. Keitel, J. Kilpatrick, & F. K. S. Leung (Eds.), *Second international handbook of mathematics education* (pp. 9–30). Dordrecht, The Netherlands: Kluwer.

Zeleke, S. (2004) Learning disabilities in mathematics: A review of the issues and children's performance across mathematical tests. *Focus on Learning Problems in Mathematics,* 26(4), 1–18.

Zevenbergen, R. (2001). Mathematics, social class, and linguistic capital: An analysis of mathematics classroom interactions. In B. Atweh, H. Forgasz, & B. Nebres (Eds.), *Sociocultural research on mathematics education. An international perspective* (pp. 201–215). Mahwah, NJ: Erlbaum.

Zevenbergen, R. (2002). Teachers' beliefs about teaching mathematics to students from socially disadvantaged backgrounds: Implications for social justice. In L. Burton (Ed.), *Which way social justice in mathematics education* (pp. 133–151). Westport, CT: Praeger.

AUTHOR NOTE

We would like to thank Sarah Lubienski and Carol Malloy for their valuable comments on an early draft of this chapter. We are also grateful to Sue Wotley for her assistance in searching the literature.

27

RESEARCH ON TECHNOLOGY IN MATHEMATICS EDUCATION

A Perspective of Constructs

Rose Mary Zbiek, M. Kathleen Heid, and Glendon W. Blume

THE PENNSYLVANIA STATE UNIVERSITY

Thomas P. Dick

OREGON STATE UNIVERSITY

Educators have witnessed explosive growth in the availability of computing technology for mathematics classrooms during the last quarter of a century, and this growth has been accompanied by unbridled enthusiasm for the potential of new technologies in the teaching and learning of mathematics (Fey et al., 1984). A parallel interest in understanding the impact of these technologies on learning, teaching, and curriculum has engulfed the research community. Perhaps the very passion that stimulates research and the growth of new technologies also accounts for why there are many interesting studies and a need for explicit ways to integrate those studies into a solid body of research that both spurs continued scholarship and informs practice. This chapter elaborates on how constructs may be useful tools to further research and enlighten classroom instruction, curriculum matters, and tool development.

The chapter begins with background issues regarding technology in mathematics education and a discussion of what constructs are and how they can be used to further develop, interpret, and apply research. It continues with discussion of constructs related to several key areas for existing and potential research: technological tools and technology-based mathematical activities, students' behavior in the context of technology, teaching issues related to technology in mathematics education, and effects of technology on mathematical curriculum content.

SETTING THE STAGE

Connecting Past and Future

In the previous *Handbook*, Kaput (1992) documented the continuing growth of electronic technology and predicted, "major limitations of computer use in the coming decades are likely to be less a result of technological limitations than a result of limited human imagination and the constraints of old habits and social structures" (p. 515). At that time, Kaput portrayed structures that conceivably underpinned the user's interaction with a broad range of computing technologies, but the structures were not yet actualized. Rather than predicting where hardware and software will go in the years to come, we consider where research has been and may go, including a ubiquitous role of technology in work since 1992 and in work that takes us back more than 15 years.

The 14 open questions that Kaput proposed in 1992 have been addressed with varying degrees of interest and of success. One complication is the extent to which research on the teaching and learning of mathematics in technological settings has been premised on the development and accessibility of sophisticated computing tools and appropriate and willing testing sites—both of which take time and resources different from what is required in many areas of mathematics education research. As a consequence, research on technology-intensive mathematics teaching and learning has only recently begun to mature into a well-articulated area of scholarship. As is true of many areas of research, further growth in this arena will depend on the development and articulation of a shared understanding of the phenomena under consideration. As research in this area has developed, researchers have noted, explicitly or implicitly, constructs that describe teaching and learning in these environments. This chapter does not propose a specific agenda for future research on technology in mathematics education but rather submits that scholarly discussion and activity should be framed around constructs that advance collective understanding of past, present, and future work. An attempt to further the development of a shared language of constructs can inform and guide research on the teaching and learning of mathematics in the context of technology.

Technical and Conceptual Activities

Advancing the collective wisdom about the role of technology in mathematics education requires careful distinctions between two different kinds of mathematical activity: technical and conceptual. The technical dimension of mathematical activity is about taking mathematical actions on mathematical objects or on representations of those objects. Procedures can then be built out of sequences of mathematical actions (or out of previously built procedures). Examples of technical mathematical activity include geometric construction and measurement, numerical computation, algebraic manipulation, graphing, graphical transformation, translation between notation systems, solving equations, creating diagrams, displaying, collecting, sorting, and so on. Conceptual mathematical activity involves understanding, communicating, and using mathematical connections, structures, and relationships. Examples of conceptual mathematical activity include finding and describing patterns (inductive reasoning), defining, conjecturing, generalizing, abstracting, connecting representations,[1] predicting, testing, proving, and refuting.

Technical mathematical activity is concerned primarily with tasks of mechanical or procedural performance, whereas conceptual mathematical activity is concerned with tasks of inquiry, articulation, and justification. The potential for student access to technology to influence both the technical and conceptual dimensions of their mathematical activity has been suggested by Borwein (2005) in his description of experimental mathematics (Borwein & Bailey, 2003) in which the computer is used for (a) gaining insight and intuition, (b) discovering new patterns and relationships, (c) graphing to expose mathematical principles, (d) testing and especially falsifying conjectures, (e) exploring a possible result to see whether it merits formal proof, (f) suggesting approaches for formal proof, (g) replacing lengthy hand derivations with tool computations, and (h) confirming analytically derived results.

For researchers interested in the impact of technological tools on mathematics learning and teaching, the distinction between the technical and conceptual dimensions of mathematical activity is both intriguing and problematic. Kaput (1992) noted that "off-loading" routine computations provides a learning efficiency in terms of compacting and enriching experiences. For example, computers or calculators can enable students to examine several graphs in a relatively short amount of time. In turn, this compacted experience allows the student to focus on the common attributes of a family of functions or on the consequences of varying a parameter. The compacted technical activity thus affords an opportunity for enriched conceptual activity. Technology that facilitates connectedness and sharing of results heightens those affordances.

As Artigue (2002) noted, there is a temptation for researchers to explicitly or implicitly treat technical and conceptual dimensions of mathematical activity as though they were fundamentally opposed types of activity. This dualistic epistemological stance tends to view technical activity as entirely mechanical and thus largely devoid of meaning, whereas meaningful learning must be derived from conceptual activity. In terms of its implications for integrating technology

[1] As many who have studied representations have noted (see Goldin & Kaput, 1996, for a discussion of representation), there is a need to distinguish internal representations from external representations. We humans have no direct access to internal representations, and we infer their existence from students' actions on external representations, which might better be termed inscriptions or written artifacts. In this text, when we use the term representation, we mean to indicate an external representation unless stated otherwise.

into mathematics classrooms, such a stance suggests the hypothesis that the primary influence of computing technology is to "free" students from all technical work so they may engage directly in conceptual activity. However, as elaborated by Lagrange (1999) and Hoyles, Noss, and Kent (2004), technical activity using computing technology may involve a combination of routine mechanical actions informed by conceptual reasoning. As such, a line cannot be drawn neatly between technical activity and conceptual activity that distinguishes clearly which is more mathematically meaningful. The underlying and perhaps synergistic relationship between technical and conceptual activities is assumed in this chapter.

Cognitive Technological Tools in Mathematics Education

A discussion of the technical and conceptual dimensions of mathematical activity informs the use of technology in mathematics education, but it does not explain why the potential influence of computer technology may be profoundly different from that of noncomputer tools such as physical manipulatives or drawing instruments. Research questions focused on the influence of technology on mathematics learning and teaching can be sharpened and refined with attention to special differences and contributions technology makes in the mathematical tasks students undertake, the kinds of mathematical activity in which they engage, and the mathematical knowledge and understandings they build.

Pea (1987) used the term "cognitive technologies" for those technologies that help "transcend the limitations of the mind . . . in thinking, learning and problem-solving activities" (p. 91). To facilitate the technical dimension of mathematical activity, a cognitive tool for mathematical activity must allow the user the means to take actions on mathematical objects or representations of those objects. To facilitate the conceptual dimension of mathematical activity, reactive visual feedback becomes critically important (e.g., Hillel, Kieran, & Gurtner, 1989). A cognitive tool must react in response to the user by providing clearly observable evidence of the consequences of the user's actions "at the surface of the screen" (Balacheff & Kaput, 1996; Balacheff & Sutherland, 1994).

The use of the term cognitive tool in this chapter is intended to include a broad range of technologies used for either the technical or conceptual dimensions of mathematical activity. Distinctions among types of cognitive tools may vary among researchers, and tools having a wide array of uses often do not fit cleanly into any given categorization scheme. For example, although the term microworld is used extensively in the literature on technology in mathematics education, it is not at all clear that authors share a common definition. Microworlds can vary widely in complexity, ranging from simple systems of virtual manipulatives having a narrow set of objects and actions with limited rules and consequences to exploratory statistical worlds and dynamical geometry systems having a rich array of objects and actions as well as flexible rules that allow for complicated sequences of actions to be captured for repetition or animation.

Balacheff and Kaput (1996) drew a distinction between a microworld and a simulation, the latter of which they described as providing a virtual model of some real-world phenomena, including parameter inputs that can be entered or controlled by the user. The resulting effects of varying the parameter values are then observable by the user. In some cases, a microworld might also be considered a simulation (provided its domain of phenomenology provides a suitable virtual model of some real-world phenomena). The use of the term cognitive tool in this chapter is intended to be inclusive of both microworlds and simulations as well as representational toolkits (Dick, 1992) such as computer algebra systems (CAS), graphing calculators, and spreadsheets that provide a variety of computational tools for dealing with symbolic expressions, graphs, and numeric data. Such tools can include direct access to real-world physical phenomena, either through data collection probes—through interfaces such as the Microcomputer or Calculator Based Laboratory (MBL or CBL)—or through "drivers" of external devices (such as remotely controlled robots). In each case, these tools help transcend the limitations of the mind.

In addition to cognitive tools, other forms of technology are designed to facilitate mathematics learning, teaching, and teacher development. Many of these technologies are cognitive in nature, but they are not mathematical tools. For example, intelligent cognitive tutors, such as the geometry and algebra tutors developed at Carnegie Mellon University (Anderson, Boyle, Corbett, & Lewis, 1990; Ritter, Haverty, Koedinger, Hadley, & Corbett, 2007), provide an alternative learning situation—as tutors, not as tools (even though students have access to a collection of tools within such tutors). Video formats, including the videodisc technology of *The Adventures of Jasper Woodbury* (Zech et al., 1998), allow students to encounter and pursue problems situated in real-world contexts. In these cases, a primary goal of the technology is problem presentation and branching, not use as a problem-solving tool. Online textbooks, online courses, and Internet communities such as the Math Forum (Renninger, Weimar, & Klotz,

1998) offer avenues to serve and connect students and teachers in new ways and in distant settings. Collections of video cases of teaching, including those connected with the Third International Mathematics and Science Study (TIMSS; TIMSS Video Mathematics Research Group, 2003), offer opportunities for teachers and researchers to make teaching an object of study (Cestari, Santagata, & Hood, 2004). Although these and other forms of technology are part of the growing literature in mathematics teaching and learning and could also be used to make the case for constructs, we intentionally drew examples for this chapter from the substantial body of work done regarding cognitive tools to articulate the argument for the usefulness of constructs to conceptualize studies, to connect existing empirical studies, to unfold theories, and to provide a language with which to talk about practice informed by empirical findings and theoretical insights.

What Are Constructs?

What do we mean by *constructs*? *The American Heritage Dictionary* (Pickett, 2000) defines construct as a "concept, model, or schematic idea" or "a concrete image or idea" (pp. 394–395). Our constructs are concepts that have explanatory power regarding technology and the teaching and learning of mathematics. The constructs we have selected for this chapter are ones that have specific applications to mathematics, that have an empirical basis, and that help one understand relationships among tool, activity, students, teacher, and curriculum content.

The constructs on which we will focus are those that are abstracted from research, those that arise from within emerging theories, and those that have arisen from rich descriptions of practice. Although some are specific to particular types of technology or arose in the data analysis of a particular study, the constructs are essentially explanatory tools that transcend a variety of venues. Each construct shows promise in many different settings, including different types of technology, different mathematical content domains, and different groups of learners. Each construct may start to define or redefine dimensions or stages or capture an underlying component of a larger phenomenon.

Our purpose is to raise configurations of promising constructs to the consciousness of the mathematics education community and to suggest ways in which these constructs have been and can be useful to researchers. Constructs can help frame and influence research and enhance the coherence and connectedness of findings. Use of particular constructs can help refine research questions by drawing attention to structure, relationships, and features related to the roles of technology in the mathematical tasks and activities in which students engage, to the behaviors in which students engage as they work on mathematical tasks in the presence of technology, to the roles of teachers in technology-intensive mathematics education, and to the nature of mathematical content that is learned in technology-intensive curricular worlds. A construct can serve in qualitative and quantitative settings in a variety of ways, including as an independent variable, a dependent variable, a covariate, an explanatory factor, or as the object of study. In addition to their usefulness in research, the chosen constructs are informative for classroom practice and software development.

The rest of this chapter identifies particularly useful constructs in looking at how tools affect aspects of the teaching and learning of mathematics. As suggested by the thick lines added as one moves from the left-hand side of Figure 27.1 to the right-hand side of that

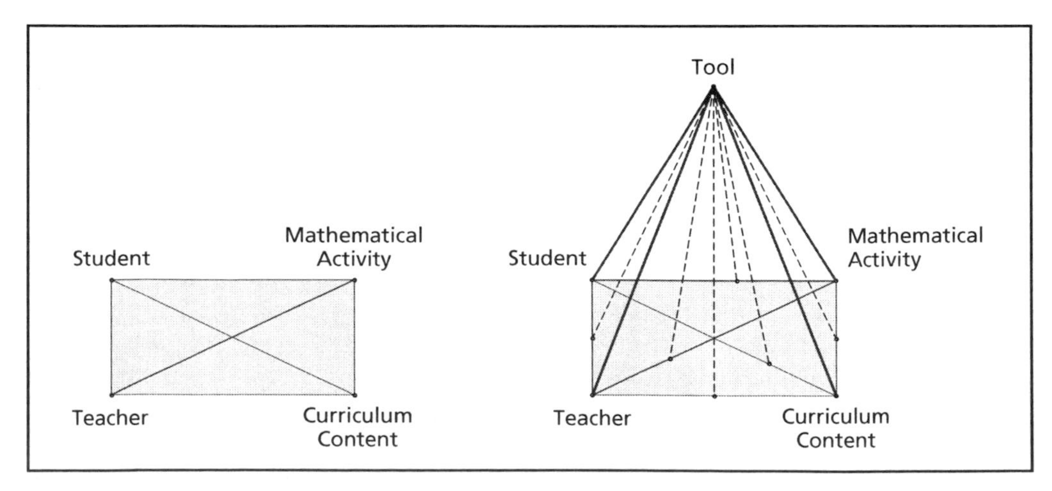

Figure 27.1 Mediation relationships among technology, student, teacher, mathematical activity, and curriculum content.

figure, each of student, mathematical activity, teacher, and curriculum content is mediated by the use of and configuration of the tool. Tool use and configuration also mediate the relationships between any two of the four areas, as denoted by the dotted lines in the right-hand side of Figure 27.1. The remainder of the chapter is organized around the four areas. In the discussion of each area, we identify constructs that address how technology mediates that area and how technology mediates relationships of that area to other aspects of the learning and teaching process. Each section articulates both how these constructs enrich an understanding of existing and potential research and how selected constructs are interconnected. At the end of the chapter, we identify seven specific ways in which constructs inform research on technology in the teaching and learning of mathematics and illustrate each way with constructs discussed in the chapter.

TOOLS AND MATHEMATICAL ACTIVITY

Certain characteristics of cognitive tools can serve as useful constructs for both mathematics education researchers and tool designers. The three constructs discussed here—externalized representation, mathematical fidelity and cognitive fidelity—relate to the ways cognitive tools can provide special opportunities or impediments for learning in the context of mathematical activity. Special attention is drawn to two types of fidelity of the externalized representations of mathematical objects provided by or accessed by the cognitive tool and the technical actions on and linkages to those representations available to the user. Mathematical fidelity refers to the faithfulness of the tool in reflecting the mathematical properties, conventions, and behaviors (as would be understood or expected by the mathematical community). Cognitive fidelity refers to the faithfulness of the tool in reflecting the learner's thought processes or strategic choices while engaged in mathematical activity.

Externalization of Representations—Dynamic Actions and Linkages

Cognitive tools play a special role in mathematical activity by externalizing representations (Heid, 1997). Through externalized, though limited, surrogates for a student's internal mental representations displayed on the surface of the screen, *externalized representations* become visible phenomena that can be shared and discussed with others (e.g., other learners or the teacher). By bringing such representations literally to the surface, a cognitive tool can allow for unique opportunities for exposing cognitive conflicts. "As mathematical thinking in the face of technology is examined, it will be important not to lose sight of the degree of effectiveness of each application of technology in facilitating students' externalization of representations" (Heid, 1997, p. 8).

Once an external mathematical representation has been created or accessed, the cognitive tool can act as a "user agent" (Kaput, 1992) to perform specific mathematical actions or procedures at the command of the student. Looking at externalization through this idea of user agency suggests research questions that center on a role given to the student—that of tool manager or director. For what kinds of mathematical tasks do students choose to use the technology? How do students use the technology to carry out these tasks? For example, this pair of questions comprised a major strand of a report synthesizing the research findings on handheld graphing technology (Burrill et al., 2002). The authors indicated that this technology was used primarily as a computational and visualization tool and for tasks involving moving between external representational forms of functions, but they also noted that student choices in using the technology were heavily mediated by the teacher. The researchers reported finding few studies (e.g., Ruthven, 1990) that examined what students might do differently because of the access to the handheld graphing technology.

Actions taken on externalized representations with a cognitive tool differ in important ways from actions taken with manipulatives or physical tools. The actions taken with a physical tool are less constrained and are not necessarily as mathematically meaningful as actions on a cognitive tool. For example, a student could stack base-10 blocks vertically to make a toy tower or use a compass point to poke holes in a piece of paper. The extent to which actions with a physical tool are mathematically meaningful is at the discretion of the user. Furthermore, even when an action with a physical tool has consequences with mathematical meaning, that meaning may go unnoticed or be misinterpreted by the student. Physical tools do not automatically react to a user's action with feedback. The student must work to extract feedback from interactions with a physical tool, perhaps with assistance or direction from a teacher.

In its role as a user agent, a cognitive tool can constrain the possible actions on an external representation to be ones that are potentially mathematically meaningful, and it can enforce mathematical rules of behavior of the objects on which it acts. The cognitive tool can also respond to the actions chosen by the student by automatically making potentially mathematically meaningful consequences of the actions more overtly apparent. Of course, the question of whether the user of a cognitive tool recognizes or correctly interprets an action or its consequences as mathematically meaningful remains, but these constraints and the reactive feedback provided lend greater support for sense-making by the student. Thus, unlike the physical tool, the cognitive tool provides a "constraint-support system" (Kaput, 1992) for mathematical activity. Kaput claimed that the computational contribution of a cognitive tool becomes most apparent in the area of multiple representations when it is used to establish a "hot link," that is, a dynamic connection between two representations such that an action taken in one representation is automatically reflected in the other. For example, the dilation of a geometric figure in a dynamical geometry program might be accomplished through a "point, click, and drag" action by the user. If there is a hot link to a visually displayed numerical measurement of some attribute of the figure, such as area or perimeter, then changes in the displayed measurement value are dynamically updated continuously as the figure is dilated. Such a hot link achieves a particularly high degree of immediacy in the reactive feedback provided to the user.

Because at least two representations are involved, Kaput (1992) noted that the directionality of a hot link should be considered. In the geometric example mentioned above, the described hot link is unidirectional—the user action is taken on the geometric figure with two related, observable consequences: change in its size and change in the related numerical measurement. A distinctly different hot link is created if, instead, the user action is taken on the numerical measurement (for example, by typing in a specific value in a display field) with the resulting change in value triggering a change in the object by dilation (or some other transformation parameterized by the measurement value). Of course, a hot link could be bidirectional if it offers the option of taking action in either of the two linked representations with the consequences automatically reflected in both. Bidirectionality would, of course, depend on univalence of the mapping in both directions.

In this discussion, we have made the case that cognitive tools provide special environments offering not only access to external representations, but also specific opportunities and "rules of engagement" for the learner's mathematical activity with those representations. In the section on students and mathematical activity later in this chapter, we introduce research constructs that draw sharper distinctions among the ways in which a learner may act within and interact with a technological environment. In the present section, our focus is on characteristics of the cognitive tool itself. For the mathematics education researcher to make sense of the learner's mathematical experiences in such an environment, we suggest that it is important to attend to the faithfulness of those representations to the mathematics accepted by the wider community and to the cognitive intentions of the user.

Mathematical Fidelity

In order to function effectively as a representation of a mathematical "object," the characteristics of a technology-generated external representation must be faithful to the underlying mathematical properties of that object. The actions that can be taken on that representation, and the consequential behavior resulting from those actions, should reflect accurately the expected mathematical characteristics and behavior. The degree to which this is achieved is a measure of the *mathematical fidelity* of the tool (Dick, 2007).[2] The desirability for a high degree of mathematical fidelity may seem obvious, but ambiguities of context and technological limitations may make it hard to achieve. The desirability for "user friendliness" may compete with faithfulness to mathematical structure for the tool designer's priorities.

Dick (2007) used the term mathematical fidelity to point out that the mathematics of the tool does not always represent mathematics as it is understood by the mathematics community. This construct draws attention to the notion that the mathematics students experience in working with technology may not be that which they are intended to experience. There is a range of ways that the disparities accentuated by mathematical fidelity may play out in technology-based mathematics teaching and learning. These ways include differences in the underlying mathematics of seemingly similar computer programs or differences between the mathematics that is intended in a visual display and the possible different interpretations of that display.

Consider, for example, typical calculator graphs of the sine function with increasingly small periods. Figure 27.2 shows a screen capture representation of the graph of such a function. This technology-generated graph of $f(x) = \sin(100x)$ suggests a periodicity and pattern of behavior that does not reflect those of the function rule.

[2] Later in this chapter, we discuss a different construct of *mathematical concordance* to account for mismatches among the mathematics of the tool, the teacher, the mathematics of the task, and the mathematics of the student.

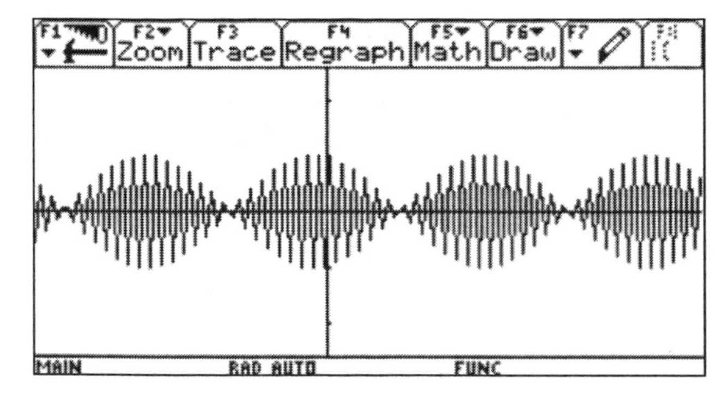

Figure 27.2 Screen capture representation of the graph of the function with rule $f(x) = \sin(100x)$.

While the term itself has not been used widely by researchers to date, the idea behind mathematical fidelity appears repeatedly, particularly in describing phenomena at the interface between the learner and the tool. Dick (2007) noted three areas in which a lack of mathematical fidelity can emerge. First, a lack of mathematical fidelity can emerge through discrepancies between tool and mathematical syntax conventions. For example, the common convention of hierarchy of operations for arithmetic operations may be violated for implied multiplication in the interests of ease of user entry of expressions. Second, lack of mathematical fidelity can emerge through underspecification in mathematical structure. For example, the tool may allow "functions" to be defined by algebraic expressions without specified domains. This can lead to such anomalies as the graph of $f(x)=|\ln(x)|$ (absolute value of the natural logarithm of the real variable x) suggesting a larger domain of real numbers for this function than the domain of real numbers for $f(x)=\ln(x)$. Finally, lack of mathematical fidelity can emerge through limitations in representing continuous phenomena with discrete structures and finite precision numerical computations.

Syntax discrepancies or underspecification of structure have potential for generating confusion for learners and teachers, but it is unclear whether they pose mere nuisances or more serious impediments. When they arise in the midst of classroom activity, it is plausible that the impact on learners may be largely mediated by the teacher's reactions and interpretations and how such events shape the attitude of the teacher toward technology. Difficulties dealing with the limitations due to discrete structures and finite precision (called "discretisation processes" by Artigue, 2002) have received attention from researchers, especially in the study of function representations and limits. In a study of students exploring asymptotic behavior of rational functions using the Function Supposer, Yerushalmy (1997) noted the "discrepancy between technology as a support for visual perception and the manipulation of mental objects, but a numerical tool that cannot support the concept of 'approaching infinity'" (p. 3). Yerushalmy pointed out that the graphical investigation of horizontal asymptotes is inherently thwarted by the impossibility of exploring the entirety of an infinite domain—at the threshold of the precision limitations of the machine, a horizontal asymptote cannot be distinguished from a horizontal tangent. One of the students in Yerushalmy's study came up with an example that illustrates the difficulty of distinguishing a removable discontinuity from a vertical asymptote when relying on visual evidence: The values of

$$f(x)=\frac{10^{12}-x^2}{10^6-x}$$

near $x=10^6$ are very large (visually suggesting a vertical asymptote), but

$$\lim_{x\to 10^6} f(x)=2\cdot 10^6.$$

Even steep portions of continuous function graphs (such as that of $f(x)=x^2$) could visually suggest vertical asymptotes to students.

Hart (1991) conducted extensive task-based interviews with students using graphing technology with a multiple-representation-based calculus curriculum. Several of these tasks were designed specifically to create situations in which a machine-generated representation (usually graphic or numeric) might suggest information contradictory to a student-generated representation (usually algebraic). These tasks included (a) numerical investigations of limits such as

$$\lim_{x\to 0}(1+0.5x)^{1/x} \text{ or } \lim_{x\to 0}\frac{1}{1+2^{1/x}}$$

in which substitution of a sequence of small numerical values for x (0.001, 0.000001, 0.0000000000001) would produce conflicting information due to round-off precision limitations; (b) graphical investigations of functions such as $f(x)=x^3(x-1)^4$ and its derivative $f'(x)=x^2(x-1)^3(7x-3)$, in which the graphs at typical scaling could hide critical point information; and (c) graphical consideration of definite integrals such as $\int_0^{3\pi/4}\sec^2(x)dx$ in which a misapplication of the Fundamental Theorem of Calculus would yield a result conflicting with the graph of the function over the interval of integration. Hart examined how the student arrived at a resolution of these apparent conflicts. In many cases, the source of the contradictory information was essentially a breach of mathematical fidelity in the machine's representation due to either precision limitations or visual anomalies related to the discrete screen representation of a continuous graph.

Hart found that students' resolutions of such situations were influenced by their confidence in their own algebraic skills, with students having stronger algebra skills less likely to consider a machine-generated graph as authoritatively correct. Interestingly, Hart also found that students with strong algebraic skills using a traditional curriculum without technology were more reluctant to use graphical representations even when the representations provided much more accessible information than a symbolic formula.

One issue centering on mathematical fidelity arises when the mathematics underlying particular tools is not evident to the user. Having to program into the software some behavior not defined by the mathematical system being modeled, software engineers sometimes inadvertently create their own mathematical systems. In dynamical geometry tools, for example, software designers make decisions about the behavior of "arbitrarily placed" points when the figure on which they are placed is dragged (Goldenberg, Scher, & Feurzeig, 2006). Different programmers make different decisions and produce ostensibly similar microworlds that act differently in selected cases because they are governed by different rules. The mathematics of the tool can become the object of study resulting in new constructs such as that of *Dynamic Geometry* as an area of mathematics. Colette and Jean-Marie Laborde (2007, p. 44) described activity in this new area of mathematics:

> Instead of considering these problems solely as design problems, the researchers involved in the Cabri project recognized these problems as theoretical problems about a new kind of geometry, a dynamical geometry with its own objects and relations. A new field of research about the fundamentals of such a geometry was established. In particular, the theoretical problems pertaining to the axiomatic development of Dynamic Geometry remain under discussion at present. Contributions are being made, some of which were discussed specifically in a vibrant European meeting in Germany in December, 2000. (Gawlick & Henn, 2001, as cited in Laborde & Laborde, 2007)

Dynamical geometry software creates new opportunities to model physical or mechanical motion governed by geometrical constraints (e.g., gear linkages, rolling objects). The simulation of the physical motion can be modeled in a dynamic drawing of the construction as the user drags a point or other element on screen. González-López (2001) studied "the fidelity of the construction by comparing the moving drawings with the real mechanism's performance" (p. 130). González-López echoed the concern of Hölzl (1996, as cited in González-López, 2001) that the demands on dragging (direct manipulation via mouse or stylus device) to provide a faithful simulation go beyond geometry: "The geometrical laws supporting the motion of pictures have no direct link with the physical motion laws nor the continuity of an analytic model" (p. 130).

As technology becomes an increasingly integrated part of school mathematics, careful analysis of issues of mathematical fidelity will be needed. This type of research will necessitate intense collaboration involving mathematicians, computer scientists, and mathematics education researchers.

Cognitive Fidelity

In the context of intelligent tutoring systems, Beeson (1989) has used the term *cognitive fidelity* to refer to the degree to which the computer's method of solution resembles a person's method of solution. Dick (2007) has adopted the term in a somewhat wider context to include the degree to which the (cognitive) tool actions explicitly reflect the user's cognitive actions. Hence, if the external representations afforded by a cognitive tool are meant to provide a glimpse into the mental representations of the learner, then the cognitive fidelity of the tool reflects the faithfulness of the match between the two. In turn, a tool having a high degree of cognitive fidelity holds particular promise to researchers in making visible the thinking of the user.

Olive, Steffe, and others have exploited the idea of a constraint-support system construed to support cognitive fidelity in their research on children's development of rational number concepts through the design of a collection of computer microworlds known collectively as TIMA (Tools for Interactive Mathematical Activity; see Biddlecomb, 1994; Olive, 2000; Olive & Steffe, 1994; Steffe & Olive, 2002). These cognitive tools offer on-screen manipulatives analogous to counters, sticks, and fraction bars. The possible actions on these objects that are afforded to the user are explicitly tied to the development of mental numerical schemes identified in extensive empirical research of children's thinking. These actions include unitizing, uniting, fragmenting, segmenting, partitioning, replicating, iterating, disembedding, and measuring. Hence, the TIMA technology is designed to provide cognitively faithful representations and actions.

Olive and Lobato (2007) made the case that a cognitive tool such as TIMA can allow cognitively faithful actions more easily than physical tools:

> One way that technology can enhance the learning of rational number concepts is through the use of com-

puter tools that allow students to enact psychological operations that are difficult to perform with physical materials. In order to establish a relation between a part and a whole in a fractional situation, the child needs to mentally disembed the part from the whole. With physical materials it is not possible to remove a part from the whole without destroying the original whole. With static pictures the part is either embedded in the whole or is drawn separate from the whole. In the former case the child has to mentally unitize one part of the whole while maintaining the unity of the whole and compare these two abstracted units. In the latter case the child has to compare the separate units while imagining that one is embedded in the other. Using a computer tool that provides the child with the ability to dynamically pull a part out of a partitioned whole while leaving the whole intact, the child can enact the disembedding operation that is necessary to make the part-to-whole comparison. This example of one way in which technology can enhance learning draws upon the psychological analysis of rational numbers.

Cognitive fidelity is a particularly important consideration for researchers. By providing action choices to the learner that faithfully reflect potential cognitive choices, tools such as the TIMA technology can provide to researchers more powerful evidence of patterns in children's thinking. In turn, an improved understanding of children's thinking can better inform continuing development of the tools.

An Illustration Involving Parameter Manipulation

Further research into the relationship between cognitive fidelity and the effectiveness of hot links among external representations used as learning aids can offer insights not only to mathematics education researchers but also to instructional designers and engineers of cognitive tools. To appreciate some of the complex issues involved and how the constructs discussed here may be helpful, one might consider a common type of hot-link activity that exploits technology: parameter manipulation with dynamic consequences in another representation. Dynamic virtual simulations of physical phenomena often offer exactly this kind of hot link.

Parameter manipulation is also used widely in investigating families of functions using graphing technology. We will use tool-design decisions about parameter manipulations to illustrate how issues of mathematical and cognitive fidelity can raise research questions. In such an activity, the learner manipulates a representation of the value of a parameter in the symbolic expression for a function—such as the value of the parameter a in the family of functions defined by $f(x) = \sin(ax)$—and observes the resulting graphical representation of that function. The learning goal is usually an explicit recognition of the mathematical connection between the parameter and some salient characteristic of the function (such as the period). Although simultaneous visual access to both the symbolic form and the graph of the function seems naturally desirable in a technological implementation of this activity (say with a split screen or two adjacent windows onscreen), there are timing issues related to the parameter changes. Specifically, if a student types in a new parameter value, does the graph change while the student is typing or only after the student signals completion of the change (perhaps by clicking or pressing an ENTER button)? In either case, the screen world may temporarily present conflicting mathematical information to the user (such as an "old" graph simultaneously displayed with a "new" symbolic function formula), a breach of mathematical fidelity.

A tool designer might decide to introduce a "slider bar" as a virtual manipulative controlling the value of the parameter. In this setting, the student need not type in a parameter value. Instead, the student moves the slider to change a number-line representation of the value of the parameter, and the graphical representation is hot-linked to transform continuously with the move of the slider bar. Changing the slider position now directly controls both the shape of the graph and the numerical value of the parameter. The intent behind most slider bars is to enable the user to run through many individual parameter values in a short time, effectively animating the visual consequences of the rapidly changing parameter values. However, it is plausible that the learner's dominant cognitive perception of the user action taken is now more kinesthetic (moving a slider) than mathematical (changing a parameter value). In terms of the cognitive fidelity, the slider bar is not simply a convenient alternative to typing in the parameter value. Rather, it is a third external representation—a virtual manipulative that competes for the user's attention with the symbolic and graphical representations. Does the physical sense of moving a slider obscure rather than enhance the desired cognition of a connection between the parameter value and a salient visual consequence? Such a question is an open one with relevance to both the mathematics education researcher and the tool designer.

The discussion here of some of the issues that arise in this particular case of parameter manipulation is intended to suggest just a few of the complexities of the interplay between a cognitive tool and the learner. The constructs of mathematical fidelity and cognitive fidelity can serve not only to provide a useful vocabulary to cognitive scientists for examining some of those complexities but may also highlight tool characteristics that provide insights about learning to both instructional designers and tool developers. For the mathematics education researcher, these notions of fidelity can provide a lens for interpreting or making sense of the learner's experiences within the special environment for mathematical activity provided by a cognitive tool. These fidelity constructs are ones in which cognitive-tool designers might have a keen interest, for their engineering and programming decisions bear directly on both. In addition, curriculum developers and instructional designers who assume that learners have access to a cognitive tool also have a special stake in considering the implications these fidelity constructs have for their intended learning outcomes. Researchers are in a unique position to aid these other stakeholders by using fidelity constructs to highlight issues at the interface between the learner and the machine and by more clearly describing what features of the learning environment a cognitive tool affords.[3]

STUDENT–TOOL RELATIONSHIP

When a student is positioned to engage in mathematical activity in the context of technology a central question is, How will he or she use the technology? A response to this question requires both identification of the point at which the student will choose to bring the technology into play in the process of carrying out the mathematical activity and the determination of the extent to which the technology will function effectively to help the student to accomplish his or her mathematical goal. The constructs of instrumental genesis and technique have productively focused conversation centered on student–tool activities and have concentrated on the relationship between the student and the tool.

Development of Student–Tool Relationships

Instrumental genesis is one of the constructs that have helped frame the conversation about the student–tool relationship. It relates the notion of "instrument" and derives from ergonomic theories emanating from outside of mainstream mathematics education. Verillon and Rabardel (1995) differentiated between an artifact and an instrument, pointing out the need to consider not only the capacity of the artifacts with which one is working and the tasks in which that person is engaged, but also the relationship between the person and the artifact. The notion of instrument, then, is a psychological one, not a description of a material artifact. The artifact does not automatically assume the role of user-agent (with the computer as a vehicle for accomplishing the purpose of the user, Kaput, 1992); the user needs to develop an understanding of the capacity of the artifact and to develop a relationship with the artifact. As Verillon and Rabardel stated, "The instrument does not exist in itself, it becomes an instrument when the subject has been able to appropriate it for himself and has integrated it with his activity" (p. 84). That is, it is not until this relationship is developed that the artifact becomes what Kaput (1992) described as a user-agent. Artigue (2002) noted the individual and social nature of the development of an instrument:

> Thus an instrument is a mixed entity, part artefact, part cognitive schemes which make it an instrument. For a given individual, the artefact at the outset does not have an instrumental value. It becomes an instrument through a process, called instrumental genesis, involving the construction of personal schemes or, more generally, the appropriation of social pre-existing schemes. (p. 250)

As the organizing construct for this conversation about instruments, instrumental genesis is the process of an artifact becoming an instrument (or the process of developing meaningful ways to use the artifact). It is through instrumented action that artifacts evolve into instruments. The question remains, How, specifically, do artifacts such as calculators or computer software become mathematical instruments—tools that the user can employ for his or her own mathematical purposes? This process is not merely a matter of the user becoming acquainted with stan-

[3] A relatively new paradigm of educational research methodology may prove especially appropriate in an arena that attracts the interests of these different communities. *Design experiments* (Brown, 1992; Collins, 1992) employ scientifically based research approaches in complex natural learning settings (such as mathematics classrooms), especially those mediated by some technological innovation (such as cognitive tools). The aim of a design experiment is to explore how the technology affects both student learning and educational practice. Design experiments can inform and provide insights into (a) student cognition and learning, (b) the structure of the instructional environment, and (c) design improvements in the technology itself. This kind of research methodology begs for partnerships and collaborations among educational practitioners, researchers, and technological engineers. However, such partnerships are hard to establish, given that the respective goals of these different communities are not necessarily shared, and the tension that exists between the traditions of scientifically based research approaches and those of naturalistic classroom observation (see McCandliss, Kalchman, & Bryant, 2003).

dard uses of the tool; neither the instrument nor the user is unaffected by the development of the instrumental genesis. As described by Guin and Trouche (1999), "the subject has to develop the instrumental genesis and efficient procedures in order to manipulate the artefact. During this interaction process, he or she acquires knowledge which may lead to a different use of it" (p. 201). Thus, instrumental genesis includes both the user shaping the tool for her or his purposes (instrumentalization) and the user's understanding being shaped by the tool (instrumentation). The core of instrumental genesis in mathematics education is understanding the mathematics of the technology and being able to use it for one's own purposes.

The construct of instrumental genesis is helpful to researchers in examining the role of technology in learning. It explains how technology does not have the same automatic power for all users and how its intelligent use requires both conceptual and technical knowledge. The construct of instrumental genesis was foundational to Drijvers' (2003) report of his investigation of how the CAS (in this case, the Texas Instruments TI-89 calculator) could be used to contribute to higher level understanding of the concept of parameter by 9th- and 10th-grade students. Using instrumental genesis, Drijvers accounted for the role of the CAS in students' learning as a mediator between the physical artifact of the technology and the mental actions of the students. He identified an inventory of obstacles students encountered in developing their relationships with the CAS for the purpose of solving the problems he posed; his analysis of that list of obstacles revealed both conceptual and technical difficulties. For example, one of the obstacles students encountered in the instrumentation process was difficulty in entering expressions into the CAS. Entering expressions requires a conceptual appreciation for the structure of the expressions being entered (e.g., recognizing the need for parentheses in order to avoid entering incorrect expressions) as well as technical knowledge of the notations and conventions of the CAS (e.g., the CAS's treatment of implicit multiplication, such that *ab* is interpreted as a variable name rather than the product of *a* and *b*, and its convention of responding with "$\sqrt{}$(" when the square root key is depressed, requiring the user to enter the ending parenthesis). That the development of full capacity in using a CAS requires both conceptual and technical understanding is not surprising, but the construct of instrumental genesis helped focus attention on this important perspective on learning to use a CAS.

Aspects of Mathematics in Cognitive Tool Use

Attention by researchers to the development of an instrument from an artifact calls for a closer examination of the capacities of the instrument that are to be developed. Thus, concern about instrumental genesis is naturally accompanied by concern about *technique*, the nature of technical capacity that goes beyond rote application of procedures. Examination of this interplay between the constructs of instrumental genesis and technique results in a deeper understanding of both constructs. Guin and Trouche (1999) used the construct of instrumental genesis in their investigation of technique used by 15- to 16-year-olds and 17- to 18-year-olds in CAS environments. Their data helped the researchers identify two phases of instrumental genesis with a CAS calculator. The first phase was a discovery phase in which students were performing actions that helped them detect the function of each key (in this phase students were strongly dependent on the computer, often failing to recognize other relevant information). Students in this phase made almost no use of "understanding tools" (e.g., interpretation, inference, comparison, coordination). In the second phase the students cut down on the number of commands they used, organized their actions using fewer commands, and coordinated their use of those commands. They developed a tendency to seek and use relevant strategies. One of the products of the Guin and Trouche (1999) work is the development of profiles that characterized students' behavior in using the CAS to solve mathematics problems. With their attention focused on particular aspects of instrumental genesis—instrumentation, instrumentalization, and technique—researchers have been better able to articulate the ways in which students develop in their abilities to use technologies productively (students' techniques) as well as identify conceptual and technical obstacles to that development. Being able to characterize instrumental genesis, of course, does not guarantee the ability to develop strong capabilities in instrumented action. Having brought the construct of instrumental genesis both into her work with the teachers in an experimental CAS setting and into the design of the curriculum, Artigue (2002) cautioned, "Nevertheless, the students' instrumental knowledge remained fragile as did the mathematical knowledge they began to build in the field of analysis" (p. 262).

The construct of instrumental genesis can be extended in mathematics education beyond the development of cognitive tools. Although much of the literature on instrumental genesis centers on physically or electronically present tools and technologies, it is equally viable to think of representations as artifacts that function to a greater or lesser extent as instruments. The questions are parallel. How does a representation become an instrument? Is there a predictable trajectory from representation as artifact to representation as instrument? How students act on multiple technologically available representations is a function of both the stage of their personal instrumental genesis and their understanding of the artifact as a representation of some mathematical object. The core of a student's use of external representations is the extent to which the artifact is not a replacement for the object but serves as a representation to the student of the abstract mathematical object. As Guin and Trouche (1999) pointed out, "It is precisely because mathematical objects are not in the tangible world that the differentiation between a mathematical object and its representation is at the heart of the learning process" (p. 207).

A quintessential and prototypical example that points out the importance of distinguishing between representation as the entity of interest and representation as an encryption that particularizes a mathematical object plays out through *figure* and *drawing*. Used to characterize how students think about geometric displays in dynamical geometry sketches, these constructs, figure and drawing, point out the difference between thinking of the referent (drawing as the visible inscription on the computer screen) and thinking of what is being represented (figure as the theoretical object). This "figure/drawing" distinction is not, of course, limited to technological environments; geometry teachers have long worried about students basing their school geometry proofs on some nonessential feature of the accompanying diagram, and students have long wondered why the distinction matters. The confusion is exacerbated in dynamical geometry environments with the temptation to think of the seemingly endless set of displays under dragging as equivalent to the theoretical object rather than representations of that object. With the powerful range of visual representations available in technological environments, constructs like figure and drawing are helpful in focusing research on mathematical concepts and techniques instead of on less meaningful treatments of procedures. The general constructs of instrumental genesis and technique and the prototype constructs of figure and drawing suggest some overarching perspectives on students' relationships to tools. The next section characterizes students' mathematical activity in the presence of cognitive tools.

STUDENTS AND MATHEMATICAL ACTIVITY

Drawing on theory and research on cognitive mathematical tools, we have identified several constructs that are useful when studying what students do as they use technology in learning mathematics. When students use cognitive tools as they engage in mathematical activity they engage in that activity with goals in mind, and in the course of taking actions in pursuit of those goals, they exhibit certain observable behaviors that can be related to their goals. This section will examine several constructs related to students' use of technology in their mathematical activity: expressive activity and exploratory activity and students' work method when using technology. Expressive activity and exploratory activity are a pair of related constructs that are connected to the nature of students' goals, whereas students' work method when using technology is related to behaviors students exhibit while using technology to engage in mathematical activity.

Goals of Student Activity

Students may have a variety of goals as they work with technology, goals that depend in part on the cognitive demand of the mathematical task. For example, if the cognitive demand of the task is of a purely technical nature, the student's goal in using technology might be to use the tool to carry out a computational algorithm, producing a numeric or symbolic result. If, however, the cognitive demand of the task is of a conceptual nature, the student's goal might be to use the tool to create multiple measures of some attribute from which a generalization might result. So, tasks with quite different cognitive levels might be associated with very different goals. Not surprisingly, students' differing goals often are associated with different types of activity when using technology. For example, if they use cognitive tools to create a representation of a mathematical object (e.g., function, geometric figure) or to construct a game or tool, their mathematical activity may differ substantially from what it would be if they were working with an already-created tool or an already-created external representation of a mathematical object. In particular, two distinct types of activity often are apparent: *exploratory* activity and *expressive* activity. These constructs derive from but also differ somewhat from Bliss and Ogborn's (1989) exploratory mathematical modeling (working with a model created by someone else) and expressive mathematical modeling (creating one's own model). Doerr and Pratt (in press) described how exploratory modeling and expressive modeling differ and how they can provide different insights into student learning.[4]

Neither exploratory activity nor expressive activity is a singular entity; each can be thought of as being represented by a continuum. Exploratory activity might range from structured or prescribed activity (e.g., "Press this key. What do you get? Why does that happen?"), to a somewhat less structured exploratory activity (e.g., "Explore this relationship using a triangle, a quadrilateral, a pentagon"), to much less structured activity (e.g., using a specified tool or approach, students create their own examples of a particular mathematical object or instances of a mathematical relationship). Expressive activity might range from students' selection of a numeric or symbolic computation to be performed and a tool to carry out that computation, to creation of a graph for the solution of an equation, to much more open-ended activity such as creating a game for another student, an activity in which students act as a *bricoleur* or *bricoleuse*—one who builds new structures or objects (Papert, 1993). Expressive activity entails the creation and use of external representations (see the section titled *Externalization of Representations–Dynamic Actions and Linkages*) and therefore may be more conducive than exploratory activity to the development of representational fluency (as described in a subsequent section with that title).

When students are given a procedure to carry out, they are engaging in exploratory activity; however, when students decide which procedures to use they are engaging in expressive activity, albeit a somewhat restricted expressive activity in some instances. It is important to note that both exploratory activity and expressive activity can occur in what often are described as "explorations." One type of exploratory activity is "guided exploration," in which the student's goal is to produce a predetermined result that was chosen by the teacher. In contrast, less guided exploration, in which the student selects some aspect(s) of what to investigate and the approach used to investigate it, is characteristic of expressive activity. When students use technology to help answer a question posed by the teacher using a process suggested by the teacher, their (exploratory) activity is quite different from expressive activity in which they attempt to answer a question of their choosing with their choice of process. One sees quite different outcomes when studying what students do when they engage in these two forms of "explorations." One also sees that the nature of students' exploratory or expressive work depends on both the task and the activity. For example, suppose a teacher, textbook, or peer suggests a task and a particular method to accomplish the task. The student who follows the suggested path to address the given task likely is working in an exploratory mode. The student who creates an original way to address the given task may be working in an expressive mode. The student who follows the suggested path but generates a new task within that path also is working in an expressive way.

One type of setting in which expressive activity results might be one in which a student's goal is simply to "play"[5] with the tool, "trying out" various actions that the tool can perform on a particular mathematical object simply to see what kind of results the tool will produce. In an expressive activity such as unstructured play, students can challenge themselves to determine a tool's capabilities and limitations, and they can begin to develop an understanding of how the tool can accomplish what they intend so they can begin the process of converting the tool into an instrument. From the standpoint of instrumental genesis (Guin & Trouche, 1999)—a construct addressed in depth in the *Student–Tool Relationship* section—this play serves to develop in students an intense, personal, and purposeful relationship with the tool, namely, the psychological component that, when paired with the artifact, comprises the instrument.

The perceived benefits from students' initial play with a technological tool derive primarily from theory (e.g., the instrumental approach) rather than from empirical studies that explicitly examined play. However, some research studies that have investigated students' actions when using technology have found that unguided play may not be optimal. Pratt and Ainley (1997) found not only that elementary students produced drawings more often than constructions when given minimal guidance in their introduction to a dynamical geometry tool but also that those drawings did not lead to insights about powerful mathematical ideas that were embedded in the Cabri microworld in

[4] Doerr (2001) and Doerr and Pratt (in press) made the case that in working with a model created by someone else (exploratory modeling in the sense of Bliss & Ogborn, 1989) students may develop facility with an expanded set of representations and develop conceptual understanding by exploring the consequences of their assumptions. When students create their own models (expressive modeling) there is an opportunity to observe particular aspects of their conceptions, because they make explicit those conceptions as they select objects, represent them, and interpret and validate outcomes from their models. Different insights into learning result from observing what occurs when one does what one is directed to do with a tool as opposed to when one initiates what is to be done with the tool.

[5] This type of play differs from the play that occurs when playing a game, in which the structure of the game provides direction to the student.

which they were working. Glass and Deckert (2001) suggested that, rather than allowing students' play to be relatively unguided, it be structured by tasks that focus students' attention on relevant mathematical notions. The structuring of students' exploratory activity provided by or facilitated by cognitive tools can focus students on mathematical characteristics and facilitate symbolic descriptions (Clements & Battista, 2001). Play with cognitive tools may involve expressive activity, giving students some freedom to choose tasks, but it also requires some focusing or orchestration by the teacher. Some additional issues related to play that research might address concern the role of play in the instrumentation process; the effects of different amounts and types of structured, exploratory play and unstructured, expressive play; and the effects of different types of guidance that a teacher might offer as the student–tool relationship develops.

Often many paths are available to a student when using a tool in an exploration related to a mathematical task. This multiplicity of paths leads to the "play paradox" described by Hoyles and Noss (1992) and Noss and Hoyles (1996): Because many cognitive tools offer to students such a wide variety of approaches to solving problems, students might not encounter, in the course of their explorations of a problem with the tool, the particular mathematical ideas that were identified as goals by their teacher or by the developers of the curriculum materials. Students often have many possible ways to proceed when using technological tools, but giving students the freedom to select their own approaches raises the possibility that some of the options available to them enable them to bypass the ideas that their teacher intends for them to encounter in the course of their activity. Noss and Hoyles (1996) contended that when one is engaged in play (of a game), one attends to what one experiences in the game, not necessarily to the rules that govern the play of the game, and they noted that

> if the teacher brings a new idea to the student's attention from *outside* the activity, then he or she is no longer playing: yet, if it is not imported, he or she might never encounter the idea. Play confers meaning to an activity but blurs the specificity of the intended meaning. (p. 71)

This blurring of specificity might (or might not) result in a change in the level of abstraction. During play the student might engage in a higher level of mathematics than that intended by the teacher, activity that might be related to the student's work method (described in the section on *Students' Technology-Specific Behaviors*).

If the teacher simply brings a new idea to the student's attention during play, that intervention might substantially change the activity, from being expressive to being exploratory. Noss and Hoyles (1996) suggested somewhat different responses to the play paradox.

> Our resolution of this paradox, which is of course only provisional, is that our pedagogical interventions were centered upon encouraging pupils to *reflect* on the task at hand. Of course, this is often quite heavy-handed in comparison to the laissez-faire position which some (including, at other times, ourselves) have advocated. (p. 46)

Noss and Hoyles also noted that off-computer discussions were designed to "help children *explicitly* to confront mathematical notions and distinguish them from everyday ones" (p. 54).

Arguments that arise in relation to the play paradox can be viewed as arguments about the exploratory–expressive distinction. At times, the play paradox may result when students treat a task as one involving expressive activity (trying a variety of approaches to see what they produce) while their teacher is expecting exploratory activity (exploring a particular domain using a particular process). This discrepancy can be useful in explaining why outcomes from students' technology use may differ from those that the teacher expects and that the curriculum is structured to promote.

Despite the potential detrimental effects of play to undermine teachers' and curriculum developers' goals as suggested by the play paradox, students' use of technology in a relatively unstructured way also can be quite productive. Harel (1990) and Harel and Papert (1991) provided an example of a very open-ended type of expressive activity that involved students taking on the role of *bricoleur* or *bricoleuse*, in which their goal was to design or produce a tool for other learners. This research centers on the notion of constructionism, which is based on the principle that learning results from a student's construction of artifacts, and that such construction can promote the development of mathematical ideas:

> Learners are particularly likely to make new ideas when they are actively engaged in making some type of external artifact—be it a robot, a poem, a sand castle, or a computer program—which they can reflect upon and share with others. (Harel & Papert, 1991, p. 1)

The work of Harel (1990) and Kafai (1995, 1996) suggests that children learn in several ways through designing a game or instructional software for younger students. They learn through design, for example, attempting to integrate the subject matter into a game;

they also learn through programming, which requires reformulation of their knowledge; and they learn through constructing their own mathematical representations. Kafai (1995) identified two types of content integration in the games children created: extrinsic, in which the designer focused more on the game than on the subject matter, and intrinsic, in which the subject matter was more closely integrated with the game. Kafai noted that in several different experiments most children's games employed extrinsic integration, for example, stopping the play at some point to ask the player a content-related question, rather than integrating the content into the game's context. According to Kafai (1995), a challenge for researchers is to determine how to "engage students in more dialogue about making intrinsically integrated educational games" (p. 298). It is tempting for researchers to assume that with sophisticated technology, students could easily create a game in which mathematics is an integral part; however, students might view them as quite separate. Creation of games that intrinsically embody the mathematics requires students to know different mathematics and understand it in a different way from what is required when playing a game and engaging with mathematics are viewed as being sequential, rather than integrated, enterprises.

We next consider how research might profit from the constructs of exploratory activity and expressive activity. Consider research that reports the use of technology to "investigate" some mathematical idea. To better account for outcomes, instead of characterizing students' activity merely as an investigation using technology, researchers need to analyze students' activity when using technology both in terms of the exploratory–expressive distinction as well as on one of the individual continua, either of exploratory activity or expressive activity. This should provide additional explanatory power to studies that previously have considered "exploration," "investigation," or "inquiry" with technology as a singular entity. By using the distinctions that these two constructs provide, researchers can characterize students' activity more accurately and be able to relate differences in outcomes to the type of activity in which students were engaged, avoiding the possibility of conflicting results that are unexplainable without such a distinction.

Even in response to a particular task, students' goals for using a technological tool may differ greatly, and those goals may be reflected in either exploratory or expressive activity. Researchers who have studied students' use of the dragging feature of dynamical geometry tools have observed that students may have different goals when employing dragging. Arzarello et al. (1998) identified three types of dragging, each with quite different goals (also see Hollebrands, Laborde, & Sträßer, 2006; Rivera, 2005). One type is wandering dragging, a somewhat random type of dragging in which the student's goal is to search for regularities or interesting results that occur when some object is dragged. For example, a student who uses wandering dragging to drag a vertex of a triangle from which an altitude is constructed might observe something about the location of the point representing the foot of the altitude (see Figure 27.3).

In the second type, *lieu muet* dragging (dragging in which the student tacitly or explicitly maintains some condition), the student's goal is to preserve some regularity in the drawing, for example, keeping the measures of one angle and one side of a triangle constant as a vertex is dragged. Figure 27.4 illustrates this type of dragging.

The third type, dragging test (dragging to test a hypothesis), has a different goal, namely, to determine whether a conjecture (the student's or someone else's) is true. For example, in Figure 27.5 a student might test the hypothesis that when $\angle ACB$ is obtuse, the perpendicular from A lies in the exterior of ΔABC by dragging vertex A in such a way as to keep $\angle ACB$ obtuse.

Wandering dragging involves a somewhat unconstrained search for results: What interesting result might one notice when examining many examples of some type of figure? *Lieu muet* dragging also involves a search for results, but with some particular condition constraining those results: What is the outcome when

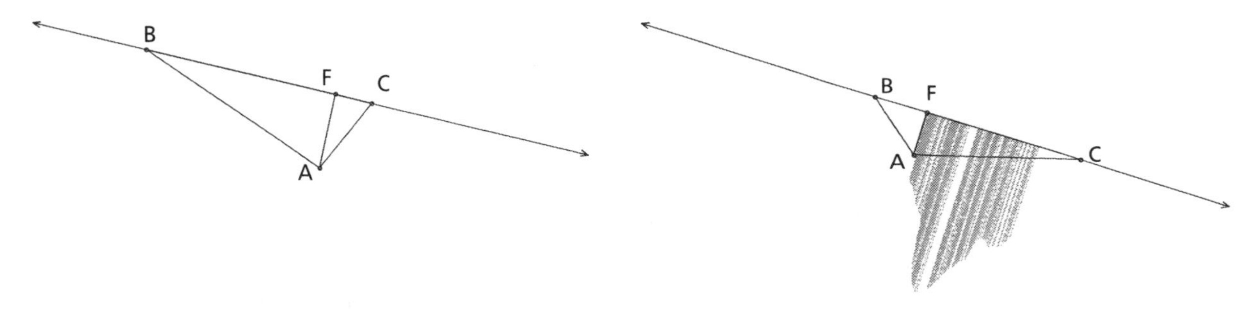

Figure 27.3 A student who drags point A might note what happens to point F.

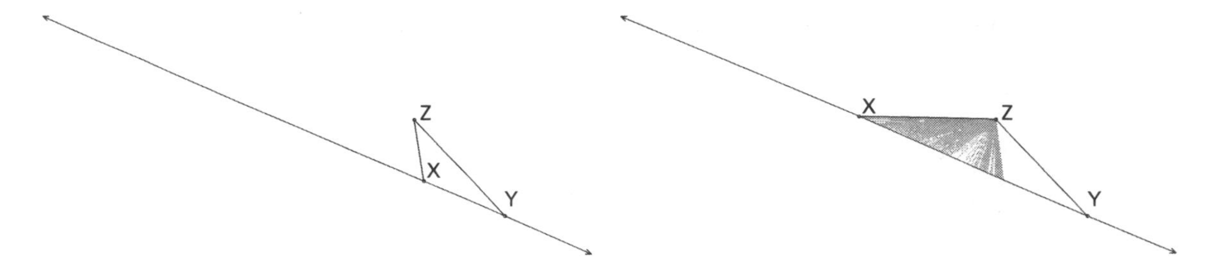

Figure 27.4 A student might drag point X along line XY, keeping m∠XYZ and the length of segment YZ constant.

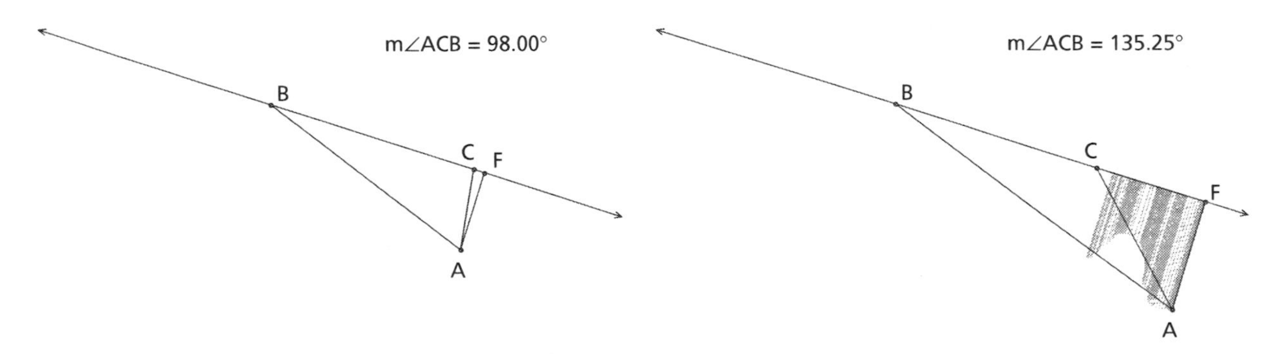

Figure 27.5 A student might drag point A to test a hypothesis about the location of point F.

some aspect of a figure remains invariant? Dragging to test a hypothesis involves a search, not for results, but for confirmation (or disconfirmation) of some particular result: Does this result hold true for many different figures created from a given set of conditions?

These types of dragging can include both exploratory and expressive activity. A student using dragging to test a hypothesis might be doing so in response to a teacher's prompt to drag one vertex of a triangle in a given sketch (e.g., "Try an acute triangle, a right triangle, and an obtuse triangle" to determine whether a particular result holds for each). Such use of technology illustrates exploratory activity. Or, the students' activity might be a much more open-ended wandering dragging that involves the creation of a figure and dragging of its parts to search for some invariant that could lead to a conjecture about such figures. The expressive nature of such activity is reflected in the student's selection of the figure, the dragging to be carried out, and the conclusions that can be drawn. That expressive activity results in part from the task, the curriculum materials, and the teacher's questions and prompts.

These dragging types are important for research because they identify qualitatively different meanings that might underlie a single action made possible by a technological tool and a particular goal. The exploratory–expressive constructs might help a researcher to explain certain learning outcomes that result from student use of one of these types of dragging rather than another and, again, might help the researcher to explain potentially conflicting results that are masked by the lack of distinction among the types of dragging activities in which students were engaged. Second, relating students' goals and their uses of a tool's capabilities can help to explain very different behaviors that might be observed as students use a tool. Other differences in students' behaviors when using technological tools are the focus of the section that follows.

Students' Technology-Specific Behaviors

When using technological tools, students' goals influence their behaviors, and because not all students' goals are necessarily the same, they do not all exhibit similar behaviors. As observed by Ball and Stacey (2005), data from Pierce (2002, as cited in Ball & Stacey, 2005) suggest that some students are always judicious users of technological tools, some students persist in using technological tools capriciously and inappropriately, and a large number of middle-ground students can be educated to work judiciously. Studying the behaviors students exhibit when working with technological tools can lead to insights about the appropriateness of their use of those tools and about their understanding of mathematics.

A particularly useful idea in characterizing students' behaviors when using technological tools is the construct of *work method.* Guin and Trouche (1999) identified five work methods, or characterizations of behaviors, exhibited by 17- to 18-year-old students who used symbolic calculators: a random work method, a mechanical work method, a resourceful work method, a rational work method, and a theoretical work method. For each, they indicated the primary source of students' understanding (e.g., investigation, comparison, or inference), the source or tool(s) students used to obtain information (e.g., calculator, paper and pencil, or both calculator and paper and pencil), and students' method(s) of proof (e.g., accumulation of evidence, analogy, or demonstration based on inference). Trouche (2005b) offered several additional components of work method relating to time (e.g., time needed for calculator work) and the student's control of her or his activity (e.g., command process, defined as the student's conscious attitude to consider the information available from all sources and to seek mathematical consistency between those sources). Guin and Trouche also noted that students generally did not operate exclusively with a single work method, and that a student's behaviors often evolved during the instrumentation process, reflecting different work methods at different times.

Trouche (2005b) used a task involving limits of polynomial functions to illustrate contrasting work methods. Students were given the following symbolic rule for a function:

$$P(x) = 0.03x^4 - 300.5003x^3 + 5004.002x^2 \\ -10009.99x - 100100$$

and were asked to "determine its limit at $+\infty$" and "determine a calculator window which confirms your result" (Trouche, 2005b, p. 205). A student who used a random (also referred to by Trouche as automatistic) work method was unable to analyze the graph that appeared in the standard graphing-calculator window (which did not include the function's right-most relative minimum) or to determine what might be a more informative window, created a calculator table of values for $P(x)$ for values of x from 1 to 7, and concluded that the limit seemed to be $-\infty$. In demonstrating a mechanical work method—a calculator-restricted work method—a student employed a variety of calculator commands: The Trace command led to location of points that lay outside of the standard window, Zoom commands facilitated searches, and the student obtained the desired result exclusively on the calculator, without reference to its mathematical basis and without recording the work. A student who used a resourceful work method sought coherence when confronted with conflicting results. The student first looked for confirmation of an assertion, based on previously learned theorems, that $\lim_{x \to +\infty} P(x) = +\infty$. When the student used a graphic representation of the function, it suggested that the function was decreasing rapidly, even for large values of x. This contradiction led the student to the conclusion that the limit would be $-\infty$, because when there was a great discrepancy between the coefficient of the polynomial's highest degree term and a much larger coefficient of the term with the next-to-largest power, the term with the larger coefficient would dominate. The work of a student using a rational work method entailed factoring out x^4 and finding the limit of each factor using theorems on limit sums and products, then attempting to use first and second derivatives to provide information that would be useful for determining an appropriate calculator window (but not finishing due to lack of time). In contrast to the proof method (demonstration) of the student who used the rational work method, a student who used the theoretical work method established results primarily by analogy, identifying the function as a polynomial function, invoking the theorem concerning its limit at $+\infty$ being that of its highest degree term, and using the overall shape of the graph of a fourth-degree polynomial to make a sketch of the graph.

Students who exhibited the random (or automatistic) work method generally employed trial and error, but without verification of results produced by the CAS. These students apparently had a goal that involved finding some calculator operation that would produce a result (not necessarily known to the student to be a correct result) for the task at hand. Other researchers have identified a number of behaviors similar to those of students who exhibited Guin and Trouche's random work method. These behaviors seem to result when students search for some appropriate action with a tool. However, each of these behaviors lacks a mathematical focus. For example, Guin and Trouche (1999) observed in weaker students aged 15 to 16 "avoidance strategies" such as random trials and "zapping" to other commands in the same menu, similar to what Ball and Stacey (2005) reported. Guin and Trouche noted that calculator use does not automatically lead to questioning and reflection:

> No doubt this behaviour is different from the activity in the paper/pencil environment where trying something else already requires careful thought. The main difference between the two environments is precisely that which occurs with the calculator, students lose consciousness of the task and there is little mathematical work in their activity. (p. 213)

Trouche (2005a) noted several other behaviors in which a mathematical focus is lacking in students' work. Trouche reported that Defouad (as cited in Trouche, 2005a) identified zapping, in which students quickly change the graph window without analyzing information from the graph; oscillation between several techniques or strategies; and overchecking that uses more of the calculator's capabilities than is necessary. Trouche (2005a) also reported that in his previous work (Trouche, 1997, as cited in Trouche, 2005a) he noted the phenomenon of automatic transportation, in which students entered all of the problem data into the calculator, with the goal of finding a command that could produce the desired result directly.

Many of these student behaviors involve searching for a command that will generate some result. The goals of these "hunting" behaviors are similar to the goal of curve fitting described by Zbiek (1998)—finding, without using the desired mathematical analysis of the problem situation, some curve that "fits" the data. These goals that entail searching but lack a focus on pertinent mathematical features also appear in Guin and Trouche (1999): "Weaker students often do not question results; activities do not necessarily lead to reflective work, but may instead lead to behaviour she [Artigue, 1995] has called 'fishing behaviour'" (p. 200).

Guin and Trouche's mechanical work method resulted primarily from simple manipulations and led to reasoning based on an accumulation of consistent results from the tool. This seems to reflect a goal that involves finding regularities across multiple instances rather than identifying an underlying mathematical structure. This is similar to what Noss and Hoyles (1996) termed "pattern spotting," when students viewed patterns between numbers or geometric figures as equally valid regardless of those patterns' underlying mathematical structures. Students who engage in wandering dragging may be exhibiting a mechanical work method: They produce a collection of tool-generated results, from which they might note some regularity (without necessarily identifying the mathematical basis for that regularity).

In reference to students' use of spreadsheets, Friedlander and Tabach (2001) found that some students extended a table to thousands of rows to answer a question that could have been answered by using a position formula (specifying the value in a cell in terms of *n*, its position in the sequence) rather than the recursive formula that they had used to extend the table (specifying the value in one cell in terms of the value in the preceding cell). Despite the efficacy of such an approach, the authors contended that students who used this form of "formula dragging" abused the power of a spreadsheet by constructing large tables in lieu of using what the authors deemed to be a more efficient symbolic approach (e.g., solving an equation), and they noted that tasks with large numbers brought this issue to the fore for students who wanted to use the symbolic expression rather than the numeric values to obtain a result. Students who use a mechanical work method base their results on an accumulation of tool-generated instances rather than on analysis of underlying mathematical relationships.

In other work methods identified by Guin and Trouche (1999), mathematical analyses played a more prominent role. In the resourceful work method students used a wide range of solution strategies and a variety of sources to generate information (e.g., calculator, paper and pencil, theoretical analyses), sometimes basing results on observations and other times basing them on theoretical analyses. The rational work method entailed little CAS use, and inferences played a primary role in reasoning. In the theoretical work method, understanding derived primarily from interpretation rather than from investigation (as in the mechanical work method) or comparison (as in the resourceful work method); theoretical knowledge, rather than the calculator (as in the mechanical work method) or paper and pencil (as in the rational work method), was the primary tool for information; and reasoning was essentially based on analogy.

Researchers who have attempted to identify productive and unproductive behaviors when students use technology often have focused on inappropriate behaviors similar to those described in conjunction with the random work method, many of which appear to result when students search (sometimes randomly) for appropriate actions. What is needed is research that identifies constructs that are associated with the development of judicious use of technology. An initial step toward identification of those constructs might be empirical studies that provide descriptions of students' successful use of technology (e.g., Dick, 1992; Heid, Hollebrands, & Iseri, 2002), studies that focus on observations of students' behaviors when using technological tools, and studies that identify successful strategies for changing undesirable behaviors. As more is learned about what causes students to use certain work methods, teachers might help students make more judicious use of technology—to use more mathematically based and more productive work methods—causing students whose behaviors are passive, random, or lacking in thoughtfulness to become more judicious users of technology and redirecting them toward behaviors that are based, not so much on the results produced by the tool, but on reflection about the underlying mathematics.

Because the work-method construct consists of a collection of elements that characterize aspects of students' understanding, coordination of information, tool use, and approach to proof, it provides a composite description of the nature of the mathematical basis on which a student's work with technological tools is grounded. This can be useful to researchers in linking changes over time in students' work methods as they use technological tools to the process of instrumentation, in identifying extreme behaviors that may account for unexpected outcomes when using technology, and as a way to relate students' understanding to their behaviors when using technological tools. A study of work methods may also reflect what patterns of privileging occur in the classroom (see the section on the Relationship Between Technology and Practice for additional discussion of the privileging construct).

As noted previously, identifying students' type of activity and behaviors when using technological tools is important, and from a research standpoint it is desirable to use the type of activity in which they engage (exploratory or expressive) and their corresponding behaviors (work method) to infer what students are thinking. An important feature of some technological tools that facilitates this inference is their ability to make students' thinking easier to infer from the actions they take with the tool. In fact, in some instances tools have been designed explicitly to reflect students' thinking by providing choices that mirror students' mental actions. For example, Tools for Interactive Mathematical Activity "provide children with possibilities for enacting their mathematical operations with whole numbers and fractions. They also provide the teacher/researcher with opportunities to provoke perturbations in children's mathematical schemes and observe children's mathematical thinking in action" (Olive & Lobato, in press). In a slightly different way Logo makes students' thinking explicit by enabling students to express geometric ideas in the form of a program that represents an "instantiation of students' expression of geometric ideas" (Clements & Battista, 2001, p. 145). Noss and Hoyles (1996) described programming in general as "a tool for expressing and articulating ideas" (p. 57). They noted that Illich (1973) described such tools as convivial tools—tools that allow users to express their meaning in action.

Students' mathematical behavior is, of course, greatly influenced by the ways in which teachers choose to engage them in mathematical activity. The next section takes up the issue of teaching practice in technological environments.

RELATIONSHIP BETWEEN TECHNOLOGY AND PRACTICE

The relationship between technology and a particular teacher's practice is not always predictable. Two constructs are useful in understanding how teachers involve technology in their work. The degree of match between a particular cognitive technology and a teacher's practice and beliefs underlies the construct of pedagogical fidelity. The construct of privileging centers on a teacher's implicit and explicit choices, including but not limited to choices about technology use.

Match Between Technology and Practice

A teacher's readiness to use a particular form of technology and the nature of how a teacher's use of the technology unfolds center around how the teacher's practice and the nature of that technology align. Dick (2006) assumed that "students learn mathematics by taking mathematical actions . . . on mathematical objects . . . , observing the mathematical consequences of those actions, and reflecting on their meanings" (p. 334). He then described the pedagogical faithfulness of a technological tool as the degree to which the student sees use of the tool as something that furthers this learning. Dick positioned his discussion of pedagogical faithfulness under the more general umbrella of *pedagogical fidelity*. We see pedagogical fidelity more generally as the extent to which teachers (as well as students) believe that a tool allows students to act mathematically in ways that correspond to the nature of mathematical learning that underlies a teacher's practice, regardless of whether the teacher's view of learning is consistent with Dick's assumption that learning requires reflection on observations of the mathematical consequences of mathematical actions on mathematical objects.

Although pedagogical fidelity has not been an explicit research topic, the construct can be used to explain seemingly disconnected or convoluted sets of findings and observations from studies of teachers' thinking, planning, and use of technology in mathematics classrooms. As an example, we consider several studies (Lumb, Monaghan, & Mulligan, 2000; Ruthven & Hennessy, 2002; Zbiek, 1995) that address relationships between mathematical exploration or investigation and classroom use of technology.

In a study of secondary mathematics teachers from seven schools, Ruthven and Hennessy (2002) noted the promotion of investigation as one factor to which teachers attributed successful classroom use of spreadsheets, graphing utilities, geometry programs, and databases. Their observation might be interpreted as indicating that each of these teachers valued some form of investigation as part of their view of learning and attributed success to the extent to which the tools were faithful facilitators of this valued mathematical activity. In this study, it appears that teachers experienced high pedagogical fidelity because technology supported the investigations they valued for all students and was particularly helpful for learners whom they viewed as "low-attaining students."

Reports of other studies relate situations in which we wonder whether a key challenge for the teacher was a low degree of pedagogical fidelity. Zbiek (1995) described LeAnne as a teacher who valued a form of investigation in mathematics, but who also wanted students to use particular mathematical procedures in their work and to have a relatively high degree of commonality in how they answered questions. The computer algebra system used in her classroom supported the mathematical investigation she valued, but it also allowed students to have freedom to pursue different methods and ideas in their investigations. The extent to which this freedom resulted in a mismatch between what LeAnne wanted in her classroom and what the tool supported suggests that a low degree of pedagogical fidelity may have been a key factor in LeAnne's decisions to alter tasks in ways that curtailed students' opportunities to investigate mathematical phenomena. A similar observation of a teacher seeing the extent to which a computer algebra system allows students many options is found in Lumb et al.'s (2000) description of Steve's view of materials. As a teacher of Advanced level or A-level 16- and 17-year-old students, Steve began with worksheets to help students learn to use Derive. As students became sufficiently able users of the software, Steve continued to use the worksheets and did not move to using the textbook. Among the many factors that may have played into Steve's decision to continue with worksheets is pedagogical fidelity. The computer algebra system offered many options and Steve's decision may have arisen from "a desire to 'lead' students to go down a particular route they believe is beneficial to learning" (p. 236).

In these studies and others (e.g., Bottino & Furinghetti, 1996; Mariotti & Bussi, 1998), we see classroom events influenced by teachers' views of how technology (and curriculum materials) facilitates or impedes mathematical investigations as a form of student mathematical activity that enhances learning, apparently influenced by what happens in instructional settings and how teachers view the appropriateness and usefulness of the technology in their classrooms. The usefulness of pedagogical fidelity in understanding the relationship between technology and practice extends beyond studies of classroom instruction. Becker and colleagues (Becker & Riel, 2000; Ravitz, Becker, & Wong, 2000) surveyed 4,083 U.S. Grades 4–12 teachers (not all mathematics teachers) chosen by a national probability sample of schools and purposive samples of schools with greatest per-capita computer technology and of schools with substantial participation in educational reform efforts. The research team identified the extent to which each teacher embraced one or some combination of two models of instruction: traditional instruction or constructivist-compatible instruction. Ravitz et al. (1998) carefully articulated what they meant by these two forms of instruction:

> *Traditional Transmission Instruction* is based on a theory of learning that suggests that students will learn facts, concepts, and understandings by absorbing the content of their teacher's explanations or by reading explanations from a text and answering related questions. Skills (procedural knowledge) are mastered through guided and repetitive practice of each skill in sequence, in a systematic and highly prescribed fashion, and done largely independent of complex applications in which those skills might play some role.
>
> *Constructivist-Compatible Instruction* is based on a theory of learning that suggests that understanding arises only through prolonged engagement of the learner in relating new ideas and explanations to the learner's own prior beliefs. A corollary of that assertion is that the capacity to employ procedural knowledge (skills) comes only from experience in working with concrete problems that provide experience in deciding how and when to call upon each of a diverse set of skills. (p. 3)

Teachers were also categorized with respect to the extent to which they reported professional involvement (e.g., working substantially with teachers from other schools, presenting at workshops, attending conferences, publishing, teaching college courses). Based on the self-report data, Becker and colleagues concluded that teachers who were more involved professionally were more likely than other teachers to report constructivist-compatible instruction and greater use of technology. Although these results might suggest that the constructivist-compatible instruction description captures key aspects of practice that suggest high pedagogical fidelity with the technology used, we caution that such a claim requires further investigation. The sample of teachers was not limited to teachers of mathematics, and the technology involved ranged greatly, including but not limited to skill games, electronic mail, browsers, word processors, simulations, and spreadsheets.

The importance of pedagogical fidelity also may be seen in its relationship to other constructs. One example to illustrate this point involves a potential relationship between pedagogical fidelity and the instrumentation and instrumentalization aspects of instrumental genesis. Instrumentation involves the ways in which a student's mathematics is influenced by the technology. In cases in which the tool supports the desired mathematical activities, there may be a high potential for the student to develop rich instrumentation schemes. In contrast, when there is not a high degree of pedagogical fidelity, the student may need to adapt the tool in nontrivial ways to meet the mathematical activity that is consistent with the teacher's view of learning. The student's instrumentalization, or adaptations of the tool to the mathematics, may be well supported and the student may thrive in this situation, or instrumentation may not be supported and the student may struggle to achieve learning goals.

The extent to which instrumentation and instrumentalization are supported in the absence of strong pedagogical fidelity may depend greatly on the teacher's instrumental orchestrations. As described by Trouche (2005b), instrumental orchestration refers "to an organization of the artifactual environment, that an institution (here the schooling institution) designs and puts in place, with the main objective of *assisting* the *instrumental genesis of* individuals (here students)" (p. 210). Careful attention in research work to the orchestrations that teachers establish and how these orchestrations reflect the pedagogical fidelity of the tools may inform practice as well as help researchers and theoreticians to unravel the complexity of teaching with technology. In addition, articulating orchestrations when describing classroom settings may be one way to add some clarity to what is meant by commonly used but uninformative labels, such as "experimental class," "technology group," and "nontraditional classroom."

When students are engaged in expressive activity or exploratory activity, their teacher might take on different roles (teacher roles are addressed more fully in a subsequent section titled *Incorporating Technology Into Practice*). In exploratory activity students might be involved primarily in doing, with the teacher functioning as manager of the task to be done. When students are engaged in expressive activity, the teacher may function as the manager of reflection, ensuring that students are creating something on which it is worth reflecting and focusing that reflection on the mathematical ideas involved.

Beyond describing pedagogical fidelity in particular cases and using the construct to synthesize studies, describe research settings, and connect constructs, there is a question of whether and how one might change the degree of pedagogical fidelity. Empirical evidence suggests such changes in teachers' perceptions of the fit between the technology and their view of learning and goals for students' mathematical activity. Graham and Thomas (2000) developed an early-algebra module that used a "press–see–explain" pattern. After pressing keys and seeing results, students routinely were asked to explain the results. The positive reaction to technology by teachers in this study may reflect how the press–see–explain pattern gave these teachers a specific pedagogical strategy to elicit the mathematical explanations they valued. This strategy formed a bridge between technology use by students and evidence of learning as valued by teachers. The teachers seemingly saw this fit between the calculator use and their classroom activity goals, which suggested use of the press–see–explain pattern might have influenced teachers' views of calculator use. Pedagogical fidelity would be enhanced if, for example, teachers moved from seeing calculator use as detracting from learning goals or being good only for checking answers to seeing a press–see–explain form of calculator use that contributed to classroom mathematics goals.

Pedagogical Choices

Beyond the overarching choice to use technology at all, teachers make daily choices about how to use technology in their classrooms. Kendal and Stacey (1999, 2000, 2001) adapted the notion of *privileging* from Wertsch (1990) to describe how teachers, intentionally or unintentionally, use frequently or place a priority on certain things in their practice. Teachers' objects of privileging included types of representation, skills or concepts, and by-hand or by-technology methods. Through written tests and interviews, Kendal and Stacey elaborated a connection between calculus classroom practice and student performance with evidence that students with the same scores on a common calculus test but taught by different teachers differed in their understanding and performance in ways that reflected their teachers' classroom patterns of privileging. One might also ask whether teachers may privilege—or be thought by their students to privilege—certain mathematical activities (e.g., exploration over exposition, production of examples over articulation of definition, illustration over proof) and what relationships, if any, privileging of mathematical activities has to student understanding and performance.

One form of privileging is embodied in Buchberger's (1989) White Box/Black Box Principle for Using Symbolic Computation Software in Math Education. Buchberger argued that some areas of mathematics, such as common integration rules, are "trivialized" in the sense that "there is a (feasible, efficient, tractable) algorithm that can solve any instance of a problem from this area" (p. 10). He then noted a critical question given the power of symbolic software: "*Should math students learn area X of mathematics when this area has been trivialized?*" (p. 11). His response to this question is the Principle:

> I think it is totally inappropriate to answer such a question by a strict "yes" or "no." Rather, the answer depends on the stage of teaching area *X*.
>
> - *In the stage where area X is new* to the students, the use of a symbolic software system realizing the algorithms of area *X* as black boxes would be a disaster. . . . Students have to study the area thoroughly, i.e., they should study problems, basic concepts, theorems, proofs, algorithms based on the theorems, examples, hand calculations.
> - In the *stage where area X has been thoroughly studied*, when hand calculations for simple examples become routine and hand calculations for complex examples become intractable, students should be allowed and encouraged to use the respective algorithms available in the symbolic software systems. (Buchberger, 1989, p. 13)

The principle obviously promotes privileging of symbolic software use only when the simple problems are routine and complex problems are intractable. Less obviously, the principle embodies privileging of many facets of mathematics (e.g., problems, concepts, theorems, proofs, algorithms) and of mathematics that not only is going to be introduced with all of these facets but also used beyond its introduction. Coupled with the examples offered by Kendal and Stacey, Buchberger's White Box/Black Box Principle is a reminder that privileging may have various layers, some of which are less obvious and all of which might influence student learning and achievement.

The white box/black box notion in this principle has been borrowed by many others to convey two extremes regarding whether students are aware of the mathematics they ask the technology to do (white box) or are using the technology without expectation of knowing the mathematical working of the tool (black box). Several authors (e.g., Pimm, 1995) have argued that black-box treatment of technology may be detrimental to students. Lagrange (1999) suggested that students for whom the technical aspects of symbolic work are masked by CAS use may not be able to develop rich technique, a construct discussed earlier in this chapter in *Aspects of Mathematics in Cognitive Tool Use*. In contrast, several authors have contended that some seemingly black-box uses of technology may be productive. Heid (1988) described how college students were able to engage in mathematical investigations and develop conceptual understanding of calculus topics when manipulation details were delegated primarily to the equivalent of CAS. Her students were able to engage in these investigations without having first mastered by hand the routines that were being executed by the symbolic manipulation program. Doerr and Zangor (2000) noted that the black-box use of a graphing calculator did not impede secondary school students' encounters with mathematical reasoning. They concluded that the teacher's questioning drew students' attention from the workings of the tool to the mathematical justifications of the work, suggesting that this black-box use supported justification activity.

Lurking beneath white-box/black-box debates are potentially different privileging schemes and issues of pedagogical fidelity. For example, where one stands on the white box/black box issue may reflect the extent to which one privileges by-hand computation. If one believes that knowing the details of the manipulations that could be delegated to the tool is critical mathematical activity for students' learning of mathematics, the privileging patterns intrinsic to the White Box/Black Box Principle strengthen the pedagogical fidelity of the CAS. Privileging, particularly when driven by an adherence to the White Box/Black Box Principle, may also help to explain students' differing work methods (a construct described in the section on *Students' Technology-Specific Behaviors*). Interesting research questions arise when one considers how patterns of privileging might unfold differently in two classrooms, both of which claim to adhere to Buchberger's principle. To what extent might privileging connections among basic concepts, hand calculations, theorems, and proofs correspond to students' use of resourceful work methods? To what extent might students in classrooms characterized by the absence of privileging these connections but otherwise following the White Box/Black Box Principle be more likely to resort to random work methods or mechanical work methods than students in classrooms that privilege the connections?

We believe privileging also may apply to what technology offers as well as to what teachers do. For example, Goldenberg and Cuoco (1998) discussed geometry environments in terms of the mathematical concept of function. An interesting point they made regards how students may see continuity and dynamic dependency in dynamical geometry versus seeing function as algorithm in a static geometry environment. Dynamical geometry privileges a view of function rooted in dragging, or "continuous real-time transformation" (Goldenberg & Cuoco, 1998, p. 351). Goldenberg and Cuoco contrasted this view of function in dynamic settings with function as constructive algorithm in static geometry environments (e.g., Geometric Supposer). This observation suggests the need to consider what mathematical views software privileges as well as what teachers (and curriculum) privilege. In short, the match between what the software privileges and what the teacher privileges may be a key aspect of pedagogical fidelity.

Incorporating Technology Into Practice

In addition to constructs to use in articulation of choices teachers make and actions they take, other constructs help to describe other aspects of how teachers go about their work and grow personally and professionally as technology users. One might look at the *teacher role*, a construct that involves describing and contrasting various roles that teachers perform in the classroom. Zbiek and Hollebrands (in press) described 11 roles based on those identified and used by Fraser and colleagues (1987), Farrell (1996), and Heid, Sheets, and Matras (1990). As examples of how teacher role may be useful, we will use two roles (Counselor and Technical Assistant) and their definitions as given by Zbiek and Hollebrands. The Counselor role is one in which the teacher is familiar with the mathematical task at hand and provides mathematical assistance when students ask for it. As a Technical Assistant, the teacher helps students with hardware and software difficulties.

Teacher role is a useful construct to compare technology-present and non-technology-present classroom work of teachers and their students. In particular, several studies have suggested that teachers acted in a Counselor role in technology-using settings more than they or other teachers did in nontechnology settings. On the basis of observations using particular programs rather than mathematics tools, Fraser and colleagues (1987) noted that the technology changed the dynamics of mathematics classrooms with 12- to 14-year-old students by taking on particular roles normally assumed by the teacher while the teacher moved into roles infrequently held when the technology was not in use. In particular, the researchers noted that, when the computer was in use, teachers would join with students in what we call a Counselor role, which seemed to be absent in most teachers' practice. Building from the roles used by Fraser and colleagues, Farrell (1996) compared the roles taken on by six teachers when using graphing calculators in a technology-using precalculus course to the roles taken on by those teachers in the same classes when the technology was not in use. Each 5-minute section of videotaped lessons was coded according to whether the teacher ever acted in a particular role and whether any calculator use happened, among other things. On the basis of 283 coded segments, Farrell concluded that teachers functioned as Counselors more frequently when technology was in use (43% of all technology-using 5-minute segments) than when technology was not in use (19% of all non-technology-using segments). The increased prevalence of a Counselor role may not be limited to secondary school settings, and it may be perceived by instructors as well as by observers. When surveyed by Rochowicz (1996), calculus instructors reported taking on a Counselor role when computers or calculators were used in their calculus classrooms in engineering and pre-engineering schools. This set of studies by Rochowicz, Farrell, and Fraser and colleagues suggests that bringing technology into a classroom corresponds to increased instances of the Counselor role.

Teacher role also may be used to describe tensions in a teacher's practice. Despite the seeming association of technology with the likelihood of teachers working in a Counselor role, the Counselor role may be in conflict with teachers' expectations and competing roles. Timmerman (1998/1999) described how Susan, an elementary school teacher, struggled with a desire to take on a Counselor role while using TIMA software (Olive & Steffe, 1994) with Grade 4 students given access to only one computer. Susan was accustomed to all students being engaged in the same task in a teacher-controlled environment. Having two students working together without the teacher's input while the rest of the class was involved in other activity did not allow her control and perhaps did not allow her to be in a Counselor role. Piliero (1994) described a teacher's tension between wanting to serve a Counselor role but more often serving as a Technical Assistant, reacting to students' technical issues as they were using Function Probe (Confrey, 1991, as cited in Piliero, 1994) in a precalculus class. Teacher roles provide a way to capture the tensions that might be useful in identifying differences between teachers' perceptions and practice—differences that might influence or reflect difficulties in implementing technology-intensive curricula.

In addition to ways to describe teachers' roles within the classroom and tensions they feel, ways are needed to describe how technology-rich practice develops. Teacher role may be useful in capturing teachers' experiences as they become fluent users of cognitive tools as pedagogical tools, but they would not explain the total journey of learning to teach with technology. Beaudin and Bowers (1997) introduced the PURIA model for CAS incorporation that Zbiek and Hollebrands (in press) expanded to account for teachers' growth in the use of any technological tool. The model notes that teachers engage in one or more modes of technology use: personal Play (use without mathematical aims), personal Use (use with clear mathematical goals), Recommendation (recommending or trying technology with small groups of students), Implementation (use of the technology in the classroom), and Assessment (careful attention to the impact of technology use on student learning). Zbiek and Hollebrands argued that PURIA is consistent with evidence found in the reports of prior research. Although it is early to see empirical evidence to support or challenge the value of PURIA as a framework for supporting teachers' growth as practitioners in technology-rich classrooms, the extent to which the classifications provide a way to talk about decades of research on teacher technology use is promising.

PURIA's promise as a way to describe teachers' growth in teaching mathematics with technology includes the extent to which it fits with other constructs related to technology and mathematics teaching and learning. The Play and Use modes allow for teachers' instrumental genesis as they become proficient users of the cognitive tool. As teachers move among Recommend, Implement, and Assess modes, a different type of genesis occurs as teachers grow to see the technology as a pedagogical instrument. Within this genesis of the pedagogical instrument, pedagogical instrumentalization arises from teachers' desires to fit the technology with existing or emerging privileging patterns and teacher roles. The ease with which pedagogical instrument genesis occurs may reflect the degree of pedagogical fidelity.

TECHNOLOGY AND CURRICULUM

The effects of technology on the teaching and learning of mathematics is a function of the curriculum in which it is embedded—how the technology is used, with what mathematical activities students engage, what curriculum materials are used, and upon what content the curriculum is focused. We have identified several central constructs that capture the opportunities for change in curriculum facilitated by technology: representational fluency, mathematical concordance, amplifiers, reorganizers, microprocedures, and macroprocedures.

Representational Fluency

One of the reasons offered for the importance of incorporating technology in mathematics instruction is the access technology can offer to content that had not previously been included in the school curriculum. This access is facilitated through a range of technological capabilities particularly suited to mathematics instruction. Primary among those technological capabilities is the capacity to accommodate multiple representations, hot-linked and interactive, and a key construct related to the availability of these representations is *representational fluency*. Researchers have used the phrase representational fluency to refer to the interaction between student and representation, usually indicating an ability to move between and among representations, carrying the meaning of an entity from one representation to another and accruing additional information about the entity from the second representation. Entailed in representational fluency is the ability to draw meaning from different representations of the same mathematical entity and to generalize (Becker & Rivera, n.d.) across different representations. Sandoval, Bell, Coleman, Enyedy, and Suthers (2000) provided a comprehensive definition of representational fluency—one that serves well in capturing the phenomenon in the context of multirepresentational techno-mathematical environments:

> We view *representational fluency* as being able to interpret and construct various disciplinary representations, and to be able to move between representations appropriately. This includes knowing what particular representations are able to illustrate or explain, and to be able to use representations as justifications for other claims. This also includes an ability to link multiple representations in meaningful ways. (p. 6)

In this chapter, the term representational fluency is construed more broadly than the translation between representations. It includes, as suggested by Sandoval and colleagues, interacting with each of the representations in a meaningful manner. For our purposes, representational fluency includes the ability to translate across representations, the ability to draw meaning about a mathematical entity from different representations of that mathematical entity, and the ability to generalize across different representations. Our operational definition aligns with the notion of representational versatility that Hong and Thomas (2002) took "to include both fluency of translation, and the ability to interact procedurally and conceptually with individual representations" (p. 1002).

Representational Fluency and Mathematics Learning

The study of mathematics centers on the study of abstract concepts, and individuals' access to those concepts is mediated[6] through externally present representations of those concepts. These externally present representations are organized in representational systems—sets consisting of external representations, the entities they represent, and the rules that govern transformations of those representations. Each representation system has the potential for revealing certain properties of the represented quantity (and for concealing others), and a set of representations with the potential for foregrounding different aspects of a mathematical entity ostensibly affords students opportunities to consider those entities in a broad range of settings. Availability of representations, however, is insufficient to ensure that students will take advantage of their availability; representational fluency is the issue. Kaput (1989) described the intimate relationship between representational fluency and mathematical meaning, citing translations between mathematical representation systems and translations between mathematical representations and nonmathematical systems (e.g., physical systems, natural language, pictures) as two of four primary sources of mathematical meaning (p. 168). Through the use, interpretation, and coordination of multiple (external) representations teachers and students communicate about mathematical entities; when a broad range of representations is coupled with representational fluency, the opportunity for building rich meaning grows. Technology expands the repertoire of representations available to students for the mathematical entities they study as well as the opportunities to carry meaning between representations; dynamic, interactive technology is fertile ground for the development and use of representational fluency.

The importance of representational fluency is in its capacity to enhance mathematical understanding, and researchers have provided evidence that mathematical understanding emerges when students actively engage in making connections between multiple representational forms. Hall, Kibler, Wenger, and Truxaw (1989) observed how important situational representation was to the success of students who had mastered manipulation of symbolic representations, years later, when they were asked to develop answers to standard word problems. Koedinger and Tabachneck (1994) noted the importance of informal representations in college students' solutions of word problems, and Tabachneck and colleagues (Tabachneck, Koedinger, & Nathan, 1994) observed that merely having access to representations did not ensure student success with word problems. Although their study focused on the strategies students used in solving word problems, it can be interpreted to suggest that success in solving word problems using multiple representations depends on students' abilities to use the representations interactively, that is, it depends on their representational fluency. Students differ in the extent to which they use representations interactively (that is, in their representational fluency), and these differences are linked to their success in solving verbally stated problems (Tabachneck et al., 1994). Representations are said to be used interactively when an intermediate representation provides feedback that informs the original representation. One with representational fluency can turn to another representation when he or she encounters a local difficulty and return to the original representation once the difficulty is overcome. Researchers (Gluck, 1999/2000; Koedinger, 2001; Koedinger & Anderson, 1998) have identified ways in which technology can encourage the development of representational fluency. For example, coaching students, at opportune times, in appropriate use of the available technological representations in thinking about labels for quantities can assist them in making connections to more concrete situations, and technology-assisted support for thinking about numerical relationships can help students make connections to arithmetic knowledge and facilitate the transition to fluent use of algebraic symbols.

Representational Fluency and Technological Environments

Because technology has the potential for broadening the representational horizon, consideration of representation is fundamental to the development of educational technologies that support problem solving, learning, and learner-centered design issues (Underwood et al., 2005, take this position in their consideration of the development of applets). Researchers (Hoadley & Kirby, 2004) see the thoughtful incorporation of representations as essential to the development of technology:

> Thus, representations do a lot of work in the learning context. . . . Seeing the representations that form part of the educational environment and trying to optimize those representations is an important part of the design and instruction. In particular, computers give us unprecedented ability to create, manipulate, and share representations of a wide variety of types. Understanding how they do their work, and how we may design them, is vital. (p. 2)

Enhancement of representational fluency is an affordance, not an automatic result, of the use of tech-

[6] We consider computer tools to be one of Vygotsky's "psychological tools" or "signs," and hence a means of mediation.

nology. Depending on how it is configured and used, technology can function as cognitive support for the development of representational fluency. Technology like Geometer's Sketchpad or Cabri that dynamically coordinates the actions taken by the user with the representations that appear on the screen has the potential for helping the user to draw meaning about a mathematical entity from its representations if the user reflects on the connection between his or her action and the change in representation. Technology with hot-linked representations that automatically and simultaneously update in reaction to a user action provide the potential for enhancing the user's ability to translate and generalize across representations of the same mathematical entity. The redundancy in information offered by multiple-linked representations has the potential to enhance the user's ability to draw meaning about a mathematical entity from different representations of that mathematical entity. As Kaput (1989) pointed out, "We can have our cake and eat it too, in the sense of being able to trade on the accessibility and strengths of different representations without being limited by the weakness of any particular one" (p. 179). Kaput explained further the potential of technology for enhancing representational fluency but warned that the success of the enhancement depends on the choice of representation and the configuration of the learning environment:

> Appropriate experience in a multiple, linked representation environment may provide webs of referential meaning missing from much of school mathematics and may also generate the cognitive control structures required to traverse those webs and tap the real power of mathematics as a personal intellectual resource. However, one must choose such representations carefully and, even more importantly, ensure that the learning environment supports rich sets of actions that will expose underlying invariances and thus enable the student to weave a flexible and enduring web of mathematical meaning. (Kaput, 1989, p. 180)

Gifford (2005) echoed that point of view as he pointed out that technological environments alone are unlikely to enhance representational fluency and that characteristics of curricula that are likely to enhance representational fluency are those characteristics that are inherent in the dynamic interactive representations present in many mathematics-specific technologies:

> Research . . . suggest[s] that lessons that afford students access to increasingly complex mathematical representations, or that require students to negotiate their way between and among multiple representations of the same mathematical concepts, procedures, or problem-solving strategies are much more likely to increase representational fluency than conventional mathematics print-based curricula materials. (p. 3)

Representational Fluency as a Construct for Research

The promise of representational fluency as a construct lies in its potential for providing a focused representational perspective to research on mathematics learning and teaching. In particular, representational fluency provides a lens researchers can productively use in investigations of mathematical activity and of students' mathematical thinking.

Representational fluency as a lens to examine the mathematical activity in classrooms. When viewed as a unit, the ability to translate across representations, the ability to draw meaning about a mathematical entity from different representations of that mathematical entity, and the ability to generalize across different representations can be examined as they affect mathematical activity. Researchers have addressed the effects of representational fluency on mathematical inquiry and on the foci of mathematical curricula.

Research studies have suggested ways in which representational fluency can catalyze mathematical inquiry. For high school precalculus students' investigations of asymptotes in a qualitative technological environment, work in one representation precipitated work in a parallel representation; one student's graphical description of a function with a nonhorizontal slanted linear asymptote ("There is an imaginary line that divides the function in two," Yerushalmy, 1997, p. 18) led her group to a discussion of the form of symbolic function rules that would produce such results. Yerushalmy observed, "The learner's action on an unexplored object presented in one representation (the graph produced by the software) leads to inquiry about the processes underlying that object explored through a parallel representation (the symbolic procedure)" (p. 22). The representational fluency present in the group because of individuals' collective knowledge of representations allowed the group to traverse a path from graphical to symbolic in order to pursue their conjecture. This notion of representational fluency of a group was also studied by Schwartz (1995) who analyzed how students make shared use of multiple representations when they work together. One wonders how the representational fluency of a group relates to the representational fluency of individuals in the group. There may be a connection to expressive activity and exploratory activity (described in the section, Students and Mathematical Activity). In a group, those who contribute most to the discussion may be shaping the group's representational fluency by engaging in expressive activity; those who watch and interpret may be engaged in exploratory activity. So, representational fluency can be a function of either expressive or exploratory activity, and group members may develop different levels of representational fluency depending on the nature of their expressive or exploratory participation in the group's activity.

The potential that representational fluency offers in the context of multirepresentational settings affords the opportunity for curricula to focus on central mathematical objects and processes and for researchers to study the development of an understanding of these central objects. Chazan (1999) observed that whereas the focus of his teaching in an introductory algebra class in a traditional setting centered on procedures of a litany of solving equations and simplifying expressions, his introductory algebra class in a multirepresentational technological setting focused on central mathematical objects of function and variable. The promise of this functions approach to algebra rests squarely on students' development of representational fluency.

One curriculum constructed as a multirepresentational technology-intensive, functions-based approach to introductory algebra is the Computer-Intensive Algebra (CIA) curriculum (Fey et al., 1991). Analysis of the effects of this newly configured content has been highly influenced by the construct of representational fluency. Studies of implementations of CIA in 8th- and 9th-grade classes demonstrated that a technology-intensive, functions-based introduction to algebra could be constructed to productively enable students to develop an understanding of the concepts of algebra while they used computing tools to move between and among representations and to execute routine procedures (Heid, 1996a). Understanding of fundamental concepts of introductory algebra seemed to be evidenced through representational fluency, and curricula that capitalized on multirepresentational technology provided a platform for activation of that fluency. A range of research studies have documented that in the CIA classes students demonstrated representational fluency as they used computers as tools in solving realistic problems (Heid, Sheets, Matras, & Menasian, 1988; Heid & Zbiek, 1993). That this development of representational fluency may have been a function of the curriculum rather than of the ability level of the students was suggested in a study that compared 9th-grade CIA students with 11th-grade students in an advanced mathematics class. This research reported that students in the 9th-grade CIA classes developed greater flexibility in reasoning with multiple embodiments of mathematical functions than did a group of comparable aptitude that had 2 years of additional exposure in traditional mathematics courses beyond high school algebra (Sheets, 1993).

Representational fluency as a lens to examine how students think about mathematics. Attention to representational fluency has also enabled researchers to observe the ways students call on representations as they develop strategies for solving problems. The availability of multiple representations and the rapidity with which they can be generated provides students with options when they encounter surprising results, and some researchers have suggested that at times those options allow students to avoid interpreting unfamiliar results (Tabachneck et al., 1994).

These situations can easily arise in the context of work with CAS, in that there is a constant need for CAS users to interpret results, and that need elevates particularly when the results of computations are unexpected. In the context of a teaching experiment with 6th- and 7th-grade students in their first formal exposure to work with symbolic algebra (Heid et al., 2002), Kevin, a very bright 7th-grade student, encountered an unexpected result. In the course of creating a function rule to match a given set of characteristics, Kevin noticed that the CAS was giving him a positive value for $\left|\sqrt{x+5}\right|$ when $x = -10$. Not having previously encountered

$$|a+bi| = \sqrt{a^2+b^2}$$

as the modulus of a complex number, Kevin did not venture further on trying to interpret the result and revised his rule to eliminate the expression, $\left|\sqrt{x+5}\right|$. Because of the ease of changing the particular symbolic representation and the capability of testing the new rule immediately by graphing it, Kevin was able to quickly generate an acceptable rule. With the multiple representations available in technological environments, students are afforded the opportunity to pursue any of a range of alternative strategies, at times enabling them to steer clear of unfamiliar content. Kevin's encounter with contradictory information was not unlike that of Trouche's (2005b) student whose resourceful work method was described previously in this chapter. Although Kevin's resolution of the conflict led to a correct result, Trouche's student adopted a new (incorrect) theorem to explain the conflict between a previously learned result concerning limits of polynomial functions and the result suggested by the values in a table of function values. Perhaps the availability of multiple representations contributes to students' adoption of a resourceful work method—one that employs multiple approaches—or adoption of a resourceful work method increases the likelihood that students will call on multiple representations.

The availability of multirepresentational technology has opened new curricular horizons in mathematics classrooms, and representational fluency is key to thinking about the effects of those new curricula. Kaput (1994), in his SimCalc project, developed ways to use multirepresentational technology to introduce the ideas of calculus to young learners. Stroup (2002) posited that the cognitive constructs on which younger learners can build as they engage in qualitative calculus experiences differ considerably from the cognitive structures of standard calculus. He noted that the overlap of the standard calculus and the qualitative calculus "allow[s] us to reason powerfully about change and to engage successfully the dynamism of our experience," and suggested that the technology enables a "*dynamics revolution* in learning" (p. 206). The potential for dynamic, interactive, hot-linked, and connected multirepresentational technology to enable the observation of students' representational fluency allows researchers to focus on how students (including younger students) can think differently about traditional topics.

Not only has the potential for representational fluency accelerated the introduction of traditional topics, it has allowed researchers to examine students' (and teachers') understandings of concepts that have not been a traditional part of the curriculum. In some cases, these studies have revealed general ways of understanding that appear to be critical in these new environments. For example, the dynamic linked-representational statistical tool, Fathom, allows students to simulate random sampling and examine graphs of the resulting distributions of statistics as the samples are generated. A study examining students' understanding of sampling distribution revealed the complexity of making distinctions among mathematical representations and calling up and using those distinctions at appropriate times (Heid, Perkinson, Peters, & Fratto, 2005). Understanding of sampling distribution required students not only to make careful distinctions involving a sampling distribution, proper subsets of that distribution, a single sample, and a population, but to properly select and apply probabilistic or deterministic reasoning to those concepts—all in the context of representations that appeared very similar. In this case, representational fluency required holding on to the mathematical/statistical meaning of each representation. This ability to make and manage intricately related mathematical distinctions is reflective of representational fluency at a high level of sophistication.

The study of representational fluency can be tightly targeted. Cobb (1999), for example, used purposefully produced tools with constrained capacities so that he and his colleagues could observe particular facets of students' reasoning. The minitools focused 7th-grade students' attention on characteristics of distributions, and the researchers observed students creating a new concept of "hills" to describe features of the data sets students graphed.

The promise of representational fluency as a construct. With a constantly expanding array of newly conceived representations and with current ready access to dynamic, interactive, hot-linked, and connected multirepresentational technology, mathematics teachers and students are becoming increasingly aware of the promise of representations in mathematics teaching and learning. Key to learning, however, is not solely the availability of externally present representations but the actions taken on those representations. That is, it is not so much that the representations exist but that the student interacts in meaningful ways with those representations. Representational fluency as a construct focuses attention on these meaningful mathematical actions.

Representational fluency can be thought of as meaningful and fluent interaction with representations. Researchers can investigate students' representational fluency as outcomes, as conditions, or as stages of development. A mathematical entity can be viewed through any of its myriad representations, and students' understandings of an entity are revealed in their interactions with representations. Researchers can use representational fluency in developing an understanding of how students think about traditional school mathematics or about mathematics newly available to schools because of technology. The lens of representational fluency is also an important tool for revealing the nature of students' understandings of fundamental mathematical ideas. Students who move from one representation of a mathematical entity to another while holding on to the meaning of each of the representations are exhibiting a certain depth of understanding of that mathematical entity. Without the multirepresentational capacity of technology, researchers would not have as broad an access to observing student interaction with representations.

Researchers have offered two somewhat opposing points of view regarding the role of representational fluency in mathematical problem solving. Some (Heid et al., 2002; Tabachneck et al., 1994) have pointed to representational fluency as allowing an end-run for students as they encounter a result they cannot explain. Students with a high degree of representational fluency can quickly generate an alternative representation and avoid having to interpret an unexpected result. Other researchers have credited representational fluency with allowing students to focus on the mathematical entities being represented instead of on a singular representation of that entity (see Chazan, 1999). Because of the availability and ease of production of multiple technology-based representations of the same mathematical entity, research using this construct can focus on the overall meaning that students credit to a range of representations. By examining how students use fundamental concepts in their work within a range of representations, research using the construct of representational fluency as a lens can provide evidence of the depth of students' understandings of the concepts themselves. Using either perspective on the role of representational fluency, researchers can structure their inquiries into students' mathematical thinking.

Finally, as a construct, representational fluency directs researchers' attention to a capability that may be linked to students' levels of mathematical understanding and allows researchers to examine the extent to which outward manifestations of capabilities in dealing with representations are indicators of deep understanding. A central set of research questions occurs at the interface of representational fluency and other constructs. A productive line of research, for example, might be the relationship between representational fluency and work method or between representational fluency and privileging. Is a student who adopts a resourceful (multiapproach) work method likely to develop a higher level of representational fluency than a student who uses a rational work method (deriving results not from instances but from theorems)? Do particular work methods hide students' representational fluency? Are the students of teachers who privilege multiple representation types or multiple technologies likely to develop more sophisticated representational fluency than the students of teachers who privilege only one representation or technology? These, and a host of other related questions, can help to structure informative empirical studies.

Mathematical Concordance

New content with technological tools can be either intended or unintended. In some cases, technological representations of mathematics are mathematically misleading, and we have referred to the construct of mathematical fidelity to describe this mismatch. But the lack of alignment between the characteristics of a technology-generated representation and the mathematical characteristics and behavior of the represented object are not the only sites for mathematical misalignment. Lack of alignment of mathematics can occur in any of the range of sources for the mathematics that arises in technology-intensive learning environments, including the teacher's mathematics, the student's mathematics, the text's mathematics, and the tool's mathematics. We capture the compatibility between any two elements of the range of different embodiments of mathematics that arise in technology-intensive environments with the term *mathematical concordance.* We will describe the degree of mathematical concordance as a continuum. Mathematical concordance is different from mathematical fidelity in that mathematical fidelity is about a standard, whereas mathematical concordance is about degree of alignment without reference to a standard.

Researchers have, for example, investigated concordance between the mathematics in which students engage and the mathematics the teacher intended in designing the technological task. At times this low level of concordance is manifested in students' not substantially engaging in the intended mathematical task. In their description of research on the relationship between what was intended to be learned and what was actually learned in the context of 13-year-olds' work on mathematics within a Logo microworld, Hoyles and Noss (1992) observed the task of having the Logo turtle sketch a larger House similar to a given one revert into the task of creating a larger house that looked similar to the given House. Whereas the researchers' and instructors' goal was to engage the students in thinking about proportions, the students' task was to build a bigger House without paying attention to ratios. The students' mathematics circumvented the mathematics intended by the teacher. Incidents like this having drawn their attention to this mathematical discord, the researchers noted in their discussion of the play paradox the lack of mathematical concordance:

> The issue we want to pin-point here is what we have termed the *play paradox.* This derives from the following contradictory situation: on the one hand, the computer offers pupils rich and diverse ways of exploring and solving a problem and building upon rather fragile intuitions of the mathematical domain. Yet in exploiting this diversity, pupils are as likely to avoid encountering the mathematical nuggets so carefully planted by their teachers (although of course they may encounter other—unintended—ones). (pp. 45–46)

Dugdale (2007) observed a related phenomenon at the beginning of her work in the development of the program Green Globs. The goal of Green Globs was to write equations whose graphs pass through a pseudo-randomly generated set of globs on a Cartesian plane. Although the program was intended to engage students in thinking about the graphs of different families of functions, students subverted this goal on an early version of the program when they discovered functions such as $y = 10\sin(10x)$ to be single functions whose compact periodicity ensured that they would hit every glob. In this case there was a low level of concordance between the intended mathematics of the task and the mathematics in which students engaged. Although not fully engaged in the mathematics of the task as intended, the students' mathematics was related to the intended mathematics.

Sutherland and Balacheff (1999) provided a salient example of a low level of mathematical concordance in their report on teaching experiments that revealed "the tension between the student's constructions (which are tied to the phenomenology of the screen, and the student's previous knowledge) and the knowledge which the teacher intends the student to learn" (p. 21). Their example (see Figure 27.6) occurs in the context of a student working on the following problem in a Cabri-Geométrè environment (the authors cited the source of the problem as Capponi & Laborde, 1995, Cabri-classe, Sheet 4-10).

Construct a triangle ABC. Construct a point P and its symmetrical point P1 about A. Construct the symmetrical point P2 of P1 about B, construct the symmetrical point P3 of P2 about C. Move P.

What can be said about the figure when P3 and P are coincident?

Construct the point I, the midpoint of [PP3].

What can be said about the point I when P is moved? Explain.

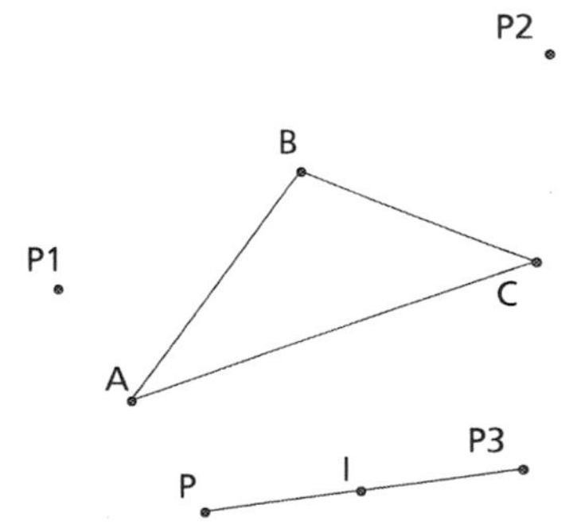

P2
B
P1
C
A
P3
P
I

Figure 27.6 Capponi & Laborde's (1995) geometry task.

As intended by the teacher and the task, the student in the study noticed that, as P is moved to different parts of the plane, I is stationary and that ABCI is a parallelogram. The teacher understood that because ABCI is a parallelogram and because A, B, and C are fixed, then I is also fixed. The student focused on establishing that ABCI is a parallelogram but did not see the connection to I being stationary. When the student, with scaffolding provided by the teacher, arrived at a proof that ABCI is a parallelogram, the teacher was satisfied that the student had solved the problem, despite the fact that the student never established why I is stationary. Identification of the low level of mathematical concordance in this example highlights the importance of the teacher's awareness of the extent to which the students' mathematical goals match those of the teacher. The researchers accentuated the importance of the teacher's recognition of level of mathematical concordance and pointed out that degree of mathematical concordance may be more visible to them in technological environments, arguing that

> the teacher is a crucial mediator between individually constructed knowing and socially constructed knowledge. Whatever the medium, students can always construct meanings which are different from those which the teacher intends, the situation is no more complex with computational media than it is with paper and pencil media. However we suggest that with computational media the meanings which students construct are potentially more transparent to teachers if they learn to make sense of the phenomena of the screen. Moreover student/teacher interactions are also potentially more transparent to researchers when students work with computer-based environments. (p. 23)

As the researchers indicated, it is important that the teacher recognized the degree of mathematical concordance and accurately interpreted the student's mathematics in order to steer the student's attention in a fruitful direction.

Although mathematical concordance is certainly not confined to learning in technological environments, the insertion into the student–task–teacher mix of an additional party (the computer) that has its own rules of operation can increase the probability of lack of agreement on mathematical goals. Filtering one's research on mathematics learning in technological environments through the construct of mathematical concordance may provide a more complete account of technology-based learning of mathematics. Of course, divergent mathematical interpretations do not always arise from one of them being incorrect. Sometimes a low level of concordance arises from variant interpretations of the images on the screen. Goldenberg (1988) pointed out how one could provide numerous interpretations for the nature of transformations of graphs as they appear on computer or calculator screens. For example, he pointed out the difficulty of interpreting the transformation of the graph of $f_1(x) = x^2 - 2$ to the graph of $f_2(x) = 2x^2 - 2$; three different possible transformations (horizontal scaling transformation, vertical transformation, and rotational transformation) are suggested in a diagram that appeared in Goldenberg's article (see Figure 27.7).

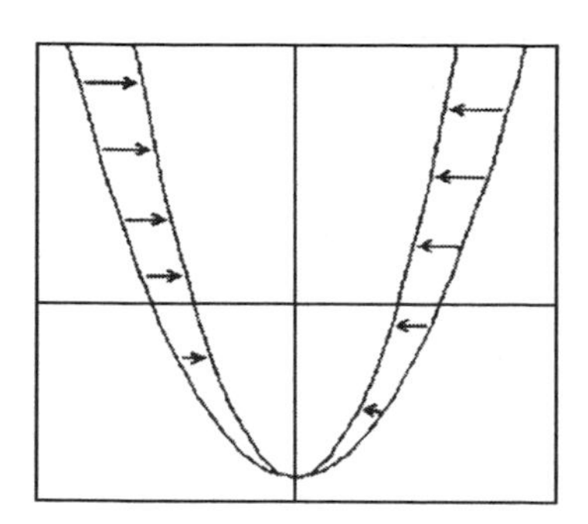
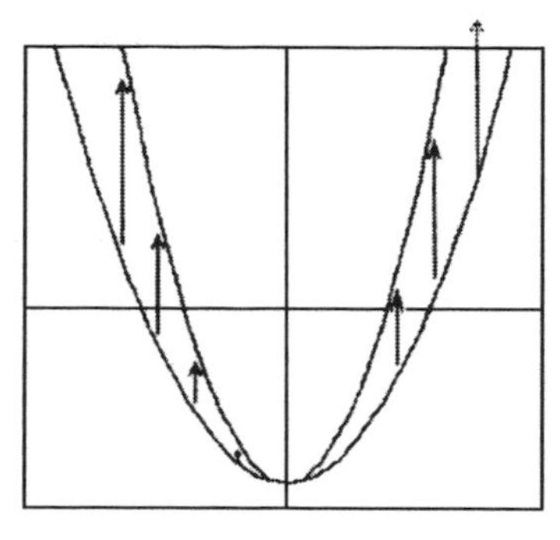
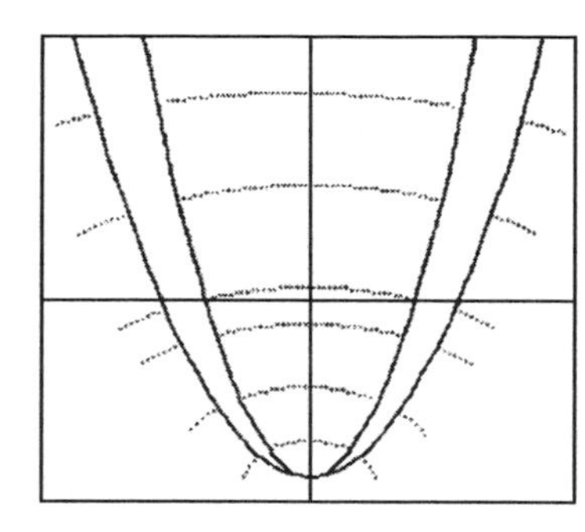

Figure 27.7 Visual descriptions of three different transformations of a graph (Goldenberg, 1988, p. 158). *Note.* Reprinted from *Journal of Mathematical Behavior, 7,* E. P. Goldenberg, Mathematics, metaphors, and human factors: Mathematical, technical, and pedagogical challenges in the educational use of graphical representations of functions, pp. 135–173, copyright 1988, with permission from Elsevier.)

At times the teacher's mathematics will appear to be different partly because the teacher is operating in a new environment and taking some shortcuts. An example of this was reported by Heid and colleagues (Heid, 1996b; Heid, Zbiek, & Blume, 1994) when they recounted how a teacher's (Sara's) weak concept of function notation coupled with her minimal understanding of the computer algebra program she was using led to her systematic misuse and misinterpretation of mathematical notation for function. Within Sara's tasks was the need to evaluate expressions with function notation, such as $f(4)$ given the function f with rule $f(x)=3x^2-10x+5$. She was using a CAS that required the user to enter in one window an expression to be evaluated and to go to another window to redefine a function. Being somewhat unsteady in her use of function notation, being tentative about the working of the computer program, and being accustomed to using the window to redefine functions, whenever Sara needed to find the value of a function, f, for a given input value, a, she redefined the function by inserting the input value into the function rule in the redefinition window instead of issuing a command to evaluate $f(a)$. For example, to evaluate $f(4)$ given $f(x)=3x^2-10x+5$, Sara input the expression, $3(4)^2-10(4)+5$, in the redefine window. There was no conflict generated by this action in as much as her action resulted in the numerical value she needed. Her action, however, changed the function rule to a constant function with a value of $f(4)$ and resulted in her losing the potential to use the function rule stored in the computer (now newly defined) in the remainder of the task. Her students followed her strategy with resulting confusion when subsequent use of the function rule stored in the computer yielded unexpected results. What Sara saw as an evaluation tool was designed as a redefinition tool. In this case, the low level of concordance between the mathematics of the teacher and the mathematics of the tool had unanticipated, unfortunate consequences for her students.

The construct of mathematical concordance can serve to remind researchers that, in analyzing the ways that teachers and students interact with cognitive tools, it is helpful to consider the mathematics of the tool, the mathematics of the teacher, the mathematics the teacher intended through particular technology-based activities, the mathematics that the student engaged in as a result of the technology-based activity arranged by the teacher, and the mathematics that is learned. The construct of mathematical concordance reminds researchers of the importance of the matches and mismatches among these different understandings of mathematics in technological environments.

Amplifiers and Reorganizers: New Emphasis and New Order

In addition to creating technology-intensive curricula that affect the content of school mathematical activity, researchers have developed ways of thinking about technology-based curricula that suggest alternative organizations and emphases in mathematics instruction. Each of these new configurations is accompanied by research on the effects of that vision of curriculum. An overall way to think about and classify these new approaches was suggested by Pea (1985), who used the constructs of *amplifiers* and *reorganizers* to describe possible curricular roles of technology. Pea pointed out that certain uses of technology are amplifiers in that they accept the goals of the current curriculum and work to achieve those goals better. Others are reorganizers in that they change the goals of the curriculum by replacing some things, adding others, and reordering still others. Heid (1997) found these constructs useful in characterizing and differentiating among the ways in which technology had been used in reports of curriculum research to change the order, emphasis, or content of the mathematics curriculum. Early research on the effects of technology-intensive curricula started in the context of the reform of college calculus courses, and the range of curricular arrangements used in these studies can be characterized by the constructs of amplifier and reorganizer. Technology use in the "calculus reform" projects fell somewhere along a continuum of reorganizers in that, when technology was used, it generally had the effect of changing the content of the calculus curriculum in some significant way. Technology was instrumental in restructuring courses around applications, in making a multirepresentational approach feasible, and in helping focus calculus courses on fundamental concepts. Myriad dissertations and other studies focused on analyzing the effects of these restructured technology-intensive curricula (e.g. Beckmann, 1988/1989; Bookman, 2000; Chappell & Killpatrick, 2003; Cooley, 1996; Cunningham, 1992; Ellison, 1994; Estes, 1990; Girard, 2002/2003; Hart, 1991; Hawker, 1987; Heid, 1988; Judson, 1988, 1990; Meel, 1996; Melin-Conejeros, 1992/1993; Palmiter, 1991; Park, 1993; Park & Travers, 1996; Parks, 1995/1996; Porzio, 1995; Roddick, 1998; Schrock, 1990; Wells, 1996). Similar distinctions can be made about research on curricular approaches to the teaching and learning of algebra at the school level. By taking into account the reorganizer–amplifier distinction, researchers can isolate important structural differences in the uses of technology in the teaching and learning of mathematics.

Sequencing and Emphasis: Microprocedures and Macroprocedures

Technology served as a vehicle to reorganize the curriculum in many of the calculus reform projects of the 1990s, and researchers studying those curricula treated the reorganized curricula as a significant contextual attribute. The advent of symbolic algebra software made possible attempts to have students use computing technology to execute routine procedures, which led to questions about the role of traditional routine procedures without the calculator in mathematics curricula. One category of concerns centered on the sequence of treatments of procedures, examining whether it was better for students to learn the by-hand procedures before using technology to produce the results or whether it made more sense to have students learn to use the technology to produce results before learning by-hand procedures to produce those results.

A pair of constructs that is useful in conceptualizing the use of computer algebra systems in mathematics instruction is that of *microprocedures* and *macroprocedures*. For example, Heid (1988) investigated the effects on students' conceptual and procedural understandings of a conceptually oriented curriculum that deferred development of routine skills (microprocedures) until the last few weeks of a 15-week applied calculus course and that focused student attention during the first part of the course on the general procedures (macroprocedures) and concepts of calculus. Heid (2003) used the constructs of macroprocedure and microprocedure to describe the design of curricula like these: "Mathematical procedures can be viewed as macroprocedures whose major procedural steps are component microprocedures" (p. 39). Computing technology makes the enlistment of macroprocedures (and the deferment of by-hand execution of microprocedures) possible with the considerable capability of technology to allow the user to chunk mathematical work in packages of various sizes. Lagrange (2000) and Artigue (2002) have raised issues about such resequencing efforts, pointing to the difficulties of what the French have termed the conceptual–technical cut. Ruthven (2002), however, pointed out that the macro-micro strategy in the Heid study deferred only routine skills and not the more general notion of technique promoted by the French researchers:

> In its own terms, then, the experimental teaching approach studied by Heid did indeed succeed in 'resequencing skills and concepts.' It deferred routinisation of the customarily taught skills of symbolic manipulation until the final phase of the applied calculus course, while the attention given to a broader range of problems and representations in the innovative main phase supported development of a richer conceptual system. Equally, however, in the terms of the French theory, this conceptual development grew out of new techniques constituted in response to this broader range of tasks, and from greater opportunities for the theoretical elaboration of these techniques. At the same time, standard elements emerged from these new tasks, characteristic of the types of problem posed and the forms of representation employed, creating a new corpus of skills distinct from those officially recognised. (p. 284)

The constructs of microprocedures and macroprocedures draw on notions of subprocedures and superprocedures, constructs proposed by Davis (1984) to describe sequences of instructions in mathematics from a cognitive science perspective. Microprocedures and macroprocedures, however, are intended to encompass more than sequences of instructions and are broad enough to take into account the wider notions of technical expertise and knowledge suggested in the French notion of technique. It is the notion of controlling the grain size of the mathematical activity that is encompassed in the constructs of macroprocedures and microprocedures. Certain configurations of technology put that control in the hands of the teacher and the learner.

Constructs related to technology's effects on the content of school mathematics have potential for affecting research on learning, teaching, and curricular alternatives. Representational fluency provides a lens for understanding mathematical activity and students' mathematical thinking. Mathematical concordance provides a way to think about the extent to which the tool, the teacher, the students, and the curriculum address the same mathematics. The constructs of amplifier and reorganizer provide a global categorization scheme for the curricular content choices that technology enables. The constructs of macroprocedure and microprocedure address the relationship between ways of learning mathematics and ways of organizing mathematics curricula. Representational fluency and mathematical concordance provide targets for attention in research that examines students' thinking and interactions in technological environments. Amplifier, reorganizer, macroprocedure, and microprocedure provide ways for researchers to investigate technology-intensive curricular approaches.

SPECIFIC CONTRIBUTIONS OF CONSTRUCTS

We have focused this chapter on identifying and elaborating on constructs that have promise for informing research on technology in the teaching and learning of mathematics. In that endeavor we have recognized the importance of constructs to refine current directions for research, to suggest new directions, to identify promising variables, to assist research in dealing with the complexity of technology-based mathematics teaching and learning, and to characterize how learning is different (and how it is the same) in technological and non-technological environments.

Refining current directions for research. Consideration of constructs can sharpen one's research focus. It helps researchers deal with complexity by identifying critical features of the phenomena they are observing. For example, the construct of instrumental genesis points to the importance, in examining the effects of technology in mathematics learning, of documenting not only what a student does but the student's patterns of technological action with the tool as well as the student-tool relationship.

Suggesting new directions for research. Constructs help identify targets for new research endeavors as well as ways to organize and interpret existing research. For example, because technology-based approaches to mathematics curricula are remarkably varied, it is helpful in generalizing their effects to look for commonalities. Even though researchers may not have identified their technological approaches as amplifiers or reorganizers, applying these constructs to their reports of curricular research helps in recognizing a critical difference in technology-intensive mathematics curricula. Focus on constructs also raises issues about missing areas of research. Realizing the plethora of constructs relating to inappropriate uses of technology, for example, researchers may be motivated to study and develop characterizations of judicious use of technology.

Identifying promising variables. In mathematics, defining an entity brings it into being. In mathematics education, identifying a construct brings the phenomenon or particular distinctions related to the phenomenon into awareness. Constructs give researchers a finer grained descriptive language for variables in the teaching and learning process—the same variable might function as an independent variable, a dependent variable, or an intervening variable. For example, student achievement could be analyzed as a function of students' dragging behaviors, students' dragging behavior could be analyzed as a function of the type of mathematical activity in which the students are engaged, or students' dragging behavior could be viewed as an intervening variable between the mathematical activity and student achievement.

Dealing with the complexity of technology-based mathematics teaching and learning. Technology mediates (and complicates) the complex interplay among teacher, learner, tasks, and content. The insertion of technological tools (with their own independent rules of operation) as mediators in the teacher–student–task mix increases the likelihood of a mismatch among mathematical or pedagogical goals. Identifying constructs related to these goals helps focus research on important relationships related to these goals. Identifying constructs related to technology use with respect to teachers, students, activity, and content can help researchers handle the complexity. Recognizing the importance of constructs such as pedagogical fidelity and privileging helps to capture important features of teaching in technological contexts. These features can then be examined more closely. Some genres of constructs are particularly helpful in dealing with complexity. Fidelity constructs enable focus on the compatibility of goals. The insertion of an additional party into the teacher–student–task mix that has its own rules of operations increases the probability of some degree of mismatch among mathematical goals. This highlights the importance of determining the degree of agreement among different understandings of mathematics in technological worlds. It offers the opportunity for a more complete account of technology-based learning of mathematics. Constructs suggest promising configurations of phenomena of interest in research endeavors to explain the impact of technology in the teaching and learning of mathematics. For example, having identified roles of technology, pedagogical fidelity, and instrumental genesis as constructs of importance, one might be drawn to consider elaborating how roles taken by technology or by the teacher in a technological mathematics setting relate to pedagogical fidelity or instrumental orchestration.

Characterizing how learning is different (and how it is the same) in technological and nontechnological environments. Constructs can shed light on the particular facets of learning mathematics that are different in technological contexts. They accentuate the differences technology makes in learning different types of mathematics (conceptual and technical). For example, one might research the role of cognitive tools, perhaps as embodied in intelligent tutors, in learning skills or developing techniques and the role of multirepresentational software in developing an understanding of mathematical concepts. Constructs can identify specific differences in the technology itself that make a difference in mathematics learning. For example, highlighting features of mathematical technology that characterize the representations they offer may lead to research on the effects of those features. Research using the constructs that describe representations can better illuminate the depth of students' understandings of the concepts themselves.

Constructs can promote boundary crossing. Some constructs that arise outside of mathematics education can also be useful in analyzing the teaching and learning of mathematics, and some constructs that arise in technology-related research can also be informative in nontechnological settings. For example, the construct of instrumental genesis, drawn from outside of mathematics education, turns out to be particularly useful in thinking about technology use in mathematics teaching and learning. Having developed the application of instrumental genesis to computing-technology use in mathematics classrooms, it is natural to apply the construct to other uses of tools (say, manipulatives) in mathematics classrooms.

Consideration of constructs can be informative to teachers and tool developers as well as to researchers. Teachers aware of the notion of privileging may have more of a tendency to examine their own teaching for this bias. Having identified the constructs of mathematical fidelity and pedagogical fidelity, tool developers may have more of a tendency to attend to these features as they build tools, and potential users may have more of a tendency to consider these features in their choice and use of tools.

CONCLUSION

The identification of constructs important to the teaching and learning of mathematics in technological environments paves the way for a comprehensive and connected approach to research in this area. It is our hope that by drawing attention to these constructs we will assist in the formation, framing, and situation of that research. Focusing on the constructs we have identified will assist researchers in explaining not only the effect of technology on the student, teacher, mathematical activity, and curriculum content, but also the effects of technology on the interaction among student, teacher, mathematical activity, and curriculum content in environments that are characterized by intensive use of cognitive mathematics technologies.

REFERENCES

Anderson, J. R., Boyle, C. F., Corbett, A. T., & Lewis, M. W. (1990). Cognitive modeling and intelligent tutoring, *Artificial Intelligence, 42*(1), 7–49.

Artigue, M. (1995). Une approche didactique de l'intégration des EIAO [A pedagogical approach to the integration of EIAO]. In D. Guin, J.-F. Nicaud, & D. Py (Eds.), *Environnements interactifs d'apprentissage avec ordinateur* (pp. 17–29). Paris: Eyrolles.

Artigue, M. (2002). Learning mathematics in a CAS environment: The genesis of a reflection about instrumentation and the dialectics between technical and conceptual work. *International Journal of Computers for Mathematical Learning, 7*, 245–274.

Arzarello, F., Micheletti, C., Olivero, F., Robutti, O., Paola, D., & Gallino, G. (1998). Dragging in Cabri and modalities of transition from conjectures to proofs in geometry. In A. Olivier & K. Newstead (Eds.), *Proceedings of the 22nd conference of the International Group for the Psychology of Mathematics Education* (Vol. 2, pp. 32–39). South Africa: University of Stellenbosch.

Balacheff, N., & Kaput, J. J. (1996). Computer-based learning environments in mathematics. In A. J. Bishop, K. Clements, C. Keitel, J. Kilpatrick, & C. Laborde (Eds.), *International Handbook of Mathematics Education* (pp. 469–501). Dordrecht, The Netherlands: Kluwer.

Balacheff, N., & Sutherland, R. (1994). Epistemological domain of validity of microworlds. The case of Logo and Cabri-géomètre. In R. Lewis & P. Mindelsohn (Eds.), *Lessons from learning* (pp. 137–150). Amsterdam: Elsevier.

Ball, L., & Stacey, K. (2005). Teaching strategies for developing judicious technology use. In W. J. Masalski & P. C. Elliott (Eds.), *Technology-supported mathematics learning environments, 2005 Yearbook of the National Council of Teachers of Mathematics* (pp. 3–15). Reston, VA: National Council of Teachers of Mathematics.

Beaudin, M. & Bowers, D. (1997). Logistics for facilitating CAS instruction. In J. Berry, J. Monaghan, M. Kronfellner, & B. Kutzler (Eds.), *The state of computer algebra in mathematics education* (pp. 126–135). Lancashire, England: Chartwell-York.

Becker, H. J., & Riel, M. M. (2000, December). *Teacher professional engagement and constructivist-compatible computer use.* (Teaching, Learning, and Computing: 1998 National Survey. Report No. 7). University of California, Irvine and University of Minnesota, Center for Research on Information Technology and Organization. Retrieved July 11, 2006 from http://www.crito.uci.edu/tlc/html/findings.html

Becker, J. R. & Rivera, F. (n.d.). Establishing and justifying algebraic generalization at the sixth grade level. Retrieved June 13, 2006 from 130.65.82.235/rivera/nsfcareer.fa05/RR_becker.rtf

Beckmann, C. E. (1989). Effect of computer graphics use on student understanding of calculus concepts (Doctoral dissertation, Western Michigan University, 1988). *Dissertation Abstracts International, 50*(05), 1974.

Beeson, M. (1989). Logic and computation in MATHPERT: An expert system for learning mathematics. In E. Kaltofen & S. Watt (Eds.), *Computers and mathematics* (pp. 202–214). New York: Springer-Verlag.

Biddlecomb, B. D. (1994). Theory-based development of computer microworlds. *Journal of Research in Childhood Education, 8,* 87–98.

Bliss, J., & Ogborn, J. (1989). Tools for exploratory learning. *Journal of Computer Assisted Learning, 5,* 37–50.

Bookman, J. (2000). Program evaluation and undergraduate mathematics renewal: The impact of calculus reform on student performance in subsequent courses. In S. Ganter (Ed.), *Calculus renewal: Issues for undergraduate mathematics education in the next decade* (pp. 91–102). New York: Plenum Press.

Borwein, J. M. (2005). The experimental mathematician: The pleasure of discovery and the role of proof. *International Journal of Computers for Mathematical Learning, 10,* 75–108.

Borwein, J. M., & Bailey, D. H. (2003). *Mathematics by experiment: Plausible reasoning in the 21st Century.* Natick, MA: AK Peters.

Bottino, R., & Furinghetti, F. (1996). Emerging of teachers' conceptions of new subjects inserted in mathematics programs: The case of informatics. *Educational Studies in Mathematics, 30,* 109–134.

Brown, A. L. (1992). Design experiments: Theoretical and methodological challenges in creating complex interventions. *Journal of the Learning Sciences, 2,* 141–178.

Buchberger, B. (1989). Should students learn integration rules? *SIGSAM Bulletin, 24*(1), 10–17.

Burrill, G., Allison, J., Breaux, G., Kastberg, S. Leatham, K., & Sanchez, W. (2002). *Handheld graphing technology in secondary mathematics: Research findings and implications for classroom practice.* Dallas, TX: Texas Instruments.

Capponi, B., & Laborde, C. (1995). *Cabri-classe: Apprendre la géométrie avec un logiciel* [Cabri-classe: Learning geometry with software]. Argenteuil, France: Editions Archimède.

Cestari, M. L., Santagata, R., & Hood, G. (2004). Teachers learning from videos. In B. Clarke, D. M. Clarke, G. Emanuelson, B. Johansson, D. B. Lambdin, F. K. Lester, et al. (Eds.), *International perspectives on learning and teaching mathematics* (pp. 489–502). Göteborg, Sweden: National Center for Mathematics Education.

Chappell, K. & Killpatrick, K. (2003). Effects of concept-based instruction on students' conceptual understanding and procedural knowledge of calculus. *Primus, 13,* 17–37.

Chazan, D. (1999). On teachers' mathematical knowledge and student exploration: A personal story about teaching a technologically supported approach to school algebra. *International Journal for Computers in Mathematical Learning, 4,* 121–149.

Clements, D. H., & Battista, M. T. (with Sarama, J.). (2001). *Journal for Research in Mathematics Education Monograph Number 10: Logo and Geometry.* Reston, VA: National Council of Teachers of Mathematics.

Cobb, P. (1999). Individual and collective mathematical development: The case of statistical data analysis. *Mathematical Thinking and Learning, 1,* 5–43.

Collins, A. (1992). Toward a design science of education. In E. Scanlon & T. O'Shea (Eds.), *New directions in educational technology* (pp. 15–22). Berlin: Springer-Verlag.

Cooley, L. A. (1996). Evaluating the effects on conceptual understanding and achievement of enhancing an introductory calculus course with a computer algebra system (Doctoral dissertation, New York University, 1995). *Dissertation Abstracts International, 56*(10), 3869.

Cunningham, R. F. (1992). The effects of achievement of using computer software to reduce hand-generated symbolic manipulation in freshman calculus (Doctoral dissertation, Temple University, 1991). *Dissertation Abstracts International, 52*(07), 2448.

Davis, R. B. (1984). *Learning mathematics: The cognitive science approach to mathematics education.* Norwood, NJ: Ablex.

Dick, T. (1992). Super calculators: Implications for calculus curriculum, instruction, and assessment. In J. Fey & C. Hirsch (Eds.), *Calculators in mathematics education* (pp. 145–157). Reston VA: National Council of Teachers of Mathematics.

Dick, T. (2007). Keeping the faith: Fidelity in technological tools for mathematics education. In G. W. Blume & M. K. Heid (Eds.), *Research on technology and the teaching and learning of mathematics: Syntheses, cases, and perspectives. Vol. 2: Cases and perspectives* (pp. 333–339). Greenwich, CT: Information Age.

Doerr, H. (2001). Learning algebra with technology: The affordances and constraints of two environments. In H. Chick, K. Stacey, J. Vincent, & J. Vincent (Eds.), *Proceedings of the 12th study conference of the International Commission on Mathematical Instruction: The future of the teaching and learning of algebra* (Vol. 1, pp. 199–206). Melbourne, Victoria, Australia: The University of Melbourne.

Doerr, H. M., & Pratt, D. (2007). The learning of mathematics and mathematical modeling in the context of technology. In M. K. Heid & G. W. Blume (Eds.), *Research on technology and the teaching and learning of mathematics: Syntheses, cases, and perspectives. Vol. 1: Research syntheses.* Greenwich, CT: Information Age.

Doerr, H., & Zangor, R. (2000). Creating meaning for and with the graphing calculator. *Educational Studies in Mathematics, 41,* 143–163.

Drijvers, P. H. M. (2003). *Learning algebra in a computer algebra environment: Design research on the understanding of the concept of parameter* (Doctoral dissertation, Utrecht University, 2003).

Utrecht, The Netherlands: CD-β Press, Center for Science and Mathematics Education.

Dugdale, S. (2007). From network to microcomputers and fractions to functions: Continuity in software research and design. In G. W. Blume & M. K. Heid (Eds.), *Research on technology and the teaching and learning of mathematics: Syntheses, cases, and perspectives. Vol. 2: Cases and perspectives* (pp. 89–112). Greenwich, CT: Information Age.

Ellison, M. (1994). The effect of computer and calculator graphics on students' ability to mentally construct calculus concepts (Doctoral dissertation, University of Minnesota, 1993). *Dissertation Abstracts International, 54*(11), 4020.

Estes, K. A. (1990). Graphics technologies as instructional tools in applied calculus: Impact on instructor, students, and conceptual and procedural achievement (Doctoral dissertation, University of South Florida, 1990). *Dissertation Abstracts International, 51*(04), 1147.

Farrell, A. (1996). Roles and behaviors in technology-integrated precalculus classrooms. *Journal of Mathematical Behavior, 15*, 35–53.

Fey, J. T., Atchison, W. F., Good, R. A., Heid, M. K., Johnson, J., Kantowski, M. G., et al. (1984). *Computing and mathematics: The impact on secondary school curricula.* Reston, VA: National Council of Teachers of Mathematics.

Fey, J. T., Heid, M. K., Good, R., Blume, G. W., Sheets, C., & Zbiek, R. M. (1991). *Computer-intensive algebra.* College Park, MD: University of Maryland.

Fraser, R., Burkhardt, H., Coupland, J., Phillips, R., Pimm, D., & Ridgway, J. (1987). Learning activities and classroom roles with and without computers. *The Journal of Mathematical Behavior, 6*, 305–338.

Friedlander, A., & Tabach, M. (2001). Developing a curriculum of beginning algebra in a spreadsheet environment. In H. Chick, K. Stacey, J. Vincent, & J. Vincent (Eds.), *Proceedings of the 12th study conference of the International Commission on Mathematical Instruction: The future of the teaching and learning of algebra* (Vol. 1, pp. 252–257). Melbourne, Victoria, Australia: The University of Melbourne.

Gifford, B. R. (2005). *Learning conductor mathematics: A community source distributed learning environment.* Berkeley, CA: The Distributed Learning Workshop.

Girard, N. R. (2003). Students' representational approaches to solving calculus problems: Examining the role of graphing calculators (Doctoral dissertation, The University of Pittsburgh, 2002). *Dissertation Abstracts International, 63*(10), 3502.

Glass, B., & Deckert, W. (2001). Making better use of computer tools in geometry. *Mathematics Teacher, 94*, 224–229.

Gluck, K. A. (2000). Eye movements and algebra tutoring (Doctoral dissertation, Carnegie Mellon University, 1999). *Dissertation Abstracts International, 61*(03), 1664.

Goldenberg, E. P. (1988). Mathematics, metaphors, and human factors: Mathematical, technical, and pedagogical challenges in the educational use of graphical representations of functions. *Journal of Mathematical Behavior, 7*, 135–173.

Goldenberg, E. P., & Cuoco, A. A. (1998). What is dynamic geometry? In R. Lehrer & D. Chazan (Eds.), *Designing learning environments for developing understanding of geometry and space* (pp. 351–368). Mahwah, NJ: Erlbaum.

Goldenberg, E. P., Scher, D., & Feurzeig, N. (2006). What lies behind dynamic interactive geometry software? In G. W. Blume & M. K. Heid (Eds.), *Research on technology and the teaching and learning of mathematics: Syntheses, cases, and perspectives. Vol. 2: Cases and perspectives* (pp. 53–87). Greenwich, CT: Information Age.

Goldin, G. A., & Kaput, J. J. (1996). A joint perspective on representation in learning and doing mathematics. In L. P. Steffe, P. Nesher, P. Cobb, G. A. Goldin, & B. Greer (Eds.), *Theories of mathematical learning* (pp. 397–430). Mahwah, NJ: Erlbaum.

González-López, M. J. (2001). Using dynamic geometry software to simulate physical motion. *International Journal of Computers for Mathematical Learning, 6*, 127–142.

Graham, A., & Thomas, M. (2000). Building a versatile understanding of algebraic variables with a graphic calculator. *Educational Studies in Mathematics, 41*, 265–282.

Guin, D., & Trouche, L. (1999). The complex process of converting tools into mathematical instruments: The case of calculators. *International Journal of Computers for Mathematical Learning, 3*, 195–227.

Hall, R., Kibler, D., Wenger, E., & Truxaw, C. (1989). Exploring the episodic structure of algebra story problems. *Cognition and Instruction, 6*, 223–283.

Harel, I. (1990). Children as software designers: A constructionist approach for learning mathematics. *Journal of Mathematical Behavior, 9*, 3–93.

Harel, I., & Papert, S. (1991). *Constructionism.* Norwood, NJ: Ablex.

Hart, D. (1991). Building concept images: Supercalculators and students' use of multiple representations in calculus (Doctoral dissertation, Oregon State University, 1991). *Dissertation Abstracts International 52*(12), 4254.

Hawker, C. (1987). The effects of replacing some manual skills with computer algebra manipulations on student performance in business calculus. *Dissertation Abstracts International, 47*(08), 2934.

Heid, M. K. (1988). Resequencing skills and concepts in applied calculus using the computer as a tool. *Journal for Research in Mathematics Education, 19*, 3–25.

Heid, M. K. (1996). A technology-intensive functional approach to the emergence of algebraic thinking. In C. Kieran, N. Bednarz, & L. Lee (Eds.), *Approaches to algebra: Perspectives for research and teaching* (pp. 239–255). Dordrecht, The Netherlands: Kluwer.

Heid, M. K. (1996, April). *One teacher's understanding of mathematics and technology and its relationship to students' use of a computer algebra system.* Paper presented at the 1996 annual meeting of the American Educational Research Association, New York.

Heid, M. K. (1997). The technological revolution and the reform of school mathematics. *American Journal of Education, 106*(1), 5–61.

Heid, M. K. (2003). Theories for thinking about the use of CAS in teaching and learning mathematics. In J. T. Fey, A. Cuoco, C. Kieran, L. McMullin, & R. M. Zbiek (Eds.), *Computer algebra systems in secondary school mathematics education* (pp. 33–52). Reston, VA: National Council of Teachers of Mathematics.

Heid, M. K., Hollebrands, K., & Iseri, L. (2002). Reasoning, justification, and proof, with examples from technological environments. *Mathematics Teacher, 95*, 210–216.

Heid, M. K., Perkinson, D. B., Peters, S. A., & Fratto, C. L. (2005). Making and managing distinctions: The case of sampling distributions. In G. M. Lloyd, M. Wilson, J. L. M. Wilkins, & S. L. Behm (Eds.), *Proceedings of the 27th annual meeting of the North American Chapter of the International Group for*

Psychology of Mathematics Education—North American Chapter [CD]. Blacksburg, VA: Virginia Polytechnic Institute and State University.

Heid, M. K., Sheets, C., & Matras, M. A. (1990). Computer-enhanced algebra: New roles and challenges for teachers and students. In T. J. Cooney (Ed.), *Teaching and learning mathematics in the 1990s, 1990 Yearbook of the National Council of Teachers of Mathematics* (pp. 194-204). Reston, VA: National Council of Teachers of Mathematics.

Heid, M. K., Sheets, C., Matras, M., & Menasian, J. (1988, April). *Classroom and computer lab interaction in a computer-intensive environment.* Paper presented at the annual meeting of the American Educational Research Association, New Orleans, LA.

Heid, M. K., & Zbiek, R. M. (1993). Nature of understanding of mathematical modeling by beginning algebra students engaged in a technology-intensive conceptually based algebra course. In J. R. Becker & B. J. Pence (Eds.), *Proceedings of the 15th annual meeting of the North American Chapter of the International Group for the Psychology of Mathematics Education* (Vol. 1, pp. 128–134). San Jose, CA: Center for Mathematics and Computer Science Education, San Jose State University.

Heid, M. K., Zbiek, R. M., & Blume, G. (1994). Case studies on empowering secondary mathematics teachers in computer-intensive environments. In D. Kirshner (Ed.), *Proceedings of the sixteenth annual meeting of the North American Chapter of the International Group for the Psychology of Mathematics Education* (Vol. 2, pp. 203–209). Baton Rouge: Louisiana State University.

Hillel, J., Kieran, C., & Gurtner, J. (1989). Solving structured geometric tasks on the computer: The role of feedback in generating strategies. *Educational Studies in Mathematics, 20,* 1–39.

Hoadley, C., & Kirby, J. A. (2004, June). *Socially relevant representations in interfaces for learning.* Paper presented at the International Conference of the Learning Sciences. Santa Monica, CA. Retrieved June 13, 2006 from www.tophe.net/papers/Hoadley-Kirby-icls04.pdf

Hollebrands, K., Laborde, C., & Sträßer, R. (2007). Technology and the learning of geometry at the secondary level. In M. K. Heid & G. W. Blume (Eds.), *Research on technology and the teaching and learning of mathematics: Syntheses, cases, and perspectives. Vol. 1: Research syntheses.* Charlotte, NC: Information Age.

Hong, Y. Y., & Thomas, M. O. J. (2002). Representational versatility and linear algebraic equations. *Proceedings of the International Conference on Computers in Education* (Vol. 2, pp. 1002–1006), Auckland, New Zealand: ICCE.

Hoyles, C., & Noss, R. (1992). A pedagogy for mathematical microworlds. *Educational Studies in Mathematics, 23,* 31–57.

Hoyles, C., Noss, R., & Kent, P. (2004). On the integration of digital technologies into mathematics classrooms. *International Journal of Computers for Mathematical Learning, 9,* 309–326.

Illich, I. (1973). *Tools for conviviality.* New York: Harper and Row.

Judson, P. T. (1988). Effects of modified sequencing of skills and applications in introductory calculus (Doctoral dissertation, The University of Texas at Austin, 1988). *Dissertation Abstracts International, 49*(06), 1397.

Judson, P. T. (1990). Elementary business calculus with computer algebra. *Journal of Mathematical Behavior, 9,* 153–157.

Kafai, Y. B. (1995). *Minds in play: Computer game design as a context for children's learning.* Hillsdale, NJ: Erlbaum.

Kafai, Y. B. (1996). Learning design by making games. In Y. B. Kafai & M. Resnick (Eds.), *Constructionism in practice: Designing, thinking, and learning in a digital world* (pp. 71–96). Hillsdale, NJ: Erlbaum.

Kaput, J. J. (1989). Linking representations in the symbol systems of algebra. In S. Wagner & C. Kieran (Eds.), *Research issues in the learning and teaching of algebra* (pp. 167–194). Reston, VA: National Council of Teachers of Mathematics.

Kaput, J. (1992). Technology and mathematics education. In D. Grouws (Ed.), *Handbook of research on mathematics teaching and learning* (pp. 515–556). New York: Macmillan.

Kaput, J. (1994). Democratizing access to calculus: New routes to old roots. In A. Schoenfeld (Ed.), *Mathematical thinking and problem solving* (pp. 77–156). Hillsdale, NJ: Erlbaum.

Kendal, M., & Stacey, K. (1999). Varieties of teacher privileging for teaching calculus with computer algebra systems. *International Journal of Computer Algebra in Mathematics Education, 6,* 233–247.

Kendal, M., & Stacey, K. (2000). Acquiring the concept of derivative: Teaching and learning with multiple representations and CAS. In T. Nakahara & M. Koyama (Eds.), *Proceedings of the 24th annual conference of the International Group for the Psychology of Mathematics Education* (Vol. 3, pp. 127–134). Hiroshima, Japan: Hiroshima University.

Kendal, M., & Stacey, K. (2001). The impact of teacher privileging on learning differentiation with technology. *International Journal of Computers for Mathematical Learning, 6,* 143–165.

Koedinger, K. R. (2001). Cognitive tutors as modeling tool and instructional model. In K. D. Forbus & P. J. Feltovich (Eds.), *Smart machines in education: The coming revolution in educational technology* (pp. 145–168). Menlo Park, CA: AAAI/MIT Press.

Koedinger, K. R., & Anderson, J. R. (1998). Illustrating principled design: The early evolution of a cognitive tutor for algebra symbolization. *Interactive Learning Environments, 5,* 161–180.

Koedinger, K. R., & Tabachneck, H. J. M. (1994, April). *Two strategies are better than one: Multiple strategy use in word problem solving.* Paper presented at the annual meeting of the American Educational Research Association, New Orleans, LA.

Laborde, C., & Laborde, J.-M. (2007). The development of a dynamical geometry environment: Cabri-géomètre. In G. W. Blume & M. K. Heid (Eds.), *Research on technology and the teaching and learning of mathematics: Syntheses, cases, and perspectives. Vol. 2: Cases and perspectives* (pp. 31-52). Greenwich, CT: Information Age.

Lagrange, J.-B. (1999). Complex calculators in the classroom: Theoretical and practical reflections on teaching pre-calculus. *International Journal of Computers for Mathematical Learning, 4,* 51–81.

Lagrange, J.-B. (2000). L'intégration d'instruments informatiques dans l'enseignement: Une approche par les techniques [The integration of technological tools in instruction: A techniques approach]. *Educational Studies in Mathematics, 43,* 1–30.

Lumb, S., Monaghan, J., & Mulligan, S. (2000). Issues arising when teachers make extensive use of computer algebra. *International Journal of Computer Algebra in Mathematics Education, 7,* 223–240.

Mariotti, M., & Bussi, M. (1998). From drawing to construction: Teacher's mediation within the Cabri environment. In A. Olivier & K. Newstead (Eds.), *Proceedings of the 22nd confer-*

ence of the International Group for the Psychology of Mathematics Education (Vol. 3, pp. 247–255). Stellenbosch, South Africa: Faculty of Education.

McCandliss, B., Kalchman, M., & Bryant, P. (2003). Design experiments and laboratory approaches to learning: Steps toward collaborative exchange. *Educational Researcher, 32*(1), 14–16.

Meel, D. E. (1996). A comparative study of honor students' understandings of central calculus concepts as a result of completing a Calculus & Mathematica® or a traditional calculus curriculum (Doctoral dissertation, University of Pittsburgh, 1995). *Dissertation Abstracts International, 57*(01), 142.

Melin-Conejeros, J. (1993). The effect of using a computer algebra system in a mathematics laboratory on the achievement and attitude of calculus students (Doctoral dissertation, University of Iowa, 1992). *Dissertation Abstracts International, 53*(07), 2283.

Noss, R., & Hoyles, C. (1996). *Windows on mathematical meanings: Learning cultures and computers.* Dordrecht, The Netherlands: Kluwer.

Olive, J. (2000). Computer tools for interactive mathematical activity in the elementary school. *International Journal of Computers for Mathematics Learning, 5,* 241–262.

Olive, J., & Lobato, J. (in press). The learning of rational number concepts using technology. In M. K. Heid & G. W. Blume (Eds.), *Research on technology and the teaching and learning of mathematics: Syntheses, cases, and perspectives. Vol. 1: Research syntheses.* Charlotte, NC: Information Age.

Olive, J., & Steffe, L. P. (1994). *TIMA: Bars* [Computer software]. Acton, MA: William K. Bradford.

Palmiter, J. R. (1991). Effects of computer algebra systems on concept and skill acquisition in calculus. *Journal for Research in Mathematics Education, 22,* 151–156.

Papert, S. (1993). *The children's machine: Rethinking school in the age of the computer.* New York: Basic Books.

Park, K. (1993). A comparative study of the traditional calculus course versus the Calculus & Mathematica course (Doctoral dissertation, University of Illinois at Urbana-Champaign, 1993). *Dissertation Abstracts International, 54*(01), 119.

Park, K, & Travers, K. J. (1996). A comparative study of a computer-based and a standard college first-year calculus course. *CBMS Issues in Mathematics Education, 6,* 155–176.

Parks, V. W. (1996). Impact of a laboratory approach supported by 'Mathematica' on the conceptualization of limit in a first calculus course (Doctoral dissertation, Georgia State University, 1995). *Dissertation Abstracts International 56*(10), 3872.

Pea, R. D. (1987). Cognitive technologies for mathematics education. In A. H. Schoenfeld (Ed.), *Cognitive science and mathematics education* (pp. 89–122). Hilldale, NJ: Erlbaum.

Pickett, J. P. (Ed.). (2000). *The American heritage dictionary of the English language* (4th ed.). Boston: Houghton Mifflin.

Piliero, S. C. (1994). The effects of a problem-based curriculum, multi-representational software, and teacher development on the knowledge, beliefs and practices of a secondary mathematics teacher (Doctoral dissertation, Cornell University, 1994). *Dissertation Abstracts International, 55*(5), 1215.

Pimm, D. (1995). *Symbols and meanings in school mathematics.* London: Routledge.

Porzio, D. T. (1995). The effects of differing technological approaches to calculus on students' use and understanding of multiple representations when solving problems (Doctoral dissertation, The Ohio State University, 1994). *Dissertation Abstracts International, 55*(10), 3128.

Pratt, D., & Ainley, J. (1997). The construction of meanings for geometric construction: Two contrasting cases. *International Journal of Computers for Mathematical Learning, 1,* 293–322.

Ravitz, J. L., Becker, H. J., & Wong, Y. (2000). *Constructivist-compatible beliefs and practices among U.S. teachers.* (Teaching, Learning, and Computing: 1998 National Survey. Report No. 4). University of California, Irvine and University of Minnesota, Center for Research on Information Technology and Organization. Retrieved July 11, 2006, from http://www.crito.uci.edu/tlc/html/findings.html.

Renninger, K. A., Weimar, S. A., & Klotz, E. A. (1998). Teachers and students investigating and communicating about geometry: The Math Forum. In R. Lehrer & D. Chazan (Eds.), *Designing learning environments for developing understanding of geometry and space* (pp. 465–488). Mahwah, NJ: Erlbaum.

Ritter, S., Haverty, L., Koedinger, K. R., Hadley, W., & Corbett, A. T. (2007). Integrating intelligent software tutors with the mathematics classroom. In G. W. Blume & M. K. Heid (Eds.), *Research on technology and the teaching and learning of mathematics: Syntheses, cases, and perspectives. Vol. 2: Cases and perspectives* (pp. 157–181). Greenwich, CT: Information Age.

Rivera, F. D. (2005). An anthropological account of the emergence of mathematical proof and related processes in technology-based environments. In W. J. Masalski & P. C. Elliott (Eds.), *Technology-supported mathematics learning environments, 2005 Yearbook of the National Council of Teachers of Mathematics* (pp. 125–136). Reston, VA: National Council of Teachers of Mathematics.

Rochowicz, J. A. (1996). The impact of using computers on calculus instruction: Various perceptions. *Journal of Computers in Mathematics and Science Teaching, 15,* 423–435.

Roddick, C. D. (1998). A comparison study of students from two calculus sequences on their achievement in calculus-dependent courses (Doctoral dissertation, The Ohio State University, 1997). *Dissertation Abstracts International 58*(07), 2577.

Ruthven, K. (1990). The influence of graphic calculator use on translation from graphic to symbolic forms. *Educational Studies in Mathematics, 21,* 431–450.

Ruthven, K. (2002). Instrumenting mathematical activity: Reflections on key studies of the educational use of computer algebra systems. *International Journal of Computers for Mathematical Learning, 7,* 275–291.

Ruthven, K., & Hennessy, S. (2002). A practitioner model of the use of computer-based tools and resources to support mathematics teaching and learning. *Educational Studies in Mathematics, 49,* 47–88.

Sandoval, W. A., Bell, P., Coleman, E., Enyedy, N., & Suthers, D. (2000, April). *Designing knowledge representations for epistemic practices in science learning.* Paper presented at the annual meeting of the American Educational Research Association, New Orleans, LA.

Schrock, C. S. (1990). Calculus and computing: An exploratory study to examine the effectiveness of using a computer algebra system to develop increased conceptual understanding in a first semester calculus course (Doctoral dissertation, Kansas State University, 1989). *Dissertation Abstracts International, 50*(07), 1926.

Schwartz, D. L. (1995). The emergence of abstract representations in dyad problem solving. *Journal of the Learning Sciences, 4*, 321–354.

Sheets, C. (1993). Effects of computer learning and problem-solving tools on the development of secondary school students' understanding of mathematical functions (Doctoral dissertation, University of Maryland, 1993). *Dissertation Abstracts International, 54*(5), 1714.

Steffe, L. P., & Olive, J. (2002). Design and use of computer tools for interactive mathematical activity (TIMA). *Journal of Educational Computing Research, 27*, 55–76.

Stroup, W. M. (2002). Understanding qualitative calculus: A structural synthesis of learning research. *International Journal of Computers for Mathematical Learning, 7*, 167–215.

Sutherland, R., & Balacheff, N. (1999). Didactical complexity of computational environments for the learning of mathematics. *International Journal of Computers for Mathematical Learning, 4*, 1–26.

Tabachneck, H. J. M., Koedinger, K. R., & Nathan, M. J. (1994). Toward a theoretical account of strategy use and sense-making in mathematics problem solving. In *Proceedings of the 16th annual conference of the Cognitive Science Society* (pp. 836–841). Hillsdale, NJ: Erlbaum.

Timmerman, M. (1999). Learning in the context of a technology-enriched mathematics teacher education course: Two case studies of elementary teachers' conceptions of mathematics, mathematics teaching and learning, and the teaching of mathematics with technology (Doctoral dissertation, The Pennsylvania State University, 1998). *Dissertation Abstracts International, 59*(8), 2896.

TIMSS Video Mathematics Research Group. (2003). Understanding and improving mathematics teaching: Highlights from the TIMSS 1999 Video Study. *Phi Delta Kappan, 84*, 768–775.

Trouche, L. (2005a). An instrumental approach to mathematics learning in symbolic calculator environments. In D. Guin, K. Ruthven, & L. Trouche (Eds.), *The didactical challenge of symbolic calculators: Turning a computational device into a mathematical instrument* (pp. 137–162). New York: Springer.

Trouche, L. (2005b). Instrumental genesis, individual and social aspects. In D. Guin, K. Ruthven, & L. Trouche (Eds.), *The didactical challenge of symbolic calculators: Turning a computational device into a mathematical instrument* (pp. 197–230). New York: Springer.

Underwood, J. S., Hoadley, C., Lee, H. S., Hollebrands, K., DiGiano, C., & Renninger, K. A. (2005). IDEA: Identifying Design Principles in Educational Applets. *Educational Technology Research and Development, 53*, 99–112.

Verillon, P. & Rabardel, P. (1995). Cognition and artifacts: A contribution to the study of though[t] in relation to instrumented activity. *European Journal of Psychology in Education 9*, 77–101.

Wells, P. J. (1996). Conceptual understanding of major topics in first semester calculus: A study of three types of calculus courses at the University of Kentucky (Doctoral dissertation, University of Kentucky, 1995). *Dissertation Abstracts International, 56*(09), 3493.

Wertsch, J. (1990). *The voice of rationality in a sociocultural approach to mind.* Cambridge, MA: Cambridge University Press.

Yerushalmy, M. (1997). Reaching the unreachable: Technology and the semantics of asymptotes. *International Journal of Computers for Mathematical Learning, 2*, 1–25.

Zbiek, R. M. (1995). Her math, their math: An in-service teacher's growing understanding of mathematics and technology and her secondary school students' algebra experience. In D. Owens, M. K. Reed, & G. M. Millsaps (Eds.), *Proceedings of the 17th annual meeting of the North American Chapter of the International Group for the Psychology of Mathematics Education* (pp. 214-220). (ERIC Document Reproduction Service No. ED389610)

Zbiek, R. M. (1998). Prospective teachers' use of computing tools to develop and validate functions as mathematical models. *Journal for Research in Mathematics Education, 29*, 184–201.

Zbiek, R. M., & Hollebrands, K. (in press). A research-informed view of the process of incorporating mathematics technology into classroom practice by inservice and prospective teachers. In M. K. Heid & G. W. Blume (Eds.), *Research on technology and the teaching and learning of mathematics: Syntheses, cases, and perspectives. Vol. 1: Research syntheses.* Charlotte, NC: Information Age.

Zech, L., Vye, N. J., Bransford, J. D., Goldman, S. R., Barron, B. J., Schwartz, D. L., et al. (1998). An introduction to geometry through anchored instruction. In R. Lehrer & D. Chazan (Eds.), *Designing learning environments for developing understanding of geometry and space* (pp. 439–464). Mahwah, NJ: Erlbaum.

AUTHOR NOTE

The authors gratefully acknowledge the helpful comments and suggestions offered by George Bright, Sharon Dugdale, and Frank Lester on an earlier version of this chapter. We also gratefully acknowledge the support of the National Science Foundation (NSF Grant ESI 0087447). However, the ideas presented here are those of the authors and not necessarily those of the reviewers or of NSF.

28

ENGINEERING CHANGE IN MATHEMATICS EDUCATION

Research, Policy, and Practice

William F. Tate

WASHINGTON UNIVERSITY IN ST. LOUIS

Celia Rousseau

UNIVERSITY OF MEMPHIS

In 1985, the Mid-Atlantic Equity Center published *Mathematics and Science: Critical Filters for the Future,* addressing in part mathematics education, school leadership, and academic opportunity for traditionally underserved students. In her monograph, DeAnna Banks Beane argued, "The success of many intervention programs demonstrates that there are no permanent barriers to minority student achievement in science and mathematics. However, the data tell us that the longer we wait to intervene, the more invincible the barriers become" (p. 1). Her remarks are a reminder of the challenges and opportunities in school mathematics requiring clarification and associated strategies for change and improvement. This chapter represents an effort to build upon and extend beyond the literature on school mathematics as discussed in *Mathematics and Science: Critical Filters for the Future.* The purpose of this chapter is to provide a review of mathematics education research and relevant policy literature that can inform school leadership attempting to build effective learning environments for students traditionally underserved in school mathematics.

The political and educational landscape in school mathematics has changed in important ways since 1985. Three significant changes are discussed here. The first change is the introduction of mathematics standards to the education community, specifically the 1989 release of the *Curriculum and Evaluation Standards for School Mathematics* published by the National Council of Teachers of Mathematics (NCTM). This document was a part of a series of mathematics standards documents produced by the Council (NCTM, 1991, 1995, 2000). The role of standards in educational practice and policy-making has gained traction, and today dominates discourse related to school mathematics. The NCTM Standards documents and related educational policy developments have resulted in the rapid evolution of standards-based language. In the post *Curriculum and Evaluation Standards for School Mathematics* era (1990–2002), the word *standard* produced 26,843 documents in the ERIC database. Although the citations were not all directly related to school mathematics, the point here is that standards-based language permeates the education terrain. Most states have standards for school mathematics that signal to local school districts goals for instruction and desired student outcomes.

A second change is a movement calling for educational leadership to more directly address issues of learning and teaching in schools (Rowan, 1995). Significant changes are taking place in the ways the constructs of teaching and learning are now being defined by researchers, practitioners, and policymakers. During the 1980s and 1990s, cognitive models of teaching and learning were formulated and tested, and many small-scale efforts to transition from the predominant behaviorist models of instructional theory occurred. This research and development has implications for understanding best practice in the design of educational goals, implementation of instructional practices, and development of assessment techniques. Thus, instructional leadership requires a deep understanding of research and development. Many state and federal policies require school district instructional leadership to document the effectiveness or research-base undergirding local change strategies. This represents a new demand on those charged with district-wide and school-level improvements. This chapter is designed for instructional leaders facing today's research-focused managerial demands.

A third and related change in the educational landscape is the *No Child Left Behind Act of 2001* (NCLB). The NCLB Act calls for a new level of Title I accountability by requiring each state to implement accountability systems covering all public schools and public school students. These systems must be based on rigorous state standards in mathematics, annual testing for all students in Grades 3–8, and annual statewide progress objectives ensuring that all groups of students reach established levels of mathematical proficiency within 12 years. Assessment findings and state progress objectives must be broken out by poverty, race, ethnicity, disability, and limited English proficiency to ensure that no group is left behind. School districts and schools that fail to make adequate yearly progress (AYP) toward statewide proficiency goals will, over time, be subject to improvement, corrective action, and restructuring measures aimed at getting them back on course to meet state standards. Schools that meet or exceed AYP objectives or close achievement gaps will be eligible for State Academic Achievement Awards.

Setting rigorous state standards and related accountability models is placing significant pressure on school districts to rethink past practice and to look for effective and sound strategies to support the teaching and learning of mathematics. This chapter is designed to assist teachers, administrators, and community supporters in their efforts to incorporate research-based strategies into the school mathematics program. Additionally, and equally as important, the chapter is an effort to provide mathematics education researchers with a synthetic review of factors that influence students' learning of mathematics.

ENGINEERING CHANGE: AN OVERVIEW

How can teachers be more productive teachers of school mathematics? How can students' learning of school mathematics be accelerated? These are difficult questions; yet Hamming provided insight into one appropriate response.

> Knowledge and productivity are like compound interest. Given two people of approximately the same ability and one person who works 10% more than the other, the latter will more than twice outproduce the former. The more you know, the more you learn; the more you learn, the more you can do; the more you can do, the more the opportunity—it is very much like compound interest. I don't want to give you a rate, but it is a very high rate. Given two people with exactly the same ability, the one person who manages day in and day out to get in one more hour of thinking will be tremendously more productive over a lifetime.[1]

The teaching and learning process is embedded in a complex web of schools, school districts, communities, and state governance systems that each play a role in expanding students' opportunity to learn and think about mathematics. Some have criticized the mathematics education community for failing to adequately articulate how access and opportunity to learn mathematics can be expanded to traditionally underserved students (Apple, 1992; Hilliard, 1991; Meyer, 1991). NCTM has recognized this criticism. Recent standards documents produced by the NCTM have called for a focus on equity. For example, the *Principles and Standards for School Mathematics* (NCTM, 2000) stated the following:

> The vision of equity in mathematics education challenges a pervasive societal belief in North America that only some students are capable of learning mathematics. This belief, in contrast to the equally pervasive view that all students can and should learn to read and write in English, leads to low expectations for too many students. Low expectations are especially problematic because students who live in poverty, students who are not native speakers of English, students with disabilities, females, and many nonwhite students have traditionally been far more likely than their counterparts in other demographic groups to be the victims of low expectations. Expectations must be

[1] This quote is taken from a transcription of the Bell Communications Research Seminar, March 7, 1986.

raised—mathematics can and must be learned by all students. (pp. 12–13)

High expectations for all students are a new challenge in school mathematics education. Past reform efforts in mathematics education were designed for more select groups. For example, in the post-*Brown* era, the "new math" reform movement sought to improve mathematics education in the United States, as it was thought that good scientific education was a vital component of a strong national defense program and a robust economy (Kliebard, 1987). Initiated in response to the launch of Sputnik by the Soviet Union, this mathematics reform effort designed to address the nation's scientific crisis did little to address the problems of students of color in urban and rural areas of the United States (Garcia, 1995; Neito, 1995; Tate, 1997). Many responsible for the reform effort stated that their programs should be limited to "college-capable" students (Devault & Weaver, 1970; Kliebard, 1987; NCTM, 1959). The code words *college capable* were a signifier to the educational establishment that only a select few communities and students were appropriate for the reform activities. This is not to say that urban, rural, and poor communities were completely denied opportunities to participate in this reform effort. Instead, these opportunities were limited and insufficient for the curricular and pedagogical changes called for within the reform movement. Thus, for many students—particularly African American and Hispanic students—the late 1950s and 1960s are best characterized as an era of benign neglect with respect to opportunity to learn challenging, high-level mathematics (Tate, 1996).

The focus of mathematics curriculum and pedagogy has evolved in a cyclic fashion. In the late 1960s and early 1970s, a different mathematics movement, "back to basics," emerged, which focused primarily on elementary and middle schools (NCTM, 1980). This movement was a product of policy directives conceived to address equality of educational opportunity through compensatory education. The back-to-basics effort called for instruction in a narrow set of rudimentary mathematics procedures and facts, often to the exclusion of conceptually rich tasks and advanced mathematical ideas.

Members of the National Council of Supervisors of Mathematics (NCSM) were concerned about the effect this would have on the teaching and learning of mathematics appropriate to the needs in a modern society. A NCSM 1977 position paper urged that we move forward, not "back" to the basics. Not included in the back-to-basics movement were 10 important areas of mathematics students would find essential as adults: problem solving; applying mathematics to everyday situations; alertness to reasonableness of results; estimation and approximation; appropriate computational skills; geometry; measurement; reading, interpreting, and constructing tables, charts and graphs; using mathematics to predict; and computer literacy. The NCSM position paper was widely influential in school mathematics circles; however, the back-to-basics movement had a pronounced impact on the learning opportunities of low-income urban schools (Strickland & Ascher, 1992).

On the positive side, the basic skills effort resulted in limited gains on narrowly defined aspects of school mathematics for traditionally underserved student demographic groups (Secada, 1992). It served as an existence proof that when teachers and administrators agreed on and supported a common goal in mathematics, students would learn the content. However, as the vision of mathematics education has shifted from largely rudimentary notions to a more challenging standard, the limitations of past pedagogical and school organizational support systems are apparent. The National Research Council (2001) stated the following:

> To many people, school mathematics is virtually a phenomenon of nature. It seems timeless, set in stone—hard to change and perhaps not needing to change. But the school mathematics education of yesterday, which had a practical basis, is no longer viable. Rote learning of arithmetic procedures no longer has the clear value it once had. The widespread availability of technological tools for computation means that people are less dependent on their own powers of computation. At the same time, people are much more exposed to numbers and quantitative ideas and so need to deal with mathematics on a higher level than they did just 20 years ago. Too few U.S. students, however, leave elementary and middle school with adequate mathematical knowledge, skill, and confidence for anyone to be satisfied that all is well in school mathematics. (p. 407)

One response to the current state of affairs in school mathematics has been the rapid development and adoption of state-level mathematics standards. Typically, an accountability model is associated with the mathematics content standards to provide indicators of student progress. However, situated in the time between the adoption of mathematics standards and the application of accountability models are important aspects of the educational process. There is an unstated assumption that standards and accountability models are only part of the solution strategy for school improvement in mathematics education. The assumption is that armed with quantitative data, local leadership—teachers, mathematics coordinators and supervisors, principals, assistant superintendents, superintendents, and school board members—will proactively respond to data. The way in which local school leadership responds to system data has a profound consequence for students' opportunity to learn. Today's school leadership is being called to engineer school improvement and the classroom learning setting. This chapter is written with the hope that it will help the reader understand how research-based strategies can support the engineering of positive change to the structures supporting the teaching and learning of mathematics in educational settings.

The engineer as a metaphor representing a change agent requires a brief explanation. To some, the engineer may seem to be synonymous with the scientist.[2] The distinction between a scientist and an engineer is partially clarified by examining two activities related to the preparation of each professional—analysis and design. In science classes, students are required to solve problems, observe phenomena in lab settings, record observations, and perform calculations. This process is the essence of analysis. In engineering classes, the instruction often stresses the importance of design. The difference between analysis and design can be described in the following way: If only one solution to a problem exists and discovering it merely entails putting together pieces of discrete information, the activity is probably analysis (Horenstein, 2002). In comparison, if more than one solution exists, and if determining a reasonable path demands being creative, making choices, performing tests, iterating, and evaluating, then the activity is design. Design often includes analysis; however, it also must involve at least one of these latter components. Horenstein (2002) offered the following example to further clarify the difference between analysis and design:

> [A] remote-controlled buoy is located off the coast of California and is maintained by the U.S. National Oceanic and Atmospheric Administration (NOAA). It provides 24-hour data to mariners, the Coast Guard, and weather forecasters. Processing the data stream from this buoy, posting it on the Internet, and using information to forecast the weather are examples of analysis. Deciding *how* to build the buoy so that it meets the needs of NOAA is an example of design. (p. 29)

Administrator and teacher leadership charged with addressing our nation's school mathematics challenges must decide how to build effective programs. Clearly, there is more than one solution to our school mathematics problem, and designing appropriate solutions will require creativity, hard choices, performance tests, iterative action, and evaluation. Like engineers, mathematics educators must study access and opportunity-to-learn issues in great depth and then design an intervention—"learn to build." In contrast, most scientists construct instruments to measure and study phenomena of interest—they "build to learn." This chapter is dedicated to those interested in "learning to build" outstanding school mathematics programs.

The next two sections provide an examination of the challenges facing school mathematics change agents. The second section documents changes in U.S. mathematics achievement by reviewing population trends and national achievement trend studies. A focus of this section is to determine achievement trends of various racial-ethnic and socioeconomic groups. The third section examines opportunity-to-learn (OTL) factors that have the potential to positively influence the learning of mathematics. The intent in this section is to offer possible building blocks to support the engineering of positive change in school mathematics and to review the work of some scholars who have designed school mathematics improvement models based on important OTL factors. The fourth section provides a closer look at research-based cases of successful mathematics programs. This section will highlight both classroom and organizational components that are present in high performing school mathematics programs. The fifth section is a review of two large-scale change efforts in school mathematics and related evaluation information. The sixth and final section is a brief review of the engineering perspective—learning to build—and its importance for school mathematics improvement.

LEARNING TO BUILD: THE PROBLEM DEFINED

Who are the children in the classrooms of today and citizenry of tomorrow? One challenge posed in recent calls for mathematics reform is to improve the academic performance of all students, and particularly students traditionally underserved in mathematics classrooms. For instance, the *National Education Goals Report: Building a Nation of Learners* (National Education Goals Panel, 1995) called for the mathematics performance of all students at the elementary and

[2] In fact, Hurd (1997) argued that the paradigmatic boundaries of science are shifting toward a science guided by the coaction of science and technology, perceived as an integrated system. Further, he indicated, many in the science community speculate that engineering education may be the best preparation for the natural sciences. Yet, this speculation suggests there are distinct paradigmatic differences.

secondary levels to increase significantly in every quartile and for the distribution of minority students in each quartile to better reflect the student population as a whole. Addressing this challenge requires some insight into the population trends of various demographic groups. A particular focus of this section is population trends over the past 30 years.

The student race/ethnicity population trends have changed dramatically since the 1985 release of *Mathematics and Science: Critical Filters for the Future*. Table 28.1 provides insight into this trend.

Table 28.1 Percentage Distribution of Public School Students in K–12, by Race: Fall 1972 and 2003

Race/Ethnicity	Fall 1972	Fall 2003
White	78%	58%
Black	15%	16%
Hispanic	6%	19%
Other	1%	7%

Note: From U.S. Department of Education, 2005.

The information in Table 28.1 requires additional explanation. The U.S. school-age population declined between 1980 and 1990 but became more diverse. The United States General Accounting Office (GAO, 1993) reported in 1990 about 44.4 million school-age children (ages 5–17), a decline of more than 2.3 million, or 5.8% since 1980. In 1992, the percentages of male and female students 5–18 years old enrolled in school were 51.4% and 48.6%, respectively (National Science Foundation, 1994).

Changes in the racial-ethnic characteristics of the U.S. population have been a part of American life since the first European settlements. However, only in recent decades has the population in the United States become less, rather than more, White. The racial-ethnic diversity of the country is much greater now than at any previous period in history and seems on course to become progressively more diverse for some time to come (Riche & Pollard, 1992; Vernez, 1992). This diversity is reflected in recent trends of school-age children.

During the 1980s, the White school-age population declined by more than 4 million children, or about 12%, and the number of African American children decreased by about 250,000, or about 4%. In contrast, the number of Hispanic school-age children increased by 1.25 million, or 57%, and the number of Asian children rose by over 600,000—an 87% increase. In 1990, White children made up less than 70% of the total school-age children, down from about 75% in 1980 (GAO, 1993).

Similar to the total school-age population, poor children are now more racially and ethnically diverse. On the 1990 census, an individual or family would be categorized as poor if its annual before-tax cash income was below the corresponding poverty threshold for a family of that size. On the 1990 census, the poverty cutoff for a family of four was a 1989 income of $12,674. During the 1980s, the number of poor school-age children increased by 6% from about 7.2 million to 7.6 million (GAO, 1993). The national poverty rate for school-age children rose from 15.3% to 17.1%. The number of poor Hispanic and Asian children grew by almost 600,000 while the number of poor White children declined, and the African American school-age population in poverty remained relatively constant.

Despite the decline in poor White children, they continued to make up more than 40% of all poor school-age children in 1990, but this percentage changed dramatically by geographic region. White children represented over two thirds of all rural poor children and approximately one third of the urban school-age population in poverty. Yet, regardless of region, African American children experienced the *highest rates* of school-age poverty, from almost 41% in rural areas to 34% in urban areas.

Three other traditionally underserved demographic groups—immigrant households, linguistically isolated (LI) households, and limited English proficiency (LEP) households—each contributed about 5% of all school-age poverty children (GAO, 1993).[3] Many of the children were categorized into more than one of these groups. When adjusted for overlap, these three groups totaled nearly 4 million children—more than 9% of all school-age children. More than 30% of these 4 million children were also classified as poor.

[3] Some definitions of terms are required here. LI children are in households where no persons aged 14 or older speak "only English" and no persons aged 14 or older who speak a language other than English speak English "very well." There is no accepted definition of LEP. The term generally refers to students who have difficulty with speaking, writing, or reading English. The GAO (1993) defined LEP children as all persons aged 5 to 17 living in families whose members the Census reported as speaking English "well," "not well," or "not at all." It should be noted that there is considerable variation in actual English-speaking ability among those classified in the "speaks English well" category.

Current demographic trends should be examined in light of mathematics achievement trends. As the demographic context of the United States changes rapidly, how well is our system of education performing in school mathematics across demographic groups?

PROFICIENCY TRENDS IN MATHEMATICS

The purpose of this section is to document changes in mathematics achievement by examining national trend studies to better understand the status of the United States education system. The discussion of national trend data is offered for two related reasons. The first reason is to clearly describe the student achievement problem. The trend studies reviewed in this section are in part a reflection of past practice in school mathematics. Thus, the mathematics trends are linked to limitations of the implemented curriculum, pedagogy, and school organizational strategies. A second reason to discuss national mathematics trend studies is to describe the measures used to determine mathematics achievement and to interpret the findings with a focus on engineering change.

The trend studies should be examined with several concepts in mind. Miller (1995) argued that three intertwined concepts should be taken into consideration when attempting to build effective strategies to accelerate minority student performance on the basis of academic achievement data:

1. Generally, differences in academic achievement patterns among racial/ethnic groups reflect the fact that the variation in family resources is greater than the variation in school resources. His analysis of achievement patterns and resource allocations confirms that most high-SES students receive several times more resources than most low-SES students receive, and much of this resource gap is a function of family resources rather than school resources.
2. Demographic group educational advancement is an intergenerational process. From this perspective, education-related family resources are school resources that have accumulated across multiple generations. On average, investments in the current generation of African American, Hispanic, and American Indian children in the form of intergenerationally accrued education-relevant family resources are significantly less than comparable investments in White and Asian children.
3. Educational attainment is a function of the quality of education-relevant opportunity structure over several generations. The pace of educational advancement depends on multiple generations of children attending good schools.

Miller (1995) stated the following about these three interrelated concepts:

> Current variations in education-relevant family resources are heavily a function of variations in the historical opportunity structure experienced by generations of racial/ethnic groups. At the same time, the quality of the contemporary opportunity structure is crucial to the further evolution of family resource variation patterns. The nation's ability to accelerate the intergenerational advancement process for minorities may be decisively shaped by its capacity to *engineer* [my emphasis] a more favorable opportunity structure for them in the years ahead as well as to supplement family and school resources for those groups at a level commensurate with their actual needs. (pp. 339–340)

The first step in engineering change is problem identification. One goal of recent federal (NCLB Act) legislation and state policy focused on mathematics standards and accountability is to document the achievement level of traditionally underserved students' yearly progress and to provide performance trends at a local level. The theory of action of most standards-based reform initiated at state and federal levels of governance suggests that, armed with quantitative data on how students perform against standards, school leadership will react by making instructional changes required to improve student performance. According to the National Research Council (1999a),

> Research on early implementation of standards-based systems shows, however, that many schools lack an understanding of the changes that are needed and lack the capacity to make them. The link between assessment and instruction needs to be made strong and explicit (p. 5).

Why do schools lack an understanding of administrative changes that are needed to improve student performance on specifically designated tests? One or all of the following problems may hamper many school leaders:

- Failure to disaggregate and organize data by race, class, language proficiency, or other relevant demographic variables

- Failure to align local content standards with external performance standards associated with the designated testing system
- Failure to align the testing cycle and fiscal planning

One reason many schools lack the insight to make appropriate instructional changes is related to how they organize and analyze data. Although many states, schools districts, and schools disaggregate data to help provide a more accurate picture of student performance, many educational leaders do not have insight into student mathematical performance by demographic group. This is problematic in that student achievement patterns and trends are potentially overlooked; thus, opportunities for instructional intervention are lost, and future student performance is hampered. Further, lack of clarity about the relationship between content standards and performance standards can result in the implementation of curriculum that is not consistent with outcome measures being employed (NRC, 1999a). Thus, any discussion of achievement trends should be coupled with a clear description of what is being measured. Moreover, the discussion of trends must occur in a time frame that allows for immediate intervention. The timing of tests and the administrative planning cycle further complicate the possibility of intervention. In many states, the test results are produced after fiscal planning has taken place in school districts. This disconnect makes it difficult to plan appropriate interventions for the upcoming school year.

Racial-Ethnic Trends[4]

Rapid growth of the school-age population and changing discourses about racial categories have made it more difficult to classify racial-ethnic, immigrant, and language groups. For example, within the Hispanic, Asian, and African American populations, distinct subgroups have formed, and many have requested unique demographic characterizations. Most national trend analysis of mathematics performance is not conducted at this level of detail. This limitation stated a review of this literature remains instructive for evaluating national trend direction in school mathematics.

NAEP Trends

The National Assessment of Educational Progress (NAEP) trend assessment is largely a basic skills examination. To measure performance trends, subsets of the same items have been a part of successive assessments. Some items have been included in each examination. This practice means that findings from nine NAEP trend assessments provide insight into how students' mathematics proficiency has changed from 1973–1999. NAEP mathematics proficiency scores are available for 1973, 1978, 1982, 1986, 1990, 1992, 1994, 1996, 1999, and 2003.[5] Tests are administered to a sampling of students across the United States at Ages 9, 13, and 17. The scale scores, which range from 0 to 500, provide a common metric for determining levels of proficiency across assessments and demographic characteristics. NAEP scores reflect student performance at five levels on the scale:

- Level 150—Basic Arithmetic Facts
- Level 200—Beginning Skills and Understanding
- Level 250—Basic Operations and Beginning Problem Solving
- Level 300—Moderately Complex Procedures and Reasoning
- Level 350—Multi-Step Problem Solving and Algebra

The performance-level categories were developed for the 1973 assessment and have continued to be used through the 1999 assessment. However, the language associated with these categories has evolved and changed over this time period. Thus, the "engineer" charged with making decisions about curriculum, teaching, and other relevant educational inputs should be aware that this trend analysis may use language consistent with today's standards-based discourse (NCTM, 1989, 2000). However, the test items may not reflect the problem-solving and reasoning descriptions found in more recent standards documents and state content and performance assessment documents. With this limitation noted, the NAEP trend analysis is a valuable gauge of student performance progress over time. Table 28.2 provides a summary of NAEP racial-ethnic trends in mathematics performance from 1973–1999.

[4] The trend studies reviewed in this chapter are limited to select national-level analyses that provide insight into student mathematics performance across demographic groups. National studies that did not disaggregate data by demographic group are not included. Moreover, no state-level trend studies or international studies (TIMMS) are included. The period from 1985 to 1999 is a particular focus of this trend analysis summary. This report continues the 1985 effort of DeAnna Banks Beane.

[5] The 2003 NAEP scores are not included in this discussion. At the time of publication, these findings were not included in the trend study.

Table 28.2 NAEP Trends in Average Mathematics Scale Scores by Race/Ethnicity

Race/Ethnicity	Year	Age 9 Scores	Age 13 Scores	Age 17 Scores
White	1999	238.8 (0.9)	283.1 (0.8)	314.8 (1.1)
	1996	236.9 (1.0)	281.2 (0.9)	313.4 (1.4)
	1994	236.8 (1.0)	280.8 (0.9)	312.3 (1.1)
	1992	235.1 (0.8)*	278.9 (0.9)*	311.9 (0.8)*
	1990	235.2 (0.8)*	276.3 (1.1)*	309.5 (1.0)*
	1986	226.9 (1.1)*	273.3 (1.3)*	307.5 (1.0)*
	1982	224.0 (1.1)*	274.4 (1.0)*	303.7 (0.9)*
	1978	224.1 (0.9)*	271.6 (0.8)*	305.9 (0.9)*
	1973	225.0 (1.0)*	274.0 (0.9)*	310.0 (1.1)*
Black	1999	210.9 (1.6)	251.0 (2.6)	283.3 (1.5)
	1996	211.6 (1.4)	252.1 (1.3)	286.4 (1.7)
	1994	212.1 (1.6)	251.5 (3.5)	285.5 (1.8)
	1992	208.0 (2.0)	250.2 (1.9)	285.8 (2.2)
	1990	208.4 (2.2)	249.1 (2.3)	288.5 (2.8)
	1986	201.6 (1.6)*	249.2 (2.3)	278.6 (2.1)
	1982	194.9 (1.6)*	240.4 (1.6)*	271.8 (1.2)*
	1978	192.4 (1.1)*	229.6 (1.9)*	268.4 (1.3)*
	1973	190.0 (1.8)*	228.0 (1.9)*	270.0 (1.3)*
Hispanic	1999	212.9 (1.9)	259.2 (1.7)	292.7 (2.5)
	1996	214.7 (1.7)	255.7 (1.6)	292.0 (2.1)
	1994	209.9 (2.3)	256.0 (1.9)	290.8 (3.7)
	1992	211.9 (2.3)	259.3 (1.8)	292.2 (2.6)
	1990	213.8 (2.1)	254.6 (1.8)	283.5 (2.9)*
	1986	205.4 (2.1)*	254.3 (2.9)	283.1 (2.9)*
	1982	204.0 (1.3)*	252.4 (1.7)*	276.7 (1.8)*
	1978	202.9 (2.2)*	238.0 (2.0)*	276.3 (2.3)*
	1973	202.0 (2.4)*	239.0 (2.2)*	277.0 (2.2)*

Note: Standard errors of the scale scores appear in parentheses.

* Significantly different from 1999. From NAEP 1999 Trends in Academic Progress, NCES (2000).

The racial-ethnic mathematics scores as measured by the NAEP long-term trend assessment improved for all racial-ethnic subgroups from 1973–1999. The scores for Black and Hispanic students are less consistent than those for White students and demonstrate more abrupt changes. However, the samples of Black and Hispanic students are smaller than that of White students. Smaller samples typically have more variability. Overall, the NAEP trend assessment indicates that all three racial-ethnic groups have experienced positive growth in mathematics proficiency. However, no group by Age 17 was performing on average at the highest student performance level. This finding is a concern given that the performance levels are more closely aligned with a basic skills mathematics curriculum.

NELS Trends

The National Education Longitudinal Study of 1988 (NELS:88) included a nationally representative sample of over 10,000 students, followed from 8th grade (1988) through 12th grade (1992) in nearly 800 high schools nationwide. The schools in the study include public, Catholic, and other private schools and represent a range of enrollment, religious affiliations, geographic settings, school social compositions, as well as various levels of restructuring activity (Newmann & Wehlage, 1995). The NELS:88 mathematics tests were constructed to measure both high-level and low-level skills at three points in time: 1988, 1990, and 1992. Thus, students in the sample were assessed in mathematics at Grades 8, 10, and 12, respectively. The difficulty levels of the first and second follow-up mathematics tests were adapted to the students' performance levels in the previous administration. Each mathematics test had 40 items. Eighty-one items were used in all forms of the test. The different forms of the test were equated using item response theory (IRT) so the various forms of the test could be equated with a common metric. Units on these tests refer to the number of items answered correctly, after the IRT procedures were used to score the tests and to assign all students on the same scale.

Green, Dugoni, Ingels, and Camburn (1995) reported findings from the NELS:88 second follow-up data set that included mathematics achievement results of high school seniors in 1992. The 1992 NELS:88 second follow-up examination items represent items typically characterized as traditional, basic skills curriculum. Five levels of mathematics proficiency were defined in the study. Green et al. found that African American and Hispanic students were less likely than White and Asian students to demonstrate advanced proficiencies (Levels 4 and 5) on the standardized test of mathematics (12% and 20% compared with 39% and 45%, respectively). Further, 50% of the African American and 42% of Hispanic students were categorized at Level 1 or below. In comparison, 14% of the Asian and 21% of White students performed at Level 1 or below.

Rasinski, Ingels, Rock, and Pollack (1993) compared mathematics scores for sophomores in the 1980 High School and Beyond (HS& B) study and the 1990 NELS:88 (a follow-up study conducted in 1990) by using an IRT scaling procedure that linked the two assessment instruments. The HS&B sophomore cohort mathematics test administered in 1980 consisted of 38 test items and required students to complete the examination in 21 minutes. The test items were quantitative comparisons that required students to mark which of two quantities is greater, indicate their equality, or note a lack of sufficient information to determine a relationship between the quantities. The 1990 NELS:88 first follow-up mathematics test contained 40 items to be completed in 30 minutes. The test items assessed advanced skills of comprehension and simple mathematical application skills. The items included geometric figures, graphs, word problems, and quantitative comparisons (as in the HS&B). Consistent with the HS&B, a multiple-choice format was used in this follow-up test. To compare the performance of the 1980 HS&B sophomore cohort and the 1990 NELS:88 sophomores, 16 quantitative comparisons from the HS&B were included in the 1990 NELS:88 mathematics assessment. Thus, the findings from this study should be viewed as a comparative analysis of a narrow scope of the mathematics content. All racial and ethnic groups with the exception of Asian students made statistically significant gains in mathematics performance on the test. In each administration of the test, Asian students on average were the highest performing of the four demographic groups. African American and Hispanic students gained more than Asian and White students in this comparison.

Racial-Ethnic Trend Analysis Summary

The NAEP trend analysis indicates improvement between 1973 and 1999 in all racial-ethnic groups at each age level. During this period, African American and Hispanic students made larger gains than did White students; thus, the performance gap on this assessment between White students and the other two demographic groups closed slightly. The 1980 HS&B and the 1990 NELS:88 sophomore cohort study reported a similar result: African Americans, Hispanics, and Whites made statistically significant gains in mathematics achievement. Further, the gains made by African American and Hispanic students were larger than those of White students.

The NAEP trend analysis and the 1990 NELS:88 sophomore cohort study indicate that the mathematics performance on basic skills items over the past 20 years has improved for the largest racial-ethnic demographic groups in the United States. However, no racial-ethnic demographic group has consistently produced scores that are associated with the highest levels of performance being measured by the NAEP trend analysis.

Socio-Economic Trends

The concept of social class is derived from multiple academic domains and literatures. Most conceptions of social class are linked to the economic roots of class and to varying degrees couple class with political and cultural indicators. The traditional practice in school mathematics achievement data is to organize a hierarchy of classes—working class, lower-middle class, middle class, and so on. This hierarchical framework objectifies high, middle, and low positions on some metric, such as socioeconomic status (SES) in which "Parents' Education" or "Family Income" is a proxy for class. The limitations of this practice are discussed elsewhere (Grant & Sleeter, 1986; Knapp & Woolverton, 1995; Secada, 1992). However, for the purpose of understanding SES trends in mathematics achievement, a proxy like "Parents' Education" is instructive. One major limitation of this proxy—and others like it—is that school administrators cannot intervene directly on this variable. However, the importance of family resources must not be ignored in a discussion of achievement (Miller, 1995).

NAEP Trends

From 1978 to 1999, the National Assessment of Educational Progress provided trends in average mathematics proficiency by the highest level of education that students reported for either parent. A summary of the trends in average mathematics scale scores for students at three age levels by parents' highest level of education is provided in Table 28.3.

Students who indicated their parents had less than a high school education have exhibited overall gains in average mathematics proficiency since 1978 across all ages. For students who reported their parents' highest education level was high school graduation, the average proficiency trend has generally improved at Ages 9 and 17. The performance of 13-year-olds in this group was relatively the same during this time period. For students with a parent who graduated from college, only 9-year-olds had an average score in 1999 that was significantly higher than in 1978.

NELS Trends

Rasinski and colleagues' (1993) comparison of sophomore cohorts from the 1980 HS&B study and the 1990 NELS:88 follow-up study documented a consistent pattern of positive gains within SES groups during this period and a difference that is related to student SES. Four SES categories were created by framing the socioeconomic status composite into SES high quartile, SES high middle half, SES low middle half, and SES low quartile. The findings suggest that the highest quartile improved more than the lowest quartile; however, approximately 12% of the lowest quartile in 1990 was missing math test scores, whereas nearly all the 1980 lowest quartile reported mathematics scores. The researchers speculated that the lowest quartile gain could be biased downward as a result of the missing data. The missing data make any interpretation of differential gain between quartiles difficult. However, within each data set—i.e., HS&B 1980 and

Table 28.3 NAEP Trends in Average Mathematics Scale Scores by Parents' Highest Level of Education

Parents' Level of Education	Test Year	Age 9 Scores	Age 13 Scores	Age 17 Scores
Less than High School	1999	213.5 (2.8)	256.2 (2.8)	289.2 (1.8)
	1996	219.8 (3.3)	253.7 (2.4)	280.5 (2.4)*
	1994	210.0 (3.0)	254.5 (2.1)	283.7 (2.4)
	1992	216.7 (2.2)	255.5 (1.0)	285.5 (2.3)
	1990	210.4 (2.3)	253.4 (1.8)	285.4 (2.2)
	1986	200.6 (2.5)*	252.3 (2.3)	279.3 (2.3)*
	1982	199.0 (1.7)*	251.0 (1.4)	279.3 (1.0)*
	1978	200.3 (1.5)*	244.7 (1.2)*	279.6 (1.2)*
Graduated High School	1999	224.4 (1.7)	264.0 (1.1)	299.1 (1.6)
	1996	221.2 (1.7)	266.8 (1.1)	297.3 (2.4)
	1994	225.3 (1.3)	265.7 (1.1)	295.3 (1.1)
	1992	222.0 (1.5)	263.2 (1.2)	297.6 (1.7)
	1990	226.2 (1.2)	262.6 (1.2)	293.7 (0.9)*
	1986	218.4 (1.6)*	262.7 (1.2)	293.1 (1.0)*
	1982	218.3 (1.1)*	262.9 (0.8)	293.4 (0.8)*
	1978	219.2 (1.1)*	263.1 (1.0)	293.9 (0.8)*
Graduated College	1999	239.7 (0.8)	285.8 (1.0)	285.2 (3.9)
	1996	239.7 (1.4)	282.9 (1.2)	287.3 (4.0)
	1994	237.8 (0.8)	284.9 (1.2)	282.7 (3.8)
	1992	236.2 (1.0)*	282.8 (1.0)*	290.2 (3.9)
	1990	237.6 (1.3)	280.4 (1.0)*	276.8 (2.8)
	1986	231.3 (1.1)*	279.9 (1.4)*	280.6 (2.4)
	1982	228.8 (1.5)*	282.3 (1.5)	271.7 (1.8)*
	1978	231.3 (1.1)*	283.8 (1.2)	275.7 (1.9)*

Note: Standard errors of the scale scores appear in parentheses.

* Significantly different from 1999. From NAEP 1999 Trends in Academic Progress, NCES (2000).

1990 NELS:88 Follow-up—SES status is clearly related to mathematics performance.

Green et al. (1995) reported findings from the 1992 NELS:88 second follow-up survey of seniors. In one analysis, Green and colleagues compared achievement across racial and ethnic groups while controlling for SES. The two lowest proficiency levels—below basic and Level 1—and the two highest proficiency levels—Levels 4 and 5—are contrasted. The data indicate that achievement differences exist even when the effects of socioeconomic status are held constant. For example, this study reported that significant differences existed between Whites' and African Americans' test performance within each SES category. Also, there were significant differences between White and Hispanic seniors in the high SES group. The percentage differences among racial and ethnic groups were generally larger in the higher SES groups. There was one exception: Differences in Asian and White seniors' performance were not significant.

SES Trend Analysis Summary

The studies reviewed in this section should be considered with population trends in mind. Clearly, poverty is more severely concentrated among Hispanic and African American children than it is among Whites. Across the various studies of mathematics achievement, a strong relationship between SES and mathematics achievement was present. These studies indicate a need to improve the mathematics achievement of low-SES students as a whole, and even more pressing is the need to raise the mathematics achievement of low-SES minority students. In light of these findings and population trends, the need for intervention in the two geographic regions with the highest poverty levels—urban and rural communities—is apparent.

OPPORTUNITY TO LEARN FACTORS: TIME, QUALITY, AND DESIGN

A close look at the achievement trends reviewed in the prior section suggests that student demographic background is strongly related to mathematics achievement. This is important to know; however, demographic background is out of the control of the teacher, instructional supervisor, school board member, and other school personnel. An educator interested in improved student performance in mathematics must focus on the variables associated with learning mathematics that can be influenced by specific action and intervention. Yet, as Miller (1995) argued, ignoring family resource variables and SES factors could limit the progress of any reform effort. For students living in poverty, the nature and extent of their in-school experience in many cases must rival the combined in-school and out-of-school academic experiences of resource-rich families, if the goal is closing the achievement gap without penalty to any group. This is not a well-accepted position, however the logic of the argument tacitly undergirds the achievement gap debate. One response to current student underperformance is to examine how opportunity-to-learn variables might inform the design of active intervention on student learning.

Opportunity-to-Learn (OTL) as an important construct influencing—and possibly explaining—the impact of instruction was introduced during the 1960s. Carroll (1963) included OTL as one of five critical constructs in his model of school learning. He defined OTL as the amount of time allocated to the learner for the learning of a specific task. If, for instance, the task assigned a student is to understand the concept of place value, opportunity-to-learn is simply the amount of time the student has available to learn what place value is.

In Carroll's (1963) model, opportunity-to-learn is contrasted with the amount of time the student requires to learn a principle or concept. This latter construct is largely related to the student's aptitude in a concept domain. Thus, whereas teachers have some control over the time available for student learning, they have little control over the time required for student learning. Carroll also contrasted OTL with the amount of time the student actually spends engaged in the learning process. The latter variable, often referred to as time-on-task or engaged time, is thought to be affected by the perseverance of the student and the quality of the teaching. In Carroll's model, OTL represents the maximum value for engaged time.

In contrast to Carroll, Husén (1967) organized OTL in terms of the relationship between the mathematics content taught to the student and mathematics content assessed by achievement tests. In Husén's model, OTL is the overlap of mathematics taught and mathematics tested. Simply stated, the greater the overlap, the greater the opportunity-to-learn.[6]

[6] Carroll's (1963) and Husén's (1967) opportunity-to-learn models have two important differences. First, whereas Carroll's model describes OTL as an instructional variable (under the control of teachers), Husén's model frames OTL as a measurement variable. Second, Carroll describes OTL as a continuous variable, whereas Husén designed OTL as a dichotomous variable. The most important concern from Carroll's perspective is how much time the student has to learn a specific concept. The most important concern from Husén's perspective is whether or not a student has been provided with quality instruction relative to the concepts included on achievement tests.

Scholars, school leaders, and government agencies have used various combinations of Carroll's and Husén's models to design their own frameworks of opportunity-to-learn (National Governors' Association, 1993; Robitaille & Travers, 1992; Winfield, 1987, 1993). However, Stevens (1993a) identified four variables related to teacher instructional practice and student learning that consistently emerge in these interpretations. In this section, two of the variables are combined; thus, the following three variables form an opportunity-to-learn framework:

1. *Content exposure and coverage variables* measure the amount of time students spend on a topic (time-on-task) and the depth of instruction provided. These variables also measure whether or not students cover critical subject matter for a specific grade or discipline.
2. *Content emphasis variables* affect the selection of topics within the implemented curriculum and the selection of students for basic skills instruction or for higher order skills instruction.
3. *Quality of instructional delivery variables* reveals how classroom pedagogical strategies affect students' academic achievement.

The purpose of these OTL variables is to determine whether or not students are provided sufficient access to learn the mathematics curriculum expected for their grade level and age. According to Stevens (1993b), the OTL variables are "deceptively simple" (p. 234). In general, research in this area examines one variable at a time; however, the OTL conceptual framework developed by Stevens (1993a, 1993b) encourages teachers, administrators, and researchers to examine the interaction of all three variables simultaneously (see Table 28.4).

This theoretical framework will remain a theory, rather than an active change strategy for most teachers, unless their work is part of a coherent "design" that allows them to take advantage of what is known about opportunity to learn. Two very important variables that emerge from the OTL literature are *time* and *quality*. Time and quality are critical variables because they can be altered with interventions. Thus, time and quality variables derived from the OTL literature form a basis for the construction of school design strategies aimed to improve learning. For purposes of management and leadership, design is critical.

Think of "design" as an innovative portfolio of strategies that will provide students appropriate content exposure, content coverage, content emphasis, and quality instructional delivery. The term *design* is used here to describe how school personnel can construct and package opportunities to learn. Those responsible for the education of children need to be challenged to accept a greater level of responsibility for how teaching and learning is organized. Every educator—teachers, principals, superintendents, and school board members—should have a clear understanding of how the school system and, more specifically, how each school is designed to improve student performance in mathematics. Too many educators fail to see the limitation of long-standing design principles. Still others fail to recognize existing design principles. Some may question the need for a transparent opportunity-to-learn design. However, not having a design is a design for failure. Each state has a measurement system to gauge student performance. These systems are transparent. Similarly, every school and school district should have a learning design that

Table 28.4 Opportunity to Learn: A Theoretical Framework Derived from International Assessments and Research Studies to Examine Students' Access to Intended Curriculum

Variable/Related Study	Definition
Content exposure and coverage (Leinhardt & Seewald, 1981; Leinhardt, 1983; Brophy & Good, 1986; Winfield, 1987, 1993; Suter, 2000)	Teacher arranges class so that there is time-on-task for students. Teacher arranges adequate time for students to learn subject matter and to cover adequately a specific topic. Teacher arranges the curriculum to overlap test content.
Content emphasis (Floden, Porter, Schmidt, Freeman, & Schwille, 1981; LeMahieu & Leinhardt, 1985; McDonnell, Burstein, Catterall, Ormseth, & Moody, 1990; Oakes, 1990; Porter, 1989, 1993; Stevens 1993b; Suter, 2000)	Teacher chooses content from the curriculum to teach. Teacher chooses the dominant level to teach the curriculum (recall, higher order skills). Teacher chooses which skills to teach and which skills to highlight with different groups of students (ability grouping and tracking).
Quality of instructional delivery (Brophy & Good, 1986; Stevens, 1993b; Stevenson & Stigler, 1992)	Teacher uses different pedagogical strategies to meet the learner's needs. Teacher has understanding of the subject matter.

is transparent, open to ongoing monitoring, assessment, and revision.

The appropriate design and management of OTL variables is central to the improvement of school mathematics for many students. The remainder of this section will be devoted to the role of time, quality, and design as they relate to student OTL with traditionally underserved student groups.

Time and School Mathematics

Policies and practices that influence content coverage and time on task in school mathematics are pivotal to the improvement of student performance in the domain. The purpose of studying these opportunity-to-learn variables is to determine whether or not students are provided sufficient time to learn the mathematics curriculum expected for their grade level and age. One very basic principle related to time should be transparent in every classroom. Significant time should be dedicated to mathematics instruction each school day. Further, appropriate time should be allotted to ensure students develop understanding of key concepts and procedures.[7] Many factors can influence whether or not this basic principle is followed. In this section, several factors related to time and school mathematics will be reviewed.

Course-Taking

Two of the most powerful predictors of school mathematics achievement in large-scale assessments of mathematics have been (a) increased time on task in high-level mathematics and (b) the number of courses taken in mathematics. Generally, these two predictors are interrelated. Evidence indicates that African American, Hispanic, and low-SES students are less likely to be enrolled in higher level mathematics courses than middle-class White students (Secada, 1992). Further, White students on national assessments of mathematics achievement consistently outperform African American and Hispanic students. Thus, it is not shocking that a positive relationship between mathematics achievement and course taking exists across measurement systems (e.g., NAEP, SAT, and ACT).

Course-taking options in the United States are organized in a technology that takes on two forms—curricular and ability tracking. Many comprehensive high schools offer a wide range of mathematics courses linked to various work-related opportunities. No student could experience all of the coursework, so schools design technologies to regulate the selection process. To this end, students in most high schools are sorted into a curricular track involving a specific course sequence and, ultimately, different opportunities to learn mathematics. Generally, three curricular tracks—college preparation, vocational, and general education—are offered within most traditional high schools. The college preparation track clearly has higher status and provides greater opportunity to learn more demanding mathematics. Curricular tracking has serious implications for student opportunity to learn mathematics.

Similarly, ability tracking is a technology used to sort students into curriculum experiences.[8] This mechanism for sorting provides different levels of instruction to students across two tracks based on perceived ability. This version of sorting is more difficult to recognize because course labeling can disguise the practice. For example, schools may offer two different courses in geometry. Both may have the same title; however, the mathematics covered in each course may differ in dramatic ways. Another sorting strategy is to offer students different entry points into the college-preparatory coursework at different times (e.g., freshmen year versus junior year). The organizational structure of the school may recognize many tracks or just a few; schools may or may not link tracks to a block of courses or to mathematics only; and schools may have loosely or tightly coupled curricular and ability tracking. Additionally, students may or may not have the option to move across tracks. The opportunity to negotiate new curricular possibilities is an important equity consideration.

Tracking is a serious challenge to mathematics achievement and opportunities to learn mathematics. In theory, tracking as a technology is designed to benefit all students. However, evidence strongly suggests that this goal is not being accomplished (Hoffer, Rasinski, & Moore, 1995). Instead, research studies have indicated that even when tracking systems have positive effects, those effects are more closely associated with those students assigned to high-status tracks (Oakes, 1990; Rock & Pollack, 1995).

One possible solution to the differential opportunities to learn across tracks is to constrain the curriculum options in mathematics at the second-

---

[7] Sufficient and appropriate time to learn the mathematics curriculum should be a data-driven decision. Certain mathematical concepts are more difficult to understand. System-wide as well as classroom-based assessments data can inform the process. Both assessment formats are informative with respect to determining the amount of time to devote to a concept.

[8] We use the term *ability grouping* because this is consistent with the literature on tracking. However, a more appropriate term is *perceived ability*.

ary level. Currently, African Americans and Hispanic students are overrepresented in vocational programs and low-track options. Lee, Croninger, and Smith (1997) found that students learn more mathematics in schools that offer them a narrow curriculum composed of college-preparatory academic courses. This research is suggestive, rather than definitive.

A word of caution: "Course-taking patterns are an important indicator of system quality." Many students may be enrolled in low-track mathematics courses due in part to prior experiences in elementary and middle school mathematics. Merely mandating a narrower curriculum consisting of college prep mathematics will not address the endemic quality problem of the preK–8 mathematics program. Thus, it is imperative that curriculum constraints toward the college prep model at the secondary level occur in tandem with a close examination of the preK–8 effort.

One state-level change strategy to improve elementary and middle school mathematics is to align the mathematics curriculum with state assessments. This model has implications for time and school mathematics. The next section examines this strategy.

Assessment Practices

The mathematics curriculum in many school districts is aligned with mathematics standards adopted or derived from state or national curricular frameworks. The standards-based reform of mathematics education is often part of a larger systemic change effort that includes academic standards in the core disciplines by grade, holding all students to the same standards, statewide assessments closely linked to the standards, accountability systems with varying levels of consequences for results, computerized feedback systems, and data for continuous improvement (NRC, 1999a). State-level assessment systems and most national testing proposals call for students to be tested in mathematics and reading (NRC, 1999b). This practice has implications for content coverage and time on task in mathematics classrooms in urban school districts and other school systems with large percentages of traditionally underserved students.

Students' opportunities to learn mathematics are influenced by the assessment policies of the school district. Assessment policy often influences the nature of pedagogy in a classroom. The influence of standardized tests—and, more recently, state-mandated testing—is arguably greater in high-minority classrooms. In a nationwide survey, teachers of high-minority classrooms reported test-specific instructional practices more often than teachers of low-minority classrooms (Madaus, West, Harmon, Lomax, & Viator, 1992). For example, in high-minority classrooms, about 60% of the teachers reported teaching test-taking skills, teaching topics known to be on the test, increasing emphasis on tested topics, and starting test preparation more than a month before the examination. These practices were reported significantly less often in low-minority classrooms. Moreover, mathematics teachers with high-minority classes indicated more pressure from school district officials to improve test scores than teachers with low minority classes.

Today, school districts across the country use testing technology as a mechanism to measure school and student progress. However, the role of testing technology is much greater than measurement concerns. Tests do change or at least influence teaching behavior. Many districts are ignoring best practice related to assessment and school mathematics. Two recommendations related to school mathematics and student assessment performance are listed below:

1. Design a curriculum, select quality instructional materials, align curriculum and instructional materials, and then use aligned instructional materials all year. Testing systems are intended to measure the quality of a school's instructional program. Avoid spending significant time on test preparation. If the combination of the curriculum, instructional materials, and teaching fall short of school district goals, then these factors must be reviewed and improved upon.
2. Use state and classroom assessment data as a way to build a solid instructional program linked directly to student thinking in the content domain.

Fiscal Adequacy

Limited course-taking options and narrow assessment practices are compounded by problems of fiscal inadequacy and resource distribution. The Council of Great City Schools [CGCS] (1992) calculated that the average-per-pupil expenditure in 1990–1991 was $5,200 in large urban school districts compared with $6,073 in suburban public school systems. Although both types of school systems allocated about 62% of their budgets to classroom instruction, urban schools spent about $506 less per child on instruction. Although this study does not use current data related to fiscal resources, it reflects a growing fiscal disparity between urban school systems and some suburban systems and illustrates an important point. How money is spent should be examined carefully. For example, the Commissioner of Education of New York State reported, "The more advantaged districts [in New York State], spend over $3,000 more per student and pay their teachers $20,000 more annually. Students in more advantaged districts are substantially more likely than students in less advantaged districts to perform with distinction on Regents examinations, and they are more than twice as likely to plan to attend four-year colleges" (New York State Department of Education, 2002, pp. vi–vii). The CGCS (2003) calculated that the New York Public Schools would need $12,537 per pupil to have the resources equivalent to the highest achieving school districts in the state. The fiscal support undergirding instructional practice has implications for meeting new and more challenging demands in mathematics education.

Over the past decade, the average-per-pupil expenditure has constantly increased for urban and suburban school districts. Yet, as Cohen, Raudenbush, and Ball (2003) proposed, rather than focus on fiscal resources as the center of research and policy-making, teaching and learning should be centered and questions of adequate fiscal resources should derive from carefully planned instructional programs. The call for "Mathematics for All" or "Algebra for All" associated with many state content standards proclamations has placed new demands on urban school systems to prepare larger numbers of students in content traditionally reserved for a small percentage of students. Never before has there been a greater need to extend the amount of time students have with mathematics content that is aligned to state curriculum guides and appropriate tests.

Unfortunately, the old saying "time is money" is directly applicable to the implementation of design strategies capable of providing students more time on task in mathematics. Some considerations related to extending time for students in mathematics are listed below:

- Preschool availability;
- Early intervention programs for low performing schools;
- Extended school day opportunities;
- After- and before-school tutorial programs;
- Saturday school;
- Summer school enrichment for all students (not just remediation);
- Community college/university programs;
- Longer school day or expanded year;
- Enrichment and mentoring programs; and
- More individualized or small group instruction.

Each of the strategies listed is integral to a standards-based approach to educational policy-making. These strategies require a sound vision that is directly linked to fiscal policy. State standards provide an opportunity to plan for success. A simple planning strategy includes (a) adopting a set of mathematics standards, (b) identifying resources needed to achieve the standards (including time-related strategies), (c) formulating a long-term plan that aligns the standards and resources, (d) developing the plan before spending money, and (e) adopting the necessary structural changes to maximize cost-effectiveness. Planning for more engaged time in mathematics is a purposeful act that should be aligned with fiscal management. A school district's portfolio of mathematics practices and interventions should be clearly aligned to uniform content goals and fiscal management. Too often, districts fail to produce aligned practices and fiscal policy. Yet, a portfolio of aligned practices, interventions, and fiscal policy is the essence of a district's learning design.

Quality and School Mathematics

Most observers should recognize that the call for more demanding standards in mathematics is a signal for not only what students must know but also what teachers must understand and school systems must support. High standards in school mathematics demand quality instruction and supporting infrastructure. The purpose of this section is to examine quality factors that influence mathematics instruction.

The OTL literature defines *quality* as classroom pedagogical strategies that affect students' academic achievement. In this case, quality is defined as those pedagogical strategies that positively influence student achievement in school mathematics. Before discussing quality factors, a baseline review detailing what is typical with respect to mathematics pedagogy is helpful.

Traditionally, mathematics pedagogy has emphasized whole-class lectures with teachers modeling one strategy for solving a problem and students passively listening to the explanation. Generally, the lecture is followed by students working alone on a large set of problems that reflect the lecture topic (Fey, 1981; Porter, 1989; Stodolsky, 1988). The purpose of the lecture and problem set is to prepare students to produce correct responses to narrowly defined problems. This pedagogical strategy is often coupled with curricular or ability grouping, with many African American and Hispanic students selected to participate in compensatory mathematics programs that focus on the mastery of low-level computational skills (Strickland & Ascher, 1992). These phenomena are so "normal" in many schools, they have become cultural artifacts. The achievement trends as a result of this model of instruction were reviewed earlier in this section.

In contrast, high-quality mathematics programs generally deviate in important ways from the "normal" approaches to mathematics instruction and classroom practice. A comparison of mathematics teachers in higher and lower performing schools conducted by the North Central Regional Educational Laboratory (NCREL, 2000) revealed important quality factors related to instruction. The findings are summarized in Table 28.5.

Note that the NCREL findings must be understood in light of the centrality of students' *mathematical reasoning* in higher performing schools. Higher performing schools and teachers provide a learning environment that supports sustained engagement on rigorous mathematical tasks. Teaching as characterized in the higher performing schools is complex and demanding. In contrast, the teaching in lower performing schools is routine and limited with respect to teacher-student discourse patterns. Further, instructional practices in lower performing schools do not center on students' mathematical understandings and thinking. The characteristics found in teachers working in higher performing schools can be supported in other schools by administrative planning and instructional leadership with the following specific actions:

- Provide professional development that prepares teachers to focus on mathematical understandings and reasoning.
- Provide ongoing professional development focused on content, effective instruction, and student thinking in the content domain.
- Design a curriculum that provides sufficient exposure to difficult concepts.
- Develop programs to address the impact of student/teacher mobility in low-performing schools.

Each of these quality factors will be discussed in greater detail. The focus of the discussion will center on why these factors are key support mechanisms for achieving the quality teaching characteristics indicated in the NCREL study.

Quality Professional Development

What are the "best practices" related to the professional development of mathematics teachers? Every year, school districts sponsor thousands of professional learning opportunities for teachers. There has been a gradual shift in thinking about professional development in many sectors including education and the corporate world (Meister, 1998). A summary of recent shifts in emphasis related to professional development is provided in Table 28.6.

Are shifts in thinking about professional development (as reflected in Table 28.6) consistent with research in mathematics education and teacher learning? What works?

Garet, Porter, Desimone, Birman, and Yoon (2001) conducted the first large-scale empirical comparison of effects of different characteristics of professional development on teachers' learning. The study used a national probability sample of 1,027 mathematics and science teachers. The results confirm and extend the

Table 28.5 Characteristics of Classrooms and Teachers in Higher and Lower Performing Schools

Higher Performing Schools	Lower Performing Schools
Teachers and students participate in two-way conversations about mathematical ideas.	Conversations tend to be one-way: The teachers tell information to students or look for answers and move on.
Classes exhibit the characteristics of learning communities. There are norms in place so students and teachers are learning together.	Classes have few learning community characteristics. Individuals are more disconnected.
Teachers push for mathematical meaning behind the task.	Teachers lead math tasks; however, meaning-oriented discussion is missing.
Teachers have high expectations that all will learn. They review concepts often, explain ideas thoroughly, invite student thinking, assess student competence, and reteach when necessary.	The expectation is that other sources of help will fill in gaps for struggling students.
Teachers build continuity in the mathematical domain from day to day.	Little continuity is built into mathematical content from day to day.
Students are comfortable with classroom routines and expectations and take initiative in their progress. (They know where to find enrichment materials when finished with an assignment and get started on their own.)	Classroom routines are teacher initiated rather than student initiated. Teachers often remind students of expectations.

Note: From NCREL, 2000, p. 18

Table 28.6 Professional Development Paradigm Shift from Staff Training to Learning

	Old Training Paradigm	Learning Paradigm
Location	Central office	On demand—anywhere
Content	Upgrade math skills	Build core workplace competencies
Methodology	Lecture	Action learning
Audience	Individual teachers	Intact teams of teachers, principals, other staff
Faculty	External university professor/consultants	Internal senior-level district staff and a consortium of university professors and consultants
Frequency	One-time service	Continuous learning process
Goal	Build teacher's inventory of skills	Solve real education issues and improve classroom teaching

Source: Adapted from Meister (1998).

literature on "best practice" in several ways. The study confirms past literature in that the research indicates sustained and intensive professional development is more likely to influence teacher learning, as reported by teachers, than shorter professional development. Also, the research indicates that professional development that focuses on academic work (content), provides teachers opportunities for "hands-on" work (active learning), and is integrated into the daily life of the school (coherence) is more likely to result in enhanced knowledge and skills.

Garet and Associates (2001) extended what is known about professional development and confirmed speculation in the following manner:

> Our results provide support for previous speculation about the importance of collective participation and the coherence of professional development activities. Activities that are linked to teachers' other experiences, aligned with other reform efforts, and encouraging of professional communication among teachers appear to support change in teaching practice, even after the effects of enhanced knowledge and skills are taken into account. Such coherence has been hypothesized as important, but with little direct empirical support in the literature to date. Similarly our data provide empirical support that the collective participation of groups of teachers from the same school, subject, or grade is related both to coherence and active learning opportunities, which in turn are related to improvements in teacher knowledge and skill and changes in classroom practice. (p. 936)

This study suggests that if those who are concerned about education are serious about improving the quality of teaching in mathematics classrooms, they need to support and invest in professional learning opportunities for teachers that foster enhanced instructional practice. A major challenge to the kind of professional development outlined in this study is cost. It is very important that sufficient resources be in place to support a quality professional development model.

Quality Curriculum

The need for a demanding mathematics curriculum aligned with high-quality instructional materials is intuitively obvious. Unfortunately, there is often confusion about the relationship between curriculum and instructional materials. Many systems purchase instructional materials and then treat them like a curriculum. Teachers strive to teach the book from cover to cover with little reflection about the curriculum. In other school districts, curriculum guides are designed, distributed at one-day workshops, placed in school storage, and never used again.

There is serious need for quality district-level curriculum guides in mathematics. In many states, the state-level curriculum framework offers little guidance related to focus; instead, litanies of discrete topics are listed. At the school-district level, curriculum quality can be achieved if the following recommendations for developing and implementing guides are taken seriously:

- Focus on mastery objectives only.
- Reduce the scope of coverage.
- Provide and support the development of more cognitively demanding enrichment materials.
- Allow for variations in completion time and instructional strategy.
- Provide quality instructional materials to schools in a timely fashion.
- Educate principals by focusing their learning opportunities on the relationship between the

curriculum guides, district achievement goals, and test materials.

In the world of high-stakes testing, there is tremendous pressure, real or perceived, to teach to the test. A high-quality curriculum guide that demonstrates an alignment between the instructional materials (including the enrichment materials) and assessment tasks is more likely to result in students' experiencing a coherent and cognitively demanding mathematics classroom than pure reliance on test guidelines.

Mobility and Mathematics

How do schools address the challenges to quality mathematics instruction presented by student and teacher mobility in low-performing schools? High mobility causes a great deal of stress on campus officials attempting to serve these students. In the context of high mobility, a quality curriculum guide that standardizes the curriculum and instructional materials is vital. While schools and classes may deviate on pacing, teachers have a reasonable opportunity to meet individual needs if the curriculum guide has narrowed the coverage and focused on mastery objectives. Further, it is very important that individual student data is transmissible to the new school setting. This will give the teacher an opportunity to construct a data-driven program of study for the student.

Teacher mobility in low-performing schools also is a major problem. Often, new teachers to a system are sent to low-performing schools. The result is not surprising. These teachers either leave the profession or get seniority and transfer to another school. This pattern is consistent and endemic. The result: Low-performing schools are constantly staffed by less experienced teachers and, in many situations, by teachers with emergency teaching certificates.

The time has come for new models of operating in these schools. Questions related to how to retain teachers in low-performing schools require empirical evidence. Here are a couple of speculations on the issue. Retention of teachers in challenging settings may be linked to instructional leadership. Good principals create learning communities that support teachers and students.

Another potential strategy involves the recruitment process. Perhaps with incentives cohorts of well-established teachers can be recruited to low-performing schools. The emphasis is on *cohort.* The goal in this strategy is to embed a core group of excellent teachers in the school setting to influence and mentor the less experienced teachers.

Clearly, it is a disservice to new teachers and students to place novice teachers in the most challenging settings. Remedies to the mobility problem will require major rethinking in the areas of human resource management and fiscal management. Moreover, mobility issues are often compounded with other system challenges like cultural factors and student language background.

Culture and Mathematics Learning

Today, many calls for equity in mathematics education borrow from opportunity-to-learn constructs found in national and international testing programs. In fact, OTL constructs are foundational in this section. These constructs frame equity largely as the overlap of content taught and content tested. The overlap of content taught and content tested is a serious policy concern. Moreover, opportunity-to-learn constructs have additional explanatory power if aligned with the cultural factors that influence students' mathematics learning (Tate, 1995). Research suggests that developers of equity-related policies in mathematics education should carefully consider incorporating recommendations found in the *Professional Standards for Teaching Mathematics* (NCTM, 1991), which call for mathematics pedagogy to build on (a) how students' linguistic, ethnic, racial, gender, and socioeconomic backgrounds influence their learning; (b) the role of mathematics in society and culture; (c) the contribution of various cultures to the advancement of mathematics; (d) the relationship of school mathematics to other subjects; and (e) the realistic application of mathematics to authentic contexts (see e.g., Ladson-Billings, 1994; Meyerson, 2002; Moses & Cobb, 2001; Nelson-Barber & Estrin, 1995; Rousseau & Tate, 2003; Secada, 1996).

The first NCTM recommendation calls for understanding how demographic group membership may be linked to the learning of mathematics. This recommendation is consistent with *No Child Left Behind* legislation that requires a national accounting of student performance in mathematics by demographic group. However, gathering achievement data by demographic group is very different than reflecting on race-related achievement patterns. Rousseau and Tate (2003) found that mathematics teachers in their study were reluctant to reflect on race and student performance in mathematics. Instead, some teachers in their study indicated they were color-blind and did not notice race or attend to matters of race-related patterns of student achievement. Further, many of the teachers were unwilling to link poor student performance to their teaching or other school-related factors. Teachers tended to blame the students and their families. This kind of practice represents a unique challenge for instructional leadership attempting to engineer school mathematics improvement. The challenge suggests a need for professional practice among teachers and school leaders that differs radically from traditional formats. The need for study groups composed of teachers and school leadership is clear. These groups must foster trust and openly communicate about data and students; in particular, the challenge of discussing race and culture must be met (e.g., Tatum, 1992).

Other cultural factors in school mathematics are related to quality design and school change. Many of these cultural factors largely deal with the aim of school mathematics. More specifically, they deal with the nature and extent to which school mathematics curriculum is linked to the liberal arts tradition of reasoning and inquiry in contexts broader than the problems and concepts found in the discipline of mathematics. Should school mathematics include investigations of how mathematics is used in society and culture? For example, how relevant is political numeracy? Mathematics is part of many aspects of a democracy and can inform the reasoning associated with policy formation and policy analysis. Is this appropriate for middle school or high school students?

Some teachers have embarked on student-led integrated problem-solving investigations that include mathematics, statistics, legal analysis, multimedia techniques, scientific method, and connections to other disciplines (Tate, 1995). The approach is consistent with calls for authenticity in mathematics instruction (Meyerson, 2002). This strategy is designed to build on students' interest and to provide a liberal arts approach to the middle school and high school experience. However, the liberal arts approach may not be consistent with the current demands of high-stakes testing environments. Any disconnect between the liberal arts perspective of schooling and mathematics education is worthy of discussion by teachers, instructional leaders, and policymakers.

Similarly, for some elementary instructional leaders, providing an integrated learning experience that connects mathematics, science, and reading is desirable. This kind of integrated approach has many merits including efficient use of time and building on best practice in early childhood education (Bredekamp & Copple, 1997). If the culture of testing, specifically test preparation activities, substitute for real learning experiences and best practice, then long-term skills like student reasoning ability may be sacrificed.

Language and Mathematics Learning

In a policy analysis of urban students acquiring English and learning mathematics in the context of reform, Secada (1996) raised the following two questions: "Should their [urban school] efforts at reforming school mathematics specifically address the status of students acquiring English? Or should urban schools assume that these students' needs will be addressed under the broader aegis of reform?" (p. 422). Secada argued that failure to consider the specific learning needs of students acquiring English would be a mistake. He maintains that educators might usefully examine common learning processes that cut across language learning and mathematics learning. Two potential areas of analyses include (a) psychological processes that are common to understanding language and mathematics and (b) sociolinguistic and cultural processes that support the creation of discourse communities in school including how sense making takes place and is validated in these communities (e.g., Kintsch & Greeno, 1985; Lampert, 1990). Secada (1996) described a potential scope of work for educators related to bilingual education and school mathematics:

> Newly developing models for teaching mathematics should be scrutinized for their applicability to bilingual learners and adapted as necessary. The limitations of evolving ways to teach mathematics (Lampert, 1990) is a reason to question, but not reject, the developing visions for teaching mathematics (NCTM, 1991). Maybe, with some adjustments—specifically inviting these students to add their thoughts, encouraging them to use their native languages and asking others to translate, slowing down the fast-pace tempo of the classroom, creating an atmosphere in which language variation in the community of discourse is an accepted fact of life—these methods can apply to bilingual learners. (p. 440)

Secada's remarks concerning bilingual learners and school mathematics focused on the importance of modifying instructional time and appropriate instructional accommodations—both critical OTL variables. Time and quality factors permeate the discussion of the research-based cases of the next section.

RESEARCH-BASED CASES OF SCHOOL MATHEMATICS REFORM[9]

In every mathematics reform effort, significant time should be devoted to information gathering and group study of other similar change efforts. Learn as much as possible about related design. The intent of this section is to review a select set of research-based

[9] Additional information about these cases can be found in Tate and Rousseau (2002) and the January 1996 issue of *Urban Education*. There is some overlap in the evidence presented in this chapter and the Tate and Rousseau article. The cases are presented in this chapter to highlight the importance of time and quality factors.

cases to serve as a model for future information-gathering activities.

The group of studies reviewed in this section was included for three reasons. First, the studies were part of large multi-year projects focused on classroom-based research. These studies provide insight into how time, quality, and design interact to produce positive academic results in school mathematics. In each of the projects reviewed, student performance in mathematics improved over time. Second, each project at some point examined equity-related concerns and looked to intervene in school settings where student proficiency in mathematics was underdeveloped. Finally, each project was included because participants engaged in an effort to reform school mathematics in a manner that was consistent with the teaching practices and curricular goals found in the National Council of Teachers of Mathematics (NCTM) reform documents or state/local mathematics standards. This is not an endorsement of the NCTM Standards but a review of research with the potential to shed some light on the standards debate. The three projects included in this review each have comprehensive research and evaluation components including data on student advancement and other educationally relevant indicators of progress. The research and evaluation aspects of these projects are important because of the rapid advancements of state standards. Massell (1994) reported 41 states have adopted mathematics standards that at least in part are consistent with the NCTM standards series. A brief history of this series is warranted.

In 1980, NCTM, a professional organization of mathematics teachers, supervisors, and college professors, released *An Agenda for Action*, which described a 10-year reform process. A central goal of *An Agenda for Action* was to move the focus of school mathematics from a strictly basic skills curriculum to a more balanced approach that included more demanding mathematics content and appropriate pedagogy to implement this content. Subsequently, but not as a direct result of *An Agenda for Action*, NCTM sponsored the development of the *Curriculum and Evaluation Standards for School Mathematics* (1989), the *Professional Standards for Teaching Mathematics* (1991), the *Assessment Standards for School Mathematics* (1995), and most recently, *Principles and Standards for School Mathematics* (2000). These documents were a product of extensive literature reviews and a series of technical reports that described key themes and ideas in school mathematics. This series of reform documents and the movement to reform school mathematics are important from an equity perspective. Past reform efforts have failed to significantly improve opportunity to learn mathematics for African American, Hispanic, and low-SES students (Tate, 1996). Thus, a close review of more recent reform efforts is part of the process of learning to build.

Cognitively Guided Instruction (CGI)

Researchers at the University of Wisconsin developed Cognitively Guided Instruction (CGI). The CGI foundation was in part established on Carpenter and Moser's (1983) analysis of young children's learning of addition and subtraction. Subsequently, other research was conducted to understand how teacher knowledge of children's thinking would affect teachers' pedagogical actions and student learning (Carpenter, Fennema, Peterson, & Cary, 1988; Carpenter, Fennema, Peterson, Chiang, & Loef, 1989). This research suggested that knowledge extracted from studies of learners' thinking can be used by teachers to strategically influence students' learning. The CGI research program supports the argument that knowledge of students' thinking, when integrated, robust, and a part of the established curriculum, can affect the teaching and learning of mathematics (Fennema & Franke, 1992). Carpenter, Fennema, Franke, Levi, and Empson (1999) described the CGI design process and model as follows:

> Our research has been cyclic. We started with explicit knowledge about the development of children's mathematical thinking, which we used as a context to study teachers' knowledge of students' mathematical thinking and the way teachers might use knowledge of students' thinking in making instructional decisions. We found that although teachers had a great deal of intuitive knowledge about children's' mathematical thinking, it is fragmented and, as a consequence, generally did not play an important role in most teachers' decision making. If teachers were to be expected to plan instruction based on knowledge of students' thinking, they need some coherent basis for making instructional decisions. We designed CGI to help teachers construct conceptual maps of the development of children's mathematical thinking in specific content domains. (p. 105)

CGI is not associated with particular instructional materials. Moreover, CGI does not have an explicit equity component, nor is it targeted at a particular group of students. However, it has been successfully implemented in classroom settings with diverse student groups.

For example, Carey, Fennema, Carpenter, and Franke (1995) described CGI classrooms in a predominantly African American school district. Twenty-two 1st-grade teachers from 11 schools in Prince George's County, Maryland, an urban school setting bordering Washington, D.C., participated in a research project organized to evaluate the efficacy of CGI with African American students. The student demographics in the classrooms of the study exceeded 70% African American. Further, 7 of the 11 schools participated in Chapter 1, a federally funded program of Title I of the Elementary and Secondary Act, a strong indicator of high concentrations of low-income students in a school. The teachers who participated in the study attended a 2-week summer in-service program that was followed with 5 full-day professional development days offered during the academic year. The researchers documented a change in the teachers' implemented mathematics curriculum, with a greater focus on problem solving beyond that typically associated with the first-grade curriculum. The teachers also displayed an ability to take advantage of student thinking about important mathematical ideas, ultimately building on student understanding to establish new knowledge of school mathematics.

Villasenor and Kepner (1991) reported on the implementation of CGI in a minority context. The study was carried out with 12 treatment classes and 12 control classes in which the percentage of non-White populations ranged from 57% to 99%. The CGI group performed significantly better on a 14-item word problem posttest, an interview on word problems, and an interview on number facts. The CGI students also used advanced strategies significantly more often than non-CGI students on both problem solving and number facts. Peterson, Fennema, and Carpenter (1991) argued that "Villasenor's results are important because they provide concrete evidence for the effectiveness of the CGI approach with a disadvantaged population of students" (p. 78).

The CGI studies suggest that an important set of quality factors related to mathematics instruction is how well teachers (a) understand the structure of a specific mathematical concept, (b) understand students' thinking about the particular mathematical idea, and (c) implement instructional strategies that build on this knowledge of student thinking. As a set, these quality factors are powerful indicators of good instruction with a strong relationship to student learning and performance on outcome measures.

Project IMPACT

Project IMPACT "is a school-based teacher enhancement model for elementary (K–5) mathematics instruction designed to foster student understanding and to support teacher change in predominantly minority schools" (Campbell, 1996, p. 449). Six schools were involved in the original study (three treatment and three control). The model involved (a) a summer in-service program, (b) an on-site mathematics specialist in each school, (c) manipulative resources for each classroom, and (d) teacher planning and instructional problem solving during a common grade-level planning period each week. The focus of the model was on instructional approaches consistent with a cognitive perspective on learning, emphasizing interaction and collaboration rather than the typical direct instruction approach.

Unlike CGI, Project IMPACT focused specifically on teaching for understanding in urban schools. Thus, content addressing "teaching mathematics in culturally diverse classrooms" was included in the program's summer in-service. Supported by campus-based mathematics specialists, instructional change occurred in most treatment classrooms, particularly where the instructional leadership by the principal encouraged and embraced the reform process. The students in these schools were assessed in the middle and at the end of each school year. Campbell (1996) summarized the results:

> The influence of the IMPACT treatment on student achievement was not immediate. The students in the IMPACT treatment schools did not evidence statistically significant higher achievement, as compared to the students in the comparable-site schools, until the middle of second grade; however, once established, this mathematics differential continued through second and third grade. (p. 463)

White (1997), in her dissertation, examined the nature of questioning in four 3rd-grade classrooms both before and after the teachers went through the Project IMPACT summer in-service program. The study documented the question-response pattern, the cognitive level of the question (low or high), and the race and gender of the students who responded. White found that students' educational experiences, as reflected in classroom questioning, differed both between and, in some cases, within classes. Two teachers, Ms. Davis and Ms. Tyler, were fairly equitable in their distribution of questions.[10] "They posed questions to all students across questioning patterns and cognitive levels" (White, 1997, p. 300).

In Ms. Atkins's class, however, the distribution was more skewed. Overall, females answered the majority of the questions. Yet, a look at the different cognitive levels reveals racial patterns as well. White and Asian females answered most of the high-level questions. Black and Hispanic female students were asked a relatively low number of high-level questions. According to

[10] The names were changed to protect the identities of the teachers.

White (1997), the origin of this disparity lies in Ms. Atkins's perceptions of students' academic ability and her own discomfort with mathematics. Ms. Atkins wanted to ask high-level questions, but her own lack of understanding caused her to call only on students whom she thought would give the correct answer. Thus, only the students perceived to be of high ability were selected to answer high-level questions. A similar pattern of focusing only on the students who were perceived to have the greatest mathematical understanding was found in the class of the fourth teacher, Ms. Smith.

This very detailed study of question and response patterns is important for at least two reasons. First, it documents a partial success story for Project IMPACT in terms of improving equity in classrooms. Two of the four teachers appeared to change their practices as a result of their participation in the initial IMPACT summer in-service and ongoing campus-level assistance. Both Ms. Davis and Ms. Tyler were more equitable in their distribution of questions after the in-service than they had been before.

This study is also important because it suggests the need to look closely at teachers' explanations for their actions in order to more fully understand what is happening in the classroom. For example, the case of Ms. Atkins indicates that teachers' inequitable actions can originate from a variety of sources, including inadequate content knowledge.

Project IMPACT suggests that another quality indicator related to mathematics instruction is the relationship among teachers' knowledge of mathematics, teachers' understanding of student thinking about mathematics, and teachers' understanding of race/gender interactions in classroom settings. Project IMPACT is consistent with other research programs that indicate the importance of treating cultural background as a resource for learning (Rousseau & Tate, 2003). For example, Knapp (1995) found that teachers in high-poverty schools who placed the greatest emphasis on meaning in their mathematics instruction made two significant shifts in their thinking about learners. First, they viewed learners as active participants in learning, and second, the teachers used cultural dimensions of instruction to sustain engaged time with academic work. This also is an important lesson from Project IMPACT.

QUASAR

QUASAR is described as "an educational reform project aimed at fostering and studying the development and implementation of enhanced mathematics instructional programs for students attending middle schools in economically disadvantaged communities" (Silver & Stein, 1996, p. 476). One purpose of the project was to help students develop a meaningful understanding of mathematical ideas through engagement with challenging mathematical tasks. The QUASAR project supported teachers and administrators in six urban middle schools. Each school site worked with a resource partner—typically, mathematics educators from local universities—to improve the school's mathematics instructional program with a focus on mathematical understanding, thinking, reasoning, and problem solving. The site teams operated independently in the design and implementation of its curriculum plan, professional development, and other features of its instructional program. There were regular interactions among representatives from all QUASAR sites. Moreover, each site-based team benefited from financial support, technical assistance, and advice from the QUASAR staff housed at the Learning Research and Development Center at the University of Pittsburgh.

Silver and Stein (1996) described three different analyses used to assess the effectiveness of instruction in QUASAR sites. Unlike the CGI and IMPACT studies, there was no control group in the QUASAR study. One method used to determine the impact of QUASAR was the examination of changes in student performance over time. The results from the first 3 years of the project indicated that "students developed an increased capacity for mathematical reasoning, problem solving, and communication during that time period" (Silver & Stein, 1996, p. 505). A second method of evaluation used a variety of tasks from the National Assessment of Educational Progress (NAEP) as pseudo-control groups (Silver & Lane, 1995). The QUASAR students were given items from the 1992 eighth-grade NAEP. The results were compared to those of NAEP's national sample and disadvantaged urban sample. The findings from the analysis of student performance on the nine open-ended tasks were very informative about the effectiveness of QUASAR. QUASAR students performed at least as well as the national sample on seven of the nine tasks. Silver and Lane (1995) noted that this is an important result, in light of the fact that the national sample had significantly outperformed the disadvantaged urban sample on all nine tasks.

> The findings clearly suggest that the mathematics performance gap between more and less affluent students has been significantly reduced for students attending the QUASAR schools. Thus, the performance of QUASAR's students is far greater than would have been expected, given their demographic similarity to NAEP's disadvantaged urban sample, and one can infer that the instruction at QUASAR has a beneficial impact on students' mathematical performance. (p. 62)

A third method of evaluation examined outcomes other than achievement, considering whether QUASAR instruction was linked to increased access and success in algebra coursework. Silver and Stein (1996) reported that students from QUASAR schools were both qualifying for and passing algebra in ninth grade at substantially higher rates than before QUASAR.

The QUASAR project reinforces the importance of students' engaged time with cognitively demanding mathematical concepts. Sustained engaged time with quality mathematics tasks resulted in improved student performance on a wide range of indicators.

Research Case Summary

These research cases have three overlapping themes worthy of note. First, high-quality mathematics was at the center of each effort. More specifically, each program called for mathematics instruction using mathematical tasks not typically associated with the lower educational expectations often found in large urban school districts. In each program, mathematical proficiency consisted of a balance between conceptual understanding and procedural proficiency. Too often instruction is based on extreme positions, rather than a balanced approach with increasing cognitive demand.

A second common feature of each program was the important role of classroom-based assessment designed to better understand student thinking about mathematical ideas. These studies of mathematics teaching support the idea that teachers' knowledge of students' reasoning when integrated with a balanced mathematics curriculum can positively affect the teaching and learning of traditionally underserved students. The assessments used in the research-based cases of mathematics reform were designed to support the learning process and to identify areas in which further instruction is needed. The measures included direct observations of children during classroom activities, evaluation of student work, and asking questions in class. The portfolio of measures in the research-based case studies differed from the measures used to gauge national trends. The latter measures are designed to inform the public about trends in student performance or the effectiveness of large-scale educational programs. The research-based cases included standardized tests, selected NAEP items, and classroom-based activities.

A third common feature of each case was the presence of a strong mathematics professional development program. Two key features of these programs were teacher learning opportunities in the areas of school mathematics content and how children's mathematical knowledge develops in the content domains, including what knowledge students were likely to bring with them to school.

These three common themes represent important quality factors. They reinforce how the combination of quality curriculum, cognitive-based assessment tools, and integrated professional development are central to school mathematics design.

TAKING THE DESIGN TO SCALE

The three cases reviewed in the prior section provide important lessons regarding the design of effective programs. However, they are relatively small in scale. Arguably, CGI has had the most widespread impact of the three programs. Hundreds of teachers from various parts of the country have participated in CGI professional development sessions. Although teacher participation in CGI has been linked to student achievement, CGI was not designed to address OTL issues on a systemic basis. Thus, in order to more fully understand how the OTL variables described in the previous sections might come together in the design of a systemic reform effort, it is important to look at examples of design taken to scale. More specifically, each example represents an effort to reform mathematics education in a school district or set of school districts.

The Urban Systemic Initiative

In 1993, the National Science Foundation began the Urban Systemic Initiative (USI) program. The purpose of the program has been to catalyze large-scale systemic change directed toward improving the science and mathematics achievement of students in urban schools. In the 1998–1999 school year, 21 USI sites served over 4.5 million students or 10% of American students in public schools. The students in these school districts represent some of the poorest in the United States, with 69% of the students eligible to receive free or reduced lunch. In addition, 12% are students with limited English proficiency (Kim, Crasco, Blank, & Smithson, 2001).

Six "drivers" serve as the basis for the urban systemic reform program. The drivers have an interdisciplinary philosophical foundation including principles derived from organizational theory, policy studies, ethics, and cognitive science. In practice, the drivers were designed to address mathematics, science, technology, and engineering. However, in this chapter, the focus will be on the role of the drivers in school mathematics. The six drivers for the National Science Foundation educational system reform are

1. Implement standards-based curricula as represented in instructional practice and student assessment system-wide.
2. Develop a coherent and consistent policy portfolio that supports high-quality mathematics and science education.
3. Converge all resources that are directly or tangentially linked to the support of mathematics and science education into a singular program of upgrades and improvement in mathematics and science for all students.
4. Build broad-based support from parents, policymakers, higher education community, business and commerce, philanthropic organizations, and other segments of the community for the mathematics and science programming and the critical decisions related to system-wide programmatic efforts.
5. Accumulate a broad and comprehensive evidentiary base of student performance indicators—for example, achievement test scores, course-taking patterns, college admissions rates, college majors, and Advanced Placement tests.
6. Improve the achievement of all students, including those traditionally underserved (Kim et al., 2001).

The drivers are consistent with the time and quality variables discussed earlier in this chapter. For example, the USI sites have emphasized implementing quality curriculum and instruction. According to Kim et al. (2001), nearly all of the USI sites have decreased their emphasis on more traditional lecture and demonstration methods and moved towards the types of high-quality instructional practices described earlier. Similarly, most of the USI schools have implemented instructional materials that are aligned with the NCTM Standards documents, including 89% of USI elementary schools and 90% of USI middle schools (Kim et al., 2001). In many cases, these instructional materials were developed with funding from NSF (e.g., *Everyday Mathematics*; *Investigations in Number, Data, and Space*; *Connected Mathematics Project*). In addition, the adoption of these instructional materials has been connected to the revision of district curriculum guides.

Policy changes have also impacted OTL in USI sites. According to Kim et al. (2001), 13 USI sites reported the elimination of tracking and all remedial and non-college preparatory mathematics. In addition, mathematics requirements for high school graduation have increased in several USI districts since the start of the initiative—changing either from 2 years to 3 years (Chicago, Detroit, Phoenix, Cleveland, Columbus, and Fresno) or from 3 years to 4 years of required high school mathematics (Dallas and New Orleans). Moreover, in 12 of the 21 districts, the graduation requirements for the district exceeded the state graduation requirements (Kim et al., 2001).

Reform efforts in USI sites also have begun to address fiscal adequacy issues through the coordination of multiple funding sources, including Title I, Eisenhower, and other federal, state, and local programs. According to Kim et al. (2001), a total of $547 million dollars in additional funding was leveraged by the urban systemic districts during the 1999 fiscal year. Moreover, other resources have been brought to the reform effort through partnerships with several different types of entities, including higher education institutions and the private sector. For example, partnerships with higher education institutions have provided the Memphis USI with professional development for its teachers as well as academic support and enrichment activities for students in the Memphis City Schools (Kim et al., 2001).

USI sites have put a significant emphasis on providing quality professional development to their teachers, investing an average of 49% of USI funds on professional development activities. The specific nature of mathematics professional development activities varies across sites, but these activities generally focus on content knowledge, curriculum implementation, assessment, and new methods of teaching. According to Kim et al. (2001), approximately 80 to 90% of USI teachers have been actively involved in professional development, and many report changing their classroom practice as a result.

One of the outcomes of interest with respect to measuring the impact of the USI program is enrollment in gate-keeping and higher level mathematics courses. Changes in enrollment in the USI sites appear to indicate success in getting more students to higher levels of the mathematics pipeline. For example, during the 1993–1994 school year, 158,673 students, or 30% of the total number of 9th–12th-grade students in the first cohort of schools (including Baltimore, Chicago, Dallas, Detroit, El Paso, Miami-Dade, New York, and Phoenix) were enrolled in gate-keeping and higher level mathematics courses. By the 1998–1999 school year, the number of high school students from the first cohort enrolled in these courses had increased to 253,458, or 46% of the total number of students (Kim et al., 2001). In addition, increases in enrollment in Algebra I or above for 8th-grade students have occurred across racial groups.

In the first cohort of sites, African American 8th-grade student enrollment in Algebra I or above increased 58% from the 1993–1994 school year to the 1998–1999 school year. Hispanic 8th-grade student enrollment in Algebra I or above increased 59% during the same time period. For White students, there was a 26% increase in enrollment in 8th-grade Algebra or beyond. In addition to increases in enrollment rates for all groups, the disparities between groups have decreased. In the first cohort of USI sites, there was a 21% decrease in the enrollment disparities between White and African American students in gate-keeping and higher level mathematics courses between the 1993–1994 and 1998–1999 school years. Similar trends can be seen in Hispanic-to-White disparities (Kim et al., 2001).

Another important indicator of the impact of the USI program is student standardized test scores. All seven of the sites in the first cohort have seen an overall improvement in student test scores between the 1993–1994 and 1998–1999 school years. In addition, five of the seven sites in the first cohort have shown reductions in the minority-to-White test score gap. For example, at the start of the USI program in New York, the average passing rate for African American and Hispanic students on the CAT-5 was 26%, compared with 59% for White students. In the 1998–1999 school year, the average passing rate for African American and Hispanic students was 42%, compared with 72% for White students. The scores of both minority and White students increased, and the minority-to-White test score gap decreased (Kim et al., 2001).

Thus, in many ways, the USI program brings together many of the OTL factors described earlier into a design to improve achievement and equity for students who have traditionally been least well-served by public schools. The description of the general USI program drivers and accomplishments gives some indication, in broad strokes, of the possibility of intervention and improvement at a systemic level. However, what this description of the USI program lacks is an understanding of how the various time and quality factors interact to produce positive changes in student achievement. For this reason, we offer another example of the systemic reform of mathematics education.

Mathematics: Application and Reasoning Skills (MARS)

The MARS[11] project was a collaboration between the Baltimore City Public School System (BCPSS) and the University of Maryland. The goal of the project was to "develop, implement, and evaluate a model for systemic reform in elementary mathematics" (Campbell, Bowden, Kramer, & Yakimowski 2003). The project, which lasted from 1996 to 2002, involved 106 elementary schools in BCPSS. Among the several different components of the MARS project, we will focus on three: (a) curriculum, (b) instructional support, and (c) professional development.

One of the key elements of the MARS project involved the redesign of the BCPSS curriculum guide and the subsequent adoption of instructional materials that supported the objectives outlined in the curriculum guide. In addition to this attention to quality curriculum, another element of MARS intervention involved support for teachers as they attempted to change their instructional practices. Exemplary teachers from BCPSS were released from their classroom responsibilities to serve as Instructional Support Teachers (ISTs). The ISTs were generally assigned to one or two schools and provided day-to-day on-site instructional support to teachers. They also designed professional development and organized grade-level planning meetings in their schools to help teachers understand and implement the curriculum. "The IST acted as coach, demonstration teacher, and program catalyst, supporting individual teachers' efforts to implement intended instructional strategies in the mathematics classroom" (Campbell et al., 2003, p. 4).

A third key element of the MARS project was ongoing professional development for teachers. A variety of professional development activities were provided to teachers. One of these opportunities was the Summer Institute. The Summer Institutes were from 5 to 14 days in duration and focused on (a) mathematical content knowledge of teachers; (b) teaching mathematics for understanding; (c) research on children's learning of mathematics concepts, organized with a grade-level focus; and (d) teaching mathematics in urban classrooms (Campbell et al., 2003). After the new curriculum guide and instructional materials were introduced, the Summer Institutes also served as an opportunity for teachers to learn how to implement the curriculum. In addition to Summer Institutes, the MARS project involved grade-specific Saturday Workshops during the school year that focused on the mathematics content that would be taught in the upcoming quarter and on the pedagogical strategies and materials appropriate for that content. Other professional development opportunities included af-

[11] The MARS project was supported by a Local Systemic Change (LSC) grant from NSF. Although similar to the USI Program in its systemic approach, the LSC program was narrower in scope (in the MARS example, focusing only on mathematics at the elementary grades) and was not specifically targeted at urban areas.

ter-school mini-courses addressing mathematics content and 3-hour graduate courses at the University of Maryland. During the 2001–2002 school year, 2,082 teachers (or 56% of the total number of elementary teachers) from BCPSS participated in some form of elementary mathematics professional development.

Standardized test results demonstrated significant increases in student achievement over the course of the MARS project. For example, between the 1997–1998 and 2000–2001 school year, the second-grade composite scores on the Terra Nova standardized test increased approximately 20 national percentile points. Similar gains were made in Grades 3 and 5. Fourth-grade gains were slightly less and first-grade gains slightly more than the other grades over the same time period (Campbell et al., 2003).

What makes MARS significant for this discussion is not simply the indication that systemic reform can impact student achievement, although that is certainly an important result. Campbell and colleagues (2003) have also examined the impact of the various components of the system, thus allowing greater insight into the influence of the time and quality factors. One of the components of the project that made a difference in student achievement was professional development. Teachers' participation in professional development that was high quality was associated with higher student achievement on the Terra Nova, at levels that were statistically significant. Further, results indicate that the effect was cumulative. For each additional high-quality professional development activity attended corresponding gains in student achievement were evident.

However, not all of the professional development opportunities offered through MARS were determined to be of high quality. Campbell and her colleagues (2003) found that the characteristics of high-quality professional development were (a) the professional development session was taught by either an IST, a well-qualified master teacher with 3 or more years of involvement in the project, or a partner from the University of Maryland; (b) the professional development integrated mathematics content and reform pedagogy; (c) if the professional development occurred following the revision of the curriculum guide, it focused on the guide and adopted textbooks; (d) if the professional development was a workshop, it was organized by grade level; and (e) if the professional development was a workshop, it focused on upcoming objectives in the mathematics curriculum, addressing both mathematics content and pedagogy (Campbell et al., 2003). These characteristics point to the interaction of quality professional development with other quality factors, specifically quality curriculum. For, Campbell and colleagues (2003) found that even high-quality professional development failed to have a positive effect on student achievement unless it integrated the textbook resources and curriculum guide. Thus, the opportunity for teachers to learn about the curriculum that they were to teach appears to have been an important component of high-quality professional development in the Baltimore case.

Another important element of the MARS intervention was the instructional support provided to the schools. However, as with professional development, not all cases of instructional support were equally effective. Moreover, differences in relative effectiveness seem to have been related to both time and quality factors. A comparison of student achievement in 2000–2001 between schools with on-site instructional support and those without support revealed no significant difference in student achievement on the Terra Nova. Yet, significant differences were evident when high-quality support persons—teachers with a strong understanding of mathematics content and the standards-based curriculum and pedagogy—were assigned to a school for at least 4 years. According to Campbell and associates (2003), "less knowledgeable resources for school-based professional support or knowledgeable resources with less time for teachers to have access to that support were not as effective" (p. 48).

In summary, the MARS example adds to our understanding of systemic reform by highlighting the connections among OTL variables. Although the MARS example points to several of the same characteristics of high-quality professional development as the studies cited previously (e.g., targeted at a grade-level, focused on content and pedagogy), it also indicates a connection between quality curriculum and quality professional development. Moreover, the results suggest that high-quality instructional support of sufficient duration can impact student achievement. Thus, the example of MARS demonstrates how several of the OTL time and quality variables can be brought together to design a systemic program that is successful in raising student achievement.

It's Time to Design

Rigorous mathematics standards preK–12 require thoughtful action and planning. Moreover, the building blocks for engineering a mathematical revolution in any school are time, quality, and design. These three pillars of OTL are foundational for the improvement of the teaching and learning of mathematics in school settings. There are many paths for organizing and implementing change in school mathematics; however, the failure to consider time and quality factors and design issues carefully is a recipe for lost educational opportunity. Jere Confrey and colleagues at the *Systemic Research Center for Education in Mathematics, Science, and Technology* (SYRCE) designed Figure 28.1 to serve as a conceptual model to inform the organizational design of mathematics teaching and learning in schools.

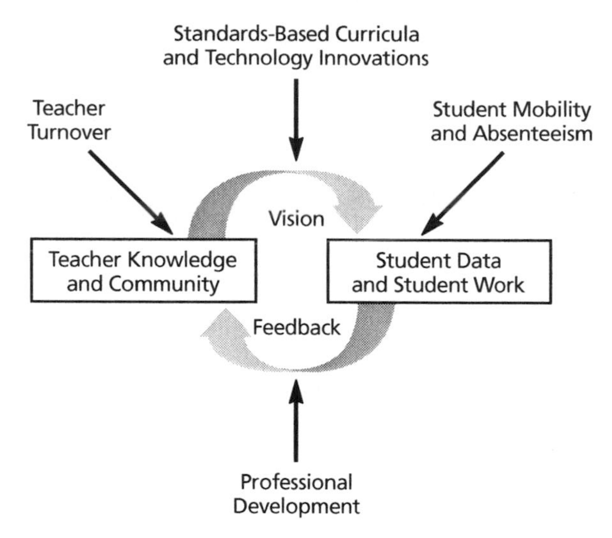

Figure 28.1 SYRCE conceptual model. *Source:* J. Confrey (personal communication, November 2004)

Figure 28.1 is included because it serves as a reminder of key opportunity-to-learn factors and how they interact in our systems of education. As an engineer, it is important to keep in mind a broader conceptual model of school change. This is especially important in light of the day-to-day realities of teaching, administration, and political challenges that face educators. The many events and distractions, such as leadership changes and financial shortfalls, that occur in schools and school districts only become real problems when they influence these three foundational pillars of learning and teaching. Unfortunately, instability and interim leadership are rife everywhere in public schooling. Recall, engineers "learn to build." Both learning and building require stable, long-term, and insightful leadership. The building process for the engineer is supported by the following model development sequence:

- Model Construction
- Model Exploration
- Model Application
- Model Revision

Models are a language for describing patterns, patterns that can be observed and tested in the real world of schools. Thus, the process of model development emphasizes understanding school factors that influence opportunity to learn mathematics. The model development sequence requires additional clarification.

Model Construction

OTL requires a clear vision and set of learner goals. Many school districts accomplish this part of model construction by adopting all or portions of state mathematics standards and in combination with local system objectives create specific district-level goals for what all students should know and be able to do. District mathematics goals are important quality indicators for teachers, administrators, parents, students, and the broader community. Thus, multiple communication strategies for each constituency should be part of the model's design. Each constituency should be provided samples of student work. The work should exemplify system-wide expectations and illustrate student work products that meet the district's mathematics standards. Setting high standards and communicating a clear vision are only part of the model construction process. Ensuring that all students have access to a quality curriculum also is part of the process.

A number of steps are vital to initiating a quality curriculum. Each school should design an academic plan based on local mathematics standards and an associated accountability structure. The process includes

1. Communicating the importance of consistent application of curricular programs and standards.
2. Eliminating courses and academic experiences lacking the rigor of mathematics standards or inconsistent with developmentally appropriate mathematics practice.
3. Providing adequate time in the curricular design to focus on core subjects.
4. Building a timely framework for monitoring student progress using data.
5. Providing teacher and administrator professional development focused on classroom strategies for assessing student learning of mathematics standards.

Vision, curriculum quality, and accountability systems are vital to constructing a model. However, policy and resource alignment also contribute to a comprehensive model design. It is difficult to imagine a sustained change strategy that does not include front-end alignment of school policies and resources to support rigorous standards. For example, ongoing mathematics professional development should be part of the model construction process. In addition, effective student support programs should be prepared and ready to implement in response to updates on student assessments and mobility information. Too often, responses to these types of data are not part of the upfront planning process. Consequently, many school districts find themselves engaged in triage mode, piecing together programs in real-time, rather than implementing planned data-driven interventions. Resources aligned with an appropriate vision, quality curriculum, and accountability systems are part of the technical core of the model construction process. The technical core is vitally important, but not enough. Ultimately, people explore the model and make decisions.

Model Exploration

Model construction includes planning for reflective examination of student, teacher, and school-level progress. Embedded in the model construction process are key elements of model exploration. The constructed model should include rich opportunities to explore the progress of the organization in relation to its goals. Yet, model exploration is not merely an instrumental activity; instead, exploration is cultural and people driven, with teachers and administrators central to the evaluative process.

As Figure 28.1 indicates, teacher community and knowledge are important components of the feedback loop. Teachers interpret student performance and assessments, and in turn they should rethink their practice and that of the school in conjunction with other instructional leaders. This process is greatly enhanced where community norms are established and teacher turnover is low. Teacher community and norms are linked to the model application process. The importance of teacher collective practice in school mathematics cannot be overstated (see, e.g., Gutierrez, 1996). As Campbell (1996) noted, “it may be unreasonable to expect sustained and reflective reform in isolated classrooms across urban settings. It is not unreasonable to address reform in urban schools where teachers and administrators are working together to develop a shared purpose and meaning” (p. 453).

Both administrators and teachers require training to support their community efforts to build a student-centered, data-driven organization. Classroom strategies for assessing student learning should be central to the common professional development. The community-building and professional-development activities also must be combined with an aligned monitoring process that includes

1. Timely and usable data on student progress.
2. Opportunities for mid-course corrections based on data.
3. Disaggregated student achievement data and mobility information.
4. Recognition and reward for positive results.

Model exploration potentially offers important benefits to schools and school districts. First, model exploration can provide highly visible evidence related to opportunity to learn and support administrators and teachers seeking successful exemplars of effective mathematics practice (Skrla & Scheurich, 2001). Second, incremental success in student achievement and teacher effectiveness can lead to higher expectations and goals for academic achievement of all demographic groups.

Model Application

One can focus on the design of the model and matters of model exploration without attending to the classroom practices and support systems associated with instruction. Most school systems begin the academic year with some kind of school improvement plan, and although the quality of these plans vary widely, plans for change models exist. Further, it is quite common to find some level of model exploration in most school systems. Although, the quality factors associated with the exploration are often limited. Yet, the most serious challenge facing leadership is model application or implementation. Many instructional programs are adopted, distributed, and discarded each year.

A well-thought-out plan is required to gain any implementation traction. A key factor for success in this area is a fine-tuned curriculum guide that clearly delineates important content aligned to appropriate instructional materials. Further, research-based professional development for teams of teachers and instructional leaders is vital. The work of the teacher collective enterprise and significant time engaged with high-quality professional development are central to the implementation process. And quality student assessments—classroom-based as well as more summative assessments—are linked to collective and individual reflection by members of the teacher community. Quality curriculum, professional development, collective practice, and the other aspects of the model design are a support system for creating a successful teaching and learning process.

Teachers are key to model application. Clearly, mathematics teachers at all levels, kindergarten through college, are central to the improvement of mathematics education. If professional development is to a make a difference to students in the classroom, it must be teacher-focused and student-centered. Stigler and Hiebert (1999) wrote:

> Improving something as complex and culturally embedded as teaching requires the efforts of all players, including students, parents and politicians. But teachers must be the primary driving force behind change. They are in the best position to understand the prob-

lems that students face and to generate possible solutions. (p. 135)

Successful model application is largely classroom based and includes

1. Teachers and students participating in two-way conversations about mathematics.
2. Teachers pushing for both mathematical meaning behind quality curricular tasks and procedural fluency.
3. Teachers daily building instruction on student thinking and continuity in the mathematics domain.
4. Teachers and administrators organizing classes and support systems to ensure adequate time for students to learn the mathematics content.
5. Teachers and administrators maintaining high expectations for all learners.

These five components of instruction are foundational to model application and opportunity to learn. Although these components are stable in a broader sense, at times the design is not quite adequate to achieve system goals or more ambitious goals require a revised model.

Model Revision

Careful reflection on the progress being made is central to future progress. How is the model helping or hindering teachers and students? Each phase of the model development sequence should be reviewed, and critical discussions about fine-tuning or model abandonment must be held. Revision is not merely an end-of-the-year activity. Model revision is a continuous process that should be part of the design consideration. To aid in both the design and model revision process, the Appendix includes an *Engineering Change Assessment Instrument.* This instrument can be used to help keep track of school and school district progress on key opportunity-to-learn factors.

Taking a model-based approach to the mathematics change process involves constructing a change strategy, exploring the qualities and feasibility of the strategy, applying or implementing the strategy, and based on the data continuously fine-tuning the strategy. A close examination of many school district strategic plans will reveal many goals and targets related to the state accountability systems. Often, what is absent is a coherent change strategy or model for improvement in mathematics that takes seriously the time, quality, and design considerations reviewed in this document. Few school districts explore the qualities or feasibility of their change effort. Rather, many buy into a model without considering the conditions and constraints that exist in the system. For example, this is evident when major initiatives to improve mathematics, reading, and science at the elementary level are occurring simultaneously. The point is not that three reform strategies cannot occur together; rather, failure to coordinate the changes is a common strategic flaw. Many reasonable change strategies are destroyed as a result of the failure to consider the feasibility of the model.

Engineering Progress: Limiting Conditions

Limiting conditions are conditions that materially affect the appraisal process and, as a consequence, the value conclusion. For example, having no electrical power in a school building is a limiting condition, as it prohibits the use of computers and other electronic equipment. This notion can be applied to mathematics education. What limiting conditions exist that can affect student performance on mathematics outcome measures and, as a consequence, the public's perception of school quality? A brief examination of recent legislation will provide some insight into this question.

Current educational policy and law, more specifically NCLB, calls for educators to carefully examine student achievement by demographic group. Moreover, schools and administrators are being held accountable for improved student performance. This is a radical departure from past educational practice. In fact, this law represents a major shift in federal discourse related to matters of equality. According to Crenshaw (1988), a legal scholar, there are two visions of equality present in anti-discrimination legislation and discourse. One view of equality, which she referred to as the restrictive view, "treats equality as a process, downplaying the significance of actual outcomes" (p. 1341). This section highlighted important processes related to student learning. Although critical to engineering positive progress, the opportunity-to-learn recommendations and other research-based lessons are limited. Specifically, they represent building blocks. However, someone must build the building. Thus, a potential limiting condition is related to people, more specifically to educators willing to do the hard work associated with building mathematical minds.

In the case of school mathematics, the "product" is optimal student learning as measured by state-mandated tests—the intended outcome. Hence, current law as reflected in NCLB is more consistent with another view of equality—the expansive view. Professor Crenshaw (1988) stated this second view of equality, the expansive vision, "stresses equality as a result and looks to real consequences" (p. 1341). The NCLB legislation calls for educators to reflect seriously on student outcomes by demographic group. This expansive view suggests that equal treatment of students is not equitable if it leads to differential outcomes. This perspective conflicts with the worldview held by many educators that "equal treatment" of all students is optimal. If educators assume a "one-size-fits-all" approach to classroom practice, without careful reflection and planning for individual as well as collective student learning, the product is likely to be unequally distributed opportunities to learn and continued underperformance of traditionally underserved students. In the case of NCLB this "equal treatment" worldview and its associated ideological perspective—color blindness—are limiting conditions.

Many educators who view the "equal treatment" position as a well-meaning and fair perspective assume color blindness as a political or ideological stance. Part of the problem with color blindness is that it ignores students and their performance. Irvine (1990) stated, "by ignoring students' most obvious physical characteristic, race, . . . teachers are also disregarding students' unique cultural behaviors, beliefs, and perceptions—important factors that teachers should incorporate, not eliminate, in their instructional strategies" (p. 54). Today, the color-blind method of mathematics education creates barriers to true equality by erecting barriers in school mathematics such as

1. Persistent tracking,
2. Fewer opportunities for African American and Hispanic students to learn from the best qualified teachers,
3. Less access to technology, and
4. Cultural discontinuity between school mathematics and the family life of diverse student groups (National Science Board, 1991; Oakes, 1990; Piller, 1992; Stanic, 1991).

The first three of these barriers to equity, which are quantifiable, are considered acceptable indicators of unequal educational conditions. In contrast, this fourth barrier is subtle and difficult to identify and measure in everyday schooling. The matter of family life and school mathematics warrants additional discussion.

Less frequently, educators have explored the experiences of stakeholders other than teachers in the process of school mathematics reform (see Graue & Smith, 1996). This is a serious limiting condition. In particular, there is limited information available on how parents

- Perceive their children's mathematics instruction,
- Interpret their children's performance in light of mathematics standards and state testing, or
- View their role in the mathematics education process.

As Graue and Smith (1996) noted, despite parental presence in many aspects of educational reform rhetoric, researchers of school mathematics practice and design have shown little interest in parents.[12] Ethically, in the context of high-stakes testing and reformed practice, this is a condition that must be addressed by the focused efforts of scholars, school-based educators, and community-based organizations. Matters of ethics represent the final limiting conditions.

The pressures of high-stakes testing, public disclosure of testing results, and the real possibility of job loss or demotions have placed a new and heavy burden on professional educators. This burden will cause some to carefully design change strategies and create exciting learning environments for students. For others, the score-high mentality will create new ethical dilemmas. Reports have emerged of schools and districts removing large numbers of students from opportunities to test, rather than creating appropriate opportunities to learn. The ethical challenges are endless when the stakes are high and very real. As with good engineers, educators must factor ethics into their thinking, everyday planning, and ultimately their design strategy. Failure to do so will endanger the building process.

In pursuit of an engineering strategy in mathematics education, the importance of dedicated educators cannot be underestimated. Many educators are excellent, but some are not. The following questions highlight the traits of those ready to improve school mathematics:

- Does the educator listen to new ideas with an open mind?
- Does the educator consider a variety of solution methods associated with student learning before choosing a design approach?
- Is the educator content with determining a learning design on the basis of trial and error?
- Does the teacher use phases such as, "I need to understand why students learn mathematics with this approach" and "Let's consider all possibilities."

If educators are eager to listen, open to a variety of educational solutions, never content with just trial-

[12] One exception is Family Math (www.lhs.berkeley.edu/equals/FMnetwork.htm).

and-error methods, and pressed to know why a method works with students, they represent the type of teachers and instructional leaders who can engineer changes in mathematics education. These kinds of educators are foundational for "Learning to Build."

REFERENCES

Apple, M. (1992). Do the standards go far enough? Power, policy, and practice in mathematics education. *Journal for Research in Mathematics Education, 23*, 412–431.

Beane, D. B. (1985). *Mathematics and science: Critical filters for the future.* Bethesda, MD: Mid-Atlantic Equity Consortium.

Bredekamp, S. & Copple, C. (Eds.). (1997). *Developmentally appropriate practice in early childhood programs.* Washington, DC: National Association for the Education of Young Children.

Brophy, J., & Good, T. (1986). Teacher behavior and student achievement. In M. Wittrock (Ed.), *Handbook of research on teaching* (pp. 328–375). New York: Macmillan.

Campbell, P. (1996). Empowering children and teachers in the elementary mathematics classrooms of urban schools. *Urban Education, 30*(6), 449–475.

Campbell, P. F., Bowden, A. R., Kramer, S. C., & Yakimowski, M. E. (2003). *Mathematics and reasoning skills* (No. ESI 95-54186). College Park, Maryland: University of Maryland, MARS Project.

Carpenter, T. P., Fennema, E., Franke, M. L., Levi L., & Empson S. (1999). *Children's mathematics: Cognitively guided instruction.* Portsmouth, NH: Heinemann.

Carpenter, T. P., Fennema, E., Peterson, P., & Carey, D. (1988). Teachers' pedagogical content knowledge of students' problem solving in elementary arithmetic. *Journal for Research in Mathematics Education, 19*, 385–401.

Carpenter, T. P., Fennema, E., Peterson, P., Chiang, C. P., & Loef, M. (1989). Using knowledge of children's mathematics thinking in classroom teaching: An experimental study. *American Educational Research Journal, 26*, 499–532.

Carpenter, T. P., & Moser, J. M. (1983). The acquisition of addition and subtraction concepts. In R. Lesh & M. Landau (Eds.), *The acquisition of mathematics concepts and processes* (pp. 7–14). New York: Academic Press.

Carroll, J. B. (1963). A model of school learning. *Teachers College Record, 64*, 723–733.

Carey, D., Fennema, E., Carpenter, T. P., & Franke, M. L. (1995). Equity and mathematics education. In W. Secada, E. Fennema, & L. Adajian (Eds.), *New directions for equity in mathematics education* (pp. 93–125). New York: Cambridge University Press.

Cohen, D. K., Raudenbush, S. W., & Ball, D. L. (2003). Resources, instruction, and research. *Educational Evaluation and Policy Analysis, 25*(2), 119–142

Council of Great City Schools. (1992). *National urban education goals: Baseline indicators, 1990–1991.* Washington, DC: Author.

Council of Great City Schools. (2003). Brief of the Council of Great City Schools, amicus curiae, *Campaign for fiscal equity v. State of New York* (NY Ct. App. 2003). Retrieved March 21, 2006 from http://www.cgcs.org/pdfs/NYBrief.pdf

Crenshaw, K. W. (1988). Race, reform, and retrenchment: Transformation and legitimation in antidiscrimination law. *Harvard Law Review*, 101, 1331–1387.

Devault, M. V., & Weaver, J. F. (1970). Forces and issues related to curriculum and instruction, K–6. In A. F. Coxford & P. S. Jones (Eds.), *A history of mathematics education in the United States and Canada* (pp. 92–152). Washington, DC: National Council of Teachers of Mathematics.

Fennema, E., & Franke, M. L. (1992). Teachers' knowledge and its impact. In D. A. Grouws (Ed.), *Handbook of research on mathematics teaching and learning* (pp. 147–164). New York: Macmillan.

Fey, J. T. (1981). *Mathematics teaching today: Perspectives from three national surveys.* Reston, VA: National Council of Teachers of Mathematics.

Floden, R., Porter, A. C., Schmidt, W., Freeman, D., & Schwille, J. (1981). Responses to curriculum pressures: A policy-capturing study of teachers' decisions about context. *Journal of Educational Psychology, 73*, 129–141.

Garcia, E. E. (1995). Educating Mexican-American students: Past treatment and recent developments in theory, research, policy, and practice. In J. A. Banks & C. A. McGee Banks (Eds.), *Handbook of research in multicultural education* (pp. 372–387). New York: Macmillan.

Garet, M. S., Porter, A. C., Desimone, L., Birman, B. F., & Yoon, K. S. (2001). What makes professional development effective? Results from a national sample of teachers. *American Educational Research Journal, 38*(4), 915–945.

General Accounting Office. (1993). *School age demographics: Recent trends pose new educational challenges* (HRD-93-105BR). Washington, DC: United States General Accounting Office.

Grant, C. A., & Sleeter, C. E. (1986). Race, class, and gender in education research: An argument for integrative analysis. *Review of Educational Research, 56*, 195–211.

Graue, M. E., & Smith, S. Z. (1996). Parents and mathematics education reform: Voicing the authority of assessment. *Urban Education, 30*(4), 395–421.

Green, P. J., Dugoni, B. L., Ingels, S. J., & Camburn, E. (1995). *A profile of the American high school senior in 1992.* Washington, DC: U.S. Department of Education.

Gutierrez, R. (1996). Practices, beliefs, and cultures of high school mathematics departments: Understanding their influence on student advancement. *Journal of Curriculum Studies, 28*(5), 495–529.

Hilliard, A. (1991). Do we have the will to educate all children? *Educational Leadership, 49*, 31–36.

Hoffer, T. B., Rasinski, K. A., & Moore, W. (1995). *Social background differences in high school mathematics and science coursetaking and achievement* (NCES No. 95–206). Washington, DC: U. S. Department of Education.

Horenstein, M. N. (2002). *Design concepts for engineers.* Upper Saddle River, NJ: Prentice Hall.

Hurd, P. D. (1997). *Inventing science education for the new millennium.* New York: Teachers College Press.

Husén, T. (Ed.). (1967). *International study of achievement in mathematics: A comparison of twelve countries.* New York: Wiley.

Irvine, J. (1990). *Black students and school failure: Policies, practices, and prescriptions.* Westport, CT: Praeger.

Kim, J. J., Crasco, L. M., Blank, R. K., & Smithson, J. (2001). *Survey results of urban school classroom practices in mathematics and science: 2000 report.* Norwood, MA: Systemic Research.

Kinstch, W., & Greeno, J. G. (1985). Understanding and solving arithmetic word problems. *Psychology Review, 92*, 109–129.

Kliebard, H. M. (1987). *The struggle for the American curriculum 1893–1958.* New York: Routledge & Kegan Paul.

Knapp, M. (1995). *Teaching for meaning in high poverty classrooms.* New York: Teachers College Press.

Knapp, M. S., & Woolverton, S. (1995). Social class and schooling. In J. A. Banks & C. A. McGee Banks (Eds.), *Handbook of research on multicultural education* (pp. 548–569). New York: Macmillan.

Ladson-Billings, G. (1994). *The dreamkeepers: Successful teachers of African American students.* San Francisco: Jossey-Bass.

Lampert, M. (1990). Connecting inventions with conventions. In L. P. Steffe & T. Wood (Eds.), *Transforming children's mathematics education* (pp. 253–265). Hillsdale, NJ: Erlbaum.

Lee, V., Croninger, R., & Smith, J. (1997). Course-taking, equity, and mathematics learning: Testing the constrained curriculum hypothesis in U.S. secondary schools. *Educational Evaluation and Policy Analysis, 19,* 99–121.

Leinhardt, G. (1983). Overlap: Testing whether it is taught. In G. F. Madaus (Ed.), *The courts, validity, and minimum competency testing* (pp. 153–170). Boston: Kluweer-Nijhoff.

Leinhardt, G., & Seewald, A. (1981). Overlap: What's tested, what's taught? *Journal of Educational Measurement, 18*(2), 85–96.

LeMahieu, P., & Leinhardt, G. (1985). Overlap: Influencing what's taught, a process model of teachers' content selection. *Journal of Classroom Interaction, 21*(1), 2–11.

Madaus, G. F., West, M. M., Harmon, M. C., Lomax, R. G., & Viator, K. A. (1992). *The influence of testing on teaching math and science in grades 4–12.* Boston: Boston College, Center for the Study of Testing, Evaluation, and Educational Policy.

Massell, D. (1994). Setting standards in mathematics and social studies. *Education and Urban Society, 26,* 118–140.

McDonnell, L., Burstein, L., Catterall, J., Ormseth, T., & Moody, D. (1990). *Discovering what schools really teach: Designing improved coursework indicators.* Santa Monica, CA: RAND.

Meister, J. C. (1998). *Corporate universities: Lessons in building a world-class work force.* New York: McGraw-Hill.

Meyer, M. R. (1991). Equity: The missing element of recent agendas for mathematics education. *Peabody Journal of Education, 66*(2), 6–21.

Meyerson, M. I. (2002). *Political numeracy: Mathematical perspectives on our chaotic constitution.* New York: W. W. Norton & Company.

Miller, L. S. (1995). *An American imperative: Accelerating minority educational advancement.* New Haven, CT: Yale University Press.

Moses, R. P., & Cobb, C. E. (2001). *Radical equations: Math and literacy and civil rights.* Boston: Beacon Press.

National Center for Educational Statistics. (2000). *NAEP 1999 trends in academic progress* (NCES No. 2000–469). Washington, DC: Author.

National Council of Supervisors of Mathematics. (1977, January). Position paper on basic skills. Retrieved March 21, 2006, from http://mathforum.org/ncsm/NCSMPublications/position.html#basmathskills.

National Council of Teachers of Mathematics. (1959). The secondary mathematics curriculum. *Mathematics Teacher, 52,* 389–417.

National Council of Teachers of Mathematics. (1980). *An agenda for action: Recommendations for school mathematics of the 1980's.* Reston, VA: Author.

National Council of Teachers of Mathematics. (1989). *Curriculum and evaluation standards for school mathematics.* Reston, VA: Author.

National Council of Teachers of Mathematics. (1991). *Professional standards for teaching mathematics.* Reston, VA: Author.

National Council of Teachers of Mathematics. (1995). *Assessment standards for school mathematics.* Reston, VA: Author.

National Council of Teachers of Mathematics. (2000). *Principles and standards for school mathematics.* Reston, VA: Author.

National Education Goals Panel. (1995). *The national education goals report: Building a nation of learners.* Washington, DC: Author.

National Governors' Association. (1993). *The debate on opportunity-to-learn standards.* Washington, DC: Author.

National Research Council. (1999a). *Testing, teaching, and learning.* Washington, DC: Author.

National Research Council. (1999b). *High stakes: Testing for tracking, promotion, and graduation.* Washington, DC: Author.

National Research Council. (2001). *Adding it up: Helping children learn mathematics.* Washington, DC: National Academy Press.

National Science Board. (1991). *Science and engineering indicators* (NSB No. 91-1). Washington, DC: U.S. Government Printing Office.

National Science Foundation (1994). *Women, minorities, and persons with disabilities in science and engineering: 1994.* Arlington, VA: Author.

Neito, S. (1995). A history of the education of Puerto Rican students in U. S. mainland schools: "Losers," "Outsiders," or "Leaders"? In J. A. Banks & C. A. McGee Banks (Eds.), *Handbook of research in multicultural education* (pp. 388–411). New York: Macmillan.

Nelson-Barber, S., & Estrin, E. T. (1995). Bringing Native American perspectives to mathematics and science teaching. *Theory into Practice, 34*(3), 174–185.

New York State Department of Education (2002). *New York, The state of learning.* Albany, NY: The State University of New York and New York State Department of Education

Newmann, F. M., & Wehlage, G. G. (1995). *Successful school restructuring.* Madison: Wisconsin Center for Education Research.

North Central Regional Educational Laboratory. (2000, February). *A study of the differences between higher- and lower- performing Indiana schools in reading and mathematics.* Naperville, IL: Author.

Oakes, J. (1990). Opportunities, achievement, and choice: Women and minority students in science and mathematics. In C. B. Cazden (Ed.), *Review of research in education* (Vol. 16, pp. 153–222). Washington, DC: American Educational Research Association.

Peterson, P., Fennema, E., & Carpenter, T. (1991). Using children's mathematical knowledge. In B. Means, C. Chelemer, & M. Knapp (Eds.), *Teaching advanced skills to at-risk students: Views from research and practice* (pp. 68–101). San Francisco: Jossey-Bass.

Piller, C. (1992). Separate realities. *MACWORLD, 9*(9), 218–231.

Porter, A. C. (1989). A curriculum out of balance: The case of elementary school mathematics. *Educational Researcher, 18*(5), 9–15.

Porter, A. C. (1993). Defining and measuring opportunity to learn. In S. L. Traiman (Ed.), *The debate on opportunity-to-learn standards* (pp. 33–72). Washington, DC: National Governors' Association.

Rasinski, K. A., Ingels, S. J., Rock, D. A., & Pollack, J. M. (1993). *America's high school sophomores: A ten-year comparison* (NCES No. 93-087). Washington, DC: National Center for Educational Statistics.

Riche, M. F., & Pollard, K. M. (1992). *The challenge of change: What the 1990 census tells us about children.* Washington, DC: U. S. Department of Education.

Robitaille, D. F., & Travers, K. J. (1992). International studies of achievement in mathematics. In D. A. Grouws (Ed.), *Handbook of research on mathematics teaching and learning* (pp. 687–709). New York: Macmillan.

Rock, D. A., & Pollack, J. M. (1995). *Mathematics course-taking and gains in mathematics achievement* (Statistical Analysis Rep. NCES No. 95-714). Washington, DC: National Center for Educational Statistics.

Rousseau, C., & Tate, W. F. (2003). No time like the present: Reflecting on equity in school mathematics. *Theory into Practice, 42*(3), 210–216.

Rowan, B. (1995). Learning, teaching, and educational administration: Toward a research agenda. *Educational Administration Quarterly, 31*(3), 344–354.

Secada, W. G. (1992). Race, ethnicity, social class, language, and achievement in mathematics. In D. A. Grouws (Ed.), *Handbook of research on mathematics teaching and learning* (pp. 623–660). New York: Macmillan.

Secada, W. G. (1996). Urban students acquiring English and learning mathematics in the context of reform. *Urban Education, 30*(4), 422–448.

Silver, E. A., & Lane, S. (1995). Can instructional reform in urban middle schools help narrow the mathematics performance gap? Some evidence from the QUASAR project. *Research in Middle Level Education, 18*(2), 49–70.

Silver, E. A., & Stein, M. K. (1996). The QUASAR project: The "Revolution of the Possible" in mathematics instructional reform in urban middle schools. *Urban Education, 30*(4), 476–521.

Skrla, L., & Scheurich, J. J. (2001). Displacing deficient thinking in school leadership. *Education and Urban Society, 33*(3), 235–259.

Stanic, G.M.A. (1991). Social inequality, cultural discontinuity, and equity in school mathematics. *Peabody Journal of Education, 66*(2), 57–71.

Stevens, F. (1993a). *Opportunity to learn: Issues of equity for poor and minority students.* Washington, DC: National Center for Educational Statistics.

Stevens, F. (1993b). Applying an opportunity-to-learn conceptual framework to the investigation of the effects of teaching practices via secondary analyses of multiple-case-study summary data. *Journal of Negro Education, 62,* 232–248.

Stevenson, H., & Stigler, J. (1992). *The learning gap: Why our schools are failing and what can we learn from Japanese and Chinese education.* New York: Summit Books.

Stigler, J. W., & Hiebert, J. (1999). The teaching gap. New York: Free Press.

Stodolsky, S. (1988). *The subject matters: Classroom activity in mathematics and social studies.* Chicago: University of Chicago Press.

Strickland, D. S., & Ascher, C. (1992). Low-income African American children and public schooling. In P. W. Jackson (Ed.), *Handbook of research on curriculum* (pp. 609–625). New York: Macmillan.

Suter, L. E. (2000). Is student achievement immutable? Evidence from international students on schooling and student achievement. *Review of Educational Research, 70*(4), 529–545.

Tate, W. F. (1995). School mathematics and African American students: Thinking seriously about opportunity-to-learn standards. *Educational Administration Quarterly, 31,* 424–448.

Tate, W. F. (1996). Urban schools and mathematics reform: Implementing new standards. *Urban Education, 30,* 371–378.

Tate, W. F. (1997). Race-ethnicity, SES, gender, and language proficiency trends in mathematics achievement: An update. *Journal for Research in Mathematics Education, 28,* 652–679.

Tate, W. F., & Rousseau, C. (2002). Access and opportunity: The political and social context of mathematics education. In L. D. English (Ed.), *Handbook of international research in mathematics education* (pp. 271–299). London: Erlbaum.

Tatum, B. D. (1992). Talking about race, learning about racism: The application of racial identity development theory in the classroom. *Harvard Educational Review, 62,* 1–24.

U.S. Department of Education, National Center for Education Statistics. (2005). *The condition of education* (NCES 2005-094). Washington, DC: U.S. Government Printing Office.

Vernez, G. (1992). Needed: A federal role in helping communities cope with immigration. In J. B. Steinberg (Ed.), *Urban America: Policy choices for Los Angeles and the nation.* Santa Monica, CA: RAND.

Villasenor, A. & Kepner, H. (1991). Arithmetic from a problem-solving perspective: An urban implementation. *Journal for Research in Mathematics Education, 24,* 62–69.

White, D. Y. (1997). *The mathematics classroom question and response patterns of third-grade teachers in high minority population schools* (Doctoral dissertation, University of Maryland, College Park, 1997). Dissertation Abstracts International, 58-09A, 3451.

Winfield, L. F. (1987). Teachers' estimates of test content covering in class and first-grade students' reading achievement. *Elementary School Journal, 87,* 438–445.

Winfield, L. F. (1993). Investigating test content and curriculum content overlap to assess opportunity to learn. *Journal of Negro Education, 62,* 288–310.

AUTHOR NOTE

We wish to thank the following colleagues who took the time to review various drafts of this chapter: Francena Cummings, Paul Cobb, Joseph Harris, Kay Gilliland, Gerunda Hughes, Rowhea Elmesky, Jere Confrey, Marilyn Strutchens, Michele Chappell, Yuria Orihuela, Doris Hipps, Jerome Shaw, Karen King, Josephine Doline, Louise Beyers, Cindy McIntee, Marilyn Irving, DeAnnna Beane, Sheryl Denbo, and Alicia Darnesbourg. Portions of this chapter are reprinted with permission from SERC@SERVE. We want to thank Jere Confrey and the McKenzie Group for permission to use tools constructed under their respective programs of research and development. The opinions expressed in this chapter are those of the authors and do not necessarily reflect the views of the supporting agencies or reviewers. Versions of this chapter were presented at the 2003 Annual Meeting of the National Council of Teachers of Mathematics and the 2004 Annual Meeting of the American Educational Research Association.

Appendix

Opportunity-to-Learn Building Blocks	Rating Low–High 1	2	3	4	5	Description/Areas to Be Addressed
1 Establishing Essential Structures—Organizational Context						
1.1 Set high standards						
1.1.1 Develop clear and rigorous goals for what students should know and be able to do 1.1.2 Communicate expectations to teachers, administrators, parents, students, community 1.1.3 Provide samples of student work that meet standards and communicate what is expected to all constituencies						
1.2 Ensure that all students have access to a challenging curriculum						
1.2.1 Implement school-based planning and accountability processes 1.2.2 Emphasize consistent application of curricular programs and standards 1.2.3 Eliminate watered-down courses and tracking 1.2.4 Ensure that curriculum and assignments are aligned with standards 1.2.5 Focus on core subjects: language arts, reading, mathematics, science, social studies						
1.3 Align standards with assessment and monitor progress						
1.3.1 Provide timely, usable data on student progress 1.3.2 Encourage mid-course corrections based on data 1.3.3 Recognize and reward positive results 1.3.4 Disaggregate student achievement data and carefully monitor minority student data 1.3.5 Train teachers in classroom strategies for assessing student learning to determine the impact of teaching and classroom activities						
1.4 Develop programs to address the impact of student/teacher mobility in low-performing schools to include						
1.4.1 Standardization of curriculum and materials 1.4.2 Centralized storage of data 1.4.3 Transmission of individual student data 1.4.4 Staff training on impact of mobility and how to mitigate this factor (see Dept. of Defense school model) 1.4.5 Recruitment and retention of quality teachers at low-performing schools 1.4.6 Additional use of instructional/curriculum specialists						
1.5 Develop innovative strategies to						
1.5.1 Organize schools into smaller units—small schools have been shown to improve both gap and achievement levels 1.5.2 Provide quality, sustained professional development and increase its availability of schools and through technology and other innovative approaches 1.5.3 Encourage collaborative and team approaches 1.5.4 Foster collegial networks among teachers, principals, schools 1.5.5 Extend teacher and team planning time						

Opportunity-to-Learn Building Blocks	Rating Low–High 1	2	3	4	5	Description/Areas to Be Addressed
1.6 Provide needed policies and resources to support						
1.6.1 Ongoing, quality school site professional development 1.6.2 Effective, timely student support programs 1.6.3 Parent/community support initiatives 1.6.4 Rigorous standards-based teaching and learning for all students 1.6.5 Governance/organizational alignments necessary to ensure accountability, student progress, and quality teaching and learning for all students						
1.7 Foster research-based characteristics of high-performing schools						
1.7.1 Minimize bureaucracy 1.7.2 Promote effective leadership at all levels 1.7.3 Promote access and equity for all students 1.7.4 Implement rigorous standards-based programs 1.7.5 Provide significant planning time for teachers 1.7.6 Orient learning around essential skills, concepts, and inquiry 1.7.7 Foster parent involvement 1.7.8 Develop a sense of team, collaboration, and family within the school system or school 1.7.9 Develop strategies to use research on "bridging the gap" to inform decision-making and to influence the actions of legislators, teacher preparation institutions, educators, community leaders, and parents 1.7.10 Accountability for quality classroom teaching						
2 Increasing Opportunities to Learn						
2.1 Establish a process for review of subject area achievement goals quarterly, to address needed changes in subject area and grade-level expectations, standards-based teaching/learning practices, and assessment techniques						
2.2 Foster high levels of teacher engagement in "bridging the gap" process through activities that promote involvement with						
2.2.1 School or school system goals for high achievement 2.2.2 Identification of skills, practices, and knowledge for effective teaching and high achievement 2.2.3 Teacher training in working with diverse learners						
2.3 Establish a process for monitoring classroom implementation of performance and content standards to include						
2.3.1 Depth of content coverage, extent and time allocated to key concepts, and emphasis on general principles and major concepts 2.3.2 Quality of instructional delivery and teaching practices such as hands-on activities, use of technology, and student performance activities that include work products						

Opportunity-to-Learn Building Blocks	Rating Low–High 1	2	3	4	5	Description/Areas to Be Addressed
2.4 Foster implementation of new approaches to support meaningful ways to make schools/classrooms more compatible with the diverse home backgrounds of students to						
2.4.1 Reduce cultural isolation 2.4.2 Identify individual student learning styles and build on strengths students bring 2.4.3 Use modeling and interactive practices and scaffolding to guide students to complete complex tasks 2.4.4 Enhance use of group learning techniques such as cooperative and team learning, peer tutoring, and reciprocal teaching						
3 Expanding Community Connections						
3.1 Foster increased and meaningful family involvement activities						
3.1.1 Provide specific training to families in tutoring and encouraging students to meet high standards 3.1.2 Provide community and extended educational opportunities for families at schools, in projects, churches, and community centers 3.1.3 Develop family resource centers 3.1.4 Initiate or extend family participation opportunities to foster meaningful roles for families in support of school system/school goals and to encourage family involvement as a team effort in a supportive school environment 3.1.5 Communicate importance of family support in establishing homework time, study skills, and academic standards in the home						
3.2 Establish necessary collaboration between schools and other community resources to provide needed assistance to						
3.2.1 Ensure availability of support mechanisms within each school 3.2.2 Establish programs to forge connections between families and community organizations in support of student academic learning and rigorous standards 3.2.3 Establish opportunities for meaningful proactive roles for business and industry in shaping and supporting rigorous standards at district and school-site levels 3.2.4 Involve universities in "bridging the gap" initiatives that strengthen the K–16 community connections and programs						
4 Enhancing Teacher Quality and Practice						
4.1 Implement professional development programs that focus on standards-based learning aligned with assessment and with emphasis on practices that promote in-depth content knowledge, key concepts and general principles, real-life hands-on activities, application of technology, and modeling effective teaching strategies						
4.2 Implement quality teaching practices and training programs that utilize strategies for achieving high expectations for all and focus on the need to address and maximize individual students' learning styles						

Opportunity-to-Learn Building Blocks	Rating Low–High 1	2	3	4	5	Description/Areas to Be Addressed
4.3 Develop effective classroom assessment practices to enhance						
4.3.1 Use of data to make timely decisions 4.3.2 Use of a variety of classroom assessments including performance tasks and applications, surveys, interval testing, observations, standardized and criterion-referenced tests, student portfolios, teacher logs, course content, modes of instruction, time- allocated key concepts, student activities, etc.						
4.4 Implement teacher training programs that develop skills related to student behavior, especially						
4.4.1 Developing cultural sensitivity 4.4.2 Working with nonmotivated students 4.4.3 Developing effective classroom management skills						
4.5 Provide teachers with powerful, proven techniques successful in "bridging the gap" and in raising student achievement through						
4.5.1 Guiding students to create meaning and solve problems 4.5.2 Providing interactive and collaborative learning activities 4.5.3 Linking academics to community resources 4.5.4 Promoting action research in classrooms						
4.6 Establish personnel practices and programs that promote						
4.6.1 The need to invest in professional development, making available both time and resources 4.6.2 Availability of quality teaches at all schools and in all classrooms 4.6.3 Appropriately prepared and certified teachers 4.6.4 In-house teacher recruitment programs 4.6.5 Degree programs for work-force advancement 4.6.6 Collaborative programs with universities to support pre-service, in-service and community resource needs 4.6.7 Staff stability and quality in low performing schools and reduced teacher mobility 4.6.8 Assignment of high-performing teachers to neediest schools						
4.7 Communicate that research supports three essential characteristics of quality teaching						
4.7.1 In-depth knowledge of content 4.7.2 Effective teaching practices and pedagogy for all students 4.7.3 Higher levels of student achievement						
5 Creating Student Support Networks						
5.1 Create school environments that foster student resilience, provide timely needed academic, social, and psychological support mechanisms						
5.1.1 Develop clear and rigorous goals for what students should know and be able to do 5.1.2 Communicate expectations to teachers, administrators, parents, students, community						

Opportunity-to-Learn Building Blocks	Rating Low–High 1	2	3	4	5	Description/Areas to Be Addressed
5.1.3 Provide samples of student work that meet standards and communicate what is expected to all constituencies						
5.2 Extended availability of pre-school and early intervention programs to address needs of low-performing schools and school clusters in critical needs areas, K–12						
5.3 Implement innovative approaches for providing more time on task for students in low-performing schools/school clusters by introducing programs that provide						
5.3.1 Extended-school-day learning opportunities 5.3.2 After-and before-school tutorial programs 5.3.3 Saturday school 5.3.4 Summer school enrichment programs 5.3.5 Community/adult school extended-day programs 5.3.6 Community college and university programs 5.3.7 Longer school day approaches 5.3.8 Expanded school year approaches 5.3.9 Enrichment and mentoring programs 5.3.10 Efficient and productive use of time						
5.4 Increase student support programs to meet needs of students related to increased standards in mathematics through						
5.4.1 Development of supplementary learning activities and programs to promote achievement of more rigorous standards 5.4.2 Timely support for high-level advanced courses 5.4.3 Peer tutoring and adult mentoring programs for at-risk students 5.4.4 Collaborative programs established at universities, libraries, museums, community centers, and with business and industry to support and reinforce academic standards 5.4.5 Programs that support and increase the number of minority students successfully completing honors/ advance-placement courses						
5.5 Closely monitor the availability, enrollment, and content of student support programs to assure access and equity for all students, especially students from low-performing schools/ clusters, and to ensure quality of support program content						
5.6 Focus on early identification of students who are at-risk of academic failure in the classroom and implement processes that promote achievement						
5.6.1 Develop early intervention programs for all levels and subjects K–12 5.6.2 Increase emphasis on reading in early grades 5.6.3 Implement literacy programs across the curriculum K–12 5.6.4 Provide special programs for second-language students 5.6.5 Establish school-linked services/resources to meet multi-ethnic population needs						

Note: The Engineering Change Assessment Instrument is a slightly modified version of the "Bridging the Gap" tool created by the McKenzie Group for urban and rural communities attempting to conduct systemic reform. The McKenzie tool provides a robust and comprehensive framework that can help inform the change process.

29

EDUCATIONAL POLICY RESEARCH AND MATHEMATICS EDUCATION

Joan Ferrini-Mundy and Robert E. Floden

MICHIGAN STATE UNIVERSITY

In this chapter we review and synthesize research in four areas of educational policy: K–12 mathematics standards and frameworks, K–12 mathematics instructional materials and curriculum, assessment, and the mathematics teaching profession. For each area, we will discuss selected studies that illustrate the formulation, status, implementation, and effects of educational policy that relate to mathematics education. Research on policy *formulation* describes the ways in which politics, practice, and the discipline of mathematics have shaped policy decisions. Studies on the *status* of policy examine and describe variation in policy, primarily at the state level. Research on policy *implementation* has investigated the ways that policies have been interpreted, introduced into practice, and modified as faculty, curriculum developers, and others carry them out. Research on the *effects* of policy looks at the impact policies have had on mathematics teaching and learning. Within each of these four policy areas we discuss both research that has been completed and research that is needed.

We undertook an extensive, though not exhaustive, literature search as the basis for this review. The studies reviewed are eclectic in their origin, with some conducted by those who see themselves primarily as mathematics educators, and others carried out by educational policy experts, assessment experts, or others who have found the domain of mathematics education and its intersection with educational policy a fertile ground for exploration. We hope that our review will encourage and guide further research, both by mathematics educators and by those working in fields that bear on mathematics education.

THE REVIEW PROCESS

We began the review by selecting search criteria for identifying relevant literature and organizing the studies with a framework that allowed us to address, as appropriate, the four mathematics education policy areas and the four types of policy-related research. The search was based on the broad WordNet® definition of policy as "a plan of action adopted by an individual or social group" (Cognitive Science Laboratory, 2005). In education, such plans for action may be generated from a variety of sources. Examples of major federal policies include the Individuals with Disabilities Education Act (IDEA) of 1997 (U.S. Department of Education, 1997), or more recently, the No Child Left Behind Act (NCLB) of 2001 (U.S. Department of Education, 2002). States are also sources of education policy. For example, many states mandate teacher education and certification procedures or require that schools administer state-approved pupil assessments. States generate or approve instructional standards or curriculum frameworks. Many states and districts also set criteria for the selection of textbooks, uses of technology, and professional development activities for teachers.

With NCLB the federal government has begun to play a greater role in K–12 policy decisions and related decisions about teacher preparation that affect mathematics education specifically. The increasingly prominent role played by the federal government accompanies the highly visible role that has been played by national professional societies and organizations for the past two decades. In the late 1980s, for example, the National Council of Teachers of Mathematics (NCTM) embarked upon a pioneering effort to create content, teaching, and assessment standards in K–12 mathematics (see NCTM, 1989, 1991, 1995, 2000). This effort by NCTM led to policy recommendations, many of which were adopted or adapted by states and districts. In addition, NCTM's work was taken as a model for the development of standards in other school subjects and influenced a wide range of education policies and education research funding priorities. Hence policies created by professional organizations or national groups such as NCTM or Achieve are also legitimate foci for this review.

Additional sources of policy include funding programs offered by federal and state agencies and private foundations. The programs offered by the National Science Foundation (NSF), such as the Statewide Systemic Initiatives, or the Centers for Learning and Teaching, are also plans of action, and so are considered policies for consideration in our review.

In our literature search, we gave most attention to empirical studies or literature reviews that appeared in refereed sources; we gave less emphasis to the plethora of articles that offered rhetorical (often highly charged) commentary about policy. Empirical studies in mathematics education research have historically focused on core issues of teaching and learning, both in the United States and internationally. This focus is reflected in the first *Handbook of Research on Mathematics Teaching and Learning* (Grouws, 1992), in which 8 of 29 chapters focus on specific subject areas within mathematics. More recent surveys, including that of Sfard and her colleagues for the Tenth International Congress on Mathematics Education (Sfard, 2005) confirm that most mathematics education researchers address issues and problems they consider fundamental to the improvement of mathematics teaching and learning in classrooms, rather than in a broader policy arena. In recent years, this is evident in the predominance of small-scale studies using qualitative methodologies to focus deeply on specific problems about the teaching or learning of mathematics. This emphasis can be corroborated by a simple categorization of articles published in the *Journal for Research in Mathematics Education* over the past three years, where the research mode appearing in a majority of articles is qualitative.

The policies that are devised by local governments, the states, the federal government, professional associations, and funding agencies are critical to the practice of mathematics education. Fundamental research about mathematics teaching and learning alone stands little chance of influencing teaching and learning on a broad scale unless some mathematics education experts become deeply engaged in the policy arena and produce research about policy. Many crucial research questions fall within this domain and have major implications for the health and future of mathematics education. Such questions include:

- In a high accountability policy environment, what is the impact of high-stakes K–12 assessment on mathematics instruction?
- Do teacher certification policies support continued growth in teachers' development of subject matter knowledge for teaching?

These questions do not address what might currently be the most significant policy arena for mathematics education—policies related to subject matter standards at the local, state, and national level. (See Lester & Ferrini-Mundy, 2004, for a summary of a research-agenda-setting conference on this matter.) The need for research addressing policy issues suggests interesting opportunities for mathematics education researchers and for policy researchers who are willing to address questions in this arena.

POLICY AREA ONE: K–12 MATHEMATICS STANDARDS AND FRAMEWORKS

National organizations for mathematics education, especially NCTM, have worked to develop content standards over the past two decades. Prior to that time, despite periodic national reports and general recommendations about curriculum, there was little specific guidance at the national level for K–12 curriculum. International comparative research that has examined curriculum and standards internationally, such as the Third International Mathematics and Science Study (TIMSS), has highlighted the fact that the standards established by other countries are an important influence on mathematics teaching and achievement.

Standards have been a prominent policy instrument in the United States in the past two decades. In most cases, standards are designed to apply to particular categories that are important for policy. These categories include content, curriculum, performance, teaching, and opportunity to learn. Though the labels have been inconsistently applied, basic descriptions of the categories are as follows: *Content* standards specify the topics that should be covered, often with links to grade levels or bands of grade levels (e.g., content for grades 4–6). *Curriculum* standards, sometimes called a curriculum *framework,* describe the general outlines of the intended curriculum. *Performance* standards specify what students are expected to be able to do. *Teaching* standards describe the instructional approaches teachers should use to help students learn mathematics. *Opportunity to learn* standards, which were discussed but never formally developed in mathematics at the national level, would describe the resources that should be devoted to supporting instruction. (For details see Darling-Hammond, 1994; National Council on Education Standards and Testing, 1992; Oakes, 1989; Smith & O'Day, 1991; Special Study Panel on Education Indicators, 1991.)

State policy makers, with the encouragement of academic policy analysts (e.g., Smith & O'Day, 1991), have seen the formulation of standards as a key step in setting the direction for K–12 mathematics education, both as a way of communicating with practitioners, and as a way to link local and state policy. The U.S. federal government explored the idea of setting official national standards via the Educational Summit in 1989 and the Goals 2000: Educate America Act in 1994 (U.S. Department of Education, 1994), and some policy analysts continue to advocate for national standards (see Ravitch, 2005), but thus far that responsibility has been left to the states.

As noted earlier, in discussing policy research in each area of K–12 mathematics education, we divide the body of the research into four types: research on formulation, status studies, studies of implementation, and studies on policy effects or impact. In the area of standards, research on *formulation* includes studies of the ways that government agencies and other organizations have decided to develop standards and gone about constructing standards and framework documents. *Status* studies provide documentation about the state of standards development or use at the state or local level. We think of studies of *implementation* as works that examine standards documents to determine whether they are coherent, clear, specific, and consistent with other policy documents including other standards or assessments. Some documents suggest that states intended to align their standards with NCTM standards, so studies that look at alignment with NCTM can also be considered implementation studies. Studies of *policy effects* or impact measure the degree to which standards or frameworks have effected a change in teaching practice or student achievement.

Research on the Formulation of K–12 Standards and Frameworks

Policy research can illuminate the ways standards have been developed by looking at the interplay of political, social, and disciplinary factors in the construction of the particular standards adopted by a state agency or national organization, and in the adoption of standards as a policy tool. In the area of standards, where values and beliefs about what should constitute curricular emphases are central, such research can also give glimpses into how values and beliefs influence the development of standards. In the past two decades, most policy research on the formulation of K–12 mathematics education standards has focused on two cases: the series of NCTM standards documents and the evolving California *Curriculum Frameworks and Standards.*

McLeod, Stake, Schappelle, Mellissinos, and Gierl (1996) offered a retrospective account of the development of the NCTM *Curriculum and Evaluation Standards for School Mathematics* (1989). They highlight the role that declining mathematics test scores on U.S. and international assessments played in propelling key actors in the government and in professional organizations toward the creation of systematic plans for improvement. NCTM, influenced in part by the belief that weaknesses in the U.S. curriculum accounted for the poor showing in the Second International Mathematics Study, chose to focus their policy formation efforts on the K–12 curriculum. The actual proposal for creating standards came as a result of discussion at a retreat sponsored by the Conference Board of the Mathematical Sciences (CBMS) focused on a 1983 National Science Board Report (NSB, 1983) that called on professional organizations to take responsibility for improving their fields. In the face of advice from federal policy makers against using the language of standards or of national standards, NCTM decided to design the standards, which were to serve as a guide for the development of higher quality curriculum, instruction, and classroom evaluation. The authors shared a commitment to moving beyond a predominance of "'shopkeeper' arithmetic skills" (NCTM, 1989, p. 3) in the U.S. mathematics curriculum, toward an emphasis on problem solving and application.

The Reagan Administration reduced the NSF budget for K–12 mathematics and science education research shortly after the publication of *A Nation at Risk* (National Commission on Excellence in Education, 1983), in part as a result of controversy over federally-funded curriculum efforts (Dow, 1991). In this context, NCTM worked without the aid of government funding programs to produce the 1989 *Curricu-*

lum and Evaluation Standards. The NCTM effort drew upon English-language publications and information produced throughout the international community. Studies from England, the Netherlands, and Australia figured prominently in the conceptualization of the *Standards.* Also prominent was research underpinned by social constructivist views of learning. Concerns about how standards would be received in the U.S. context of local control of education and teacher autonomy were prominent in the development of the document.

Accounts of the development of *Principles and Standards for School Mathematics* (NCTM, 2000) are provided by Ferrini-Mundy and Martin (2003) and Martin and Berk (2001). Drawing on Silver (1990), Ferrini-Mundy and Martin argue that the development of *Principles and Standards* was influenced by theoretical perspectives from research, methodologies from research, and findings from research. They describe the process that NCTM implemented for obtaining input from the field about early drafts of the document, synthesizing and using the results. Martin and Berk (2001) provide an account of the cyclical relationship between research and standards development.

The research that is available about the formulation of NCTM standards documents suggests that the national and local policy context, the availability of research and theoretical perspectives, and the shared commitment by those involved in the development toward a meaningful mathematics education for all students all had impact on the content and form of the documents.

Wilson's research (2003) on the development of the California Mathematics Frameworks highlights the effects of national and state conflicts about the content of K–12 mathematics (also known as the "Math Wars"). That context led to the development of a standards document quite different from the NCTM Standards. Her historical account is based on document analysis and interviews with most of the major figures in the development. It shows how the early version of the Framework was constructed primarily by leaders in the California mathematics education community, working from contemporary ideas in that community about the nature of mathematics, what aspects of mathematics were most important for K–12 pupils, and what approaches to instruction fit best with the intended mathematical learning.

She goes on to describe how the goals and instructional approaches became the subject of controversy, with some research mathematicians joining conservative political actors in criticizing the 1998 Framework and eventually replacing the initial set of mathematics educators as the leaders in the development of later versions of the Framework. All the key players in this California story accepted the idea that standards, both the Curriculum Framework and later the standards used to guide development of K–12 state assessments, were an important device for influencing mathematics education. The story Wilson tells reveals the ways in which the particular content of the standards was influenced both by content experts (i.e., mathematics educators and research mathematicians) and by others who saw the content as a symbol for broader political issues. The California story is not unique. Similar issues and events play out in other domains where a particular policy is likely to have a high impact and a far reach, especially when the policy calls for a radical departure from the current state of affairs.

As described by Martin and Berk (2001), accountability concerns—in the form of expectations for proven or evidence-based standards and curricula — led to serious questions about the research basis for the 1989 NCTM *Curriculum and Evaluation Standards.* (See Wu, 1998.) In their review, Martin and Berk describe the many ways in which the 1989 Standards generated a flurry of research on the applicability, adaptability, and impact of NCTM's curriculum policy. In addition to the multitude of studies that focused on local efforts to understand and implement the Standards and standards-based curricula, 49 states had developed their own mathematics curriculum standards, claiming that these were based in some sense on NCTM's standards, by the mid-nineties (cf. Council of Chief State School Officers [CCSSO], 1995).

Research on the Status of K–12 Standards and Frameworks

Various types of policy analyses are conducted to summarize the status of standards development by state. Periodic reports by the Council of Chief State School Officers (see CCSSO 1997, 1998, 2001, 2002, 2005) provide current summary information by state about such things as the "status of state content standards in academic subjects." Similarly, the annual *Quality Counts* reports (Editorial Projects in Education, 2004, 2005) provide information about the status of K–12 standards, frameworks, and curricular expectations, including "grades" by state on standards and accountability. Other organizations also assign "grades" to state standards, according to various frameworks and perspectives developed by organization. Aside from specific details about where individual states are in their standards efforts, the general findings from these types of status reports are that there is great diversity in state standards, and that judgments about their quality are highly sensitive to the criteria that groups or individuals establish for judging them.

The American Federation of Teachers (AFT) has published reports about that state of state standards (see AFT, 2001, 2003) In their 2001 report, AFT concludes that, although many states were making advances in creating standards, the progress on standards-based reform was being hindered because the assessments were not necessarily aligned with standards.

Another type of status study of standards that has figured prominently in public policy discussions is international work such as that undertaken as part of the 1995 Third International Mathematics and Science Study (TIMSS). William Schmidt and colleagues conducted analyses of the K–12 mathematics standards and textbooks for students in the TIMSS focal grades—grades 4, 8, and 12—from nearly 50 countries (Schmidt, McKnight, Valverde, Houang, & Wiley, 1997). In particular, mathematics standards and textbooks from the U.S. were compared to those from countries whose students performed at the highest levels on the TIMSS eighth-grade assessment. These comparisons revealed that standards and textbooks in the high-achieving countries reflected a kind of curricular coherence not present in U.S. standards and curricula (Schmidt, McKnight, & Raizen, 1997). In particular, Schmidt et al. claim that the mathematics standards and curricula in the high achieving countries are organized according to the structure of the discipline of mathematics (Schmidt, Wang, & McKnight, 2005; Valverde & Schmidt, 2000). The analysis is summarized in Figure 29.1.

This figure reveals what these high-achieving countries collectively intended to be taught at each grade level as reflected in their national mathematics standards documents and analyzed using the TIMSS Curriculum Framework and document analysis technique (Robitaille et al., 1993; Survey of Mathematics and Science Opportunities [SMSO], 1991, 1995). An analysis of the 1989 NCTM *Curriculum and Evaluation Standards for School Mathematics* compared it with the composite curriculum from the high-achieving TIMSS countries at grade eight (Schmidt, Wang, & McKnight, 2005), and found (given the grade-band structure of the NCTM document) that coverage of all but seven of the 32 topics that are intended in the top-achieving TIMSS countries is recommended for all of grades one through eight. The authors note that in general, based on their analyses of national and state standards in mathematics and science, "Topics enter and linger, so that each grade typically devotes instructional attention to many more topics than is typical of the six high-achieving countries; in addition, each topic stays in the curriculum for more grades than in the high-achieving countries" (p. 525).

Research on the Implementation of K–12 Standards and Frameworks

One type of implementation study is the analysis of state and local standards documents, with attention to clarity, specificity, alignment, and mathematical appropriateness; these features that are contributors to interpretation and implementation. Meeting these criteria is viewed by some as important for ensuring that the policy—that is, standards—have a significant influence on practice.

The literature we identified indicates that the various state standards for K–12 mathematics are highly variable in these features of implementation. Klein et al., 2005), for example, graded the standards on clarity, definiteness, and testability. States receiving the highest overall grades were California, Indiana, and Massachusetts (Grade A) and Alabama, New Mexico, Georgia (Grade B). The national average was D and 29 states received a grade of D or F. In discussing the reasons for the poor grades, the reviewers note such factors as: too much emphasis on calculators, inadequate emphasis on basic number facts and standard algorithms, and insufficient developmental guidance for mathematical reasoning and problem solving. This analysis is the third sponsored by the Fordham Foundation (see also Braden et al., 2000; Raimi & Braden, 1998).

Ongoing analyses of state standards are underway at the NSF-funded Center for the Study of Mathematics Curriculum (CSMC) (see Reys et al., 2006). This status study, rather than grading state standards on specific quality criteria, instead is descriptive. The project was intended to "describe the emphasis and grade-level placement of particular learning expectations as presented in state grade level expectation documents and to document variations across states." (Reys et al., 2006, p. 1). They report a general lack of consensus in what is included at particular grade levels across the states. In an effort sponsored jointly by NCTM and the Association of State Supervisors of Mathematics (Jackson, 2004), experts reviewed state mathematics standards to examine whether or not there is a "de-facto national curriculum for school mathematics" (p. 1355). In the report of this work, Lott and Nishimura (2005) provide some commentary about what reviewers found to be characteristics of the least and most informative state standards. Standards that used finer grain sizes, used clear language, gave examples, and focused on "big ideas" were seen as more informative, and standards that used imprecise mathematical language, used broad or vague language, were too wordy, or used multiple objectives in single statements were less informative (p. 63). Both of these policy analysis efforts are focused on similarities and differences in standards across states,

Topic	Grade 1	2	3	4	5	6	7	8
Whole Number: Meaning	■	■	■	●	●			
Whole Number: Operations	■	■	■	■	●			
Measurement Units	▲	■	■	■	■	■	●	
Common Fractions			▲	■	■	●		
Equations & Formulas			▲	●	●	●	■	■
Data Representation & Analysis			▲	▲	●	●		▲
2-D Geometry: Basics			▲	●	●	●	■	■
2-D Geometry: Polygons & Circles				●	●	●	■	■
Measurement: Perimeter, Area & Volume				●	●	●	●	▲
Rounding & Significant Figures				●	●			
Estimating Computations				●	●	●		
Whole Numbers: Properties of Operations				▲	●			
Estimating Quantity & Size				▲	▲			
Decimal Fractions				●	■	●		
Relation of Common & Decimal Fractions				●	■	●		
Properties of Common & Decimal Fractions					●	●		
Percentages					●	●		
Proportionality Concepts					●	●	●	▲
Proportionality Problems					●	●	■	■
2-D Geometry: Coordinate Geometry					▲	▲	●	●
Geometry: Transformations						●	●	●
Negative Numbers, Integers, & Their Properties						▲	●	
Number Theory							●	▲
Exponents, Roots & Radicals							●	●
Exponents & Orders of Magnitude							▲	▲
Measurement: Estimation & Errors							▲	
Constructions Using Straightedge & Compass							■	▲
3-D Geometry							●	■
Geometry: Congruence & Similarity								■
Rational Numbers & Their Properties								▲
Patterns, Relations & Functions								▲
Proportionality: Slope & Trigonometry								▲

▲ Intended by 4 of 6
● Intended by *all* but one (5 out of 6)
■ Intended by *all* top-achieving countries

Figure 29.1 Mathematics topic coverage in high-achieving countries in TIMSS 1995. Adapted from Schmidt, Wang, and McKnight (2005).

in pursuit of the question of whether there is so much consistency that there is, in fact, a "standard" national curriculum. Both found that that there is wide variation and disparity in what individual states require in mathematics at particular grades.

Another class of studies about implementation of standards includes efforts to examine how teachers used standards in their classroom practice. In addition, we have selected some studies that discuss teachers' interpretation of the meaning of standards, given the relationship of interpretation to implementation, as well research looking at more general responses from teachers and administrators to the overall standards phenomenon.

Cohen and Ball (1990) examined early implementation of the 1985 California Mathematics Framework, which shared many features with the 1989 NCTM Standards. They found that despite efforts by the California Department of Education to provide teachers only those textbooks that were aligned to the standards and to create state assessments that were also aligned, teachers continued to view mathematics in traditional ways and struggled to interpret and implement the standards in their classrooms. They tended to interpret the policy as suggestions for topics that they should add to the list rather than as elements useful for various types of mathematical reasoning. (Also see Ferrini-Mundy & Johnson, 1994.) Teachers tended to filter the new policy through the lens of established practice rather than to respond to the new policy by altering their lens.

This notion of filtering the policy through a particular lens is closely related to the matter of interpretation of standards. Researchers in the NCTM Recognizing and Recording Reform in Mathematics Education study (see Ferrini-Mundy & Schram, 1997) found that for teachers, implementation of standards often meant attending only to the surface features of the documents, particularly to advice in the standards that was perceived to be pedagogical, rather than to the mathematical shifts and emphases recommended by NCTM standards. In a study of teachers interpreting state standards in Connecticut, Hill (2001) found that teachers were not familiar with the technical terms that standards writers used in their policy documents, concluding that: "teachers reading these documents imputed more local, and sometimes conventional, definitions to these words. As a result, state standards lost their force." (p. 289).

Similarly, Wilson and Floden (2001) found that standards-based reform meant different things to different teachers and administrators. For some teachers, the standards-based reform was hardly noticeable, "flowing into a large stream of other reforms." (p. 213). Other teachers found that standards-based reform provided clarity and language for thinking about their practice. Some teachers who have a history of improving professional practice on their own viewed the standards as stifling. Given the varieties of standards-based reform that teachers encountered, the researchers consistently found teachers and administrators seeking to take a "balanced" approach, which mostly meant a commitment to mixing old with new, and using their professional judgment.

Policy research often finds that implementation is undercut by alignment failure (Webb, 1997). In some cases, such alignment failures are due to assessments that are inconsistent with alignments in curriculum and instruction. In other cases failures are due to deviations in instruction that undermine alignments in curriculum and assessment. In many cases implementation studies find misalignments (of varying degrees of severity) between all three areas: curriculum, instruction, and assessment (La Marca, 2001).

Webb (1999) analyzed mathematics and science standards in four states for their alignment with assessments in those states, using the following criteria: categorical concurrence, range of knowledge concurrence, balance of representation, and depth of knowledge consistency. Although his purpose was primarily to develop and refine methodology for such alignment analysis, he found that the standards and assessments were not consistently well-aligned, signaling that one of the central theoretical assumptions about the standards movement—that is, alignment—is not necessarily in place in implementation. With NCLB requiring this alignment, future status and implementation studies may reveal different trends.

Research on implementation has led to an increased understanding of the role alignment plays in successful reform (Porter, 2002; Smith & O'Day, 1991) and meaningful interpretation of assessment results (Messick, 1989). A knowledge brief produced by WestEd (Ananda, 2003) discusses issues of alignment in the context of NCLB and the attendant emphasis on assessment results as the basis for rewards and sanctions for schools, teachers, and students. The NCLB policy on alignment indicates that "states may select and design their assessment of their own choosing" provided that state assessments are aligned with state academic (i.e., performance) standards and subject-area content standards for each grade level. According to Ananda, the "myriad assessment-and-accountability provisions are pushing states and districts to address many alignment issues that have often received short shrift" (p. 1).

Research on the Effects of K–12 Standards and Frameworks

To think about the effects of the policy tool of K–12 mathematics standards, it is helpful to consider the framework for understanding the influence of standards put forth in the National Research Council (NRC, 2002a) report on that topic. This report notes that standards may influence K–12 practice through three channels: curriculum, teacher development, and assessment and accountability. With respect to curriculum, standards may shape the development of curriculum materials and documents (e.g., district scope and sequence charts), which then influence classroom instruction. With respect to teacher development, standards may be used as a focus for professional development, which may affect teachers' learning by persuading teachers to change their ideas about what is important to teach, or by motivating teachers to pursue opportunities to learn about mathematics content or pedagogy. By leading to changes in teachers' knowledge, skills, and dispositions, standards may thereby have an influence on K–12 teaching and learning. With respect to assessment and accountability, standards may be a basis for such systems, thus giving teachers, students, and others reasons for classroom practice changes that would lead to high performance on the assessments.

Policy studies sometimes focus on a subset of these channels. At other times they look only at the effects on teaching and learning, without trying to understand the channel though which the influence has come. One line of work has focused on standards as part of systemic reform, where assessments and accountability are prominent features but curriculum and teacher development also play a part. Another line of work has focused on the teacher development channel, looking at the ways in which teachers interpret standards. Some investigators have done systematic comparisons of teaching practices with the operating set of standards. The correspondence to standards, or lack thereof, is seen as an indication of the influence of standards.

Central themes that emerged from these studies are that in the early stages of implementing standards, the absence of existing instructional materials that embodied the ideas of the standards led many teachers to adopt superficial aspects of the reforms (Ferrini-Mundy & Johnson, 1997). Teachers may believe they have changed their practice to fit a set of standards, and may talk the language of the standards, without making changes that authors of the standards saw as most important (Spillane & Zeuli, 1999). Those who are most likely to make the changes to practice that are expected by the policy are teachers who have an opportunity to collaborate with colleagues in their efforts to understand the expected changes and enact them in their classrooms (Ferrini-Mundy & Johnson, 1997). The effects of policy are also greater when teachers have deep conceptual understanding of the mathematics they are expected to teach (Lloyd & Wilson, 1998) or participate in professional development activities that promote the development of such understanding (Spillane & Zeuli, 1999). Wilson and Floden (2001) found that, despite the fact that, for some teachers, the clarity of standards and aligned accountability systems proved a catalyst for creating a more coherent practice, most teaching remained more familiar than new, and more ordinary than challenging. Recent work by Jacobs et al. (2006), in their analysis of TIMSS 1999 video study, indicates that "typical mathematics teaching (in the U.S.), in both 1995 and 1999, is more like the kind of traditional teaching reported for most of the past century." (p. 6).

Observations: Research about K–12 Mathematics Standards and Frameworks

We found very little empirical work that has documented and described the processes of standards development and formulation. Given the range of values and commitments that the diverse constituencies (mathematics educators, teachers, mathematicians, administrators) who develop standards bring to the task, and the complexity and differences of viewpoint reported anecdotally by many who have worked on state and national standards, additional research about how standards are created might be useful in assisting new groups coming to the work of creating state, district, or national standards and frameworks. There is much descriptive material available about the status of standards and frameworks, and a trend toward more analytic examinations of state documents. Again, in that work we see the strong influence of differing values and priorities in the choices of criteria for evaluation. Research about implementation and effects of the policies embodied in standards and frameworks is relatively sparse, and does not provide compelling evidence about the impact of standards in mathematics on teachers' classroom practice. The complexity of undertaking research about implementation and effects is acknowledged in the NRC report about studying the effects of standards (NRC, 2002a), and the framework provided in that document may be a useful tool to guide future efforts in this area.

POLICY AREA TWO: K–12 MATHEMATICS INSTRUCTIONAL MATERIALS AND CURRICULUM

We discuss three areas of policy that relate to instructional materials and curriculum in K–12 mathematics: federally-funded programs to develop curriculum and instructional materials; policies that govern the selection of curriculum materials in states, districts, and schools; and policies about course-taking and curricular opportunities to learn mathematics in schools.

The NSF has played a leading role in supporting the development of mathematics instructional materials through its funding policies, beginning in 1952 with its support of the University of Illinois Committee on School Mathematics. In the "new math" era alone, spanning the period from 1957–1970, NSF invested more than $500 million in improvement projects for mathematics and science education. Between 1989 and 2002, the NSF invested $93 million to fund the development of 13 comprehensive mathematics curricula (NRC, 2004, p. 1). Three of these curricula were at the elementary school level, five were at the middle school level, and five were at the secondary school level (NRC, 2004, pp. 19–20). Between 1995 and 2000, several of these major projects received another round of funding to create second editions.

The selection of instructional materials by school districts is governed by a mix of policies. Both at the federal and state level, policies have included the requirement that curricula or instructional materials used in school districts seeking federal funding should be research-based. In some states, such as California, this requirement was present as early as 1999 (cf. http://www.signetwork.org/SIG2000/California.htm). Title IX (Section 9101) of NCLB (NCLB, 2001, Title IX, Sec. 9101, No. 37) see http://www.ed.gov/policy/elsec/leg/esea02/pg107.html#sec9101) requires the use of materials that are scientifically based, and defines scientifically based research as "research that involves the application of rigorous, systematic, and objective procedures to obtain reliable and valid knowledge relevant to education activities and programs." In order to receive funding from the U.S. Department of Education's Comprehensive School Reform (CSR) program for 2002 and beyond, schools must implement a program that "uses proven strategies and methods for learning, teaching, and school management based on scientifically based research and effective practices, and used successfully in multiple schools . . . and that has been found through scientifically based research to significantly improve student academic achievement, or has shown strong evidence that it will." (http://www.ed.gov/programs/compreform/legislation.html).

New policy relevant to instructional materials in mathematics could emerge from the efforts of the recently appointed National Mathematics Panel. The Panel has been charged to "advise the President and the Secretary of Education on the best use of scientifically based research to advance the teaching and learning of mathematics . . . examine and summarize the scientific evidence related to the teaching and learning of mathematics, with a specific focus on preparation for and success in learning algebra . . . issue two reports containing policy recommendations on how to improve mathematics achievement for all students (http://www.ed.gov/about/bdscomm/list/mathpanel/factsheet.html).

In addition to policies at the federal level, a number of states have specific policies about selection of instructional materials (see Seeley, 2003, for a historical summary of mathematics textbook adoption in the U.S.).

Finally, mathematics course-taking policies, such as requiring all students to take algebra in the eighth grade, requiring that algebra two must be completed for high school graduation, or requiring all students to take four years of mathematics in high school, are now common in states and school districts.

The Advanced Placement (AP) Program, developed in 1955 by the College Board, is one specific area of the high school curriculum that is governed by a national policy-like process. Topic outlines for AP courses are generated largely by surveying colleges and universities. AP course outlines (and assessments) are designed to mirror the content of typical college-level introductory courses. The courses are intended to be an acceptable substitute for the introductory courses offered at more than 2,000 colleges and universities (see http://apcentral.collegeboard.com/program/0,,150-0-0-0,00.html, and National Research Council, 2002a).

There has been some policy support for the AP program in the form of U.S. Department of Education funding—the *Advanced Placement Incentive (API) Program* (http://www.ed.gov/programs/apincent/awards.html). These grants are intended to help schools comply with the mandates of the Elementary and Secondary Education Act—specifically the "Access to High Standards Act" within NCLB (see http://www.ed.gov/policy/elsec/leg/esea02/pg14.html for details). API grants support initiatives to promote pre-AP and AP programs for low-income middle and high school students. Since 2002, the USDE has awarded API program grants to 14 state departments of education as well as to local districts in 10 other states and one U.S. territory. While the California, Texas, and New York Departments of Education have not received API grants, multiple districts within these states and several consortia that include these and other states have also received grants.[1]

Research on the Formulation of K–12 Instructional Materials and Curriculum Policies

The formulation of NSF policy to develop K–12 curriculum materials in the post-Sputnik era has been well documented (e.g., Dow, 1991; Lappan & Wanko, 2003). The idea that newer, more modern instructional materials would be crucial to maintaining the nation's scientific and technological superiority was promulgated by leading mathematics educators such as Edward Begle and Robert Davis, and groups of mathematicians and scientists were at the forefront in the development of the materials. Less has been written about the process of formulating the NSF plans to

[1] The U.S. Department of Education's full list of *Advanced Placement Incentive (API) Program* grantees is accessible at http://www.ed.gov/programs/apincent/awards.html.

fund comprehensive curricula at all three grade levels in the late eighties and early nineties. Language from one of these solicitations indicates the agency's commitment to advancing the standards movement as it had been defined by NCTM:

> Because of the focus on problems of mathematics education in our schools, educators have been working toward a plan for improvement. Coming from a long process of deliberation and consensus, the *NCTM Curriculum and Evaluation Standards* and *Everybody Counts* document first steps toward this plan. These and other recent documents address issues such as broadening the content of the current mathematics curriculum, shifting emphasis from computational to problem solving skills, and recognizing the cognitive factors which enter into teaching and learning. They provide a sound philosophical basis for changes in content, method, and assessment. They signal high current interest in mathematics education providing both a foundation and an opportunity for improvement. (NSF, 1989, p. 4)

We were unable to locate any scholarly accounts of the processes of formulating "approved textbook" lists in various states, or about the formulation of K–12 course and credit policies.

Research on the Status of K–12 Instructional Materials and Curriculum Policies

We are aware of one study summarizing funding policies for K–12 instructional materials development. A review of the mathematics education portfolio in NSF's Education and Human Resources directorate (Ferrini-Mundy, Earle, & King, 2005; NSF, 2004) found an imbalance in the cycle of innovation, where there was little evidence of cumulativity across instructional materials projects, research about learning, and implementation in mathematics education projects resulting from solicitations issued between 1990 and 2003. Of the 133 separate solicitations issued by the Directorate in that period, 10 were for instructional materials development, and the investment in materials development represented only about 5% of the total spending on mathematics education during that period.

Efforts to gauge the quality of instructional materials developed under federal policies can be seen as status studies. The U.S. Department of Education engaged in a process to determine "exemplary" instructional materials (U.S. Department of Education, 1999), and identified five of the NSF-funded materials as exemplary. These were: *Cognitive Tutor Algebra, College Preparatory Mathematics, Connected Mathematics, Core-Plus Mathematics Project,* and the *Interactive Mathematics Program.* This process resulted in considerable controversy, to be discussed subsequently.

Finn and Ravitch (2004, p. 3) have provided some status research on textbook selection and adoption policies. They report that 21 states across the U.S. have implemented state-wide textbook adoption policies that govern the selection and adoption of textbooks by schools and districts.[2] Among these, 10 states choose textbooks from which districts must select, 7 states provide an approved list of textbooks from which districts can choose, and 3 states use a combination of these strategies that involves mandating texts for certain grade levels or courses but allowing districts to choose from an adoption list in other grades and courses. Similar textbook adoption policies exist at the district level in many cases where there is no state-wide policy. While the specific procedures for selecting texts for adoption or adoptions lists vary, most districts adopt textbooks on a five to seven year cycle and the process generally includes the use of some kind of committee review process. Some states, such as Texas, generate a list of nonconforming textbooks that districts and schools are expected to avoid. According to Finn and Ravitch (2004) and several other sources, the majority of texts that receive approval are published by four corporations: Pearson, Houghton Mifflin/McDougal Littell, Harcourt/Holt, and McGraw Hill, reflecting recent consolidation of publishing companies into a small set.

Status research is available concerning K–12 curricular and opportunity to learn policies. This includes policy recommendations for mathematics course-taking. NCTM recommends that all students study four years of mathematics in high school (NCTM, 2000, p. 371), increased from a recommendation of three years in the 1989 Standards (NCTM, 1989, p. 124). In 42 states and the District of Columbia, course requirements for high school graduation are set by officials at the state level. In the remaining states, this responsibility is delegated to local school boards.

A number of national reports also have argued for a four-year requirement, with specific advice about the nature of the mathematics that students should study. For instance, Achieve, Inc. recommends that state policy makers "require all students to take a common college and work-preparatory curriculum in math . . . align academic standards in high school with the knowledge and skills required for college and workplace success . . . pursue accelerated options for

[2] For maps of adoption states see the AAP web site and the Fordham Institute site at http://www.publishers.org/school/schoolarticle.cfm?SchoolArticleID=32 and http://www.edexcellence.net/institute/publication/publication.cfm?id=335, respectively.

earning postsecondary credit while in high school; and monitor results through postsecondary education and use the data to help improve the rigor of course offerings and instruction in high school." (Achieve, 2004b, p. 3). Increasingly, states and districts are requiring high school exit examinations, and these exams typically include mathematics as a core area.

Achieve (2004b) documented the number and types of mathematics courses that students are required to complete for graduation. There are 13 states that require only two years of mathematics study. Twenty-four states and the District of Columbia require three years, and only five states (Alabama, Arkansas, Mississippi, South Carolina and West Virginia) require all students to complete four mathematics courses for graduation. Nearly half of the states (22) do not specify which mathematics courses students need to take. Of those that do, only Arkansas, Indiana and Texas now or soon will require Algebra I, Geometry, and Algebra II (Achieve, 2004b, pp. 11–12).

Descriptive data about the status of the AP program is made available regularly by the College Board. Originally, the program served only top students from a few high schools. As of 2002, approximately 62 percent of U.S. high schools offered AP courses (NRC, 2002b). Among the eight separate AP courses in mathematics subjects, the most prevalent and popular courses are Calculus AB (first semester college equivalent), Calculus BC (second semester college equivalent), and Statistics. According to the College Board's summary report on nation-wide AP exam participation, nearly 300,000 eleventh and twelfth graders took one or more exams in a mathematics subject in 2004 (http://www.collegeboard.com/student/testing/ap/exgrd_sum/2004.html).

Research on the Implementation of K–12 Instructional Materials and Curriculum Policies

There has been some research about the implementation of the NSF instructional materials funding policy and its products in the past two decades. Most of these analyses are alignment studies comparing the instructional materials developed under the policy to various frameworks that are either mentioned explicitly in the solicitations or that are of interest in the context of the policy. For example, the American Association for the Advancement of Science (AAAS) Project 2061 (http://www.project2061.org/) analyzed middle grades mathematics materials (both NSF-funded and commercial) with respect to the AAAS Benchmarks for Science Literacy (AAAS, 1993). The systematic alignment study process that they implemented involved examining the materials with respect to six benchmarks: "a concept benchmark dealing with fractions and operations on them; a skill benchmark dealing with equivalent forms of numbers; a concept and a skill benchmark dealing with properties of shapes and computations of circumference, area, and volume; and two concept benchmarks dealing with graphing and equations." (AAAS, 2000, http://www.project2061.org/publications/textbook/mgmth/report/part1.htm). The process also involves looking at how instructional issues are represented in the materials. AAAS concludes "It is clear that reform principles for mathematics education are beginning to have an impact on textbooks and on the developers and publishers who create them." (AAAS, 2000, http://www.project2061.org/publications/textbook/mgmth/report/part1b.htm).

One contribution of such research and evaluation processes has been the design of tools and procedures for measuring alignment, which proves to be a rather difficult process. In TIMSS, Schmidt and his colleagues developed tools and methodologies for comparing instructional materials from the U.S. with those used in other countries. (see Schmidt et al., 1997). In an analysis of several of the NSF-supported mathematics materials, Schmidt et al. (2001) claimed that in many cases these materials lacked the rigor and high standards found in instructional materials from high-achieving nations in TIMSS, and that they also perpetuated the U.S. "mile wide, inch deep" problem of including too many topics in too much of a spiral curriculum. Schmidt argues that a coherent mathematics curriculum should follow the logical structure of the discipline, and that topics should be started and completed, rather than revisited without increase in depth over many years in the school curriculum.

We located systematic accounts by authors of their work in designing instructional materials and consider these descriptions of the implementation of NSF instructional materials policy. Some studies are reported in a special issue of the *American Journal of Education*. Alper, Fendel, Fraser, and Resek (1997) discuss the principles that served as the foundation for their development of the *Interactive Mathematics Program*, one of the earliest of the NSF-funded high school curricula. These principles were: "Students must feel at home in the curriculum; Students must be personally validated as they learn; Students must be actively involved in their learning; Students need a reason for doing problems" (pp. 150–151). They then provide illustrations from their curriculum materials of how each of these principles was instantiated. Usiskin (1997) provides an historical account of the role of application in school mathematics curriculum, including explanation of the ways that applications are handled in the University of Chicago School Mathematics Project, a program developed simultaneously with the NSF-funded materials. He notes that goals for what students should understand about applications are made as explicit about goals for what students should know about skills. Accounts such as these that provide insights into how developers of instructional materials chose particular ideas from standards documents to frame their work, and are useful as an indicator of how policy goals are interpreted.

There is scholarly activity that relates to the implementation of policies about requirements of using research-based curricula. As states and school districts struggle to find out if the instructional materials they wish to use are research based, and/or have been validated as being effective through research, some resources are emerging. The What Works Clearinghouse (WWC) (http://www.whatworks.ed.gov/whoweare/overview.html) was established in 2002 by the U.S. Department of Education's Institute of Education Sciences to provide educators, policymakers, researchers, and the public with a central and trusted source of scientific evidence of what works in education. The WWC aims to:

> promote informed education decision making through a set of easily accessible databases and user-friendly reports that provide education consumers with high-quality reviews of the effectiveness of replicable educational interventions (programs, products, practices, and policies) that intend to improve student outcomes. (What Works Clearinghouse, What We Do, Overview)

In fall 2004, the WWC released a report on middle grades mathematics materials, which cited only ten studies that met their standards for inclusion. These studies were randomized controlled trials or studies that used quasi-experimental designs. Based on these studies, the report concluded that only five sets of instructional materials had been studied with research designs that met the WWC standards, and the findings about their effectiveness were equivocal.

The only research we located describing the implementation of state textbook adoption policies is focused on the process in California.[3] Wilson (2003) reports on a highly politicized and controversial process around California's textbook adoption policy in the 1990s, pitting teachers who supported the reforms against state officials concerned with basic skills and procedures, all surrounded by suspicions about undue influence of the commercial publishing industry. Jacob and Akers (2000) provide an account of the California mathematics education policy process during the 1995–1998 period, where a Curriculum Commission, serving in an advisory role to the California Board of Education, chose to enforce a part of the state education code that called for instructional materials to "'incorporate principles of instruction reflective of current and confirmed research" (see Jacob & Akers, 2000, p. 4). They argue that this stipulation eliminated a number of "reformist" (NSF-supported) curricula from the approved list.

There is a more robust body of research describing implementation of course- taking and opportunity-to-learn policies. In a recent Council of Chief State School Officers (CCSSO) report (Blank & Langesen, 2005), some data are included regarding algebra in grade eight. NAEP data from 2003 indicate that about 27% of eighth graders across the U.S. report studying algebra in eighth grade, up only slightly from 1996. They also report on course-taking patterns in high school:

> As of 2004, 50 percent of U.S. high school graduates took Four Years of high school math. Rates of high school math enrollment vary widely by state. Data as of 2003-04 school year show that nine states had over 50 percent of high school students take Trigonometry or Pre-Calculus by their graduation, which indicates the percent of students taking four years of high school math (with Algebra 1 the first year). In the U.S., 72 percent of graduates took Three Years of high school math. As of the 2003-04 school year, 17 states had over 70 percent of students take Algebra 2 or Integrated Math 3 by graduation, and seven states had over 80 percent at this level. (p. 3). In addition, "In 2003, 27 percent of grade 8 students took Algebra 1 for high school credit, which represents an increase of 3 percentage points from the rate in 1996. In 9 states over 30 percent of students reported taking Algebra 1 in grade 8 as of 2003; and a majority of states had over 25 percent taking Algebra. (p. 17)

Data such as these indicate that policy recommendation calling for higher expectations and earlier introductions of algebra are being implemented.

In another study of high school mathematics course-taking patterns, Kher, Schmidt, Houang, and Wiley (2005) found that the number of routes through high school mathematics course options is surprisingly high and that what constitutes four years of high school mathematics can look radically different for students of similar backgrounds in a single school.

Our search for state and district policies concerning advanced placement courses did not reveal any that require participation in AP mathematics as a condition of high school graduation. However, a

[3] The textbook policies of California, Texas, and Florida drive the production of the textbook publishing industry (Finn & Ravitch, 2004).

number of states and districts have engaged in highly organized efforts to substantially increase students' participation in the AP program. One example of this is an initiative in Illinois that promotes AP courses to parents and students, provides support for students' academic achievement, and provides supplies and materials for coursework. Other aspects of the initiative include training teachers and administrators in AP philosophy and methodology, encouraging teaming between high school and middle school teachers, supporting the mentoring of new AP teachers by experienced AP teachers, and promoting teachers' professional development by funding their attendance at institutes conducted by the College Board (see http://www.isbe.net/curriculum/pdf/AP_grant-awards_release.pdf for more information).

To support expanded participation in AP programs, in 2002 the College Board began holding an annual *National Advanced Placement Program® Equity Colloquium* to assist educators in their efforts to ensure that teachers and students across the socioeconomic spectrum have equitable access to AP courses. These supports—and, according to Klopfenstein (2004), many pressures—have led to hundreds of schools (many of them small, rural, or low income) to implement AP programs in the past several years. The Dallas-based AP incentive program successfully uses incentives, including financial supports for teachers and for students based on exam results, reports increased percentages of minority students enrolling in AP courses, taking AP exams, receiving passing grades on AP exams, and applying for college scholarships. (O'Donnell, 2006).

Research on the Effects of K–12 Instructional Materials and Curriculum Policies

The main type of research available about the effects of NSF instructional materials policies is in the form of evaluations of the materials produced with NSF funding. An NRC study committee was charged to evaluate the quality of the 13 mathematics curricula produced with NSF funding beginning in the late 1980s, together with 6 commercially produced curricula, and produced the report, *On Evaluating Curricular Effectiveness: Judging the Quality of K–12 Mathematics Evaluations* (NRC, 2004). After reviewing 147 research and evaluation studies, the committee concluded that

> the corpus of evaluation studies as a whole across the 19 programs studied does not permit one to determine the effectiveness of individual programs with a high degree of certainty, due to the restricted number of studies for any particular curriculum, limitations in the array of methods used, and the uneven quality of the studies. (p. 3)

These findings complicate further the recent challenges that curricular choices be based on the available research base about those materials.

The most recent wave of NSF-funded instructional materials also has had another tangible outcome in that it has elicited a strong response from mathematicians and others who are critical of these materials for a variety of reasons. In the fall of 1999 a group of more than 200 mathematicians and others signed an open letter to then-Secretary of Education Richard Riley that was published in the Washington Post (November 18, 1999). The letter (see www.mathematicallycorrect.com/riley.htm) called for the U.S. Department of Education to withdraw its endorsement of the ten K–12 mathematics programs—five endorsed as exemplary and five as promising (http://www.enc.org/ed/exemplary/), on the grounds that:

> It is not likely that the mainstream views of practicing mathematicians and scientists were shared by those who designed the criteria for selection of "*exemplary*" and "*promising*" mathematics curricula.... Even before the endorsements by the Department of Education were announced, mathematicians and scientists from leading universities had already expressed opposition to several of the programs listed above and had pointed out serious mathematical shortcomings in them.... [W]e believe that it is premature for the United States Government to recommend these ten mathematics programs to schools throughout the nation. We respectfully urge you to withdraw the entire list of "*exemplary*" and "*promising*" mathematics curricula, for further consideration, and to announce that withdrawal to the public. We further urge you to include well-respected mathematicians in any future evaluation of mathematics curricula conducted by the U.S. Department of Education. Until such a review has been made, we recommend that school districts not take the words "*exemplary*" and "*promising*" in their dictionary meanings, and exercise caution in choosing mathematics programs. (http://www.mathematically-correct.com/riley.htm)

The endorsement was not withdrawn.

There is a considerable amount of anecdotal discussion about the effects of the textbook adoption policies, but again, we found little actual research on this topic. Representatives of commercial textbook companies respond to these adoption policies by sometimes creating special editions of texts that are designed to meet the standards of a given state (e.g., California, Texas), and have claimed that the reason that U.S. textbooks are "a mile wide and an inch deep" is in part because publishers strive to include the union of all topics in all of the state standards, particularly in the "textbook adoption" states (C. Dillender, personal communication, Oct. 1, 2003). Delfattore (1992) confirms this view, noting that, because publishers must demonstrate specifically how their textbooks align with the state's standards or curriculum, this requirement "puts publishers on notice that their books had better conform as closely as possible to the curricula of those states" (p. 123).

Similarly, we found little research describing the effects of course-taking and opportunity-to-learn policies. It is early to know the results of the trend toward requiring algebra at the eighth grade, though a number of interesting questions are suggested in RAND Mathematics Study Panel (2003): How do algebra course-taking policies affect the subsequent mathematics course-taking patterns and achievement of urban and minority students? Should algebra be offered at the eighth grade? What is the impact on high school graduation, on placement in college mathematics, and on entry to technical careers in the workforce? Do students who have studied algebra at the eighth grade continue on to reach advanced placement mathematics courses?

We found few studies that looked directly at questions such as these. There is some interest in the impact of participation in AP mathematics courses and high school calculus experience on later course-taking patterns, college placement, or achievement. (e.g., Ferrini-Mundy & Gaudard, 1992; Morgan & Maneckshana, 2000; Morgan & Ramist, 1998; NRC, 2002b). Several studies have indicated that students who take AP courses, even if they do not take or pass the exam, achieve higher college grades than non-AP in the courses into which they place (e.g., Morgan & Crone, 1993; Morgan & Ramist, 1998). However, the advantage of AP participation may diminish as students move to more advanced courses. This is not altogether surprising if one considers the AP experience as acceleration to college-level work. The advantage of early exposure is likely to diminish as students with AP experience enter their second year of college.

Observations: Research about K–12 Mathematics Instructional Materials and Curriculum Policies

It would be very helpful to future policy development to have more research about how policies are formulated, particularly for policy related to mathematics instructional materials development and selection, areas that have emerged as a highly controversial. Such research would likely reveal that the policies guiding these areas are not necessarily grounded in research findings about what might lead to effective instructional materials in mathematics. More research about formulation also would highlight areas where policy makers are working without an available base of research, and might encourage mathematics education scholars to undertake studies in areas that are central for policy development.

Research about implementation and impact does reveal that, even when policies seem well thought-out and systematic, the resulting implementation and products can lead to substantial disagreement and debate. One implication might be that the policy-making components should include wider diversity and breadth in terms of the stakeholders, with more intentional inclusion of mathematicians, mathematics educators, assessment experts, and K–12 practitioners.

POLICY AREA THREE: ASSESSMENT IN K–12 MATHEMATICS

In the U.S., assessment systems have long been a part of federal and state policy systems. The two most prominent current assessment policies are a cluster of policies associated with the federal No Child Left Behind (NCLB) legislation and periodic national assessments in mathematics, some conducted as part of the National Assessment of Educational Progress (NAEP) and others as part of international comparative studies. Some states and districts have additional assessment policies, including an increasing use of mathematics testing as a requirement for high school graduation.

With the NCLB, the federal government instituted a major new federal assessment policy, requiring states to assess all students in mathematics each year in grades 3–8 (starting in 2005–2006) and once during grades 10–12. Although the law requires testing that produces scores for individual students, the accountability requirements are specified at the school level, rather than at the level of teachers or students. Larger organizational units—districts and states—are also held accountable, with a requirement that they make adequate yearly progress toward the 2013–14 goal of getting all students to the state-determined level of academic proficiency. The requirement for adequately yearly progress also specifies that adequate progress be made for subgroups of students specified by race and income. The assessment initiative builds on state policies already in place.

The intent of the federal NCLB assessment requirements is to improve the achievement of all students by holding schools accountable, including accountability for the achievement of designated student subgroups. The legislation's implicit theory of action is that requiring the publication of assessment results and tying those results to changes in schools and options for parental choice will lead educators to work toward improving achievement.

In addition to requirements for state assessments, federal policy makers have created a system for national student assessments in mathematics—the National Assessment of Educational Progress (NAEP)—and have supported U.S. participation in international assessments of student achievement, including TIMSS (renamed after TIMSS 1995 as the *Trends in International Mathematics and Science Studies*) and the *Program for International Student Assessment* (PISA). Beginning with the 2002–2003 school year, federal policy for NAEP has strengthened, now making state receipt of Title I funds contingent on participation in the NAEP program component that reports results at the state level.[4] The goals of NAEP are to compare student achievement in states and other jurisdictions and to track changes in achievement of fourth-, eighth-, and twelfth-graders over time in mathematics, reading, writing, science, and other content domains (National Center for Education Statistics [NCES], 2005).

Similarly, the federal government supports U.S. participation in international assessments, including assessments of mathematics achievement, to provide a basis of comparison—across countries and over time.

> With the emergence and growth of the global economy, policymakers and educators have turned to international comparisons to assess how well national systems of education are performing. These comparisons shed light on a host of policy issues, from access to education and equity of resources to the quality of school outputs. They provide policymakers with benchmarks to assess their systems' performances, and to identify potential strategies to improve student achievement and system outputs. (http://nces.ed.gov/surveys/international/IntlIndicators/)

Other state and district assessment policies vary, with purposes that include providing public information, determining the distribution of resources or interventions, and serving as tools for accountability or for promoting changes in practice.

In short, federal and state policy makers have put in place a variety of required student assessments, intended to provide information on student achievement, with possibilities for comparisons across schools, among countries, and over time. NCLB has required states to link assessments to state standards, a connection often, but not universally, made previously. The NCLB assessment requirements are, according to the U.S. Department of Education, intended to generate information educators and policy makers can use to improve student achievement.

> The tests will give teachers and principals information about how each child is performing and help them to diagnose and meet the needs of each student. They will also give policymakers and leaders at the state and local levels critical information about which schools and school districts are succeeding and why, so this success may be expanded and any failures addressed. (U.S. Department of Education)

Some assessment policies are tied to specific consequences, such as government intervention or parental choice; others are intended to spur change by making assessment results public. Mathematics is a core subject in such assessments.

Research on the Formulation of Policy for K–12 Assessment

As the stated purposes for NCLB testing and NAEP illustrate, assessment policies are established with multiple purposes in mind. Haertel (1999) distinguishes three main purposes for large-scale assessment: providing information for accountability, evaluation, and comparison; focusing public attention; and changing education practices (p. 6). The stated purpose of NCLB assessment highlights the purpose of changing practice, suggesting that change will be spurred by the public attention and accountability provisions built into the overall NCLB legislation.

Policy analysts, drawing on their own experiences as the policy was developed, have described aspects of the development of the assessment policies within NCLB. Goertz (2001) illustrates the accounts given, which indicate that NCLB modified the approach taken in previous federal legislation by requiring states to put in place more uniform and extensive policies for assessment, with federal mandates for improvements in test performance. NCLB mandated that education agencies seeking funds under the legislation (e.g., states and local districts) select at least one assessment in each of mathematics, language arts, and science for common use by all public schools in their purview for measuring the achievement of all students. McDonnell (2005) argues that the NCLB assessment policy

[4] The NAEP Program includes the Main NAEP and Long-term Trend NAEP. Content tested on the Main NAEP is revised every ten years. Every two years Main NAEP reports mathematics results at the national level years for grades 4, 8 and 12 and results at the state level for grades 4 and 8. The content of the Long-term Trend NAEP has remained "relatively fixed" (NCES, 2005, p. 4), with results to be reported every four years for 9-, 13-, and 17-year olds.

can be seen as part of an evolution in policy over two decades, as research showed that assessments could be an effective way to create the changes in education practice that were needed to move student achievement closer to the high standards that were being set. For the recent standards and accountability movements, the seminal paper by Smith and O'Day (1991) is often cited as a summary of the rationale for using a combination of standards and assessments to promote change in education.

Porter and his colleagues have further developed a framework for specification of assessment policies designed to increase student achievement (e.g., Porter, Chester, & Schlesinger, 2004).

Some accounts, however, see the increasing use of assessments as serving less noble purposes, describing the increased use of testing as misguided or even as a covert ploy for maintaining social stratification (Apple, 2006; Oakes, 1985). These critiques lump assessments of mathematics in with a generally increased policy emphasis on assessment and accountability.

The political history behind recent changes in assessment policy in California has been reported in more detail than is usually found in reports on the formulation of mathematics assessment policies (see Jackson, 1997a, 1997b; Wilson, 2003). In concert with its Mathematics Framework, the California Department of Education had developed an assessment, called the California Learning Assessment System (CLAS), which incorporated open response items in mathematics. As part of the broader controversy about the California approach to mathematics, the assessment came under fire.

Research on the Status of Policy for K–12 Assessment

Documentation about state level assessment policies is included in several periodic reports produced by CCSSO, the AFT, *Education Week*, and the NSF. One aspect of status reports are descriptions of which government units conduct mathematics assessments, at which grade levels, with what frequency. For instance, *Quality Counts* 2005 (Editorial Projects in Education, 2005), includes a state-by-state summary of assessments administered by grade and by subject. For example, Arkansas, Georgia, Utah and the District of Columbia report a mathematics assessment in grade one; there are 13 states that do not assess mathematics in grade four.

Sometimes status reports go further, including descriptions of the mathematics content that is assessed. For high school graduation tests, for example, The Center on Education Policy (2004) reported that in 2004, 27 states had a high school exit exam. Achieve, Inc. (2004a) provides a more substantive analysis in *Do graduation tests measure up: A closer look at state high school exit exams.* The report indicates that about half of the states require such tests as a condition for earning a high school diploma. They have conducted an analysis of the mathematical demands of high school exit exams in the states of Florida, Maryland, Massachusetts, New Jersey, Ohio, and Texas, and draw conclusions such as:

> The questions on these exams reflect material that most students study by early in their high school careers. In math, the tests place a heavier emphasis on prealgebra and basic geometry and measurement concepts than on concepts associated with Algebra I and high school geometry.
>
> . . . The "cut scores" required to pass the tests reflect modest expectations. To pass the math tests, students in these states need to successfully answer questions that, on average, cover material students in most other countries study in 7th or 8th grade. . . .
>
> The tests measure only a fraction of the knowledge and skills that colleges and employers say are essential. Achieve, Inc. (2004a, pp. 1–2)

These analyses of the status of assessment policies show both that the prevalence of assessments is increasing and that the expectations for student performance remain modest. If the intent of mathematics assessment policies is to increase student achievement, the goal seems to raise the floor, rather than to push to high levels.

Research on the Implementation of Policy for K–12 Assessment

If the logic model for assessment policies is that making assessment results public and connecting them to incentives will push the system to bring more students up to the standards, a key aspect of implementation is having the assessments align with the standards. Alignment between assessments and standards has been the subject of frequent study in the last two decades.

In the past decade, several systems have been used for examining alignment between mathematics standards and assessments (e.g., Schmidt et al., 2001; Smithson & Porter, 2004; Webb, 1999). Each of the systems includes dimensions for classifying the content of both assessments and standards, which are used by raters to describe the distribution of content in an assessment or set of standards. These studies often have found a surprising disjuncture between the content specified in the standards and the content included on the assessment, even when the standards and assessment were adopted within a single state. In describing example studies of alignment, Porter (2002) notes that a study designed to look at the alignment between state standards and assessments in seventh grade mathematics found that, across four states, assessments were as closely aligned to standards in other states as they were to the state's own standards. Although Porter notes that, because tests represent only a sample of intended content, perfect alignment should not be expected, this study suggests that state standards may not be sufficiently specific to allow an assessment to be tightly aligned with them. "Another possibility is that states have much more work to do to bring their assessments into alignment with their standards (Webb, 1999). This finding seems probable, and it is one about which U.S. Department of Education officials are expressing concern" (p. 6).

The misalignment between standards and assessments is more severe for more challenging content areas. A recent summary of alignment studies produced by the American Educational Research Association (2003) concludes that many tests fail to measure adequately the cognitive complexity or "depth of knowledge" described in state standards and that:

> For example, in an analysis of 8th-grade math standards and tests in one state, Porter found that the test questions concentrated on length and perimeter, area and volume, and the performance of routine procedures. But the tests generally neglected more complex topics and cognitive demands included in the standards, such as angles or solving nonroutine problems. Similarly, in an analysis of math standards in three states, Webb found that nearly all of the states failed to measure adequately the "depth of knowledge" described in their standards. (pp. 2, 4)

An alignment study carried out by Achieve concluded that when states use existing norm-referenced tests for their state assessment systems, "these tests do not measure the content of the standards well" (Achieve, Inc., 2002, pp. 8–9).

Research on the Effects of Policy for K–12 Assessment

Research does not yet give a clear picture of the effects of mathematics assessment policies. Varying conclusions are being drawn, with publications sometimes leading to sequences of responses and rejoinders.

Early commentaries on testing policies sometimes criticized the effects that testing programs had on the curriculum, claiming that tests focused on low-level skills, leading teachers to an equally low-level curriculum. But those commentaries seldom cited solid empirical evidence. As measurement experts later noted, research was needed to determine the broad effects of testing (Mehrens, 1998; Porter, Floden, Freeman, Schmidt, & Schwille, 1988).

Subsequent research has shown that effects of testing policies are varied, depending in part on local context and capacity. The conclusion is consistent with research on a broad range of education policies (e.g., Berman & McLaughlin, 1997). Firestone, Schorr, and Monfils (2004) studied the effects of mathematics and science testing policies in New Jersey. The mandated state tests included both multiple-choice and short open-ended questions, intended to measure a range of pupil knowledge, from computational skills to problem solving. Its orientation seemed likely to encourage teaching that included, but went beyond, simple mastery of algorithms.

Firestone and his colleagues found that the influence of the mandated test varied across schools and districts. They saw changes in instruction, many of which might be called, "teaching to the test," but which ran the gamut from drilling on test mechanics to asking probing questions, and trying to get students to explain the reasoning behind their answers. Explanations of differences in effects of testing policies included availability of resources, local political context, and educators' interest in and knowledge about education reform. This research indicates that attempts to determine the typical effect of an assessment policy may be frustrated by variation in what any one policy will produce.

When scholars make broad claims, they often get responses from others who support opposing claims. One recent example is the exchange that began with Amrein and Berliner's (2002) publication of an analysis of the data on the consequences of the implementation of high-stakes state testing policies. Using nationally-normed assessments (e.g., SAT, ACT, NAEP), the investigators looked for changes in student achievement in 18 states which began high-stakes testing polices at varying points in the past two decades. They concluded that, though results on the high stakes exam itself typically increased when higher stakes were attached, the results did not transfer to other measures of academic achievement. Hence, these researchers concluded that apparent positive effects were due to exclusion of students from the high-stakes tests or to narrow forms of training that were unrelated to the intended content of schooling.

Rosenshine (2003) responded with an alternative analysis that led to the conclusion that states with "clear" high-stakes testing policies did have higher scores on the NAEP.

> The results in Table 1 show the average NAEP increases in the "clear" high-stakes states were much higher than the increases in the comparison states. In 8th grade mathematics and in 4th grade reading the mean increase for the clear high-stakes states was double the increase for the states without consequences. The effect sizes for the comparisons were .35 for 4th grade math, .79 for 8th grade math, and .61 for 4th grade reading. (http://epaa.asu.edu/epaa/v11n24/, paragraph 7)

His analysis differed in comparing states with clear high-stakes testing to those without, rather than looking at individual states across time, as Amrein and Berliner did. Amrein-Beardsley and Berliner (2003) published a rejoinder, finding a positive effect of high-stakes testing in 4th-grade NAEP scores, but not elsewhere.

This exchange illustrates the ongoing debates about the impact of state and national assessment policies. The sources of disagreement are multiple. Some focus on methodology: What comparison group should be used? Can a single setting (e.g., state or district) serve as its own control over time? How can the effects of a testing policy be disentangled from other policies, some changing, some not? What is an appropriate outcome measure? How can controls be built in for changes in the population tested?

Other disagreements come from differences in values attached to outcomes. Is a narrowing of the content taught (to match the test) a desired or disparaged outcome? Are increases in average achievement the most important aspect of improvement, or is narrowing the gap between rich and poor, or Black and White, the most important goal?

A few studies have attempted to consider the impact of high school exit exams on other variables. For instance, Marchant and Paulson (2005) analyzed the relationship of high school graduation exams to graduation rates and SAT scores, through secondary analyses of College Board and NCES data. They found that, "states requiring graduation exams had lower graduation rates and lower SAT scores. Individually, students from states requiring a graduation exam performed more poorly on the SAT than did students from states not requiring an exam." (p. 1). This result, coupled with the Achieve result about low expectations on these exams, suggests that this policy is not necessarily having the effect of restoring graduation as a valid indicator of preparedness.

Most scholars seem to believe that instituting mathematics testing policies will have effects, but the field seems far from reaching agreement on just what those effects will be, and whether they are desirable.

Observations: Research about Assessment Policy in K–12 Mathematics

Assessment policies in current mathematics education are among the most prominent and prescriptive of all subject areas. The federal policy emphasis on assessment has increased with No Child Left Behind, extending the trajectory along which federal policy makers have been pushing for higher achievement outcomes for two decades. Federal policies have not dictated the specific content to be assessed, but through requirements for broad student participation and connections to NAEP, national policies press states to push achievement higher, and in the general direction represented by the content of NAEP. Analyses of the status and implementation of assessment policies reveal upward movement in expectations, though not yet to levels that match the rhetoric of calling for mastery of challenging content by all students.

Research on the impact of assessment policies shows effects, particularly when these policies are aligned with other policies. Researchers debate whether all the effects observed are desirable, though the general weight of evidence is that assessment is having effects roughly along the lines policy makers are advocating.

POLICY AREA FOUR: THE MATHEMATICS TEACHING PROFESSION

The profession of mathematics teaching is governed and affected by policy conditions of various types at the local, state, and federal levels. In this section we discuss policy relevant to four areas that affect mathematics teaching: entry into the teaching profession (including initial licensure, accreditation, teacher preparation programs); recruitment, retention, and teacher induction; continuing licensure, credentials, and teacher quality requirements; and teacher professional development. Entry into the teaching profession is governed by a complex and largely state-driven set of policies for certification and licensure. These policies include "state-regulated initial certification of new teachers, state-required teacher tests for initial licensing, and voluntary professional accreditation of teacher preparation programs and institutions." (Cochran-Smith & Zeichner, 2005, p. 26). After a provisional or initial license to teach is granted, continuing licensure depends on meeting additional requirements, such as passing further tests, completing additional education, or getting a favorable rating from an employing school. The specific requirements, tests, and passing scores vary across states. Many states also require districts to provide some type of induction support to beginning teachers as they enter the career. In addition, there are policies associated with teacher induction, advanced credentialing and teacher professional development that we will discuss in this section.

The education of prospective teachers is governed by state and national accreditation policies that apply at the program and institutional level. The National Council on Accreditation of Teacher Education (NCATE) and the Teacher Education Accreditation Council (TEAC) are the two national accrediting agencies for teacher education programs. NCATE's process includes a specialty program area review component which, in the case of mathematics, is overseen jointly by NCATE and NCTM. State program approval processes vary, but generally involve a self-study and state-administered review.

Other policies address the induction and retention of beginning teachers. The Interstate New Teacher Assessment and Support Consortium (INTASC) is a "consortium of state education agencies and national educational organizations dedicated to the reform of the preparation, licensing, and on-going professional development of teachers." (CCSSO, 2006, p. 1). INTASC has developed core standards for what beginning teachers should know, and also has created model performance assessments for teachers, thus providing policy guidance to state agencies. INTASC has standards specific to mathematics (INTASC, 1995), which call for a view of mathematics teaching consistent with the NCTM Standards. In particular, they note:

> They will be expected to provide challenging mathematics to students at very different levels of mathematical understanding and to adequately address the full range of student questions that arise from learning diversity. Teachers of mathematics at every grade level will be expected to teach more mathematics and more challenging mathematics to their students, and to be knowledgeable about the curricula that precede and follow the mathematics they teach. (p. 6)

Local and state policies have been developed about teacher placement in areas of high need (e.g., policies that allow teachers lacking adequate credentials in mathematics to teach mathematics if properly qualified teachers cannot be hired, often tied to district-level alternative certification programs) and policies about teacher retention (e.g., induction programs.) The problem of attrition among beginning teachers has been well documented (see Grissmer & Kirby, 1987, 1997; Lortie, 1975; Veenman, 1984); these attrition issues may be relatively more serious in mathematics, particularly in urban areas.

Periodically, federal programs have been implemented to increase the size of the mathematics and science teaching force. The most notable is the "Troops to Teachers" (TTT) program, established in 1994 by the U.S. Department of Defense as an alternative certification program.

Among the influential federal policies over the past several decades relative to teaching and teachers (see Earley, 2000), the most current and visible are the policies related to improving teacher quality, recruitment, and preparation in the 1998 Amendments to the Higher Education Act (HEA)[5] and those related to ensuring elementary and secondary students' access to highly qualified teachers in NCLB. Such federal policies may be influenced by some of the national reports in recent years that have addressed issues about the teaching profession generally (e.g., National Commission on Teaching and America's Future [NCTAF], 1996), as well as issues in mathematics teacher education (e.g., CBMS, 2001; National Commission on Mathematics and Science Teaching for the 21st Century, 2000; Mathematical Sciences Education Board, 1996; NRC, 2001a, 2001b). Such reports typically include strong recommendations for policy. NCTM's 1991 *Professional Standards for Teaching Mathematics* offer specific standards for policymakers, such as

> Policymakers in government, business, and industry should take an active role in supporting mathematics education by accepting responsibility for
>
> - supporting decisions made by the mathematics education professional community that set directions for mathematics curriculum, instruction, evaluation, and school practice. (p. 179)

In the mathematics domain, most recent recommendations have focused primarily on the need to strengthen content preparation of teachers.

In response to issues of teacher quality, NCLB requires that all public school teachers of core academic subjects (which includes mathematics) must be "highly qualified" by the end of the 2005-06 school year, with consequences if this is not achieved. (U.S. Department of Education, 2001, pp. 40–41) To be highly qualified, teachers must have a bachelor's degree and full state certification or licensure, and must prove

---

[5] HEA was enacted in 1965. The most recent reauthorization by the full Congress was in 1998 (see P.L. 105–244). The 1998 reauthorization was set to expire in 2003 but has been extended several times in response to difficulties achieving agreement on the provisions of the reauthorization.

that they know each subject they teach (U.S. Department of Education, 2001). In addition NCLB requires states to measure the extent to which all students, particularly minority and disadvantaged students, are being taught by highly qualified teachers. States are expected to adopt goals and plans to ensure that all teachers are highly qualified and must publicly report their plans and progress in meeting these goals. NCLB allows states to develop an additional way for current teachers to demonstrate subject-matter competency and meet highly qualified teacher requirements. Proof may consist of a combination of teaching experience, professional development, and knowledge in the subject garnered over time in the profession.

Beyond state credentialing processes, one primary vehicle for advanced teacher credentialing is the National Board for Professional Teaching Standards (NBPTS). Founded in 1993–94, the NBPTS awards advanced certificates for experienced teachers who successfully complete its required assessments. The NBPTS is a non-profit organization, governed by a board of directors, the majority of whom are K–12 teachers. The significance of NPBTS certification varies across states, with some states linking it to pay increases and others simply treating it as information that local districts can use as they see fit. NBPTS offers two certificates in mathematics: adolescence and young adulthood (NBPTS, 2001) and middle childhood through early adolescence (NBPTS, 1998).

In contrast with the strong state policy control, national guidance, and federal control over entry into the profession, teacher preparation programs, induction, and teacher quality, there is very little state-level regulation about the quality and content of the professional development that is expected for the renewal of teaching licenses (Hirsch, Koppich, & Knapp, 2001, cited in Desimone, Garet, Birman, Porter, & Suk Yoon, 2003). School districts, states, professional organizations, universities, and private organizations all are engaged in the business of offering professional development for teachers. The policy environment in this domain is similarly scattered.

The policies that probably have had the most significant impact on teacher professional development in mathematics are federally funded programs for teacher improvement. For example, the National Defense Education Act (NDEA) (PL 85-864) was the main source of funding for institutes for teachers of mathematics and science in the post-Sputnik era, from the mid 1950s until the late 1970s. On the heels of this, the Education for Economic Security Act (EESA) (PL-98-377) was passed in 1984, charging the NSF to provide professional development for teachers of mathematics and science. From 1994–2002, the Eisenhower Professional Development Program—a part of Title II of the Elementary and Secondary Education Act of 1994—was "the federal government's largest investment solely focused on developing the knowledge and skills of classroom teachers in mathematics and science" (Desimone et al., 2003, p. 615).

More recently, the NSF has supported teacher professional development through a variety of programs, including Teacher Enhancement, Systemic Initiatives, and Mathematics and Science Partnerships. And, the U.S. Department of Education provides funding to states for mathematics and science partnerships that offer professional development for teachers.

In summary, the policies affecting the mathematics teaching profession include policies about initial licensure and accreditation of teacher preparation programs; recruitment, retention, and induction; continuing licensure and teacher quality, and professional development. As in other areas, the policies in this arena are generally not specific to mathematics, and there is little mathematics education-specific research that has examined the formulation, status, implementation, and effects of these policies. However, some of the more general research may have important implications for mathematics education. Wilson and Youngs (2005) summarize the historical and political issues surrounding what they posit as a central question in this area: do accountability processes lead to higher teacher quality? (p. 592). They go on to call for rigorous research in this area as a start in addressing the question. We would add that research particular to mathematics education issues, within the more general issues about the teaching profession, would also enlighten the policy discussion.

Research about the Formulation of Policies about the Mathematics Teaching Profession

We found few studies about the formulation of policy relative to the mathematics teaching profession. Although there are historical studies of the overall evolution of requirements (e.g., Sedlak, 1989), and the recommendations for mathematics teacher preparation (e.g., Ferrini-Mundy & Graham, 2003), we do not include those here, because they do not directly address how policies were formed.

There is some available research on policy formulation relative to the teaching profession more generally. Porter, Youngs, and Odden (2001) summarized the processes of developing criteria and standards used in several key teacher assessments, which represent policies governing entry into the profession. As a starting point for the development of Praxis III, researchers at Educational Testing Service (ETS) established a conceptual guide to describe teaching (Dwyer & Villegas, 1993, cited in Porter et al., 2001). This document included analyses of the job of teaching, reviews of research about teaching, and analyses of state standards for licensure.

Earley (2000) provides an account of some aspects of the formulation of the 1998 Higher Education Act (HEA) legislation. She concludes that the reauthorization was strongly influenced by the 1996 NCTAF report. The policy introduced new accountability and reporting measures for states and institutions of higher education; Earley speculates that the provisions in the HEA were included in part as accountability measures responding to assertions that "teacher education is a culprit in school reform failures." (p. 33.) As we saw in research about the formulation of standards, it is clear that the context in which policies are constructed can be quite influential in the shaping of the policies.

Porter et al. (2001) contrast the approach used by ETS in developing Praxis III with the approach to the development with that used by INTASC and NBPTS. INTASC also has conducted careful documentation of its development processes (Moss, 1998, referenced in Porter et al., 2001; Porter et al., 2001, p. 276). INTASC convened groups of experienced professionals in specific subject areas to design standards. They provide a detailed description of how the NBPTS standards vetting process occurs, where various layers of committees and experts review the proposed standards and accompanying instruments. The distinction noted by the authors in these two approaches to policy formulation—an analytic, job-demand-based conceptual framework, versus consensus standards developed by practitioners—is interesting. Mathematics teaching-specific policy recommendations generally seem to be in the form of standards developed by an expert practitioner consensus process.

Research about the Status of Policies about the Mathematics Teaching Profession

Various organizations provide periodic information about changes and general features of state-level policies concerning teacher entry, induction, continuing licensure and professional development. It is often difficult to get detailed information about the mathematics-specific elements of the policies.

For many years, the professional association of state officials with responsibility for teacher certification compiled the requirements for every state. This organization, the National Association of State Directors of Teacher Education and Certification (NASDTEC), produces paper copies of these regulations (e.g., NASDTEC, 1997, 2000). A version of the regulations in a uniform format is also published by a private company dedicated to this work (Teacher Certification Publications; http://home.earthlink.net/%7eteachercertification/index.html). The Web has eased the compilation of these regulations, which can now be accessed through links at the web sites of the Education Commission of the States (ECS) (http://www.ecs.org/) and the National Center for Education Information (NCEI) (http://www.ncei.com/resources.html).

Some policy analysts work to pull out what they see as critical features of these descriptions, such as by compiling summaries of recent changes in teacher certification policies, or describing general trends in state policies. Hirsch et al., (2001), for example, write that

> Certification and teacher testing are the principal vehicles to date for promoting standards for teaching. Multi-tiered certification systems are becoming increasingly popular ways to demonstrate the growth of teachers' competence over time. While a growing number of states are establishing teacher assessment systems, states are encountering complicated issues regarding the role, validity, and technical quality of teacher assessments. For example, states have yet to evolve cost-effective means of assessing actual classroom performance—or, even more problematic, predicting future performance, in the case of assessments done at the time of initial licensing. (p. 6)

These descriptive studies show some of the common features of teacher certification requirements, such as completion of a bachelor's degree, completing coursework in subjects to be taught, and, for most types of certification, completing some form of practice teaching. Some form of teacher testing is required in 42 states (NASDTEC, 2000, cited in Wilson & Youngs, 2005, p. 596.) Youngs, Odden, and Porter (2003) provide status information about state policy in the area of teacher licensure for the 2001–02 year. They were interested in the extent to which states were using performance assessment, and found that only 18% of states were doing so.

In response to teacher shortages, states often create and implement policy alternatives to university-based teacher certification routes. Feistritzer and Chester (2003) report that most states have some version of an alternative certification program for teachers, and since 1985 about 200,000 people have been certified to teach through alternative routes. Editorial Projects in Education (2004) reports that in 29 states, candidates in alternative routes are required to demonstrate subject matter expertise.

We found little status research or policy analysis describing the status or characteristics of mathematics teacher education programs. The Association of Mathematics Teacher Educators conducted a survey, (Karp, 2005) to better understand the work of mathematics teacher educators in universities and colleges across the country. They collected survey responses from 427 institutions, and learned that for prospective teachers of grades K–2, most institutions required between 0 and 6 credits of mathematics. In addition, the majority of institutions required between 0 and 3 credits in mathematics teaching methods, and at about 2/3 of the institutions there were 0–3 mathematics education faculty. It is difficult to assemble a comprehensive national picture of the requirements, staffing, and resources involved in the preparation of teachers of K–12 mathematics.

At the state and district levels, beginning teacher induction policies have emerged as a response to the problems of teacher quality and teacher retention. There is no federal policy about teacher induction programs or support for beginning teachers, but Britton, Paine, Pimm, and Raizen (2003) claim that more than thirty states do have policies in this area. Many of these require the mentoring of new teachers. Some school districts also have policies and related programs for new teacher induction in response to state policies. Smith and Ingersoll (2004), on the basis of data from the NCES "Schools and Staffing Survey" (SASS, report increases in the numbers of teachers participating in induction programs, from 51% in 1993–94 to 79% in 1999–2000.

The status studies also show the great variation in the specifics of state requirements. In addition to the "regular" certificate, many states offer others types of certificate, such as "emergency" certificates for cases in which a school is unable to find a certified teacher for some position. These "emergency" certificates have lower requirements and are intended to be valid for only a short time, until a regularly certified teacher can be found.

Research on Implementation of Policies about the Mathematics Teaching Profession

Most studies of policy implementation in the teaching profession are conducted to look at general trends and issues, rather than mathematics-specific matters. Here we highlight some studies that we feel are particularly relevant, and discuss how the trends that they identify may have implications for mathematics.

Some of the status studies described in the previous section characterize the written state policies about what is needed to teach, but as with other policies, actual practice and implementation may depart from the written policy. Ideally, if the policies requiring a "regular" certificate were thoroughly implemented, the vast majority of teachers would meet those requirements. Policy studies of the certification status of a state's teaching force give a picture of how well that intention is being met. State teacher certification is not always required for teaching in independent schools. Even within the public schools, however, policy studies have shown that many students are taught by teachers who hold emergency certificates, or whose primary certification is not for the subject area. Darling-Hammond has addressed the prevalence of what she calls "substandard" teacher licenses, noting substantial differences across states. "In Wisconsin and eleven other states, . . . no new elementary or secondary teachers were hired without a license in their field in 1994. By contrast, in Louisiana, 31% of new entrants were unlicensed and another 15% were hired on substandard licenses. At least six other states allowed 20% or more of new public school teachers to be hired without a license in their field" (1999, p. 15).

Some studies look at these issues specifically in mathematics. The National Science Board (2006) reports that 80% of public high school mathematics teachers were fully certified in mathematics in 2002, while 20% were either not fully certified or certified in another field. During the past two decades, there has been a steady problem of recruiting and retaining qualified teachers of mathematics to teach in urban and/or high need districts. In 1999–2000, 69% of U.S. students in grades 5–8, and 31% of students in grades 9–12, were taught mathematics by teachers with no major or certification in the subject (NCES, 2004, p. 10). Situations such as these, which also affect urban and rural youth especially severely, are often a consequence of difficulty recruiting and retaining qualified teachers. Using data from the NCES' (1999) SASS, Ingersoll (1999b) found that in 1994, 27% of public secondary school teachers assigned to teach mathematics lacked a certificate to teach mathematics. This was a larger percentage than in any other core content area. Out-of-field teaching is associated with school characteristics, with more out-of-field teaching in high-poverty schools and in smaller secondary schools. Ingersoll makes explicit that out-of-field teaching can be seen as a failure to implement certification policy.

Some policy studies have addressed the extent to which the intent of state or national program standards and teacher examinations is implemented in the actual requirements for coursework in programs, and in assessment scoring. For example, some research has examined whether teacher examinations and procedures for determining pass rates actually uphold requirements. Analysts at the Education Trust (Mitchell & Barth, 1999) have examined the subject matter content of teacher certification tests from the two major suppliers, ETS and National Evaluation Systems (NES). They selected a sample of state certification tests in English, mathematics, and science, and engaged their staff and a panel of expert consultants in rating each of the items. For secondary mathematics, this study found that most content on the teacher tests was material from the high school curriculum. The balance of "routine" and more complex items varied across tests, with the test used in Massachusetts having the most challenging items. From this analysis, the authors concluded that the tests were not sufficiently advanced, because "the tests assess mostly tenth to eleventh grade level content. Nothing in either the Praxis II tests or the NES tests probes the intellectual substance of college mathematics that should equip high school teachers with robust backgrounds for dealing with the myriad ideas that will emerge from discussion with students" (Mitchell & Barthe, 1999, p. 10). In addition, many states require teachers to get only half the items correct. The authors conclude that implemented policies for teacher certification do not reflect the rigor of the policies' intensions.

Once teachers are hired, there are challenges in retaining them, in the profession and in particular in high-need schools (Ingersoll, 1999a, 1999b). Retention of teachers of mathematics, particularly in urban settings, is especially difficult. Research reveals a variety of reasons for teacher mobility and attrition; there is some disagreement in the research about how prominently salary figures in this context. Strong factors in teacher attrition include poor working conditions (Ingersoll & Smith, 2003) and issues related to student race and achievement (Hanushek, Kain, & Rivkin, 2001). Neuhaus (2003) reports on a state-funded program in California to increase the proportion of fully certified teachers in low-performing schools. The Teaching as a Priority Program (TAP) involved incentives such as signing bonuses and supplementary funds available for mathematics and literacy instruction. Findings indicate that these kinds of policy incentives might not be effective in inducing teachers to stay in low-performing schools.

In the U.S., one of the most sweeping and heavily resourced statewide induction policy-program combinations is the California Beginning Teacher Support and Assessment System (BTSA), which encourages school districts and other agencies to offer induction programs, and which provides state funding for such programs. (from http://www.btsa.ca.gov/, retrieved January 19, 2005).

In the case of teacher induction, it may be possible to improve the implementation of policies by looking at international practices, where induction policies are in place in a number of countries. The policies tend to be more elaborate and structurally infused within both the national culture and the education system than those in the U.S.; (see Britton et al., 2003, for a report of a study of induction in Shanghai, Switzerland, New Zealand, France, and Japan). In this volume, the researchers describe the experience of the *stagiaire* (first year teacher) in mathematics and the elaborate and highly prescribed induction system in France, which focuses on subject matter. For instance, they recount the centrality for the new teacher of "'le cours'—the formal, mathematical content to be organised and taught." (p. 238). The researchers observed an emphasis on oral fluency in mathematics and careful selection and evaluation of preparatory mathematical tasks as part of the content of the induction experience in mathematics. Induction policies in the U.S. tend not to be subject-specific, and thus their implementation is also general.

Evaluation studies that are aimed specifically at whether funded programs have met the program goals as specified in the "policy" of the funding agency are a type of research that could be construed as studying policy implementation. For example, in a study of district-level implementation of mathematics inservice programs for teachers, as part of the national evaluation of the Eisenhower Professional Development program, Desimone, Porter, Birman, Garet, and Suk Yoon (2002) examined "the degree to which the [PD] activity has a content focus—that is, the degree to which the activity is focused on improving and deepening teachers' content knowledge in mathematics and science" (p. 1274), a goal of the program. They found that "districts that align professional development with standards and assessments are more likely to offer reform types of activities and are more likely to engage in continuous improvement efforts" (p. 1292).

Some professional development funding programs require or encourage involvement of higher education faculty and disciplinary faculty specifically. Desimone et al. (2002) in the Eisenhower Professional Development Program evaluation, found that "postsecondary institution projects in education departments have significantly more coordination with districts than postsecondary institution projects in mathematics or science departments" (p. 633), and that "projects in mathematics/science departments have high content focus regardless of the type of postsecondary institution they are in, whereas projects in education and other departments have a high content focus only if they are in research/doctoral universities." (p. 636). They concluded that the quality of professional development experiences would be strengthened by close coordination between districts and postsecondary institutions. Howey (2000) was also referenced, saying that encouraging K–12 and university faculty to work together will require new policies that offer incentives and working mechanisms for such collaborations.

In summary, studies of policy implementation suggest that there is a distance between the idealized version of the policy and its applied version in practice, particularly in the area of teacher certification. This becomes important in understanding research that examines the effects of policies, because policies are not always implemented with fidelity.

Research on the Effects of Policies Related to the Mathematics Teaching Profession

The fundamental questions in this area center around whether particular policies, especially those related to accountability, credentialing, and program components as specified in state, national, or federal policy, have an impact. The question is complicated not only by the mediating effects of the fidelity of implementation, but also by determining the most suitable outcome variables. These include teacher retention, teacher effectiveness, and ultimately, student performance. However, little research has examined the impact of policy on pupil performance.

In the past decade, scholars have vigorously debated what effect requirements for teacher certification have on the effectiveness of teaching. Sometimes the debates focus on teacher certification per se, comparing teachers at the secondary level with regular certification in the subject area to teachers lacking such certification. At other times, the debates focus on the requirements for teacher certification, especially coursework and test performance, looking for associations between those requirements and pupil performance.

Linda Darling-Hammond argues that teacher certification is associated with teaching effectiveness (e.g., Darling-Hammond, 2000a, 2000b; Darling-Hammond, Berry, & Thoreson, 2001). Those on the other side of the debate include Ballou and Podgursky (2000) and Walsh (2001). Wilson and Youngs' (2005) careful review of rigorous empirical research on the effects of teacher certification concludes that "the trend is toward favoring certified teachers . . . particularly in the area of mathematics" (p. 614). The studies they reviewed used teachers' learning and knowledge, teaching practice, and/or student learning as outcomes. They note that the great variation in requirements for certification and how researchers define certification make it difficult to draw strong general conclusions about its effects.

Research also has examined the associations between particular aspects of certification and related teacher education programs and the learning of teachers' pupils. A recent review of the effects of taking subject matter courses found that "Only in secondary school mathematics do the studies provide a clear answer to questions about connection between amount of teachers' subject matter study and pupil achievement" (Floden & Meniketti, 2005, p. 269). Though research indicates that there is an effect of completing some mathematics coursework, the research does not go so far as to support the requirement for a major in mathematics (or its equivalent), as is built into many state certification requirements and has historically been the norm (see Ferrini-Mundy & Graham, 2003). "For secondary school mathematics, one study showed that the benefit of additional subject matter courses declined after the first five courses, suggesting that, even here, the relationship between subject matter study and student achievement is not as simple as 'more is better'" (Floden & Meniketti, 2005, pp. 269–270).

Some research also offers support for completing coursework on the teaching of mathematics, another typical part of the requirements for mathematics teacher certification. Monk (1994) found that undergraduate mathematics pedagogy courses contribute more to performance gains of teachers' pupils than do undergraduate mathematics courses, and that the influence of advanced mathematics courses beyond a threshold set about five may be negligible.

Passing teacher tests is now part of the certification requirements in most states. Little evidence exists, however, about the effects of this requirement on teacher effectiveness. Wilson and Youngs' literature review (2005) found that much of the research focused on tests that are no longer used and that the literature as a whole gave little information about the connection between performance on teacher tests and pupil achievement.

For advanced certification, however, two recent studies have found an association between passing the NBPTS examination and student achievement. A study in North Carolina (Goldhaber & Anthony, 2004) and one in Florida (Cavalluzzo, 2004) both have found that pupils of NBPTS-certified teachers learn more than pupils of teachers in comparison groups. The North Carolina study found this result for elementary school mathematics (with a smaller, positive effect in reading) and the Florida study for 9th and 10th grade mathematics. Sanders, Ashton, and Wright (2005) undertook a comparison of National Board Certified Teachers' (NBCT) students' performance growth in mathematics to the performance growth of students of other teachers. They found that "students of NBCTs did not have significantly better rates of academic progress than students of other teachers." (p. 2).

Research on the impact of alternative route programs and induction programs does not typically single out mathematics as a focus, but some of the general findings may have interesting implications for mathematics education. Alternative route programs vary widely in their design and implementation. Those that assume strong subject matter background, such as Troops to Teachers or Teach for America, tend to include preparation that is focused on more general professional practice issues. The more effective alternative route programs are those that include some of the main characteristics of university-based teacher education programs, such as significant support and mentoring, clinical experiences, etc. (see Feistritzer & Chester, 2003; NCTAF, 2003; Wilson, Floden, & Ferrini-Mundy, 2001). Some research has suggested that teachers prepared through alternative routes have reached higher educational achievements than their counterparts who have been certified through traditional processes (Shen, 1999). In his 2003 review of research, Allen concludes that there is only limited support for the conclusion that alternative-route completers can be as effectively as traditionally prepared peers. Thus the research is inconclusive.

Most induction policies require mentoring—that is, systematic support offered by experienced teachers to beginning teachers. (see Feiman-Nemser, Schwille, Carver, & Yusko, 1999); others also include requirements for participation in professional development. Some researchers are concerned that the focus on mentoring is too narrow and fails to sufficiently problematize the challenges faced by the new teacher. Britton et al. (2003) note that,

> "Mentoring" has come to stand for an automatic remedy to a problem that tends to remain unexamined. It is variously seen as insufficient experience of the exigencies of teaching, inadequate information about local practices and customs in the particular school or incomplete knowledge of various sorts required for teaching. . . . Share the common presumption that the novice arrives lacking something. . . the assistance of a mentor—is somehow to make up for this deficiency. (p. 1)

Smith and Ingersoll (2004) found that when mentoring is coupled with other, subject-specific supports, retention is improved. They comment that "Of all beginners who entered teaching in the 1999–2000 school year, 16% received none of the aforementioned induction or mentoring supports in their first year. Their predicted probability of turnover at the end of the first year was 40%. Twenty-two percent of beginning teachers received three induction components: a helpful mentor from their same field, common planning time with other teachers in their subject area, and regularly scheduled collaboration with other teachers on issues of instruction. Their turnover probability was 28%" (p. 6).

There has been a considerable amount of research on the impact of induction and mentoring programs on teachers' attitudes about their practice, again without specific attention to mathematics. The California Educational Research Cooperative undertook a study of the impact of the BTSA program in 1995–96, reported in Mitchell, Scott, Takahashi, and Hendrick, 1997. About 45% of the teacher respondents reported that their support providers worked in the same subject area and at the same grade level. Overall, the evaluation found that "teachers felt that the program was most beneficial in helping them plan instruction, while support providers, site administrators, and BTSA staff members felt the program was most valuable in helping teachers to organize and manage their classroom" (Porter et al., 2001, p. 284). Bartell and Ownby (1994, summarized in Porter et al., 2001) conducted a study of the implementation of local BTSA programs from 1992–1994, focusing mainly on the nature of the interactions of mentors and teachers, without specific attention to subject matter. BTSA projects at the local levels were charged with taking the lead on this work in 1994. In their account of the various processes of review, the recommendations that came from the field, and the revisions, Porter et al. do not mention any particular focus as having emerged on issues of the place of subject matter, or more specifically, mathematics, in the development of new teachers.

Strong (2005) conducted a review of research about induction programs. He comments:

> Of those [studies] that compare different kinds of programs, it appears that comprehensive support such as that advocated by Wong (2004) and identified as having enhanced 'packages' of components by Smith and Ingersoll (2004) are most likely to have an effect on turnover than more basic mentoring programs. (p. 192)

In summary, research that has examined the effects of alternative route and induction programs suggests that programs which are more comprehensive and include subject-specific elements may be more likely to impact teachers' decisions to remain in the classroom. Johnson and Kardos (2002) have found that new teachers are interested in assistance with curriculum implementation and advice from colleagues in their subject areas. Research that is subject-specific and that makes the link to teaching practice and to student performance is very much needed.

A problem with research about the effects of professional development policy has been the difficulty in linking to student achievement. An analysis released by the U.S. General Accounting Office (GAO) (1984) examined the impact of the NSF Mathematics and Science Teacher Institutes of the '70s and '80s, and found no evidence to connect teachers' participation in the institutes to the subsequent performance of their students in mathematics and science. They report, "as a group, these studies fail to show any relationship between teacher knowledge and the knowledge gain of their students" (GAO, 1984, p. 33, as cited in Earley & Schneider, 1996, p. 308). More recent work suggests that mathematics professional development that is curriculum-based can have an impact on the achievement of teachers' pupils (Cohen & Hill, 2001), as can intervention that is multi-dimensional and involves in-class coaching (Balfanz, MacIver, & Byrnes, 2006). In her review of studies that examined the impact of professional development by considering evidence of student learning, Kennedy (2002) concludes that "programs that focus on subject matter knowledge and on student learning of particular subject matter are likely to have larger positive effects on student learning than are programs that focus mainly on teaching behaviors" (Kennedy, 2002, p. 25).

Observations: Research about Mathematics Teaching Profession Policies

Teachers play a central role in mathematics education, so policy makers and policy researchers are keenly interested in understanding how policies affect teacher recruitment, preparation, selection, retention, and continuing education. In short, they are interested in policies intended to affect the quality of mathematics teachers. Policy research across subject areas has also identified teacher quality policies as an area of high priority and interest.

Given that strong interest, the amount of informative research on mathematics teacher quality policies is disappointing. The status of teacher quality policies across states has been documented, and periodically graded in publications such as *Quality Counts*. But the grades are based more on the beliefs of the report authors than on research about policy implementation and effects. The scholarly journals have published competing analyses about the policies such as requirements for certification, creation of alternate routes for teacher preparation, and requirements for teacher education program approval. Such studies usually report results for mathematics, along with other subject areas. The scholarly debates remain unsettled, in part due to the methodological challenges in conducting research on these policies.

Those reviewing the teacher quality policy literature (e.g., Cochran-Smith & Zeichner, 2005) offer recommendations for strengthening this area of research. They suggest, for example, building more comprehensive data bases, developing consistent definitions of key terms, paying more attention to impact, and situating research in relevant theoretical frameworks. These recommendations all apply to research on mathematics education teacher quality policies. At present, research in this area remains thin.

SUMMARY AND CONCLUSIONS

This review of research on mathematics education policy has themes that echo those of the broader education policy research literature: variability attributable to the decentralized governance of U.S. education; differences between formal and implemented policies; policy formulation influenced by many factors, with research one among many; and difficulty in tracing changes in pupil learning to particular policies. Policy research in mathematics education does, however, have distinctive features: a longer history with standards and associated assessments; more prominent disputes about desired outcomes; and a relatively large body of research, compared to policy research on other subject areas.

The results for research on policy formulation, status, implementation, and effects reveal the diversity across U.S. states and governmental agencies. Research on mathematics education policy formulation finds policies largely shaped by beliefs and values in the overall education context. In the past quarter-century, for example, standards have been widely promoted. Research studies show how the emphasis on standards has come in large part from the rhetorical appeal of holding high expectations. In mathematics education, research also has shown how important the inter-group (i.e., among mathematics educators, research mathematicians, parents, politicians) differences in value attached to particular areas of mathematics have been in affecting standards development and acceptance.

Research on policy formulation has, though, also documented ways in which research results have shaped policy. Current assessment policies, for example, were shaped by research showing how previous policies failed to produce the changes policy makers desired. As standards have been revised, research results have also been consulted.

Research on the status of mathematics education policies has been motivated by the variability across political contexts. This research, often sponsored by organizations outside higher education (e.g., the Council of Chief State School Officers, the American Federation of Teachers), is of considerable interest to policy makers who attend to policy trends. Research shows many states or other governments adopting similar policies, with variation in policy details. Anecdotal evidence suggests that policy makers base their own decisions in part on their understandings of what their counterparts are doing. Some status studies add evaluative judgments to the descriptions of policies, giving "grades" to standards or the requirements for certification as a mathematics teacher. Those grades, particularly when published prominently, are also likely to influence continuing policy developments. Such evaluative studies might be considered studies of policy implementation. Once again, differences in values play an important role in the research: a narrowly specified standard might be seen as admirably specific or disastrously confining.

Implementation studies in mathematics education have shown the slippage between intended policy and implemented policy found in most policy research. Some research shows, for example, that assessments intended to be "challenging" include a large proportion of low-level content. Assessment policies may intend to match tests to standards, but the tests are more like each other than like the standards. Nevertheless, the studies often show some effects, generally in the intended direction. Revisions in assessments move them toward greater challenge or alignment, though not as far as policy makers generally intended.

Likewise, the research on effects of mathematics education policies finds policy impact muted by complexity of the phenomena and disputed on value grounds. The effects of policies are documented better for mathematics education than for other school subjects. Effects of teacher quality policies, for example, are usually found primarily in mathematics. Analysts continue to debate what research shows about these policies, due in part to disputes about which effects matter most (e.g., making the curriculum or increasing test scores) and about how key policy variables (e.g., alternate route, full certification) should be defined.

Research shows differences in development and focus across policy areas. Policies around standards were adopted earlier in mathematics than in other subject areas, so research is now examining the second and third revisions of these policies. After initial general enthusiasm, the initial standards policies were criticized on grounds of form, content, and process. Later versions of policies reflected responses to those criticisms; policy research on mathematics education standards policies also developed, with more refined questions and methods. The result has been a clearer picture of these policies which, though varied, have been shown to have effects.

Policy research on mathematics instructional materials is less well developed. The policies themselves have been less consistent over time and across contexts. The research on these policies is thin, especially in characterizing the intended content and effectiveness in improving student learning. Many of the materials are developed by private companies who may wish to protect their proprietary interests.

In contrast, assessment policies have been widely studied, with mathematics often chosen as the focal subject area. Studies consider assessment policy in conjunction with standards and accountability mechanisms. This policy area shows evidence of research informing further policy development, though other considerations also play a major role in setting policy. Disputes about the desired outcomes lead to disagreements about what direction policy should take, but assessment policy seems likely to grow in importance in the coming years.

Policy makers of every political persuasion agree that teacher quality is a key policy area. Many studies have described, and often judged, the status of teacher quality policies across states. Research on formulation, implementation, and effects is scarcer. Many obstacles impede research on teacher quality policies: key variables are imprecisely defined, myriad factors other than these polices affect their implementation and effects, and the organizations that mediate the policies (e.g., institutions of higher education) are themselves varied and notoriously difficult to influence. The work in mathematics education is, again, more informative than that in other subject areas, but the common conclusion reviewers draw is that the research base is thin.

The literature describes cases, especially around standards and assessments, where research has informed mathematics education policy. Policy will never be based solely on research; values, professional judgment and political expediency will always have important roles, and are not always in line with results of research.

Better connections among research, policy, and practice can be built, however. Such connections can increase the chances that policies achieve their intended effects and that mathematics educators understand the ways in which they are affected by, and might be able to shape policies. Mathematics educators could also gain a better understanding of how policies and policy research are connected to goals to which various actors aspire.

Policy that affects mathematics education, when it has involved discipline-specific experts, relied primarily on mathematics educators in the period following the New Math, up until the mid 1990s, and thus tended to reflect the shared ideals and values of that community. In the past 10 years there has been a shift to widen the influence to include research mathematicians, whose ideals feature very strong commitment to the centrality of subject matter. Education policy researchers often take mathematics as the subject matter context for their investigations, but sometimes have a limited understanding of the ideals and values of both mathematics educators and research mathematicians. Greater collaboration among mathematics educators, mathematicians, and policy experts would be helpful in building a body of policy-relevant mathematics education research.

REFERENCES

Achieve, Inc. (2002). *Staying on course: Standards-based reform in America's schools: Progress and prospects.* Washington, DC: Achieve, Inc.

Achieve, Inc. (2004a). *Do graduation tests measure up? A closer look at state high school exit exams.* Washington, DC: Author.

Achieve, Inc. (2004b). *The expectations gap: A 50-state review of high school graduation requirements.* Washington, DC: Author.

Alper, L., Fendel, D., Fraser, S., & Resek, D. (1997). Designing a high school mathematics curriculum for all students. *American Journal of Education, 106*(1), 148–178.

American Association for the Advancement of Science. (1993). *Benchmarks for science literacy.* Washington DC: Author

American Association for the Advancement of Science (2000). *Project 2061 textbook evaluations.* Washington, DC: Author. Retrieved October 30, 2006, from http://www.project2061.org/publications/textbook/default.htm.

American Educational Research Association. (2003). Standards and tests: Keeping them aligned. *Research Points, 1*(1).

American Federation of Teachers. (2001). *Making standards matter 2001.* Washington, DC: Author.

American Federation of Teachers. (2003). *Where we stand: Standards-based assessment and accountability.* Washington, DC: Author.

Amrein, A., & Berliner, D. C. (2002). High-stakes testing, uncertainty, and student learning. *Education Policy Analysis Archives, 10*(18). Retrieved July 7, 2006, from http://epaa.asu.edu/epaa/v10n18/

Amrein-Beardsley, A., & Berliner, D.C. (2003). Re-analysis of NAEP math and reading scores in states with and without high-stakes tests: Response to Rosenshine. *Education Policy Analysis Archives, 11*(25). Retrieved July 20, 2006 from http://epaa.asu.edu/epaa/v11n25/

Ananda, S. (2003). *Rethinking issues of alignment under No Child Left Behind.* San Francisco: WestEd.

Apple, M. W. (2006). What can we learn from Texas about No Child Left Behind? *Educational Policy, 20*(3), 551–560.

Balfanz, R., MacIver, D. J., & Byrnes, V. (2006). The implementation and impact of evidence-based mathematics reform in high-poverty middle schools: A multi-site, multi-year study. *Journal for Research in Mathematics Education, 37*(1), 33–64.

Ballou, D., & Podgursky, M. (2000). Reforming teacher preparation and licensing: What does the evidence show? *Teachers College Record, 101*(1), 5–26.

Bartell, C.A., & Ownby, L. (1994). *Report on the implementation of the beginning teacher support and assessment program, 1992–1994.* Sacramento: California Commission on Teacher Credentialing and California Department of Education.

Berman, P., & McLaughlin, M. W. (1997). *Federal programs supporting educational change: Factors affecting implementation and continuation* (Vol. 7). Santa Monica, CA: RAND.

Blank, R. K., & Langesen, D. (2005) *State indicators of science and mathematics education 2005.*Washington, D.C.: Council of Chief State School Officers.

Braden, L., Finn, C. E. J., Lerner, L. S., Munroe, S., Petrilli, M. J., Raimi, R. A., et al. (2000). *The state of state standards 2000.* Washington, DC: The Thomas B. Fordham Foundation

Britton, E., Paine, L., Pimm, D., & Raizen, S.A. (Eds.). (2003). *Comprehensive teacher induction: Systems for early career learning.* Dordrecht, The Netherlands: Kluwer.

Cavalluzzo, L. C. (2004). *Is national board certification an effective signal of teacher quality?* Alexandria, VA: The CNA Corporation.

Center on Education Policy (2004). *State high school exit exams: A maturing reform.* Washington, DC: Author.

Cochran-Smith, M., & Zeichner, K. (Eds.). (2005). *Studying teacher education: The report of the AERA Panel on Research and Teacher Education.* Mahwah, NJ: Erlbaum.

Cognitive Science Laboratory. (2005). *WordNet 2.1.* Princeton, NJ: Princeton University. Retrieved October 30, 2006, from http://wordnet.princeton.edu/perl/webn.

Cohen, D., & Ball, D. (1990). Policy and practice: An overview. *Educational Evaluation and Policy Analysis, 12*(3), 233–239.

Cohen, D. K., & Hill, H. C. (2001). *Learning policy: When state education reform works.* New Haven, CT: Yale University Press.

Conference Board of the Mathematical Sciences (2001). *The mathematical education of teachers.* Washington, DC: American Mathematical Society and Mathematical Association of America.

Council of Chief State School Officers. (1995). *State education accountability reports and indicator reports: Status of reports across the states.* Washington, DC: Author.

Council of Chief State School Officers. (1997). *Mathematics and science content standards and curriculum frameworks: States progress on development and implementation.* Washington, DC: Author.

Council of Chief State School Officers. (1998). *Key state education policies on K–12 education: Standards, graduation, assessment, teacher licensure, time and attendance - A 50-state report.* Washington, DC: Author.

Council of Chief State School Officers. (2001). *State standards and state assessment systems: A guide to alignment.* Washington, DC: Author.

Council of Chief State School Officers. (2002). *Key state policies on PK–12 education: 2002.* Washington, DC: Author.

Council of Chief State School Officers. (2005). *Key state education policies in PK–12 education: 2004.* Washington, DC: Author.

Council of Chief State School Officers. (2006). *Interstate New Teacher Assessment and Support Consortium (INTASC)* Washington, DC: Author.

Darling-Hammond, L. (1994). National standards and assessments: Will they improve education? *American Journal of Education, 102*(4), 478–510.

Darling-Hammond, L. (1999). *Teacher quality and student achievement: A review of state policy evidence.* Seattle: University of Washington, Center for the Study of Teaching and Policy.

Darling-Hammond, L. (2000a). Teacher quality and student achievement: A review of state policy evidence. *Educational Policy Analysis Archives, 8*(1).

Darling-Hammmond, L. (2000b). Reforming teacher preparation and licensing: Debating the evidence. *Teachers College Record, 102*(1), 28–56.

Darling-Hammond, L., Berry, B., & Thoreson, A. (2001). Does teacher certification matter? Evaluating the evidence. *Educational Evaluation and Policy Analysis, 23*(1), 57–77.

Delfattore, J. (1992). *What Johnny shouldn't read: Textbook censorship in America.* New Haven, CT: Yale University Press.

Desimone, L., Garet, M. S., Birman, B. F., Porter, A., & Suk Yoon, K. (2003). Improving teachers' in-service professional development in mathematics and science: The role of postsecondary institutions. *Educational Policy, 17*(5), 613–649.

Desimone, L., Porter, A. C., Birman, B. F., Garet, M. S., & Suk Yoon, K. (2002). How do district management and implementation strategies relate to the quality of the professional development that districts provide to teachers? *Teachers College Record, 104*(7), 1265–1312.

Dow, P. B. (1991). *Schoolhouse politics: Lessons from the Sputnik era.* Cambridge, MA: Harvard University Press.

Dwyer, C. A., & Villegas, A. M. (1993). *Guiding conceptions and assessment principles for the Praxis series: Professional assessments for beginning teachers.* Princeton, NJ: Educational Testing Service.

Earley, P. M. (2000). Finding the culprit: Federal policy and teacher education. *Educational Policy, 14*(1), 25–39.

Earley, P. M., & Schneider, E. (1996). Federal policy and teacher education. In J. Sikula, T. Buttery, & E. Guyton (Eds.), *Handbook of research on teacher education* (2nd ed., pp. 306–319). New York: Simon & Schuster.

Editorial Projects in Education. (2004, January). *Quality counts 2004—Count me in: Special education in an era of standards. Education Week.*

Editorial Projects in Education (2005). *Quality counts 2005—No small change: Targeting money towards student performance. Education Week.*

Feiman-Nemser, S., Schwille, S., Carver, C., & Yusko, B. (1999). *A conceptual review of literature on new teacher induction.* East Lansing, MI: National Partnership for Excellence and Accountability in Teaching, Michigan State University College of Education.

Feistritzer, C. E., & Chester, D. T. (2003). *Alternative teacher certification: A State-by-state analysis, 2003.* Washington, DC: National Center for Education Information.

Ferrini-Mundy, J., Earle, J., & King, K. (2005). *Relevance, performance and quality in Mathematics education research and development: Implications from The National Science Foundation mathematics education portfolio review project.* Manuscript in preparation.

Ferrini-Mundy, J., & Gaudard, M. (1992). Secondary school calculus: Preparation or pitfall in the study of college calculus? *Journal for Research in Mathematics Education, 23*(1), 56–71.

Ferrini-Mundy, J., & Graham, K. J. (2003). The education of mathematics teachers in the United States after World War II: Goals, programs, and practices. In G. Stanic & J. Kilpatrick (Eds.), *A history of school mathematics.* (Vol. 2, pp. 1193–1308.) Reston, VA: National Council of Teachers of Mathematics.

Ferrini-Mundy, J., & Johnson, L. P. (1994). Implementing the Curriculum and Evaluation Standards: Recognizing and recording reform in mathematics: New questions, many answers. *Mathematics Teacher, 87*(3), 190–193.

Ferrini-Mundy, J., & Johnson, L. P. (1997). Building the case for standards-based reform. In D. M. Bartels & J. O. Sandler (Eds.), *Implementing science education reform: Are we making an impact?* (pp. 157–186). Washington, DC: American Association for the Advancement of Science.

Ferrini-Mundy, J., & Martin, G. (2003). The role of research in shaping NCTM's Principles and Standards. In J. Kilpatrick, D. Shifter, & G. Martin (Eds.), *A research companion to Principles and Standards for School Mathematics.* Reston, VA: National Council of Teachers of Mathematics.

Ferrini-Mundy, J., & Schram, T. (Eds.). (1997). *The Recognizing and Recording Reform in Mathematics Education Project: Insights, issues, and implications.* [JRME Monograph Series, No. 8]. Reston, VA: National Council of Teachers of Mathematics.

Finn, C. E., & Ravitch, D. (2004). *The mad, mad world of textbook adoption.* Washington, DC: Fordham Institute.

Firestone, W. A., Schorr, R. Y., & Monfils, L. F. (2004). *The ambiguity of teaching to the test: Standards, assessment, and educational reform.* Mahwah, NJ: Erlbaum.

Floden, R. E., & Meniketti, M. (2005). Research on the effects of coursework in the arts and sciences and in the foundations of education. In M. Cochran-Smith & K. M. Zeichner (Eds.), *Studying teacher education: The Report of AERA Panel on Research and Teacher Education* (pp. 261–308). Mahwah, NJ: Erlbaum.

Goertz, M. (2001). *The state context for implementing the No Child Left Behind Act,* Queenstown, MD: The Aspen Institute.

Goldhaber, D. D., & Anthony, E. (2004). *Can teacher quality be effectively assessed?* Washington, DC: Urban Institute.

Grissmer, D. W., & Kirby, S. N. (1987). *Teacher attrition: The uphill climb to staff the nation's schools* (Report No. R-3512-CSTP). Santa Monica, CA: The RAND Corporation.

Grissmer, D. W., & Kirby, S. N. (1997) Teacher turnover and teacher quality. *Teachers College Record, 99*(1), 45–56.

Grouws, D. (Ed.). (1992). *Handbook of research on mathematics teaching and learning.* New York: Macmillan.

Haertel, E. H. (1999). Validity arguments for high-stakes testing: In search of the evidence. *Educational Measurement: Issues and practice, 18*(4), 5–9.

Hanushek, E. A., Kain, J. F., & Rivkin, S. G. (2001). *Why public schools lose teachers.* Greensboro, NC: Smith Richardson Foundation.

Hill, H. C. (2001). Policy is not enough: Language and the interpretation of state standards. *American Educational Research Journal, 38*(2), 289–318.

Hirsch, E., Koppich, J. E., & Knapp, M. S. (2001, February). *Revisiting what states are doing to improve the quality of teaching: An update on patterns and trends.* Seattle: University of Washington, Center for the Study of Teaching and Policy.

Howey, K. (2000). *A review of challenges and innovations in the preparation of teachers for urban contexts: Implications for state policy.* Washington, DC: Office of Educational Research and Improvement, U.S. Department of Education.

Ingersoll, R. M. (1999a). *Teacher turnover, teacher shortages, and the organization of schools. A CTP working paper.* Seattle: University of Washington, Center for the Study of Teaching and Policy.

Ingersoll, R. M. (1999b). The problem of underqualified teachers in American secondary schools. *Educational Researcher, 28*(2), 26–37.

Ingersoll, R. & Smith, T. (2003). The wrong solution to the teacher shortage. *Educational Leadership, 60*(8), 30–33.

INTASC. (1995). *Model standards in mathematics for beginning teacher licensing & development: A resource for state dialogue.* Washington DC: Council of Chief State School Officers.

Jackson, A. (1997a). The math wars: California battles it out over mathematics education (Pt. 1). *Notices of the American Mathematical Society, 44*(6), 695–702.

Jackson, A. (1997b). The math wars: California battles it out over mathematics education (Pt. 2). *Notices of the American Mathematical Society, 44*(7), 817–823.

Jackson, A. (2004). Mathematicians' group to provide advice on math standards. *Notices of the American Mathematical Society, 51*(11), 1355–1356.

Jacob, B., & Akers, J. (2000) "Research based" mathematics education policy: The case of California 1995–1998. *International Journal for Mathematics Teaching and Learning.* Retrieved October 31, 2006, from http://www.internep.org/.

Jacobs, J. K., Hiebert, J., Givvin, K. B., Hollingsworth, H., Garnier, H., & Wearne, D. (2006). Does eighth-grade mathematics teaching in the United States align with the NCTM Standards? Results from the TIMSS 1995 and 1999 Video Studies. *Journal for Research in Mathematics Education, 37*(1), 5–32.

Johnson, S.M., & Kardos, S.M. (2002). Keeping new teachers in mind. *Educational Leadership, 59*(6), 12–16

Karp, K. (2005) *AMTE Survey.* Paper presented at the Michigan State University Education Policy Forum, Alexandria, VA.

Kher, N., Schmidt, W., Houang, R., & Wiley, D. E. (April, 2005). *The Heinz 57 curriculum: When more may be less.* Paper presented at the annual meeting of the American Educational Research Association, Montreal, Quebec, Canada.

Klein, D., Braams, B., Parker, T., Quirk, W., Schmidt, W., & Wilson, S. (2005). *State of the state math standards.* Washington, DC: Fordham Foundation.

Klopfenstein, K. (2004). The advanced placement expansion of the 1990s: How did traditionally underserved students fare? *Education Policy Analysis Archives, 12*(68). Retrieved July 6, 2006, from http://epaa.asu.edu/epaa/v12n68/

La Marca, P. M. (2001). Alignment of standards and assessments as an accountability criterion. *Practical Assessment, Research, and Evaluation, 7*(21). Retrieved July 21, 2006, from http://PAREonline.net/getvn.asp?v=7&n=21

Lappan, G., & Wanko, J. (2003). The changing roles and priorities of the federal government in mathematics education in the United States. In G. Stanic & J. Kilpatrick (Eds.), *A history of school mathematics.* pp. 897–930. Reston, VA: National Council of Teachers of Mathematics.

Lester, F. Jr., & Ferrini-Mundy, J. (Eds.). (2004). *Proceedings of the NCTM research catalyst conference, September, 2003.* Reston, VA: National Council of Teachers of Mathematics.

Lloyd, G. M., & Wilson, M. (1998). Supporting innovation: The impact of a teacher's conceptions of functions on his implementation of a reform curriculum. *Journal for Research in Mathematics Education, 29*(3), 248–274.

Lortie, D. C. (1975). *Schoolteacher: A sociological study.* Chicago: The University of Chicago Press.

Lott, J., & Nishimura, K. (Eds.). (2005). *Standards and curriculum: A view from the nation.* Reston, VA: National Council of Teachers of Mathematics.

Marchant, G. J., & Paulson, S. E. (2005). The relationship of high school graduation exams to graduation rates and SAT scores. *Education Policy Analysis Archives, 13*(6). Retrieved July 6, 2006, from http://epaa.asu.edu/epaa/v13n6/.

Martin, G., & Berk, D. (2001). The cyclical relationship between research and standards: The case of Principles and Standards for School Mathematics. *School Science and Mathematics, 101,* 328–339.

Mathematical Sciences Education Board. (1996). *The preparation of teachers of mathematics: Considerations and challenges—A letter report.* Washington, DC: Mathematical Sciences Education Board & National Research Council.

Mathematically Correct. (1999). *An open letter to United States Secretary of Education, Richard Riley.* Washington, DC: Washington Post. Retrieved July 21, 2006, from http://www.mathematicallycorrect.com/riley.htm.

McDonnell, L. M. (2005). No Child Left Behind and the federal role in education: Evolution or revolution? *Peabody Journal of Education, 80*(2), 19–38.

McLeod, D., Stake, R., Schappelle, B., Mellissinos, M., & Gierl, M. J. (1996). Setting the standards: NCTM's role in the reform of mathematics education. In S. A. Raizen & E. D. Britton (Eds.), *Bold ventures: U.S. innovations in science and mathematics education* (Vol. 3, pp. 12–132). Dordrecht, The Netherlands: Kluwer.

Mehrens, W. A. (1998). Consequences of assessment: What is the evidence? *Education Policy Analysis Archives, 6*(13).

Messick, S. (1989). Validity. In R. L. Linn (Ed.), *Educational measurement* (3rd ed., pp. 13–103). New York: American Council on Education and MacMillan Publishing.

Mitchell, R., & Barthe, P. (1999). How teacher licensing tests fall short. *Thinking K–16, 3*(1), 3–19.

Mitchell, D., Scott, L., Takahashi, S., & Hendrick, I. (1997, June). *The California Beginning Teacher Support and Assessment program: A statewide evaluation study.* Riverside, CA: California Educational Research Cooperative.

Monk, D. H. (1994). Subject matter preparation of secondary mathematics and science teachers and student achievement. *Economics of Education Review, 13*(2), 125–145.

Morgan, R., & Crone, C. (1993). *Advanced placement examinees at the University of California: An examination of the freshman year courses and grades of examinees in biology, calculus, and chemistry.* (Statistical Report 93–210). Princeton, NJ: Educational Testing Service.

Morgan, R., & Maneckshana, B. (2000). *AP students in college: An investigation of their course-taking patterns and college majors.* (Statistical Report 2000-09). Princeton, NJ: Educational Testing Service.

Morgan, R., & Ramist, L. (1998). Advanced placement students in college: An investigation of course grades at 21 colleges. (Statistical Report 98–13). Princeton, NJ: Educational Testing Service.

Moss, P. A. (April, 1998). *Response to Porter, Odden, & Youngs.* Paper presented at the annual meeting of the American Educational Research Association, San Diego, CA.

National Association of State Directors of Teacher Education and Certification. (1997). *Manual on certification and preparation of educational personnel in the United States and Canada, 1997–98.* Mashpee, MA: Author.

National Board for Professional Teaching Standards. (1998). *Middle childhood through early adolescence/mathematics standards (for teachers of students ages 7–15).* Washington, DC: Author.

National Board for Professional Teaching Standards. (2001). *Middle childhood through early adolescence/mathematics standards (for teachers of students ages 14–18+).* Washington, DC: Author.

National Center for Education Statistics. (1999). *SASS—Schools and Staffing Survey.* Washington, DC: U.S. Department of Education.

National Center for Education Statistics (2004). *Qualifications of the public school teacher workforce: Prevalence of out-of-field teaching 1987–88 to 1999–2000.* (No. NCES 2002-603 Revised). Washington, DC: U.S. Department of Education, Institute of Education Sciences. Retrieved October 30, 2006, from http://nces.ed.gov/pubs2002/2002603.pdf.

National Center for Education Statistics (2005). *International Education Indicators.* Washington, DC: U.S. Department of Education.

National Commission on Excellence in Education. (1983). *A nation at risk: The imperative for educational reform.* Washington, DC: U.S. Government Printing Office.

National Commission on Mathematics and Science Teaching for the 21st Century. (2000). *Before it's too late: A report to the nation from the National Commission on Mathematics and Science Teaching for the 21st Century.* Washington, DC: U.S. Department of Education.

National Commission on Teaching and America's Future. (1996). *What matters most: Teaching for America's future.* New York: Author.

National Commission on Teaching and America's Future. (2003). *No dream denied: A pledge to America's children.* New York: Author.

National Council on Education Standards and Testing. (1992). *Raising standards for American education. A report to Congress, the Secretary of Education, the National Education Goals Panel, and the American people.* Washington, DC: Author.

National Council of Teachers of Mathematics. (1989). *Curriculum and evaluation standards for school mathematics.* Reston, VA: Author

National Council of Teachers of Mathematics. (1991). *Professional teaching standards for school mathematics.* Reston, VA: Author.

National Council of Teachers of Mathematics. (1995). *Assessment standards for school mathematics.* Reston, VA: Author.

National Council of Teachers of Mathematics. (2000). *Principles and standards for school mathematics.* Reston, VA: Author.

National Research Council. (2001a). *Educating teachers of science, mathematics, and technology: New practices for the new millennium.* Washington, DC: National Academy Press.

National Research Council. (2001b). *Knowing and learning mathematics for teaching.* Washington, DC: National Academy Press.

National Research Council (2002a). *Investigating the influence of standards.* Washington, DC: National Academy Press.

National Research Council (2002b). *Learning and understanding: Improving advanced study of mathematics and science in U.S. high schools.* Committee on Programs for Advanced Study of Mathematics and Science in American High Schools. J. P. Gollub, M. W. Bertenthal, J. B. Labov, & P. C. Curtis Jr., Editors.. Center for Education, Division of Behavioral and Social Sciences and Education. Washington, DC: National Academy Press.

National Research Council. (2004). *On evaluating curricular effectiveness: Judging the quality of K–12 mathematics evaluations.* Committee for a Review of the Evaluation Data on the Effectiveness of NSF-Supported and Commercially Generated Mathematics Curriculum Materials. Washington, DC: The National Academies Press.

National Research Council, Committee on Research in Education. (2005). *Advancing scientific research in education.* Washington, DC: National Academies Press.

National Science Board. (1983). *Educating Americans for the 21st century.* Washington, DC: U.S. Government Printing Office.

National Science Board. (2006). *Science and engineering indicators 2006* (Vol. 1, NSB 06-01; Vol. 2, NSB 06-01A). Arlington, VA: National Science Foundation.

National Science Foundation. (1989). *Materials for middle school mathematics instruction.* Arlington, VA: Author.

National Science Foundation. Division of Research Evaluation and Communication. (2004). *The mathematics education portfolio brief* (No. NSF 05-03). Arlington, VA: Author.

Neuhaus, R (2003). *A preliminary report on the effects of the TAP program on the attraction and retention of fully credentialed teachers in API 1–5 schools.* (Planning, Assessment, and Research Division Publication No. 135): Los Angeles: Los Angeles Unified School District, Program Evaluation and Research Branch.

Oakes, J. (1985). *Keeping track: How school structure inequality.* New Haven, CT: Yale University Press.

Oakes, J. (1989). What educational indicators? The case for assessing the school context. *Educational Evaluation and Policy Analysis, 11*(2), 181–199.

O'Donnell, P. (2006). *Rising above the gathering storm: Employing America for a brighter economic future.* Statement for the Subcommittee on Education and Early Childhood Development, Committee on Health, Education, Labor and Pensions. Washington, DC. Retrieved October 30, 2006, from http://help.senate.gov/Hearings/2006_03_01_a/ODonnell.pdf

Porter, A. C. (2002). Measuring the content of instruction: Uses in research and practice. *Educational Researcher, 31*(7), 3–14.

Porter, A. C., Chester, M. D., & Schlesinger, M. D. (2004). Framework for an effective assessment and accountability program: The Philadelphia example. *Teachers College Record, 106*(6), 1358–1400

Porter, A., Floden, R., Freeman, D., Schmidt, W., & Schwille, J. (1988). Content determinants in elementary school mathematics. In D. A. Grouws, T. J. Cooney, & D. Jones (Eds.), *Effective mathematics teaching* (pp. 96–113). Reston, VA: National Council of Teachers of Mathematics.

Porter, A. C., Youngs, P., & Odden, A. (2001). Advances in teacher assessments and their uses. In V. Richardson (Ed.), *Handbook of research on teaching* (4th ed., pp. 259–297). Washington, DC: American Educational Research Association.

Raimi, R. A., & Braden, L. S. (1998). *State mathematics standards: An appraisal of math standards in 46 states, the District of Columbia, and Japan.* Washington, DC: Thomas B. Fordham Report.

RAND Mathematics Study Panel. (2003). *Mathematical proficiency for all students: Toward a strategic research and development program in mathematics education.* Santa Monica, CA: RAND Education.

Ravitch, D. (2005, June 20th). Ethnomathematics: Even math education is being politicized. *Wall Street Journal, p.* A14

Reys, B.J., Dingman, S., Olson, T.A, Sutter, A., Teuscher, D., Chval, K., et al. (2006). *The intended mathematics curriculum as represented in state-level curriculum standards: Consensus and confusion?* Columbia, MO: Center for the Study of Mathematics Curriculum, College of Education, University of Missouri-Columbia.

Robitaille, D. F., Schmidt, W. H., Raizen, S. A., McKnight, C. C., Britton, E., & Nicol, C. (1993). *Curriculum Frameworks for Mathematics and Science (TIMSS Monograph No. 1).*Vancouver, Canada: Pacific Educational Press.

Rosenshine, B. (2003, August 4). High-stakes testing: Another analysis. *Education Policy Analysis Archives, 11*(24). Retrieved August 9, 2005, from http://epaa.asu.edu/epaa/v11n24/.

Sanders, W. L., Ashton, J. J., & Wright, S. P. (2005). *Comparison of the effects of NBPTS certified teachers with other teachers on the rate of student academic progress.* Arlington, VA: National Board for Professional Teaching Standards.

Schmidt, W. H., McKnight, C., Houang, R. T., Wang, H., Wiley, D. E., Cogan, L. S., et al. (2001). *Why schools matter: A cross-national comparison of curriculum and learning.* San Francisco: Jossey-Bass.

Schmidt, W. H., McKnight, C. C., & Raizen, S. A. (1997). *A splintered vision: An investigation of U.S. science and mathematics education.* Dordrecht, The Netherlands: Kluwer.

Schmidt, W.H., McKnight, C.C., Valverde, G.A., Houang, R.T., & Wiley, D.E. (1997). *Many visions, many aims. Vol 1: A cross-national investigation of curricular intentions in school mathematics.* Dordrecht, The Netherlands: Kluwer.

Schmidt, W. H., Wang, H. C., & McKnight, C. C., (2005). Curriculum coherence: An examination of U.S. mathematics and science content standards from an international perspective. *Journal of Curriculum Studies, 37*(5), 525–559.

Sedlak, M. W. (1989). Let us go and buy a school master. In D. Warren (Ed.), *American teachers: History of a profession at work* (pp. 257–290). New York: Macmillan.

Seeley, C. L. (2003). Mathematics textbook adoption in the United States. In G. M. A. Stanic & J. Kilpatrick (Eds.), *A history of school mathematics* (Vol. 2, pp. 957–988). Reston, VA: National Council of Teachers of Mathematics.

Sfard, A. (2005). What could be more practical than good research? On mutual relations between research and practice of mathematics education. *Educational Studies in Mathematics, 58*(3), 393–413.

Shen, J. (1999) Alternative certification: Math and science teachers. *Educational horizons, 78*(1), 44–48.

Silver, E. A. (1990). Contributions of research to practice: Applying findings, methods, and perspectives. In T. J. Cooney (Ed.), *Teaching and learning mathematics in the 1990s* (pp. 1–11). Reston, VA, National Council of Teachers of Mathematics.

Smith, M. S., & O'Day, J. (1991). Systemic school reform. In S. H. Fuhrman & B. Malen (Eds.), *The politics of curriculum and testing: The 1990 yearbook of the Politics of Education Association* (pp. 233–267). London: Falmer Press.

Smith, T., & Ingersoll, R. (2004). What are the effects of induction and mentoring on beginning teacher turnover? *American Educational Research Journal, 41*(2), 681–714.

Smithson, J. L., & Porter, A. C. (2004). From policy to practice: The evolution of one approach to describing and using curriculum data. In M. Wilson (Ed.), *Towards Coherence Between Classroom Assessment and Accountability.* The 2004 yearbook of the National Society for the Study of Education. Chicago: University of Chicago Press

Special Study Panel on Education Indicators. (1991). *Education counts: An indicator system to monitor the nation's educational health.* Washington, DC: U.S. Department of Education, National Center for Education Statistics.

Spillane, J. P., & Zeuli, J. S. (1999). Reform and teaching: Exploring patterns of practice in the context of national and state mathematics reforms. *Educational Evaluation and Policy Analysis, 21*(1), 1–27.

Strong, M. (2005). Teacher induction, mentoring, and retention: A summary of the research. *The New Educator, 1*(3), 181–198.

Survey of Mathematics and Science Opportunities. (1995). *General topic trace mapping: Data Collection and processing* (Curriculum Analysis Technical Report Series No. 5). East Lansing, MI: U.S. TIMSS National Research Center, College of Education, Michigan State University.

U.S. Department of Education. (1994). *Goals 2000: Educate America Act.* Washington, DC: Author. Retrieved December 28, 2005, from http://www.ed.gov/legislation/GOALS2000/TheAct/index.html

U.S. Department of Education. (1997). *Individuals with Disabilities Education Act (IDEA) of 1997.* Washington, DC: Author. Retrieved November 9, 2005, from http://www.ed.gov/offices/OSERS/Policy/IDEA/the_law.html

U.S. Department of Education. (2002). *The No Child Left Behind Act of 2001.* Washington, DC: Author. Retrieved December 27, 2005, from http://www.ed.gov/policy/elsec/leg/esea02/index.html

U.S. Department of Education, Mathematics and Science Expert Panel. (1999). *Exemplary and promising mathematics programs,* Washington, DC: Author.

U.S. Department of Education. (n.d.). *Stronger accountability: Testing for results.* Retrieved July 21, 2006, http://www.ed.gov/nclb/accountability/ayp/testingforresults.html

U.S. General Accounting Office (1984) *New directions for federal programs to aid mathematics and science teaching.* Washington, DC: Author.

Usiskin, Z. (1997). Applications in the secondary school mathematics curriculum: A generation of change. *American Journal of Education, 106*(1), Reforming the Third R: Changing the School Mathematics Curriculum. (Nov. 1997), pp. 62–84.

Valverde, G. A., & Schmidt, W. H. (2000). Greater expectations: Learning from other nations in the quest for "world-class standards" in U.S. school mathematics and science. *Journal of Curriculum Studies, 32*(5), 651–687.

Veenman, S. (1984). Perceived problems of beginning teachers. *Review of Educational Research, 54*(2), 143–178.

Walsh, K. (2001). *Teacher certification reconsidered: Stumbling for quality.* Baltimore: Abell Foundation.

Webb, N. L. (1997). *Criteria for alignment of expectations and assessment in mathematics and science education.* (Research monograph, No. 6). Washington, DC: Council of Chief State School Officers.

Webb, N. L. (1999). *Alignment of science and mathematics standards and assessments in four states* (Research Monograph No. 18). Madison: National Institute for Science Education, University of Wisconsin, Madison.

What Works Clearinghouse, Institute for Education Sciences. Retrieved October 30, 2006, from http://www.whatworks.ed.gov/whoweare/overview.html.

Wilson, S. M. (2003). *California dreaming: Reforming mathematics education.* New Haven, CT: Yale University Press.

Wilson, S. M., & Floden, R. (2001). Hedging bets: Standards-based reform in classrooms. In S. H. Fuhrman (Ed.), *From the capitol to the classroom: Standards-based reform in the states* (pp. 193–216). Chicago: University of Chicago Press.

Wilson, S. M., Floden, R. E., & Ferrini-Mundy, J. (2001). *Teacher preparation research: Current knowledge, gaps, and recommendations* (Document R-01-3). Seattle: University of Washington, Center for the Study of Teaching and Policy.

Wilson, S. M., & Youngs, P. (2005). Research on accountability processes in teacher education. In M. Cochran-Smith & K. M. Zeichner (Eds.), *Studying teacher education: The Report of AERA Panel on Research and Teacher Education* (pp. 591–643). Mahwah, NJ: Erlbaum.

Wong, H. (2004). Producing educational leaders through induction programs. *Kappa Delta Pi Record, 40*(3), 106–111.

Wu, H. (1998). *Some observations on the 1997 battle of the two standards in the California math war.* Berkeley: University of California at Berkeley, Department of Mathematics.

Youngs, P., Odden, A., & Porter, A. (2003). State policy related to teacher licensure. *Educational Policy, 17,* 217–236.

AUTHOR NOTE

We appreciate the advice provided by Andrew Porter and Iris Weiss who reviewed early outlines and drafts of this chapter. We acknowledge the support of Jean Beland, Glenda Breaux, Kuo-Liang Chang, and Aaron Mosier in the preparation of this manuscript.

30

MATHEMATICS CONTENT SPECIFICATION IN THE AGE OF ASSESSMENT

Norman L. Webb

WISCONSIN CENTER FOR EDUCATION RESEARCH

We live in the age of assessment. At no other time in the history of education in the United States, and for that matter in other countries, have so many people, resources, and time been devoted to measuring students' knowledge and performance in mathematics. Assessment has come to be viewed by policy makers and others to be more a lever for reform than merely an indication of the effectiveness of reform. The loosely coupled assessment system pervades every classroom in every school in the nation, at least in grades 3–8 and high school.

The three other assessment chapters (by Wiliam, de Lange, and Wilson) in this handbook discuss formative assessments at the classroom level, and with regard to large-scale assessments and high stakes assessments. Collectively, their chapters address different purposes for assessments along with a number of other considerations for assessments. This chapter attends to a common issue, not directly addressed by the other assessment authors, that pertains to any purpose of assessment and the use of assessment by researchers. The mathematical content specification for an assessment needs to attend to the purpose for the assessment, but also needs to relate to the qualities of the assessment including its validity for making inferences from the results and its alignment to instruction and the curriculum. This chapter describes an approach for specifying the content for an assessment. The content specification criteria discussed here apply to assessments developed by classroom teachers, to large-scale assessments, and particularly to assessments used by researchers.

An assessment is one instrument frequently used by researchers to measure student knowledge and learning. The mathematical content measured by an assessment will greatly influence the inferences and conclusions drawn from the results. If the researcher is trying to determine students' knowledge of a mathematical construct, such as proportional reasoning, then the items that are included on the assessment can determine what inferences can be made. If the items only measure students' computation of percentages, then the results would produce information on a very limited part of the proportional reasoning domain. An assessment on proportional reasoning could include items measuring students' computation of rates and ratios, along with reasoning with ratios and proportional relationships (e.g., similar figures and rates). The complexity of the items included on the instrument also could vary, from requiring simple recall of information to solving nonroutine problems. How to construct, or select items, for an assessment designed to meet research needs and requirements requires consideration of a number of factors. Content specification is one of the most important. Of course, other factors also are important, including psychometrics—e.g., item difficulty, point-biserial correlations, differential item functioning (DIF) analysis, item format, and the length of the assessment.

Researchers who evaluate curriculum are faced with the issue of using assessments that are fair to all of the mathematics programs included in the analysis. An innovative curriculum can target content areas not normally included in traditional mathematical programs. One strategy is to use an existing instrument, such as a state assessment or national test. However, these instruments will very likely not capture all that students are learning in a given curriculum, including both traditional and innovative programs. Content analysis is one technique that can be used to reveal what content is covered in any assessment and how that coverage compares to the content emphases of the curricula under investigation. Knowing how content has been specified for existing assessment instruments can be very helpful in determining whether the assessment is appropriate for a given application.

In the age of assessment, research is being increasingly reported at the policy level. Student performance on state assessments is being compared to state scores on the National Assessment of Educational Progress (NAEP). Countries and states are being compared on international tests, such as the Third International Mathematics and Science Study (TIMSS) and *Programma for International Student Assessment* (PISA). Achievement gaps between different groups within a population are being studied and reported. Only rarely do such policy studies or other research studies use the mathematical content on the assessments as a basis for drawing conclusions. Sometimes such studies may report on mathematical topics included on the assessments, but do not report on the range of content assessed under these topics, or the complexity of the content measured. A language for content specification can facilitate such a discussion by helping researchers focus on what mathematics is being assessed on different instruments, to what depth that mathematics is being assessed, and how the emphasis on different mathematical topics varies among instruments.

Assessment has assumed increasing importance to schools and teachers with the advent of the *No Child Left Behind Act* in 2003 (see Chapter 24). Assessment results are being used to make decisions on the effectiveness of schools and to identify "failing" schools. And, beginning in 2005–2006, mathematics and reading are the two content areas in which states are to assess every student in grades 3–8 and once in high school each year. These tests are to be aligned to state standards. States that do not meet all of the requirements, as judged by peer review, are subject to the loss of millions of dollars of Title I funding. As a consequence, the annual state mathematics assessments have become very high stakes tests.

Schools faced with sanctions for poor results on the annual state assessments are incorporating benchmarks or other periodic assessments as one means of monitoring students' progress in learning what is specified in state standards. These benchmark tests, which are administered one to six times during the school year, are to measure whether students learn the materials covered within a six-week period, or the extent to which students are progressing on content tested on the state assessment. Along with monitoring students' progress, teachers are encouraged to use formative assessments (see Chapter 23) to reveal more about how students are understanding mathematics and what instructional interventions students need to further their understanding.

International comparisons of mathematics performance also have increased in the past decade. Whereas, nearly 20 years elapsed between the First and Second International Mathematics and Science Studies sponsored by the International Association for the Evaluation of Educational Achievement (IEA), assessments for the Third International Mathematics and Science Survey (TIMSS) were administered in grades 4 and 8 three times in eight years (1995, 1999, and 2003). This time span overlapped the administration of the Organisation for Economic Co-operation and Development's (OECD) *Programma for International Student Assessment* (PISA) administered in 2000 and 2003 in a number of nations (see Chapter 25).

Policy makers and politicians are turning more often to assessments as a means for strengthening education and making schools, teachers, and students more accountable for student learning. Although large-scale assessments can be expensive to develop, administer, and score, the cost to mandate an assessment as a graduation requirement is far less than most of the alternatives—for example, teacher professional development, curriculum development, and reduced teacher/student ratios. But in the haste to comply with federally required timelines or state legislation to have every student be assessed by an end-of-year assessment or that students should pass an assessment to graduate from high school, the timeline to develop such assessments is circumvented or compressed beyond what best practice for assessment development requires. In particular, two years is a minimum time for developing a large-scale test because creation of such a test requires developing a test blueprint, selecting or developing at least twice the number of tasks that will be needed, piloting these tasks, discarding or refining the tasks, creating a test form, field testing the test form, conducting bias and alignment studies, refining the tasks and the test form, and, finally, field testing the test at that time during the school year at which the test will be administered. However, in the United States, state assessment directors are being given less than one year to come up with tests that are to be administered statewide. This pressure comes in part from the number of assessments that need to be developed by states because of increasing the number of grades being assessed from three to seven and the number of different groups that are to be assessed including the regular population, English as a second language students, and students with disabilities. This pressure also is imposed by unrealistic expectations of state legislators who too quickly require students to pass a test for graduation with the intent of raising expectations that will lead to improve learning by students.

The push to "politicize" assessments by holding schools and teachers more accountable for the results on the assessments and the urgency to impose such accountability systems on states and schools has resulted in less attention given to general principles that are generally applicable to any form of assessment.[1] A paper-and-pencil test is one of a variety of assessment forms. Others include making observations, analyzing student work samples, and interviewing students. Strong pressures on teachers to meet accountability requirements and to avoid sanctions can result in the assessed curriculum becoming the enacted curriculum (Porter & Smithson, 2001a, 2001b). Under such pressures, teachers can be forced to make their classroom assessments (and even their classroom instruction) conform more with those used for the accountability system (generally, multiple-choice items) and have their instruction restricted to what is measured on the large-scale high stakes assessments. What can happen under such pressure is that teachers may give less attention to how individual students are developing their mathematical understanding, may not use assessment techniques that reveal students' thinking, may make decisions on too little information (e.g., one or two assessment items), may not be able to rectify different results from different sources of information, and may disregard the mathematics that is being assessed and how this mathematics relates to the mathematics students should be learning. In short, the use of such high-stakes assessments can have very unfortunate effects on teaching practices.

Assessment is a process for making inferences about what students know and can do related to a content domain. In theory, any assessment produces only a sample of what a student knows and can do. Unless the content domain, a body of knowledge, is very narrow, any assessment is too limited to measure all possible ways students may be able to demonstrate their knowledge of the content domain. Results from an assessment then produce some information about what a student knows. How much confidence an educator can have in this information for inferring what a student knows about a content domain will depend on qualities of the assessment, such as reliability and validity. A clear statement of the content domain is necessary to establish content validity for an assessment. Without specifying the content in some detail, making inferences about what a student knows or can do in relation to a content domain is very difficult. Clear content specifications are an essential component for assessment. In what follows, I discuss specific criteria that can be used in specifying content for an assessment.

CONTENT SPECIFICATION

Central to any form of assessment is specifying what knowledge of mathematics or what mathematical skills should be measured (Webb, 2006). Content specification is an issue at each level of education, from that of a teacher inquiring about what a student knows and can do to that of assessment developers creating a national or international test. At each level, a number of choices have to be made to decide what assessment tasks to include in order to produce both reliable and valid information about a student, or a group of students. Frequently, inferences are made about what students know and can do that are beyond the capacity of the assessment instrument to provide. In part, this is because those making the decisions do not fully understand what content knowledge is being assessed and how the assessment instrument was constructed. An item analysis of a large-scale assessment can result in teachers making instructional decisions about students based solely on their responses to one item, which is too restricted a sample to indicate much beyond the results of the individual item. States also try to make multiple uses of a single assessment, such as measuring the expected standards for K–12 and predicting performance in higher education. How the content is specified for different assessments can be very revealing about those inferences that can and cannot be made from students' results.

Content specifications need to be shaped by the structure of the content knowledge being assessed. Mathematics has domains of knowledge and inquiry that are interconnected, but generally are referred to as separate areas of study; for example, school math-

[1] Here assessment is used in its broadest sense as gathering information about student learning to make decisions.

ematics is generally broken down into topics such as number, geometry, measurement, algebra, data analysis, probability, and statistics. One starting point for specifying content for a mathematics assessment is to identify what the major mathematical topics are that should be represented on the assessment and what emphasis should be given to each. But these are broad topics that need further delineation. Not only do assessment developers need to know more about what subtopics are to be assessed under each topic, but they need other information, such as the level of complexity that should be represented on the assessment, the form of the assessment tasks (multiple-choice, short-answer, constructed-response), and whether items should measure isolated skills, or assess how students have integrated their understanding across skills. Also, assessment developers should know whether the results from the assessment will be reported by topic, as a total score, or in some other way. Even a teacher who wants to determine what a student knows about a very isolated topic, such as addition of whole numbers, needs to specify in some way the domain of knowledge the student is to know and what the indicators are that the student either has this knowledge, or is progressing toward acquisition of this knowledge. Content specifications then need to be based on the structure of the content and an understanding of how students acquire greater sophistication in mathematics. That is, content specifications should be theory based along with assessment in general (Webb, 1992).

For a large-scale assessment, content specifications take the form of a test blueprint or assessment framework. For classroom assessment, content specifications are presented more informally, but should be tied to the content standards and expectations, as well as to a theory of learning. For both large-scale and classroom assessment, understanding how students learn and how students demonstrate their learning should influence the specification of content for an assessment (National Research Council, 2001).

In the late 1940s, Ralph Tyler introduced into curriculum evaluation the concept of considering multiple dimensions when making a judgment about curriculum effectiveness (Tyler, 1949). In 1956, Benjamin Bloom, a student of Tyler, and his colleagues produced a taxonomy for the purpose of specifying behavioral levels that could be used to evaluate the sophistication of educational goals and assessments (Bloom, Engelhart, Furst, Hill, & Krathwohl, 1956). This was at the time that behavioral psychology was a dominant view of learning and the mathematics education community embraced Bloom's Taxonomy in designing assessments and curriculum. The content-by-behavior assessment matrix of the National Longitudinal Study of Mathematical Abilities (NLSMA), a project of the School Mathematics Study Group, was greatly influenced by Bloom's work. This matrix listed mathematical topics as the rows (number, geometry, measurement, and algebra) and behavioral levels as columns (computation, knowledge, application, and problem solving). Assessment content is then specified by the percent of the items that should correspond to each cell (number by computation, geometry by application, etc.). But with the rise of cognitive psychology, greater attention began to be given to how students develop relationships among mathematical ideas and how their mental schemas form through mental construction, rather than by rote stimulus and response. Content-by-behavior matrices became less frequently used in structuring assessments. For example, the writers of the National Council of Teachers of Mathematics (NCTM) *Curriculum and Evaluation Standards for School Mathematics* (NCTM, 1989) avoided specifying a content-by-behavior matrix in the assessment component, but rather focused on different types of knowledge. The argument against using a content-by-behavior matrix was that mathematical knowledge should not be pigeonholed into discrete cells (comprehension—number, application—geometry, etc.), but rather should be assessed by activities that extend across cells. That is, an assessment item or task may require a student to use number and geometry, along with recall of facts and strategic thinking. Or, assessments may be structured as a cluster of tasks all building on a single context. Similarly, the National Assessment of Educational Progress framework for the 1990–2003 incorporated the *Curriculum and Evaluation Standards* and used a content-by-abilities matrix in its assessment framework (National Assessment Governing Board, 1996). In this framework five content strands was one dimension while three mathematical abilities served as the second dimension (see Table 30.1).

With the advent of systemic reform in the 1990s and the need to consider the relationships among components of the educational system, the concept of alignment grew in importance. Systemic reform is based on the premise that in order to produce lasting increases in student performance it is necessary to change a multiplicity of components simultaneously, rather than focusing on only one component such as professional development, or curriculum, or assessment. As such, alignment can be defined as the degree to which expectations and assessments are in agreement and serve in conjunction with one another to guide an education system toward students learning what they are expected to know and do (Webb, 1997).

Table 30.1 Content-by-Abilities Matrix Used for the 1990–2003 National Assessment of Educational Progress (NAEP) Mathematics Framework

Content Strands	Conceptual understanding	Procedural knowledge	Problem solving
Number sense, properties, and operations			
Measurement			
Geometry and spatial sense			
Data analysis, statistics, and probability			
Algebra and functions			

The notion of alignment was central to systemic reform and became a critical requirement in the reauthorization of Title I in 1994, requiring states to show that assessments used in monitoring the progress of students served by Title I are aligned with standards. The national *No Child Left Behind Act* requires that all states develop challenging academic content and student achievement standards and tests that are aligned to these standards in the core academic areas of reading/language arts, mathematics, and science (U.S. Department of Education [USDE], 2003). In the establishment of *No Child Left Behind* legislation, alignment of assessments and standards, as judged by a peer-review process, was instituted as one of the key requirements states had to meet to get approval by the USDE.

In addition to the stature it has been given by the USDE, the alignment of expectations for student learning with assessments for measuring students' attainment of these expectations is an essential attribute for an effective standards-based education system. Alignment is a quality of the relationship between expectations and assessments and not an attribute of any one of these two system components. Alignment describes the match between expectations and assessments that can be legitimately improved by changing either student expectations or the assessments. As a relationship between two or more system components, alignment is determined by using the multiple criteria described in detail in a National Institute for Science Education (NISE) research monograph, *Criteria for Alignment of Expectations and Assessments in Mathematics and Science Education* (Webb, 1997).

In standards-based education, alignment criteria are additional considerations for assessment frameworks. Test developers at all levels need to consider the coherence of the system and how the assessments relate to standards, curriculum, and instruction. Five criteria help specify attributes of assessments that are considered aligned with standards and other system components:

Categorical Concurrence
Depth-of-Knowledge Consistency
Range-of-Knowledge Correspondence
Balance of Representation
Structure-of-Knowledge Comparability

Categorical Concurrence

An important aspect of alignment between content standards and assessments is whether both address the same content categories. The categorical-concurrence criterion provides a very general indication of alignment if both documents incorporate the same content. The criterion of categorical concurrence between domains and an assessment is met if the same or consistent categories of content appear in both documents. This criterion is judged by determining whether an assessment includes items measuring content from each content standard. The number of items that should be included on an assessment to produce some information about a student's performance on a content standard or domain of knowledge is primarily a question of reliability. How many items should a test include to produce reliable inferences from the student's responses about the student's proficiency on a content standard or domain of knowledge? Other important factors for consideration, such as breadth of coverage with a sufficient level of complexity, are judged by the other alignment criteria.

The minimum number of items or tasks that should be used to make judgments about a student's knowledge and proficiency of a content standard or domain of knowledge depends on the level of precision of the judgments. In general, the greater the number of items, the more precise will be the judgments. However, tests and other means for gathering information on student knowledge are constrained by time and other factors. An argument can be made for six items as being the minimum acceptable level for meeting the *Categorical Concurrence* alignment criterion if an assessment is to produce reliable information about what a student knows within a content standard. The number of items, six, is based on estimating the number of items that could produce a reasonably reliable subscale for estimating students' mastery of content on that subscale. Of course, many factors have to be considered in determining what a reasonable number is, including the reliability of the subscale, the mean score, and cutoff score for determining mastery. Using a procedure developed by Subkoviak (1988) and assuming that the cutoff score is the mean and that the reliability of one item is .1, it is estimated that six items would produce an agreement coefficient of at least .63. This indicates that about 63% of the group would be consistently classified as masters (a designated number of items scored correctly as determined by some process to represent a satisfactory level of attainment) or nonmasters, if two equivalent test administrations were employed. The agreement coefficient would increase if the cutoff score were increased to one standard deviation from the mean to .77 and, with a cutoff score of 1.5 standard deviations from the mean, to .88. Usually states do not report student results by content domains, or require students to achieve a specified cutoff score on subscales related to a domain. If a state did do this, then the state would seek a higher agreement coefficient than .63. But, six items are assumed as a minimum for an assessment measuring content knowledge related to a content standard or domain and as a basis for making some decisions about students' knowledge of that domain. If the mean for six items is 3 and one standard deviation is one item, then a cutoff score set at 4 would produce an agreement coefficient of .77. Any fewer items with a mean of one-half of the items would require a cutoff that would only allow a student to miss one item. This would be a very stringent requirement, considering a normal standard error of measurement on the subscale (Downing & Haladyna, 2006).

A minimum number of tasks or amount of information also applies to classroom assessments. Teachers, through probing a student's thinking, can gain insight into the student's conception of a mathematical idea or how the student may be drawing a relationship among mathematical ideas. However, for a teacher to make a generalization about a student's conception to all instances of the mathematical idea or to a domain of knowledge, the teacher will need to observe or interact with the student on a number of occasions and with different examples of the same mathematical idea or ideas to assure that any conclusions reached about the student's knowledge are valid. A teacher will have increasing confidence about a student's mathematical knowledge and performance with additional observations in different settings.

Depth-of-Knowledge Consistency

Content standards and assessments can be aligned not only on the category of content covered by each, but also on the basis of the complexity of knowledge required by each. Depth-of-knowledge consistency between content standards and assessment indicates alignment if what is elicited from students on the assessment is as demanding as what students are expected to know and do as stated in the standards. Four depth-of-knowledge levels are defined to compare the content complexity of content standards with the complexity of an assessment. The definitions were influenced by the Bloom Taxonomy and other schemes for considering content complexity, but are not the same. Bloom's Taxonomy was created at a time when behaviorism was a dominant psychological theory and it strongly reflects this perspective in portraying levels of knowledge. The depth-of-knowledge levels are designed to be used in content analyses and to portray content complexity that is related to cognitive processing (Webb, 1997). The judgment is based on a content analysis rather than on a psychological analysis. Thus, the cognitive processing is inferred from an analysis of the structure of content and the knowledge that students have at a given grade level. For the depth-of-knowledge levels to be verified as cognitive levels would require significant research in cognitive laboratories to determine whether, in fact, these levels represent meaningful degrees in mental functioning. As the definition now exists, there clearly is some relationship between cognition and content complexity, but the main use of the definitions is to compare content descriptions and to use this analysis to make some judgment of the comparability between the complexity of the expectations for student learning and what is required by the assessment.

Interpreting and assigning depth-of-knowledge levels to both standards within domains and assessment tasks is based on the degree to which the standards expect and the tasks require students to process mathematical ideas, from very routine algorithms to non-routine problem solving (Webb, 1999). The descriptions of the levels for mathematics are:

Level 1 (*Recall*) includes the recall of information such as a fact, definition, term, or a simple procedure, as well as performing a simple algorithm or applying a formula. That is, in mathematics, a one-step, well defined, and straight algorithmic procedure should be included at this lowest level. Other key words that signify Level 1 include "identify," "recall," "recognize," "use," and "measure." Verbs such as "describe" and "explain" could be classified at different levels, depending on what is to be described and explained.

Level 2 (*Skill/Concept*) includes the engagement of some mental processing beyond a habitual response. A Level 2 assessment item requires students to make some decisions as to how to approach the problem or activity, whereas Level 1 requires students to demonstrate a rote response, perform a well-known algorithm, follow a set procedure (like a recipe), or perform a clearly defined series of steps. Keywords that generally distinguish a Level 2 item include "classify," "organize," "estimate," "make observations," "collect and display data," and "compare data." These actions imply more than one step. For example, to compare data requires first identifying characteristics of the objects or phenomenon and then grouping or ordering the objects. Some action verbs, such as "explain," "describe," or "interpret," could be classified at different levels depending on the object of the action. For example, interpreting information from a simple graph, requiring reading information from the graph, also is at Level 2. Interpreting information from a complex graph that requires some decisions on what features of the graph need to be considered and how information from the graph can be aggregated is at Level 3. Level 2 activities are not limited only to number skills, but can involve other sorts of skills (e.g., visualization skills and probability skills). Other Level 2 activities include noticing and describing non-trivial patterns, explaining the purpose and use of experimental procedures; carrying out experimental procedures; specifying an algebraic equation and solution; making observations and collecting data; classifying, organizing, and comparing data; determining congruent figures; and organizing and displaying data in tables, graphs, and charts.

Level 3 (*Strategic Thinking*) requires reasoning, planning, using evidence, and a higher level of thinking than the previous two levels. In most instances, requiring students to explain their thinking is at Level 3, as are activities that require students to make conjectures. The cognitive demands at Level 3 are complex and abstract. The complexity does not result from the fact that there are multiple answers, a possibility for both Levels 1 and 2, but because the task requires more demanding reasoning. An activity, however, that has more than one possible answer and requires students to justify the response they give would most likely be at Level 3. Other Level 3 activities include drawing conclusions from observations; citing evidence and developing a logical argument for concepts; explaining phenomena in terms of concepts; and using concepts to solve problems.

Level 4 (*Extended Thinking*) requires complex reasoning, planning, developing, and thinking, most likely over an extended period of time. The extended time period is not a distinguishing factor if the required work is only repetitive and does not require applying significant conceptual understanding and higher-order thinking. For example, if a student has to take the water temperature from a river each day for a month and then construct a graph using the data gathered, this would be classified at Level 2. However, if the student is to conduct a river study that requires taking into consideration a number of variables, this would be at Level 4. At Level 4, the cognitive demands of the task should require analyzing and making inferences by considering a number of sources. Work could entail developing an original model for representing a problem with a number of facets. Students should be required to make several connections—to relate ideas within the content area or among content areas—and to select one approach among many alternatives on how the situation should be solved, in order to be at this highest level. Level 4 activities include developing and proving conjectures; designing and conducting experiments; making connections between a finding and related concepts and phenomena; combining and synthesizing ideas into new concepts; and critiquing experimental designs.

Depth-of-knowledge of an assessment task is different from the difficulty of that task. Difficulty of an item is a psychometric term that is represented by the percent of students who answer the item correctly. The difficulty of an item is related to depth-of-knowledge, but is also determined by other factors, such as the students' opportunity to learn and repetitive actions. It is conceivable that an item at depth-of-knowledge (DOK) Level 1, such as "Recall the first 20 digits of pi," is difficult, whereas an item that requires strategic reasoning, but of a type for which students have had extensive instruction, is actually less difficult.

Depth-of-knowledge is as relevant to classroom assessment as to large-scale assessment. For teachers to make judgments about students' mathematical knowledge requires them to assure that the tasks used for students to demonstrate their knowledge is at an appropriate depth-of-knowledge level. Within a standards-based system, the appropriate level is defined by the content standards. In other systems, the appropriate level could be defined by the expected rigor, or factors used to determine which students have the requirements for advancement or certification.

Range-of-Knowledge Correspondence

For assessments to be considered aligned with content standards, the assessments need to cover an adequate breadth of content, as expressed in the standards, in addition to addressing the categories and depth-of-knowledge. The range-of-knowledge correspondence criterion is used to judge whether a comparable span of knowledge expected of students by a standard is the same as, or corresponds to, the span of knowledge that students need in order to correctly answer the assessment items. The criterion for correspondence between span of knowledge for a content standard and an assessment considers the proportion of more refined delineations of content that are addressed on the assessment. Most content standards are further described by objectives or benchmarks. The range-of-knowledge correspondence criterion then considers the proportion of the objectives or benchmarks under a standard that has at least one corresponding item. In alignment studies, an assessment is considered to have attained a minimum acceptable level for range-of-knowledge correspondence if it has items corresponding to at least 50% of the objectives under a standard.

The concept of range applies to the relation between an assessment and other statements of content besides content standards. In any specification of a content domain of knowledge—e.g., operations with whole numbers, problems requiring proportional reasoning, or mathematical proof—the content can be further described by more specific content subdomains. For example, operations with whole numbers can be divided into addition with whole numbers, subtraction, multiplication, and division. The 2005 mathematics framework for the National Assessment of Educational Progress (NAEP) identifies five main content areas:

- number properties and operations,
- measurement,
- geometry,
- data analysis and probability, and
- algebra.

Each of these content areas is further defined by more refined content topics. For example, number properties and operations include:

- represent numbers,
- order numbers,
- compute with numbers,
- make estimates appropriate to given situations,
- use ratios and proportional reasoning, and
- apply number properties and operations to solve real-world and mathematical problems.

An acceptable range for NAEP would exist if the assessment included items from all six of these areas under Number Properties and Operations because the purpose of NAEP is to produce information on the performance of students in the states and the nation as a whole on the full spectrum of mathematical content. Creators of more limited assessments, constrained by the time available for testing and the number of possible items, have to make choices as to the number of the subtopics that can be addressed by an assessment. What proportion of the subtopics should be the minimal acceptable level for range will then depend on how student results will be reported, the purpose of the assessment, the number of subtopics, and the priorities assigned to topics. A test developed by a teacher, for example, to determine what students learned from a unit of instruction may only need to have items that related to number computations and estimation.

Range-of-knowledge correspondence is more difficult to attain if the content expectations are partitioned among a greater number of domains and a larger number of objectives. If a content standard or topic is further defined by a long list of very detailed and narrow content topics, then a greater number of items or more compound items (those that assess more than one mathematical idea) are needed to address an adequate proportion of these very detailed topics for an assessment to be considered as having sufficient range as expressed by the expectations. Very detailed statements of standards have been a problem for some states attempting to show that their assessments meet the alignment criteria of range. Because states are required to show that their assessments measure an adequate breadth of content as expressed by the state's content standards, this is a greater challenge if the five or six mathematics standards are delineated in a long list of objectives. Some states had developed standards in the mid 1990s to serve as curriculum guides for teachers. The intent of these curriculum standards was to give teachers as much detail as possible to help them make instructional decisions for students. The *No Child Left Behind* mandate, however, requires states to be held accountable for assuring that their assessments cover an adequate breadth of content, as expressed in their standards. As such, the curriculum standards are being used as accountability documents, not simply as curriculum guides. Standards serving as an accountability document do not require the detail in specifying what content students should be able to know and do. As an accountability document, standards can articulate in general terms the expected knowledge and performance, while leaving the detailed specifications for other documents. Moreover, standards used for accountability should contain sufficient detail to guide policy development, assessment, and curriculum. Then the more detailed description of content, and precisely what students should know and do, can be explained at the required level of specificity in curriculum frameworks or assessment blueprints.

Balance of Representation

In addition to comparable depth and breadth of knowledge, aligned content standards and assessments require that knowledge be distributed equally in both. The range-of-knowledge correspondence criterion only considers the number of objectives or subtopics within a content standard or domain; it does not take into consideration how the assessment items or activities are distributed among these standards. The balance-of-representation criterion is used to indicate the degree to which one objective or subtopic is given more emphasis on the assessment than another and whether this is appropriate.

Not all mathematical content should be given the same emphasis across the grades. Number and operations generally would have greater emphasis than other topics in the earlier grades, while algebra and geometry would increase in emphasis in the higher grades. Also, within these broad mathematical areas, content topics will vary in importance, a factor that should be reflected on an assessment designed to measure students' mathematical knowledge as expected by curriculum standards or by other forms of expectations. Generally, assessment frameworks, such as NAEP, will specify how items should be distributed among the major topics or content areas (see Table 30.2). Similar attention should be given to constructing a classroom assessment or evaluating a student's progress in learning. Not all mathematical content and ideas have the same degree of importance or the same number of attributes to warrant greater emphasis or a greater number of assessment tasks. Assessments should reflect these variations in importance and complexity and have appropriate balance of representation as reflected in the content standards and expectations.

Structure-of-Knowledge Comparability

The underlying conception of mathematics knowledge in expectations and assessments should be in agreement. For example, if standards indicate that students should understand mathematics "as an integrated whole" (National Council of Teachers of Mathematics, 1989), then the assessment activities should be directed toward the same ends. Both expectations and assessments should embody similar requirements for how students are to draw connections among ideas. Assessment of knowledge only as fragmented skills, for example, would not be consistent with the conception of how students should understand mathematics through the relationship among ideas as expressed in NCTM's *Standards* documents (1989, 1995, 2000).

The preliminary draft of the Illinois Academic Standards (Illinois State Board of Education, 1996) stated, "Solving problems is at the heart of mathematics. Mathematics is a collection of concepts and skills; it is also a means of investigation, reasoning, and communicating" (p. 25). One of five applications of learning in this document was "Making Academic Connections: Recognize and apply connections of important information and ideas within and among academic learning areas" (p. 26). The depiction of mathematics in these statements implies that students should learn a collection

Table 30.2 Distribution of Mathematics Questions as Specified by the NAEP 2005 Mathematics Framework (National Assessment Governing Board, 2004)

<table>
<tr><th>Content Areas</th><th>Grade 4 Target</th><th>Grade 8 Target</th><th>Grade 12 Target</th></tr>
<tr><td>Number properties and operations</td><td>40</td><td>20</td><td>10</td></tr>
<tr><td>Measurement</td><td>20</td><td>15</td><td rowspan="2">30*</td></tr>
<tr><td>Geometry</td><td>15</td><td>20</td></tr>
<tr><td>Data analysis and probability</td><td>10</td><td>15</td><td>25</td></tr>
<tr><td>Algebra</td><td>15</td><td>30</td><td>35</td></tr>
</table>

* Measurement and geometry are combined for grade 12.

of concepts and skills along with learning how to solve problems and applying information to other fields. An assessment system aligned with a comparable structure of knowledge as represented by these expectations would have to gather information on students' understanding of specific concepts and skills, their ability to solve problems, and their application of mathematics to other academic learning areas.

Judging the structure of knowledge comparability, along with many of the other criteria, requires considering the full assessment system. Expectations and assessment are fully aligned according to this criterion if the structure of mathematical understanding expressed in expectations is represented in the assessment system. In the case of judging whether its assessment system is aligned with the Illinois Academic Standards, some part of the state system should gather information on how students are able to identify appropriate applications of mathematical ideas both within mathematics and in other fields. For example, students could be asked to develop a mathematical model of a physics experiment. There would be less alignment with the Illinois Academic Standards if little or no part of the assessment system asked students to demonstrate knowledge of how they are able to draw relationships among mathematical ideas and among mathematical ideas that have applications to other fields. The standards and assessments will show very low alignment in the structure of knowledge if students are only asked to demonstrate their knowledge of mathematics by recalling or applying isolated concepts, procedures, and skills.

As with the other alignment criteria, the structure-of-knowledge comparability applies to classroom assessments as well as to large-scale assessments. In determining what mathematics students know and can do, teachers need to consider the interaction among mathematical ideas and how students should make sense of these. In a standards-based system, the curriculum standards should communicate or imply how students are to understand the structure of mathematical ideas. When using other forms that indicate expectations, such as a curriculum document or textbook, teachers may need to probe deeper in order to understand the underlying connections among mathematical ideas and how students are to build a more sophisticated understanding of mathematics.

Structure-of-knowledge comparability also is related to the theory of learning that is implied in the curriculum and the expectations for learning. Such a theory of learning needs to clarify how students' knowledge of the content area develops, how their knowledge deepens through forming relationships among the concepts and skills in the content area, and the pathways students can use to achieve competence in the content area (Webb, 2006). A fully aligned assessment will represent the underlying theory of learning and produce information that is comparable to this theory and how students are to know mathematics.

Beware of Using a Single Assessment for Multiple Purposes

The five alignment criteria I have discussed apply to all forms of assessments as discussed by de Lange, Wiliam, and Wilson (in this volume). Alignment is critical in a high-stakes standards-based system, as discussed by Wilson. In other high stakes environments such as selecting students for post-secondary education or in competitions, alignment of assessment content to standards or curriculum is less a concern than having reliable assessments with adequate predictive validity. Assessments used to select students should have psychometric qualities that are sufficient for use in verifying that the decisions made are accurate. For such tests, the criteria of range and depth of knowledge are of less concern than including items that will distinguish those students who have a greater probability of being successful in higher education from those who have a lower probability of being successful, or those who are experts from novices (Chi, Glaser, & Rees, 1982). The depth and breadth of items on the assessment is of concern only to the degree that these attributes predict what students will be required to know and do for future studies or to identify those who have specialized knowledge.

Some state assessment programs are confronting the issue of using one assessment for multiple purposes. State legislatures and policy makers (e.g., Michigan and Maine) are mandating that the high school graduation assessment be comparable to a college entry assessment, such as the College Board SAT or American College Testing ACT. Their rationale is that the K–12 curriculum in the state should prepare students sufficiently so they are ready for higher education. The high school graduation test should look forward to preparation for higher education rather than looking backwards to coverage of the curriculum. The preparation required for higher education is not always considered the same as what should constitute a quality K–12 education. A K–12 education should prepare students for higher education, but it must also prepare students to go into the work force and to become good citizens. For example, a strong background in algebraic rigor is important preparation for the study of mathematics in higher education, whereas a quality K–12 mathematics program may include data analysis, probability, and applying mathematical knowledge in context, as discussed by de Lange in this volume.

The state of Michigan conducted a study of the alignment of ACT and SAT assessments with its high school content standards to determine whether either of these assessments would be considered to be aligned with the content standards (Webb, 2005). If one of these assessments used to predict success in higher education was considered aligned as required by the *No Child Left Behind Act,* then the state could serve two purposes with the one assessment. The study indicated that the Michigan Educational Assessment Program (MEAP) assessment (51 items) had the best alignment with the state's challenging curriculum standards. Both the ACT (60 items) and the SAT (54 items) were found to overemphasize solution of algebraic equations, simplifying expressions, and determining geometric properties to a greater extent than is needed to sufficiently provide a measure of students' attainment of the stated expectations. Data analysis and probability were not considered sufficiently emphasized on the ACT and SAT in relationship to the Michigan content standards. Since both the ACT and the SAT have proven records in making higher education entry decisions, modifying either of these was not an option. The only other option was to supplement each of these assessments with items in the content areas with insufficient coverage in relationship to Michigan's content standards. Assessments are generally designed for specific purposes. When the results of an assessment are used for other than the intended purpose, there is an increased chance that invalid inferences will be made from the assessment results.

SUMMARY

In the age of assessment, gathering information to make decisions about students is pervasive in all aspects of the educational system. Critical to producing reliable and valid information about students' mathematical knowledge is having clear specifications about what students should know and be able to. Clear content specifications apply to all forms of assessments including large-scale assessments, high-stakes assessments, and formative assessments. In constructing or selecting an assessment, different criteria can be used to express the content students should know and be able to do, or how the content is measured by an assessment aligned to content standards or other statements of expectations. One criterion is that the assessment should include a sufficient number of tasks related to each topic so that reliable decisions can be made by the student's knowledge of each topic (*Categorical Concurrence*). The assessment also should include tasks at a sufficient level of complexity (*Depth-of-Knowledge Consistency*) and that adequately cover the content (*Range-of-Knowledge Correspondence*). Because not all areas within a content domain have equal importance, the specification of content should include some indication of what emphasis should be given to different content areas within a content domain, and the distribution of items on the assessment should reflect this relative emphasis (*Balance of Representation*). In mathematics, rarely are students to learn only isolated facts, skills, and concepts; rather, they are expected to learn how mathematical ideas relate to each other and how to use these ideas in concert with each other. How the mathematical ideas within a content domain are structured and how students are to integrate these ideas with ideas in other content domains should be evident in expectations for student learning and in how assessments are constructed (*Structure-of-Knowledge Comparability*).

These criteria for content specifications are only a few factors that need to be considered in developing an assessment. Other factors include the underlying theory of learning, item type, length of the assessment, purpose, and how the results are to be reported. Another important consideration is that policy makers who mandate assessments, teachers, and consumers for assessment results all need to have some basic assessment literacy about what can and cannot be concluded from the scores of different types of assessments and what is necessary to develop a good assessment for a specific purpose. Advancement through the age of assessment requires taking responsibility for the appropriate use of assessment results, research on how assessments can be more effectively utilized by teachers to further student learning, and less invasive ways to gather information from students. Critical to all of this is a clear understanding and specification of the mathematics to be assessed.

Researchers have a special responsibility for attending to the specification of content on any assessment instrument they use. Reporting of research findings should include some indication of what content knowledge was assessed and how representative the content measured by an instrument is of the content from the targeted domain. This chapter has identified a number of dimensions for specifying content for assesment. In the age of assessment with the proliferation of tests, it is easy to find assessment results at all levels of the educational system. However, researchers who use results from existing assessments and their own instruments are challenged to insure that they are clear about what content knowledge and sklls are being measured and that they have attended to the different dimension for specifyng content.

REFERENCES

Bloom, B. S. Englehart, M.D., Furst, E.J., Hill, W.H., & Krathwohl, D.R. (1956). *Taxonomy of educational objectives: The classification of educational goals. Handbook I: Cognitive domain.* New York: David McKay Co.

Chi, M. T. H., Glaser, R., & Rees, E. (1982). Expertise in problem-solving. In R. J. Sternberg (Ed.), *Advances in the psychology of human intelligence* (Volume 1),(pp. 7-76). Hillsdale, NJ: Erlbaum.

Downing & T. M. Haladyna (Eds.). (2006). *Handbook of test development.* Mahwah, NJ: Erlbaum.

Illinois State Board of Education. (1996). *Illinois Learning Standards (preliminary draft).* Springfield, IL: Author.

National Assessment Governing Board. (1996). *Mathematics framework for the 1996 National Assessment of Educational Progress.* Washington, DC: Author.

National Assessment Governing Board. (2004). *Mathematics framework for the 2005 National Assessment of Educational Progress.* Washington, DC: U. S. Department of Education. Retrieved July 29, 2006, from http://www.nagb.org/pubs/m_framework_05/toc.html.

National Council of Teachers of Mathematics. (1989). *Curriculum and evaluation standards for school mathematics.* Reston, VA: Author.

National Council of Teachers of Mathematics. (1995). *Assessment standards for school mathematics.* Reston, VA: Author.

National Council of Teachers of Mathematics. (2000). *Principles and standards for school mathematics.* Reston, VA: Author.

National Research Council. (2001). *Knowing what students know: The science and design of educational assessment.* Committee on the Foundations of Assessment. J. Pellegrino, N. Chudowsky, & R. Glaser (Eds.), Board on Testing and Assessment, Center for Education. Division of Behavioral and Social Sciences and Education. Washington, DC: National Academy Press.

Porter, A. C., & Smithson, J. L. (2001a). Are content standards being implemented in the classroom? A methodology and some tentative answers. In S. H. Fuhrman (Ed.), *From the capitol to the classroom: Standards-based reform in the states* (pp. 60-80). Chicago: National Society for the Study of Education, University of Chicago Press.

Porter, A. C., & Smithson, J. L. (2001b). Defining, developing, and using curriculum indicators. Consortium for Policy Research in Education (CPRE) Research Report Series RR-048. Philadelphia: CPRE, University of Pennsylvania, Graduate School of Education.

Subkoviak, M. J. (1988). A practitioner's guide to computation and interpretation of reliability indices for mastery tests. *Journal of Educational Measurement, 25*(1), 47-55.

Tyler, R. W. (1949). *Basic principles of curriculum and instruction.* Chicago: University of Chicago Press.

U.S. Department of Education. (2003). *The school improvement knowledge base: Standards and assessments, Non-regulatory guidance.* Washington, DC: Author.

Webb, N. L. (1992). Assessment of students' knowledge of mathematics: Steps toward a theory. In D. A. Grouws (Ed.), *Handbook of research on mathematics teaching and learning* (pp. 661-683). New York: Macmillan.

Webb, N. L. (1997). *Criteria for alignment of expectations and assessments in mathematics and science education.* Council of Chief State School Officers and National Institute for Science Education Research Monograph No. 6. Madison: University of Wisconsin, Wisconsin Center for Education Research.

Webb, N. L. (1999). *Alignment of science and mathematics standards and assessments in four states.* Council of Chief State School Officers and National Institute for Science Education, Madison, WI: University of Wisconsin, Wisconsin Center for Education Research.

Webb, N. L. (2005). *Alignment analysis of mathematics standards and assessments, Michigan, high school.* A report prepared for the Michigan Department of Education, Lansing, Michigan.

Webb, N. L. (2006). Identifying content for assessing student achievement. Chapter 8. In S. M. Downing & T. M. Haladyna (Eds.), *Handbook of test development.* Mahwah, NJ: Erlbaum.

AUTHOR NOTE

The author would like to acknowledge the support of the National Science Foundation and the United States Department of Education to develop the alignment criteria. This work was done in cooperation with the Technical Issues in Large-Scale Assessment, a collaborative of the Council of Chief State School Officers. In addition, Maria Cormier, Rob Ely, Meredith Alt, and Brian Vesperman made important contributions to the development and dissemination of the depth-of-knowledge levels.

31

REFLECTIONS ON THE STATE OF AND TRENDS IN RESEARCH ON MATHEMATICS TEACHING AND LEARNING

From Here to Utopia

Mogens Niss

ROSKILDE UNIVERSITY, DENMARK

INTRODUCTION: IMAGINE UTOPIA

Why do we do research on the teaching and learning of mathematics? Although there are several different answers to this question depending on the degrees to which—and the ways in which—respondents are involved in the research enterprise, there is, I think, one ultimate answer that lies underneath all others: We do research on the teaching and learning of mathematics because there are far too many students of mathematics, from kindergarten to university, who get much less out of their mathematical education than would be desirable for them and for society. We believe that they could learn much more, and in much better ways, if the conditions and circumstances for teaching and learning were different. Simply put, at the end of the day research in mathematics education is done because we want to know why students do not get enough out of their mathematical education, and what we can do to remedy the situation.

Now, let us imagine that Utopia will have dawned on mathematics education research in, say, 2050, in the sense that the discipline will, by then, be able to provide final, exhaustive and effective answers to all important questions in our field. What would this mean, and what would the imagined situation look like, more specifically? By referring to this situation as "Utopia," it has already been admitted that it will never occur (and if it did, mathematics education researchers would be out of business, except—perhaps—as custodians of eternal treasures!). Nevertheless, pondering on the possible characteristics and potential implications of Utopia may be a means for reflection on where our field is today, and on where we are, or should be, heading in the future. Of course, the entire endeavour presupposes that by 2050 there will still be such a thing as explicit mathematics teaching and learning for at least some categories of students who need our services. In other words, it is a basic assumption for the following deliberations that the teaching and learning of mathematics will not have been integrated away into the teaching and learning of other subjects or been hidden in technological systems focusing on something else. Another basic assumption is that there will still be such a thing as institutions devoted to education (i.e., schools and universities and suchlike).

I submit that Utopia would mean that we would know and understand—and be able to document it—how to plan, orchestrate and carry out mathematics teaching so as to ensure that every student accomplishes an optimal learning of mathematics in accordance with his or her needs, wishes, capacity, and potential. On closer inspection, reaching Utopia would involve (at least) the following.

First of all, we would know and understand *why we should teach mathematics* to various categories of students in society, both from static and dynamic perspectives related to the development of society, acknowledging that the answer(s) to this question might be different for different student categories in different societies. This knowledge would be based on well founded and documented surveys of the location of mathematical insight and competencies in various quarters of society, thus providing convincing answers to the derived question: *who in society needs what mathematical insight and competencies, and for what purposes?*, where "society" is to be understood in a multi-faceted sense.

We would know and understand, in concrete terms, exactly what it means *to master mathematics* in all its dimensions and facets. More specifically, we would know, in an exhaustive and complete way, what the fundamental constituents of such mastery are, at an *overarching level* across age, grades, and institutions, and across mathematical topics. We would further know what it means to master different topics such as numbers, algebra, geometry, calculus, analysis, probability etc., as well as the interplay between topics. And we would know what it means and takes to be able to put different kinds of mathematics to use for extra-mathematical purposes in a variety of contexts and situations. We would further know what it means to be able to make mathematical inferences and to justify or assess mathematical statements and claims, as we would know what it is to pose and solve mathematical problems, and so forth and so on.

Then we would also have a complete knowledge and understanding of *how learning of mathematics takes place*, again in all the dimensions and facets of mathematics, for different categories of individuals and groups as a function of personal, social, societal, cultural, and contextual variables. This would also include knowing possible *limits* to mathematical learning for different sorts of individuals, for example as a function of biological, motivational and attitudinal factors. Moreover, we would know what paths mathematical learning can take, and what factors determine which paths are actually taken by different students under different circumstances.

We would know the life situations, interests, preoccupations, and motivations of different categories of students so well that their *beliefs, affects, and attitudes to mathematics and its study*, and the impact of these on their learning approach and behavior, would be known and understood by us. We would further know how we could influence (if desirable) these beliefs, affects, and attitudes in ways that pay due respect to students' personal, social and cultural integrity.

Furthermore, we would have a complete knowledge and understanding of *how teaching of mathematics can assist* different individuals and groups of learners in accomplishing optimal learning of mathematics according to their needs and potentials. This would include knowing the ways, approaches, methods and techniques that help bringing about such learning, as well as the boundary conditions and limitations of their use and effect, as it would imply knowing the circumstances under which an approach or a method should not be put to use, as well as the effects of different approaches or methods under given circumstances. This would further include complete knowledge and understanding of what various kinds of learning materials and teaching aids (e.g., textbooks, ICT systems, concrete materials), and student activities can offer to the teaching and learning of mathematics, and how they can be utilized in optimal ways.

In addition, we would know and understand how the teaching and learning of mathematics can be viewed in relation to the teaching and learning of *other subjects*. What significant similarities (and differences) are there between mathematics and other educational subjects in these respects? This would also imply that we knew to what extent, and how, mathematics teaching and learning can gain from interaction with other subjects, both in terms of content and in terms of organized co-operation between subjects.

Assuming that live teachers will still be needed in Utopia as key players in the teaching of mathematics, we would have a complete overview and understanding of *what competencies different kinds of mathematics teachers, from pre-school to university, should possess* so as to be able to design, plan, orchestrate and carry out optimal teaching of mathematics. We would know *how to prepare* such teachers through pre-service education, and how to *maintain and develop their professional identity and competencies* by means of in-service activities of various sorts. We would further know and understand teachers' beliefs, affects and attitudes towards mathematics and its teaching and learning, and the influence of these on students' learning outcomes.

In Utopia we would have complete knowledge of the *ways in which students' mathematical competencies and achievements can be assessed* in comprehensive and effective ways. This implies knowing the characteristics and qualities of a full array of assessment modes and instruments, including their administration and operation, again as a function of the boundary conditions, contexts and situations at issue. It would further include knowing and understanding when to implement what modes and instruments for what aims.

Finally, we would know exactly *what policies*, conducted at different decisional and administrational levels, involving individuals, institutions, organizations, and national or international authorities, *would be efficient* so as to reach the objectives that follow from the above-mentioned points.

When it comes to the *meta-level*, we would have a complete overview and understanding of *mathematics education research as a discipline*, including its characteristic research questions, philosophical underpinnings, theoretical constructs, methodologies, techniques, and results, and including different schools of thought within, and approaches to, the field. We would know how, and why, this discipline resembles, and differs from, specific other disciplines, as we would know and understand the ways in which mathematics education research draws upon and is supported by other relevant disciplines, as well as the impact of this on the research enterprise.

THE NATURE OF THIS CHAPTER

The majority of the chapters in this handbook undertake the task of surveying and reviewing the development of mathematics education research, with regard to selected issues and topics, since the publication of the first Handbook, edited by Douglas Grouws, in 1992 (Grouws, 1992). The developments considered are to do with perspectives on, approaches to, and results of research, in order to portray what new land has been reclaimed by our field during the last decade and a half.

The nature of this, final, chapter is somewhat different. It might have been possible to concentrate on providing an overarching synthesis of the development of the field, a "synthesis of syntheses," based on the "progress reports" constituted by the individual chapters. Although I shall, in fact, do something in that direction, I have chosen instead to position the current state and the "rate" of development of our field in relation to "Utopia." How far away from Utopia are we, what progress has been made during the last fifteen years to move in that direction, and what—if anything—is missing in the course we have set to get there?

OBSERVATIONS CONCERNING TRENDS REFLECTED IN THE PREVIOUS CHAPTERS

This section presents and discusses four observations of a somewhat general nature made during my reading of the previous chapters of this Handbook.

First Observation: Three Kinds of Widening Perspectives of Research

The *first observation* that emerges from the previous chapters in this volume is the continual widening of the perspectives of research, within, by and large, any area of mathematics education.

Since the major take-off of mathematics education research in the late 1960's, our field has experienced a *widening of research perspectives* in three different respects.

Firstly, researchers have revealed, and continue to reveal, that the issues being discussed and the phenomena, objects and *problématiques* being investigated in our field are much more complex than we used to think. The set of factors that influence a given phenomenon or entity in significant ways, is much larger than acknowledged in earlier research. So, researchers insist, we have to include such additional factors in our research and pay due respect to their actual or potential roles. One example of this is the influence of students' metacognition and beliefs on their problem solving behavior (see the chapters by Schoenfeld, and Lesh & Zawojewski in this volume). Another example, represented in several chapters in this handbook (e.g., Cobb's and Artigue, Batanero, & Kent's, and Battista's), is the shift away from primarily taking into account the cognition of the individual learner towards studying him or her as situated in different kinds of socio-cultural surroundings and within different institutional practices: classroom, school or university, communities of practice, (sub)culture, society at large, or region. Moreover, the teaching and learning of mathematics is no longer seen as a function of student characteristics and teacher knowledge only. Teachers' beliefs and affect are important factors as well, as is reflected in Philipp's chapter on this very theme. The fact that mathematics teaching takes place in societal institutions under the influence of economic, political, ideological, and organizational forces, suggests that research should include national and local issues of policy and change. In this handbook two chapters (by Ferrini-Mundy & Floden, and by Tate & Rousseau) are devoted to such issues. Societies' focus on accountability and high-stakes testing, and on international comparisons such as TIMSS and PISA of student achievements in different countries, are now factors that have a direct impact on mathematics teaching and learning. Hence they should be studied by researchers, as depicted in the chapters by de Lange and Wilson.

One consequence of the growing complexity of our objects and subjects of study is that the empirical and theoretical frameworks and methodologies adopted in research have to account for the resulting new and emergent factors and their interplay. This is what happens when Lesh and Zawojewski propose to include studies of the problem solving of people adapting mathematics for use in everyday environments and in work-places that require heavy use of mathematics. It is also what happens when Artigue, Batanero and Kent in their chapter talk about historico-epistemological and semiotic approaches to complement cognitive studies of tertiary mathematics education (e.g., focusing on linear algebra) and when they look at what neuroscience and embodied cognition have to offer to post-secondary mathematics education research. Thus, *the investigational frameworks themselves become more numerous and increasingly complex.* This development, in turn, gives way to the borrowing of constructs, frameworks and methodologies from an increasing number of other fields, such as neuroscience, psychology, education, philosophy, history, sociology, statistics, psychometrics, political science, economics, anthropology, ethnography, in addition, of course, to mathematics and all the disciplines and areas in which mathematics plays non-negligible parts.

In itself, the recognition that the teaching and learning of mathematics, and hence mathematics education research, are utterly complex is not new.

In the somewhat informal publication "Theory of Mathematics Education (TME)" (Steiner, et al, 1984; see also Steiner, 1985), which is a report partly of Topic Area 5 at ICME-5 in Adelaide, 1984, and partly of a subsequent mini-conference, the chief organiser Hans-Georg Steiner in his opening paper (pp. 16–32) made the following statement, which can almost be seen as a programmatic declaration for the establishment and development of a theory of mathematics education:

> Mathematics education is a field whose domains of reference and actions are characterized by an *extreme complexity*: the complex phenomenon "mathematics" in its historical and actual development and its interrelation with other sciences, areas of practice, technology, and culture; the complex structure of teaching and schooling within our society; the highly differentiated conditions and factors in the learner's individual cognitive and social development etc. In this connection the great variety of different groups of people involved in the total process plays an important role and represents another specific aspect to the given complexity.
>
> Within the whole system several *sub-systems* have evolved. They do not always operate sufficiently well, especially they often lack mutual interconnection and cooperation. With respect to certain aspects and tasks mathematics education itself as a discipline and professional field is one of these sub-systems. On the other hand, it also is the only scientific field to be concerned with the total system.
>
> A *systems approach* with its self-referent tasks can be understood as an organizing *meta-paradigm* for mathematics education. It seems to be a necessity in order to cope with complexity at large, but also because of the fact that the *systems character* shows up in each particular problem in the field. (p. 16)

In his paper, and in his summary of it, Steiner (1984) warns against ". . . non-scientific *short-range pragmatism* " as well as against "strong *reduction of complexity* favoring special aspects or research paradigms . . ." (i.e., against undue reduction of complexity) (p. 7). What we have seen in mathematics education research over the last couple of decades is that this complexity is now being taken seriously, not only at the rhetorical level but, more importantly, in actual practice. As indicated above, the chapters in this handbook demonstrate how this is taking place.

Secondly, along with the fact that mathematics teaching has developed in scope so as to address new categories of learners at all educational levels in increasingly varied environments and subject to more and more diverse conditions, *new phenomena, entities and problems* of mathematics education have been placed on the agenda of research. While mathematics education research in former times mainly focused on learners in primary and secondary school, and to some extent on their teachers, the scope has widened to such an extent that, today, pre-school children (as reflected in the chapter by Clements & Sarama), several sorts of tertiary students, also in fields to which mathematics is a service subject rather than a core subject, (as is the case with engineering, which is dealt with in a special section in the chapter on post-secondary mathematics education (by Artigue, Batanero, & Kent)) are now subjects of research. The same is true of adults returning to education, out-of school mathematics learning (see Presmeg's chapter), children playing computer games, popularization of mathematics in informal settings including ICT, media and museums, and so on. As mathematics teaching is often delivered indirectly —in fields in which mathematics is integrated in ways that make it invisible or at least implicit—or is delivered by means of ICT systems in which it is hidden, new research items have been added to the agenda. Also, the teaching and learning of new mathematical topics are being studied. This Handbook has made an attempt at placing topics born to model features of the real world, such as probability and statistics under the aegis of mathematics education research. The chapters on statistics (by Shaugnessy) and probablity (by Jones,

Langrall, & Mooney), and the statistics and probability section of the post-secondary education chapter are reflections of this. These are all instances of a different type of widening of research perspectives.

The addition of new items to the research agenda in combination with the endeavour to pay due respect to the complexity of the teaching and learning of mathematics, has led many mathematics educators to look at issues and problems that are actually outside of mathematics education proper but belong to general psychological, educational, or sociological spheres that have some bearing on mathematics education. Schoenfeld's chapter in large parts deals with the nature of empirical research in general, with examples drawn from pharmacological research as well as from mathematics education. Schoenfeld proposes a general schematization for conducting empirical research, starting from a real world situation, of which a conceptual model is constructed, and some of whose aspects are then captured in a representation system. Analyses are performed within the representation system, whose outcomes are interpreted within the conceptual model so as to allow for inferences to be made about the original real world situation. In his chapter, Cobb considers issues belonging to general epistemology, where he ends up making a plea for Putnam's pragmatic realism, and issues belonging to psychology, where he compares experimental psychology, cognitive psychology, socio-cultural theory and distributed cognition. The APOS theory (Action-Process-Object-Schema), by Dubinsky and colleagues, is actually a neo-Piagetian theory of general cognition, even though it is meant to be oriented towards mathematical cognition. Lots of research issues pertaining to curriculum reform and change (see the chapter by Ferrini-Mundy & Floden, and by Tate & Rousseau), to teacher's professional development (Sowder), to equity (see the Diversity in Mathematics Education chapter, and Bishop & Forgasz's chapter) are actually of a general nature, not specific to mathematics, as is also reflected in the literature reviewed in the respective chapters. In summary—and this is the third instance of widened perspectives to be mentioned here—we are witnessing that *phenomena, entities and problems beyond mathematics education* are now being dealt with by mathematics education researchers.

Second Observation: Constructing Theoretical Frameworks

While the previous observation included a widening of the theoretical and empirical investigational frameworks adopted in mathematics education research, the *second observation* focuses on mathematics educators' actual struggle to establish such frameworks. It is evident from the chapters in this Handbook that although there is certainly no convergence to be seen in the frameworks employed, many of the authors attempt to establish one as far as their own chapter is concerned—often a framework in whose development they have (had) a key role themselves. Cobb argues for a deliberately permissive approach to investigational frameworks that he calls "theorizing as *bricolage*." As mentioned above, Schoenfeld, in addition to revisiting his own construct "metacognition," proposes a framework for empirical research that focuses on "Pasteur's quadrant," which combines a quest for fundamental understanding with considerations of use. This framework is also meant to dissolve what he sees as an unfortunate dichotomy between quantitative and qualitative studies. Harel and Sowder base their chapter on the notion of "proof scheme," a construct which they have developed over the last decade or so. Lesh and Zawojewski propose "a models and modelling perspective" on problem solving which differs from traditional perspectives by placing traditional problem solving as a subset of "applied problem solving as modelling activity," rather than the other way round. In the chapter on post-secondary education, the authors outline Godino and Batanero's attempt, with colleagues, to create a framework incorporating ontological, anthropological and semiotic ideas while focusing on the relationship between students' personal meaning with respect to a mathematical concept and the intended institutional meaning,.

This suggests that even for a particular theme or topic, like problem solving, or proof and proving, it is not a settled matter what framework to adopt. Once again an overarching, unifying framework, agreed upon by all researchers, is not in sight. In several chapters (see Cobb, Lesh & Zawojewski, and Artigue, Batanero, & Kent) considerable effort is spent on surveying and reviewing available frameworks, without necessarily advocating or constructing a particular one.

To this point I have talked of an investigational framework as a somewhat undefined notion. What does it actually mean? It is not a well-defined notion but I propose that we take it to consist of at least three components: (a) A perspective on the issues and questions to be investigated; (b) A set of theoretical constructs (e.g,. concepts, notions, and assumptions) more or less sharply defined, coined to capture essential entities of significance to the issues and questions in focus of the framework; and (c) A set of preferred methods considered suitable for the investigation of the issues and questions of interest, with particular regard to the ways in which the theoretical constructs come into play. For a research approach to deserve the name "investigational framework," it is not necessary that the three components are explicitly introduced. The important thing is that they are actually present in the approach. I should make it clear at this point that to me the term "investigational framework" is not identical with the term "theory," which in my view is a more elaborate construct to which I shall return later.

Third Observation: Political Pressure on Mathematics Education Research

My *third observation* is that several Handbook chapters (Cobb, Ferrini-Mundy & Floden, Schoenfeld) touch upon demands from political and administrative quarters for evidence-based research and accountability in education in general and in mathematics education in particular. In the USA this demand is very evident indeed, but in other countries, such as the UK and Denmark, similar developments may have been encountered and the movement is undoubtedly gaining momentum in a world dominated by neo-libertarian approaches to economy and politics. The thrust of the requirements inherent in this demand is that educational research should provide solid evidence, "proof," of its claims, in particular of claims that lead to recommendations concerning subject matter content, processes, curricula, materials, assessment instruments, teaching approaches, and so forth. Unless such evidence is produced claims should be counted as unwarranted, and recommendations put forward on the basis of them should be discarded as "political views" disguised in scholarly or scientific gowns. In the U.S., the demands are accompanied by rather specific politically rooted requirements of the methodologies to be adopted by "acceptable" research, namely that such research should emulate clinical trials in pharmacology.

The vast majority of mathematics education researchers are likely to discard such requirements as being, in and of themselves, political in nature but disguised as a concern for scholarly health. The requirements are further discarded as expressions of a rather primitivistic reduction of the complexity of the practice and research of the teaching and learning of mathematics, a reduction which—as Schoenfeld attempts to demonstrate in his chapter on research method—also is too primitivistic to capture the pharmacological research, which by advocates of the evidence-based demand is meant to serve as a role model for educational research.

Although it is, on the one hand, annoying for seriously working mathematics education researchers to be met with potentially destructive requirements to their very enterprise, requirements that are even given strength by being invoked in public funding of research, there are two other sides of the coin (you have to think of the coin as a rather thick one, so that the edge can be counted as a third side!). Firstly, the fact that politicians, administrators, and lobbyists find it worthwhile to take position vis-à-vis mathematics education research suggests that they take such research seriously, even if many of them counteract it. In a not too distant past it was almost out of the question that research was in play to underpin recommendations for practice. Many reactions against attempts to substantiate recommendations by reference to research seem to be based on dissatisfaction with the recommendations themselves rather than with the underpinning of them. However, it is not only politicians and lobbyists with particular perspectives and interests who are beginning to take mathematics education research and its outcomes seriously. Practising teachers, teacher trainers and others do the same but usually in more embracing ways, even though they often ask for more specific results that can guide them in their everyday work in concrete ways. All this being said—and here comes the third side of the coin—although it is indeed unreasonable to ask for evidence-based research in the pharmacological sense only, it should be admitted that the nature of our results are not always easy to understand and use, because they are multi-faceted and need substantial interpretation, and that we are much better at demonstrating what does not work and at providing explanations of why, than the opposite. Knowing what not to do does assist our delimination of what to do, but not quite in ways that can produce operational guidelines for tomorrow's practice.

Fourth Observation: The U.S. is the Focus of Attention

The *fourth observation* is probably one that can primarily be made by non-U.S. readers of the chapters in this Handbook. While it is certainly true that the bulk of the content of the previous chapters is completely international in scope and significance, it is also the case that some of the issues addressed, the perspectives adopted, and large parts of the literature reviewed, have distinct U.S. flavors to them. There are two reasons for this.

The first, which is obvious and natural, was also valid for the first Handbook, namely that a handbook crafted within the framework of the NCTM must pay particular attention to matters pertaining to and relevant for the education system(s) in the U.S., not the least so since important developments concerning the teaching and learning of mathematics in the U.S. and worldwide are spurred by the NCTM as an organisation and by quite a few of its members, including the authors of this volume. While the preoccupation with matters American is easy to explain, it is less obvious why many U.S. researchers are not so inclined to consider literature from other countries, even though there are hosts of non-American research publications available in English that provide insights which can shed light on numerous issues of interest to U.S. researchers. With noticeable exceptions this bias is also reflected in the chapters in this volume. Allow an outsider to note that there is a fair amount of truth in the final paragraph of Judith Sowder's chapter, when she writes "... we in the United States are too often insular and not open to answers found in other countries. Thus, more than one border needs to be crossed while we, researchers, administrators, policy makers, and teachers, seek to better educate teachers for the purpose of improving student learning." I believe that we would all benefit from the cross-fertilization that would arise if more U.S. researchers paid as much attention to non-U.S. research as non-U.S. researchers pay attention to research from the United States.

The other reason, which deserves some consideration, is of a different nature. Numerous chapters in this volume make explicit reference to the current politico-ideological situation in the U.S., and the ways in which it exerts an impact on mathematics education in general and on research in particular. As the "Math Wars" in California, Florida (see, for instance, the sections on these wars in the chapter by the Diversity in Mathematics Education Group) and elsewhere have shown, together with the present views and policies of the current national administration, mathematics education has become subject of a strongly politicized discourse. This seems to be rather unique in history. Although some of the heated debates and controversies over the so-called New Mathematics movement in the '50s and '60s made their way into the media and sometimes also reached national political levels in some countries (e.g., Norway) the discussion by and large remained an internal one amongst mathematics educators and mathematicians, accompanied by some spill-over into the public domain.

How can we interpret the entering of mathematics education into political agendas? Is this a singular phenomenon, local in time and space, related to the U.S. government 2001–2008? I don't think so. While governments in democracies come and go (and the latter is at least as important as the former), the political interest in the teaching and learning of mathematics goes beyond transient governments. It is deeply rooted in the societal underground. This is not the place to offer a more detailed analysis of the relationships between mathematics education, society, and politics, but allow me to refer to a chapter, "Goals of Mathematics Teaching," that I rote for another handbook, the *International Handbook of Mathematics Education* (Niss, 1996). In that chapter, I proposed that society provides mathematics education to its citizen for one or more of the following three fundamental reasons:

- in order to contribute to the *technological and socio-economic development* of society at large, either as such or in competition with other societies/countries;
- in order to contribute to *society's political, ideological and cultural maintenance and development*, either as such or in competition with other societies/countries;
- in order to *provide individuals with prerequisites which may help them to cope with life* in the various spheres in which they live: education or occupation; private life; social life; life as a citizen. (p. 13)

In countries that offer public mathematics education to all its children and youth, analyses suggest that the predominant reason is the first of these. The other two are important as well, but are subsidiary reasons adopted by some countries in certain eras. In principle, these three reasons are neither in concert nor in conflict with one another. However, in a given country at a given time the specific consequences drawn from them may be either in concert or in conflict. This happens if, for instance, the teaching time allocated to pursuing goals derived from the reasons favors one over the others, like if most activities were orchestrated to pursue goals related to the third reason.

I submit that what we are witnessing, as far as the political interest in mathematics education is concerned, is a struggle between the consequences of these reasons. The current U.S. administration seems to primarily insist on goals derived from the first reason, with some subsidiary attention being paid to the second one, while goals related to the third reason only deserve attention to the extent they are not in conflict with the primary ones. At the same time, many mathematics educators tend to emphasise goals related to the third reason, which is to do with empowering people to live in society, while maintaining that there is no intrinsic conflict between this and paying due attention to the other two (cf. the Handbook chapters by Bishop & Forgasz, and by the Diversity in Mathematics Education Group).

Potential conflicts between societal quarters advocating one reason over the others are not of a superficial nature. They are to do with fundamental views and interests concerning the direction and ways in which society should move and develop. Let me mention just one point for illustration.

It is fair to claim that since World War II most societies in the world have actively sought to make mathematics education available to all citizens (for example, see Sowder, Bishop & Forgasz, and the chapter by the Diversity in Mathematics Education Group), first as regards primary education, then with respect to secondary education, and recently in more and more countries with respect to tertiary education also. What is important here is not the extent to which countries have actually been successful in this endeavor, but the very endeavor itself. The basic reason for this is, I submit, the technological and socio-economic reason. Now, in recent years we have been witnessing that the question has been raised, in some quarters in highly developed countries, whether it is really justified, from the point of view of societal needs and the needs of individuals in their everyday life, to subject everyone to a substantial dose of mathematics teaching. The fact that, today, so much mathematics of utilitarian societal relevance is embedded in technological systems in implicit ways, has been taken by some to suggest that perhaps not every citizen needs mathematics beyond a certain minimum, which might well be decreasing, for coping with his or her occupational and private life. Of course, any society needs a minority of high-level specialists with advanced mathematical education, but why, the argument goes, should this fact be used to bother the majority with painful mathematics beyond their needs and natural capacity, and why should those who are unhappy with the study of mathematics be allowed to prevent or jeopardize the development of front-line mathematical competencies with those who can and should benefit from demanding mathematics teaching? To counter this position, some mathematics educators might argue that although this line of reasoning might look seductive if a narrow utility argument based on the technological and socio-economic reason is adopted, a completely different conclusion would follow if focus were on equipping individuals with prerequisites for becoming knowledgeable, concerned, analytically inclined citizens with the capability to exert well-founded criticism of those in power in society. This would amount to pleading for goals related to the third of the fundamental reasons.

Such issues and their relatives will not disappear with transient governments but will remain with us in the future. I would even venture to predict that fights over such issues are likely to be intensified in the next couple of decades. Against this background, it will be interesting to see how a third NCTM research handbook will describe and analyse the "mathematico-political" situation in another decade and a half.

These deliberations suggest that USA, as far as political influence on mathematics education is concerned, is at present experiencing movements and developments that are going to spread to the rest of the world in the future, whence the preoccupation of several authors of this Handbook with this influence as it manifests itself in the U.S.

REFLECTIONS ON THE STATE OF AND PROSPECTS FOR MATHEMATICS EDUCATION RESEARCH

I am now ready to take a general look at how far we have come in approaching Utopia, if the distance is to be judged on the basis of the chapters in this Handbook. My overall conclusion is "not so far." This being said, I hasten to add that on average a lot of progress has been made since the first Handbook appeared in 1992. Below, I shall try to outline the state of affairs as I interpret it.

Why Mathematics Education? For Whom?

To what extent do we fully know and understand *why we should teach mathematics to various categories of students in society*, and to what extent do we know and understand *who in society needs what mathematical insights and competencies, and for what purposes?* These issues are not key points in this Handbook, but aspects of them are briefly touched upon in some of the chapters, in particular the ones that deal with diversity and equity, in the engineering section of the chapter by Artigue, Batanero, and Kent, and in the chapter by Lesh and Zawojewski with its numerous remarks concerning problem-solving practices in professions that make heavy used of context-embedded mathematics.

It may seem pretty natural that a handbook on research on the teaching and learning of mathematics does not focus much on issues of "Why mathematics education? For whom?," because research tends to enter the stage only when mathematics education is actually being given/offered to certain groups of students. Nonetheless, well-founded answers to these questions are crucial to all aspects of practice of and research on the teaching and learning of mathematics. Most mathematics educators are convinced that it is essential to provide some mathematical education to everyone in society, although they are likely to disagree on what mathematics different kinds of people need and on how they need it. They would probably subscribe to the arguments related to some or all of the three reasons indicated in the previous section, in addition, perhaps, to some others. Mathematics educators often have vivid images of where, how, and by whom mathematics is being used in society's work-places and in other disciplines and subjects, of what role mathematical competence plays for people as private and social individuals and as citizens, and of what role mathematics has in the development of society and culture from historical or contemporary perspectives. These images may well contain different sub-images with regard to different mathematical topics or processes.

Such images and convictions, which are often sound and strong, mostly rely on individual experience and impressions. Their soundness and strength notwithstanding, these images and convictions only rarely have a solid foundation and documentation in terms of research. This has at least three significant consequences. Firstly, a well-founded—I hesitate to write "evidence-based"—solid knowledge of the actual and potential nature, location, and distribution of mathematical knowledge and competence in different segments of different types of societies would help us orchestrate and tune teaching to different kinds of students much better than today. Secondly, we would be in a much better position to justify our claims about the relevance of mathematics education to various categories of learners, to people who have not accumulated similar piles of impressions and experiences to the ones that we, mathematics educators, have accumulated as individuals or as a community. Thirdly, in addition to providing better justification of our claims such well-documented knowledge would help us become better at communicating with others about the whole enterprise of mathematics teaching and learning.

Thus, I submit that we are lacking comprehensive overviews of mathematics in different kinds of societies and cultures and an insight into the educational consequences that may or should be drawn from such overviews. In other words, we are lacking a well-founded *socio-cultural ethnography of mathematics.* It is true that research within the ethno-mathematics movement (see Presmeg's chapter, and, for a specific example, Gerdes, 1996) has provided us with parts of such an ethnography as regards some "developing" countries and as regards particular cultural or ethnic groups in "developed" countries. It is also true that there are many illuminating studies of situated knowledge and communities of practice focusing on the discrepancy between in- and out-of-school contexts, like the famous study by Carraher, Carraher & Schliemann (1985) and Nuñez, Schliemann, & Carraher (1993) of Brazilian street vendors. Studies of the mathematics of different professions, for example, banking and nursing (Hoyles, Noss, & Pozzi, 2001; Noss & Hoyles, 1996; Noss, Hoyles, & Pozzi, 2002) and engineering (Kent & Noss, 2003), have begun to appear, but are still rather sporadic and far from systematic across significant societal practices. Much more needs to be done in these directions if we are to come closer to Utopia, and the current mostly static, "snapshot," studies ought to be complemented by dynamic, "moving picture," studies of change and evolution.

In addition to studies of vocational and professional practices, and studies of mathematics in arts and crafts, studies are lacking that uncover the mathematical competencies involved in different kinds of citizenship. What does it take, for instance, to be able to understand, judge and take position towards specific civic and political issues, for example, concerning economy, environment, climate, planning, resource exploitation, epidemics, and demography? Many teaching materials contain examples referring to such issues, but they are usually tailored to fit into instructional sequences, and are normally not authentic in a large scale sense. What I am proposing is that studies within the socio-cultural ethnography of mathematics be extended to chart and survey aspects of citizenship on which mathematics has a bearing.

What Does It Mean to Master Mathematics?

The next issue to be addressed is to what extent we know and understand exactly *what it means to master mathematics* in all its dimensions and facets, *inside mathematics as a discipline and in extra-mathematical situations and contexts,* both at an overarching level (across age, grades, and institutions, and across mathematical topics), and with special regard to different topics. The general answer to this question is that we know and understand much more today than fifteen years ago, but that there is still a non-negligible distance to Utopia.

Marked progress has been made when it comes to understanding and describing what it means to master various mathematical topics. Essential parts of this progress are captured by the topics chapters in this Handbook, on whole number concepts and operations (by Verschaffel, Greer, & De Corte), on rational numbers and proportional reasoning (Lamon), on algebra learning (by Carraher & Schliemann, and Kieran, respectively), on geometric and spatial thinking (by Battista), on probability (by Jones, Langrall, & Mooney), and on statistics (by Shaughnessy and by Artigue, Batanero, & Kent at the tertiary level). Topics belonging to post-secondary levels such as calculus and linear algebra are touched upon in the last-mentioned chapter, but are not dealt with as topics per se. One common feature of current portraits of the mastery of these topics is a much more sophisticated understanding of the interwoven variety of facets and components of each topic than was found one-and-a-half decade ago. Another common feature is that the mastery of one topic is conceptually and procedurally related to the mastery of other topics. Most importantly, perhaps, today mastery of a topic is portrayed as being intimately related to multi-faceted links to a variety of semantic environments that allow for interpretation and sense-making of concepts, rules, procedures, facts, and results.

Similar pictures can be drawn of the mastery of mathematical topics that are not so well represented in this Handbook such as discrete mathematics, analysis, and abstract algebra and geometry, which predominantly pertain to the tertiary level. Here the body of research is much more limited than for numbers, arithmetic, elementary algebra, and geometry, but the volume is growing, in particular as regards mathematical analysis and linear algebra.

What we are still largely lacking is an in-depth (profound rather than detailed) charting of how the mastery of a given topic is related to the mastery of other topics, and of how mastery of topics is related to mastery of problem solving, modelling, and proof and proving, respectively. Here, Lesh and Zawojewski in their chapter argue that there is no such thing as context- and situation-independent problem solving in mathematics, and that in general mathematical problem solving has not made any real progress during the last 10–15 years. Rather, they suggest, problem solving is developed along with the learning of mathematical concepts in rich environments that "ask for" or lend themselves to modelling. While it is pretty clear that there is not much evidence of subject-free problem solving, it seems to me that it remains an unsettled matter to what extent it is possible to identify and characterize generic problem solving within mathematics but across different domains.

There are certainly strong connections between problem solving and modelling. However, they are not, I submit, just two sides of the same coin, because much mathematical modelling concerning extra-mathematical domains is being performed for reasons different from solving problems in the usual sense of that word, and because a lot of intra-mathematical problem solving is being performed without regard to anything extra-mathematical. There is a fairly substantial research literature on the mastery of mathematical modelling that ought to be taken into consideration in that context but which does not figure prominently in this Handbook.

During the last one or two decades mathematics education researchers have attempted to identify and characterize mastery of mathematics in terms that are not locked into particular mathematical topics and do not pertain to particular educational levels. One example of this endeavour can be found in the *Curriculum and Evaluation Standards* of the National Council of Teachers of Mathematics (NCTM, 1989) and in the *Principles and Standards for School Mathematics* (NCTM, 2000). The 1989 *Standards* listed—in addition to grade-specific mathematical topics—four characteristics of mathematics, meant to help foster students' "mathematical empowerment": "Mathematics as Problem Solving," "Mathematics as Communication," "Mathematics as Reasoning," and "Mathematical Connections." In the same spirit, the 2000 *Standards* provided a similar, but slightly expanded list: "Problem Solving," "Reasoning and Proof," "Communication," "Connections," and "Representation." More recently, "Adding it Up" (National Research Council, 2001), edited by Kilpatrick, Swafford, and Findell, considered five strands of overarching *mathematical proficiency*: "conceptual understanding," "procedural fluency," "strategic competence," "adaptive reasoning," and "productive disposition."

I and colleagues in Denmark have worked since the late 1990's on developing the notion of a system of overarching *mathematical competencies* that are meant to capture what it means to master mathematics across topics and educational levels. I propose eight such competencies: "mathematical thinking competency," "mathematical problem solving competency," "mathematical modelling competency," "mathematical reasoning competency," "mathematical representations competency," "symbols and formalisms competency," "communications competency," and "aids and tools competency." (For a brief outline, see Niss, 2004a). This system of competencies also serves to underpin the mathematics framework of the OECD/PISA international comparisons of student achievements (OECD, 2003).

It appears that we have not yet come very far in establishing a comprehensive, consistent, and robust characterization of mathematical mastery in general terms, substantiated by a massive empirical corroboration. This is a point where lots of research is needed if we are ever to arrive in Utopia. But to quote the late Bristish prime minister Winston Churchill, "This is not the end. It is not even the beginning of the end. But it is the end of the beginning."

How Does Mathematics Learning Take Place?

To obtain complete knowledge and understanding of *how learning of mathematics takes place* is arguably the most formidable task in mathematics education research. How far have we come in accomplishing this task? It would be an impossible endeavour to attempt to answer this question in any kind of detail, at the end of a volume in which several hundred pages have been spent on surveying and reviewing the state of the art with regard to this very question. However, are there any aggregated conclusions to draw?

There are. On average a lot of progress has been made towards a deeper and more sophisticated insight into the mazes of mathematical learning, both in general and with respect to particular domains and processes. It is remarkable, however, that some of the chapters (e.g., Lamon's, and Lesh & Zawojewski's) talk with some pessimism about stagnation in research achievements over the last decade or so, although they do see promising new germs that are likely to open new avenues for coming to grips with how learning of certain topics or processes takes place or, rather, may take place. This is even more remarkable when contrasted with the much more optimistic tones found in other chapters (e.g., Verschaffel, Greer, & De Corte's and Harel & Sowder's) which report major progress since the last Handbook. The same is true of the learning of probability, where cognitive models of probabilistic reasoning have shed light on the learning of phenomena and concepts such as chance, random process, sample space, and probability measurement since the previous Handbook. Also the chapter by Zbiek, Heid, Blume, and Dick report on major progress in our understanding of how technology can facilitate learning in a multitude of ways, in particular when it comes to representation of mathematical objects, phenomena, and processes. Some chapters lie in between, for examples Battista's, when he reports on solid progress as regards the cognitive side of geometric and spatial reasoning—except for the role of what he calls technological enhancements offered by computer environments—but points to a lack of research that can account for the effects of affects and socio-cultural factors on students' learning.

This suggests that progress is not uniform across domains. Research can run out of steam within one domain while gaining momentum within another. It would be interesting to investigate the mechanisms behind such developments. It is certainly not the case that the domains in which research is reported to have stagnated have already reached Utopia. There is a long way to go for that to happen in all aspects of mathematics learning. Nevertheless, we have come much closer to knowing and understanding how students' mathematics learning is construed, and the conditions under which it takes place. In one sentence: Mathematics learning is utterly complex, much more so than we thought. Through a multitude of studies over the last fifteen years research has been able to move beyond the mere recognition of complexity towards uncovering its scope and components. However, huge efforts remain in order to achieve major steps towards a justified reduction of complexity, the ultimate scientific goal of all scientific research.

What Are Students' Beliefs, Affects, and Attitudes to Mathematics?

In this Handbook, students' beliefs, affects, and attitudes to mathematics and its study have not been dealt with as an independent theme. Rather this theme has been parcelled out to be addressed, more or less sporadically or in passing (with some exceptions, e.g., Battista's, Lesh & Zawojewski's, and Schoenfeld's chapters), by the other chapters. This implies that there is not much of an overview to reflect on. Internationally, however, there is an amount of research worth considering—new research included (see, for instance, Leder, Pehkonen, & Törner, 2002)—but it may be that there has been some slowing-down in such research during the last decade.

One might hypothesize that the relative slowing down of research on beliefs, affects, attitudes, and motivation with regard to mathematics is a reflection of doubts being raised about how well-defined and well-delineated the basic notions are, and how clearly they can be disentangled from cognition in mathematics education. An attempt at defining key concepts is made by Philipp in his chapter on teachers' beliefs and affect. On the other hand, all mathematics educators, be they practitioners or researchers, seem to agree that there are affective factors that exercise a strong influence on students' learning. This suggests that there is a need for investing a considerable amount of effort on sorting things out, both conceptually and empirically, before Utopia is in sight in this respect.

How to Teach Mathematics So As to Ensure Optimal Learning Outcomes for Students?

One of my favorite quotations is from the Polish-American mathematician and mathematical physicist Mark Kac, who is claimed to have once said "I can't define my wife, but I can recognize her when I see her." This dictum (which may be apocryphal as I haven't been able to locate it anywhere) points to important limits to formalization of human experience. There are many things that we know without being able to document them in a formalized way, while obeying certain rules and procedures. This is often the case when human beings' extraordinary pattern recognition is involved.

The Kac quotation, or to whomever it should be attributed, came to my mind when reading Hiebert and Grouws' chapter on the effects of classroom mathematics teaching on students' learning. One of their section headings bears a strong resemblance with the quotation: "A claim that appears obvious is strikingly difficult to specify." In several places in their chapter they return to statements like "teaching matters . . . but *how* is not easy to document," and "documenting the effects of teaching on learning is methodologically difficult." That goes for teaching in general and for mathematics education in particular. When broken down with regard to effective teaching for learning of particular topics or processes, these difficulties prevail, in some respects in an amplified form. Things get even more complicated when we want to investigate the specific influence of curricula and teaching materials, including textbooks, on students' learning, separating such influence from other instructional variables (see Stein, Remillard, & Smith's chapter).

Against this background, to review—here—our current knowledge and understanding of how teaching of mathematics can assist the optimal learning with different individuals and groups of learners is far beyond reach. Once again, we can ask whether there are aggregate conclusions to consider, across the chapters. And once again, there are.

Considerable progress has been made when it comes to devising, designing, orchestrating, and implementing mathematics teaching and student activities that seem to promote students' learning in forceful ways. But as indicated, it is difficult to document that the desirable learning outcomes result from teaching rather than from other variables in play. Characteristic of such teaching is that it pays due and articulate respect to the variety of factors that may influence student learning. In other words successful teaching makes a point of explicitly identifying, monitoring and controlling such factors as part of the teaching-learning environment. Moreover, research findings suggest that for students to obtain desirable learning outcomes it is important that they are engaged in activities where they have to "struggle" (in a productive sense of that word) with important mathematics, preferably in collaborative environments that challenge their work in constructive and supportive ways. Delicate balances are to be struck here, as "struggles" and "challenges" can so easily become obstacles rather than promoters of learning, if they are in mismatch with students' situations and prerequisites. Striking such a balance is one objective in The Mathematical Task Framework, developed by Stein and her colleagues and referred to in several chapters in this Handbook.

Since the mid 1970s the role of information and communication technology (pocket calculators, computers, software systems) has been discussed intensively among mathematics educators. In the beginning the topic of the discussion was rather simplistic: Good or bad? And little solid evidence was provided for either position. Today we know and understand, much more than we did fifteen years ago, and in much more subtle ways, how technology can (and cannot) be put to use as a crucial instrument for teaching (see Zbiek, Heid, Blume, & Dick's chapter). This includes a deeper insight into the significance of the teaching and learning enviroments in which ICT is put to use and the conditions that govern them. One overall finding is that ICT as such can be both rather harmful and highly helpful to the teaching and learning of mathematics. The outcome is almost completely determined by the the didactic and pedagogical design and implementation carried out by the teacher. Another finding (see Artigue, Batanero, & Kent's chapter) is that technology can no longer be seen as just a tool for mathematics education, an add-on. Rather we should consider technology as something that profundly shapes and changes the very ways in which we work with and do mathematics.

We are not only making progress as regards understanding how teaching can contribute to learning. We have made considerable progress in characterizing and understanding what happens in the classroom, and in teaching practices. As regards American classrooms and teachers, the chapter by Franke, Kazemi, & Battey, is an eye-opener. Such research was only in its infancy fifteen years ago, and new panoramas are likely to be added to our knowledge in the years to come.

We are still far from being in a position where we can design and implement teaching that—guaranteed—leads to optimal learning outcomes for all the students we teach. The problem is not only to document the effects. Even though we have made remarkable progress in the design and implementation of successful and effective teaching for many categories of students, we have only limited successes in really assisting weak learners. So, here too, the way to Utopia is long.

How Should the Teaching and Learning of Mathematics and of Other Subjects Be Related?

On this issue the present Handbook is—as was the previous one—largely silent. It would of course be unreasonable to criticize the book (i.e., the editor!) for this. The book has to limit itself to what is seen as its primary priorities. Besides, while there was a growing interest in the 1970s and 1980s in linking the teaching of mathematics to other subjects, mainly in order to provide avenues for import-export activities, which, from the point of view of mathematics, would provide fuel for work on mathematical applications and modelling, this interest was never really accompanied by research.

This is a pity. I submit that important insights would be gained and lessons to learned, also for the benefit of mathematics education research, by studies that compare and contrast mathematics and other teaching subjects. We already know and understand, in the "folk wisdom sense" of Mark Kac, a lot of the differences and similarities in play between mathematics and other fields. But making them subject of research studies would make us much better at identifying and characterizing those features and factors that are tied to teaching and learning in general, or focused on certain types of subjects, and those that are specific to mathematics. Once such features and factors were identified and located we could invest special effort on uncovering the ways in which the features and factors, with respect to which mathematics differs from other subjects, influence the teaching and learning of mathematics. That, in turn, would help us understand when and why mathematics becomes difficult (or easy, for that matter) for students, and what we can do to assist them in overcoming these difficulties.

From the point of view of Utopia we can never claim to have arrived there until we are able to pinpoint the ways in which all important aspects of mathematics teaching and learning are similar to, respectively different from, the corresponding aspects of teaching and learning of other subjects. I therefore propose that such a comparative research programme be undertaken as one way to bring us closer to Utopia.

What Competencies Should Mathematics Teachers Possess and How Should They Be Developed?

There was a time, not so long ago, when the prevalent conception of teachers of mathematics could be condensed—and caricatured—as follows. A teacher of mathematics is a person who knows some mathematics, i.e. some concepts, facts, results, rules, methods, and procedures, on the one hand, and who knows how to teach students, in general terms, on the other hand. Thus the competencies a mathematics teacher should possess were of two kinds, subject matter knowledge, and general pedagogical skills. Satisfactory and efficient teaching was expected to flow from the happy mating of these two kinds of competencies.

Today we know much better. We know that general pedagogy—or education—should not be kept separate from subject matter knowledge. We know that mathematical knowledge for teaching (see the chapter by Hill, Sleep, Lewis, & Ball) is an essential form of knowledge that is not identical to knowledge of mathematics as a theoretical edifice. We further know that teachers, in addition to possessing solid mathematical competencies, should also possess mathematics-specific "didactic competencies," such as "curriculum competency," "teaching competency," "uncovering of learning competency," "assessment competency," "collaboration competency," and "professional development competency" (see Niss, 2004a). We know that teachers' beliefs and affects towards the nature and role of mathematics, mathematical activities, and the teaching and learning of mathematics, as well as towards students and their potentials for learning, are crucial elements in their luggage (see Philipp's chapter). And we know that although teachers' initial preparation does matter a lot for the ways in which they are initiated to their professional career, continual professional development focused on the teaching and learning of mathematics as an integrated entity, and the building of a professional identity in collaboration with different sorts of colleagues—peers, mentors, researchers—are even more critical to the way and extent to which they can develop as teachers who can help students learn mathematics to the best of their capacities.

An impressive amount of research over the last twenty years, on all aspects of teachers' pre-service preparation, in-service development, and professional identity, in several countries, has brought us to where we are today. Sfard (2005) even talks of the present era in mathematics education research as the "era of the teacher." Despite this progress, there are still many things that need to be done before we reach Utopia. For, there is no end to what it takes to be(come) a Utopian teacher, a fact that makes such a person more ideal than real. Against that background, it is important to identify the components that are particularly instrumental in the preparation and development of mathematics teachers, so as to reduce the infinite complexity of shaping the protagonists in the teaching and learning of mathematics.

How to Assess Students' Mathematical Competencies and Achievements?

Assessment of students has always been a key element in mathematics education. Nevertheless research has demonstrated that the assessment instruments typically employed in mathematics classrooms and in high-stakes testing are often so limited in scope that they can only assess a few aspects and components of students' knowlegde and skills. And the ones that are left out tend to be the ones involved in aggregate and higher order competencies. Since "what you assess is what you see," this typically gives rise to a mismatch between the goals pursued in teaching and learning of mathematics and the assessment modes and instruments adopted. How far have we come in changing this picture so that we can become able to assess the full range of mathematical competencies in manners that are comprehensive, valid and reliable?

As is the case with teaching, there are several approaches to assessment that seem to work and are valid, if judged on an experiential basis. But it is difficult to convincingly demonstrate that they work, i.e. it is not always easy to achieve reliability.

Nevertheless, substantial progress has been made during the last decade and a half, as regards devising, designing and testing of assessment instruments and as regards research on assessment practices (see Wiliam's chapter in this Handbook). Along with such research, sharpened conceptual frameworks have been developed as tools to capture the multiple different purposes of assessment and the dimensions of the assessment modes and instruments. Research has also made progress when it comes to investigating the consequences and effects of high-stakes testing (see Wilson's chapter).

However, only very little progress has been made as regards the assessment of essential ingredients in mathematical competencies, such as asking questions, conjecturing, posing problems, constructing argument, including formal proofs, making use of and switching between representations, communication, and suchlike. Not only is research lacking, assessment instruments are largely lacking as well. Much assessment and testing is still focused on students' solving of already formulated problems. This shows that as far as assessment is concerned there is indeed a long way to Utopia.

What Policies Are Efficient to Achieve the Goals that Follow from Answers to the Previous Questions?

Issues of policy are, almost by definiton, determined by local, including national, rather than global conditions and circumstances. It is therefore not so easy to draw many overarching lessons from the highly informative overviews of research made available to us by Ferrini-Mundy and Floden, and by Tate and Rousseau in their chapters. Nor is it so clear, as in the other respect dealt with here, what constitutes Utopia with respect to policy. For in Utopia, the political atmospheres and environments prevalent in today's world, are not likely to be present. The reader might infer from this statement that I do not consider the current state of affairs, within which policies have to operate, ideal. There seems to me to be a non-negligible amount of room for political improvement in a number of places, but I shall refrain from specifying the directions in which such improvement should go.

Instead, I shall focus on one characteristic trait of policy in mathematics education in the U.S., a trait that cannot but impress a non-U.S. observer. I am thinking of the role of professional assocations and organizations in mathematics education and mathematics, without which the situation would have been completely different, and, to be sure, worse, than is the case. The ways in which professional associations like NCTM, MAA, AMS, AAAS, ASI, among others, and committees created by them, have proactively worked to pave the way for reform in numerous aspects of the teaching and learning of mathematics, and have taken the lead in instigating research on them, is an inspiration for those of us who live in countries where change and reform is predominantly a top-down process, with political and administrative authorities holding the reins. In many of our countries, the role played by professional associations is often a reactive one, for instance limited to providing comments on reform proposals "sent down" from above. There is no reason why such associations should not take up similar activities as in the U.S. Although it is not so clear what exactly Utopia is, such activities would bring us closer.

What is the Nature of Mathematics Education Research as a Discipline?

Several chapters in this volume focus on fundamental aspects of mathematics education as a research discipline. This is true of the three "foundation" chapters by Cobb (on philosophy), Silver and Herbst (on theory), and Schoenfeld (on method), as well as of the present one. Moreover, the far majority of the other chapters actually contain sections in which the authors grapple with the nature of our field. This shows that authors feel it necessary to address this, fairly general issue, in order for them to portray and position their particular issue, topic, or theme within the larger research landscape.

In other words, researchers on the teaching and learning of mathematics are pretty much preoccupied with its foundations and characteristics. This would not be the case had the nature of our field been a settled matter. Most research mathematicians, physicists, and historians do not include sections on foundations and scholarly identity in their publications on Lie groups, viscous liquids, or the French revolution. They work in fields with clear foundational identities, not necessarily in a philosophical, but in pragmatic and paradigmatic sense.

Mathematics education research does not have a clear identity. Well, it does when it comes to the objects of our investigations, the teaching and learning of mathematics in all its dimensions and facets. But when it comes to the goals of our research, opinions begin to differ, albeit not necessarily very much. Are we doing research exclusively in order to obtain intellectual elucidation of certain phenomena and processes, are we striving at ameliorating teaching and learning of mathematics for given groups of students, are we striving at empowering people to cope with life in a changing, complex world, or are we striving at improving society at large? When it comes to the nature of the activities we undertake to study our objects and pursue our goals, opinions really begin to differ.

Let us just take a look at one, "simple," specific question that divides opinions among mathematics education researchers. The question is "Where does mathematics education research belong? To the social sciences, to the humanities, or to the sciences (including mathematics)? Somewhere else?" I do certainly not intend to answer this question here (well, I do: it belongs "somewhere else"), but the fact that researchers in our field come up with conflicting answers to it—although the majority would probably answer: "the social sciences," and not just for pragmatic sociological or institutional reasons—suggests that there is a long way to go before we have a complete overview and understanding of the nature of our field in all its dimensions and facets, which is what Utopia is all about.

HOW DO WE GET CLOSER TO UTOPIA?

I shall finish by addressing three general issues that in my view deserve particular attention for research on mathematics teaching and learning to move substantially forward.

Concepts and Definitions in Research on the Teaching and Learning of Mathematics

The French mathematician Henri Poincaré (1908) in a well known article *L'avenir des Mathématiques* (reprinted as the chapter 'The future of mathematics' in "Science and Method"), made the often quoted statement that mathematics is the art of giving the same name to different things ("Je ne sais si j'ai déja dit quelque part que la Mathématique est l'art de donner le même nom à des choses differentes"). This was meant to be opposed to the situation in poetry which may be characterised as the art of giving different names to the same thing. Anna Sfard has proposed, on several occasions (e,g., Sfard, 2005, p. 409f; see also the remarks by Silver and Herbst in this Handbook concerning international conversations about research in mathematics education), that there is a tendency for mathematics education research, too, to be the art of giving different names to the same thing. She makes a call for "conceptual accountability" (Sfard, 2005, p. 419). I share this view.

More generally, it seems to me that key concepts in research on the teaching and learning of mathematics are often used without being defined at all (e.g. "theory" or "theoretical," see the next section), are vaguely defined (e.g. "problem," "method," "proof"), or are defined differently by different researchers (e.g., "modeling," "reasoning," "representation," "belief"). Moreover, as Sfard has pointed out, even though definitions are being provided in scholarly publications, these definitions are seldom operational in the sense that there is an effective way to check whether a given entity actually falls under the definition or not.

Of course, as mathematics education research is usually reported in writing that makes heavy use of ordinary language, with technical terms interspersed, it is inconceivable that every single concept and term would be clearly and unambiguously defined, because such an endeavour would inevitably lead to an infinite regress. However, the fact that the task of defining key concepts, terms and notions as clearly as possible is extremely difficult should not be taken as an argument against trying our very best. Such a counter-argument would be similar to saying that "as a perfectly aseptic environment is impossible, one might as well conduct surgery in a sewer," as the economist Robert Solow is claimed to have once said.

I would like to propose that committees of experts in our field be established to undertake the task of surveying and reviewing current usages attached to key concepts and terms relevant to their areas of expertise, in order to come up with draft definitions to be discussed in the mathematics education research community at large prior to revision, followed—hopefully—by agreed-upon definitions. If such work were coordinated across committees, preferably by the International Commission on Mathematical Instruction (ICMI), otherwise by other relevant bodies, we might achieve the mounting of a rather authoritative "Dictionary of concept and terms in mathematics education research," to be revised at regular intervals. Although coordination would eventually be needed, the initiation of such work would not necessarily have to await coordinated and concerted efforts by central bodies. Small-scale initiatives to pave the way would be valuable in their own right.

The Concept and Role of Theory in Mathematics Education Research

The notion of "theory" is crucial in any scholarly or scientific discipline, including research on the teaching and learning of mathematics. Schoenfeld, in his chapter, argues that any research methodology is fundamentally theory-laden, even though the theory may well be implicit. Although the words "theory" and "theoretical" come up in more and more research publications, it is neither clear what "theory" is actually is supposed to mean, nor what foundations theories have and what parts they play in mathematics education. The absence of clear definitions of "theory" and "theoretical" in many publications that invoke these terms implies that they tend to become undefined notions in our discourses, thus laden with whatever connotations authors and readers bring to the interpretation of the research publications.

As is stated several times in this Handbook, there is no such thing as a well-established unified or unifying "theory of mathematics education" which is supported by a majority of mathematics education researchers. On the contrary, different groups of researchers represent different, if not opposing, schools of thought. This suggests that a lot of work needs to be done concerning the origins, foundations, developments, roles, and uses of theories in mathematics education research. We need work of a descriptive nature, focusing on uncovering and analysing the actual state of affairs, as well as work of a prescriptive, normative nature, focusing on what a theoretical underpinning of research in our field should look like. At the very least we should begin to discuss these issues. An excellent beginning has been made by Silver and Herbst in their chapter in this handbook. Their placing of "theory" in mathematics education as a mediator between the vertices of a triangle composed of "problems," "research," and "practice" seems very fruitful and worth further exploration.

I shall risk my skin by proposing the following definition of the concept of theory, not just within mathematics education research, but in general (see also Niss, 2006). The point is not (so much) to plea for acceptance of this particular definition but rather to stimulate further discussion of what a definition should look like.

A *theory* is a system of concepts and claims with certain properties, namely

- A theory consists of an *organized network of concepts* (including ideas, notions, distinctions, terms, etc.) *and claims* about some extensive domain, or a class of domains, consisting of objects, processes, situations, and phenomena.
- In a theory, the *concepts are linked in a connected hierarchy* (oftentimes, but not necessarily, of a logical or proto-logical nature), in which a certain set of concepts, taken to be basic, are used as building blocks in the formation of the other concepts.
- In a theory, the *claims are either* basic hypotheses, assumptions, or axioms, taken as *fundamental* (i.e., not subject to discussion within the boundaries of the theory itself), *or* statements obtained from the fundamental claims by means of *formal* (including deductive) *or material* (i.e., experiential or experimental with regard to the domain(s) of the theory) *derivation.*

Theories differ—greatly—with respect to the origin and nature of the *domains* (or classes of domains)—and the entities that populate them—to which they refer. They further differ with respect to the origin and nature of the *concepts* and conceptual *networks and hierarchies* involved, and regarding the fundamental and derived *claims* that are made. Theories also differ with respect to the kinds of methods by which claims are *derived* and *justified.* Finally, theories differ as far as their degrees of *stability, coherence,* and domain *coverage* are concerned.

In research, theories may serve different *purposes.* (A given theory, can, of course, serve several purposes at the same time.) One—crucial—purpose of a theory is to provide *explanation* of some observed phenomenon. Explaining a phenomenon by way of a theory means that the occurrence of the phenomenon can be obtained as a claim in the theory under specified conditions and circumstances, the fulfilment of which are substantiated as part of the explanation. A closely related purpose is to provide *predictions* of the (possible) occurrence of certain phenomena as a claim resulting from the (possible) fulfilment of the preconditions of their occurrence. Prediction will sometimes, but not necessarily always, rely on explanation of causes and mechanisms, while explanation will sometimes, but not necessarily always, give rise to predictions. Another purpose is to provide *guidance for action or behavior* by employing knowledge of substantiated claims within a theory as a means to plan and implement action or behaviors so as to achieve desirable outcomes.

A fourth purpose of a theory, and perhaps the most important one as regards research on mathematics education, is to provide *a structured set of lenses*, through which aspects or parts of the world can be approached, observed, studied, analysed or interpreted. This takes place by selection of the elements to be considered important in the context, by focusing on certain features, issues, or problems; by adopting and utilising particular perspectives, and by providing a *methodology* for answering questions concerning the domain(s) considered.

A related, but not identical, purpose is to provide *a safeguard against unscientific approaches* to a problem, issue or theme, like, for example, haphazard and inconsistent choices with regard to terminology, research methodology, and interpretation of results. This purpose, which is a prominent one in mathematics education research, is pursued by articulating underlying assumptions and choices and making them subject to discussion; by situating one's research within some framework; and by explicating its characteristics vis-à-vis possible alternatives. A concise way of expressing this purpose is due to the late David Wheeler (personal communication) when he said "the purpose is simply to keep us honest."

Any mathematics educator has experienced criticism of our field from sceptical if not hostile colleagues in other fields (especially mathematics, psychology and general education) with regard to the foundations of our research and its results. Researchers in the humanities and the social sciences often have encountered similar criticism. This gives rise to the sixth and final purpose of theory in research: to provide (some) *protection against attacks from outside.*

In research on the teaching and learning of mathematics all these purposes seem to be in play, with varying emphases for different researchers. However, as the specific *role* of theory in a given piece of research is only seldom made explicit, let alone discussed, it is not so easy to detect this role empirically. This is, of course, the case with the rather numerous research papers which do not explicitly invoke or employ any theory at all. But it is also the case with publications that do invoke a theory without going beyond the mere invocation as such (Lerman & Tsatsaroni, 2003). This happens pretty often, when some theoretical framework is being referred to in the beginning or at the end of the publication without having any presence in between. In such cases one might be tempted to suppose that the main purpose in play is the last-mentioned one, unless it is just a matter of convention or ritual: there must be some mentioning of theory in a research publication. When a theory is actually employed in a piece of research, not just invoked, its role can be to serve as an overarching framework, to organise a set of observations and interpretations of singular but related phenomena into a coherent whole, to provide distinctions and terminology to the research, or to offer a research methodology. A number of other, more specific, roles have been identified in the chapter by Silver and Herbst. In mathematics education research, one role of theory seems to be predominant, namely that as a purveyor of sets of *hypotheses* put forward to account for observed phenomena. This relates to the overarching purpose of providing explanation. The fact that we usually encounter loosely organised sets of, typically a few, hypotheses rather than elaborate edifices of tightly organized networks of concepts and claims (as required in my definition of theory) is a reflection of the weak theoretical underpinning of our field.

It is characteristic that the majority of the theories actually invoked (but not necessarily employed) in mathematics education research are borrowed from other fields and disciplines. In addition to mathematics, we encounter statistics, psychometrics, philosophy, neuroscience, psychology, pedagogy, education, linguistics, semiotics, sociology, anthropology, political science, and history. There are also examples of what has been named "home-grown theories" developed more or less within the discipline of mathematics education research itself. Without engaging, here, in a discussion of the extent to which they are theories in the sense defined above, let me just mention, for example, APOS, Action-Process-Object-Schema (see, e.g. the chapter by Artigue, Batanero, & Kent), the distinction between concept definition and concept image, the process-object duality and its ramifications into encapsulation, reification, and procepts; the French theories of didactical situations (Brousseau), of didactical transposition (Chevallard), and conceptual fields (Vergnaud). American theories of socio-mathematical norms (Cobb and Yackel), teachers' mathematical knowledge for teaching (Ball and colleagues), and metacognition (Schoenfeld). All of these are treated in previous chapters.

The collection of theories put to use in mathematics education does not seem to be satisfactory or sufficient to serve as the foundation for research on the teaching and learning of mathematics. The theories imported from other fields are of too general a nature to be transposed to offer a sufficient pool of specific results and concrete methodologies, so as to provide complete guidelines for research in our field. The home-grown theories currently available are of too limited scopes to provide a comprehensive coverage of our field.

In Utopia, what would a full-fledged theory of mathematics education look like? I realise that Cobb, in the vein of his advocacy for "theorizing as *bricolage*" in his chapter in this Handbook, would probably deem this question at best irrelevant, at worst impossible. Similarly, Silver and Herbst, in their chapter, make a plea for what they call "an ecumenical approach," while also seeing benefits in contemplating the possibility of a "grand theory of mathematics education." Nevertheless, I think the question is worth taking seriously, in order to envision a potential role model for theory development. Without in any way pretending to be able to answer the question, a theory of mathematics education would, I would like to submit, contain at last the following sub-theories, somewhat similar to what Silver and Herbst call "middle-ranged theories," linked together as a coherent and consistent system:

- a sub-theory of *mathematics as a discipline and a subject* in all its dimensions, including its nature, role, and exercise in *society and culture*;
- a sub-theory of *individuals' and groups' affective notions*, experiences, emotions, attitudes, and perspectives with regard to their actual and potential encounters with mathematics, and the outcomes thereof;
- a sub-theory of *individuals' and groups' cognitive notions*, beliefs, experiences, and perceptions with regard to their actual and potential encounters with mathematics, and the outcomes thereof;
- a sub-theory of *the teaching of mathematics* seen within all its institutional, societal, national, international, cultural and historical contexts;
- a sub-theory of *teachers of mathematics*, as individuals and as communities, including their personal and educational backgrounds and professional identities and development.

All sub-theories have to account for *situating* their objects, processes, situations, and phenomena *in all the contexts and environments that influence them*, be they scientific, biological, anthropological, linguistic, philosophical, economic, sociological, political, or ideological. Each sub-theory has to live up to the general requirements of a scientific/scholarly theory, for example, the ones outlined above, including accounting for the ways in which its claims are obtained and justified. The sub-theories *cannot just be juxtaposed*. They have to be integrated into a coherent and consistent whole, alone because mathematics is a constituent component in all of them.

For each of the five domains pointed to here it is probably possible to create several meaningful sub-theories. When combined this gives rise to a potential of several competing "grand" theories of our field. So we shall presumably never have just one complete theory of mathematics education. However, based on our accomplishments in the past three decades we have the potential to really begin to undertake the task of moving towards such overarching theories.

Although it may be true that the search for a grand theory of mathematics education is futile, at least at the present state of development in our field, what would not be futile would be to undertake a meta-research programme that we might call *theory archaeology*. By this I mean a systematic effort to identify, uncover and analyse the alleged as well as the actual roles of "theory," "theoretical framework," and "theorizing" encountered in a wide selection of research publications in mathematics education research. Such a programme would help us transcend the rather unfortunate state of affairs in our field where terms derived from "theory" are ubiquitous without, in many cases, providing more coverage than did the emperor's new clothes. At least such an endeavour would assist in keeping us honest, to reiterate, at a meta-level, David Wheeler's dictum.

Past Insights Should Not Be Forgotten

From time to time researchers in our field deplore its apparent lack of accumulation. There are several causes for this deficiency. Here I shall concentrate on one of them.

We are not good enough at looking back at what our predecessors have accomplished and to take advantage of it in our current thinking. Our colleagues of the past are the ones that first broke what later became the roads on which we walk today. Their insights were no less deep than ours, but were shaped by experience, intuition and reflection in territories that were much more virgin than the territories at our disposal. These territories have later been cultivated step by step by means of research, much of which was forseen and requested by our "ancestors." But thirty to forty years ago, research, in the modern sense, was scarce. The approaches of our predecessors are sometimes pejoratively referred to as "opinionating." Yet quite a few of their opinions were instrumental in pointing out what was worth investigating. We have a bad habit of discarding writings of the past, unless they are singular either because they are early instances of what later became mainstraim developments, or because they can serve as especially articulate examples of the naïve and undeveloped past that we, thank God, have left long ago (was it yesterday?).

I plead for revisiting past accomplishments, not in order to honor our predecessors, although quite a few of them do indeed deserve it, but because it is in the best interest of our field to revisit the contributions of the past. We need a scholarly researched history of mathematics education research. There do exist excellent contributions to such a history (Kilpatrick, 1992, focusing primarily on the U.S.), but much more is needed, and of an international scope. Being aware of the history of mathematics education research would not only help us find forgotten, or even hidden, treasures, but would also make us less susceptible to the fads and fashion waves that haunt our field from time to time. The observations and comments made at the conclusion of my plenary lecture at ICME-9, Tokyo-Makuhari 2000 (Niss, 2004b), unfortunately do not seem to be outdated today, 6–7 years later. Allow me to quote myself :

> As I see it, there is such a huge body of highly valuable outcomes of excellent research that need to be surveyed, structured, brought together and synthesized. It appears to be one of the weaknesses of our profession that many of us, myself included, tend to write and speak too much and read and contemplate too little. I would conjecture that if more systematic revisiting and synthesizing of research were carried out we would realise that "we know more than we know that we know." (p. 53)

The other chapters of the present Handbook (re)present major steps forward to doing just that, identifying, digesting, surveying and synthesizing significant contributions to mathematics education research, so as to make us more conscious of what we know. Yet, the chapters also show that what we know is still highly diversified and complex, in the sense that it is not easy to produce brief and clear conclusive statements that are both correct, focused and concrete.

CONCLUSION

By its very nature, this chapter presents personal observations and reflections on the field of mathematics education as a research domain, based on all the previous chapters of this Handbook. What would Utopia—of course a point at infinity—in our field look like? Where are we today? What progress have we made during the last fifteen years? What could/should we do in order for us to make substantial new progress in our endeavour to approach Utopia? These are the questions that preoccupied me—and hopefully the readership as well—in the sections above.

Unavoidably, my observations and reflections are premised on my background, experiences, values, opinions, and taste. But I am profiting massively from the work of my colleagues, the authors of this Handbook. Needless to say, it has not been possible in this chapter to do justice to all the preceding ones. I want to conclude by expressing my gratitude to and admiration for the editor and the authors of this Handbook. By their efforts to identify, survey, and review fundamental parts and aspects of mathematics education research they have done invaluable service to our field and to our community, and not least to its newcomers, all of whom would be well advised to seriously study the relevant chapters of this volume at an early stage of their careers. In addition to bringing us up to date with significant accomplishments over the last couple of decades, the authors have succeeded in demonstrating that our field is not only alive and kicking but amidst a vibrant development of reclaiming promising new land. The construction of roads to Utopia is well under way.

REFERENCES

Carraher, T.N., Carraher, D.W., & Schliemann, A.D. (1985). Mathematics in the streets and schools. *Bristish Journal of Educational Psychology, 3,* 21–29.

Gerdes, P. (1996). *Femmes et géometrie en Afrique Australe.* Paris: L'Harmattan.

Grouws, D. (Ed.). (1992). *Handbook of research on mathematics teaching and learning.* New York: MacMillan.

Hoyles, C., Noss, R., & Pozzi, S. (2001): Proportional reasoning in nursing practice. *Journal for Research in Mathematics Education, 32*(1), 4–27.

Kent, P. & Noss, R. (2003). *Mathematics in the university education of engineers (A report to The Ove Arup Foundation).* London: Ove Arup Foundation. (http:www.theovearupfoundation.org/arupfoundation/pages/ViewContent.cfm?RowID=25)

Kilpatrick, J. (1992). A history of research in mathematics education. In D. Grouws (Ed.), *Handbook of research on mathematics teaching and learning* (pp. 3–38). New York: MacMillan.

Leder, G., Pehkonen, E., Törner, G. (Eds.). (2002). *Beliefs: A hidden variable in mathematics education?* Dordrecht, The Netherlands: Kluwer.

Lerman, S. & Tsatsaroni, A. (2003). A sociological description of changes in the intellectual field of mathematics education research: Implications for the identities of academics. http://myweb.lsbu.ac.uk/~lerman/ESRCProjectHOMEPAGE.html

National Council of Teachers of Mathematics. (1989). *Curriculum and evaluation standards for school mathematics.* Reston, VA: Author.

National Council of Teachers of Mathematics. (2000). *Principles and standards for school mathematics.* Reston, VA: Author.

National Research Council. (2001). *Adding it up: Helping children learn mathematics.* Washington, DC: National Academy Press.

Niss, M. (1996). Goals of mathematics teaching, In A. J. Bishop, K. Clements, C. Keitel, J. Kilpatrick, & C. Laborde. (Eds.),

International handbook of mathematics education (Vol. I, pp. 11–47). Dordrecht, The Netherlands: Kluwer.

Niss, M. (2004a). The Danish "KOM" Project and possible consequences for teacher education. In R. Strässer, G. Brandell, B. Grevholm, & O. Helenius. (Eds.), *Educating for the future. Proceedings of an International Symposium on Mathematics Teacher Education,* (pp. 179–190). Gothenburg, Sweden: The Royal Swedish Academy of the Sciences, and National Center for Mathematics Education, University of Gothenburg.

Niss, M. (2004b). Key issues and trends in research on mathematical education. In H. Fujita, Y. Hashimoto, B. R. Hodgson, P. Y. Lee, S. Lerman, & T. Sawada (Eds.), *Proceedings of the Ninth International Congress on Mathematical Education* (pp. 37–57). Boston: Kluwer.

Niss, M. (2006). The concept and role of theory in mathematics education. Proceedings of NORMA05: Relating practice and research in mathematics education—Fourth Nordic Conference on Mathematics Education, Trondheim, Norway, 2nd–6th September, 2005.

Noss, R., & Hoyles, C. (1996). The visibility of meanings: Modelling the mathematics of banking. *International Journal of Computers for Mathematical Learning, 1*(1), 3–31.

Noss, R., Hoyles, C., & Pozzi, S. (2002). Abstraction in expertise: A study of nurses' conception of concentration. *Journal for Research in Mathematics Education, 33,* 204–229.

Nuñez, T., Schliemann, A.D., & Carraher, D.W. (1993). *Street mathematics and school mathematics.* Cambridge, UK: Cambridge University Press.

Organisation for Economic Co-operation and Development. (2003). *The PISA 2003 assessment framework: Mathematics, reading, science and problem solving knowledge and skills.* Paris: Author.

Poincaré, H. (1908). L'avenir des mathématiques. *Revue générale des sciences pures et appliqués* [The future of mathematics. *General journal of pure and applied sciences*], *19,* 930–939.

Sfard, A. (2005). What could be more practical than good research?. *Educational Studies in Mathematics, 58*(3), 393–413.

Steiner, H.-G. (1984). Theory of Mathematics Education (TME). In Steiner et al. 1984, see below. pp. 16–41.

Steiner, H.-G. (1985). Theory of mathematics education (TME): An introduction. *For the Learning of Mathematics, 9*(2), 24–33.

Steiner, H.-G., Balacheff, N., Mason, J., Steinbring, H., Steffe, L.P., Brousseau, G., Cooney, T.J., & Christiansen, B. (1984, November). *Theory of mathematics education (TME).* Occasional paper 54. Bielefeld, Germany: Institut für Didaktik der Mathematik der Universität Bielefeld.

AUTHOR NOTE

I would like to extend my sincere thanks to the editor, Frank K. Lester, Jr., for having given me the unique opportunity to reflect on the chapters in this Handbook. Although this chapter has had no reviewers in the classical sense, its final version has greatly benefitted from his comments and suggestions.

About the Contributors

Michèle Artigue earned a Ph.D. in Mathematical Logic and began her career doing research in that area. She progressively moved toward doing research in mathematics education, thanks to her involvement in the activities of research, innovation and teacher training developed at the Institute of Research in Mathematics Education at the University Paris 7. In this field, beyond theoretical contributions on the relationships between epistemology and didactics, and on didactical engineering, her main research topics have been the teaching and learning of mathematics at university level, and especially the didactics of calculus and analysis, and the integration of computer technologies into mathematics education. She is currently full professor in the Mathematics Department at the University Paris 7 and director of the master program in didactics at this University. She has many editorial and scientific responsibilities, and since 1998 has been vice-president of the International Commission on Mathematics Instruction.

Deborah Loewenberg Ball is William H. Payne Collegiate Professor and Dean of the School of Education, at the University of Michigan. Ball's work draws on her many years of experience as an elementary classroom teacher. Her research focuses on mathematics instruction and on interventions designed to improve its quality and effectiveness. Her research groups study the nature of the mathematical knowledge needed for teaching and develop survey measures that have made possible analyses of the relations among teachers' mathematical knowledge, the quality of their teaching, and their students' performance. Of particular interest in this research is instructional practice that can intervene on significant patterns of educational inequality in mathematics education. Currently a member of the National Math Panel, Ball served on the Glenn Commission on Improving Mathematics Education for the 21st century, chaired the Rand Study Panel on programmatic research in mathematics education, co-chaired the international Study on the Professional Education of Teachers of Mathematics sponsored by the International Commission on Mathematical Instruction, and she is a trustee of the Mathematical Sciences Research Institute in Berkeley.

Carmen Batanero is Senior Lecturer of Mathematics Education at the University of Granada, Spain, where she teaches mathematics education and statistics education, and supervises doctoral dissertations on statistical education. Before joining this Department in 1988, she worked in the Department of Statistics where she taught probability and biostatistics courses for 11 years and completed her Ph.D. in Statistics (Stochastic Point Processes and Systems Reliability). Her primary research interest is statistics education, with special emphasis in the teaching of statistical inference. She is also interested in the theoretical and methodological bases for research on mathematics education and the education of mathematics teachers to teach statistics. Her publications include research papers, chapters in some international books, and books directed to teachers. She was co-editor of *Statistics Education Research Journal* (2002–2003), President of the International Association for Statistical Education (IASE) (2001–2003), and member of the International Commission for Mathematics Instruction (ICMI) Executive Committee (2003–2006). She also helped to establish the stochastic group at PME (Psychology of Mathematics Education). She has served on the International Programme Committee of several international conferences and is currently coordinating a Joint ICMI /IASE Study on Statistics Education.

Dan Battey is an Assistant Professor of Mathematics Education at Arizona State University, where he has been on the faculty since 2005. Previously, he was a post-doctoral fellow in Diversity in Mathematics Education at the UCLA Graduate School of Education & Information Studies. His work centers on engaging urban elementary school teachers in opportunities to learn within and from their practice of teaching mathematics in a way that sustains and generates change. Drawing on research of students' mathematical knowledge, he studies teacher knowledge, identity, and practice as well as equity issues within the context of urban schools.

Michael T. Battista is Professor of Mathematics Education in the College of Education at Michigan State University. Most of his research focuses on the learning and teaching of geometry and geometric measurement, and on the use of technology in mathematics teaching. He has served on the editorial panel for the *Journal for Research in Mathematics Education*, is one of the authors of the geometry modules for the *Investigations in Number, Data, and Space* curriculum, is the geometry author for the grades 3-6 *Navigations* problem solving and reasoning books, is on the editorial panel for NCTM's 71st yearbook on geometry (2009), and is serving on NCTM's *Research Committee*. He has also served as a workshop co-leader in the Mathematical Association of America's *Preparing Mathematicians to Educate Teachers* program. Currently he is directing two NSF grants, the first to develop a *Cognition Based Assessment* system for elementary mathematics, and the second to investigate how elementary teachers learn, make sense of, and use research on students' mathematical thinking. At Michigan State, he is co-leader of the mathematics group for the Carnegie Foundation funded *Teachers for a New Era* project, and he is faculty leader for secondary mathematics in the secondary teacher education program.

Alan J. Bishop is Emeritus Professor of Education at Monash University, Melbourne, Australia where he has been since 1992. He took his PhD at Hull University in UK, then worked as Lecturer in mathematics education at Cambridge University, UK, from 1969–1992. His first area of research concerned visualisation, spatial ability and geometry learning, but in the 1970s he became more interested in socio-cultural aspects of mathematics education. He was Editor of the journal *Educational Studies in Mathematics* from 1978-90, has been an Advisory Editor to that journal since then, and is Managing Editor of the book series *Mathematics Education Library* published by Springer. He was Chief Editor of the two international handbooks of mathematics education published by Kluwer in 1996 and 2003, and has published numerous books, chapters and papers.

Glendon W. Blume is Professor of Education (Mathematics Education) at the Pennsylvania State University, where he has been a member of the faculty since 1986. His primary research interests focus on the areas of technology in mathematics teaching and learning and on secondary mathematics teachers' understanding of mathematics. He has been a co-Principal Investigator or Faculty Research Associate on grants funded by the National Science Foundation for over 15 years. He has served on the editorial panel for the 2005 Yearbook of the National Council of Teachers of Mathematics and has served on or chaired several committees of the Council. He has been a co-editor of the Yearbooks of the Pennsylvania Council of Teachers of Mathematics since 1989, and he has co-authored four book chapters that provide interpretive reports based on data from the National Assessment of Educational Progress from 1992 through 2003.

David W. Carraher is Senior Scientist at TERC, in Cambridge, MA, where he directs the TERC-Tufts *Early Algebra, Early Arithmetic Project* (www.earlyalgebra.terc.edu). He has been the Principal Investigator in NSF-funded longitudinal studies of the development, implementation, and evaluation of Early Algebra lessons in grades 2 to 5. He has published widely about learning mathematics in and out of school, rational number, and early algebra, and is the author of prize-winning mathematics education software. He is the director of research of the NSF-MSP project, *The Fulcrum Institute for Education in Science*, another collaboration between TERC and Tufts University. As Professor of Psychology in Brazil, he carried out influential research on everyday mathematics in and out of school. He chaired the scientific program committee of PME-95 (Psychology of Mathematics Education), edited, with R. Nemirovsky, a CD-based *JRME* Monograph about the *Uses of Video in Mathematics Education Research*, and co-edited, with Jim Kaput and Maria Blanton, *Algebra in the Early Grades*, published by Erlbaum in 2007. In 2006 he won the Premier Concours de Photos Numériques, Ste. Suzanne, Mayenne, France.

Douglas H. Clements is Chancellor's Professor of Education at University of Buffalo, SUNY. His primary research interests lie in the areas of the learning and teaching of geometry, computer applications in mathematics education, the early development of mathematical ideas, and the effects of social interactions on learning. His most recent interests are in creating, using, and evaluating a research for research-based curriculum and in taking successful curriculum to scale using technologies and learning trajectories. He has published over 90 refereed research studies, 6 books, 50 chapters, and 250 additional publications. He has directed several projects funded by the National Science Foundation and the U.S. Education Department's Institute of Educational Sciences, including curriculum development, investigations of children's development of geometric and spatial ideas, and experiments using large-scale randomized trials. His most recent projects include studies of the implementation of a research-based intervention including the *Building Blocks* preschool curriculum, which was developed with NSF funding. He was chair of the Editorial Panel of NCTM's research journal, the *Journal for Research in Mathematics Education*.

Paul Cobb is Professor of Mathematics Education at Vanderbilt University, where he has been a member of the faculty since 1992. He teaches courses in elementary mathematics methods, design research methodology, and the institutional setting of mathematics teaching and learning. His research interests focus on instructional design, the classroom microculture, and the broader institutional setting of mathematics teaching and learning. He is an elected member of the National Academy of Education and received the Hans Freudenthal Medal for cumulative research program over the prior ten years from the International Commission on Mathematics Instruction (ICMI) in 2005. He an Invited Fellow of the Center for Advanced Studies in the Behavioral Sciences and received the award for outstanding article published in the *Journal for Research in Mathematics Education* in 1996. He currently serves on the editorial boards of seven journals in mathematics education and the learning sciences.

Erik De Corte is Emeritus Professor of Educational Psychology and member of the Center for Instructional Psychology & Technology at the University of Leuven, Belgium. His major research interest is to contribute to the development of theories of learning from instruction and the design of powerful learning environments, focusing thereby on learning, teaching, and assessment of thinking and problem solving, especially in mathematics. He was the first President (1985–1989) of the European Association for Research on Learning and Instruction (EARLI), and the founding editor of the EARLI journal *Learning and Instruction* (1990–1993). From 1987 until 2002 he was associate editor of the *International Journal of Educational Research.* He co-edited (with the late F.E. Weinert) the *International Encyclopedia of Developmental and Instructional Psychology* (1996). In 1997 he received the EARLI "Oeuvre Award for Outstanding Contributions to the Science of Learning and Instruction." He was President of the International Academy of Education from 1998 until 2006. Respectively in 2000 and 2003 he was conferred the doctorate honoris causa of the Rand Afrikaans University, Johannesburg, and the University of the Free State, Bloemfontein, South Africa. During the academic year 2005–2006 he stayed as a Fellow at the Center for Advanced Study in the Behavioral Sciences at Stanford.

Jan de Lange is a Full Professor in Mathematics Education in the Department of Mathematics of the Faculty of Science at Utrecht University, The Netherlands. He has worked at the institute that is now known as the Freudenthal Institute since 1976 and has been leading this institute since. His interest in issues in mathematics education ranges from pre-primary to tertiary, from modeling and applications to assessment, from software to the web, and from educational design to implementation problems. Recently he has explored how cognitive psychology and educational neurosciences can, in some kind of complementarity, contribute to the theory of learning mathematics. He has been a member of the Mathematical Sciences Education Board at the National Research Council, Washington D.C., on the National Committee for TIMSS, on the international TIMSS-R committee, is Chair of the International Mathematics Experts Group of the PISA study of OECD (Paris), a core member of the Numeracy and Literacy Commission of OECD, and member of a content validity commission of NAEP. He has been PI on many national and international projects, published several hundred articles, and delivered lectures in more than sixty countries. He is visiting Professor at the University of Wisconsin at Madison, and Director of FI-us at the University of Colorado at Boulder.

Thomas P. Dick is Professor of Mathematics and Coordinator of Collegiate Mathematics Education at Oregon State University, where he has been a member of the faculty since 1986 and directs both the Mathematics Learning Center and the Math Excel program. His primary research interests lie in the areas of advanced mathematical thinking, mathematical discourse, and the influences of technology on mathematics teaching and learning, especially calculus. He currently serves on the editorial panel for the *Journal for Research in Mathematics Education* (2004–2007) and as an associate editor for *School Science and Mathematics.* He is a former chair of the Committee for Research in Undergraduate Mathematics Education (American Mathematical Society/Mathematical Association of America) and a former chair of the College Board's Advanced Placement Calculus Development Committee. He is an advisor on several projects involving the use of technological tools in mathematics instruction.

The Diversity in Mathematics Education Center for Learning and Teaching (DiME) is a consortium of three universities and three school districts that focuses on issues of equity in mathematics education. Consortium partners include the University of Wisconsin-Madison, the University of California at Los Angeles, and the University of California at Berkeley. About a dozen faculty and 30 graduate fellows and post doctoral fellows participate in DiME. DiME seeks to create new knowledge and research, new resources and tools, and a new generation of researchers and instructional leaders who are capable of making significant progress on issues of equity and diversity in mathematics education. Among the issues addressed by DiME research are (a) how opportunities for students from non-dominant cultures to learn mathematics with understanding are supported and how they are limited in schools and classrooms; (b) how teachers can be supported to engage in culturally relevant pedagogy, which builds on students' cultural histories and engages them in using mathematics to address issues of social justice; and (c) how mathematics learning is mediated at multiple levels by issues of race and power. The chapter on Race, Culture and Mathematics Education is based on a center-wide initiative to address various explanations for differential mathematics achievement. Building on that work, an editorial panel was selected to take the lead in writing the handbook chapter. The panel consisted of the following DiME members. **Vanessa R. Pitts Bannister** is an Assistant Professor of Mathematics Education at the Virginia Polytechnic Institute and State University located in Blacksburg, VA. She is interested in the relationships between teachers' pedagogy and knowledge of equity/diversity issues, curriculum, subject matter and student thinking. **Tonya Gau Bartell** is an Assistant Professor of Mathematics Education at the University of Delaware. Her primary research interest lies in examining how teachers learn to teach mathematics for social justice, including building understanding of how teachers might actually practice critical pedagogy and how teacher educators might help in-service teachers do so. **Dan Battey** is an Assistant Professor of Mathematics Education at Arizona State University. His research centers on teacher learning and equity issues in mathematics education, particularly in urban schools. **Victoria M. Hand** is an Assistant Professor of Mathematics Education at the University of Wisconsin at Madison. Her research investigates how particular pedagogical approaches, tools, and activity structures in mathematics classrooms afford and constrain students' negotiation of their sociocultural practices and identities. **Joi Spencer** is an Assistant Professor of Mathematics Education at the University of San Diego. Her research is interested in how discourse around race shapes the learning opportunities of African American children in mathematics. DiME Fellows also involved in writing the chapter include Filiberto Barajas, Rozy Brar, Kyndall Brown, Indigo Esmonde, Mary Q. Foote, Charles Hammond, Carolee Koehn, Mara G. Landers, Mariana Levin, Shiuli Mukhopadhyay, Ann Ryu, Marian Slaughter, Anita A. Wager. DiME directors Thomas Carpenter, Megan Franke, and Alan Schoenfeld provided oversight and guidance for the writing of the chapter.

Joan Ferrini-Mundy is a University Distinguished Professor in Mathematics Education at Michigan State University, where she has been Associate Dean and Director for Science and Mathematics Education in the College of Natural Science since 1999. Her experiences include two years as a Visiting Scientist at the National Science Foundation, and four years as the Director of the Mathematical Sciences Education Board at the National Research Council. For 18 years, Ferrini-Mundy was a member of the Mathematics Department at the University of New Hampshire. She has been active with the National Council of Teachers of Mathematics, serving on its Board of Directors from 1993 to 1996, and as chair of the Writing Group for the 2000 *Principles and Standards for School Mathematics.* Ferrini-Mundy has received numerous federal, state, and private foundation grants for projects in research and teacher education. Currently she is PI for an NSF Comprehensive Mathematics and Science Partnership project involving about 60 school districts in Michigan and Ohio. She is also serving as Project Director for an initiative with the Michigan Department of Education to review and improve policies relative to teacher preparation. Her research interests include mathematics teacher knowledge and K–12 mathematics program improvement.

Robert E. Floden is Professor of Teacher Education, Measurement & Quantitative Methods, and Educational Psychology at the Michigan State University College of Education, where he has been a faculty member since 1977. His research has examined the effects of education policies on teaching and learning, with a special emphasis on roles of preservice teacher preparation and professional development. He is currently co-PI of Michigan State University's *Teachers for a New Era* initiative and co-PI on a project developing measures of teachers' mathematical knowledge for teaching algebra. From 1987 to 1989 Floden spent time at the Universität Tübingen as an Alexander von Humbolt Fellow. He was Features Editor of *Educational Researcher* from 1993 to 1995 and Editor of *Review of Research in Education* from 2003 to 2005. Floden served as president of the Philosophy of Education Society in 2004-5 and received the Margaret B. Lindsey Award for Distinguished Research in Teacher Education from the American Association of Colleges for Teacher Education in 2006. Floden's work has been published in the *Handbook of Research on Teaching*, the *Handbook of Research on Teacher Education*, and many journals.

Dr. Helen J. Forgasz is a Senior Lecturer in the Faculty of Education, Monash University (Australia), where she has been a member of the academic staff for four years. Her research interests in mathematics education include: gender equity issues, the affective domain, technology use for mathematics learning, and learning settings. From 2003–2006, Helen was the editor of the *Mathematics Education Research Journal* and is currently a member of the editorial boards of *Educational Studies in Mathematics* and the *Journal for Mathematics Teacher Education.* She is a member of the International Committee for the International Group for Psychology in Mathematics Education, and has served on the executive committee of the Mathematics Education Research Group of Australasia as Vice President (Research) 2001–2003 and Vice President (Conferences) 1999–2001. Helen has presented many papers at national and international conferences, and has published widely in scholarly and professional books and journals.

Megan Loef Franke is an associate professor in the Graduate School of Education and Information Studies at UCLA and the Director of Center X: Where Research and Practice Intersect for Urban School Professionals. Dr. Franke's research focuses on understanding teacher learning and professional development, particularly as it pertains to the teaching of mathematics. She uses her developing understanding of these issues to consider issues of urban school reform and social justice. Over the past 20 years Dr. Franke has worked with the Cognitively Guided Project to support and learn from teachers as they attempt to make use of the development of children's mathematical thinking. She has published in the *Journal for Research in Mathematics Education*, *Teaching and Teacher Education*, *American Educational Research Journal*, and *Journal for Mathematics Teacher Education*.

Brian Greer is currently an independent scholar, having held positions in the School of Psychology, Queen's University, Belfast, 1969–2000, and Department of Mathematics and Statistics, San Diego State University, 2000–2003. Major research and theoretical interests have included multiplicative reasoning, word problem solving, and the relationships between cognitive psychology and mathematics education. His current focus is on the cultural, historical, and political contexts of mathematics education

Douglas A. Grouws is Professor of Mathematics Education and William T. Kemper Fellow at the University of Missouri. He has a long history of research and scholarship in the mathematics education field. He was a member of the National Academy of Science committee that produced the recent report, *On Evaluating Curricular Effectiveness: Judging the Quality of K–12 Mathematics Evaluations*. He is author or co-author of three books including the editorship of the first *Handbook of Research on Mathematics Teaching and Learning*. A lifetime member of the National Council of Teachers of Mathematics, he currently serves on NCTM's task force on the interpretation of the results of the NAEP assessments in mathematics. He is co-Principal Investigator at the Center for the Study of Mathematics Curriculum (CSMC) and he currently directs two National Science Foundation funded projects: Mathematics Through Technology (MTT) and Comparing Options in Secondary Mathematics: Investigating Curriculum (COSMIC).

Guershon Harel is Professor at the Mathematics Department at the University of California, San Diego. Previously he served as Associate Editor of the *American Mathematical Monthly*, co-editor of the *Research in Collegiate Mathematics Education Series*, and Chair of the Editorial Board of the *Journal for Research in Mathematics Education*. Harel has research interest in cognition and epistemology of mathematics and their application in mathematics curricula and the education of mathematics teachers. Until the mid 1990s, Harel's research interest revolved around the notions of multiplicative conceptual field and advanced mathematical thinking, with particular attention to the concept of function, proof, and the learning and teaching of linear algebra. He is co-editor of two books in these areas: *The Development of Multiplicative Reasoning in the Learning of Mathematics* and *The Concept of Function: Aspects of Epistemology and Pedagogy*. In the mid 1990s, he began to center his attention on the learning and teaching of proof, and since then he has been working collaboratively in this area with Larry Sowder.

M. Kathleen Heid is Distinguished Professor of Education and Professor of Mathematics Education at The Pennsylvania State University, where she has been a member of the faculty since 1984. Her research has centered on the impact of technology on the teaching and learning of mathematics, particularly algebra and calculus, and on mathematical knowledge of secondary teachers. She served on the Board of Directors of the National Council of Teachers of Mathematics from 2003 to 2006, served on the Board of Governors for the Mathematical Association of America from 1998 to 2003, and is currently Editor-Elect of the *Journal for Research in Mathematics Education*. She has co-directed several federally funded mathematics education projects centered on creating computer-intensive curricula for secondary schools and preparing teachers to teach mathematics in technology-intensive environments, she has served on the editorial boards of several research journals, and she is co-editor of *Research on Technology in the Teaching and Learning of Mathematics: Syntheses, Cases, and Perspectives*. She is currently co-PI for the NSF-funded Mid-Atlantic Center for Mathematics Teaching and Learning.

Patricio Herbst is an Associate Professor of Education at the University of Michigan, where he has been a member of the faculty since 1999. His primary research interest is on mathematics instruction in classrooms, particularly in relation to the teachers' management of students' reasoning and proving in high school geometry, and on the design and use of multimodal representations of teaching to elicit teachers' perspectives on different kinds of instruction. He has served on the Executive Board of the AERA Special Interest Group on Research in Mathematics Education and is on the editorial boards of *Journal of Mathematics Teacher Education, Educational Studies in Mathematics, and Recherches en Didactique des Mathématiques*.

James Hiebert is the Robert J. Barkley Professor of Education at the University of Delaware, where he teaches in programs of teacher preparation, professional development, and doctoral studies. His professional interests focus on mathematics teaching and learning in classrooms. He has edited books on students' mathematics learning and co-authored the books *Making Sense: Teaching and Learning Mathematics with Understanding* and *The Teaching Gap: Best Ideas from the World's Teachers for Improving Education in the Classroom*. He served on the National Research Council committee that produced *Adding It Up* and *Helping Children Learn Mathematics*, was director of the mathematics portion of the TIMSS 1999 Video Study, and is a PI on the NSF-funded Mid-Atlantic Center for Teaching and Learning Mathematics. He has served on the editorial boards for *American Educational Research Journal, Cognition and Instruction, Elementary School Journal, Journal for Research in Mathematics Education, Journal of Educational Psychology, Journal of Mathematical Behavior,* and *Mathematics Thinking and Learning.*

Heather C. Hill is an assistant professor and associate research scientist at the University of Michigan. Her primary work focuses on developing measures of mathematical knowledge for teaching, and using these measures to evaluate public policies and programs intended to improve teachers' understanding of this mathematics. Her other interests include the measurement of instruction more broadly, instructional improvement efforts in mathematics, and the role that language plays in the implementation of public policy. She received a Ph.D. in political science from the University of Michigan in 2000 for work analyzing the implementation of public policies in law enforcement and education. She has served as section chair for AERA division L (politics and policy), and on the editorial board of *Journal for Research in Mathematics Education.* She is the co-author, with David K. Cohen, of *Learning Policy: When State Education Reform Works* (Yale Press, 2001).

Graham A. Jones, is an Emeritus Professor at both Illinois State University, where he was a faculty member from 1991–2002, and Griffith University, Australia, where he was Pro-Vice-Chancellor from 1986–1991. His major research activities have focused on the formulation and validation of cognitive models in probability, statistics, and number sense, and the use of these models in teaching and learning at the elementary and middle school levels. He has also investigated teacher knowledge in geometry, probability, and statistics during teacher enhancement projects. His publications include numerous research articles and books in these areas of interest and he has recently edited the book, *Exploring Probability in School: Challenges for Teaching and Learning* (Springer). From 1979 until 1984 he was the first president of the *Mathematics Education Research Group of Australasia* (1979–1984), and has served for varying periods of time on the editorial boards of the *Journal for Research in Mathematics Education, Mathematics Thinking and Learning, Mathematics Education Research Journal* (Australia) and *The Mathematics Educator* (Singapore). For the past 20 years he has regularly been involved in coordination roles, in teacher development and elementary education, at International Congresses in Mathematical Education.

Elham Kazemi is Associate Professor of Mathematics Education at the University of Washington, where she has been on the faculty since 1999. Her research interests include studying the development of intellectual community among teachers and students in elementary mathematics. In particular, she has studied how teachers learn about children's reasoning in classroom practice and how teachers' collaborative inquiry creates new cultures of learning within schools. Her work with teachers is informed by her study of children's social and intellectual experiences in classrooms, and in recent years comparing children's experience across mathematics and literacy. Over the last six years, she has been an investigator in several federally-funded grants designing and studying professional education, especially for teacher leaders. Her works appears in places such as *Elementary School Journal, Theory into Practice, Journal of Mathematical Behavior,* and *Journal for Mathematics Teacher Education.*

Phillip Kent is a Research Officer in mathematics education at the School of Mathematics, Science and Technology, University of London Institute of Education. His research specializes in mathematics in workplaces and in professional practice, particularly in relation to the evolution of working practice due to information technology, and how new mathematical technologies may change the learning and use of mathematics in workplaces. He has done nationally-funded research projects and published widely in this area. He is a member of the London Knowledge Lab, an inter-disciplinary research institute of the University of London. Before joining the Institute of Education in 2001, he was at Imperial College, University of London, where he gained a PhD in Applied Mathematics in 1992, and worked for eight years in undergraduate curriculum development projects.

Carolyn Kieran is Professor of Mathematics Education at the Université du Québec à Montréal, where she has been a faculty member of the Department of Mathematics since 1983. Her primary research interest is the learning and teaching of algebra, with a particular focus on the roles played by computing technology. She served as President of the International Group for the Psychology of Mathematics Education from 1992 to 1995 and as a member of the Board of Directors of the National Council of Teachers of Mathematics from 2001 to 2004. Past contributions also include the Mathematics Learning Study committee and its project volume *Adding It Up*, the International Program Committee for the 12th ICMI Study on Algebra, the chair of the editorial panel of the *Journal for Research in Mathematics Education*, the editorial board of *A History of School Mathematics*, and the vice presidency of the Canadian Mathematics Education Study Group. Recent publications comprise chapters in the *Handbook of Research on the Psychology of Mathematics Education*, and *The Future of the Teaching* *and Learning of Algebra*, and an article in the *International Journal of Computers for Mathematical Learning*.

Susan J. Lamon is a professor in the Department of Mathematics, Statistics, and Computer Science at Marquette University, where she has been on the faculty since 1989. Her research employs children's and adults' thinking to investigate the nature and development of proportional reasoning, its relationships to rational number knowledge, and its enhancement through the process of mathematical modelling.

Her research has led to the development and testing of classroom materials designed to facilitate rational number knowledge and proportional reasoning in elementary through university-level mathematics. In addition to serving on the executive committee of the International Committee on Teaching Mathematical Modelling and Applications (ICTMA), she reviews articles related to rational numbers and proportional reasoning for 10 major journals, serves on advisory boards for several federally-funded projects, and consults internationally with Ministries of Education to improve mathematics curricula.

Cynthia W. Langrall is a Professor in the Mathematics Department at Illinois State University. Her research interests are in the development of elementary and middle school students' probabilistic and statistical reasoning. Publications in these areas include several book chapters and articles in *Mathematical Thinking and Learning*, *Journal for Research in Mathematics Education*, and *Educational Studies in Mathematics*. She is also interested in research on teaching and teacher professional development and works with teachers in settings around the country to enhance the mathematics learning of all students. At Illinois State, her teaching is centered on theory and research courses for graduate students and mathematics content and methods courses for elementary and middle school prospective teachers.

Richard Lesh, Rudy Distinguished Professor of Learning Sciences at Indiana University, has been the Director of the Center for Research on Learning & Technology and is the Chair of the Learning Sciences Program in the Indiana University's School of Education. More than 250 journal articles and books reflect his international reputation concerning learning and problem solving in mathematics education, and research and assessment design in mathematics and science education. His projects, books, and software have been directed toward audiences ranging from kindergarten through college, as well as from teachers to researchers. He has been a Dean for Research at Northwestern University and Purdue University, a Principal Research Scientists at the Educational Testing Service in Princeton, and a Program Officer for the Research Division at the Education Directorate for the National Science Foundation. Current or recent publications include: *The Handbook of Design Research in Mathematics & Science Education* (Kelly & Lesh, 2000), *Foundations for the Future in Mathematics Education* (Lesh, Hamilton & Kaput, in press), and *Beyond Constructivism: Models & Modeling Perspectives on Mathematics Problem Solving, Learning & Teaching* (Lesh & Doerr, 2003). He also is the associate editor for the *International Journal for Mathematical Thinking & Learning*.

Frank K. Lester, Jr. is Chancellor's Professor of Education, Martha Lea and Bill Armstrong Chair of Teacher Education, and Professor of Mathematics Education and Cognitive Science at Indiana University, where he has been a member of the faculty since 1972. His primary research interests lie in the areas of mathematical problem solving and metacognition, especially at the elementary and middle school levels. During the past 10 years he has become increasingly interested in the study of classroom assessment of problem solving and higher-order thinking in mathematics, as well as the theoretical and philosophical foundations for research on mathematics learning, teaching, and curriculum. From 1991 until 1996 he was editor of the *Journal for Research in Mathematics Education* and from 1999–2002 he was a member of the Board of Directors of the National Council of Teachers of Mathematics. He also served on the International Steering Committee for International Group for Psychology in Mathematics Education from 1988–1992. He has been on the advisory boards of several federally-funded projects and on the editorial boards of various journals, including *International Journal of Research on Mathematics, Science, and Technology Education*, *Journal of Educational Psychology*, *Mathematics Education Research Journal*, and *Mathematics Thinking and Learning*.

Jennifer Lewis is a doctoral student in mathematics education and teacher education at the University of Michigan, having been a teacher in elementary and middle schools. Her research focuses on teacher education efforts designed to improve mathematics instruction. She is also interested in research that examines education across the professions.

Edward S. Mooney is an associate professor of mathematics education at Illinois State University. His research has focused on middle school students' understanding of statistics. He has authored or coauthored articles in *Journal of Education for Students Placed At Risk*, *Mathematical Thinking and Learning*, and *Teaching Statistics*.

Mogens Niss is Professor of Mathematics and Mathematics Education. He was trained as a pure mathematician (specializing in topological measure theory) at the University of Copenhagen, where he spent the first years of his academic career. In 1972 he joined the founding staff of Roskilde University, where he has remained ever since. This caused a reorientation of his research interests, which today include the teaching and learning of mathematical applications and modeling, the justification problem in mathematics education from societal perspectives, assessment in mathematics education, and the nature and role of mathematics education as an academic field. Within these fields he has published numerous papers and books. He is a member of the editorial board of several journals. During the years 1987–1998 he was a member of the Executive Committee of The International Commission on Mathematical Instruction, of which he was the Secretary General, 1991–1998. He has served on a multitude of national and international committees, commissions, and governing boards, often as a chair. He was the Chair of the International Programme Committee for ICME-10, held in Copenhagen in 2004. Also, he was the Director of the Danish KOM project, 2000–2002, and the Director of the Danish National Graduate School of Mathematics and Science Education.

Randy Philipp is professor of mathematics education in the School of Teacher Education at San Diego State University and a member and former associate director of the university's Center for Research in Mathematics and Science Education. His research focus is on the relationships among prospective and practicing mathematics teachers' knowledge, beliefs, and practices. He directed the federally-funded research and development project *Integrating Mathematics and Pedagogy* (IMAP). In addition to publications, video products, and a web-based beliefs survey, a Children's Mathematical Thinking course was developed and institutionalized through IMAP. He and Vicki Jacobs are currently co-directing federally-funded research to map a trajectory for the evolution of elementary school teachers engaged in sustained professional development. He coauthored *Middle-Grade Mathematics Teachers' Mathematical Knowledge and Its Relationship to Instruction* with Judy Sowder, Barbara Armstrong, and Bonnie Schappelle. He has served as associate editor of the *Journal for Research in Mathematics Education*, co-chair of the American Educational Research Association (AERA) Special Interest Group for Research in Mathematics Education, and chair of the AERA annual program for Division C, Section 2. In 2006 he became the first recipient of the Association of Mathematics Teacher Educators' Award for Excellence in Teaching.

Norma Presmeg has been a Professor in the Mathematics Department at Illinois State University since 2000. Prior to this appointment she served on the faculty of Florida State University for ten years, after emigrating from South Africa, where she served at the University of Durban-Westville from 1986 to 1990. Following from her doctoral research at Cambridge University (1982–1985), her research interests over the last 25 years have involved the potential and pitfalls of visualization in the teaching and learning of mathematics, equity and the role of culture in this teaching and learning, the significance of mathematical prototypes, metaphor and metonymy, and more recently, the use of semiotics as a theoretical lens for research in mathematics education. She is currently an editor of *Educational Studies in Mathematics*, and book review editor of *Journal for Research in Mathematics Education.* She served as Co-Chair of the Special Interest Group: Research in Mathematics Education of the *American Educational Research Association* from 2000 to 2002, and she served on the International Steering Committee for the International Group for the Psychology of Mathematics Education (PME) from 1997 to 2001. She was a plenary speaker at the 30th Annual Meeting of PME in Prague, July 2006.

Janine Remillard is an Associate Professor at the University of Pennsylvania and chair of the Foundations and Practices of Education Division. Her research interests include teachers' use of and learning from mathematics curriculum materials and mathematics teaching and learning in urban schools and communities. She is currently a co-P.I. of *MetroMath: The Center for Mathematics in America's Cities*, an NSF-funded Center for Learning and Teaching and a research associate with the Center for the Study of Mathematics Curricula. Her work has been published in *Review of Educational Research, Curriculum Inquiry, Journal for Research in Mathematics Education, Elementary School Journal, Journal of Mathematics Teacher Education, the School Community Journal, Urban Review* and. *Mathematics Thinking and Learning.* She co-editor of the forthcoming volume: Teachers' Use of Mathematics curriculum Materials: Research Perspectives on the Relationship between Teachers and Curriculum Materials.

Celia K. Rousseau is an assistant professor in the Department of Instruction and Curriculum Leadership at the University of Memphis. Her primary research interests lie in the areas of equity, critical race theory, and mathematics education in urban schools. She has authored or coauthored several articles and book chapters on these topics. She recently co-edited a book entitled: *Critical Race Theory in Education: All God's Children got a Song.* In addition, she has been involved in several projects in the Memphis area related to the professional development of secondary mathematics teachers.

Julie Sarama is Associate Professor of mathematics education at the University at Buffalo, SUNY. She conducts research on the implementation and effects of her own software environments in mathematics classrooms, young children's development of mathematical concepts and competencies, implementation and scale-up of educational reform, and professional development. She is co-author of the award-winning *Turtle Math*, as well of over 40 refereed articles, 16 chapters, 7 units of the *Investigations in Number, Data, and Space* project, and the research-based *Building Blocks* curriculum. She also has co-authored 1 book, 20 software titles and many additional curriculum, journal, and magazine publications. She is current directing several projects funded by the National Science Foundation and the U.S. Education Department's Institute of Educational Sciences (IES), including two large-scale research projects investigating the implementation of educational reform and the professional development that is a necessary component of large-scale innovations. She has taught secondary mathematics and computer science, gifted math at the middle school level, preschool and kindergarten mathematics enrichment classes, and mathematics methods and content courses for elementary to secondary teachers.

Analúcia D. Schliemann is Professor of Education at Tufts University. She studied in Brazil, France, and England, where she obtained her Ph.D. in Developmental Psychology, and she was a Fulbright Scholar at LRDC, University of Pittsburgh. In Brazil, she worked for 20 years at the Federal University of Pernambuco. Her work on everyday mathematics has been published in English, Portuguese, Spanish, and Japanese. Some of her studies appear in *Street Mathematics and School Mathematics* (with T. Nunes and D. Carraher, New York: Cambridge University Press, 1993). At Tufts (since 1994), her research focuses on children's early algebraic reasoning and learning through classroom activities designed to explore the algebraic character of arithmetic in elementary school (see www.earlyalgebra.terc.edu). She co-authored, with D. Carraher and B. Brizuela, *Bringing Out the Algebraic Character of Arithmetic*, published by Erlbaum in 2007. She is equally at home in Theories of Learning and Development, Mathematics Education, Developmental Psychology, and Cross-Cultural Psychology.

Alan Schoenfeld holds the Elizabeth and Edward Conner Chair in Education and is Affiliated Professor of Mathematics at the University of California, Berkeley. Schoenfeld has served as president of the American Educational Research Association and as vice president of the National Academy of Education. He is a Fellow of the American Association for the Advancement of Science and a Laureate of Kappa Delta Pi. Schoenfeld obtained his Ph.D. in mathematics from Stanford University in 1973. Soon afterwards he began conducting research on mathematical thinking, teaching and learning. Major foci of his work have included problem solving, assessment, constructing analytic models of teaching, and issues of diversity; he has an ongoing concern with finding productive mechanisms for systemic change and for deepening the connections between educational research and practice. Schoenfeld was one of the founding editors of *Research in Collegiate Mathematics Education*, and has served as associate editor of *Cognition and Instruction*. In all of his work, issues of research methods have been a central concern.

J. Michael Shaughnessy has taught and researched in mathematics and mathematics education for over 30 years. Much of his career has been devoted to the professional development of pre-service and in-service mathematics teachers. He is currently a Professor of Mathematics Education in the Department of Mathematics and Statistics at Portland State University, Portland, OR where he has worked for the last 15 years. His principal research interests lie in the areas of teaching and learning probability and statistics, and teaching and learning geometry. He was a member of the Board of Directors of the National Council of Teachers of Mathematics from 2001–2004, and served as co-chair of the Board for the Special Interest Group for Research in Mathematics Education of the American Educational Research Association from 2005–2007. He has served on the editorial boards of a number of journals, including the *Journal for Research in Mathematics Education, Mathematics Thinking and Learning,* and *Mathematics Education Research Journal.*

Edward A. Silver is the William A. Brownell Collegiate Professor of Education, and Professor of Mathematics at the University of Michigan. He also currently serves as Associate Dean for Academic Affairs in the University of Michigan's School of Education. His scholarly interests include the study of mathematical thinking, especially mathematical problem solving and problem posing; the design and analysis of intellectually engaging and equitable mathematics instruction for students; innovative methods of assessing and reporting mathematics achievement; and effective models for enhancing the knowledge of teachers of mathematics. He has published numerous articles, chapters, and books on these topics. For much of the 1990s Silver directed the QUASAR project, an ambitious design experiment, which stimulated and studied efforts to improve mathematics instruction in urban middle schools. He is a co-PI of the NSF-funded Center for Proficiency in Teaching Mathematics. He served as editor of the *Journal for Research in Mathematics Education* from 2000–2004. He has been on the advisory boards of several federally-funded projects and on the editorial boards of various journals, including the *American Educational Research Journal*, *Cognition and Instruction*, and *Mathematical Thinking and Learning*. He was the 2004 recipient of the Award for Outstanding Contributions of Educational Research to Practice from the American Educational Research Association.

Laurie Sleep is a doctoral student in mathematics education at the University of Michigan. Her research interests build upon her experience as an elementary teacher and include designing and studying ways to help prospective teachers develop mathematical knowledge for teaching. She teaches both mathematics methods and content courses for prospective teachers and was named a University of Michigan Outstanding Graduate Student Instructor in 2006. Sleep holds a B.A. in business-economics, a master's degree in education from UCLA, and a master's degree in mathematics from the University of Michigan.

Margaret Schwan Smith is an Associate Professor in the Department of Instruction and Learning in the School of Education and a Research Scientist at the Learning Research and Development Center, both at the University of Pittsburgh. She works with pre-service elementary, middle, and high school mathematics teachers at the University of Pittsburgh, with doctoral students in mathematics education who are interested in becoming teacher educators, and with practicing middle and high school mathematics teachers and coaches locally and nationally through several funded projects. Over the past decade she has been developing research-based materials for use in the professional development of mathematics teachers and studying what teachers learn from the professional development in which they engage. She is currently a member of the Board of Directors of the National Council of Teachers of Mathematics (2006–2009). In 2006 she was selected to receive the Chancellor's Distinguished Teaching Award given annually to honor outstanding faculty at the University of Pittsburgh.

Judith Sowder is Professor Emerita of Mathematics and Statistics at San Diego State University. For many years her research focused on children's development of number sense, but more recently on the instructional effects of teachers' mathematical knowledge at the elementary and middle school level. She served as editor of the *Journal for Research in Mathematics Education* from 1996 to 2000 and was elected to serve on the National Council of Teachers of Mathematics Board of Directors from 2000 to 2003. In 1996 she received the San Diego State University Alumni Association Award for Outstanding Teaching, Work and Contributions to the University and Community, and in 2000 the Lifetime Achievement Award from NCTM. She served on the International Committee (steering committee) for International Group for Psychology in Mathematics Education from 1992–1996. She has received research funding for several research projects from the National Science Foundation and the U. S. Department of Education, and a grant from QUALCOMM to develop a Professional Development Collaborative between San Diego State University and surrounding school districts. She continues to serve on advisory boards for several projects, and is currently coauthoring a mathematics textbook for elementary and middle school teachers.

Larry Sowder is Professor Emeritus at San Diego State University, where he has been a faculty member since 1986. Earlier, he served as a high-school mathematics department chairman and physics teacher in northern Indiana before enjoying both graduate school at the University of Wisconsin and then a seventeen year stint at Northern Illinois University. His research interests fall under a problem-solving rubric, ranging from an earlier focus on grades 4-8 students' work with applications of mathematics as embodied in the humble story problem, through the learning and teaching of mathematical proof devising. His collaborations with Guershon Harel have given him an enlightening and delightful professional development experience.

Mary Kay Stein holds a joint appointment as Professor in the School of Education and Senior Scientist at the Learning Research and Development Center, both at the University of Pittsburgh. Over the past decade, her research has transitioned from an exclusive focus on classroom-based mathematics teaching and learning to research that seeks to understand how institutional, interpersonal and policy contexts shape teachers' practice. Her work has been published in the *Journal for Research in Mathematics Education*, the *American Educational Research Journal, Teachers College Record*, *Urban Education*, and the *Harvard Educational Review*. She is the lead author of a widely used casebook for mathematics professional development, *Implementing Standards-Based Mathematics Instruction* and co-author of a book on educational reform in San Diego (*Reform as Learning*). Over the past several years, Dr. Stein has been a principal or co-principal investigator of a number of grants, including two each from the National Science Foundation, the Spencer Foundation, and the MacArthur Foundation. Dr. Stein has served on several national panels including the National Academy of Education's Panel on Strengthening the Capacity of Research to Impact Policy and Practice, and NCTM's Standards Impact Research Group.

William F. Tate is the Edward Mallinckrodt Distinguished University Professor in Arts & Sciences at Washington University. Professor Tate holds Arts and Sciences academic appointments in Education, American Culture Studies, Urban Studies, and Applied Statistics and Computation. He also is chair of the Department of the Education. Additionally, he is a participating faculty member in the School of Medicine's program in Audiology and Communication Sciences. Other responsibilities include serving as the principal investigator and project director of the St. Louis Center for Inquiry in Science Teaching and Learning. The center is a multidisciplinary effort to build sustainable models of human resource development in the sciences and to integrate these models into the cultural resources of an urban community. He has authored scores of scholarly journal articles, book chapters, edited volumes, monographs, and textbooks focused on human resource development in mathematics, science, and technology education, urban studies, and race and American education. His most recent book project is a co-edited volume (with Gloria Ladson-Billings) entitled: *Education Research in the Public Interest: Social Justice, Action, and Policy*. He is president-elect of the American Educational Research Association.

Lieven Verschaffel has been a full professor in the faculty of Psychology and Educational Sciences of the University of Leuven since 2000. From 1979–2000, he held several research positions at the Fund for Scientific Research, Flanders (Belgium). His major research interest is psychology of (elementary) mathematics education. During the past 20 years he has been editor or co-editor of several scientific journals on research on learning and instruction in general and mathematics education in particular. He was editor of the major Dutch education journal *Pedagogische Studiën*, assistant editor of *Learning and Instruction*, and associate editor of *Research Dialogue in Learning and Instruction*. Currently, he is a member of the editorial board of *Learning and Instruction, Educational Research Review, Mathematical Thinking and Learning, Educational Studies in Mathematics* and *Adults Learning Mathematics*, Since 2005 he has been co-editor of the book series *New Directions in Mathematics and Science Education* (Sense Publishers). He is a much-sought-after speaker at international conferences and has written chapters for handbooks about his area of research. Currently, he is coordinator of the international scientific network "Design, development and implementation of powerful learning environments" sponsored by the Fund for Scientific Research–Flanders, and of the Concerted Research Action "Adaptive expertise in mathematics education," sponsored by the Research Fund of the University of Leuven.

Norman L. Webb is a senior research scientist for the Wisconsin Center for Education Research of the University of Wisconsin-Madison. His current principle area of research is evaluation, assessment, and alignment. He has served as the principal investigator on a number of projects including the Adding Value project supporting the evaluation of NSF's Mathematics Science Partnerships, the Study of Systemic Reform in Milwaukee Public Schools, and the Systemic Reform Study Team of the National Institute for Sciences. He is the leader of the evaluation team for the Center for Integrating Teaching and Learning and an evaluator for SCALE. He has directed evaluations of curriculum and professional development projects. His academic training was in mathematics and mathematics education. He chaired the evaluation working group, one of four groups who wrote the Curriculum and Evaluation Standards for School Mathematics for the National Council of Teachers of Mathematics. He co-authored with Thomas Romberg, *Reforming Mathematics Education in America's Cities.* He edited the NCTM 1993 yearbook on classroom assessment and wrote a chapter, "Assessment of Students' Knowledge of Mathematics: Steps Toward a Theory," for the 1992 *Handbook of Research on Mathematics Teaching and Learning.*

Dylan Wiliam is Deputy Director of the Institute of Education, University of London, where he is also Professor of Educational Assessment. After a first degree in mathematics and physics he taught in urban schools in London for 7 years, earning further degrees in mathematics and mathematics education. In 1984 he joined King's College London where he led the mathematics teacher education program. Between 1989 and 1991 he coordinated the development of large-scale assessments for the national curriculum of England and Wales. After his return to King's, he completed his PhD, dealing with the psychometric issues involved in creating systems of age-independent levels of attainment. From 1996 to 2001 he was the Dean of the School of Education at King's College London, and from 2001 to 2003, he served as Assistant Principal of the College. Over the last 10 years, principally in collaboration with Paul Black, he has explored the potential for assessment to support learning (sometimes called assessment *for* learning, or formative assessment). From 2003 to 2006, he directed the Learning and Teaching Research Center at ETS, Princeton, NJ, where he led the development of a range of professional development materials for helping teachers use assessment to improve their teaching.

Linda Dager Wilson has worked in the field of mathematics education research since 1993, specializing in assessment issues, following 13 years as a high school mathematics teacher and a PhD in mathematics education from the University of Wisconsin. She was the assistant project director of the *Assessment Standards for School Mathematics*, published by the National Council of Teachers of Mathematics in 1995. While on the faculty at the University of Delaware, she served in Washington DC as a mathematics coordinator for the Voluntary National Test in Mathematics. Since coming to Washington in 1999, she has worked on several national projects, including the 2005 mathematics framework for the National Assessment of Educational Progress and more recent updates for grade twelve. While at the Center for the Study of Assessment Validity and Evaluation at the University of Maryland, she developed methods for increasing access to mathematics assessment items for all learners. She recently chaired a task force on large scale assessment for NCTM. For the past three years she has worked as a consultant at Project 2061, taking primary responsibility for the development, scoring, and analysis of the student assessments in a research project on the teaching of middle grades mathematics.

Judith S. Zawojewski has been an Associate Professor of Mathematics Education at Illinois Institute of Technology since 2001. She has long been interested in the teaching and learning of problem solving and modeling in the context of classroom practice. Her commitment to link theory and practice is reflected in her publications, including co-authorship in a series of books for the National Council of Teachers of Mathematics: *Navigating through Problem Solving and Reasoning in Grade 3* (2004), *Grade 4* (2005, *Grade 5* (2006), and *Grade 6* (in press) and research chapters in *Beyond Constructivism: A Models and Modeling Perspective on Problem Solving, Learning and Instruction in Mathematics and Science Education* (2003), *Results from the Sixth Mathematics Assessment of the National Assessment of Educational Progress* (1997), and *Results from the Seventh Mathematics Assessment of the National Assessment of Educational Progress* (2000). Dr. Zawojewski has served as a reviewer for various research and practice-based journals, chaired the editorial panel for *Mathematics Teaching in the Middle School* from 1994–1997 and served on the editorial board of *School Science and Mathematics* from 2001–2006. She has also held leadership positions and served on advisory boards for federally-funded research and development projects.

Rose Mary Zbiek is Associate Professor of Mathematics Education at The Pennsylvania State University. One strand of her research on technology addresses the impact of mathematical tools on teaching secondary school mathematics, particularly involving the use of computer algebra systems. Another strand of her work involves the use of technology in mathematical modeling, with a particular focus on developing understanding of curricular mathematics. Her current interests center on mathematical understandings of prospective secondary mathematics teachers and how these understandings synergistically relate to emerging classroom practice. She currently is Chair of the International Committee for Computer Algebra in Mathematics Education and recently completed her term as treasurer and board member of the American Education Research Association's Special Interest Group for Research in Mathematics Education. She served on the National Council of Teachers of Mathematics writing team for the *Curriculum Focal Points for Prekindergarten through Grade 8 Mathematics*. She is series editor of the National Council of Teachers of Mathematics' forthcoming Essential Understandings Project, which explicates key mathematical understandings needed by teachers of mathematics in Pre-Kindergarten through Grade 12.

AUTHOR INDEX

A

B

C

D

E

F

G

H

I

J

K

L

M

N

O

P

Q

R

S

T

U

V

W

X

Y

Z

SUBJECT INDEX

A

B

C

D

E

F

G

H

I

J

K

L

M

N

O

P

Q

R

S

T

U

V

W

Y